Lecture Notes in Mechanical Engineering

Lecture Notes in Mechanical Engineering (LNME) publishes the latest developments in Mechanical Engineering - quickly, informally and with high quality. Original research reported in proceedings and post-proceedings represents the core of LNME. Volumes published in LNME embrace all aspects, subfields and new challenges of mechanical engineering. Topics in the series include:

- Engineering Design
- Machinery and Machine Elements
- Mechanical Structures and Stress Analysis
- Automotive Engineering
- Engine Technology
- Aerospace Technology and Astronautics
- Nanotechnology and Microengineering
- Control, Robotics, Mechatronics
- MEMS
- Theoretical and Applied Mechanics
- Dynamical Systems, Control
- Fluid Mechanics
- Engineering Thermodynamics, Heat and Mass Transfer
- Manufacturing
- Precision Engineering, Instrumentation, Measurement
- Materials Engineering
- Tribology and Surface Technology

To submit a proposal or request further information, please contact the Springer Editor in your country:

China: Li Shen at li.shen@springer.com
India: Dr. Akash Chakraborty at akash.chakraborty@springernature.com
Rest of Asia, Australia, New Zealand: Swati Meherishi at swati.meherishi@springer.com
All other countries: Dr. Leontina Di Cecco at Leontina.dicecco@springer.com

To submit a proposal for a monograph, please check our Springer Tracts in Mechanical Engineering at http://www.springer.com/series/11693 or contact Leontina.dicecco@springer.com

Indexed by SCOPUS. The books of the series are submitted for indexing to Web of Science.

More information about this series at http://www.springer.com/series/11236

Magd Abdel Wahab
Editor

Proceedings of the 13th International Conference on Damage Assessment of Structures

DAMAS 2019, 9–10 July, 2019, Porto, Portugal

Springer

Editor
Magd Abdel Wahab
Laboratory Soete
Ghent University
Ghent, Belgium

ISSN 2195-4356 ISSN 2195-4364 (electronic)
Lecture Notes in Mechanical Engineering
ISBN 978-981-13-8330-4 ISBN 978-981-13-8331-1 (eBook)
https://doi.org/10.1007/978-981-13-8331-1

This Springer imprint is published by the registered company Springer Nature Singapore Pte Ltd.
The registered company address is: 152 Beach Road, #21-01/04 Gateway East, Singapore 189721, Singapore

Preface

This volume contains the Proceedings of the 13th International Conference on Damage Assessment of Structures, DAMAS 2019, Porto, Portugal, 9–10 July 2019. DAMAS has a long history of almost 24 years. The first DAMAS conference took place in 1995 (Pescara, Italy), followed by a biannual meeting in 1997 (Sheffield, UK), 1999 (Dublin, Ireland), 2001 (Cardiff, UK), 2003 (Southampton, UK), 2005 (Gdansk, Poland), 2007 (Torino, Italy), 2009 (Beijing, China), 2011 (Oxford, UK), 2013 (Dublin, Ireland), 2015 (Ghent, Belgium) and 2017 (Kitakyushu, Japan). The thirteenth edition of DAMAS conference series, DAMAS 2019, is held in the Faculty of Engineering of the University of Porto (FEUP), Portugal.

The conference is established as a major international forum for research topics relevant to damage assessment of engineering structures and systems including numerical simulations, signal processing of sensor measurements and theoretical techniques, as well as, experimental case studies. The presentations of DAMAS 2019 are divided into four main sessions, namely (1) Structural Health and Condition Monitoring, (2) Damage in Civil Engineering, (3) Damage in Machineries and (4) Damage in Composite Materials.

The Organising Committee is grateful to keynote speakers, Prof. Keith Worden, Department of Mechanical Engineering, The University of Sheffield, UK, for his presentation entitled 'Graphical Methods for Structural Health Monitoring' and Prof. Huajiang Ouyang, School of Engineering, University of Liverpool, Liverpool, UK, for his presentation entitled 'Data-driven output only structural damage localisation'.

Special thanks go to members of the Scientific Committee of DAMAS 2019 for reviewing the articles published in this volume and for judging their scientific merits. Based on the comments of reviewers and the scientific merits of the submitted manuscripts, the articles were accepted for publication in the conference proceedings and for presentation at the conference venue. The accepted papers are of a very high scientific quality and contribute to advancement of knowledge in all research topics relevant to DAMAS conference.

Finally, the Organising Committee would like to thank all authors, who have contributed to this volume and presented their research work at DAMAS 2019.

Magd Abdel Wahab
Chairman of DAMAS 2019

Organising Committee

Chairman

Magd Abdel Wahab	Ghent University, Belgium

International Organisation Committee

Nao-Aki Noda	Kyushu Institute of Technology, Japan
Huajiang Ouyang	University of Liverpool, UK
Vadim Silberschmidt	University of Loughborough, UK

International Scientific Committee

F. Aymerich	University of Cagliari, Italy
S. K. Bhalla	Indian Institute of Technology Delhi, India
F. Casciati	University of Pavia, Italy
R. Ceravolo	Politecnico di Torino, Italy
F. Chu	Tsinghua University, China
G. De Roeck	KU Leuven, Belgium
L. Faravelli	University of Pavia, Italy
S. Fassois	Patras University, Greece
M. Friswell	University of Swansea, UK
G.-R. Gillich	Eftimie Murgu University of Resita, Romania
J.-C. Golinval	University of Liege, Belgium
A. Gonzalez	University College Dublin, Ireland
K. Holford	University of Cardiff, UK
S. Khatir	Ghent University, Belgium
M. P. Limongelli	Politecnico di Milano, Italy
G. Manson	University of Sheffield, UK
T. McCarthy	University of Wollongong, Australia
N.-A. Noda	Kyushu Institute of Technology, Japan

Contents

Structural Health and Condition Monitoring

Machine Learning Techniques for Structural Health Monitoring

Bibin Kurian and Ranjith Liyanapathirana(✉)

Western Sydney University, Penrith, NSW 2751, Australia
r.liyanapathirana@uws.edu.au

Abstract. Structural Health Monitoring has become a hot topic in recent decades as it provides engineers with sufficient information regarding the damages on civil infrastructure by analysing data obtained from the monitoring sensors installed in the structures. Commonly, the process of implementing a damage identification strategy for aerospace, civil and mechanical engineering infrastructure is referred to as Structural Health Monitoring (SHM). The development of smart sensors and real-time communication technologies via Wireless Sensor Networks (WSN) has empowered the advancement in SHM. Recently, statistical time series models have been widely used for structural damage detection due to the sensitivity of the model coefficients and residual errors to the damages in the structure. Increasingly Machine Learning (ML) algorithms are employed for damage detection tasks. This research sheds light on the methodologies to predict the structural damage on concrete structures with the help of sensor technology by effectively combining data science and ML strategies. Experimental test results publicly available are used, where the tests have been performed with varying stiffness and mass conditions with the assumption that these sources of variability are representative of changing operational and environmental conditions in addition to changes caused by damage. To enhance the accuracy of damage detection, instead of the traditional time series analysis, ML is used for learning from prior experience. To detect the existence and location of the damage in the structure, we use supervised learning, and for measuring the severity of the damage, unsupervised learning is used. Accuracy results are obtained with three well-known ML algorithms (KNN- k Nearest Neighbour, SVM-support vector machine and RFC random forest classifier). In this study, the Random Forest Classifier algorithm generated good predictions on damaged and undamaged conditions with good accuracy, when compared to the KNN algorithm and Support Vector Machine algorithm under the supervised mode of machine learning. The utilisation of sensor technology effectively combined with aspects of Artificial Intelligence (AI) such as Machine Learning has the potential to implement a more efficient SHM system.

Keywords: Damage detection algorithms · Structural health monitoring · Machine learning · Non-destructive testing · Time-series analysis · Data science

M. A. Wahab (Ed.): DAMAS 2019, LNME, pp. 3–24, 2020.
https://doi.org/10.1007/978-981-13-8331-1_1

1 Introduction

The method of employing damage detection techniques in various engineering fields such as aerospace, mechanical and civil is called Structural Health Monitoring (SHM) or Non-destructive Testing and Evaluation (NDE). The terminology 'damage' in effect can be defined as the variations subject to the geometric characteristics or materialistic properties of the system, comprising the modifications to the boundary conditions and system connectivity, which can create adverse effects on the performance of the systems. There exist different kinds of Non-Destructive Evaluation (NDE) methodologies and tools for localised monitoring of such systems [1]. At present, the known damage detection approaches include experimental methods of visual or localised nature. The use of acoustic signals, ultrasonic methods, radiographs, magnetic field methods, thermal field methods or Eddy current methods are some of the conventional techniques employed in damage detection. The limitations of these methods are the prior knowledge about the location where damage occurs is required, and also the detected area should be easily accessible. Apart from these drawbacks, the methods mentioned above can identify only the damages on the surface of the structures. Thus methodologies including the examination of vibration characteristics of the structural systems, which can be applied to complex areas were introduced to overcome these limitations [2].

On the other hand, during the past three decades of research in this field, attempts to determine the damages in a structural system as a whole were carried out. There has been a tremendous increase in the number of research projects conducted during the recent decade in the field of SHM. The increased attention in SHM of structures has been due to the importance of life-safety factors as well as cost-effective benefits on civil and mechanical structural systems [1]. The goal of this research is to study, by way of analysing existing experimental results readily available on the internet, the structural damage in the presence of operational and environmental variations using vibration-based damage identification procedures, with the help of machine learning algorithms that aim to improve the damage detection efficiency.

2 Literature Review

2.1 Structural Health Monitoring

Reviews based on examining the techniques to measure structural vibration have been found from earlier times. The attention in monitoring civil, mechanical and aerospace structures to understand and detect the damage has been prevalent in the structural engineering field [1, 2]. Structural Health Monitoring (SHM) using vibration analysis is based on five levels, which are detection, localisation, classification, assessment, and prediction [3]. Linear and non-linear types are the two major classifications meant for structural damage. Even after the occurrence of damage, a linear-elastic structure will exist as the same, where the modal properties and variations due to geometric or materialistic changes can be demonstrated using a linear equation.

On the other hand, non-linear damage occurs when an initially linear-elastic model or structure changes into a non-linear manner after the occurrence of damage. For instance, development of a crack from wear and tear may consequently open and close under vibrating in its normal operating circumstances is a suitable example for non-linear damage. A reliable damage detection technique will be relevant to both types of damages in general [2].

As mentioned before, the process of SHM involves different steps. At first, the system is monitored over time, using a set of sensors and observations are inferred based on periodic samples of measurements of dynamic responses obtained from the same sensors. Extraction of the features is the next stage, where the features which can bring about damages are being extracted from these observed measurements. Later a statistical analysis is performed on these extracted damage sensitive features to assess the current conditions and health of the structural system. In case of long term structural monitoring scenarios, the output of such statistical process is updated regularly, in order to obtain information that substantiates capacity of the structure t smoothly o function when it is subjected to ageing and deterioration resulting from various environmental conditions.

Moreover, when the structure undergoes adverse impacts due to the occurrence of events like earthquakes or heavy loading, SHM is an emergency aid to validate the structure's functional reliability. The collapse of the North Carolina Bridge [4] in the USA is an incident that paved the way for focusing on the structural health monitoring techniques inspired from the aspect of life-safety. Also, the advancement in wireless sensor networks (WSN) has influenced the SHM technology, which facilitates the wireless transmission of the monitored parameters, mainly featuring the remote access of the SHM systems.

2.2 Damage Detection Techniques

Centred on the works of Rytter as referred to in [1], the author groups the process of SHM into five phases:

- Identification of damage occurred in a structure
- Localisation of damage
- Identification of the damage type
- Quantification of damage severity
- Estimation of the remaining service life of the structure.

Doebling [2] presents a recent thorough review of the vibration-based damage identification methods. He has proposed several diverse methods for the identification and localisation of damage with the aid of vibration response measuring methods. However, for real-time applications, these methodologies are subjected to certain limitations, which make them ineffective for detection of the damage at the initial stages. Thus to overcome the drawbacks of such methods, researchers have introduced statistical analyses of the structure through time series analysis.

The authors Farrar and Warden [1] have proposed a damage detection method by combining five processes which are closely interrelated to each other. This method includes:

- Structural Health Monitoring (SHM)
- Condition Monitoring (CM)
- Non-Destructive Evaluation (NDE)
- Statistical Process Control (SPC)
- Damage Prognosis (DM).

Among these methods, SHM is generally applied to aircraft structures and buildings to detect real-time damages in the whole structural system. CM is similar to SHM, but it is deployed in assessing the damages into SHM, but addresses damage identification in rotating mechanical systems and reciprocating machinery, which are used specifically in manufacturing and power generation plants [6]. NDE is a localised offline method carried out once the damage has been traced. Apart from these, NDE is utilized in monitoring some prefabricated structures such as rails and pressure containers. For performing NDE, the general methods include the use of acoustic waves, X-rays and microscopic observatory techniques [7]. NDE is implemented and applied to smaller areas of a structure where the location of the potential damage is identified, which make it mainly useful only for analysing the characteristic of the detected damage, provided prior information regarding the damaged spot [6]. Converse to the structure-based method, SPC is regarded as process-oriented technique, where a set of sensors are used in order to monitor the variations in the carried out processes, which can provide information related to the structural damages. Damage prognosis is a different technique used after the detection of the damage, in order to estimate the remaining functional lifetime of a structural system [1].

2.3 SHM Methods in Civil Structures

The examination of the structural health of surviving structures like buildings and bridges after disastrous events such as hurricanes and earthquakes is generally time-consuming and expensive as most of the critical structural members and links are being masked under architectural surfaces. In the case of acute structures such as major bridges, power plants, hospitals, military centres and incineration plants, it is inevitable to evaluate and reassure the structural health as soon as possible [8]. Moreover, the information on such critical and catastrophic collapses should be broadcasted to the emergency services at the earliest, in order to save peoples life. Mostly, such emergency alert messages are delayed because of the weather conditions, and unreachability to the damaged locations due to the obstacles from the collapse. In most of the situations, the collapses occurring in the near future are not noticeable from outside of the structure [8].

Even though in the field of aerospace and mechanical industries, SHM has been extensively utilized, it was in the recent only attention was given to the civil infrastructures, as a result of their gradual weakening due to blast loading variations, for example. Various categories of SHM procedures have been developed for analysing the structural states. Vibration-based, strain-based, electrical impedance-based probability-based and statistical based methods [9]. In the last decade, a significant amount of research has been carried out theoretically and experimentally. Side by side, efforts have been taken to develop systems for typical seismic event damage monitoring, such as earthquakes scenarios. Those systems integrate a sensing and processing unit, a mechanism for transmission and acquisition of data and damage verifying systems together. Innovative developments in micro-scale sensors, wireless and GPS technologies and systems capable of increased calculation abilities have effectively contributed to solving the hindrances in SHM systems.

SHM is generally concentrated on damage identification, that is whether the damage has occurred or not as well as prediction of the extent of the damage. Apart from these, additional information regarding the damage, such as geometrical properties, configuration, networking, shape are not considered or considerably simplified at the time of categorising the damage sensitive characteristic features for SHM [7]. However, such parameters can be used well enough to estimate the depth of damage occurred. At present, there are several NDE techniques to detect the structural damages, but at the same time, those include expensive visual processes or experimental analysis confined to particular areas in the structure, including ultrasonic techniques, magnetic field procedures, X-ray technology and Eddy-current methods and so on. Non-destructive methods are subjected to limitations, as these require the knowledge about the damaged site in advance, provided it should be readily accessible [10]. A damage identification scheme that evaluates the non-linear features of the structure to determine the damage could be very much useful in the sense as most structural damages bring some non-linear changes in the structure [10]. To lessen the costly human intervention and complex safeguarding SHM judgements in damage characterisation processes, computational analysis with the help of algorithms are required.

2.4 Time Series Analysis

When we record some processes which change with time, we get a time series. For instance, the variation of share market prices in a short period recorded as stock exchange census is a perfect example of time series analysis. Those recordings can be observations of either continuous or discrete variable or events. In SHM, mainly discrete set of observations that changes with time are monitored using time series analysis method; those observed values are obtained at equal and regular intervals of time [5, 11]. When we analyse data at different time slots, it results in distinct problems. The sampling data is always dependent on time, and this put a limit to the use of random samples for standard statistical methodologies. Such a case, make the use of time series analysis important in statistical modelling techniques [12]. Earlier, physical and environmental science problems were solved using time series analysis. Scrutinising the recorded information plotted across time is the initial stage in any time series analysis execution [12]. By primarily analysing the recorded history of acceleration in a structure over time, the source of damage is traced. At first, a data normalisation procedure is carried out, and this process compares the detected or recorded signal with a reference signal which is closer to that of under undamaged condition of the structure [4]. For detecting and locating the damage, the most widely used statistical time series analysis method is Auto Regressive-Auto-Regressive with exogenous input (AR-ARX). The structural damage is identified or recognised based on the fact that, if the model developed through the statistical prediction of undamaged time series data measurements could not replicate or estimate the newly generated time series, then the current structure is said to be damaged [13].

2.5 Machine Learning Approach in SHM

The researcher Arthur Samuel as in [14], contributed one of the best definitions for Machine Learning as a tool to offer the computers the ability to learn without programming explicitly. It is one of the most widely used algorithms now to monitor the health of infrastructures, which includes two different types of learning methods such as supervised and unsupervised learning. When the information from both damaged and undamaged scenario is accessible, the pattern recognition in statistical modelling can come under the supervised learning method category [1]. For instance, Group classification and regression analysis are examples that fall under the supervised learning algorithm. When the information or data regarding the damaged scenario is not available, the unsupervised learning algorithm is applied in SHM systems. Novelty detection or outlier detection is the prime category of algorithms executed in the application of unsupervised learning methods. However, all algorithms above make use of the statistical modelling of findings derived from enhancing the damage detection process in SHM [1]. In order to solve the issue of pattern recognition, several machine learning algorithms are utilized. The authors Farrar and Worden [1] used neural networks, SVM (Support Vector Machines) and genetic algorithms (GA). Gaussian Classifiers, SVM, Random Forests and Adaboost algorithms are mentioned by the author William Nick and others in their research on SHM using machine learning techniques [15].

In the development of an agent-based SHM, acoustic emission signals are used, where signals are categorized according to source processes, which are accompanied by significant crack developments in the structural systems. The agent-based system can replace complex communication and computational monitoring techniques, which can respond to the circumstances immediately by proposing an appropriate set of techniques. The need to respond quickly necessitates the system to have a selection of classifiers with different features so that it can utilise a combination of those to apply for distinct situations. As mentioned earlier, unsupervised learning can detect and identify the spot of damage, whereas a supervised method can contribute information about the severity and nature of the damage [15].

2.6 Existing Machine Learning Methodologies

There are different types of learning systems, based on the aspect of learning. These include supervised learning, unsupervised learning and reinforcement learning, which will be described in detail later in this chapter. Firstly, the way of learning is essential and based on this, different strategies and algorithms are developed and following are some of the ways that a machine or learner transforms the knowledge acquired.

- Learning by being programmed: The system learns and performs the task based on the program or code implemented by an external entity into the system [16]. Here the system is capable of working strictly within the coded ways unless the external entity modifies the program.

- Learning by memorisation: The data or facts stored in various format is used by the system to learn the environment or task. The system is not capable of learning or performing aspects beyond the information fed in its memory.
- Learning from examples: This is the most established method of learning, where the system learns or understands the process from previous outputs or past experiences of the working environment of a task. The previous experiences and performing conditions serve as an example to the machine which enables it to predict future outputs. The examples fed to the system can be favourable examples or counterexamples so that the system can make decisions and predict accordingly. This is considered as a supervised method of learning as there are some specific outputs or previous experiences for some specific inputs given to the system. Learning from examples is regarded to be a more efficient method to make valid inferences than learning from instructions or by being programmed type [17].
- Learning from observations: Here the system uses the observations to discover and form theories, analyses and classify the information gained through observations and so on to learn and arrive at inferences [16]. Learning from observation is an example of unsupervised learning as there are no specific set of inputs and corresponding outputs, i.e. the system learns without supervision or an external entity who act as a teacher.

3 Methodology

This section discusses various methods planned in accomplishing the development of an effective Structural Health Monitoring system with the help of Artificial Intelligence (AI) techniques like Machine Learning, to improve the damage detection in SHM systems. The proposed methodology for this research is using experimental analysis and applied research, which tries to analyse the findings (available on the internet for use by the research community) of an experiment conducted at the Engineering Institute (EI) located at Los Alamos National Laboratory (LANL), USA, in partnership with the Laboratory for Concrete Technology and Structural Behaviour (LABEST) of the Faculty of Engineering of the University of Porto (FEUP), Portugal on structural health monitoring, where time series analysis is carried out to detect the damages in civil infrastructure. Also, the project's primary focus is to apply machine learning techniques in order to improve the damage detection process. For this, an existing database with results of damage detection from the experiment as mentioned above using time series analysis is utilised. A set of training data is to be created from the stated database to perform machine learning (ML) methods such as supervised and unsupervised learning techniques.

3.1 Experimental Setup at LANL [10]

According to [10], a three-storey building prototype shown in Fig. 1 is created for the experiment, and linear and non-linear impacts are introduced repeatedly with the help of a bumper mechanism. The three-storey test structure is made from unistrut columns and aluminium floor plates [10]. Floors consist of 0.5 in. thick plates of aluminium with two-bolt connections to brackets on the unistrut column, with an adjustable mechanism for the floor heights. An aluminium plate of 1.5 in. thick forms the base, and support brackets for the columns were also bolted to this plate. All bolted connections were stiffened to a torque of 25 Nm in the undamaged state. At the bottom of the base plate, in order to introduce free horizontal displacement, four firestone air-mount isolators were fixed. In order to maintain the level of the base with the shaker, isolators were assembled on aluminium blocks and sheet of wood. The isolators were inflated to 10 psi. The shaker was connected to the structure of a 6 in. long stinger with a diameter of 0.375 in., which is connected to a hole in the base plate at a mid-height. For the proper excitation of torsional and translational motion, the shaker was connected 3.75 in. apart from the corner of the 24 in. side of the structure.

In this experimental setup [10], the damage is introduced by relaxing the preloads applied by the bolts at the joints of the building structure. A "healthy" joint is held together by bolts that are torqued to a value of 25 Nm. In order to test the sensitivity of the damage detection ability, multiple levels of damages are introduced in the structure. There are 24 piezoelectric accelerometers with two at each joint and one linked to the plate and other attached to the column. The configuration in [10] enables the detection of relative motion of the column with respect to the floor. Each accelerometer is having a minimum sensitivity of 1 V/g. In order to facilitate the digitising of the accelerometer and to force the analog signals from the transducer, a data acquisition system which is accessible from a laptop or PC is used.

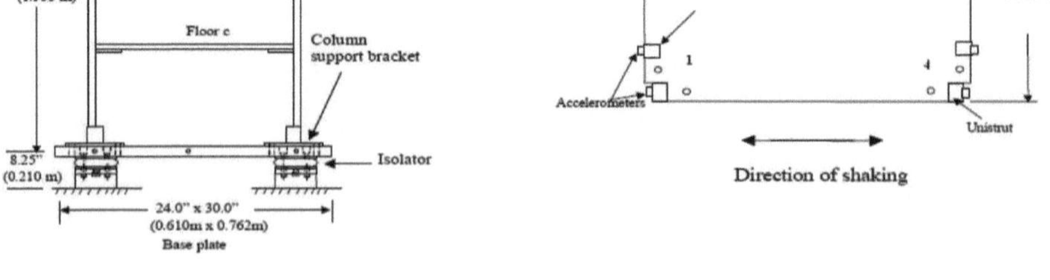

Fig. 1. Side view and top view of the assembled three-storey structure in LANL Lab [10]

3.2 Steps in Conducting the Test

The test structure mentioned above is analysed using an SHM damage detection process which is the focus of this study. SHM involves the following four progressions based on a statistical pattern recognition paradigm [18]:

- Operational evaluation
- Data Acquisition
- Feature extraction
- Statistical modelling for feature classification.

Operational evaluation involves assessing the parameters that facilitate the computing of the possible extends of damage occurrence and extent of damage in the deployment of Structural health monitoring systems. In real cases, monitoring of structures requires consideration of operational and environmental effects. Temperature changes sometimes generate more significant changes in parameters than the effect of damage in the early state. Thus it becomes difficult to distinguish between the two causes [19]. This stage also constitutes the modifications in SHM characteristics to be streamlined to monitor the distinct features of the damage identification system. The data acquisition step includes the selection of application specific processes such as data collection, processing, transmission and storage and so on, along with the excitation scenario that comprises the information regarding the type, number and location of the sensors assembled in the structure. These aspects will not add in determining the existence of damage or measuring the damage, whereas the data acquisition and sensing activities compare and measure the response to the excitation mechanisms or to the loads the structure subjected from the environment and operating conditions. The sensor outputs, which are associated with the nature of damage to be detected and the underlying sensor technology utilised can provide information about the occurrence and location of the damage [20].

The next stages Feature extraction involves analysing the damage sensitive features from the recorded sensor output measurements to predict the damage conditions of the structure. That is to differentiate damaged and undamaged structure using SHM. The extracted Damage Sensitive Features (DSF) are classified based on the statistical modelling techniques utilised with the help of algorithms such as machine learning to evaluate the damaged condition in detail [20]. Theoretically, when the extent of damage increases, DSF that holds the changes of structural properties [21] will increase, will be converted in some logical manner. The damage detection and localisation approach based on the residual forces method was used successfully by some researchers to locate structural damage based on an analytical model for damage identification [22].

3.3 Supervised ML Algorithms Used

As we know, machines are generally designed to perform tasks effortlessly. Humans are capable of classifying and recognising various aspects, which a machine cannot do in general. In some cases, humans can perform some tasks like speech or handwriting recognition without efforts. However, it might be difficult for them to explain how they recognise the differences. Machine learning algorithms are strategical techniques designed to fill the gap of perceiving or understanding how the task was carried out [23]. A machine or computer can learn or understand the aspects behind a task such as recognition or classifying through machine learning algorithms. Different individuals perform the same tasks differently according to the situations and their capacity of intelligence. Similarly, based on the application and the scenarios, many machine learning algorithms can be implemented through a computer to perform simple to complex tasks beyond human limits. In this project, we mainly focused on supervised learning algorithms. Following are some commonly used algorithms that come under supervised mode in a machine learning algorithm.

- KNN classifier
- Support Vector Machine
- K-Means Clustering
- Random Forest
- Neural Networks
- Bayesian.

Among these, for improving the damage detection method in structural health monitoring, we utilised three algorithms, namely KNN classifier, Support Vector Machine and Random Forest as described below. ML algorithm classification in MATLAB is shown in Fig. 2.

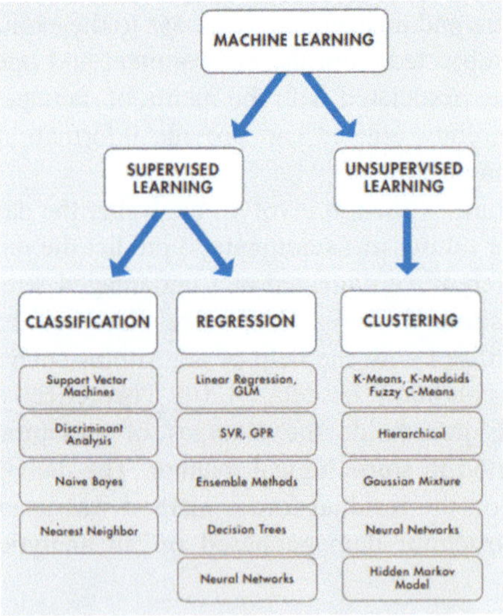

Fig. 2. ML algorithm classification

K-Nearest-Neighbor Classifier (KNN). KNN is one of the most widely used supervised learning algorithms for pattern recognition. It is considered to be one of the modest learning algorithms in machine learning when no prior knowledge of the learning environment is available or when the available knowledge is at a minimum level. The K-Nearest Neighbour algorithm works on the principle that objects or examples in a training sample that are closer to each other have similar characteristic features [24]. The nearest neighbour rule is shown in Fig. 3.

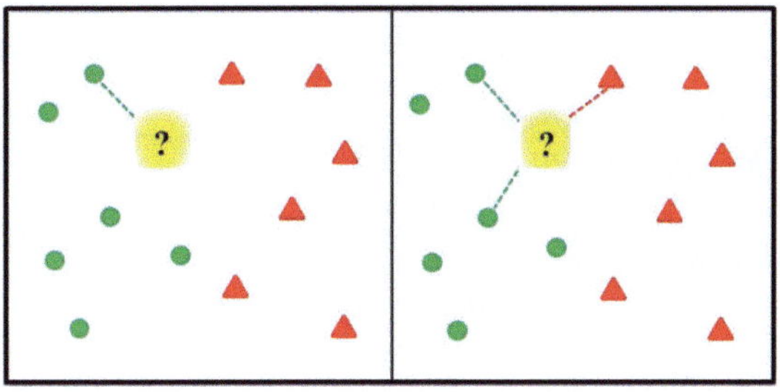

Fig. 3. Nearest neighbour rule

In KNN, the classification of examples occurs in accordance with the class of their 'k' closest neighbouring examples.

For an optimum value of k, this conventional classifier independent of specific parameters shows good performance levels in a learning environment. According to the k-nearest rule of the neighbourhood, a data sample is assigned a class label with the most frequent occurrence among the k-nearest test samples or examples. When there are two or more similar classes, the sample is assigned class labels that possess minimum average distance to it [25]. For example, if a bird looks like a duck make sounds like a duck and walk like a duck, and then it is probably a duck only. Although there are different means to evaluate k-nearest neighbour, the most regarded method is the classification based on estimating the Euclidian distance because of its ease in use, better productivity and efficiency.

The Euclidian distance between two vectors X_i and X_j can be calculated as shown in Eq. (1) [25], where $X_i = X_i^1, X_i^2, X_i^3 \ldots X_i^n$ and $X_j = X_j^1, X_j^2, X_j^3 \ldots X_j^n$.

The difference, Euclidian distance,

$$\mathrm{D} = \sqrt{\sum\nolimits_{k}^{n=1}} (xik - xjk)2 \tag{1}$$

KNN classifier is also called by the name instance-based classifier, as it can perform in learning environments where unknown instances can be classified based on the neighbour distance function. KNN algorithm is powerful and easy to implement, while at the same time, it is not considered to be suitable for data sets with larger dimensions. It is also known as a memory-based classifier as it necessitates the storage of all training examples in the memory of the learner at the time of running the algorithm.

The nearest neighbour rule, in its simplest form, is when the value of 'k' equals 1. Depending on the value of 'k', each sample is compared to find similarity or closeness with 'k' surrounding samples. In short, when k = 1, each sample is classified by comparing it with one nearest sample. When k = 4, the individual samples undergo comparison with the nearest four samples in and hence the unknown one is classified accordingly. Following is a simple example of a KNN algorithm [26].

for *all the unknown samples unknown(i)*
for *all the known samples known(j)*
estimate the Euclidian distance between unknown(i) and known(j)
end for

find the k smallest distance
locate the equivalent samples

known(j1), …, known(jk)
assign unknown(i) to the most frequent class
end for

Figure 3 illustrates the nearest neighbour rule for k = 1 and k = 4, where samples are labelled into two different classes (red and green). As shown in Fig. 3, in the second case, when k = 4, out of four samples, three of them fall into the same class and only one is under a different class. Estimating the neighbourhood value k is crucial as different values of k can produce different probable outcomes in classification [26].

If the k value is minimal, the query sample may fail to find any nearest neighbour as it could get into some mislabelled noisy data points. On the other hand, a large k value may result in overlapping of invalid samples from other classes, which in turn makes the classification technique much tougher. Thus, the KNN classifier performance is evaluated based on the selection of the parameter 'k'.

Support Vector Machines (SVM). SVM is one of the most widely utilised machine learning algorithms in the supervised mode for applications such as pattern classification, forecasting and decision-making tasks. SVMs are deployed by making use of randomised training set instances or samples which are categorised in advance [27]. SVM was the first algorithm which evolved from a kernel based system, where samples from input environment are transformed into a multidimensional feature space with the help of a kernel function by creating a hyperplane that separates the training data samples for the classification process. Some of the primarily used kernel functions [27] to convert the input space into corresponding feature space includes Linear functions, Polynomial functions and Radial Basis Functions (RBF).

As mentioned previously, the SVM model uses a hyperplane to separate or classify two or more classes of data samples, where the focus is to maximise the underlying margin between the data samples [28]. For a data sample of lower orders, with two classes, the hyperplane is a straight line (as seen in Fig. 4) that separates the two classes with a marginal distance, whereas in a training set with multi-dimensional samples, the hyperplane becomes a plane as the name suggests. The primary goal of an SVM classifier is to figure out an optimal hyperplane between the classes, where a hyperplane is said to be optimal if it is capable enough to separate the classes with the largest possible margin. The data samples that lie along the marginal line are termed as support vectors, as they serve as the means to classify the dissimilar classes.

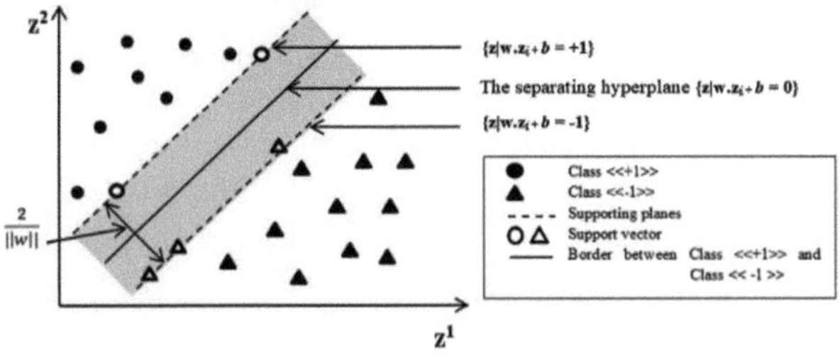

Fig. 4. Hyperplane illustration

SVM is a quadratic programming algorithm in convex nature, as it optimises the parameters globally rather than performing time-consuming local optimisations, which is done in the case a concave function [27]. Hence SVM optimises the linear parameters in a training set on a global basis to arrive at inferences at a faster rate. There are two major parameters set to SVM classifier, which includes smoothness parameter γ and penalty parameter C in a radial-based-function [27]. The smoothness parameter decides the functional mapping of space inputs into multidimensional features, whereas the model complexity and minimising the error in fitting the classifier is influenced by the penalty parameter. Support vector machine algorithm finds its application in areas like image classification, handwriting recognition, face detection, pedestrian detection and diagnostics of tumours from scanned results. SVM is found to produce good results on a smaller set of training data sets, whereas for a huge data sample it consumes more time, thereby making the computational cost high. Moreover, SVM is regarded as a less effective method in environments were overlapping of classes occurs which makes the dataset noisier.

Random Forest Classifier Algorithm. A Random Forest Classifier (RFC) is an ensemble algorithm under the supervised method of machine learning, where the term ensemble indicates that the samples are analysed as a whole or a group rather than viewing them individually. The fundamental classifiers used in RFC algorithms are decision trees, where many decision trees are produced to perform classification. The aspect of random sampling is carried out in two stages [29].

Firstly, for bootstrap samples, which are used for introductory processing of algorithms, are sampled randomly, and secondly, input attributes for individual decision trees are also sampled randomly. The major aspect of random error generalisation in RFC is based on the correspondence of individual decision trees with the base tree used for classification [29].

RFC is regarded to be well suited for a large number of datasets. RFC also can manage the missing data samples while processing a huge dataset by effectively estimating them, by evaluating the generalisation errors, where the generation of decision trees occurs concurrently in a progressive manner. RFC model is deployed in a wide range of areas, which includes, texts classification and extraction, arranging books in a library according to various specified categories, statistical analysis in stock markets, diagnosing various medical conditions such as heart failures, cancer and tumours from scanned outcomes, liver failures and so on.

The decision tree structure is inspired or framed based on the features of a normal tree in nature, which consists of a root, branches and leaves. Similarly, a decision tree starts with a root and moves in the downward direction, which is divided into circular nodes interconnected with branches or segments and nodes are ended or terminated at leaf nodes [30]. A node indicates some features, where the branches indicate the range of information. The branches or segments that contain the range of values serve as the separation points for a group of samples with specific characteristics. Depending on the data samples and learning environment, the number of branches evolving from a node can be different. Pre-classified data samples are used to generate Decision trees; the classification of samples into the different class is primarily based on some characteristic features that divide the data into the finest levels. The process of tree generation continues until all the samples in a subset fall into the same class.

The decision trees are built based on the parameter k_n which indicates the number of points to be estimated on the leaf nodes. The value n denotes the training set size, which determines the minimum size of a leaf node.

In the next section, we consider the implementation of the above methods in SHM.

3.4 Experimental Evaluation and Implementation of ML in Structural Health Monitoring

To implement the ML in SHM, as mentioned in the introductory section, the database from the Los Alamos National Laboratory (LANL) resulted from the testing of a three-storey building bookshelf-like structure is used in this work. The test database includes a large number of files produced through the time series analysis experiment. Out of the 24 channels, which represent the outputs from the 24 accelerometers located, two each in the four corners of each storey of the three-storey bookshelf structure gives out a resulting data set of size 26 × 8192 data samples, where 2 channels were found to be of extra information, which is mentioned briefly in the following training data set preparation stage.

3.5 Training Data Set Preparation

Here, the damage sensitive features are extracted from the collected database, in order to do statistical analysis by implementing the ML technique. There are separate data sets for the two cases, Damaged and Undamaged condition at 8192 test instants with input voltages 2 V, 5 V and 7 V applied to the accelerometers to check and ensure the repeatability of the experiment. Among these, for preparing the training set, a data set corresponding to the 2-V application on accelerometers is used. The data set features were extracted carefully as there were some wrong outputs as the channel nine did not work properly and channel nine sensor outputs were recorded later on channel 26. Also, channel 25 represented time stamps for test instances, which is considered to be insignificant while preparing the training set. So channel 25 was removed, and channel 9 was replaced with channel 26 data in the data cleaning process. For implementing a predictive model, the features needed to be classified with class labels. Damaged and Undamaged data sets were combined to form a training data set of 25 × 16,384 with the class labelling shown in Table 1 below.

Table 1. Labelling classes

Class label	Description
DA	Damaged condition
UD	Undamaged condition

Thus a combined data set matrix of 25 × 16,384 briefly illustrates the following information:

25 columns: Column (1–24) represents 24 sensor output channel & column 25 for Class labels.

16384 rows: Represents test instances. Rows (1-8192) indicate Undamaged test results and (8193–16384) with Damaged ones

3.6 Training and Applying ML

Python programming language is used to implement the ML and corresponding Python codes used are mentioned in each step of training and modelling. The sample input and its responses are represented by vector X and Y, respectively, where X includes the input accelerometer sensor outputs, and Y is the response represented as Damaged and Undamaged. Out of the total data samples, as in most ML algorithm implementations, 80% of the data is generally used for training purposes, and 20% for testing the model, i.e., the training and testing split ratio is 0.20. The seed value represents a random number generator factor, which initiates the training according to the split size ratio.

```
X = array[:, 0:23]
Y = array[:, 24]
test_train_split_size = 0.20
seed = 20
X_train, X_test, Y_train, Y_test = model_selection.\
   train_test_split(X,Y, test_size=test_train_split_size, random_state=seed)
return X_train, X_test, Y_train, Y_test
```

There are three machine learning algorithms used to test the trained data, which includes K-Nearest Neighbour (KNN), Support Vector Machine (SVM), and Random Forest Classifier (RFC). The following are the steps in Python that define each classifier.

```
def KNN_test(X_train, X_test, Y_train, Y_test):
    knn = KNeighborsClassifier()
    knn.fit(X_train, Y_train)
    predictions = knn.predict(X_test)

def SVM_SVC_test(X_train, X_test, Y_train, Y_test):
    svm = SVC()
    svm.fit(X_train, Y_train)
    predictions = svm.predict(X_test)

def RF_test(X_train, X_test, Y_train, Y_test):
    rf = RandomForestClassifier(n_estimators=700)
    rf.fit(X_train, Y_train)
    predictions = rf.predict(X_test)
```

After setting up the classification model, the fitting of the model is done. For instance, we use the svm.fit () function to generate fitted values from the past data samples by extending them, in order to make predictions for the unknown samples. In the evaluation of the models, the accuracy of each classifier model is estimated with the help of confusion matrix and classification report which are explained in detail in the results section next.

4 Results and Discussion

4.1 Confusion Matrix

A confusion matrix is generated in each modelling, which summarises the prediction results in a matrix form. The data set values used for testing is represented in count values, and it is indicating the confused stage of the model while making predictions. A confusion matrix is generated in the following format using Python code:

Confusion_Matrix = confusion_matrix (Y_test, predictions)

[[TP FN]
[FP TN]]

The confusion matrix generated for SVM is given below as a reference:

Confusion Matrix

[[1061 539]
[88 1589]]

The row of a confusion matrix signifies the predicted class, and the column of the matrix indicates the actual class. The elements or count value in a confusion matrix are:

- TP: True Positive values; Actual class was True, predicted correctly as True (Damaged predicted as Damaged)
- FN: False Negative values; Actual class was True, predicted wrongly as False (Damaged predicted as Undamaged)
- FP: False Positive values; Actual class was False, predicted wrongly as True (Undamaged predicted as Damaged)
- TN: True Negative values; Actual class was False, predicted correctly as False (Undamaged predicted as Undamaged).

The efficiency of the model increases as the number of False Negatives (FN) and False Positives (FP) reduces. So a good predictive model classifier makes good predictions by avoiding FN & FP.

4.2 Classification Reports

Classification report is another means that evaluates the performance of a supervised classifier. The classification reports for three ML algorithms used in this project are described below

Classification report is obtained by using the below mentioned Python code

```
Classification_Report = classification_report (Y_test, predictions)
```

A classification report includes the following parameters:

Precision. Precision is defined as the ability of a classifier not to predict any false value of a class as True. In other words, how precisely the model predicts the true class values as a true and false class as false. In an SHM context, if we consider the precision for a damaged class for Random Forest Classifier 0.94, it means the model successfully predicted 94% of damaged conditions correctly as damaged itself, out of the total number of damaged predictions made. The 'precision' factor can also be calculated manually from the confusion matrix as follows:

For RFC Confusion Matrix:

[[1445 155]
[91 1586]]

Precision is also described as the ratio of true positives to the overall number of positive predictions made.

$$\begin{aligned}\mathbf{Precision} &= \mathrm{TP}/\left(\mathrm{TP} + \mathrm{FP}\right)\\ &= 1445/\left(1445+91\right)\\ &= \mathbf{0.94}\end{aligned}$$

Recall. The Recall factor is the ratio of true positives to the overall number of predictions made in that specific class. It is also called the *sensitivity* of the classifier- that is the ability to find all positive predictions of a class. The recall for a damaged class in RFC is computed as follows.

$$\begin{aligned}\textbf{Recall} &= \mathrm{TP}/\,(\mathrm{TP}\ +\ \mathrm{FN})\\ &= 1445/\,(1445+155)\\ &= \mathbf{0.90}\end{aligned}$$

f1 Score. This value is derived using the precision and recall values, where a calculated weighted harmonic average or mean of precision and recall gives the f1 score. For an ideal classifier, the f1 score will be 1. The value of f1 in its worst or minimum state is 0. f1 score is used to evaluate the efficiency of a model, similar to the accuracy of the classifier.

$$\textbf{f1} = 2 * (\mathrm{Recall}\ *\ \mathrm{Precision})\,/\,(\mathrm{Recall}\ +\ \mathrm{Precision})$$

Support: This gives the number of true values belonging to the particular class. The total numbers of support for both classes together form the testing data (20% of the entire dataset), obtained by splitting the data set to train and test the models.

KNN, SVM and RFC classification reports are shown in Tables 2, 3, and 4, respectively.

Table 2. KNN classification report

KNN	Precision	Recall	f1-score	Support
Damaged	0.98	0.72	0.83	1600
Undamaged	0.79	0.98	0.88	1677
Average/Total	0.88	0.86	0.85	3277

Table 3. SVM classification report

SVM	Precision	Recall	f1-score	Support
Damaged	0.92	0.66	0.77	1600
Undamaged	0.75	0.95	0.84	1677
Average/Total	0.83	0.81	0.80	3277

Table 4. RFC classification report

RFC	Precision	Recall	f1-score	Support
Damaged	0.94	0.90	0.92	1600
Undamaged	0.91	0.95	0.93	1677
Average/Total	0.93	0.92	0.92	3277

4.3 Results Interpretation

The results obtained from the ML classifier models are interpreted to conclude. For this, the precision-recall curve, Damage Prediction comparison charts and Accuracy charts are used.

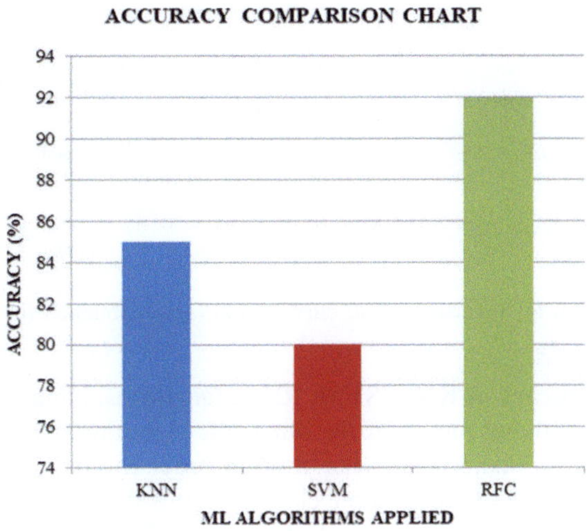

Fig. 5. Comparison of accuracy of ML algorithms.

Comparison of ML algorithms – based on predictions. The precision-recall curve explains how good the classifier model is. For an ideal or best classifier, the precision and recall value will be the one as shown in Fig. 6(a). A precision-recall curve is obtained by connecting the points marked according to each classifier with the value best value of precision and recall, i.e., 1. When comparing the obtained precision-recall curve as shown in Fig. 6(b) for the three ML algorithm model classifiers, Random Forest Classifier is found to be the most efficient classifier than the Support Vector Machine and K-Nearest Neighbour. KNN shows a moderate performance and SVM shows less performance when compared to KNN and RFC.

RFC made a good number of predictions among all the classifiers. Here the number of true positives, where exact damaged predictions were accounted for 1445 out of the total number of observations used for testing the model. Similarly, several true negative predictions, that is undamaged predicted exactly as undamaged was more found in KNN classifier. SVM predictions were intermediate between both KNN and RFC. False positive predictions were found the minimum in the case of KNN as only 27 numbers of undamaged observations were predicted as damaged. Also, false negatives were seen as the minimum for RFC, where the number of damaged samples predicted as undamaged was only 155. Overall, Random Forest Classifier is considered to be a good classifier from the above interpretation.

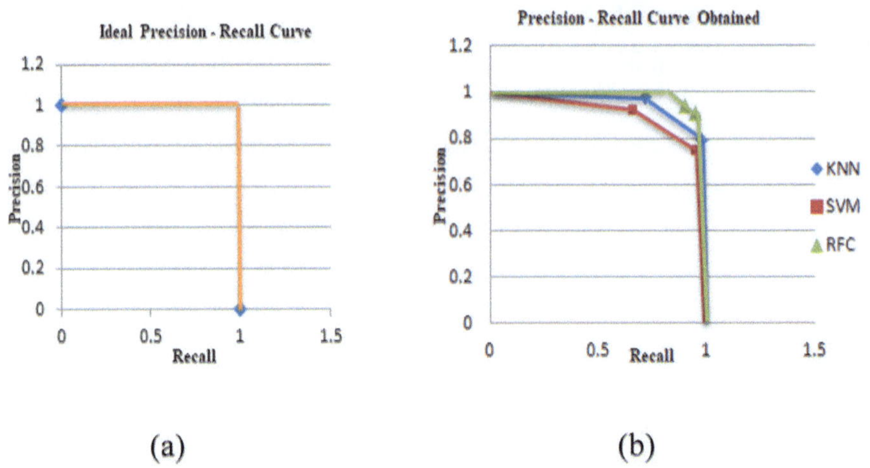

Fig. 6. Precision-recall curve: Ideal (a) v^s Obtained (b).

Comparison of accuracy of KNN, SVM, RFC algorithms. Figure 5 above shows the graph that compares the accuracy of the three ML algorithm classifiers used for this study. Accuracy is the major parameter used in order to compare and evaluate the performance or efficiency of one or more classifiers. It is the ratio of correct predictions to the total number of predictions made by a classifier. Accuracy is expressed in percentage or decimal formats. In Python, the accuracy is generated using the following code.

$$\text{Accuracy} = \text{accuracy\_score}\ (\text{Y\_test},\ \text{predictions})$$

Accuracy is calculated theoretically by using the following formula:

$$\text{Accuracy} = \text{TP} + \text{TN} / (\text{TP} + \text{TN} + \text{FP} + \text{FN})$$

Among the above three classifiers, the RFC gives the highest percentage of accuracy of 0.92 or 92%, where, SVM and KNN showed an accuracy of 80% and 85%, respectively.

5 Summary and Conclusion

In this paper, we have discussed some aspects of Structural Health Monitoring (SHM), primarily the implementation of Machine Learning in SHM in an attempt to improve the damage detection techniques. With the help of relevant literature that describes SHM, we discussed different damage detection methods and methods adopted on civil structures to monitor those. A brief explanation of Machine Learning (ML) techniques and its classification is provided. An experimental setup available in the literature was described in detail in the methodology along with a thorough explanation and implementation of three supervised learning classifiers KNN, SVM and RFC. An existing database from the experiment conducted on a simulated three-storey building model at the Los Alamos National Laboratory (LANL) in partnership with the Laboratory for Concrete Technology and Structural Behaviour (LABEST) of the Faculty of Engineering of the University of Porto (FEUP) on SHM was used to demonstrate the use of ML. Python programming language was used to implement the ML algorithms, and results were analysed using Excel and MATLAB software. According to the results obtained among the three ML classifier models used for damage detection, Random Forest Classifier algorithm generated good predictions on damaged and undamaged conditions with good accuracy, when compared to K-Nearest Neighbours algorithm and Support Vector Machine algorithm under the supervised mode of machine learning.

References

1. Farrar, C.R., Worden, K.: An introduction to structural health monitoring. Philos. Trans. R. Soc. A: Math. Phys. Eng. Sci. **365**(1851), 303–315 (2007)
2. Doebling, S.W., Farrar, C.R., Prime, M.B., Shevitz, D.W.: Damage Identification and Health Monitoring of Structural and Mechanical Systems from Changes in Their Vibration Characteristics: A Literature Review, p. 133. Los Alamos National Laboratory, USA (1996). LA-13070-MS
3. Khatir, S., Dekemele, K., Loccufier, M., Khatir, T., Wahab, M.A.: Crack identification method in beam-like structures using changes in experimentally measured frequencies and particle swarm optimization. Comptes Rendus Mécanique **346**(2), 110–120 (2018)
4. Hoon, S., Charles, R.F.: Damage diagnosis using time series analysis of vibration signals. Smart Mater. Struct. **10**(3), 446–519 (2001)
5. Rosales, M.J., Liyanapathirana, R.: Data-driven innovations in structural health monitoring, In: Proceedings, DAMAS 2017, Kitakyushu, Japan (2017)
6. Dervilis, N.: A machine learning approach to structural health monitoring with a view towards wind turbines. Doctoral dissertation, University of Sheffield (2013)
7. Gillich, G.R., Abdel Wahab, M., Yan, R., Araújo dos Santos, J.V.: Damage models and assessment methods. Shock. Vib. **2016**, 1 (2016)
8. Kiremidjian, A.S., Straser, E.G., Meng, T.H., Law, K., Soon, H.: Structural damage monitoring for civil structures. In: Proceedings of SPIE, 3489, Figure 2, pp. 93–100 (1997)
9. Zhou, Y.L., Abdel Wahab, M.: Damage detection using vibration data and dynamic transmissibility ensemble with auto-associative neural network. Mechanics **23**(5), 688–695 (2017)
10. Fasel, T.R., Sohn, H., Farrar, C.R.: Application of frequency domain ARX models and extreme value statistics to damage detection. In: Smart Structures and Materials, Proceedings of SPIE, vol. SPIE 5057, pp. 145–156 (2003)
11. Hamilton, J.D., Avent, S.R.: Time series analysis. Water Resour. Res. **3**(3), 1–85 (2005)
12. Hamilton, J.D.: Time Series Analysis. Princeton University Press, Princeton, NJ (1994)
13. Jayawardhana, M., Zhu, X., Liyanapathirana, R.: An experimental study on distributed damage detection algorithms for structural health monitoring. J. Phys: Conf. Ser. **305**, 12068 (2011)

14. Vitola, J., Tibaduiza, D., Anaya, M., Pozo, F.: Structural damage detection and classification based on machine learning algorithms. In: 8th European Workshop on Structural Health Monitoring, pp. 5–8 (2016)
15. Nick, W., Asamene, K., Bullock, G., Esterline, A., Sundaresan, M.: A study of machine learning techniques for detecting and classifying structural damage. Int. J. Mach. Learn. Comput. **5**(4), 313–338 (2015)
16. Ayodele, T.: Machine Learning Overview. https://doi.org/10.5772/9374 (2010)
17. Jaime, G., Carbonell, R.S.: Machine learning: a historical and methodological analysis. Assoc. Adv. Artif. Intell. **4**(3), 1–10 (1983)
18. Farrar, C.R., Cornwell, P.J., Doebling, S.W., Prime, M.B.: Structural Health Monitoring Studies of the Alamosa Canyon and I-40 Bridges, vol. LA-13635-M, pp. 1–170 (2000)
19. Gillich, G.R., Furdui, H., Wahab, M.A., Korka, Z.I.: A robust damage detection method based on multi-modal analysis in variable temperature conditions. Mech. Syst. Signal Process. **115**, 361–379 (2019)
20. Figueiredo, E., Park, G., Figueiras, J., Farrar, C.R., Worden, K.: Structural Health Monitoring Algorithm Comparisons Using Standard Datasets. Los Alamos National Laboratory Report LA-14393 (2009)
21. Jayawardhana, M., Zhu, X., Liyanapathirana, R.: Damage detection of reinforced concrete structures based on the Wiener filter. Aust. J. Struct. Eng. **14**, 57–70 (2013)
22. Behtani, A., Bouazzouni, A., Khatir, S., Tiachacht, S., Zhou, Y.L., Wahab, M.A.: Damage localization and quantification of composite stratified beam structures using residual force method. J. Phys.: Conf. Ser. **842**(1), 12028 (2017). IOP Publishing
23. Mohammed, M., Badruddin Khan, M., Bashier, E.: Machine Learning: Algorithms and Applications. CRC Press, Boca Raton (2016)
24. Khamis, H.S., Cheruiyot, K.W., Kimani, S.: Application of k-nearest neighbour classification in medical data mining. Int. J. Inf. Commun. Technol. Res. **4**(4) (2014)
25. Kataria, A., Singh, M.D.: A review of data classification using k-nearest neighbour algorithm. Int. J. Emerg. Technol. Adv. Eng. **3**(6), 354–360 (2013)
26. Imandoust, S.B., Bolandraftar, M.: Application of k-nearest neighbour (knn) approach for predicting economic events: theoretical background. Int. J. Eng. Res. Appl. **3**(5), 605–610 (2013)
27. Chao, C.F., Horng, M.H.: The construction of support vector machine classifier using the firefly algorithm. Comput. Intell. Neurosci. **2015**, 2 (2015)
28. Demidova, L., Nikulchev, E., Sokolova, Y.: Big data classification using the SVM classifiers with the modified particle swarm optimization and the svm ensembles. Int. J. Adv. Comput. Sci. Appl. (IJACSA) **7**(5), 294–312 (2016)
29. Kulkarni, V.Y., Sinha, P.K.: Effective learning and classification using random forest algorithm. Int. J. Eng. Innov. Technol. (IJEIT) **3**(11) (2014)
30. Ali, J., Khan, R., Ahmad, N., Maqsood, I.: Random forests and decision trees. IJCSI Int. J. Comput. Sci. Issues **9**(5), 272–278 (2012)

Bridge Monitoring Using Geophones: Test and Comparison with Interferometric Radar

Gabriele Corsi[1], Ferdinando Frediani[1], Lapo Miccinesi[2(✉)], Michelangelo Micheloni[3], and Massimiliano Pieraccini[2]

[1] Move Solutions, Move S.r.l., Milan, Italy
[2] Department of Information Engineering, University of Florence, Florence, Italy
lapo.miccinesi@unifi.it
[3] Studio Micheloni S.R.L., Florence, Italy

Abstract. More than 30% of European highway bridges present structural criticalities. Their continuous health monitoring is a priority. Conventional sensors are accurate and reliable, but they are often too expensive for continuous monitoring. Possible low-cost alternative equipment are the geophones, that are able to detect the displacement of structures by integrating in time their response. They can be easily installed and can provide continuous health monitoring by transmitting the data to a remote server.

The aim of this paper is to assess the capability of a geophone sensor to provide continuous, accurate and reliable data about the dynamic loads of a bridge. In order to validate its performance, it has been experimentally compared with an interferometric radar. As preliminary test the geophone has been fixed to a horizontal steel plate. The radar has been positioned under the steel plate in order to detect the same displacement component. The steel plate has been excited with controlled pulses. Finally a network of geophone sensors, provided with its control and transmission electronics, has been installed on a well-known bridge in Florence, Italy (the "Amerigo Vespucci" bridge). The interferometric radar has been installed under the deck close to an abutment. The obtained results both in controlled environment and in the in-field test are in good agreement, although the geophone appears less sensible to impulsive stimulus than the radar.

Keywords: Geophone sensor · Vespucci bridge · Interferometric radar · Bridge monitoring

1 Introduction

The recent Morandi bridge disaster (Genova, Italy) increased attention on large structures health problems. Indeed about 30% of European highway bridges present structural criticalities [1]. Moreover the age of many bridges is more than 60 years, as an America study of 2013 highlights [2]. In this context continuous monitoring of dynamic properties of bridges is a priority and a challenge for civil and electronic engineers.

M. A. Wahab (Ed.): DAMAS 2019, LNME, pp. 25–34, 2020.
https://doi.org/10.1007/978-981-13-8331-1_2

The conventional sensors (radar, total station, accelerometer) are accurate and reliable, but often they are too expensive for continuous monitoring. Furthermore a sensor network can be more appropriate to monitor the whole structure [3–6].

The aim of this paper is to assess the capability of a geophone sensor to provide continuous, accurate and reliable data about the dynamic loads of a bridge. This device is able to detect the displacement of structures by integrating in time their response and it can operate in a network by sending the measurement results of different devices to a remote server.

In detail a particular device made with an analog compensated geophone has been created by the joint venture between Move Srl and Studio Micheloni Srl.

The performance of the geophone sensor has been experimentally compared with an interferometric radar [7–10] in a controlled environment and during a test on Vespucci bridge, Florence, Italy.

2 The Geophone Sensor

A block scheme of the geophone sensor is shown in Fig. 1. The geophone is able to measure the speed-changing of a structure in the gravity direction. Geophone is a passive sensor based on an inductance. It can measure speed (dx/dt) by measuring the voltage (V) in according to Eq. 1.

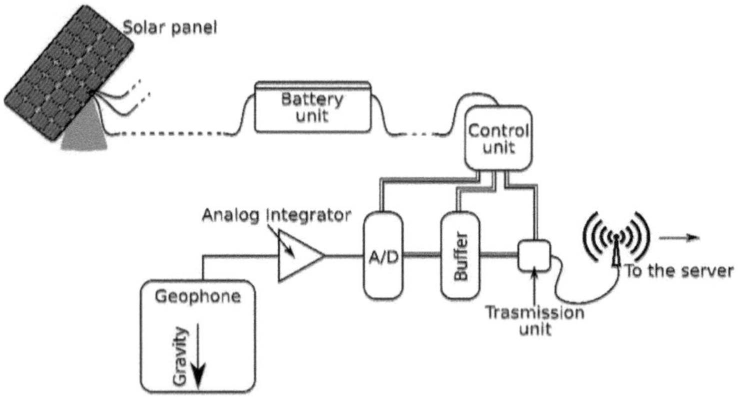

Fig. 1. Block scheme of geophone sensor

$$V = S_g \frac{dx}{dt} \tag{1}$$

Where S_g is the sensitivity, $[V/(m/s)]$, and it is the proportional constant between voltage and speed and it depends on the specific device. For the geophone selected the sensitivity is $16.5V/(m/s)$ [11].

The response of geophone (V) is integrated using an analog integrator in order to obtain the displacement. The analog integrator compensated the geophone response in order to obtain a flat band for the displacement form 1 Hz to 20 Hz. The displacement is digitized by a 10-bit A/D converter.

The control unit provides a 100 Hz clock to A/D converter and it manages the buffer and the communication with the server.

The device is able to send data to a remote server via LoRaWAN network or via USB cable. A battery unit provides the power supply for each device and it ensures an operating time of one/two weeks. The battery unit can be recharged by a solar panel for operating indefinitely.

The device can work in debug-mode or threshold-mode. In the debug-mode, the device is able to provide a continuous measurement and it sends all recorded data via USB. The debug-mode can be used for laboratory test or in a controlled environment. The threshold-mode is used for in-field operations. In this case the control unit checks the last value of the measurement and if it is larger than a fixed threshold, it sends a packet with 40 s data (20 s before and 20 s after threshold) to a remote server. The threshold can be adjusted for different applications. The server is able to manage thousands of devices installed also in different structures and the final user can access to the main data of the structure of interest.

The performance of geophone sensor has been compared with an interferometric radar. A radar is a remote sensor able to detect targets in its field of view by sending and receiving an electromagnetic wave. The interferometric radar is able to measure small displacements by measuring the differences of phase between the sending and receiving electromagnetic wave. The radar operated a Continuous Wave – Step Frequency signal in K_u band and it allows to detect the natural frequencies in the band from DC to more than 100 Hz, by applying the Fourier transform to the displacement [7, 8]. The response of interferometric radar is flat in all bandwidth. Table 1 reports a brief comparison between the performance of both sensors.

Table 1. Comparison between performance of geophone and interferometric radar.

	Geophone sensor	Interferometric radar
$f_{Acquisiton}$	100 Hz	(25 – 300) Hz
f_{min}	1 Hz	DC
f_{max}	20 Hz	> 100 Hz
Acquisition time	40 s	> 1 h

3 The Amerigo Vespucci Bridge, Florence, Italy

The Vespucci bridge was built in 1957 after WWII. The bridge replaces an older bridge destroyed by German mines and its design win a national contest for its originality and innovation. During 2017 some strong deteriorations were found on the structure, with particular reference to erosion under one pier (Fig. 2). For this reason, in 2018 the

Municipality of Florence commissioned an extensive monitoring campaign and possible restoration works.

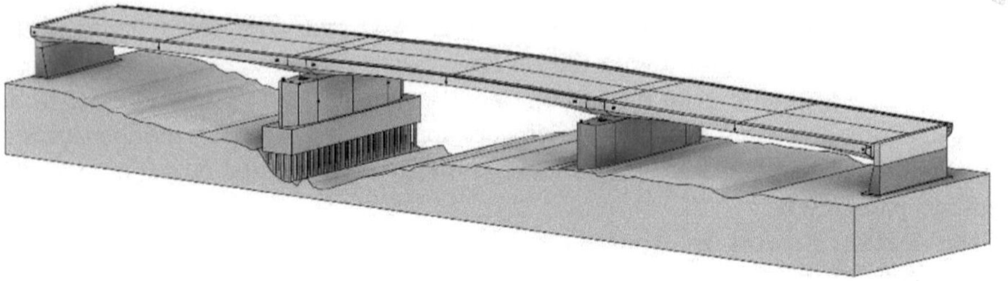

Fig. 2. Design of Vespucci bridge with erosion under left pier

The bridge was built with prestressed reinforced concrete and it is composed by three spans over two piers that are holding a four-carriage roadway. Each span is a flat arch designed to do not interfere with the skyline of the old town.

In order to reinforce the arches a counterweight system was inserted the structures. As Fig. 3 shows a comb of cables was included in each arch and it is plugged inside the piers. This comb stiffs the structure during the static load and changes the dynamic properties of the bridge and it introduces slow relaxing movements.

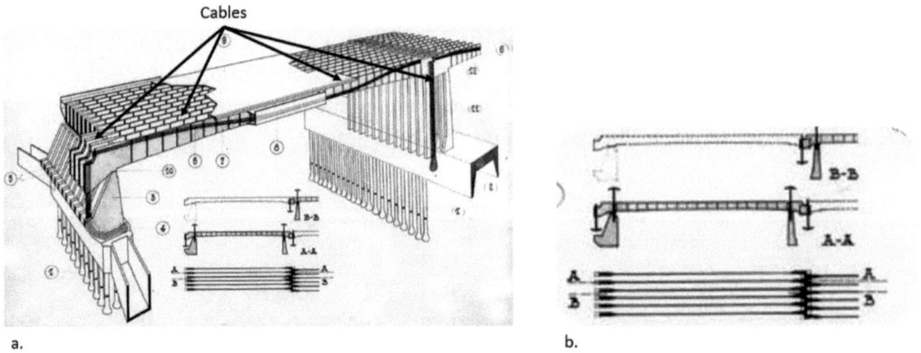

Fig. 3. Design of counterweight system of Vespucci bridge (a). Detail of comb of cables (b) [12].

Given this particular and original configuration of the bridge structure, it is very important to adopt an accurate monitoring system, which can correctly capture the dynamic behavior of the bridge.

4 Experimental Results

The geophone sensor and the interferometric radar were preliminary compared in a controlled environment and finally during a dynamic test of Vespucci bridge, Florence, Italy.

4.1 Test in Controlled Environment

As preliminary test a geophone sensor in debug-mode has been fixed on a steel plate (Fig. 4). The steel plate was 2 m long, 0.5 m large and 2.5 mm height. It was cantilevered at 3 m height over a wall. The interferometric radar was positioned under the steel plate in order to detect the same displacement component of the geophone. The steel plate vibrated exited by short pulse stimulus.

Fig. 4. Experimental setup of controlled environment

Figure 5 reports the comparison between the displacements measured by both sensors. Radar appears to be more sensitive to the impulsive stimulus than geophone, indeed the two displacements become more similar after 1 s by the stimulus.

The natural frequency of steel plate was measured with both sensors by calculating the Fast Fourier Transform of the displacement with a padding factor equal to 100. Furthermore, random decrement technique (RDT) [13, 14] has been applied to displacement in order to retrieve the natural spectrum, but the results were not improved. The comparison between the spectra is showed in Fig. 6. The frequency measured was (2.131 ± 0.007) Hz for both sensors.

4.2 In-Field Test: Vespucci Bridge, Florence, Italy

A network of six geophones, operating in threshold-mode, was installed under the spans of Vespucci bridge in order to provide continuous monitoring during the renovations works. When the works will be finished, the sensors will allow to monitor the

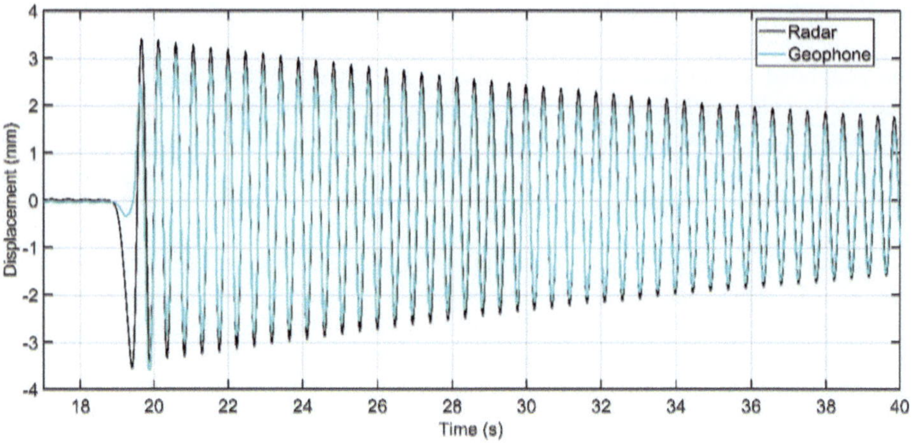

Fig. 5. Comparison between displacements measured by interferometric radar (Black) and by geophone system (Cyan).

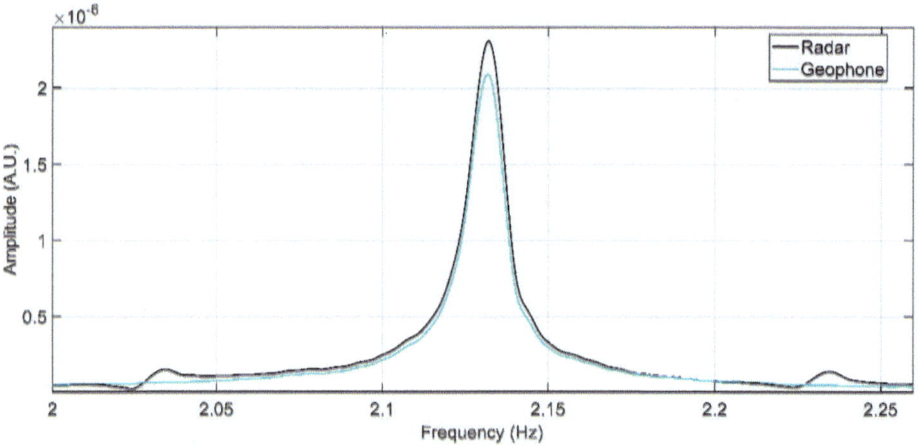

Fig. 6. Comparison between frequencies measured with interferometric radar (Black) and geophone system (Cyan).

bridge under the road live load. One single geophone has been used for the comparison with the radar.

With reference to Fig. 7 the geophone was located under the south span on the right side of carriageway. A standard target (corner reflector) for the interferometric radar was fixed close to the geophone. The radar was installed on a concrete platform 3.15 m under the bridge. A 10000 kg truck was exploited as stimulus to test the bridge. The truck went up on a 0.2 m step. When the track dropped from the step, the bridge was excited by an impulsive stimulus.

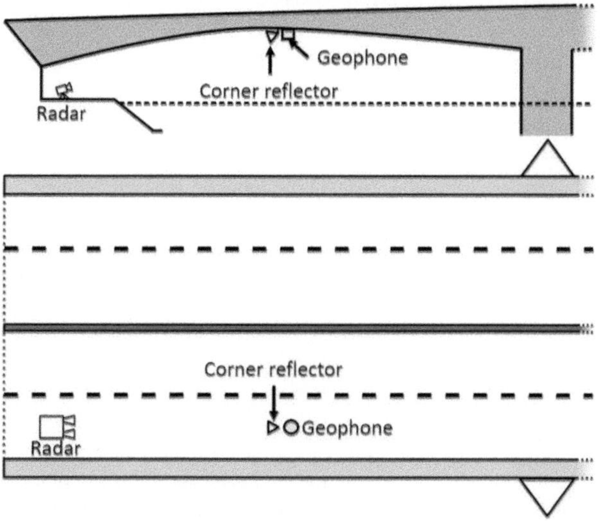

Fig. 7. Measurement setup during the dynamic test of Vespucci bridge.

The displacement measured by the radar is shown in Fig. 8. This displacement has been projected on vertical axis to be comparable with the geophone measurement. The comparison between geophone and radar is shown in Fig. 9. As Figs. 8 and 9 show, the radar has been able to detect the slow movement of the bridge. The approaching of the truck excited the modes relative to the counterweight system that induced this slow movement.

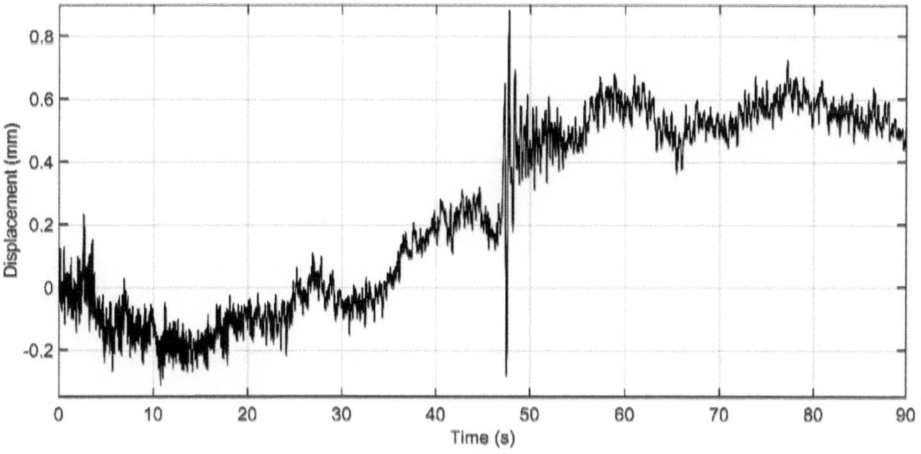

Fig. 8. Displacement measured by interferometric radar.

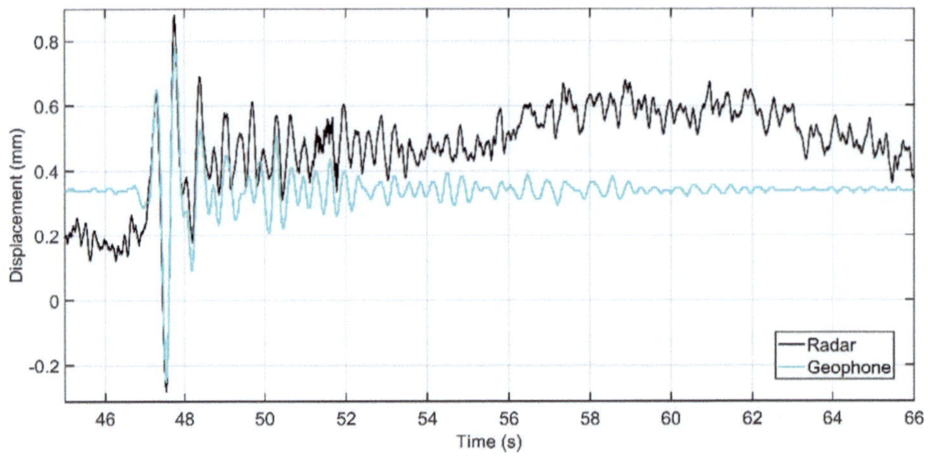

Fig. 9. Comparison between displacements measured by the interferometric radar (Black) and the geophone system (Cyan) zoomed on the stimulus by using their own bandwidths.

It is important to note that this comparison was performed by using their own bandwidths of both sensors. Comparison by using the same band can be more appropriate. Therefore, a high-pass filter ($f_{LOW} = 0.5\,\mathrm{Hz}$) was applied to both the measurements (Fig. 10). The two plots appear to be in good agreement.

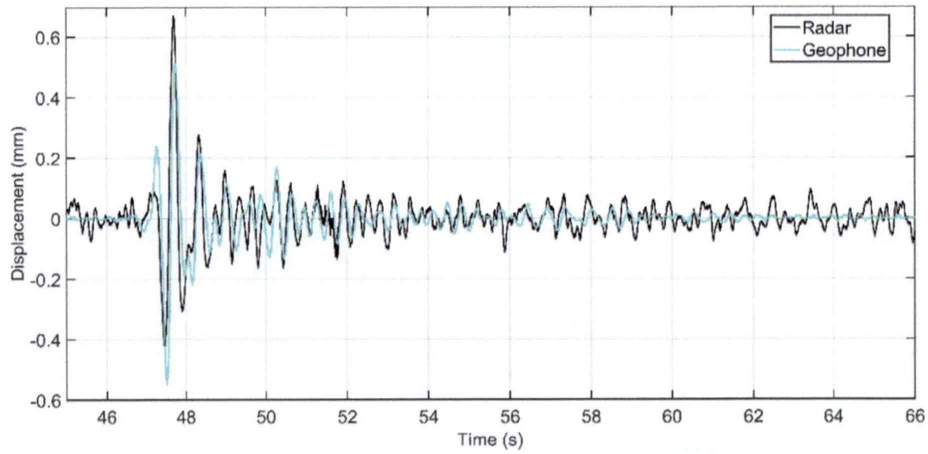

Fig. 10. Comparison between displacements in a common bandwidth of interferometric radar (Black) and geophone system (Cyan).

The spectra of the two sensors are shown in Fig. 11. The frequencies detected by radar have been $f_1 = (2.37 \pm 0.05)\,\mathrm{Hz}$, $f_2 = (3.12 \pm 0.05)\,\mathrm{Hz}$. The frequencies detected by geophone have been $f_1 = (2.40 \pm 0.05)\,\mathrm{Hz}$, $f_2 = (3.11 \pm 0.05)\,\mathrm{Hz}$ and

$f_3 = (2.78 \pm 0.05)$ Hz. The frequencies f_1 and f_2 are in good agreement, while f_3 is not detected by interferometric radar.

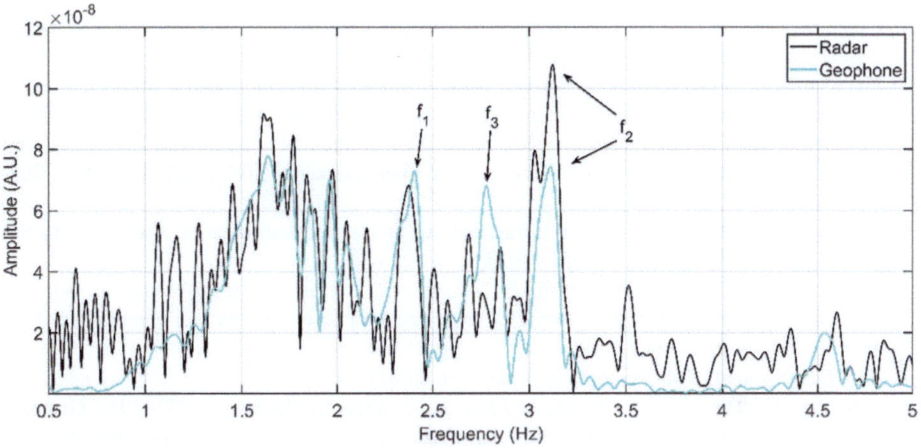

Fig. 11. Natural frequencies of Vespucci bridge measured with radar (Black) and geophone (Cyan)

5 Conclusions

A geophone sensor has been tested and its performance has been experimental compared with an interferometric radar. The tests were performed both in a controlled environment and during the dynamic test of the Vespucci bridge, Florence, Italy. The displacements measured are in good agreement for the two sensors both in controlled environment and during an in-field test, as well as the natural frequencies.

The geophone appears to be less sensitive to the impulsive stimulus. However, it can be used in most cases and it can provide continuous, and accurate monitoring of large structures.

References

1. Technical Committee 11 Bridges and Other Structures: Reliability-Based Assessment of Highway Bridges (1999), web site
2. American Road & Transportation Builders Association: Over 54,000 American Bridges Structurally Deficient, Analysis of New Federal Data Shows (2018), web site
3. Chae, M.J., Yoo, H.S., Kim, J.Y., Cho, M.Y.: Development of a wireless sensor network system for suspension bridge health monitoring. Autom. Constr. **21**(1), 237–252 (2012)
4. Kim, S., et al.: Health monitoring of civil infrastructures using wireless sensor networks. 2007 6th International Symposium on Information Processing in Sensor Networks, pp. 254–263. Cambridge, MA (2007)

5. Basharat, A., Catbas, N., Shah, M.: A framework for intelligent sensor network with video camera for structural health monitoring of bridges. Third IEEE International Conference on Pervasive Computing and Communications Workshops, pp. 385–389. Kauai Island, HI (2005)
6. Guan, S., Bridge, J.A., Li, C., DeMello, N.J.: Smart radar sensor network for bridge displacement monitoring. J. Bridge Eng. **24**(1), January (2019)
7. Pieraccini, M., Fratini, M., Parrini, F., Atzeni, C.: Dynamic monitoring of bridges using high-speed coherent radar. IEEE Trans. Geosci. Remote Sens. **44**(11), 3284–3288 (2006)
8. Pieraccini, M., Parrini, F., Fratini, M., Atzeni, C., Spinelli, P., Micheloni, M.: Static and dynamic testing of bridges through microwave interferometry. NDT and E Int. **40**(3), 208–214 (2007)
9. Pieraccini, M.: Monitoring of civil infrastructures by interferometric radar: a review. The Sci. World J. **2013**, Article ID 786961 (2013)
10. Pieraccini, M., Fratini, M., Parrini, F., Atzeni, C., Bartoli, G.: Interferometric radar vs. accelerometer for dynamic monitoring of large structures: an experimental comparison. NDT & E Int. **41**(4), 258–264 (2008)
11. ION Geophysical: SM-7 Geophone datasheet (2001)
12. Fabbrizzi, F.: The Vespucci bridge in Florence. A street on the river, 1953–1957 Giuseppe Giorgio Gori, Enzo Gori, Ernesto Nelli, Riccardo Morandi. In: Fupress, Firenze architettura (1), 88–95 (2018)
13. Ibrahim, S.R.: Random decrement technique for modal identification of structures. J. Spacecraft and Rockets **14**(11), 696–700 (1977)
14. Cheynet, E.: Modal parameters identification from ambient vibrations https://www.mathworks.com/matlabcentral/fileexchange/55557-modal-parameters-identification-from-ambient-vibrations (2019), MATLAB Central File Exchange. Retrieved March 22, 2019

A Damage Detection Method by Transient Damping Feature Based on Monitoring Data

Ayaho Miyamoto(✉)

Yamaguchi University, Ube, Japan
miya818@yamaguchi-u.ac.jp

Abstract. This paper introduces the details of a transient damping feature extracting method based on the bridge monitoring data with the aid of the frequency slice wavelet transform (FSWT) which is a time-frequency space analysis tool and its application to damage detection of the bridges. The proposed method via the state representation methodology (SRM) algorithm has been verified by the state probably density distribution using the vibration signal from a laboratory bridge monitoring system. As a result, the system state vector as a non-parametric state variable will be able to apply to assess the state change (damage) of the bridge structures by expression of the state probably density distribution.

Keywords: Bridge health monitoring · Vibration signal · Transient damping · System state representation (SRM) · Damage detection · State probably density distribution

1 Introduction

Lifetime engineering conjunction with the structural health monitoring system (SHMS) for the civil infrastructures, such as bridges, highway networks, is becoming one of the most important issues in all over the world [1, 2]. Also it becomes an important thing that how to detect damages from such huge number of monitoring data, that is a big challenge for analysis to discover the damage information of a target structure. The author has been developing a new concept of the state representation methodology (SRM) combined with the frequency slice wavelet transform (FSWT), which is a new time–frequency space analysis tool, is proposed for assessing a bridge condition based on bridge monitoring data [3–5]. It presents a general idea for non-parametric description of system state, and includes how to define system state or record the variable of system state, and express the system state by a state representation equation (SRE). The SRE can be approximated similarly by support vector machines (SVM) methods [6]. Before using the SRM, it is necessary that the system features need to be extracted from the system responses. Consequently, a new time-frequency space analysis tool called FSWT is introduced. The FSWT has many new properties in

A. Miyamoto—Visiting Prof., Swiss Federal Institute of Technology (EPFL), Lausanne, Switzerland.

M. A. Wahab (Ed.): DAMAS 2019, LNME, pp. 35–50, 2020.
https://doi.org/10.1007/978-981-13-8331-1_3

contrast of traditional wavelet [6, 7]. The FSWT analysis will powerfully reveal the characteristics of vibration signal. This paper proposes a transient damping as a new concept by applying a modal damping function (MDF) to the analytical results thus obtained, amplitude envelope ($x = Ae^{-\alpha t}$) coefficients, α are extracted, and feature vectors are constructed. With the aid of FSWT analysis, a new feature extracting method about the transient damping is introduced for SRM assessment method and is applied to the vibration response obtained from a laboratory bridge monitoring system. Experiments show the SRM features are sensitive and stable when its condition is changed. Figure 1 shows the procedure for detecting damage by expressing a damage-induced change with an SRM-based probabilistic model. In the SRM application, a non-parametric condition state variable (system state; feature), ζ is defined as $\zeta \in [0, 1]$ or $\zeta \in (0, 1)$, which is a function of its state space:

$$\zeta = f(\lambda, \omega) = 1 \tag{1}$$

where, λ is the system structure alias parameters vector, ω is the system response alias features vector, ζ is called the system state, that is, $\zeta = 1$ means that the system is in normal (no-damage) state, otherwise $\zeta \neq 1$ means that the system is changing (deteriorating).

In here, the determination of the state variable depends on the system structure parameters and the system response features. The main response feature of a system is its frequency spectrum and modal parameters. However, it is usually difficult to know each modal frequency exactly. At the same time, the response of a complex system

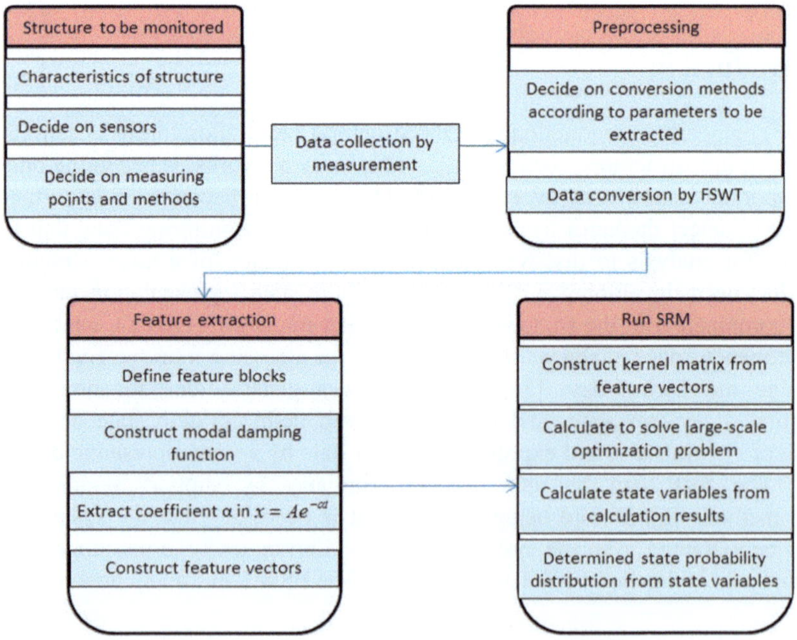

Fig. 1. Flow of damage detection using SRM.

always have quite many close modal frequencies and high frequencies, and so it is very difficult to separate them from the time and frequency domains. Therefore, we have to give up the exact feature vectors instead with the approximation or new feature vectors. Naturally, we expect that the new features can reveal the characteristics of the original system. For example, the damping or damping (ratio) vector of each modal frequency is an important feature. Therefore, the SRE can be expanded as,

$$\zeta = f(\lambda, \omega, \zeta) \tag{2}$$

where, ξ is the damping (ratio) alias vector of each modal frequency.

This paper introduces the details of a transient damping feature extracting method based on the bridge monitoring data with the aid of the frequency slice wavelet transform (FSWT) which is a time-frequency space analysis tool and its application to damage detection of the bridges. The proposed method via the state representation methodology (SRM) algorithm has been verified by the state probably density distribution using the vibration signal from a laboratory bridge monitoring system. As a result, the system state vector as a non-parametric state variable will be able to apply to assess the state change (damage) of the bridge structures by expression of the state probably density distribution.

2 FSWT Analysis Tool and Applications

A new time–frequency space analysis tool which called the frequency slice wavelet transform (FSWT) analysis tool has been developed as a pre-processing for damage detection based on a large amount of monitoring (vibration) data. The FSWT analysis makes it possible not only to accurately extract peaks in the time–frequency space but also to be superior to the other methods in resistance to noise.

Suppose $\hat{f}(u)$ and $\hat{p}(\omega)$ are the Fourier transform (FT) of the functions, $f(t)$ and $p(t)$, respectively. For any $f(t) \in L^2(R)$, the frequency slice wavelet transform (FSWT) is defined directly in the frequency domain as:

$$W_f(t, \omega, \kappa) = \frac{1}{2\pi} \int_{-\infty}^{+\infty} \hat{f}(u) \hat{p}^*(\kappa \frac{u - \omega}{\omega}) e^{iut} du \tag{3}$$

where, the scale κ is a constant or a function of ω, t and u, and the star '*' means the conjugate of a function. Here, $\hat{p}(\omega)$ is called a frequency slice function (FSF). More details of the FSWT analysis can be referred in Refs. [3, 8].

As a simple example, Fig. 2(a)–(h) show the results of FSWT analysis tool application to the monitoring (vibration) data (acceleration) by the impact hummer tests on a simply supported girder (laboratory) bridge model with three main girders [5] as shown in Fig. 3, which is the minimum number of girders needed to obtain necessary information such as load distribution characteristics in the direction perpendicular to the bridge axis for explaining the feature extract algorithm. In here, the left top, Fig. 2 (a) shows an original acceleration signal, and the right top, Fig. 2(b) shows its Fourier (FFT) spectrum. Figure 2(c) and (d) are the 2D and 3D maps of FSWT analysis for the original signal, respectively. Figure 2(e), (f) and (g) are the FSWT reconstructed signal (three segmented signals), and the right bottom Fig. 2(h) shows their synthesized signal. From the comparison of these results with FFT analysis, it is found that the FSWT analysis tool will be able to reveal clearly the characteristics (damping, etc.) of the measured (monitoring) signal in a time-frequency space.

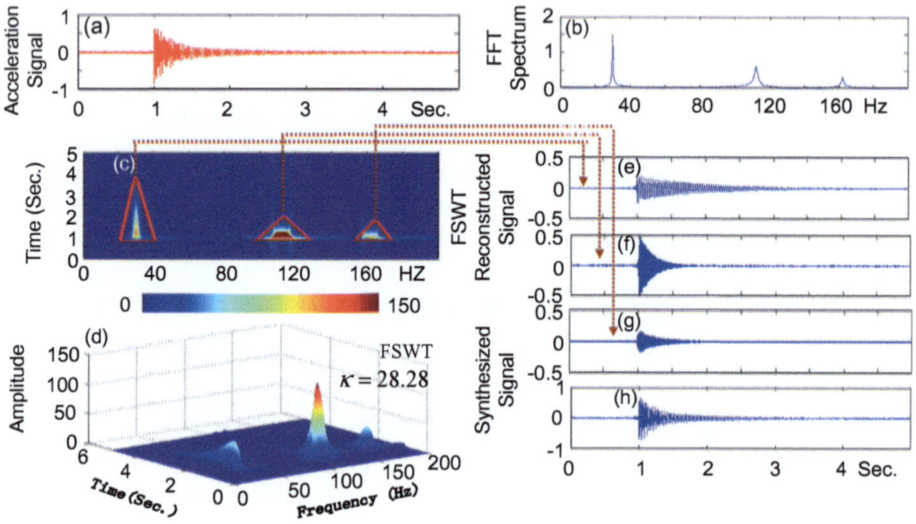

Fig. 2. Example of time–frequency space analysis results obtained by FSWT analysis.

Fig. 3. General view of a three-girder (laboratory) bridge model for impact hammer tests.

As another example, Fig. 4(a)–(i) show the time-frequency images of FSWT analysis tool application to the original signal (acceleration) from a double (twice) impact hammer test on the laboratory bridge model, compared with FFT spectrum. In here, the left top, Fig. 4(a) shows the original acceleration signal of double impact, and the right top, Fig. 4(d) shows its FFT spectrum. Figure 4(b) and (c) are 2D and 3D map of FSWT analysis. Figure 4(e)–(h) are the reconstructed signals, and Fig. 4(i) is their synthesized signal.

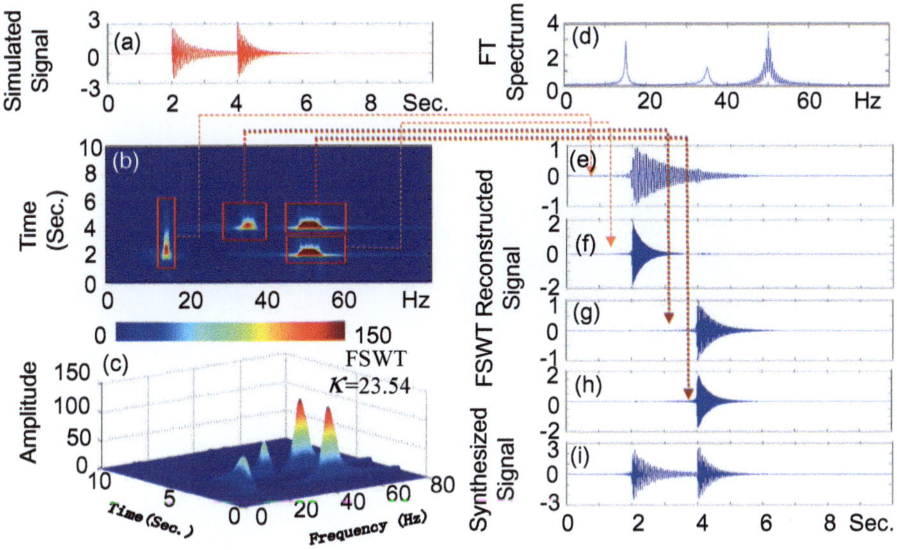

Fig. 4. Example of FSWT time–frequency space analysis results for twice impact hammer test data and their comparison.

Based on these results, it will be summarized as follows:

(1) FSWT analysis result has four separated peaks obviously as shown in Fig. 4(b) and (c),
(2) FSWT is able to show that two groups of components are at the same impacting time with different frequency, and two components are in same frequency at the different time,
(3) On the other hand, the Fourier(FFT) spectrum can only show the different frequency signals.

Furthermore, for discussing the noise filtering effects in time and frequency domain of FSWT analysis, Fig. 5(a)–(e) show the results of FSWT analysis tool application to the monitoring data by the single-girder moving load tests [5] which simulated a vehicle is made to move on a simply supported girder. In here, Fig. 5(a) and (b) show an original signal including noise and its Fourier (FFT) spectrum, respectively. And Fig. 5(c) and (d) are 2D and 3D maps of FSWT analysis. Figure 5(e) is the reconstructed signal by parts of FSWT analysis which are marked with the red rectangles in Fig. 5(c). As the results, by using FSWT analysis tool, it will be able to make very clearly visualization in 2D and 3D maps that there are four peaks at the different time and in near frequencies, even some noises can be seen on the FFT spectrum. Especially, the reconstructed signal in time domain shown in Fig. 5(e) is almost the perfect for reducing noises and obtaining their high accurate modal parameters.

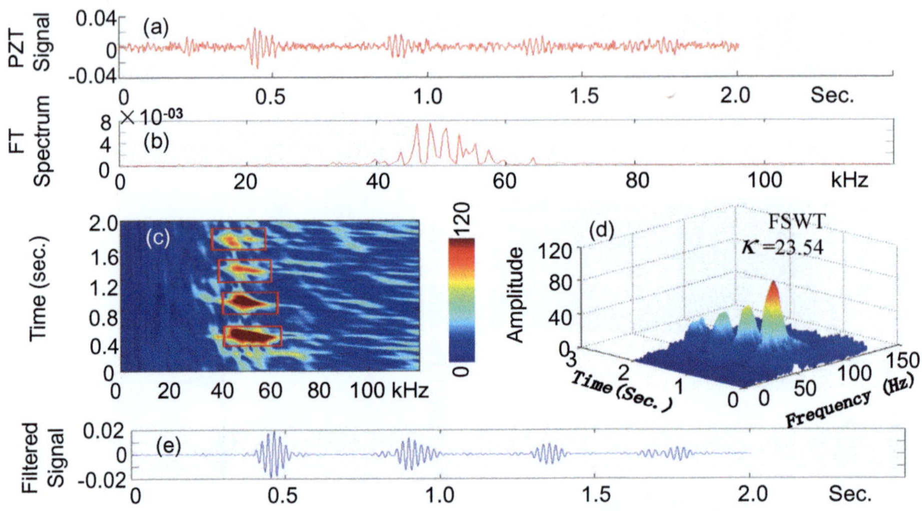

Fig. 5. Example of FSWT time–frequency space analysis results for moving load test data including noise and their comparison.

Finally, because the application results of FSWT analysis tool which defined as Eq. (3) in the frequency domain depend on the frequency slice function (FSF), $\hat{p}(\omega)$, we need to verify that which type of FSF has good advantage for extracting the time-frequency resolution and amplitude response. Figure 6(a)–(e) show the comparison of FSWT analysis results in time–frequency space using a sample signal as shown in Fig. 6(a) with the four FSF functions. In here, it is found that Fig. 6(b) and (c) exist the short time-frequency smearing, and Fig. 6(d) has the frequency band energy leakage problem because of an infinite support width function. Then, from the better localization property of FSF point of view, we applied the Gaussian function, $\hat{p}(\omega) = e^{-\frac{1}{2}\omega^2}$ to the FSWT analysis tool in this study.

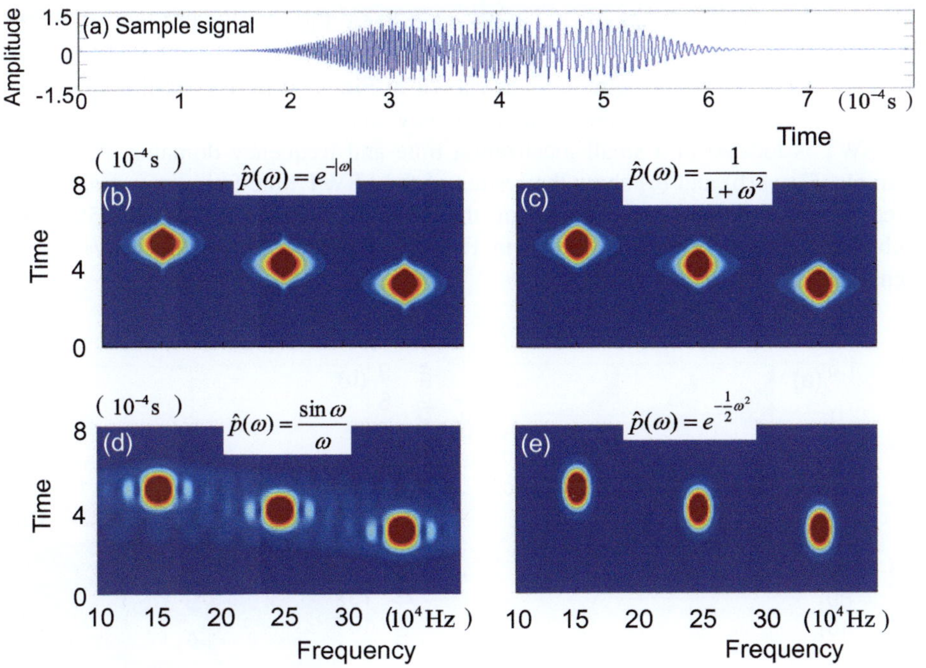

Fig. 6. Comparison of the frequency slice function (FSF) effects on FSWT time–frequency analysis.

3 Modal Transient Damping Feature Analysis and Feature Extract Algorithm

This session focuses on the modal signal separation and damping parameter identification using in damage detection based on the results of FSWT analysis tool application to the structural health monitoring data (big data). On the main application of FSWT analysis to damage detection in the bridges, an accurate estimation method is necessary to develop both in time and frequency domains for single modal damping identification (damping ratio of each modal frequency). In here, with the aid of FSWT, a new feature extracting method about transient damping, α is introduced for SRM assessment (damage detection) method and is applied to the vibration response obtained from a laboratory bridge monitoring system.

3.1 FSWT Envelope Analysis

The features of a damping vibration signal include its frequency and damping ratio. Figure 7(a) shows a simulated sample damping vibration signal (acceleration) which has been discussed in Session 2. The FFT spectrum shown in Fig. 7(b) is easy to know the frequency feature of it. However, it is usually not easy to get the damping ratio exactly because in general a damping vibration signal always includes some interference frequencies and noise. On the other hand, Fig. 7(c) and (d) reveal that the FSWT analysis makes three clear separated peaks with different time domain. And the envelope curves of it are quite clear as shown in Fig. 7(c) by FSWT time-frequency transform. A result of the FSWT time-frequency analysis ensures that the distribution of FSWT is located in a small localization time and frequency domains. As another example, Fig. 8(b) and (c) show the results of the FSWT time-frequency analysis and three transient damping envelopes of the different frequencies for a 25% noise including vibration signal as shown in Fig. 8(a) (compared with the above sample signal).

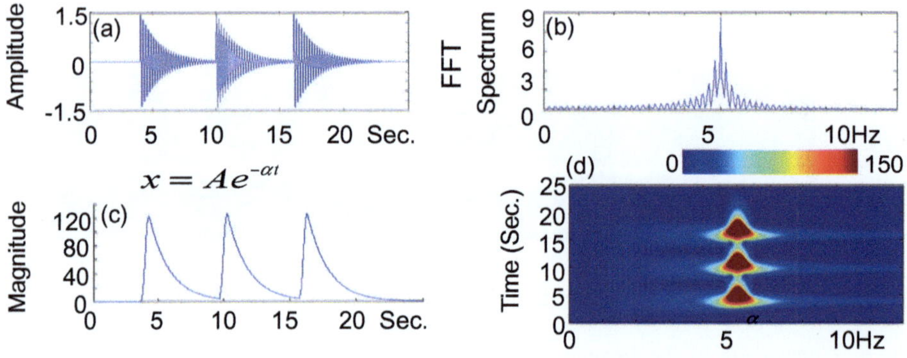

Fig. 7. Example of FSWT time–frequency analysis result for sample vibration data and its envelope curve.

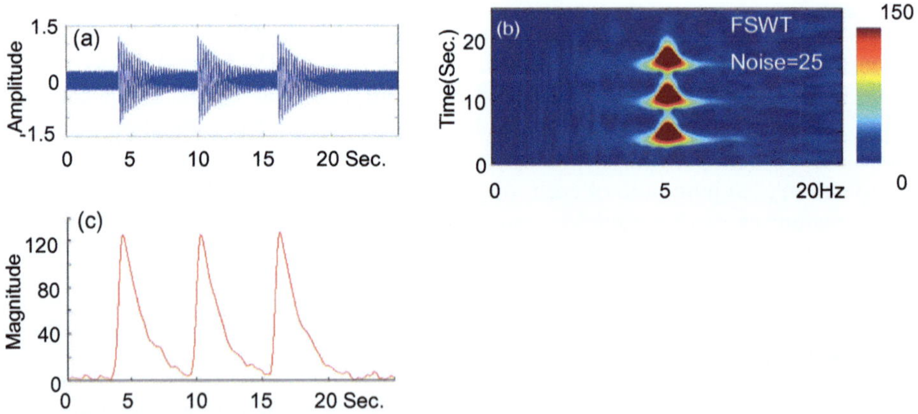

Fig. 8. Example of FSWT time–frequency analysis for sample vibration data including 25% noise and its envelope curve.

In order to simplify the computation of transient damping features and ensure their stabilities, a new damping method via FSWT analysis and statistical methods is presented in the following:

If $x = Ae^{-\alpha t}$, when $t \in [0, T] \subset (0, +\infty)$, let

$$S(x^n) = \frac{1}{T}\int_0^T x^n dt, \ n = 0.5,\ 1,\ 2,\ 4 \tag{4}$$

It is obvious that $S(x^n)$ are easily computed from the time data x. Then the damping parameter, α can be calculated by:

$$\alpha = \frac{4\sqrt{S^2(\sqrt{x})S(x^2) - S^3(x)}}{(S^2(\sqrt{x})\sqrt{S(x)})/T} = MDF(x) \tag{5}$$

$$\bar{\alpha} = \frac{1}{M}\sum\nolimits_{i=1}^{M} \alpha_i \tag{6}$$

According to Eq. (5), a statistic formula is established to compute the modal damping. It is significant to note that $x = Ae^{-\alpha t}$ is an envelope curve of a damping vibration signal. The modal damping function (*MDF*) is commonly denoted as,

$$\alpha = MDF\ \ (x) \tag{7}$$

Here, the *MDF* can be used any other equations [5], and also to get the damping parameter from the damping vibration signal.

3.2 Damping Feature Extract Algorithm

In a real bridge system, the system response may be stochastic. The system must make some responses when it is excited by environment conditions such as cars, winds, earthquake etc. Therefore, some trigger conditions ever exist. We call the time of a trigger condition as the Time Trigger Lines (TTLs). Now we take a FSWT image as shown in Fig. 2(c) (marked with red triangles) as an example to explain the feature extract algorithm.

The steps for the feature extraction algorithm are as follows:

STEP 1: Compute the FSWT expression for each sensor response, and denote it by $W_i(t, \omega, \alpha),\ i = 1, 2, \ldots N_s$, where N_s is the number of sensors.

STEP 2: Take the maxima of $|W_i(t, \omega, \kappa)|$, i.e. $|W|_M = \max\limits_{t \geq 0, \omega \geq 0, i} |W_i(t, \omega, \kappa)|$ and record $(t_M, \omega_M, |W|_M)$ as a trigger condition.

STEP 3: Define the frequency feature sampling line (FSL) and time feature sampling line (TSL), and record M_f, the total number of FSLs, and M_t, the total number of TSLs. The small block is called a feature block (FB), as shown in Fig. 9.

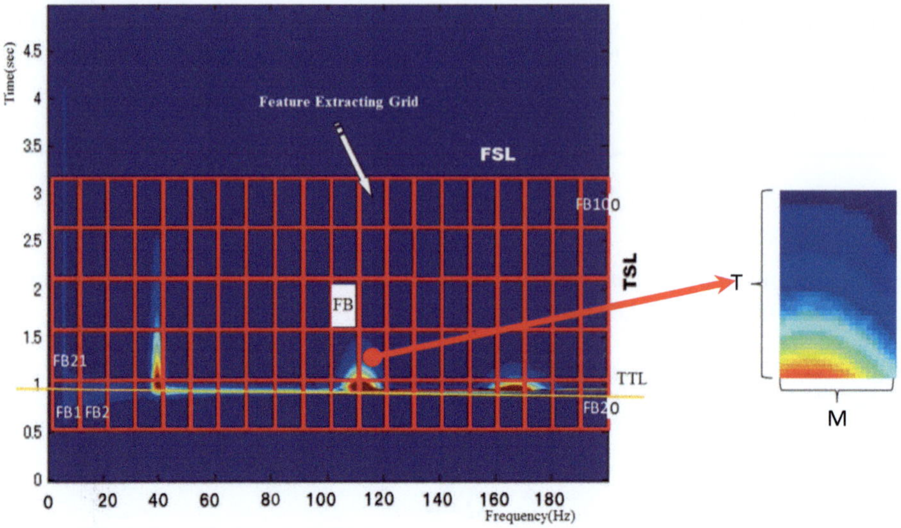

Fig. 9. Example of blocks in time–frequency space defined for impact hammer testing (feature extracting grid).

STEP 4: Compute the modal damping feature at each feature block as

$$\alpha_{ipq} = MDF(|W_i(t,\omega,\kappa)|)|_{(t,\omega)\in \mathrm{FB}_{pq}},$$
$$i = 1,2,\ldots N_s, p = 1,2\ldots M_f, q = 1,2\ldots M_t \quad (8)$$

STEP 5: Define: $V_i = (W_{M1}, \alpha_{ipg})\ i = 1,2,\ldots N_s, p = 1,2\ldots M_f, q = 1,2\ldots M_t,$ where,

$$W_{ipq} = E(|W_i(t,\omega,\kappa)|)|_{(t,\omega)\in\ \mathbf{FB}_{pq}} \quad (9)$$

where $E(\cdot)$ means the average value, and V_i is the time-frequency feature (TFF) vectors of the system.

4 Damage Detection Results and Discussions

Feature quantities are extracted by applying Eq. (4) to the feature blocks in time-frequency space defined as Fig. 9, and the results are substituted into Eq. (5) (*MDF*(*x*)). The coefficient (damping parameter), α in the amplitude envelope obtained by arbitrary domain FSWT result was calculated at multiple locations, and the coefficients thus determined were used as feature vectors for damage detection. The calculation of Eq. (4) is easy, and the calculated value, $S(x^n)$ thus obtained is substituted into Eq. (5) to calculate the coefficient α. In this study, the value of α for each feature block calculated from the measurement data obtained from the experiments carried out by the laboratory bridge monitoring system for all combinations mentioned earlier was used as an input, and a state probability distribution model (probabilistic model) was constructed by using SRM.

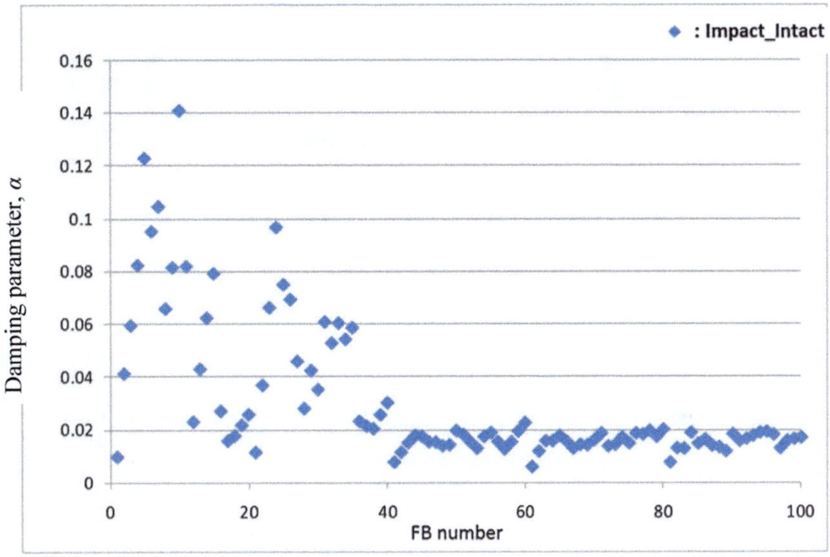

Fig. 10. Example of damping parameter extraction from the normal state (no-damage) bridge model by impact hammer test.

Figure 10 shows an example of calculated block-by-block values of transient damping parameter, α for the laboratory bridge model in a sound (no-damage) condition obtained from the impact hammer test. These results indicate that feature quantities in a block increase when there is a peak frequency (see Fig. 2(c)) in the block which represents the damping characteristics of a multi-modal signal by a frequency slice algorithm.

Figure 11 shows the overall logics of SRM condition assessment (damage detection) based on FSWT analysis and a concept of the transient damping. There are three main steps: (1) Compute signal feature and solve feature model, (2) Construct the state variables, and (3) Make the statistics of the state variables. According to the operation logics, we can realize the final aim of SRM, namely state probability distribution of the state variables, ζ for damage detection. In here, three patterns of the SRM scale parameter σ, namely, $\sigma = 1.0$, $\sigma = 1/2$ and $\sigma = 1/8$ [5, 9], were examined in evaluating the effects of different scale parameter values on damage detection results. Also, SRM-based state variable values determined by following the steps shown in Fig. 11 were compared to examine how the state probability distribution changes in the no-damage girder cases and damaged girder cases, respectively.

Figure 12 shows the FSWT analysis results for observing at three locations; C1 $\sim$ 3 girders [5] in the normal state. The first three main modes are: f_1 = 29.3 Hz,

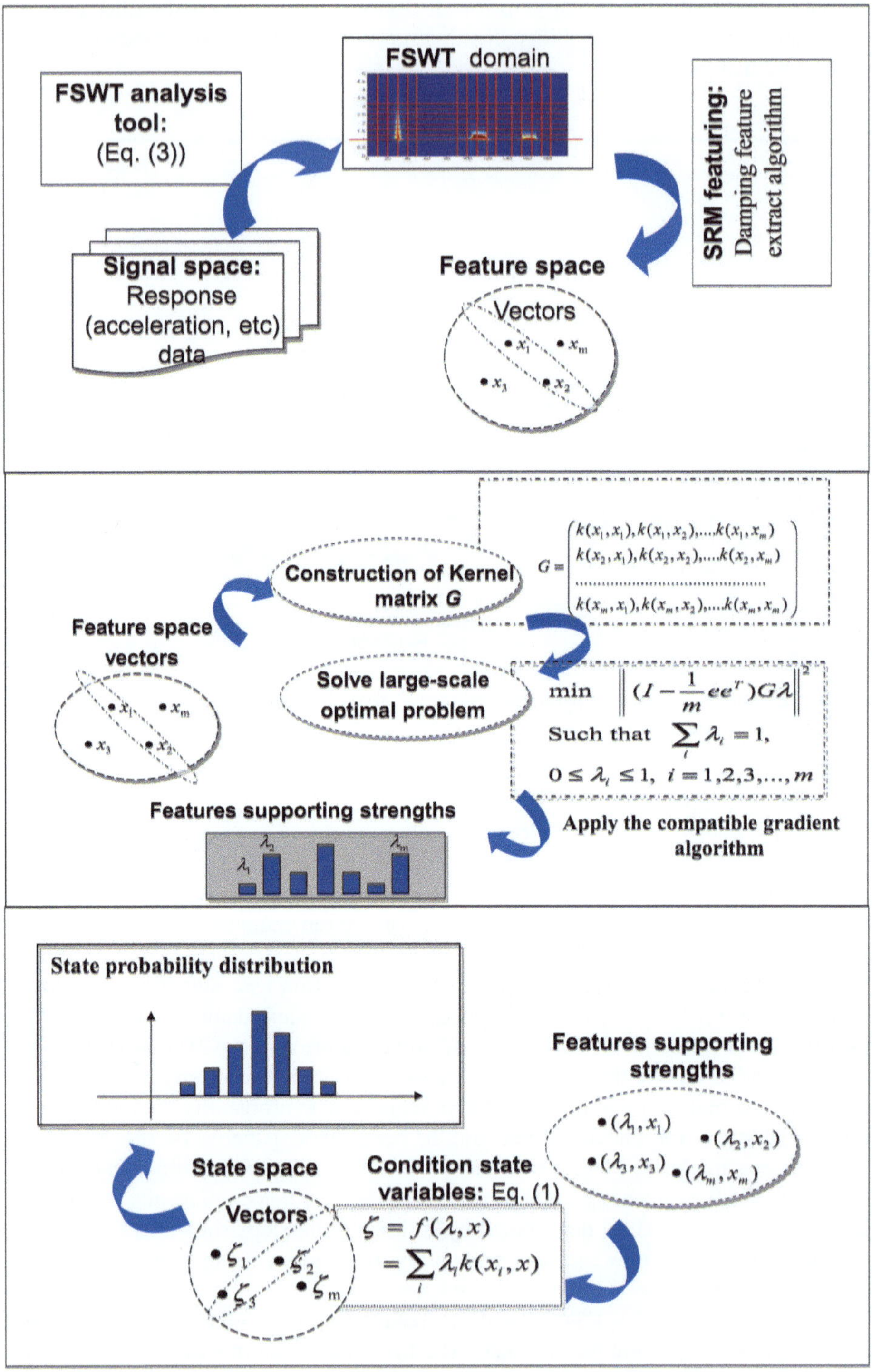

Fig. 11. Operation logics of SRM condition assessment (damage detection).

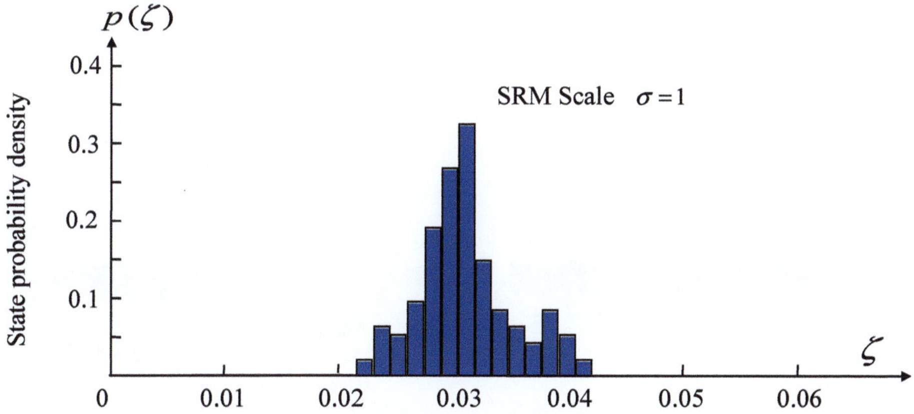

Fig. 12. Example of state probability density distribution for normal state (no-damage).

f_2 = 112.3 Hz, and f_3 = 164.0 Hz (see Fig. 2), which are almost same for all observing sensors. In fact, we cannot easily distinguish the differences among 990 groups [5] of datum. Based on the above featuring methods and SRM techniques, we can build a state function, $\xi = f(\lambda, x)$ by SRM methods, where x is a feature vector extracted by the "Feature Extract Algorithm" discussed in Session 3. Finally, we can get the state probability density distribution as shown in Fig. 12. From Fig. 12, it is found that the system state value is concentrated at nearby ζ = 0.03.

Next, in order to show that how to change the state probability density distribution after deterioration (damaged state) of the bridge, a serious damage was introduced artificially in the bottom flange at C1 (edge) girder (see Fig. 3) of the laboratory bridge model as a specific example. Figure 13(a)–(e) show, the FSWT time–frequency analysis results for damaged state at C1 (edge) girder and its state probability density distribution. From the comparison of these results with the normal state (no-damaged) shown in Fig. 2, it is found that the vibration mode was then changed, therefore the responses shown in Fig. 13(d), except the first mode f_1 = 29.3 Hz, have additional different modes obviously. Some new modal frequencies appear at the nearby of frequency 112.3 Hz and 164.0 Hz, and they might not be the same in the different observing positions, and the response magnitudes of modal 29.3 Hz are greater than that of normal state. By using the same state function, $\zeta = f(\lambda, x)$ that is built in normal test phase, in the same way, we can get the state probability density distribution as shown in Fig. 13(e). This means that the system state value, ζ is shifted greatly by flange deterioration of C1 (edge) girder (stiffness reduction) from ζ = 0.03 (peak value of normal state) to nearby ζ = 0.0 (peak value of damaged state).

Finally, based on the SRM algorithm as shown in Fig. 11, by using future extracting grids, namely, a small block as feature block (FB) as shown in Fig. 9, we will be able to get the distribution of state variable ζ, that is the SRM state probability distribution, as shown in Fig. 14. Figure 14 shows the results of comparison between C2 (center) girder damaged and no-damaged girders on the SRM state probability density distribution. It is clear that damages like girder stiffness reduction tend to not

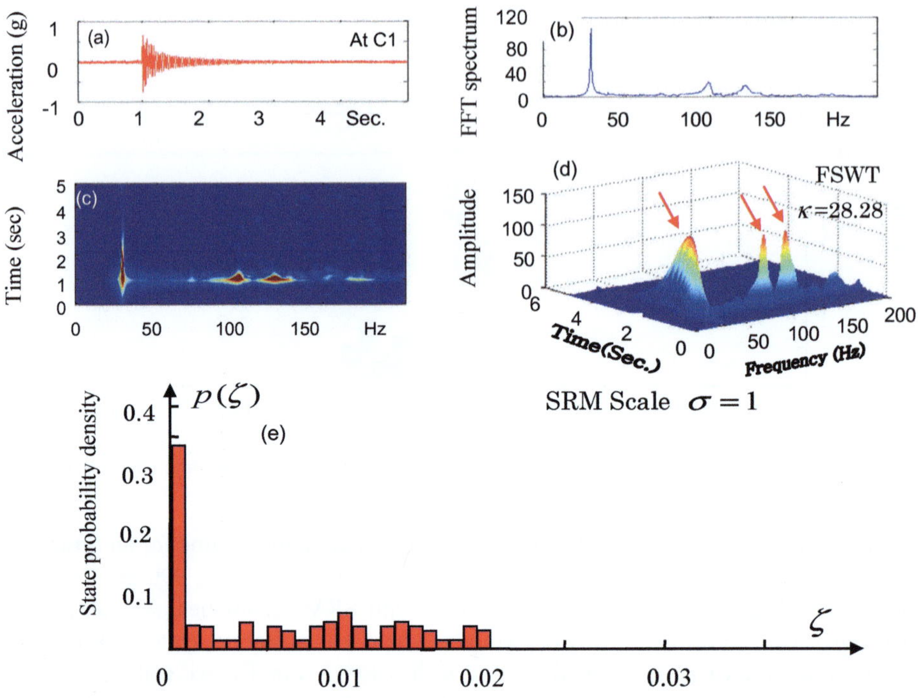

Fig. 13. Example of FSWT time–frequency analysis results for damaged state at C1 (edge) girder and its state probability density distribution.

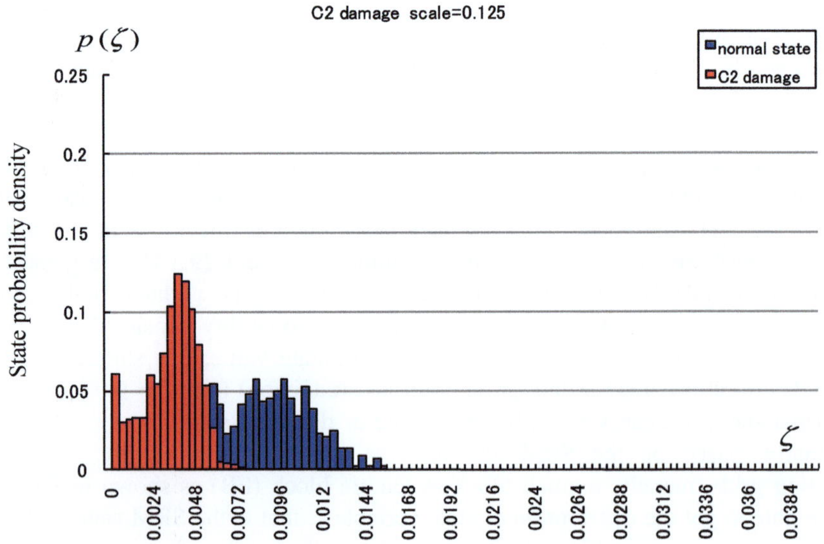

Fig. 14. Comparison between C2 (central) girder damaged and no-damaged (normal) girders on SRM state probability density distribution.

only move away the peak value of SRM state probability distribution from their normal (original) state (ex. $\zeta \fallingdotseq 0.0096 \rightarrow 0.0036$ for C2 (central) girder damage) but also change the parameters related to state variable. Then, it is found that based on these distributions, we will be able to recognize the condition changes between the current state and previous state (normal state) in a deteriorating bridge. However, more discussions and applications are still needed to be improved for practical application to bridge condition assessment.

5 Conclusions

The feature extracting method from a large amount of continually collected measurement data by the SHM system is an important problem [10] in the SRM condition assessment (damage detection). This paper therefore proposes a new basic concept of transient damping parameter. At the same time, a new time-frequency transform, called FSWT is introduced as the pre-processing for damage detection in this study. By the aid of FSWT analysis, a new feature extracting method using the transient damping parameter is proposed for applying to practical use of the SRM assessment method. The featuring algorithm is also applied to the vibration responses obtained from the laboratory bridge monitoring system. Experiments show the SRM features are sensitive and stable with the system changes when its condition has been changed.

The results obtained from this study are as follows:

(1) A new time–frequency space analysis tool which called the frequency slice wavelet transform (FSWT) analysis tool has been developed as a pre-processing for damage detection based on a large amount of monitoring (vibration) data. The FSWT analysis makes it possible to not only accurately extract peaks in the time–frequency space but also have high noise resistance for SRM condition assessment which is a new idea for a non-parametric description of system state.

(2) A feature extracting method of the transient damping parameter, α as the state variables with aid of FSWT analysis results was proposed by the state representation methodology (SRM) which is applied to assess the system condition. Through verification tests by the laboratory bridge monitoring system, the SRM will be able to become an useful method for bridge condition assessment in structural health monitoring (SHM) system.

References

1. Miyamoto, A: Usage management of civil structures. In: Boller, C., Chang, F., Fujino, Y. (eds.) Encyclopedia of Structural Health Monitoring, vol. 4, pp. 1635–1671. Wiley, Oxford, UK (2009)
2. Yanev, B.: Bridge Management. Wiley, Hoboken, USA (2007)
3. Yan, Z.-H., Miyamoto, A., Jiang, Z.-W.: Frequency slice wavelet transform for transient vibration response analysis. Mech. Syst. Signal Process. **23**(5), 1474–1489 (2009)

4. Yan, Z.-H., Miyamoto, A.: State Representation Methodology (SRM) and Its Application to Bridge Condition Assessment, Lifetime Management Book No. 9, The Research Center for Environmental Safety, Yamaguchi University (2009)
5. Miyamoto, A., Yabe, A., Yan, Z.-H.: Chapter 11: A new damage detection method for bridge condition assessment in structural health monitoring (SHM). In: Yurish, S.Y. (ed.) Advances in Signal Processing: Reviews, IFSA Book Series, vol. 1, pp. 407–479. International Frequency Sensor Association (IFSA), Barcelona, Spain (2018)
6. Cristianini, N., Shawe-Tayor, J.: An Introduction to Support Vector Machines and Other Kernel-based Learning Methods, Cambridge University Press, Cambridge, UK (2000)
7. Daubechies, I.: Ten Lecture on Wavelets, CBMS-NSF regional conference series in applied mathematics, vol. 61. SIAM, Philadelphia, USA (1992)
8. Yan, Z., et al.: An overall theoretical description of frequency slice wavelet transform. Mech. Syst. Sign. Proces. (2009). https://doi.org/10.1016/j.ymssp.2009.07.002
9. Miyamoto, A., Brühwiler, E.: Damage detection method based on state representation methodology (SRM). In: Proceedings of the 4th Conference on Smart Monitoring, Assessment and Rehabilitation of Civil Structures (SMAR 2017), Zurich, Switzerland, Paper No. 15, pp. 1–10, 2017
10. Masson, G., Cogan, S., Bouhaddi, N., Lombard, J.P., Bonini, J.: Parameterized reduced models for efficient optimization of structural dynamic behavior. AIAA-2002-1392, American Institute of Aeronautics in Astronautics, 2002

Finite Element Modelling and Damage Detection of Seam Weld

Xiuming Yang[1,2], Huajiang Ouyang[2(✉)], Xinglin Guo[1], and Dongsheng Li[3]

[1] State Key Laboratory of Structural Analysis for Industrial Equipment, Dalian University of Technology, Dalian 116024, People's Republic of China
Xiuming.yang@liverpool.ac.uk, xlguo@dlut.edu.cn
[2] University of Liverpool, Liverpool L69 3BX, UK
{Xiuming.yang, h.ouyang}@liverpool.ac.uk
[3] Guangdong Engineering Center for Structure Safety and Health Monitoring, Department of Civil and Environmental Engineering, Shantou University, Guangdong 515063, China
lids@stu.edu.cn

Abstract. Seam welds are widely used in assembled structures for connecting components. However, the dynamic effects of a seam weld are often difficult to characterise in numerical models for several reasons: (1) it is often not wise to build a fine mesh on the seam line which will add considerable computational cost for a structure with many welds, (2) the mechanical properties of weld materials are not well known; (3) sometimes some geometric information about welds is not known beforehand. In this work, the finite element model of a welding connection part is developed by employing CSEAM element in NASTRAN and its feasibility for representing a seam weld is investigated. Based on this result, a damage detection method by updating the properties of the built CSEAM elements is also proposed for welding quality assurance. The damage takes the form of a gap in the weld which causes a sharp change of model strain energy at the edges of the gap for certain vibration modes. Specifically, the model strain energy shape is used as the objective function. A Kriging model is introduced for efficiency and simulation of a T-shaped welded plate structure to demonstrate the effectiveness of this method.

Keywords: Seam weld · Model updating · Kriging model · Model strain energy

1 Introduction

Assembled structures are widely used in civil and mechanical engineering. Different structural members are produced independently and then connected together by special joining techniques. The joint formed has an important influence on the overall dynamical characteristics of the structure. Thus the accurate representation of the joints in the Finite Element (FE) models has significant research value [1].

Welding is one of the most commonly used joining techniques, whose FE modelling has drawn great attention in the past decades [2]. At the early stage, special

M. A. Wahab (Ed.): DAMAS 2019, LNME, pp. 51–62, 2020.
https://doi.org/10.1007/978-981-13-8331-1_4

elements are designed for connection, like partly rigid beam element used in frame structures [3–5]. Then, with the development of Computer-aided engineering software, many useful connectors are available in structural analysis software packages to represent welded joints in the FE model, like RBE2, ACM2 and CWELD [6]. The last one is specially designed for spot weld modelling in Nastran [7] and has been wildly used and studied by many researchers. References [8, 9] show that after reliable model updating, CWELD elements can successfully represent the laser spot weld joints in a top-hat structure. Further research by Abu Husain et al. about damage identification [10] and uncertainty analysis [11] on that kind of structure also benefits from the superiority of the CWELD element.

Unlike the mature application of spot weld modelling, a seam weld which creates a weld seam line to connect parts together has not attracted much research, even though it is also widely used. Most research about FE modelling of seam welded joints does not focus on reflecting the dynamical characteristics, but on their deformation [12], rotational stiffness [13], fatigue capacity [14, 15] and residual stresses [16]. In the field of structural dynamics, Zahari et al. [17, 18] studied the modelling of friction stir weld joints. But their model is simplistic and thus cannot be used generally. Chee [19] and Rahman [20] presented common finite element modelling of T-shaped structures connected by fillet welds. Their focus was on the impact of the types of elements used in plates while qualified welding is completed. The joint part modeled by a shell or a solid element is not flexible enough to represent the damage in a weld. In most cases, a seam weld provides a firm connection. But it may still suffer from cracks, underfill, burn through, incomplete fusion, long services, welders' faults and so on. Therefore, an effective modelling method of seam welded joints that can properly reflect the welding quality and can be further applied in damage identification of the welding part is required in modern industry.

Besides different joint modelling methods, model updating is applied in most of the publications mentioned above. In actual applications, geometric and material properties of the welding part are often assumed but they may slowly vary over time or their accurate values are not known beforehand. One remedy is to update these models by minimizing the differences between the predicted results and the measured experimental results by optimization methods. After that, the welding part is regarded as properly modelled. More details about model updating methods are given in [21].

Model updating is also an effective means for damage identification, especially when an initial model is available. Usually, the stiffness of the models before and after damage is compared and the reduction of the stiffness of the later indicates presence of damage. Vibration-based damage identification make use of dynamic responses such as frequency, mode shape, frequency response function, mode shape curvature and modal strain energy (MSE), any of which can be used as a damage index. Among them, MSE is adopted in this research as it was found to be more sensitive to damage by Alvandi and Cremona [22]. Several works about MSE-based damage identification are listed as follows:

Doebling et al. [23] used MSE as a mode selection criteria as a basis for other methods. Stubbs et al. [24] proposed a method using the decrease of MSE as an index for damage detection and successfully applied it on the I-40 bridge. Cornwell et al. [25] generalized that method to plate-like structures. In these two papers, MSE was calculated by the curvature of the measured mode shapes. On the other hand, Shi et al. [26–28] developed a similar method in which the MSE calculation was based on the FE model. In addition to localization, he also explored a damage quantification method by updating the MSE of suspected damaged elements and derived the sensitivity formulas for optimization. Recently, some improvement of that method has been made by Li et al. [29] and Moradipour et al. [30, 31]. Artificial Intelligence algorithms were also introduced in this method by Seyedpoor [32] and Kaveh et al. [33, 34].

In this paper, the CSEAM element of Nastran is adopted to model the seam weld joint of a T-shape plate structure. The advantages and disadvantages of this element are discussed and its application in damage identification is shown through an FE simulation. In the simulation, the damage is in the form of an unwelded gap in the weld seam line and model updating method based on MSE is applied to detect the gap. The resulting optimization problem is solved by a genetic algorithm while Kriging model is added for computational efficiency.

2 Seam Welded Joint Modelling

The simplest way to join two plates (modelled by shell elements) of a T-shaped structure together is to delete extra overlapping points at the connection parts, as shown in Fig. 2a, Sect. 4. Thus, the two parts are assumed to be melt together representing the weld is firm enough. But in this case, the joint of the model is not adjustable and cannot be used for further calibration.

A smarter approach is to include extra shell or solid elements and delete overlapping points between them and the two plates. In this way, the quality of the weld can be changed through modifying the parameters of these elements. Model updating methods could be applied to make the model more accurate and reliable.

However, all of these methods mentioned above are limited by the mesh distribution of the two plates. They can hardly work with misalignment of the two meshes especially when the two element sizes are different. In order to solve more general problems, in this research the CSEAM element of Nastran is explored. This type of element is specially designed for modelling seam welds and can easily overcome the misalignment problem.

Demonstration about how a CSEAM element works is shown in Fig. 1. At first, two pairs of shell elements are built in Patran (Fig. 1a). Then, two hex elements are added between those two pairs and each node of the hex elements is connected to the four nodes of the corresponding shell elements by RBE3 elements (Fig. 1b). After this, the relative displacement between two connected shell elements is limited and a CSEAM element is created as such an element set.

Two shell elements can easily be connected by a CSEAM element whether they are parallel or vertical. It is not necessary for the CSEAM element to take the same length as the shell elements. Thus it can easily join two plates with different meshes together. Obviously, this element can not only be used for a tee joint, but also a lap joint, corner joint and edge joint. But it is not included in any pre- or post-processing software packages, which makes it inaccessible.

(a)

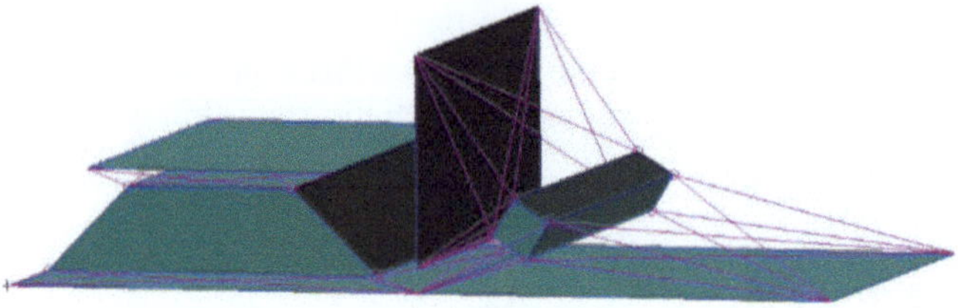

(b)

Fig. 1. Demonstration about how CSEAM element is built in Nastran: (a) Semi-finished structure in Patran; (b) CSEAM elements rebuilt in Patran.

3 Basic Theory

3.1 Model Updating Method

When an FE model is built, the discrepancies between the measured and predicted responses, like natural frequencies and mode shapes, are unavoidable. And the model can be improved by systematically adjusting the structural parameters to minimise these discrepancies. Such a procedure is called model updating.

In seam weld joint modelling, even though the welds are well produced, model updating should still be applied to reflect the real firmness. Usually the stiffness of the whole seam element is selected as a design parameter and the minimisation is carried out via a residual-based objective function:

$$\min \sum_{i=1}^{m} W_i \left(\frac{\omega_i^{\mathrm{a}}}{\omega_i^{\mathrm{e}}} - 1 \right)^2 \tag{1}$$

where $\boldsymbol{\omega}_i^{\mathrm{a}}$ and $\boldsymbol{\omega}_i^{e}$ are the predicted and measured frequencies of the ith mode, respectively. $\mathbf{W}_i$ indicates the weighting coefficient of the residual of that mode.

Measured frequencies are enough for updating the intact model. However, in this research, the focus is the damage identification of the weld joint. As mentioned before, the damage is introduced as a gap within the weld seam representing missing connection. This time the mode shape must be used to localize the gap.

Raw mode shape data are not very sensitive to damage. Many methods took advantage of their derivatives, for example, mode shape curvature, flexibility matrix and MSE were used and compared in reference [22]. The MSE is adopted here, whose theory and improvement are shown as follows:

Energy stored in the *j*th element at mode *i* before and after the occurrence of damage is defined as:

$$MSE_{ij} = \boldsymbol{\phi}_i^{\mathrm{T}} \mathbf{K}_j \boldsymbol{\phi}_i,\ MSE_{ij}^{\mathrm{d}} = \boldsymbol{\phi}_i^{\mathrm{dT}} \mathbf{K}_j \boldsymbol{\phi}_i^{\mathrm{d}} \tag{2}$$

where $\boldsymbol{\phi}_i$ is the mode shape vector, $\mathbf{K}_j$ is the global stiffness matrix of the *j*th element. $(*)^{\mathrm{d}}$ represents the damaged states.

In traditional MSE methods, the elemental modal strain energy change ratio is used as a good indicator for damage localization and is defined as:

$$MSECR_{ij} = \frac{\left| MSE_{ij}^{\mathrm{d}} - MSE_{ij} \right|}{MSE_{ij}} \tag{3}$$

But in this case, damage exists in the weld part whose mode shapes cannot be measured directly. Thus the measured MSEs of joint elements are unavailable and Eq. (3) is unusable.

To solve this problem, in this paper a new MSE-based model updating method is proposed for seam weld joint damage identification. This method utilizes the MSE of the shell elements of the horizontal plate connected by the weld joint to localize the gap. Assuming the vertical plate is in bending, the MSE of the shell elements of the horizontal plate will increase and decrease sharply across the seam line at the edge of the gap. If the theoretical model predicted MSEs are closest to the experimental ones, it can be regarded as properly updated and the reduction of the design parameter shows the location of the gap. The objective function to maximise this MSE shape similarities is defined as:

$$\min f = -\frac{\mathbf{MSEa}^{\mathrm{T}} \cdot \mathbf{MSEe}}{|\mathbf{MSEa}| \cdot |\mathbf{MSEe}|} \tag{4}$$

where a '-' is added to turn the maximum problem into a minimum problem. **MSEa** and **MSEe** are theoretical and experimental MSE vectors of the shell elements along the weld seam line for a certain mode, respectively. One row of shell elements nearest and parallel to the weld is enough to localize a gap in the weld.

3.2 Kriging Surrogate Model

To solve the optimization problem in Eq. (4), it is not wise to use sensitivity based method as formulas will be complex and it is time-consuming to extract global stiffness and mass matrices from a structural analysis software. Instead, one of the famous evolutionary algorithms, genetic algorithm (GA), is applied.

GA is commonly used to generate a high-quality solution to optimization problems by using bio-inspired operators such as mutation, crossover and selection. It has a higher probability of identifying a global optimum solution than the gradient-based approach. Even if the process of computing gradient is impossible, it can still be applied.

Direct utilization of GA can easily solve the problem. However, GA usually requires many calculations of the objective function to get a result. Each of them needs to invoke Nastran to get the dynamic responses of the structure with the changed parameters. The whole process incurs a high computational workload.

To decrease the time taken for Nastran to calculate responses, a Kriging model is first established and then inserted into the GA process. A Kriging model is a surrogate model based on a stochastic process. It maps the input parameters to the corresponding responses mathematically which can be written as:

$$y(\mathbf{x}_i) = \mathbf{f}^{\mathrm{T}}(\mathbf{x}_i)\boldsymbol{\beta} + z(\mathbf{x}_i),\ i = 1, 2, \cdots, n \tag{5}$$

where $\mathbf{f}(\mathbf{x})$ is a polynomial vector of the sample $\mathbf{x}$, $\boldsymbol{\beta}$ is the vector of the linear regression coefficients to be estimated and $z(\mathbf{x})$ represents errors and is assumed to be a stochastic process that follows a normal distribution of $N(0, \sigma^2)$ with a zero mean and standard deviation σ.

After the initial sampling of the design parameters (input) and the corresponding objective function values (output), a surrogate model is built for fast calculating output at random input points instead of invoking Nastran. In this way, much time saving is made when applying GA.

In the event that the initial sampled Kriging model is not precise enough to get proper results, new samples should be added to make the model more reliable especially in the area around the minimum. A Kriging model can predict the response of a new point and its mean square error. If the distance between the new point and the existing sample points is longer, this error is larger. A balance about the value and error must be reasonably considered to find the most likely minimum point. Such a point will be added to the samples and the Kriging model will be renewed until convergence.

This method is called efficient global optimization. For more details and formulas of this methods and Kriging model, the readers are referred to references [35, 36]. The procedure to implement this method to solve optimization problems is concluded as follows:

Step 1: Generate initial sample points of the updating parameters.

Step 2: Run the FE analysis program to calculate the objective function output vector of the sample points and construct the initial Kriging Model.

Step 3: Find the the most likely minimum point of the current Kriging model by GA and add it into the set of the sample points.

Step 4: Calculate the response output of the new sample point and reconstruct the Kriging model by the all sample points and response output.

Step 5: Check whether the procedure has converged. If so, then stop and the sample point leading to the minimum objective function value is the updating result. Otherwise, go back to step 3 and continue adding new sample points

4 Numerical Simulation

In this section, a finite element simulation is used to test the performance and robustness of the presented method. As shown in Fig. 2, two plates are connected together to simulate a tee weld joint configuration with free edge conditions.

(a)

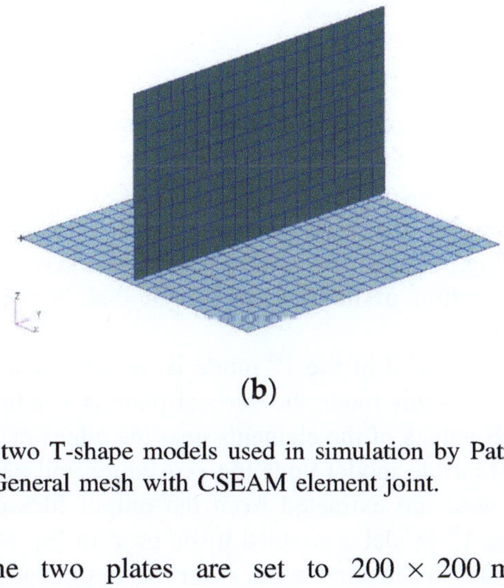

(b)

Fig. 2. Illustrations of two T-shape models used in simulation by Patran: (a) Fine mesh with merged-node joint (b) General mesh with CSEAM element joint.

The sizes of the two plates are set to 200 × 200 mm (horizontal) and 200 × 120 mm (vertical) with 6 mm thickness. They share the same material properties: Young's modulus of 70 GPa, mass density of 2769 kg/m^3 and Poisson's ratio of 0.33. They are welded across the middle line of the horizontal plate. Two different models are constructed with shell elements.

In the 1st model shown in Fig. 2a, very fine mesh (element size is set to 1 mm) is distributed onto both plates. Their connected parts, edge and middle line, are overlapped. Welding connection is constructed by merging the 2 nodes at the same location as mentioned in Sect. 2. The purple line shows the nodes that have been merged. Also, a gap is left in the middle of the weld seam representing the part of the welding failure (damage). The responses of this model are regarded as experimental responses.

In the 2nd model shown in Fig. 2b, the element size is set to 10 mm for general use. There is a small distance between the two plates due to thickness of the horizontal plate. 20 CSEAM elements are created on both sides of the vertical plate to join the two plates together element by element, which cannot be seen in Patran. The material properties of these elements are set the same as the plates initially. Then their Young's modulus (E) is updated based on the first five frequencies of the 1st model without damage by Eq. (1). The updated E is 28 GPa and the improvement in frequencies is shown in Table 1.

Table 1. Natural frequency improvement after updating.

Mode	1st model (Hz)	Original 2nd model (Hz)	Updated 2nd model (Hz)
1	401.3	442.1	398.9
2	459.8	463.1	456.3
3	642.6	668.0	651.2
4	680.3	694.5	656.9
5	897.5	920.7	909.8
Average error		3.9%	1.5%

The proposed method is applied to identify the damage gap in the 1st model. The left edge point and length of the gap is chosen as updating parameters a and b, respectively. They both change from 0 to 20 with the constraint: $a + b \leq 20$ (unit: cm). As the gap represents the welding fault, the connection stiffness is 0 in the gap. During the updating process, E of each CSEAM element is multiplied by a factor which equals the proportion of its element length that is not covered by the gap represented by a and b.

In this simulation, the MSE of the 1st mode is used for damage identification. As shown in Figs. 3 and 4, in this mode the vertical plate is bending laterally across the seam line and the MSE values of the elements near the edges of the gap are extremely high, which indicates that the model updating procedure will get correct results.

All of the MSEs used are extracted from the output files of Nastran while 100 elements' MSEs of the 1st model are added to be used in Eq. (4) with the MSE of 1 element of the 2nd model in the same area. After the optimization, the updated a and b are compared with their actual values in Table 2. It is clear that the damage gap can be identified correctly with a tiny error by this method. This error is caused by inevitable modelling error which influences the updating result of the gap length more.

To check the robustness of the proposed method, 30% white Gaussian noise is added to the experimental MSE of the 1st model. The contaminated MSE of the horizontal plate is shown in Fig. 4 and the updating result by these data is also listed in Table 2. This time the error of the result is larger caused by noise, but is still acceptable for damage localisation.

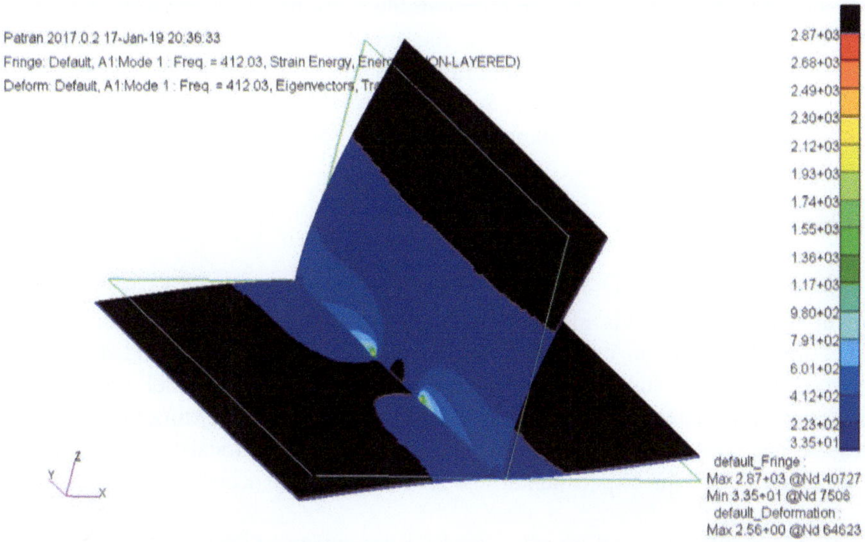

Fig. 3. Mode shape of Mode 1 of the 1st model and each element's MSE shown in color.

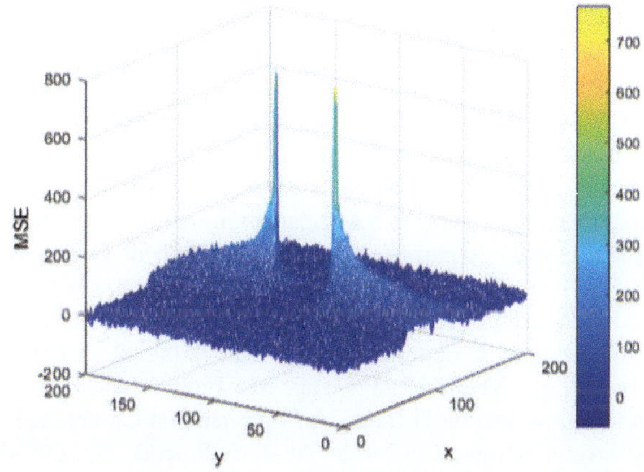

Fig. 4. Contaminated MSE of the horizontal plate of the 1st model in mode 1.

Table 2. Actual location and updating results of the gap.

Parameters	Actual value	Updating results without noise	Error	Updating results with noise	Error
a	7.9	8.06	2%	8.14	3%
b	4.5	4.2	6.7%	4.0	11%

It should be noted that although considerable noise has been added to represent the real response, in practice the error of the measured MSE could be even larger. This is because the angular displacements are difficult to measure and play an important role in Eq. (2) especially for shell elements. So research about how to effectively measure and calculate MSE and the noise influence based on it should be further explored.

5 Conclusions

In this paper, the finite element modelling of seam weld joint based on CSEAM element is presented. This element performs well in connecting two shell element meshes together and can be conveniently used in model updating and damage identification. A Kriging surrogate model updating method using model strain energy is then proposed to detect and localise the damage caused by a welding fault in the form of a gap. The simulation results show that this method can localize the gap correctly and is robust when the experimental test results are fairly accurate.

Acknowledgements. The authors are grateful for the support of the National Natural Science Foundation of China (51578107, and 51778103), and STU Scientific Research Foundation for Talents (Grant No. NTF18012).

References

1. Van Belle, L., Brandolisio, D., Deckers, E., Jonckheere, S., Claeys, C., Pluymers, B., Desmet, W.: Experimental validation of numerical structural dynamic models for metal plate joining techniques. J. Vib. Control **24**(15), 3348–3369 (2018)
2. Zahari, S.N., Zakaria, A.A.R., Sani, M.S.M., Zaman, I.: A review on model updating of joint structure for dynamic analysis purpose. In: MATEC Web of Conferences, pp. 1–6, EDP Sciences (2016)
3. Ahmadian, H., Mottershead, J.E., Friswell, M.I., Soc Exptl, M.: Joint Modelling for Finite Element Model Updating, Soc Experimental Mechanics Inc, Bethel (1996)
4. Horton, B., Gurgenci, H., Veidt, M., Friswell, M.: Finite element model updating of the welded joints in a hollow section H-frame. In: International Conference on Applications of Modal Analysis: recent Advances in Modal Analysis Practice, pp. 309–335 (1999)
5. Horton, B., Gurgenci, H., Veidt, M., Friswell, M.I.: Finite Element model updating of a welded space frame. In: Proceedings of IMAC, pp. 529–535 (2000)
6. MSC.Software, MSC Nastran 2017 Quick Reference Guide (2016)
7. Fang, J., Hoff, C., Holman, B., Mueller, F., Wallerstein, D.: Weld Modeling with MSC. In: Nastran Proceedings 2nd MSC Worldwide (2000)
8. Husain, N.A., Khodaparast, H.H., Snaylam, A., James, S., Dearden, G., Ouyang, H.: Finite-element modelling and updating of laser spot weld joints in a top-hat structure for dynamic analysis. Proc. Inst. Mech. Eng. Part C-J. Eng. Mech. Eng. Sci. **224**(C4), 851–861 (2010)
9. Rani, M.N.A., Kasolang, S., Mohd, M.H.O., Yunus, A., Wan Iskandar Mirza, W.I.I., Ouyang, H.: Finite element modelling and modal based updating for the dynamic behaviour of a laser spot welded structure. In: 23rd International Congress on Sound and Vibration, ICSV 2016, July 10, 2016–July 14, 2016, pp. 1–8, International Institute of Acoustics and Vibrations (2016)

10. Abu Husain, N., Snaylam, A., Khodaparast, H.H., James, S., Dearden, G., Ouyang, H.: FE model updating for damage detection - application to a welded structure. In: Key Engineering Materials, vol. 413–414, pp. 393–400, Trans Tech Publications Ltd (2009)
11. Abu Husain, N., Khodaparast, H.H., Ouyang, H.J.: Parameter selection and stochastic model updating using perturbation methods with parameter weighting matrix assignment. Mech. Syst. Signal Proc. **32**, 135–152 (2012)
12. Zeng, P., Gao, Y., Lei, L.P.: Local equivalent welding element to predict the welding deformations of plate-type structures. Sci. China Ser. E-Technol. Sci. **51**(9), 1502–1506 (2008)
13. Garifullin, M., Bronzova, M., Jokinen, T., Heinisuo, M., Kovacic, B.: Effect of fillet welds on initial rotational stiffness of welded tubular joints. In: Alpatov, S., Prentkovskis, O., Sterling, R.L., Kaliampakos, D. (eds.), Procedia Engineering, vol. 165, pp. 1643–1650. Elsevier Science Bv (2016)
14. Aygül, M.: Fatigue analysis of welded structures using the finite element method, Ph. D. Thesis, Chalmers University of Technology, Gothenburg, Sweden (2012)
15. Fayard, J.-L., Bignonnet, A., Van, K.D.: Fatigue Design of Welded Thin Sheet Structures, vol. 22, pp. 145–152. Elsevier (1997)
16. Peric, M., Tonkovic, Z., Rodic, A., Surjak, M., Garasic, I., Boras, I., Svaic, S.: Numerical analysis and experimental investigation of welding residual stresses and distortions in a T-joint fillet weld. Mater. Des. **53**, 1052–1063 (2014)
17. Zahari, S.N., Sani, M.S.M., Abu Husain, N., Ishak, M., Zaman, I.: Dynamic analysis of friction stir welding joints in dissimilar material plate structure. J. Teknol. **78**(6–9), 57–65 (2016)
18. Zahari, S.N., Sani, M.S.M., Ishak, M.: Finite element modelling and updating of friction stir welding (FSW) joint for vibration analysis. In: Ghani, S.A.C., Hamzah, W.A.W., Alias, A. (eds.), MATEC Web of Conferences, vol. 90, EDP Sciences (2017)
19. Chee, N.C., Bakar, A.R.A.: Finite element modeling of arc welded joints. Jurnal Mekanikal **23**(1), 15–30 (2007)
20. Rahman, R.A., Zubair, M., Amin, N.: Finite element modeling, correlation and model updating of stiffened plate. Jurnal Mekanikal **16**, 91–106 (2003)
21. Mottershead, J.E., Link, M., Friswell, M.I.: The sensitivity method in finite element model updating: a tutorial. Mech. Syst. Signal Proc. **25**(7), 2275–2296 (2011)
22. Alvandi, A., Cremona, C.: Assessment of vibration-based damage identification techniques. J. Sound Vib. **292**(1–2), 179–202 (2006)
23. Doebling, S.W., Hemez, F.M., Peterson, L.D., Farhat, C.: Improved damage location accuracy using strain energy-based mode selection criteria. Aiaa J. **35**(4), 693–699 (1997)
24. Stubbs, N., Kim, J.T., Farrar, C.R.: Soc Exptl, M.: Field verification of a nondestructive damage localization and severity estimation algorithm. In: Proceedings of the 13th International Modal Analysis Conference, vols 1 and 2, Soc Experimental Mechanics Inc, Bethel, pp. 210–218 (1995)
25. Cornwell, P., Doebling, S.W., Farrar, C.R.: Application of the strain energy damage detection method to platelike structures. J. Sound Vib. **224**(2), 359–374 (1999)
26. Shi, Z.Y., Law, S.S.: Structural damage localization from modal strain energy change. J. Sound Vib. **218**(5), 825–844 (1998)
27. Shi, Z.Y., Law, S.S., Zhang, L.M.: Improved damage quantification from elemental modal strain energy change. J. Eng. Mech.-ASCE **128**(5), 521–529 (2002)
28. Shi, Z.Y., Law, S.S., Zhang, L.M.: Structural damage detection from modal strain energy change. J. Eng. Mech.-ASCE, **126**(12), 1216–1223 (2000)

29. Li, Y.C., Wang, S.Q., Zhang, M., Zheng, C.M.: An improved modal strain energy method for damage detection in offshore platform structures. J. Mar. Sci. Appl. **15**(2), 182–192 (2016)
30. Moradipour, P., Chan, T.H.T., Gallage, C.: An improved modal strain energy method for structural damage detection, 2D simulation. Struct. Eng. Mech. **54**(1), 105–119 (2015)
31. Moradipour, P., Chan, T.H.T., Gallage, C.: Benchmark studies for bridge health monitoring using an improved modal strain energy method. In: Procedia Engineering **188**, 194–200 (2017)
32. Seyedpoor, S.M.: A two stage method for structural damage detection using a modal strain energy based index and particle swarm optimization. Int. J. Non-Linear Mech. **47**(1), 1–8 (2012)
33. Kaveh, A., Zolghadr, A.: Cyclical parthenogenesis algorithm for guided modal strain energy based structural damage detection. Appl. Soft. Comput. **57**, 250–264 (2017)
34. Kaveh, A., Zolghadr, A.: Guided modal strain energy-based approach for structural damage identification using tug-of-war optimization algorithm. J. Comput. Civil. Eng. **31**(4), 12 (2017)
35. Jin S.S., Jung H.J.: Sequential surrogate modeling for efficient finite element model updating, Comput. Struct. **168**, 30–45 (2016)
36. Yang, X., Guo, X., Ouyang, H., Li, D.: A kriging model based finite element model updating method for damage detection. Appl. Sci. **7**(10), 1039 (2017)

Detection of Multiple Cracks Using an Energy Method Applied to the Concept of Equivalent Healthy Beam

Gilbert-Rainer Gillich[1(✉)], Alexandra Teodora Aman[1], M. Abdel Wahab[2(✉)], and Cristian Tufisi[1]

[1] Department of Mechanical Engineering, "Eftimie Murgu" University of Resita, P-ta Traian Vuia 1-4, 320085 Resita, Romania
gr.gillich@uem.ro

[2] Soete Laboratory, Faculty of Engineering and Architecture, Ghent University, 9052 Zwijnaarde, Belgium
magd.abdelwahab@ugent.be

Abstract. We introduce herein the concept of global stiffness of a damaged beam, and the Equivalent Healthy Beam (EHB) as a damage model. This model simplifies the process of assessing multiple cracks in beams. It consists of a thinner beam with constant cross-section, but with unchanged mass, which stores in each vibration mode identical energy as the damaged beam. Hence, the model has similar frequencies as for the damaged beam. We estimate the amount of the stored energy from the beam deflection under dead mass. Each additional crack provokes a deflection increase, and consequently the beam stores less energy. The resulted EHB becomes therefore thinner. We performed simulations for beams with transversal cracks which are present on the opposite beam faces. By iteratively removing one crack relative to the other, we obtained the values of the deflections and the eigenfrequencies for the first vibration modes. Comparing the separate frequency drop due to cracks with the effect of the concomitantly acting cracks, we demonstrated the superposition principle applies successfully for most locations of the multiple cracks. As an exception, for nearby-located cracks the principle does not apply. In such a case, a bigger frequency drop is noticed.

Keywords: Damage detection · Vibration · Multi-cracked beam · Frequency shift · Static deflection · Equivalent healthy beam

1 Introduction

Monitoring of structures to assess their integrity has become a current concern of researchers and practitioners. There are currently many damage detection methods that are based on the link between changes occurring in the structure and certain modal parameter shifts [1–5]. The faults most often considered by researchers who develop non-destructive control methods are single transverse cracks. For this kind of damage, the researchers distinguish between open and breathing cracks, which are considered to influence differently the dynamic behavior of the structures [6, 7].

M. A. Wahab (Ed.): DAMAS 2019, LNME, pp. 63–78, 2020.
https://doi.org/10.1007/978-981-13-8331-1_5

Approaches in which two or more cracks affect the structure at the same time are also presented in the literature. However, the number of works in which multiple cracks were detected, compared to those showing the case of detecting single cracks, is very small. The most common approach of beams with multiple cracks considers the structure split into more segments, which are linked by massless torsional springs [8–11]. These springs represent the beam slices weakened by the cracks. The models constructed by using this approach lead to algebraic systems with many equations and unknowns, which allow by laborious calculation the determination of the natural frequencies and mode shapes. The location of the damages is found either from the analysis of the frequency shift [12] or from the modal shape alteration [13]. A different approach was presented in Ref. [14], where an energy method was employed. The frequencies of the damaged beam were calculated by subtracting the contribution of the spring(s) to the potential energy stored in the beam. The case of precise localization of multiple damages by involving the Functional Model Based Method was presented in Ref. [15], in which the damage was simulated by additional masses. By exploring the literature, we found just vibration-based damage detection methods dedicated to assessing multiple cracks that dealt with open cracks. All authors were stating this aspect or considering it implicitly, without declaring it. The reason that only open cracks were taken into account was the difficulty of the torsion spring to efficiently model breathing cracks. The nonlinearity brought by the breathing crack was considered to be responsible for the difficulties of applying the superposition principle to beams with such damages.

The use of the bi-harmonic oscillator in modeling the behavior of beams with a breathing crack was analyzed in Ref. [7], and the limitations of this model were revealed. In prior research [16, 17] the authors and collaborators have shown the necessity of a model with two degrees of freedom to completely model the beam with a breathing crack. It results from the additional rotation occurred due to the crack if it is open. We also found the frequencies in the open and closed stage were similar, i.e. the half periods associated with the open stage had the same length in time as those of the closed stage [18]. Therefore, if just the natural frequencies are targeted, there is no difference between the open and breathing cracks, which cause the same energy drop. Consequently, we develop a model for the cracked beam which has the same frequency as the real cracked beam. The model consists of a beam with a constant cross-section; therefore it is called the EHB. Employing this concept, we demonstrate in the following that the superposition is applicable in the case a beam with multiple cracks, except the situation of nearby-located cracks.

2 The EHB Model and Its Use in Predicting Frequency Changes

Let us consider two cantilever beams, the first one has a crack and the second one is thinner than the original but having a uniform cross-section. Both beams attain the same free end deflection if subjected to the same load. According to Castigliano's theorem, the two beams store the same amount of energy. This means that, if the force is released, they will vibrate with the same frequency.

If the crack is of open type, the amplitudes are similar, while for breathing cracks in the closed stage, the amplitude is lower than in the open stage (see Fig. 1). However, if the system is conservative, the total energy remains unchanged during vibration and so the frequencies in the two stages of the breathing crack evolution.

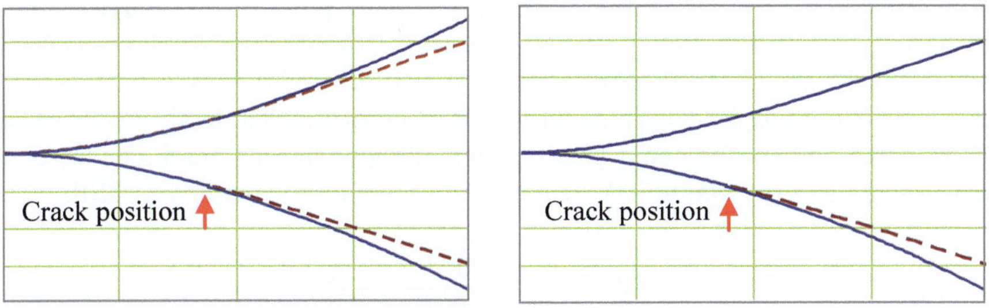

Fig. 1. Maximum displacement for the beam with open crack (left image) respectively closed crack (right image) located near the beam centre. The continuous line represents the cracked beam and the dashed line represents the healthy beam.

We have shown in [19] that the relation between the deflection of the free beam end, δ, and the natural frequency, f_i, of any bending mode is found from the relation:

$$f_i = \frac{\lambda_i^2}{2\pi}\sqrt{\frac{g}{8\delta}} \tag{1}$$

where λ_i is the dimensionless eigenvalue and g is the Earth gravity constant. The deflection δ for a healthy beam is calculated as:

$$\delta = \frac{\bar{m}gL^4}{8EI} \tag{2}$$

where $\bar{m}$ is the mass per unit length, L is the beam length, E is Young's modulus and I is the area moment of inertia. If this beam gets a crack with depth d at location c, its deflection at the free end increases to $\delta(c,d)$. The same deflection is achieved by a healthy beam with smaller cross-section area $A_{eq}(c,d)$ and area moment of inertia $I_{eq}(c,d)$. Because of the same amount of stored energy, the frequency of the damaged beam is identical with the frequency of the thinner healthy beam, i.e.:

$$f_i(c,d) = \frac{\lambda_i^2}{2\pi}\sqrt{\frac{g}{8\delta(c,d)}} \tag{3}$$

The relation between the natural frequencies of the healthy and damaged beam results from Eqs. (1) and (3), and depends just on the deflections in the two states

$$f_i(c,d) = f_i\sqrt{\frac{\delta}{\delta(c,d)}} \tag{4}$$

This dependency is expected, since the deflection is a measure of the stored energy that clearly influences the beam's natural frequencies. Because the ratio between the deflection of the healthy and damaged beams does not include terms that dependent on the end supports, Eq. (4) is applicable for any boundary conditions.

If the crack is located at the position where the curvature achieves maxima, i.e. the fixed end has $c = 0$ for the cantilever beam, the crack produces the biggest effect in terms of deflection increase and frequency decrease, respectively. This is what we call the *equivalent healthy beam* (EHB). It has the lowest flexural rigidity $EI(0,d)$ among all healthy beams used to model the beam with a crack of a given depth. In the sense of this continuous model, the beam gets a *global stiffness* and the crack gets associated with a damage severity coefficient. The damage severity γ is defined as [20]:

$$\gamma(0,d) = \frac{\sqrt{\delta(0,d)} - \sqrt{\delta}}{\sqrt{\delta(0,d)}} \tag{5}$$

The above relation was proven to work also for laminated composites [21]. From Eqs. (1) and (4) one can find the frequency drop Δf_i due to a crack located at the point of maximum of the curvature as:

$$\Delta f_i(0,d) = f_i - f_i(0,d) = f_i\left[1 - \sqrt{\frac{\delta}{\delta(0,d)}}\right] = f_i\frac{\sqrt{\delta(0,d)} - \sqrt{\delta}}{\sqrt{\delta(0,d)}} = f_i \cdot \gamma(0,d) \tag{6}$$

Our previous researches [22–24] revealed the dependency between the crack position and the global stiffness decrease, which is obviously followed by a frequency drop. The mathematical relation describing this dependency is:

$$\gamma(c,d) = \frac{\sqrt{\delta(c,d)} - \sqrt{\delta}}{\sqrt{\delta(c,d)}} = \frac{\sqrt{\delta(0,d)} - \sqrt{\delta}}{\sqrt{\delta(0,d)}}\left[\bar{\phi}_i''(c)\right]^2 = \gamma(0,d)\left[\bar{\phi}_i''(c)\right]^2 \tag{7}$$

where $\bar{\phi}_i''(c)$ is the normalized modal curvature. The term $\bar{\phi}_i''(c)$ defines the bending moment (or modal curvature) evolution along the beam for the different vibration modes i, and is the unit at the location where these features achieve the maximum. For the general case $c \neq 0$, the frequency drop can be calculated from the relation:

$$\Delta f_i(c,d) = f_i - f_i(c,d) = f_i\left[1 - \sqrt{\frac{\delta}{\delta(c,d)}}\right] = f_i\frac{\sqrt{\delta(c,d)} - \sqrt{\delta}}{\sqrt{\delta(c,d)}} \tag{8}$$

In a compacted form, Eq. (8) becomes:

$$\Delta f_i(c,d) = f_i \cdot \gamma(c,d)\left[\bar{\phi}_i''(c)\right]^2 \tag{9}$$

This is a simple relation to derive the frequency drop for beams with any boundary conditions, either for breathing or for open cracks. For the latter case, when loss of mass is noticed, this should be considered in the frequency drop prediction [25].

It is advantageous to evaluate the frequency drop after normalization, i.e. after dividing it by the frequency of the healthy beam in the given mode. This facilitates a comparison between different vibration modes. The relative frequency shift (RFS), is calculated as:

$$RFS_i = \Delta\bar{f}_i(c,d) = \frac{\Delta f_i(c,d)}{f_i} = \gamma(c,d)\left[\bar{\phi}_i''(c)\right]^2 \tag{10}$$

Equations (9) and (10) permit predicting the frequency drop that occurs due to a crack. The thickness of the equivalent healthy beam h_{eq} (0,*d*) is found from the relations expressing the frequency of the healthy beam and the frequency of the cracked beam (modeled as an EHB with constant thickness), respectively, as:

$$f_i = \frac{\lambda_i^2}{2\pi}\sqrt{\frac{EI}{\rho A L^4}},\; f_i(0,d) = \frac{\lambda_i^2}{2\pi}\sqrt{\frac{EI_{eq}(0,d)}{\rho A_{eq}(0,d)L^4}} \tag{11}$$

The frequency ratio found from the relations given in Eq. (11) is:

$$\frac{f_i(0,d)}{f_i} = \sqrt{\frac{h_{eq}^2(0,d)}{h^2}} = \frac{h_{eq}(0,d)}{h} \tag{12}$$

which shows that the thickness of the equivalent healthy beam is in inverse relation to the frequency drop, i.e.:

$$h_{eq}(0,d) = \frac{f_i(0,d)}{f_i}h = [1-\gamma(0,d)]h \tag{13}$$

Consequently, the EHB is a beam with a constant cross-section that has the parameters similar to the healthy beam, except for the thickness that is reduced in accordance with Eq. (13). This model reflects the cracked beam behavior in terms of frequencies and considers the crack severity, thus the crack is positioned to produce the most important frequency drop possible. For the cantilever beam, this crack position is at the fixed end. It has to be mentioned that, for a given damage depth, the thickness h_{eq}(0,*d*) of the EHB is the same, irrespective to the boundary conditions.

Obviously, if the damage is located in a position *c* different from the fixed end, the thickness of the beam having a continuous cross-section and the similar frequency as the damaged one increases because the effect of the crack is less significant. In this case, the thickness also differs in function of the vibration mode number *i*.

This damaged beam model, which we call *Mode and Location Adjusted Equivalent Healthy Beam* (ML-EHB) is defined by the beam thickness $h_i(c,d)$, which is:

$$h_i(c,d) = \frac{f_i(c,d)}{f_i}h = \left\{1-\gamma(0,d)\left[\bar{\phi}_i''(c)\right]^2\right\}h \tag{14}$$

To calculate the effect of the crack location it is necessary to employ the square of the normalized modal curvature to diminish the frequency drop. From Eqs. (7) and (8), we obtain the frequency drop due to a crack located in a certain position:

$$\Delta f_i(c,d) = f_i - f_i(c,d) = f_i\left[1 - \sqrt{\frac{\delta}{\delta(c,d)}}\right] = f_i\frac{\sqrt{\delta(c,d)} - \sqrt{\delta}}{\sqrt{\delta(c,d)}} \tag{15}$$

Consequently, the eigenfrequencies of the cracked beam are:

$$f_i(c,d) = f_i\left[1 - \frac{\sqrt{\delta(c,d)} - \sqrt{\delta}}{\sqrt{\delta(c,d)}}\right] = f_i\left\{1 - \gamma(0,d)\left[\bar{\phi}_i''(c)\right]^2\right\} \tag{16}$$

The equivalent healthy beam in association with the modal curvature can be used to generate a many damage scenarios. By employing these scenarios, we can predict the frequency drop produced by the considered crack to the beam at any location along the beam. Therefore, damage assessment can be made as an inverse problem. In the following sections, we demonstrate the concept of the equivalent healthy beam as well as the application of the superposition principle for any kind of transverse cracks which are far enough from each other generally works. We also found out the distance range for the cracks for which the superposition principle is not applicable.

3 Numerical Simulation

In this section, we present the analysis carried out by the means of the finite element method (FEM), using ANSYS program, with the main goal to prove the application of the superposition for frequency prediction if two or more cracks affect the beam's vibrational behavior. The analyzed specimen is a steel beam with initially constant cross-section, which is fixed at the left end and free at the right end. The dimensions and the main material characteristics are presented in Table 1.

Table 1. Dimensions and physical-mechanical properties of the cantilever beam.

Length L [mm]	Width B [mm]	Thickness h [mm]	Mass density ρ [kg/mm^3]	Young modulus E [N/m^2]	Poisson ratio ν [-]
1000	50	5	7850	$2 \cdot 10^{11}$	0.3

The considered damages are described as transversal breathing cracks of depth $d = 1$ mm. One crack is positioned on the upper face of the beam at distance $c_{U1} =$ 160 mm from the fixed end, whilst the second crack act on the opposite face. First, the crack located on the bottom face is placed at distances $c_{B1} = 100$ mm, $c_{B2} = 300$ mm and $c_{B3} = 700$ mm. Afterward, it is iteratively removed with a step of $s = 0.5$ mm in the range $c = 150 \div 170$ mm. A schematic of the cracked beam is presented in Fig. 2.

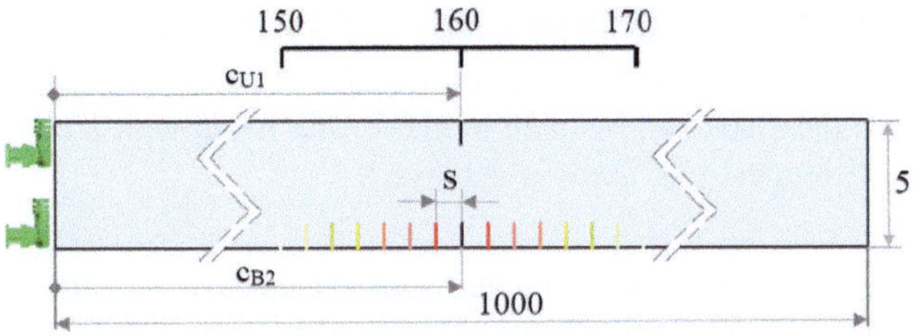

Fig. 2. Schematic representation of the crack positions on the damaged beam

In order to obtain accurate results a fine mesh was applied to all cases, using hexahedral elements of 1 mm size, resulting in a number of 251488 elements, 1171495 nodes for the intact cantilever beam. Obviously, for the cracked beams the number of elements increases in conformity with the crack position.

3.1 Validation of the EHB Concept

The first step is to prove the validity of the EHB and the subsequent ML-EHB concept. To this aim, we suppose the existence of just one crack positioned at the upper face, which is located at distance $c = 6$ mm. For this cracked beam, as well as for the healthy beam, we determine with the modal analysis module of ANSYS the eigenfrequencies of the first five bending vibration modes. Then, we calculated the thickness $h_{eq}(0,1)$ of the EHB using Eq. (13) from the frequency ratio. The necessary parameters and the result are given in Table 2.

Table 2. The equivalent healthy beam thickness calculated from the frequency ratio

Mode no.	Eigenfrequency		Beam thickness	
	healthy beam f_1^{FEM}[Hz]	crack at fixed end $f_1^{FEM}(0,1)$ [Hz]	healthy beam h [mm]	EHB $h_{eq}(0,1)$ [mm]
1	4.0898	4.0775	5	4.9849

Next, we find the eigenfrequencies for the first five bending vibration modes by performing simulations for the EHB with the thickness given in Table 2. The achieved results were compared with those obtained from the FEM analysis for the beam with a crack near the fixed end. The accomplished eigenfrequencies and the resulted errors are presented in Table 3. The small frequency changes produced by the crack are always a problem when performing damage detection method. To be able to measure the eigenfrequencies with precision of three or four digits, we developed an advanced method that included procedures for excitation, acquisition and signal processing [26–28].

The results obtained for the EHB model and for the cracked beam are in good agreement, the absolute values of the errors for all five modes being less then 0.05%. This demonstrates the EHB is a realistic model of the beam with a crack placed at the position where the biggest curvature is achieved.

Table 3. Frequencies for the healthy beam, the cracked beam and the EHB with the thickness obtained from calculus employing the frequency ratio

Mode no.	Freq. - healthy beam f_i^{FEM} [Hz]	Freq. - EHB $f_i^{FEM}(0,1)$ [Hz]	Freq. - cracked beam $f_i^{FEM}(0,1)$ [Hz]	Error ε_i [%]
1	4.0898	4.0776	4.0775	−0.00221
2	25.6265	25.5495	25.552	0.00978
3	71.7547	71.5391	71.552	0.01802
4	140.627	140.205	140.2463	0.02944
5	232.520	231.823	231.9143	0.03936

To prove that the damage severity can be used to calculate the thickness $h_{eq}(0,1)$ of the EHB using Eq. (13), we perform static analysis for the healthy and damaged beam. With this data, we calculated the damage severity employing Eq. (5). This feature is finally used to define the EHB thickness that is used in the simulation. All data accomplished by simulation and calculus is given in Table 4. One can observe a thickness $h_{eq}(0,1)$ close to that obtained if employing the frequency ratio, the difference being less then 0.1%.

Table 4. Equivalent healthy beam thickness calculated by employing the severity

Mode no.	Deflection at the beam free end		Severity $\gamma(0,1)$	Beam thickness	
	Healthy beam δ [mm]	Cracked beam $\delta(0,1)$ [mm]		Healthy beam h [mm]	EHB $h_{eq}(0,1)$ [mm]
1	22.948	23.132	0.003985	5	4.9800

The frequencies predicted using the severity calculated with Eq. (16) are listed in Table 5. These values are precise, the errors being less than 0.15%. This study also validates Eq. (5) since it successfully makes the link between the thicknesses of the healthy beam and that of the EHB model.

Table 5. Frequencies for the cracked beam obtained from direct simulation respectively simulation for the EHB with the thickness obtained from calculus by employing the damage severity

Mode no.	Frequencies - cracked beam $f_i^{FEM}(0,1)$ [Hz]	Frequencies - EHB $f_i^{calc}(0,1)$ [Hz]	Error ε_i [%]
1	4.0775	4.0735	0.0980
2	25.552	25.523	0.1100
3	71.552	71.468	0.1173
4	140.246	140.067	0.1277
5	231.914	231.594	0.1380

3.2 Validation of the ML-EHB Model

In the previous subsection, the position of the crack was near the fixed end. If the crack is located elsewhere, its effect on the frequency drop is diminished because the stress in that region is lower than that at the fixed end. Equation (14) shows the way how the normalized modal curvatures $\bar{\phi}_i''(160)$ are used to calculate the thickness $h_i(160,1)$ of the ML-EHB for the crack location $c = 160$ mm and different vibration modes. To prove the validity of the Eq. (14) proposed in this paper, we calculate the thickness $h_i(160,1)$ from the thickness of the healthy beam $h = 5$ mm, the damage severity $\gamma(0,1) = 0.003985$ and the curvature for different vibration modes at distance $c = 160$ mm (listed in Table 6). For the beams with such thickness we made simulations to find the eigenfrequencies. These are finally compared with the frequencies found by simulation for the crack location $c = 160$ mm and the errors are calculated.

Table 6. Frequencies for the cracked beam respectively for the ML-EHB with the thickness obtained by calculus

Mode no.	Curvature $\bar{\phi}_i''(160)$	Thickness $h_i(160,1)$ [mm]	Freq. ML-EHB $f_i^{FEM}(160,1)$ [Hz]	Freq. cracked beam $f_i^{FEM}(160,1)$ [Hz]	Error ε_i [%]
1	0.779883	4.987881	4.080	4.081	0.0245
2	0.243933	4.998814	25.620	25.621	0.0039
3	0.183067	4.999332	71.745	71.747	0.0027
4	0.506523	4.994888	140.483	140.513	0.0216
5	0.659706	4.991328	232.118	232.192	0.0320

From Table 6, one can observe the small resulted errors. This permitted us to consider that the ML-EHB is suitable for modeling the dynamic behavior of the cracked beams in order to find their eigenfrequencies.

3.3 Application of the Superposition Principle

In the previous subsections, the concept of the EHB and the ML-EHB model were validated. Now, we can say that a beam with a crack on any location acts, from the point of view of frequency responses, as a healthy beam with reduced cross-section. This motivated us considering the superposition principle, the research presented in this subsection aiming to demonstrate the application of the supposition method.

Tests made for cracks located relatively away from each other

We consider the beam with two cracks acting simultaneously on the beam. The crack on the upper face is located at distance $c_{U1} = 160$ mm from the fixed end, whilst the crack located on the bottom face act at locations $c_{B1} = 100$ mm, $c_{B2} = 300$ and $c_{B3} = 700$ mm.

We first perform the modal analysis for one crack acting at a time and then calculate the frequency drop for each damage case. The results are presented in Table 7.

Table 7. Frequency drop for the beam with one crack obtained by simulation

Mode no.	Freq. drop $\Delta f_i^{FEM}(100, 1)$ [Hz]	Freq. drop $\Delta f_i^{FEM}(160, 1)$ [Hz]	Freq. drop $\Delta f_i^{FEM}(300, 1)$ [Hz]	Freq. drop $\Delta f_i^{FEM}(700, 1)$ [Hz]
1	0.011849	0.0088	0.005536	0.000185
2	0.027535	0.005	0.0092	0.02732
3	0.014705	0.007	0.11942	0.15905
4	0.00112	0.11423	0.08839	0.10201
5	0.07277	0.32845	0.04451	0.04285

Afterward, we consider the simultaneous action of the upper crack and a bottom crack and performed the three modal analyses. Finally, we compare the frequencies obtained if two cracks act at a time with the frequency of the healthy beam from which we subtract the correspondent frequency drops. The results are presented in Table 8.

Table 8. Frequencies obtained from direct simulation and by superposing the crack effect

Mode no.	Freq. with cracks at c = 100 mm and c = 160 mm [Hz]		Freq. with cracks at c = 300 mm and c = 160 mm [Hz]		Freq. with cracks at c = 700 mm and c = 160 mm [Hz]	
	Simulation	Superposing	Simulation	Superposing	Simulation	Superposing
1	4.0682	4.0691	4.0745	4.0754	4.0799	4.0808
2	25.592	25.593	25.610	25.611	25.592	25.593
3	71.731	71.732	71.627	71.627	71.588	71.587
4	140.494	140.512	140.408	140.425	140.394	140.411
5	232.069	232.119	232.094	232.147	232.100	232.149

One can observe that the frequencies deduced in the two different ways agree with each other, so the superposition can be use to find the frequency drop caused by two or more cracks. This facilitates creating damage scenarios to predict the frequencies changes in the case of multiple cracks, with clear application in damage detection.

Tests made for closely located cracks

The effect of damage is linked to the energy stored in the slice where the damage is located because here the state of stress is disturbed due to the crack. It was shown that, if the disturbance of state of stress does not interfere, the superposition can be applied to estimate the frequency changes. The question is what happens if the cracks are close to one another and the disturbance in the state of stress interferes? In this subsection, we analyze the case of two cracks located on opposite faces of the beam; the bottom crack is replaced step-by-step as shown in Fig. 2.

To have a global picture on the phenomenon, we represented graphically the frequency evolution for the 41 damage cases. The first data presented in Fig. 3 regards the frequency of the healthy beam (black dashed line), which is obviously the highest value in all modal representations. Then, we represent the frequency of the beam with the upper crack located at $c_{U1} = 160$ mm and the frequency of the ML-EHB with the thickness $h_i(160{,}0)$ calculated by employing the damage severity and curvatures. The first is represented by a continuous green line, the latter with dark violet circles. Both frequencies are obtained by modal analysis. One can observe that these frequencies, deduced in two different ways, agree for all modes.

Furthermore, we consider the effect of the second damage, which is placed on 41 positions, affecting one by one the real beam which has already a crack at $c_{U1} = 160$ mm.

The frequencies obtained from the modal analysis by considering the simultaneous effect of two cracks are plotted with blue circles linked by a continuous blue line. In addition, we derive the effect of each crack and apply the superposition method. The frequencies obtained in this way are plotted with red dashed lines. Finally, we took the ML-EHB the upper crack located at $c_{U1} = 160$ mm as reference and calculated the effect of the bottom cracks on it by employing Eq. (16). The results, obtained for 11 crack locations are represented with brown squares.

Analyzing how the superposition works, from the Fig. 3, we can conclude that in a narrow range where the state of stress due to the two cracks interferes, the actual effect on the beam frequency is greater if the sum of the individual effects. The closer the cracks are, the bigger the frequency decrease is. For the studied case, this range is around 10 mm. This means that the superposition is applicable if the distance between the cracks is bigger then 5 mm. On the other hand, the frequency of the beam with one crack (the upper crack in this case) can be used as a benchmark, similar to the healthy beam. If several cracks are present on the beam, the frequency can be calculated step by step for each individual crack starting from the cumulated effect of the previously considered cracks.

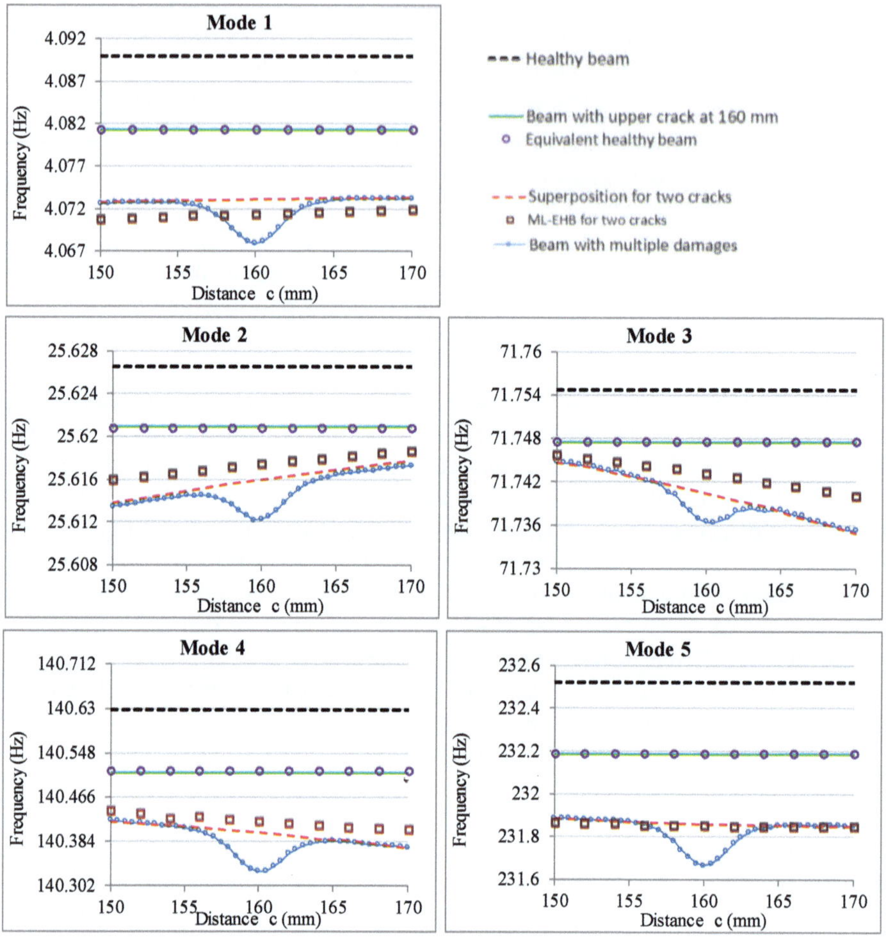

Fig. 3. Frequency evolution with the crack location

4 Discussion on the Superposition Principle

It is interesting to see why the superposition fails if the cracks are close one to another. To this aim, we analyze the stress state disturbances for the upper crack located at c_U = 160 mm and the bottom crack removed by 0.5 mm in the range c_B = 150 ÷ 170 mm. Figure 4 shows the state of stress for the vibration mode one. The slices on which the state of stress is disturbed by the cracks are marked with a dashed rectangle.

From the fist image in Fig. 4, where the distance between the cracks is Δc_1 = 10 mm, one can observe that the state of stress due to the two cracks do not interfere. For this case the superposition works, as it can be seen in Fig. 3 for all vibration modes. It happens also for the case of the bottom crack located at 154 mm from the fixed end, shown in the second image in Fig. 4, where Δc_2 = 6 mm.

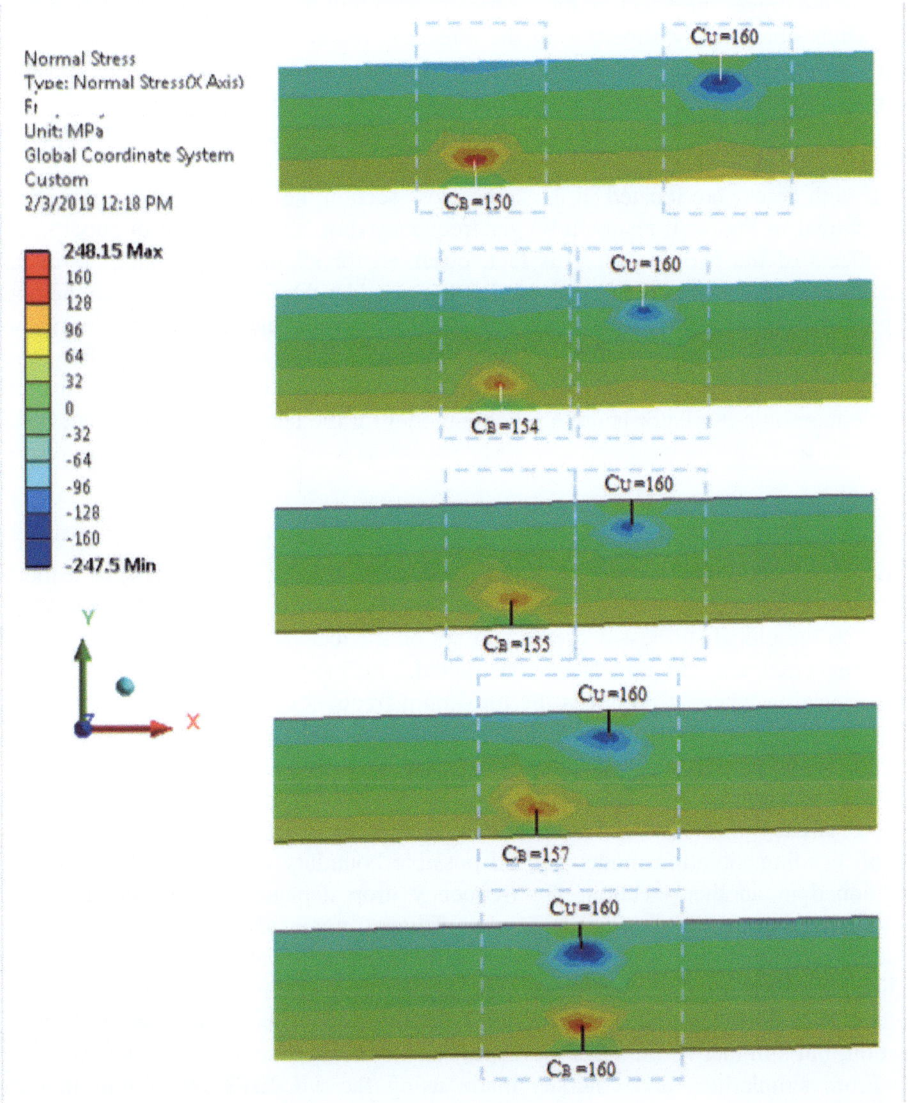

Fig. 4. Interference of the state of stress disturbances for several particular distances between the two cracks

If decreasing the distance between the cracks to $\Delta c_3 = 5$ mm, one can observe in the third image in Fig. 4 the slices on which the stress state is disturbed are in contact. Correlating this aspect with the frequency shifts represented in Fig. 3, it can be deduced this is the limit from which on the superposition will not work anymore. From here on, the effect of the two cracks on the frequency drop exceed the sum of the effects of the two cracks, as shown in Fig. 3. The interference of the state of stress is illustrated in the fourth image in Fig. 4, obtained for and $\Delta c_4 = 3$ mm. From now on, the closer the damage the bigger the effect on the frequency drop will be. The biggest deviation from the results obtained by superposing the effects is found for the cracks located at the same distance from the fixed end, hence $\Delta c_5 = 0$ mm. After passing the position $c_B = 160$ mm, the bottom crack is moving away from the upper crack and the frequency drop decreases consequently until the superposition method can be applied. This happens at $c_B = 165$ mm for all vibration modes, see Fig. 3.

If both cracks are located in the same cross-section, i.e. $c_B = c_B = 160$ mm in our case shown in Fig. 4, it results a bigger frequency drop as in the case of superposing the effects of the two cracks. This fact, observed for all modes in Fig. 3, is in disagreement with the breathing crack model that consider the frequency drop in the open stage to be proportional to the beam stiffness decrease in this stage. Considering two open stages for a similar breathing crack positioned on opposite faces, the frequency drop should be double, which is not the case. Therefore, we can conclude that the simple harmonic oscillator is not a reliable model for the beam with a breathing crack.

5 Conclusion

The paper introduces the concept of the Equivalent Healthy Beam (EHB) as a model for beams with cracks, whether they are of open or breathing type. The EHB model is based on the global stiffness decrease and not on the local stiffness decrease, on which the actual cracked beam models are constructed. This model always considers the crack at the location where it produces the maximum frequency drop, i.e. where the maximum bending moment is achieved, and makes a direct link between the frequency drop and the beam deflection occurred due to a crack. Because the EHB model is a slimmer beam as the original but with constant cross-section, its behavior can be described by simple, already known mathematical relations. As a particularity, the EHB is the same for all bending vibration modes and all possible boundary conditions. If the crack is positioned in another location, the frequency drop depends also on the boundary conditions and the vibration mode number. To take in consideration these elements, a model generalizing the EHB was designed. This is the so-called Mode and Location Adjusted Equivalent Healthy Beam (ML-EHB), which considers in addition to the deflections also the modal curvature. This model is also a beam with uniform cross-section, but slimmer as the EHB in respect to the local value of the modal curvature.

From simulations and calculus made using the ML-EHB, we found that the superposition can be applied if more cracks that are far enough from each other affect the beam. If the distance between the defects located on opposite beam faces is small, the frequency decreases more than the superposition indicates. For the analyzed beam, this distance was found to be 5 mm, similar with the beam thickness. In future research we will focus on cracks located on the same face of the beam and will consider also other boundary conditions.

References

1. Fan, W., Qiao, P.: Vibration-based damage identification methods: a review and comparative study. Struct. Health Monit. Int. J. **10**(1), 83–111 (2011)
2. Rezvani, K., Maia, N.M.M., Sabour, M.H.: A comparison of some methods for structural damage detection. Sci. Iran. **25**(3B), 1312–1322 (2018)
3. Khatir, S., Dekemele, K., Loccufier, M., Khatir, T., Abdel Wahab, M.: Crack identification method in beam-like structures using changes in experimentally measured frequencies and Particle Swarm Optimization. Comptes Rendus Mécanique **346**(2), 110–120 (2018)
4. Ostachowicz, W.M., Krawczuk, M.: Analysis of the effect of cracks on the natural frequencies of a cantilever beam. J. Sound Vib. **150**, 191–201 (1991)
5. Zenzen, R., Belaidi, I., Khatir, S., Abdel Wahab, M.: A damage identification technique for beam-like and truss structures based on FRF and Bat Algorithm. Comptes Rendus Mécanique **346**(12), 1253–1266 (2018)
6. Rivola, A., White, P.R.: Bispectral analysis of the bilinear oscillator with application to the detection of fatigue cracks. J. Sound Vib. **216**(5), 889–910 (1998)
7. Bovsunovsky, A., Surace, C.: Non-linearities in the vibrations of elastic structures with a closing crack: A state of the art review. Mech. Syst. Signal Process. **62–63**, 129–148 (2015)
8. Vigneshwaran, K., Behera, R.K.: Vibration analysis of a simply supported beam with multiple breathing cracks. Procedia Eng. **86**, 835–842 (2014)
9. Labib, A., Kennedy, D., Featherston, C.: Free vibration analysis of beams and frames with multiple cracks for damage detection. J. Sound Vib. **333**, 4991–5003 (2014)
10. Mazanoglu, K., Sabuncu, M.: A frequency based algorithm for identification of single and double cracked beams via a statistical approach used in experiment. Mech. Syst. Signal Process. **30**, 168–185 (2012)
11. Moezi, S.A., Zakeri, E., Zare, A.: Structural single and multiple crack detection in cantilever beams using a hybrid Cuckoo-Nelder-Mead optimization method. Mech. Syst. Signal Process. **99**, 805–831 (2018)
12. Patil, D.P., Maiti, S.K.: Detection of multiple cracks using frequency measurements. Eng. Fract. Mech. **70**(12), 1553–1572 (2003)
13. Altunışık, A.C., Okur, F.Y., Kahya, V.: Modal parameter identification and vibration based damage detection of a multiple cracked cantilever beam. Eng. Fail. Anal. **79**, 154–170 (2017)
14. Mazanoglu, K., Yesilyurt, I., Sabuncu, M.: Vibration analysis of multiple-cracked non-uniform beams. J. Sound Vib. **320**, 977–989 (2009)
15. Sakaris, C.S., Sakellariou, J.S., Fassois, S.D.: Vibration–based multi–site damage precise localization via the Functional Model Based Method. Procedia Eng. **199**, 2072–2077 (2017)
16. Gillich, G.R., Abdel Wahab, M., Praisach, Z.I., Ntakpe, J.L.: The influence of transversal crack geometry on the frequency changes of beams. In: Proceedings of International Conference on Noise and Vibration Engineering (ISMA2014) and International Conference on Uncertainty in Structural Dynamics (USD2014), pp. 485–498. Leuven (2014)
17. Gillich, G.R., Praisach, Z.I.: Detection and quantitative assessment of damages in beam structures using frequency and stiffness changes. Key Eng. Mater. **569**, 1013–1020 (2013)
18. Nitescu, C., Gillich, G.R., Abdel Wahab, M., Manescu, T., Korka, Z.I.: Damage severity estimation from the global stiffness decrease. J. Phys: Conf. Ser. **842**, 012034 (2017)
19. Gillich, G.R., Tufoi, M., Korka, Z.I., Stanciu, E., Petrica, A.: The relations between deflection, stored energy and natural frequencies, with application in damage detection. Rom. J. Acoust. Vib. **13**(2), 87–93 (2016)

20. Praisach, Z.I., Gillich, G.R., Protocsil, C., Muntean, F.: Evaluation of crack depth in beams for known damage location based on vibration modes analysis. Appl. Mech. Mater. **430**, 90–94 (2013)
21. Gillich, G.R., Praisach, Z.I., Abdel Wahab, M., Vasile, O.: Localization of transversal cracks in sandwich beams and evaluation of their severity. Shock Vib. **2014**, 607125 (2014)
22. Gillich, G.R., Praisach, Z.I.: Modal identification and damage detection in beam-like structures using the power spectrum and time–frequency analysis. Signal Process. **96**(A), 29–44 (2014)
23. Gillich, G.R., Birdeanu, E.D., Gillich, N., Amariei, D., Iancu, V., Jurcau, C.S.: Detection of damages in simple elements. In: Annals of DAAAM for 2009 & Proceedings of the 20th International DAAAM Symposium, vol. 20(1), pp. 623–624 (2009)
24. Gillich, G.R., Praisach, Z.I., Negru, I.: Damages influence on dynamic behaviour of composite structures reinforced with continuous fibers. Materiale Plastice **49**(3), 186–191 (2012)
25. Gillich, G.R., Praisach, Z.I., Wahab, M.A., Gillich, N., Mituletu, I.C., Nitescu, C.: Free vibration of a perfectly clamped-free beam with stepwise eccentric distributed masses. Shock. Vib. **2016**, 2086274 (2016)
26. Mituletu, I.C., Gillich, G.R., Maia, N.M.M.: A method for an accurate estimation of natural frequencies using swept-sine acoustic excitation. Mech. Syst. Signal Process. **116**, 693–709 (2019)
27. Gillich, G.R., Maia, N., Mituletu, I.C., Praisach, Z.I., Tufoi, M., Negru, I.: Early structural damage assessment by using an improved frequency evaluation algorithm. Lat. Am. J. Solids Struct. **12**(12), 2311–2329 (2015)
28. Gillich, G.R., Mituletu, I.C., Negru, I., Tufoi, M., Iancu, V., Muntean, F.: A method to enhance frequency readability for early damage detection. J. Vib. Eng. Technol. **3**(5), 637–652 (2015)

Study Regarding the Effect of Crack Branching on the Eigenfrequencies of Beams

Cristian Tufisi, Gilbert-Rainer Gillich(✉), Codruta Oana Hamat, and Tiberiu Manescu

Department of Mechanical Engineering, Universitatea "Eftimie Murgu" din Resita, P-ta Traian Vuia 1-4, 320085 Resita, Romania
gr.gillich@uem.ro

Abstract. The paper presents a study regarding the vibration behavior of Euler-Bernoulli beams which have cracks with complex shapes. The aim is to show that diverse orientations of the crack branches occurring in a given region of the structure produce different changes of the eigenfrequencies for the different vibration modes, dependent on the position and propagation angle of the crack. The research is conducted employing the finite element analyses, which are used to describe the dynamic response of a structure affected by a transversal open crack followed by two branches, each oriented at a different angle. These cracks are traditionally referred to as Y-shaped crack. We calculated the relative frequency shifts for the first six transverse vibration modes and extracted the damage signature for the considered branched crack location. The damage signatures can be used as patterns in damage detection procedures, transforming the procedure of detecting the crack position and severity from the eigenfrequency changes in an inverse problem.

Keywords: Damage detection · Beam-like structure · Branched crack · Damage signature · Frequency shift

1 Introduction

Damages reduce the capacity of the beams to store energy because the slices where damage is present are subject to stiffness decrease. As a consequence, the eigenfrequencies of damaged beams decrease [1–4]. The frequency decrease depends on the reduction of the cross-sectional area, hence on the crack depth, but also on the crack position [5–7]. For transverse cracks, either open or breathing, mathematical relations which permit predicting the frequency drop if the crack depth and position are known to exist and are widely presented in the literature [8–11]. Most mathematical relations were derived empirically, from the fracture mechanics theory, and are applicable just for particular cases [12, 13]. Our research has been focused on finding a mathematical relationship with a large degree of generality, and we have succeeded in creating a relationship that can be applied to any beam-like structure if it is subject to a transverse crack [14–16].

Our recent research is focused on cracks with Y-shaped branches [17], which have increased complexity and therefore more parameters that influence the frequency

M. A. Wahab (Ed.): DAMAS 2019, LNME, pp. 79–91, 2020.
https://doi.org/10.1007/978-981-13-8331-1_6

changes. This makes approaching such structures more difficult. The investigation presented in this paper is a numerical study destined to find the dynamic response of beams with branched cracks, in particular, to define the effect of the position of the transverse crack component relative to the longitudinal crack component.

2 Materials and Methods

2.1 The Beam Used for FEM Simulation

To find the effect of a branched crack on the eigenfrequencies of beam-like structures, we select for this analysis a cantilever beam made of steel, fixed on the left end and free at the other, for which we indicated the dimensions in Table 1.

Table 1. The geometry of the healthy beam

Length L [mm]	Width b [mm]	Thickness h [mm]	Cross-sectional area A [mm^2]	Moment of inertia I [mm]
1000	50	5	250	52083.33

The finite element model of the cantilever is shown in Fig. 1. To ensure the proper constraints, at the left end we applied the *fixed support constraint* and let *free* the other one. For the static analysis we defined a gravitational acceleration of 9.81 m/s^2.

Fig. 1. Healthy beam with the applied boundary conditions and a zoom to highlight the mesh.

The relevant physical-mechanical properties, extracted from the ANSYS library for the *structural steel*, are presented in Table 2.

Table 2. Physical and mechanical properties of the structural steel used in the study

Mass density [kg/m^3]	Young modulus [N/m^2]	Poisson ratio [-]	Yield strength [MPa]	Min. elongation [%]
7850	$2 \cdot 10^{11}$	0.3	355	20

A fine mesh, including hexahedral elements with the maximum size of an edge of 2 mm, was generated. After meshing, the healthy beam consists in 43587 elements and 231639 nodes. For the damaged beam, because of the crack, the higher number of elements is requested.

2.2 The Branched Crack

To create diverse damage scenarios, we have considered a branched crack which has the transverse component located at a distance $c = 160$ mm from the fixed beam end. In the first case, the crack is modeled with a transverse component of depth $a = 1.5$ mm, followed by a branch of length $l = 1.5$ mm that forms an angle $\alpha_0 = 90°$ with the other crack component. This is a so-called L-shaped crack, which has the horizontal branch oriented to the right. We denote this crack as L_R. For the second case we considered a T crack, modeled with a transverse component that has the same depth a as L_R crack, but followed by two branches of equal length l spread in the horizontal direction.

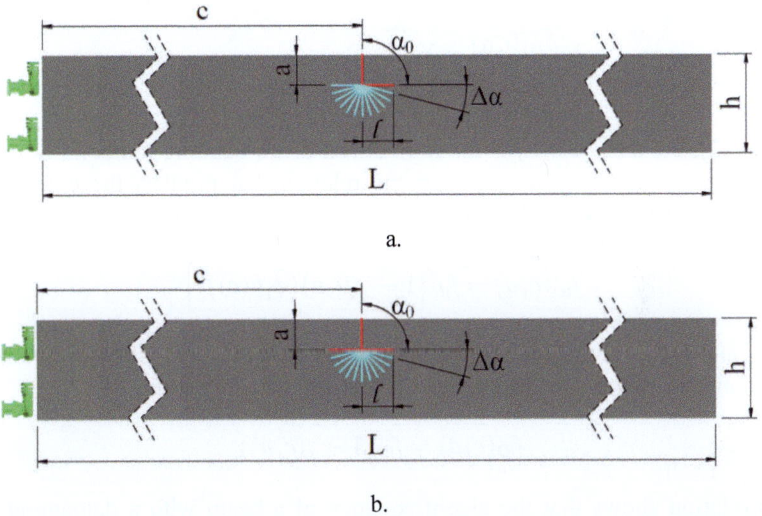

Fig. 2. Branched crack geometry: (a) L – shaped crack, (b) T – shaped crack. The initial branches are marked with a red line and the iterative positions for the rotated right branch with blue lines.

In order to evaluate the effects of the orientation of the crack branches on the beam's eigenfrequencies, we have iteratively removed the right branch of the crack in clockwise direction with a step $\Delta\alpha = 15°$ for both damage cases. Therefore, the right branch has different angles $\alpha_k = \alpha + k\Delta\alpha$, as presented in Fig. 2a for the L_R crack and Fig. 2b for the T crack.

2.3 Theoretical Background and the Test Procedure

At first, we aim finding the beam's free end deflections under dead mass and the eigenfrequencies for the first six transverse modes by means of the ANSYS simulation software. We considered for this analysis the healthy beam and the beam subjected to all described damage scenarios.

Next, we calculate the damage pseudo-severity $\gamma(c, a)$, which is found from the deflection δ_U of the healthy beam and the deflection $\delta_D(c, a)$ of the beam with a crack of depth a which is located at distance c from the fixed end [18], by employing the mathematical relation:

$$\gamma(c,a) = \frac{\sqrt{\delta_D(c,a)} - \sqrt{\delta_U}}{\sqrt{\delta_D(c,a)}} \tag{1}$$

In earlier studies [19], we have shown that the relationship between the pseudo-severity $\gamma(c,a)$ and the severity $\gamma(0,a)$ calculated for the crack located on the beam segment which support the highest bending moment (the fixed end, in the case of a cantilever beam) is expressed as:

$$\gamma(c,a) = \frac{\sqrt{\delta_D(0,a)} - \sqrt{\delta_U}}{\sqrt{\delta_D(0,a)}} \left(\bar{\phi}_i''(c)\right)^2 = \gamma(0,a)\left(\bar{\phi}_i''(c)\right)^2 \tag{2}$$

In Eq. (2), the term $\bar{\phi}_i''(c)$ is the normalized beam bending moment, or curvature, at the location c. Let us now consider the relation that predicts the eigenfrequency evolution with the crack depth a and location c, which is [20]:

$$f_{iD}(c,a) = f_{iU}\left[1 - \gamma(0,a)\left(\bar{\phi}_i''(c)\right)^2\right] \tag{3}$$

where f_{iD} and f_{iU} are the eigenfrequencies of the *i-th* transverse vibration mode for the damaged and healthy beam. From Eq. (2) and Eq. (3), we obtain:

$$f_{iD}(c,a) = f_{iU}(1 - \gamma(c,a)) \tag{4}$$

This relation shows that the eigenfrequency of a beam with a damage at a given position $c \in (0, L]$ can be directly found if the deflections in the healthy and damaged states are known. After simple substitutions, the eigenfrequencies of the damaged beam result as:

$$f_{iD}(c,a) = f_{iU}\sqrt{\frac{\delta_U}{\delta_D(c,a)}} = \kappa(c,a) \cdot f_{iU} \tag{5}$$

where the $\kappa(c,a) = \sqrt{\delta_U}\left(\sqrt{\delta_D(c,a)}\right)^{-1}$.

Since the deflections for the healthy and the damaged beams are known from simulation, we can calculate the coefficients $\kappa(c, a)$ for the proposed damage scenarios.

With these coefficients and the frequencies of the healthy beam, we can predict the frequencies of the damaged beam for all considered vibration modes by employing Eq. (4). These frequencies, obtained from simulation and calculus (which we call the hybrid method), are compared with the frequencies obtained directly from simulation in order to proof the reliability of the deduced Eq. (4) for the case of branched cracks. This equation is important for damage detection because it makes possible quantifying the frequency changes occurred due to a crack.

To have comparable values, it is convenient to work with normalized frequency values, which means the frequency of the healthy beam for any mode is one. The normalized frequency shift for a transverse vibration mode is nominated Relative Frequency Shift (RFS). The RSF for the *i*-th vibration mode is calculated, according to [21], with the mathematical relation:

$$RFS_i(c,a) = \frac{f_{iU} - f_{iD}(c,a)}{f_{iU}} = \frac{\sqrt{\delta_D(c,a)} - \sqrt{\delta_U}}{\sqrt{\delta_D(c,a)}} = \gamma(0,a)\left(\bar{\phi}_i''(c)\right)^2 \qquad (6)$$

From Eq. (6), one can deduce that the RFSs of the damaged beam can be found just knowing the deflection of the healthy beam, the deflection of the beam with a crack at the fixed end, and the modal curvature of the damaged beam slice in healthy state. All these values can be found *a priori*, so a database can be crated for any imaginable damage scenarios. These are finally compared with the measurement results to locate the crack and quantify its dimensions [22]. The second aspect of our study is focused on finding if the RSFs for branched cracks are similar to these for transverse cracks, and how the severity is affected when the crack propagates in various directions. This is made by comparing the RSFs for different values of the angle α_k.

The succession of steps performed for these tests is following:

1. We find from FEM simulation the healthy beam's free end deflection under dead mass;
2. We find from FEM simulation the beam's free end deflection under dead mass for the L_R and T cracks located near the fixed beam's end;
3. We find from FEM simulation the beam's free end deflection under dead mass for the L_R and T cracks located at c = 160 mm;
4. We calculate the damage severity with the results obtained at points 1 and 2 and by involving Eq. (1) applied for the particular case c = 0 mm;
5. We calculate the damage pseudo-severity with the results obtained at points 1 and 3 and by involving Eq. (1) applied for the particular case c = 160 mm;
6. We test the reliability of Eq. (2) by comparing the pseudo-severities with the severities adjusted with the modal curvatures.
7. We find from FEM simulation the eigenfrequencies for the healthy beam;
8. We find from FEM simulation the eigenfrequencies for the beam with the L_R and T cracks located at c = 160 mm and all subsequent cracks;
9. We calculate the RSFs for the damage scenarios and compare the results to observe the changes occurred due to the branch angle alteration.

3 Results and Discussion

By static simulations, we find the deflections for the healthy beam and the beam with cracks having L_{R90} and T_{90} shapes, i.e. $\alpha = 90°$, which have the location of the transverse component at $c = 6$ mm respectively $c = 160$ mm. The results obtained by Fem simulation are presented in Table 3. For these deflections, we calculate the severities and the pseudo-severities; the results are also presented in Table 3.

Table 3. The free end deflections of the beam under dead mass and the resulted severities.

Crack type	Deflection - healthy beam δ_U	Deflection - crack at 6 mm $\delta_D(6,1.5)$	Severity $\gamma(6,1.5)$	Deflection crack at 160 mm $\delta_D(160,1.5)$	Pseudo-severity $\gamma(160,1.5)$
L_{R90}	22.948	23.585	0.013597	23.328	0.008178
T_{90}	22.948	23.835	0.018784	23.483	0.011457

Knowing that the normalized modal curvature is $\bar{\phi}_i''(160) = 0.7808$ we apply Eq. (2) to find out if it is possible to calculate the pseudo-severity at a given position from the severity and modal curvatures. By applying Eq. (2) for the achieved results, hence using the hybrid method, we found following pseudo-severities:

- for the L_{R90} crack $\gamma_{LR90}(160,1.5) = 0.008289$
- for the T_{90} crack $\gamma_{T90}(160,1.5) = 0.011451$

which are very close to these obtained directly from simulation, the errors being less than 1.5%. This justifies considering Eq. (2) applicable for complex-shaped cracks.

From the modal analysis we found the eigenfrequencies for the healthy beam and the beam with the cracks described in previous section. The results are presented in Fig. 3. From this figure one can observe that in all cases the crack produced a frequency drop. It can also be noticed that the L-shaped crack produces quite similar frequency drop as the T-shaped crack, with the observation that for the latter a slightly bigger frequency drop was obtained. This is justified by the supplementary left crack branch. However, the contribution of this additional branch was proved to be low.

We draw attention to the two particular positions of the branch of the crack we rotate. The first refers to the L_R crack that has $\alpha_k = 180°$, which is actually a transverse crack with the depth 3 mm. Similarly, it is the case of the T crack, which is a transverse crack with the depth 3 mm and a longitudinal branch. For this particular branch angle the highest frequency drop is obtained for both crack types. The second case is the angle $\alpha_k = 270°$, where the two crack are similar, actually being a L-shaped crack with the longitudinal branch oriented to the left. Here, the same frequencies are obtained for both crack types. With the frequency values plotted in Fig. 3 we found the evolution of the RFSs with the branch rotation α_k. The obtained values are used to plot Fig. 4. The particular cases L_R with $\alpha_k = 90°$ and $\alpha_k = 180°$ respectively T crack with $\alpha_k = 90°$ and $\alpha_k = 180°$ are explicitly presented in Tables 4 and 5.

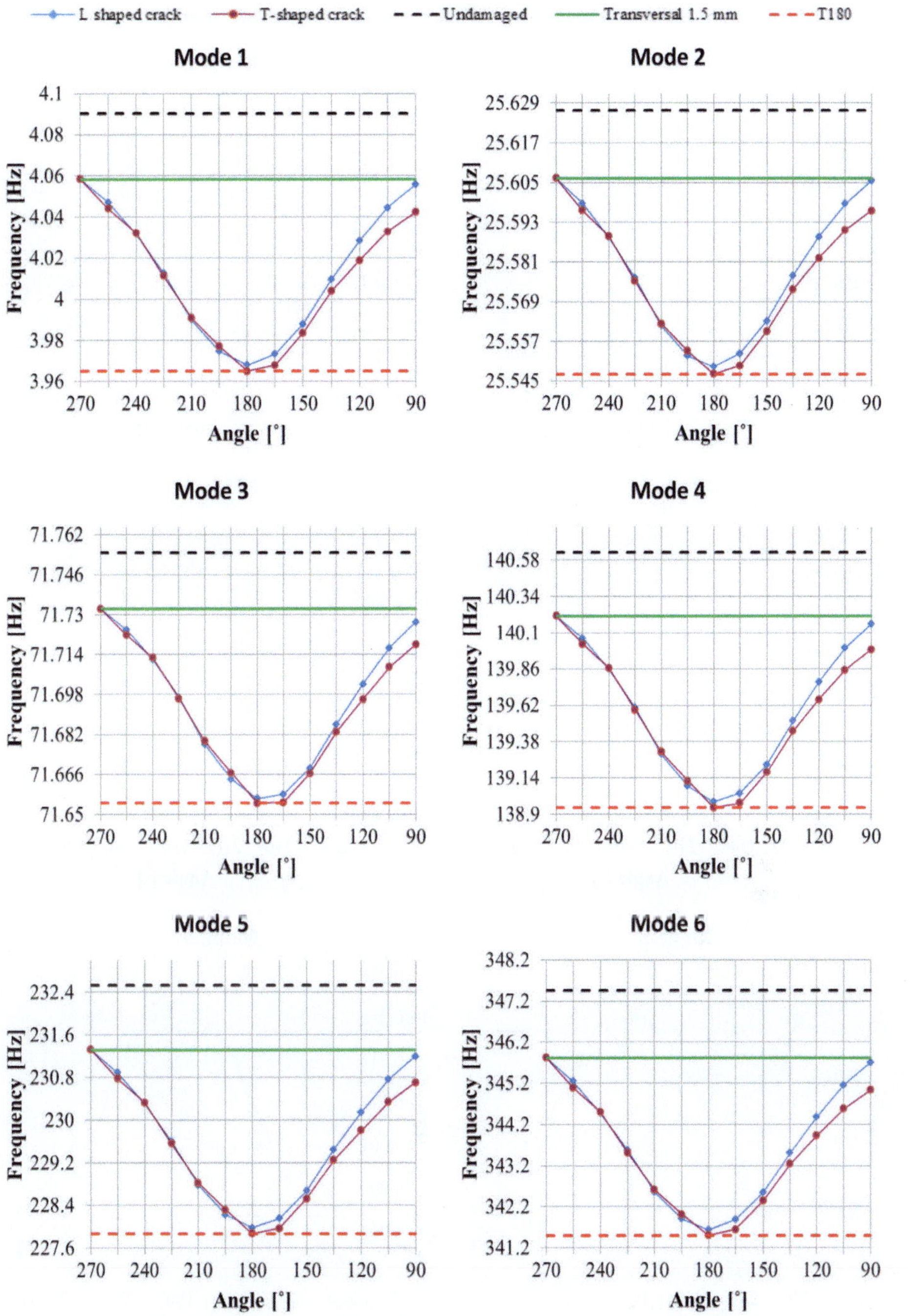

Fig. 3. Eigenfrequency evolution with the crack branch rotation.

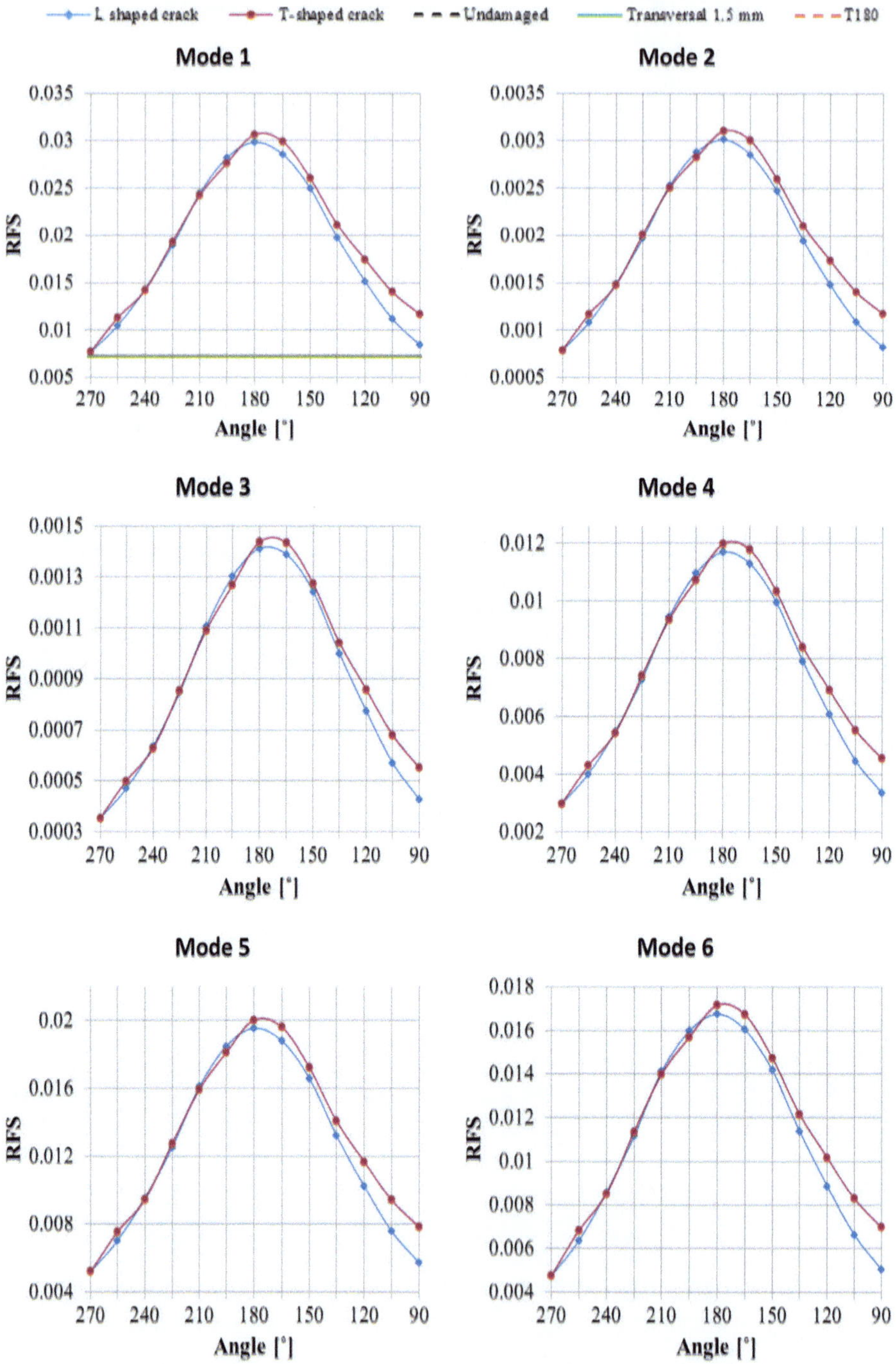

Fig. 4. Relative Frequency Shift evolution with the crack branch rotation.

Table 4. Comparison of frequencies obtained directly from FEM and calculated involving the hybrid method for L-shaped crack

L_{90} analytical	L_{90} FEM	Percent diff.	L_{180} analytical	L_{180} FEM	Percent diff.
4.056551	4.055576	0.02%	3.971244	3.968029	0.08%
25.41697	25.60549	0.74%	24.88246	25.54929	2.68%
71.16788	71.72701	0.79%	69.67125	71.65629	2.85%
139.4775	140.1564	0.49%	136.5444	138.9817	1.79%
230.6192	231.191	0.25%	225.7694	227.9752	0.98%
344.6111	345.6971	0.32%	337.3641	341.6345	1.27%

Table 5. Comparison of frequencies obtained directly from FEM and calculated involving the hybrid method for T-shaped crack

T_{90} analytical	T_{90} FEM	Percent diff.	T_{180} analytical	T_{180} FEM	Percent diff.
4.043141	4.042322	0.02%	3.967903	3.964728	0.08%
25.33295	25.59642	1.04%	24.86153	25.54704	0.03%
70.93262	71.71796	1.11%	69.61264	71.65453	0.03%
139.0164	139.9872	0.70%	136.4295	138.9426	0.02%
229.8569	230.7005	0.37%	225.5795	227.8656	0.01%
343.4719	345.0312	0.45%	337.0803	341.4947	0.01%

The differences in the tables above, resulted by comparing the eigenfrequencies achieved by the hybrid and the FEM method, are quite similar. In general, we found differences less than 1.2%, excepting the L_{180} crack these are up to 2.85%.

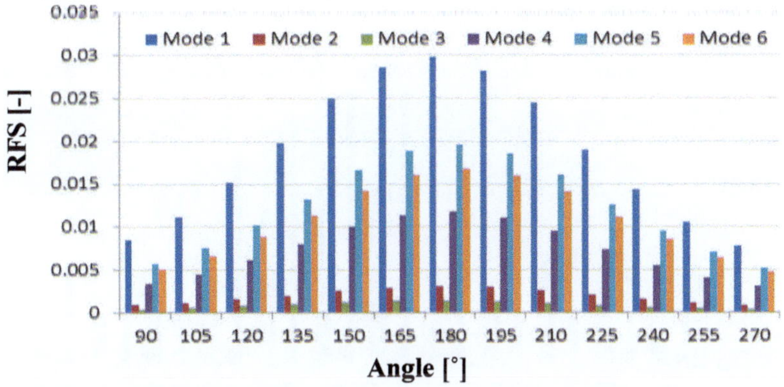

Fig. 5. The Relative Frequency Shifts for the beam with an L-shaped crack.

We calculate the RFSs for the beam with an L_R crack, for all positions of the rotated branch, and represent the achieved values in Fig. 5. One can observe that the pattern for this crack type does not change, since the ratios between the RFS amplitudes are the same irrespective to the angle α_k. This is even better illustrated in Fig. 6, where the Damage Location Coefficients are calculated by dividing the values of the RFS for a given case to the biggest value in the series [22]. Here, the RFS for mode one is this value, therefore it achieves in Fig. 6 always the value one. This property makes finding the crack location independent of its severity estimation.

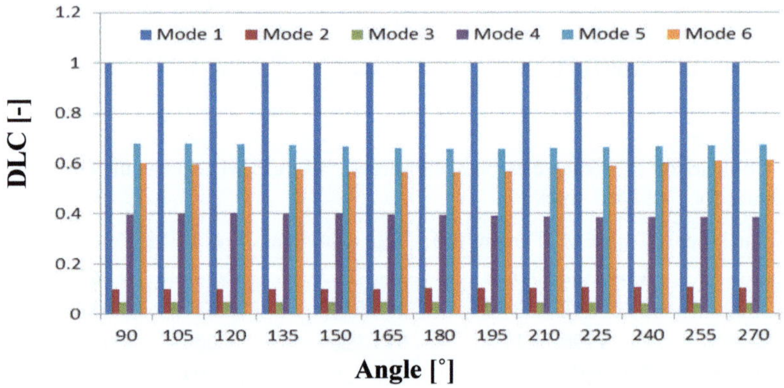

Fig. 6. The Damage Location Coefficients for the beam with an L-shaped crack.

However, the absolute amplitudes of the RFS depend on the branch orientation because different severities are obtained for different angles α_k. If the crack position is found from the DLCs, by dividing one of the RFSs to the corresponding DLC one obtain the damage severity. This should be preferably made for the biggest RFS, to avoid small numbers which are most susceptible to introduce errors (Fig. 7).

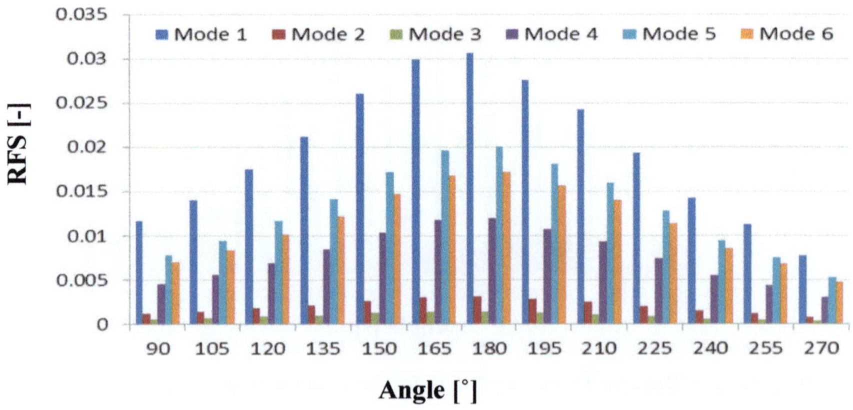

Fig. 7. The Relative Frequency Shifts for the beam with a T-shaped crack.

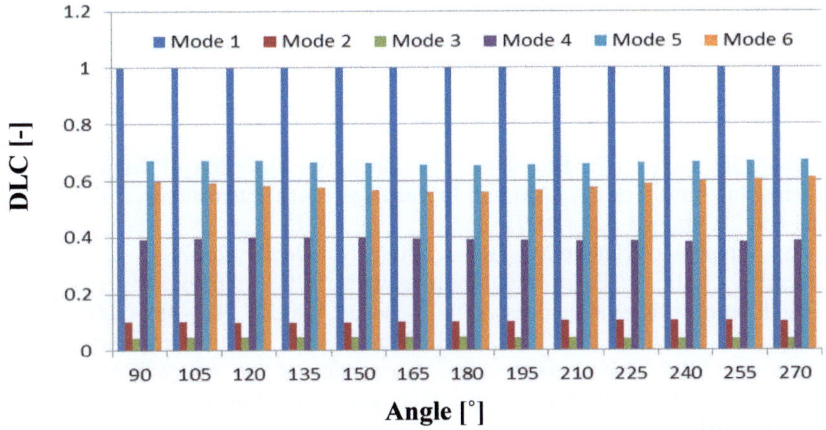

Fig. 8. The Damage Location Coefficients for the beam with a T-shaped crack.

The same comments as for the L-shaped crack are valid for the T-shaped crack too, see Figs. 7 and 8. Comparing Fig. 6 with Fig. 8 clearly results that the DLCs are similar, which shows that the DLCs are in connection whit the damage location only. Moreover, the DLCs are similar to that of a transverse crack. This is proved by the crack L_{180}, which is actually a transverse crack with depth $a + l = 3$ mm.

4 Conclusion

The paper analyzes how cracks with oblique components affect the dynamical behavior of beams. The aim was to find out if it is possible to extend the use of the earlier derived mathematical relations predicting the frequency changes due to transverse cracks to cracks with a more complex shape.

By calculating the severity and pseudo-severity for two types of cracks, we found the relation between these two features is still given by the normalized modal curvature found for the healthy beam at the slice which will be affected. Comparing the results obtained directly from FEM simulation which these obtained involving the hybrid method, we found errors less than 1.3%. This confirms that, the relation contrived for transverse cracks is generally applicable, irrespective of the direction of crack propagation.

We also found out that the only parameter that needs to be considered when looking for the crack position is the location at which the end of the crack emerges from the beam. The crack evolution inside the beam, if it is present in a relatively narrow beam segment, is less important and has no relevant influence on the frequency changes. This is also found from the Relative Frequency Shifts calculated for the two crack types and considering the various orientation of the oblique branch differ only as amplitudes, the pattern being the same. Moreover, the normalized RFSs, namely the Damage Location Coefficients are independent of the damage severity, therefore being perfect tool to identify the crack position on the beam.

References

1. Yang, Z.B., Radzienski, M., Kudela, P., Ostachowicz, W.: Scale-wavenumber domain filtering method for curvature modal damage detection. Compos. Struct. **154**, 396–409 (2016)
2. Gillich, G.R., Gillich, N., Birdeanu, E.D., Iancu, V.: Detection of damages in simple elements. In: Annals of DAAAM and Proceedings of the International DAAAM Symposium, vol. 20, pp. 623–624 (2009)
3. Rao, S.: Vibration of continuous systems. Wiley, New Jersey (2007)
4. Caddemi, S., Calio, I.: Exact closed-form solution for the vibration modes of the Euler-Bernoulli beam with multiple open cracks. J. Sound Vib. **327**, 473–489 (2009)
5. Praisach, Z.I., Minda, P.F., Gillich, G.-R., Minda, A.A.: Relative frequency shift curves fitting using FEM modal analyses. In: Proceedings of the 4th WSEAS International Conference on Finite Differences - Finite Elements - Finite Volumes - Boundary Elements, pp. 82–87 (2011)
6. Zhang, K., Yan, X.: Multi-cracks identification method for cantilever beam structure with variable cross-sections based on measured natural frequency changes. J. Sound Vib. **387**, 53–65 (2017)
7. Gillich, G.-R., Praisach, Z.-I., Onchis-Moaca, D., Gillich, N.: How to correlate vibration measurements with FEM results to locate damages in beams. In: Proceedings of the 4th WSEAS International Conference on Finite Differences - Finite Elements - Finite Volumes - Boundary Elements, pp. 76–81 (2011)
8. Sinha, J.K., Friswell, M.I., Edwards, S.: Simplified models for the location of cracks in beam structures using measured vibration data. J. Sound Vib. **251**(1), 13–38 (2002)
9. Gillich, G.R., Praisach, Z.I., Onchis, M.D.: About the effectiveness of damage detection methods based on vibration measurements. In: International Conference on Engineering Mechanics, Structures, Engineering Geology, International Conference on Geography and Geology – Proceedings, pp. 214–219 (2010)
10. Song, Y.Z., Bowen, C.R., Kim, A.H., Nassehi, A., Padget, J., Gathercole, N.: Virtual visual sensors and their application in structural health monitoring. Struct. Health Monit. **13**(3), 251–264 (2014)
11. Rizos, P.F., Aspragathos, N., Dimarogonas, A.D.: Identification of crack location and magnitude in a cantilever beam from the vibration modes. J. Sound Vib. **138**(3), 381–388 (1990)
12. Ostachowicz, W.M., Krawczuk, C.: Analysis of the effect of cracks on the natural frequencies of a cantilever beam. J. Sound Vib. **150**(2), 191–201 (1991)
13. Chondros, T.J., Dimarogonas, A.D., Yao, J.: A continuous cracked beam vibration theory. J. Sound Vib. **215**(1), 17–34 (1998)
14. Gillich, G.R., Mituletu, I.C., Praisach, Z.I., Negru, I., Tufoi, M.: Method to enhance the frequency readability for detecting incipient structural damage. Iran. J. Sci. Technol., Trans. Mech. Eng. **41**(3), 233–242 (2017)
15. Gillich, G.R., Negru, I., Praisach, Z.I.: Damages influence on dynamic behaviour of composite structures reinforced with continuous fibers. Mater. Plast. **49**(3), 186–191 (2012)
16. Gillich, G.R., Maia, N.M.M., Mituletu, I.C., Praisach, Z.I., Tufoi, M., Negru, I.: Early structural damage assessment by using an improved frequency evaluation algorithm. Lat. Am. J. Solids Struct. **12**(12), 2311–2329 (2015)

17. Gillich, G.R., Praisach, Z.I., Hamat, C., Gillich, N., Ntakpe, J.L.: Crack localization in L-shaped frames. In: Herisanu, N., Marinca, V. (eds.) Acoustics and Vibration of Mechanical Structures - AVMS-2017. Springer Proceedings in Physics, vol. 198, pp. 315–322. Springer, Cham (2018)
18. Gillich, G.R., Tufoi, M., Korka, Z.I., Stanciu, E., Petrica, A.: The relations between deflection, stored energy and natural frequencies, with application in damage detection. Rom. J. Acoust. Vib. **13**(2), 87–93 (2016)
19. Nitescu, C., Gillich, G.R., Abdel Wahab, M., Manescu, T., Korka, Z.I.: Damage severity estimation from the global stiffness decrease. J. Phys.: Conf. Ser. **842**(1), art. 012034 (2017)
20. Gillich, G.R., Praisach, Z.I., Iavornic, M.C.: Reliable method to detect and assess damages in beams based on frequency changes. In: Proceedings of the ASME 2012 International Design Engineering Technical Conferences and Computers and Information in Engineering Conference IDETC/CIE 2012, pp. 3806–3810 (2010)
21. Gillich, G.R., Praisach, Z.I., Iancu, V., Furdui, H., Negru, I.: Natural frequency changes due to severe corrosion in metallic structures. Strojniški vestnik - J. Mech. Eng. **61**(12), 721–730 (2015)
22. Gillich, G.R., Abdel Wahab, M., Praisach, Z.I., Ntakpe, J.L.: The influence of transversal crack geometry on the frequency changes of beams. In: International Conference on Noise and Vibration Engineering ISMA2014, ID 666 (2014)

The Application of Spatial Filtration for Damage Detection in Structures with Multiple Poles

Krzysztof Mendrok(✉)

Department of Robotics and Mechatronics, AGH University of Science and Technology, al. Mickiewicza 30, 30-059 Krakow, Poland
mendrok@agh.edu.pl

Abstract. For many years, modal analysis and its results has been used to detect and locate damage. Also modal filtration, which is related to modal analysis, was recently applied for this purposes. Especially the method, which involves the detection of peaks that appear on the badly filtered output characteristics of the spatial filter. This poor filtration results from the fact that the spatial filter is tuned for the object in the reference (undamaged) state, and the filtered characteristics are recorded on the structure in the current (possibly damaged) state. Modal filter coefficients are the coordinates of a reciprocal modal vector which main property is orthogonality for all modal vectors except the one for which filter has been set. When the damage occurs, the modal vectors change, and the orthogonality of the reciprocal modal vector ceases to occur. Then the modal filter does not work properly. There are therefore additional peaks on the filtered characteristics and their presence is a symptom of damage.

The question arises how the method will work for the system, which contained double or multiple poles. In such a case, badly filtered peaks could be located at the same frequency as the main peak resulting from the natural frequency to which the filter was set. This can lead to failure in damage detection. Therefore, appropriate simulations have been carried out, the results of which will be presented in the article.

Keywords: Modal filter · Damage detection · Multiple poles

1 Introduction

Many methods of damage detection and location are based on the analysis of changes in modal model parameter. Starting from changes in natural frequency [1], through changes in damping coefficients, [2] to various modifications of the shape of modes [3–6]. The modal filter is a tool related to modal analysis [7, 8]. This is a kind of spatial filter that removes all structural components from the vibration response of the system except the one to which the filter is set. This filtration is carried out in accordance with the formula (1):

M. A. Wahab (Ed.): DAMAS 2019, LNME, pp. 92–101, 2020.
https://doi.org/10.1007/978-981-13-8331-1_7

$$\eta_r(\omega) = \psi_r^T \cdot \{x(\omega)\} \tag{1}$$

where: ψ_r – r-th reciprocal modal vector
$x(\omega)$ – vector of system responses.

Further, the modal coordinates η_r might by scaled to the known input, and the FRF is calculated with all peaks, except r – th, filtered out.

The modal filter was originally developed for vibrations control [9]. Soon a number of other applications were found, such as identification of excitation forces, comparison of modal models, and health monitoring of the structure [8].

For the first time, the modal filter, and in principle its adaptive [10] version, was described in the context of monitoring in the work of Shelley and Slater in 1993 [11]. The solution was designed to track changes in selected object modes. Thanks to that it was possible to control object vibrations and detect changes in its structural parameters. These changes could be further used to detect damage. The authors have developed several variants of an adaptive modal filter. Operating in domain of time or frequency, with knowledge of excitation forces or without. An attempt to use an adaptive modal filter to monitor the structural integrity of selected elements of a wind turbine tower was described in [12].

Another application of the modal filter to detect damage was presented by El-Ouafi Bahlous et al. [13]. In this approach, modal filtration was only an element of data preparation. Based on the data registered in the reference state, collected in the current state and the FE model of the object, residual functions ware calculated. The functions had a normal distribution with a mean value equal zero for systems without damage and different from zero for systems with structural changes. The generalized log-likelihood ratio test was used to verify the statistical parameters of the residues.

Also Deraemaeker and Preumont in 2006 used modal filtration for damage detection [14]. The described method consisted in filtering the frequency characteristics of the object in the current state with a modal filter tuned to the data collected in the reference state. The method showed great potential for practical application, because it had such advantages as: the simplicity of the algorithm, low demand for computing power, work on measurement data without the need to identify the modal model at every diagnosis and a certain resistance to changes in external conditions. In addition, its results were easy to interpret.

The method was further developed to allow its use for operational data [15, 16], It was also extended to allow the localization of the damage [17, 18] and filtration of harmonic components [19]. An industrial monitoring system based on this concept has also been developed. For this purpose, a series of simulation and experimental studies of the method's behavior in various conditions and for various objects were carried out [20–22].

In this work, further tests of the effectiveness and robustness of the method were carried out. This time its effectiveness for multi-pole systems was verified. A detailed description of the method and the motivation to undertake this type of experiment was presented. Created simulation model and obtained results were described.

2 Goal of the Research

In this section the explanation what is the main purpose of the research and why it is important will be placed.

To apply method described at the end of Chapter 1, first the modal analysis of the object in the reference (undamaged state) has to be performed. As a result of the analysis, modal parameters are identified: natural frequencies ω_n, modal damping coefficients ξ_n and modal vectors Φ_n. Next, on the basis of this modal model, reciprocal modal vectors Ψ_n (modal filter coefficients) are calculated. Further operation of the method is shown in Fig. 1 (for a reference object) and Fig. 2 (for an object with damage).

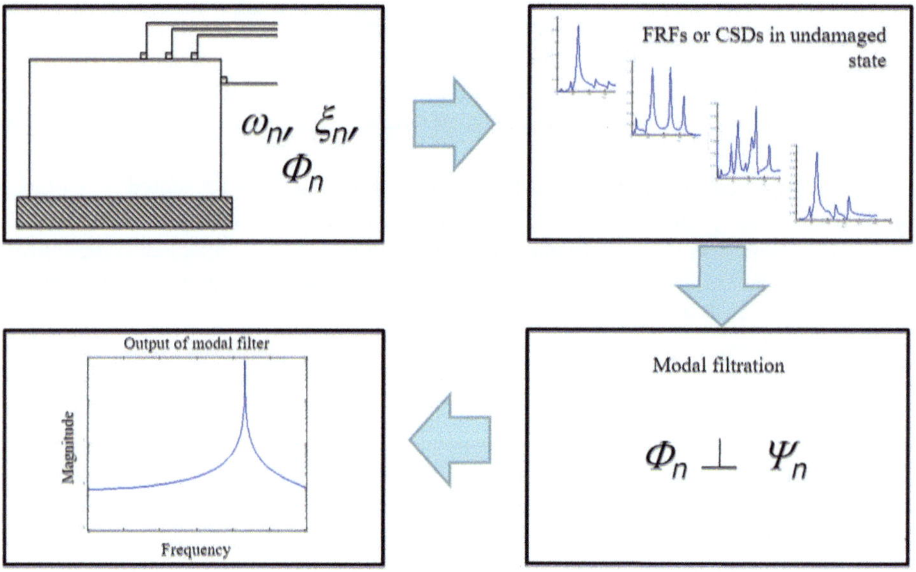

Fig. 1. Method operation for undamaged system

The upper left block in Fig. 1 shows the measurement of the object's vibrations in the undamaged state (modal parameters with index n - for which modal filter coefficients are determined). Then (right upper block) the frequency characteristics of the object are calculated (FRFs or CSDs) and are subjected to modal filtration (bottom right block). Due to the fact that modal parameters are identical to those for which filter coefficients have been determined (there is orthogonality between the relevant modal vectors and the selected reciprocal modal vector), the filtration takes place correctly and the output shows a characteristic with only one maximum - corresponding to the natural frequency to which the filter was tuned.

In the damaged case, a local drop of stiffness appears in the object, which is reflected by the change in the stiffness matrix of the object's mathematical model. It further results in different values of modal parameters (indexed with letter d), and

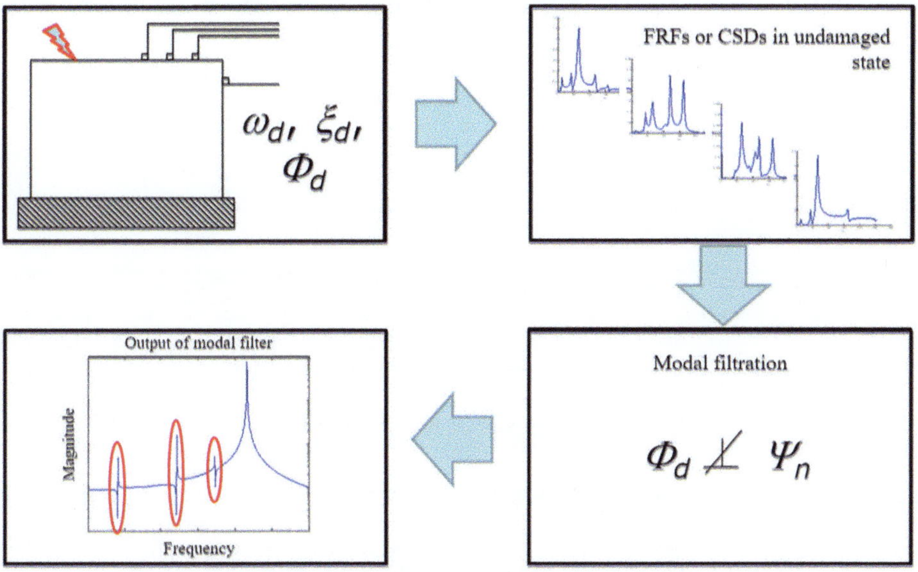

Fig. 2. Method operation for damaged system

changed shapes of recorded time histories and estimated frequency characteristics. The key issue here is the change of modal vectors, which are no longer orthogonal to the selected reciprocal modal vector. Lack of the orthogonality spoils the quality of the filtration and additional peaks appear on the output.

Recently, the author carried out a modal analysis of the symmetrical system, which contained double poles. It prompted him to consider how the damage detection method would work for such an object. In such a case, badly filtered peaks could be located at the same frequency as the main peak resulting from the natural frequency to which the filter was set. This can lead to failure in damage detection. Therefore, appropriate simulations have been carried out, the results of which are presented in the next sections.

The following research plan was adopted: first, a system with 6 degrees of freedom was created with three poles in the same frequency and the method was tested in such conditions. Then an analysis was made to answer the question: how far the poles must be, for the method to work.

3 Simulation Model

For further simulations, a model with 6 degrees of freedom was developed. It is shown in Fig. 3. Its physical parameters are gathered in Table 1.

The following notation was used: the stiffness between mass i and j – k_{ij}, the damping coefficient between mass i and j – c_{ij}. Viscous damping was assumed for the model, because the modal filter works only for the normal modal vectors.

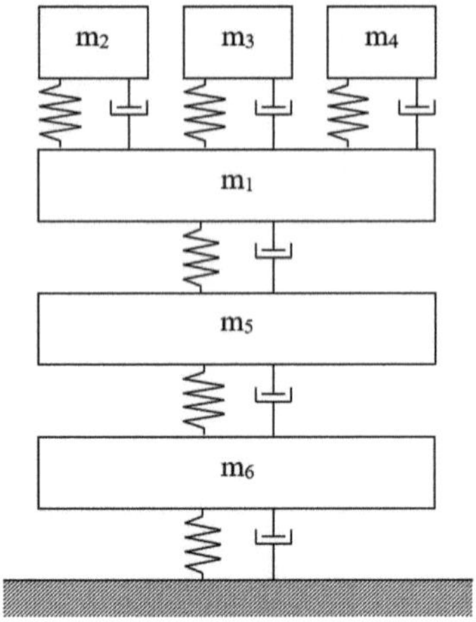

Fig. 3. Scheme of the simulation model

Table 1. Physical and modal parameters of the simulation model

Mass [kg]	Damping coeff. [N s/m]	Stiffness coeff. [N/m]	Natural frequ. [Hz]
$m_1 = 1$; $m_2 = 11$; $m_3 = 11$; $m_4 = 11$; $m_5 = 1$; $m_6 = 1$;	$c_{12} = 3$; $c_{13} = 3$; $c_{14} = 3$; $c_{15} = 5$; $c_{56} = 5$; $c_{60} = 9$;	$k_{12} = 900$; $k_{13} = 900$; $k_{14} = 900$; $k_{15} = 70000$; $k_{56} = 60000$; $k_{60} = 60000$;	$\omega_1 = 8.5$; $\omega_2 = 9$; $\omega_3 = 9$; $\omega_4 = 116$; $\omega_5 = 319$; $\omega_6 = 455$

In order to calculate the modal parameters of the model, the eigenvalue problem was solved assuming zero initial conditions for displacements and velocities. As a result, six eigenvalues and eigenvectors were obtained. On their basis, natural frequencies (presented also in Table 1) and mode shapes were determined.

As one can notice the physical parameters were set in the way to obtain three natural frequencies almost at the same frequency. Two of them were identical due to the symmetry of the system. The third one is distant by 0.5 Hz, which with the assumed frequency resolution of the characteristics equal to 0.5 Hz practically creates a system with a triple pole.

4 Obtained Results

In this section the results of structural changes detection for the model with triple pole will be presented. Also analysis how far (in terms of natural frequency) one from another the poles should be located to allow the method work properly.

4.1 Results for the System with Triple Pole

Six simulations were carried out. In each of them, the stiffness coefficient of subsequent springs was lowered by 5%. Next, the outputs of modal filters set for subsequent mode shapes were observed. In total, 36 sets of results were obtained. The most interesting were cases of damage in springs k_{12}, k_{13} and k_{14}, when the modal filter was set to one of the first three mode shapes. In these simulations, the effect, that was anticipated, appeared, i.e. the damage was not detected. In tests, where the modal filter was set to higher natural frequencies, the method worked correctly. Also cases of damage in other springs did not cause problems in the operation of the method. Figures 4 and 5 present the results of modal filtration for damaged spring k_{13} and filter set for 1st and 2nd mode shape respectively (results for 3rd mode shape are the same). In Fig. 6, just for comparison, selected case in which the method worked correctly is presented.

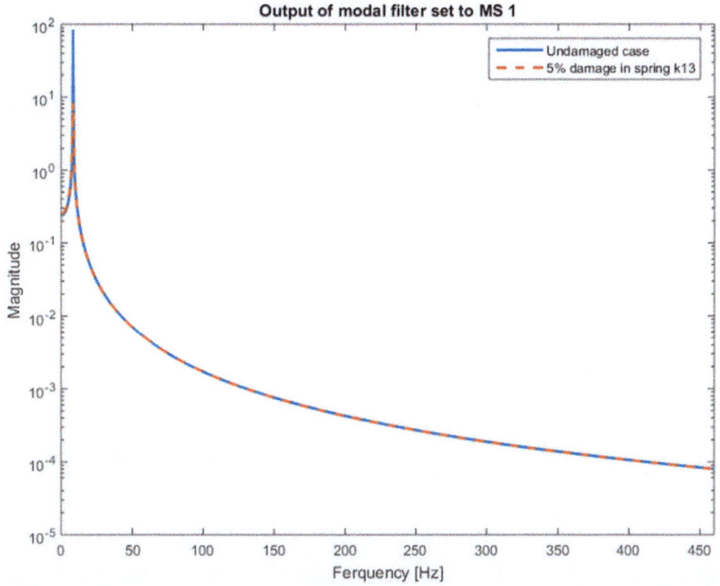

Fig. 4. Results of modal filtration for damage in spring k_{13} and modal filter tuned to mode shape no. 1

It is visible that there are cases where the method based on modal filtration cannot detect damage located in the system with specific structure, with multiple poles.

Of course, this does not disqualify the method, because as shown in Fig. 6, the damage is detected by a modal filter set to other mode shape.

4.2 Analysis of Required Pole Distance in the Frequency Domain

In the next step, the physical parameters of the prepared model were slightly changed, to move apart the poles of triple pole (change their natural frequency). Only the stiffness

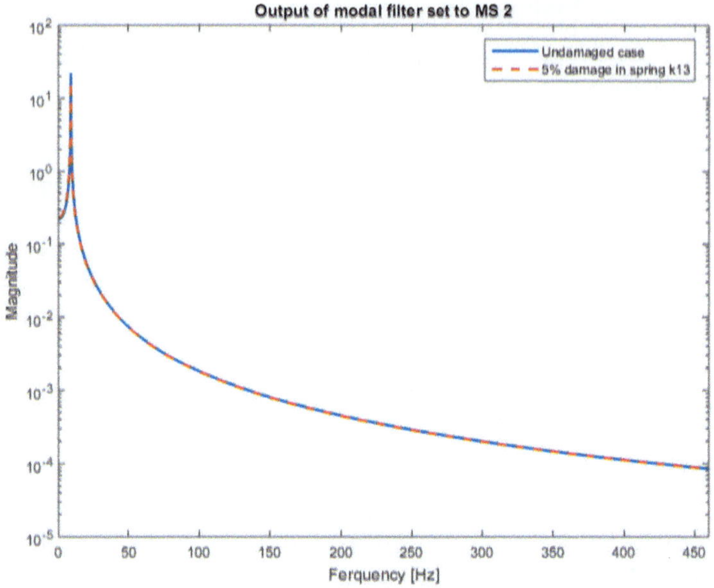

Fig. 5. Results of modal filtration for damage in spring k_{13} and modal filter tuned to mode shape no. 2

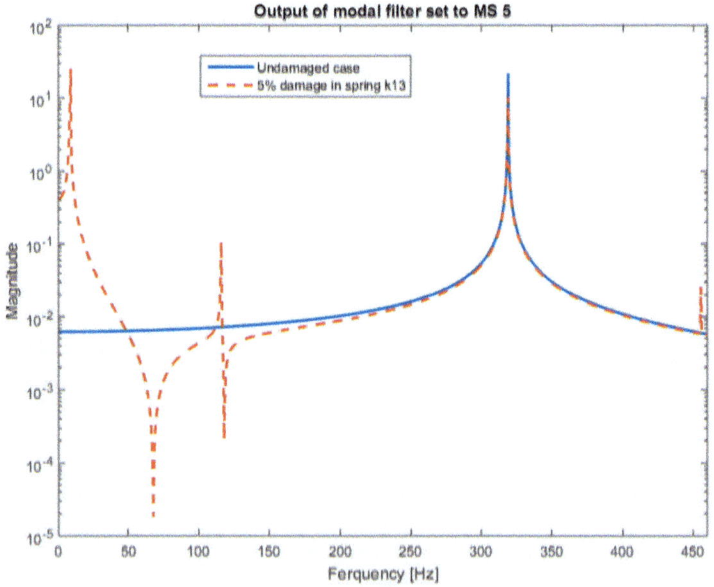

Fig. 6. Results of modal filtration for damage in spring k_{13} and modal filter tuned to mode shape no. 5

coefficients of the springs k_{12}, k_{13} and k_{14} were operated. Because the frequency resolution of the characteristics was 0.5 Hz, the same value for the step of the natural frequency changes was adopted. Three tests were carried out, i.e. the multiple pole was split into three poles that were distant from each other by 1, 1.5 and 2.2 Hz. Table 2 presents the changed values of stiffness coefficients and the corresponding three natural frequencies of interest. Figure 7 shows the results of modal filtration for subsequent simulations (including the triple pole case). The filter was set to the 3rd mode shape and the spring k_{13} was damaged (similar as in the previous section of the article). The frequency band was limited to 20 Hz in order to better visualize area of interest.

Table 2. Physical and modal parameters of the model in consecutive simulations

Simulation 1 – poles distant by 1 Hz		Simulation 2 – poles distant by 1.5 Hz		Simulation 3 – poles distant by 2.2 Hz	
Stiff. coeff. [N/m]	NF [Hz]	Stiff. coeff. [N/m]	NF [Hz]	Stiff. coeff. [N/m]	NF [Hz]
k_{12} = 1270; k_{13} = 1500; k_{14} = 1730;	ω_1 = 10.75; ω_2 = 11.9; ω_3 = 12.9;	k_{12} = 1150; k_{13} = 1500; k_{14} = 1850;	ω_1 = 10.3; ω_2 = 11.8; ω_3 = 13.3;	k_{12} = 900; k_{13} = 1500; k_{14} = 2100;	ω_1 = 9.5; ω_2 = 11.8; ω_3 = 14;

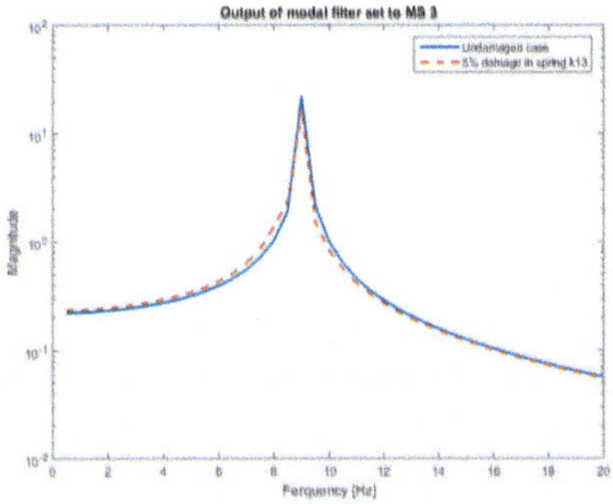

Triple pole

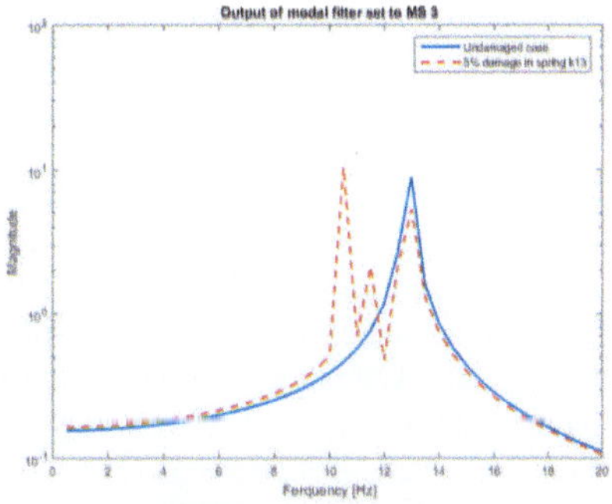

Poles distant by 1 Hz

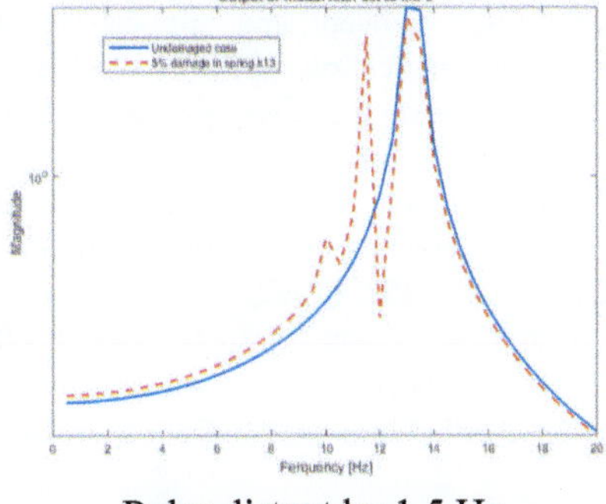

Poles distant by 1.5 Hz

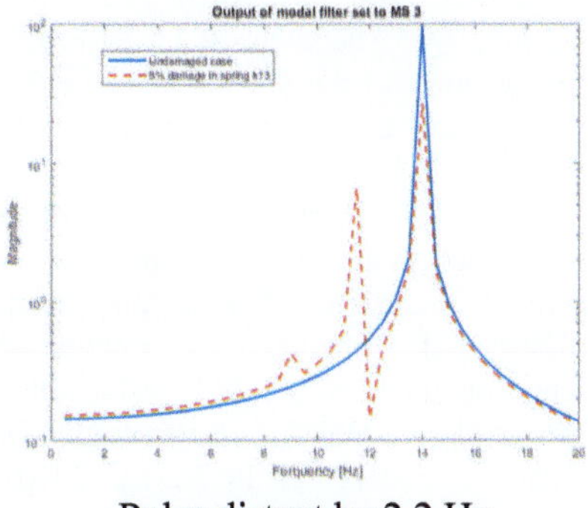

Poles distant by 2.2 Hz

Fig. 7. Results of modal filtration for consecutive simulations

In each of the obtained results it was possible to detect the damage - additional maxima appeared on the modal filter outputs. The conducted simulations showed that the difference in the natural frequency of neighboring poles should be twice as large as the frequency resolution of the characteristics, so that the method allows detection of damage.

5 Summary

As demonstrated in the article, there are cases where the damage detection method based on modal filtration of the object's frequency characteristics may not work correctly. However, performed simulations proved that it does not disqualify the method. The appropriate analysis of the dynamics of the tested system and the use of many modal filters for detection allows to eliminate the risk of undetected damage.

Acknowledgement. The author would like to acknowledge support from 16.16.130.942-KRIM subvention.

References

1. Ruotolo, R., Surace, C.: Damage assessment of multiple cracked beams: numerical results and experimental validation. J. Sound Vib. **206**(4), 567–588 (1997)
2. Kawiecki, G.: Modal damping measurement for damage detection. Smart Mater. Struct. **10** (3), 466 (2001)
3. Ahmadian, H., Mottershead, J.E., Friswell, M.I.: Damage location indicators from substructure mode shapes. Inverse Probl. Eng. **8**(4), 309–323 (2000)
4. Abdel Wahab, M.M., De Roeck, G.: Damage detection in bridges using modal curvatures: application to a real damage scenario. J. Sound Vib. **226**(2), 217–235 (1999)
5. Rucka, M., Wilde, K.: Application of continuous wavelet transform in vibration based damage detection method for beams and plates. J. Sound Vib. **297**(3–5), 536–550 (2006)
6. Shi, Z.Y., Law, S.S., Zhang, L.M.: Structural damage detection from modal strain energy change. J. Eng. Mech. **126**(12), 1216–1223 (2000)
7. Meirovitch, L., Baruh, H.: The implementation of modal filters for control of structures. J. Guid. Control. Dyn. **8**(6), 706–716 (1985)
8. Zhang, Q., Allemang, R.J., Brown, D.L.: Modal filter: concept and applications. In: Proceedings of the 8th International Modal Analysis Conference, pp. 487–496 (1990)
9. Meirovitch, L., Baruh, H.: Control of self-adjoint distributed-parameter systems. J. Guid, Control (1982)
10. Shelley, S.J., Freudinger, L.C., Allemang, R.J.: Development of an On-line parameter estimation system using the discrete modal filter. In: Proceedings of the 10th International Modal Analysis Conference, pp. 173–183, San Diego, CA, USA (1992)
11. Slater, G.L., Shelley, S.J.: Health monitoring of flexible structures using modal filter concepts. Proc. SPIE **1917**, 997–1008 (1993)
12. Mendrok, K., Uhl, T.: Health monitoring of off-shore wind energy power plant with use of adaptive modal filter. In: ISMA (2010)

13. El-Ouafi Bahlous, S., Abdelghani, M., Smaoui, H., El-Borgi, S.: A modal filtering and statistical approach for damage detection and diagnosis in structures using ambient vibrations measurements. Journal Vib. Control **13**(3), 281–308 (2007)
14. Deraemaeker, A., Preumont, A.: Vibration based damage detection using large array sensors and spatial filters. Mech. Syst. Signal Process. **20**(7), 1615–1630 (2006)
15. Mendrok, K., Uhl, T.: The application of modal filters for damage detection. Smart Struct. Syst. **6**(2), 115–133 (2010)
16. Mendrok, K., Kurowski, P.: Operational modal filter and its applications. Arch. Appl. Mech. **83**(4), 509–519 (2013)
17. Mendrok, K., Uhl, T.: Modal filtration for damage detection and localization. In: Structural Health Monitoring 2008: Proceedings of the Fourth European Workshop (2008)
18. Mendrok, K., Uhl, T.: Experimental verification of the damage localization procedure based on modal filtering. Struct. Heal. Monit. **10**(2), 157–171 (2011)
19. Mendrok, K., Wójcicki, J., Uhl, T.: An application of operational deflection shapes and spatial filtration for damage detection. Smart Struct. Syst. **16**(6), 1049–1068 (2015)
20. Mendrok, K., Uhl, T.: Numerical tests of a damage detection procedure based on modal filtration. In: Proceedings of the 5th European Workshop on Structural Health Monitoring (2010)
21. Mendrok, K., Uhl, T., Maj, W., Packo, P.: SHM system based on modal filtration. Key Eng. Mater. **518**, 289–297 (2012)
22. Mendrok, K., Maj, W., Uhl, T.: Industrial tests of the SHM system based on modal filtration. In: Proceedings of the 6th European Workshop on Structural Health Monitoring (2012)

Experimental Validation of Damage Indices Based on Complex Modes for Damage Detection in Vibrating Structures

F. Iezzi[1(✉)], C. Valente[1], and F. Brancaleoni[2]

[1] Department of Engineering and Geology, University "G. D'Annunzio" of Chieti-Pescara, Viale Pindaro 42, 65127 Pescara, Italy
fabrizio.iezzi@unich.it

[2] Department of Architecture, Roma Tre University, Largo G.B. Marzi 10, 00153 Rome, Italy

Abstract. Commonly, the damage detection techniques for structures prone to earthquakes are based on the changes of modal parameters between different structural states, namely before and after a seismic event. Recently, in this context, damage indices based on measures of the imaginary part of complex mode shapes turned out to be an interesting alternative as compared to the direct use of the modal parameters. In fact, numerical simulations proved that the increase in mode shapes complexity comes from the energy dissipated along with damage occurrence. In literature, several indices have been suggested on purpose. However, such indices, for being successful in damage detection, should possess at least two properties with respect to the damage severity: monotony, to guarantee solution uniqueness, and sensitivity, to cope with incipient damage. The present work aims to investigate experimentally the behavior of one of these indices that has been proved the most effective by previous numerical studies. To this end, the results provided by a laboratory physical model subjected to base motion have been used. The damage severity was graded by stepping the amplitude of the base motion so that quasi-linear conditions of the overall structure response can be preserved. The model response was processed via the joint use of the Empirical Mode Decomposition and the Complex Plane Representation methods. The resulting damage index was used to validate the numerical results of previous studies; it was found that numerical and experimental values of the damage index are well comparable.

Keywords: Damage indices · Complex modes · Dynamic experimental tests

1 Introduction

Damage detection is an important research area of structural engineering. In the past, it has received considerable attention, especially in the dynamic field, from the pioneering work in [1]. This subject, although well established, still rises interest particularly for civil structures. In fact, the recent worldwide seismic events have made timely and important the development of new techniques for the structural damage detection.

M. A. Wahab (Ed.): DAMAS 2019, LNME, pp. 102–123, 2020.
https://doi.org/10.1007/978-981-13-8331-1_8

In the seismic field, the analysis of the dynamic behavior of a structure can be used to assess the suffered damage. To this end, the use of modal parameters, or their appropriate functions, shows effective as long as the structure preserves a quasi-linear behavior. In fact, changes in the physical properties of a structure (mass, stiffness, energy dissipation) are reflected in as much changes of the modal parameters (frequencies, mode shapes, damping).

There exist a significant number of techniques based on changes in modal parameters and targeted to damage assessment [2]. Changes in natural frequencies have been the topic of numerous research studies due to the ease of their identification [3]. These studies revealed that the frequency changes are sensitive to damage but hardly allow its localization and further that detectable changes could require high damage levels. On the contrary, the mode shapes are far more effective to localize the damage than frequencies [4], but they rarely suffice alone for damage quantification issues. Damping has seldom been used for damage assessment. This is due to the existence of multiple damping mechanisms and the large scatter in estimating damping values [5, 6]. A review of the literature, however, suggests that damping may prove, in some cases, to be more advantageous than detection schemes based on frequencies and mode shapes. The use of damping for damage detection is mentioned in several applications [7–15].

Recently, the damping has been used in conjunction with the effects induced on the mode shapes in energy dissipating structures [16–20]. The fundamental hypothesis is that undamaged structures exhibit proportional damping, yielding real natural mode shapes. Structural damage implies a loss of stiffness and an increase of dissipated energy during vibration with respect to the undamaged situation: this is associated with non-proportional damping. The natural mode shapes become complex and their imaginary part can be used for damage evaluation.

A number of indices has been proposed and analyzed to provide an effective measure of the mode shapes complexity [16–21]. A group of indices requires the knowledge of the damping matrix of the structure; whereas another group of indices requires the knowledge of the mode shapes. From an experimental point of view, the first group has only a theoretical value; the practical applicability is therefore restricted to the second group of indices. However, to get successful indices, two properties should be fulfilled: monotony and sensitivity both related to uniqueness aspects.

The present work aims to provide an experimental validation of the effectiveness of the indices based on the measure of complexity of the mode shapes. The analyzed data are collected from the seismic response of a framed r/c laboratory structure mounted on a shaking table. The intensity of the base motion is progressively raised in order to increase the damage severity.

The paper is organized as follows. Initially, the assumed relation between structural damage and modal complexity is presented. Then, the damage indices are briefly introduced and divided according to their measurability. Subsequently, the adopted framework for the identification of complex mode shapes, based on the joint and sequential use of the EMD method [22, 23] and of the CPR method [24, 25], is applied to experimental data.

Based on the above study, the most representative index is selected to be used in the procedure of damage detection. The results are finally discussed and issues related to monotonicity, sensitivity, stability and repeatability of the damage index are also addressed.

In conclusion, it deserves to underline that the data relevant to laboratory tests were kindly provided by the Italian Civil Protection Department and in particular by the Seismic and Volcanic Risk Office of Rome, which performed the experimental tests, together with the ENEA (Italy) and Basilicata University (Italy), within the TREMA project (Technologies for the Reduction of seismic Effects on Architectural Manufacts) [26].

2 Relation Between Structural Damage and Modal Complexity

The discrete form of the equation of motion of a structure with linear behavior endowed with viscous damping is:

$$\mathbf{M}\ddot{\mathbf{x}}(t) + \mathbf{C}\dot{\mathbf{x}}(t) + \mathbf{K}\mathbf{x}(t) = \mathbf{P}(t) \tag{1}$$

where t is the time variable; $\mathbf{x}$, $\dot{\mathbf{x}}$, and $\ddot{\mathbf{x}}$ are, respectively, the displacement, velocity and acceleration vectors; in turn $\mathbf{M}$, $\mathbf{C}$ and $\mathbf{K}$ are, respectively, the mass, damping and stiffness matrices and $\mathbf{P}$ is the load vector.

Even though the matrices $\mathbf{M}$ and $\mathbf{K}$ are diagonalizable, $\mathbf{C}$ can be diagonalizable or not depending whether the damping is proportional or non-proportional (that is $\mathbf{C}$ is a combination or not of $\mathbf{M}$ and $\mathbf{K}$). In the first case, the mode shapes are real and their components share the same phase; in the second case, the mode shapes are complex and their components have different phases [27]. Further, the more $\mathbf{C}$ is non-proportional, the greater the imaginary part of the mode shapes is, that is to say the complexity of the mode shapes increases along with the increase of the $\mathbf{C}$ non-proportionality.

A theoretical test to check if the mode shapes are real or complex (i.e. if $\mathbf{C}$ is proportional or not) is the simultaneous validity of the three relations below [28]:

$$\begin{aligned} \mathbf{K}\mathbf{M}^{-1}\mathbf{C} &= \mathbf{C}\mathbf{M}^{-1}\mathbf{K} \\ \mathbf{M}\mathbf{K}^{-1}\mathbf{C} &= \mathbf{C}\mathbf{K}^{-1}\mathbf{M} \\ \mathbf{M}\mathbf{C}^{-1}\mathbf{K} &= \mathbf{K}\mathbf{C}^{-1}\mathbf{M} \end{aligned} \tag{2}$$

whose applicability can be extended also to the non-viscous damping case [29].

In actual structures, the damage is often related to energy dissipation of hysteretic type as a consequence of local or diffused plasticity. In these conditions, the discrete form of the equation of motion (1) takes on the non-linear form:

$$\mathbf{M}\ddot{\mathbf{x}}(t) + \mathbf{H}[\mathbf{x}(t)] = \mathbf{P}(t) \tag{3}$$

in which $\mathbf{H}(\mathbf{x}) = \mathrm{d}\mathbf{F}(\mathbf{x})/\mathrm{d}\mathbf{x}$ is the matrix of the instantaneous non-linear stiffness that depends on the reaction force $\mathrm{d}\mathbf{F}(\mathbf{x})$. The increase of plasticity (i.e. of damage) implies simultaneously the stiffness reduction and the energy dissipation increase. Both these effects are accounted for through $\mathbf{H}(\mathbf{x})$ in Eq. (3). Each effect contributes to make non-proportional the damping and, hence, to make complex the mode shapes.

For simplicity, it is assumed that these two elementary effects are uncoupled. This assumption corresponds to linearize the Eq. (3) and allows to examine separately the consequences due to each single effect:

$$\mathbf{M}\ddot{\mathbf{x}}(t) + \mathbf{C}_{\mathbf{eq}}\dot{\mathbf{x}}(t) + \mathbf{K}_{\mathbf{s}}\mathbf{x}(t) = \mathbf{P}(t) \tag{4}$$

The matrices $\mathbf{C}_{\mathbf{eq}}$ and $\mathbf{K}_{\mathbf{s}}$ in Eq. (4) assume, respectively, the meaning of matrix of equivalent damping and matrix of secant stiffness. As the damage progresses, the effects induced by $\mathbf{C}_{\mathbf{eq}}$ and $\mathbf{K}_{\mathbf{s}}$ increase and the same occurs to the modal complexity. Previous studies [20] have shown that the effects related to $\mathbf{C}_{\mathbf{eq}}$, i.e. to the energy dissipation, are far more sizeable than those related to $\mathbf{K}_{\mathbf{s}}$, i.e. to the stiffness, consequently these latter will not be considered further in the present context of damage detection.

3 Damage Indices Based on Modal Complexity

A number of indices has been proposed in the literature to quantify the damping non-proportionality and consequently the modal complexity, quantity that can be used to detect the damage in the structures [16–21]. These indices can be divided in two groups depending on their experimental identifiability: unmeasurable indices and measurable indices. The unmeasurable indices require the knowledge of the damping matrix of the structure that is not directly identifiable from experimental tests; therefore, they have essentially a theoretical value. On the contrary, the measurable indices require the knowledge of the mode shapes that can be easily derived from experimental tests; therefore, they are valuable from a practical point of view. In view of the above, the unmeasurable indices quantify directly the damping non-proportionality, but indirectly the mode shapes complexity. The opposite happens for the measurable indices since they estimate directly the mode shape complexity, but indirectly the damping non-proportionality. Generally, it turns out that the values of the unmeasured indices are higher than that of the measurable ones for a given damage level since the structural damage induces first the damping non-proportionality (primary damage effect) and then the mode shapes complexity (secondary damage effect) [18].

In any case, however, the indices of both type provide for damage detection and quantification, but do not allow for localization because of their scalar nature. Their effectiveness is therefore related to the capacity in providing prompt and robust damage detection. The experimental verification of the existence of a monotonic relation between damage and its effects (energy dissipation, damping non-proportionality and modal complexity) is of fundamental importance for uniqueness issues and the safe application of the damage indices as damage measures.

In this regard, it is important to point out that problems involving high modal density [21, 24] or modal data incompleteness [18], even if cause an over-content of complexity in the mode shapes, do not affect the monotonous trend of the damage indices. Some of these indices, selected on the basis of their sensitiveness to the mode shapes complexity, are briefly discussed below for both cases of unmeasurable and measurable indices.

3.1 Unmeasurable Indices

The unmeasurable indices refer to viscously damped structures according to $\mathbf{C_{eq}}$ in Eq. (4). For the general case of non-viscous damping, see [30].

A first set of unmeasurable indices (I_{NM1}) is based on a partitioning of the damping matrix into a diagonal matrix and a non-diagonal matrix of which appropriate ratios are computed. In one case, the indices derive from the ratio between the sum of the non-diagonal matrix components and the sum of the components of the diagonal one [31–33]. In another case, the indices are functions of the determinant of the ratio between the non-diagonal matrix and the diagonal one [33] and are related to the components average of the non-diagonal matrix [31].

A second set of unmeasurable indices (I_{NM2}) is given by appropriate functions of the difference between $\mathbf{KM}^{-1}\mathbf{C}$ and $\mathbf{CM}^{-1}\mathbf{K}$ of Eq. (2). They measure the amount of non-correlation of the non-proportional damping matrix with respect to the mass and stiffness matrices [34].

A third set of unmeasurable indices (I_{NM3}) makes use of the maximum and minimum eigenvalues of the damping matrix [35].

3.2 Measurable Indices

A first set of measurable indices (I_{M1}) is of geometric nature and is based on the concept of the modal polygon (i.e. the polygon formed in the complex plane by joining the components of a mode shape). These indices are functions of the modal polygon area normalized by the maximum possible area. This maximum area can be set according to different choices: it can be the area of an equilateral polygon formed by the mode shape components of unit length [33] or it can be the area of the circle whose radius has the largest amplitude among those of the mode shape components [36].

A second set of measurable indices (I_{M2}) is based on the phase of the mode shapes. In [33], the indices are function of the difference between the maximum and minimum phases of the mode shape components. In [36], the ratio between the phase of a mode shape (the average of the phases of its components) and the phase of the mode shape endowed with the maximum value of the imaginary part (corresponding to a phase angle equal to 90°) is considered.

A third set of measurable indices (I_{M3}) comes from the concept of collinearity between the real and the imaginary part of a mode shape. In [37] an appropriate function of the ratio between these two parts is considered.

A fourth set of measurable indices (I_{M4}) arises from the measure of the degree scatter of a mode shape into the complex plane. In [37] an average of the imaginary part of the mode shape is used.

A fifth set of measurable indices (I_{M5}) weighs the importance of the modal complexity quantity. In [38] the ratio between the length of the imaginary part and the overall length of the mode shape is assumed.

3.2.1 Complex Mode Shapes Rotation

It is important to remind that the experimental modal analysis may lead to the identifications of mode shapes endowed with an over-content of complexity that should be removed prior the computation of the damage indices I_M.

Two techniques are available in the literature on purpose. Both of them try to minimize the imaginary part of the mode shapes. This is done by rotating the generic mode shape in such a way to obtain the closest alignment with the real axis [37] or with the real part of the mode shape [29]. Actually, only the rotation provided in [37] is experimentally applicable, and will therefore be used in the following, since the formulation proposed in [29] needs the use of the structural matrices that are hardly identifiable in practical situations.

The effects of the minimization of the imaginary part of the mode shape on three different measurable indices (I_{M1}, I_{M2} and I_{M3}) are shown in Fig. 1 according to a case study worked out in [20].

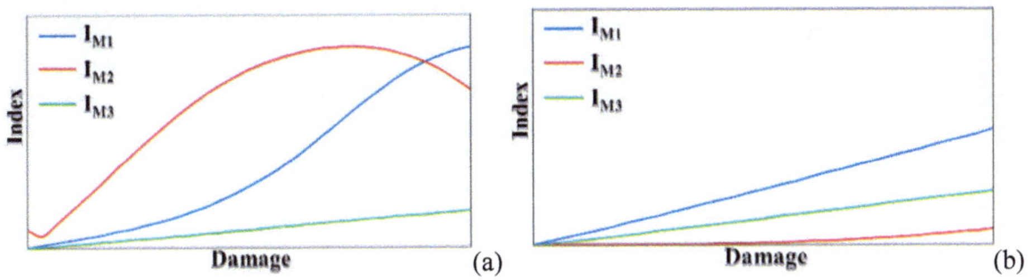

Fig. 1. Trends of indices I_{M1}, I_{M2}, I_{M3}: (a) prior the mode shapes rotation (mode 3); after the mode shapes rotation (mode 3).

The comparison between Fig. 1a and Fig. 1b shows that the indices are useless for damage detection without the preliminary rotation of the mode shapes in the complex plane. In fact, due to the fictitious imaginary contributions, the indices may not be monotonous, as it happens for I_{M2}, and may have variable sensitivity against damage, as it happens for I_{M1}. In addition, it is observed that the rotation of the mode shapes entails two further effects: modifies the relative importance among the different indices and reduces their absolute values.

The above results refer to a particular mode shape, but remain valid in general as shown in Fig. 2 in which the index I_{M1} is reported before and after the rotation of the first three mode shapes.

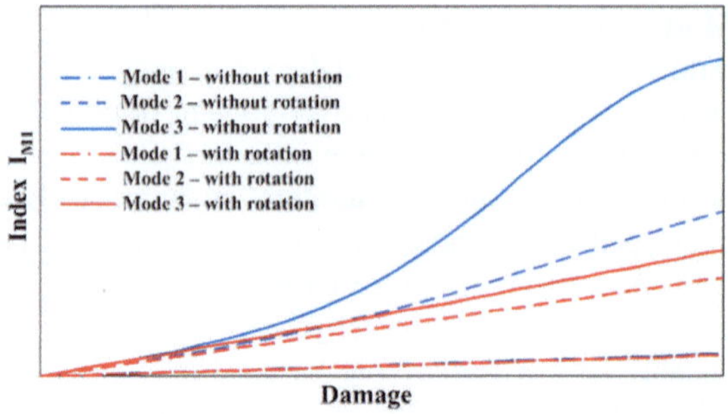

Fig. 2. I_{M1} index vs damage for the first three modes, before and after the mode shapes rotation.

4 Case Study and Methodology

The structural model, the experimental tests and the vibration data referred hereafter were kindly made available by the Italian Civil Protection Department and in particular by the Seismic and Volcanic Risk Office of Rome, which performed the experimental tests together with the ENEA (Italy) and the Basilicata University (Italy), within the TREMA project (Technologies for the Reduction of seismic Effects on Architectural Manufacts) [26].

4.1 The Structural Model

The structural model is a three dimensional reinforced concrete frame characterized by FRP strengthened beam-column joints. The model scale is 1:4 and its geometry is given in Fig. 3 according to two sections in the X and Y directions respectively. The actual physical model prepared for shaking table tests is shown in Fig. 4.

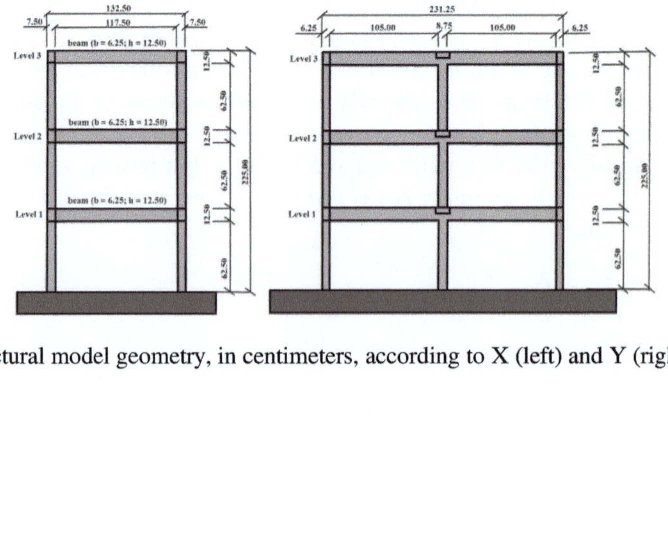

Fig. 3. Structural model geometry, in centimeters, according to X (left) and Y (right) directions.

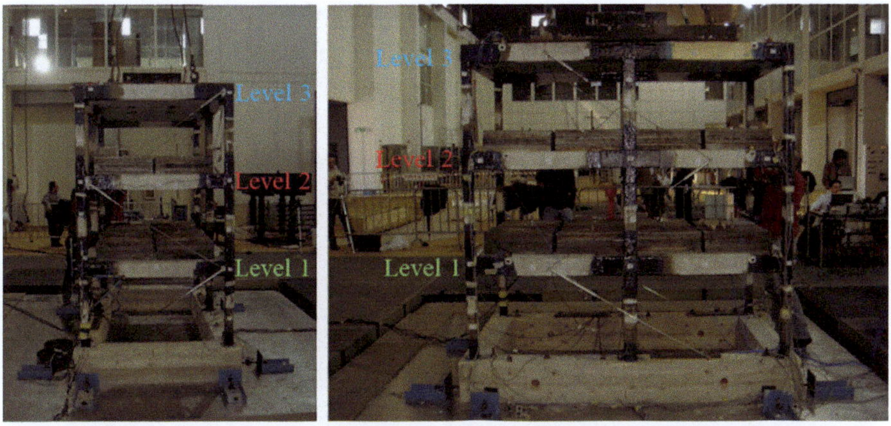

Fig. 4. Images of the structural model prepared for shaking table tests.

4.2 Dynamic Tests

The shaking table tests were carried out at ENEA-Casaccia (Rome, Italy) in the framework of the above said TREMA project. The 4 × 4 m, 6 dofs shaking table facility has a frequency working range [0; 50 Hz] with a maximum peak ground acceleration (PGA) of 3 g being "g" the gravity acceleration.

The seismic action is provided by the Colfiorito registration Fig. 5, that belongs to the earthquake that hit the center Italy (Umbria and Marche regions) in September 27th 1997.

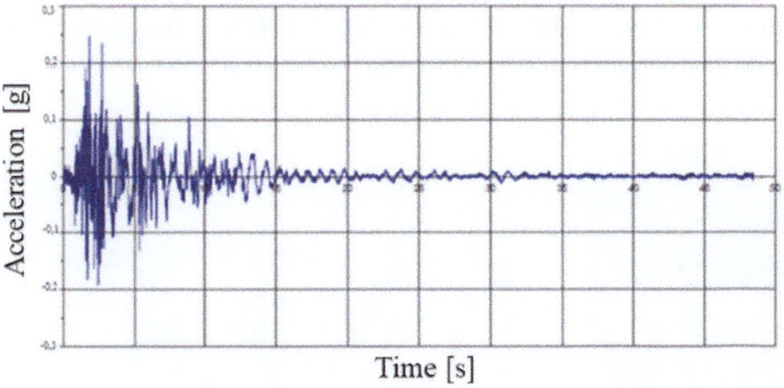

Fig. 5. Colfiorito earthquake (September 27th 1997), West-Est, component.

The Colfiorito earthquake was scaled to seven PGA steps (0,1 g; 0,4 g; 0,5 g; 0,6 g; 0,7 g; 0,8 g; 0,9 g) and as many tests were performed according to a discrete gradually increasing seismic action capable to progressively damaging the structural model, Fig. 6. In particular, this figure shows the damage suffered at the columns base for a PGA value of 0,6 g and 0,9 g respectively. For PGA intensity of 0,6 g the columns base already present a severe damage characterized by a high reduction of the resistant section and significant deformations of the longitudinal reinforcement bars. The same damage features, slightly increased, are observed for a PGA intensity of 0,9 g.

Fig. 6. Damage at columns base for PGA = 0,6 g (left) and PGA = 0,9 g (right).

Finally, dynamic tests at each PGA step were replicated three times to check the response stability, Fig. 7.

The dynamic behavior of the model was monitored through accelerometers placed at each of the three structure levels, Fig. 7a. The corresponding acceleration time histories are given in Fig. 7b according to step 1 (PGA = 0,1 g) for the three levels (Levels 1–3) and the three replications (Tests 1–3).

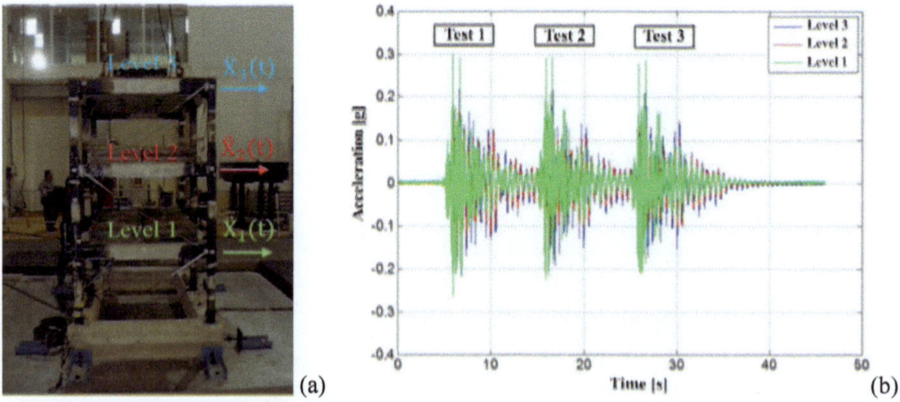

Fig. 7. (a): Measurement stations (3 Levels); (b): Acceleration time histories (3 Levels) for three repeated tests (Tests 1 to 3), PGA = 0,1 g.

4.3 Damage Index

The purpose of the present work is to demonstrate the effectiveness of the measurable indices for damage detection in actual structures. To this end and for definiteness, the most performant index among all the measurable indices is adopted. This is the so called "modal imaginary ratio" that weighs the imaginary part with respect to the overall length of the mode shape [38]:

$$I = \frac{\|\mathrm{Im}(\boldsymbol{\Psi}_r)\|}{\||\boldsymbol{\Psi}_r|\|} \tag{5}$$

where $\boldsymbol{\Psi}_r$ is the r-th mode shape rotated in the complex plane according to [37]; $\mathrm{Im}(\boldsymbol{\Psi}_r)$ stands for the imaginary part of $\boldsymbol{\Psi}_r$, whereas $\| . \|$ is the Euclidean 2-norm operator and $| . |$ is intended as the componentwise absolute value.

In the case of proportional damping, regardless the energy dissipation entity, the mode shapes are real and the index I is zero. On the contrary, in the case of non-proportional damping, the index I increases along with the dissipated energy and tends to 0,70 when the imaginary part is equal to the real one. Indicatively, this value can be considered as an upper limit for the index I itself. However, in order to provide a more readable measure, the index I is expressed in normalized form in the interval percentage [0; 70].

4.4 Complex Mode Shapes Identification

4.4.1 Identification Procedure

The index I of Eq. (5) is used for damage detection purposes. In particular, the damage is estimated through the distance of the values attained by the index I between a structural state and the reference one. In this respect, for a successful damage detection, the index I should be an increasing monotonic function of some measure of the seismic intensity as for instance the PGA.

In view of the above, the preliminary step of the procedure is the identification of the complex mode shapes in order to compute the index I through the Eq. (5). In the present work, the identification of the complex mode shapes is based on the joint and sequential use of the EMD method [22, 23] and of the CPR method [24, 25].

The CPR method is an output-only technique that works in the time domain and is particularly effective for the identification of modal parameters of general damped structures. The formulation exploits the Hilbert Transform (HT) potential [25] in solving problems involving (quasi)mono-harmonic signals [24]. Therefore, in order to apply the CPR method to identify the complex mode shapes, it is necessary to decompose in advance the multi-harmonic output signals (acceleration time histories) in their individual mono-harmonic components. This is accomplished by the use of the EMD method that is a time domain technique capable to decompose a multi-component signal into a set of mono-harmonic signals: the intrinsic mode functions (IMFs) [22, 23].

The results of the application of the EMD method to a sample recorded multi-component acceleration time history (case PGA = 0,1 g - Test 1), Fig. 8, are given in

Figs. 9, 10 and 11 where the first, second and third (quasi)mono-harmonic components (ordered from the lowest to the highest frequency) are shown for each of the three structural model levels.

The CPR method is then applied recursively to the first, second and third (quasi) mono-harmonic components to get an estimate of the complex mode shapes.

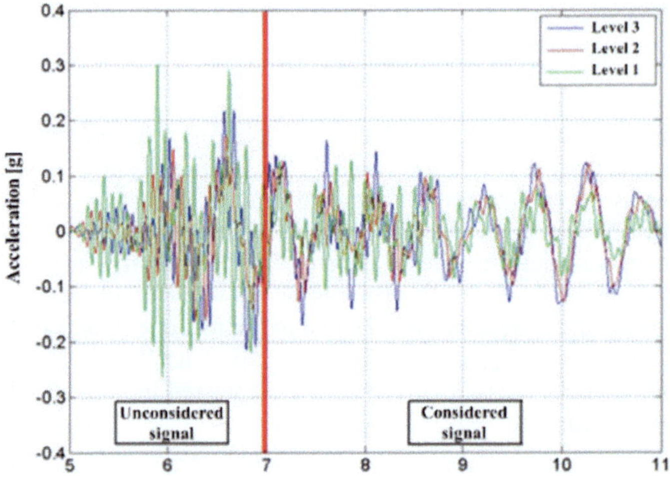

Fig. 8. Sample acceleration time history (case: PGA = 0,1 g – Test 1).

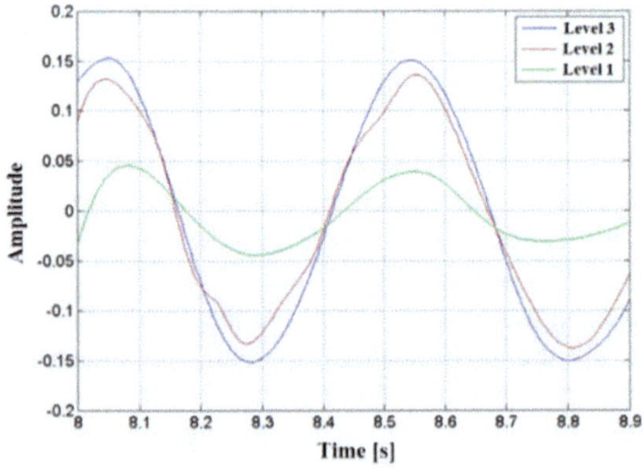

Fig. 9. First component of the multi-harmonic response acceleration of Fig. 8 (case: PGA = 0,1 g – Test 1).

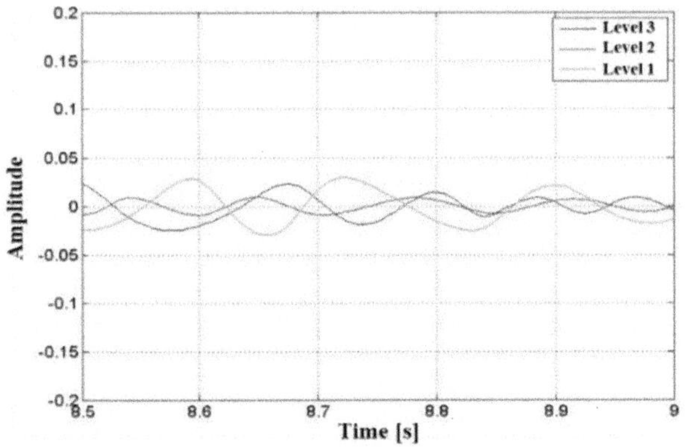

Fig. 10. Second component of the multi-harmonic response acceleration of Fig. 8 (case: PGA = 0,1 g – Test 1).

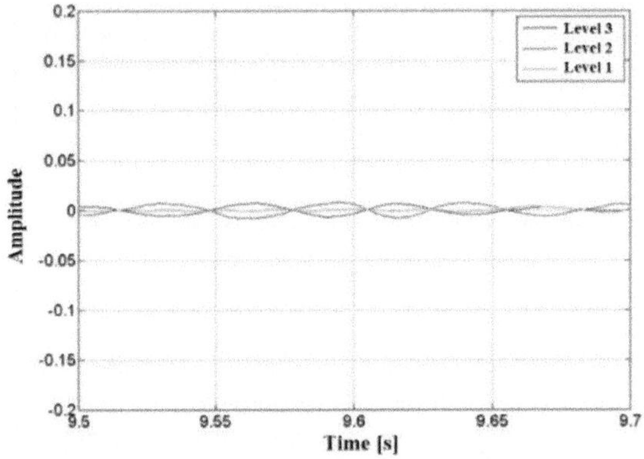

Fig. 11. Third component of the multi-harmonic response acceleration of Fig. 8 (case: PGA = 0,1 g – Test 1).

4.4.2 Identified Complex Mode Shapes

The identified complex mode shapes, after their rotation according to [37], are reported in Table 1 for Test 1, in Table 2 for Test 2 and in Table 3 for Test 3. Each table contains, for each analyzed PGA step (0,1 g; 0,4 g; 0,5 g; 0,6 g; 0,7 g; 0,8 g; 0,9 g), the three identified complex mode shapes (Mode 1, Mode 2, Mode 3) defined according their modal components at the three level (Level 1, Level 2, Level 3) of the tested frame, Fig. 7.

It is important to observe that the results of the three tests should be comparable since they are three replications of the same PGA step. The scatter entity in the comparison can be used to assess the results accuracy and robustness. However, these topics are not the main concern of the present work and will not be therefore discussed further.

Table 1. Identified complex mode shapes - Test 1.

PGA	Level	Mode 1	Mode 2	Mode 3
0,1 g	3	+ 1,00 + 0,01i	− 0,42 − 0,03i	− 0,72 − 0,02i
	2	+ 0,86 − 0,01i	− 0,06 + 0,11i	+ 0,99 − 0,01i
	1	+ 0,23 + 0,02i	+ 1,00 − 0,01i	− 0,10 + 0,07i
0,4 g	3	+ 1,00 + 0,01i	− 0,72 − 0,19i	− 0,40 − 0,01i
	2	+ 0,76 + 0,01i	+ 0,71 + 0,09i	+ 0,99 − 0,02i
	1	+ 0,34 − 0,03i	+ 0,98 − 0,19i	− 0,28 − 0,08i
0,5 g	3	+ 1,00 + 0,00i	+ 0,99 − 0,16i	− 0,94 + 0,13i
	2	− 0,29 − 0,03i	− 0,68 − 0,09i	+ 0,99 + 0,04i
	1	− 0,08 + 0,09i	− 0,41 − 0,27i	− 0,16 − 0,45i
0,6 g	3	+ 0,35 − 0,03i	+ 0,96 − 0,27i	− 0,87 + 0,19i
	2	+ 0,54 + 0,14i	− 0,87 − 0,37i	+ 1,00 − 0,01i
	1	+ 0,99 − 0,07i	− 0,88 − 0,00i	− 0,39 − 0,44i
0,7 g	3	− 0,98 − 0,21i	+ 0,88 + 0,04i	− 0,96 + 0,28i
	2	− 0,04 + 0,17i	− 0,67 + 0,39i	+ 0,99 + 0,03i
	1	+ 0,82 − 0,21i	− 0,97 − 0,23i	− 0,50 − 0,46i
0,8 g	3	− 0,95 + 0,32i	+ 0,61 + 0,06i	− 0,89 + 0,31i
	2	+ 0,89 + 0,29i	+ 0,14 − 0,40i	+ 0,99 + 0,05i
	1	+ 0,93 − 0,01i	− 1,00 − 0,01i	− 0,34 − 0,60i
0,9 g	3	+ 0,78 − 0,38i	+ 0,60 − 0,22i	+ 0,89 − 0,26i
	2	+ 0,97 + 0,25i	+ 0,13 + 0,22i	+ 1,00 − 0,04i
	1	+ 0,60 + 0,01i	− 0,89 + 0,45i	+ 0,50 + 0,69i

Table 2. Identified complex mode shapes - Test 2.

PGA	Level	Mode 1	Mode 2	Mode 3
0,1 g	3	+ 0,99 − 0,02i	− 0,54 + 0,01i	+ 0,99 + 0,02i
	2	+ 0,50 + 0,02i	+ 0,49 + 0,12i	− 0,77 − 0,10i
	1	+ 0,15 + 0,05i	+ 0,99 − 0,05i	− 0,40 + 0,24i
0,4 g	3	+ 0,97 + 0,03i	− 0,10 − 0,13i	− 0,30 + 0,10i
	2	+ 0,99 − 0,05i	+ 0,99 + 0,03i	+ 0,99 − 0,06i
	1	− 0,24 − 0,12i	+ 0,62 − 0,07i	− 0,34 − 0,28i
0,5 g	3	+ 0,02 + 0,14i	− 0,11 − 0,01i	+ 0,14 − 0,13i
	2	+ 0,19 + 0,02i	− 0,26 + 0,22i	+ 0,99 + 0,13i
	1	+ 1,00 − 0,01i	+ 0,99 + 0,05i	− 0,47 + 0,27i
0,6 g	3	+ 0,63 + 0,17i	− 0,34 − 0,16i	− 0,76 + 0,21i
	2	− 0,21 + 0,07i	− 0,35 + 0,22i	+ 0,99 − 0,15i
	1	− 0,99 + 0,10i	+ 0,99 + 0,02i	− 0,77 − 0,46i
0,7 g	3	+ 0,15 − 0,28i	+ 0,40 + 0,26i	− 0,59 − 0,26i
	2	+ 0,20 + 0,09i	− 0,05 − 0,02i	+ 0,01 + 0,30i
	1	+ 0,99 + 0,02i	− 0,99 + 0,17i	+ 0,99 − 0,13i
0,8 g	3	− 0,29 + 0,31i	− 0,20 + 0,28i	+ 0,99 − 0,08i
	2	+ 0,43 + 0,15i	+ 0,99 + 0,11i	− 0,85 − 0,27i
	1	+ 0,99 + 0,02i	+ 0,27 − 0,29i	− 0,27 + 0,48i
0,9 g	3	+ 0,96 + 0,26i	− 0,13 + 0,83i	− 0,90 + 0,29i
	2	+ 0,33 − 0,33i	+ 0,55 − 0,03i	+ 1,00 + 0,01i
	1	− 0,91 + 0,23i	− 0,99 − 0,16i	− 0,39 − 0,65i

Table 3. Identified complex mode shapes - Test 3.

PGA	Level	Mode 1	Mode 2	Mode 3
0,1 g	3	+ 0,74 – 0,06i	+ 0,09 – 0,16i	– 0,40 + 0,22i
	2	+ 0,99 + 0,04i	– 0,50 + 0,10i	+ 0,99 + 0,08i
	1	– 0,00 + 0,00i	+ 0,99 + 0,06i	– 0,07 – 0,03i
0,4 g	3	– 0,50 – 0,09i	– 0,27 – 0,11i	+ 0,78 + 0,21i
	2	+ 0,38 – 0,08i	+ 0,80 + 0,06i	– 0,98 + 0,19i
	1	+ 0,99 – 0,08i	– 0,99 + 0,11i	+ 0,00 + 0,00i
0,5 g	3	– 0,99 + 0,10i	+ 0,33 – 0,17i	– 0,62 + 0,15i
	2	+ 0,64 + 0,10i	+ 0,78 + 0,22i	+ 0,99 – 0,09i
	1	– 0,04 – 0,21i	+ 0,99 – 0,14i	– 0,63 – 0,32i
0,6 g	3	+ 1,00 – 0,04i	– 0,13 + 0,24i	– 0,76 + 0,23i
	2	– 0,01 – 0,00i	– 0,06 + 0,10i	+ 0,88 – 0,13i
	1	+ 0,21 + 0,20i	+ 0,99 + 0,03i	– 0,94 – 0,34i
0,7 g	3	+ 0,13 – 0,25i	– 0,92 + 0,14i	– 0,53 + 0,33i
	2	+ 0,22 + 0,02i	+ 0,99 + 0,11i	+ 0,99 + 0,16i
	1	+ 0,63 + 0,03i	– 0,01 – 0,40i	– 0,00 + 0,00i
0,8 g	3	+ 0,99 – 0,12i	+ 0,98 + 0,20i	– 0,75 + 0,31i
	2	– 0,13 + 0,12i	– 0,44 – 0,04i	+ 1,00 – 0,00i
	1	– 0,63 – 0,23i	– 0,56 + 0,46i	– 0,29 – 0,81i
0,9 g	3	+ 0,65 + 0,25i	– 0,47 – 0,58i	– 0,82 + 0,43i
	2	– 0,01 + 0,05i	– 0,34 + 0,57i	+ 0,94 – 0,03i
	1	– 0,98 + 0,21i	+ 0,99 – 0,05i	– 0,45 – 0,89i

5 Results

The damage index I is computed by inserting the complex mode shapes of Tables 1, 2 and 3 into Eq. 5. The trend of the index I against the seismic intensity is then analyzed for the purpose of damage detection. Issues concerning sensitivity, stability and repeatability of the index I are also addressed.

5.1 Damage Detection

The Figs. 12, 13 and 14 report, for a selected mode shape, the normalized damage index I against the seismic intensity (PGA) respectively for the three replications named Test 1, 2 and 3.

By the inspection of the Figs. 12, 13 and 14, a preliminary observation shows that the I-PGA curves are quite similar for the three replications: Tests 1, 2 and 3. Further, it is apparent that, regardless the mode shape considered, the damage index I proves to be a monotonic increasing function of the seismic intensity (PGA). This is a fundamental property of the index I for a successful damage detection. In fact, this property guarantees the solution uniqueness particularly in those cases in which the damage presence is inferred by the measure of the distance between the present structural state and a reference one.

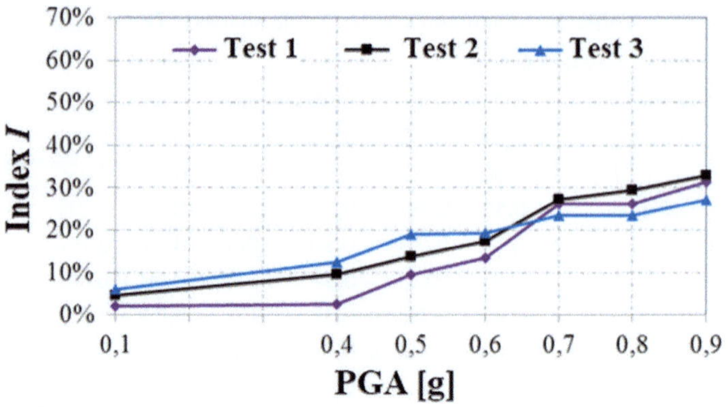

Fig. 12. Mode 1: damage index *I* vs. PGA for Tests 1, 2 and 3.

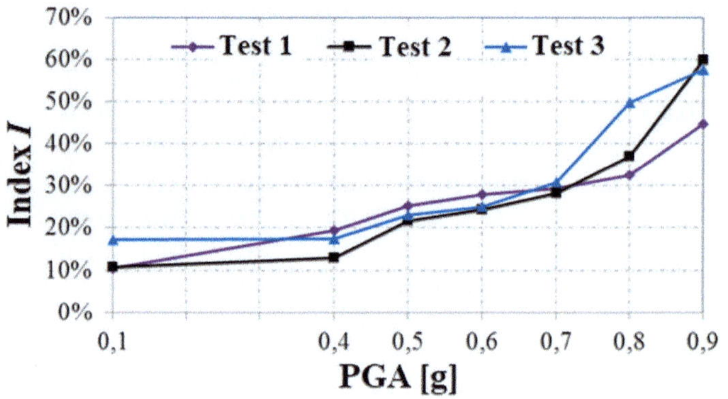

Fig. 13. Mode 2: damage index *I* vs. PGA for Tests 1, 2 and 3.

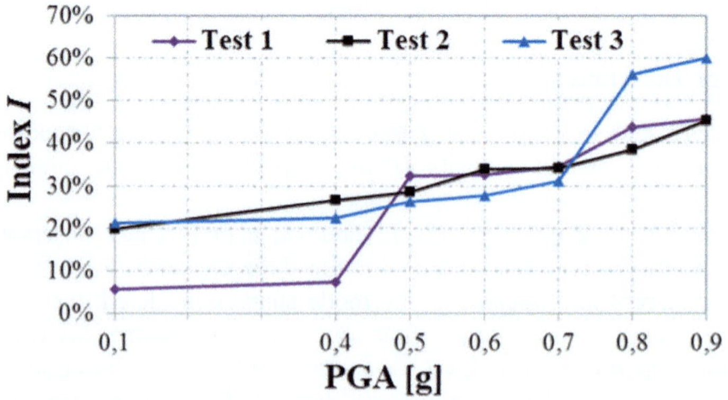

Fig. 14. Mode 3: damage index *I* vs. PGA for Tests 1, 2 and 3.

Finally, the monotonic increasing character of the index *I*, for being effectively used in damage detection, is confirmed by Fig. 15 where the variation of the index *I*, averaged over the three replications (Tests 1, 2 and 3), is reported for the three Modes 1, 2 and 3.

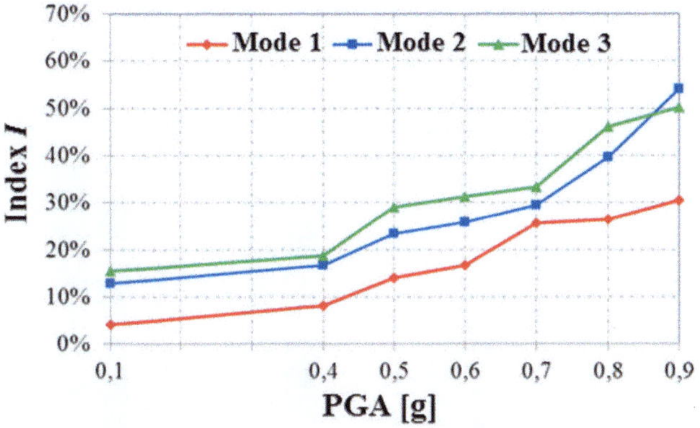

Fig. 15. Average damage index *I* vs. PGA for Modes 1, 2 and 3.

5.2 Damage Index Sensitivity

The sensitivity of the damage index is defined as the rate of change of *I* corresponding to a unit change of the seismic intensity PGA. As a first approximation, resort is made to a linear regression of the data so that the sensitivity equals the slope of the interpolating straight line. This result provides for an averaged estimate of the sensitivity all over the PGA range considered, Fig. 16. The adopted approximation holds true in the case of viscoelastic type energy dissipation [20], but it remains effective in the case of structures experiencing plasticization as those under consideration characterized by hysteretic type energy dissipation [16].

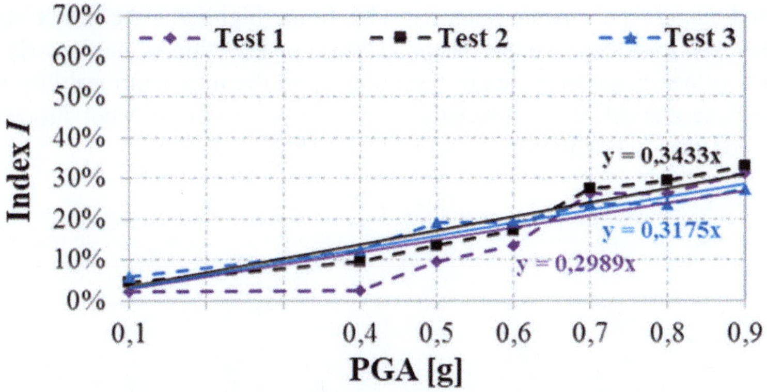

Fig. 16. Linearization of *I* - PGA relationship for Mode 1 and Tests 1, 2 and 3.

The Fig. 17 summarizes the sensitivity of the damage index I for the three mode shapes (Modes 1–3) and the three tests (Tests 1–3). From the figure, it is possible to observe that the sensitivity increases with the increase of the mode shape order and also with the replicated tests (damage accumulation). The use of higher order mode shapes is then preferable for damage detection purposes even if the sensitivity related to the first mode (approx. 1/3) still appears significant to avoid false estimates.

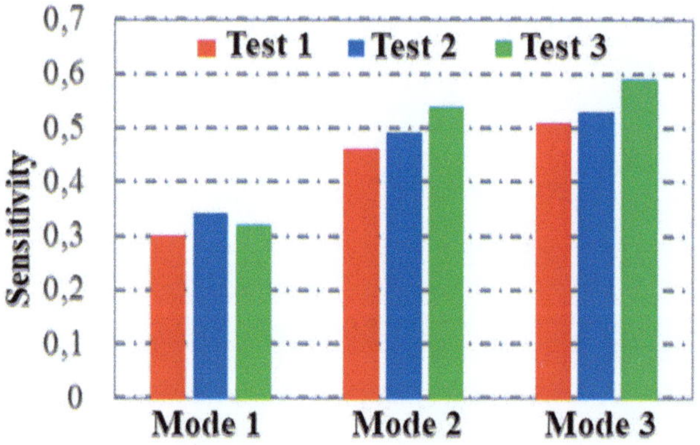

Fig. 17. Sensitivity of I for the three modes and the three replications (Tests 1, 2 and 3).

5.3 Damage Index Stability

The stability S of the damage index I is herein assessed using the ratio:

$$S = \frac{I_{max} - I_{min}}{I_{avg}} \tag{6}$$

where I_{max}, I_{min} and I_{avg} are respectively the maximum, minimum and average values of I among the three replicated Tests 1, 2 and 3 for each of the three mode shapes considered. The results are shown in Fig. 18 where the stability of the damage index I is reported against the seismic intensity. This figure highlights the scarce stability of I regardless the considered mode shape. Unfortunately, this is particularly true for lower PGA values range in which one should desire the maximum stability to reduce the risk of false identification especially in the case of small incipient structural damages. The contrary happens in the case of higher PGA values.

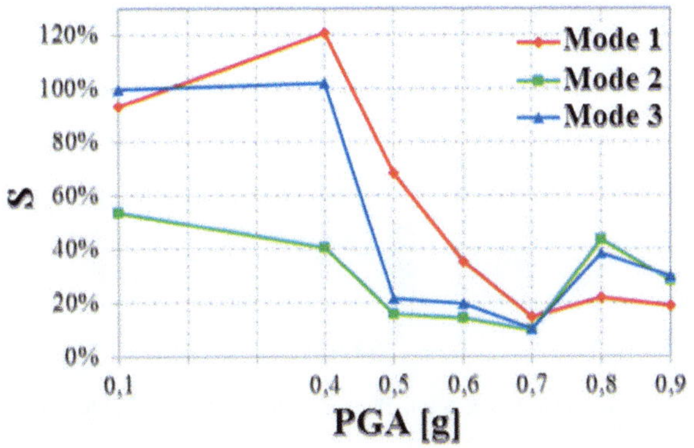

Fig. 18. Stability of damage index *I*.

5.4 Damage Index Repeatability

The repeatability of the damage index is studied by comparing, mode by mode, the values attained by *I* in the three replicated tests (Tests 1, 2 and 3).

The inspection of Figs. 19, 20 and 21 helps on purpose. In order to get perfect repeatability of the results the three histograms relevant to each mode shape should overlap. In view of this and as a general comment, it can be concluded that a reasonably good level of overlapping exists for all the considered cases.

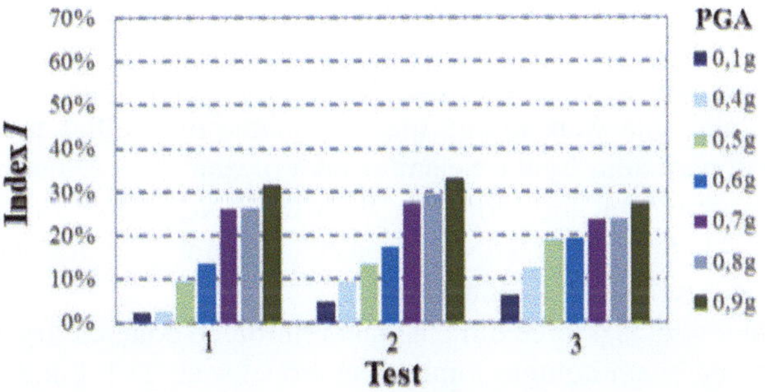

Fig. 19. Mode 1: damage index *I* vs. PGA for the three replicated Tests 1, 2 and 3.

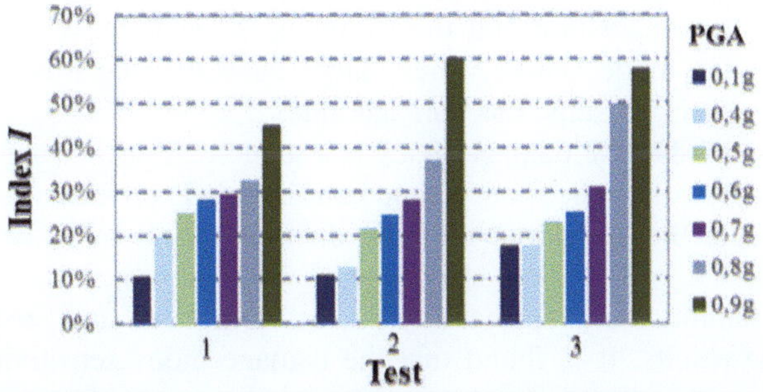

Fig. 20. Mode 2: damage index *I* vs. PGA for the three replicated Tests 1, 2 and 3.

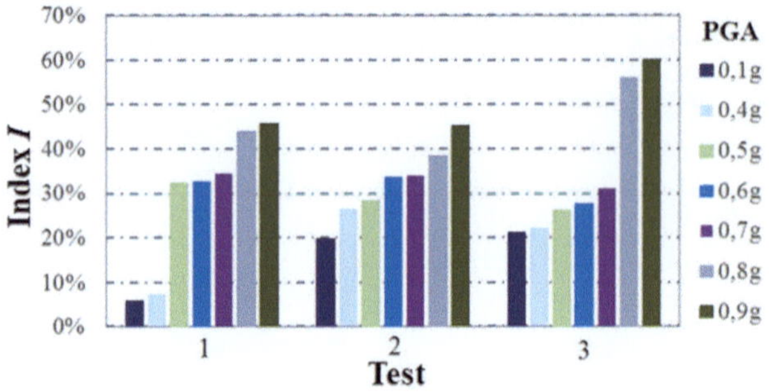

Fig. 21. Mode 3: damage index I vs. PGA for the three replicated Tests 1, 2 and 3.

6 Conclusions

The present work is targeted to validate experimentally the effectiveness of damage indices constructed on measures of the imaginary part of complex mode shapes.

The basic assumption is that the increase in mode shapes complexity comes from the energy dissipated along with damage occurrence. There exist indeed two type of indices. The unmeasurable indices require the knowledge of the damping matrix of the structure that is not directly identifiable from experimental tests; therefore, they have essentially a theoretical value. On the contrary, the measurable indices require the knowledge of the mode shapes that can be easily derived from experimental tests; therefore, they are valuable from a practical point of view. Among the measurable indices the so called "modal imaginary ratio", that proved the most effective by previous numerical studies, is adopted in the present work. Such damage index weighs the imaginary part with respect to the overall length of the mode shape.

The experimental data were kindly made available by the Italian Civil Protection Department and in particular by the Seismic and Volcanic Risk Office of Rome, which performed the experimental tests together with the ENEA (Italy) and the Basilicata University (Italy), within the TREMA project (Technologies for the Reduction of seismic Effects on Architectural Manufacts).

The structural model is a three dimensional reinforced concrete frame characterized by FRP strengthened beam-column joints. The model scale is 1:4 and it is placed on a shaking table so that the structural damage is caused and increased by stepping the amplitude of the base motion that replicates the Colfiorito earthquake.

The acceleration time histories have been processed via the joint use of the Empirical Mode Decomposition and the Complex Plane Representation methods to identify the complex mode shapes. These latter have then be used to compute the "modal imaginary ratio" assumed as damage index.

The results show that the damage index is a monotonic increasing function of the damage severity regardless the mode shape considered. Therefore, the damage detection is possible via the comparison of the damage index value between different structural states.

Sensitivity, stability and repeatability issues have been analyzed to assess the robustness of the results. It is found that the damage index sensitivity increases, in general, with the mode shape order. However, the first mode shape already leads to significant values for the damage index. Therefore, any mode shapes is suitable for the index computation.

As concerns the stability issue, it turned out a scarce stability of the index regardless the considered mode shape. Unfortunately, this is particularly true for lower PGA values range in which one should desire the maximum stability to reduce the risk of false identification especially in the case of small incipient structural damages. On the contrary, as concerns the repeatability issue, it is found that a reasonably good level of matching exists for all the considered replications.

A final comment concerns the good correspondence between the present experimental results and the numerical results obtained in previous works by the authors. This outcome can be considered as a validation of the numerical results.

Acknowledgements. The support of the Seismic and Volcanic Risk Office of the Italian Department of Civil Protection in providing the experimental data is gratefully acknowledged.

References

1. Cawley, P., Adams, R.D.: The location of defects in structures from measurements of natural frequencies. J. Strain Anal. **14**, 49–57 (1979)
2. Gunes, B., Gunes, O.: Structural health monitoring and damage assessment part I: a critical review of approaches and methods. Int. J. Phys. Sci. **8**(34), 1694–1702 (2013)
3. Salawu, O.S.: Detection of structural damage through changes in frequency: a review. Eng. Struct. **19**(9), 718–723 (1997)
4. Shi, Z.Y., Law, S.S., Zhang, L.M.: Damage localization by directly using incomplete mode shapes. J. Eng. Mech. **126**(6), 656–660 (2000)
5. Rainieri, C., Fabbrocino, G., Cosenza, E.: Some remarks on experimental estimation of damping for seismic design of civil constructions. Shock Vibr. **17**, 383–395 (2010)
6. Rainieri, C., Fabbrocino, G., Cosenza, E.: Structural damping: theoretical models and experimental evaluations (in Italian). In: Proceedings of 13th ANIDIS Conference, Bologna, Italy (2009)
7. Bovsunovkii, A.P.: On the efficiency of using damping characteristics of structural components for damage diagnostics. Strength Mater. **34**(6), 560–569 (2002)
8. Zonta, D., Modena, C., Bursi, O.S.: Analysis of dispersive phenomena in damaged structures. In: Proceedings of European COST F3 Conference on System Identification and Structural Health Monitoring, pp. 801–810. Madrid, Spain (2000)
9. Modena, C., Sonda, D., Zonta, D.: Damage localization in reinforced concrete structures by using damping measurements. Key Eng. Mater. **167–168**, 132–141 (1999)
10. Lage, Y., Cachão, H., Reis, L., Fonte, M., de Freitas, M., Ribeiro, A.: A damage parameter for HCF and VHCF based on hysteretic damping. Int. J. Fatigue **62**, 2–9 (2014)
11. Zhang, Z., Hartwig, G.: Relation of damping and fatigue damage of unidirectional fiber composites. Int. J. Fatigue **24**, 713–718 (2002)
12. Montalvão, D., Ribeiro, A.M.R., Duarte-Silva, J.: A method for the localization of damage in CFR plate using damping. Mech. Syst. Signal Process. **23**, 1846–1854 (2009)
13. Kawiecki, G.: Modal damping measurement for damage detection. Smart Mater. Struct. **10**, 466–471 (2001)

14. Egba, E.I.: Detection of structural damage in building using changes in modal damping mechanism. Int. J. Eng. Manag. Sci. **3**(3), 250–255 (2012)
15. Curadelli, R.O., Riera, J.D., Ambrosini, D., Amani, M.G.: Damage detection by means of structural damping identification. Eng. Struct. **30**, 3497–3504 (2008)
16. Iezzi, F., Valente, C.: Mode shapes complexity for damage identification of structures experiencing plasticization. In: Proceedings of 9th International Conference on Computational Methods, ScienTech Publisher, Paper ID 3030, Rome, Italy (2018)
17. Lofrano, E., Paolone, A., Ruta, G., Taglioni, A.: Perturbation damage indicators based on complex modes. Procedia Eng **199**, 1949–1954 (2017)
18. Iezzi, F.: Structural damage identification through modal complextiy (in Italian). Ph.D. Thesis, University "G. d'Annunzio" of Chieti-Pescara, Italy (2016)
19. Iezzi, F., Valente, C., Zuccarino, L.: The measure of the modal complexity as indicator of structural damage (in Italian). In: Proceedings of 14th ANIDIS Conference, L'Aquila, Italy (2015)
20. Iezzi, F., Spina, D., Valente, C.: Damage assessment through changes in mode shapes due to non-proportional damping. J. Phys.: Conf. Ser. **628**, 012019 (2015)
21. Iezzi, F., Valente, C.: Modal density influence on modal complexity quantification in dynamic systems. Procedia Eng. **199**, 942–947 (2017)
22. Attoh-Okine, N., Barner, K., Bentil, D., Zhang, R.: The empirical mode decomposition and the Hilbert-Huang transform. EURASIP J. Adv. Signal Process. **2008**, 1–2 (2008)
23. Rato, R.T., Ortigueira, M.D., Batista, A.G.: On the HHT, its problems, and some solutions. Mech. Syst. Signal Process. **22**, 1374–1394 (2008)
24. Gabriele, S., Iezzi, F., Spina, D., Valente, C.: The effects of modal density in system identification using the Hilbert transform. In: Proceedings of IEEE Workshop on Environmental, Energy and Structural Monitoring Systems, Naples, Italy (2014)
25. Ahan, S.L.: Hilbert Transforms in Signal Processing. Artech House, Boston, USA (1996)
26. Dolce, M., Moroni, C., Nigro, D., Ponzo, F.C., Goretti, A., Spina, D., Lamonaca, B., Giordano, F., De Canio, G., Rainieri, N., Marnetto, R.: TREMA project: experimental evaluation of seismic performance of a RC ¼ scaled model upgraded with FRP. In: Proceedings of the fib 2nd International Congress, Naples, Italy (2006)
27. Craig, R.R., Kurdila, A.J.: Fundamentals of Structural Dynamics, 2nd edn. Wiley, Hoboken, NJ (2006)
28. Adhikari, S.: Damping modelling and identification using generalized proportional damping. In: Proceedings of 23rd International Modal Analysis Conference, Orlando, FL, USA (2005)
29. Adhikari, S.: Optimal complex modes and an index of damping non-proportionality. Mech. Syst. Signal Process. **18**(1), 1–27 (2004)
30. Adhikari, S.: Structural Dynamic Analysis with Generalized Damping Models: Identification. Wiley, New York (2014)
31. Liu, K., Kujath, M.R., Zheng, W.: Evaluation of damping non-proportionality using identified modal information. Mech. Syst. Signal Process. **15**(1), 227–242 (2001)
32. Tentor, L.B., Wicks, A.L., Cudney, H.H.: Characterization of nonproportional damping from complex modes and eigenvalues. In: Proceedings of 14th International Modal Analysis Conference, pp. 1628–1631 (1996)
33. Prater, G., Singh, R.: Quantification of the extent of nonproportional viscous damping in discrete vibratory systems. J. Sound Vib. **104**(1), 109–125 (1986)
34. Nair, S.S., Sing, R.: Examination of the validity of proportional damping approximations with two further numerical indices. J. Sound Vib. **104**(2), 348–350 (1986)
35. Tong, M., Liang, Z., Lee, G.C.: An index of damping non-proportionality for discrete vibrating systems. J. Sound Vib. **174**(1), 37–55 (1994)

36. Imregun, M., Ewins, D.J.: Complex modes – origins and limits. In: Proceedings of 13th International Modal Analysis Conference, pp. 496–506, Nashville, TN, USA (1995)
37. Liu, K., Kujath, M.R., Zheng, W.: Quantification of non-proportionality of damping in discrete vibratory systems. Comput. Struct. **77**(5), 557–569 (2000)
38. Valente, C., Spina, D., Gabriele, S., De Leonardis, A.M.: The complex plane representation method for structural damage detection. In: Proceedings of 10th International Conference on Computational Structures Technology, Valencia, Spain (2010)

Application of Modulation Transfer Effect to Damage Detection

Jakub Górski and Andrzej Klepka(✉)

Department of Robotics and Mechatronics, AGH University of Science and Technology, Al. Mickiewicza 30, 30-059 Krakow, Poland
klepka@agh.edu.pl

Abstract. The paper presents investigation of modulation transfer effect. This effect - originally observed in the early nineteen thirties in Luxembourg and in Gorky for radio waves propagation in ionosphere - was manifested by modulation transfer of a weaker wave in presence of a strong amplitude-modulated wave. The research are focused for both, application to damage detection and analysis of possible sources of modulation transfer. The different nonlinear model are analyzed to find the potential source of non-linearities responsible for sidebands transfer form low to high frequency. The models include hysteretic stiffness and quadratic damping. In experimental part of the work, modulated low-frequency signal was used as excitation of cracked beam. Simultaneously the high frequency acoustic wave has been introduced to the structure. This combination of excitations induced a modulation transfer from low-frequency to high-frequency wave in presence of structural damage. Surface bonded piezoceramic transducer and electromagnetic shaker were used to excite the structure. Laser vibrometry was used to acquire the response of the structure. The experimental work presented focuses on the analysis of modulation intensities and damage-related nonlinearities. The paper demonstrates that the method can be used for fatigue damage detection.

Keywords: Modulation transfer · Damage detection

1 Introduction

Recent years an increasing interest in nonlinear ultrasonic techniques have been observed. The methods based on different nonlinear effects have been developed for damage detection and localization. In this group, techniques based on higher harmonics generation [1, 2], slow dynamics effect [3], frequency shifting and mixing [4–6] and many others can be found. The main assumption of all methods of nonlinear acoustics is that the non-linear effect observed in response signal are damage related. For example, the modulation and the appearance of higher harmonics in the response signal for a damaged structure excited simultaneously in low and high frequency can be caused by: closing and opening action of damage (e.g. fatigue crack) or/and relative motions of damage interfaces [7]. This causes the stiffness characteristics to become non-linear. In the case of closing and opening action different stiffness applies for tension and compression phase and turns stress-strain relation in bi-linear mode

M. A. Wahab (Ed.): DAMAS 2019, LNME, pp. 124–134, 2020.
https://doi.org/10.1007/978-981-13-8331-1_9

(Fig. 1a). For relative motion of damage interfaces (sliding action) two cases are predicted: micro-slip mode and stick and slip model [8]. For the first the forces acting on surfaces in contact are to week to broke friction between asperities (Fig. 1b). It is causes that slope of the stress-strain relation change twice per cycle. For stick and slip action (when the friction forces are broken) the hysteretic stiffness characteristic is predicted (Fig. 1c). The examples of theoretical stress-strain characteristics for different crack mode presents Fig. 1.

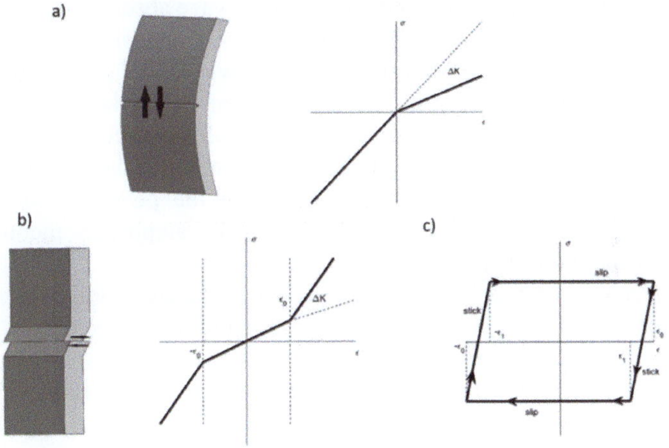

Fig. 1. Stress-strain characteristic for different crack mode: (a) bi-linear for closing-opening action, (b) piecewise linear for micro-slick and (c) hysteretic for stick and slip action.

The results of different stress-strain relation is different frequency content of the signal response for particular case. For symmetrical characteristics all harmonics are predicted. For non-symmetrical relation only odd harmonics appear in the signal spectra. Additionally, in case of damaged structure the modulation phenomena with different sidebands patterns can occur. The modulation intensity parameter are very often used for damage assessment of structures. Figure 2 present examples of spectra with different sidebands pattern for damaged and undamaged structure.

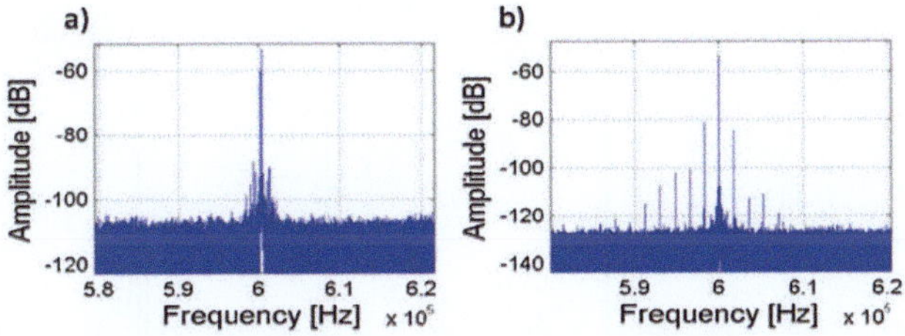

Fig. 2. Examples of ultrasonic response power spectra for nonlinear acoustics test: (a) undamaged structure, (b) damaged structure.

2 Modulation Transfer

The cross modulation method is a nonlinear acoustic technique based on the Luxembourg-Gorky (L-G) effect. Originally, this effect was observed in the early nineteen thirties in Luxembourg and in Gorky for radio waves. It was manifested by modulation transfer of a weaker wave in presence of a strong amplitude-modulated wave. The hypothesis developed by Bailey et al. [9] assumed that modulation transfer was caused by variable ionosphere absorption. It was inducted by the amplitude-modulated stronger wave. It is well known that radio wave absorption of the ionosphere is determined by the conductance related to numbers of collisions between electrons, molecules and ions. In the case of absence of radio waves, the number of collisions is proportional to velocity of electrons. When the intensity of radio wave field is significant, velocities of the electrons or number of collisions change (when the field intensity increases, the conductance of the gas decreases). These changes are time-dependent. Weak radio waves, propagating though the perturbed ionosphere area absorbed in varying degrees. In other words, they are amplitude- modulated by the frequency of strong radio waves.

Very similar phenomena have been observed in damaged solids exhibiting nonlinear behavior by Zaitsev et al. [10]. When the structure with is excited by two types of waves: strong amplitude-modulated wave (pumping wave) and a weak mono-harmonic wave (probing wave) (F) – the modulation of pumping wave is transferred to the probing wave.

Figure 3 presents modulation transfer effect. When the structure is undamaged, the spectrum of the response signal contains component related to the pumping and

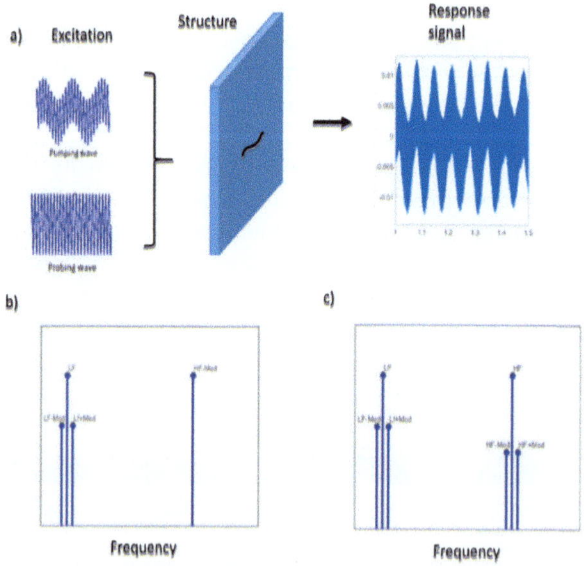

Fig. 3. Example illustrating the modulation transfer effect (a), spectrum of undamaged structure (b), spectrum of damaged structure.

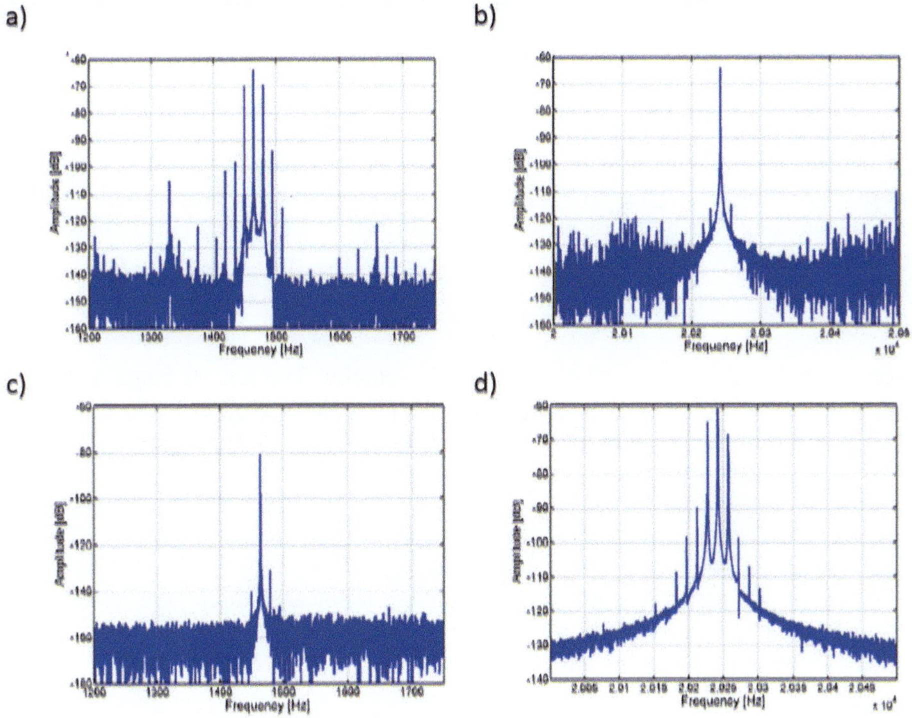

Fig. 4. Examples of ultrasonic response power spectra for modulation transfer test for undamaged and damaged specimen: (a) the spectrum zoomed around pumping wave (undamaged), (b) the spectrum zoomed around probing wave (undamaged), (c) the spectrum zoomed around pumping wave (damaged), (d) the spectrum zoomed around probing wave (damaged) [11].

probing wave (Fig. 3b). When the structure is damaged the modulation sidebands with frequency corresponding to the modulation of pumping wave appear around the high frequency probing wave (Fig. 3c). This means that modulation transfer has taken place form pumping to probing wave. This approach was used for damage detection for different types of structures [6, 11, 12]. The example of spectra based on the response signals of damaged and undamaged real structures presents Fig. 4.

The physical phenomena behind this effect are not clearly confirmed. This article focuses on an attempt to define a modulation transfer model and an experimental verification of the proposed approach.

3 Nonlinear Dissipation Energy Models

Many items in literature, describing the different models assumed nonlinear damage related effects, can be found. From this review [13] it can be suspected that phenomena related with Luxemburg-Gorky effect have fundaments in material absorption and energy dissipation. Therefore in this study, three single degree of freedom (SDOF) models with different nonlinear characteristics have been developed.

3.1 Quadratic Damping

The first model contains most basic form of polynomial damping. It is often used to describe fluid flow through orifice in automotive damper and to characterize the loading force on offshore structures. In literature it is known as Morison's equation [14] and can be described as follow:

$$F(t) = c_1\dot{x} + c_2\dot{x}|\dot{x}| \tag{1}$$

where c_1, c_2 are damping coefficients and $\dot{x}$ is velocity. The Eq. (1) contain absolute value to ensure damping force is always opposite to velocity. In model presented in this subchapter, the absolute value was removed according to [15]. The equation of motion for single degree of freedom (SDOF) model has been assumed as

$$m\ddot{x} + c_1\dot{x} + c_2\dot{x}^2 + kx = F(t) \tag{2}$$

Where m is mass, k is stiffness coefficient, c_1 is linear damping coefficient, c_2 is damping coefficient responsible for cross-modulation and $F(t)$ is excitation. Table 1 contains values of parameters for this model.

Table 1. Numerical values of quadratic damping model

m [kg]	c_1 [Ns/m]	c_2 [Ns^2/m^2]	k [N/m]
1	$1.8 \cdot 10^5$	$7.2 \cdot 10^6$	$4.63 \cdot 10^7$

3.2 Hysteresis Stiffness

The second model contains non-functional nonlinearity. The hysteresis nonlinearity in stiffness characteristic was applied to produce dynamic behavior related to surfaces in contact (fatigue crack) [13]. In this case equation of motion for SDOF is described as

$$m\ddot{x} + c\dot{x} + K_h(t) = F(t) \tag{3}$$

where c is the damping factor, m is the mass, $F(t)$ is force excitation and $K_h(t)$ is hysteresis stiffness based on mathematical model described in [7]. The stiffness in this case is described by

$$K(t) = 0.5 \cdot K_s\{1 - sign(\dot{x}(t)) \cdot sign[x(t) + x_1 sign(\dot{x}(t))]\} \tag{4}$$

$$K_h(t) = K(t)x(t) + K(t)x_1 sign(\dot{x}) + 0.5K_s(x_0 - x_1)sign(\dot{x}) \tag{5}$$

where K_s is stiffness, x_0, x_1 are instantaneous extreme values of arguments for which the value of stiffness changes. The numerical values of the parameters are presented in Table 2.

Table 2. Numerical values of hysteresis stiffness model.

m [kg]	c_1 [Ns/m]	k [N/m]
1	$1.8 \cdot 10^5$	$4.63 \cdot 10^7$

3.3 Quadratic Damping with Frequency Dependent Component

The third model has similar equation of motion to the model described in subsection 3.1. Based on work performed by Zaitsev et al. [16] that energy dissipation is frequency dependent. Thus the model gained additional frequency-dependent component $c_3(\omega)$. The restring force for assumed model can be written as

$$F(t) = c_1\dot{x} + c_2\dot{x}^2 + c_3(\omega)\dot{x}|\dot{x}| \tag{6}$$

Then, the equation of motion for SDOF system can be described as

$$m\ddot{x} + c_1\dot{x} + c_2\dot{x}^2 + c_3(\omega)\dot{x}|\dot{x}| + kx = F(t) \tag{7}$$

where m is mass, k is stiffness coefficient, c_1 is linear damping coefficient, c_2 is quadratic damping coefficient, $c_3(\omega)$ is frequency-dependent damping coefficient and $F(t)$ is excitation force. In this model the values of parameters are the same as subsection 3.1 except of $c_3(\omega)$, which is presented in Fig. 5.

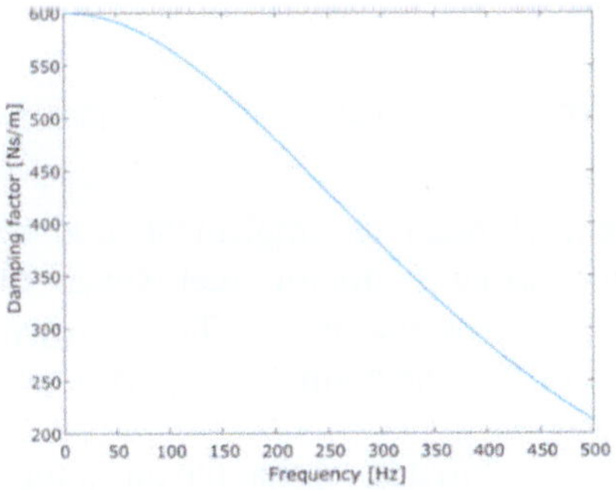

Fig. 5. Function of damping coefficient $c_3(\omega)$.

4 Model Comparison

For given parameters, three models presented in previous chapter was simulated. The models were excited by the same excitation force, described by equation

$$F(t) = F_{pm}(1 + m_{int}cos(2\pi f_s t))\cos(2\pi f_{pm}t) + F_{pr}\cos(2\pi f_{pr}t) \tag{8}$$

where F_{pm}, f_{pm} are amplitude and frequency of pumping wave excitation, F_{pr}, f_{pr} are amplitude and frequency of probing wave excitation, m_{int} is modulation intensity and f_s is frequency of modulating signal. The selected values are listed in Table 3. The normalized power spectrum of force signal is presented in Fig. 6.

Table 3. Parameters of force excitation.

F_{pm} [N]	f_{pm} [Hz]	F_{pr} [N]	f_{pr} [Hz]	m_{int}	f_s [Hz]
15	1083	0,05	25000	0,9	43

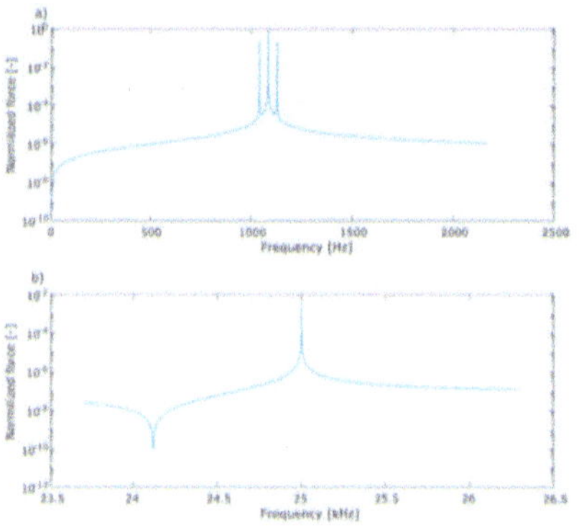

Fig. 6. Spectrum of force excitation signal for: (a) pumping wave, (b) probing wave.

The simulation were performed on implemented models in Simulink™. The defined problem was solved using the fourth order Runge–Kutta method with a fixed step length of Δt = 4 μs and total time of 5 s. The velocity response for all models were collected, and for each signal the normalized spectrum was calculated. The results are presented in Figs. 7 and 8.

It can be noticed from Fig. 7(a) and (c) that the quadratic damping model generate even and odd harmonics which is characteristic for the symmetric type of non-linearity. Compared to force excitation spectrum presented in Fig. 6, the additional sidebands are present for all visible harmonics. Additionally, components associated with modulation frequency 43 Hz are also visible on both spectrums.

In Fig. 7(b) the spectrum for model with hysteresis is presented. This type of nonlinearity is responsible for generation only odd harmonics [7].

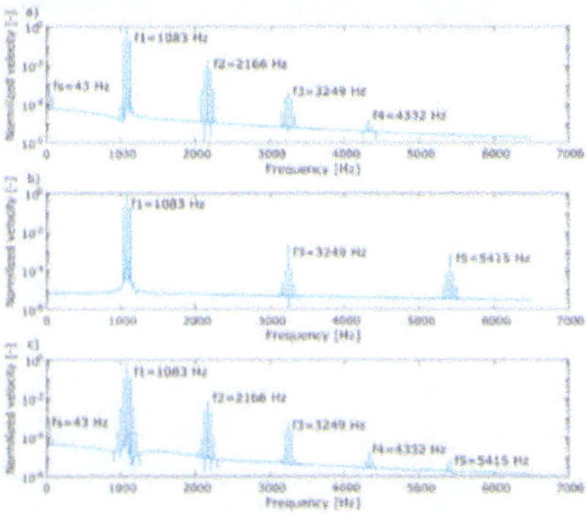

Fig. 7. Normalized velocity spectrum zoomed at low frequencies for model with: (a) quadratic damping, (b) hysteresis, (c) quadratic damping frequency dependent.

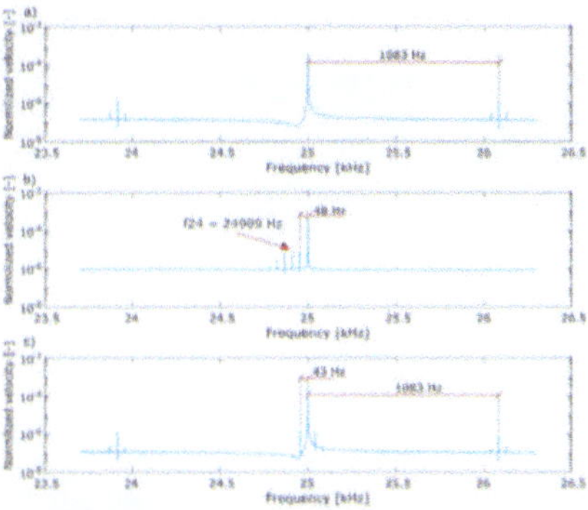

Fig. 8. Normalized velocity spectrum zoomed at high frequencies for model with: (a) quadratic damping, (b) hysteresis, (c) quadratic damping frequency dependent.

The Fig. 8 present power spectrum zoomed at pumping wave. It can be seen from Fig. 8(b) the hysteresis nonlinearity has no effect on modulation transfer. Visible sideband around probing wave are not related to modulation frequency (43 Hz) but with harmonics of modulated frequency (48 Hz). However for models with quadratic damping presented in Fig. 8(a) the cross-modulation is clearly visible. For frequency dependent damping, presented in Fig. 8(c), the modulation transfer effect can be observed.

5 Experimental Verification

To verify responses of models presented in Sect. 4, the experimental test were conducted. As specimen, a damaged slender beam manufactured from aluminum grade 6082 T6 was used. The test sample of 300 × 20 × 10 mm dimensions was cut from an aluminum sheet using a water jet cutter.

One end of the cracked aluminum beam was clamped using the VMC-6P pneumatic clamp. The 5 s, ultrasonic, harmonic signal with frequency 25 kHz was used as the high-frequency excitation. The amplitude of this signal was equal to 4 μN. The high-frequency signal was generated by Agilent 33522A signal generator and introduced to the beam through a low-profile, surface-bonded PI Ceramics PIC155 piezoceramic transducer. Simultaneously, the specimen was excited by the low-frequency amplitude-modulated signal. The amplitude of carrier wave was equal to 12 mN and its frequency was chosen to 1083 Hz. The modulation signal was sine wave of frequency 43 Hz and modulation intensity equals to 0.9. This signal was also generated by Agilent 33522A and introduced to the monitored structures through the electrodynamic shaker TMS K2007E01 with built-in amplifier and PCB 288D01 impedance head.

Both excitation signals were introduced to cantilever beam in different sides of crack. The PSV Polytec - 400 SLDV laser vibrometer was used to acquire velocity response at one position on the monitored beam. A schematic diagram illustrating the experimental arrangements is given in Fig. 9.

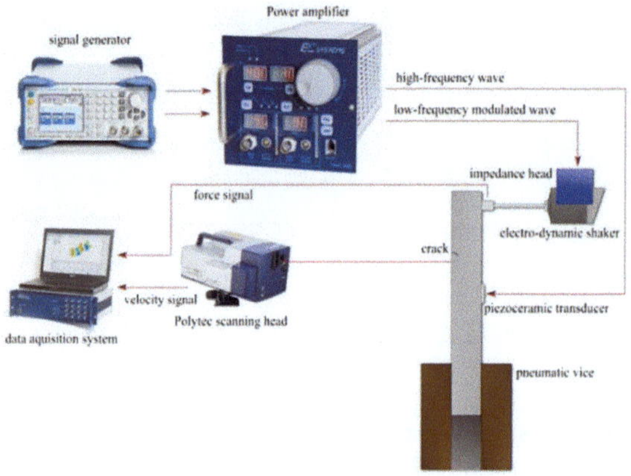

Fig. 9. Diagram illustrating experimental arrangement.

For registered velocity signal, the spectrum was calculated, then normalized and presented in Fig. 10. The Fig. 10a presents velocity spectrum zoomed at low frequencies, near the pumping wave. The structure response contains odd and even harmonics, similar to models with quadratic damping presented in Fig. 6a and c. There is also visible multiplication of sidebands in harmonic of pumping wave. Due to presence of noise in marked spectrum region, it cannot be unambiguously determined whether the modulation-related components appear in the response spectrum.

The Fig. 10b reveals response velocity spectrum zoomed at probing wave. The modulation transfer is clearly visible. The probing wave is modulated by pumping wave with its modulation sidebands. Velocity spectrum also presents modulation transfer typical to Luxemburg-Gorky effect. In reference to models spectrum near probing wave showed in Fig. 8, both quadratic damping models reveal cross-modulation. However only model with additional frequency dependent component

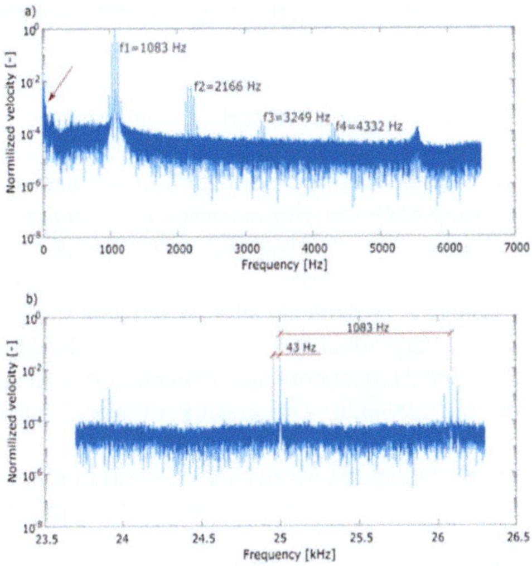

Fig. 10. Normalized velocity spectrum for damaged beam zooming at: (a) pumping wave, (b) probing wave.

reveals modulation transfer. From the presented experimental results, the model presented in subsection 3.3, represents the real structure with damage the most accurately.

6 Conclusions

The study shows modulation transfer effect associated with damage in structure. The three SDOF models with different non-linear characteristics were developed and investigated. The presented works confirmed that the modulation transfer phenomena in structures is related to damping characteristics and can be modeled. The simulation results were compared with experimental data to validate assumed damping model.

Acknowledgement. The work presented in this paper was performed within the scope of the statutory works of the Faculty of Mechanical Engineering and Robotics AGH

References

1. Broda, D., Pieczonka, L., Hiwarkar, V., Staszewski, W.J., Silberschmidt, V.V.: Generation of higher harmonics in longitudinal vibration of beams with breathing cracks. J. Sound Vib. **381**, 206–219 (2016). https://doi.org/10.1016/j.jsv.2016.06.025
2. Novak, A., Bentahar, M., Tournat, V., El Guerjouma, R., Simon, L.: Nonlinear acoustic characterization of micro-damaged materials through higher harmonic resonance analysis. NDT E Int. **45**, 1–8 (2012). https://doi.org/10.1016/j.ndteint.2011.09.006

3. Bentahar, M., El Aqra, H., El Guerjouma, R., Griffa, M., Scalerandi, M.: Hysteretic elasticity in damaged concrete: quantitative analysis of slow and fast dynamics. Phys. Rev. B - Condens. Matter Mater. Phys. (2006). https://doi.org/10.1103/physrevb.73.014116
4. Wang, L., Lie, S.T., Zhang, Y.: Damage detection using frequency shift path. Mech. Syst. Signal Process. **66–67**, 298–313 (2016). https://doi.org/10.1016/j.ymssp.2015.06.028
5. Klepka, A., Staszewski, W.J., di Maio, D., Scarpa, F.: Impact damage detection in composite chiral sandwich panels using nonlinear vibro-acoustic modulations. Smart Mater. Struct. **22**, 084011 (2013). https://doi.org/10.1088/0964-1726/22/8/084011
6. Aymerich, F., Staszewski, W.J.: Experimental study of impact-damage detection in composite laminates using a cross-modulation vibro-acoustic technique. Struct. Heal. Monit. **9**, 541–553 (2010). https://doi.org/10.1177/1475921710365433
7. Pecorari, C., Solodov, I.: Nonclassical nonlinear dynamics of solid surfaces in partial contact for NDE applications, pp. 309–326. Universality of Nonclassical Nonlinearity (2006). https://doi.org/10.1007/978-0-387-35851-2_19
8. Klepka, A., Dziedziech, K., Spytek, J., Mrówka, J., Górski, J.: Experimental investigation of hysteretic stiffness related effects in contact-type nonlinearity. Nonlinear Dyn. (2018). https://doi.org/10.1007/s11071-018-4641-z
9. Bailey, A.V., Martyn, D.F.: The influence of electric waves on the ionospherc. Philos. Mag. **18**, 369–386 (1934). https://doi.org/10.1080/14786443409462506
10. Zaitsev, V., Gusev, V., Castagnede, B.: Luxemburg-gorky effect retooled for elastic waves: a mechanism and experimental evidence. Phys. Rev. Lett. **89**, 105502 (2002). https://doi.org/10.1103/physrevlett.89.105502
11. Trojniar, T., Klepka, A., Pieczonka, L., Staszewski, W.J.: Fatigue crack detection using nonlinear vibro-acoustic cross-modulations based on the Luxemburg-Gorky effect. SPIE Smart Struct. Mater. Nondestruct. Eval. Heal. Monit. **9064**, 90641F–90641F (2014). https://doi.org/10.1117/12.2046471
12. Zaitsev, V., Nazarov, V., Gusev, V., Castagnede, B.: Novel nonlinear-modulation acoustic technique for crack detection. In: NDT and E International. pp. 184–194 (2006)
13. Broda, D., Staszewski, W.J., Martowicz, A., Silberschmidt, V.V.: Modelling of nonlinear crack–wave interactions for damage detection based on ultrasound—A review. J. Sound Vib. **333**, 1097–1118 (2014). https://doi.org/10.1016/j.jsv.2013.09.033
14. Morison, J.R., Johnson, J.W., Schaaf, S.A.: The force exerted by surface waves on piles. J. Pet. Technol. (1950). https://doi.org/10.2118/950149-g
15. Chaparro, L.: Signals and Systems Using MATLAB® (2011)
16. Fillinger, L., Zaitsev, V.Y., Gusev, V., Castagnède, B.: Nonlinear relaxational absorption/transparency for acoustic waves due to thermoelastic effect. Acta Acust. united with Acust. **92**, 24–34 (2006)

Application of Wavelet Analysis for Crack Localization and Quantification in Beams Using Static Deflections

Qiaoyu Ma and Mario Solís(✉)

Escuela Técnica Superior de Ingeniería, Universidad de Sevilla, Camino de los Descubrimientos, 41092 Sevilla, Spain
msolis@us.es

Abstract. Wavelet analysis has been proven to be an efficient tool for identifying singularities in signals, such as the effect of damage in structural deflections. This paper establishes a new approach applying this technique to identify cracks in beams using static measurements. The deflection difference of the beam before and after damage is a piecewise polynomial with discontinuities at crack locations. The crack positions can be identified at apexes of the continuous wavelet transform coefficients. At damage locations, a damage index can be defined from the y-intercept of the linear regression between the logarithms of wavelet coefficients and their corresponding scales. By normalizing itself to the internal bending moment at the damage location, the damage index becomes damage location independent. Through a numerical model, a reference map between the crack depth and the damage index can be established and further used for damage severity assessment.

Keywords: Wavelet analysis · Crack depth estimation · Trend estimate filter

1 Introduction

Since the first application of wavelet analysis in crack detection in beams [1], numerous studies of this subject have been conducted in the past two decades. Compared to other digital differentiator filters, the wavelet function has the advantage in terms of simplicity due to its characteristics [2]. At the early stage, the applications of wavelet theory for damage detection were mainly on the dynamic response of the structure in time domain. Liew and Wang [3] first applied the wavelet to the dynamic response of a beam at a certain time in space domain. Later, due to the ability of detecting singularities in a continuous signal, the focus of wavelet analysis in crack detection shifts to space domain, i.e. deflection based for static responses [4–11] and mode shape based for dynamic responses [12–21]. Moreover, Zhu and Law [22] applied the wavelet analysis to

M. A. Wahab (Ed.): DAMAS 2019, LNME, pp. 135–149, 2020.
https://doi.org/10.1007/978-981-13-8331-1_10

the operational deflection of a single measurement on a simply supported beam with multiple cracks subjected to a moving load. Meanwhile, issues related to the wavelet analysis have been studied comprehensively, such as the selection of mother wavelet and scale, boundary distortion using the finite length signal, sampling effects, crack size sensitivity, mode order sensitivity, et al. [5,23–28]. The experimental noise issue in the measurements in practical applications have been addressed as well, such as by Solís et al. [29] through using a roving mass and weighted parameter based on Signal to Noise ratio and by Cao et al. [30] via the Teager-Energy Operator.

Further, researchers have extended the use of the wavelet transform from locating the cracks to quantifying the extents. Pakrashi et al. [31] suggest that the absolute value of the wavelet coefficient of the mode shapes at the damage location can be used as a damage calibration factor. The authors also proposed to use Kurtosis analysis for crack estimation [32]. Umesha et al. [8] studied the wavelet analysis to the static deflection of a fixed-fixed beam. A generalized curve which is the envelope plot of the wavelet coefficient maximum at damage point is proposed as a reference map for crack severity calibration. However, using the absolute value of the wavelet coefficient has the drawback that it not only varies from the crack location but also depends on the scale used in the analysis. To resolve the location dependency, Andreaus and Casini [10] applied the wavelet analysis to the static deflection difference and proposed a new location independent damage index which is the ratio between the wavelet coefficient at the damage location and its corresponding curvature of the undamaged state. To avoid the scale dependency, Andreaus et al. [11] also proposed a new damage locating factor, normalized wavelet coefficient of several scales, which can also be used to estimate the damage severity by comparing the experimental results to the numerical values.

Another scale independent damage quantification method is to use the characteristics of the irregularities in the signal (deflection or mode shape) as damage index. Mallat and Hwang [33] points out that at the crack location, the local maximum wavelet coefficient is proportional to the scales in the logarithmic form. Hong et al. [34] first suggested to use the Lipschitz exponent which is the slope of the linear relation to estimate the crack severity. It was found out later that the intersection of the y-axis can be used as a damage quantification index as well [35–37]. In addition, Zhu et al. [37] proposed a new damage locating index which can eliminate the boundary effect.

In this paper, a damage locating index which uses the Continuous Wavelet Transform (CWT) information at different scales, and a damage severity index which is independent from both crack location and wavelet scales are proposed by applying the wavelet analysis to the static deflection difference of the damaged and undamaged beam. To examine the performance of the methodology, Gaussian wavelet with 2 vanishing moments (Gaus2) is used as the mother wavelet. A linear trend filter is applied to reduce the experimental noise effect and determine the number of cracks. The correlation between crack depth and the damage index is obtained through a numerical model.

In the following sections, first the theoretical background of the continuous wavelet analysis is introduced. Secondly, the crack locating and severity indices are presented along with a numerical example. Then, the experimental tests and results are provided with a guide of the noise filter application. Lastly, the conclusions are drawn.

2 Continuous Wavelet Analysis

A function $\psi(x)$ is said to be wavelet if and only if its Fourier transform $\hat{\psi}(\omega)$ satisfies

$$\int_0^{+\infty} \frac{|\hat{\psi}(\omega)|^2}{|\omega|} \mathrm{d}\omega = \int_{-\infty}^{0} \frac{|\hat{\psi}(\omega)|^2}{|\omega|} \mathrm{d}\omega < +\infty \tag{1}$$

It means that function $\psi(x)$ has a zero mean and finite length (compact support)

$$\int_{-\infty}^{+\infty} \psi(x) \mathrm{d}x = 0 \tag{2}$$

The real or complex function $\psi(x)$ is used to create a family of wavelets $\psi_{u,s}(x)$, defined as

$$\psi_{u,s}(x) = \frac{1}{\sqrt{s}} \psi\left(\frac{x-u}{s}\right) \tag{3}$$

where real number s and u are the scale and translation parameters respectively. The family of wavelet functions is a dilated or stretched version of the mother wavelet $\psi(x)$.

For a given signal $f(x)$, where x is time or space, the Continuous Wavelet Transform (CWT) is obtained integrating the product of the signal function and the wavelet function

$$Wf(u,s) = \frac{1}{\sqrt{s}} \int_{-\infty}^{+\infty} f(x) \psi^* \left(\frac{x-u}{s}\right) \mathrm{d}x \tag{4}$$

where $\psi^*(x)$ is the complex conjugate of the wavelet function. $Wf(u,s)$ is called the CWT coefficient for wavelet $\psi_{u,s}(x)$ and it measures the variation of the signal in the vicinity of u whose size is proportional to s.

In detection of singularities of deflections, the vanishing moments has an important influence. A wavelet has n vanishing moments if the following equation is carried out

$$\int_{-\infty}^{+\infty} x^k \psi(x) \mathrm{d}x = 0, \quad k = 0, 1, 2, ..., n-1 \tag{5}$$

Therefore, a wavelet with n vanishing moments is orthogonal to polynomials of degree $n-1$.

3 Static Based Crack Identification Problem

For a beam under some time invariant static loads, denote the static deflection difference of the beam between the damaged and undamaged states by v. It is known that the deflection difference v is caused by the bending moment applied at the damage location [38,39]. At the vicinity of the damage location, the cracks generate a region where the curvature of the beam changes rapidly. In practice with discrete data, the problem can be transformed into identifying the discontinuities in the deflection difference measurements.

3.1 Damage Locating Index

If the deflection has a singularity at certain point u, that means it is not differentiable at u, then the CWT coefficient at that point will have locally maximum values (apex values) for every scale. Small scales provide higher resolution on locating singularities but is more sensitive to noise in practice while high scales are more robust to noise but provide lower location resolution. Hence, to maximize the information provided by different scales, a damage locating index (LI) defined as the sum of the normalized CWT coefficients at different scales is proposed (Eq. (6)). The CWT coefficient is normalized to its maximum value for each scale (Eq. (7)). The normalized CWT coefficient, Wf_n, gives the same weight to each scale in the summation. The potential damage location is taken where the local maximum value appears, i.e. $\partial \mathrm{LI}(u)/\partial u = 0$.

$$\mathrm{LI}(u) = \sum^{s} Wf_n(u, s) \tag{6}$$

$$Wf_n(u, s) = \frac{Wf(u, s)}{\max\limits_{u}(Wf(u, s))} \tag{7}$$

3.2 Damage Severity Index

Mallat and Hwang [33] implies the connection between two indicators which can characterize the local regularity of functions, the wavelet transform and Lipschitz exponents. At the local maximum point, u_0, the wavelet coefficient satisfies

$$|Wf(u_0, s)| \leq As^{\alpha} \tag{8}$$

where A is a constant and α is the Lipschitz exponent. By normalizing the wavelet transform coefficient at the damage location to its corresponding damaged bending moment, M_0, and taking the logarithmic value on both sides of the equation, one has

$$\log_2 \left| \frac{Wf(u_0, s)}{M_0} \right| \leq \log_2 \left(\frac{A}{M_0} \right) + \alpha \log_2(s) \tag{9}$$

The damage index (DI) is taken as the y-axis interception

$$\mathrm{DI} = \log_2\left(\frac{A}{M_0}\right) \tag{10}$$

A reference correlation map between the crack depth and DI can be established through a numerical model and used for quantification purpose in practical applications.

3.3 Numerical Example

A simple numerical example of a simply supported beam is provided in this section to demonstrate the procedure of crack identification methodology. The crack-typed damage is idealized by a massless lumped rotational spring. The target beam modeled in ANSYS has a dimension of 1200 (length) × 800 (width) × 20 (height) mm and Young's modulus $E = 210$ GPa. Four rotational springs with rotational stiffness K_1, K_2, K_3 and K_4, respectively, are used to model four cracks and their stiffness values are listed in Table 1. Hereupon, all location notations are defined from the left end of the beam. For demonstration, the deflection different of the beam is generated by simultaneously implementing four pairs of unit self-equivalent bending moments on the springs respectively. The scheme of the beam is depicted in Fig. 1.

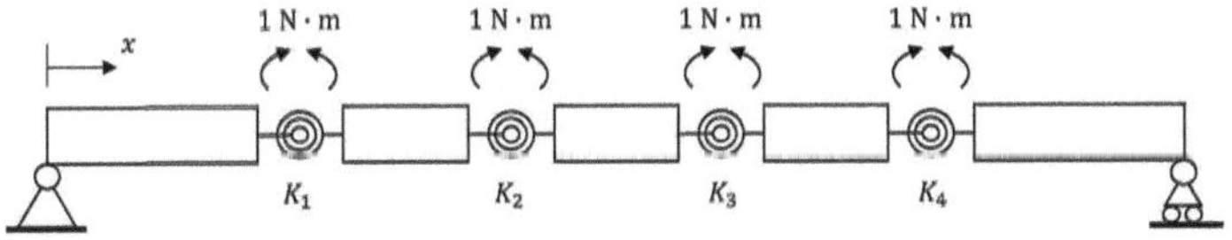

Fig. 1. Sketch of the 1D beam model with four springs as damage.

Table 1. The target beam model information

Spring label	1	2	3	4
Location, x (mm)	425	525	625	775
Stiffness, K (N m/rad)	2.00e[7]	4.12e[5]	1.73e[5]	1.35e[6]
Severity[a] (%)	10	35	50	20

[a]Crack depth to beam height ratio based on Ostachowicz's model [40]

A total number of 241 points are taken and the deflection is plotted in Fig. 2(a). The wavelet analysis using Gaus2 wavelet is applied to the deflection difference. The scale is taken to be from 1 to 4. The normalized CWT coefficients Wf_n at each scale are shown in Fig. 2(b) and the damage locating index is shown in Fig. 3(a). Locations where $\partial LI/\partial u = 0$ are associated with the damages at 425 mm, 525 mm 625 mm and 775 mm. At the damage locations, the linear relation between the local maximum CWT coefficients and the scales in logarithmic form can be seen clearly in Fig. 3(b). The DI of each crack is obtained and listed in Table 2.

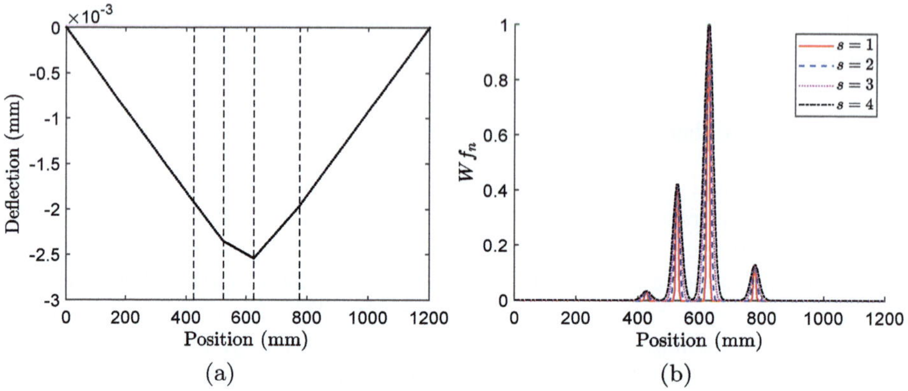

Fig. 2. The target model: (a) deflection difference (the dashed lines mark the damage locations); (b) the normalized wavelet transform coefficient of the beam.

Table 2. The damage index of the targeted damage.

Spring label	1	2	3	4
DI	−21.4	−17.4	−16.1	−19.1

A beam with the same dimension and material properties with a single spring situated at the midspan is used to establish the reference map to quantify the damage severity. The linear relation between the logarithmic wavelet coefficient and scales at the damage location of different severities is shown in Fig. 4. The DI of the reference model is listed in Table 3. By matching the values in Table 2 to the reference values in Table 3, the crack severities can be accurately evaluated in noise free condition.

Table 3. The reference values of damage index for different severities.

Severity (%)	10	20	35	50
DI	−21.0	−19.1	−17.3	−16.1

The is worthy to note that other mother wavelet can be use in place of Gaus2, though the selected one should be symmetrical and has a vanishing moment not less than 2 for statically determinate beams and 4 for statically indeterminate beams.

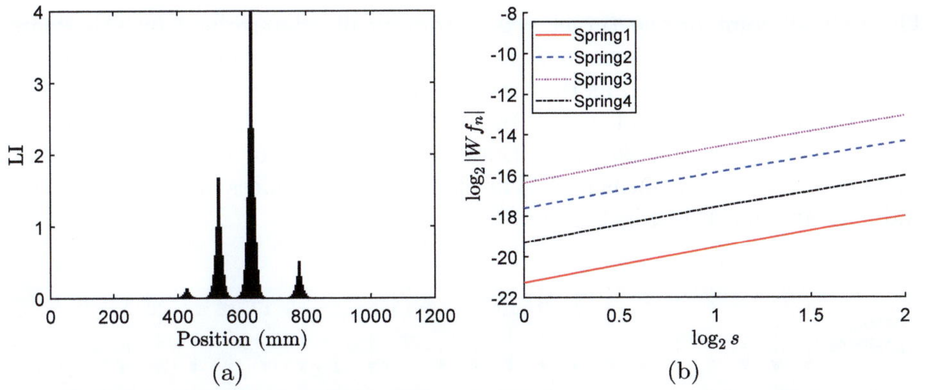

Fig. 3. The target model: (a) the location index of the beam; (b) the CWT coefficients versus scales in logarithmic at damage locations.

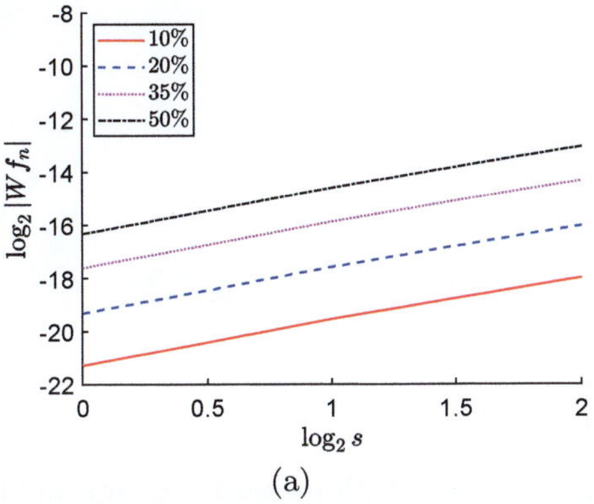

Fig. 4. The CWT coefficients versus scales in logarithmic of various damage severities with spring located at 600 mm (reference model).

4 Experimental Tests

An experimental test of a simply supported beam was conducted to examined the performance of the methodology. The tested beam is 1200 mm long, 800 mm wide and 20 mm high. The material of the beam is steel with 210 GPa Young's modulus (Fig. 5(a)). A concentrated load of 120 kg was hung at the bottom of the beam at 21 different equally distributed positions on the beam (Fig. 5(b)). The sum of the deflections of all loads were used as the deflection base. An artificial notch crack was cut at 425 mm (Fig. 5(c)). Two different severities were considered in the study, 20% (4 mm) and 50% (10 mm).

The deflections of the beam were measured by a Digital Imagine Correlation (DIC) system from GOM company (Fig. 5(d)). The measurement target is a 10 mm diameter black circle with a 5 mm diameter white circle in the center.

The central point of the target was automatically determined by the Pontos system, part of the DIC system. For each test, 20 images were captured and the average values were taken as the target coordinates. To retain the merit of the DIC system, the whole side of the beam was covered by the targets which were placed in the forms of two lines. Both the deflections of the undamaged and damaged beams were measured. The deflection differences with crack of 20% and 50% are shown in Fig. 6.

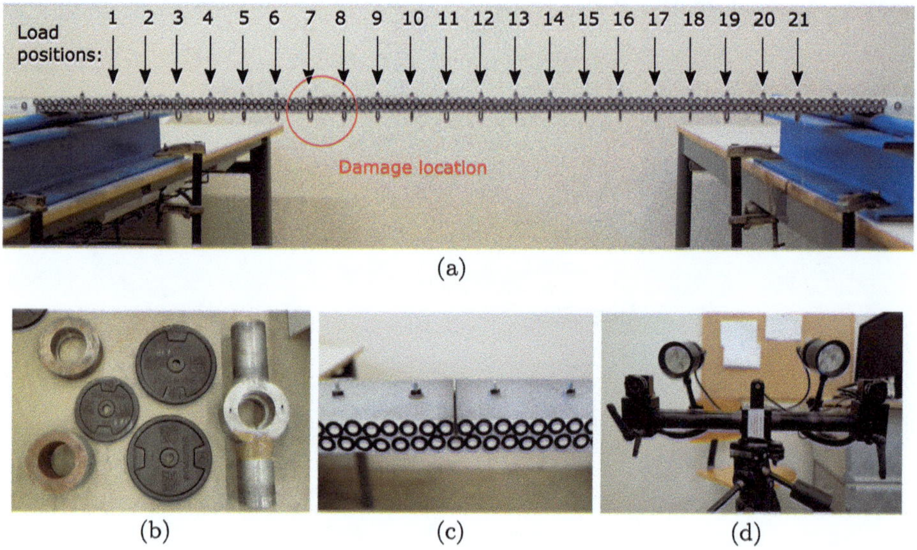

Fig. 5. (a) The experimental beam with measuring points and load positions; (b) the concentrated load; (c) the artificial crack; (d) the camera of the DIC system.

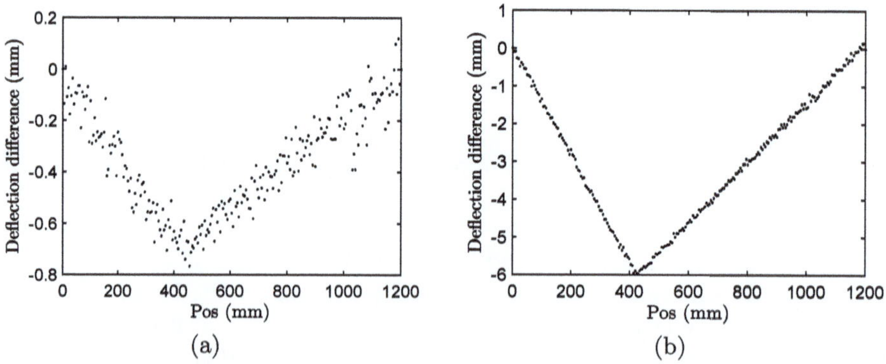

Fig. 6. The deflection difference of the damaged beam with (a) 20% (4 mm) crack; (b) 50% (10 mm) crack.

The deflection difference with 20% crack is about one tenth of the one with 50% crack. For static tests, the measurement noise effect is more severe for small damage than large severity. A denoising filter is applied to reduce the noise level.

4.1 Noise Filtering

Since the static deflection difference of a simply supported beam is linear outside of the damage region or piecewise linear in the spring model, a linear trend estimating tool named $l1$ Trend Filtering [41] is used for denoising purpose in practice. Previous studies [39,42] have shown that the $l1$ Trend Filtering can efficiently and automatically determine the locations as well as the number of the kinks. The objective function to be minimized is

$$(1/2)\sum_{k=1}^{241}(v_k - v_k^{l1})^2 + \lambda\sum_{k=2}^{240}|v_{k-1}^{l1} - 2v_k^{l1} + v_{k+1}^{l1}| \tag{11}$$

where v_k is the kth measurement and v_k^{l1} is its estimate. λ is a weighting parameter which controls the trade off between the size of the residual and the 'smoothness' of the estimate. It is obvious that as λ changes from 0 to ∞, the estimate changes from the input data inself to its linear regression fit ultimately. The size of the residual $(v_k - v_k^{l1})^2$ increases with respect to λ. Therefore, the selection of λ is essential in this denoising process. If λ is too small, the estimate is overfitted while if λ is too high, the estimate becomes underfitted. There exists a range of λ values where the estimates can be considered valid, which is named the optimal range of λ.

To evaluate the reliability of the estimates for each λ, the norm of the residual (R), between the estimate and the data, measured by Eq. (12) and the change rate of R with respect to λ ($\mathrm{d}R/\mathrm{d}\lambda$) calculated by Eq. (13) are used, where $i = 1, 2, \cdots, n$. By trying a series of λ values, the norms of the residual as well as its change rate are plot in Fig. 7.

$$R = ||v_k - v_k^{l1}|| \tag{12}$$

$$\frac{\mathrm{d}R}{\mathrm{d}\lambda} \approx \frac{R(\lambda_{i+1}) - R(\lambda_i)}{\lambda_{i+1} - \lambda_i} \tag{13}$$

It can be seen that as λ increases from 2 (2^1) to 64 (2^6), the norm of the residual rises up at a relative fast pace but the change rate quickly drops down. The change rate reaches to the minimum at $\lambda = 13$ for 20% damage (Fig. 7(a)) and at $\lambda = 9$ for 50% damage (Fig. 7(b)). As λ gets higher, the change rate increases slowly, which means that the estimate is less sensitive to the change of λ. Therefore, the λ value correponding to the minimum point in the change rate is taken as the lower bound of the optimal region, λ_L and estimates based on λ values smaller than λ_L are considered overfitted. However, there is no clear point on the chart that reveals at which point the underfitting occurs. Intuitively, the norm of the residual is a good standard to determine the point where underfitting

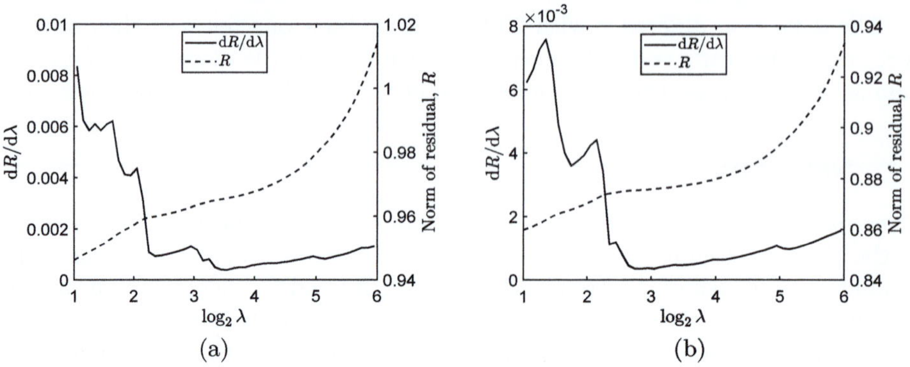

Fig. 7. The change rate of the residuals (the solid line) and the norms of the residuals (the dashed line) with different λ of (a) 20% (4 mm) crack; (b) 50% (10 mm) crack.

occurs. In this case study, it is found that by keeping the norm of residual $R(\lambda)$ below $1.05R(\lambda_L)$ can prevent underfitting. Hence, the corresponding value of λ is taken as the upper bound, λ_H. The lower and upper bound values of λ for the two experimental cases are listed in Table 4. The corresponding $l1$ Trend Filtering results are shown in Fig. 8. Estimates with λ values within this optimal range can be considered valid.

Table 4. The optimal region of λ.

Damage severity (%)	Lower bound λ_L	Upper bound λ_H
20%	11	64
50%	8	56

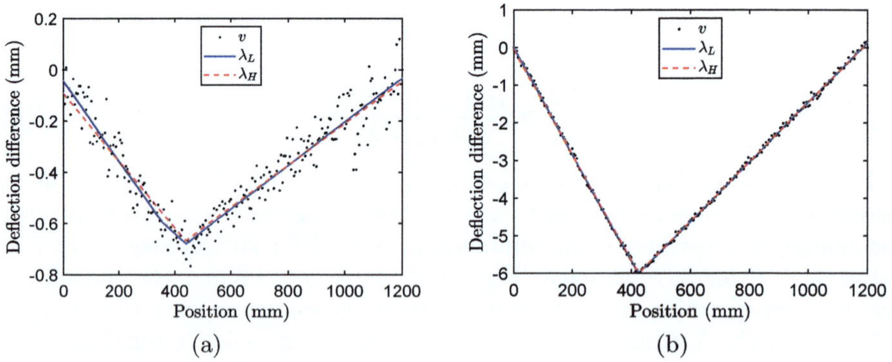

Fig. 8. The $l1$ Trend Filtering estimates with lower and upper bond λ values (a) 20% (4 mm) crack; (b) 50% (10 mm) crack.

4.2 Crack Localization Results

The $l1$ Trend Filtering leads to different kinks numbers and positions with different λ values. For the experimental study, all integer values of λ within the optimal range were taken into account. Therefore, the locating index (LI) in Eq. (6) should be written as

$$\mathrm{LI}^*(u) = \sum_{i=1}^{n} \left[\sum_{j=1}^{m} Wf_n(u, s_j) \right]_{\lambda_i} \tag{14}$$

where n is the number of λ ($n = 51$) and m is the number of scales used in the wavelet transform. In this experimental study, the scales are taken from 2 to 8 ($m = 7$). From the damage locating index (LI^*) in Fig. 9, the damage locations associated with $\partial LI^*/\partial u = 0$ are predicted at 440 mm for the beam with 20% crack and at 430 mm for the beam with 50% crack. In both cases, the damage location is accurately found.

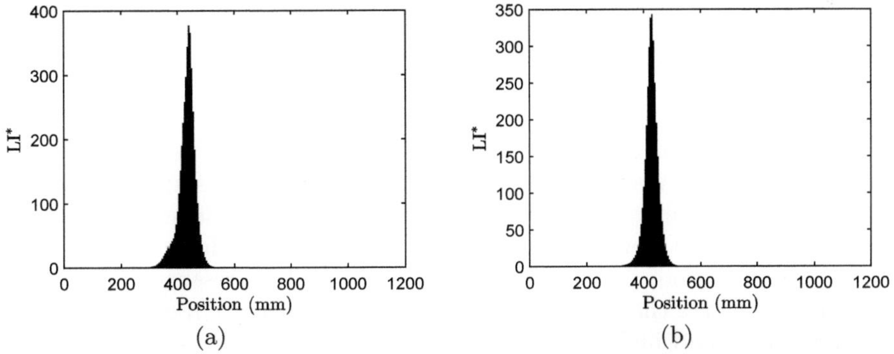

Fig. 9. The Locating Index (LI*) of the beam with (a) 20% (4 mm) crack; (b) 50% (10 mm) crack (the actual crack is situated at 425 mm).

4.3 Crack Quantification Results

A 3D numerical model beam of the same dimension was built in ANSYS with the same material properties in order to establish the DI to crack depth correlation. In this model, the load is applied at 500 mm and the crack is set at 725 mm. The measurement points are taken at the same locations as in the experimental test. The reference DI values of crack severities from 5% to 60% are obtained (Fig. 10).

The damage indices at the predicted location can be obtain for all evaluated λ values. The corresponding crack severities can be assessed with the reference map. The average crack depth values of all considered λ are listed in Table 5. The estimate crack depth of 20% is 1.0 mm lower than the actual crack depth while the prediction of 50% is 0.5 mm lower than the actual value.

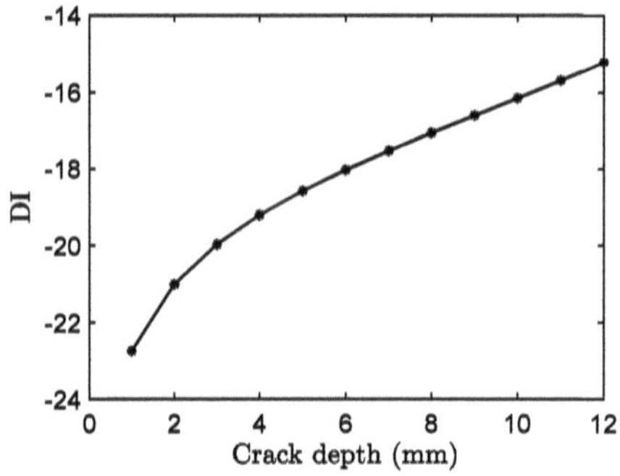

Fig. 10. The damage index (DI) reference map of the beam.

Table 5. The estimated crack depths of the experimental tests.

Damage severity (%)	20%	50%
Crack depth (mm)	3.0	9.5
Actual depth (mm)	4.0	10.0

5 Conclusion

This paper addresses the issue of crack identification in beams using static deflection measurements. A new damage locating index as well as an location independent damage severity index based on the wavelet analysis are presented. In practice, an effective trend estimating tool is applied to reduce the noise influence. A practical guide of the application of this tool is presented. The experimental results shows that the proposed methodology can accurately locate both low severe crack and high severe crack. It is worthy to note that the proposed methodology can be applied to identify multiple cracks scenarios.

Acknowledgments. This work was supported by the Consejería de Economía, Innovación, Ciencia y Empleo of Andalucía (Spain) under project P12-TEP-2546 and the Spanish Ministry of Economy and Competitiveness (Ministerio de Economía y Competitividad, Secretaría de Estado de Investigación, Desarrollo e Innovación) through research project BIA2016-43085-P. The financial support is gratefully acknowledged.

References

1. Surace, C., Ruotolo, R.: Crack detection of a beam using the wavelet transform. In: Proceedings of the 12th International Modal Analysis Conference, pp. 1141–1167 (1994)
2. Messina, A.: Detecting damage in beams through digital differentiator filters and continuous wavelet transforms. J. Sound Vib. **272**(1–2), 385–412 (2004). https://doi.org/10.1016/j.jsv.2003.03.009
3. Liew, K.M., Wang, Q.: Application of wavelet theory for crack identification in structures. J. Eng. Mech. **124**(2), 152–157 (1998)

4. Wang, Q., Deng, X.: Damage detection with spatial wavelets. Int. J. Solids Struct. **36**(23), 3443–3468 (1999). https://doi.org/10.1016/S0020-7683(98)00152-8
5. Quek, S.T., Wang, Q., Zhang, L., Ang, K.K.: Sensitivity analysis of crack detection in beams by wavelet technique. Int. J. Mech. Sci. **43**(12), 2899–2910 (2001). https://doi.org/10.1016/S0020-7403(01)00064-9
6. Rucka, M., Wilde, K.: Crack identification using wavelets on experimental static deflection profiles. Eng. Struct. **28**, 279–288 (2006). https://doi.org/10.1016/j.engstruct.2005.07.009
7. Spanos, P.D., Failla, G., Santini, A., Pappatico, M.: Damage detection in Euler-Bernoulli beams via spatial wavelet analysis. Struct. Control Monit. **13**(1), 472–487 (2006). https://doi.org/10.1002/stc.118
8. Umesha, P., Ravichandran, R., Sivasubramanian, K.: Crack detection and quantification in beams using wavelets. Comput. Aided Civ. Infrastruct. Eng. **24**(8), 593–607 (2009). https://doi.org/10.1111/j.1467-8667.2009.00618.x
9. Wu, N., Wang, Q.: Experimental studies on damage detection of beam structures with wavelet transform. Int. J. Eng. Sci. **49**(3), 253–261 (2011). https://doi.org/10.1016/j.ijengsci.2010.12.004
10. Andreaus, U., Casini, P.: Identification of multiple open and fatigue cracks in beam-like structures using wavelets on deflection signals. Contin. Mech. Thermodyn. **28**(1–2), 361–378 (2016). https://doi.org/10.1007/s00161-015-0435-4
11. Andreaus, U., Baragatti, P., Casini, P., Iacoviello, D.: Experimental damage evaluation of open and fatigue cracks of multi-cracked beams by using wavelet transform of static response via image analysis. Struct. Control Health Monit. **24**(4), 1–16 (2017). https://doi.org/10.1002/stc.1902
12. Chang, C.C., Chen, L.W.: Detection of the location and size of cracks in the multiple cracked beam by spatial wavelet based approach. Mech. Syst. Signal Process. **19**(1), 139–155 (2005). https://doi.org/10.1016/j.ymssp.2003.11.001
13. Rucka, M., Wilde, K.: Application of continuous wavelet transform in vibration based damage detection method for beams and plates. J. Sound Vib. **297**, 536–550 (2006). https://doi.org/10.1016/j.jsv.2006.04.015
14. Castro, E., García-Hernandez, M.T., Gallego, A.: Damage detection in rods by means of the wavelet analysis of vibrations: influence of the mode order. J. Sound Vib. **296**(4–5), 1028–1038 (2006). https://doi.org/10.1016/j.jsv.2006.02.026
15. Zhong, S., Oyadiji, S.O.: Detection of cracks in simply-supported beams by continuous wavelet transform of reconstructed modal data. Comput. Struct. **89**(1–2), 127–148 (2011). https://doi.org/10.1016/j.compstruc.2010.08.008
16. Jiang, X., Ma, Z.J., Ren, W.X.: Crack detection from the slope of the mode shape using complex continuous wavelet transform. Comput. Civ. Infrastruct. Eng. **27**(3), 187–201 (2012). https://doi.org/10.1111/j.1467-8667.2011.00734.x
17. Solís, M., Algaba, M., Galvín, P.: Continuous wavelet analysis of mode shapes differences for damage detection. Mech. Syst. Signal Process. **40**(2), 645–666 (2013). https://doi.org/10.1016/j.ymssp.2013.06.006
18. Xu, Y.F., Zhu, W.D., Liu, J., Shao, Y.M.: Identification of embedded horizontal cracks in beams using measured mode shapes. J. Sound. Vib. **333**(23), 6273–6294 (2014). https://doi.org/10.1016/j.jsv.2014.04.046
19. Cao, M.S., Xu, W., Ren, W.X., Ostachowicz, W., Sha, G.G., Pan, L.X.: A concept of complex-wavelet modal curvature for detecting multiple cracks in beams under noisy conditions. Mech. Syst. Signal Process. **76–77**, 555–575 (2016). https://doi.org/10.1016/j.ymssp.2016.01.012

20. Janeliukstis, R., Rucevskis, S., Wesolowski, M., Chate, A.: Multiple damage identification in beam structure based on wavelet transform. Procedia Eng. **172**, 426–432 (2017). https://doi.org/10.1016/j.proeng.2017.02.023
21. Ghanbari Mardasi, A., Wu, N., Wu, C.: Experimental study on the crack detection with optimized spatial wavelet analysis and windowing. Mech. Syst. Signal Process. **104**, 619–630 (2018). https://doi.org/10.1016/j.ymssp.2017.11.039
22. Zhu, X.Q., Law, S.S.: Wavelet-based crack identification of bridge beam from operational deflection time history. Int. J. Solids Struct. **43**(7–8), 2299–2317 (2006). https://doi.org/10.1016/j.ijsolstr.2005.07.024
23. Quek, S.T., Wang, Q., Zhang, L., Ong, K.H.: Practical issues in the detection of damage in beams using wavelet. Smart Mater. Struct. **10**, 1009–1017 (2001)
24. Gentile, A., Messina, A.: On the continuous wavelet transforms applied to discrete vibrational data for detecting open cracks in damaged beams. Int. J. Solids Struct. **40**(2), 295–315 (2003). https://doi.org/10.1016/S0020-7683(02)00548-6
25. Messina, A.: Refinements of damage detection methods based on wavelet analysis of dynamical shapes. Int. J. Solids Struct. **45**(14–15), 4068–4097 (2008). https://doi.org/10.1016/j.ijsolstr.2008.02.015
26. Montanari, L., Basu, B., Spagnoli, A., Broderick, B.M.: A padding method to reduce edge effects for enhanced damage identification using wavelet analysis. Mech. Syst. Signal Process. **52–53**(1), 264–277 (2015). https://doi.org/10.1016/j.ymssp.2014.06.014
27. Swamy, S., Reddy, D.M., Prakash, G.J.: Damage detection and identification in beam structure using modal data and wavelets. World J. Model. Simul. **13**(1), 52–65 (2017)
28. Janeliukstis, R., Rucevskis, S., Wesolowski, M., Chate, A.: Experimental structural damage localization in beam structure using spatial continuous wavelet transform and mode shape curvature methods. Meas.: J. Int. Meas. Confed. **102**, 253–270 (2017). https://doi.org/10.1016/j.measurement.2017.02.005
29. Solís, M., Ma, Q., Galvín, P.: Damage detection in beams from modal and wavelet analysis using a stationary roving mass and noise estimation. Strain **54**, e12266 (2018). https://doi.org/10.1111/str.12266
30. Cao, M., Radzieński, M., Xu, W., Ostachowicz, W.: Identification of multiple damage in beams based on robust curvature mode shapes. Mech. Syst. Signal Process. **46**(2), 468–480 (2014). https://doi.org/10.1016/j.ymssp.2014.01.004
31. Pakrashi, V., O'Connor, A., Basu, B.: A study on the effects of damage models and wavelet bases for damage identification and calibration in beams. Comput. Civ. Infrastruct. Eng. **22**(8), 555–569 (2007). https://doi.org/10.1111/j.1467-8667.2007.00510.x
32. Pakrashi, V., Basu, B., O'Connor, A.: Structural damage detection and calibration using a wavelet-Kurtosis technique. Eng. Struct. **29**(9), 2097–2108 (2007). https://doi.org/10.1016/j.engstruct.2006.10.013
33. Mallat, S., Hwang, W.L.: Singularity detection and processing with wavelets. IEEE Trans. Inf. Theory **38**(2), 617–643 (1992). https://doi.org/10.1109/18.119727
34. Hong, J.C., Kim, Y.Y., Lee, H.C., Lee, Y.W.: Damage detection using the Lipschitz exponent estimated by the wavelet transform: applications to vibration modes of a beam. Int. J. Solids Struct. **39**(7), 1803–1816 (2002). https://doi.org/10.1016/S0020-7683(01)00279-7
35. Douka, E., Loutridis, S., Trochidis, A.: Crack identification in beams using wavelet analysis. Int. J. Solids Struct. **40**(13–14), 3557–3569 (2003). https://doi.org/10.1016/S0020-7683(03)00147-1

36. Loutridis, S., Douka, E., Trochidis, A.: Crack identification in double-cracked beams using wavelet analysis. J. Sound Vib. **277**(4–5), 1025–1039 (2004). https://doi.org/10.1016/j.jsv.2003.09.035
37. Zhu, L.F., Ke, L.L., Zhu, X.Q., Xiang, Y., Wang, Y.S.: Crack identification of functionally graded beams using continuous wavelet transform. Compos. Struct. **210**, 473–485 (2018)
38. Gudmundson, P.: The dynamic behaviour of slender structures with cross-sectional cracks. J. Mech. Phys. Solids **31**(4), 329–345 (1983). https://doi.org/10.1016/0022-5096(83)90003-0
39. Ma, Q., Solís, M.: Damage localization and quantification in beams from slope discontinuities in static deflections. Smart Struct. Syst. **22**(3), 291–302 (2018)
40. Ostachowicz, W.M., Krawczuk, M.: Analysis of the effect of cracks on the natural frequencies of a cantilever beam. J. Sound Vib. **150**(2), 191–201 (1991). https://doi.org/10.1016/0022-460X(91)90615-Q
41. Kim, S.J., Koh, K., Boyd, S., Gorinevsky, D.: l_1 trend filtering. SIAM Rev. **51**(2), 339–360 (2009). https://doi.org/10.1137/070690274
42. Ma, Q., Solís, M.: Damage localization and quantification in simply supported beams using static test data. J. Phys.: Conf. Ser. **842**, 012007 (2017). https://doi.org/10.1088/1742-6596/842/1/012007

Using Enhanced Cepstral Analysis for Structural Health Monitoring

M. Ferraris[1], M. Civera[1(✉)], R. Ceravolo[1], C. Surace[1], and R. Betti[2]

[1] Politecnico Di Torino, C. Duca Degli Abruzzi, 24, 10129 Turin, Italy
marco.civera@polito.it
[2] Columbia University, New York, NY 10027, USA

Abstract. Mel-Frequency Cepstral Coefficients (MFCCs) have been proved to be viable to detect damage-induced shifts in the frequency content of structures subjected to external forces. Nevertheless, the Melodic (Mel) Scale is a perceptual feature, roughly based on human perception and originally formulated to resemble the biological mechanisms of the auditory apparatus. Thus, its straight application to non-auditive acquisition systems may be misleading. However, the intrinsic basilar assumption – that is, a nonlinear, non-constant spacing between filters in the frequency domain – is promising for the time-frequency analysis of structural dynamic behaviour. Here, an ensemble of related techniques, developed departing from the original approach, is described; different features, filter types and filter spacing have been investigated. Candidates are compared to each other and benchmarked against the previous approaches present in Literature on the challenging Case Study of the Fossano Bell Tower, considering realistic damage scenarios.

Keywords: Structural health monitoring · Damage detection · Cepstral analysis

1 Introduction

The aim of this research is to optimise the Mel-Frequency Cepstral Coefficients (MFCCs) features, which have been already proved to be effective for Structural Health Monitoring (SHM) purposes [1–3]. Specifically, three new variants are here introduced; they represent slight variations respect to the several formulations proposed in [2] and have been compared to them on a benchmark Case Study, a Finite Element (FE) model of the Santa Maria and San Giovenale Cathedral Bell Tower (Fossano, Italy) [4]. Some interesting improvements were found.

The rest of the paper is organised as follow: in Sect. 2, the theory of Mel-Frequency Cepstral Coefficients is briefly described. Section 3 presents the case study. The proposed variations are specified in Sect. 4. Section 5 reports the Results, and Sect. 6 the Conclusions.

M. A. Wahab (Ed.): DAMAS 2019, LNME, pp. 150–165, 2020.
https://doi.org/10.1007/978-981-13-8331-1_11

2 Theoretical Background

It is well-known that the structural dynamical response of a given structure is highly affected by damage-induced stiffness reduction, which is especially appreciable in the frequency domain [5]. This can be easily extended to the *quefrency* domain [6], where the MFCCs are calculated. Some hints are here recalled; a more detailed description of these techniques, plus others not investigated here, can be found in the book of Beigi [7], applied in the context of Speech Processing. A review of cepstral analysis applications to mechanical systems is also available in [8].

2.1 Cepstrum, Mel-Scale and Mel-Frequency Cepstral Coefficients

The "*cepstrum*" of a signal, as originally defined by [6], is computed as the inverse Fourier transform of the logarithm of its estimated spectrum. Some slightly-modified definitions exist; for the purpose of this research, the *power cepstrum* is the most convenient. This latter can be defined in the frequency domain as

$$\hat{f} = \left[\frac{1}{2\pi}\int_{-\pi}^{\pi} \log\left(|H(\omega)|^2\right) e^{j\omega t} dt\right]^2 \tag{1}$$

For a pulsation ω; here, $H(\omega)$ represents the Discrete Fourier Transform (DFT) of function f. The so-displayed response shows simultaneously properties from both time and frequency domain, derived by the frequency-warping step. It is noteworthy that, due to direct and inverse transform, the resulting unit of is expressed in terms of seconds, even if $\hat{f}$ is not any more a time series, again due to the frequency warping process; more on the subject can be found in [9].

Mel-Frequency Cepstral Coefficients (MFCCs) are the main focus of this discussion. The procedural steps for their extraction are as follow. Firstly, some pre-processing is performed, to decompose the signal into more stationary frames. Then, the discrete power spectrum is computed in the frequency domain. After frequency warping, the logarithm of the re-ordered spectrum in the melody-frequency scale (Mel-scale) undergoes an L-points inverse Discrete Cosine Transform (iDCT) [10]. The final result is an array of the Cepstral Coefficients (CCs), $\boldsymbol{c} \in \mathbb{R}^{L\times 1}$, computed from the melodic log-spectra, i.e.,

$$c[l] = \sum\nolimits_{m=0}^{M-1} a_m \log(H_{\mathrm{mel}}[m]) \cos\left(\frac{\pi(2l+1)m}{2M}\right) \quad l = 0, 1, \ldots, L-1 \tag{2}$$

For M critical bands and $L-1$ CCs. Amplitude a_m equals $1/M$ if $m = 0$ and $2/M$ elsewhere; $H_{\mathrm{mel}}[m]$ indicates the m-th point of the mel-spectrum. Single steps will be further described in the following Sections; for better comprehension, a flow chart is provided in Fig. 1. The preference for iDCT over standard inverse DFT is due to technical reasons, specifically, a better adherence to the Karhunen-Loève Transform.

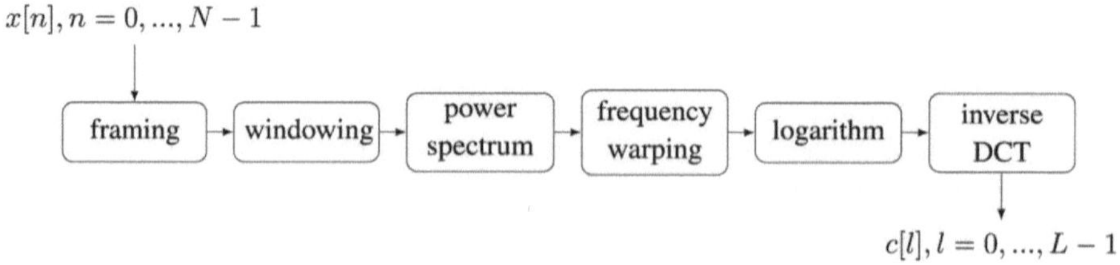

Fig. 1. MFCCs extraction procedure [2].

Therefore, the core of MFCCs is, apart from the power cepstrum itself, the Melodic Scale. This nonlinear, perceptual scale is defined by the *Mel*, with one unity equal to one-thousandth of the pitch ($\wp$) of a simple tone with a frequency of 1000 Hz and an amplitude of 40 dB above the auditory threshold [11, 12]. Several semi-empirical equations have been proposed to express the Hertz-Mel relationship (most common definition is the one by Fant [13]; a comparative study between it and other choices was done by [14]). A number of closely-related alternatives and variants for MFCCs exist; the essence remains almost unaltered, nevertheless. Basically, the frequency warping can be performed slightly differently, to obtain some other sort of nonlinear, Mel-like scale. In this ambit, the main competitor to the Melodic Scale is the Bark Scale [15], for which the most used approximated Hertz-Bark relationship is defined in [16]. As the Mel Scale, the Bark Scale is based on the auditory system; 24 empirically-defined critical bands have been proposed for it. An approximation of their equivalent Mel and Hz values, plus details over the extraction of Bark-Frequency Cepstral Coefficients (BFCCs) may be found, e.g., in [17]. The Bark-scale has been tested here as a alternative for both the previous and the novel Mel-based proposals.

3 Case Study

The bell tower of the Santa Maria and San Giovenale Cathedral investigated here (Fig. 2) is located in the town of Fossano, Piedmont, Italy. Built between 1389 and 1420, it represents an important case of architectural heritage subjected to seismic risk and highly damaged by an earthquake in 1887. The structure was the object of several safety measures and interventions, because of its structural vulnerability, especially from the point of view of the masonry quality, which was confirmed to be poor after dynamic and coring tests. The tower has been widely investigated and monitored by the Earthquake Engineering & Dynamics Group of Politecnico di Torino, specifically after the last interventions in 2012.

The clock tower is 35 m high from the ground level. The maximum thickness of the masonry walls is 1.5 m. Above the main body of the structure, there is an octagonal belfry with masonry walls 0.5 m thick; considering this last element, the bell tower reaches 46 m height, with a roof made of wood at the top of it. Three slabs are present: the first one is a masonry vault at level 9.9 m from the ground floor; the second one is at level 28.2 m and it is made of wood; the last one separates the main structure from the belfry at level 35 m and it is a composite slab, with a lower web of arched beams and an upper layer made of wood.

Fig. 2. The Santa Maria and San Giovenale Cathedral in Fossano (Italy) and its bell tower. Autofrettage interventions (executed with 11 orders of ties) are clearly visible on the external walls.

3.1 Finite Element Model

A Finite Element (FE) Model of the current structure was developed and calibrated, based on the results of ambient vibration tests [4]. In more details, the FEM was updated with acceleration data from a campaign of periodic monitoring to fine-tune the material properties of the lesioned masonry walls; material properties are here reported in Table 1, defined according to the storey level (0: ground, 1: intermediate, 2: top, and 3: belfry; further details available in [4]). This FEM will be assumed as the *baseline* condition with respect to the new damage scenarios that were considered in this study.

A Rayleigh's viscous damping model was assumed, with $\alpha = 0.0993$ and $\beta = 0.0180$ (from the experimentally-found first two flexural natural frequencies, $f_1 = 1.297$ Hz and $f_2 = 4.253$ Hz). Density and Poisson ratio were assumed as uniform on the whole structure ($\rho = 2000\ kg/\mathrm{m}^3$, $\nu = 0.3$). Acceleration time histories were calculated at the 10 nodes and along the directions of the 20 channels indicated in Fig. 3, corresponding to the actually-deployed sensors arrangement. For simulation, a sampling frequency of $f_s = 100$ Hz was set. Rectangular 8-node structural shell elements were used everywhere, with six degrees of freedom per node; mesh size was 0.5×0.5 m. Nonlinear 3D spring elements were used to represent the distributed connections with the Cathedral.

Table 1. Young's moduli of the fine-tuned Fossano bell tower baseline FEM

Storey level	E [MPa]
0	$2.690\ 10^3$
1	$1.320\ 10^3$
2	$1.250\ 10^3$
3	$2.470\ 10^3$

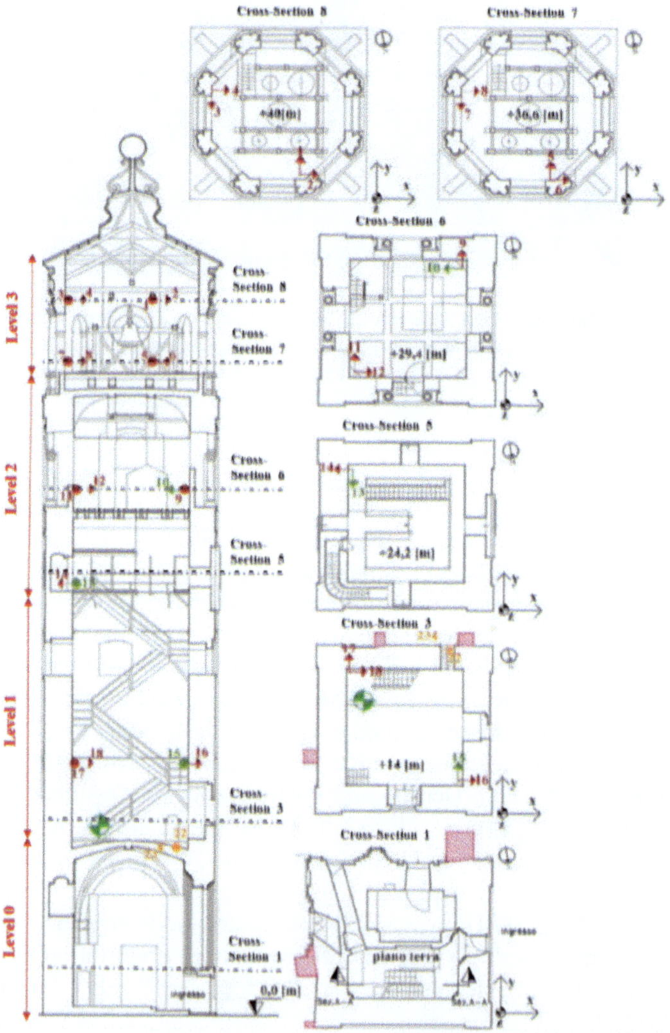

Fig. 3. Structural scheme of the bell tower. The four portions with homogeneous elastic properties are reported in the figure. Location and direction of the 20 acquisition channels are also reported.

3.2 Damage Scenarios

A total of 14 damage scenarios have been considered, as reported in Table 2. The first four (01–04) are simpler cases, with a general reduction of the Young's modulus along one of the four storeys on which the bell tower FE model was originally subdivided (see Fig. 3). Cases 05–10 represent more realistic scenarios, as often encountered in bell tower after seismic damage, as explained in detail in Fig. 4. Case 11 presents a global stiffness reduction for the whole structure, while the scenarios 12–14 involve the linking with the adjacent Cathedral. Four scenarios were also included to simulate realistic fluctuations of the material properties, which are too small to be damage-related, in order to determine the propensity of each method to cause false alarms. These cases are defined by a reduction/increase of the modulus equal or less than 1%.

Table 2. Healthy and damaged scenarios

Case	Name	Description
00	Baseline	As is, no Young's Modulus reductions
01	1st Damaged	5.00% reduction of the *E* modulus at Level 0 (all four façades)
02	2nd Damaged	5.00% reduction of the *E* modulus at Level 1 (all four façades)
03	3rd Damaged	5.00% reduction of the *E* modulus at Level 2 (all four façades)
04	4th Damaged	5.00% reduction of the *E* modulus at Level 3 (all four façades)
05–10	5th–10th Damaged	15.00% local reduction of the *E* modulus at specific, realistic points (as reported in Fig. 4)
11	11th Damaged	10.00% global reduction of the *E* modulus of all levels, all façades
12	12th Damaged	50.0% reduction of spring stiffness at the linking with the Cathedral, *x* and *y* directions
13	13th Damaged	50.0% reduction of spring stiffness at the linking with the Cathedral, *x*-direction only
14	14th Damaged	50.0% reduction of spring stiffness at the linking with the Cathedral, *y*-direction only
15	1st Baseline with fluctuations	+1.00% global increase of the *E* modulus of all levels, all façades
16	2nd Baseline with fluctuations	+0.25% global increase of the *E* modulus of all levels, all façades
17	3rd Baseline with fluctuations	−1.00% global reduction of the *E* modulus of all levels, all façades
18	4th Baseline with fluctuations	−0.25% global reduction of the *E* modulus of all levels, all façades

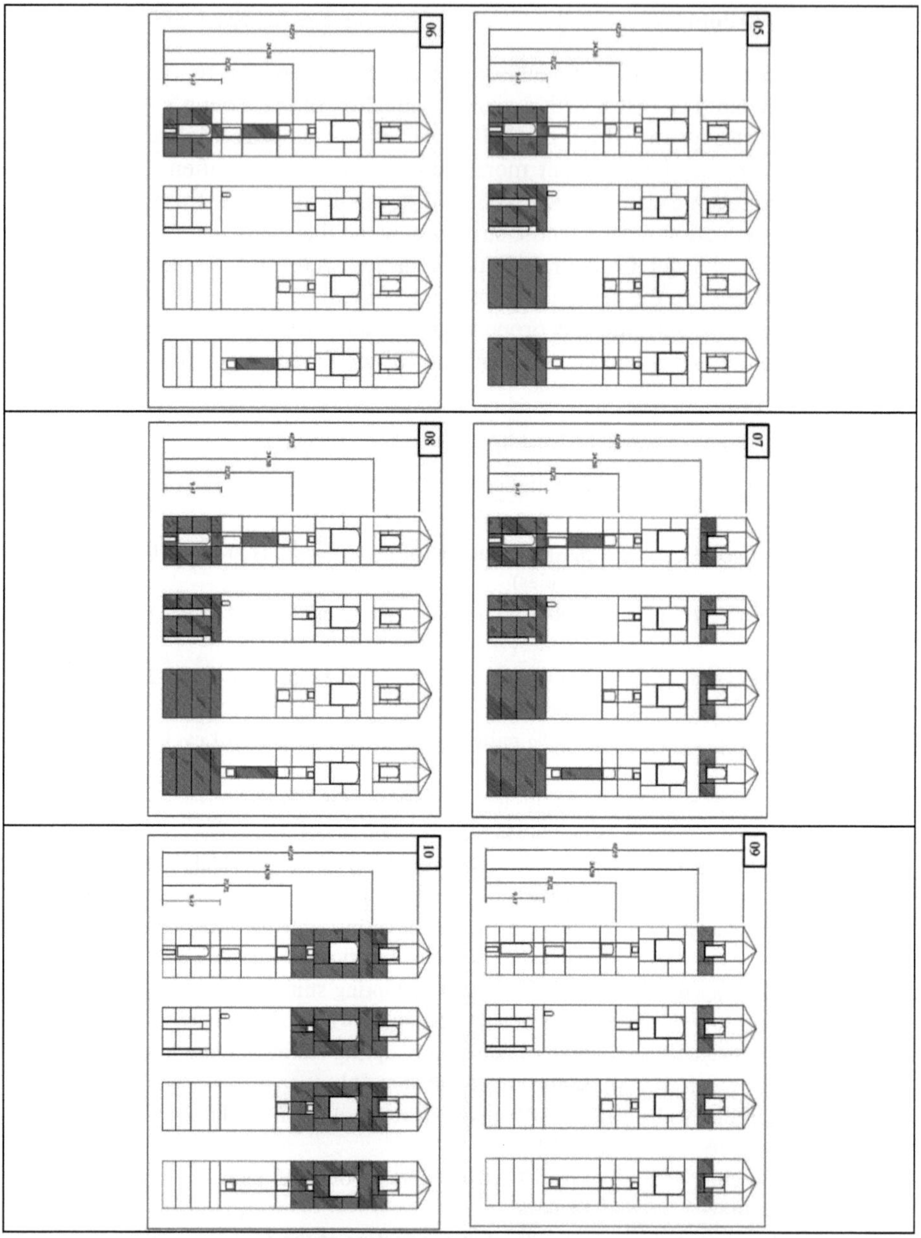

Fig. 4. Damage scenarios 05–10, based on observations on actual similar structures damaged by earthquakes; areas with reduced Young's moduli are portrayed in grey.

3.3 Seismic Parameters and Input Definition

The dynamic response of the several FE models was computed by applying to them a set of seismic strong motions. 300 spectrum-compatible earthquakes were artificially produced by means of the SIMQKE (SIMulation of earthQuaKE ground motions) software [18], in accordance with the international seismic standards. Seismic intensity parameters correspond to a return period of $T_R = 475$ years, for a rated life of $V_N = 50$ years, with $C_U = 1.0$ for a Category 2 structure, and a probability of exceedance $P_{V_R} = 0.9$ for the Limit State considered (here, the Limit State of safeguard of life).

For the specific topographic and stratigraphic categories of the structure, considering also that the bell tower interacts with the adjacent cathedral and other factors, the corresponding maximum ground acceleration was evaluated as $a_g = 0.109$ g. The spectrum-compatible earthquakes were all randomly created starting from 10 different combinations of the remaining parameters, using realistic values and in compliance with the Eurocodes and the Italian law; for instance, the damping coefficient was left free to vary in the range 0.05–0.07. The seismic input duration was set to 35 s, which again is realistic for the seismicity of the surrounding area.

4 Damage Detection Algorithm

The damage detection algorithm utilised here is the one proposed and implemented by [2]. The algorithm builds a normality model from cepstral data taken from the baseline structure and then define a threshold from it, resorting to F-distribution and the Squared Mahalanobis Distance (SMD; [19]). Once trained, the machine is tested on a mixture of time series taken from both damaged and baseline scenarios. The actual algorithm would be too long to be fully reported here; the reader is redirected to the original paper for a complete dissertation of it. The main difference between that work and this paper regards the feature extraction process, especially regarding the filtering.

For any given filter type, a filterbank is defined by two parameters: filters' shape and spacing. Filterbank spacing is a matter of data discretisation; Mel Scale, Bark Scale and similar profit from non-linearly uniform spacing to better emphasise the interesting components of the signal, such as damage-related behaviours.

The damage detection algorithm proposed by Balsamo et al. [2] relied on MFCCs and similar features; five variants were tested: (1) the standard Mel Scale; (2) a Mel Scale with linear cut-off set on one-quarter of the sampling frequency f_s; (3) Mel Scale with $f_{\text{cut-off}} = \frac{1}{8} f_s$; (4) standard Logarithmic Scale; and (5) a linearly-spaced filterbank. These have been used here in this work as a benchmark against the three original definitions proposed here and similarly derived from slight modifications of the MFCC concept, changing the corner frequency $f_{\text{cut-off}}$. Thus, the Mel-modified - Hertz relationship equation can be defined:

$$f_{mel-modified} = C \log\left(1 + \frac{f_{Hz}}{f_{\text{cut-off}}}\right) \quad \text{with } C = \frac{1000}{\log\left(1 + \frac{1000}{f_{\text{cut-off}}}\right)} \tag{4}$$

This writing (reported for base 10 for convenience, but convertible for natural logarithm if needed) maintains the Mel-scale property of having 1000 Mel corresponding to 1000 Hz but allows to change the constant C accordingly to a cut-off frequency different from the standard value of 700 Hz, similar to what tested in [2]. Here, rather than a fraction of f_s, some structure-tailored approaches were pursued; $M = 13$ bands were considered in all cases. These three approaches were collectively named the *Modified Mel-Frequency-like Cepstral Coefficients* (MMFCCs) methods.

Firstly, from the Power Spectral Density (PSD) of the baseline model, it was computed than 90% of the energy was included in between 0 and 1.8 Hz circa. Thus, this higher bound was set as the linear-nonlinear cut-off; here it was named the Energy COntent-based MMFCCs (ECO-MMFCCs).

Secondly, from the modal analysis of the pristine model, the obtained first 10 frequencies were used as centre frequencies for the first 10 filters; the last three filters were arbitrarily centred at 15, 30 and $f_s/2 = 50$ Hz, resulting in the following array of centre frequencies:

$$f_c = [1.297\ 1.392\ 3.490\ 4.253\ 4.520\ 6.469\ 7.107\ 7.157\ 7.401\ 7.736\ 15\ 30\ 50]$$

This option is referred to as the Modal Frequency-centred MMFCCs (MO-MMFCCs).

Thirdly, the symmetric triangular filters were linearly spaced with a 1-Hz constant bandwidth but restricted to the low frequencies, in an arbitrary range ($f_{low} = 0$ Hz and $f_{high} = 13$ Hz). This was called LOw-Frequency Restricted MMFCCs (LO- MMFCCs).

4.1 Signal Pre-processing

Pre-processing was performed for signal enhancement. Respect to the speech signal, the dynamical response of a structure has the advantage of being (in normal conditions) much less time-variant and non-stationary. Its heteroskedasticity, while not always negligible [20] – especially in case of damage occurring during the recording – generally is not comparable to human vocal emissions, where changes in the vibrational behaviour happen constantly and within milliseconds. Hence, *framing* procedure is almost certain to produce stationary snapshots, for the values of shifting period Ξ and frame size K commonly used in Speech Processing (i.e.., $\Xi = 350$ and $K = 700$ for a 50% frame overlap). Nevertheless, the Augmented Dickey-Fuller (ADF) test of stationarity was performed on any frame to avoid all doubt. Then, any l-th frame was windowed to suppress potential riddle effects; a Hamming window was specifically used for this aim. For the given f_s and duration, all the numerically-simulated time histories were made up by 3501 elements; this resulted in 9 overlapping frames.

5 Results

All proposed techniques were tested on the scenarios reported in Table 2, using 10 different seismic inputs for any one of them. For the reader's convenience, results are reported separately in the next three sub-paragraphs. Firstly, baseline and basic damage scenarios (cases 00 to 04) are compared. This is intended to show the algorithms capability to discern between no damage and very simple damage configurations. Then, the proposed variants are compared to the more realistic scenarios (cases 05 to 14). Lastly, the scenarios with small fluctuations respect to the baseline (cases 15 to 18) are compared to verify the robustness of the several methodologies to false alarms. The first Cepstral Coefficient, $c[0]$, has always been discarded, as known to be too sensitive to DC component and input-specific effects [1, 2]; hence, the feature vector dimension c is reduced to L-1.

5.1 Baseline and Level-Uniform Damage Cases

Figure 5 shows the best results for the first 4 damage scenarios, compared to the ones obtained by using the standard definition of the MFCCs. As can be inferred from Table 3, all the three newly proposed Mel-Modified Scales have relatively low Type 1 errors, performing slightly better than the previous approaches. ECO-MMFCCs performed as well as LO-MMFCCs but with a relatively larger incorrect value of Damage Index for the two mislabelled cases. The Damage Index values inside any given damage case are quite similar; that is a valuable proof of the robustness of the method. For comparison, the same algorithm, when fed with the Bark-scale based CCs, wrongly labelled as damaged 60% of the baseline inputs.

Table 3. Results for cases 00–04

Previous methods [2]	Type 1 errors [%]	Type 2 errors [%]	Novel proposals	Type 1 errors [%]	Type 2 errors [%]
MFCCs	60.0	0.0	BFCCs	60.0	0.0
$f_{\text{cut-off}} = \frac{1}{4} f_s$;	30.0	0.0	ECO-MMFCCs	20.0	0.0
$f_{\text{cut-off}} = \frac{1}{8} f_s$;	30.0	0.0	**MO-MMFCCs**	***10.0***	***0.0***
Log-Scale	20.0	0.0	LO-MMFCCs	20.0	0.0
Linear	60.0	0.0			

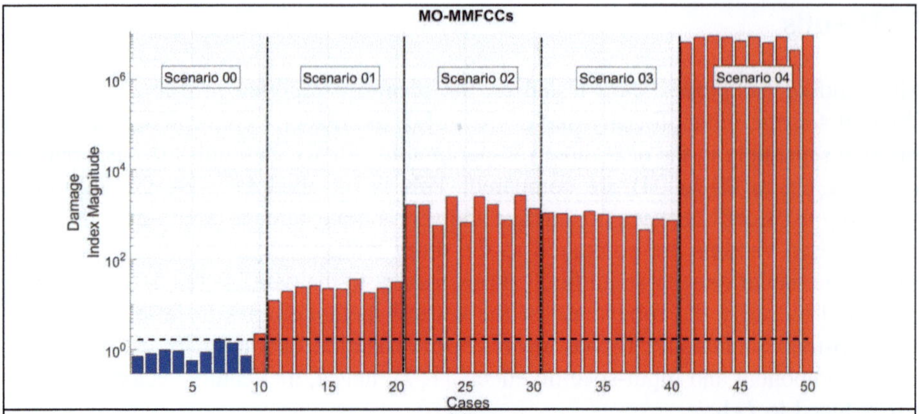

(a)

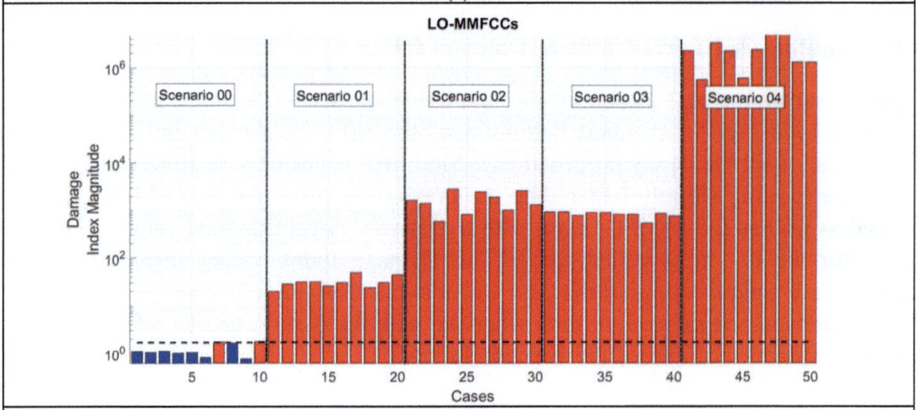

(b)

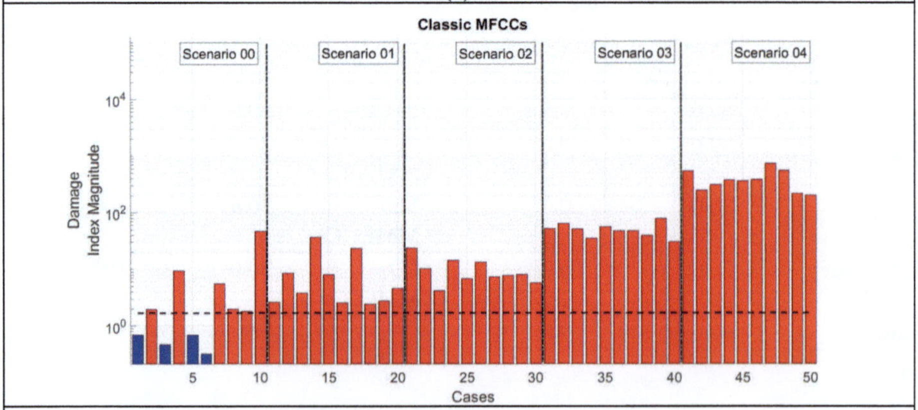

(c)

Fig. 5. Damage indices for the best (MO-MMFCCs) (a) and second to best algorithms (LO-MMFCCs) (b) compared to the standard MFCCs definition (c) scenarios 00-04 (storey-level damage).

5.2 Realistic Damage Cases

Figure 6 shows the results for the more realistic 10 damage. All of the tested methods were able to correctly spot the damage for all scenarios (results are reported in Table 4). The algorithms also recognised substantial gaps between different scenarios, which reflect the difference of damage severity; the trends between different cases into the same scenario are instead relatively flat, some specific cases apart. This may prove the viability of using CCs also for damage estimation.

Table 4. Results for cases 05–14

Previous methods [2]	Type 2 errors [%]	Novel proposals	Type 2 errors [%]
MFCCs	0.0	BFCCs	0.0
$f_{\text{cut-off}} = \frac{1}{4} f_s$;	0.0	ECO-MMFCCs	0.0
$f_{\text{cut-off}} = \frac{1}{8} f_s$;	0.0	MO-MMFCCs	0.0
Log-Scale	0.0	LO-MMFCCs	0.0
Linear	0.0		

5.3 Cases with Small Fluctuations from the Baseline

Even if the Type 1 Error ratio is greater for all the MMFCCs proposed, it must be remarked that all three the methods seem to always label $\pm 1\%$ variations of the Young's modulus as damaged, and to label $\pm 0.25\%$ deviation of the same as baseline most of the times (see Fig. 7a for instance; a similar behaviour was observed for the other two methods). This phenomenon could be justified by the fact that a relatively small difference between the pristine and the modified mechanical properties of the model's material cause a greater variation in the power spectrum of the structure, which is wrongly seen as damage. The error ratio of the previous MFCCs alternatives is instead much more uniformly spread (Fig. 7c and d), meaning that the algorithm is confused on how to properly address the different cases. A similar behaviour was encountered in the log-scale, while the linear scale and the Bark scale both behaved as the MFCCs with the sampling-dependent cut-off. Thus, since the actual definition of '*small fluctuation*' from the baseline dynamic response of the structure was arbitrary, it may be possible to tune the algorithm in order to recognise a specific threshold of interest (Table 5).

Table 5. Results for cases 05–14

Previous methods [2]	Type 1 errors [%]	Novel proposals	Type 1 errors [%]
MFCCs	***45.0***	BFCCs	47.5
$\boldsymbol{f_{\text{cut-off}}} = \frac{1}{4} \boldsymbol{f_s}$;	***45.0***	ECO-MMFCCs	62.5
$\boldsymbol{f_{\text{cut-off}}} = \frac{1}{8} \boldsymbol{f_s}$;	***45.0***	MO-MMFCCs	62.5
Log-Scale	65.0	LO-MMFCCs	65.0
Linear	45.0		

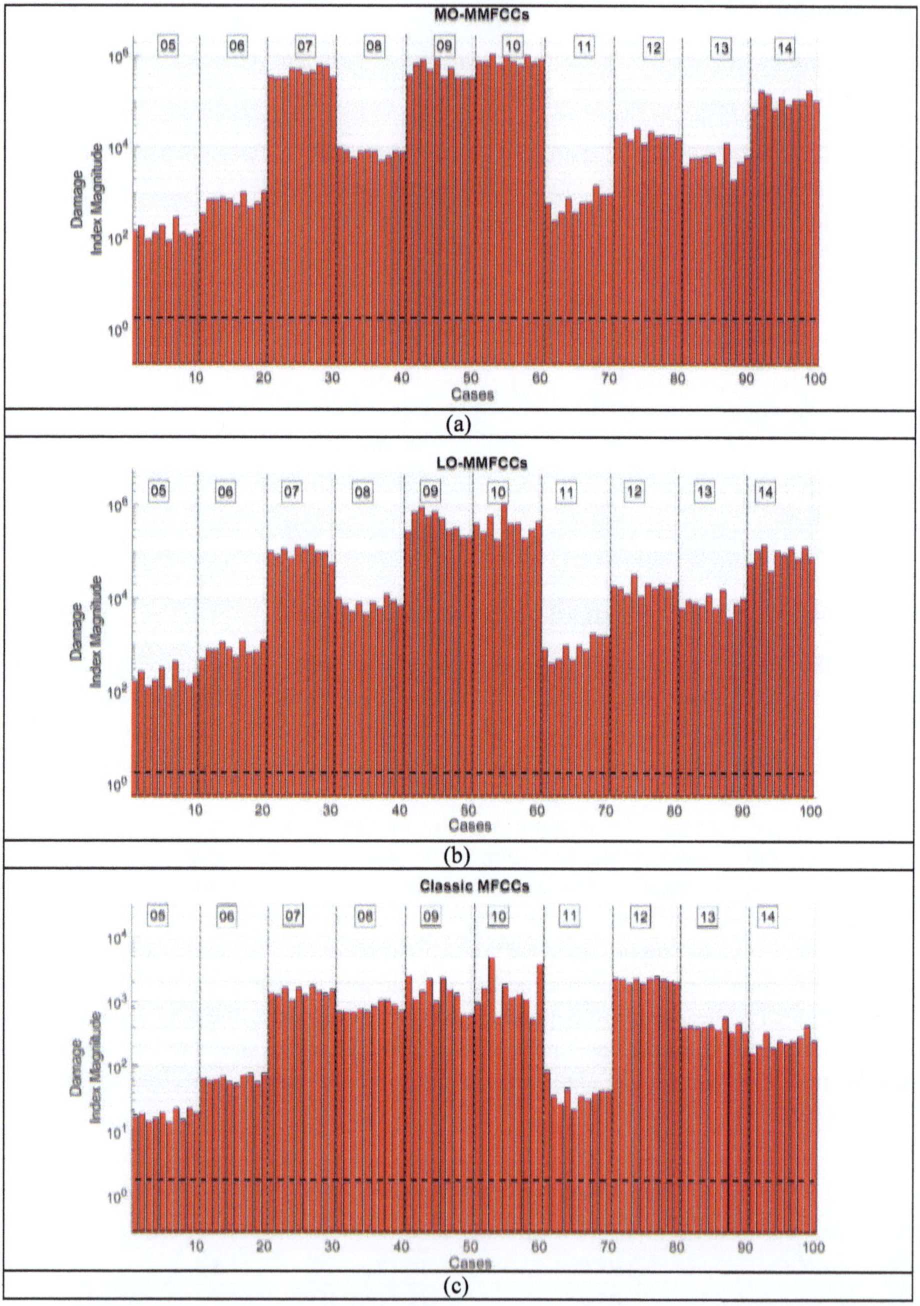

Fig. 6. Damage indices for the MO-MMFCCs (a) the LO-MMFCCs (b) and the standard MFCCs definition (c) scenarios 05–14 (realistic damage).

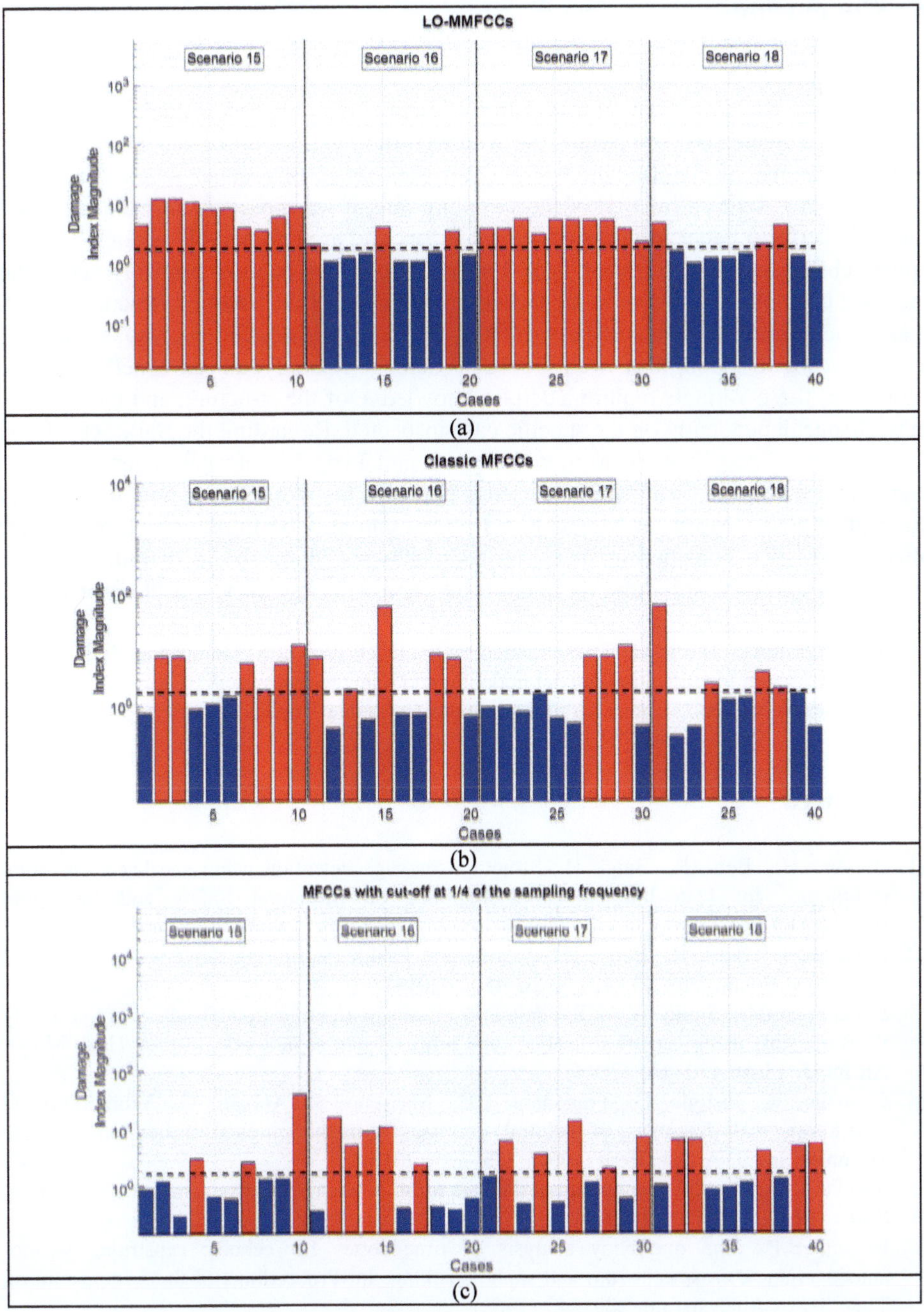

Fig. 7. Damage indices for the MO-MMFCCs (a), the LO-MMFCCs (a) the standard MFCCs definition (b) and the MFCCs with $f_{\text{cut-off}} = \frac{1}{4} f_s$ (c) [2]; scenarios 15–18.

6 Conclusions

The main aim of this research was to investigate three proposed options to improve the MFCC-based SHM concept, plus to compare them to a similar perceptual scale, known as the Bark Scale. By comparing the Mel-Modified Scales tuned on the actual frequency response of the structure to the other Mel-based approaches where the cut-off frequency was set as a fraction of the sampling rate, it was observed that the proposed methods perform better in terms of Type 1 errors and are comparable (even if with an appreciable increase) in terms of Type 2 errors. These may be most probably due to the pros and cons of being fitted on the frequency response of the baseline structure, which makes them more accurate to detect shifts in the frequency domain but also less robust, as they cannot distinguish well between damage-related and unrelated changes. Moreover, these variants require a certain knowledge of the structure, and their results may change dependently on the specific case inspected. Regarding the Bark Scale CCs, they performed poorly in all cases, with Type 1 and Type 2 error rates generally higher than any previous or novel MFCC-based proposal. This work shows how the original concept of SHM based on MFCC-like features can still be improved; several other alternatives, also considering much broader changes in the basic definition of the cepstral coefficients, are currently being tested and will be reported in the near future.

Acknowledgments. The research was funded by the INTE project by Compagnia di San Paolo (2017–18) "System Identification, model updating and damage assessment of heritage buildings and structures", partners: Politecnico di Torino and Columbia University (USA).

References

1. Balsamo, L., Betti, R., Beigi, H.: Structural damage detection using speaker recognition techniques. In: 11th International Conference on Structural Safety and Reliability (ICOSSAR) (2013)
2. Balsamo, L., Betti, R., Beigi, H.: A structural health monitoring strategy using cepstral features. J. Sound Vib. **333**(19), 4526–4542 (2014)
3. Dackermann, U., Smith, W.A., Randall, R.B.: Damage identification based on response-only measurements using cepstrum analysis and artificial neural networks. Struct. Heal. Monit. An Int. J. **13**(4), 430–444 (2014)
4. Ceravolo, R., Pistone, G., Fragonara, L.Z., Massetto, S., Abbiati, G.: Vibration-based monitoring and diagnosis of cultural heritage: a methodological discussion in three examples. Int. J. Archit. Herit. **10**(4), 375–395 (2016)
5. Salawu, O.S.: Detection of structural damage through changes in frequency: a review. Eng. Struct. **19**(9), 718–723 (1997)
6. Bogert, B.P.: The quefrency analysis of time series for echoes; cepstrum, pseudo-autocovariance, cross-cepstrum and saphe cracking. In: Proceedings of the Symposium on Time Series Analysis, pp. 209–243 (1963)
7. Beigi, H.: Fundamentals of Speaker Recognition. Springer, Heidelberg (2016)
8. Randall, R.B.: A history of cepstrum analysis and its application to mechanical problems. Mech. Syst. Signal Process. **97**, 3–19 (2017)

9. Childers, D.G., Skinner, D.P., Kemerait, R.C.: The cepstrum: a guide to processing. Proc. IEEE **65**(10), 1428–1443 (1977)
10. Ahmed, N., Natarajan, T., Rao, K.R.: Discrete cosine transform. IEEE Trans. Comput. **C-23** (1), 90–93 (1974)
11. Stevens, S.S., Volkmann, J., Newman, E.B.: A scale for the measurement of the psychological magnitude pitch. J. Acoust. Soc. Am. **8**(3), 185–190 (1937)
12. Stevens, S.S., Volkmann, J.: The relation of pitch to frequency: a revised scale. Am. J. Psychol. **53**(3), 329 (1940)
13. Fant, G.: Acoustic Theory of Speech Production: With Calculations Based on X-Ray Studies of Russian Articulations. De Gruyter Mouton, Berlin (1970)
14. Umesh, S., Cohen, L., Nelson, D.: Fitting the mel scale. In: 1999 IEEE International Conference on Acoustics, Speech, and Signal Processing. Proceedings. ICASSP99 (Cat. No. 99CH36258), vol. 1, pp. 217–220. IEEE (1999)
15. Zwicker, E.: Subdivision of the audible frequency range into critical bands (Frequenzgruppen). J. Acoust. Soc. Am. **33**(2), 248 (1961)
16. Traunmüller, H.: Analytical expressions for the tonotopic sensory scale. J. Acoust. Soc. Am. **88**(1), 97–100 (1990)
17. Herrera, A., Del rio, F.: Frequency bark cepstral coefficients extraction for speech analysis by synthesis. J. Acoust. Soc. Am. **128**(4), 2290 (2010)
18. Gasparini, D., Vanmarcke, E.H.: SIMQKE: A Program for Artificial Motion Generation. Department of Civil Engineering, Massachusetts Institute of Technology, Cambridge, MA (1990)
19. Hardin, J., Rocke, D.M.: The distribution of robust distances. J. Comput. Graph. Stat. **14**(4), 928–946 (2005)
20. Cross, E.J., Worden, K., Chen, Q.: Cointegration: a novel approach for the removal of environmental trends in structural health monitoring data. Proc. R. Soc. A Math. Phys. Eng. Sci. **467**(2133), 2712–2732 (2011)

Operational Modal Analysis of Y25 Bogie via Stochastic Subspace Identification for the Condition Monitoring of Primary Suspension Systems

Fulong Liu[2], Jiongqi Wang[1,2(✉)], Miaoshuo Li[2], Fengshou Gu[2], and Andrew D. Ball[2]

[1] School of Mathematics and Big Data, Foshan University, Foshan 528000, People's Republic of China
wjq_gfkd@163.com
[2] Centre for Efficiency and Performance Engineering (CEPE), School of Computing and Engineering, University of Huddersfield, Huddersfield HD1 3DH, UK
{fulong.liu, li.miaoshuo, f.gu, a.ball}@hud.ac.uk

Abstract. Railway vehicle suspension systems are vital to the vehicle safety and ride comfort, which is further driven by high speed operations. Condition Monitoring (CM) based online measurement is an efficient and achievable method to ensure the suspension systems working under normal function. In this paper, a potential method, which can achieve online CM of railway vehicle primary suspension, denoted as Average Correlation Signals based Stochastic Subspace Identification (ACS-SSI) was explored through simulation and experimental studies. Particularly, the dynamic performance of an Y25 bogie were investigated under the operational condition and the main focus was on the modes related to the suspension system. Firstly, ACS-SSI was presented briefly. Then, the employed test rig, an advanced dynamic test cell in the Institute of Railway Research (IRR) at University of Huddersfield, was introduced and the theoretical modal parameters of the tested bogie associating with the primary suspension system were calculated based on a multi rigid body model in the SIMPACK. The theoretical natural frequencies of bounce, roll and pitch modes are 11.07 Hz, 13.93 Hz and 15.19 Hz, respectively. Finally, ACS-SSI was adopted to identify modal parameters of the bogie using the collected responses on the four corners of the bogie frame. The pitch mode was identified successfully, which can illustrate the condition of the suspension system. Therefore, it can draw the conclusion that ACS-SSI has the potential to achieve suspension online monitoring.

Keywords: Operational modal analysis (OMA) · Railway vehicle suspension system · Stochastic Subspace Identification (SSI) · Condition monitoring (CM)

M. A. Wahab (Ed.): DAMAS 2019, LNME, pp. 166–181, 2020.
https://doi.org/10.1007/978-981-13-8331-1_12

1 Introduction

Railway vehicle suspension systems are responsible for smoothing out the ride to ensure the safety and comfortability. Hence, the components of suspension systems are vital to a vehicle, especially for the high speed railway vehicles. As a result, it is meaningful to monitor the condition of suspension systems. Condition Monitoring (CM) is an efficient and achievable way to ensure the reliability of suspension systems, therefore, numerous scholars have focused on developing suitable approaches to monitor the suspension systems [1, 2].

So far, loads of methods have been explored by different scholars to fulfil the aim of monitoring suspension systems. These methods can be categorised into two groups, track-side and on-board, according to sensors' locations. The track-side methods install sensors around the track to monitor the wheel profile, wheel impact load, and bogie performance [3, 4]. This kind of methods can only detect the target components at the selected positions along the track and obviously, online monitoring is impossible. Whereas on-board methods have the possibility to achieve real-time monitoring due to the sensors are installed on the vehicle. Compared with track-side approaches, on-board technique has plenty of advantages so it has attracted much more attentions in recent years.

Traditionally, on-board monitoring systems are focused on the detection of track faults, but the interest of employing on-board methods to monitor the vehicle dynamic performance has a considerable increase in recent years, particularly for the suspension system monitoring. As stated in [2], the on-board monitoring methods can be divided into two categories: model-based and signal-based.

Model-based methods are based on system's mathematical model to develop the relationship between excitations and responses. The most common model-based methods are on the basis of Kalman filter (KF). A KF was applied in [5, 6] to detect the vertical damper and vertical spring faults. The KF can also be applied to wheel-rail profile estimation and low adhesion detection [7]. In addition, the Rao-Blackwellised particle filter (RBPF) is widely used in the suspension system monitoring. In [8], RBPF was employed to estimate the suspension parameters for condition monitoring and the results were promising. Similarly the recursive least square algorithm was adopted in [9] for suspension monitoring by estimating the stiffness and damping of suspension systems. However, all of the model-based methods have to build an accurate vehicle dynamic model to obtain an acceptable result. It is often difficult to develop such a model on account of nonlinearities of suspension components, such as the dampers.

Another choice to monitor the suspension system is through signal-based methods. The signal-based methods illustrate the condition of suspension by directly analysing the collected responses. These methods can be divided into four main categories: time-domain, frequency-domain, time-frequency domain and correlation analysis. A full vehicle model was developed in [10] via SIMPACK to generate numerical signals when the suspension system under various scenarios, and to detect the difference through Principle Components Analysis (PCA), a time-domain method. Moreover, the same author applied three different techniques, Dempster-Shafer (D-S) evidence theory, Fisher Discrimination Analysis (FDA) and Support Vector Machine (SVM), to isolate suspension faults via four fault features in time-domain and three fault features in frequency-domain [11]. For the time-frequency domain techniques, the railway wheel-flat was investigated through a 1/5th scaled bogie in [12]. The signals were analysed by three commonly time-frequency analysis methods, Short-Time Fourier Transform (STFT), Smoothed Pseudo Wigner-Ville Transform (SPWVT) and Wavelet Transform (WT). In [13] and [14], cross-correlation analysis was adopted to detect the railway vehicle suspension on the basis that the suspension faults would result in imbalance into the vehicle dynamic systems. Nevertheless, a pre-built database including different fault sceneries is needed for the application of signal-based method, which restricted the application of such method.

On the other hand, Structure Health Monitoring (SHM) through modal identification has the merits of no need of accurate mathematical model or pre-built database. Therefore, modal analysis has become a popular approach for structure health monitoring in recent decades, especially the Operational Modal Analysis methods [15]. Stochastic Subspace Identification (SSI) is one of the most popular and outstanding OMA technique [16]. It has been widely used to identify the modal parameters of civil structures and detect the damage [16–18] under the assumption of white noise excitations. Moreover, on account of the effectiveness of SSI, it was extended by different scholars to apply in other field. For instance, a modified data-driven SSI method was proposed in [19, 20] to satisfy the harmonic excitation conditions resulting from the rotation of wind turbine. Besides, a method denoted as Average Correlation Signal based SSI (ACS-SSI) was developed on the basis of Covariance-driven SSI to suppress nonstationary situations.

Particularly, ACS-SSI was proposed in [21] to identify the modal parameters of a chassis frame of a heavy-duty dump vehicle under normal operational condition, which has to suppress the high noise and nonstationary effects. The modal parameters of the chassis frame were identified successfully during the heavy-duty working on the construction road where could cause high noise and extreme nonstationary signals. This result illustrated the effectiveness of ACS-SSI and therefore, this method was explored in a further step by applying it in vehicle suspension monitoring [22, 23] through a car experimental investigation and a SIMPACK bogie model numerical study. The suspension faults were also identified successfully through ACS-SSI method, which presented that ACS-SSI has the potential capability to identify dynamic characteristics of vehicle suspension systems. Hence, this study is planning to employ ACS-SSI to identify the modal parameters of Y25 bogie during it is running on a full-size roller rig for railway vehicle primary suspension system monitoring, which will contain much more serious responses compared with previous study.

This paper is organized as follows: Sect. 2 will introduce the ACS-SSI in brief; Sect. 3 will present the test rig and experimental set-up and a SIMPACK model for theoretical modal parameters calculation of Y25 bogie; Sect. 4 will present experimental results and discussion; then, some conclusions will withdraw in Sect. 5.

2 ACS-SSI Algorithm

Stochastic Subspace Identification (SSI) is one of the most commonly used output-only technique for OMA [24] because of its robust mathematical derivation. The SSI method typically derivate from the state space description of n degree of freedoms linear system. The purpose of SSI is to obtain the state space matrix **A** through the measured responses from l sensors. For the reference-based Cov-SSI, the output measurements from l sensors are gathered into a Hankel matrix with $2i$ block rows and j columns, and divided into a past reference and a future part. Then, a covariance matrix between all outputs and the reference channels can be calculated using the past reference part $\mathbf{Y}_p^{ref}$ multiple the future part $\mathbf{Y}_f$. The results are gathered in a block Toeplitz matrix $\mathbf{T}_{1|i}^{ref}$ and the Singular Value Decomposition (SVD) will be performed on it to obtain the observability and extended controllability matrices. The next step is to calculate the state matrix **A** by decomposing a shifted block Toeplitz matrix $\mathbf{T}_{2|i+1}^{ref}$. Consequently, the modal parameters can be extracted from state matrix **A**. The details of Cov-SSI can be found in [16].

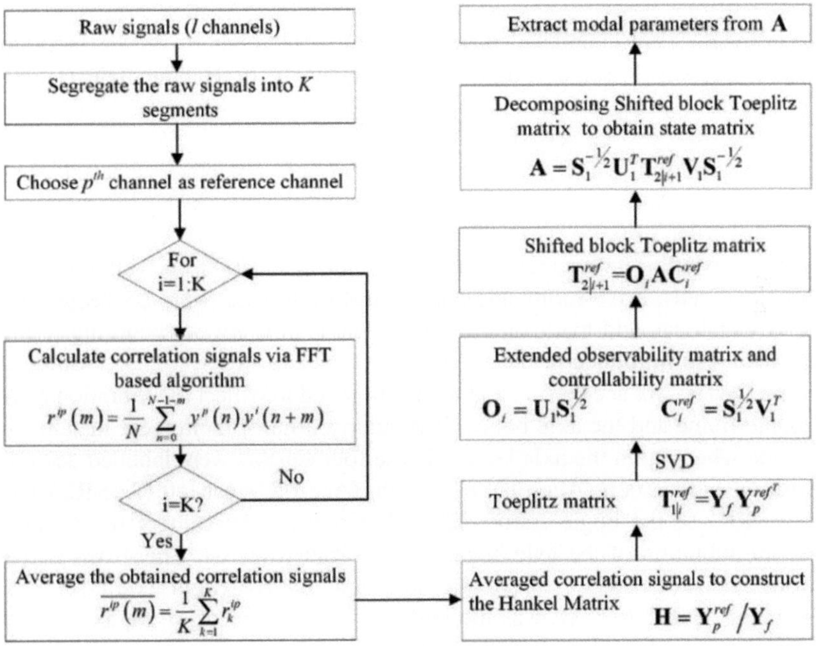

Fig. 1. Flow chart of ACS-SSI

It is well known that Cov-SSI is under the fundamental assumption that system inputs are white noise. However, this assumption is difficulty to be satisfied in real engineering applications and therefore, the ACS-SSI was proposed in [21] to reduce the deficiency of Cov-SSI.

The whole process of ACS-SSI was presented in a flow chart, shown as Fig. 1. Compared with Cov-SSI, it can be seen that a signal pre-process procedure is added to the ACS-SSI, which included four main steps. Firstly, divide the raw signals into several segments which can be achieved by dividing the data obtained from a long time test or directly obtain several data segments from several time tests. Secondly, choose a channel with high signal to noise ratio (SNR) as the reference channel. Thirdly, calculate the correlation signals of each segments and fourthly, average the obtained correlation signal segments. Then, the averaged correlation signals are adopted as the inputs to construct the Hankel matrix and the following steps are same as the Cov-SSI procedure.

The advantages of the pre-process procedure are twofold: first of all, as referred earlier, it is effective to suppress the noise and reduce the nonstationary responses caused by the color noise excitations and nonlinearity of the system; secondly, this pre-process procedure can improve the calculation speed due to the signal length are shortened by the average step, which is significant to fulfil system on-line monitoring.

3 Test Rig and Theoretical Modal Parameters

3.1 Test Rig Introduction and Experiment Setup

The experiments were conducted based on a full-size roller rig in the laboratory of Institute of Railway Research (IRR) at the University of Huddersfield (UoH). The detail of this test rig can be found in [25]. An Y25 bogie was installed on the roller rig, shown as Fig. 2. It can be seen from Fig. 2 that only one wheelset of the bogie was fixed on the rail drum, which was marked as front one. The other wheelset marked as rear one and it was laid on the track which was built in the test bed. Moreover, it can be observed that a load mass was fixed on the bogie through the central bowl and secondary suspension. The weight of the load mass is 2000 kg. Additionally, two hydraulic load cells, drive room side and control room side, were connected with the load mass which can add loads on the bogie. The load is added according to the load profile which will introduce in the following.

During the test, six accelerometers were employed to collect the vibration responses of the bogie frame and the axle boxes. Particularly, four of them were installed on the bogie frame where upon the axle boxes. These four sensors were labelled according to their position as front-left (Acc-FL), front-right (Acc-FR), rear-left (Acc-RL) and rear-right (Acc-RR) of the bogie frame. The other two accelerometers were adopted to measure the vibration of the axle boxes on the front wheelset. These two accelerometers were labelled as Acc-AL (left axle box) and Acc-AR (right axle box). It can be seen from Fig. 2 that Acc-AL was mounted on the attachment of the axle box due to the space between the bogie frame and the axle box was not enough to install the sensor. The attachment attached on the axle box with four bolts so they have the same vibration response. Acc-AR was installed in the same way.

In order to illustrate the experiment set-up clearly, a schematic diagram of the test system was draw and presented in Fig. 3. Moreover, the position of sensor were illustrated in a further step by a schematic of the top view of the bogie, shown as Fig. 3 (b). It is worth to notice that the rail drum is composed by four parts of radium rail and therefore, the rail drum has four joints which will result in period pulse excitations.

Fig. 2. Photograph of the test rig and sensor installation position

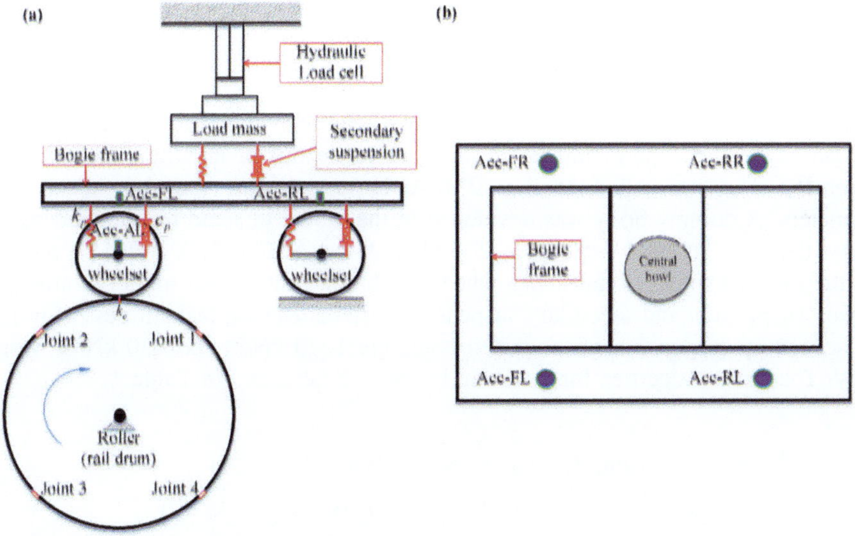

Fig. 3. (a) Schematic diagram of test rig, and (b) schematic diagram of sensor position

The model of these four sensors installed on the bogie frame is CA-YD-185 which and the model of the other two sensors installed on the axle boxes is YMC121A20. Both types of sensors have a wide frequency range, which is from 0.5 Hz to 5,000 Hz, and the sensitivity of CA-YD-185 and YMC121A20 are around 5.0 mV/(m/s2) and 2.0 mV/(m/s2), respectively.

In addition, a 16-channel data acquisition system was adopted. Its model is YE6232B. The sampling frequency of this data acquisition system can reach 96,000 Hz. During this test, the sampling frequency was set at 1000 Hz in this test considering the resonance frequency of primary suspension systems. A photograph of employed sensors and data acquisition system was presented in Fig. 4.

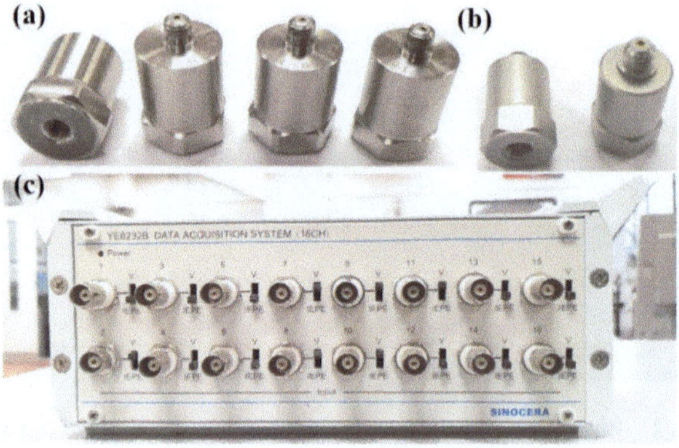

Fig. 4. (a) CA-YD-185 accelerometers; (b) YMC121A20 accelerometers; (c) YE6232B data acquisition system

3.2 Theoretical Modal Analysis

SIMPACK is a powerful software to conduct dynamic analysis of railway vehicle [5, 23]. Based on the experiment set-up, an Y25 bogie model including the secondary suspension was developed in the SIMPACK to calculate the theoretical modal parameters. A dummy body was developed in the model to stand for the load mass in reality, therefore its mass was set as 2000 kg, shown as Fig. 5(a). It can be observed that the model mainly includes two wheelsets, bogie frame, load mass dummy body, primary suspension and secondary suspension system. The vertical stiffness of primary and secondary suspension of the tested bogie are 1220 kN/m and 430 kN/m, respectively. The mass properties for the main bodies are tabulated in Table 1.

Table 1. Main body mass properties [8, 23]

Body	M(kg)	I_{xx}(kg.m^2)	I_{yy}(kg.m^2)	I_{zz}(kg.m^2)
Load mass	2000	0	0	0
Bogie frame	2615	1722	1476	3076
Wheelset	1200	740	74	740

All of the theoretical modal parameters of the Y25 bogie can be calculated through the developed SIMPACK model. Because this study is focused on the condition monitoring of primary suspension, therefore only the modes related to the primary suspension are presented, shown as Fig. 5(b1, 2, 3). The three modes are bounce, roll and pitch of which resonance frequencies are 11.08 Hz, 13.92 Hz and 15.19 Hz, respectively. It can be seen that the resonance frequencies of the primary suspension systems are around 11–16 Hz.

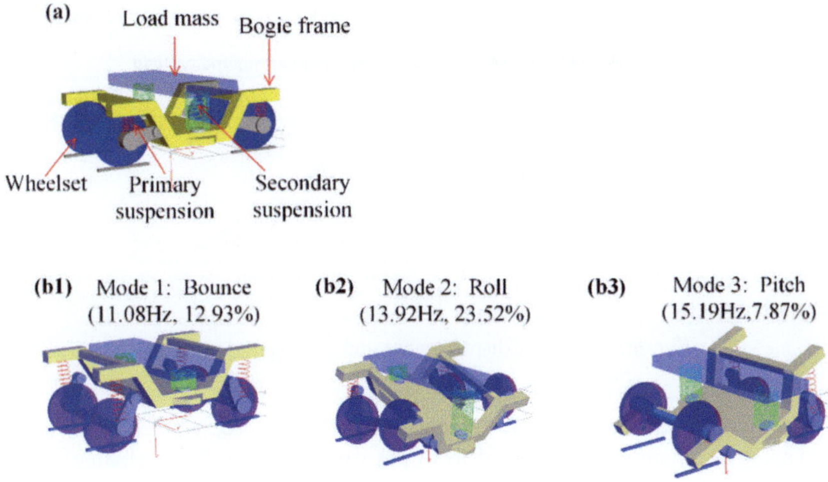

Fig. 5. (a) Developed SIMPACK model; ($b_{1,\ 2,\ 3}$) theoretical bounce, roll and pitch modes of the bogie frame

4 Experiment Results and Discussion

4.1 Raw Signal Characteristics

Load Profile of the Hydraulic Driven Load Cell

The hydraulic load cell has two load strategies to add loads on the load mass: the first strategy is setting a load value as the inputs and the load cell will move until the load value achieved; the second strategy is controlling the displacement of the load cell. In this test, the second strategy was employed and the load profiles (displacement) of the load cells are shown as Fig. 6. At the same time, the real-time loads of the load cells are recorded and presented in the same figure.

It can be seen from Fig. 6 that both load cells were keeping at original position (568 mm) in the first 78 s. Then, their displacements were increased with a constant speed to 628 mm in 60 s and decreased with the same speed in the following 60 s. Finally, their displacements were kept at the original position until the rotation speed of rail drum slowed to zero.

As can be seen from Fig. 6 that the corresponding loads of the load cells were increased along with the displacement increasing. It has to note that the Y25 bogie became a time-varying system when the loads were added on the load mass by the load

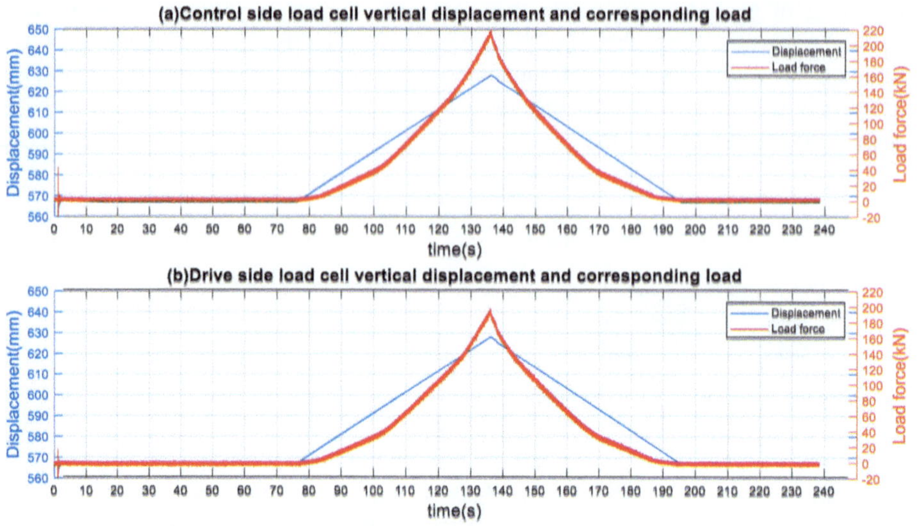

Fig. 6. Load cell displacement and corresponding load

cells. Moreover, it can be seen that the slope of the load force was changed at around 105 s. The slope change was resulted from the suspension stiffness change due to the inner spring engaged. This character indicated that Y25 bogie was a time-varying system in a further step. As a preliminary study, the Y25 bogie was considered as a time-invariant system. Therefore, only the vibration responses from 14 s to 74 s were adopted in the following analysis. In addition, the same test was repeated six times to obtain enough data for the modal identification via ACS-SSI.

Time-Domain Analysis

As referred earlier, the test with the same load profile was repeated six times and the vibration responses from 14 s to 74 s in each test were cut off to be used as the inputs of ACS-SSI. One of the six response data sets in the first five seconds is presented in Fig. 7.

First of all, it can be seen that the vibration amplitudes of the bogie frame are smaller than the vibration amplitudes on the axle boxes due to the attenuation utility of primary suspension systems. Then, it can be seen that the amplitudes of original responses of the left axle box are larger than the right one, which lead to the amplitudes of the responses at front-left (Acc-FL) of the bogie frame is greater than the front-right (Acc-FR). In addition, the response amplitudes of the bogie frame at rear-left (Acc-RL) and rear-right (Acc-RR) is smaller than the front parts, because only the front wheelset is excited by the rail drum.

Besides, it can be observed that the responses are similar to the period impulse responses, which are quite clear for the axle boxes' responses. The reason for the impulse responses is the joints on the rail drum, shown as Fig. 2, which could result pulse excitations. It is worth to notice that the period excitation will result false modes in the modal identification process and this will be analysed in a further step in the following.

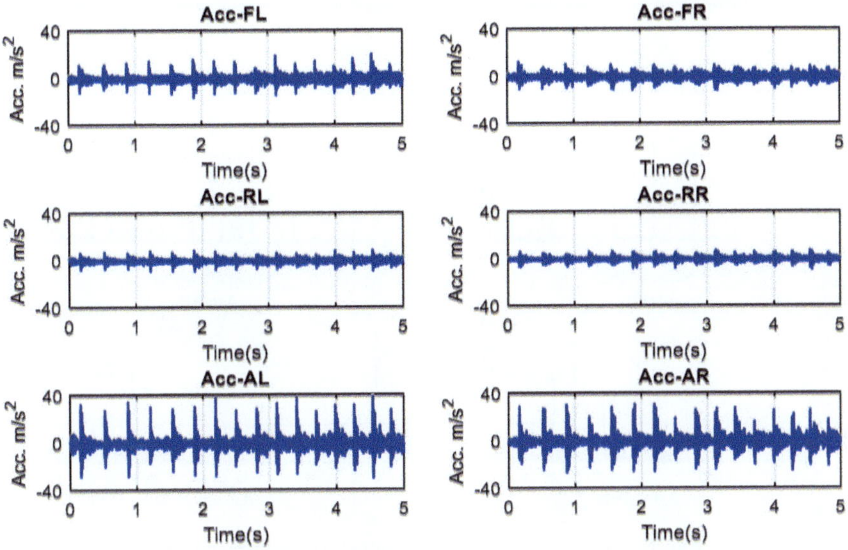

Fig. 7. One of the collected signal sets in the first five seconds

Frequency-Domain Analysis
The corresponding frequency-domain signals of the time-domain signals of the responses at the four corners of the bogie frame in Fig. 7 are presented in Fig. 8. The analysed maximum frequency is 30 Hz due to the interested frequencies are around 10 to 16 Hz according to the results obtained from theoretical modal analysis. In addition, the rotation angle of the rail drum was also recorded, one of the test is shown as Fig. 9. The actual rotation speed can be calculated based on the rotation angle, which is 54 rpm. Then the rotation frequency can be calculated according to the speed and it is 0.90 Hz. Consequently, the basic harmonic frequency will be four times of the rotation frequency on account of the four junctions on the rail drum, which will be 3.60 Hz.

As can be seen in Fig. 8, the main peaks in the frequency-domain are the harmonic frequencies caused by the drum rotation. It is apparent that they are not exactly integer multiples of 3.60 Hz, but all of them are very close to integer multiples of 3.60 Hz. It is well known that harmonic excitations will lead to false modes during modal identification process, especially when the resonance frequency is close to the harmonic excitation frequency. This is a big challenge for the modal identification and a great many of researchers have been paid attention on this problem [20, 26]. They found that one characteristic of the false mode caused by harmonic excitation is its damping ratio will be much lower, and this can be performed as the criterion to filter out the false mode.

In this paper, the harmonics were eliminated by cepstrum editing [27, 28]. The raw signals were edited in the cepstrum domain through a short-pass filter, and then recuperating to time domain signals by combing the edited cepstrum with the phase of raw signals. This step is similar to filter the signal. An example of the edited cepstrum and the filtered signals is presented in Fig. 10, moreover, the corresponding frequency

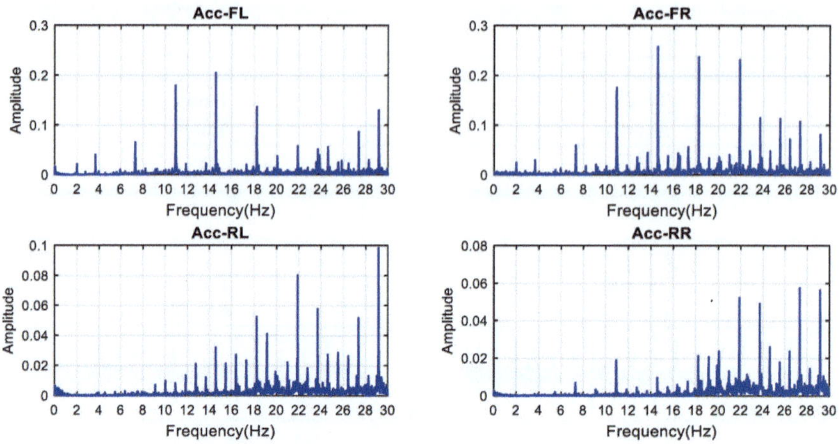

Fig. 8. An example of the frequency-domain responses

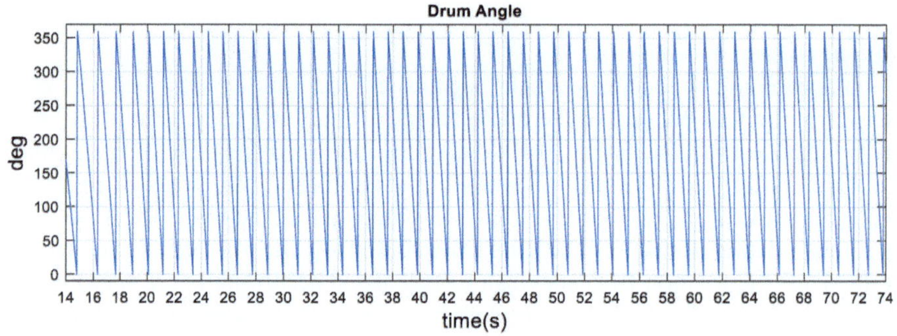

Fig. 9. Rotation degree of the rail drum

domain signals are presented in Fig. 11. It can be seen from Fig. 11 that the harmonics resulted from rotation have been removed successfully. Then, the filtered signals will be employed to do the operational modal analysis via ACS-SSI.

4.2 Modal Parameters Identified by ACS-SSI

The effectiveness of ACS-SSI for suspension monitoring has been validated through a bogie model developed in SIMPACK in [21]. In [21], the responses of the bogie frame upon the axle boxes were measured through four virtual accelerometers in the SIMPACK and then the collected signals were applied to identify the rigid modes of the bogie frame which were related to the primary suspension systems. Based on the success of this study, four accelerometers were mounted on the bogie frame where upon the axle boxes, shown as Fig. 2, to collect responses for the identification of

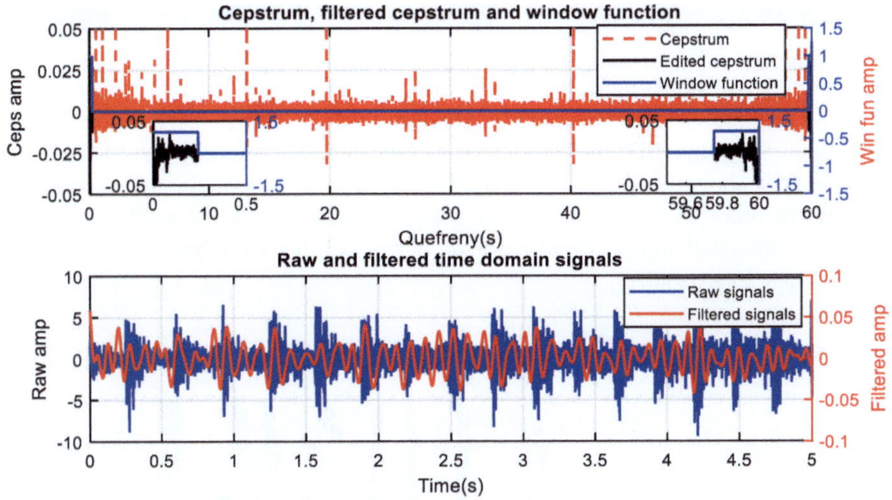

Fig. 10. Cepstrum editing and filtered signals

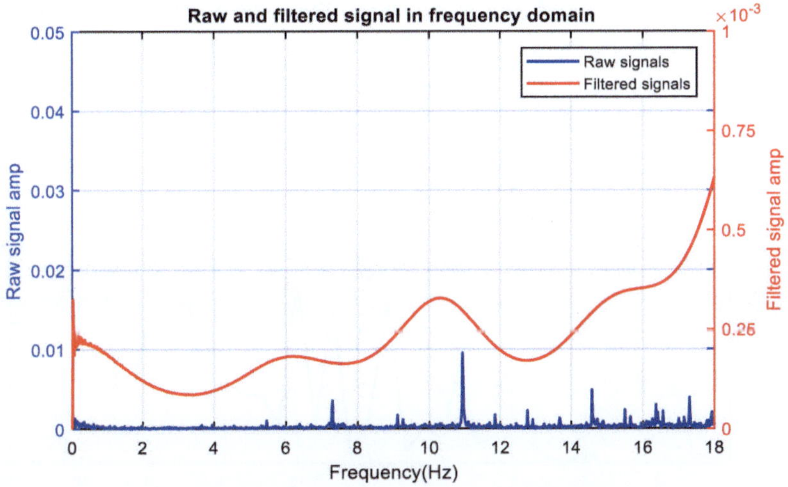

Fig. 11. FFT of the RAW and filtered signals

modal parameters of the bogie frame which are related to the primary suspension systems.

As mentioned in Sect. 3.1, the test was repeated six times to obtain enough data for ACS-SSI method. According to the flowchart of ACS-SSI method that presented in Fig. 1, the number of data segments K is equal to six in this situation. Then the correlation signals of each data segment are calculated, which will result six segments of correlation signals. The correlation signal segments are averaged and subsequently used to construct the Hankel matrix. Afterwards, the Toeplitz matrix is obtained. Later, the extended observability and controllability matrix can be obtained by SVD of the Toeplitz matrix.

Theoretically, the system order can be confirmed by SVD. However, this is untruthful because of measurements contained loads of noises [17]. Therefore, the Stabilization Diagram (SD) was employed to filter the false modes. The SD identified in this study is shown as Fig. 12. It can be seen that a lots of stable modes have been identified. In order to select the relative stable modes, the rate of stable points over orders was calculated and the result is shown in Fig. 13. It shows that six relative stable modes were selected out.

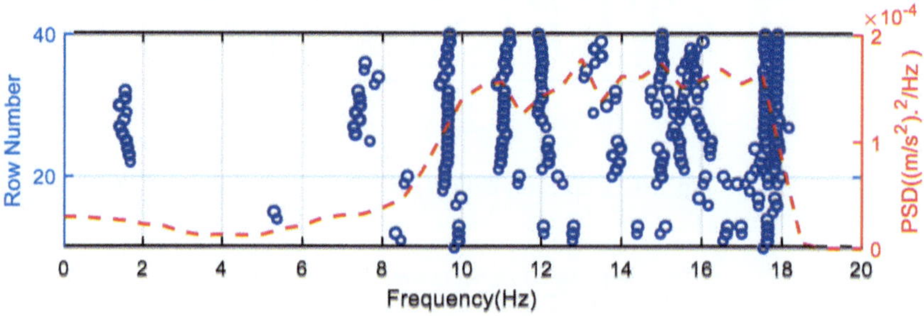

Fig. 12. Stabilization diagram

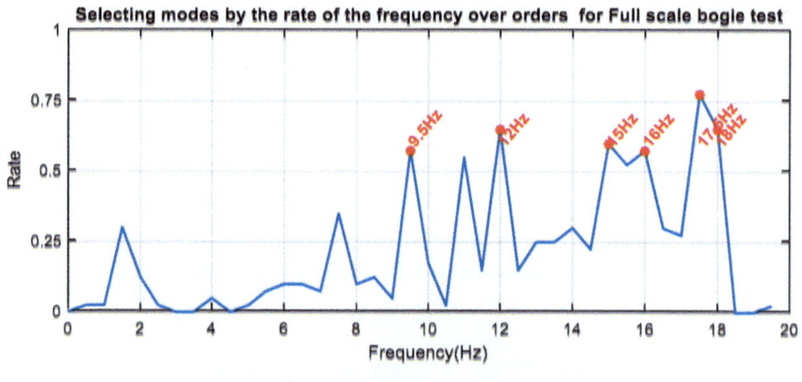

Fig. 13. Rate of identified stable modes

These selected stable modes were obtained and presented in Fig. 14. It can be seen that the mode at 15.01 Hz is pitch since the frequency, mode shape and damping ratio are agreed well with the theoretical ones. In addition, it is worth to notice that the modes at 9.61 Hz and 16.01 Hz are bounce and roll, respectively. Although their mode shapes are not ideal align with theoretical results, such results are still acceptable considering the situation of excitation and measurement noise. The other two modes at 17.57 Hz and 17.87 Hz could be false ones or the compound one since their damping ratios are quite low. Nonetheless, the goal of vehicle suspension monitoring can be fulfilled if the pitch mode was identified accurately. In other word, the identified results in this paper indicated that ACS-SSI has the potential capability for condition monitoring of railway vehicle primary suspension systems.

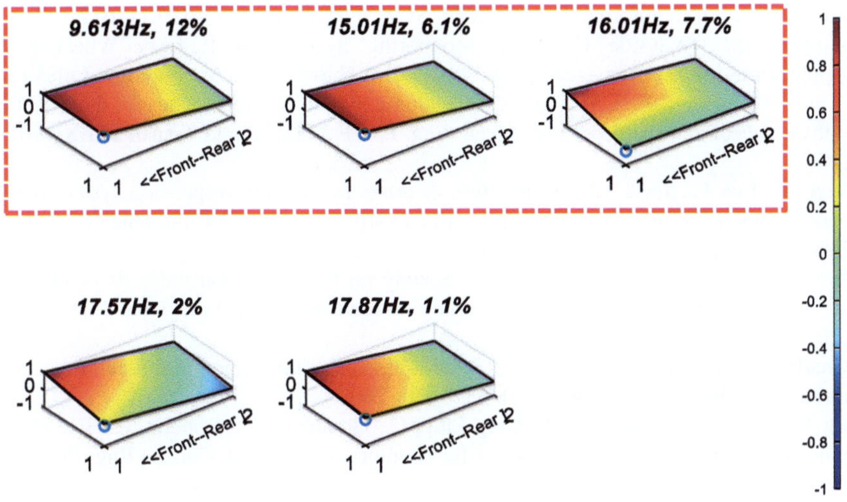

Fig. 14. Modal parameters identified by ACS-SSI method

5 Conclusions

This study focused on the dynamic characteristics identification of primary suspension systems of a Y25 bogie, which performed as a preliminary attempt to achieve online monitoring of the suspension systems. The Y25 bogie was tested on a full-size roller rig and ACS-SSI was employed to identify the modal parameters through the vibration responses on the four corners of the bogie frame where upon the axle boxes. The obtained signals were edited firstly in cepstrum to remove the harmonics caused by the rotating rail drum. Then, the pitch mode was identified accurately by ACS-SSI, which indicated the potential ability of ACS-SSI for railway vehicle suspension online monitoring. Future work will conduct experiments when the Y25 bogie suspension contains faults to verify the effectiveness of ACS-SSI in further step.

Acknowledgment. The author would like to thank the China Scholarship Council (CSC Grant No: 201608060041) and the National Natural Science Foundation of China (NSFC, Grant No: 61773021) for the sponsorship of the project carried out in this study. Moreover, the authors would like to express the appreciation to the engineer, Mr Barnaby Bryce, for his work to conduct the experiments.

References

1. Goodall, R.M., Roberts, C.: Concepts and techniques for railway condition monitoring. In: 2006 IET International Conference on Railway Condition Monitoring, pp. 90–95 (2006)
2. Li, C., Luo, S., Cole, C., Spiryagin, M.: An overview: modern techniques for railway vehicle on-board health monitoring systems. Veh. Syst. Dyn. **55**(7), 1045–1070 (2017)
3. Asplund, M.: Wayside Condition Monitoring System for Railway Wheel Profiles: Applications and Performance Assessment. Luleå University of Technology, Luleå (2016)
4. Alemi, A., Corman, F., Lodewijks, G.: Condition monitoring approaches for the detection of railway wheel defects. Proc. Inst. Mech. Eng. Part F J. Rail Rapid Transit **231**(8), 961–981 (2017)
5. Wei, X., Liu, H., Jia, L.: Fault detection of urban rail vehicle suspension system based on acceleration measurements. In: 2012 IEEE/ASME International Conference on Advanced Intelligent Mechatronics (AIM), pp. 1129–1134 (2012)
6. Wei, X., Jia, L., Liu, H.: A comparative study on fault detection methods of rail vehicle suspension systems based on acceleration measurements. Veh. Syst. Dyn. **51**(5), 700–720 (2013)
7. Charles, G., Goodall, R., Dixon, R.: Model-based condition monitoring at the wheel–rail interface. Veh. Syst. Dyn. **46**(sup1), 415–430 (2008)
8. Li, P., Goodall, R., Weston, P., Seng Ling, C., Goodman, C., Roberts, C.: Estimation of railway vehicle suspension parameters for condition monitoring. Control Eng. Pract. **15**(1), 43–55 (2007)
9. Liu, X.Y., Alfi, S., Bruni, S.: An efficient recursive least square-based condition monitoring approach for a rail vehicle suspension system. Veh. Syst. Dyn. **54**(6), 814–830 (2016)
10. Wei, X., Guo, Y., Jia, L., Liu, H.: Fault detection of rail vehicle suspension system based on CPCA. In: 2013 Conference on Control and Fault-Tolerant Systems (SysTol), pp. 700–705 (2013)
11. Wei, X., Jia, L., Guo, K., Wu, S.: On fault isolation for rail vehicle suspension systems. Veh. Syst. Dyn. **52**(6), 847–873 (2014)
12. Liang, B., Iwnicki, S.D., Zhao, Y., Crosbee, D.: Railway wheel-flat and rail surface defect modelling and analysis by time–frequency techniques. Veh. Syst. Dyn. **51**(9), 1403–1421 (2013)
13. Mei, T.X., Ding, X.J.: A model-less technique for the fault detection of rail vehicle suspensions. Veh. Syst. Dyn. **46**(sup1), 277–287 (2008)
14. Mei, T.X., Ding, X.J.: Condition monitoring of rail vehicle suspensions based on changes in system dynamic interactions. Veh. Syst. Dyn. **47**(9), 1167–1181 (2009)
15. Sohn, H., et al.: A Review of Structural Health Monitoring Literature: 1996–2001. Los Alamos National Lab, New Mexico (2003)
16. Peeters, B., De roeck, G.: Reference-based stochastic subspace identification for output-only modal analysis. Mech. Syst. Signal Process. **13**(6), 855–878 (1999)
17. Reynders, E., Pintelon, R., De Roeck, G.: Uncertainty bounds on modal parameters obtained from stochastic subspace identification. Mech. Syst. Signal Process. **22**(4), 948–969 (2008)
18. Magalhães, F., Cunha, Á., Caetano, E.: Online automatic identification of the modal parameters of a long span arch bridge. Mech. Syst. Signal Process. **23**(2), 316–329 (2009)
19. Dong, X., Lian, J., Yang, M., Wang, H.: Operational modal identification of offshore wind turbine structure based on modified stochastic subspace identification method considering harmonic interference. J. Renew. Sustain. Energy **6**(3), 033128 (2014)

20. Dong, X., Lian, J., Wang, H., Yu, T., Zhao, Y.: Structural vibration monitoring and operational modal analysis of offshore wind turbine structure. Ocean Eng. **150**, 280–297 (2018)
21. Chen, Z., Wang, T., Gu, F., Zhang, R.: Characterizing the dynamic response of a chassis frame in a heavy-duty dump vehicle based on an improved stochastic system identification. Shock Vib. **2015** (2015)
22. Liu, F., Gu, F., Zhao, Y., Ball, A.: A validation study of ACS-SSI for online condition monitoring of vehicle suspension systems. Vibroeng. Procedia **10**, 369–375 (2016)
23. Liu, F., Gu, F., Ball, A., Zhao, Y., Peng, B.: The validation of an ACS-SSI based online condition monitoring for railway vehicle suspension systems using a SIMPACK model. In: Proceedings of 23rd International Conference on Automation Computing, 7–8 September 2017, University of Hudders, October 2017
24. Tondreau, G., Deraemaeker, A.: Numerical and experimental analysis of uncertainty on modal parameters estimated with the stochastic subspace method. J. Sound Vib. **333**(18), 4376–4401 (2014)
25. Facilities: University of Huddersfield. https://research.hud.ac.uk/institutes-centres/institutes/../irr/facilities/. Accessed 22 Jan 2019
26. Li, W., Vu, V.-H., Liu, Z., Thomas, M., Hazel, B.: Extraction of modal parameters for identification of time-varying systems using data-driven stochastic subspace identification. J. Vib. Control **24**(20), 4781–4796 (2018)
27. Randall, R.B.: A history of cepstrum analysis and its application to mechanical problems. Mech. Syst. Signal Process. **97**, 3–19 (2017)
28. Peeters, C., Guillaume, P., Helsen, J.: A comparison of cepstral editing methods as signal pre-processing techniques for vibration-based bearing fault detection. Mech. Syst. Signal Process. **91**, 354–381 (2017)

A Comparative Study on Data Manipulation in PCA-Based Structural Health Monitoring Systems for Removing Environmental and Operational Variations

Callum Roberts[1(✉)], David Garcia[1], and Dmitri Tcherniak[2]

[1] Department of Mechanical and Aerospace Engineering, University of Strathclyde, Glasgow, UK
callum.roberts@strath.ac.uk

[2] Brüel & Kjær Sound & Vibration Measurement, Skodsborgvej 307, 2850 Nærum, Denmark

Abstract. Vibration-based structural health monitoring (VSHM) methodologies provide a robust, data-driven, system for damage diagnosis. However, there are a few challenges that are currently being investigated to ensure the systems are more reliable for decision-making. The features selected from the vibration responses are not only sensitive to damage but also to environmental and operational variations (EOV). This paper aims to investigate the use of a principal component analysis (PCA) based system for VSHM. In particular, the aim is to compare different approaches, using the same dataset, to explore the effect that data manipulation has on the damage detection capabilities of such a system when it is corrupted by EOV. The data that was used for this study was first taken from a simulated five degree of freedom spring-mass-damper system and secondly from an in-operation Vestas V27 wind turbine with damaged and undamaged scenarios. The simulated system was subjected to varying temperatures and involved four states; one healthy state followed by three states with increasing damage, represented by the reduction of a spring stiffness. Each combination of data manipulation was compared to determine their performance and limitations on removing EOV for reliable damage diagnosis.

Keywords: Environmental and operational variations · Wind turbine · Damage diagnosis · Principal component analysis · Correlation factors

1 Introduction

A method becoming more common for detecting damage within structures is vibration-based structural health monitoring (VSHM). VSHM encompasses a wide variety of different methods and has been around for a number of years. More recently, VSHM has been moving more towards data-driven methods, as opposed to model-based, for damage detection. Several data-driven methods already exist and are being continually developed for this purpose, for example: singular spectrum analysis [1], regression analysis [2] and principal component analysis [3]. However, when most of these

M. A. Wahab (Ed.): DAMAS 2019, LNME, pp. 182–198, 2020.
https://doi.org/10.1007/978-981-13-8331-1_13

methods are applied, they do not directly account for environmental and operational variations (EOV) which usually inhibit a system's ability to detect damage.

Therefore, there is an expanding area of interest in the field of structural health monitoring to advance current methods or develop a new one, which can purge a response of EOV without discarding any damage sensitive features. It is well-established that the features of a structural response that are used to detect damage are almost always also affected by changing temperature [4]. It is also known that the changing temperature can have a greater effect on the stiffness of the structure than the damage itself, especially in large structures. This is because the damage is a local change whereas the EOV causes a change in the global system [5]. Thus the EOV can mask any signs of the damage in the first place. Thus, removing the influence of EOV is an important development which must be completed before these methods will see implementation in real-world structures.

There have been several different approaches suggested to remove the EOV, including cointegration, factor analysis and PCA, to name a few. Cointegration is a method taken from the world of econometrics and linearly combines responses to create a stationary residual which is representative of the healthy state of the structure [6]. Any deviation from the stationarity of the healthy condition then indicates damage in the structure. Common trends are excluded as part of the cointegration process, including those that have arisen as a result of EOV [6]. Further improvements have been implemented to cointegration by introducing a regime-switching system, but it still falls short of real-world implementation [7].

PCA is a versatile method, and the one which this paper will focus on, and is used across many disciplines for applications such as facial recognition [8] and monitoring land-cover [9]. In a PCA-based system, the EOV features are considered to be different from the ones present due to damage in the structure and can therefore be separated and discarded [10]. PCA often exists alongside other methods. For example, a Gaussian Mixture Model can be used to linearize the system before PCA is used to process the data [11]. More commonly, PCA is used in conjunction with a novelty detection index such as the Mahalanobis Squared distance [10, 12], Q-statistic or T^2-statistic [3]. These indices are capable of detecting whether a structure is operating under its normal condition or whether there are any anomalies present.

An important, but perhaps sometimes overlooked, factor affecting the results of PCA-based analysis is the way in which the data is handled. There appears to be no consensus in the literature as to which data manipulation method is best for the application of PCA. In some cases [5, 13], it is not made explicitly clear in what form the data exists. With regards to normalisation, there are several ways to do it. In some applications, data compression only for example, no normalisation is done and everything is used in its raw form [14]. In other studies, the data has been mean centered [11]. However, the most commonly used method mean centers the data as well as dividing by the standard deviation to give a mean of zero and standard deviation of one [15, 16]. The method chosen in the literature varies depending on the particular application but, as demonstrated by Yan, et al. [10], the choice is not trivial.

As well as there being different ways of normalising the data, there are also different ways of manipulating the data before the PCA process. In the case where there is only a single dataset, the matrix exists as the number of measurements by the number of sensors and this is how it will be inputted into PCA, subject to normalisation. The complications arise when there are multiple realisations that must be used for a single PCA process. One method that has been adopted is to take the covariance of each realisation and then place the values from the upper triangle of the matrix in a single column [12]. The column related to each realisation is then put in a matrix, where the dimensions are the number of measurements by the number of realisations, and PCA performed on it. Another option is to arrange the data into a 3D array where each 2D matrix contains all the measurements from one sensor. The array is then "unfolded" to create a larger 2D matrix which then becomes the input for PCA [3]. The latter adds complications as there are different methods for normalising this arrangement which affect the end results of the PCA process [17].

The aim of this paper is to test and compare different combinations of normalisation and arrangement to see if there is quantitatively an optimum combination for the specific application of damage detection. This will be done by testing the combinations on data taken from a simulated five degree of freedom spring-mass damper system, which has been subjected to EOV, and then verified using taken from an in-operation Vestas V27 wind turbine. The rest of the paper is set out as follows. The different ways in which the data was normalised and arranged will be set out followed by a description of the PCA method. Case studies will then be presented on the data that was used in the study. The results from the analysis will be shown followed by a discussion of their implications. Finally, the paper will draw a conclusion from the results obtained in the study.

2 Methodology

2.1 Data Pre-processing

In order to simplify the explanations, the data dimensions have been defined in the following way; m is the number of measurements, n is the number of sensors and r is the number of realisations. Each realisation is made up of samples taken from a single test. A single $m \times n$ matrix is shown in Eq. 1.

$$\boldsymbol{X}_k = \begin{pmatrix} x_{11} & \cdots & x_{1j} & \cdots & x_{1n} \\ \cdots & \cdots & \cdots & \cdots & \cdots \\ x_{i1} & \cdots & x_{ij} & \cdots & x_{in} \\ \cdots & \cdots & \cdots & \cdots & \cdots \\ x_{m1} & \cdots & x_{mj} & \cdots & x_{mn} \end{pmatrix} = \begin{pmatrix} \boldsymbol{v}_1 & \cdots & \boldsymbol{v}_j & \cdots & \boldsymbol{v}_n \end{pmatrix}, \tag{1}$$

where $\boldsymbol{X}_k$ contains all the samples from realisation k, x_{ij} is the ith value of the sample from sensor j and $\boldsymbol{v}_j$ is the vector representation of column j.

Raw Data. Raw data requires no processing prior to the application of the PCA process. This form is rarely used but has been included in this study for the sake of completeness.

Mean Centered Data. The first normalisation method used was mean centering. This method is more widely used than raw data but it does not account for large differences in the magnitude of the measured data from sensor to sensor. Mean centering takes the mean from each column of data, from the same sensor, and subtracts it from each value in the column to give an overall mean of zero as shown in Eq. 2:

$$\bar{x}_{ij} = x_{ij} - \mu_{\boldsymbol{v}_j}, \tag{2}$$

where $\mu_{\boldsymbol{v}_j}$ is the mean value of the column vector $\boldsymbol{v}_j$.

Normalised Data. The most common method used in the application of PCA is a full normalisation process, in which the data is first mean centered and then each value divided by the corresponding standard deviation of each sensor reading. The values are calculated in the method set out in Eq. 3. From here forward, the result of this process will be referred to as "normalised data".

$$\bar{\bar{x}}_{ij} = \frac{x_{ij} - \mu_{\boldsymbol{v}_j}}{\sigma_{\boldsymbol{v}_j}}, \tag{3}$$

where $\sigma_{\boldsymbol{v}_j}$ represents the standard deviation of the column vector $\boldsymbol{v}_j$.

2.2 Feature Extraction

Two different data arrangements were considered for this study which will be covered in detail in the following sections.

Horizontal Stack. The first approach used for the analysis considers the datasets stacked in a horizontal orientation. Thus, the r sets of $m \times n$ data matrices are arranged in such a way that it becomes an $m \times (n \times r)$ matrix as the input for the PCA as shown in Eq. 4. This method utilises a large number of principal components, thus making it possible to cover a large percentage of the variance in the data without incorporating the principal components associated with the EOV and noise.

$$\boldsymbol{X} = (\boldsymbol{X}_1 \quad \ldots \quad \boldsymbol{X}_k \quad \ldots \quad \boldsymbol{X}_r) \tag{4}$$

Correlation Factor. This method is typically used to correlate measurements between different sensor locations and will be used as an alternative way of arranging data. For each realisation r, the $m \times n$ data matrix is pre-multiplied by its transverse giving a covariance matrix with dimension $n \times n$. The values from the upper triangle of the matrix are then rearranged into a column vector, $p \times 1$. The input for PCA then becomes the horizontal arrangement of these vectors giving a dimension of $p \times r$. A more complete explanation of this method can be found in the literature [18].

2.3 Damage Detection

Principal Component Analysis. As discussed earlier, PCA is a powerful statistical tool which has seen application in a wide number of fields, including VSHM. The PCA method is primarily used for dimensionality reduction but in the case of damage detection, it can also be used for feature extraction [3, 12]. The dimensionality reduction aspect makes it appropriate for the application of VSHM as there is often a large number of variables but a simpler underlying relationship. The basic concept of PCA is to project the data into a new set of coordinates whilst only keeping selected parts of the signal. The main aim of this method for damage detection is for the PCA method to highlight any differences between a base model and new data being received. This transformation projects the original data, $\boldsymbol{X}$, using a set of principal components, $\boldsymbol{P}$, to obtain a new set of principal scores, $\boldsymbol{T}$, as in Eq. 5.

$$\boldsymbol{T} = \boldsymbol{XP} \tag{5}$$

Calculating the eigenvectors and eigenvalues of the covariance matrix of the original dataset, $\boldsymbol{X}$, gives the principal components, $\boldsymbol{P}$, and their corresponding variance respectively. The principal components can then be ordered based on the amount of variance they account for. The number of principal components can then be reduced based upon how much of the original data is desired to be kept. Noise and other pollutants are typically accounted for in the components with smaller variance and thus it is desirable to discard this information [3, 10]. The original data can then be projected again using Eq. 5 and the new set of principal components to obtain a new set of principal scores. A full explanation of the method can be found in Mujica et al. [3].

Mahalanobis Squared Distance. A base model can then be created using the new principal components and principal scores associated with a known healthy condition. Any new data can then be compared to the so-called reference state by projecting the new data using the same set of principal components. The difference between the reference state and new state can then be determined using a damage index. For this study, the Mahalanobis Squared Distance was used as it has been commonly used in damage detection applications [12, 19]. The Mahalanobis Squared Distance compares each new measurement, $\boldsymbol{y}$, to the mean value of the healthy state, $\bar{\boldsymbol{x}}$, as in Eq. 6 where $\boldsymbol{\Lambda}$ represents the eigenvalues of the covariance matrix.

$$\boldsymbol{D}^2 = (\boldsymbol{y} - \bar{\boldsymbol{x}})\boldsymbol{P}\boldsymbol{\Lambda}^{-1}\boldsymbol{P}^T(\boldsymbol{y} - \bar{\boldsymbol{x}})^T \tag{6}$$

The threshold value chosen for this study has simply been chosen as the value that covers 95% of the variance of the data from the training data in an exclusive manner. This assumes that 5% of the observations from the training set are considered to be outliers. There are more involved methods for determining a more efficient threshold value but this paper simply aims to compare different ways of manipulating data and not optimize the process. The classification is being treated in an equally simply manner. Any damage data above the threshold has been said to be correctly classified and below incorrectly. The opposite is true for the testing dataset.

3 Case Studies

3.1 Simulated System

In order to test the method, a simulated system was created. The system consisted of five masses connected by springs and dampers as shown in Fig. 1. The end masses were both connected to a ground for a total of six springs, six dampers and five degrees of freedom (DoF). Each mass and damper were given the same value of 2 kg and 0.01 Ns^2/m respectively. The system was given a set of initial conditions, which were all zero apart from a small displacement to the fifth mass, and then left to vibrate freely. In order to emulate the EOV, the temperature was varied as shown in Fig. 2. The simulation was repeated many times in order to obtain multiple realisations for a setup with a varying temperature. Table 1 shows the number of realisations that were used for each of the states of the simulated system. During each realisation, the system was considered to have a constant temperature, and thus constant values for the spring constants. The vertical lines in Fig. 2 indicate the points at which the different damage scenarios were introduced.

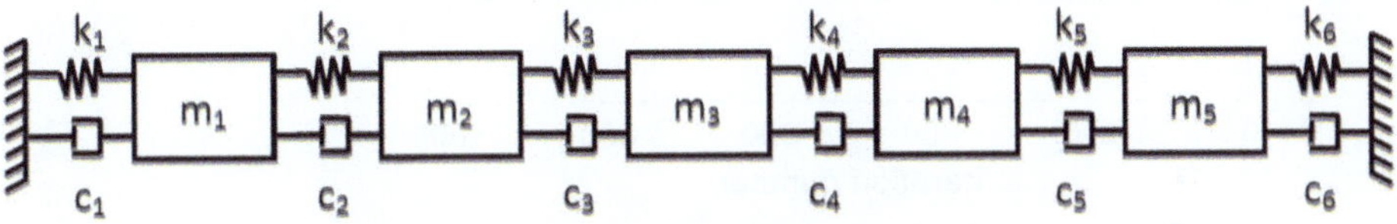

Fig. 1. Five DoF spring-mass-damper system

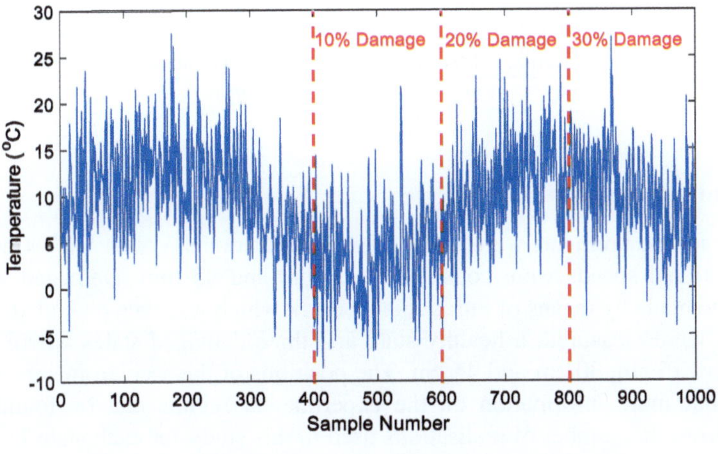

Fig. 2. Temperature profile applied to simulated system

The varying temperature affects the dynamics of the system and alters the response since the spring constants are sensitive to changes in temperature. The spring constants were the same in each spring and calculated using the simple linear model shown in Eq. 7, a simplified version of a form previously used in [7]. The effect that the changing temperature had on the natural frequencies of the system can be seen in Fig. 3. The damage was simulated by reducing k_3 by 10%, 20% and 30%.

$$k_1 = k_2 = k_3 = k_4 = k_5 = k_6 = \begin{cases} -0.15 \times T + 4, T < 0 \\ -0.05 \times T + 4, T \geq 0 \end{cases} \quad (7)$$

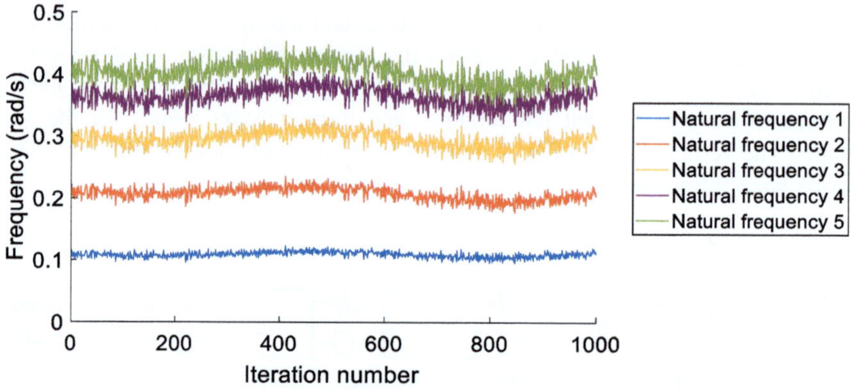

Fig. 3. Variation in natural frequencies for simulated system

Table 1. Number of realisations used in simulated system

	Undamaged		Damaged (Stiffness reduction of k_3)		
	Training	Testing	10% damage	20% damage	30% damage
Simulated data	250	150	200	200	200

3.2 Wind Turbine Data

Acceleration measurements were taken from an operational Vestas V27 wind turbine. Two operational speeds were considered, 32 rpm and 43 rpm. Damaged was introduced to the blade by means of introducing a crack which was then extended. A total of four states were measured: a healthy state and three damaged states where the crack lengths were 15 cm, 30 cm and 45 cm. The positions of the accelerometers are shown in Fig. 4 but more information on the experimental regime can be found in [12]. Table 2 shows the number of realisations used in this study for each state for the wind turbine data.

Table 2. Number of realisations used for wind turbine data

	Undamaged		Damaged (crack length)		
	Training	Testing	15 cm	30 cm	45 cm
32 rpm	500	200	66	117	105
43 rpm	500	200	158	193	132

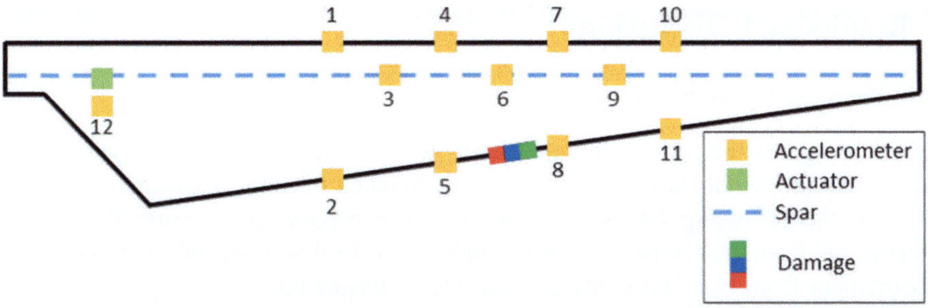

Fig. 4. Wind turbine blade layout with accelerometer positions and associated number

Since the wind turbine in the experiment was situated outside, it was exposed to a wide variety of weather conditions. One condition was the changing temperature which the simulated system aimed to emulate. Figure 5 shows how the temperature varied during the experimental regime when the turbine was operating at 32 rpm and 43 rpm.

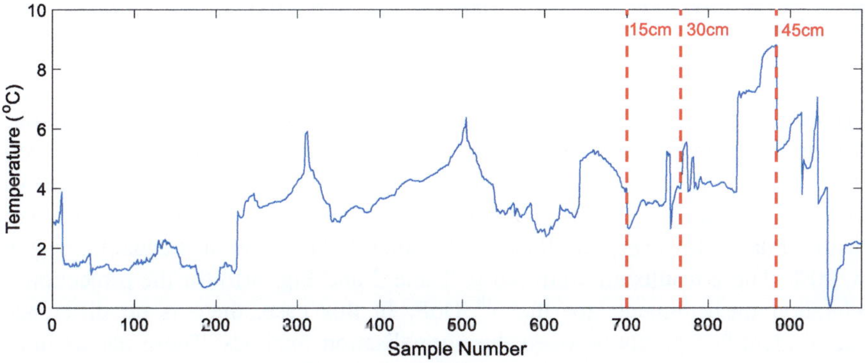

(a)

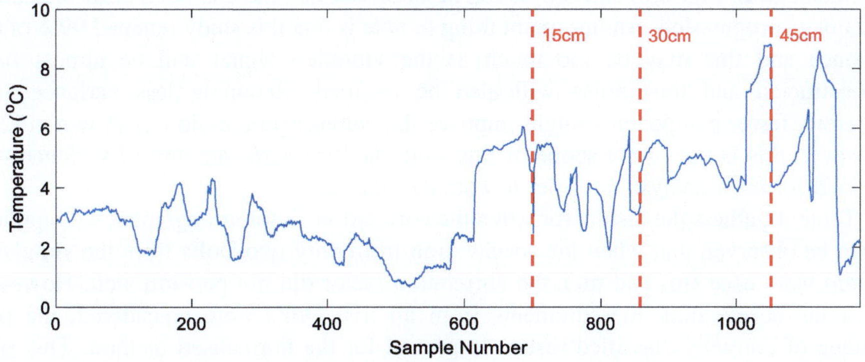

(b)

Fig. 5. Temperature profile for (a) 32 rpm (b) 43 rpm experimental regimes

4 Results and Discussion

4.1 Simulated System

Table 3 compares the results obtained from running the various normalisation procedures in the horizontal stack arrangement on data taken from the simulated system. The values in the following tables correspond to the percentage of correctly classified observations from each type of data normalisation. In this case, only the acceleration measurement from two DoFs, m_4 and m_5, were considered.

Table 3. Percentage of correctly classified results for projected acceleration responses from m_4 and m_5 when two acceleration measurements were used to obtain the PCA space in the horizontal stack arrangement (*cf.* Figure 6)

	Horizontal stack – using only DoF 4			Horizontal stack – using only DoF 5		
	Raw	Mean centered	Normalised	Raw	Mean centered	Normalised
Healthy Testing	91	91	90	89	89	89
10%	100	100	100	100	100	100
20%	100	100	100	100	100	100
30%	100	100	100	100	100	100

Table 3 and Fig. 6(a) show that projecting acceleration measurements from the 4th DoF using the principal components from the training set gives good damage detection. With regards to the normalisation procedure, it does not make much difference which method is used, with only a slightly lower detected rate in the testing region for the normalised data. However, for all cases, the correct classification is equal to, or greater than, 90%. These results are mirrored in Table 3 and Fig. 6(b) for the projection with acceleration measurements for the 5th DoF. In this case, there is no difference in correctly identified points between the normalisation methods. These results indicate that there is no best method to apply when conducting PCA in the horizontal stack arrangement. In Fig. 6(a) and 6(b) it can be observed that there is not a clear indication of damage progression. An important thing to note is that this study retained 99% of the variance and this may be too much as the vibration signal will be almost fully reconstructed and thus noise will also be retained. Retaining less variance and, therefore, fewer components might improve the damage progression as shown in [20]. However, this is out of the scope of this study and the same amount of variance was retained for each analysis in order to compare the results.

Table 4 gathers the results for when the correlation factor arrangement was applied. It can be observed that when the acceleration from only two DoFs from the simulated system were used (m_4 and m_5), the correlation factor did not perform well. However, when the acceleration measurements from all five DoFs were considered, the percentage of correctly classified results improved for the normalised method. This suggests that the normalised data is the best form for the application in the correlation factor, especially when the results are compared to those of the mean centered and raw

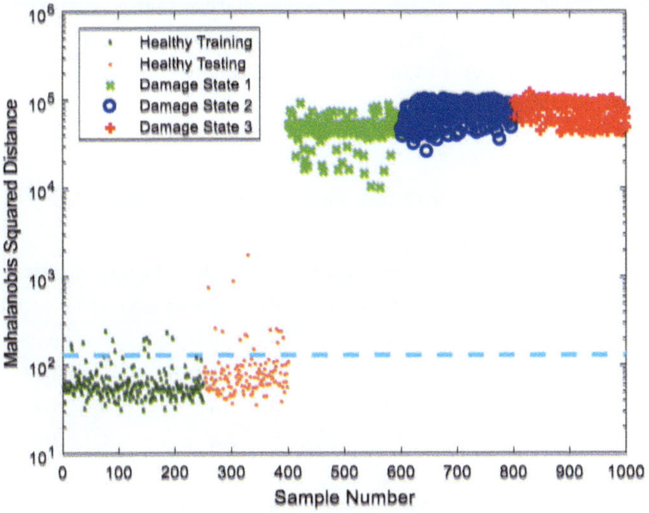

(a)

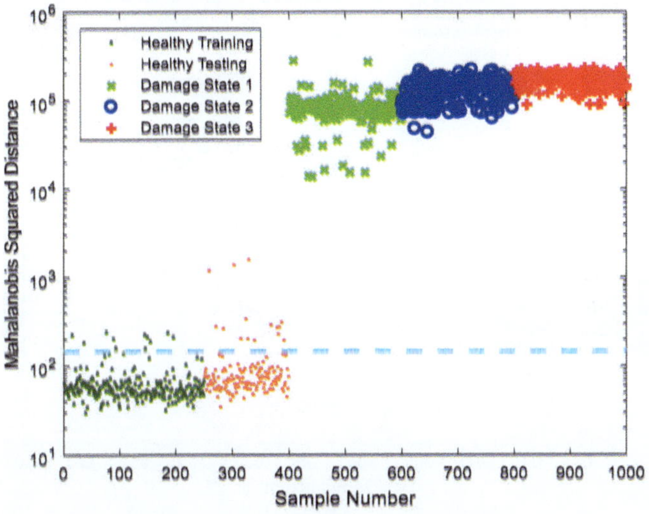

(b)

Fig. 6. Damage indices representation when only acceleration responses from (a) m_4 and (b) m_5 were projected on the PCA space. Two acceleration measurements were used to obtain the PCA space in the horizontal stack arrangement data. Furthermore, the results suggest that the correlation factor performs better when a greater number of acceleration signals are used. From Fig. 7 it can also be seen that the correlation factor exhibits clearer evidence of damage progression when compared with Fig. 6 for the horizontal stack.

Table 4. Normalised results for correlation factor arrangement conducted with two and five DoFs

	Correlation factor – using 2 DoF			Correlation factor – using 5 DoF		
	Raw	Mean centered	Normalised	Raw	Mean centered	Normalised
Testing	94	94	94	93	93	95
10%	8	8	7	19	19	100
20%	4	4	4	4	4	100
30%	2	3	2	5	4	100

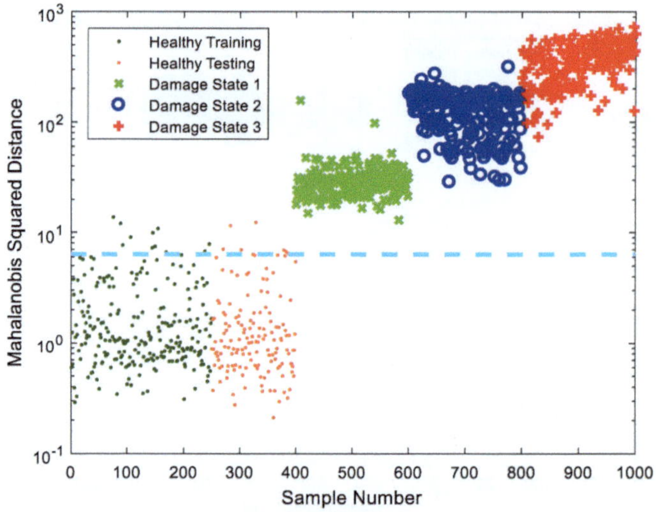

Fig. 7. Damage indices representation of normalised simulated data using acceleration measurements from all five DoFs in the correlation factor arrangement

4.2 Wind Turbine Data

Tables 5 and 6 show the percentage of correct classifications for the wind turbine data at 32 rpm and 43 rpm respectively. The measurements from eight accelerometers were used from the wind turbine, namely the four on the leading edge and four on the trailing edge of the blade. For the horizontal stack, accelerometer number 8 was chosen to be projected because it is the one closest to the damage.

Table 5. Percentage of correctly classified results for the horizontal stack and correlation arrangements for wind turbine data at 32 rpm (*cf.* Figure 8)

	Horizontal stack – projecting Acc 8			Correlation factor		
	Raw	Mean centered	Normalised	Raw	Mean centered	Normalised
Testing	3	2	0	94	98	96
15 cm	100	100	100	2	12	32
30 cm	100	100	100	14	60	100
45 cm	100	100	100	17	74	100

Table 6. Percentage of correctly classified results for the horizontal stack and correlation arrangements for wind turbine data at 43 rpm (*cf.* Figure 9)

	Horizontal stack – projecting Acc 8			Correlation factor		
	Raw	Mean centered	Normalised	Raw	Mean centered	Normalised
Testing	0	0	0	97	96	93
15 cm	100	100	100	11	13	48
30 cm	100	100	100	12	58	98
45 cm	100	100	100	28	88	100

The results from Tables 5 and 6 for the correlation factor at both 32 rpm and 43 rpm show that data pre-processing has a significant impact on the number of observations correctly classified. As with the simulated system, the normalised data performs the best but, unlike the simulated system, it struggles to detect the lesser damage. Comparing the two data arrangements, the horizontal stack does not perform well, since the observations from the undamaged blade are not correctly detected. Figures 8 and 9 indicate that the correlation factor gives better evidence of damage progression.

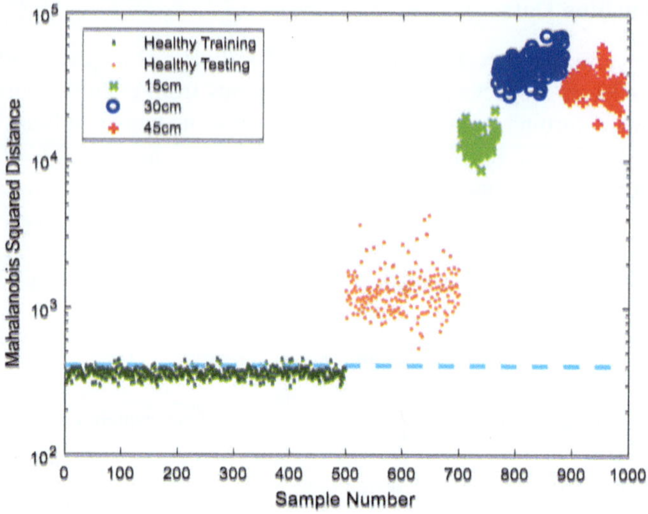

(a)

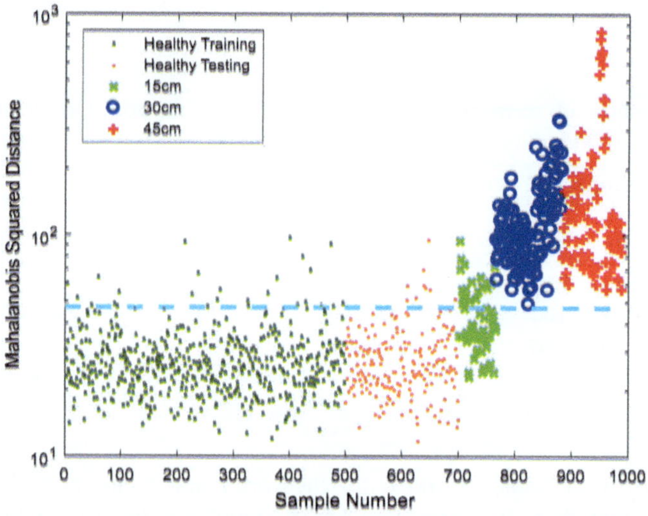

(b)

Fig. 8. Damage indices representation of the normalised wind turbine data at 32 rpm in the (a) Horizontal stack (b) Correlation factor arrangement

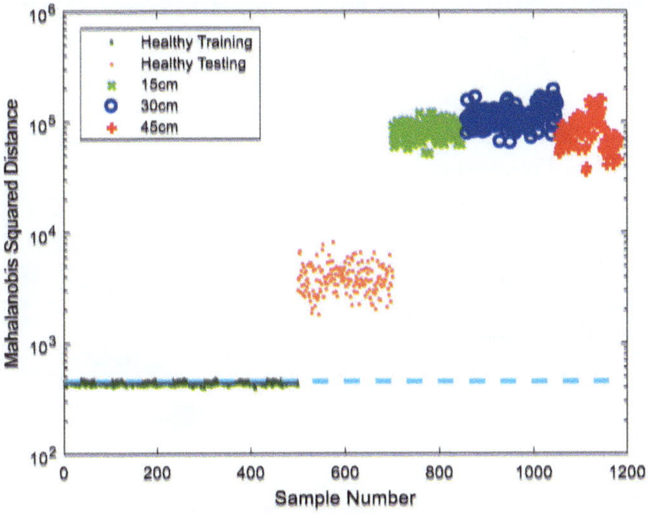

(a)

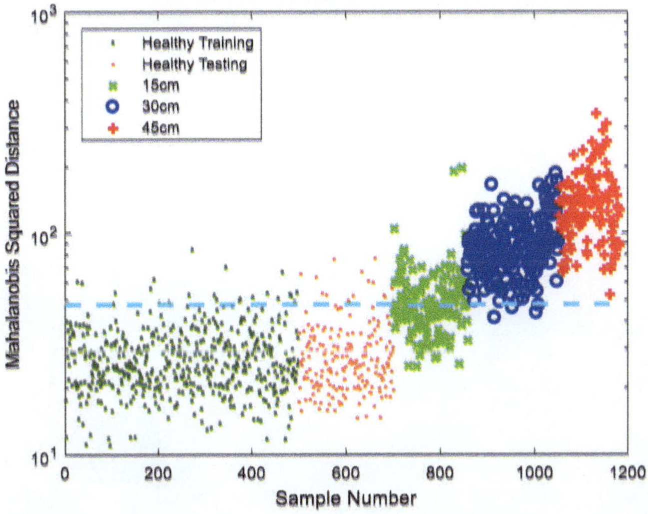

(b)

Fig. 9. Damage indices representation of the normalised wind turbine data at 43 rpm in the (a) Horizontal stack (b) Correlation factor arrangement

Table 7. Percentage of correctly classified results for the horizontal stack and correlation arrangements for filtered wind turbine data at 32 rpm (*cf.* Figure 10)

	Horizontal stack – projecting acc. 8			Correlation factor		
	Raw	Mean centered	Normalised	Raw	Mean centered	Normalised
Testing	83	83	74	97	96	97
15 cm	100	100	100	0	6	100
30 cm	100	100	100	18	47	100
45 cm	100	100	100	14	71	100

Previous studies have shown that by applying a bandpass filter to the data the detection rates can be improved [12]. Applying the filter used in the previous study produced the results in Table 7. Immediately it can be seen that the horizontal stack results improve. By observing Fig. 10(a), there is a noticeable shift in the testing data when compared to the unfiltered data in Fig. 8(a). A similar level of improvement can also be seen between Fig. 8(b) and 10(b).

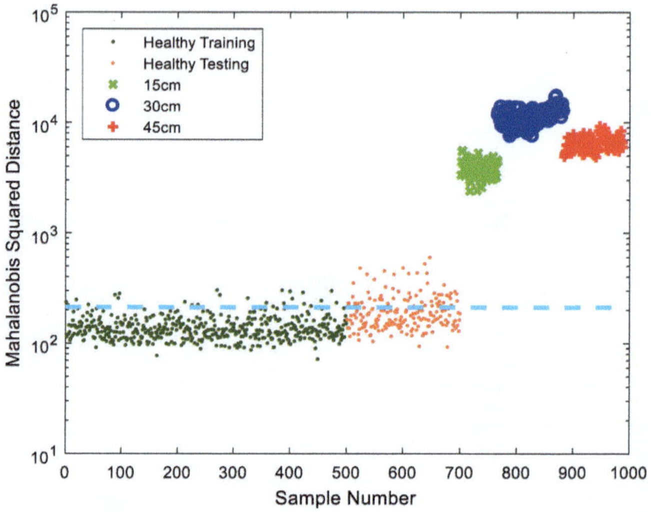

(a)

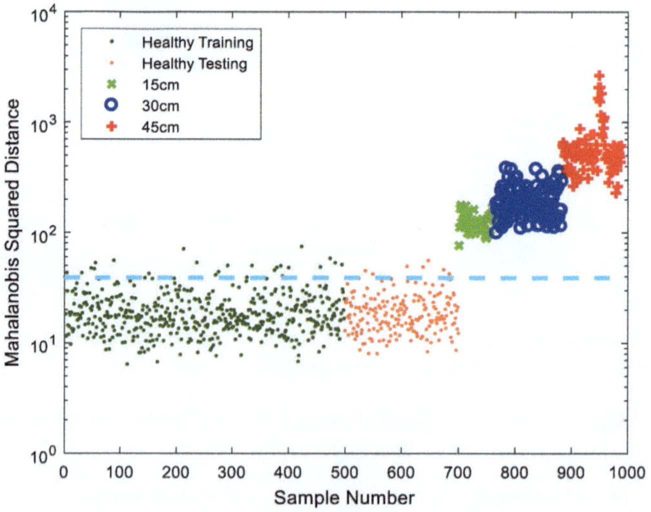

(b)

Fig. 10. Damaged indices representation of the normalised filtered wind turbine data at 32 rpm in the (a) Horizontal stack (b) Correlation factor arrangement

As with the unfiltered and simulated data, the filtered data performs optimally when the data is normalised. Furthermore, the correlation factor performs better overall than the horizontal stack as well as showing evidence of damage progression. The horizontal stack performs worst when the data is normalised, in line with the results from the unfiltered and simulated data.

A factor that has an impact on the results is the choice of accelerometers. This is especially true for the wind turbine data, as has been shown in [18]. As this study was not aimed at optimizing the detection rate, a set of accelerometers were chosen which had been proven to work. Since each signal from an accelerometer also holds a substantial amount of data, it is important to choose which part of the signal should be used to achieve damage detection. In the case of the wind turbine, the data range was always chosen to be immediately after the actuator hit, in line with [12, 18], where the excitation in the blade has a significant amplitude.

5 Conclusion

In this study, several combinations of data normalisation and arrangement were compared to determine if there was quantitatively an optimal normalisation method to use for damage detection in structures subjected to varying environmental and operational variations. The data normalisation methods used were raw data, mean centered data and normalised data as detailed in the manuscript. The data arrangements that were used were a horizontal stack and the correlation factor.

In terms of optimizing the detection rates, normalising the data was shown to give the best results for both simulated data and real data for the correlation factor but had the opposite effect on the horizontal stack. Furthermore, damage progression was present for only the correlation factor and not the horizontal stack. Overall, the correlation factor arrangement performed better than the horizontal stack if there were a suitable number of signals considered.

It should be noted that the horizontal stack arrangement considered only measurements for a single sensor, whereas at least two measurements are required for the correlation factor. This creates a bias towards the correlation factor being the best arrangement in this study. It is clear from the results that if previous knowledge is available, the implementation of a filter could significantly improve the damage detection by considering only the frequencies in a range of interest.

Acknowledgements. The authors would like to acknowledge The Carnegie Trust for the Universities of Scotland for their support of this research project.

References

1. Garcia, D., Trendafilova, I.: A multivariate data analysis approach towards vibration analysis and vibration-based damage assessment: application for delamination detection in a composite beam. J. Sound Vib. **25**(10), 7036–7050 (2014)
2. Dervilis, N., Worden, K., Cross, E.: On robust regression analysis as a means of exploring environmental and operational conditions for SHM data. J. Sound Vib. **347**(7), 279–296 (2015)

3. Mujica, L., Rodellar, J., Fernandez, A., Guemes, A.: Q-statistic and T2-statistic PCA-based measures for damage assessment in structures. Struct. Health Monit. **10**(5), 539–553 (2010)
4. Peeters, B., Maeck, J., De Roeck, G.: Vibration-based damage detection in civil engineering: excitation sources and temperature effects. Smart Mater. Struct. **10**(3), 518–527 (2001)
5. Kojidi, S.M., Dohler, M., Bernal, D., Liu, Y.: Linear Projection Techniques in Damage Detection Under a Changing Environment, pp. 325–332. Springer, New York (2013)
6. Cross, E.J., Worden, K., Chen, Q.: Cointegration: a novel approach for the removal of environmental trends in structural health monitoring data. Proc. R. Soc. **467**(2133) (2011). https://doi.org/10.1098/rspa.2011.0023
7. Shi, H., Worden, K., Cross, E.J.: A regime-switching cointegration approach for removing environmental and operational variations in structural health monitoring. Mech. Syst. Signal Process. **103**, 381–397 (2018)
8. Kim, K.I., Jung, K., Kim, H.J.: Face recognition using kernel principal component analysis. IEEE Signal Process. Lett. **9**(2), 40–42 (2002)
9. Byrne, G., Crapper, P., Mayo, K.: Monitoring land-cover change by principal component analysis of multitemporal landset data. Remote Sens. Environ. **10**(3), 175–184 (1980)
10. Yan, A., Kerschen, G., De Boe, P., Golinval, J.: Structural damage diagnosis under varying environmental conditions: part I: A linear analysis. Mech. Syst. Signal Process. **19**(4), 847–864 (2005)
11. Wah, W.S., Chen, Y.-T., Roberts, G.W., Elamin, A.: Damage detection of structures subject to nonlinear effects of changing environmental conditions. Procedia Eng. **188**, 248–255 (2017)
12. Ulriksen, M.D., Tcherniak, D., Damkilde, L.: Damage detection in an operating Vestas V27 wind. In: 2015 IEEE Workshop on Environmental, Energy, and Structural Monitoring Systems (EESMS) Proceedings, Trento, IEEE (2015)
13. Gharibnezhad, F., Mujica, L., Rodellar, J., Fritzen, C.: Damage detection using principal component analysis based on wavelet ridges. Key Eng. Mater. **569**, 916–923 (2013)
14. Park, S., Lee, J.-J., Yun, C.-B., Inman, D.J.: Electro-mechanical impedance-based wireless structural health monitoring using PCA-data compression and k-means clustering algorithms. J. Intell. Mater. Syst. Struct. **19**(4), 509–520 (2007)
15. Tang, J.: Frequency response based damage detection using principal component analysis. In: 2005 IEEE International Conference on Information Acquisition, Hong Kong, IEEE (2006)
16. Gomez Gonzalez, A., Fassois, S.: A supervised vibration-based statistical methodology for damage detection under varying environmental conditions & its laboratory assessment with a scale wind turbine blade. J. Sound Vib. **366**, 484–500 (2016)
17. Westerhuis, J.A., Kourti, T., MacGregor, J.F.: Comparing alternative approaches for multivariate statistical analysis of batch process data. Chemometrics **13**(3–4), 397–413 (1999)
18. Tcherniak, D., Molgaard, L.: Active vibration-based structural health monitoring system for wind turbine blade: demonstration on an operating Vestas V27 wind turbine. Struct. Health Monit. **16**(5), 536–550 (2017)
19. Cross, E., Manson, G., Worden, K., Pierce, S.: Features for damage detection with insensitivity to environmental and operational variations. Proc. R. Soc. A: Math. Phys. Eng. Sci. **467**(2133), 4098–4122 (2012)
20. García, D., Tcherniak, D.: An experimental study on the data-driven structural health monitoring of large wind turbine blades using a single accelerometer and actuator. Mech. Syst. Signal Process. **127**, 102–119 (2019, in press)

Damage Localisation in Thin Plates Using the Inverse Finite Element Method

Rinto Roy[1], Marco Gherlone[1], and Cecilia Surace[2(✉)]

[1] Department of Mechanical and Aerospace Engineering, Politecnico di Torino, Turin, Italy
{rinto.roy,marco.gherlone}@polito.it

[2] Department of Structural, Geotechnical and Building Engineering, Politecnico di Torino, Turin, Italy
cecilia.surace@polito.it

Abstract. This work investigates the use of the inverse Finite Element Method (iFEM) for damage localisation by reconstructing the damaged strain field of the structure. The iFEM is a proven technique for reconstructing the displacement field of a structure using surface strain measurements from discrete locations on the structure. This new approach divides a large structure into square grid cells with strain sensors located only at the boundaries of these cells. Presence of a crack in a cell causes perturbations in the strain measurements at the boundary. The iFEM reconstructed strain field will show a higher strain magnitude in those cells affected by damage in comparison to healthy, unaffected cells of the structure. Using such a relative rather than absolute measure of strain for damage localisation helps to reduce the number of strain sensors required for the Structural Health Monitoring (SHM) of large structures.

Keywords: Structural Health Monitoring · Inverse Finite Element Method · Shape sensing

1 Introduction

Structural health monitoring is seen as a key technology for the maintenance of various commercial and industrial structures. An ideal SHM system would be capable of providing real-time assessment of the integrity of a structure under any operational conditions, making use of data measured by a small number of sensors located on the structure, thus reducing the amount of human intervention and providing significant benefits in terms of reduced maintenance time and cost and an improved measure of safety for the entire system. Although it has been the focus of research for the past four decades, very few of these technologies are currently being routinely applied for the maintenance of commercial and industrial structures.

Of the large body of research that has gone into developing various SHM methodologies, the vast majority of these methods are based on the modal parameters of the structure [1]. Methods based on modal parameters normally rely on dynamic data of the structure measured while the structure is subject to vibrations and is a representation of the health status of the entire structure. A category of these damage

M. A. Wahab (Ed.): DAMAS 2019, LNME, pp. 199–212, 2020.
https://doi.org/10.1007/978-981-13-8331-1_14

detection techniques is based on the analysis of changes in mode shapes (alternatively operational deflection shapes) or their derivatives.

It is known that, in fact a crack or a delamination causes a discontinuity in the rotation and in the curvature, i.e. in the first and second derivatives of the mode shapes. Techniques of this kind proposed by the scientific community are numerous and each has advantages and disadvantages. A number of these have been proposed by the authors of this article for one-dimensional [2–5] and two-dimensional [6–8] structures. In general, common limitations of each of these techniques include the fact that a very dense sensor grid is required, and that they are highly sensitive to the presence of measurement noise, although new measurement tools such as Scanning Laser Doppler Vibrometers and video-cameras are enabling the noise-sensitivity issue to be tackled.

The modal parameters can also be used as part of an optimisation problem, to update a damaged FEM model of the structure to localise and quantify the damage [9]. Such techniques and also those based on the analysis of the natural frequencies, suffer from the limitation that the natural frequencies are affected not only by the damage but also by different operational and environmental factors. To separate the effects of operational and environmental conditions from structural changes due to damage, several authors have implemented data-normalisation procedures such as look-up tables, regression modelling, machine learning and co-integration [10–12].

As an alternative, methods based on analysing the strain or displacement field of a structure can be considered. These methods usually analyse the static behaviour of the structure and hence use strain data to describe the health of the structure at any instant of time. As strain measurements provide a picture of the local, rather than global state of the structure; it usually does not require any prior knowledge of the operational conditions of the structure. Methods such as the Boundary Element Formulation [13] and the Body Force Method [14], make use of displacement and strain measurements at various grid points of the structure for solving the inverse problem of minimising an objective function defined using various damage parameters. Such approaches are largely dependent on the measurement grids used, which can be optimised to minimise the number of sensors required [15]. Other approaches include characterising the damage as the site of violation of the strain compatibility relations [16] for a given strain field or the violation of the governing differential equations of the structure for a given displacement field [17]. But these approaches provide a highly localised detection area and often require a large number of sensors to be practically relevant.

The inverse Finite Element Method is a technique used for solving the inverse problem of reconstructing the displacement field of a structure, form surface strain measurements [18, 19]. The iFEM has primarily been used for shape sensing applications, and very little research has focused on employing the iFEM for specific SHM problems. The use of iFEM in this domain is quite appealing as the displacement reconstruction is independent of the material or loading conditions of the structure and the method is seen to be robust in the presence of measurement noise. The iFEM was successfully used in conjunction with a fibre optic sensor for locating the position of a crack in a beam specimen [21]. It was also used to develop a load-adaptive baseline for damage identification [22], where the damage location is identified by comparing this baseline strain field to experimentally measured strain at the damage location.

The present study aims to develop on the merits of the previous works by employing the iFEM for reconstructing the damage affected strain field of the structure, using a suitable strain sensing grid discretisation. This reconstructed strain field is subsequently analysed to identify the location of the crack. The paper initially discusses the general iFEM framework, and the mathematical formulations for the inverse element used for the work (Sect. 2). Section 3, introduces the new damage localisation strategy. In Sect. 4, the validity of the new method is demonstrated through various example problems of crack localisation in a thin plate specimen subjected to bi-axial loading. Different sensor discretisations and the effect of measurement uncertainties are also tested. Section 5, summarises the general conclusions and future direction of the work.

2 The Inverse Finite Element Method

This section provides a brief review of the general formulation of the iFEM [19]. The iFEM approach is based on the minimisation of a functional defined as the least-squares error between the analytic strain measures $\varepsilon_k(\{u^e\})(k = 1,\ldots,K)$ and the corresponding experimental strain measures, $\varepsilon_k^\varepsilon(k = 1,\ldots,K)$, where K is the number of independent strain measures defined based on the structural theory adopted. The analytical strain measures are obtained from the displacement field using the linear strain-displacement relations; the displacement field is interpolated within each inverse finite element by the chosen element shape functions and $\{u^e\}$ are the nodal degrees of freedom of the element. The experimental strain measures are evaluated at n discrete locations of the structure using strain sensors. For any single element, the error functional, Φ^e, is defined as the sum of the products of the least-squares error of the k-th strain component, Φ_k^e, and the corresponding weighting coefficient, w_k^e,

$$\Phi^e(\{u^e\}) \equiv \sum_k w_k^e \Phi_k^e \tag{1}$$

where, Φ_k^e is defined as

$$\Phi_k^e \equiv \frac{1}{n}\sum_{i=1}^{n}\left[\varepsilon_{k(i)}(\{u^e\}) - \varepsilon_{k(i)}^\varepsilon\right]^2 \quad (k = 1,\ldots,K) \tag{2}$$

The weighting coefficients, w_k^e, are used for providing dimensional consistency to the error functionals and also for enforcing a stronger or weaker correlation between a specific analytic and experimental strain measure. Minimisation of Φ^e with respect to $\{u^e\}$ provides the nodal degrees of freedom and the related approximated displacement and strain field in each element. Since only strain-displacement relations are used in the formulation, iFEM does not require the knowledge of any material properties or the applied loading. Thus, it is applicable to both static and dynamic conditions, without requiring inertial or damping material properties.

2.1 Plate Formulation

Considering a plate of thickness 2t, whose mid-plane is represented by the Cartesian coordinate system (x, y), the components of the displacement vector based on the framework of the Mindlin plate theory is defined as [20],

$$\begin{aligned} u_x(x, y, z) &= u + z\theta_y \\ u_y(x, y, z) &= v + z\theta_x \\ u_z(x, y, z) &= w \end{aligned} \tag{3}$$

where, the six kinematic variables, $u \equiv \{u, v, w, \theta_x, \theta_y\}^T$, are: u and v, average uniform displacements in the x and y directions, respectively; w, average transverse deflection; θ_x and θ_y, rotations of the normal about the negative x axis and positive y axis, respectively (see Fig. 1).

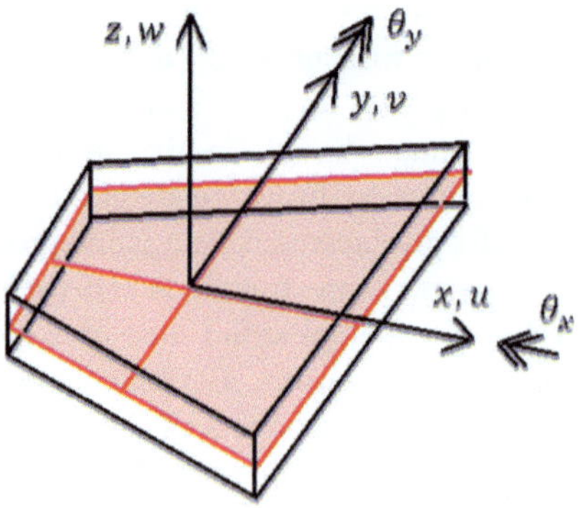

Fig. 1. Four-node quadrilateral inverse-shell element, iQS4, with the kinematic variables defined with respect to the local coordinates (x, y, z)

The strain field is identified by eight strain measures, $\{\varepsilon_k\}$ (K = 8), given by

$$\begin{aligned} e &= \{u_{,x}, v_{,y}, u_{,y} + v_{,x}\}^T = \{\varepsilon_1, \varepsilon_2, \varepsilon_3\}^T \quad \text{(Membrane Strains)} \\ k &= \{\theta_{y,x}, \theta_{x,y}, \theta_{x,x} + \theta_{y,y}\}^T = \{\varepsilon_4, \varepsilon_5, \varepsilon_6\}^T \quad \text{(Bending Curvatures)} \\ g &= \{w_{,x} + \theta_y, w_{,y} + \theta_x\}^T = \{\varepsilon_7, \varepsilon_8\}^T \quad \text{(Transverse Shear Strains)} \end{aligned} \tag{4}$$

Using the above formulation, a four-node inverse shell element, referred to as iQS4, can be developed [20]. The element is formulated with the two membrane displacements interpolated using the anisoparametric shape functions with an additional component to introduce the effect of the drilling rotation, θ_z, defined as the rotation about the positive z direction. The two out of plane rotations are also interpolated linearly; while the transverse displacement is interpolated with a quadratic polynomial. The corresponding shape functions are used for expressing the strain measures defined in Eq. 4 in terms of the nodal degrees-of-freedom and the error functional can be written using Eqs. 1 and 2.

The membrane and bending experimental strain measures, $e^{\varepsilon}_{(i)}$ and $k^{\varepsilon}_{(i)}$, can be evaluated at any mid-plane location of the plate, $(x_{(i)}, y_{(i)})$, using surface strain measurements from strain sensors embedded at the top (+) and bottom (−) surfaces of the plate,

$$\begin{aligned} e^{\varepsilon}_{(i)} &= \frac{1}{2}\left\{\left(\varepsilon^{+}_{xx} + \varepsilon^{-}_{xx}\right), \left(\varepsilon^{+}_{yy} + \varepsilon^{-}_{yy}\right), \left(\gamma^{+}_{xy} + \gamma^{-}_{xy}\right)\right\}^{T}_{(i)} \\ k^{\varepsilon}_{(i)} &= \frac{1}{2t}\left\{\left(\varepsilon^{+}_{xx} - \varepsilon^{-}_{xx}\right), \left(\varepsilon^{+}_{yy} - \varepsilon^{-}_{yy}\right), \left(\gamma^{+}_{xy} - \gamma^{-}_{xy}\right)\right\}^{T}_{(i)} \end{aligned} \tag{5}$$

As the value for the transverse shear strain measures, g, cannot be evaluated experimentally, the error functionals corresponding to the two shear strains are calculated using the equation,

$$\Phi^{e}_{k} \equiv \int_{A^e} \varepsilon^{2}_{k(i)}(u^{e})dA \quad (k = 7, 8) \tag{6}$$

where, A^e, is the area of the element. As Eq. 6, is an approximation of the value of the shear strain measures, low values are used for the weighing coefficients to enforce a weaker correlation. Also, when strain measurements from only a few strain sensors are available, some elements may be devoid of some or entirely of any strain measures. Even in such cases, any appropriately low value is used for the weighting coefficient of Eq. 1.

3 Damage Localisation Strategy

The proposed strategy for localising the position of damage in a structure using the iFEM is based on utilising two key features of the problem defined:

1. A damage on a structure will produce a local perturbation in the strain field in the vicinity of the damage and the perturbations subside the farther one moves away from the damage site.
2. The iFEM can be used for reconstructing the continuous displacement field of a structure using strain measurements from discrete points on the structure.

Using these two properties, a damage localisation strategy was developed which does not necessarily require strain data measured at extreme vicinities to the damage thus reducing the need for an unnaturally large number of sensors. As the technique developed is based on the iFEM, the strain field can be reconstructed without any knowledge of the load conditions and material properties of the structure.

3.1 Discretisation of the Structure

The damage localisation strategy developed is based on discretising the entire structure into square grid cells, with strain sensors positioned only at the boundaries of these grid cells.

Considering a plate structure defined on the R^2 coordinate space, a defect present at a location, (x_d, y_d), on the structure will produce a perturbation in the strain field, defined by, $\varepsilon^p \equiv \left\{\varepsilon^p_{xx}, \varepsilon^p_{yy}, \gamma^p_{xy}\right\}$. Considering any point on the structure, the strain tensor measured at that point,

$$\varepsilon(x,y) = \varepsilon_g(x,y) + \varepsilon^p(x,y,x_d,y_d) \tag{7}$$

where, ε_g represents the global strain field due to far field loading. The strain perturbations, ε^p, is a function of the damage location (x_d, y_d) and the measurement point (x, y) and is an inverse function of the distance between the points.

As the entire structure is discretised into square grids cells, considering any one such cell, the strain sensors present at the boundary will experience a greater magnitude of strain perturbation if there is damage present inside the cell, when compared to a case where there is no damage present within the cell.

Figure 2 presents a rough picture of the kind of strain perturbations expected at the boundaries of a cell, when a crack is located within. It shows the absolute value of strain perturbations in the normal strains at the boundary of a 1 m × 1 m grid cell, due to the presence of a 1 cm crack within the cell, when the structure is subjected to bi-axial loading.

The iFEM is used to reconstruct the global displacement field of the structure from the boundary strain measurements and the global strain field is subsequently calculated using the strain-displacement relations. The iFEM essentially acts as an interpolation

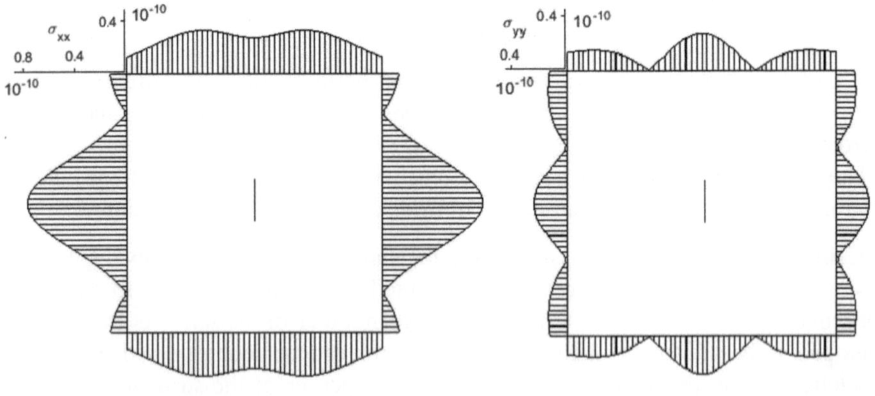

Fig. 2. The perturbations in the normal strains produced on a 1 m × 1 m enclosure due to the presence of a crack at the center

tool, based on the strain displacement relationships, that interpolates the boundary strain field over the internal surface of each grid cell. As the interpolation is based on the actual behavioural characteristics of the structure, the interpolated strains will bear more resemblance to the actual strain field generated by the damage.

The magnitude of the reconstructed strains within a damaged cell will be much higher than those in a healthy cell. This comparison of strains, between damaged and healthy grid cells is used to establish the presence and location of the damage. As the damage is localised to each grid cell and not to any single element, the size and number of grid cells on a structure will determine the precision with which the damage can be localised.

3.2 Condition for Damage Prediction

The iFEM reconstruction reproduces the 8 strain components at any point in the structure. It is desirable to condense this strain tensor information into a single variable. Thus an equivalent strain proportional to the second invariant of the deviatoric strain tensor is used [22]. Considering a plain strain measure, calculated using the membrane strains measured by a strain rosette, the equivalent strain at any location is defined as,

$$\varepsilon_{eq} = \frac{1}{\sqrt[2]{2}} \sqrt[2]{\left(\varepsilon_{xx} - \varepsilon_{yy}\right)^2 + \varepsilon_{xx}^2 + \varepsilon_{yy}^2 + 6\gamma_{xy}^2} \tag{8}$$

In an ideal scenario any damage, however small in magnitude can be detected by the strategy proposed as any damage will generate a corresponding strain perturbation however small. But in actual application, for any predefined sensor grid geometry, there exists a minimum magnitude of damage than can be successfully detected by the technique. This minimum depends on various factors such as measurement precision of the sensors and data acquisition systems, density and size of the sensor grid, magnitude of noise in the sensor data, etc.

For any particular sensor grid, if $\varepsilon_{eq,noise}$ represents the reconstructed baseline equivalent strain field of a healthy structure in the presence of practical measurement uncertainties and ε_{eq} represents the reconstructed equivalent strain at any point on the structure under similar conditions, then the presence of damage at a point (x_i, y_i) within a sensor grid cell can only be identified for those damage scenarios where,

$$\varepsilon_{eq}(x_i, y_i) > \varepsilon_{eq,noise} \tag{9}$$

The minimum magnitude of damage that can be identified within a cell is also dependent on the length of each grid cell and the sensor density on each side of the cell. It is however independent of the overall length of the entire plate. Hence the same sensor grid geometry on plates of different lengths is expected to produce the same measurement sensitivity, while a more dense or refined senor grid geometry is expected to produce an increase in measurement sensitivity.

4 Numerical Examples

The capabilities of the new technique are demonstrated by applying it to the problem of damage localisation in various plate specimens with an embedded crack and subjected to bi-axial loading. The experimental strain measurements required for the iFEM reconstruction is obtained from high fidelity FE analysis of the plate using ABAQUS 6.13. As the entire localisation methodology is based on sensing the strain perturbations caused by the presence of a crack, it is essential to capture an accurate representation of the strains due to a crack in the FE analysis. The embedded crack was modelled in ABAQUS using the seam feature which isolates the nodes present at the crack mouth. The stress concentration at the crack tip and the crack propagation is handled by calculating the J-integral around the crack tip and specifying the correct direction of the 'q'-vector. A finer mesh discretisation is applied at the vicinity of the crack, and the crack tip is modelled using an isoparametric element with a collapsed midside node.

4.1 Test Cases

The plate used for the analysis has a side length of 3.8 m × 3.8 m and a thickness of 0.0038 m. The material of the plate is considered to be an Aluminium alloy with Young's Modulus, E = 73000 MPa and Poisson's ratio, $\nu = 0.3$. It is subjected to a biaxial end loading of 10 N/m. The plate is discretised into 9 sensor grid cells, which indicates maximum of 9 possible damage locations if there is a crack present on the plate.

Central Crack Under Different Sensor Discretisation. In the first case, the plate is damaged by a crack, 25 cm in length, placed at the center of the plate as shown in Fig. 3(a). Two different sensor grids are used for crack localisation to test the effect of sensor density. The first configuration, shown in Fig. 4(a), has a high sensor density (440 strain rosettes) and correspondingly uses a fine iFEM mesh (400 elements), while the second configuration, shown in Fig. 4(b), has a relatively lower value of sensor density (152 strain rosettes) and uses a coarser iFEM mesh (49 elements).

Due to the uniform nature of the sensor grids used, in both cases the crack will be positioned at the center of the central sensor grid cell. This addresses a worst case scenario, regarding the location of a crack inside a sensor grid cell. As the crack is at the center, it is equidistant from all sides of the grid cell and will produce the smallest strain perturbation in the sensors positioned around it. In comparison, a crack positioned closer to one side of the cell than the other, would have produced a larger strain perturbation on the sensors adjacent to it, thus making it easier to identify the crack location.

Short Crack Under End Effects. The previous case presented an ideal situation where the crack is at the center of the plate and does not resemble most practical situations. The location of the crack might affect the magnitude of strain perturbations it produces. A crack located at the ends of the plate, will be susceptible to end effects which might diminish the effect of the crack. Hence a case is tested where the crack is positioned near the corner of the plate, but still positioned at the center of the sensor grid present at

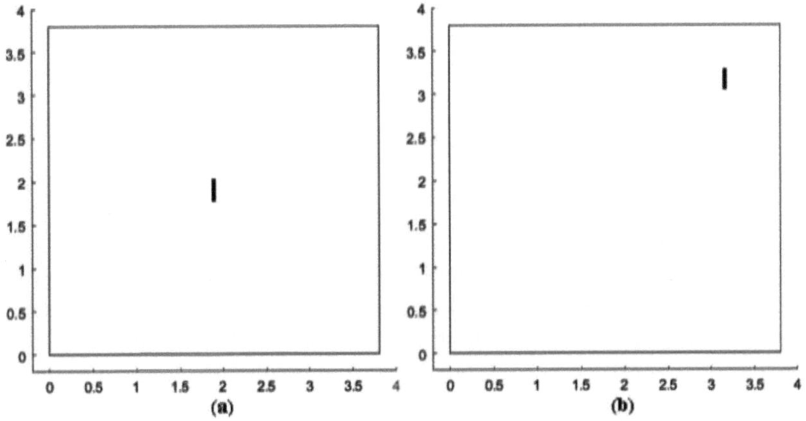

Fig. 3. Plate specimen with two crack configurations: (a) crack positioned at the center of the plate, (b) crack positioned near the edges of the plate

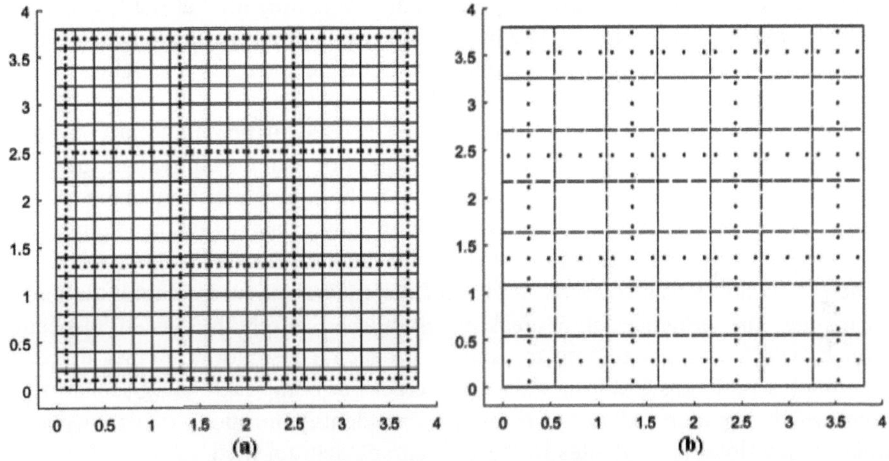

Fig. 4. The sensor and corresponding iFEM discretisation used, (a) dense mesh with 442 sensors and 400 inverse elements, (b) coarse mesh with 152 sensors and 49 inverse elements (black dots indicate the location of a strain rosette)

the corner, as shown in Fig. 3(b). This succeeds in producing the desired end-effects of the plate while also making sure that the crack is not unfairly close to any line of sensors, which might result in an easy detection. This same case is also used to demonstrate the sensitivity of the technique to cracks of smaller lengths. Hence the plate is damaged by a shorter crack, 5 cm in length and the high density sensor grid of Fig. 4(a) is used for the iFEM reconstruction.

Effect of Measurement Noise. As most experimental strain measurements are affected by noise, a damage detection technique using such data should be robust enough to produce reliable results without significant degradation in performance. The effect of measurement noise on the results of the technique is tested by corrupting the strain measurement data with an artificial value of measurement noise. The value of the noise is represented as a random percentage of the original sensor data,

$$\epsilon_{i,noise} = \left(1 + \frac{n}{100}\right)\epsilon_i. \tag{10}$$

where, n is a random number in the interval [−N, N]. The value of N indicates the maximum magnitude of noise in the sensor data.

The results of plate specimen with the 25 cm long crack, positioned at the center of the plate (Fig. 3(a)) and using the high density sensor discretisation of Fig. 4(a) is selected to test the sensitivity of the technique to measurement noise.

4.2 Results

For the iFEM analysis, the plate was discretised using the inverse shell element: 'iQS4' [20] defined in Sect. 2.1. As each quadrilateral iFEM element uses the 9-point Gaussian quadrature integration scheme, experimental strains measured at the Gauss-points of each element is used for the iFEM analysis. All Gauss-points in an element without a corresponding strain measurement associated with it is penalised by a weight of 10^{-4} in the iFEM formulation.

The iFEM reconstructed strain field is obtained by recalculating the strains only at the centroid of each element. This is based on an assumption that the equivalent strain field within an element is a constant and its value equal to the equivalent strain at the centroid. Although a more precise strain field could be obtained by recalculating the strains at each integration point within an element, that is left for implementation in future work.

The results are presented as basic plots of the equivalent strain magnitude in each element, and the presence of a crack is established purely by visual inspection. Although the cases presented here are suitable for such a basic form of crack detection, more complex problems featuring very small cracks or complex loading scenarios can be analysed using more advanced techniques to identify locations of strain concentrations, as possible damage sites in the reconstructed strain field.

Central Crack Under Different Sensor Discretisation. The iFEM strain reconstructions for a plate damaged by a central crack are shown in Fig. 5. The plot of the equivalent strain magnitude in each element (calculated using Eq. 8), clearly indicates the damage site as the location of the maximum strain concentration. For both sensor discretisations, that location is seen to be at the center of the plate which coincides with the actual location of the crack. The results show that even with sensor density of one-third the original one, the technique used is successful in identifying the presence of the crack. It also indicates that there is further scope for reducing the number of sensors used by using a suitable optimization scheme.

It is observed that there is a distinct difference in iFEM results due to the two sensor configurations used. Although both configurations were successful in identifying the presence of a crack, Fig. 5(a), with the higher mesh density gives a more detailed

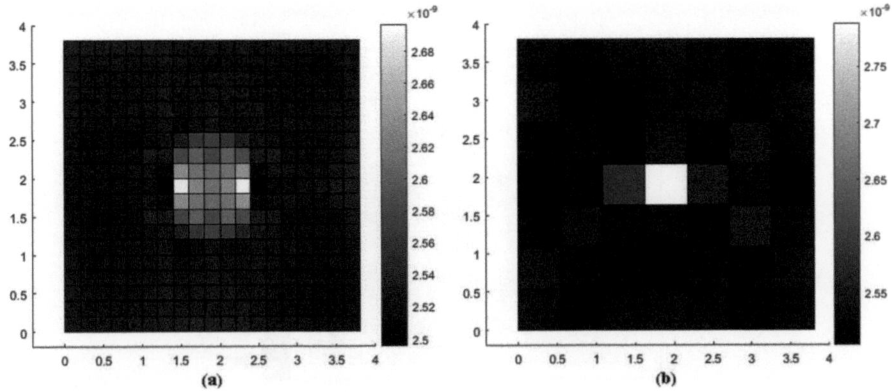

Fig. 5. iFEM reconstructed equivalent strain field for the plate specimen with the central crack, (a) using high density sensor grid, (b) using low density sensor grid

picture of the strain variations in the vicinity of the crack, while that information in lost in Fig. 5(b), due to the lower density. Although the iFEM results indicate the presence of a crack inside the central grid cell, its exact location within the cell is still unknown. A possible location of the crack inside the cell can be established by looking at the symmetry of the strain variation inside the central grid cell. The plot of Fig. 5(a) shows symmetry about both x and y-axes, providing a strong case for the presence of the crack at the center of the grid cell. It can be similarly argued that a crack present within the same cell but at a non-central location will produce asymmetries in the strain field. Hence a more detailed representation of the strain field inside the cell might be useful for future work involving characterising the features of the crack such as crack length and exact location.

Short Crack Under End Effects. The iFEM equivalent strain reconstructions for the short crack positioned near the ends of the plate is shown in Fig. 6. This scenario presents a good test case for the choice of a methodology which relies on analysing the strain field based on a relative rather than absolute measure of strain. As the length of the crack is quite short, i.e. 5 cm, the strain perturbation it produces is correspondingly quite small. That coupled with the end effects creates an added disturbance in the strain field. Because the values of strain perturbations are quite small, the reconstructed strain field shows local maximums, particularly in those elements that lie along the strain sensor grid lines. But as the damage is localised into a grid cell and not a single element, in Fig. 6, it is easy to identify the location of the crack by comparing the strain fields in each grid cell with the others.

It can be seen that the grid cell on the top-right hand corner has a higher strain distribution within the cell compared to the other cells. Hence this cell assumes a higher damage merit compared to the other cells, in this particular scenario the presence of the damage is attributed to that cell. Even if multiple damages or cracks are present inside the plate, the iFEM reconstructed strain field will produce strain concentrations in multiple cells, but as long as at least one cell remains healthy, the cells can be categorised as containing damages of different magnitudes based on a relative measure of

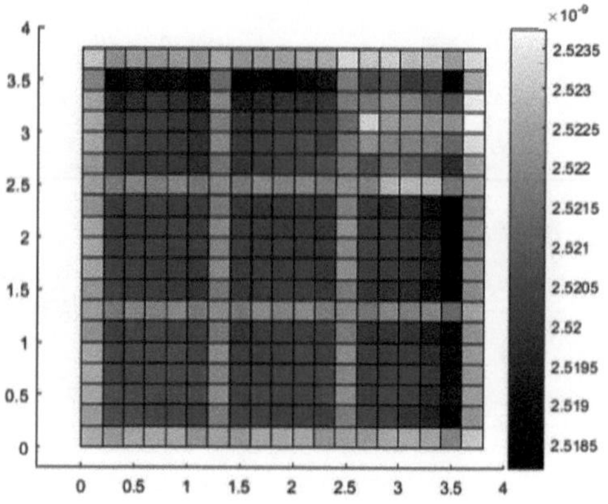

Fig. 6. iFEM reconstructed equivalent strain field for plate with a 5 cm crack located near the edge

their strain magnitude. The present scenario clearly shows that the technique is capable of detecting cracks as small as 1/20th the length of a grid cell, even under the effect of end conditions.

Effect of Measurement Noise. Two different noise magnitudes, with N = 2.5, 5, were selected for corrupting the iFEM input strain data. The iFEM reconstructed equivalent strain fields for both cases are shown in Fig. 7. Even though there is a general loss of uniformity in the strain field compared to the ideal case of Fig. 5(a), both strain fields clearly indicate a strain concentration at the center of the plate, consistent with the actual location of the crack. The figures show a number of elements with a local strain maximum or minimum throughout the strain field, but as explained previously, when analysing the strain magnitudes within each grid cell, it is easy to establish that the crack is present at the center of the plate.

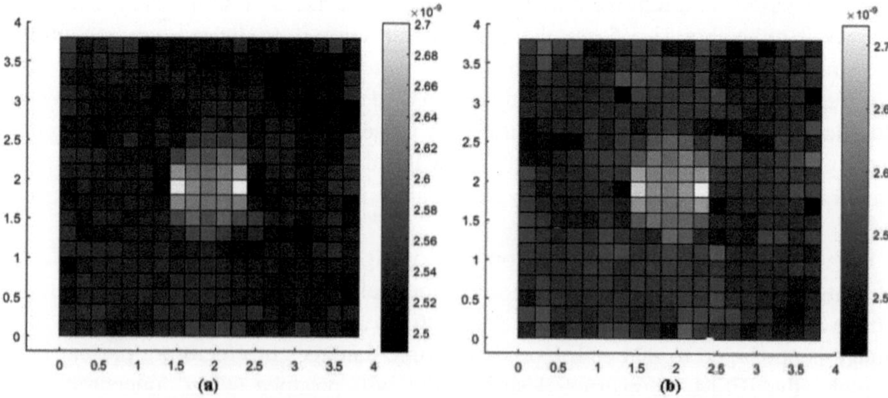

Fig. 7. iFEM reconstructed equivalent strain field for a plate with a crack at the center, (a) with a 2.5% noise level, (b) with a 5% noise level

But the plots do indicate a loss of symmetry in the strain field as results of the corruption due to noise. Hence identifying the exact location of the crack would require more manipulation of either the strain field or another method of reducing the effect of noise on the results of the iFEM reconstruction. But nevertheless, these results provide an initial validation of the robustness of the technique and its ability to produce reliable results in the presence of uncertainties.

5 Conclusions

A new methodology for damage localisation based on the inverse Finite Element Method is presented. It is based on discretising the structural domain into a square grid, with strain sensors positioned only at the boundaries of each grid cell. The iFEM is used for reconstructing the strain field based on the boundary strain measurements. The location of the damage is identified as the grid cell with the highest strain concentration based on a relative measure of the reconstructed strains. As the new methodology is based on the iFEM, it is independent of the loading conditions and the material properties of the structure.

The performance of the technique is tested for various cases of a thin plate with an embedded crack. The method was successfully used to localise a crack independent of its location on the plate. Different sensor discretisations, using a high and low density of sensors, were used to demonstrate the applicability of the technique even with a smaller number of strain measurements. Finally, the effect of measurement noise on the reconstructed strain field was tested by introducing random levels of noise data into the strain measurements. The new method was seen to produce reliable results even using such strain data, validating the robustness of the methodology proposed.

The work presented is confined to damage localisation up to the precision of a grid cell. It is meant to serve as the initial step in establishing the validity of the technique. But the methodology can be further developed to identify the exact position of the crack within the grid cell and understanding the severity of the damage by analysing the reconstructed strain field within each grid cell. Also, further optimization of the sensor discretisation within the structure can reduce the overall number sensor used for the problem. Fiber optics offer a good alternative to strain rosettes to establish the boundaries of the grid cells and future work focuses on incorporating both kinds of sensors for deriving optimal performance.

References

1. Cawley, P., Adams, R.D.: The location of defects in structures from measurements of natural frequencies. J. Strain Anal. Eng. Des. **14**(2), 49–57 (1979)
2. Surace, C., Archibald, R., Saxena, R.: On the use of the polynomial annihilation edge detection for locating cracks in beam-like structures. Comput. Struct. **114–115**, 72–83 (2013)
3. Corrado, N., Durrande, N., Gherlone, M., Hensman, J., Mattone, M., Surace, C.: Single and multiple crack localization in beam-like structures using a Gaussian process regression approach. J. Vib. Control **24**(18), 4160–4175 (2018)

4. Civera, M., Surace, C., Worden, K.: Detection of cracks in beams using treed Gaussian processes. In: Niezrecki, C. (ed.) Structural Health Monitoring & Damage Detection. Conference Proceedings of the Society for Experimental Mechanics Series, vol. 7. Springer, Cham (2017)
5. Martucci, D., Civera, M., Surace, C., Worden, K.: Novelty detection in a cantilever beam using extreme function theory. J. Phys. Conf. Ser. **1106**(1), 12–27 (2018)
6. Gherlone, M., Mattone, M., Surace, C., Tassotti, A., Tessler, A.: Novel vibration-based methods for detecting delamination damage in composite plate and shell laminates. Key Eng. Mater. **293–294**, 289–296 (2005)
7. Surace, C., Saxena, R., Gherlone, M., Darwich, H.: Damage localisation in plate like-structures using the two-dimensional polynomial annihilation edge detection. J. Sound Vib. **333**(21), 5412–5426 (2014)
8. Corrado, N., Gherlone, M., Surace, C., Hensman, J., Durrande, N.: Damage localisation in delaminated composite plates using a Gaussian process approach. Meccanica **50**, 2537 (2015)
9. Wahab, M.M.A., Roeck, G.D., Peeters, B.: Parameterization of damage in reinforce concrete structures using model updating. J. Sound Vib. **228**(4), 717–730 (1999)
10. Farrar, C.R., Worden, K.: Structural Health Monitoring: A Machine Learning Perspective. Wiley, West Sussex, U.K. (2013)
11. Cross, E.J., Worden, K., Chen, Q.: Cointegration: a novel approach for the removal of environmental trends in structural health monitoring data. Proc. R. Soc. A: Math. Phys. Eng. Sci. **467**(2133), 2712–2732 (2011)
12. Coletta, G., Miraglia, G., Pecorelli, M., Ceravolo, R., Cross, E., Surace, C., Worden, K.: Use of the cointegration strategies to remove environmental effects from data acquired on historical buildings. Eng. Struct. **183**, 1014–1026 (2019)
13. Bezerra, L.M., Saigal, S.: A boundary element formulation for the inverse elastostatics problem (IESP) of flaw detection. Int. J Numer. Meth. Eng. **36**, 2189–2202 (1993)
14. Chen, D.H., Nisitani, H.: Detection of a crack by body force method. Eng. Fract. Mech. **45** (5), 671–685 (1993)
15. Fares, N., Maloof, R.: Crack detection characterization of strain sensing grids. Int. J. Solids Struct. **35**(22), 2861–2875 (1998)
16. Wildy, S.J., Kotousov, A.G., Codrington, J.D.: A new passive defect detection technique based on the principle of strain compatibility. Smart Mater. Struct. **17**, 8 pp., 045004 (2008)
17. Wildy, S., Codrington, J.: An algorithm for identifying a crack within a measured displacement field. J. Nondestruct. Eval. **36**, 37 (2017)
18. Tessler, A., Spangler, J.L.: A variational principle for reconstruction of elastic deformations in shear deformable plates and shells. NASA/TM-2003-212445 (2003)
19. Tessler, A., Spangler, J.L.: Inverse FEM for full-field reconstruction of elastic deformations in shear deformable plates and shells. In: 2nd European Workshop on SHM, 7–9 July 2004
20. Kefal, A., Oterkus, E., Tessler, A., Spangler, J.L.: A quadrilateral inverse-shell element with drilling degrees of freedom for shape sensing and structural health monitoring. Eng. Sci. Technol. Int. J. **19**(3), 1299–1313 (2016)
21. Quach, C., Vazquez, S., Tessler, A., Moore, J., Cooper, E., Spangler, J.: Structural anomaly detection using fiber optic sensors and inverse finite element method. In: AIAA Guidance, Navigation, and Control Conference and Exhibit, p. 6357 (2005)
22. Colombo, L., Sbarufatti, C., Giglio, M.: Definition of a load adaptive baseline by inverse finite element method for structural damage identification. J. Mech. Syst. Signal Process. **120**, 584–607 (2019)

Damage Localization and Quantification in Structures Using Residual Force Indicator

M. Slimani[1(✉)], S. Tiachacht[1], S. Khatir[2], A. Behtani[1], L. Mansouri[1], A. Bouazzouni[1], and M. Abdel Wahab[2]

[1] Laboratory of Mechanics, Structure and Energetics (LMSE), Mouloud Mammeri University of Tizi-Ouzou, B.P.N°17 RP, 15000 Tizi Ouzou, Algeria
slimanimohand2006@yahoo.fr, mohand.slimani@ummto.dz
[2] Department of Electrical Energy, Metals, Mechanical Constructions and Systems, Faculty of Engineering and Architecture, Ghent University, Ghent, Belgium

Abstract. This paper presents damage identification and assessment methodology based on the changes in dynamic parameters of structures. According to continuum damage mechanics, the damage is represented by a reduction in stiffness. A Residual Force Method (RFM) and modal strain energy change ratio (*msecr*) were studied in this paper to predict the damage location. In order to demonstrate the accuracy of this method, three examples are included using free-free beam, truss structure and 3-dimensional frame structure. Single and multiple damages are considered for all structures to investigate the accuracy of these indicators. The obtained results show that RFM is capable of detecting the location and severity of diagnostic even in large-scale structures with a large number of elements.

Keywords: RFM · *Msecr* · Damage · Vibration · Plane truss · Space truss · 3D frame structure

1 Introduction

In recent years, structural damage detection has gained increasing attention from the scientific and engineering communities. A large number of proposed methods for damage identification based on static and vibration characteristics using inverse problem by Khatir et al. [1–5]. A comparison between different techniques was presented in Ref. [6], which consisted of four levels. The first level is based on damage detection, the second level is based on damage localization, the third level is based on damage assessment, and finally, the fourth level, which is the consequence of damage, predicts the remaining life of the structure. Yang and Liu [7] presented a method based on the best eigenvector concept to solve the incomplete measurement problem. Gillich et al. [8] introduced a method to identify damages in beam-like structures by analyzing the natural frequency changes of the first six transversal vibration modes. Recently, many authors used natural frequencies measurement for damage identification based on reduction in stiffness as presented in Refs [1–3, 9–11]. Pandey et al. [12] presented a technique for damage detection and localization based on the change in the structure

M. A. Wahab (Ed.): DAMAS 2019, LNME, pp. 213–224, 2020.
https://doi.org/10.1007/978-981-13-8331-1_15

flexibility matrix. Hwang and Kim [13] presented a new indicator based on frequency response for damage identification in the beam-like structure. Zenzen et al. [14] presented an approach for damage identification using frequency response for beams and truss structures. Moreover, the quantification was also investigated using inverse problem. Frequency Response Function (FRF) with optimization technique was presented also in Ref. [15]. Many researchers have proposed several applications in literature for detecting and locating the presence of damage in various types of structures. The investigation of mode shape and mode-shape-slope parameters are used by Yuen [16]. Yang [17] proposed a numerical method for structural damage detection using modal residual force criteria and matrix disassembly technique.

This work presents a RFM and *msecr* for damage detection and localization of different structures using FEM based on Matlab programming.

2 The Residual Force Method (RFM)

Structural damage identification methods based on the residual force vector are studied in Refs. [7, 18–22]. In this paper, the structural damage identification method based on the residual force vector is studied. The damage index of the jth element is here expressed as the change of the rigidity of a finite element:

$$\Delta[K]_j^e = \left([K]_j^e - [K]_{dj}^e\right) = \alpha_j [K]_j^e,\ \alpha \in [0, 1] \tag{1}$$

where $[K]_j^e$ and $[K]_{dj}^e$ are the jth element of the elementary matrix of the damaged and undamaged structure, respectively. $\Delta[K]_j^e$ represents the variation of stiffness. α indicates a loss of rigidity of jth element, i.e. for undamaged element $\alpha = 0$ and for the damaged element $\alpha = 1$. The modal residual force vector can be written as:

$$\{R\}_i = [\Delta K]\{\phi\}_{di} = [F]_i\{\alpha\} \tag{2}$$

where matrix $[F]$ is

$$[F]_{ij} = [K]_j^e[\phi]_{dij}^e \tag{3}$$

The modal residual force vector can be written as

$$\{R\}_i = ([K] - \lambda_{di}[M])\{\phi\}_{di} \tag{4}$$

λ_{di} and $\{\phi\}_{di}$ are the ith eigenvalue and eigenvector of the damaged structure, respectively.

The damage extents can be easily obtained by solving:

$$\{\alpha\}_j = [F]_{ij}^+\{R\}_i \tag{5}$$

3 Modal Strain Energy Change Ratio Indicator

From the ith mode shape of the damaged and healthy structure, the modal strain energy change ratio (*msecr*) for both structures at the *j*th element are calculated by:

$$mse_{ij}^{h} = \left(\Phi_i^h\right)^T K_j \Phi_i^h;\ mse_{ij}^{d} = \left(\Phi_i^d\right)^T K_j \Phi_i^d \tag{6}$$

where Φ_i is the ith mode shape vector and superscripts h and d denote the healthy and damaged state, respectively. K_j is the stiffness matrix of the jth element of the healthy structure. The superscript T denotes the vector transpose. Equation (6) shows that when damage occurs in elements of a system, the *mse* will change slightly in the undamaged elements, but there will be a larger change in the damaged elements. As a result, the modal strain energy change ratio (*msecr*) was proposed for damage localization. The proposed indicator is expressed by the total energy in the structure, which can be calculated by adding the *msecr*'s of all elements, then we can write:

$$msecr_j = \frac{1}{m}\sum\nolimits_{j=1}^{m} \frac{msecr_{ij}}{msecr_{ij}^{max}} \tag{7}$$

where

$$msecr_{ij} = \frac{\left|mse_{ij}^{d} - mse_{ij}^{h}\right|}{mse_{ij}^{h}};\ msecr_{ij}^{max} = \max_k\{msecr_{ik}\}$$

4 Numerical Examples

Three different examples are studied in this investigation to examine the effectiveness and the accuracy of the method based on both indicators in free vibration analysis of mechanical structures. For more accuracy, the number of modes is changed to enhance the results for *msecr* indicator as presented in following examples.

4.1 Simply Supported Beam

In the first section, we applied RFM and *msecr* for beam discretized in 20 elements as presented in Fig. 1. An experimental case has been considered, consisting of a steel beam with rectangular cross-section, of dimensions 1480 × 50 × 5 mm^3 ($L \times b \times h$). The beam was suspended by two inextensible cables, simulating "free–free" conditions. Young's modulus is 200 GPa, the density is 7850 kg/m^3 and the cross-section area is $A = b \times h$. The frequencies are presented in Table 1.

Three damage scenarios are introduced based on single and multiple damages as presented in Table 2.

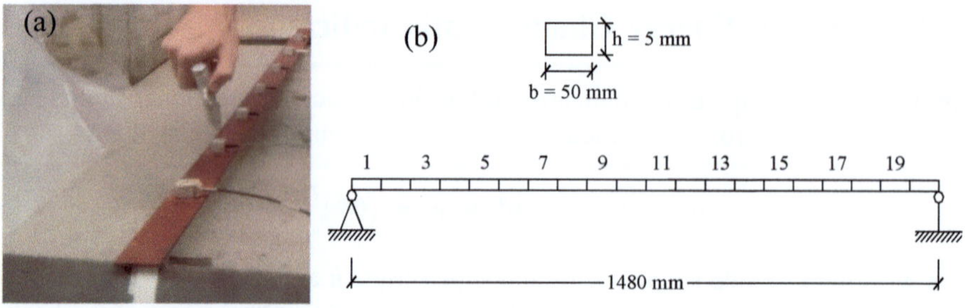

Fig. 1. (a) Excitation of the experimental beam using air pistol and (b) FE model of simply supported beam [23].

Table 1. Natural frequencies of the FE model and experimental test.

Natural frequency	Experiment [23] (Hz)	FEM (Hz)	Error	Damage (Hz)					
				Case 1	Error	Case 2	Error	Case 3	Error
1	5.10	5.08	0.23%	5.06	0.60%	5.02	1.39%	5.01	1.69%
2	20.01	20.35	−1.71%	20.18	−0.85%	20.17	−0.80%	20.02	−0.07%

Table 2. Percent of stiffness reduction of beam elements.

Case number	Element number and damage level [%]					
1	Element no.	5	–	–	–	–
	Damage	15%	–	–	–	–
2	Element no.	4	10	17	–	–
	Damage	10%	15%	10%	–	–
3	Element no.	3	7	8	11	18
	Damage	20%	10%	5%	10%	15%

The results presented in Figs. 2, 3 and 4 show that damage detection with RFM is more accurate, even in the case of several damaged elements compared with *msecr* with different modes.

4.2 25 Bar Plane Truss

In the second section, a truss structure is modeled using 2D truss elements. The finite element model consists of 25 elements and 21 DOFs. Truss elements are made from steel material with Young's modulus $E = 200$ GPa, Poisson's ratio $\nu = 0.3$, density $\rho = 7.8 \times 10^3$ kg/m^3 and cross-sectional areas are given in Table 3. The numerical model of truss structure is shown in Fig. 5.

The first ten natural frequencies of the analytical and numerical results are listed in Table 4.

Table 3. Cross-sectional area of truss members.

Number	Area (cm^2)
1–6	18
7–12	15
13–17	10
18–25	12

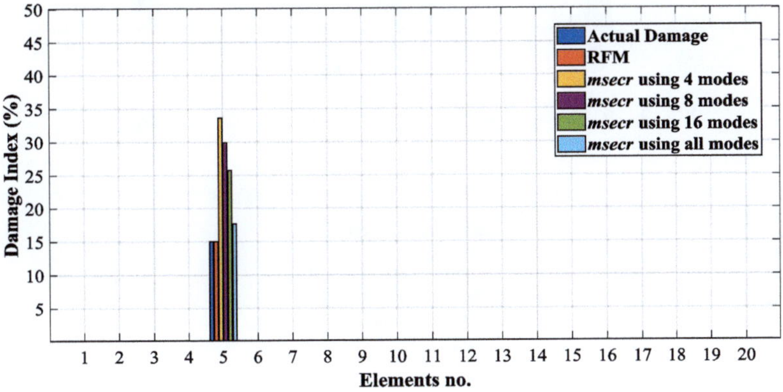

Fig. 2. A free–free simply supported beam - damage case 1.

Table 4. Natural frequencies of the analytical and numerical results.

Mode no.	FEM [24]	FEM	Error %	Damage case					
				Case 4	Error %	Case 5	Error %	Case 6	Error %
1	30.3	30.3	0.16%	30.2	0.17%	29.8	1.66%	29.6	2.42%
2	69	68.7	0.39%	68.7	0.46%	66.5	3.60%	67.5	2.19%
3	96.3	96.0	0.27%	96.0	0.30%	96.0	0.36%	95.7	0.61%
4	181.8	181.2	0.34%	181.2	0.36%	177.7	2.25%	177.9	2.17%
5	223.2	222.5	0.30%	221.6	0.73%	221.7	0.69%	215.4	3.47%
6	275.6	274.7	0.32%	274.6	0.35%	271.6	1.44%	269.8	2.09%
7	321.6	320.5	0.33%	314.0	2.35%	316.9	1.45%	320.1	0.47%
8	352	350.9	0.31%	350.8	0.35%	342.0	2.83%	346.3	1.61%
9	357.7	356.5	0.33%	356.1	0.46%	349.3	2.35%	351.3	1.80%
10	373	371.8	0.33%	371.7	0.36%	368.5	1.21%	367.9	1.37%

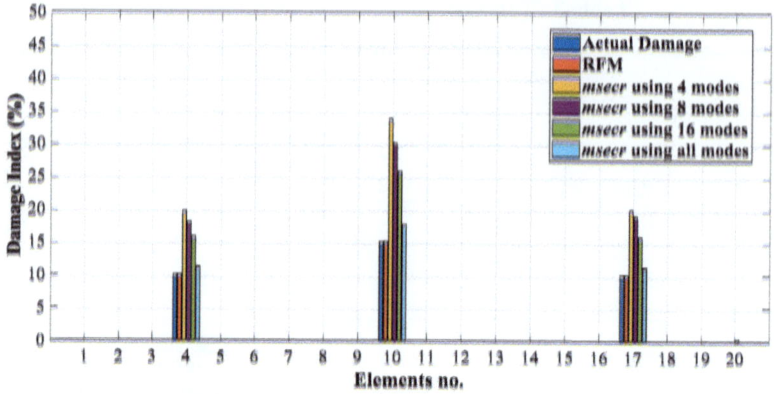

Fig. 3. A free–free simply supported beam - damage case 2.

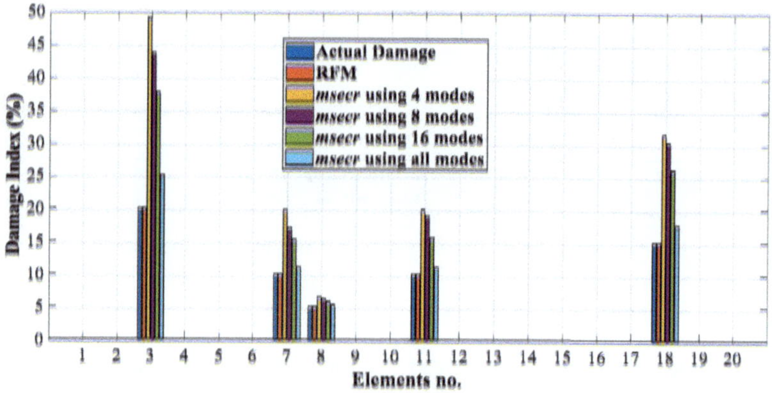

Fig. 4. A free–free simply supported beam - damage case 3.

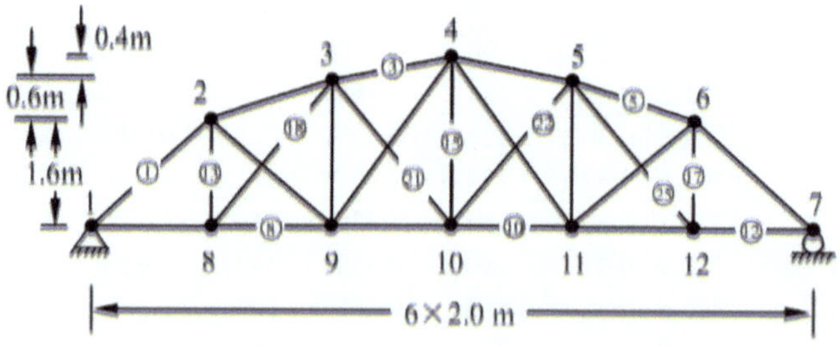

Fig. 5. The FE model of planner truss

Several damage scenarios are considered as presented in Table 5 to investigate the influence of location, severity, and number of the damaged elements on the robustness of the proposed method. The results are shown in Figs. 6, 7 and 8.

Table 5. Percent of stiffness reduction in truss elements.

Case number	Element number and percent of damage					
4	Element no.	15	–	–	–	–
	Damage	10%	–	–	–	–
5	Element no.	7	14	21	–	–
	Damage	20%	15%	15%	–	–
6	Element no.	4	9	13	20	25
	Damage	20%	20%	15%	10%	10%

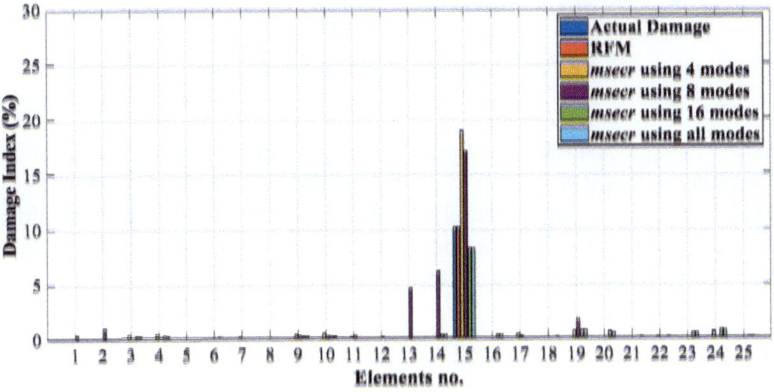

Fig. 6. A planner truss - damage case 4.

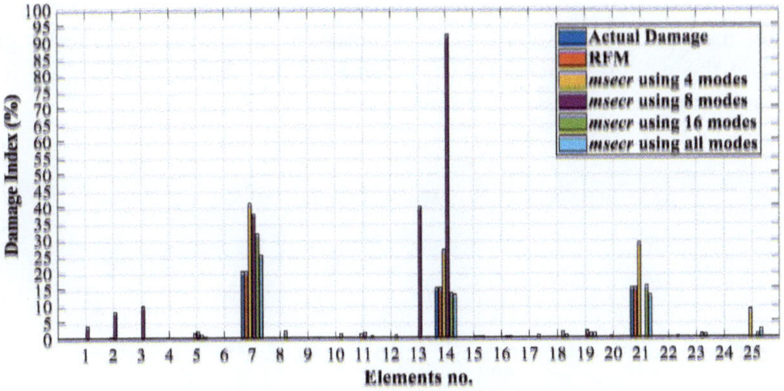

Fig. 7. A planner truss - damage case 5.

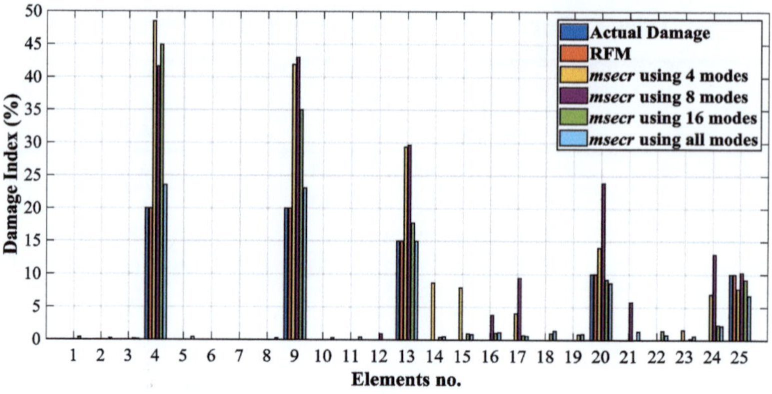

Fig. 8. A planner truss - damage case 6.

Predicted damage severities for the simulated patterns are summarized in Figs. 6, 7 and 8. It is obvious that RFM can find damage severities with high level of accuracy in case of complex structure with several damages compared with *msecr*. In addition, for the *msecr* when we increase the number of modes the results becoming better, but the CPU time is higher compared with RFM.

4.3 3D Frame Structure

A numerical mode of 3D frame structure [25], shown in Fig. 9, is used to verify the proposed indicator. The frame model is divided into ten beam elements with 6 DOFs for each node. The properties of the beam element are Young's modulus: $E = 2.1 \times 10^{11}$ N/m^2, cross-section area $S = 0.5 \times 10^{-3}$ m^2, density $\rho = 7800$ kg/m^3 and moment of inertia $I = 0.417 \times 10^{-8}$ m^4. A summary of damage scenarios is given in Table 6 and the natural frequencies of undamaged and damaged beam are listed in Table 7 and presented in Figs. 10, 11 and 12.

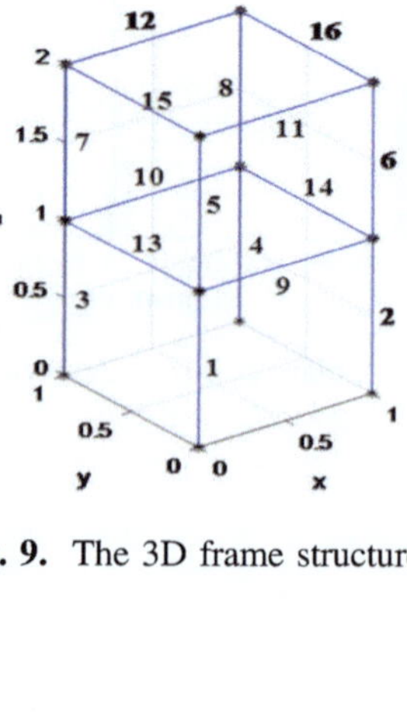

Fig. 9. The 3D frame structure.

Table 6. The 3D structure damage case investigated.

Case number	Element number and percent of damage						
7	Element no.	1	–	–	–	–	–
	Damage	10%	–	–	–	–	–
8	Element no.	9	10	–	–	–	–
	Damage	15%	15%	–	–	–	–
9	Element no.	2	3	11	12	15	16
	Damage	15%	15%	10%	10%	15%	10%

Table 7. The natural frequencies of the damaged and undamaged 3D frame structure.

$f[Hz]$	FEM	Damage					
		Case 7	Error	Case 8	Error	Case 9	Error
f_1	107.723	106.995	0.68%	105.597	1.97%	104.421	3.07%
f_2	142.545	140.875	1.17%	142.545	0.00%	137.687	3.41%
f_3	159.732	158.339	0.87%	159.438	0.18%	155.094	2.90%
f_4	435.371	434.796	0.13%	425.603	2.24%	421.585	3.17%

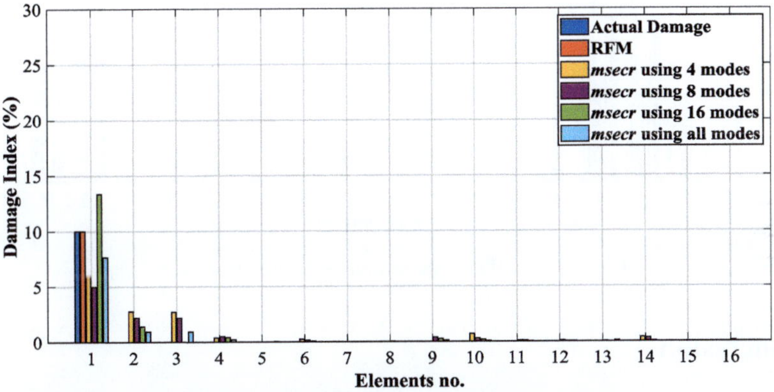

Fig. 10. A 3D frame structure - damage case 7.

In this third section, 3D frame structure was analyzed to predict the right position and severity of damaged using RFM. It can be concluded that the FRM can predict the damage accurately even in the case of a 3D frame structure.

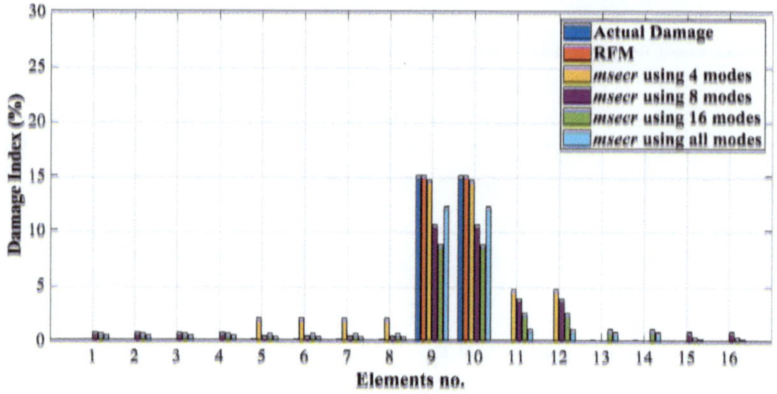

Fig. 11. A 3D frame structure - damage case 8.

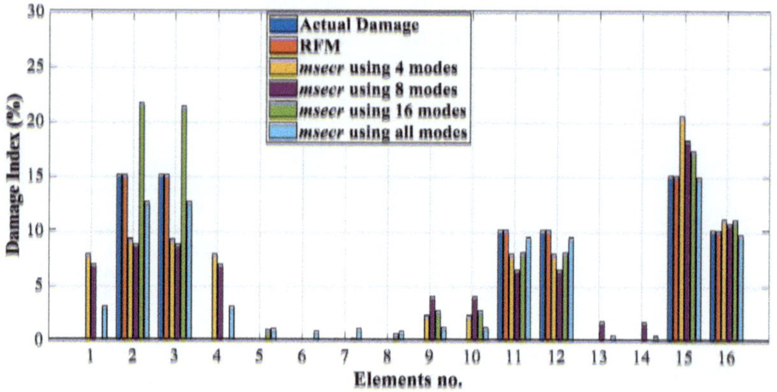

Fig. 12. A 3D frame structure - damage case 9.

5 Conclusion

In this paper, an efficient damage localization indicator, named as RFM, has been investigated using Finite Element Method (FEM) and Matlab program. Moreover, *msecr* is also used and compared with RFM. In order to verify the performance of the proposed methodology, different numerical problems with different scenarios are tested. To enhance the results of *msecr*, the variation of modes number was studied. From the results, it can be concluded that the RFM is quite efficient and robust for damage detection problems in a variety of structures and computational cost is much lower than *msecr*.

References

1. Khatir, S., et al.: Damage detection and localization in composite beam structures based on vibration analysis. Mechanics **21**(6), 472–479 (2015)
2. Khatir, S., et al.: Multiple damage detection in composite beams using particle swarm optimization and genetic algorithm. Mechanika **23**(4), 514–521 (2017)
3. Khatir, S., et al.: Crack identification method in beam-like structures using changes in experimentally measured frequencies and particle swarm optimization. Comptes Rendus Mécanique **346**(2), 110–120 (2018)
4. Benaissa, B., et al. Application of proper orthogonal decomposition and radial basis functions for crack size estimation using particle swarm optimization. J. Phys. Conf. Ser. IOP Publishing (2017)
5. Khatir, S., Wahab, M.A.: Fast simulations for solving fracture mechanics inverse problems using POD-RBF XIGA and Jaya algorithm. Eng. Fract. Mech. **205**, 285–300 (2019)
6. Rytter, A.: (1993) Vibrational based inspection of civil engineering structures. Department of Building Technology and Structural Engineering, Aalborg University
7. Yang, Q., Liu, J.: Structural damage identification based on residual force vector. J. Sound Vib. **305**(1–2), 298–307 (2007)
8. Gillich, G.-R., et al.: Early structural damage assessment by using an improved frequency evaluation algorithm. Lat. Am. J. Solids Struct. **12**(12), 2311–2329 (2015)
9. Sundermeyer, J.N., Weaver, R.: On crack identification and characterization in a beam by non-linear vibration analysis. J. Sound Vib. **183**(5), 857–871 (1995)
10. Khatir, S., et al. Genetic algorithm based objective functions comparative study for damage detection and localization in beam structures. J. Phys. Conf. Ser. IOP Publishing (2015)
11. Samir, K., et al.: Damage detection in CFRP composite beams based on vibration analysis using proper orthogonal decomposition method with radial basis functions and cuckoo search algorithm. Compos. Struct. **187**, 344–353 (2018)
12. Pandey, A., Biswas, M.: Damage detection in structures using changes in flexibility. J. Sound Vib. **169**(1), 3–17 (1994)
13. Hwang, H., Kim, C.: Damage detection in structures using a few frequency response measurements. J. Sound Vib. **270**(1–2), 1–14 (2004)
14. Zenzen, R., et al.: A damage identification technique for beam-like and truss structures based on FRF and bat algorithm. Comptes Rendus Mécanique **346**(12), 1253–1266 (2018)
15. Mohan, S., Maiti, D.K., Maity, D.: Structural damage assessment using FRF employing particle swarm optimization. Appl. Math. Comput. **219**(20), 10387–10400 (2013)
16. Yuen, M.M.F.: A numerical study of the eigenparameters of a damaged cantilever. J. Sound Vib. **103**(3), 301–310 (1985)
17. Yang, Q.: A numerical technique for structural damage detection. Appl. Math. Comput. **215** (7), 2775–2780 (2009)
18. Eun, H.C., Kim, R.J., Ahn, Y.J.: Identification of parameter matrices using residual force vector. In: Applied Mechanics and Materials. Trans Tech Publications (2013)
19. Eraky, A., et al.: Damage detection of plate-like structures based on residual force vector. HBRC J. **12**(3), 255–262 (2016)
20. Yun, G.J., et al.: A parameter subset selection method using residual force vector for detecting multiple damage locations. Struct. Control. Health Monit. Off. J. Int. Assoc. Struct. Control. Monit. Eur. Assoc. Control. Struct. **17**(1), 48–67 (2010)
21. Li, H., Lu, Z., Liu, J.: Structural damage identification based on residual force vector and response sensitivity analysis. J. Vib. Control. **22**(11), 2759–2770 (2016)

22. Zou, W.J., Zhang, Y.X., Li, C.G.: Study on truss structure damage identification base on residual force vector. In: Applied Mechanics and Materials. Trans Tech Publications (2014)
23. Shahri, A.H., Ghorbani-Tanha, A.: Damage detection via closed-form sensitivity matrix of modal kinetic energy change ratio. J. Sound Vib. **401**, 268–281 (2017)
24. Esfandiari, A., et al.: Structural model updating using frequency response function and quasi-linear sensitivity equation. J. Sound Vib. **326**(3–5), 557–573 (2009)
25. Tiachacht, S., et al.: Damage assessment in structures using combination of a modified Cornwell indicator and genetic algorithm. Eng. Struct. **177**, 421–430 (2018)

Damage Detection in Truss Structures Using Transmissibility Combined with Optimization Techniques

Roumaissa Zenzen[1(✉)], Samir Khatir[2], Idir Belaidi[1], and M. Abdel Wahab[2]

[1] Department of Mechanical Engineering, University M'hamed Bougara Boumerdes, LEMI Laboratory, 35000 Boumerdes, Algeria
Zenzen-roumaissa@hotmail.com, R.zenzen@univ-boumerdes.dz

[2] Department of Electrical Energy, Metals, Mechanical Constructions, and Systems, Faculty of Engineering and Architecture, Ghent University, 9000 Ghent, Belgium
Khatir_samir@hotmail.fr, samir.khatir@ugent.be

Abstract. The paper presents an effective approach based on Modal Assurance Criterion (MAC) formulation, transmissibility function and Particle Swarm Optimization (PSO) for damage assessment in truss structures. The Finite Element Method (FEM) is used to build the structures using Matlab. The main purpose of this study is to apply the transmissibility technique as an objective function based on MAC formulation to predict the damage location and severity. The objective function used in the proposed approach is based on transmissibly using MAC formulation (TMAC). The results show that the present methodology can reliably identify damage scenarios with higher accuracy even in case of complex structures.

Keywords: Damage identification · Transmissibly · Modal Assurance Criterion (MAC) · Particle Swarm Optimization (PSO)

1 Introduction

Recently, structural health monitoring (SHM) and vibration-based techniques have been developed and widely applied to predict the damage location in mechanical and civil engineering. Many definitions of transmissibility have been proposed. The transmissibility defined as a transformation matrix between two sets of displacements as presented in Refs [1–3]. Furthermore, the transmissibility was applied in force identification and discussed in Ref [4]. The application of transmissibility in engineering was reviewed in Ref [5]. In the last years, different methods of transmissibility have been developed. Zhou et al. [6] proposed a new transmissibility approach based mahalanobis distance with a structural forced dynamic response. Another transmissibility analysis method based on the transmissibility of nonlinear output frequency response functions (NOFRFs) for the detection and location of damage via nonlinear features in multi-degree-of-freedom (MDOF) structural systems was proposed by Zhao et al. [7]. Fan et al. [8] developed a new method of transmissibility based on wavelet

M. A. Wahab (Ed.): DAMAS 2019, LNME, pp. 225–233, 2020.
https://doi.org/10.1007/978-981-13-8331-1_16

transform for damage detection. Another approach of damage detection based on transmissibility in the frequency domain is presented by Li et al. [9] to identify the damage of shear connectors in slab-on-girder bridge structures with or without reference data from the undamaged structure. To increase the precision of damage detection many heuristics methods had been developed and combined with the damage detection methods based on vibration analysis as cuckoo search, Genetic Algorithm (GA), BAT and PSO algorithms were presented by Khatir et al. [10–12]. The combination of the methods allows obtaining the level of location and quantification of damage. Xu et al. [13] used the methods of natural frequency and mode shape to detect damage than those methods had been combined with cuckoo search and genetic algorithms for quantifying the damage. Genetic Algorithm (GA) had been coupled with three methods of damage detection using Modal Assurance Criterion (MAC) based on frequency, mode shape, and frequency with mode shape together as objective function presented by Khatir et al. [14]. In this paper, a modified MAC function based on transmissibility will be presented for damage detection combined with PSO. A planner truss structure with 25 elements is studied using the proposed approach.

2 The Transmissibility Concept

Considering linear multiple degrees of freedom (MDOF) system and dynamic equilibrium equation can be written as:

$$[\boldsymbol{M}]\{\ddot{\mathbf{x}}(t)\} + [\boldsymbol{D}]\{\dot{\boldsymbol{x}}(t)\} + [\boldsymbol{K}]\{\boldsymbol{x}(t)\} = \boldsymbol{F}(t) \tag{1}$$

where $[M]$, $[D]$ and $[K]$ represent the $n \times n$ mass, damping, and stiffness matrices, respectively. $F(t)$ is the input force vector and $x(t)$ contain the responses of each degree of freedom. To solve the differential equation, we can use Fourier or Laplace transform. The transmissibility between point i and a reference point j can be defined as:

$$\mathrm{T}_{(i,j)}(\omega) = \frac{X_i(\omega)}{X_j(\omega)} \tag{2}$$

where X_i and X_j are the complex amplitudes of the system response and ω is the frequency. To calculate the transmissibility, the method of frequency response function (FRF) is used.

$$\mathrm{T}_{(\mathrm{i,j})}(\omega) = \frac{\frac{X_i(\omega)}{F_t}(\omega)}{\frac{X_j(\omega)}{F_t}(\omega)} = \frac{H_{it}(\omega)}{H_{jt}(\omega)} \tag{3}$$

where t is the single excitation node and H is the frequency response function.

From Eq. (3), we can see that the transmissibility can be defined as the ratio of two FRFs. As the FRF represent an efficient detector of damage, we can conclude that the transmissibility also can be used for damage identification.

2.1 Objective Function

One of the most popular and sensitive tools for the quantitative comparison of modal vectors is the Modal Assurance Criterion (MAC). The modal assurance criterion is defined as a scalar constant relating the degree of consistency (linearity) between two different sets of vectors $\{\Phi^A\}$ and $\{\Phi^B\}$. Suppose that the size of Φ^A is $n \times m_A$ and size Φ^B is $n \times m_B$, which represents two collections of modal deformations corresponding to two different sets, n represents the degree of freedom, m_A and m_B are numbers of considerate modes on state A and B. the coefficient MAC is calculated over this approach:

$$\mathrm{MAC_{j,k}} = \frac{\left|\sum_{i=1}^{n} \Phi_{i,j}^{A} \Phi_{i,k}^{B}\right|^2}{\sum_{i=1}^{n}\left(\Phi_{i,j}^{A}\right)^2 \sum_{i=1}^{n}\left(\Phi_{i,k}^{B}\right)^2} \begin{pmatrix} \forall j = 1, \ldots m_A \\ \forall k = 1, \ldots m_B \end{pmatrix} \tag{4}$$

The MAC takes a value between [0–1]. Values larger than 0.9 indicate consistent correspondence, whereas small values indicate poor resemblance of the two shapes [15]. As a new approach, we replaced the mode shape in the MAC formulation by the transmissibility to present a new indicator called TMAC as the objective function:

$$\mathrm{TMAC_{j,k}} = \frac{((t_{(i,j)}^{d})^T (t_{(i,j)}^{d}))^2}{((t_{(i,j)}^{d})^T (t_{(i,j)}^{d}))((t_{(i,j)}^{u})^T (t_{(i,j)}^{u}))} \tag{5}$$

where $t_{(i,j)}^{d}$ is the transmissibility under the damaged condition, which represents the measured values by numerical or experimental, while $t_{(i,j)}^{u}$ represents the transmissibility under intact condition, which represents the calculated values by optimization techniques, where $()^T$ means transpose. And the objective function is given as follows:

$$\mathrm{Ob\_F} = \mathrm{abs}\,(\mathrm{TMAC_{j,k}} - 1) \tag{6}$$

3 Particle Swarm Optimization (PSO)

Particle Swarm Optimization (PSO) is an evolutionary computational technique developed by Kennedy and Eberhart [16]. Khatir et al. [17, 18] used PSO coupled with FEM for damage identification using dynamical data. PSO considers each individual as a particle and extends it to N dimensions. For each iteration, the particles evaluate their fitness and share the best positions with the swarm. The particle updates its position and velocity for each iteration according to its best previous computed position, known as p_{best}, and that of the best particle, known as g_{best}, found so far. The updating of the velocity, v_i, and position x_i of each particle are performed also for each iteration as presented in the following equations.

$$v_{i+1} = \omega v_i + c_1 r_1 (p_{best} - x_i) + c_2 r_2 (g_{best} - x_i) \tag{7}$$

$$x_{i+1} = x_i + v_{i+1} \tag{8}$$

where v_i is the particle velocity after the ith iteration, v_{i+1} is the particle velocity after the $(i + 1)^{th}$ iteration, x_i is the particle position after the ith iteration and x_{i+1} is the particle position after the $(i + 1)^{th}$ iteration. ω presented the inertia weight, c_1 and c_2 are the learning factors and r_1 and r_2 are random numbers that are uniform distribution between 0 and 1. The parameters used for both algorithm are presented in Table 1.

Table 1. The cross-sectional area of truss members.

Member	Cross-section area (cm^2)
1–6	18
7–12	15
13–17	10
18–25	12

4 Numerical Examples

4.1 A 25-Bar Plane Truss

In the second section, a finite element model consists of 25 elements, 12 joints, and 21 DOFs is constructed for 2D 25-bar plane truss structure as presented in Fig. 1. Truss elements are made from a steel material with Young's modulus $E = 200\,\text{GPa}$, Poisson's ratio $v = 0.3$, density $\rho = 7.8 \times 10^3\,\text{kg}\,\text{m}^{-3}$ and cross-sectional areas are given in Table 1. The excitation F is loaded at node 12. The position of the first accelerometer is at node 8 and the second one is located on node 9. The damage scenarios are presented in Table 2.

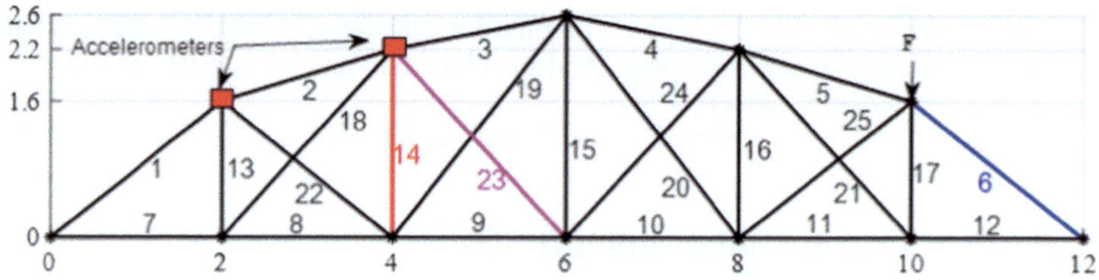

Fig. 1. The FE model of planner truss

Table 2. Damage scenarios for the planner truss structure.

Damage scenarios		Damaged elements (% Reduction in stiffness)				
D1	Element Nbr Reduction in stiffness [%]	6 50	–	–	–	–
D2	Element Nbr Reduction in stiffness [%]	6 50	14 30	–	–	–
D3	Element Nbr Reduction in stiffness [%]	6 50	14 30	23 20	–	–

In order to validate and illustrate the applicability of the proposed method based on the inverse problem, the aforementioned damage detection procedure is conducted and the results are as follows.

4.2 Damage Scenario 1 (D1)

Transmissibility T(9, 8) for the first damage scenario used as an objective function based on MAC formulation is presented in Fig. 2.

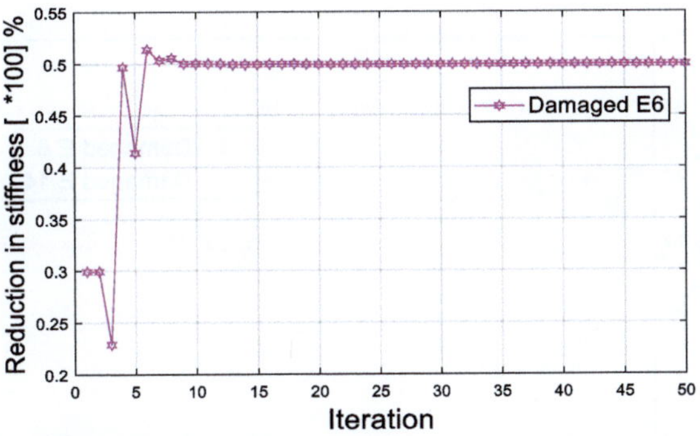

(a) Reduction in stiffness

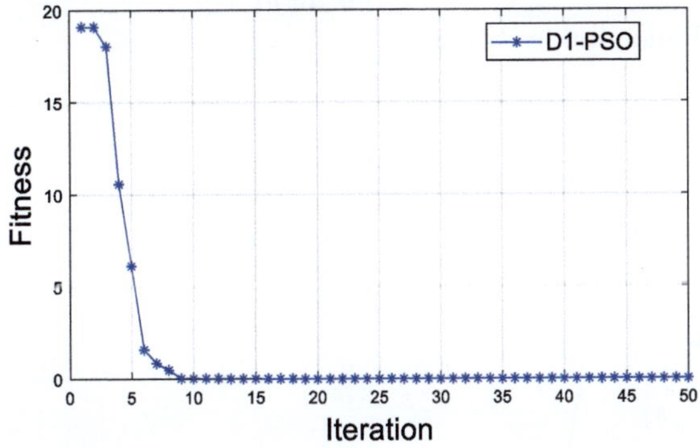

(b) Fitness

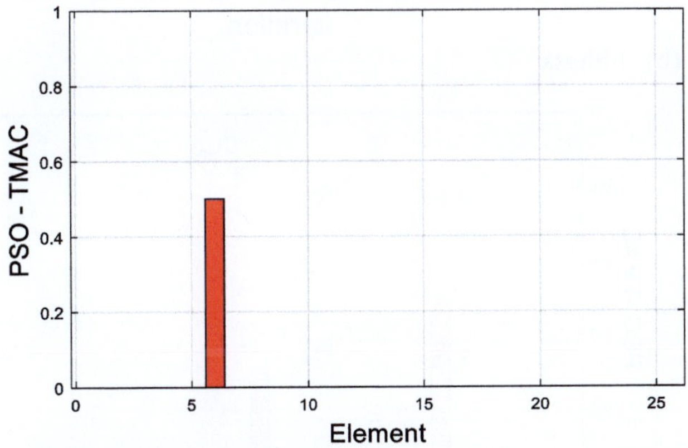

(c) Predicted element with severity

Fig. 2. PSO-TMAC for damaged scenario 1 (D1)

4.3 Damage Scenario 2 (D2)

For the second scenario, the results of damaged elements 6 and 14 with a loss of rigidity 50% and 30% respectively, are presented in Fig. 3. TMAC-1 is used as an objective function with PSO.

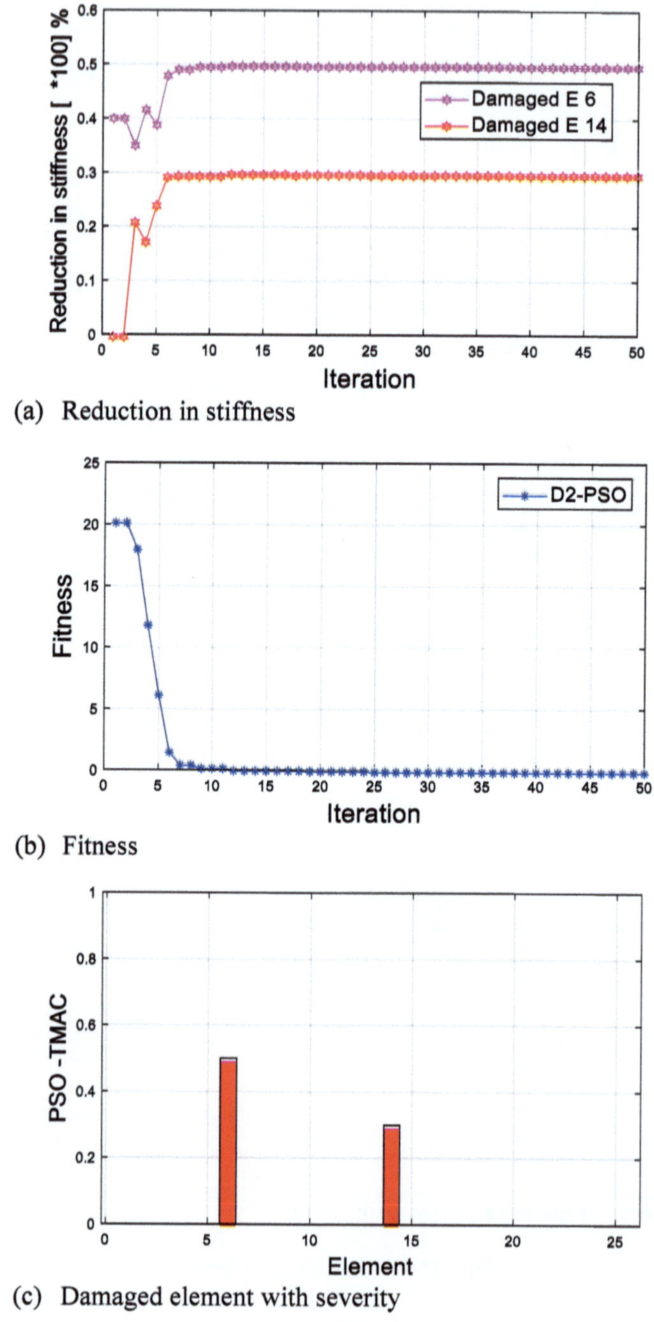

Fig. 3. PSO-TMAC for damaged scenario 2 (D2)

4.4 Damage Scenario 3 (D3)

In the last scenario, the damaged elements 6, 14 and 23 are considered with loss of rigidity 50%, 30%, and 20% respectively. The results obtained by TMAC-1 and PSO combined with FEM truss planner with 25 elements are presented in Fig. 4.

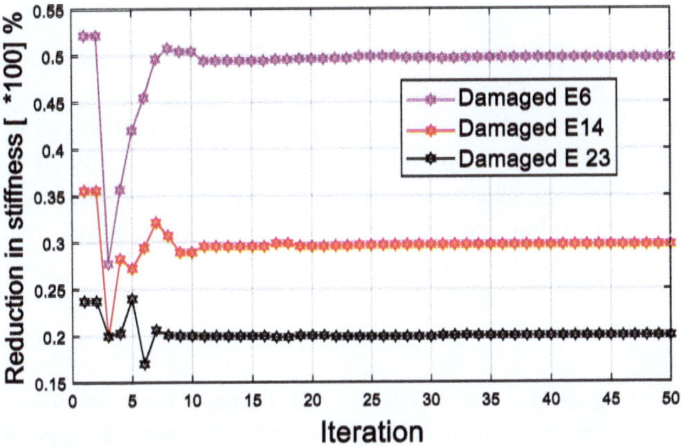

(a) Reduction in stiffness

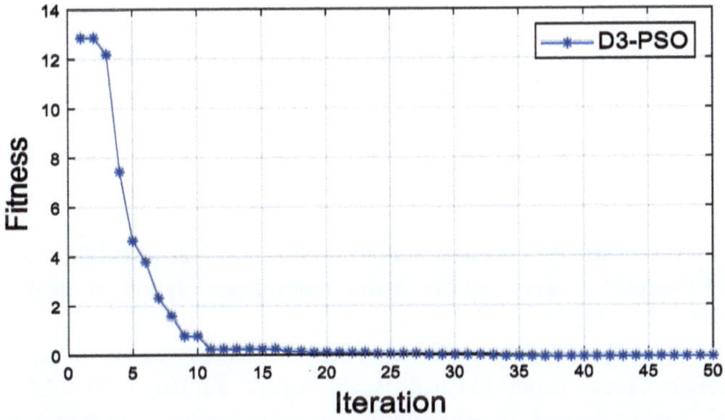

(b) Fitness

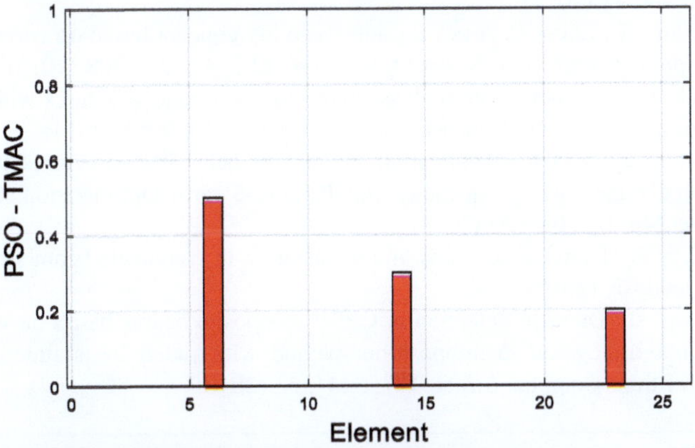

(c) Damaged elements with their severity

Fig. 4. PSO-TMAC for damaged scenario 3 (D3)

From the obtained results, it is clear that the TMAC-1 used as an objective function along with the PSO technique can detect the damage location and severity with higher accuracy even in case of several damages.

5 Conclusion

In this study, a new approach for damage identification in truss structures is investigated. Modal Assurance Criterion (MAC) formulation based on Transmissibility (TMAC) is used as an objective function. Particle Swarm Optimization (PSO) is used to solve the inverse problem to predict the location and severity of damage in truss structure with 25 elements. The results show that the present methodology can reliably identify damage with higher accuracy even for complex structures.

References

1. Zhou, Y.L., Maia, N.M., Abdel Wahab, M.: Damage detection using transmissibility compressed by principal component analysis enhanced with distance measure. J. Vib. Control **24**(10), 2001–2019 (2018)
2. Chesné, S., Deraemaeker, A.: Damage localization using transmissibility functions: a critical review. Mech. Syst. Signal Process. **38**(2), 569–584 (2013)
3. Maia, N.M., Urgueira, A.P., Almeida, R.A.: Whys and wherefores of transmissibility. In: Vibration Analysis and Control-New Trends and Developments. IntechOpen, London (2011)
4. Maia, N., Lage, Y., Neves, M.: Recent advances on force identification in structural dynamics. In: Advances in Vibration Engineering and Structural Dynamics. InTech, London (2012)
5. Buchberger, B.: Bruno Buchberger's PhD thesis 1965: an algorithm for finding the basis elements of the residue class ring of a zero dimensional polynomial ideal. J. Symb. Comput. **41**(3–4), 475–511 (2006)
6. Zhou, Y.L., et al.: Damage detection in structures using a transmissibility-based Mahalanobis distance. Struct. Control. Health Monit. **22**(10), 1209–1222 (2015)
7. Zhao, X.Y., et al.: A new transmissibility analysis method for detection and location of damage via nonlinear features in mdof structural systems. IEEE/ASME Trans. Mechatron. **20**(4), 1933–1947 (2015)
8. Fan, Z., Feng, X., Zhou, J.: A novel transmissibility concept based on wavelet transform for structural damage detection. Smart Struct. Syst. **12**(3_4), 291–308 (2013)
9. Li, J., et al.: Damage detection of shear connectors in bridge structures with transmissibility in frequency domain. Int. J. Struct. Stab. Dyn. **14**(2), 1350061 (2014)
10. Khatir, S., et al.: Crack identification method in beam-like structures using changes in experimentally measured frequencies and Particle Swarm Optimization. Comptes Rendus Mécanique **346**(2), 110–120 (2018)
11. Khatir, S., et al.: Damage detection and localization in composite beam structures based on vibration analysis (2015)
12. Samir, K., et al.: Damage detection in CFRP composite beams based on vibration analysis using proper orthogonal decomposition method with radial basis functions and cuckoo search algorithm. Compos. Struct. **187**, 344–353 (2018)

13. Xu, H., Liu, J., Lu, Z.: Structural damage identification based on cuckoo search algorithm. Adv. Struct. Eng. **19**(5), 849–859 (2016)
14. Khatir, S., Belaidi, I., Serra, R., Benaissa, B., Saada, A.A.: Genetic algorithm based objective functions comparative study for damage detection and localization in beam structures. J. Phys.: Conf. Ser. **628**(1), 012035 (2015). IOP Publishing
15. Pastor, M., Binda, M., Harčarik, T.: Modal assurance criterion. Procedia Eng. **48**, 543–548 (2012)
16. Shi, Y., Eberhart, R.C.: Empirical study of particle swarm optimization. In: Proceedings of the 1999 Congress on Evolutionary Computation, CEC 99. IEEE (1999)
17. Khatir, S., et al.: Multiple damage detection in composite beams using Particle Swarm Optimization and Genetic Algorithm. Mechanika **23**(4), 514–521 (2017)
18. Khatir, S., et al.: Crack identification method in beam-like structures using changes in experimentally measured frequencies and Particle Swarm Optimization. Comptes Rendus Mécanique **346**, 110–120 (2018)

Bolted Joint Monitoring Using the Elastic Wave Propagation

Rafał Kędra(✉) and Magdalena Rucka

Gdańsk University of Technology, 80-233 Gdańsk, Poland
{rafal.kedra,magdalena.rucka}@pg.edu.pl

Abstract. The paper presents an analysis of wave propagation through the contact zone in pretensioned bolted lap joints. A mathematical Schoenberg's model of wave propagation through the interface between two elastic media was discussed. The experiment was carried out using piezoelectric transducers. A single lap joint with different preload levels was examined. The value of the bolt load was measured by a force washer transducer. In the first step, an indicator based on the signal energy was tested. It has been shown that the visible trend is observable only for selected frequency and small range of bolt load. For this reason, advanced signal processing techniques were used for quantitative comparing the recorded signals. The presented experimental results are consistent with the theoretical prediction and they can be used to monitor the bolted joint state in a wide range of the pretension force.

Keywords: Damage detection · Wave propagation · Bolted joint

1 Introduction

An intensive development of the non-destructive testing methods based on elastic wave propagation in recent years has resulted in numerous novel techniques of determining the bolt joint status. Most of them focus on advanced signal processing using e.g. pattern recognition techniques [1], probabilistic neural networks [2] and cross-correlation method [3]. However, there is still a lack of thorough explanation of the physical nature of the elastic wave propagation through the contact zone. The mathematical models used to support a diagnostic process are relatively simple and they introduce simplifications of contact conditions [4] or they assume a perfect bonding within the contact zone [5]. This last assumption was experimentally disproved; it has been proved that the value of the wave reflection coefficient is related to the contact stresses [6].

In this work, a mathematical Schoenberg's model of contacting bodies was applied. The variability of the wave transmission coefficient in the range of contact stiffness corresponding to the stresses occurring in the contact zone of bolted joints was analyzed. Theoretical results were experimentally verified based on a simple bolt connection model.

M. A. Wahab (Ed.): DAMAS 2019, LNME, pp. 234–243, 2020.
https://doi.org/10.1007/978-981-13-8331-1_17

2 Theoretical Background

A bolted joint consists of many interacting elements. Contact zones around the bolt are formed as a result of introducing a pretension force. Their size and the contact stress distribution depend on the value of the bolt load. Furthermore, it has been experimentally proved that the value of stresses in the contact zone has a significant effect on the wave propagation character, especially on the transmission coefficient. One of the approach to modeling this type of imperfect connections is based on contact stiffness definition and it assumes the discontinuity of the displacements field. For a two-dimensional plane strain problem, it was mathematically formulated by Schoenberg [7]. Assume two elastic, isotropic and homogenous half-planes with the Young's modulus E, the Poisson's ratio ν and the mass density ρ connected continuously. The connection quality is determined by two independent parameters, i.e. the normal contact stiffness k_N and the tangential stiffness k_T. The incident wave at the boundary of media can be described by the equation:

$$\begin{bmatrix} u_1 \\ u_2 \end{bmatrix} = \begin{bmatrix} \sin\theta \\ \cos\theta \end{bmatrix} e^{i\omega x_2 \cos\theta / \alpha}, \tag{1}$$

in the case of P wave, or for SV wave by the equation:

$$\begin{bmatrix} u_1 \\ u_2 \end{bmatrix} = \begin{bmatrix} -\cos\theta \\ \sin\theta \end{bmatrix} e^{i\omega x_2 \cos\phi / \beta}, \tag{2}$$

where θ and ϕ are the angles of incidence of P and SV waves respectively, ω is the wave angular frequency. The parameters α and β are compressional and shear wave speeds:

$$\alpha = \sqrt{\frac{E(1-\nu)}{(1+\nu)(1-2\nu)\rho}}, \qquad \beta = \sqrt{\frac{E}{2(1+\nu)\rho}}. \tag{3}$$

The coefficients of reflection (R) and transmission (T) can be determined using the equation:

$$\{R_P\, R_{SV}\, T_P\, T_{SV}\}^T = A^{-1} B, \tag{4}$$

where the matrix A has a form:

$$A = \begin{bmatrix} -p & \gamma\cos\phi & p & \gamma\cos\phi \\ \gamma\cos\theta & q & \gamma\cos\theta & -q \\ -\sin\theta & -\cos\phi & \sin\theta - \frac{i\omega\gamma\cos\theta}{k_T} & -\cos\phi + \frac{i\omega q}{k_T} \\ \cos\theta & -\sin\phi & \cos\theta - \frac{i\omega p}{k_N} & \sin\phi - \frac{i\omega\gamma\cos\phi}{k_T} \end{bmatrix}, \tag{5}$$

and parameters γ, p, q are defined as follows:

$$\gamma = 2\rho\beta \sin\phi, \quad p = \rho\alpha - \gamma \sin\theta, \qquad q = \rho\beta \cos^2\phi - \frac{\gamma \sin\phi}{2}. \tag{6}$$

The vector B can be built from the elements of matrix A. It takes the form of B_P according the Eq. (7) for P wave incidence and for SV wave is described by Eq. (8) as B_{SV}:

$$B_P = \{\, -A_{11} \quad A_{21} \quad -A_{31} \quad A_{41} \,\}^T, \tag{7}$$

$$B_{SV} = \{\, A_{12} \quad -A_{22} \quad A_{32} \quad -A_{42} \,\}^T. \tag{8}$$

To analyze the variation of transmission coefficients, the ratio between the tangential stiffness and normal stiffness was set as constant according to the formula determined by Mindlin for the axisymmetric Hertzian contact [8]:

$$\frac{k_T}{k_N} = \frac{2\,(1 - \nu)}{2 - \nu}. \tag{9}$$

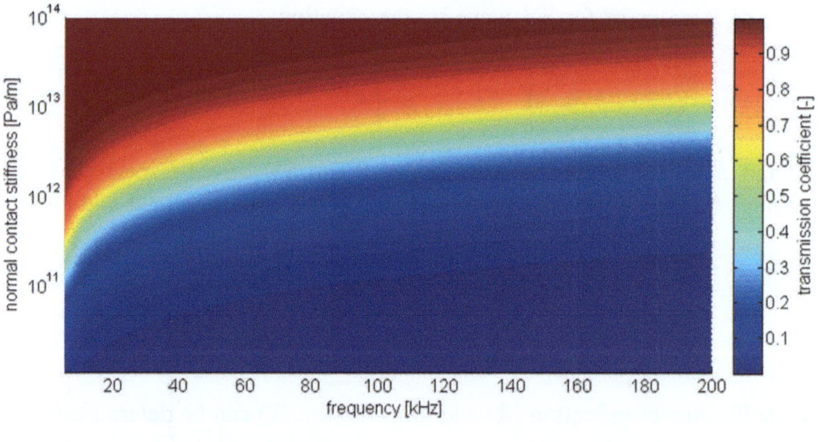

Fig. 1. The real part of P wave transmission coefficient value in the case of P wave incidence (T_{PP}) at angle 45° on steel-steel boundary.

The solution of Eq. (4) for material parameters as steel (E = 200 GPa, ν = 0.3 and ρ = 7850 kg/m^3) and an angle of incidence equal to 45° is presented in Figs. 1 and 2. It can be seen that for the P wave incidence the variation of the transmission coefficients depends on the wave frequency. In the case of coefficient T_{PP} (P wave transmission) it has asymptotic character and as the frequency grows the coefficient value increases in higher stiffness range. The SV wave transmission coefficient T_{PSV} has a local extremum, which also shifts as the frequency increases. Additionally, it should also be noted that the coefficient T_{PP} in general does not depend on the angle of incidence but in for T_{PSV} the angle of incidence changing affects in particular the maximum value of the coefficient. The variability of transmission coefficients for SV wave incidence was omitted due to their small values in the analyzed area. The range of contact stiffness, where significant changes of transmission coefficient occur, coincides with the expected range for bolt connections. There are many mathematical models and experimental methods of contact stiffness evaluation and according to majority of them for contact pressures below 200 MPa the maximum contact stiffness does not excide 1.3×10^{15} Pa/m [9].

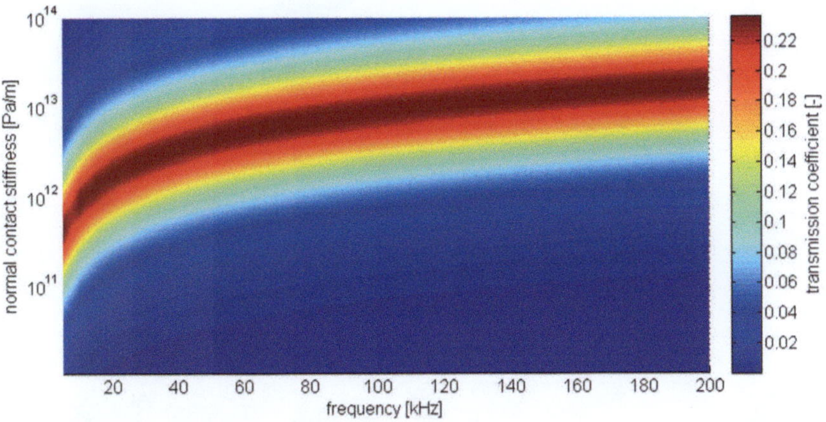

Fig. 2. The real part of SV wave transmission coefficient value in the case of P wave incidence (T_{PSV}) at angle 45° on steel-steel boundary.

3 Experimental Study

3.1 Scope of Research

The case study analysis concerns the simple model of a bolted joint. It was made of two steel plates (thickness 3 mm, length 460 mm and width 40 mm) assembled by a hexagon bolt, nut (diameter of 12 mm) and washers. Figure 3 shows the geometry of the examined connection. During tests, the joint was mounted vertically in a steel frame (Fig. 4a). The bolt tightening was realized by a torque wrench and the pretension level was determined using the force washer transducer.

The elastic waves were excited and measured using the piezoelectric transducers Noliac and the PAQ-16000D system. The configuration of transducers is shown in Fig. 4b. The actuator was placed on one side of the connection, and sensors were attached along the trace of symmetry plane at the second side. The signal used to wave excitation was a 60–120 kHz sine limited to the duration of five periods and modulated by the Hanning window.

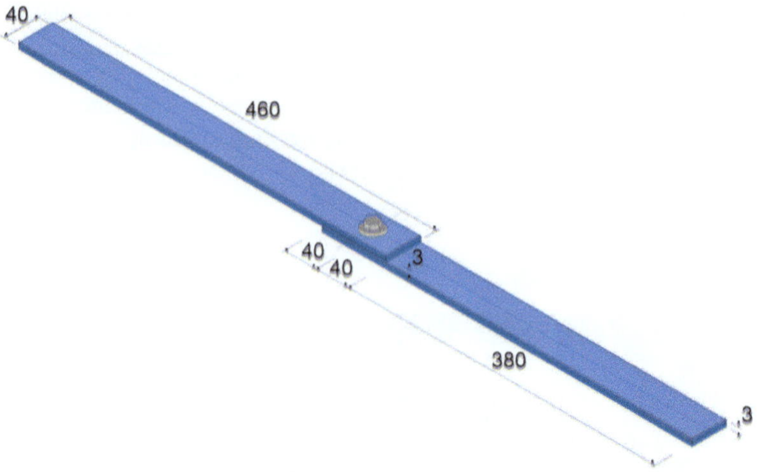

Fig. 3. The geometry of analyzed bolted joint.

a)

b)

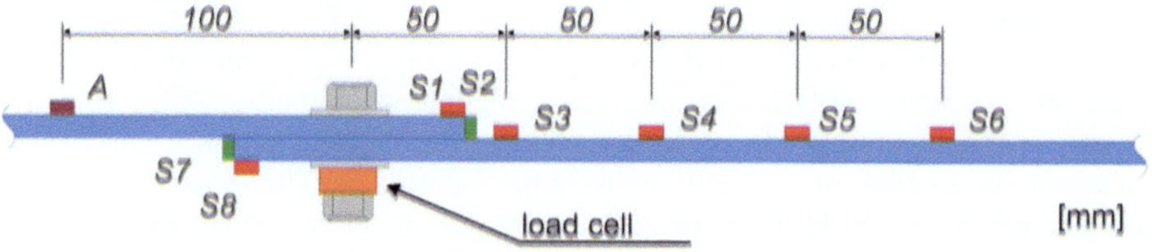

Fig. 4. Experimental set-up: (a) joint mounted in steel frame; (b) transducers arrangement.

3.2 The Results and Discussion

Based on theoretical analysis one can expect that with the increase of the bolt load value the growth of the wave amplitude can be observed. However, in the case of the analyzed connection there are several effects disturbing this variability. Firstly, in the initial period of the bolt load increase, the contact zone is growing. Secondly, waves in plate-like elements are dispersive, which results in the change of propagating wave packet shape. Finally, due to the small width of the connection in the relation to the wavelength, the propagating distribution is interfering with the reflections from the edges of the plates.

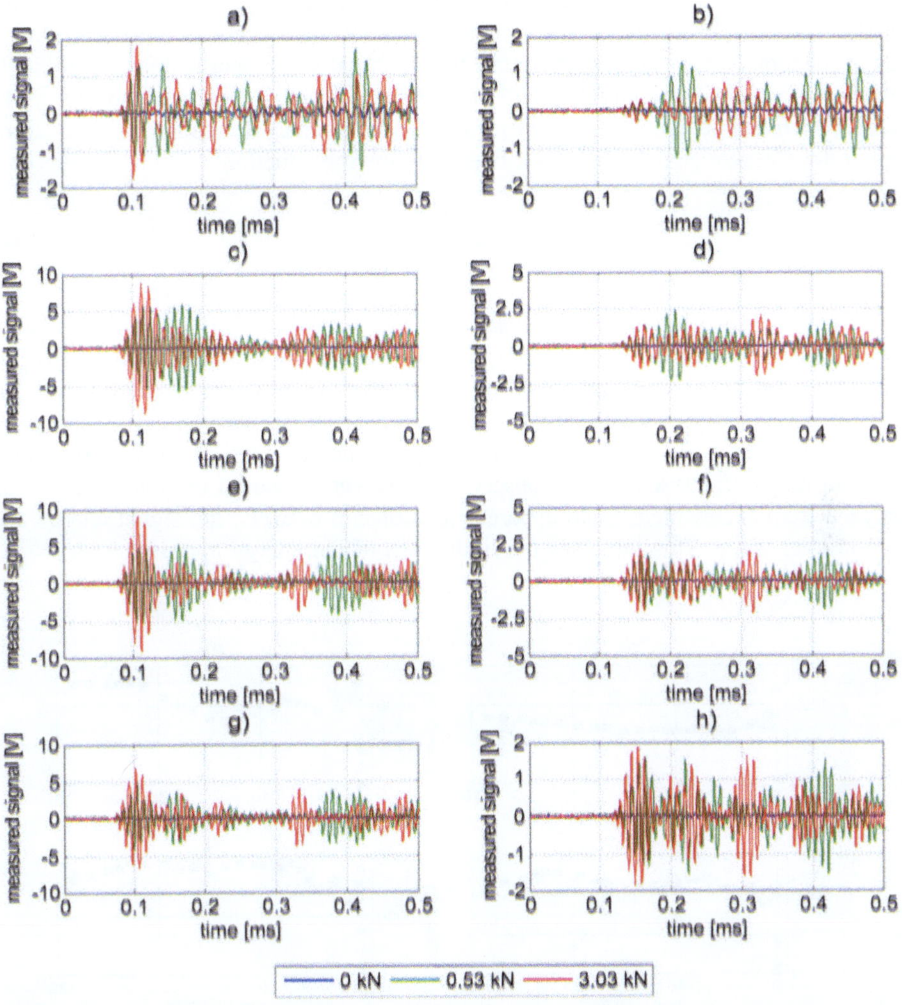

Fig. 5. The experimentally registered wave signals: (a) S4 – 60 kHz; (b) S6 – 60 kHz; (c) S4 – 80 kHz; (d) S6 – 80 kHz; (e) S4 – 100 kHz; (f) S6 – 100 kHz; (g) S4 – 120 kHz; (h) S6 – 120 kHz.

The exemplary results presented in Fig. 5 reveal no clear trend. The presented signals contain many overlapping wave packets. In general, the first one have the largest amplitude, which also increases for the growing bolt load value. For subsequent wave packets, this dependence is not maintained. The reason for this is that the first wave packet is a result of the wave propagation through the contact zone. The next ones are the result of the reflection of the wave from the edges of the plate. When the wave arrives to the contact zone, its part is transmitted to the second plate next to the washer, bolt head/nut and simultaneously it travels along the edge of the hole. The wave spreads in all directions in the bottom plate, then it is reflected from the longitudinal and transverse edges of the plate and the disturbance continues to propagate in the lower plate. Those reflections from edges due to a heterogeneous distribution of stresses in the contact zone and the lack of symmetry in real connections are superposing. For these reasons, the initial variability of the signal is the most important for diagnostic purposes. Due to many reflections appearing in the signals their qualitative comparison is very difficult. The alternative may be quantitative indicators. One of the most popular is the energy of a signal, which can be defined in both time [10] and frequency domains [11]. In the time domain for discrete signals it is described by following equation:

$$E = \sum_{i=1}^{n} x_k^2, \tag{9}$$

where x_k is the k-th value of signal and n is number of samplings points. Figure 6 presents the results of the signal energy calculation for the investigated model of the bolted lap joint. In general, after a strong increase in the initial range of bolt load, the value of signal energy stabilizes and only slight changes are visible. The exception is the excitation frequency equal to 80 kHz, where an increase of the signal energy in the bolt load range 10–20 kN can be observed. However, it should be noted that for all analyzed frequencies there is no apparent relationship between the signal energy and the bolt load value. This indicates the limitations of the energy approach.

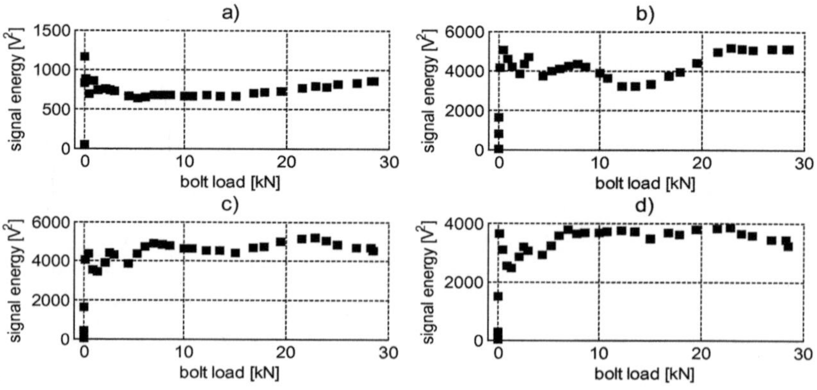

Fig. 6. The energy of signals registered by sensor S3 for connection excitation: (a) 60 kHz; (b) 80 kHz; (c) 100 kHz; (d) 120 kHz.

Table 1. The group velocity of the lamb mode A0 for the steel plate thickness 3 mm.

Frequency [kHz]	60	80	100	120
Theoretical group velocity [m/s]	2716	2877	2984	3050

Table 1 summarize the theoretically determined velocities of the antisymmetric Lamb mode A0 for the steel plate (E = 200 GPa, ν = 0.3 and ρ = 7850 kg/m^3) with a thickness of 3 mm. They allow to determine the period of signal duration, which is directly related to the wave propagation through the contact zone. In the case of the connection with a width of 40 mm, this is approximately 13–14 μs.

Due to the dispersion character of Lamb waves, the amplitude analysis was performed based on the Windowed Fourier Transform (WFT) in MATLAB. An initial part of each registered waveform was transformed into the frequency domain using WFT and its spectrogram was obtained. Then for each signal related to the particular bolt value, the spectrogram column corresponding to the excitation frequency was selected and as a results the two dimensional map of amplitude changes was created. The exemplary result obtained in this manner for the connection of a width 4 cm, sensor S3 and excitation frequency 120 kHz is presented in Fig. 7. It can be seen that in the initial time range (0.05–0.1 ms) for higher values of the bolt load, a more intense amplitude increase is observed. As mentioned above, those initials parts of signals are associated direct with wave propagation through contact zone. The Fig. 8 shows the intersections of amplitude maps at selected time points related with ends of reflection-free parts of signals. It can be seen that with the increase of the excitation frequency the maximum amplitude appears for higher value of the bolt load, which is consistent with the theoretical results (see Fig. 1).

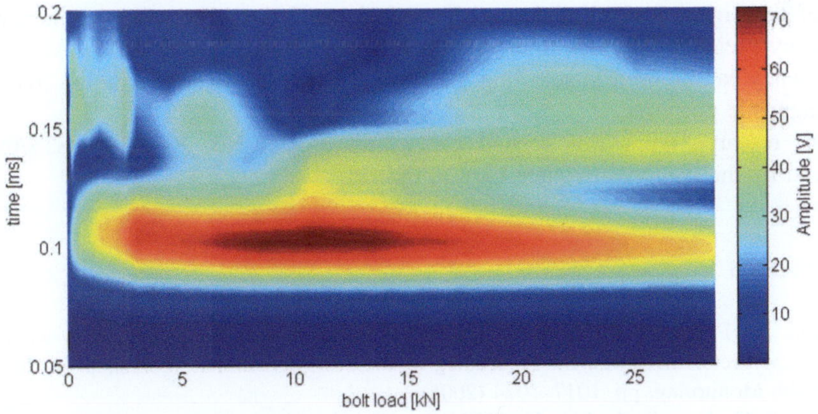

Fig. 7. The initials part of signals amplitude variability for connection of a width 4 cm and excitation frequency 120 kHz registered by sensor S3.

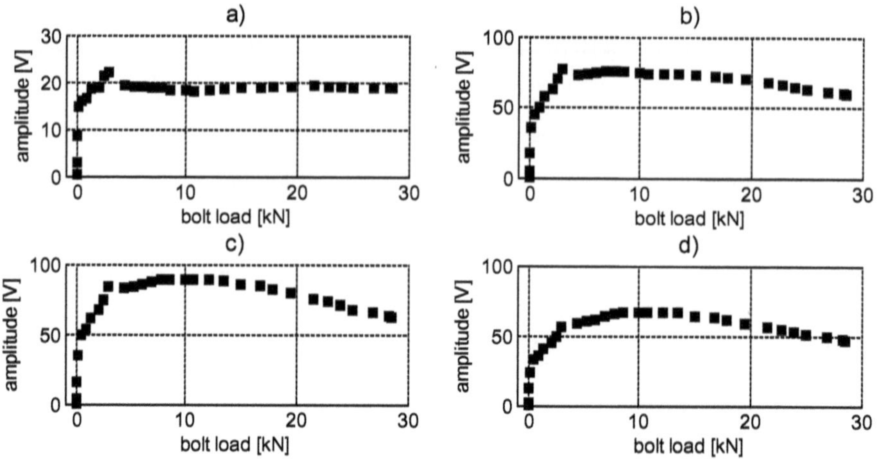

Fig. 8. The amplitude versus bolt load dependence for initial part of signal registered by sensor S3 at time point and excitation frequency: (a) 0.0945 ms, 60 kHz; (b) 0.0895 ms, 80 kHz; (c) 0.0895 ms, 100 kHz; (d) 0.0845 ms, 120 kHz.

4 Conclusion

The paper discusses the elastic wave propagation through the contact zone on the bolted connection example. The theoretical analysis was made using Schoenberg's model. In order to verify the correctness of the model, a single lap joint was tested and the influence of the bolt load value on the variability of recorded signals as well as the signal energy was analyzed. The experimental data were processed using the Windowed Fourier Transform. It has been shown that the amplitude variation of the initial part of the signal is similar to the theoretical prediction, i.e. with the increase of frequency excitation the maximum amplitude appeared for higher bolt load value. This dependency can be used in selecting the appropriate value of the excitation frequency for diagnostic purposes.

The research work was carried out within project No. 2015/19/B/ST8/00779, financed by the National Science Centre, Poland.

References

1. Mita, A., Fujimoto, A.: Active detection of loosened bolts using ultrasonic waves and support vector machines. In: Proceeding of the 5th International Workshop on Structural Health Monitoring, pp. 1017–1024 (2005)
2. Park, S.-H., Yun, C.-B., Roh, Y.: PZT-induced lamb waves and pattern recognitions for on-line health monitoring of jointed steel plates. In: Proceedings of SPIE 5765, Smart Structures and Materials 2005: Sensors and Smart Structures Technologies for Civil, Mechanical, and Aerospace Systems, pp. 364–375 (2005)

3. Ruan, J., Zhang, Z., Wang, T., Li, L., Song, G.: An anti-noise real-time cross-correlation method for bolted joint monitoring using piezoceramic transducers. Smart Struct. Syst. **16** (2), 281–294 (2015)
4. Rhee, I., Choi, E., Roh, Y.-S.: Guided wave propagation introduced by piezelectric actuator in bolted thin steel members. KSCCE J. Civ. Eng. **16**, 398–406 (2012)
5. Huda, F., Kajiwara, I., Hosoya, N., Kawamura, S.: Bolt loosening analysis and diagnosis by noncontact laser excitation vibration tests. Mech. Syst. Signal Process. **40**, 589–604 (2013)
6. Starzynski, G., Buczkowski, R.: Ultrasonic measurements of contact stiffness between rough surfaces. J. Tribol. **136**, 034503–5 (2014)
7. Schoenberg, M.: Elastic wave behavior across linear slip interfaces. J. Acoust. Soc. Am. **68**, 1516–1521 (1980)
8. Mindlin, R.D.: Compliance of elastic bodies in contact. J. Appl. Mech. **16**, 259–268 (1949)
9. Du, F., Hong, J., Xu, Y.: An acoustic model for stiffness measurement of tribological interface using ultrasound. Tribol. Int. **73**, 70–77 (2014)
10. Wang, T., Song, G., Wang, Z., Li, Y.: Proof-of-concept study of monitoring bolt connection status using a piezoelectric based active sensing method. Smart Mater. Struct. **22**, 087001 (2013)
11. Amerini, F., Meo, M.: Structural health monitoring of bolted joints using linear and nonlinear acoustic/ultrasound methods. Struct. Health Monit. **10**, 659–672 (2011)

A Generic Framework for Application of Machine Learning in Acoustic Emission-Based Damage Identification

Abhishek Kundu[1(✉)], Shirsendu Sikdar[2], Mark Eaton[1], and Rukshan Navaratne[3]

[1] Cardiff School of Engineering, Cardiff University, The Parade, Queen's Building, Cardiff CF24 3AA, UK
KunduA2@cardiff.ac.uk
[2] Institute of Fluid-Flow Machinery, Polish Academy of Sciences, 14, Fiszera Street, 80-231 Gdansk, Poland
[3] University of South Wales, Treforest Campus, Pontypridd CF37 1DL, UK
https://www.cardiff.ac.uk/people/view/364404-kundu-abhishek

Abstract. Advanced non-destructive monitoring scheme is necessary for modern-day lightweight composite structures used in aerospace industry, due to their susceptibility to barely visible damages from minor impact loads. Acoustic emission (AE) based monitoring of these structures has received significant attention in the past few years primarily due to their possibility of use in operating structures under service loads. However, localization and characterization of damages using AE is still an open area of research. The exploration of the space of signal features collected by a distributed sensor network and its reliable mapping to damage metrics (such as location, nature, intensity) is still far from conclusive. This problem becomes more critical for composite structures with complex features/geometry where the localized effects of discontinuity in geometric or mechanical properties do not make it appropriate to rely on simple signal features (such as time difference of arrival, peak amplitude, etc.) to identify damage. In this work, the AE signal features (which are spatially and temporally correlated) have been mapped to the damage properties empirically with a training dataset using meta-modeling techniques. This is used in the online monitoring phase to infer the probabilistic description of the acoustic emission source within a hierarchical Bayesian inference framework. The methodology is tested on a carbon fibre composite panel with stiffeners that is subjected to impact and dynamic fatigue loading. The study presents a generalized machine learning-based automated AE damage detection methodology which both localizes and characterizes damage under varying operational loads.

Keywords: Acoustic emission · Feature extraction · Machine learning · Damage characterization · Gaussian process

M. A. Wahab (Ed.): DAMAS 2019, LNME, pp. 244–262, 2020.
https://doi.org/10.1007/978-981-13-8331-1_18

1 Introduction

Acoustic Emission (AE) is the result of changes within a structure or material which results from mechanical phenomenon of plastic deformation, friction and rubbing or crack growth which lead to elastic wave propagation in a structure. The propagating AE waves generate minute surface-displacements when they travel along a surface. These displacements can then be converted to the voltage response by using the piezoelectric AE-sensors. These transient signals can be analysed in real-time using most modern AE systems and can also be stored and digitised to allow further signal processing. The triangulation methods can be effectively used to predict the damage source-locations in the target structure by using a distributed network of AE-sensors that detects the released energy from an AE-source, as in the case of determining the epicentre of an earthquake in seismology. A schematic representation of the AE based SHM process is described in Fig. 1 which clearly indicates that any changes in the geometry or material will affect the resultant recorded signal.

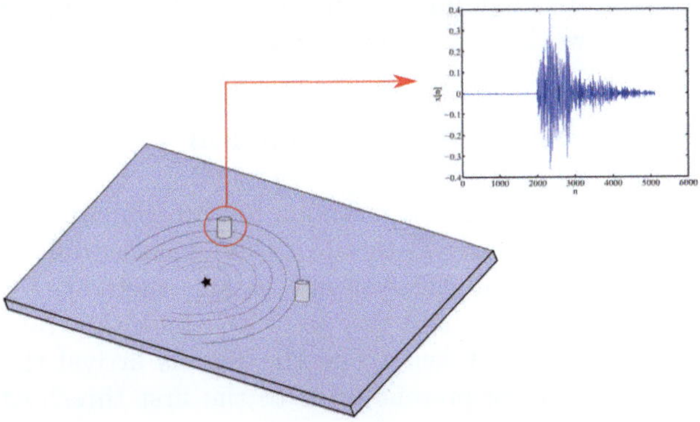

Fig. 1. AE measurement process and typical AE signal recorded at a sensor.

Accurate localization of damage-source location using AE signals in complex structures is significantly challenging as the proper assessment of boundary level uncertainties in the mathematical model is complex in nature and associated with substantial computational overhead [1]. The loss in accuracy in complex structures is mainly occurred due to the failure in accurate determination of time-of-arrival of the AE-signal and due to the overly involved in the simplified representation of the wave propagation path and velocity during source-location calculations [2,3]. The present study aims to overcome the above limitations using a novel approach which calibrates and trains the damage-model using a compound-correlation-metric among the AE-signals registered from the predefined network of AE-sensors. This is then utilized to create a predictive model which performs inverse damage characterization using the Bayesian inference framework [4,5].

This study aims to overcome the above limitations using a novel approach which trains and calibrates the damage model using a compound correlation metric between the signals recorded with the distributed sensory network. When only using the arrival time of the signal at the individual sensors (and the difference in arrival time) as the basis of training the model, a large portion of the signal data is not being utilized fully. Additionally error is incurred in the ad hoc definitions of signal threshold values which are used to calculate arrival time. Moreover, it is strongly based on the assumption of the existence of a travel path between the damage and the sensor location (this can be considered as an underlying regularization) which might often not be the case. The proposed methodology, in contrast, constructs a compressed representation of the full signal characteristics using a projected correlation metric (as discussed in Sect. 3) which utilizes the full signal characteristics to infer the source of the incoming waves. The Gaussian process based surrogate regression approach explicitly accounts for the uncertainty in lack of training data and/or the error incurred using the process described in Sect. 4. This is followed by Sect. 5 which gives the description of the test rig on which the AE source localization algorithm is applied, Sect. 6 where the main findings of the study has been presented and the conclusion and future works is included in Sect. 7.

2 Acoustic Emission Source Location

Time of arrival (TOA) based direct methods for determination of arrival time (and hence the difference in arrival time) of the signal at different sensors are widely used in all commercial AE systems. In this method, a predetermined signal threshold level is defined and the point at which the transducer voltage response exceeds this level is taken to be the signals arrival time. It is, however, possible for signal to be present prior to the first threshold crossing and attenuation of signal amplitude can mean variation in signal arrival time determination. A wide range of approaches have been developed in an attempt to improve the arrival time determination compared with the traditional threshold crossing technique. A range of frequency based techniques including filtering [6], cross-correlation [7] and wavelet transforms [8,9] have been investigated, however, statistical approaches based on 6th order statistical moments [10] and the Akaike Information Criteria (AIC) [11,12] have been shown to be more reliable. The AIC approach in particular has been demonstrated to be very robust across a range of materials and structures [13].

Standard AE location algorithms assume a single, constant, wave propagation speed, however, in composite materials the wave speed is seen to vary with propagation direction and is dependent on fibre orientations within the layup used. Several researchers have tried to address this challenge and some success has been by extending the traditional time of arrival optimization scheme to include a variable wave speed dependent on propagation direction [14,15]. Ciampa and Meo [16] adopted a novel approach whereby closely spaced sensor pairs were used to reduce the number of unknown propagation velocities in a set of simultaneous non-linear equations that describe the source position. An iterative Newton approach was adopted to solve the unknowns in the equations and therefore yield an estimate of the source position without prior knowledge of the wave speeds in the material. Despite these advances achieved in anisotropic monolithic materials, none of the approaches are capable of accounting for geometric complexities such as access holes, curvatures and thickness changes that commonly occur in industrial structures. Alternatively a mapping approach has been proposed and rigorously validated [13, 17–20] in which a structure is mapped using artificial AE sources (such as a H-N source [21, 22]) to derive an empirical relationship between the known source position and the resultant arrival times at an array of sensors. This relationship can then be used to determine the source origin for a set of measured arrival times. The empirical nature of the approach inherently accounts for all material and structural complexity such as anisotropy and geometric features.

The previously discussed approaches are deterministic and do not consider the uncertainty in the measurements and calculations performed. This is particularly relevant to their application to industrial environments where uncertainty is seen to increase, with varying operating conditions such as temperature affecting wave propagation and therefore reducing reliability. To account for the uncertainty that can be experienced in an industrial environment researchers have begun to adopt probabilistic approaches. Schumacher *et al.* [23] developed an approach based on Bayesian statistics for AE source location in a reinforced concrete beams that accounts for uncertainties and errors that exist within the measurement and calculation process. A simplified model for the concrete beam was developed, in which the mean of the wave slowness, the standard deviation of the wave slowness, the event time and the standard deviation of the observed arrival times are represented as prior probability density functions (PDF). The initial PDF of each parameter was then refined using experimental data collected from H-N sources at known positions on the beam surface. The refined model could then be used to predict the most likely position of any subsequent AE sources. The approach reduced the mean error of 22 arbitrarily located H-N sources from ∼40 mm down to ∼30 mm. Further work by Zarate *et al.* [24] developed a Bayesian framework based on a ray tracing model of AE wave propagation in liquid filled storage tanks. The approach allowed structure borne and water borne wave paths to be considered. Using a Markov Chain Monte Carlo (MCMC) method to sample the posterior distribution of the source position in x and y coordinates the most probably source position could be determined. Both of these probabilistic approaches are limited to homogeneous materials and simple geometries, i.e.direct and uninterrupted wave paths. The study of signal characteristics mapped to the damage location has been undertaken by the present authors [25, 26], whereby Bayesian inference has been used for source localization in the prediction stage. This paper is intended to highlight the main aspects of that proposed approach, especially in Sects. 3 and 4, and extend it further to the problem of damage characterization.

This main aim to the present work is to combine the Bayesian probabilistic mapping of the source of AE to the correlation characteristics of the signal collected at the distributed sensor network. This would not only capture the essential information pertaining to the phase difference and attenuation of elastic waves travelling different distances over the surface of the composite structures to reach individual sensors but also include the complex effects of boundary reflection within the signal correlation characteristics. The uncertainty due to the measurement noise and experimental errors would be explicitly accounted for in the probabilistic model and conditioned on the training data generated with H-N sources.

3 Important Signal Characteristics for Damage Identification

3.1 Compressed Cross-Correlation Signal Features for AE

The signal features recorded at different locations on the structure being investigated can be mapped to the damage characteristics to obtain a data-driven identification framework. The main challenge with this approach is that the sensors data can be affected by various ambient and operational factors, such as changes in dispersion behavior, sensor-structure coupling characteristics, amongst others. Hence a trained data-driven model which does not account for the changes in baseline environmental, operational conditions would provide erroneous damage identification with changes in the above conditions.

One way to compensate for this is to focus on sensor data in relation to each other. Specifically, the correlation measure between the signals observed at different sensor locations in a distributed sensory network is quite useful because the normalized correlation measures are invariant to measurement noise and can compensate for changes in ambient condition, such as temperature sensitivity.

If we consider n_s sensors distributed on the structure under study, the collected signal is represented as

$$\mathbf{X} = \{\mathbf{x}_i : \mathbf{x}_i = (x_{i1}, x_{i2}, \dots, x_{in})^\mathsf{T} \quad \forall i = 1, \dots, n_s\} \tag{1}$$

where each sample x_{ij} has been collected at sensor i at n discrete time points $j = 1, \dots, n$. An initiation of crack or a growth in crack size is accompanied by a packet of ultrasonic wave which propagates radially outwards from its point of origin along the plane of the composite structure. Each sensor captures signal $\mathbf{x}_i$ as show in Fig. 1. Additionally reflection from boundaries and/or other geometrical discontinuities (such as holes) result in a complicated propagation characteristics. It is expected that the accumulated data from the distributed sensory network would contain distinguishing features (statistics) which would map the detected waveforms uniquely to the source location.

The correlation matrix $\bar{\mathbf{R}}$ between the signals collected at each of the n_s sensors is defined as

$$\bar{\mathbf{R}}_\tau = \begin{bmatrix} \mathbf{R}_{11}(\tau) & \mathbf{R}_{12}(\tau) & \cdots & \mathbf{R}_{1n_s}(\tau) \\ \mathbf{R}_{21}(\tau) & \mathbf{R}_{22}(\tau) & & \mathbf{R}_{2n_s}(\tau) \\ \vdots & & \ddots & \vdots \\ \mathbf{R}_{n_s1}(\tau) & \mathbf{R}_{n_s2}(\tau) & \cdots & \mathbf{R}_{n_sn_s}(\tau) \end{bmatrix} \tag{2}$$

where each element $\mathbf{R}_{ij}(\tau)$ of the correlation matrix is defined in the continuous time domain as

$$R_{ij}(\tau) = \int_{-\infty}^{\infty} x_i(t)x_j(t+\tau)dt \tag{3}$$

for real-valued signal components. For L^1 integrable signals, i.e. $\int_{-\infty}^{\infty} |x(t)|\, dt < \infty$ the Fourier transform of the signal exists and is related to the terms in $\bar{\mathbf{R}}$ matrix. In Eq. 3, when $i = j$, the terms $\mathbf{R}_{ii}$, i.e. the diagonal terms in the matrix $\bar{\mathbf{R}}_\tau$, give the autocorrelation measure of the signals at each sensor.

For discrete time signals the cross-correlation for real-valued signals is defined as

$$\mathbf{R}_{ij}(\tau) = \mathbb{E}\left[(\mathbf{x}_i[n] - \boldsymbol{\mu}_i)(\mathbf{x}_j[n+\tau] - \boldsymbol{\mu}_j)\right] \quad \forall i,j = 1,\ldots,n_s \tag{4}$$

The cross-correlation is often normalized by the respective auto-correlation functions at zero lag such that

$$\mathbf{r}_{ij}(\tau) = \frac{\mathbf{R}_{ij}(\tau)}{\sqrt{\mathbf{R}_{ii}(0)\mathbf{R}_{jj}(0)}} \quad \text{where} \quad |\mathbf{r}_{ij}(\tau)| \leq 1 \quad \forall i,j,\tau \tag{5}$$

and we denote the normalized version of $\bar{\mathbf{R}}_\tau$ as $\bar{\mathbf{r}}_\tau$ where each element of the matrix in Eq. 2 is normalized as per Eq. 5. It is important to note that $\mathbf{R}_{ij}(\tau) = \mathbf{R}_{ji}(-\tau)$ and $\mathbf{R}_{ij}(\tau) = \mathbf{R}_{ij}(-\tau)$, hence $\mathbf{r}_\tau$ is a symmetric matrix for all τ. The normalized correlation coefficient is utilized here to balance the varying intensity of the AE source. The mapping which links training dataset to the source location assumes that the source intensity is normalized across the entire training dataset which is achieved with the normalized correlation coefficient.

The correlation matrix captures the essential information regarding the correlation of the signals captured using the distributed sensor network which contains essential information regarding the delay or phase of the arriving signals. The matrix $\bar{\mathbf{R}}_\tau$ is constructed over the interval $-\tau_s \leq \tau \leq \tau_s$ such that each of its elements $\mathbf{R}_{ij}(\tau)$ or the normalized cross-correlation $\mathbf{r}_{ij}(\tau)$ is a vector of dimension $2\tau_s + 1$. Since the correlation matrix is symmetric, only the upper triangular part is considered in a matrix representation

$$\bar{\mathbf{r}}_{\Delta_k,\tau_s} = \left[\mathbf{r}_{\tau_s,i} : \mathbf{r}_{\tau_s,i} \in \mathbb{R}^{2\tau_s+1} \text{ and } -\tau_s \leq \tau \leq \tau_s \quad \forall i \in \mathcal{I}\right] \tag{6}$$

where $\mathcal{I}$ is the set of indices associated with the ordering of the elements of the upper triangle of $\mathbf{r}_\tau$ at some instance τ and $\mathbf{r}_{\tau_s,i}$ is a vector which contains the cross-correlation measure over the interval $\pm\tau_s$. It is important to note that the subscript Δ_k of the matrix $\bar{\mathbf{r}}_{\Delta_k,\tau_s}$ $k = 1,\ldots,n_t$ denotes normalized correlation data evaluated for the k-th test point out of a total of n_t training sets. Thus the full set of training correlation data is given as

$$\bar{\mathbf{r}}_{\Delta,\tau_s} = \left[\bar{\mathbf{r}}_{\Delta_1,\tau_s}, \bar{\mathbf{r}}_{\Delta_2,\tau_s}, \ldots, \bar{\mathbf{r}}_{\Delta_{n_t},\tau_s}\right] \quad \text{where} \quad \bar{\mathbf{r}}_{\Delta_k,\tau_s} \in \mathbb{R}^{2\tau_s+1 \times n_{\mathcal{I}}}, \quad k = 1, \ldots, n_t \tag{7}$$

where $n_{\mathcal{I}}$ is the cardinality of the set $\mathcal{I}$.

The objective here is to obtain a map between the normalized correlation matrix representing the signal data for each of the n_t training sets and the corresponding coordinates of the acoustic source. However, the computational demands of modelling the large predictor matrix $\bar{\mathbf{r}}_{\Delta,\tau_s}$ is substantial and would make the subsequent real-time model prediction infeasible. Thus we seek an optimal reduced basis which can be used to represent the data for each training set such that

$$\begin{aligned} &\text{find} \quad \mathcal{T} : \mathbb{R}^{2\tau_s+1 \times n_{\mathcal{I}} n_t} \rightarrow \mathbb{R}^{n_r \times n_{\mathcal{I}} n_t} \quad \text{where} \quad \mathcal{T}(\bar{\mathbf{r}}_{\Delta_i,\tau_s}) = \boldsymbol{\alpha}_{i,\tau_s} \forall i = 1, \ldots, n_{\mathcal{I}} n_t \\ &\qquad \text{such that} \quad \Phi_{\Delta,\tau_s} = \underset{\phi_{\Delta_j,\tau_s}, \forall j}{\arg\inf} \left\| \Phi_{\Delta,\tau_s}[\boldsymbol{\alpha}_{1,\tau_s}, \ldots, \boldsymbol{\alpha}_{n_{\mathcal{I}} n_t,\tau_s}] - \bar{\mathbf{r}}_{\Delta_i,\tau_s} \right\|_F \end{aligned} \tag{8}$$

where $n_r << 2\tau_s + 1$ and $\boldsymbol{\alpha}_{i,\tau_s} \in \mathbb{R}^{n_r}$. Thus the transformation $\mathcal{T}$ expresses the normalized correlation terms using an optimal basis on which the solution would be projected and an additional orthonormality constraint of $\Phi^{\mathsf{T}}_{\Delta,\tau_s} \Phi_{\Delta,\tau_s} = \mathbf{I}$ can be imposed on the basis. This optimal basis can be informed by the left singular eigenvectors of the training correlation data contained in $\bar{\mathbf{r}}_{\Delta,\tau_s}$ or the principal eigenvectors associated with $\bar{\mathbf{r}}^{\mathsf{T}}_{\Delta,\tau_s} \bar{\mathbf{r}}_{\Delta,\tau_s}$. The compressed correlation data is be expressed in terms of this optimal basis as

$$\underline{\boldsymbol{\alpha}}_{\tau_s} = \{\underline{\boldsymbol{\alpha}}_{1,\tau_s}, \ldots, \underline{\boldsymbol{\alpha}}_{n_t,\tau_s}\} = \Phi_{\Delta,\tau_s} \quad \text{where} \quad \bar{\mathbf{r}}^{\mathsf{T}}_{\Delta,\tau_s} \quad \text{where} \quad \underline{\boldsymbol{\alpha}}_{\tau_s} \in \mathbb{R}^{n_r \times n_{\mathcal{I}} n_t} \tag{9}$$

The matrix $\underline{\boldsymbol{\alpha}}_{\tau_s}$ consists of a collection of $n_{\mathcal{I}} n_t$ vectors each of dimension n_r, thus for each of the n_t training points we get a set of $n_{\mathcal{I}}$ vectors which contains the compressed information.

Next we introduce the linear map $vec(\cdot) : \mathbb{R}^{n \times k} \rightarrow \mathbb{R}^{nk}$ which transforms a collection of k vectors each of dimension n into a single vectors of dimension nk (where each of the k vectors are stacked end to end). The matrix $\bar{\mathbf{r}}_{\Delta,\tau_s}$ is transformed as

$$\underline{\boldsymbol{\alpha}}_v = \left\{vec(\underline{\boldsymbol{\alpha}}_{1,\tau_s}), \ldots, vec(\underline{\boldsymbol{\alpha}}_{n_t,\tau_s})\right\} \quad \text{where} \quad \underline{\boldsymbol{\alpha}}_v \in \mathbb{R}^{n_r n_{\mathcal{I}} \times n_t} \tag{10}$$

such that each of the $n_r n_{\mathcal{I}}$ vector in $\underline{\boldsymbol{\alpha}}_v$ corresponds to the compressed data collected at the distributed sensor network on the structural system.

The problem then defined in terms of finding the map $\mathcal{G} : \boldsymbol{\alpha}_{v,xy} \rightarrow r_{xy}$ where r_{xy} is a vector of the Euclidean distance of AE source from each sensor in the distributed sensory network and the subscript (xy) denotes the Cartesian coordinates of the AE source. The problem is posed as the following multivariate least square regression problem

$$\gamma = \underset{\gamma \in \mathbb{R}^{n_r n_{\mathcal{I}}}}{\arg\inf} \left\| \boldsymbol{\alpha}_{v,xy} \gamma - r_{xy} \right\|_2 \tag{11}$$

However, for each training run, there is an uncertainty around the measured signal. We construct a scalar error indicator following from Eq. 11 as

$$\varepsilon_r = \left\|\hat{r}_{xy} - r_{xy}\right\| \quad \text{where} \quad \hat{r}_{xy} = \boldsymbol{\alpha}_{v,xy}\boldsymbol{\gamma} \tag{12}$$

where $\hat{r}_{xy}$ is the approximate identified location of the acoustic source associated with the training data where the known true source is at $r_x y$ distance from the sensors. Equation 12 gives the error associated with the regression map $\mathcal{G}$: $\boldsymbol{\alpha}_{v,xy} \rightarrow r_{xy}$. Thus the error indicator ε_r for the training dataset is mapped on the spatial domain using a Gaussian process surrogate, discussed in the following section, based on which the probabilistic prediction of the source location is performed based on the observed signal data at the distributed sensory network.

3.2 Discrete Scale-Time Decomposition of Cross-Correlation

The cross-correlation based signal compression can be further filtered to localize specific frequency components in the time domain. This gives the advantage of accurately identifying the arrival of specific propagation modes at different sensor locations. Hence a better understanding of the dispersion behavior is obtained by this. Additionally, the proportion of energy concentrated in different wave modes can be calculated from the scale-time decomposition of the signals.

Using discrete and continuous wavelet transform for signal characterization and feature extraction is an established field of study. We briefly discuss the fundamental aspects of the principle of wavelet transform here and its application to the cross-correlation metric. Assuming a mother wavelet of the form $\Psi_{s,\Delta}(t) = \frac{1}{\sqrt{s}}\Psi\left(\frac{t-\Delta}{s}\right)$ and applied to a signal $x(t)$ we get

$$X_i^{s,\Delta} = \int_t x_i(t)\Psi_{s,\Delta}^*(t)dt \tag{13}$$

where s and Δ are the scale and time parameters respectively. It is to be noted that the dual of this relationship exists, which enables the reconstruction of the original signal from the inverse wavelet transform. Applying the wavelet transform to the normalized cross-correlation measure (defined in Eq. 5) results in

$$\mathbf{r}_{ij}^{s,\Delta} = \int_\tau \mathbf{r}_{ij}(\tau)\Psi_{s,\Delta}^*(\tau)d\tau = \frac{1}{\sqrt{\mathbf{R}_{ii}\mathbf{R}_{jj}}}\int_t\int_\tau x_i(t)x_j(t+\tau)\Psi_{s,\Delta}^*(\tau)d\tau dt \tag{14}$$

Simplifying the expression in the Eq. 14 and applying change of variables we obtain

$$\mathbf{r}_{ij}^{s,\Delta} = \frac{1}{\sqrt{\mathbf{R}_{ii}\mathbf{R}_{jj}}}\int_t x_i(t-\Delta)X_j^{s,t}dt \tag{15}$$

This facilitates rapid calculation of the $\mathbf{r}_{ij}^{s,\Delta}$ as features using which the localization and characterization of AE sources can be made.

The proportion of energy concentrated in the propagating wave modes of the AE signals recorded at a particular sensor location (say i) is given in a scale-time sub-domain $s \in s_h$; $\Delta \in \Delta_h$ as

$$E_i^{s_h,\Delta_h} = \int_{s_h}\int_{\Delta_h} \left|X_i^{s,\Delta}\right|^2 ds\, d\Delta \tag{16}$$

where s_h and Δ_h are sub-domains defined the scale and time axis respectively. The energy concentration in a scale-time sub-domain $[s_h, \Delta_h]$ of the wavelet transformed correlation measures is given by

$$E_{ij}^{s_h,\Delta_h} = \int_{s_h}\int_{\Delta_h} \left|\mathbf{r}_{ij}^{s,\Delta}\right| ds\, d\Delta \tag{17}$$

Here $E_{ij}^{s,\Delta}$ gives a quantitative estimate of the product of energy concentration in the correlated signals. For the case when $i = j$ and $\Delta = 0$, Eq. 17 gives the energy concentration in the scale-time domain of the signal at sensor i. The quantity $E_{ij}^{s,\Delta}$ would be incorporated as features in the Gaussian process surrogate to facilitate the data-driven framework for AE source characterization in the

4 Gaussian Process Regression with Signal Features

In the previous section, the error indicator ε_r is mapped to the (x, y) coordinates of the AE source as $f_\varepsilon \colon [0, l_x] \times [0, l_y] \ni (x, y) \mapsto \varepsilon_r \in \mathbb{R}^+$. In this section, we introduce a Gaussian process model [27] to compute the map f_ε.

The coordinates (x, y) of the AE source over the domain $[l_x, l_y]$ serves as the parameter space or the input space over which the error indicator ε_r is trained. Here it is assumed that the observed error output vector $[\varepsilon_{r_1}(x, y), \ldots, \varepsilon_{r_{n_t}}(x, y)]$ are realisations of a Gaussian stochastic process with parametrized mean and covariance functions. The model structure is thus expressed as

$$\varepsilon_r(x, y) = \mathbf{h}(x, y)^\mathsf{T}\boldsymbol{\beta} + Z(x, y) \tag{18}$$

where $\mathbf{h}(x, y)$ is a vector of known functions and $\boldsymbol{\beta}$ is an unknown hyperparameter to be estimated from the data. The choice of $\mathbf{h}(\boldsymbol{\xi})$ is an active research area [28,29]. The function $Z(\cdot)$ is a stochastic process with mean zero and covariance function

$$Cov(Z((x, y)), Z(x', y')) = \sigma_z^2 C((x, y), (x', y')) \tag{19}$$

where $C(\cdot, \cdot)$ is a correlation function and σ_z^2 is the process variance, a hyperparameter that can also be estimated from the data. In order to choose a valid positive-definite correlation function, we consider products of one-dimensional correlations [30] with functional forms

$$C((x, y), (x', y')) = \exp\left\{ -b_i\left|x - x'\right|^2 - b_i\left|y - y'\right|^2\right\} \tag{20}$$

where $b_i > 0$ for all i with $x, x' \in [0, l_x]$ and $y, y' \in [0, l_y]$. The above correlation function is infinitely differentiable which is convenient when Gaussian processes are used to model not only the code output, but also its derivatives [31]. The vector of smoothness hyperparameters $\mathbf{b} = (b_1, b_2)^\mathsf{T}$ quantifies the rate at which the output varies over the spatial domain.

Let $\mathcal{D} = \{((x,y)_i, \varepsilon_r(x,y)_i) | i = 1, \ldots, n_t\}$ be a set of n_t training runs. Given this observed dataset, Bayes' theorem is used to estimate the hyperparameters as $\mathbb{P}(\beta, \sigma_\mathbf{z}^2, \mathbf{b}|\mathcal{D}) \propto \mathbb{P}(\mathcal{D}|\beta, \sigma_\mathbf{z}^2, \mathbf{b})\mathbb{P}(\beta, \sigma_\mathbf{z}^2, \mathbf{b})$, where $\mathbb{P}(\beta, \sigma_\mathbf{z}^2, \mathbf{b}|\mathcal{D})$ is the posterior probability of the hyperparameters, $\mathbb{P}(\mathcal{D}|\beta, \sigma_\mathbf{z}^2, \mathbf{b})$ is the likelihood, $\mathbb{P}(\beta, \sigma_\mathbf{z}^2, \mathbf{b})$ is the prior of the hyperparameters, and $\mathbb{P}(\mathcal{D})$ is the marginal likelihood. A detailed derivation of prior-to-posterior analysis along with the hyperparameter estimation is given in [32,33].

The assumed Gaussian process prior on the code's output implies that the posterior distribution is also a Gaussian process. Once the hyperparameters are estimated, the mean of the posterior distribution approximates the output of ε_r^* at any point (x^*, y^*) on the physical domain. The variance of the posterior distribution quantifies the uncertainty that arises from having only a limited number of observations [34]. It can be shown that the posterior distribution is a Gaussian with the posterior mean and covariance function given in [27]. The detailed analysis is not repeated here for the sake of brevity.

The error indicator can hence be evaluated at any (x, y) (as predicted by the multivariate least square regression problem in Eq. 11) and $\varepsilon(x, y)$ gives the associated uncertainty in the acoustic source location. The advantage offered by the above approach is that following the trained Gaussian process surrogate, the error around any predicted value can be computed directly from the surrogate in a computationally efficient manner. This enables real-time damage localization in in-situ operating structures.

The error $\varepsilon(x, y)$ constructed here is used in conjunction with the least square estimator (LSE) using the correlation measures (presented in Sect. 3) gives a probabilistic prediction of the AE source location conditional on the training data. At the prediction stage the reduced correlation matrix, constructed using the optimal basis Φ_{Δ,τ_s}, is mapped to the AE source and the associated error is evaluated from the trained Gaussian process surrogate. Thus a robust probabilistic estimate of the AE source location is obtained using the correlation characteristics of the signal collected from the distributed sensor network.

5 Experimental Setup

In order to develop and validate the above approach for damage localization, data was collected from a stiffened carbon fibre composite panel, representative of an aerospace structure.

5.1 Sample Details

The manufactured stiffened panel is presented in Fig. 2. The stiffeners were purchased from Easy Composites Ltd. (Staffordshire, UK) they consist of a 90° L-shaped cross-section with laminate thickness of 3 mm and cross-section dimensions of 25 × 25 mm and a length of 600 mm. The stiffeners are made from 2 × 2 twill high strength carbon fibres in an epxoy matrix with fibres aligned in the 0° and 90° directions. The skin was manufactured from 12 plies of Cytec MTM28/T800HB/200/42% 2 × 2 twill weave carbon fibre composite material with a $(0)_{12}$ layup and was cured in an autoclave in line with the manufacturers recommended cure cycle. Following curing the thickness was 2.85 mm and the skin panel was cut to 550 × 600 mm using a water cooled diamond tipped cutting wheel. The skin and stiffeners were lightly abraded and degreased in preparation for bonding using permabond ET5429 adhesive. The final overall panel dimensions were 550 × 600 mm with the stiffeners running vertically. Two aluminium dumby ribs (representative of attachment to a wing rib for a composite wing skin) were attached at 1/4 and 3/4 hight of the panel by drilling and bolting using 12 M4 bolts for each, as seen in Fig. 2(b). The top and bottom edges of the panel were potted into 20 mm deep aluminium frames using Airtech TMR2001 high temperature laminating resin to allow application of a compressive load (not considered in this paper). A 500 × 500 mm grid with 50 mm resolution was applied to the skin side of the panel (Fig. 2(a)) and was used to aid the collection of training data.

(a) Grid side (b) Stiffener side

Fig. 2. Manufactured panel (a) from stiffener side showing sensor positions and (b) from skin side showing grid used for data collection.

5.2 Panel Instrumentation

The panel was instrumented with five AE sensors as seen in Fig. 2(b), five McWade NS-3303 (300 kHz), the larger gold coloured sensors, and three Mistral Group Ltd. Nano30 sensors (300 kHz), the smaller silver sensors. Four McWade sensors are arranged in a 275 × 175 mm square with the fifth placed centrally. The Nano30 sensors are arranged in a 75 mm spaced triangular array. The sensor outputs are amplified by 40dB using a McWade PA3303 pre-amplifier for the McWade sensors and a Mistral Group Ltd. 2/4/6 (20–1200 kHz) pre-amplifier for the Nano30 sensors. A silicon adhesive (Loctite 595) was used to attach the sensors and provide a suitable acoustic couplant, the adhesive was allowed to cure for 24 hrs before any data acquisition or testing was undertaken and the correct coupling of the sensors was assessed using a H-N source [21,22]. The AE data was recorded with a Mistras Group Ltd. PCI-2 acquisition system using a 45dB threshold level. The detected signals were sampled at 5 MHz for a duration of 1000 µs: 600 µs after the threshold crossing point and with a pre-trigger of 400 µs. The time when the threshold crossing occurs for each test point has been recorded and shown in Fig. 3 and has been discussed in Sect. 6.

5.3 Data Acquisition

AE data was collected from the manufactured composite panel using a H-N source [21,22] to excite artificial AE waves. The H-N source is recognised as a standard reference source (ASTM E976) for AE testing and requires the fracture of a 0.5 mm diameter 2 H pencil lead agains the sample surface at an angle of 30°. This is facilitated using a propelling pencil fitted with a plastic rocker the rocker is placed on the sample and the pencil rotated until the lead contacts the surface and then fractures. This results in minute elastic deformation of the surface under the tip of the pencil lead and when the lead fractures the elastic energy stored in the surface is rapidly released and excites a broadband elastic stress wave. This is highly representative of the rapid release of elastic energy that occurs when cracks and fractures grow in materials and hence is why it has been adopted as a standardised reference source for AE testing. The H-N source commonly excites a larger amplitude signal than a real fracture, however, the source mechanism is still representative and it has been shown to be a suitable artificial source for training a system for the detection and analysis of AE signals from real fracture events as seen in the Delta T Mapping techniques discussed above.

The data used in this work were collected using H-N sources performed at the nodes of the grid applied to the front of the stiffened panel. Ten H-N sources were conducted at each of the grid nodes within the 500×500 mm grid shown in Fig. 2(a) and for each H-N source five AE signals were recorded and stored (one from each sensor).

6 Results and Discussion

A sample of the collected sensor data is shown in Fig. 3 where each red dot signify the time (x-axis) of arrival of the wave at a sensor (5 sensors denoted by "Ch 1" to "Ch 5") and its maximum amplitude (y-axis). A *hit* is recorded when the signal received at a particular sensor exceeds a preset threshold value. Figure 3 shows the hits recorded simultaneously at all the five sensor locations. All the hits between the vertical blue lines on each channel indicate a test performed with the H-N source at a grid point (Fig. 2(a)) on the panel. Around 10 tests have been performed at each grid point due to which there are approximately 10

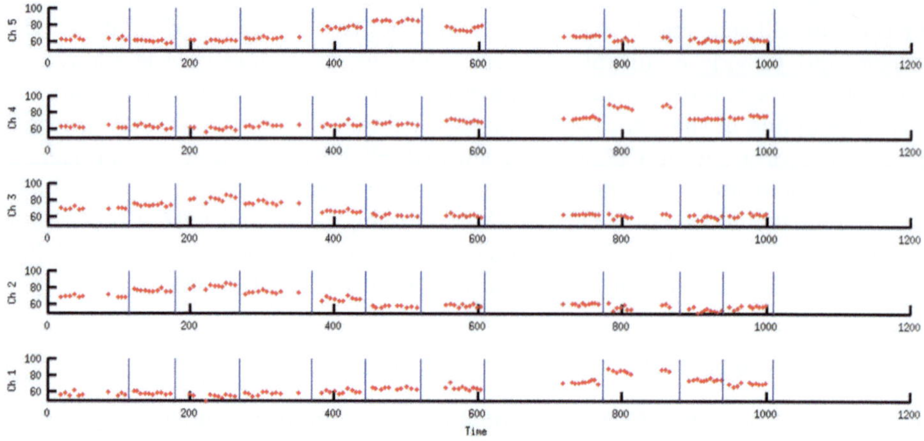

Fig. 3. Normalized correlation coefficients of the signal collected at the distributed sensor network (only the first 4 out of the total of 5 sensor signals shown) over discrete time steps $-100 \leq \tau_s \leq 100$

hits (red dots) between two consecutive blue lines. The red dots are aligned in time across all channels except for the small time difference associated with the waves travelling different distances from the source to reach each sensor. All the tests represented in Fig. 3 have been performed along the 11 test grid points at $x = 0$ (the bottommost line in the grid). This figure essentially represents only the time of arrival for the signal at the individual sensors, but the correlation measure between the sensors is shown in the next figure.

The optimal basis on which the correlation matrix is projected to obtain a compressed representation of the information collected at the sensors. A total of 30 basis functions have been used to approximate the terms of the correlation matrix. The basis functions are orthonormalized and have been calculated using the left eigenvectors of the correlation matrix using the singular value decomposition. The rapidly decaying singular value spectrum of the ensures that a good approximation is obtained with the reduced basis. The correlation matrix projected on these 30 optimal bases is termed as the reduced correlation matrix (Fig. 4).

The accuracy of the identified AE source using the LSE is shown in Fig. 5. The black dots indicate the training points on the 10×10 grid of the composite panel while the red dots indicate the approximate identified acoustic source locations obtained using the least square mapping of the reduced correlation matrix to the vector of the distance of the acoustic source from the sensor network (as discussed in Eq. 11). The error $\varepsilon_r(x, y)$ is given by the distance of the identified source locations (red dots) to the known position of AE (black squares).

This error indicator $\varepsilon_r(x, y)$ varies as per the accuracy of the fit and a Gaussian process surrogate is used to build a probabilistic error surface over the spatial domain as discussed in Sect. 4. Figure 5 gives the mean and standard

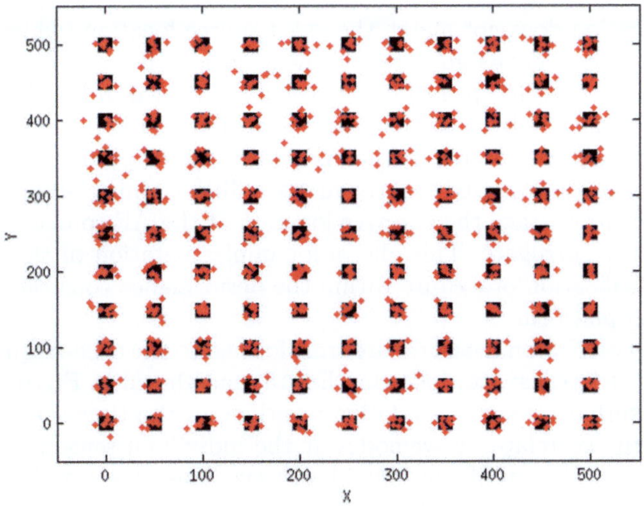

Fig. 4. The accuracy of fit of the training dataset based on the observed correlation matrix over the physical domain of the composite panel. The square blocks shown in black constitute the actual test grid while the red dots are LSE of the AE source.

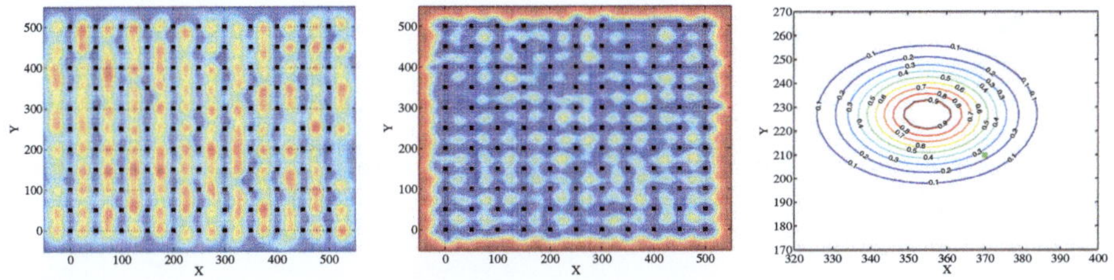

(a) Mean of the error surface (b) Standard deviation of the error surface (c) Standard deviation of the error surface

Fig. 5. The mean an standard deviation of the error response surface over the domain of the plate. The colormap of (a) varies between 0–45 mm while for the standard deviation is between 0–20 mm

deviation of the posterior error surface conditional on the observations. This gives a distribution of the accuracy of the LSE as a function of the spatial coordinates. It can be verified that the mean error is almost zero at those training locations where the accuracy of the least square method is maximum. The locations where there are no training points show a high value of mean error and correspondingly high standard deviation of error which is an expected behavior.

Figure 5(c) shows the probability distribution associated with an identified test point. The actual location of the acoustic source is shown as the green dot (although this information has not been used to train the model). The assumed Gaussian process surrogate predicts the probability distribution of the identified acoustic source location conditional on the training data and though the mean true location does not match the exact source location (which is expected due to experimental errors and measurement noise), the true source location is contained within the 95% probability envelope as shown in Fig. 5(c).

Thus effectiveness of the methodology is demonstrated in terms of the accuracy of identifying the source of the AE. The cost of producing the gaussian process surrogate is incurred mostly in an offline training stage while in the online identification stage the effective location of the AE source is derived efficiently from the surrogate. This allows for implementation of the technique in real-time identification procedure within the performance constraints of the signal processing platforms.

The cross-correlation measures are transformed to the time-frequency domain using the wavelet transform, following Eq. 15, and shown in Fig. 6. The wavelet transformed surfaces show the relative positions in the time-scale of the point of arrival of the correlated wavemodes at the individual sensors in addition to the degree of correlation of the individual wavemodes (as amplitude along the z-axis). These are used as additional features (in addition to the normalized cross-correlated metrics given in Fig. 3).

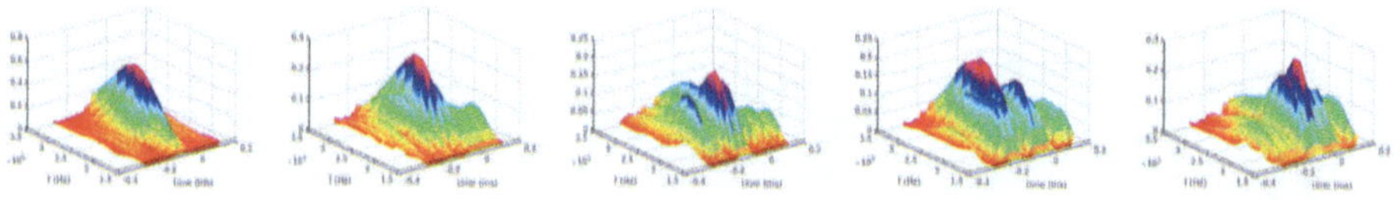

Fig. 6. The wavelet transformed surface of the cross-correlation measure between sensor pairs i.e. $\left|\mathbf{r}_{ij}^{s,\Delta}\right|$ where $i = 1$ and $j = 1, \ldots, 5$

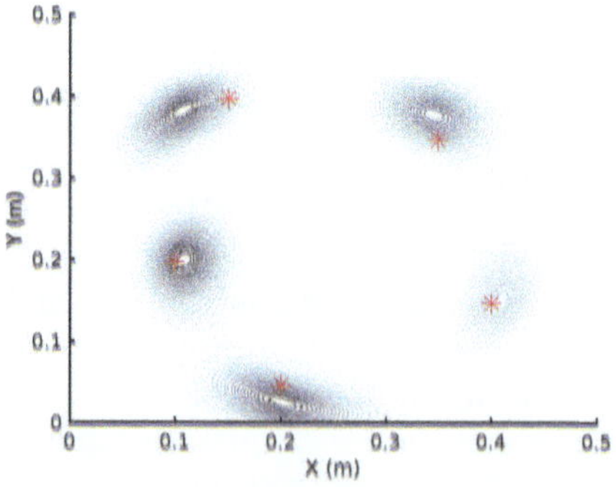

Fig. 7. The accuracy of localization of damage at multiple locations. The true AE location sources are marked with $*$ while the isoprobability contours describe the possible location of the AE sources.

Using the features shown in the cross-correlation spectrum and features from the wavelet transformed surfaces given in Fig. 6, the localization of the damage on the multilayered composite structure with H-N sources have been performed and shown in Fig. 7. This shows the location of five different AE sources using the isoprobability contours while the actual location is shown with the red marker. The figure superimposes identification performed for multiple AE source signals and the isoprobability contours signify the probability of source location based on the information collected from the 5 sensors distributed on the surface of the composite panels. The results show a good accuracy of AE source localization based on the identification features derived from the cross-correlation metrics and does not rely on any assumed shape/geometry or ad hoc regularization conditions applied for identification. Thus this has the potential to be applied in industrial scenarios as a robust identification technology for acoustic source location. The proposed methodology would also be applicable for structures operating under service loads (with the energy content concentrated in the low frequency regime) which is a fundamental advantage with AE-based SHM techniques.

Another set of tests were performed on the composite panel, when it was subject to compressive fatigue loading cycles between 0.5–5 kN (peak-to-peak) for 1000 cycles at a frequency of 1 Hz. The distributed sensory network 'listened' to damages developing in the composite structure and waveforms of the signal that were collected has been utilized for (a) source localization and (b) characterization of the nature of damage. The source localization was performed following the same procedure reported earlier in this section for H-N sources based on the wavelet transformed features of the full correlation measure of the signals at the sensors. The damage classification was performed using the k-nearest neighbours (KNN) algorithm. The KNN algorithm is a supervised, non-parametric and instance-based method used for classification.

A training dataset comprising of a mixture of H-N sources and fatigue test data was used to train a KNN model. The training data set thus comprised of signal features X_f with a categorical variable c pairs as $\mathcal{D} = \{(X_f, c)_i,\ i = 1, \ldots, n_{train}\}$. The categorical variable c takes binary values of $[0, 1]$ which correspond to the two different class of tests - the H-N source and the fatigue loading. The Mahalonobis distance was used to determine the distance in the feature space for the test data and a probabilistic estimate of the class variable associated with each test data point was determined. It has been observed that a validation accuracy of 97% was obtained for the aggregate of test data using the KNN algorithm. Thus the classification algorithm based on the signal features were successful in identifying the acoustic from H-N test data to that of fatigue loading.

The strength of the KNN algorithm in accurate classification of the test type shows the potential of the used signal features for identifying the nature of damage. The application of the proposed methodology in this paper to identify the generation of cracks in the composite panel under fatigue loading cycles shows the potential of the proposed method for an automated probabilistic identification framework to improve the accuracy and minimize the computational overhead associated with the damage detection in composites.

7 Conclusion and Future Work

The study demonstrates the effectiveness of the proposed correlation feature based identification of AE sources in applications of non-destructive testing. The effectiveness of the method stems from using direct evaluations of normalized autocorrelation functions which effectively reduces the impact of signal noise and captures the essential correlation information between the signals recorded with the distributed sensor network. A reduced set of basis vectors which compress the correlation information reduces the memory and computational overhead associated with the method. A Gaussian process surrogate is fitted on the error surface over the spatial domain to explicitly consider the error associated with deterministic evaluators. The results demonstrate the effective of the method in predicting the source location with test data on the same panel.

The promises of the study extends to a number of interesting future investigations for extending and improving the proposed methodology to address additional challenges. Some of these are

- The wavelet transformed features of the correlation metrics can be used to derive the energy distribution in propagating modes from which the characterization of the damage type (based on categorical damage models would be accomplished) can be performed.
- The portability of the trained surrogate i.e. the ability of the trained model to predict AE sources in nominally similar test panels.
- A hierarchical probabilistic model of the AE source mapped to the correlation matrix, where identification stages can be compartmentalized to correspond to various levels of refinement.

Subsequent work would also focus on using physics based model of ultrasonic wave propagation in composite panels which would provide a means of identifying the nature of damage induced in the structure which has not been addressed in this study. This would allow us to go beyond the black-box input-output mapping techniques which is expected to significantly improve the performance of the identification algorithm and provide additional information pertaining to the degradation of health of operational structures.

References

1. Kundu, A., Adhikari, S., Friswell, M.I.: Stochastic finite elements of discretely parameterized random systems on domains with boundary uncertainty. Int. J. Numer. Methods Eng. **100**(3), 183–221 (2014)
2. Miller, R., Carlos, M., Findlay, R., Godinez-Azcuaga, V., Rhodes, M., Shu, F., Wang, W.: Acoustic emission source location, 3rd edn., pp. 121–146. ASNT (2005)

3. Eaton, M., Pullin, R., Holford, K.: Towards improved damage location using acoustic emission. J. Mech. Eng. Sci. Proc. Inst. Mech. Eng. Part C **226**(9), 2141–2153 (2012)
4. Kennedy, M.C., O'Hagan, A.: Bayesian calibration of computer models. J. R. Stat. Soc. Ser. B-Stat. Methodol. **63**(3), 425–450 (2001)
5. Kundu, A., Matthies, H., Friswell, M.: Probabilistic optimization of engineering system with prescribed target design in a reduced parameter space. Comput. Methods Appl. Mech. Eng. **337**, 281–304 (2018)
6. Miller, R., Anastasopoulos, A., Carlos, M., Demeski, R., Vallen, H., Walker, J.: Acoustic Emission Signal Processing, 3rd edn., pp. 147–180. ASNT (2005)
7. Ziola, S., Gorman, M.: Source location in thin plates using cross-correlation, **90**, 2551–2556 (1991)
8. Aljets, D., Chong, A., Wilcox, S., Holford, K.: Acoustic emission source location in plate like structures using a closely arranged triangular sensor array. J. Acoust. Emiss. **28**, 85–98 (2010)
9. Hamstad, M., O'Gallagher, A., Gary, J.: A wavelet transform applied to acoustic emission signals: part 2: source location. J. Acoust. Emiss. **20**, 62–82 (2002)
10. Lokajicek, T., Klima, K.: A first arrival identification system of acoustic emission (AE) signals by means of a higher-order statistics approach. Meas. Sci. Technol. **17**, 2461–2466 (2006)
11. Akaike, H.: Markovian representation of stochastic processes and its application to the analysis of autoregressive moving average processes. Ann. Inst. Stat. Math. **26**, 363–387 (1974)
12. Kurz, J.H., Grosse, C., Reinhardt, H.W.: Strategies for reliable automatic onset time picking of acoustic emission and of ultrasound signals in concrete. Ultrasonics **43**, 538–546 (2005)
13. Pearson, M., Eaton, M., Featherston, C., Pullin, R., Holford, K.: Improved acoustic emission source location during fatigue and impact events in metallic and composite structures, pp. 1–18 (2016)
14. Kundu, T., Das, S., Jata, K.: Point of impact prediction in isotropic and anisotropic plates from the acoustic emission data. J. Acoust. Soc. Am. **122**(4), 2057–2066 (2007)
15. Kundu, T., Das, S., Martin, S., Jata, K.: Locating point of impact in anisotropic fibre reinforced composite plates. Ultrasonics **48**, 193–201 (2008)
16. Ciampa, F., Meo, M.: A new algorithm for acoustic emission localisation and flexural group velocity determination in anisotropic structures. Compos. Part A **41**(12), 1777–1786 (2010)
17. Baxter, M., Pullin, R., Holford, K., Evans, S.: Delta T source location for acoustic emission. Mech. Syst. Signal Process. **21**(3), 1512–1520 (2007)
18. Eaton, M.J., Pullin, R., Holford, K.: Acoustic emission source location in composite materials using delta T mapping. Compos. Part A **43**(6), 856–863 (2012)
19. Al-Jumaili, S., Pearson, M., Holford, K., Eaton, M., Pullin, R.: Acoustic emission source location in complex structures using full automatic delta T mapping technique **72–73**, 513–524 (2016)
20. Scholey, J., Wilcox, P., Wisnom, M., Friswell, M.: A practical technique for quantifying the performance of acoustic emission systems on plate-like structures. Ultrasonics **49**, 538–548 (2009)
21. Hsu, N.N., Breckenridge, F.R.: Characterization and calibration of acoustic emission sensors. Mater. Eval. **39**(1), 60–68 (1981)
22. ASTM.: A standard guide for determining the reproducibility of acoustic emission sensor response. American Society for Testing and Materials, **E976** (2010)

23. Schumacher, T., Straub, D., Higgins, C.: Toward a probabilistic acoustic emission source location algorithm: a Bayesian approach, **331** (2012)
24. Zarate, B., Pollock, A., Momeni, S., Ley, O.: Structural health monitoring of liquid-filled tanks: a Bayesian approach for location of acoustic emission sources, **24** (2014)
25. Kundu, A., Eaton, M.J., Al-Jumali, S., Sikdar, S., Pullin, R.: Acoustic emission based damage localization in composites structures using Bayesian identification. J. Phys.: Conf. Ser. **842** (2017)
26. Kundu, A., Sikdar, S., Eaton, M., Navaratne, R.: Probabilistic method for damage identification in multi-layered composite structures. In: Proceedings of the 9th European Workshop on Structural Health Monitoring (2018)
27. Kundu, A., DiazDelaO, F., Adhikari, S., Friswell, M.: A hybrid spectral and meta-modeling approach for the stochastic finite element analysis of structural dynamic systems. Comput. Methods Appl. Mech. Eng. **270**, 201–219 (2014)
28. Oakley, J.E., O'Hagan, A.: Probabilistic sensitivity analysis of complex models: a Bayesian approach. J. R. Stat. Soc. B **66**(3), 751–769 (2004)
29. Keane, A., Nair, P.: Computational Approaches for Aerospace Design. Wiley, Chichester (2005)
30. Sacks, J., Welch, W., Mitchell, T., Wynn, H.: Design and analysis of computer experiments. Stat. Sci. **4**(4), 409–435 (1989)
31. O'Hagan, A.: Bayesian Statistics 4, chap. In: Some Bayesian Numerical Analysis, pp. 345–363. Oxford University Press, Cambridge (1992)
32. Haylock, R., O'Hagan, A.: Bayesian statistics 5, chap. In: On Inference for Outputs of Computationally Expensive Algorithms with Uncertainty on the Inputs. Oxford University Press, Oxford (1996)
33. Oakley, J.: Eliciting Gaussian process priors for complex computer codes. Stat. **51**(1), 81–97 (2002)
34. Rougier, J.: Probabilistic inference for future climate using an ensemble of climate model evaluations. Clim. Chang. **81**(3), 247–264 (2007)

Selection of Small Sensor Arrays for Localization of Damage in Complex Assemblies Using Vibro-Acoustic Signals

Philip Becht[1,2,3](✉), Elke Deckers[1,2], Claus Claeys[1,2], Bert Pluymers[1,2], and Wim Desmet[1,2]

[1] Department of Mechanical Engineering, KU Leuven, 3001 Leuven (Heverlee), Belgium
Philip.Becht@kuleuven.be
[2] DMMS Lab, Flanders Make, 3001 Leuven (Heverlee), Belgium
[3] SIM M3 Program, Technologiepark 935, 9052 Zwijnaarde, Belgium
http://www.mech.kuleuven.be/en.pma/research/mod

Abstract. A procedure is proposed to suggest a good placement of sensors to form an array of a pre-defined size in order to localize damage in complex assemblies using the Time-Reversal MUltiple SIgnal Classification (TR-MUSIC) algorithm. The technique takes into account the amount of information that can be obtained from each possible sensor location and the difference in information provided by each possible sensor location. The proposed procedure is successfully validated on the numerical example of an aluminium framework and on the experimental example of an aluminium framework covered by a honeycomb panel.

Keywords: Selection sensor array · Time-Reversal MUSIC · Non-destructive testing and evaluation

1 Introduction

The interest to timely detect damage in a structure can be motivated either by safety [1,2] and/or by economical [3] considerations. From a safety point of view, many structures, such as e.g. airplanes are about to reach the end of the life span they were originally designed for. As it is economically not feasible to replace these old structures, they require a well defined monitoring scheme [1] in order to allow for a repair before a catastrophic event happens. Furthermore, the increasing use of novel lightweight materials with often unknown long-term degradation imposes an additional safety driven motivation to inspect the integrity of a mechanical structure [1,2].

An additional economical driver for the inspection of a product is the shift towards predictive maintenance, meaning that a component is only replaced once damage is detected, which obviously requires knowledge about its health

M. A. Wahab (Ed.): DAMAS 2019, LNME, pp. 263–282, 2020.
https://doi.org/10.1007/978-981-13-8331-1_19

status. Another example for an economical motivation is the aim to reduce the out- and throughput of damaged products through a mass production line by implementing an end-of-line or in-line inspection strategy [4].

Currently, a huge number of strategies making use of explicitly induced vibrations of a test object is available. Those can be classified in low- and high frequency methods [5]. Low-frequency methods typically operate in a frequency range, where the vibration is dominated by a small number of modes [6] and make use of the change of modal parameters, such as natural frequencies or mode shapes [5,7]. The wavelength in this frequency range is rather large as compared to the overall size of the structure and as a consequence, the modal features show only little sensitivity to small damage.

Commonly known representatives for high frequency damage detection techniques are ultrasonic bulk waves and guided waves, which operate usually in a high kHz (guided waves [8,9]) to multiple MHz range (ultrasonic bulk waves [10]). Due to this high operational frequency, the wavelength compared to the size of the test object is significantly smaller than for low-frequency techniques, which results in a higher sensitivity even to small damage. On the other hand, the classical guided waves inspection technique requires that the wave reflected from or transmitted through the damage is clearly visible in the recorded signal. This requirement restricts this strategy to simple geometries [11] and large structures with relatively small inspection regions [12]. Ultrasonic bulk waves do not suffer from this restriction, but due to the small inspection zone covered by these techniques, they rely on a fully accessible inspection zone [13]. A time consuming point-by-point inspection of the test object is the consequence.

As comes to the fore from the discussion in the previous paragraphs, neither low- nor high-frequency techniques can be used straightforward for the rapid inspection of moderately or highly complex parts and components, which is required for their application to the examples given in the beginning of this section. For this reason, Becht et al. [6,14,15] proposed to perform inspection in a mid-frequency range, where the necessary sensitivity of high-frequency methods is still present and the robustness and applicability to complex parts, characteristic for low-frequency techniques is not yet lost. In their publications, the authors use a Time-Reversal MUltiple SIgnal Classification (TR-MUSIC) algorithm to localise damage.

As demonstrated in [14], it is possible to achieve a reliable localisation of damage even if the installed number of excitations and sensors is below the theoretical limit. However, as for all vibration based techniques, also for this approach the placement of all sensors and excitations is extremely important for a reliable inspection [16]. This is the underlying motivation of this publication, proposing an algorithm leading to a good choice of sensor locations to form an array.

The paper is structured as follows: In Sect. 2, the TR-MUSIC algorithm is explained in order to facilitate the reader to follow the proposed selection algorithm for a sensor array, which is detailed in Sect. 3. This algorithm is numerically verified and experimentally validated in Sects. 4 and 5, respectively. Concluding remarks are given in Sect. 6.

2 Time-Reversal MUSIC

The TR principle states that if the response of an excitation, recorded at multiple sensor locations is time-reversed (the measured signal is re-emitted at the sensor locations reversed in time - last time-step is re-emitted first), the wave-field will converge at the source location due to a constructive superposition of the re-emitted signals of all sensors [17,18]. This principle can also be used for the localisation of a damage, regarding the damage as a scatterer and the scatterer as the source of a scattered wave-field. The damage can then be found by searching the location where the time-reversed wave-field converges. The scattered wave-field $\Psi_s(\omega, x_n, x_m)$ due to an excitation x_m, recorded at sensor x_n can be determined using Eq. (1), provided that the healthy condition $\Psi_{init}(\omega, x_n, x_m)$ is known and the current, potentially damaged structure $\Psi_{meas}(\omega, x_n, x_m)$ is measured [19]:

$$\Psi_{meas}(\omega, x_n, x_m) = \Psi_{init}(\omega, x_n, x_m) + \Psi_s(\omega, x_n, x_m). \tag{1}$$

Equation (1) holds for one rotational frequency ω. In the following, frequency dependency will be implicitly assumed if not stated otherwise.

An alternative to improve the performance of classical TR is the TR-MUSIC algorithm [20]. This algorithm requires a number of sensors N and excitations M and the scattered field measured for all combinations of N and M. This is stored in the matrix $\boldsymbol{K}$ with the element

$$K_{n,m} E_{x_m} = \Psi_s(x_n, x_m), \tag{2}$$

where E_{x_m} is the $m - th$ excitation.

It can be shown [6,14,21] that the singular value decomposition of $\boldsymbol{K}$ gives

$$\boldsymbol{K} = \boldsymbol{\Phi} \boldsymbol{\chi} \boldsymbol{\Psi}^H, \tag{3}$$

where the M singular vectors in $\boldsymbol{\Psi}$ describe the propagation from the excitation locations to the scatterer and the N singular vectors in $\boldsymbol{\Phi}$ describe the propagation from the scatterer to the sensors. The scattering event itself is considered in the diagonal matrix $\boldsymbol{\chi}$ containing the singular values. The superscript $\bullet^H$ denotes the complex conjugate transpose of a matrix. Note that this formulation explicitly also holds for extended and multiple scatterers.

Assuming that the number of sensors and excitations is higher than the number of singular values necessary to fully describe the scattered wave field R_0, there are R_0 non-zero singular values in $\boldsymbol{\chi}$, and the same number of physically meaningful singular vectors in $\boldsymbol{\Phi}$, mapping the damaged zone to the sensor array. The space spanned by these singular vectors is called the signal space S_{Φ}. In addition to the R_0 singular vectors in S_{Φ}, there are $N - R_0$ singular vectors without any physical interpretation, but with the mathematical property of being orthogonal to S_{Φ}. The space spanned by these vectors is referred to as noise space N_{Φ}.

From the TR principle, it is known that the best possible focus from the sensor side on the damage is provided by the time-reversed signal received from the damage. This knowledge is used in the definition of a column vector $\boldsymbol{g}_{\Phi,x}$, where the $m-th$ of the in total M elements in $\boldsymbol{g}_{\Phi,x}$ is a transfer function from an arbitrary point x in the inspection zone to the $m-th$ sensor. Since all the information relating to the propagation from the damage to the sensor array is given by the signal space singular vectors, it follows that $\boldsymbol{g}_{\Phi,x}$ fully lies in S_Φ if x is a point in the damaged zone. In consequence, due to the orthogonality between S_Φ and N_Φ, in this case, $\boldsymbol{g}_{\Phi,x}$ shares no information with N_Φ, which results mathematically in:

$$\boldsymbol{g}_{\Phi,x}^H \boldsymbol{\Phi}_n = 0, \text{ for } \boldsymbol{\Phi}_n \in N_\Phi, \tag{4}$$

where $\boldsymbol{\Phi}_n$ is one of the $N - R_0$ noise space singular vectors and the superscript $\bullet^H$ indicates a complex conjugate transpose. On the other hand, if x is not a point in the damaged zone, it follows that

$$\boldsymbol{g}_{\Phi,x}^H \boldsymbol{\Phi}_n \neq 0, \text{ for } \boldsymbol{\Phi}_n \in N_\Phi, \tag{5}$$

since $\boldsymbol{g}_{\Phi,x}$ shares information with the noise space.

This property is used for the definition of a pseudospectrum

$$I(x) = \left(\mid \boldsymbol{g}_{\Phi,x}^H \boldsymbol{\Phi}_n \mid\right)^{-1}, \text{ for } \boldsymbol{\Phi}_n \in N_\Phi, \tag{6}$$

which reaches its maximum if x is a point in the damaged zone.

In order to improve the robustness of the method a summation over all noise space singular vectors [22, 23] and a multiplication over the measured frequency range [24, 25] is applied. This results in the formulation of the pseudospectrum as it is used in this publication:

$$I(x) = \left(\prod_{f=1}^{N_f} \sum_{n=R_0+1}^{N} \mid \boldsymbol{g}_{\Phi,x,f}^H \boldsymbol{\Phi}_{n,f} \mid \right)^{-1}, \text{ for } \boldsymbol{\Phi}_n \in N_\Phi. \tag{7}$$

The value R_0 depends on the test object and the complexity, size and severity of the damage. In the general case of a crack or volume damage, the singular values in Eq. (3) slowly converge towards zero, but do never reach it, such that a truncation is needed to make the TR-MUSIC theory applicable to this case. Becht et al. [6, 14] showed that even when applying a very rudimentary truncation and for the case that M and/or N are smaller than R_0, the algorithm still reliably highlights (parts of) the damaged area. It is this property, which allows to keep the number of sensors and excitations low, but comes at the price that the installed sensors span only a subset of S_Φ. In this scenario, a good placement of the sensors becomes particularly important.

The same derivation introduced in this section for the pseudospectrum using the sensor array can also be derived based on the excitation array, applying the respective substitutions (e.g. $\boldsymbol{g}_\Psi$ instead of $\boldsymbol{g}_\Phi$).

3 Selection Algorithm for Sensor Array

If there are not sufficient sensors and excitations to span the complete space S_{Φ}, the matrix $\boldsymbol{K}$ changes to $\boldsymbol{K}^{meas}$, which has a lower rank than $\boldsymbol{K}$ and the theoretically desired R_0 is replaced by the really applied value R_0^{meas}. In this case, the available sensors and excitations must be placed in a way, so that they span a space, which comes as close as possible to S_{Φ} in order to have a greater chance that the rank of $\boldsymbol{K}^{meas}$ comes as close as possible to the rank of $\boldsymbol{K}$. An obvious first approach to this problem could be a simple trial and error method, where all possible sensor and excitation locations are tested for all possible damage scenarios. Certainly, for a very limited number of options, this is a promising approach, however, the number of possible sensor arrays increases rapidly, so that there are e.g. 184756 possible array configurations for an array of 10 elements and 20 possible locations to place each element. Furthermore, the performance of all array configurations would need to be evaluated for every single damage that can be imagined. This simple example illustrates that the trial and error approach quickly reaches its limits. Therefore, in the following, a procedure is proposed to find combinations of sensor locations providing a large amount of information.

In a first step, the best case, in which all possible sensor locations are used is assumed. The link between these N^{max} sensor locations and the X points in the inspection zone, at which $I(x)$ is calculated is given by the $X \times N^{max}$ matrix $\boldsymbol{G}_{\Phi,f}^{max}$, which is defined as

$$\boldsymbol{G}_{\Phi,f}^{max} = \left[\boldsymbol{g}_{\Phi,x_1,f}\; \boldsymbol{g}_{\Phi,x_2,f} \cdots \boldsymbol{g}_{\Phi,X,f}\right]^T . \tag{8}$$

In order to take into account multiple frequencies, but only those that contain a potentially valuable contribution, a frequency selection as proposed in [14] is performed and the information available in the F^{select} selected frequencies is compiled in the $XF^{select} \times N^{max}$ matrix

$$\boldsymbol{G}_{\Phi}^{max} = \left[\boldsymbol{G}_{\Phi,f_1}^{max\,T}\; \boldsymbol{G}_{\Phi,f_2}^{max\,T} \cdots \boldsymbol{G}_{\Phi,F^{select}}^{max\quad T}\right]^T . \tag{9}$$

Each of the N^{max} columns of $\boldsymbol{G}_{\Phi}^{max}$ links to exactly one possible sensor location and contains the transfer path at all frequencies considered from this sensor to all points at which $I(x)$ is calculated.

A singular value decomposition applied on $\boldsymbol{G}_{\Phi}^{max}$ results in the unitary matrices $\boldsymbol{U}$ and $\boldsymbol{V}$ (unitary: $\boldsymbol{U}\boldsymbol{U}^H = \boldsymbol{I}$, with the identity matrix $\boldsymbol{I}$) and a diagonal matrix $\boldsymbol{S}$, such that $\boldsymbol{G}_{\Phi}^{max} = \boldsymbol{U}\boldsymbol{S}\boldsymbol{V}^H$. The $n-th$ column of $\boldsymbol{G}_{\Phi}^{max}$ is then exactly

$$\boldsymbol{G}_{\Phi,n}^{max} = \boldsymbol{U}\boldsymbol{S}\boldsymbol{V}_n^H , \tag{10}$$

where $\boldsymbol{V}_n$ is the $n-th$ row of $\boldsymbol{V}$. This operation is depicted in Fig. 1.

Assuming $\boldsymbol{G}_{\Phi}^{max}$ to be full rank, the best possible approximation with a rank N, where N is the desired number of sensors, is given by

$$\boldsymbol{G}_{\Phi,n}^{max} \approx \boldsymbol{U}_{trunc}\boldsymbol{S}_{trunc}\boldsymbol{V}_{n,trunc}^H , \tag{11}$$

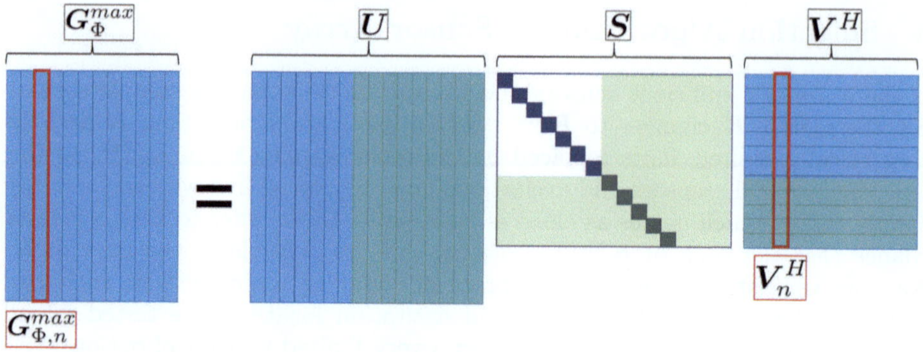

Fig. 1. Illustration of singular value decomposition and truncation corresponding to Eqs. (10) and (11). Highlighted in red: How to calculate the $n-th$ column of $\boldsymbol{G}_{\Phi}^{max}$, relating to the $n-th$ potential sensor location and in light green: The influence of the truncation.

with $\boldsymbol{V}_{n,trunc}$ the first N entries of $\boldsymbol{V}_n$, $\boldsymbol{U}_{trunc}$ the first N columns of $\boldsymbol{U}$ and $\boldsymbol{S}_{trunc}$ a diagonal matrix containing the first N diagonal elements of $\boldsymbol{S}$. This is illustrated in Fig. 1 by the light green pattern, which corresponds to the neglected, truncated elements. $\boldsymbol{V}_{n,trunc}$ conforms to the blue elements in the red rectangle marking $\boldsymbol{V}_n$. The issue with this approximation is that it lacks physical interpretation. A result could e.g. be $[0.5 \text{ sensor } n_1 \ \sqrt{3} \text{ sensor } n_2 \cdots -3 \text{ sensor } N^{max}]$, which might be the best low rank approximation of $\boldsymbol{G}_{\Phi}^{max}$, but is practically not helpful, since still all sensors would need to be installed. Therefore, the aim is to find the N sensor locations out of the N^{max} possible sensor locations that in combination result in the best possible approximation of the information in $\boldsymbol{G}_{\Phi}^{max}$.

In order to achieve this goal, the fact that each column of $\boldsymbol{G}_{\Phi}^{max}$ corresponds to one possible sensor location is used. Furthermore, the best possible low rank approximation of each column is known from Eq. (11). In the hypothetical case that the first N entries of $\boldsymbol{V}_n$ and thus all entries of $\boldsymbol{V}_{n,trunc}$ were equal to zero, the approximation of the corresponding column in $\boldsymbol{G}_{\Phi}^{max}$ would be zero as well. This can be interpreted as: The sensor represented by the respective column does not influence at all the N most important singular values and vectors. It follows that this sensor can be discarded without significant loss of information. The other extreme is that there are only entries in $\boldsymbol{V}_{n,trunc}$, but the truncated elements are equal to zero. In this case, the respective sensor would maximally contribute to the span of the N most important singular vectors and since $\boldsymbol{V}$ has the properties of a unit matrix, $\boldsymbol{V}_{n,trunc}\boldsymbol{V}_{n,trunc}^H$ would be equal to 1. Continuing along this line, the product $\boldsymbol{V}_{n,trunc}\boldsymbol{V}_{n,trunc}^H$ ranges between 0 and 1 and holds information about how much the $n-th$ sensor contributes simultaneously to the span of the N most important singular vectors. The higher the product, the higher the contribution of the respective sensor. This relation is used to define the weight [26]

$$\xi_n = \frac{\boldsymbol{V}_{n,trunc}\boldsymbol{V}_{n,trunc}^H}{N}, \tag{12}$$

computed for all N^{max} possible sensor locations.

One could try to pick the N sensor locations with largest ξ_n to construct an array. Unfortunately, this is not sufficient, as the example of two infinitesimally close sensors shows. Both would be assigned the exact same weight, although it is obvious that also in combination they do not span a two-dimensional space, since they are virtually equal to each other. In order to cope with this condition, it is proposed to follow the steps below to construct a sensor array:

1. Calculate ξ_n for all N^{max} possible sensor locations.
2. Discard sensor locations that are weighted with $\xi_n < \xi_{min}$, where ξ_{min} is a manually set threshold, which aims to prevent the placement of a sensor at a location, which is almost irrelevant in its contribution to the rank of $\boldsymbol{G}_{\Phi}^{max}$.
3. Select the sensor location corresponding to the largest weight ξ_n as first sensor location in the array.
4. Calculate the MAC-value [27] between the column of $\boldsymbol{G}_{\Phi}^{max}$ corresponding to the first selected sensor ($\boldsymbol{G}_{\Phi,\xi_n 1}^{max}$) and the column corresponding to the second largest weight ($\boldsymbol{G}_{\Phi,\xi_n 2}^{max}$):

$$MAC(\boldsymbol{G}_{\Phi,\xi_n 1}^{max}, \boldsymbol{G}_{\Phi,\xi_n 2}^{max}) = \frac{\left|\boldsymbol{G}_{\Phi,\xi_n 1}^{max}{}^H \boldsymbol{G}_{\Phi,\xi_n 2}^{max}\right|^2}{\left|\boldsymbol{G}_{\Phi,\xi_n 1}^{max}{}^H \boldsymbol{G}_{\Phi,\xi_n 1}^{max}\right| \left|\boldsymbol{G}_{\Phi,\xi_n 2}^{max}{}^H \boldsymbol{G}_{\Phi,\xi_n 2}^{max}\right|}. \tag{13}$$

5. If the calculated MAC-value is smaller than a manually set threshold MAC_{Φ}, then include the sensor location associated with the second largest weight ξ_n in the array. If the calculated MAC-value is greater than MAC_{Φ}, then go back to step 4 and test the column associated with the third largest weight ξ_n.
6. Repeat steps 4 and 5, for the third, fourth, fifth, ... largest ξ_n until either the N desired array elements are found or all adequate sensor locations have been tested. In steps 4, the MAC-value is calculated for all sensor locations already known to be in the array to the potentially new sensor location. In step 5, the potentially new sensor location is only added if all MAC-values are smaller than MAC_{Φ}.

The variable MAC_{Φ} ensures that only sensor locations are included that add sufficiently new information, while ξ_{min} can be understood as an interruption criterion for the loop depicted in the bullet points above. If the combination of MAC_{Φ} and ξ_{min} does not allow to find the desired number of sensors, either a smaller array must be used or the routine needs to be re-started with changed parameters.

4 Numerical Verification

The selection strategy developed in Sect. 3 is numerically verified at the example of a reinforced aluminium frame, as it is shown in Fig. 2. The three vertical beams are modelled with a L-shaped cross section with a wall-thickness of 4 mm and a length of 0.6 m. The two horizontal beams have a C-cross section with a wall-thickness of 3 mm and a length of 1.45 m. Horizontal and vertical beams are connected at the six intersections with a web of rigid elements, simulating a bolted connection (blue elements in Fig. 2 bottom – horizontal beams hidden in order to make connecting elements visible). Furthermore, the horizontal elements are connected to each other with rigid elements (black) in vertical and diagonal direction in order to re-enforce the structure. There are in total six different damage scenarios modelled, each of them by removing the connecting spider-web resembling the bolted connection between one horizontal and one vertical beam. The damage scenarios are numbered from 1 to 6 as shown in Fig. 2 top. $I(x)$ is computed for the $X = 6$ possible damage locations.

It is assumed that only the top of the two horizontal beams is accessible to apply sensors or excitations. Only vibration in normal direction is recorded and excited. The pseudospectrum is computed from sensor side based on an array of $N = 6$ sensors. $M = 20$ excitation locations are chosen in order to ensure that the

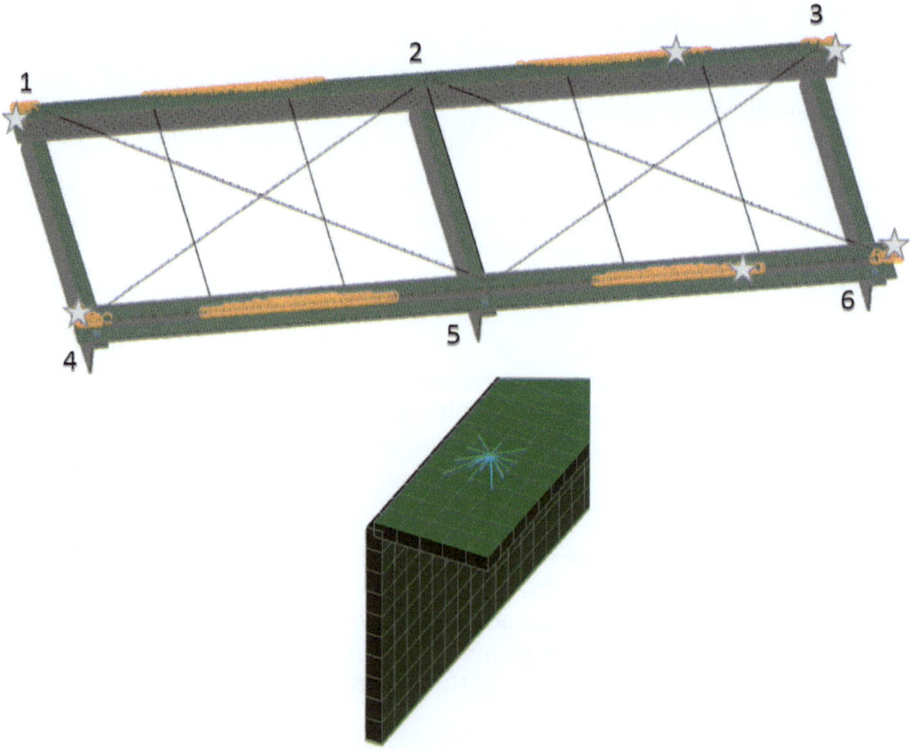

Fig. 2. FE model of a reinforced aluminium frame (top) and zoom into one representative connection between horizontal and vertical beams with hidden horizontal beam (bottom).

degradation of the detection quality results from the reduction of the number of sensors. In order to form the sensor array, $N^{max} = 500$ possible sensor locations, whose nodes are indicated by orange circles in Fig. 2, are available. This allows for more than $2.1 \cdot 10^{13}$ possible combinations of sensor arrays, which is obviously too many for a trial and error approach. The simulation includes $F = 300$ discrete frequencies equally spread over a range from 100 Hz to 250 Hz (due to the frequency selection with $MAC_{max} = 0.75$ and $\Psi_{s,min} = 0.35$ (selection procedure and parameters MAC_{min} and $\Psi_{s,min}$ are defined in [14]) roughly half the frequencies are used to calculate the pseudospectrum – slightly varying depending on the damage case).

In order to verify that a sensor array with sensors selected according to Sect. 3 ($MAC_{\Phi} = 0.75$, $\xi_{min} = 0.001$, $N = 6$) is able to find all simulated damage scenarios, this array (sensor locations indicated in Fig. 2 top by grey stars) is used to calculate the pseudospectra for all six scenarios. The results of those simulations are shown in the first column of Figs. 3 and 4, where on the x-axis the six connections are listed and on the y-axis, the corresponding $I(x)$ normalised by the largest $I(x)$ for this damage case is plotted. It can be concluded that all tested damages are indicated uniquely and correctly by this sensor array. The pseudospectrum at the highlighted location is in all cases at least a factor 10^5 larger than the pseudospectrum calculated for the other potential damage locations.

Furthermore, important for the evaluation of the proposed selection strategy is a comparison of the localisation results of the selected sensor array to the pseudospectra of other sensor arrays of the same size. Since the large number of possible sensor arrays makes it impossible to compute the pseudospectra for all of them, only a subset of 500 randomly chosen sensor arrays with $N = 6$ are tested with respect to their capability to localise the six damage scenarios. Those results are summarised in the right column of Figs. 3 and 4. For the generation of these graphs, a localisation is rated as unique, when the pseudospectrum at one connection is at least 100 times larger than the pseudospectrum at all other connections. The y-axis represents the number of sensor arrays, which uniquely indicate the damage listed on the x-axis. All arrays that do not hint uniquely at one connection are summed up in the last bar. Following this logic, e.g. the localisation of damage 1 (Fig. 3b) is uniquely and correctly performed by 207 out of the 500 randomly selected sensor arrays with $N = 6$. On top of that, there are 2 sensor arrays falsely highlighting connection 3, 26 sensor arrays falsely highlighting connection 4, 8 sensor arrays falsely highlighting connection 5 and 36 sensor arrays falsely highlighting connection 6. These in total 72 sensor arrays are the most critical cases, since they falsely, but uniquely indicate one connection. The last bar refers to the 221 sensor arrays, which do not lead to a unique identification of one damaged connection. However, since damage 1 was applied, this is still a false result. In conclusion, this means that only 207 out of 500 randomly selected sensor arrays consisting of $N = 6$ sensors are able to correctly localise damage 1.

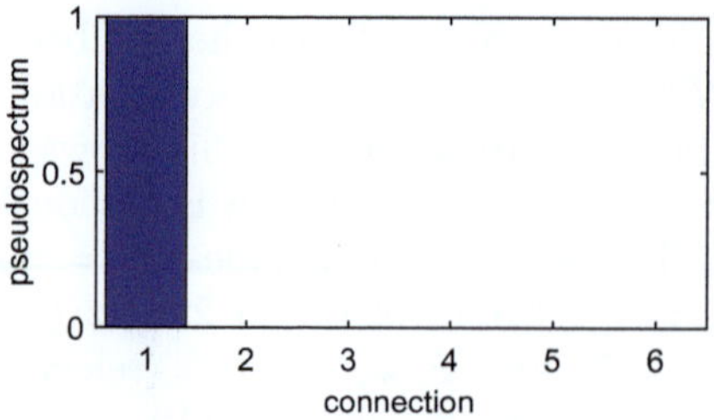

(a) Damage 1 with selected array of $N = 6$

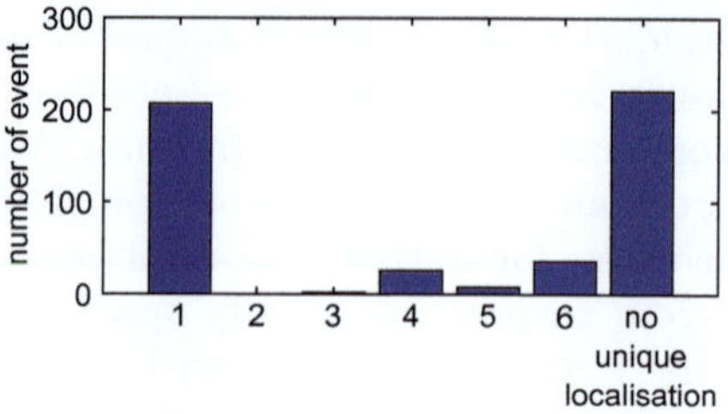

(b) Damage 1 with random arrays of $N = 6$

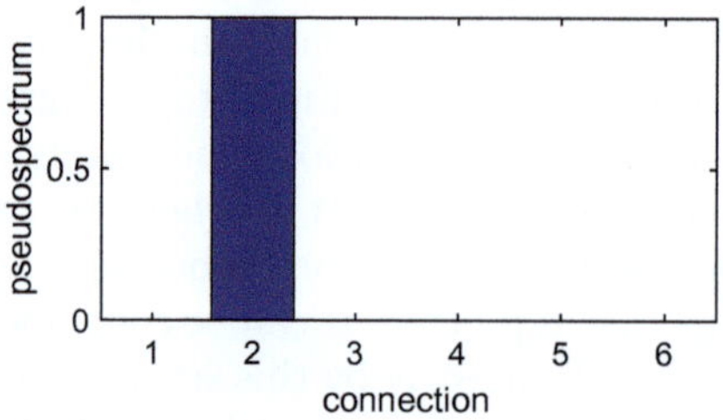

(c) Damage 2 with selected array of $N = 6$

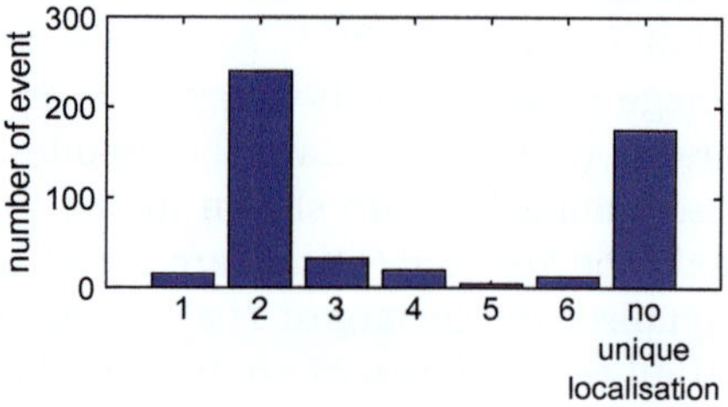

(d) Damage 2 with random arrays of $N = 6$

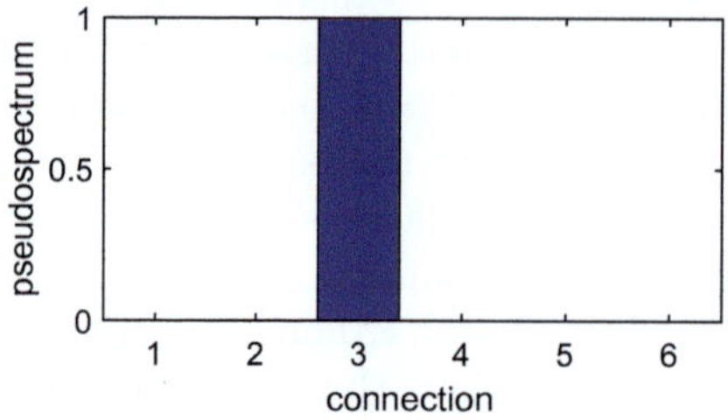

(e) Damage 3 with selected array of $N = 6$

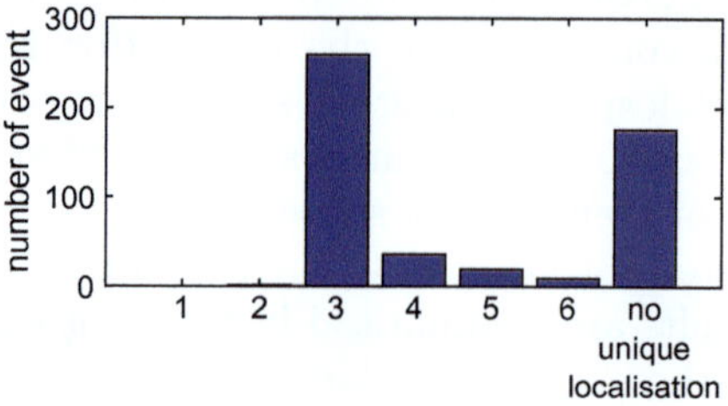

(f) Damage 3 with random arrays of $N = 6$

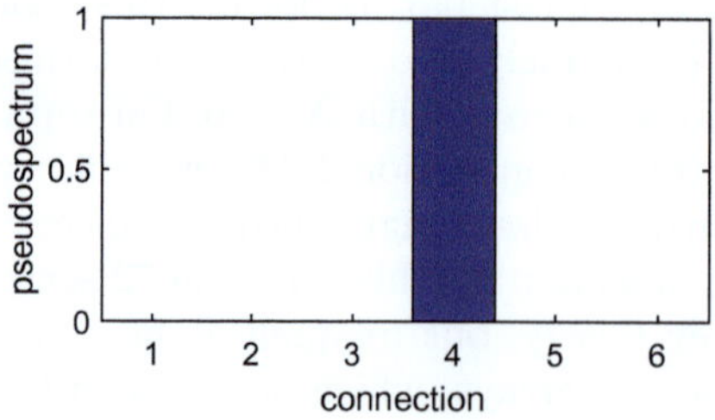

(g) Damage 4 with selected array of $N = 6$

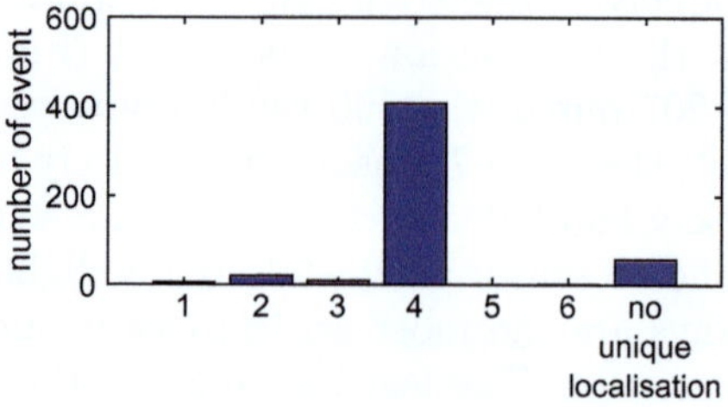

(h) Damage 4 with random arrays of $N = 6$

Fig. 3. $I(x)$ for an array selected based on the criteria in Sect. 3 (left) and for 500 randomly chosen arrays of the same size (right) for damage at connections 1 to 4 (rows).

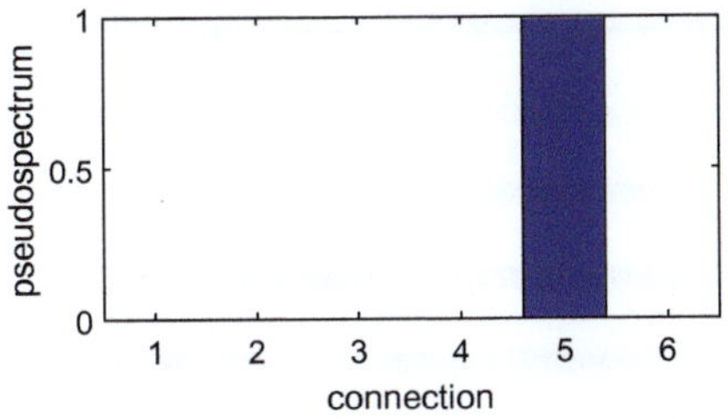

(a) Damage 5 with selected array of $N = 6$

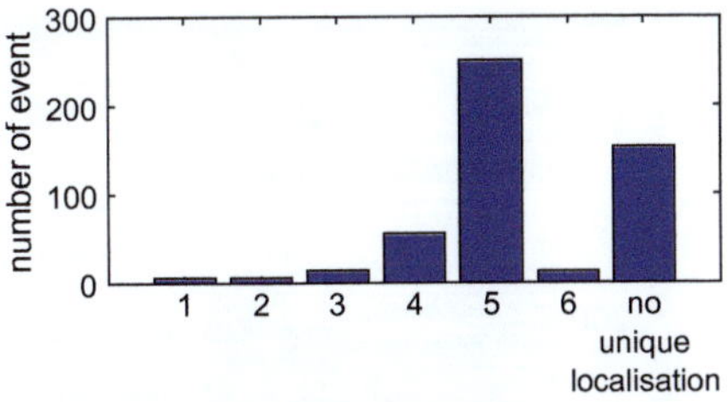

(b) Damage 5 with random arrays of $N = 6$

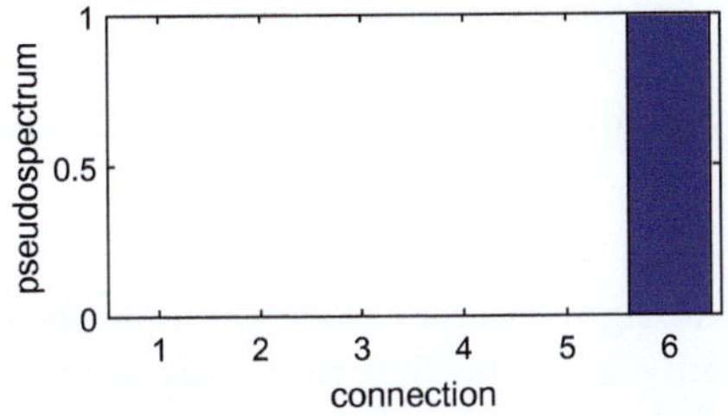

(c) Damage 6 with selected array of $N = 6$

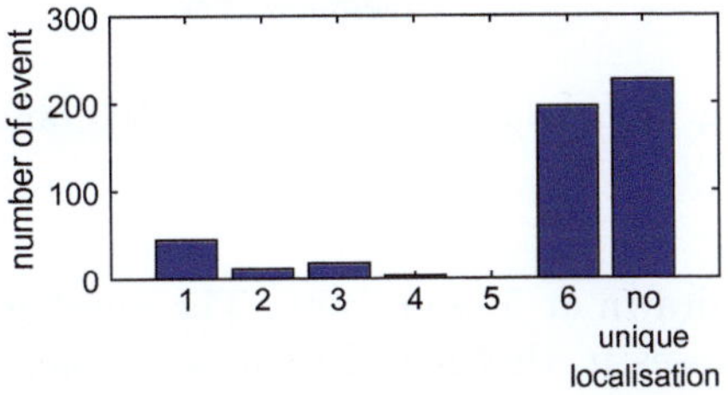

(d) Damage 6 with random arrays of $N = 6$

Fig. 4. $I(x)$ for an array selected based on the criteria in Sect. 3 (left) and for 500 randomly chosen arrays of the same size (right) for damage at connections 5 and 6 (rows).

The discussion of Fig. 3d, f, h and Fig. 4b, d can be done in a similar fashion and leads to comparable findings, being that for all damage scenarios, only a limited number of the randomly selected sensor arrays can localise the damage correctly.

So far, each randomly constructed array is evaluated separately at each damage scenario. Therefore, from the right column of Figs. 3 and 4, it cannot be derived if a sensor array, which can localise damage 1 is also able to point out a damage at connection 2 correctly. In order to evaluate this, the same 500 sensor arrays are analysed additionally, with respect to their ability to localise all six damage scenarios. The outcome is presented in Fig. 5. In the graph at the left, on the x-axis, the six damage cases are listed and the y-axis contains the 500 analysed randomly composed sensor arrays. A field is coloured black if the respective array is able to uniquely localise the damage on the x-axis, if not the field is marked white. The graph at the right represent a further condensation of the results, so that the number of the sensor combination is plotted on the x-axis (at the left it is plotted on the y-axis) and the y-axis is the sum over all damage cases of the data shown at the left. Following this logic, if an array is able to localise all 6 cases correctly, a cross is set at the corresponding sensor configuration at 6 (left y-axis). If it can localise only 5 cases correctly, a cross

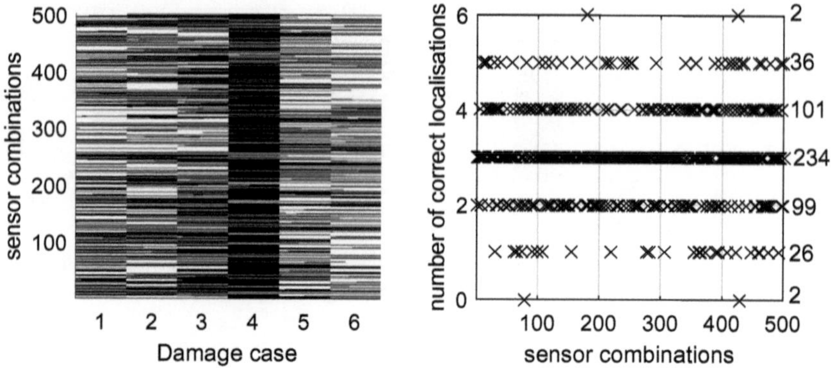

Fig. 5. Localisation results with 500 randomly composed sensor arrays of $N = 6$ sensors.

is drawn at 5, and so on. The number at the right y-axis equals the number of crosses in this row. In this example, e.g. only 2 of the 500 randomly selected sensor arrays are able to uniquely localise all damage cases, while 36 can localise 5 out of 6 damage cases correctly, 101 can localise 4 out of 6 and so on. It follows that 498 sensor-arrays are unable to properly illuminate all 6 potential damage regions and hence provide false information in at least one of the scenarios.

This comparison illustrates the importance of the selection of good sensor locations and demonstrates that sensor arrays selected based on the algorithm in Sect. 3 behave better than 498 out of 500 randomly selected sensor array of the same size.

In order to check whether there is a relation between the spread of the sensors (how distributed are the sensors over the test object?) and the success of the localisation, an index CI is defined. For the computation of CI, first the euclidean distance between two sensors $||\boldsymbol{x}_1 - \boldsymbol{x}_2||$ is calculated. Then the binary variable $cluster_{1,2}$ is defined, which equals 1 if $||\boldsymbol{x}_1 - \boldsymbol{x}_2|| > 0.3625\,\text{m}$ and 0 if $||\boldsymbol{x}_1 - \boldsymbol{x}_2|| \leq 0.3625\,\text{m}$. The value 0.3625 m equals 1/4 of the longest beam in the setup (also with varying values the conclusions do not change). Finally, the clustering index

$$CI = \sum_{n1}^{N} \sum_{n2}^{N} cluster_{n1,n2} \tag{14}$$

is computed. The greater CI, the less sensors in an array are closer than 1/4 of the longest beam away from each other. Thus, a high CI indicates little clustering, while a very low CI means that most sensors are at positions close to each other.

In Fig. 6, the 500 investigated sensor arrays are grouped according to the number of correct localisations. This results e.g. in a group of 2 sensor arrays being unable to detect any of the test scenarios, in a group of 26 array configurations localising only one damage case correctly, and so forth. The average of

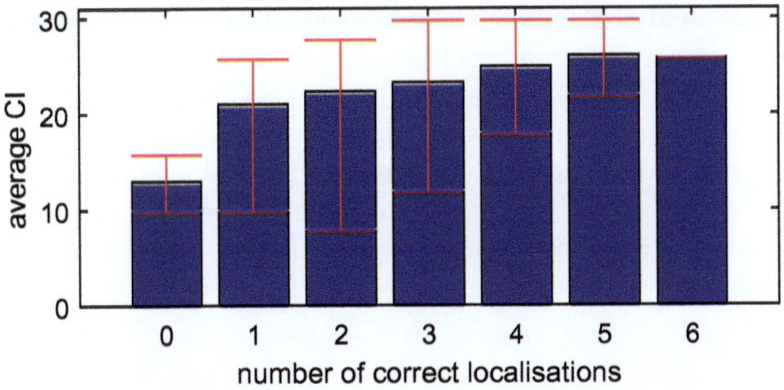

Fig. 6. Average clustering index for 500 randomly composed sensor arrays of $N = 6$ sensors – same as in Fig. 5.

CI per group is put as blue bars against the number of successful localisations. Furthermore, the maximum and the minimum CI per group are indicated by red lines.

Figure 6 shows that the average of CI correlates in tendency with the number of correct localisations of an array. However, the slope towards high numbers of correct localisations becomes rather small. Between 5 and 6 correct localisations ($CI_5 = 26.07$, $CI_6 = 26$) the slope even changes sign (approximately the slope is zero). Furthermore it should be kept in mind that the observed trend exists explicitly only for the average values indicated by the blue bars. Taking into account the smallest and largest values of CI per group, it can be seen that the spread around the average value is large. Furthermore, there are different arrays with the highest possible value of $CI = 30$ localising only 3, 4 or 5 damage cases correctly, while the arrays localising all 6 damage cases correctly show $CI = 26$. This points out that the proposed selection algorithm is superior to a sensor selection merely based on geometrical criteria.

Additional tests are performed with sensor arrays selected with changed values for MAC_Φ and ξ_{min} with the outcome that up to two sensors in the array change position. However, still all of the tested, selected arrays are able to address all damage scenarios. The fact that there are other arrays that are able to act correctly in all damage scenarios highlights the robustness of the selection algorithm to the user-set parameters MAC_Φ and ξ_{min}, but also the fact that the selected sensor array can only be regarded as a 'good' choice, but is not guaranteed to be the optimal one.

5 Experimental Validation

In this section, the developed algorithm for the selection of sensor arrays is applied to a physical experiment in order to validate its performance using real measurement data. The experimental validation is performed on an aluminium frame covered by a honeycomb panel. The frame is composed of three vertical

C-cross-sectional beams (thickness 4 mm), which are joined at each end with a horizontal beam with I-cross section. The I-cross-section is constructed by merging two C-cross-sectional beams (thickness 3 mm) with rivets. The panel covering the frame has a thickness of 10.6 mm and the in-plane dimensions are 1.4 m × 0.5 m. It is connected to the frame with bolts and double-sided tape. Each of the six connections resulting from the crossing of horizontal and vertical aluminium beams is mounted with two bolts. An overview of the test setup, as well as a detailed side view on one connection is presented in Fig. 7.

Fig. 7. Front view on the test setup (top) and side view on the bottom left connection (bottom).

Potential damage locations are, as before in the numerical example, the six connections between the horizontal and vertical aluminium beams. The number of possible damage scenarios is theoretically infinite, since the severity of a damage at multiple or only one connection can change gradually. The validation is performed at eight damage cases (subset of all theoretically possible damages). Experiments are conducted at three different discrete severity levels (one loose bolt at one connection, one removed bolt at one connection, two removed bolts at the same connection) and at three of the six potentially damaged connections (one or multiple connections damaged simultaneously). A full list of the damage scenarios is shown below [14]:

1. One loose bolt at the top central connection (invisible with the bare eye);
2. One removed bolt at the top central connection;
3. One loose bolt at the bottom left connection (invisible with the bare eye);
4. One removed bolt at the bottom left connection;
5. Two removed bolts at the bottom left connection;
6. One loose bolt at the top right connection (invisible with the bare eye);
7. One removed bolt at the top right connection;
8. One removed bolt at the top right connection and one removed bolt at the top central connection.

For the TR-MUSIC algorithm, 7 microphones and 6 accelerometers (1D, out-of plane) are placed in front of the test object, facing the honeycomb panel or directly on the aluminium frame. However, none of the sensors is in the line of sight of one of the potentially damaged connections. The sensors measure the response of a hammer impact at three points on the front-side of the frame, resulting in $N^{max} = 13$ and $M = 3$. The data was gathered for [14] and is used in this paper for the experimental validation of the algorithm derived and numerically verified in the previous sections. The needed vectors $\boldsymbol{g}_{\Phi,x}$ are measured at the healthy test setup.

Input for the selection algorithm are the parameters $MAC_{\Phi} = 0.3$ and $\xi_{min} = 0.05$ and the desired total number of sensors in the array N. It is not necessary to specify the number of accelerometers and microphones separately. MAC_{Φ} is set lower than in the previous example, which can be explained by the fact that in Sect. 4, the potential sensor locations are on continuous domains, while in the experimental case used in this section, the potential sensor locations are discrete and far away from each other. As a consequence, the columns in $\boldsymbol{G}_{\Phi}^{max}$ can become rather similar in the numerical example, while they are naturally more different in the experimental test setup. In order to obtain a selection based on MAC_{Φ}, this value needs to be adapted accordingly. The outcome for sensor arrays of different size found by the algorithm is depicted in Fig. 8. The colour code is the same as previously in Fig. 5(left), where the case 8 is also assessed correct if only one of the two damaged connections is highlighted uniquely (for further explanation see [14]).

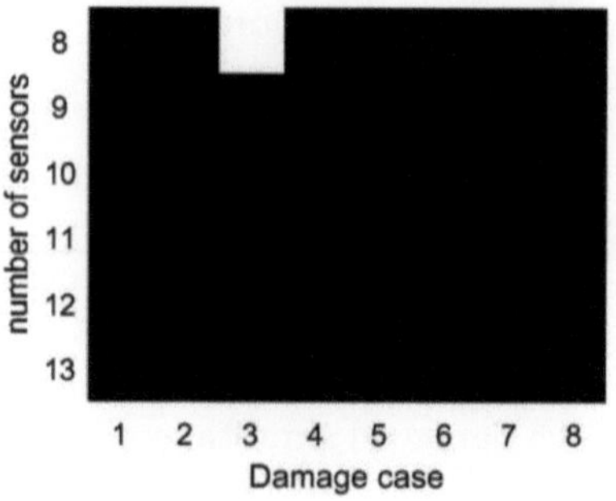

Fig. 8. Localisation results with sensor arrays of different size, selected based on the algorithm in Sect. 3 with $MAC_{\Phi} = 0.3$ and $\xi_{min} = 0.05$.

From Fig. 8, it can be observed that the selected arrays with $N = 9$ (3 microphones and 6 accelerometers) to $N = N^{max} = 13$ output the correct localisation of all cases studied. The array of size $N = 8$ (3 microphones and 5 accelerometers), resulting from the algorithm in Sect. 3, delivers a correct localisation of 7 cases, but fails in the case of a loose bolt at the bottom left connection.

An overview of the localisation results of all 715 possible array configurations with $N = 9$ and all 1287 possible array configurations with $N = 8$ for the experimental data is presented in Fig. 9. In order to be rated as successful localisation, the factor between $I(x)$ at the correctly highlighted connection and at other connections must exceed 100. In the graphs at the left, the results are presented in the style known from Fig. 5. From the graphs at the right, it follows that there are 161 sensor arrays with $N = 9$ that correctly localise all damage cases, while with $N = 8$ only 152 sensor combinations are able to localise all damage cases investigated without errors.

The comparison of Figs. 8 and 9 (top row) for $N = 9$ shows that the selected sensor array is one of the 161 sensor combinations, with which it is possible to tackle all damage cases in this study. It follows that there are 554 sensor configurations of the same size that perform worse in this example. As mentioned before, the array found with the algorithm described in Sect. 3 of size $N = 8$ gives one false localisation, while there are 152 sensor combinations with $N = 8$ that are able to correctly localise all cases studied. Since there are connections included in the inspection region, which are not part of any of the test cases, it is likely that some of the 152 sensor combinations fail at the localisation of damage at these locations. At the same time, the selected array is deliberately designed to be sensitive to all damage locations. Therefore, the 152 sensor combinations being successful in this example do not necessarily show a superior performance to the sensor array selected by the algorithm.

In conclusion, at the 8 investigated damage cases, the selected array of $N = 9$ performed better than 77% of all possible array configurations and the selected array of $N = 8$ performed better than 73% of all possible array configurations, while 11% could locate one more damage than the selected array with $N = 8$. As the design of the selected array is such that it is sensitive to all possible damage locations (not only to the 8 cases investigated), it is likely that the performance of the selected arrays even improves compared to the randomly composed ones if more damage cases are studied. However, the fact that the selected arrays outperform roughly three quarters of all possible arrays, although only a subset of all possible damage scenarios is studied, supports the value of the proposed selection algorithm.

A significant difference between the composition of successful sensor arrays and those failing at the localisation of at least one damage scenario is not obvious. In both cases shown in Fig. 9, $N = 8$ and $N = 9$, there are arrays containing between 3 and 6 microphones, which correctly localise all damage cases. Also the average distance between the sensors in the array and the tested damage locations is not correlating with the number of correct localisations. The direct application of CI as in Fig. 6 is not possible, since there are now two different types of sensors in the array, providing substantially different information, independent of their proximity to each other. Also calculating CI separately for microphones and accelerometers does not lead to a correlation between the

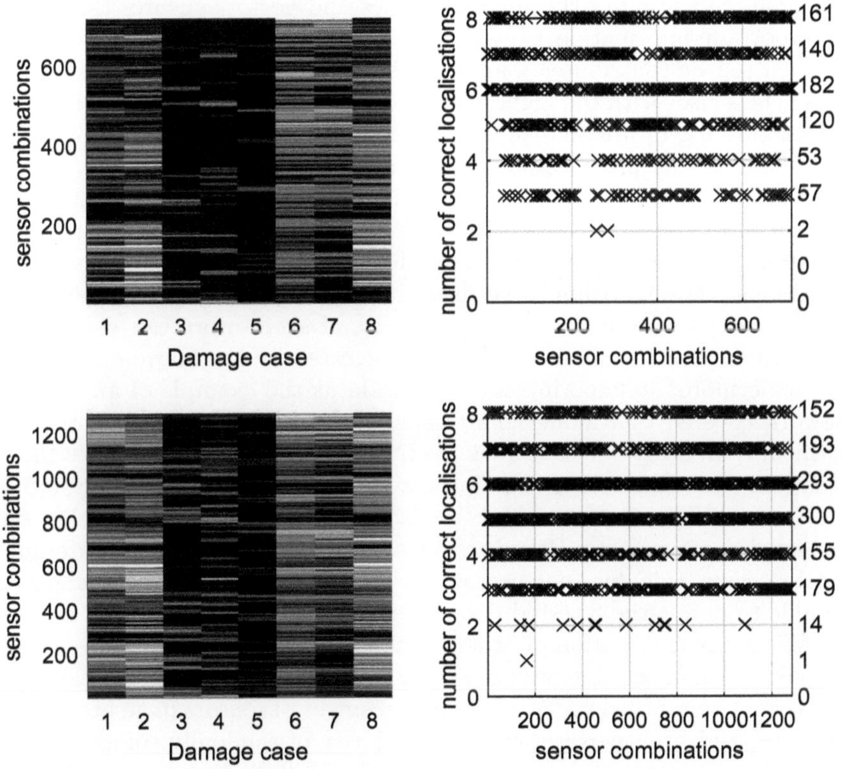

Fig. 9. Localisation results with all possible sensor arrays of $N = 9$ sensors (top) and $N = 8$ sensors (bottom).

geometrical location of the sensors and the localisation quality of the array. This comes to the fore, when analysing the microphone placement suggested for $N = 8$. Here, the microphones selected are at the bottom left, bottom right and top right of the metallic support array (see Fig. 7) and thus as far away from each other as possible in this experiment. Still the localisation of one tested scenario failed with this array.

6 Conclusion

In this publication, the TR-MUSIC algorithm is applied to localise damage in a complex assembly based on the measurement of structural and acoustic vibrations. It is known that the placement of sensors and excitations is essential for the functioning of the TR-MUSIC algorithm, especially if only a small number of sensors and excitations is used. Against this background, an algorithm is proposed, which suggests a good placement of sensors to form an array of a pre-defined size.

The algorithm makes use of a singular value decomposition of a matrix containing the transfer functions from all possible sensor locations to all inspection points. The truncation of the singular values and vectors ensures that only the most important information is kept. A weight is defined, which describes how much each sensor contributes jointly to the span of the most relevant singular vectors. The sensor with the highest weight is selected as first sensor of the array. Then, making use of a MAC-value, it is examined whether the sensor with the next highest weight provides sufficiently different information from the previously selected sensor(s). If this is the case, it is added to the sensor array. This procedure is repeated until a sensor array of the desired size is constructed.

In a numerical example of an aluminium framework with 500 possible sensor locations, the algorithm is verified. It is shown that only two out of 500 randomly composed sensor arrays consisting of six sensors can compete with the performance of the six-sensors array suggested by the above described routine. Furthermore, an experimental validation at the example of an aluminium framework covered by a honeycomb panel with 13 possible sensor locations is performed. If an array of 9 sensors is used, the suggested array provides the right localisation, while most other combination fail at least at some of the tested damage scenarios. The selected array composed of 8 sensors fails to localise one of the 8 tested damage cases. However, since there is in theory an infinite number of damage cases, this does not necessarily mean that other array configurations being able to detect the 8 tested damage cases are necessarily better, since they might fail at the localisation of other damages.

Based on a numerical as well as on an experimental example, it can be concluded that the proposed routine for the placement of sensors in an array delivers suggestions, which outperform the vast majority of randomly composed arrays. Additional research with the aim to guarantee to find the smallest functioning sensor array for a specific application case can further strengthen this approach towards the future.

Acknowledgements. The work leading to this publication has been funded by the project "DETECT-IV", which fits in the MacroModelMat(M3) research program, coordinated by Siemens (Siemens PLM software, Belgium) and funded by SIM (Strategic Initiative Materials in Flanders) and VLAIO (Flanders Innovation & Entrepreneurship Agency). Elke Deckers is a postdoctoral fellow of the Research Foundation - Flanders (FWO). The Research Fund KU Leuven is gratefully acknowledged for its support.

References

1. Farrar, C.R., Worden, K.: Structural Health Monitoring: A Machine Learning Perspective. Wiley, Hoboken (2013)
2. Giurgiutiu, V., Cuc, A.: Embedded non-destructive evaluation for structural health monitoring, damage detection, and failure prevention. Shock Vibr. Digest **37**, 83–105 (2005). https://doi.org/10.1177/0583102405052561
3. Shull, P.J.: Nondestructive Evaluation: Theory, Techniques, and Applications. CRC Press, Boca Raton (2002)
4. Hertlin, I.: Acoustic resonance analysis for safety parts in mass production. In: Proceedings of 16th World Conference on NDT in Montreal (ndt.net) (2004)
5. Zou, Y., Tong, L., Steven, G.: Vibration-based model-dependent damage (delamination) identification and health monitoring for composite structures—a review. J. Sound Vibr. **230**, 357–378 (2000). https://doi.org/10.1006/jsvi.1999.2624
6. Becht, P., van Ophem, S., Deckers, E., Claeys, C., Pluymers, B., Desmet, W.: Model based vibro-acoustic defect detection. Faculty of Engineering, KU Leuven, chap. 1, pp. 1–26 (2018)
7. Doebling, S.W., Farrar, C.R., Prime, M.B.: A summary review of vibration-based damage identification methods. Shock Vibr. Digest **30**, 91–105 (1998)
8. Ennaceur, C., Mudge, P., Gan, T.H.: Long range inspection of engineering assets using guided ultrasonic waves. In: 17th World Conference on Non-Destructive Testing (2008)
9. Rose, J.L.: Ultrasonic Guided Waves in Solid Media. Cambridge University Press, Cambridge (2014)
10. Burrascano, P., Callegari, S., Montisci, A., Ricci, M., Versaci, M.: Ultrasonic Nondestructive Evaluation Systems. Industrial Application Issues. Springer, Heidelberg (2014)
11. Alleyne, D., Cawley, P.: The long range detection of corrosion in pipes using lamb waves. In: Review of Progress in Quantitative Nondestructive Evaluation, vol. 14B, pp. 2073–2080. Springer (1995). Chap. 7: Materials' Degradation and Specific Applications
12. Su, Z., Ye, L., Lu, Y.: Guided lamb waves for identification of damage in composite structures: a review. J. Sound Vibr. **295**, 753–780 (2006). https://doi.org/10.1007/978-1-4615-1987-4_265
13. Carandente, R., Ma, J., Cawley, P.: The scattering of the fundamental torsional mode from axi-symmetric defects with varying depth profile in pipes. J. Acoust. Soc. Am. **127**, 3440–3448 (2010). https://doi.org/10.1121/1.3373406
14. Becht, P., van Ophem, S., Deckers, E., Claeys, C., Pluymers, B., Desmet, W.: Loose bolt detection in a complex assembly using a vibro-acoustic sensor array. Mech. Syst. Sig. Process. **130**, 433–451 (2019)
15. Becht, P., van Ophem, S., Deckers, E., Claeys, C., Pluymers, B., Desmet, W.: Experimental application of TR-MUSIC for loose bolt detection in complex assemblies. In: Proceedings of SHM-NDT (ndt.net) (2018)

16. Yan, Y., Cheng, L., Wu, Z., Yam, L.: Development in vibration-based structural damage detection technique. Mech. Syst. Sig. Process. **21**, 2198–2211 (2007)
17. Oestges, C., Kim, A.D., Papanicolaou, G., Paulraj, A.J.: Characterization of space-time focusing in time-reversed random fields. IEEE Trans. Antennas Propag. **53**, 283–293 (2005)
18. Roux, P., Fink, M.: Time reversal in a waveguide: study of the temporal and spatial focusing. J. Acoust. Soc. Am. **107**, 2418–2429 (2000)
19. Lehman, S.K., Devaney, A.J.: Transmission mode time-reversal super-resolution imaging. J. Acoust. Soc. Am. **113**, 2742–2753 (2003)
20. Yavuz, M.E., Teixeira, F.L.: On the sensitivity of time-reversal imaging techniques to model perturbations. IEEE Trans. Antennas Propag. **56**, 834–843 (2008)
21. Marengo, E.A., Gruber, F.K., Simonetti, F.: Time-reversal music imaging of extended targets. IEEE Trans. Image Process. **16**, 1967–1984 (2007). https://doi.org/10.1109/TIP.2007.899193
22. Lev-Ari, H., Devancy, A.: The time-reversal technique reinterpreted: subspace-based signal processing for multi-static target location. In: Proceedings of the 2000 IEEE Sensor Array and Multichannel Signal Processing Workshop, pp. 509–513. IEEE (2000)
23. Zhang, X.-Y., Tortel, H., Litman, A., Geffrin, J.-M.: An extended-dort method and its application in a cavity configuration. Inverse Prob. **28**, 115008 (2012)
24. Ruvio, G., Solimene, R., Cuccaro, A., Ammann, M.J.: Comparison of noncoherent linear breast cancer detection algorithms applied to a 2-D numerical model. IEEE Antennas Wirel. Propag. Lett. **12**, 853–856 (2013)
25. Solimene, R., Dell'Aversano, A., Leone, G.: Interferometric time reversal music for small scatterer localization. Progress Electromagn. Res. **131**, 243–258 (2012)
26. Mahoney, M.W., Drineas, P.: CUR matrix decompositions for improved data analysis. Proc. Natl. Acad. Sci. **106**, 697–702 (2009)
27. Allemang, R.J., Brown, D.L.: A correlation coefficient for modal vector analysis. In: Proceedings of the 1st International Modal Analysis Conference, Orlando, vol. 1, pp. 110–116 (1982)

Efficient Algorithm for Frequency Estimation Used in Structural Damage Detection

Gilbert-Rainer Gillich[1], Dorian Nedelcu[1(✉)], Cristian-Tatian Malin[1], Istvan Biro[2], and M. Abdel Wahab[3(✉)]

[1] Department of Mechanical Engineering, "Eftimie Murgu" University of Resita, P-Ta Traian Vuia 1-4, 320085 Resita, Romania
d.nedelcu@uem.ro
[2] Faculty of Engineering, University of Szeged, Mars Tér 7, Szeged 6724, Hungary
[3] Soete Laboratory, Faculty of Engineering and Architecture, Ghent University, 9052 Zwijnaarde, Belgium
magd.abdelwahab@ugent.be

Abstract. In damage detection processes, the accuracy of estimating the eigenfrequencies of structures is crucial because the frequencies are not highly sensitive to damage. This paper analyses the accuracy of the Discrete Fourier Transform when estimating the frequency and amplitude of sine waves, identifies its limitations and proposes an algorithm to significantly improve the attained results. Standard methods used to evaluate the eigenfrequencies fail because the results depend on the position of the spectral lines, which are related to the acquisition time. Frequently, interpolation involving the amplitude peaks displayed on several spectral lines located around the maximizer is employed to improve the frequency readability. The estimated results are improved indeed, but the achieved precision still depends on the acquisition time. We develop an algorithm that uses the maximizer of signals with different time lengths, which are obtained from the original acquired signal by cropping. The three selected maximizer are used for parabolic interpolation input data. The maximum of the regression curve represents a precise estimate of the amplitude, associated with the true frequency of the targeted harmonic component. The efficiency of the algorithm is demonstrated for harmonic and multi-harmonic signals.

Keywords: Signal processing · Accurate frequency estimation · Discrete Fourier Transform · Interpolation · Excel VBA

1 Introduction

Damage detection using modal parameters extracted from vibration signals gained the attention of numerous researchers and practitioners in the last decades [1–5]. The most common parameter is the eigenfrequency because, in the market, simple and robust equipment is available to measure it.

However, there is a problem when using this modal parameter, namely the low sensitivity of the frequency change due to the damage [6], which makes the accurate

M. A. Wahab (Ed.): DAMAS 2019, LNME, pp. 283–300, 2020.
https://doi.org/10.1007/978-981-13-8331-1_20

frequency estimation a necessity. It is also important to ensure the repeatability of the results because measurement confidence guarantees early stage damage detection.

Most beam-like structures have large intervals between their harmonics, so it is easy to discern between consecutive frequencies. Observing incipient damage is still difficult because in an early state the damage produces a small frequency drop [7]. If involving standard frequency evaluation, the frequencies of the signal components are calculated and displayed in the spectrum at equidistantly distributed lines. The position of these lines depends on the signal length so that the frequency drop is observed just if the damage grows enough to produce such shift of the frequency that determines the maximizer to move to the anterior line in the spectrum [8]. Even if no structural change occurs, it is possible that for a different signal length the maximizer moves to a neighbor spectral line. As a consequence, an apparent frequency decrease or increase can be suggested without the structure being subject of deterioration [9]. This is why there is a need for using advanced frequency estimation algorithms that provide exact information regarding the frequency even if it has the value between two spectral lines.

This paper presents a review of some actual interpolation methods used to increase the precision of the frequency estimation and proposes a new algorithm implemented in MS Excel-VBA that can precisely indicate the frequency components of a signal irrespective to the signal length taken for the analysis.

2 Review of the Main Actual Interpolation Methods

2.1 Discrete Fourier Transform

Let us consider an analog signal $x(t)$ representing the vibration response of a structure. To give computers the possibility to transmit, store, and process the signal, it must be converted in a digital signal $x(n)$. This is a sequence of N samples $(x_0, \ldots x_n, \ldots x_{N-1})$ representing the values of the analog signal captured in N equidistantly taken time intervals. The distance between two successive samples is referred to as time resolution τ and depends on the sampling rate F_S that is defined as the number of samples taken in one second. For the digital signal $x(n)$, the relation between its length t_S, number of samples and the sampling rate is:

$$t_S = (N-1)\tau = \frac{N-1}{f_S} \tag{1}$$

Discrete Fourier Transform (DFT) find a set of sinusoids, which can be added together to reconstruct the original signal $x(n)$. The period T_1 of the first sinusoid is equal to the length in time of the analyzed signal. For this period, the fundamental frequency is $f_1 = 1/T_1$. The reconstructed signal component having this frequency is displayed in the spectrum as the first spectral line.

The next sinusoid fit the signal length twice, hence $t_S = 2T_2$, so that the second component is displayed on the spectral line where the frequency is $f_2 = 2f_1$. For the general case, we can write the frequency of the k-th harmonic component $f_k = kf_1$, where k is referred to as the spectral line number. The distance between the spectral lines is constantly f_1. Therefore this value is known as the frequency resolution and is denoted with Δf. For $k = 0$, this is not sinusoid but the DC value. If the signal contains no continuous component, the amplitude value displayed at $f_0 = 0$ Hz should be null, but the DFT is not always able to ensure it. In fact, DFT creates, for each spectral line, a sequence of values:

$$X_k = \sum_{n=0}^{N-1} x_n[\cos(-2\pi kn/N) + i\sin(-2\pi kn/N)] \tag{2}$$

If the number of spectral lines is $k = 0\ldots N{-}1$, thus equal to the number of samples of the time-domain signal, it results in a number of N linearly independent equations with N unknowns and it is possible to calculate the real and imaginary parts of the coefficients X_k, as:

$$\mathrm{Re}X_k = \frac{2}{N}\sum_{n=0}^{N-1} x_n[\cos(-2\pi kn/N)] \tag{3}$$

$$\mathrm{Im}X_k = \frac{2}{N}\sum_{n=0}^{N-1} x_n[\sin(-2\pi kn/N)] \tag{4}$$

Hence, the find the absolute values of the coefficients using the relation:

$$X_k = \sqrt{(\mathrm{Re}X_k)^2 + (\mathrm{Im}X_k)^2} \tag{5}$$

DFT represented by these coefficients contains N spectral lines, from which it is sufficient to display only half of them, due to symmetry [10]. It is known that the precision achieved by frequency evaluation using DFT depends on the signal acquisition time. Two favorable cases exist: (1) it is possible to acquire a signal in an extremely long time in order to obtain a fine frequency resolution, or (2) the signal length fit a whole number of periods T for the targeted frequency. For damage detection, the fulfillment of the first condition is problematic because vibration signals acquired as a structural response are rapidly damped, especially in the case of higher frequencies [11]. Consequently, the estimated frequencies have a significant deviation from the true values. Because the period of the targeted component is not known, we can acquire the whole cycles just by accident. The problem in estimating frequencies if this condition is not met is the occurrence of spectral leakage, which introduces errors in the estimated frequencies [12]. This aspect is detailed below.

Let us consider a discrete signal with period T and frequency f, acquired in the time t_S. If the acquisition time does not contain a whole number k of cycles with length T but is a little bit longer, which is usually the case, we can write:

$$t_S = T(k + \delta) \tag{6}$$

where δ is a fraction of one cycle. It follows:

$$\frac{1}{t_S} = \frac{1}{T(k+\delta)} \tag{7}$$

Or, taking the inverse of the two fractions, results:

$$f = \Delta f(k+\delta) \tag{8}$$

Hence, the true frequency falls between two spectral lines. Some researchers prefer to work with the normalized frequency, which is obtained by dividing the frequency at the frequency resolution. For the general case, the normalized frequency is:

$$\bar{f} = \frac{f}{\Delta f} = k+\delta \tag{9}$$

From Eq. (9) it can easily be deduced that for the frequency that is the value corresponding to a spectral line, the normalized frequency value is even number of the spectral line. This is illustrated in DFT representation in Fig. 1, where on the spectral line k is plotted against the amplitude of the harmonic signal component A_k. If A_k is the biggest value in a given frequency bandwidth, it is known as the maximizer. Figure 1 shows also the two neighbor spectral lines $k - 1$ and $k + 1$ and the two amplitudes A_{k-1} and A_{k-1} displayed at these lines.

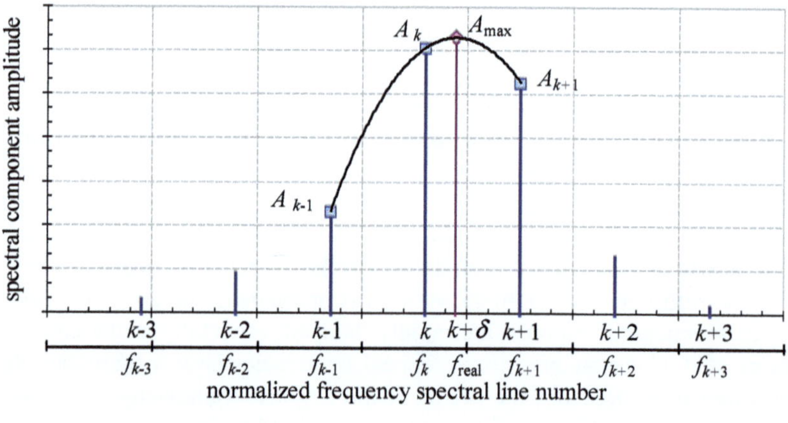

Fig. 1. DFT output for a sinusoid (frequency f = 5 Hz) by standard evaluation.

One can observe in Fig. 1 that the frequency $f_{\text{real}} \neq k\Delta f$ is not correctly estimated by the frequency associated to the maximizer A_k. Some interpolation methods were developed to find the frequency between two spectral lines. An analysis of the results achieved by using these interpolation methods is given in next subsection.

2.2 Determining the Precision of the Main Actual Interpolation Methods

So far as we know, the frequency values estimated with standard methods, e.g. DFT, provide results directly linked to the position of the spectral lines of one spectrum obtained for a single time length, which is in most cases the acquisition time. Finding f_{real} at an inter-line position relies on finding the regression curve fitting to several points obtained in the DFT. If just two points in the spectrum are considered, a method to weight their influence is employed. In all cases, finding the correction term δ ensuring a corrected frequency f_{corr} as close as possible to f_{real} is targeted. This can be made either by using at the abscissa the spectral line numbers or the frequencies. In the first case the corrected spectral line number is estimated as:

$$k_{corr} = k + \delta \tag{10}$$

Or, for the latter case the corrected frequency is found from Eq. (9).

The estimation of f_{corr} employing the correction coefficient δ is referred to as the *fine frequency estimation*, as opposed to the *coarse frequency estimation* performed by directly locating the DFT maximum [13].

Interpolation Based on Two Points in the Spectrum.

Grandke developed a method [14] that involves the peak A_k and the largest neighbor amplitude of DFT achieved from a signal windowed by a Hann window. In the spectral representation in Fig. 15 the maximizer largest neighbor is A_{k-1}. It is also possible to get the larges neighbor at the spectral line A_{k+1}. In both cases, following steps are performed to calculate the corrected frequency. First the ratios

$$\alpha^- = \frac{A_{k-1}}{A_k} \text{ or } \alpha^+ = \frac{A_{k+1}}{A_k} \tag{11}$$

are calculated I function of the bigger neighbor of the maximizer. Afterwards, the correction term is calculated as:

$$\delta^- = \frac{2\alpha^- - 1}{\alpha^- + 1} \text{ or } \delta^+ = \frac{2\alpha^+ - 1}{\alpha^+ + 1} \tag{12}$$

Finally, the frequency result as:

$$f_{corr} = (k + \delta^-)\Delta f \text{ or } f_{corr} = (k + 1 + \delta^+)\Delta f \tag{13}$$

Quinn proposed a method [15] that directly uses the DFT of the signal. Because no windowing is employed, the method is much faster as that proposed by Grandke. The method implies both neighbor amplitudes of the maximizer and request two interpolations, each of them involving just two amplitudes. The following ratios are calculated for Quinn's estimator:

$$\alpha = \frac{A_{k-1}}{A_k} \text{ and } \alpha = \frac{A_{k+1}}{A_k} \tag{14}$$

Next, the correction terms are calculated as:

$$\delta_1 = \frac{\alpha_1}{1 - \alpha_1} \text{ and } \delta_2 = -\frac{\alpha_2}{1 - \alpha_2} \tag{15}$$

Finally, the frequency result as:

$$f_{\text{corr}} = (k + \delta_{1,2})\Delta f \tag{16}$$

If $|\delta_1| < |\delta_2|$, then the correction term δ_2 is chosen in Eq. (16), else δ_1 is chosen.

Jain et al. [16] proposed an interpolation method similar to that proposed by Quinn, but the correction coefficients were calculated in a different way. For $A_{k-1} > A_{k+1}$, the correction term is calculated from the relations:

$$\alpha_1 = \frac{A_k}{A_{k-1}} \text{ and } \delta_1 = \frac{\alpha_1}{1 + \alpha_1} \tag{17}$$

and the frequency results from the relation:

$$f_{\text{corr}} = (k - 1 + \delta_1)\Delta f \tag{18}$$

If $A_{k-1} \leq A_{k+1}$, the correction term is calculated from the relations:

$$\alpha_2 = \frac{A_{k+1}}{A_k} \text{ and } \delta_2 = -\frac{\alpha_2}{1 - \alpha_2} \tag{19}$$

and the corrected frequency is derived as:

$$f_{\text{corr}} = (k + \delta_2)\Delta f \tag{20}$$

The tests are performed on a signal with the frequency $f_{\text{real}} = 4.89$ Hz, the amplitude $A_{\text{real}} = 1$ mm/s^2 and the initial time length $t_S = 1.5$ s. The original signal is generated using $N_S = 301$ samples by a sampling rate $f_S = 200$ Hz, resulting a time resolution $\Delta t = 0.005$ s. The efficiency of the interpolation methods is tested for eleven time lengths achieved by stepwise truncating the original sinusoidal signal. To obtain the ten shorter signals, $N_{S-it} = 8$ samples (i.e. $t_{S-it} = 0.05$ s) are removed at each step. The results achieved employing the three above described methods are presented in Fig. 2.

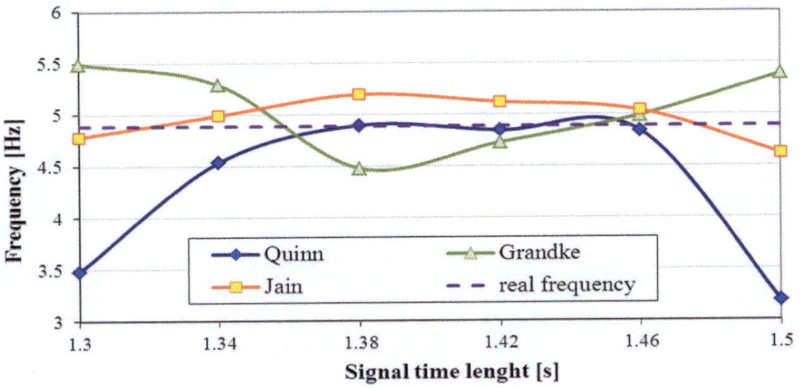

Fig. 2. Frequency estimation tests employing interpolation methods based on two points.

Interpolation Based on Three Points in the Spectrum.
The next analyzed methods involve three amplitudes at interpolation. Ding [17] proposed a barycentric method, where the correction term δ results from the relation:

$$\delta = \frac{A_{k+1} - A_{k-1}}{A_{k-1} + A_k + A_{k+1}} \tag{21}$$

Instead, Voglewede [18] proposed a correction term that involves the quadratic method. The correction term is found from the relation:

$$\delta = \frac{A_{k+1} - A_{k-1}}{2(2A_k - A_{k-1} - A_{k+1})} \tag{22}$$

A quite similar quadratic estimator of the correction term is introduced by Jacobsen in [19], which results from the relation:

$$\delta_{\text{Jac}} = \frac{A_{k+1} - A_{k-1}}{2A_k - A_{k-1} - A_{k+1}} \tag{23}$$

For all these methods, the corrected frequency is calculated as:

$$f_{\text{corr}} = (k + \delta)\Delta f \tag{24}$$

Tests are performed by involving the same signal and procedure as previously described. The results are presented in Fig. 3.

From Figs. 2 and 3, it can be seen that the errors in frequency estimation still depend on the acquisition time, i.e. the closer the time length t_S to a multiple of periods T, the higher the precision of the estimation is. However, the errors are significant and unpredictable, so this approach is not recommended to evaluate the eigenfrequencies of structures for damage detection purposes.

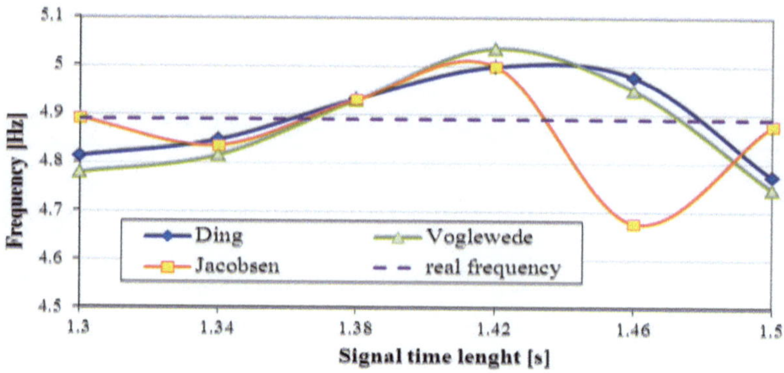

Fig. 3. Frequency estimation tests employing interpolation methods based on two points.

3 The Proposed Frequency Estimation Method

3.1 Description of the Propose Frequency Estimation Method

The weakness of the existing interpolation methods mainly consist in the fact that the points on which the interpolation is made come from the same DFT. We develop a new frequency estimation method based on three points achieved from three different DFTs that are points of a sinc function [20]. The peak values distributed of this function is close to a polynomial one, so that a second-order polynomial interpolation is made. To obtain the three points for interpolation, the method involves shortening the signal in time domain and calculating the DFT at iteration. In the following sub-sections we present the algorithm implemented in a program developed by the authors in MS-VBA, which is used for the accurate frequency estimation.

Signal Generation.

The test signal is generated involving the "SignalGeneration" sheet, but it is also possible to import real measured signals. To evaluate the accuracy of the results, in the paper we use signals generated with known frequencies. It is possible to create signals composed by up to three harmonic components whit defined amplitudes and frequencies. The operator can also impose the number of samples N_S and time resolution Δt. The signal generated or taken directly from the acquisition system is transferred to the "DFT" sheet.

Standard DFT Analysis.

After it is created or imported, the signal is transferred from to the "DFT" sheet. Here, it is represented in a chart, as that shown in Fig. 4a. By click on the "DFT calculation" button, the real and imaginary coefficients are generated according to Eqs. (3) and (4), which are further used to calculate the complex module with Eq. (5). These values are displayed in a spectral representation as that shown in Fig. 4b. In these calculations, the ratio $2/N_S$ is not considered, because we intend obtaining dissimilar amplitudes in the frequency domain representations for the truncated signals. The amplitudes attain in this way higher values for longer signals in the time domain, i.e. bigger number of samples contained in the signal.

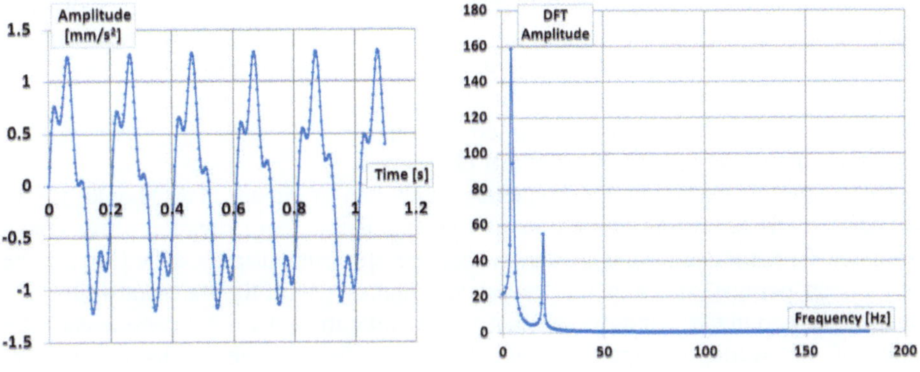

a. b.

Fig. 4. Generated harmonic signal and its standard

Setting of the Number of Peaks to be Considered in the DFTs Obtained After Signal Truncation.
After DFT is calculated, a window is displayed that request setting the number *m* of peaks to be extracted from DFTs calculated after each signal truncation. In the normal case, because the frequencies of a beam are not close together and leakage does not significantly affect the amplitudes of the neighbor harmonic components, the selected number should be the number of the components observed in DFT. But, if the structure is excited with a frequency close to the targeted frequency component, it achieves here the highest amplitudes and the number of peaks can be set as one. A procedure to excite the structure in such way is described in [20].

After entering the value of *m*, the signal is truncated by 2 samples at iteration, until the half signal length remains. DFTs are calculated for each truncated signal. From these spectra, the number of peaks set before is selected and displayed in an overlapped spectrum, as shown in Fig. 5.

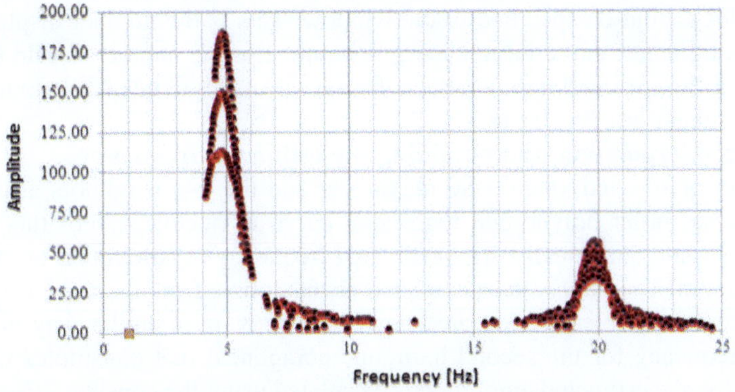

Fig. 5. Overlapped DFT for the cropped signals

Each curve in Fig. 5 represents the peak amplitudes for DFTs that correspond to a value of m. Note that, because we didn't consider the value $2/N_S$ when calculating the complex coefficients, the amplitudes of these curves differ. Dissimilar happens if considering $2/N_S$. In this case, the amplitudes of the curves are equal, but do not permit identifying the points to be used for interpolation.

Selection of the Frequency Bandwidth of Interest.

The next approach is setting the frequency bandwidth of interest. This should include the targeted frequency and frame it as narrow as possible. An indication of the coarse estimated frequency is found in the overlapped spectrum displayed in Fig. 5. This information permits selecting the lowest and highest value for the bandwidth. The values are typed in two windows which open one after the other. As a consequence, the limits of the overlapped spectrum are set, and it displays the results for just one harmonic component, as shown in Fig. 6.

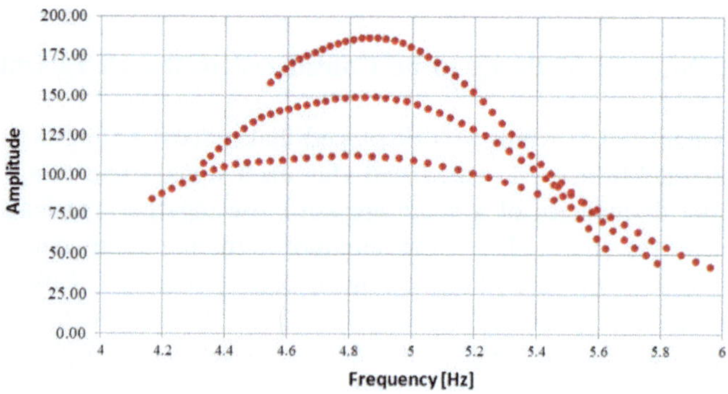

Fig. 6. Zoom on the overlapped DFT displaying the selected bandwidth.

Selection of the Number of Cycles of the Targeted Harmonic Component.

Next, a window opens, showing the number of cycles k for which the curve with the highest amplitude (see Fig. 6) is calculated. It requests setting the number of cycles for which the amplitudes are selected for interpolation. Here, the number of maximum integer cycles should be specified in the window. This is the number displayed in the window if the highest curve has a clear maximum, else the number should be reduced with one. For the specified number, the software selects all frequency-amplitude points and displays them graphically in Fig. 7

Performing interpolation and obtaining the estimated frequency.

From the set of selected values, the maximizer and the two neighbors are found, for which a second order polynomial regression curve is calculated. For this curve, the maximum if found analytically. The estimated frequency is now found as the abscissa of the maximum. The three values extracted as maximizer from the three DFTs and the maximum found by interpolation are shown in Fig. 8. In a similar way is found the estimated frequency for the second harmonic component that constitutes the original signal. Finally, the estimated amplitude is calculated using the constant $2/N_{S\text{-max}}$, and is

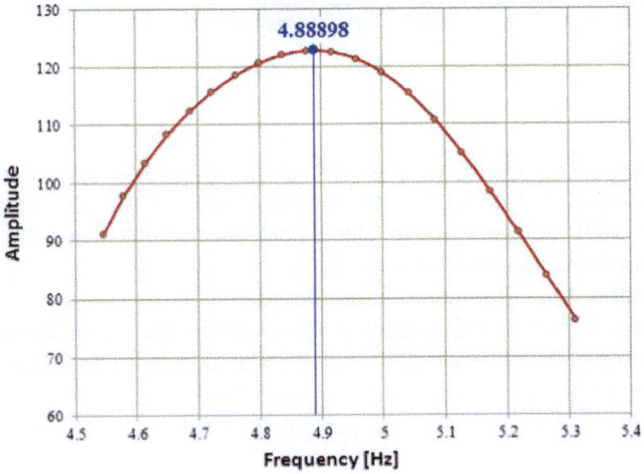

Fig. 7. Overlapped DFT for the cropped signals

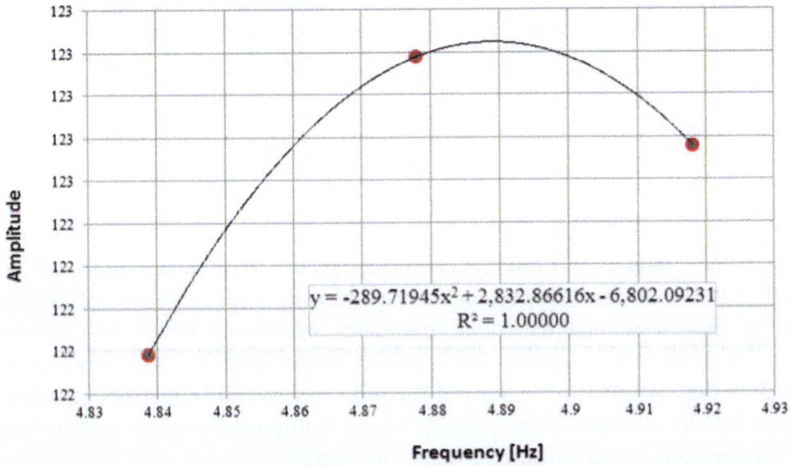

Fig. 8. The interpolation performed for three points that belong to three DFT

displayed along with the estimated amplitude. $N_{S\text{-max}}$ represents the number of samples contained in the truncated signal which ensures the maximizer in Fig. 7.

Overlapping DFTs ensure a fine frequency resolution. Because the interpolation is made by using three amplitudes from different DFTs, all close to the real amplitude, the results are expected to be accurate. However, a slight deviation to the right is noticed, because the distribution of the maximizer is not symmetric but follows a pseudo-sinc function [21].

3.2 Numerical Tests to Validate the Precision of the Frequency Estimation

To highlight the accuracy of the contrived method and the subsequent developed program, we made simulation on signals with known harmonic components. This facilitates comparing the results achieved from the frequency estimation with that of the generated frequency. The frequency estimation results are also compared with those obtained from DFT of the original signal and simulations made by applying the interpolation methods presented in Sect. 2.

Signal Containing One Harmonic Component.

The original signal with the frequency $f = 4.89$ Hz is again considered in this subsection, generated in the same conditions as the signal tested with known interpolation methods presented in Sect. 2, and for the same time lengths. The results, obtained involving the developed algorithm, are presented in Table 1.

Table 1. Data used for tests and the achieved results for the signal with $f = 4.89$ Hz.

t_S [s]	Δf [Hz]	N_S	m	f_{min}	f_{max}	k	f_{corr} [Hz]	A_{corr} [mm/s^2]
1.50	0.666667	301	1	4	6	6	4.8889816	0.9945186
1.46	0.684932	293	1	4	6	6	4.8889816	0.9945186
1.42	0.704225	285	1	4	6	6	4.8889816	0.9945186
1.38	0.724638	277	1	4	6	6	4.8889816	0.9945186
1.34	0.746269	269	1	4	6	6	4.8889816	0.9945186
1.30	0.769231	261	1	4	6	6	4.8889816	0.9945186

From Table 1 and Fig. 9 one can observe that the frequencies estimated with the proposed method do not depend on the acquisition time and are definitely more accurate as these obtained by actual frequency estimators which use interpolation methods. The error is found to be 0.02%, which totally fulfill the requirements for early damage detection. Another advantage of the proposed method consists in the fact that it makes possible to estimate the amplitude, which is impossible when current frequency estimation methods based on interpolation are used.

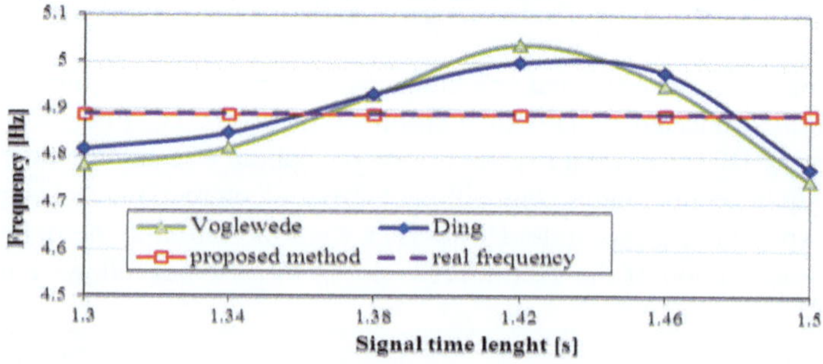

Fig. 9. Frequency estimation results achieved by employing the known methods in comparison with those achieved by the proposed method

Signal Containing Two Harmonic Components.
To show the method works if the analyzed signal contains more harmonic components, we performed tests for a signal simulating the first two out of plane vibration modes of a cantilever beam. The first component is again the sinusoidal signal with the frequency f_1 = 4.89 Hz and the amplitude $A_1 = 1\,\text{mm/s}^2$. For the second harmonic component we choose the frequency f_2 = 28.75 Hz, which is expected for the second harmonic of an Euler–Bernoulli cantilever beam. The amplitude is considered $A_2 = 0.3\,\text{mm/s}^2$. The two components are found separately.

First, the fundamental frequency is estimated. We use the same truncation strategy as for the single component signal. Accurate frequency estimation is possible, the results presented in Table 2 sustaining this conclusion. The error is still 0.02%.

Table 2. Data used for tests and the achieved results for the fundamental frequency.

t_S [s]	Δf [Hz]	N_S	m	f_{min}	f_{max}	k	f_{corr} [Hz]	A_{corr} [mm/s^2]
1.50	0.666667	301	2	4	6	6	4.888632	0.995148
1.46	0.684932	293	2	4	6	6	4.888632	0.995148
1.42	0.704225	285	2	4	6	6	4.888632	0.995148
1.38	0.724638	277	2	4	6	6	4.888632	0.995148
1.34	0.746269	269	2	4	6	6	4.888632	0.995148
1.30	0.769231	261	2	4	6	6	4.888632	0.995148

We did not succeed to find the second harmonic with the strategy described for the previous tests. This happened because the number of samples per cycle was too low. For higher frequencies, the algorithm requests a minimum number of samples per cycle, which is found to be 20 for an accurate estimation. Therefore, we reduced the time resolution from Δt = 0.005 s to Δt^* = 0.002 s and maintained unaltered the number of samples of the original signal, that are N_S = 301. The shortened signals are obtained by extracting 8 samples by iteration. The results are presented in Table 3. One can observe that the frequency is well estimated, the maximum error being 0.127 Hz. This shows that, even for signals containing more harmonics, the frequencies can be estimated well even for the harmonics with small amplitude.

Table 3. Data used for tests and the achieved results for the second harmonic.

t_S [s]	Δf [Hz]	N_S	m	f_{min}	f_{max}	k	f_{corr} [Hz]	A_{corr} [mm/s^2]
0.600	0.666666667	301	10	27	30	16	28.83255	0.319129
0.584	0.684931507	293	10	27	30	16	28.83255	0.319129
0.568	0.704225352	285	10	27	30	16	28.83255	0.319129
0.552	0.724637681	277	10	27	30	15	28.83457	0.306852
0.536	0.746268657	269	10	27	30	15	28.83457	0.306852
0.520	0.769230769	261	10	27	30	14	28.87619	0.285024

The next approach was testing if a small frequency drop, as that occurring in early damage state, can be accurately quantified. The original signal is that described above, hence it is generated with $N_S = 301$ samples and time resolution $\Delta t = 0.002$ s, resulting the signal time length $t_S = 0.6$ s. It has two harmonic components that have the frequency $f_1 = 4.89$ Hz and the amplitude $A_1 = 1\,\text{mm/s}^2$, respectively the frequency $f_2 = 28.75$ Hz and the amplitude $A_2 = 0.3\,\text{mm/s}^2$. To see if small frequency changes are observable and possible to be quantified accurately, we simulated five small frequency drops Δf_2 and obtained five reduced frequencies f_{2-R}. Afterward, we estimated the frequency for each generated signal and compared the generated frequency drop with that obtained by estimation. The absolute error is also calculated and used to appreciate the accuracy of the performed frequency estimations.

Table 4. Estimator's sensitivity analysis to frequency changes for the second harmonic.

Generated data		Estimated data				Error [%]
f_2 [Hz]	f_{2-R} [Hz]	Δf_2 [Hz]	f_2 [Hz]	f_{2-R} [Hz]	Δf_2 [Hz]	
28.75	28.73	0.02000	28.83255	28.81252	0.02003	0.32
28.75	28.71	0.04000	28.83255	28.79265	0.03990	0.32
28.75	28.69	0.06000	28.83255	28.77289	0.05966	0.32
28.75	28.6056	0.14440	28.83255	28.68969	0.14286	0.31
28.75	28.60	0.15000	28.83255	28.68412	0.14843	0.31
28.75	27.50	1.25000	28.83255	27.59033	1.24222	0.29

From Table 4 one can observe that, even for the component with the smaller amplitude, the proposed frequency estimator is able to find fine frequency changes. The accuracy permitted observing the differences between the generated frequency drops 28.6056 Hz and 2.6 Hz and even quantifying this difference. On the other hand, the bigger frequency changes are also estimated with accuracy, which is proved by the results in last row in Table 4.

Table 5. RSFs obtained from generated signals and estimations for the second harmonic.

Generated data		Estimated data				Error [%]
f_2 [Hz]	f_{2-R} [Hz]	RFS [%]	f_2 [Hz]	f_{2-R} [Hz]	RFS [%]	
28.75	28.73	0.069565	28.83255	28.81252	0.069467	0.14
28.75	28.71	0.139130	28.83255	28.79265	0.138396	0.52
28.75	28.69	0.208695	28.83255	28.77289	0.206925	0.84
28.75	28.6056	0.502260	28.83255	28.68969	0.495485	1.34
28.75	28.60	0.521739	28.83255	28.68412	0.514813	1.32
28.75	27.50	4.347826	28.83255	27.59033	4.308380	0.90

Damage detection methods developed by the authors [22–25] make use of the Relative Frequency Shift (RFS), which is calculated as:

$$\mathrm{RFS}_i = \Delta\bar{f}_i = \frac{f_i - f_{i-\mathrm{R}}}{f_i} \tag{24}$$

where f_i is the frequency of the healthy beam and $f_{i-\mathrm{R}}$ that of the damaged beam for the i-th vibration mode. Usually we express this values in percent, thus these are multiplied by 100. The RSF values for the frequency drops are presented in Table 5. It clearly results that the estimated frequency values permit finding correct RFSs.

The question is if the method works for low frequencies, being known here are the biggest problems in observing frequency changes. This happen mainly because the frequency drops due to damage in early stage, for the first vibration mode, are much smaller as the half of the frequency resolution. As a consequence, the damage occurrence is observable just until the frequency drop approaches $\Delta f/2$. For a signal with the time length $t_S = 1.5$ s results $\Delta f = 0.667$ Hz, so that a frequency drop $\Delta f_1 = 0.02$ Hz will not be observed. The change in the spectrum consists, in this case, in an amplitude alteration because another distribution on the same spectral lines results. Therefore, if the original signal's frequency $f_1 = 4.89$ Hz drops to $f_{1-\mathrm{R}} = 4.87$ Hz the change will not be noticed. By employing the proposed method to test if the frequency drop $\Delta f_1 = 0.02$ Hz is observed, we obtained the results presented in Table 6. We conclude this small change is observable and the achieved results permit calculating a reliable RFS. Another test, made for a bigger frequency drop which is still not observable by applying the standard frequency evaluation, lead to the same conclusion. The results are also presented in Table 6.

Table 6. RSFs obtained from generated signals and estimations for the first harmonic.

Generated data		Estimated data					
f_1 [Hz]	$f_{1\text{-R}}$ [Hz]	Δf_1 [Hz]	RFS [%]	f_1 [Hz]	$f_{1\text{-R}}$ [Hz]	Δf_1 [Hz]	RFS [%]
4.89	4.87	0.02	0.4089	4.88863	4.86946	0.01917	0.3921
4.89	4.79	0.10	2.0449	4.88863	4.78471	0.10391	2.1256

3.3 Recommendation

After performing a series of simulations, we found out the settings that should be imposed on the acquisition system and the developed software to allow evaluating the highest and the lowest frequency of interest. These are following:

- The length in time of the original signal should cover at least six and a half periods of the fundamental frequency, i.e. $t_S > 6.5T_1$. This condition refers to the lowest frequency and aims ensuring a symmetric distribution of the peaks in the overlapped spectrum;
- The time resolution of the original signal should ensure twenty samples for each period of the highest frequency of interest, i.e. $T_{\max} > 20\Delta t$. This condition refers to the highest frequency and aims achieving a dense overlapped spectrum around the presumed inter-spectral line;

- The number of maxima m, searched when plotting the overlapped spectrum, can be set much bigger as the presumed harmonics. This condition refers to the harmonic with low amplitude for which ensuring sufficient maxima is necessary.
- The number k of cycles selected when estimating the frequency of the original signal should be maintained in all further estimations for the damaged beam. This refers to all harmonic components and aims to guarantee the repeatability of results.
- The number k of cycles for a given component should be estimated from the signal length t_S and the coarse estimated frequency of that component and should be taken in such way to ensure sufficient time for truncation, so the condition $t_S - kT_i > 0.3\,T_i$ and $t_S - kT_i < 0.6\,T_i$ should be met. This condition refers to all harmonic components and aims ensuring conditions to keep k unchanged when lower frequencies are assessed.

It is sometimes challenging to fulfill simultaneously the first two conditions, as a substantial number of samples are needed. In such cases, the measurements should be focused on acquiring signals to be processed to evaluate low frequencies. Afterward, shorter signals with improved time resolution should address the high frequencies.

4 Conclusion

Accurate frequency estimation is crucial for the detection of damage in early stage. Current frequency estimation methods, both the standard estimation by DFT as well as estimations involving interpolation fail in detecting small frequency changes. This prevents structural changes from being observed at an early stage. To overcome this limitation, we developed a frequency estimation method and software based on it, which have shown that even a small frequency drop can be estimated.

Tests performed with this software have shown it can separately evaluate the frequencies of a signal composed by more harmonics with different amplitudes. Differences between the results obtained for the fundamental frequency when it was estimated from the harmonic signal and from the composed signal are insignificant (less than 0.0003 Hz). It was also shown that the harmonics with low amplitude from multi-tone signals can be accurately estimated. Repeatability was assured for all estimations. When simulating a slight frequency drop, the software was able to identify it. It was observed a difference even if the generated frequencies are 28.6056 Hz and 28.6 Hz. For all frequency changes we succeed to calculate the Relative Frequency Shifts, irrespective to how small the frequency drop was.

Future research will focus on the complete automation of the estimation process, which will be based on the features extracted through the coarse analysis of the signal in its incipient phase.

References

1. Ostachowicz, W.M., Krawczuk, M.: Vibration analysis of a cracked beam. Comput. Struct. **36**(2), 245–250 (1990)
2. Zenzen, R., Belaidi, I., Khatir, S., Wahab, M.A.: A damage identification technique for beam-like and truss structures based on FRF and bat algorithm. Comptes Rendus Mécanique **346**(12), 1253–1266 (2018)
3. Sakaris, C.S., Sakellariou, J.S., Fassois, S.D.: Vibration-based multi-site damage precise localization via the functional model based method. Procedia Eng. **199**, 2072–2077 (2017)
4. Dahak, M., Touat, N., Kharoubi, M.: Damage detection in beam through change in measured frequency and undamaged curvature mode shape. Inverse Prob. Sci. Eng. **27**(1), 89–114 (2019)
5. Zhou, Y.L., Abdel Wahab, M.: Cosine based and extended transmissibility damage indicators for structural damage detection. Eng. Struct. **141**, 175–183 (2017)
6. Zhao, J., DeWolf, J.T.: Sensitivity study for vibrational parameters used in damage detection. J. Struct. Eng. **125**(4), 401–416 (1999)
7. Gillich, G.R., Maia, N., Mituletu, I.C., Praisach, Z.I., Tufoi, M., Negru, I.: Early structural damage assessment by using an improved frequency evaluation algorithm. Lat. Am. J. Solids Struct. **12**(12), 2311–2329 (2015)
8. Gillich, G.R., Mituletu, I.C., Negru, I., Tufoi, M., Iancu, V., Muntean, F.: A method to enhance frequency readability for early damage detection. J. Vibr. Eng. Technol. **3**(5), 637–652 (2015)
9. Mituletu, I.C., Gillich, G.R., Maia, N.M.M.: A method for an accurate estimation of natural frequencies using swept-sine acoustic excitation. Mech. Syst. Signal Process. **116**, 693–709 (2019)
10. Smith, W.S.: The scientist and engineer's guide to digital signal processing. California Technical Publishing, San Diego (1997)
11. Gillich, G.R., Praisach, Z.I.: Modal identification and damage detection in beam-like structures using the power spectrum and time-frequency analysis. Signal Process. **96**(PART A), 29–44 (2014)
12. Gillich, G.R., Mituletu, I.C.: Signal post-processing for accurate evaluation of the natural frequencies. In: Yan, R., Chen, X., Mukhopadhyay, S. (eds.) Structural Health Monitoring, Measurement and Instrumentation, vol. 26, pp. 13–37. Springer, Cham (2017)
13. Gillich, N., Mituletu, I.C., Gillich, G.R., Chioncel, C.P., Hatiegan, C.: Frequency and magnitude estimation in voltage unbalanced power systems. In: Proceedings of the 10th International Symposium on Advanced Topics in Electrical Engineering (ATEE), pp. 1–4. IEEE, Bucharest (2017)
14. Grandke, T.: Interpolation algorithms for discrete Fourier transforms of weighted signals. IEEE Trans. Instrum. Meas. **32**, 350–355 (1983)
15. Quinn, B.G.: Estimating frequency by interpolation using fourier coefficients. IEEE Trans. Signal Process. **42**, 1264–1268 (1994)
16. Jain, V.K., Collins, W.L., Davis, D.C.: High-accuracy analog measurements via interpolated FFT. IEEE Trans. Instrum. Meas. **28**, 113–122 (1979)
17. Ding, K., Zheng, C., Yang, Z.: Frequency estimation accuracy analysis and improvement of energy barycenter correction method for discrete spectrum. J. Mech. Eng. **46**(05), 43–48 (2010)
18. Voglewede, P.: Parabola approximation for peak determination. Global DSP Mag. **3**(5), 13–17 (2004)

19. Jacobsen, E., Kootsookos, P.: Fast, accurate frequency estimators. IEEE Signal Process. Mag. **24**(3), 123–125 (2007)
20. Gillich, G.R., Mituletu, I.C., Praisach, Z.I., Negru, I., Tufoi., M.: Method to enhance the frequency readability for detecting incipient structural damage. Iranian J. Sci. Technol. Trans. Mech. Eng. **41**(3), 233–242 (2017)
21. Minda, A.A., Gillich, G.R.: Sinc function based interpolation method to accurate evaluate the natural frequencies. Analele Universitatii Eftimie Murgu. Fascicula de Inginerie **24**(1), 211–218 (2017)
22. Gillich, G.R., Praisach, Z.I., Negru, I.: Damages influence on dynamic behaviour of composite structures reinforced with continuous fibers. Mater. Plast. **49**(3), 186–191 (2012)
23. Gillich, G.R., Minda, P.F., Praisach, Z.I., Minda, A.A.: Natural frequencies of damaged beams - a new approach. Rom. J. Acoust. Vibr. **9**(2), 101–108 (2012)
24. Gillich, G.R., Abdel Wahab, M., Praisach, Z.I., Ntakpe J.L.: The influence of transversal crack geometry on the frequency changes of beams. In: Proceedings of International Conference on Noise and Vibration Engineering (ISMA2014) and International Conference on Uncertainty in Structural Dynamics (USD2014), pp. 485–498. Leuven (2014)
25. Gillich, G.R., Praisach, Z.I., Iancu, V., Furdui, H., Negru, I.: Natural frequency changes due to severe corrosion in metallic structures. Strojniški vestnik – J. Mech. Eng. **61**(12), 721–730 (2015)

Damage in Civil Engineering

Modal Property Extraction Based on Frequency Domain Stochastic Subspace Identification

Jau-Yu Chou and Chia-Ming Chang(✉)

National Taiwan University, 1, Sec. 4, Roosevelt Rd, Taipei 106, Taiwan
changcm@ntu.edu.tw

Abstract. Structural integrity can be investigated through observing modal properties (e.g., natural frequencies, mode shapes, and damping ratios) determined by system identification methods. To understand the dynamic behavior of a structure over time, automated modal tracking techniques are developed to reduce human-computer interaction. Moreover, modal tracking using earthquake responses is essential since the magnitudes of ambient vibrations are less effective for structural responses as well as for identification. Therefore, this study is focused on performance evaluation of the automated frequency-domain stochastic subspace identification (SSI) under both ambient and seismic excitations. The method first divides the measurements into sequential portions and then employs the refined frequency domain decomposition (rFDD). A peak-picking method is subsequently applied to extracting modal candidates with respect to natural frequencies in accordance with the rFDD results. The sequential portions of measurements are also used to construct the accumulated frequency-domain Hankel matrix, and the Hankel matrix is utilized for the frequency-domain SSI. Finally, the modal properties of the stable structural modes are selected from quick stabilization diagrams. In the numerical example, the proposed method is applied to a constant-stiffness structure with 500 sets of ambient excitation events in order to examine the stability and accuracy. Moreover, the proposed method is investigated for a continuously degraded structure using 500 sets of ambient excitation events of which the structure encounters an aging problem. Additionally, seismic excitation events are also examined using the proposed method. As a result, the proposed method is capable of extracting modal properties of structures under different types of excitations with high accuracy and shows consistency in modal tracking.

Keywords: Frequency-domain stochastic subspace system identification · Automated modal tracking · Refined frequency domain decomposition · Modal tracking · Accumulated frequency-domain Hankel matrix

1 Introduction

Structures may experience extreme loadings such as earthquakes or strong winds. These loadings will cause damage in structural elements that cannot be visibly observed by engineers and can be behind the interior decoration or nonstructural

M. A. Wahab (Ed.): DAMAS 2019, LNME, pp. 303–313, 2020.
https://doi.org/10.1007/978-981-13-8331-1_21

components. Moreover, structural collapse may occur if structural damage deteriorates to a certain level. Fortunately, structural health monitoring (SHM) provides a good opportunity to assess structural integrity.

In the field of SHM, modal properties such as natural frequencies, damping, and mode shapes can serve as an indicator to represent the soundness of structures [1]. In practical, operational modal analysis (OMA) explores modal properties based on structural responses. Brincker *et al.* developed the frequency domain decomposition (FDD) method, which is an OMA method using the spectral representation of measured signals based on ambient vibration [2]. Pioldi et al. extended the input from ambient vibration to seismic excitation in FDD and developed the refined frequency domain decomposition (rFDD) [3]. This method was later applied to various seismic excitations to evaluate the effectiveness of the method [4]. However, the accuracy of identification results highly depends on the density of spectral lines.

Stochastic subspace system identification (SSI) is another well-known identification technique for OMA using output-only measurements [5]. In the time domain, structural modes can be distinguished from spurious modes in the stabilization diagram generated from the SSI process [6]. Automated SSI methods were developed by combining the SSI with cluster analysis [7, 8]. In addition, SSI can also be carried out in the frequency domain. For example, McKelvey *et al.* proposed a subspace-based multivariable frequency-domain system identification method which employed the subspace system identification method in the frequency domain [9]. Cauberghe extended this method for structural health monitoring by forming a Hankel matrix using frequency-domain responses. A stabilization diagram was employed to obtain the modal properties of a structure [10]. Through the frequency-domain approach, the size of Hankel matrix and the system order while generating the stabilization diagram can be reduced, resulting in improved computational loading.

The objective of this study is to develop an automated system identification method based on the rFDD and the frequency-domain SSI. In this method, the measurements are first divided into sequential portions, and the rFDD can be carried out based on the sequential portions. A peak-picking method is then applied to the rFDD result, and the mode candidates in terms of natural frequency can be determined. The mode candidates are tested mode by mode within an effective frequency range using the frequency-domain SSI. Finally, the structural modes can be identified. A numerical study using 500 sets of ambient excitations is employed to investigate the consistency and accuracy of the proposed method. In addition, the proposed method is applied to the 500 sets of ambient responses for a continuously degraded building to evaluate modal tracking performance. Moreover, seismic responses of a structure are considered to evaluate performance of the proposed method under seismic responses. As a result, the proposed method is robust for tracking modal properties under different ambient conditions and for building with an aging problem (i.e., a continuously degraded structure). Moreover, the proposed method is capable of identifying modal properties based on seismic responses.

2 Methodology

In this study, a frequency-domain subspace system identification method with multi-Hankel matrices is utilized to identify the natural frequencies and mode shapes under ambient and seismic excitations. Figure 1 shows the flowchart of the proposed method. The method first divides the measurements into sequential portions and employs the refined frequency domain decomposition (rFDD). A peak-picking method is then applied to extracting modal candidates in terms of natural frequency in accordance with rFDD results. The sequential portions of measurements are used to construct the accumulated frequency-domain Hankel matrix, and this Hankel matrix is utilized for the frequency-domain SSI. Finally, the modal properties of the stable structural modes are selected from several quick stabilization diagrams.

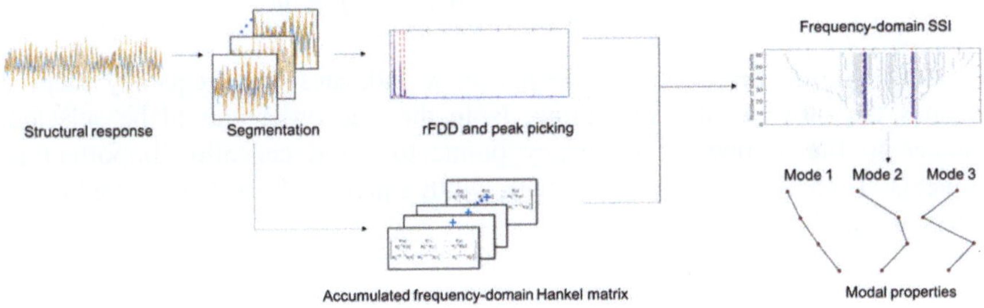

Fig. 1. Flowchart of the proposed method.

2.1 Refined Frequency Domain Decomposition (rFDD)

Consider a N-sample measurements $\mathbf{y}$ of a structure, the measurements can be divided into k portions with a measurement length as N/k. The average frequency response is calculated by

$$\mathbf{Y_f} = \frac{1}{k}\sum\nolimits_{i=1}^{k} \mathcal{F}\{\mathbf{y}_i\} \tag{1}$$

where $\mathbf{Y_f}$ is the average frequency response; $\mathcal{F}$ denotes the discrete Fourier transform. The matrix $\mathbf{C}$ is transformed from a cross-correlation matrix into the frequency domain and presented at a frequency ω as

$$\mathbf{C}(\omega) = \mathbf{Y}_\mathbf{f}^{\mathrm{T}}(\omega) \cdot \overline{\mathbf{Y}}_\mathrm{f}(\omega) \tag{2}$$

where $\overline{\mathbf{Y}}_\mathbf{f}$ is the complex conjugate of $\mathbf{Y_f}$. The matrix $\mathbf{C}$ is further been decomposed using singular value decomposition (SVD), and is given by

$$\mathbf{C}(\omega) = \mathbf{U}_l \mathbf{S}_l \mathbf{U}_l^{\mathrm{T}} \tag{3}$$

where $\mathbf{U}_l$ is a unitary complex matrix consisting of the singular vectors $\mathbf{u}_{lj}$ at a frequency ω, and $\mathbf{S}_l$ is the real diagonal matrix with the order of l. The refined frequency domain decomposition is accomplished by all of the first singular values, denoted as $\mathbf{s}$.

2.2 Peak-Picking Process

A peak-picking method is utilized to determine the peaks in s_1. These peaks are considered as mode candidates and also used in the frequency-domain SSI later. In the beginning, a smoothing technique is used to reduce the influence of noise effect in $\mathbf{s}$, the smoothing procedure can be represented as

$$\hat{\mathbf{s}}_k = \sum\nolimits_{i=-n}^{n} \mathbf{s}_{k+i}/(2n+1), \quad k > n \tag{4}$$

where $\hat{\mathbf{s}}$ represents the smoothing response; k indicates the frequency step; and n indicates the order of the smoothing. Note that the order should be selected in accordance to the number of frequency points to avoid canceling important peak components. Then, a moving-window with width equals to $2n + 1$ is applied to $\hat{\mathbf{s}}$ and the selected data is normalized by

$$\bar{\mathbf{s}}_{\mathrm{w}} = \hat{\mathbf{s}}_w / \{|\hat{\mathbf{s}}_w|\}_{\max} \tag{5}$$

where $\bar{\mathbf{s}}_{\mathrm{w}}$ is the normalized portion; the subscript "w" indicates the frequency points in the moving-window; and $|\,|$ is the absolute value indicator. The slope of the portion $\bar{\mathbf{s}}_{\mathrm{w}}$ is calculated using linear regression, and a threshold is selected to decide whether the slope value from a window is kept or eliminated. The formula is given by

$$\bar{\mathbf{s}}_{\mathrm{w}} = \alpha + \beta w$$

$$\mathbf{I}_w = \begin{cases} \beta \; if \; |\beta| \geq 0.1/n \\ 0 \; if \; |\beta| < 0.1/n \end{cases} \tag{6}$$

where α and β are the parameters obtained from the linear regression process. β will be recorded in $\mathbf{I}_w$ if the absolute slope value is larger than 10% of the largest slope (i.e., 0.1/n); otherwise, $\mathbf{I}_w$ will be set to be zero. This allows eliminating insignificant peaks

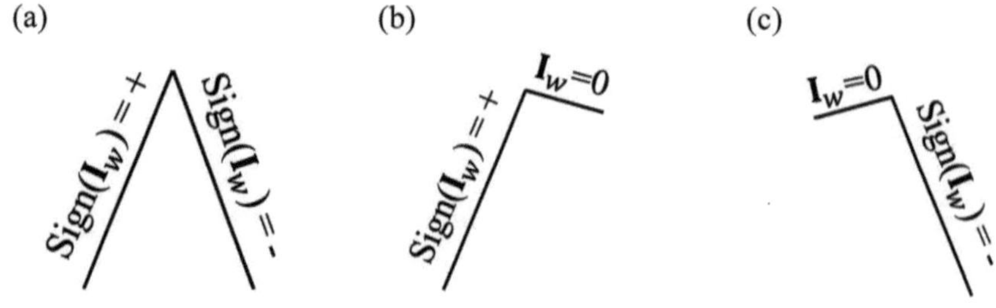

Fig. 2. Illustration of the peak-picking process: (a) positive-negative, (b) positive-zero, and (c) zero-negative situations.

in **S**. Finally, the peaks can be determined based on the signs in $\mathbf{I}_w$. As illustrated in Fig. 2, the peaks, which are the mode candidates, are selected when adjacent signs in $\mathbf{I}_w$ have patterns being positive-negative, positive-zero, or zero-negative.

2.3 Frequency-Domain Stochastic Subspace System Identification (SSI)

The frequency-domain SSI is applied to each mode candidate in terms of effective frequency range to distinguish real modes and spurious modes [11]. Reconsider the k portions of measurements in **y**, multiple frequency domain Hankel matrices can be constructed as

$$\mathcal{H}_i = \begin{bmatrix} \mathbf{Y}_i[0] & \mathbf{Y}_i[1] & \cdots & \mathbf{Y}_i[p] \\ W_{i,p}^{-1}\mathbf{Y}_i[0] & W_{i,p}^{-1}\mathbf{Y}_i[1] & \cdots & W_{i,p}^{-1}\mathbf{Y}_i[p] \\ \vdots & \vdots & \ddots & \vdots \\ W_{i,p}^{-(\frac{N}{k}-p)}\mathbf{Y}_i[0] & W_{i,p}^{-(\frac{N}{k}-p)}\mathbf{Y}_i[1] & \cdots & W_{i,p}^{-(\frac{N}{k}-p)}\mathbf{Y}_i[p] \end{bmatrix} \tag{7}$$

where $\mathcal{H}$ is the frequency-domain Hankel matrix; W_p^n is the shift phase which will not affect the identification results; p is the window length; i indicates the i-th portion. Because each Hankel matrix indicates the frequency-domain in a short time, the Hankel matrices obtained from Eq. (7) are summed such as

$$\overline{\mathcal{H}} = \mathcal{H}_1 + \mathcal{H}_2 + \cdots + \mathcal{H}_k \tag{8}$$

where $\overline{\mathcal{H}}$ is the frequency-domain Hankel matrix after accumulation. This Hankel matrix lowers some of uncertainties. The accumulated Hankel matrix is then divided into the past matrix $\overline{\mathcal{H}}_\mathrm{p}$ and the future matrix $\overline{\mathcal{H}}_\mathrm{f}$, and the subspace **O** is obtained by [12]

$$\mathbf{O} = \overline{\mathcal{H}}_\mathrm{p}/\overline{\mathcal{H}}_\mathrm{f} \cong \overline{\mathcal{H}}_\mathrm{f}\mathbf{V}_\mathrm{p,m}\mathbf{V}_\mathrm{p,m}^\mathrm{T} \rightarrow \overline{\mathcal{H}}_\mathrm{f}\mathbf{V}_\mathrm{p,m} \tag{9}$$

where $\mathbf{V}_\mathrm{p,m}^\mathrm{T}$ is neglected because the matrix can be a similarity matrix to both extended observability and controllability matrices in the SSI theory. Because the subspace, **O**, represents a product of the extend observability and controllability matrices, the system matrix and the measurement matrix can be calculated based on the two matrices, and the modal properties (i.e., natural frequency, mode shape, and damping ratio) of a specific mode can be extracted from these obtained matrices by the eigen analysis to generate the stabilization diagram. In this study, two simple criteria are used to distinguish real modes from spurious modes. One is the natural frequency error between the identified modes and mode candidates should be less than 10%; meanwhile, the modal assurance criterion (MAC) of mode shapes generated by different system orders should be higher than 0.9 [13]. By repeating the procedure to each mode candidates, the modal properties of a structure can be identified.

3 Numerical Example

A six-story, shear-type building is established to numerically investigate performance of the proposed method under ambient and seismic excitations. In this model, the mass of this building is set to 150 kg each floor, and the stiffness quantities are 80 kN/m, 76 kN/m, 72 kN/m, 68 kN/m, 64 kN/m, and 60 kN/m from the first floor to roof. The resulting natural frequencies are 0.85 Hz, 2.42 Hz, 3.86 Hz, 5.08 Hz, 6.01 Hz, and 6.70 Hz. The modal damping is 0.02. Figure 3(a) demonstrates the numerical model. The lateral absolute accelerations over all floors (i.e., AX1–AX6 in Fig. 3(a)) are utilized to evaluate the proposed method. In this study, 500 sets of structural responses under ambient vibration with peak ground acceleration (PGA) scaled to 0.05 g are first tested to explore the consistency of the proposed method. The measurements are dividing into 1, 2, and 5 portions for detailed evaluation. The number of frequency points is 4096, and the order n in Eq. (4) is 2. Figure 3(b) demonstrates the resulting mode candidates based on the rFDD and peak picking procedure in a single event. In this figure, the red-dashed lines indicate the determined mode candidates, while the blue-solid curve indicates the rFDD result. Seven mode candidates in terms of frequency selected after this procedure are 0.88 Hz, 1.17 Hz, 2.39 Hz, 4.00 Hz, 4.98 Hz, 6.10 Hz, and 6.59 Hz. These obtained mode candidates are then checked by the frequency-domain SSI. Figure 3(c) represents the identified mode shapes from the proposed method as compared to the real mode shapes. In this figure, the black lines represent the identified mode shapes, while the red-dashed lines represent the designed mode shapes. Only six modes are identified as structural modes after applying the frequency-domain SSI. The identified natural frequencies of each mode from left to right are 0.85 Hz, 2.44 Hz, 4.01 Hz, 5.00 Hz, 6.00 Hz, and 6.61 Hz, which are comparable with the designed natural frequencies. Meanwhile, the identified mode shapes meet a good agreement with the real mode shape. The result indicates high accuracy of the proposed method. Moreover, the identification results of all 500 events with different number of segmentations are shown in Fig. 4. The success rate is calculated by dividing selected results which has modal assurance criterion (MAC) values higher than 0.85. The success rate of six modes are all 100% in different number of segmentations. The low natural frequency error (VAR) and modal assurance criterion (MAC) close to 1, as compared to the real modal properties, indicate the high performance and consistency of the proposed method.

In addition, the proposed method is applied to the same 500 sets of structural responses with a continuously degraded structure in order to mimic modal tracking of real-world buildings which suffers from an aging problem. In this example, the first-story stiffness is reduced by 0.1% per event, resulting in 50% reduction of the first-story stiffness after 500 events. The number of portion measurements is 2 while evaluating the proposed method. Figure 5(a) demonstrates the identified modes and the designed modes in terms of natural frequency. In the figure, the gray dots are the real natural frequencies, and the blue stars indicate the identified natural frequencies. All modes can be successfully identified by the proposed method except for the fifth mode. This is due to the vanishing mode in the rFDD. The statistical results are shown in Fig. 5(b). The success rate of 6 modes are all over 95%, with 5 modes fully identified. The MACs are

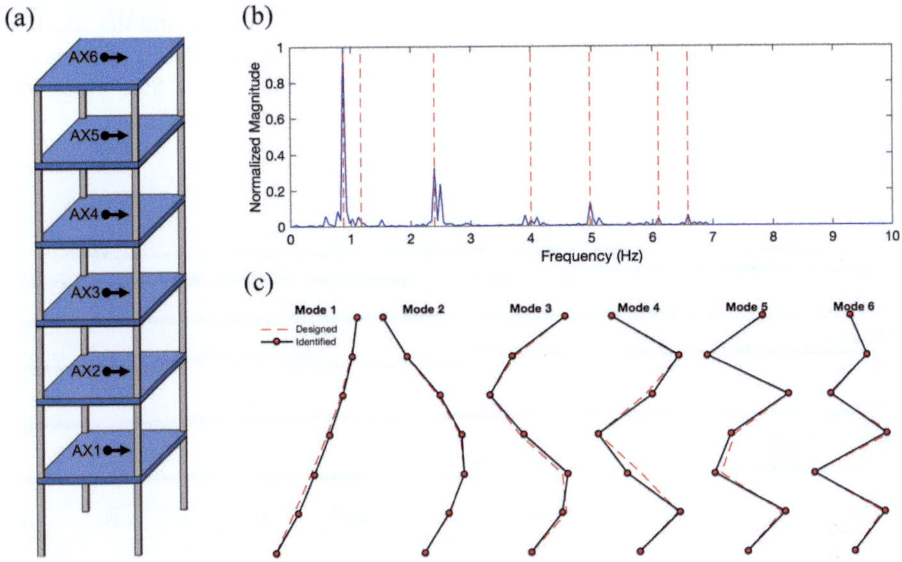

Fig. 3. (a) Illustration of the model; (b) mode candidate selection result; and (c) identified mode shapes.

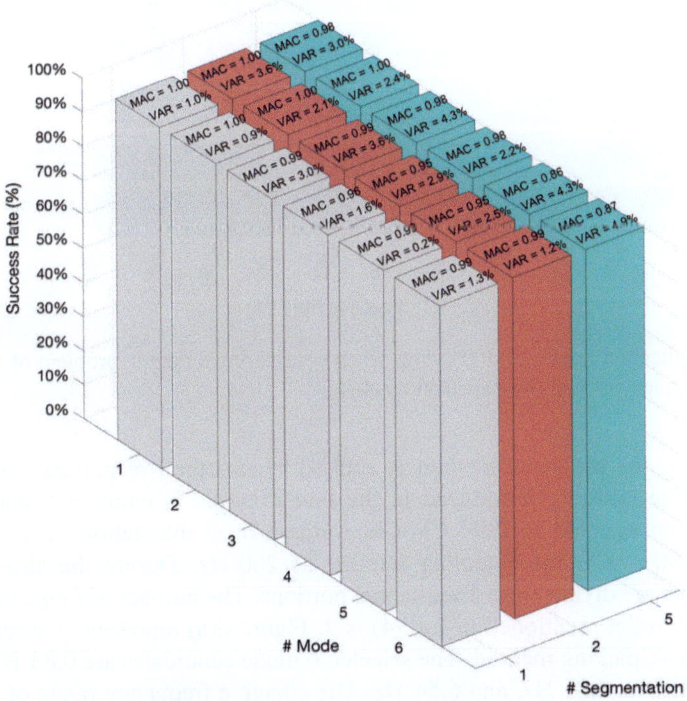

Fig. 4. Identification results of 500 events.

close to 1, indicating the consistency of the designed mode shapes and the identified mode shapes. Although the VAR (i.e., error between the real and identified natural frequencies) is higher than previous results, the error is still within 15%, which indicates the acceptable performance in modal tracking of an aging building.

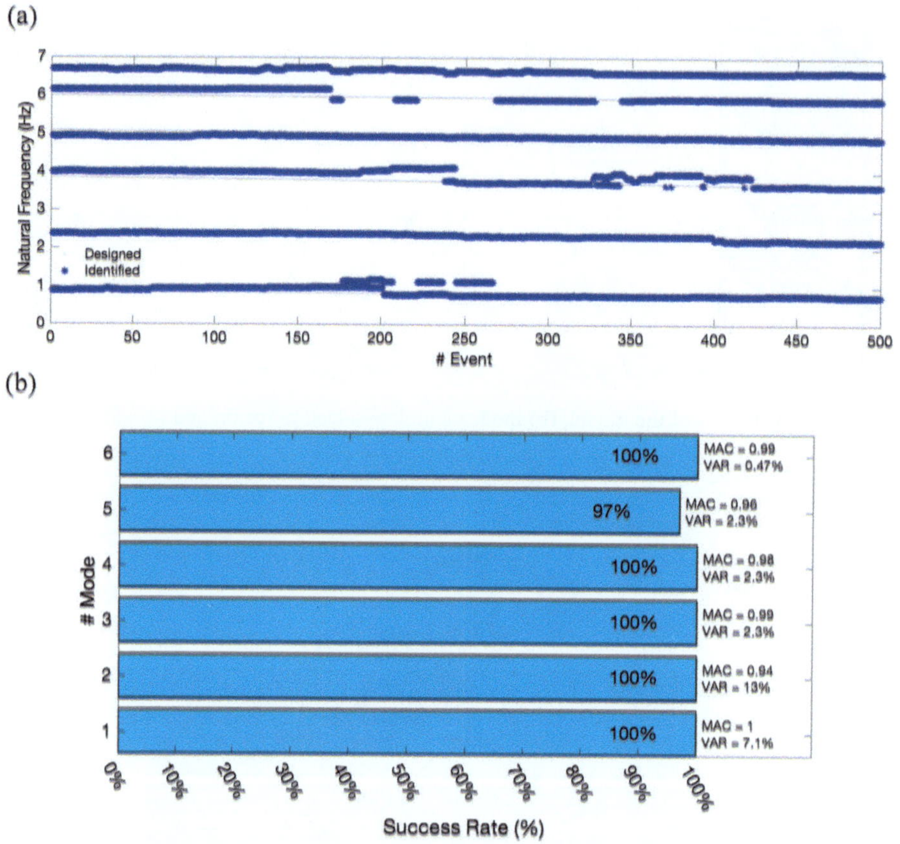

Fig. 5. Identification results of 500 events when considering an aging problem of the building: (a) natural frequencies and (b) statistical results.

Moreover, the proposed method is applied to structural responses under seismic excitation. The building introduced in the first example is employed, and the input ground motion selected is 1999 Chi-Chi earthquake at the station of TCU071. The duration is 90 s with the sampling rate set as 200 Hz. During the simulation, the measurements are divided into 5 sequential portions. The number of frequency points is 4096 and the order mentioned in Eq. (4) is 2. Figure 6(a) represents the results of the rFDD and peak-picking method. The selected 6 mode candidates are 0.83 Hz, 2.44 Hz, 3.81 Hz, 5.13 Hz, 6.01 Hz, and 6.54 Hz. The effective frequency range of each mode candidates is determined and utilized in the frequency-domain SSI. As shown in Fig. 6 (b), 6 modes are identified as structural modes. The black lines represent the identified mode shapes, while the red-dash lines represent the designed mode shapes. The high MACs and the low VARs indicate that the proposed method is capable of identifying modal properties from seismic responses of a structure with high accuracy.

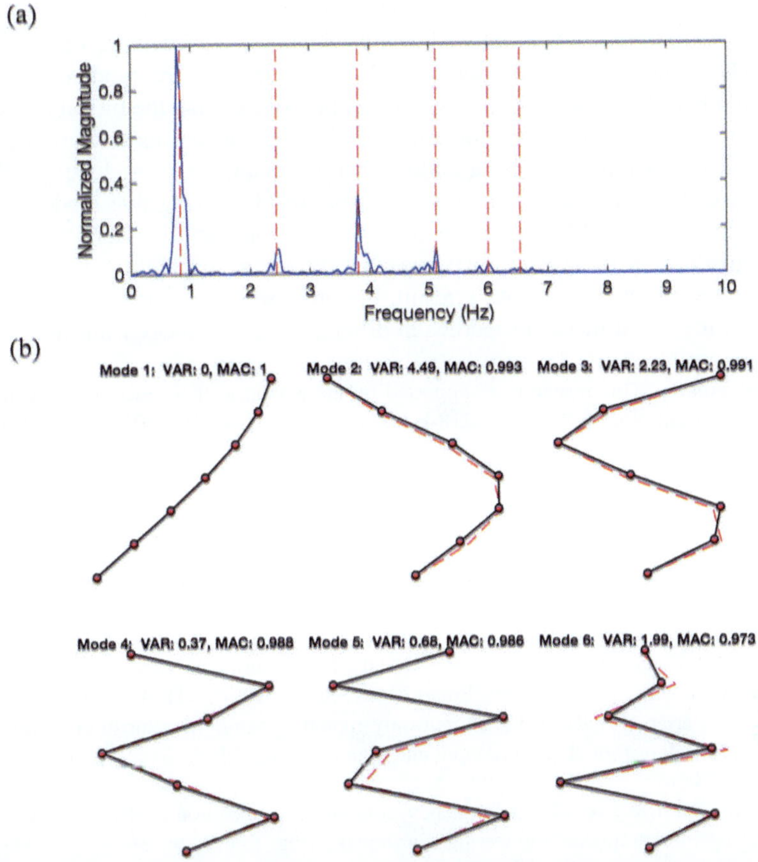

Fig. 6. Identification results of TCU071: (a) rFDD and peak-picking results and (b) comparison between the identified mode shapes and the designed mode shapes.

4 Conclusions

In this study, an automated system identification method was proposed by the integration of the refined frequency domain decomposition and the frequency-domain stochastic subspace system identification. The structural responses were first divided into sequential portions, and the refined frequency domain decomposition was carried out based on the sequential portions of measurements. A peak-picking method was applied to the rFDD results to determine mode candidates in terms of natural frequency. Then, the mode candidates were tested mode by mode with the frequency-domain stochastic subspace system identification to retain structural modes and to eliminate spurious modes. In the frequency-domain SSI, multiple frequency-domain Hankel matrices were first constructed using each portion of measurements, and an accumulated Hankel matrix was exploited. The stabilization diagram was then generated with respect to an effective frequency range of a mode candidate. Finally, the modal properties were extracted from stabilization diagrams with simple criteria.

A six-story building was employed to investigate the performance of the proposed method. The proposed method was applied to 500 sets of structural responses with different number of sequential portions. The results showed that the modal properties of the building were all successfully identified in all cases. In addition, a building with an aging problem was mimicked by a building with a continuously reducing stiffness in the first story. The proposed method was capable of identifying five modes with high accuracy and stability. Moreover, all modes were successfully identified using the proposed method from seismic responses, verifying the capability of the proposed method for seismic responses. As a result, the proposed automated system identification method identified modal properties in different circumstances with high accuracy.

Acknowledgement. This research is supported by the Ministry of Science and Technology in Taiwan under Grant No. MOST 104-2218-E-002-036 and MOST 107-3011-F-009-003.

References

1. Farrar, C.R., Doebling, S.W.: An overview modal-based damage identification method. In: International Workshop DAMAS 97 Structural Damage Assessment Using Advanced Signal Processing Procedures (1997)
2. Brinker, R., Zhang, L., Andersen, P.: Modal identification of output-only systems using frequency domain decomposition. Smart Mater. Struct. **10**(3), 441–445 (2001)
3. Pioldi, F., Ferrari, R., Rizzi, E.: Output-only modal dynamic identification of frames by a refined FDD algorithm at seismic input and high damping. Mech. Syst. Sig. Process. **68–69**, 265–291 (2016)
4. Pioldi, F., Rizzi, E.: A refined frequency domain decomposition tool for structural modal monitoring in earthquake engineering. Earthquake Eng. Eng. Vibr. **16**(3), 627–648 (2017)
5. Liu, Y.C., Loh, C.H., Ni, Y.Q.: Stochastic subspace identification for output-only modal analysis: application to super high-rise tower under abnormal loading condition. Earthquake Eng. Struct. Dynam. **42**, 477–498 (2013)
6. Peeters, B., De Roeck, G.: Stochastic system identification for operational modal analysis: a review. J. Dyn. Syst. Meas. Contr. **123**(4), 659–667 (2001)
7. Magalhães, F., Cunha, Á., Caetano, E.: Online automatic identification of the modal parameters of a long span arch bridge. Mech. Syst. Signal Process. **23**(2), 316–329 (2009)
8. Cabboi, A., Magalhães, F., Gentile, C., Cunha, Á.: Automated modal identification and tracking: application to an iron arch bridge. Struct. Control Health Monit. **24**(1), e1854 (2017)
9. McKelvey, T., Akcay, H., Ljung, L.: Subspace-based multivariable system identification from frequency response data. IEEE Trans. Autom. Control **41**(7), 960–979 (1996)
10. Cauberghe, B.: Applied frequency-domain system identification in the field of experimental and operational modal analysis. Ph.D. thesis, Vrije Universiteit Brussel (2004)

11. Chang, C.M., Chou, J.Y., Huang, S.K.: Automated extraction of dynamic characteristics using frequency-domain stochastic subspace system identification. Struct. Health Monit., under review
12. Chang, C.M., Loh, C.H.: Improved stochastic subspace identification for structural health monitoring. In: Journal of Physics: Conference Series, 11th International Conference of Damage Assessment of Structures 2015, vol. 628, p. 012010 (2015)
13. Pastor, M., Binda, M., Harčarik, T.: Modal assurance criterion. Procedia Engineering **48**, 543–548 (2012)

Damage Assessment of a Cloister Vault

A. Di Primio[1(✉)], N. Fiorini[2], D. Spina[2], C. Valente[1], and M. Vasta[1]

[1] Department of Engineering and Geology, University "G. D'Annunzio", Pescara, Italy
{alice.diprimio, claudio.valente, marcello.vasta}@unich.it

[2] Department of Civil Protection, Presidency of the Council of Ministers, Via Vitorchiano 2, Rome, Italy
{noemi.fiorini, daniele.spina}@protezionecivile.it

Abstract. Vibration-based damage identification methods are fundamental tools for the condition assessment of historical constructions prone to earthquakes. However, despite the substantial advances in the field, several issues must still be deepened to broaden the application range of such tools and to assert their effectiveness. This is particularly true for the vaulted systems considering their modal characteristics. This study deals with the damage detection of a cloister vault of the Middle Age castle sited in Bussi sul Tirino (Abruzzo, Italy), an area of moderate seismicity, which was significantly damaged by an earthquake in 2009. Due to the impossibility to have experimental measures related to the non-damaged initial state, an a priori numerical model of the undamaged vault is set up using the geometrical and mechanical properties measured on site. Ambient noise measurements were then carried out to identify the dynamic behaviour of the cloister vault in the current situation, where seismic damage is present. The a priori model is then updated on the basis of the modal parameters identified by the ambient noise measurements. The differences between the a priori and updated models are analysed to identify the presence of damage. The results found allow for discriminating the real structural damage suffered by the vault.

Keywords: Vaulted systems · Damage assessment · Environmental vibration tests

1 Introduction

Historical buildings need to be maintained, repaired and strengthened [1]. The remedial actions strongly depend on the damage experienced. The availability of robust procedures and tools for damage assessment is hence a topic of paramount importance.

The concepts for damage identification and health monitoring in the dynamic field date back of more than two decades [2]. Numerical models are often used to investigate the effects of damage, but to be fully representative they need careful calibration against experimental data [3]. Ambient vibration tests are often resorted to because of their economy, ease of execution and intensity level compatible with a safe structural

M. A. Wahab (Ed.): DAMAS 2019, LNME, pp. 314–332, 2020.
https://doi.org/10.1007/978-981-13-8331-1_22

response of the building, despite the lack of input knowledge and the necessity to use operational modal analysis techniques [4, 5].

A number of methods is available from the technical literature either self-developed [6, 7] or industrially based [4]. These latter were used to identify the modal parameters.

In the present work ambient vibration tests were performed on the vaulted system of the Bussi Castle, also known as Mediceo Castel, a historical construction. In particular, the attention was devoted to a cloister vault whose damaged conditions after the 2009 L'Aquila earthquake was studied.

For the numerical model construction, it should be taken into account that historical masonry structures have a complex geometry and are often the result of successive past interventions, while the constitutive materials can exhibit significant spatial variations in properties and internal structure. Therefore, the selection of appropriate models for structural analysis is a delicate task [8].

Many methods are presented in the literature for damage identification based on vibration signatures, but there are relatively few works discussing their application to masonry vaulted structures [9], which is the main issue of the present paper.

An a priori numerical model of the undamaged vault was set up using the geometrical and mechanical properties measured on site. Ambient noise measurements were then carried out to identify the dynamic behaviour of the cloister vault in the current situation, where seismic damage is present. The a priori model was then updated on the basis of the modal parameters identified by the ambient noise measurements. The differences between the a priori and updated models are analysed to identify the presence of damage.

2 The Bussi Castle

The Bussi Castle (known also as the Castello Mediceo) is sited in the town of Bussi sul Tirino, province of Pescara (Abruzzo, Italy). The castle was erected in the XII century by the d'Angiò family and until the XV century it was owned first by the Pietropaoli di Navelli family and then by the de Medici family. In the successive centuries, the Castel was transformed in a manor house yet preserving the aspect of a fortified building. Presently, the building belongs to the de Sanctis family and shows a fair preserved state [3] in spite of the modifications underwent during the centuries.

The present work focuses on the evaluation of the damage suffered by the cloister vault of the Castle a consequence of the L'Aquila earthquake occurred in 2009.

The considered cloister vault has almost a square plan with each side of approximately 5.50 m (Figs. 1 and 2), and a distance of 1.40 m from the shutters to the keystone (Fig. 3). The shutters rest on walls of constant depth 1.00 m, with a depth to height ratio of 0.32. Openings are present on three sides. The proximity of a door and a window on two adjacent sides creates a weak corner.

The geometry of the vault and the walls was subjected to a precise on site measurement as well as the mass density of the structural masonry and the infilling. These parameters therefore are considered "exact" and were not subjected to calibration. On the contrary, the elastic properties and the boundary conditions were subjected to updating because of their uncertainty. From the measurements, the thickness of the vault resulted to be constant and equal to 28 cm. The constitutive material, of poor mechanical properties due to the high void percentage, was characterised by the following mass density and elastic modulus: $\mu_M = 21$ kN/m^3 and E = 1500 N/mm^2. The filling was, in turn, characterised by $\mu_F = 15$ kN/m^3 (Fig. 4).

3 Crack Pattern

In order to appreciate the damage level suffered by the Bussi Castle, and particularly by the vault under consideration, some data relevant to the earthquake intensity are hereafter outlined. According to the Italian regional seismic classification, the Bussi municipality enters in seismic zone 2 with a conventional horizontal acceleration in the range 0.15 g–0.25 g. During the L'Aquila earthquake, in a zone of about 5.0 km away from the epicenter, values of the order of 0.67 g were recorded. The area of Bussi, which is 50 km away from the epicenter, underwent a shock equal to 0.06 g that is about one half of the seismic zone lower limit and one tenth of the L'Aquila earthquake. In the light of the above and the suffered damage, this means that the Bussi Castle has a high intrinsic vulnerability [3].

The crack pattern relevant to the cloister vault is shown below (Fig. 5) according to a plan view and the optical cones. Two major cracks F01 (Fig. 6) and F02 (Fig. 7) and a third minor crack F03 (Fig. 8) are observable at the vault intrados. No cracks are detectable at the top of the shutter walls.

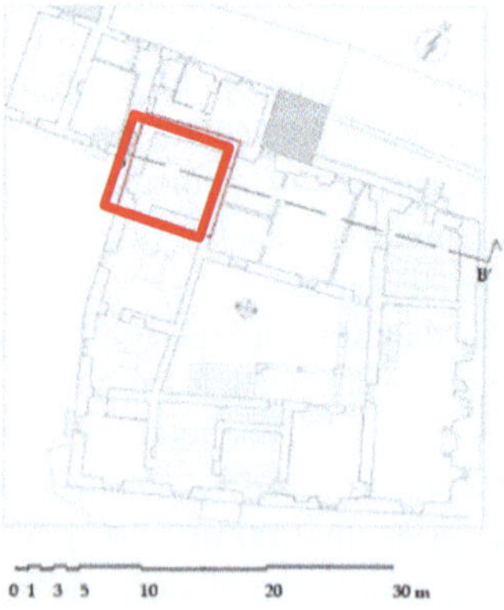

Fig. 1. Cloister vault analyzed

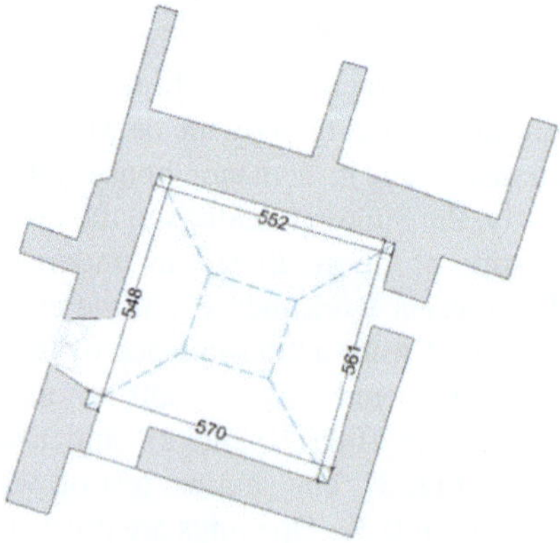

Fig. 2. Cloister vault: plan view

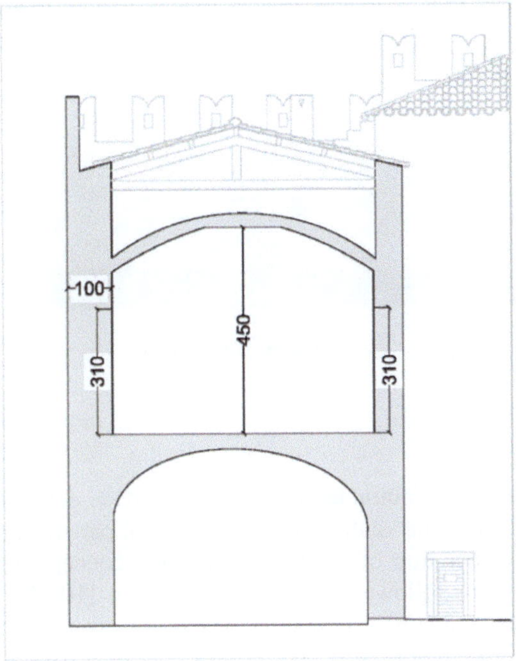

Fig. 3. Cloister vault: structural section

Fig. 4. Vault depth at keystone and constitutive disaggregated material

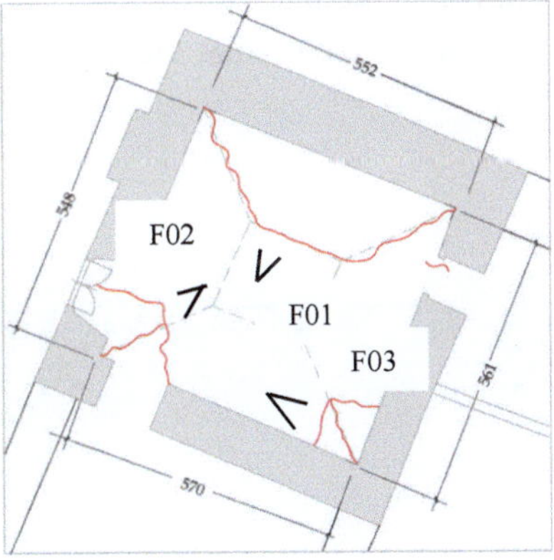

Fig. 5. Cloister vault crack pattern plane view

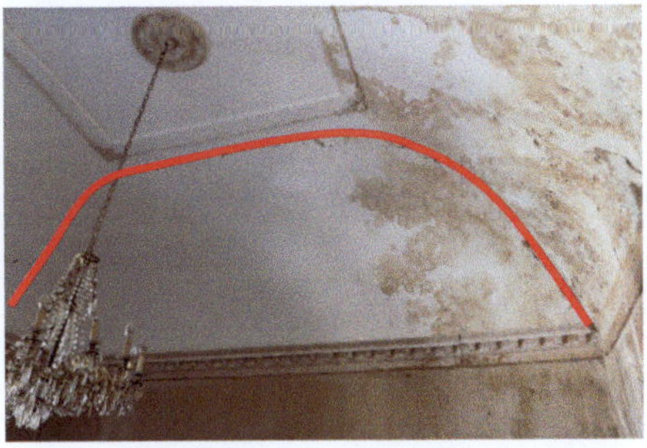

Fig. 6. Crack F01

Fig. 7. Crack F02

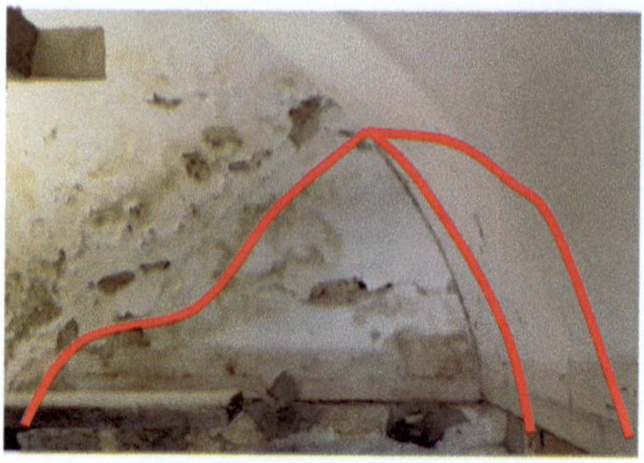

Fig. 8. Crack F03

4 Experimental Tests

The experimental tests consisted in ambient vibration tests targeted to provide data for the identification of the dynamic behavior of the vault in term of its modal parameters (frequencies and mode shapes). In consideration of the unknown input, concepts of Operational Modal Analysis (OMA) were followed for the identification [10, 11]. Preliminarily, a pre-test phase was carried out aimed at defining the optimal positioning of the instrumentation. The test configuration was as follows. A total of 31 accelerometers was used. They were placed according to two alignments I.1–I.3 (or A1–3) and I.2–I.4 (or A2–4) as shown in (Figs. 9 and 10). Each alignment has seven measurement points. The measurement point placed at the centre vault, IV.5, is shared by the two alignments. This latter and the measurement points at the shutters consist of triaxial accelerometers. The remaining measurement points consist of biaxial accelerometers oriented in such a way to record tangential and radial vibrations with respect to the arch curvature.

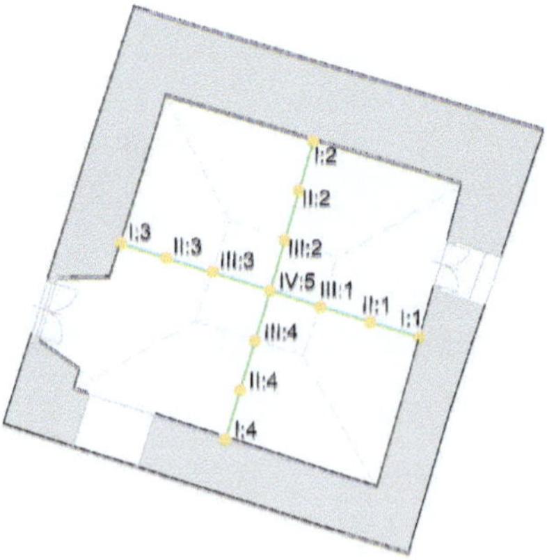

Fig. 9. Cloister vault: measur. alignments

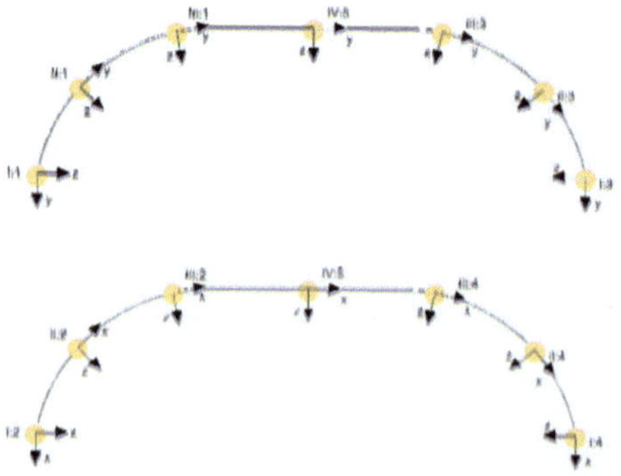

Fig. 10. Cloister vault: sensors configuration

Fig. 11. Measurement points alignment

Fig. 12. Biaxial measurement point II.4

Fig. 13. Triaxial measurement point I.1

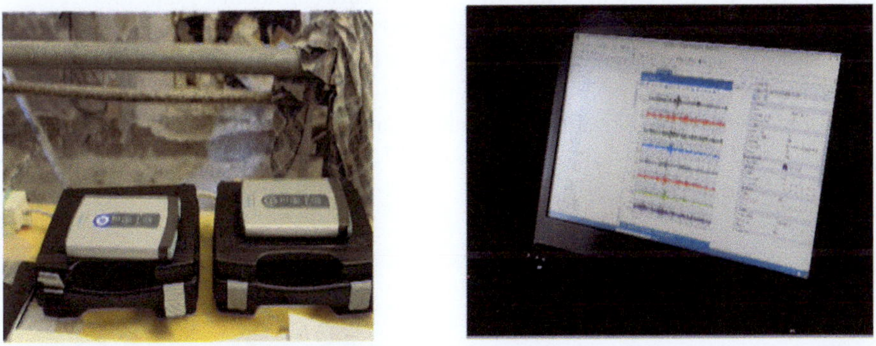

Fig. 14. Acquisition units and tablet for the management of the monitoring system

The ambient vibrations were recorded using mono-axial piezoelectric accelerometers PCB393A03, assembled as said above and characterized by a measurement range of ±5 g and a broadband resolution 0–10000 Hz with 0.00001 g rms. The sensors were connected to 24-channels LMS-SCADAS mobile recorders. The recorder is characterized by a 24-bit A/D card for a dynamic range of 150 dB. The mono-axial accelerometers were screwed on special aluminium cubes to form biaxial or triaxial sensors, the cubes were magnetically fixed to metal plates bolted to the intrados of the vault (from Figs. 11, 12, 13 and 14).

4.1 Modal Parameters Identification

The recorded signals were analyzed to extract the modal parameters of the vault in terms of frequencies, damping and mode shapes.

Among the different identification techniques available in literature [11], the LMS Test.Lab software was used. This software is based on a frequency-domain analysis through the Polymax algorithm [4]. Polymax is a modal parameter estimation method working in the frequency domain where the cross-spectra of the recorded signals are modeled as the ratio of two complex polynomials. The poles and the residual of the ratios provide frequencies, damping and mode shapes.

In order to distinguish structural modes from numerical ones, polynomials of growing order are used and the stabilization diagram is built. In the stabilization diagram a mode is define "stable" if its frequency and damping do not vary sensibly with the growth of the model order. The mode shapes are found in a second least-squares step, based on the selection of stable poles in the stabilization diagram.

In order to compare different mode shapes the Modal Assurance Criterion (MAC) [12, 13] is herein used.

$$MAC_{jk} = \frac{\left|\varnothing_{nj}^{T} * \varnothing_{sk}\right|^{2}}{\left(\varnothing_{sk}^{T} * \varnothing_{sk}\right) * \left(\varnothing_{nj}^{T} * \varnothing_{nj}\right)} \quad (1)$$

The MAC index estimates the correlation between different mode shapes, through a matrix where the cdr element is the MAC index between modes j and k.

When the MAC takes on values up to 20–30%, the modes are generally said to be not correlated (or not similar); otherwise, when the MAC is at least of the order 70–80% the modes are assumed correlated (similar).

The experimentally identified frequencies and mode shapes are reported below respectively in Table 1 and in (Figs. 15, 16, 17, 18 and 19) together with the MAC values Table 2.

Table 1. Identified frequencies

Mode	Frequencies (Hz)
1S	24.38
2S	26.88
3S	29.07
4S	36.60
5S	54.15

Below the images of the mode shapes are shown. The Cartesian rectangular coordinates are termed L, T, V to give immediate and physical interpretation of the results. The labels stand respectively for L = prevalent deformation mode in the longitudinal direction (Est-Ovest direction, see Fig. 1); T = prevalent deformation mode in the transversal direction (North-South direction, see Fig. 2); V = prevalent deformation mode in the vertical direction.

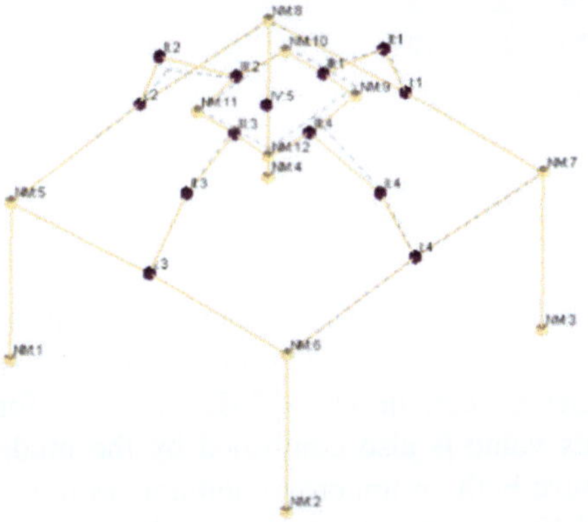

Fig. 15. Mode $1M_s\_T$

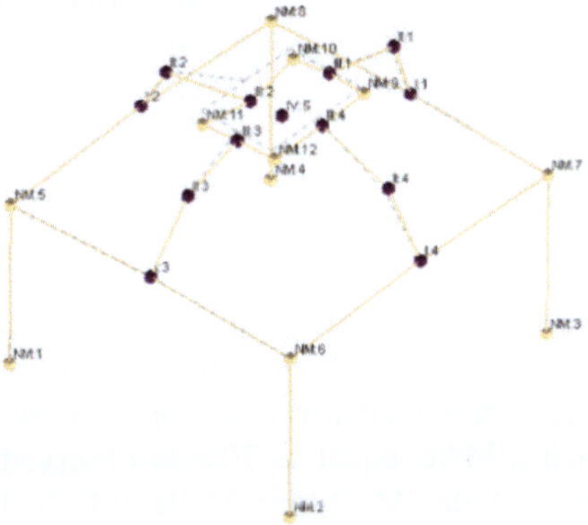

Fig. 16. Mode $2M_s\_V$

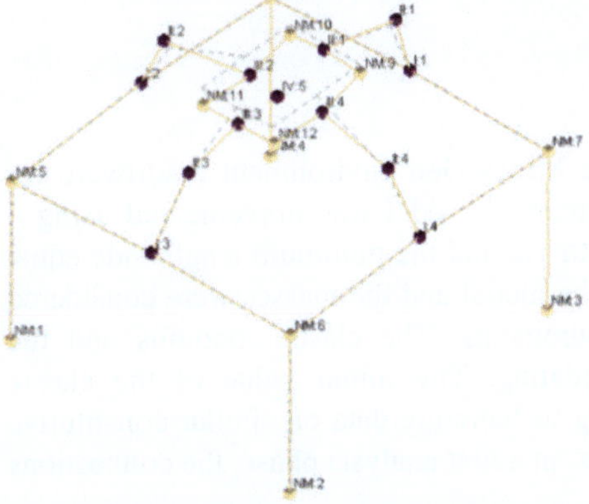

Fig. 17. Mode $3M_s\_V$

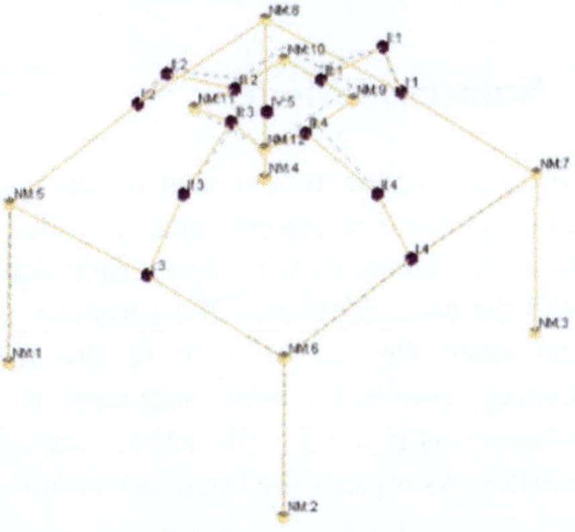

Fig. 18. Mode $4M_s\_L$

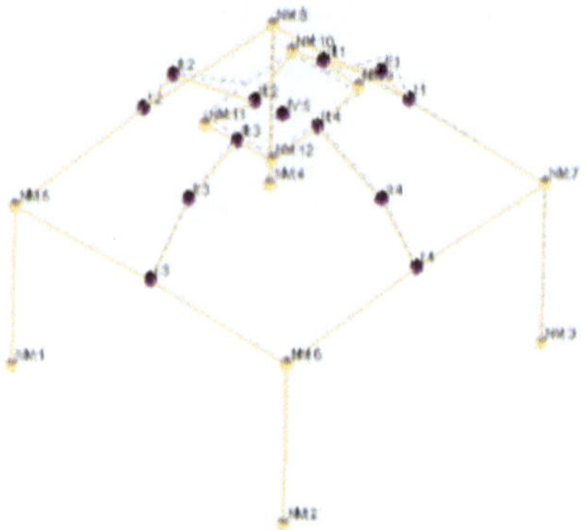

Fig. 19. Mode $5M_s\_V$

Table 2. MAC matrix – experimentally identified mode shapes

Mode	$1M_s$	$2M_s$	$3M_s$	$4M_s$	$5M_s$
$1M_s$	100	1	24	33	3
$2M_s$	1	100	70	13	12
$3M_s$	24	70	100	22	13
$4M_s$	33	13	22	100	1
$5M_s$	3	12	13	1	100

From the joint analysis of the MAC matrix, Table 2, and of the mode shapes (from Figs. 15, 16, 17, 18 and 19), it is possible to observe that the identified experimental modes are sufficiently orthogonal to each other except the cases $2M_s$ and $3M_s$, for which a MAC equal to 70% is observed. This value is also confirmed by the mode shapes Mode $2M_s\_V$ and Mode $3M_s\_V$ that have both an important component in the V direction. Off-diagonal terms of the order of 20–30% are also present and are due to the presence of non negligible vertical components that combine with the prevalent L or T components of the various mode shapes.

5 Numerical Model

The finite element model was set up in the Midas Gen environment a software for general purpose structural analyses. The numerical model was implemented using 4 nodes shell elements with aspect ratio equal to 1:2 and the minimum length side equal to 0.5 the vault thickness. The geometry of the model and the masses were considered exact since they come from in situ measurements. The elastic modulus and the boundary conditions were subjected to updating. The initial value of the elastic modulus was $E = 1.5 \cdot 106$ kN/m^2 according to literature data on similar constitutive materials. As regards the boundary conditions, in a first analysis phase, the connections of the vault on the shutter walls were considered fixed. For definiteness this model is termed "a priori" and will be labelled by the symbol MI in the following.

However, since the experimental results indicated small, yet non negligible, displacements at the connections of the vault with the shutter walls, concentrated elastic elements of appropriate stiffness were introduced at the base of the vault (Fig. 20). In particular, the rotational dofs were considered free and the stiffnesses attributed to the translational dofs were computed as follows:

$$K_V = \frac{EA}{H} \tag{2}$$

$$K_L = \frac{1}{1.2\left(\frac{H}{GA} + \frac{H^3}{nEJ}\right)} \tag{3}$$

$$K_T = \infty \tag{4}$$

where E, G, A, J, H are respectively the elastic modulus and the shear modulus, the area and the second moment of inertia of the wall cross section and the height of the wall; n is the coefficient that takes into account the degree of constraint.

The refined numerical model with respect to the "a priori" numerical model is termed "updated" and will be labelled by the symbol Mm in the following. The refinement consists in the above said modification of boundary conditions and in the calibration of the elastic modulus that was increased in the ratio 1.6 with respect to the MI model in order to get an optimal matching with the experimental results (model Ms).

The comparison between the MI and Mm models is shown below in terms of frequencies, mode shapes and MAC matrix.

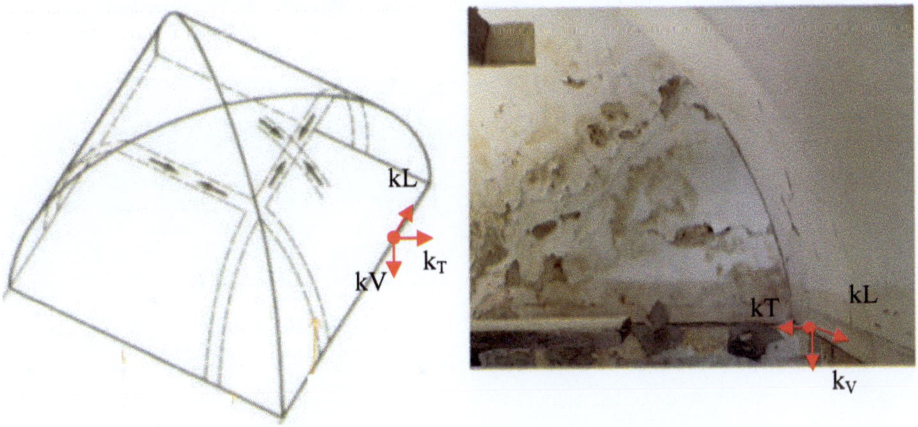

Fig. 20. Direction of action of the concentrated stiffnesses

Table 3. Frequencies: M_I–M_m comparison

Mode	Frequencies_M_I (Hz)	Frequencies_M_m (Hz)
1	23.12	24.32
2	24.48	25.77
3	28.00	28.81
4	29.13	30.51
5	35.04	42.17

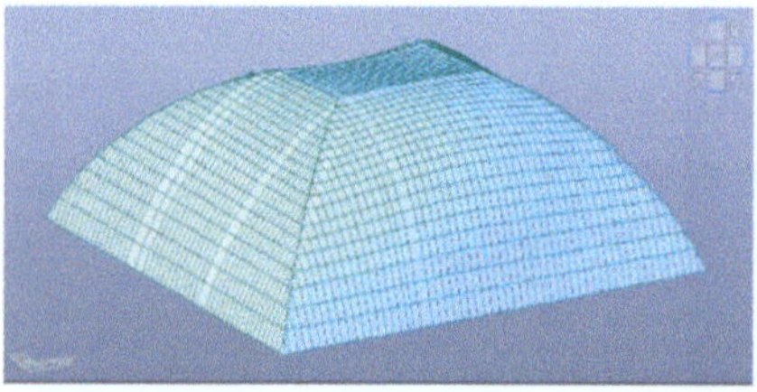

Fig. 21. Mode $1M_I$_T

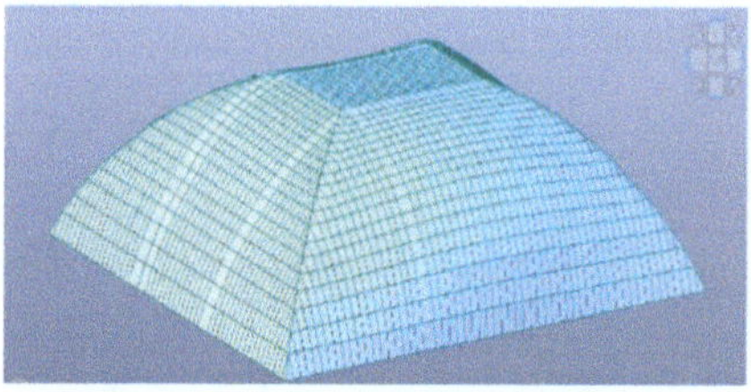

Fig. 22. Mode $1M_m$_T

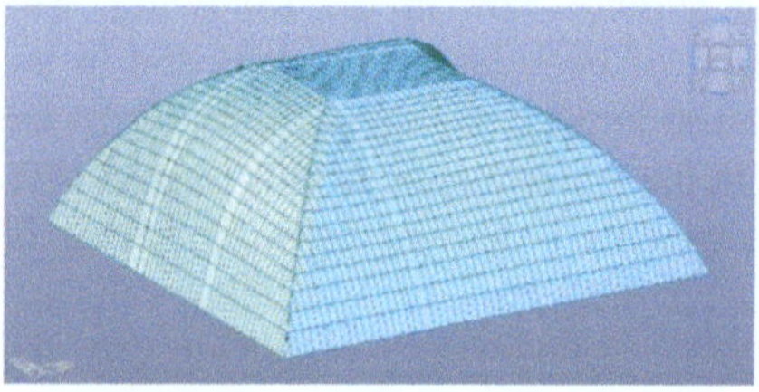

Fig. 23. Mode $2M_I$_L

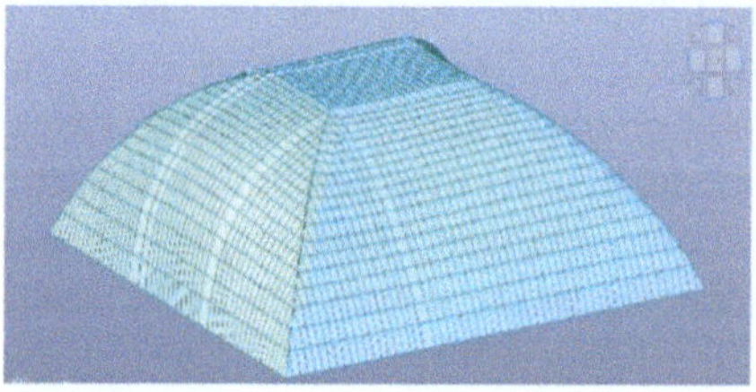

Fig. 24. Mode $2M_m$_L

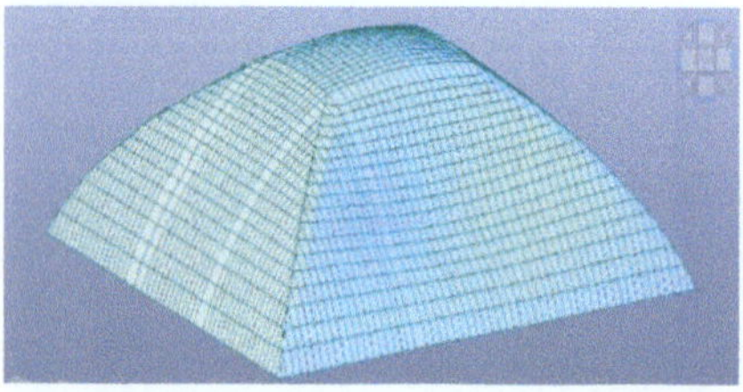

Fig. 25. Mode $3M_I$_V

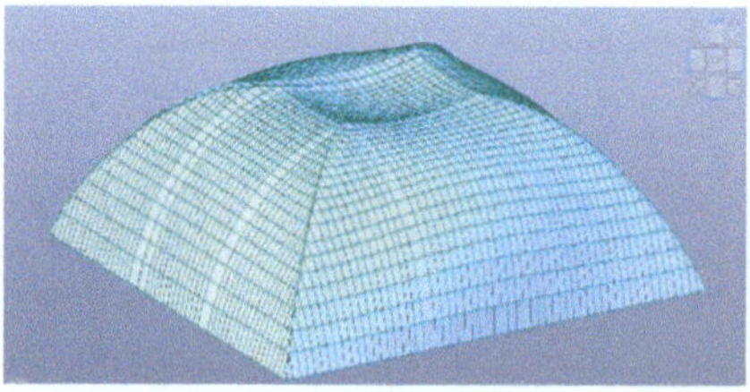

Fig. 26. Mode $3M_m$_V

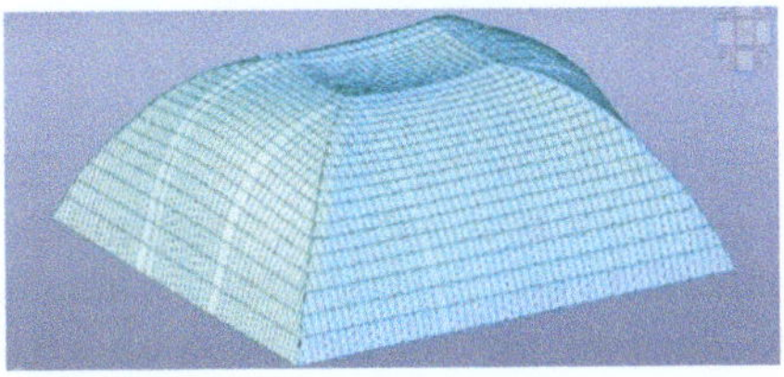

Fig. 27. Mode $4M_{I}$_V

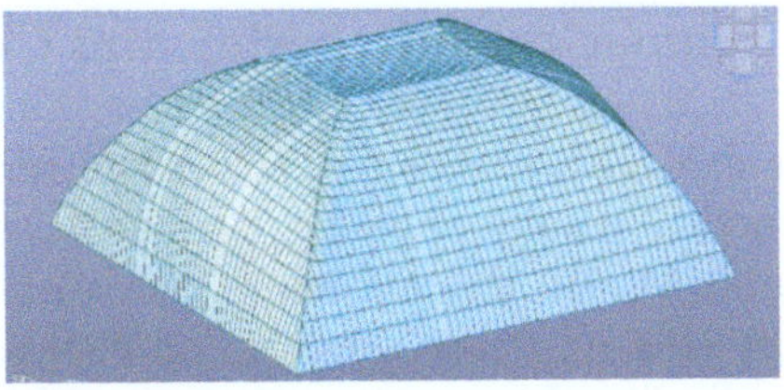

Fig. 28. Mode $4M_{m}$_V

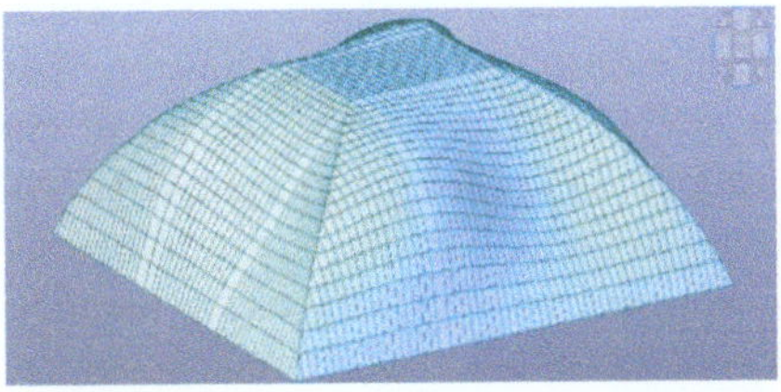

Fig. 29. Mode $5M_{I}$_Local buckled

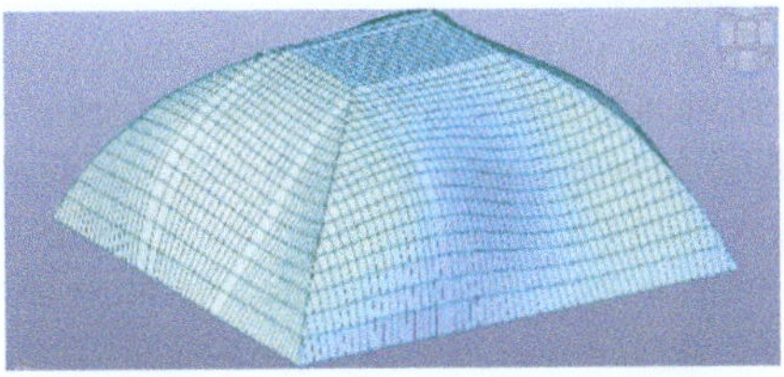

Fig. 30. Mode $5M_{m}$_Local buckled

Table 4. MAC matrix: M_{I}–M_{m} comparison

Mode	$1M_m$	$2M_m$	$3M_m$	$4M_m$	$5M_m$
$1M_I$	94	0	0	0	9
$2M_I$	0	94	0	1	0
$3M_I$	1	0	34	1	7
$4M_I$	0	1	40	85	0
$5M_I$	2	0	0	30	100

From the analysis of Tables 1 and 3 it is possible to observe that the frequencies are quite similar for the lower order modes and that they are going to worsen for the higher order modes (mode 5). In this latter case, in fact, the mode shape losses its global deformation and local buckled shapes take place.

Further, because of the stiffness releases at the boundaries of model Mm a significant increase of E is needed to restore the identified frequencies of model Ms.

As concerns the mode shapes (from Fig. 21, 22, 23, 24, 25, 26, 27, 28, 29 and 30) these remain fundamentally unchanged except the case of mode 3, almost purely vertical in model MI, feature completely lost in model Mm where mode 3 is splits into a couple of modes (mode 3 and mode 4).

The same feature is observable in the MAC matrix Table 4 that is almost purely diagonal except the case of modes 3Mm–3MI. In conclusion it can be said the mode 3 is the only one particularly sensitive to the boundary calibration.

6 Numerical – Experimental Comparison

The damage assessment is carried out according to the following procedure. First, the numerical results provided by the M_m model are compared with the experimental results M_s. If the matching between the numerical and experimental results is acceptable, in some sense to be defined, then the two models can be considered "identical" and the vault can be assumed free of damage. On the contrary, if the pairing between numerical and experimental results turns out to be defective, it is assumed that the vault experienced some damage and the discrepancies are used for damage interpretation.

In order to provide some objective measure of the observed discrepancies the MAC criterion is employed since it provides a simple scalar measure of the degree of correlation between two (eigen)vectors, i.e. mode shapes. In the following, the MAC will be given in percentage; that means if MAC = 0 the two mode shapes are "orthogonal" and completely uncorrelated, otherwise if MAC = 1 the two mode shapes are identical.

6.1 Check of Damage Presence

The check of damage presence is carried out by the comparison of the modal parameters of models M_s and M_m. The comparison of frequencies is given in Table 5, where it is observed a fairly good match of the frequencies at least for the first three modes and there is no evidence of damage.

Table 5. Frequencies: M_s–M_m comparison

Mode	Frequencies_M_s (Hz)	Frequencies_M_m (Hz)
1	24.38	24.32
2	26.88	25.77
3	29.07	28.81
4	36.60	30.51
5	54.15	42.17

Table 6. MAC matrix: M_s–M_m

Mode	$1M_m$	$2M_m$	$3M_m$	$4M_m$	$5M_m$
$1M_s$	3	1	16	30	2
$2M_s$	0	0	21	18	0
$3M_s$	0	0	41	40	1
$4M_s$	0	6	23	23	22
$5M_s$	0	2	16	9	15

Table 7. MAC matrix: M_s–M_m. Alignment A1–3

Mode	$1M_m$	$2M_m$	$3M_m$	$4M_m$	$5M_m$
$1M_s$	4	2	76	66	8
$2M_s$	1	0	90	60	0
$3M_s$	1	0	89	58	1
$4M_s$	0	2	25	18	23
$5M_s$	5	0	13	15	8

Table 8. MAC matrix: M_s–M_m. Alignment A2–4

Mode	$1M_m$	$2M_m$	$3M_m$	$4M_m$	$5M_m$
$1M_s$	3	0	11	31	0
$2M_s$	0	0	14	13	0
$3M_s$	0	0	33	42	1
$4M_s$	0	17	35	49	15
$5M_s$	0	16	32	19	20

Considering the MAC matrix of Table 6 it is noted that the first two numerical mode shapes $1M_m$ and $2M_m$ are orthogonal to all of the identified experimental mode shapes $1M_s$–$5M_s$ and hence these modes are highly uncorrelated. On the contrary, the mode shapes $3M_m$ and $4M_m$, characterized by an important translational V component, present a weak correlation with all mode shapes $1M_s$–$5M_s$. This happens because of the non negligible vertical component present in all of the mode shapes $1M_s$–$5M_s$. However, the observed lack of correlation would lead to either one of the conclusions: there is a considerable damage to the vault or the numerical model is inadequate to represent the vault vibrations. In order to better appreciate these aspects, the two main sensors alignments A2–4 and A1–3 (Figs. 9 and 10), were investigated separately, Tables 7 and 8. This analysis allows to understand if an alignment contributes negatively to the MAC index. To this end, the components orthogonal to an alignment were removed from the consideration so that only the in-plane components of each alignment have been processed.

It is found that the basic features highlighted by the complete MAC matrix are maintained also in the case of the two alignments. However, for modes $3M_m$ and $4M_m$, the MAC values of the A2–4 alignment worsen whereas those of the A1–3 alignment improve. Therefore, it can be said that there is evidence of some elements of disturbance in the direction A2–4 along with presumably a crack is localized.

6.2 Damage Evaluation

In view of the above findings, a new numerical model M_d has been set up to implement the in situ observed crack pattern (Figs. 5, 6, 7 and 8). The damage is simulated by a reduction of the elastic modulus in the finite elements interested by the crack lines (Fig. 31). The reduced value is equal to $E_r = 0.1E$.

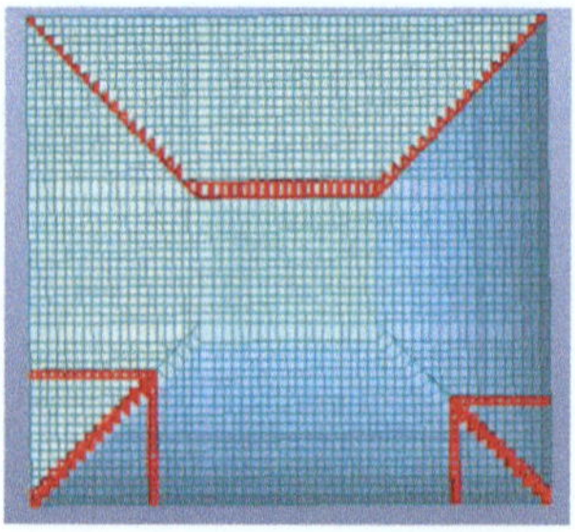

Fig. 31. Numerical model M_d: localization of crack lines.

Fig. 32. Mode $1M_{s}$_T

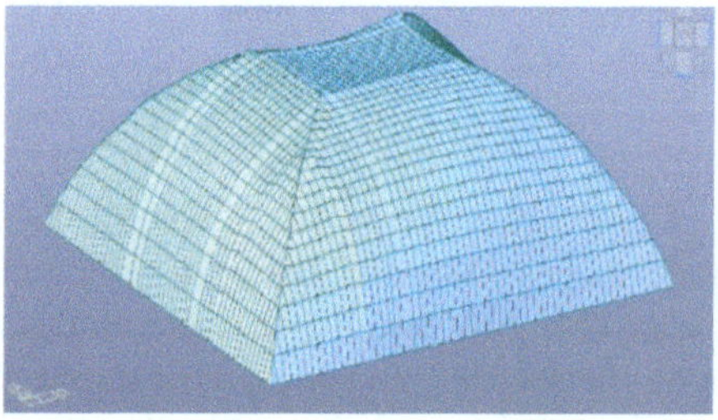

Fig. 33. Mode $1M_{d}$_T

Fig. 34. Mode $2M_{s}$_V

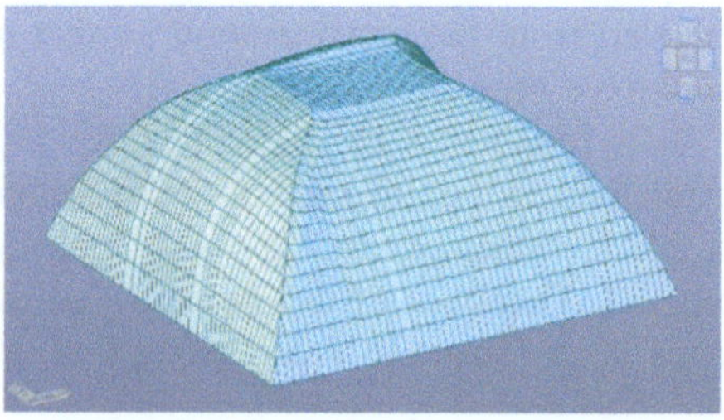

Fig. 35. Mode $2M_{d}$_L

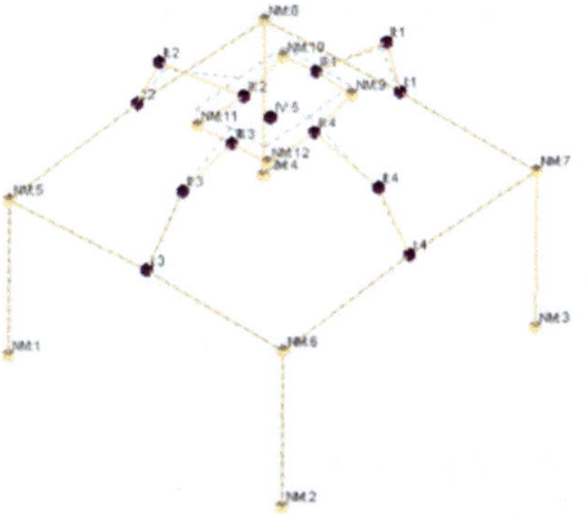

Fig. 36. Mode $3M_s$_V

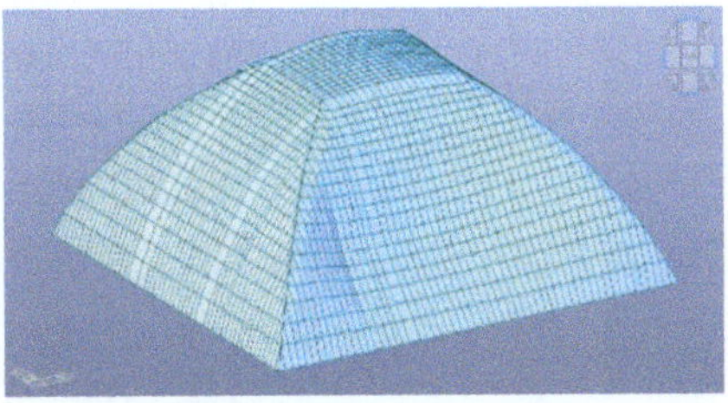

Fig. 37. Mode $3M_d$_V

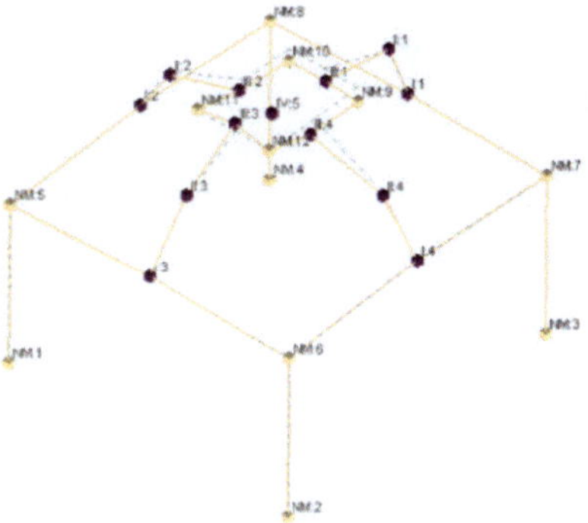

Fig. 38. Mode $4M_s$_L

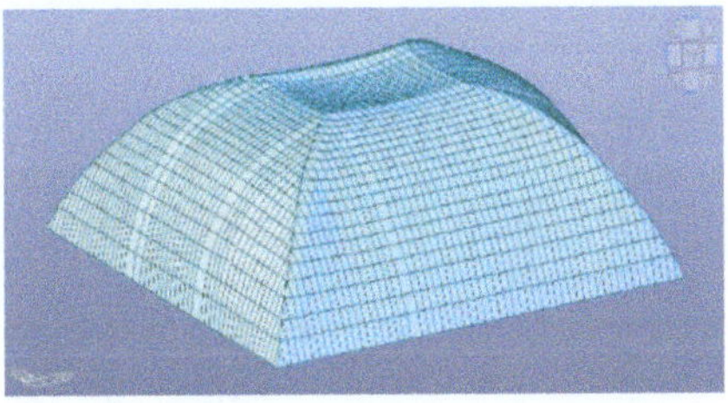

Fig. 39. Mode $4M_d$_V

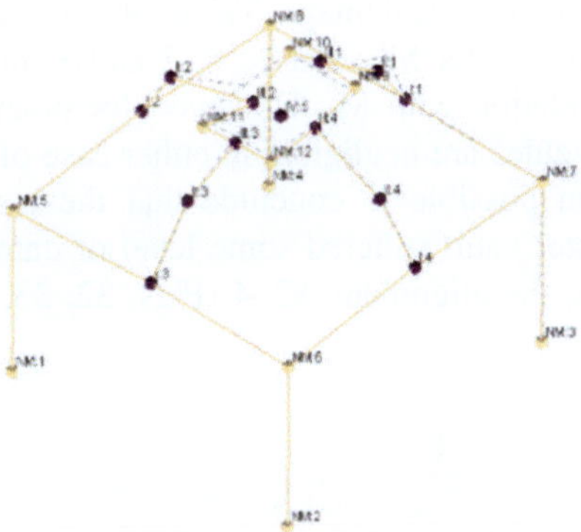

Fig. 40. Mode $5M_s$_V

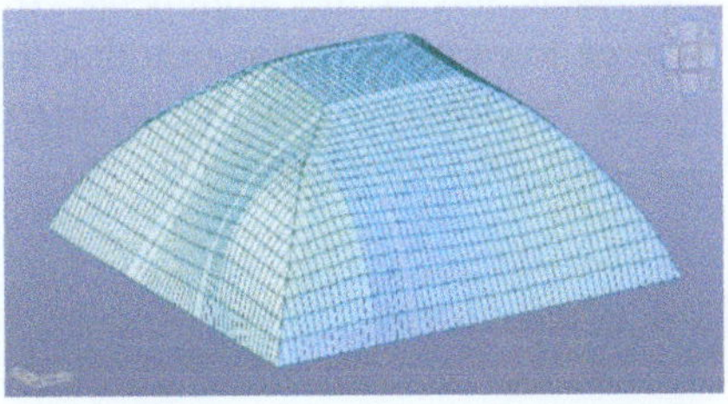

Fig. 41. Mode $5M_d$_ Torsional

Table 9. MAC matrix M_s–M_d

Mode	$1M_d$	$2M_d$	$3M_d$	$4M_d$	$5M_d$
$1M_s$	9	0	12	21	9
$2M_s$	2	0	18	50	51
$3M_s$	5	0	35	73	67
$4M_s$	0	0	19	35	20
$5M_s$	1	2	16	26	24

Table 10. MAC matrix: M_s–M_d. Alignment A1–3

Mode	$1M_d$	$2M_d$	$3M_d$	$4M_d$	$5M_d$
$1M_s$	22	2	75	72	51
$2M_s$	23	1	90	74	87
$3M_s$	27	0	89	72	88
$4M_s$	13	0	25	19	9
$5M_s$	1	1	14	13	20

Table 11. MAC matrix: M_s–M_d. Alignment A2–4

Mode	$1M_d$	$2M_d$	$3M_d$	$4M_d$	$5M_d$
$1M_s$	10	0	7	15	8
$2M_s$	0	7	13	50	53
$3M_s$	4	14	30	86	75
$4M_s$	3	20	33	78	56
$5M_s$	3	23	46	85	84

The results obtained by the model M_d embodying the damage, Tables 9, 10 and 11, ameliorate the previous results obtained by the model M_m, Tables 6, 7 and 8, in the sense that they show a higher degree of correlation with M_s. However, for practical purposes, the observed changes in the MAC values are negligible in either case of the entire vault and of the alignments. It is then possible to conclude that the results converge towards the indication that the cloister vault suffered some level of damage and that the damage is mainly localized along the alignment A2–4 (Figs. 32, 33, 34, 35, 36, 37, 38, 39, 40, and 41).

7 Conclusion

Ambient vibration tests were carried out on the cloister vault of the Bussi Castle, province of Pescara (Abruzzo, Italy) with the purpose to assess the damage suffered during the so called L'Aquila earthquake occurred in 2009.

Due to the unavailability of experimental measures prior the seismic damage, a numerical model of the undamaged vault was first set up using the geometry and mass density measured in site and then updated by calibrating the stiffness of the constitutive material and the boundaries.

The differences between the updated model and the experimental results are analysed to identify the presence of damage.

The damage assessment is carried out according to the following procedure. First, the numerical results provided by the updated model are compared with the experimental results. If the matching between the numerical and experimental results is acceptable, then the two models can be considered "identical" and the vault can be assumed free of damage. On the contrary, if the pairing between numerical and experimental results turns out to be defective, it is assumed that the vault experienced some damage and the discrepancies are used for damage interpretation.

In order to provide an objective measure of the observed discrepancies the MAC criterion is employed since it provides a simple scalar measure of the degree of correlation between two mode shapes.

The results obtained by the numerical model embodying the damage, ameliorate the previous results obtained by the updated sound model, in the sense that they show a higher degree of correlation with the experimental results. The MAC values were computed for both the entire vault and the two main alignments to investigate whether some out of plane sensors contribute negatively to the MAC index. It is found that the basic features highlighted by the complete MAC matrix are maintained also in the case of the two alignments. However, for some modes the MAC values of the A2–4 alignment worsen whereas those of the A1–3 alignment improve. Therefore, it can be said that there is evidence of some elements of disturbance in the direction A2–4 along with presumably a crack is localized.

References

1. Calık, I., Bayraktar, A., Turker, T., Karadeniz, H.: Structural dynamic identification of a damaged and restored masonry vault using ambient vibrations. Measurement **55**, 462–472 (2014)
2. Doebling, S.W., Farrar, C.R., Prime, M.B., Shevitz, D.: Damage identification and health monitoring of structural and mechanical systems from changes in their vibration characteristics: a literature review. Los Alamos National Laboratory, Los Alamos, NM (1996)
3. Di Primio, A., Vasta, M., Valente, C., Spina, D.: Monitoring and damage assessment of the Bussi castle vaulted system. In: 110th International Masonry Conference, Milan, Italy, pp. 1–11, 9–11 July 2018
4. Peteers, B., Van der Auweraer, H.: Polymax: a revolution in operational modal analysis. In: Proceedings of 1st International Operational Modal Analysis Conference, Copenhagen, Denmark, pp. 13–24, 26–27 April 2005
5. Ramos, L.F., Marques, L., Lourenco, P.B., De Roeck, G., Campos-Costa, A., Roque, J.: Monitoring historical masonry structures with operational modal analysis: two case studies. Mech. Syst. Signal Process. **24**, 1291–1305 (2010)
6. Deraemaeker, A., Reynders, E., Roeck, G.D., Kullaa, J.: Vibration-based structural health monitoring using output-only measurements under changing environment. Mech. Syst. Signal Process. **22**(1), 34–56 (2008)

7. Acunzo, G., Fiorini, N., Mori, F., Spina, D.: Modal mass estimation from ambient vibrations measurement: a method for civil buildings. Mech. Syst. Signal Process. **98**(1), 580–593 (2018)
8. Ramos, L.F., De Roeck, G., Lourenço, P.B., Campos-Costa, A.: Damage identification on arched masonry structures using ambient and random impact vibrations. Eng. Struct. **32**(1), 146–162 (2009)
9. Lourenço, P.B.: Computations of historical masonry constructions. Prog. Struct. Mat. Eng. **4** (3), 301–319 (2002)
10. Gentile, C., Saisi, A.: Ambient vibration testing of historic masonry towers for structural identification and damage assessment. Constr. Build. Mater. **21**(6), 1311–1321 (2007)
11. Ranieri, C., Fabbrocino, G.: Operational Modal Analysis of Civil Engineering Structures: AN Introduction and Guide for Applications. Springer, Berlin (2014)
12. Ewins, D.J.: Modal Testing: Theory, Practice and Application, 2nd edn. Research Studies Press Ltd., Baldock, Hertfordshire, England (2000)
13. Allemang, R.J.: The modal assurance criterion–twenty years of use and abuse. Sound Vibr. **37**(8), 14–23 (2003)

Ensemble Technique for Machine Learning with Application to Monitoring of Heritage Structures

Giorgia Coletta(✉), Gaetano Miraglia, Rosario Ceravolo, and Cecilia Surace

Department of Structural, Geotechnical and Building Engineering, Politecnico Di Torino, Corso Duca Degli Abruzzi 24, 10129 Turin, Italy
giorgia.coletta@polito.it

Abstract. Recent studies have demonstrated the effectiveness of machine learning techniques in the context of Structural Health Monitoring (SHM), where they can be applied to distinguish operational and environmental changes of dynamic features from those related to the evolution of damage. For instance, the combination of these techniques with the cointegration, a theory usually employed in econometric studies, has led to promising results, even in the detection of damage in complex monumental buildings. Several algorithms, including Support Vector Machine and Relevance Vector Machine, can be used for implementation of the multivariate regression required in the method. The choice of the algorithm to apply and the parameters to set can drastically influence the results and can lead to a wrong perception of the structural health. This paper proposes a combinatorial selection of machine learning algorithms and their settings to define the most performing among several optimal results. The ranking problem is solved by using the Plackett–Luce (PL) model-based strategy. The final aim is to obtain the damage indicator as indifferent as possible to harmless environmental and operational variations but still sensitive to structural changes, suitable to investigate the dynamic response of a structure. Data recorded by the dynamic monitoring system installed on the Sanctuary of Vicoforte, which contains the largest masonry oval dome in the world, and a calibrated finite elements model of this structure are used to simulate a damaged condition, in order to demonstrate the proposed strategy for SHM.

Keywords: Structural Health Monitoring · Plackett–Luce · Nonlinear cointegration · Support Vector Machine · Relevant Vector Machine · Dynamic monitoring system · Sanctuary of Vicoforte

1 Introduction

In recent decades, the protection of the architectural heritage has increasingly attracted the attention of the community, which is interested in preserving as best as possible the masterpieces inherited from the past. To date, structural dynamic monitoring is one of the most advanced approaches to asset protection, which represents a meeting point between architecture and engineering thanks to the minimum invasiveness and high

M. A. Wahab (Ed.): DAMAS 2019, LNME, pp. 333–349, 2020.
https://doi.org/10.1007/978-981-13-8331-1_23

efficiency. Dynamically monitoring a structure means constantly monitoring its health and having the possibility to wisely program interventions when strictly necessary. [1] This comply with the "minimum intervention" principle and leads to economic savings.

However, to be truly effective, a monitoring system must be associated with specific techniques to recognize the appearance of damage within the data collected by the sensors.

Recent studies have shown the power of the cointegration property within the Structural Health Monitoring (SHM), a property generally used in econometric field [2]. Cointegration has proved to be useful in detecting damage on different kinds of structures, including historical buildings, and in distinguishing dangerous structural evolutions from harmless effects of Environmental and Operational Variations (EOVs). In [3] this property has been successfully associated with machine learning regression techniques, i.e. Support Vector [4] and Relevant Vector Machines [5] (SVM and RVM).

Choosing to work with machine learning regression techniques when dealing with historical buildings proves to be a very valid strategy, as they are structurally complex due to original design and alterations suffered over time. However, the definition of some settings required from these algorithms is crucial to implement a good damage detection system, able to "ignore" environmental and operational variations while remaining sensitive to dynamic changes deriving from damage.

In this paper an ensemble technique is proposed for the definition of the best performing regression model in order to define a reliable damage indicator. Ensemble techniques, or simply *ensembles*, are one of the most effective machine learning methodologies today [6]. They combine different algorithms to build a model that produce a unified prediction. In general, ensembles can be classified as *sequential* (e.g. gradient boosting) and *parallel* (e.g. decision trees) or as *homogeneous* (bagging of decision trees) and *heterogeneous*, i.e. the competition come about distinct types of algorithms (e.g. neural network and decision tree) [6]. In this paper a heterogeneous ensemble is proposed.

Several regression models can give rise to a function that satisfies the specific conditions of a damage indicator. However, finding a regression function as close as possible to measurements leads to low model residual and consequently a restricted "threshold" of damage. This would lead to a very sensitive damage indicator, theoretically capable of identifying even the slightest damage. This challenge is addressed by introducing the concept of the Rank Aggregation (RA) problem [7].

The Sanctuary of Vicoforte, Italy, is a perfect case study to test this method. This imposing structure, covered by the largest oval masonry dome in the world, has been subjected over the years to differentials yielding due to the inhomogeneous foundation soil. The cracks showed at the end of 80's encouraged a strong reinforcement intervention and the installation of a static and subsequently dynamic monitoring system [8]. In this study, data from the permanent dynamic monitoring system are used [9, 10]. Moreover, a calibrated finite element model [11, 12] of the Sanctuary is employed in order to simulate a damaged condition.

Section 2 provides an overview of the cointegration method and the two machine learning algorithms chosen for regression in order to demonstrate the contribution given by the Plackett–Luce theory, described in Sect. 3. Section 4 presents the case study of the Vicoforte Sanctuary, whose experimental data are integrated with those of its FE model. Finally, the results are discussed in the final section, in Sect. 5.

2 Cointegration

The idea of cointegration arises from the econometric context in which it is used to project out data components related to non-stationary long-term trends. In general, econometrics experts are interested in knowing if a relationship between economic time-series exists and in estimating its parameters; within SHM, engineers look for relationship between dynamic variables that is damage sensitive, without being disturbed by EOVs, so that it can be used as a “structure health indicator”.

Cross et al. [2] proved the effectiveness of cointegration in damage detection, using ε, a regression model residual, as cointegration relationship. The basic idea is to fit a regression model on frequency trend over time, estimating ε and use it as a damage indicator if it proves to be stationary (i.e. the cointegration relationship exists). In the next section, the general theory of cointegration is examined.

2.1 Theory of Cointegration

To better understand the theory of cointegration, some basic concepts need to be explicit. The purpose of the section is just to give a univocal definition, useful to understand the implemented processes.

Time series are the values of a given variable over time. They are tabulated or plotted as chronologically ordered numbers or data points. In this method, the variables “frequencies” are considered as time series. In the case of the Sanctuary of Vicoforte, the frequencies have been identified hourly over 4 months period through an automatic procedure [10].

A time series can be stationary or not. A stationary time series is one whose statistical properties such as mean, variance, autocorrelation, etc. are all constant over time. In this application stationarity is a very important property owned by the structural health indicator: in an ideal case, the parameter stays stationary when structures are healthy, under all environmental and operational conditions, while it changes mean, variance, etc. in the occurrence of damage.

A stationary time series is characterized by an integration order *I(0)*. A non-stationary time series is said to be integrated of order d if its d-th difference is stationary. For instance, given a time series $x(t) \sim I(1)$, it must be differentiated only once in order to obtain a stationary process.

Now, having recalled these concepts, it is possible to give a definition of cointegration related to the proposed damage detection method: two or more non-stationary time series are said to be cointegrated if a combination of them is stationary. Then, given x_t and y_t, non-stationary time series, they are said to be cointegrated if a vector β exists such that u_t is stationary:

$$u_t = y_t - \beta x_t \tag{1}$$

The vector β is referred to as “cointegrating vector”. Notice that the previous definition applies to a “linear” cointegration as the relationship (1) is linear. The theory, however, can be simply extended to the non-linear field, considering different functions, with order greater than 1. In this way the cointegrating relationship assumes the form:

$$u_t = y_t - f(x_t) \tag{2}$$

where $f(x_t)$ is a non linear function of x_t.

In order to apply the previously discussed concepts, one needs to establish the integration order of the time series under analysis: this is performed using the Augmented Dickey–Fuller (ADF) test.

2.2 The Augmented Dickey–Fuller (ADF) Test

The Augmented Dickey Fuller (ADF) is a test used in statistics and econometrics, or for some interesting application of mathematics and computer science to economic data.

Through a comparison of the ADF statistical with some relevant critical values, the test reveals if a time series is stationary (the null hypothesis is rejected) or not (the null hypothesis is accepted) and, in this last case, how many times one must difference the time series to make it stationary [13]. In fact, the test can be repeated until the rejection of the null hypothesis is obtained, so that the integration order is deducted.

The ADF test is fundamental for the analysis, in order to establish if a cointegration relationship among variables exists. It is necessary to perform the ADF test primarily on the time series and then on the model residual.

2.3 Regression Models: Machine Learners and Kernel Function

To build a map between two (or more) numerical series it is necessary to perform a regression, which in this study is performed through two machine learners: SVM and RVM.

In the case of supervised learning, the input consists in one or more independent variables (or *predictors*) and a dependent variable (or *response variable*). In output, the regression model of the response variable is obtained and it is possible to analyze its residual, i.e. the difference between the observed value (in input) and the estimated value (in output).

Machine learning algorithms model a function that "imitates" the dynamic behavior of the *healthy structure*, under several environmental and operational conditions. When the structure is damaged, the actual data diverge from the regression function, which continues to simulate a healthy behavior of the structure, and from this gap, the cointegration method can detect the appearance of the damage.

The model residual plays the role of the cointegration relationship. It respects the cointegration definitions as it depends on the observed value of the dependent variable (i.e. y_t, referred to Eq. (2)) and the predicted value of the same variable, which is estimated by means of the independent variables and is therefore their combination ($f(x_t)$ in Eq. (2)).

Both regression techniques used here, allow to implement regression not necessarily linear. Then, the possibility to choose the *kernel function* puts the user in a condition to choose which is the best in reproducing the dependent variable behavior, among a certain number of optimal results (not unique).

For the case study of the Sanctuary, the model of Plackett–Luce is proposed to select the best regression function and algorithm among eight optimal results that are SVM and RVM algorithms with the same four Kernel Function: Linear, Gaussian, Polynomial of 2nd and 3rd orders.

2.3.1 SVM

Support Vector Machines (SVM) are supervised-learning models combined with a specific class of learning algorithms (Statistical Learning Theory) used both for classification and regression analysis [4]. The use of kernels, the possibility of making convex the optimization function (absence of local minima), the sparseness of the solution and the capacity in controlling the decision boundary by focusing on significant data points (the *support vectors*) feature an SVM regression model. The purpose of an SV regression is to find a function that deviates from observed measurements by a value no greater than the so called *insensitivity parameter,* for each point (inside the training set), and at the same time is as flat as possible. This is one of the hyperparameters that regulate the algorithm, together with the *trade-off parameter* (or *box-constraint*) that controls the penalty applied to avoiding overfitting. In fact, through the addiction of structural constraints, SVM minimizes the overfit by maximizing the margin around the discriminant line.

2.3.2 RVM

The Relevance Vector Machine (RVM) is another kernel-based approach for classification and regression introduced by Tipping in 2000 [5]. The SVM displays an excellent generalization property but it is affected by some limitations: (i) the predictions are not probabilistic; (ii) there is an exigency to tune the trade-off and the insensitivity parameters; (iii) only "Mercer" kernel functions can be employed; (iv) although relatively sparse, SVM makes liberal use of kernel functions, the requisite number of which grows steeply with the size of the training set.

Also the RVM algorithm is characterized by the sparseness capacity and achieve an excellent predictive performance comparable to SVM; however, it does not suffer from any of listed disadvantages. It makes use of a Bayesian approach to learning, where a prior over the weights is introduced. The hypothesis consists in the choice of the most probable weights configuration, which derives from an iterative process on the data set. The posterior distribution of many of the weights is often peaked around zero: it leads to a much sparser model, dependent on a smaller subset of kernel functions than SVM, and an even lower risk of overfitting. The remained non-zero weights are the *'relevance vectors'*, in deference to the principle of Automatic Relevance Determination (ARD), on which the presented approach is based, i.e. the procedure is very effective to discern the basis functions that are "relevant" for good predictions.

3 The Plackett–Luce Model

A popular method for modeling rankings of a finite collection of I items is the Plackett–Luce model. It was born from independent studies by Plackett (1975) and Luce (1959). The Luce Choice Axiom is a general axiom governing the choice probabilities of a population of choosers. In [7] the axiom is illustrated as follows: suppose that the set of items is $\{A, B, C, D\}$, and that the corresponding probabilities of choosing from this set are p_A, p_B, p_C, p_D. Now, considering a subset $\{A, C\}$, the choice probabilities are q_A and q_C. Then Luce's choice axiom states that $q_A/q_C = p_A/p_C$. It means that the choice probability ratio between two items is independent of any other items in the set (B and D). Then, consider having a set of items, and a set of choice probabilities that satisfy Luce's axiom. Image to pick one item at a time out of the set, according to the choice probabilities. Such samples give a total ordering of items, which can be thought as a sample from a distribution over all possible orderings. The form of such a distribution was considered by Plackett. The Plackett–Luce model can be used when each observation provides a complete ranking of all items, but can be extended to a partial ranking case, when not all the items are considered.

The aim of the Plackett–Luce model applied to this study, is to define the most probable ranking among all possible rankings, and then identify the first element of the list. In this case, the items "in competition" are 8 regression models of the frequencies trend of a structure. They represent the I candidates. Those who "judge", by attributing votes w_k to the candidates, are the K voters and are represented by the frequencies measured. The score is the probability of picking the candidate i among the other candidates. The Plackett–Luce model in this context can be illustrated by a simple analogy. A group of designers represents the I candidates. They are assigned the task of representing some people, the K voters, who will judge the drawings. The whole process can be divided into 3 phases:

- Phase 1: individual votes. Each voter judges the drawing with his portrait. This means that the k-th voter sees and judges the i-th candidate only for the drawing that represent him, regardless of the $k-1$ other works, i.e. the drawings representing other voters produced by that candidate.
  The choice of linking regression models to designers is not a coincidence: in fact, the models are asked to reproduce the frequencies measurements, just as the designers have to reproduce the voters in the analogy. The scores are defined as:

$$w_{k,i} = \frac{w^*_{k,i}}{\sum_{i=1}^{I} w^*_{k,i}} \tag{3}$$

where:

$$w^*_{k,i} = e^{-\left(\frac{\beta_{k,i}}{\sigma_k}\right)^2} \tag{4}$$

And:

$$\beta_{k,i} = \frac{|z_{k,e} - z_{k,i}|}{|z_{k,e}| + 1} \quad (5)$$

$\beta_{k,i}$ is the absolute normalized difference between the experimental and numerical measurements of frequency. $w^{*}_{k,i}$ moves the optimization from a minimization to a maximization problem, through a Gaussian function with 0-mean; σ_k is the standard deviation of the k-th scatter over the regression models.

- Phase 2: individual judgment on permutations. At this point, all the possible rankings (or permutations) of the designers are defined. The numbers of these possible rankings is $P = I!$. Each voter is called to give a personal judgment on each list, which, with no doubt, refers to the scores he gave to each designer (phase 1). In Matlab® environment, the lists are obtained through the *perms* command, which returns a matrix $C \in \mathbb{N}^{PxR}$ with $R = I$. The **C** matrix contains on each row a permutation and on each column the "rank position". An excerpt of the **C** matrix is shown in Table 1.

Table 1. Permutations matrix

p	r							
	1st	2nd	3rd	4th	5th	6th	7th	8th
1°	RVM polyl 3	RVM poly 2	RVM gauss	RVM linear	SVM poly 3	SVM poly 2	SVM gauss	SVM linear
2°	RVM polyl 3	RVM poly 2	RVM gauss	RVM linear	SVM poly 3	SVM poly 2	SVM linear	SVM gauss
3°	RVM polyl 3	RVM poly 2	RVM gauss	RVM linear	SVM poly 3	SVM gauss	SVM poly 2	SVM linear
4°	RVM polyl 3	RVM poly 2	RVM gauss	RVM gauss	SVM poly 3	SVM gauss	SVM linear	SVM poly 2
…	…							
40320°	SVM linear	SVM gauss	SVM poly 2	SVM poly 3	RVM linear	RVM gauss	RVM poly 2	RVM polyl 3

The individual judgment expresses the "distance", in probabilistic terms, between the p-th list considered and the personal (virtual) ranking of the voter, defined as:

$$f_{p,k} = \prod\nolimits_{r=1}^{R} \frac{w_{k,C_{p,r}}}{\sum\nolimits_{s=r}^{R} w_{k,C_{p,s}}} \quad (6)$$

- Phase 3: collective judgment on permutations. In the last phase, the voters consult to establish the most quoted list. Once the reference list has been chosen, the best designer, the worst and all intermediate ones are defined accordingly.

The collective judgment mentioned in the analogy is the probability of a list of being observed. It is calculated as:

$$f_p = \prod_{k=1}^{K} f_{p,k} = \prod_{k=1}^{K} \left(\prod_{r=1}^{R} \frac{w_{k,C_{pr}}}{\sum_{s=r}^{R} w_{k,C_{ps}}} \right) \tag{7}$$

The list characterized by the greater value of f_p is named *reference list*, and the best candidate can be derived from it.

4 Case Study: The Sanctuary of Vicoforte

The Sanctuary of Vicoforte is used to test the performance of the cointegration method with the application of the Plackett–Luce model. It is a historical religious structure located in Piedmont, Italy, which is covered by the largest masonry oval dome (see Fig. 1): axes 37,2 m and 24,89 m, height 74 m.

Fig. 1. The Sanctuary of Vicoforte with its oval dome.

As mentioned above, the structure is equipped with both static and permanent dynamic monitoring systems [8, 9].

The permanent dynamic monitoring of the lantern-dome-drum system of the Sanctuary started in December 2015. The spatial configuration of the 12 uni-axial piezoelectric accelerometers derives from the employment of the optimal sensor placement algorithms to a dynamically calibrated FE model of the structure. It can been seen in Fig. 2.

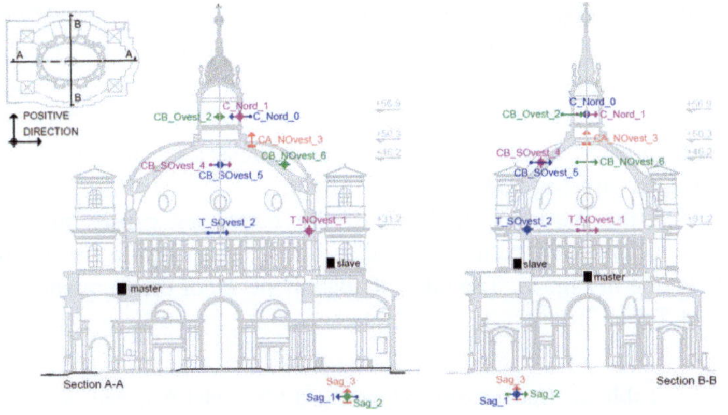

Fig. 2. Location of accelerometers: in plane (left) and in section (right)

The accelerations recorded by the sensors are collected and the frequencies of the structure are extracted from them through the Stochastic Subspace Identification (SSI) procedure [10]. Unavoidably, under general operational conditions, not all the frequencies are always identified. Since the cointegration analysis requires continuous time series, only the frequencies identified in most acceleration datasets were considered.

Higher frequencies have been neglected since previous tests [3] have attested that f_2 is enough to build a good regression model on f_1 (corresponding to the first bending mode in both directions, respectively). In fact, predictions made on the first two frequencies gave rise to equivalent damage indicators to those that consider higher frequencies too, with the advantage of a computational saving. This is probably due to the strong correlation between f_1 and f_2, consequence of the elliptical shape of the dome, which entails slightly different frequencies value and related trends. Figure 3 shows the trends of the first two frequencies of the Sanctuary.

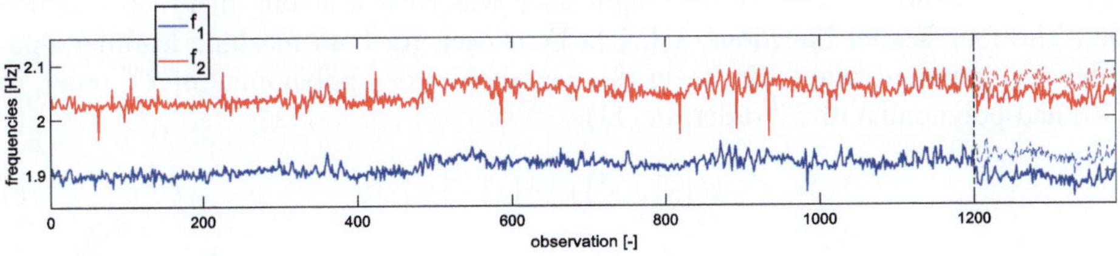

Fig. 3. Trends of the first two frequencies, 1st bending mode in X and Y directions

The data plotted in Fig. 3 have already been analyzed with the cointegration method and it has been stated that no damage has occurred or propagated during the considered monitoring period [3]. However, in order to test the effectiveness of the method, data from a damaged condition needed. Then, a virtual damage on the calibrate

FE model of the Sanctuary was induced: the damage appearance was set from 1200th observation onwards, as it can been seen in Fig. 3 (continuous line). This operation is believable thanks to the use of a thermo-mechanically calibrated FE model of the Sanctuary that was available from previous researches [11, 12]. The damage was created by reducing the elastic modulus of the buttresses by 40% in the zones characterized by higher stresses under the self-weight.

Reduction of the Young's modulus of the buttresses causes an abatement of structural frequencies from the undamaged to damaged condition (respectively dotted and continuous line in Fig. 3). The rates of abatement, *D*, of the frequencies considered are reported in Table 2:

Table 2. Undamaged and damaged FE model frequencies

Mode	f_{un} *(Hz)*	$f_{da,b}$ *(Hz)*	*D (%)*
f_1 (1st bending Y)	1.925	1.885	2.08
f_2 (1st bending X)	2.109	2.079	1.42

The first step in the analysis is to appropriately select the set of training data: it is convenient to choose a range of observations in which events, although harmless, can cause large variations in frequency trends. In this way, if similar events happen, the model succeeds in recognizing them. Accordingly, data from observation 280–700 were used as training data set.

To make sure that the cointegration approach can be used, the order of integration of the frequency trends was investigated. The ADF test firstly performed on the frequencies and secondly, after applying the difference operator on them, established an integration order *I(1)*. Thus, the 8 above mentioned regression models are fitted. The procedure requires also to set the hyper-parameters. For SVM models the trade-off parameter (or Box-Constraint), the insensitivity parameter and the kernel length scale have been set via an optimization process implemented in the Matlab® environment, which results in the regression with the minimum estimated cross-validation loss; in RVM regressions, the best kernel length scale was chosen among many tests carried out. The four Kernel Functions, $k(\{x\},\{x'\})$, chosen for both machine learning algorithms are reported below: linear, in (8), gaussian, in (9), polynomial of 2nd order, in (10) and polynomial of 3rd order, in (11).

$$k(\{x\},\{x'\}) = \{x\} \cdot \{x'\} \tag{8}$$

$$k(\{x\},\{x'\}) = exp\left(-\frac{1}{\sigma^2} < \{x\} - \{x'\}, \{x\} - \{x'\} > \right) \tag{9}$$

$$k(\{x\},\{x'\}) = (1 + \{x\} \cdot \{x'\})^2 \tag{10}$$

$$k(\{x\},\{x'\}) = (1 + \{x\} \cdot \{x'\})^3 \tag{11}$$

Where $\{x\}$ and $\{x'\}$ are the independent and the dependent variables, in this case study represented by f_2 and f_1 respectively.

The 8 regression models are plotted in Figs. 4, 5, 6, 7, 8, 9, 10 and 11: in detail, in the first subplot the identified and the predicted frequency, f_1 and $\overline{f_1}$, respectively, are compared, while the second one reports the model residual $\varepsilon = f_1 - \overline{f_1}$.

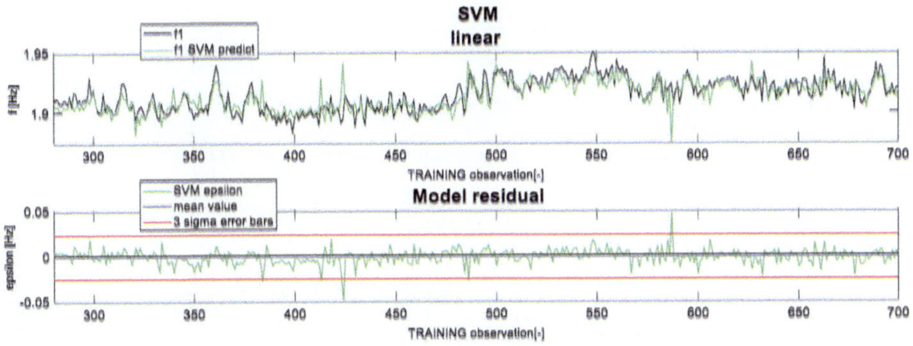

Fig. 4. SVM linear regression model and its model residual

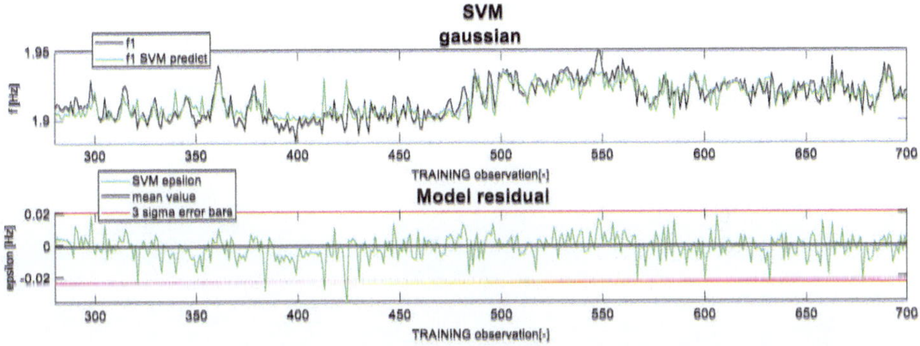

Fig. 5. SVM gaussian regression model and its model residual

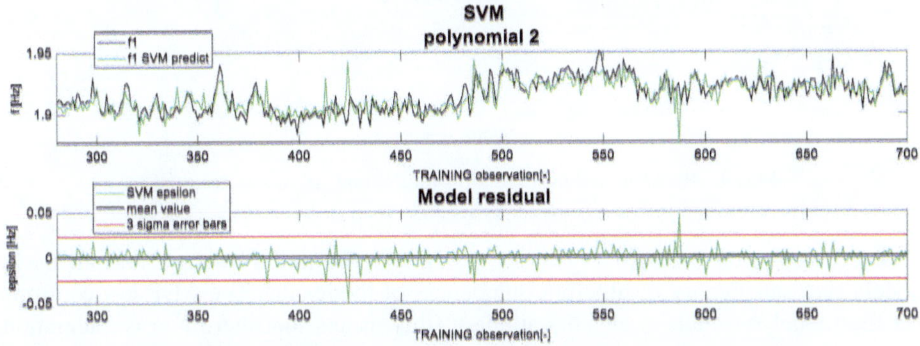

Fig. 6. SVM polynomial 2 regression model and its model residual

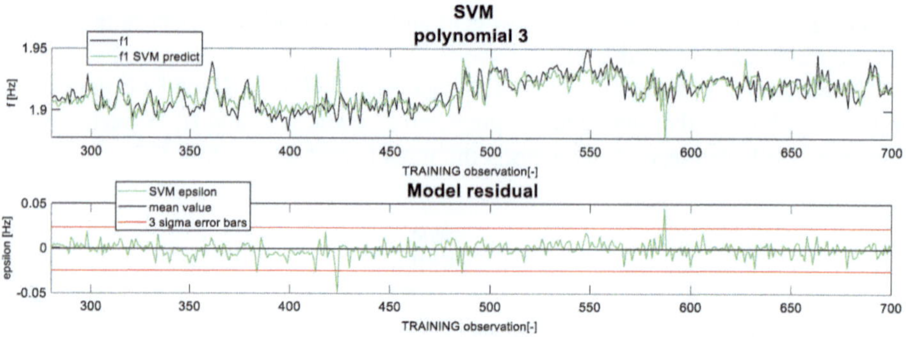

Fig. 7. SVM polynomial 3 regression model and its model residual

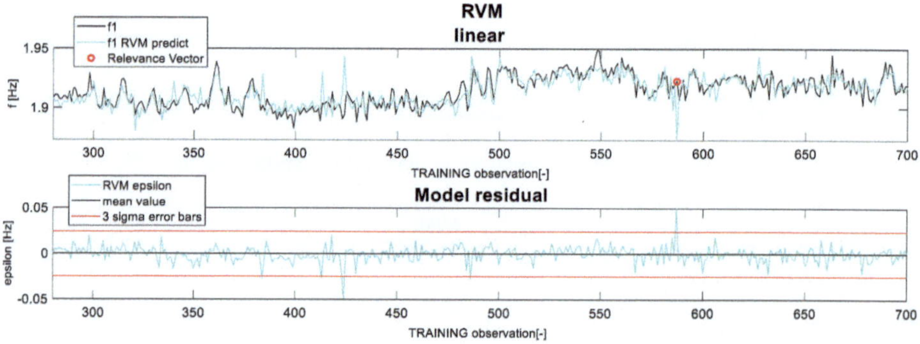

Fig. 8. RVM linear regression model and its model residual

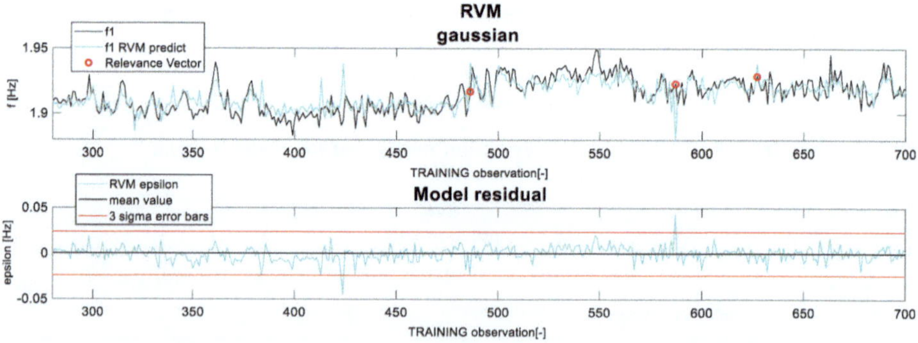

Fig. 9. RVM gaussian regression model and its model residual

The comparison between Figs. 4, 5, 6, 7, 8, 9, 10 and 11 shows that all the 8 models give satisfactory results (not uniqueness of the result). The ADF test confirms that the model residuals ε are all stationary (this means that there is a cointegration relationship between the frequencies).

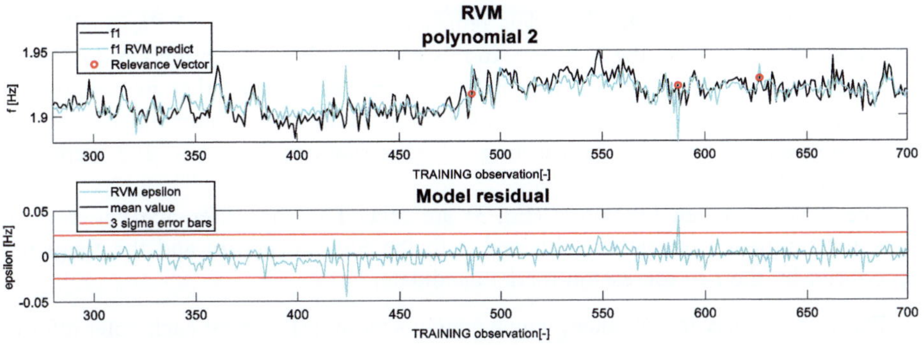

Fig. 10. RVM polynomial 2 regression model and its model residual

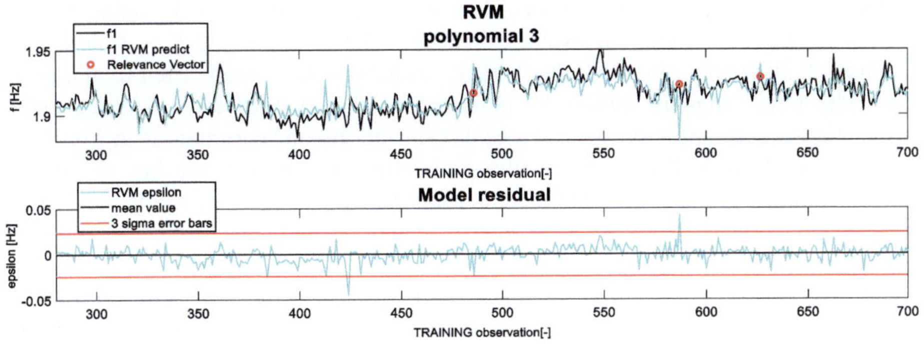

Fig. 11. RVM polynomial 3 regression model and its model residual

In the context of *statistical process control* (SPC) or *statistical quality control,* a kind of control chart can be carried up by plotting two functions of μ, the average of ε, and σ, the standard deviation of ε, in the training set: $\mu + k\sigma$ and $\mu - k\sigma$ [14]. The parameter k is chosen in order to contain a large percentage of ε, when the structure is in good condition. These two lines are the upper and lower control limits, UCL and LCL respectively, and act as a damage threshold. In this case, $k = 3$ was chosen so as more than 98,5% of training data is contained within the limits control, for all the regression models (see Fig. 12).

4.1 Application of the Plackett–Luce Model

Defined the 8 regression models that lead to efficient damage indicators, the Plackett–Luce method is applied to define their most probable ranking, from the most to the least performing. The reference list is a useful tool for choosing the kernel function and the algorithm to be implemented, in the attempt to design a fully automatic dynamic monitoring system. Through this and without needing an expert user, one could automatically know the health status of the structure.

As mentioned above, the 8 models represent the I candidates. A "testing" set of measured frequencies are the K voters. In this case $K = 105$, which amount to 25% of the observations used to train the models, is the number of voters that are unknown to the algorithms. In fact, given a set of samples, in literature it is recommended to use 4/5 for training and to keep 1/5 as a testing set. Observations from 1 to 105 were considered as voters.

The "scores" given by voters (Eq. 3) are related to the difference between the measured frequency (or rather, obtained with the SSI procedure) and the frequency reproduced by the *i-th* regression model candidate.

Phase 1: a matrix $\boldsymbol{B}$ is defined, which contains the $\beta_{k,i}$ (Eq. 5) of each voter referred to each candidate. The values within B proportionally depend on the absolute value of the difference between the measured and the predicted frequency, i.e. ε. This matrix is used to calculate w_{ki}* through Eq. 4, in which σ_k is the standard deviation of the *k-th* scatter over the 8 regression models. The last step consists in calculating the w_{ki} through the Eq. 3. All this values are collected in the $\boldsymbol{W}$ matrix, of $K \times I$ elements.

Phase 2: all the possible candidate's rankings have been defined. Their total number is $I!$, which results in in 40320, all collected on the rows of the **C** matrix, called permutations matrix, as is shown in Table 1. Considering one permutation at a time, $f_{p,k}$ is calculated through Eq. 7, that is the probability that each voter has to select a candidate, having already selected a certain numbers ($r\!-\!1$) of other candidates. These values were collected in a $K \times P$ matrix, where $P = 8!$, the number of possible permutations.

Phase 3: through Eq. 7, the reference list is defined as the one with the greatest probability of being observed. In this list, the 8 regression models are ordered and used to reproduce the trend in time of f_1, from the most to least performing. The reference list obtained is shown below (Table 3).

Table 3. Reference list.

p	r							
	1st	2nd	3rd	4th	5th	6th	7th	8th
20158°	RVM linear	SVM linear	SVM gauss	SVM poly 2	SVM poly 3	RVM poly 2	RVM gauss	RVM poly 3

The first ranking algorithm is the RVM with linear Kernel function. This does not imply that the other regression models are not suitable to fit the first frequency trend, but rather that the RVM linear model is the closest to the optimal solution, the one which best reproduces the experimental data among the 8 candidates. However, it must be noted that by changing the testing set the reference list could change. This indicates that the candidates are very "competitive": all tend to the optimal solution and there is not an algorithm and/or a function that is the best or the worse in every way. This variability does not diminish the effectiveness of the Plackett–Luce method: a future extension of the technique could consider several testing set.

The fact that the linear functions (RVM and SVM) best reproduce the 1st frequency, based on the records of the 2nd, leads to some reflections about the dynamic behaviour of the Sanctuary. As already mentioned, f_1 and f_2 correspond to the 1st bending mode according to the Y and X directions, which are respectively the directions of the principal axis of the oval dome. The result obtained shows that these two frequencies react in a fairly linearly proportional way to environmental and operational variations, likely due to the almost symmetry of the dome. It is reasonable to think that f_1 and f_2 should coincide if the dome were circular. In Fig. 12, the RVM linear function which relates the two variables is plotted.

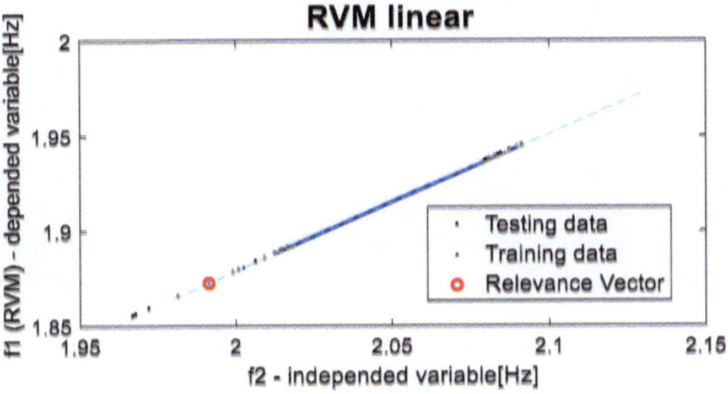

Fig. 12. Kernel linear function between f2 and f1

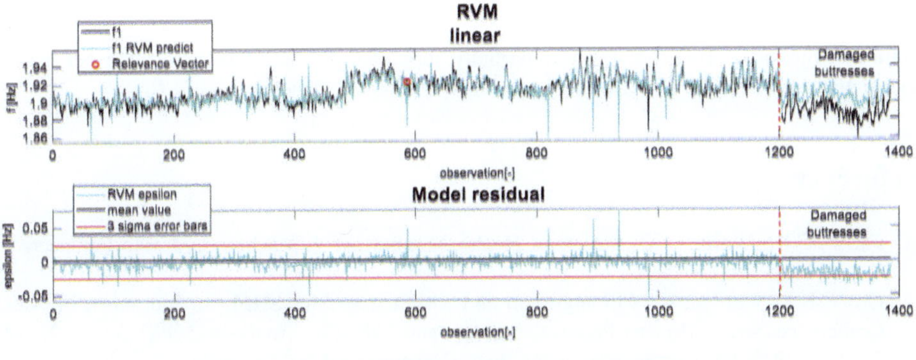

Fig. 13. RVM linear regression model on the entire data set

In Fig. 13 the best regression model is proved on the entire 4-months monitoring period (including training and testing sets) in which a damage was simulated (Fig. 3). The diagrams highlight the stationarity of the model residual until the structure condition remains unchanged from the learning (1–1200 observations): almost all the new measurements are within the control limits. When damage appears, a shift in the average residual is evident, which makes it no longer stationary. This kind of graph is indicated in the SPC literature as an *X-chart* [14]. The best regression model gives satisfactory results in the damage investigation.

5 Conclusion

The present paper introduces an original application of the Plackett–Luce model as a tool for choosing the kernel function and the algorithm which gives rise to a reliable damage indicator.

This application consists in an optimization of a previously presented method [3] in which it was necessary to establish a priori, or by trial, the regression function and the algorithm.

The Plackett–Luce model allows sorting different algorithms kinds and settings. In the presented case, 8 configurations already tested and considered optimal in detecting damage have been compared in order to select the best.

For the assumed testing data set, the linear function is found to be best in linking the first two natural frequencies of the Sanctuary, although it is the less "flexible". This is probably due to the oval symmetry of the majestic dome. In future works, other testing sets will be investigated, also in terms of the variability of the reference list connected to them, in order to increase the reliability of the method.

The Plackett–Luce model applied with the cointegration method technique resulted to be a valid tool in choosing the best regression model for different types of structures. In fact, the versatility is one of the most appreciable advantages of cointegration, which is maintained even with PL model application. Rather, it can be considered a disadvantage that the Plackett–Luce model returns the model ranking output without giving any indication of its quality: this means that the highest model in rank may still not be suitable to create a reliable damage indicator. Consequently, the regression models should be carefully analysed and "approved" as candidates before being subject to the PL procedure.

Acknowledgments. This research was partially supported by the Amministrazione del Santuario di Vicoforte and the Fondazione Cassa di Risparmio di Cuneo.

References

1. Farrar, C.R., Worden, K.: Structural Health Monitoring: A Machine Learning Perspective. Wiley, Hoboken (2013)
2. Cross, E.J., Worden, K.: Cointegration and why it works for SHM. In: Journal of Physics: Conference Series Modern Practice in Stress and Vibration Analysis (MPSVA) 2012, vol. 382, pp. 291–296. Curran Associates, Inc., Glasgow (2012)
3. Coletta, G., Miraglia, G., Pecorelli, M., Ceravolo, R., Cross, E., Surace, C., Worden, K.: Use of the cointegration strategies to remove environmental effects from data acquired on historical buildings. Eng. Struct. **183**, 1014–1026 (2019)
4. Boser, B., Guyon, I., Vapnik, V.: A training algorithm for optimal margin classifiers. In: Proceedings of 5th Annual ACM Workshop on Computational Learning Theory 1992 on Proceedings, pp. 144–152. Pittsburgh, PA, USA (1992)
5. Tipping, M.E.: Sparse Bayesian learning and the relevance vector machine. J. Mach. Learn. Res. **1**, 211–244 (2001)
6. Theobald, O.: Machine Learning for Absolute Beginners. 2nd ed. New York (2017)

7. Guiver, J., Snelson, E.: Bayesian inference for Plackett-Luce ranking models. In: Proceedings of the 26th Annual International Conference on Machine Learning, pp. 377–384. ACM (2009)
8. Ceravolo, R., De Lucia, G., Pecorelli, M.L., Zanotti Fragonara, L.: Monitoring of historical buildings: project of a dynamic monitoring system for the world's largest elliptical dome. In: Environmental, Energy and Structural Monitoring Systems (EESMS) 2015, pp. 113–118. Trento (2015)
9. Chiorino, M.A., Ceravolo, R., Spadafora, A., Zanotti Fragonara, L., Abbiati, G.: Dynamic characterization of complex masonry structures: the sanctuary of Vicoforte. In: International Journal of Architectural Heritage, pp. 296–314 (2011)
10. Pecorelli, M.L., Ceravolo, R., Epicoco, R.: An automatic modal identification procedure for the permanent dynamic monitoring of the sanctuary of Vicoforte. Int. J. Archit. Herit., 1–15 (2018)
11. Ceravolo, R., Chiorino, M., De Lucia, G., Grasso, G., Pecorelli, M.L.: Thermomechanical model updating of the world's largest oval dome. In: Van Balen, K., Verstrynge, E. (eds.) Structural Analysis of Historical Constructions - Anamnesis, diagnosis, therapy. 10th International Conference on Structural Analysis of Historical Constructions–Anamnesis, Diagnosis, Therapy, Controls 2016, pp. 1153–1160. Taylor & Francis Group (2016)
12. Ceravolo, R., De Lucia, G., Lenticchia, E., Miraglia, G.: Use of combinatorial optimisation strategies for model updating of monitored buildings. In: Proceedings of 10th International Masonry Conference, Milan, Italy (2018)
13. Fuller, W.: Introduction to Statistical Time Series, 2nd edn. Wiley, New York (1996)
14. Montgomery, D.C.: Introduction to Statistical Quality Control, 3rd edn. Wiley, New York (1996)

Sensitivity to Damage of the Forced Frequencies of a Simply Supported Beam Subjected to a Moving Quarter-Car

Arturo González, Miguel Casero(✉), and Kun Feng

School of Civil Engineering,
University College Dublin, Dublin D04 V1W8, Ireland
miguel.caseroflorez@ucd.ie

Abstract. The vibration of bridges under operational conditions can be measured via accelerometers to extract their dynamic features. These features can then be monitored in time, although only a reduced number of cause-effect scenarios can be verified on the field. Therefore, theoretical models of the bridge are often employed for covering a wider range of scenarios. For instance, a variety of damage conditions can be introduced in a calibrated bridge model to obtain the associated frequencies, which can be subsequently compared to frequencies measured on-site for assessing the bridge condition. It must be noted that these frequencies may be influenced by factors other than damage, i.e., environmental effects due to temperature changes and operational effects due to traffic. During the forced vibration of a bridge caused by a moving vehicle, the frequencies governing the bridge response depend on the mass and stiffness ratios of the vehicle to the bridge. Therefore, records in free vibration are usually preferred or alternatively, the influence of operational loads is removed from forced vibration records before assessing whether damage has occurred or not. This paper shows that forced vibration stores relevant information about damage beyond the frequency changes derived from free vibration. Eigenvalue analysis is employed to investigate how forced frequencies change with the positions of a crossing vehicle and damage. The vehicle is modelled using a quarter-car and the bridge as a simply supported finite element beam, where damage is introduced via localized stiffness losses.

Keywords: Vehicle-Bridge interaction · Damage detection · Forced vibration

1 Introduction

Structural Health Monitoring (SHM) is becoming increasingly important in several research disciplines, such as civil engineering, aeronautics, and mechanical engineering [1]. The condition of bridges, aircrafts, and machines can be monitored by changes to the dynamic properties of a structural system, i.e., natural frequencies, mode shapes and damping ratios. However, it must be noted that these changes may or may not be related to damage, i.e., frequencies will also change due to environmental and operational effects. Methods based on mode shapes include mode shape curvature [2], modal strain energy [3], changes in dynamic flexibility [4] and others, which are typically able

M. A. Wahab (Ed.): DAMAS 2019, LNME, pp. 350–362, 2020.
https://doi.org/10.1007/978-981-13-8331-1_24

to locate and quantify damage at the cost of installing a large number of sensors on the structure. On the other hand, methods based on monitoring frequencies are not so demanding in the number of sensors or energy consumption, but their capabilities are commonly limited to detecting the presence of damage, without locating or quantifying it. In the case of a bridge, the installation of only one sensor (i.e., an accelerometer) can be sufficient to capture the frequency information being sought. Frequencies can be derived from an acceleration signal by various signal processing techniques, including the Fast Fourier Transform (FFT), the Wavelet Transform (WT) and the Hilbert-Huang Transform (HHT) amongst others [5].

Compared to other structures, bridges are characterized by an operational traffic load. The latter consists of a vehicle or fleet of vehicles applying spatial- and time-varying forces to the bridge that they are crossing. Bridge and vehicle are two structural systems that, when treated in isolation from each other, have their own masses, stiffness and associated frequencies. However, when the vehicle is on the bridge, the two isolated systems become one unique system, encompassing all masses and stiffness. It is true that there is usually a disproportion between the magnitude of mass and stiffness of vehicles and bridge, which leads to very close frequencies of the bridge in free and forced vibration. Nonetheless, when heavy traffic is involved or the values of bridge and vehicle frequencies are close, the differences in dynamic behaviour between forced and free vibration of the bridge may not be negligible. The significance of these differences has been numerically investigated by [6–10], and experimentally confirmed by [11, 12].

This paper aims to carry out a preliminary assessment of how significant the impact of a vehicle is on the forced frequencies of a bridge compared to damage. Forced frequencies are calculated varying both the position of a localized stiffness loss and the position of a vehicle on the bridge. For this purpose, eigenvalue analysis is applied to the coupled system composed of a simply supported finite element beam and a quarter-car model. Section 2 describes the numerical model employed to obtain the bridge frequencies for healthy and damaged scenarios with (i.e., forced vibration) and without (i.e., free vibration) the presence of a vehicle. Section 3 shows how eigenvalues vary with the position of the quarter-car on the beam model and with the position of damage. Section 4 compares the relative changes in frequency between the different damaged scenarios and the healthy condition for free vibration and for forced vibration considering each position of the vehicle on the bridge. Finally, conclusions and recommendations for further research are presented in Sect. 5.

2 Bridge and Vehicle Models

A 20 m long simply supported beam discretized into 200 finite beam elements (0.1 m long) serves as a simplified planar model of a short/medium span road bridge. The bridge is assumed to have a uniform solid rectangular cross-section with a depth of 1 m, a width of 15.1122 m, a density of 2500 kg/m^3 and Young's modulus of 35 GPa. Table 1 provides other properties associated with this model. The values of natural frequencies are rounded to two decimal places.

The vehicle is modelled as a quarter-car with two degrees of freedom related to the vertical movement of the two masses composing the model. The largest and smallest

Table 1. Bridge properties.

Parameter	Symbol	Value	Unit
2nd moment of area	I	1.25	m^4
Mass per unit length	m	37500	kg/m
1st natural frequency	$f_{b,1}$	4.24	Hz
2nd natural frequency	$f_{b,2}$	16.97	Hz
3rd natural frequency	$f_{b,3}$	38.17	Hz

masses represent the body and axle masses respectively, and they are connected through a spring-dashpot system, which simulates the suspension stiffness and damping. The connection between vehicle and bridge models is made through a second spring, which simulates the tire stiffness. Two different configurations of the vehicle are tested, namely vehicles A and B, whose properties are listed in Table 2.

Table 2. Vehicle properties.

Parameter	Symbol	Value		Unit
		Vehicle A	Vehicle B	
Body mass	m_{body}	4250	7000	kg
Axle mass	m_{axle}	750	3000	kg
Suspension stiffness	k_s	1.8×10^6	6×10^5	N/m
Suspension damping	c_s	5000	2000	Ns/m
Tire stiffness	k_t	10^6	4×10^6	N/m
1st frequency	$f_{v,1}$	1.89	1.37	Hz
2nd frequency	$f_{v,2}$	10.09	6.26	Hz

Vehicles A and B use typical values for steel and air suspensions respectively. Gross vehicle weights of 5 tonnes for Vehicle A and 10 tonnes for Vehicle B are adopted. It must be noted that Vehicle B has the 2nd frequency closer to the 1st frequency of the bridge than the frequencies of Vehicle A. This fact is of significance given that previous research [7, 8] has concluded that when analyzing the effect of a vehicle on the forced frequencies of a bridge, not only the vehicle to bridge mass ratio is important, but also the proximity of the vehicle and bridge frequencies.

3 Eigenvalue Analysis of Coupled Bridge and Vehicle Models

As previously mentioned, forced vibration of the bridge caused by the presence of vehicles occurs at frequencies that may significantly differ from those in free vibration. Figure 1 shows the variation of the first three bridge frequencies for different locations of Vehicles A and B on the bridge. The quarter-car is placed on one of the nodes of the finite element beam model, the stiffness and mass matrices of the coupled system are computed, and the eigenvalue problem is solved. This calculation is repeated to obtain the eigenvalues for every possible position of the quarter-car on the beam. These eigenvalues provide the varying forced frequencies of the combined system as the vehicle moves along the bridge.

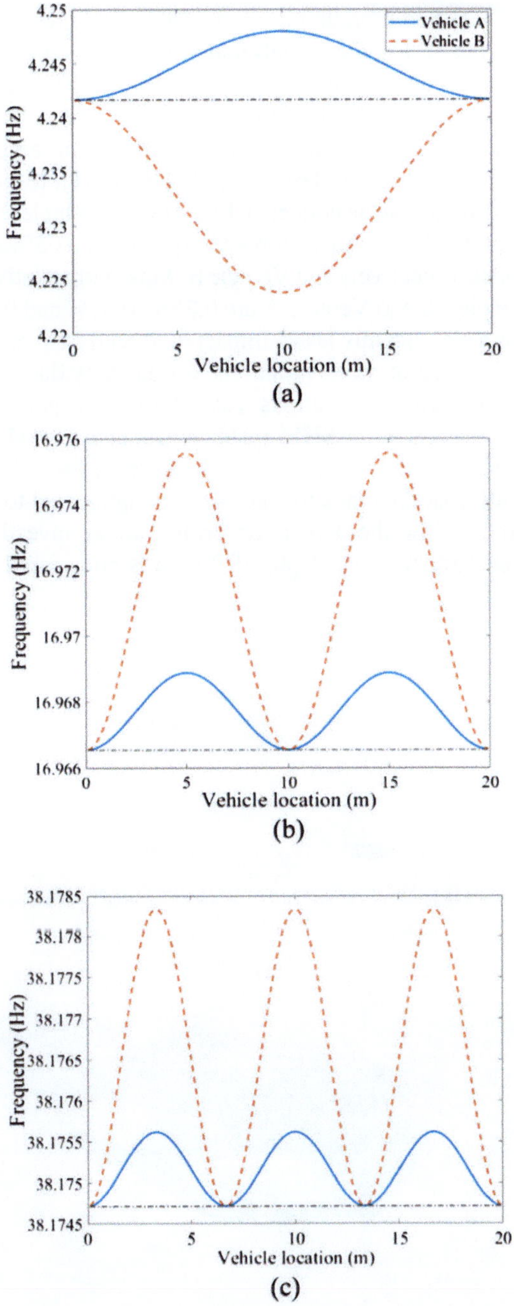

Fig. 1. Bridge forced frequencies due to the vehicle: (a) 1st frequency, (b) 2nd frequency and (c) 3rd frequency.

Bridge frequencies in free vibration given in Table 1 are represented by horizontal dashed-dotted lines in Fig. 1. For the models under investigation, the forced frequency is equal or larger than the frequency in free vibration associated with a specific mode of vibration. For the three modes of vibration, maximum forced frequencies occur for locations of the vehicle exactly matching points of maximum amplitude of the mode shape. A maximum change in forced frequency of 1.37% with respect to free vibration corresponding to the 1st mode can be noticed in Fig. 1(a) for Vehicle B placed at mid-span. Smaller maximum relative changes in the forced frequency of 0.05% and 0.01% are found for the 2nd and 3rd modes respectively and Vehicle B. Maximum relative changes in 1st, 2nd and 3rd forced frequencies due to Vehicle A are 0.26%, 0.01% and 0.002% respectively. Hence, Vehicle A has a significantly lesser impact on forced frequencies than Vehicle B.

To place the significance of these variations in frequency due to vehicle position in context, they are compared to variations caused by damage, whose location and characterization is the main goal for SHM systems. Damage is modelled here following Sinha's approach [13]: a crack of a given depth causes a loss of stiffness that varies linearly from the location of the crack to zero over a length equal to 1.5 times the depth of the bridge. Therefore, for the 1 m deep bridge under investigation, the loss of stiffness extends over 3 m. A crack depth of 20 cm is considered here (i.e., crack to

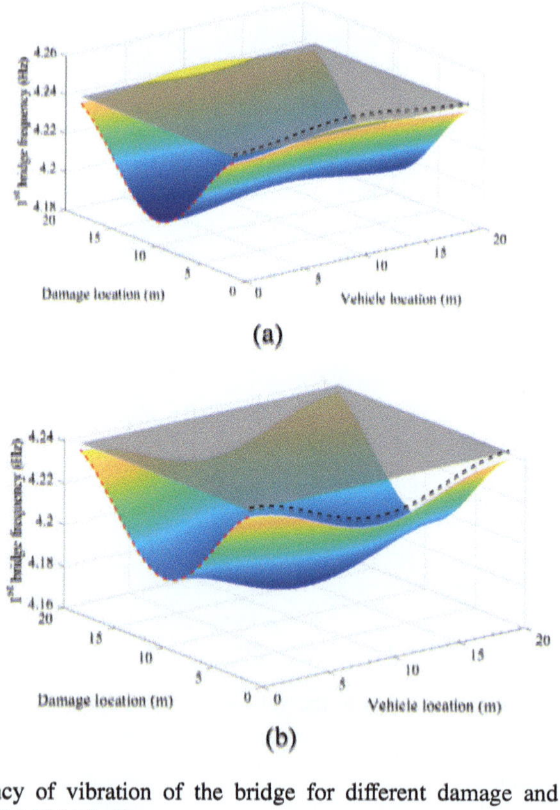

Fig. 2. 1st frequency of vibration of the bridge for different damage and vehicle locations: (a) Vehicle A and (b) Vehicle B.

beam depth ratio of 0.2), which for a rectangular cross-section, it is equivalent to a 48.8% stiffness loss at the location of the crack. Figures 2, 3 and 4 show the combined effect of damage and vehicle for all possible locations on the bridge, on the 1st, 2nd and 3rd bridge frequencies, respectively. The original bridge frequencies in free vibration given in Table 1 are represented by horizontal planes. The damage location closest to the supports starts at 1.5 m from the support to allow for the 1.5 m affected by damage to be fully located inside the bridge. In these figures, the dashed lines represent the isolated effects of vehicle and damage, whereas the continuous surfaces represent the combined effect of both on the first three frequencies of vibration of the bridge.

It is noted that for Vehicle B and the 1st frequency of the bridge (Fig. 2(b)), the frequency variation due to the vehicle itself is higher than that of damage unless the vehicle is very close to the supports. Overall, the vehicle tends to stiffen the bridge (i.e., bridge frequencies increase), while damage makes the bridge more flexible (i.e., bridge frequencies decrease). Hence, in Fig. 2(b), Vehicle B causes an increase in the 1st bridge frequency that outweighs the decrease in 1st frequency caused by damage. Therefore, damage may be concealed by the vehicle effect. The same is true for Vehicle B and the 2nd frequency (Fig. 3(b)), when damage is located near mid-span (minimal effect of damage on 2nd frequency) and the vehicle is positioned near ¼ or ¾

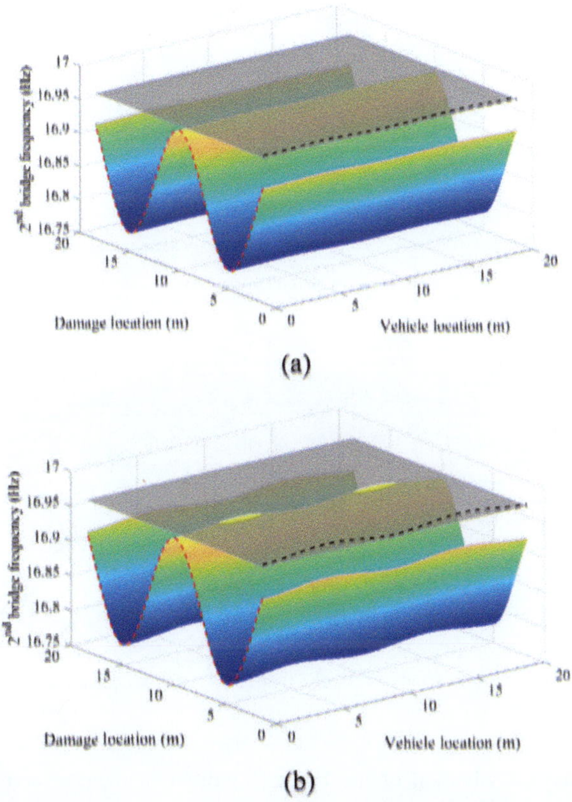

Fig. 3. 2nd frequency of vibration of the bridge for different damage and vehicle locations: (a) Vehicle A and (b) Vehicle B.

of the span (maximal effect of the vehicle position on forced frequency), and for Vehicle A and the 1st frequency (Fig. 2(a)) due to a matching of minimal damage effect (near the supports) with maximal vehicular effect (near mid-span). Changes in frequency due to damage that are hindered by those due to the position of the vehicle can be visualized by the surface rising above the horizontal plane.

In the case of the 3rd frequency (Fig. 4), the surface never surpasses the horizontal plane. This means that the impact of the vehicle on the frequency of the bridge is never larger than that of damage. However, it could be argued that this is partially due to a lack of vehicle frequencies in the proximity of the 3rd bridge frequency. Finally, given that the sensitivity to damage by any mode of vibration differs depending on the location of the damage, the use of three frequencies would minimize the risk of blind spots associated with a specific location and mode. For example, the 3rd frequency appears to exhibit relatively larger variations for damages closer to the supports than the 1st and 2nd frequencies.

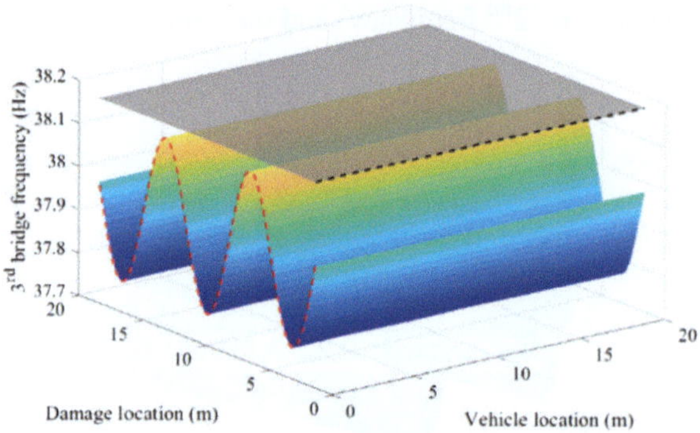

(a)

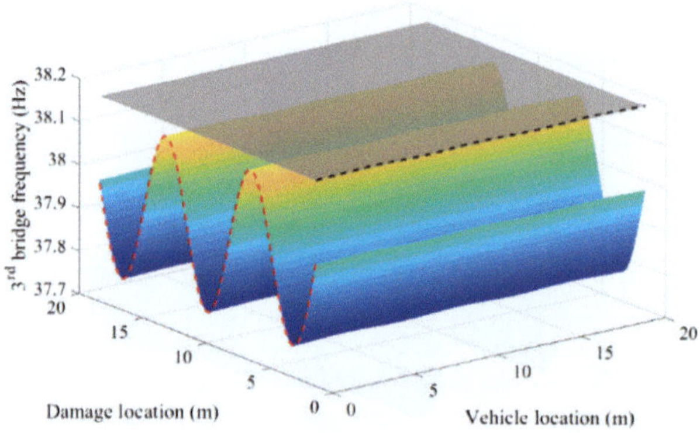

(b)

Fig. 4. 3rd frequency of vibration of the bridge for different damage and vehicle locations: (a) Vehicle A and (b) Vehicle B.

4 Comparison of Relative Changes in Frequency

The previous section has established that for a given set of vehicle parameters and damage severity, the variation of bridge frequencies depends largely on the positions of damage and vehicle. This section analyses the relative changes in frequency for the first three modes of vibration of the bridge. Relative changes in free and forced frequency are defined by Eqs. (1) and (2) respectively.

$$\Delta f_{free,i} = 100 \cdot \frac{f_{d,i} - f_{h,i}}{f_{h,i}} \tag{1}$$

$$\Delta f_{forced,i} = 100 \cdot \frac{f_{d+v,i} - f_{h+v,i}}{f_{h+v,i}} \tag{2}$$

where $i = 1, 2, 3$ indicates the mode under consideration; $f_{d,i}$ represents the frequency in free vibration of the damaged bridge, $f_{h,i}$ is the frequency in free vibration of the healthy bridge, and $f_{d+v,i}$ and $f_{h+v,i}$ are the forced frequencies of the damaged and healthy bridges respectively when the vehicle is on the bridge. While $\Delta f_{free,i}$ yields a unique value for a given damage location, $\Delta f_{forced,i}$ takes a different value for each position of the vehicle, even for a fixed damage location.

Figure 5 shows the relative change in the 1st bridge frequency for every possible location of the vehicle. Each sub-figure corresponds to a different damage location, separated by intervals of 1 m in the first half of the bridge (from damage at 2 m to damage at 10 m). The variation due to damage in the second half of the bridge is symmetric with respect to mid-span and it is omitted to avoid duplication. Dashed horizontal lines represent the variation in free vibration ($\Delta f_{free,1}$) and curved lines represent the variation in forced vibration ($\Delta f_{forced,1}$). The variation due to Vehicle A is plotted using a dashed-dotted line, and the variation due to Vehicle B is shown with a continuous line. $|\Delta f_{forced,1}|$ has a smaller absolute value than $|\Delta f_{free,1}|$ for any vehicle position or damage location. Vehicular locations corresponding to the smallest relative change in frequency (i.e., a maximum of the curve given that the latter is defined by negative numbers) are marked with vertical dashed-dotted and continuous lines for Vehicles A and B, respectively. It can be seen how the maximum point of the $\Delta f_{forced,1}$ curve (i.e., smallest relative change in forced frequency) in sub-Fig. 5(b) and (c) moves from the support towards mid-span from top to bottom figures; namely, the maximum of the curve follows the same direction as the location of the damage. Additionally, the maximum point of the $\Delta f_{forced,1}$ curve tends to happen when the vehicle is close to the position of damage, even more for locations of damage close to mid-span, where the amplitude of the 1st mode is maximum. The error in locating damage using the maximum point of the $\Delta f_{forced,1}$ curve as a reference is larger for Fig. 5(a), with locations of damage very close to the supports, than for Fig. 5(b) and (c). In terms of the magnitude of the change, for damage located at 2 m from the left support (Fig. 5(a)), the relative changes in frequency, both for free and forced vibration, are in the range of −0.037% to −0.039%. Compared to those values, damage located at mid-span (Fig. 5(c)) leads to relative changes in frequency that are almost ten times higher, with values in the range of −0.35% to −0.37%. It is also noticeable how the variation in forced frequency by

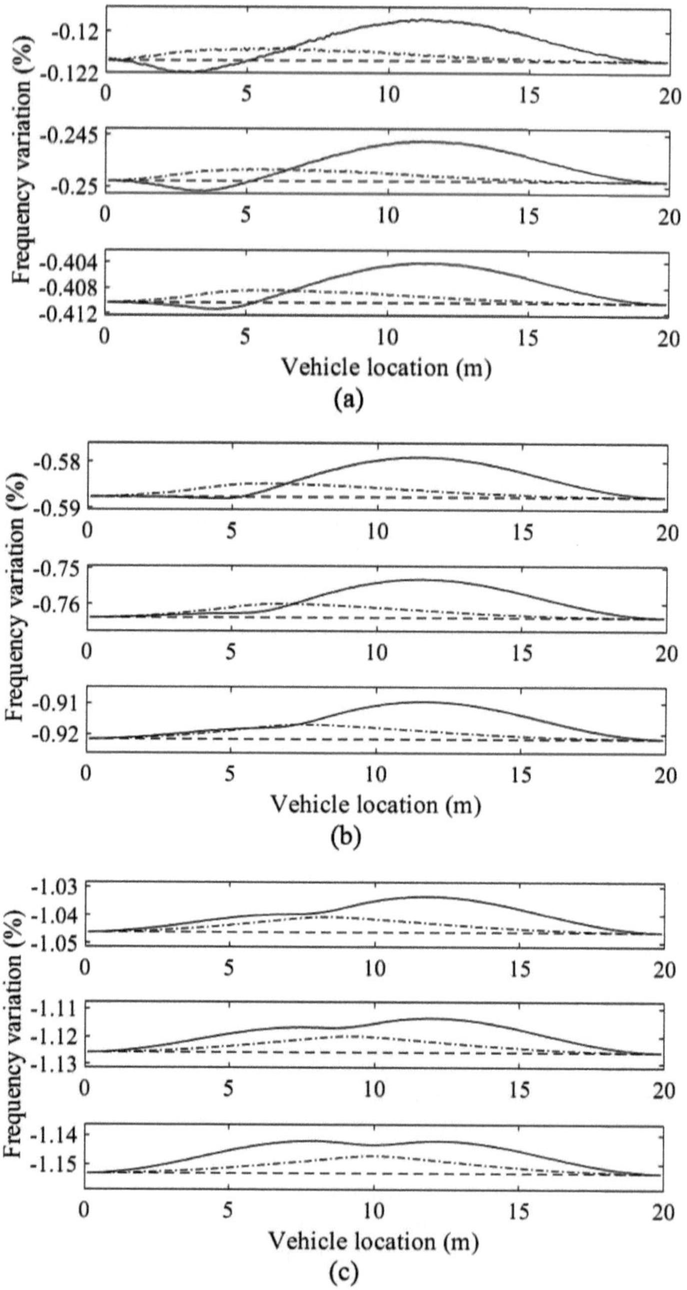

Fig. 5. Relative changes in 1st bridge frequency vs vehicle position when damage is located at (a) 2 m, 3 m and 4 m (top to bottom); (b) 5 m, 6 m and 7 m (top to bottom); and (c) 8 m, 9 m and 10 m (top to bottom).

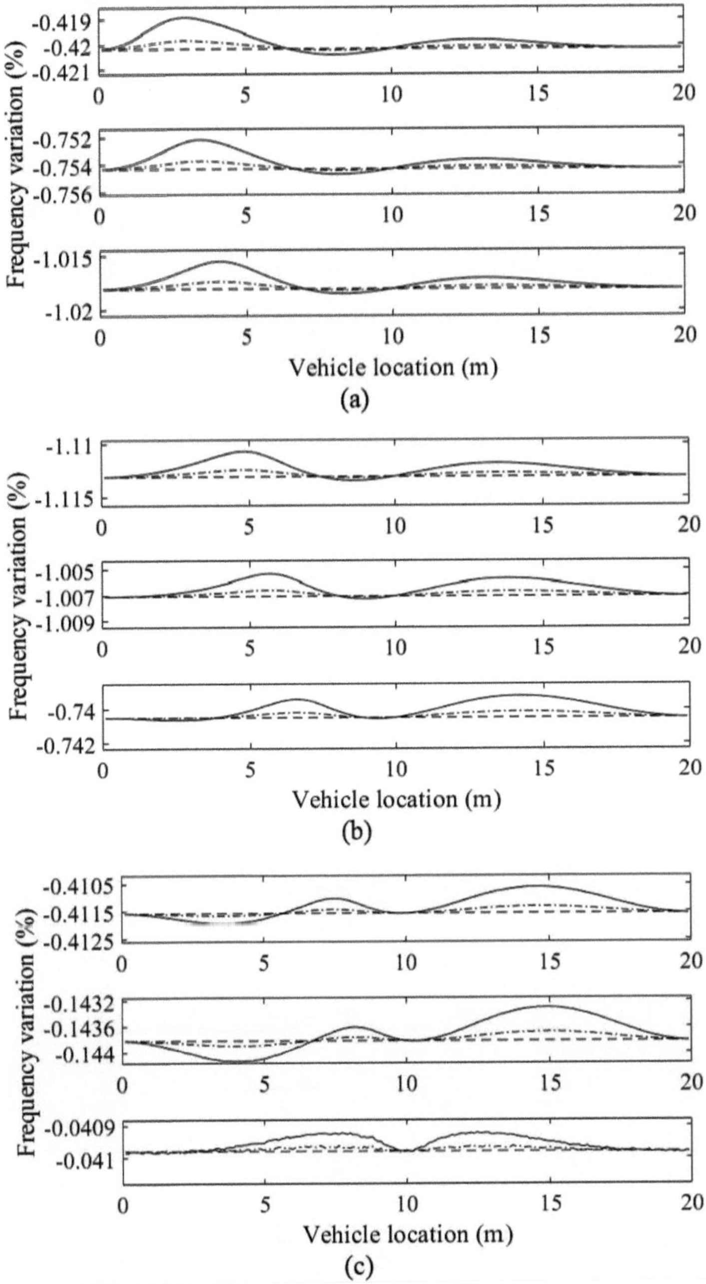

Fig. 6. Relative changes in 2nd bridge frequency vs vehicle position when damage is located at (a) 2 m, 3 m and 4 m (top to bottom); (b) 5 m, 6 m and 7 m (top to bottom); and (c) 8 m, 9 m and 10 m (top to bottom).

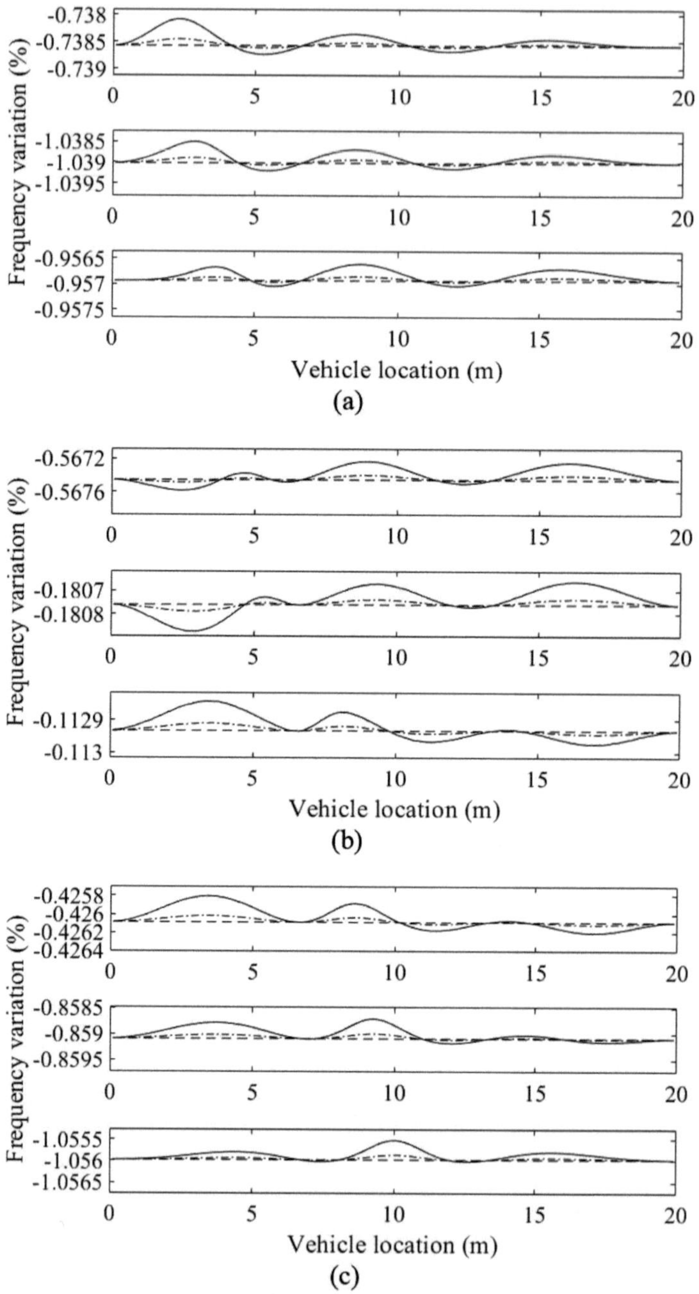

Fig. 7. Relative changes in 3rd bridge frequency vs vehicle position when damage is located at (a) 2 m, 3 m and 4 m (top to bottom); (b) 5 m, 6 m and 7 m (top to bottom); and (c) 8 m, 9 m and 10 m (top to bottom).

Vehicle A is closer to free vibration than the variation by Vehicle B. The curve corresponding to Vehicle A is flatter than the curve of Vehicle B, hence rendering the location of damage more challenging for Vehicle A than for Vehicle B.

Figure 6 shows relative changes in frequency for the 2nd mode and Fig. 7 does the same for the 3rd mode. Unlike Figs. 5, 6 and 7 show that $|\Delta f_{forced,i}|$ is sometimes larger in absolute value than $|\Delta f_{free,i}|$. In agreement with Fig. 5, the variation of relative change in forced frequency due to Vehicle A remains closer to $\Delta f_{free,i}$ for any mode or position, in comparison to Vehicle B. Vehicle B appears then to be the best-suited vehicle for locating damage. In the case of $\Delta f_{forced,2}$ (Fig. 6), the relative change in frequency at mid-span is always equal or almost equal to the value in free vibration, while noticeable peaks are found for positions of the vehicle near ¼ or ¾ of the span. The magnitude of the highest relative changes in the 2nd frequency is in the order of −0.36%, similar to those found for the 1st frequency (Fig. 5(c)), but the maximum of the curve takes place at a different damage location, i.e., at around 5 m from the left support (Fig. 6(b)). It becomes clear that when using the 2nd frequency, the damage is more accurately located near the points of maximum amplitude of the 2nd mode of vibration. For the 3rd mode, maximum points of the $\Delta f_{forced,3}$ curves correspond to damages located at 3 m (Fig. 7(a)) and 10 m (Fig. 7(c)) from the left support, with similar magnitudes as for the first two frequencies (−0.33% to −0.34%). In summary, forced frequencies are most sensitive to damage for locations near the modal points of maximum amplitude, and as the locations of damage get further away from these modal points, $\Delta f_{forced,i}$ peaks decrease in magnitude and are not as reliable as a reference for locating damage.

5 Conclusions

This paper has built on previous research about forced frequencies of a bridge due to a moving vehicle to assess how they are influenced by damage. While free vibration records provide one single value of frequency for each mode of vibration, frequencies obtained from forced vibration vary with each position of the vehicle on the bridge. A change of frequency in free vibration indicates a difference in mechanical properties or boundary conditions of the bridge between two points in time, but it gives no indication of the location of potential damage. The simplified bridge and vehicle models employed here have served to illustrate how the variation of bridge frequencies with the position of the vehicle can be used to detect and to locate damage. This variation has been more pronounced when using the vehicle with higher mass and frequencies closer to the bridge frequency. A mode of vibration can be more sensitive to damage than others depending on the location of damage. When the damage has been close to the points of maximum amplitude of the mode being considered, a peak has formed in the curves of relative changes in frequency for locations of the vehicle in the proximity of damage. The results have shown potential for development of new damage detection algorithms that aim to exploit the damage features contained in the forced vibration signal, for application to traditional SHM systems when few or null records in free vibration are available, or to modern SHM technologies based on energy-harvesting sensors or drones with a limited flying duration, i.e., drones charging bridge sensors and downloading vibration data from them for a short period of time to optimize energy efficiency. Nevertheless, this research is at an early stage, and the promising results achieved by eigenvalue analysis need to be tested in a real-life situation by capturing small changes of forced frequency with sufficient accuracy in time-frequency domain.

Acknowledgements. This research has received funding from Science Foundation Ireland (SFI)'s US-Ireland R&D partnership programme under the proposal id. 16/US/I3277 titled MARS-Fly.

References

1. Chen, S., Laefer, D., Mangina, E.: State of technology review of civilian UAVs. Recent Pat. Eng. **10**(3), 160–174 (2016). https://doi.org/10.2174/1872212110666160712230039
2. Feng, D., Feng, M.Q.: Output-only damage detection using vehicle-induced displacement response and mode shape curvature index. Struct. Control Health Monit. **23**(8), 1088–1107 (2016). https://doi.org/10.1002/stc.1829
3. Shi, Z.Y., Law, S.S., Zhang, L.M.: Structural damage localization from modal strain energy change. J. Sound Vib. **218**(5), 825–844 (1998). https://doi.org/10.1006/jsvi.1998.1878
4. Pandey, A.K., Biswas, M.: Damage detection in structures using changes in flexibility. J. Sound Vib. **169**(1), 3–17 (1994). https://doi.org/10.1006/jsvi.1994.1002
5. Casero, M., González, A., Feng, K.: Extraction of dynamic features from short acceleration data bursts: a review. In: 10th International Conference on Short and Medium Span Bridges Proceedings, p. 246 (1–10). CSCE, Quebec City (2018)
6. Yang, J., Chen, Y., Xiang, Y., Jia, X.L.: Free and forced vibration of cracked inhomogeneous beams under an axial force and a moving load. J. Sound Vib. **312**(1–2), 166–181 (2008). https://doi.org/10.1016/j.jsv.2007.10.034
7. Yang, Y.B., Cheng, M.C., Chang, K.C.: Frequency variation in vehicle-bridge interaction systems. Int. J. Struct. Stab. Dyn. **13**(2), 1350019 (2013). https://doi.org/10.1142/S0219455413500193
8. Cantero, D., OBrien, E.J.: The non-stationarity of apparent bridge natural frequencies during vehicle crossing events. FME Trans. **41**(4), 279–284 (2013)
9. González, A., Covián, E., Casero, M.: Impact of superimposed and truck live load on modal characteristics of short-span bridges. In: 6th International Operational Modal Analysis Conference Proceedings, pp. 119–128. Gijón, Spain (2015)
10. Cantero, D., Rønnquist, A.: Numerical evaluation of modal properties change of railway bridges during train passage. In: 10th International Conference on Structural Dynamics Proceedings, pp. 2931–2936. Procedia Engineering, Rome (2017). https://doi.org/10.1016/j.proeng.2017.09.345
11. Cantero, D., Hester, D., Brownjohn, J.: Evolution of bridge frequencies and modes of vibration during truck passage. Eng. Struct. **152**, 452–464 (2017). https://doi.org/10.1016/j.engstruct.2017.09.039
12. Cantero, D., McGetrick, P., Kim, C.W., OBrien, E.J.: Experimental monitoring of bridge frequency evolution during the passage of vehicles with different suspension properties. Eng. Struct. **187**, 209–219 (2019). https://doi.org/10.1016/j.engstruct.2019.02.065
13. Sinha, J.K., Friswell, M.I., Edwards, S.: Simplified models for the location of cracks in beam structures using measured vibration data. J. Sound Vib. **251**(1), 13–38 (2002). https://doi.org/10.1006/jsvi.2001.3978

Physical and Virtual Implementation of Closed-Loop Designs for Model Updating

M. S. Jensen[1(✉)], T. N. Hansen[1], M. D. Ulriksen[1], and D. Bernal[2]

[1] Department of Civil Engineering, Aalborg University, 6700 Esbjerg, Denmark
mj13@student.aau.dk, tnha14@student.aau.dk, mdu@civil.aau.dk
[2] Center for Digital Signal Processing, Northeastern University, Boston, MA 02115, USA
D.Bernal@northeastern.edu

Abstract. A recently proposed virtual implementation of output feedback based on signal processing eliminates the practical overhead associated with physical operation in closed loop. Additionally, the virtual implementation facilitates realization of multiple closed-loop systems from a single test in open loop, allows for complex gains, and removes the constraint of closed-loop stability. Care must, however, be exercised in the design of the closed-loop systems, as the errors in these are governed by the intrinsic approximations in the open-loop identification. The present paper offers an examination of this item when the closed-loop systems are designed for parameter estimation in updating of numerical models of structural systems. The differences between physical realization and the proposed virtual implementation are discussed, and the pivotal points outlined are demonstrated in the context of a numerical examination with a structural system.

Keywords: Model updating · Output feedback · Virtual implementation · Parameter estimation · Closed-loop eigenstructure

1 Introduction

Model updating through parameter estimation is a well-known discipline used in many different application areas [1]. Within structural and mechanical engineering, model updating is commonly used to calibrate a numerical model of a physical structure for use in structural design, control, health monitoring, response prediction, and so forth [2]. In this context, a typical updating approach is to minimize the discrepancy between the poles from the numerical model, $\mathscr{M}$, and target poles estimated from the physical system, $\mathscr{P}$. Here, an obvious issue is that no unique solution exists for the inverse problem when the dimensionality of $\mathscr{M}$ is larger than the number of identifiable poles from $\mathscr{P}$ [3,4]. This will often be the case, at least when using poles as targets, in structural and mechanical engineering applications, where a limited amount of poles can be identified.

M. A. Wahab (Ed.): DAMAS 2019, LNME, pp. 363–371, 2020.
https://doi.org/10.1007/978-981-13-8331-1_25

Several approaches have been discussed to increase the pole target space. Examples include (1) testing the structure in different configurations by adding known perturbations [5,6] and (2) using output feedback to design and test multiple closed-loop systems [3,4,7]. With reference to the second option, it has recently been shown that the practical overhead associated with closed-loop testing can be eliminated by use of a virtual approach, in which multiple closed-loop systems can be computed based on a single open-loop realization [8–11]. The virtual approach also removes the restrictive stability constraint [7], enables the use of complex gains because the control forces do not have to be physically delivered [11], and allows an increase in the target space based on a single closed-loop system [12]. The scope of the present paper is to examine, in terms of model updating, the basic applicability of the virtual approach by comparing it to the physical counterpart, where it is noted that the latter refers to real-time closed-loop testing.

The remainder of this paper is organized as follows: the fundamentals of output feedback and the virtual approach are briefly described in Sect. 2. The implementation of output feedback—including gain computation and selection—for model updating is outlined in Sect. 3. In Sect. 4, numerical examples are presented to demonstrate the performance of the virtual and physical approaches, and lastly, in Sect. 5, some concluding remarks close the paper.

2 Output Feedback

Let $\mathscr{P}$ be described as a linear, time-invariant system in discrete time with the direct transmission term being zero or subtracted from the measurements, then

$$x(k+1) = A_d x(k) + B_d u(k) \tag{1a}$$
$$y(k) = Cx(k), \tag{1b}$$

where $x(k) \in \mathbb{R}^{n\times 1}$ is the state, $u(k) \in \mathbb{R}^{r\times 1}$ the control input, $y(k) \in \mathbb{R}^{m\times 1}$ the output while $A_d \in \mathbb{R}^{n\times n}$, $B_d \in \mathbb{R}^{n\times r}$, and $C \in \mathbb{R}^{m\times n}$ are the system matrices. In this paper, it is assumed that $\{A_d, B_d\}$ is controllable and $\{A_d, C\}$ is observable.

Considering dynamic output feedback, the control input, $u(k)$, is the output of a discrete-time, finite-dimensional linear time-invariant system driven by $y(k)$, which is formulated as

$$x_f(k+1) = A_f x_f(k) + B_f y(k) \tag{2a}$$
$$u(k) = C_f x_f(k) + D_f y(k) + v(k) \tag{2b}$$

for some excitation v and coefficient matrices $A_f \in \mathbb{C}^{q\times q}, B_f \in \mathbb{C}^{q\times m}, C_f \in \mathbb{C}^{r\times q}$, and $D_f \in \mathbb{C}^{r\times m}$. Augmenting Eq. (2a) with Eq. (1a), using Eqs. (1b) and (2b), yield

$$\begin{Bmatrix} x(k+1) \\ x_f(k+1) \end{Bmatrix} = \begin{bmatrix} A_d + B_d D_f C & B_d C_f \\ B_f C & A_f \end{bmatrix} \begin{bmatrix} x(k) \\ x_f(k) \end{bmatrix} + \begin{bmatrix} B_d \\ 0 \end{bmatrix} v(k), \tag{3}$$

which is referred to as the compensator. As seen in Eq. (3), letting A_f, B_f and C_f equal 0 yields static output feedback, and the compensator can thus be adapted for both static and dynamic output feedback.

Let $\tilde{x}(k) = \{x(k)^T \; x_f(k)^T\}^T$, then Eq. (3) can, as shown in [9], be rewritten as

$$\tilde{x}(k+1) = \tilde{A}\tilde{x}(k) + \tilde{B}\tilde{u}(k) \tag{4}$$

with the output

$$\tilde{u}(k) = \tilde{G}\tilde{C}\tilde{x}(k) + \begin{Bmatrix} 0 \\ v(k) \end{Bmatrix} \tag{5}$$

and

$$\tilde{A} = \begin{bmatrix} A_d & 0 \\ 0 & 0 \end{bmatrix}, \quad \tilde{B} = \begin{bmatrix} 0 & B_d \\ I & 0 \end{bmatrix}, \quad \tilde{C} = \begin{bmatrix} 0 & I \\ C & 0 \end{bmatrix}, \quad \tilde{G} = \begin{bmatrix} A_f & B_f \\ C_f & D_f \end{bmatrix}. \tag{6}$$

As shown, the compensator can be viewed as a static output feedback system with the control law in Eq. (5) and the system matrices $\tilde{A} \in \mathbb{R}^{(n+q)\times(n+q)}$, $\tilde{B} \in \mathbb{R}^{(n+q)\times(r+q)}$, $\tilde{C} \in \mathbb{R}^{(m+q)\times(n+q)}$ and $\tilde{G} \in \mathbb{C}^{(r+q)\times(m+q)}$.

The virtual implementation of the compensator is achieved by using the relation between the open- and closed-loop transfer matrices, $H(z)$ and $\mathscr{H}(z)$, which in a system governed by positive static output feedback with the gain G is defined by [13]

$$\mathscr{H}(z) = (I - H(z)G)^{-1}H(z), \tag{7}$$

from which it follows that the closed-loop system can be identified from the open-loop realization. In order to incorporate the compensator, allowing for dynamic feedback, an open-loop transfer matrix is defined as [9]

$$H_C(z) = \begin{bmatrix} \frac{1}{z}I & 0 \\ 0 & H(z) \end{bmatrix}, \tag{8}$$

which complies with the dimensions of the compensator model. The compensator transfer matrix is found by substituting Eq. (8) into Eq. (7), yielding

$$\tilde{H}(z) = \begin{bmatrix} I - \frac{1}{z}A_f & -\frac{1}{z}B_f \\ -H(z)C_f & I - H(z)D_f \end{bmatrix}^{-1} \begin{bmatrix} \frac{1}{z}I & 0 \\ 0 & H(z) \end{bmatrix}. \tag{9}$$

By use of different gains, it is, in principle, possible to generate as many compensators as required from just a single open-loop realization. Worth of explicit note is that when the identification of the open-loop system is conducted in the frequency domain, the implementation of the virtual compensator follows directly. If, however, the system is identified in the time domain, one must transform to the z-domain in order to calculate Eq. (9). An approach for this is provided in [8], and it is based on mapping observer Markov parameters to $H(z)$. Furthermore, it should be noted that the inverse z-transformation filters unstable poles, hence eliminating the stability constraint.

3 Model Updating Using Closed-Loop Systems

The parameters to be estimated are gathered in $\theta \in \mathbb{R}^s$, and the updating is formulated as the following constrained optimization problem:

$$\begin{aligned} \underset{\theta \in \mathbb{R}^s}{\text{argmin}} \quad & ||\Lambda_{\mathscr{M}}(\theta) - \Lambda_{\mathscr{P}}|| \\ \text{subject to} \quad & \forall i \in [1, \text{s}] : \alpha_i \leq \theta_i \leq \beta_i, \end{aligned} \tag{10}$$

where $\alpha_i, \beta_i \in \mathbb{R}$ are lower and upper bounds on θ_i and

$$\Lambda_{\mathscr{P}} = [\Lambda_{\mathscr{P}_1} \ \cdots \ \Lambda_{\mathscr{P}p}] \in \mathbb{C}^{v \times p} \tag{11}$$

$$\Lambda_{\mathscr{M}}(\theta) = [\Lambda_{\mathscr{M}_1}(\theta) \ \cdots \ \Lambda_{\mathscr{M}p}(\theta)] \in \mathbb{C}^{v \times p} \tag{12}$$

are, respectively, v target poles and the corresponding model-predicted poles for each of the p gains. The idea is to ensure that the system to be solved in the optimization scheme is (over)determined, which is done by stacking the columns of $\Lambda_{\mathscr{P}}$ and $\Lambda_{\mathscr{M}}(\theta)$ into vectors with $vp \geq s$ rows.

There are several ways to design the required gains, such as optimizing a cost function with specific goals [7,14] or, as is currently being explored [10,11], by generating random matrices. Here, we choose to simply generate gains as random real matrices using scheme 1 [11]. The real scalars $\bar{a}$ and $\bar{b}$ should be selected such that reasonable pole shifts occur while still, since the physical approach is included for comparison, retaining system stability.

One approach to investigate the error in the realization of the closed-loop poles, using the virtual dynamic approach with the gain $\tilde{G}$, is through the poles' sensitivity with respect to some parameter, g, of the gain, that is,

$$\frac{\partial \lambda_j}{\partial g} = \psi_j^T \frac{\partial \mathscr{A}_{cs}}{\partial g} \phi_j = \psi_j^T \tilde{B} \frac{\partial \tilde{G}}{\partial g} \tilde{C} \phi_j. \tag{13}$$

Scheme 1. Generation of $\tilde{p}$ gains with $\bar{a}, \bar{b} \in \mathbb{R}$

---

for $i = 1 : \tilde{p}$ **do**

Define $A_f^i = \bar{a} R^i$ where $R^i \in \mathbb{R}^{q \times q}$ with $\Re\left(R_{kl}^i\right) \sim \mathscr{N}(0,1)$

Define $B_f^i = \bar{a} P^i$ where $P^i \in \mathbb{R}^{q \times m}$ with $\Re\left(P_{kl}^i\right) \sim \mathscr{N}(0,1)$

Define $C_f^i = \bar{b} Q^i$ where $Q^i \in \mathbb{R}^{r \times q}$ with $\Re\left(Q_{kl}^i\right) \sim \mathscr{N}(0,1)$

Define $D_f^i = \bar{b} D^i$ where $D^i \in \mathbb{R}^{r \times m}$ with $\Re\left(D_{kl}^i\right) \sim \mathscr{N}(0,1)$

Define $\tilde{G}_i = \begin{bmatrix} A_f^i & B_f^i \\ C_f^i & D_f^i \end{bmatrix}$

end for

Here, ψ_j and ϕ_j are the jth left and right eigenvectors of the compensator state matrix, $\mathscr{A}_{cs}$, presented in Eq. (3). In order to omit gains that cause undue error, we use the heuristic gain selection procedure proposed in [11], which circumvents calculation of the gain derivative. In particular, we use the initial model of the closed-loop system to define the metric

$$\gamma_i = \|\Psi_{\mathscr{M}_i}\tilde{B}\|\|\tilde{C}\Phi_{\mathscr{M}_i}\|, \tag{14}$$

where $\Psi_{\mathscr{M}_i}$ and $\Phi_{\mathscr{M}_i}$ are the left and right eigenvectors associated with the model-based poles using gain i, $\Lambda_{\mathscr{M}_i}$. Using this metric, the gain configurations yielding the lowest values are selected for use in the updating. Worth of explicit note is that since we base the metric on the initial models, pole path linearity is implicitly assumed.

4 Numerical Examples

We consider a 10-DOF shear building model and a square plate model consisting of 16 elements in the context of model updating for damage characterization. For both examples, it is assumed that there is no pre-existing knowledge regarding the location of the damage, thus θ contains, respectively, all the inter-story stiffness values in the shear building example and all the moduli of elasticity in the plate example.

In the examples, we will use the terms *simulation model* and *virtual* and *physical nominal models* in order to refer to the model used for simulations and the numerical models to be updated. In both examples, the simulation model is drawn from the manifold containing the virtual and physical nominal models, which implies that in the absence of noise there is a set of parameters for which the models to be updated coincide with the simulation model. This will, of course, not be realizable in practice. The virtual and physical nominal models are updated using the "fmincon" algorithm in MATLAB® to solve the optimization problem in Eq. (10). The required system identification is carried out using the Eigensystem Realization Algorithm [15], where the output in both examples are contaminated with 2% white Gaussian noise.

Scheme 1 is used to generate 100 gains for dynamic output feedback with $q = 2$, where the metric described in Sect. 3 is used to choose the 10 gains yielding the lowest value. Furthermore, the scheme is used to generate 10 gains that provide system stability for static output feedback. Four poles are selected from each gain configuration to form the target vector $\lambda_{\mathscr{P}} \in \mathbb{C}^{40}$, where the corresponding poles from the nominal models, $\lambda_{\mathscr{M}}(\theta) \in \mathbb{C}^{40}$, are taken as the ones yielding the lowest discrepancy to the identified target poles.

4.1 10-DOF Shear Building

The shear building illustrated in Fig. 1 is equipped with 3 displacement sensors and 1 actuator, and a 20% stiffness perturbation is introduced in the 6th floor.

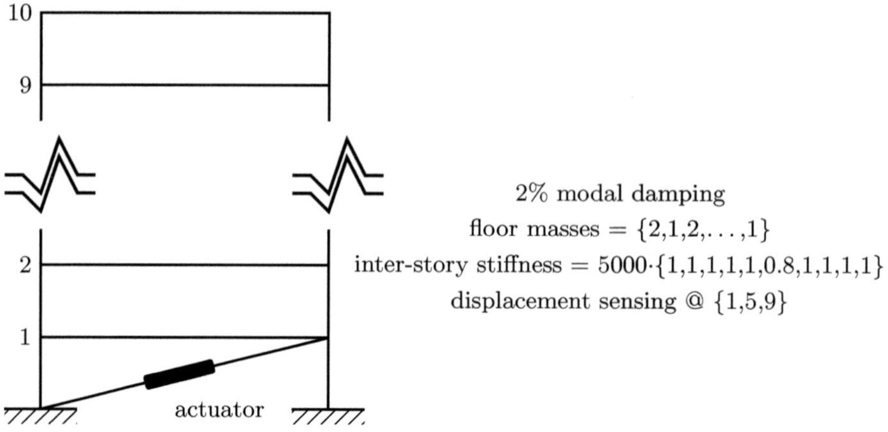

Fig. 1. Shear building simulation model with stiffness perturbation in the 6th floor, where the inter-story stiffness and the floor masses are in any consistent set of units.

In the simulation, the structure is excited with white noise in the first floor, and the resulting displacements are measured with a sampling frequency of 100 Hz at the 1st, 5th and 9th floor for a duration of 5 min.

The model updating scheme provides the results illustrated in Fig. 2, where the converged stiffness estimates, $\hat{k}_i$, of the updated model are normalized with respect to the true value of the stiffness components. The results show, qualitatively, that the performance of the virtual implementation is comparable to that of the physical. The parameters are estimated with a maximum absolute error of 0.42% and 0.49% using the virtual approach with static and dynamic feedback, respectively, and 0.48% using the physical approach. The mean absolute percentage error is 0.25% for the physical approach and, respectively, 0.19% and 0.24% for virtual static and dynamic feedback.

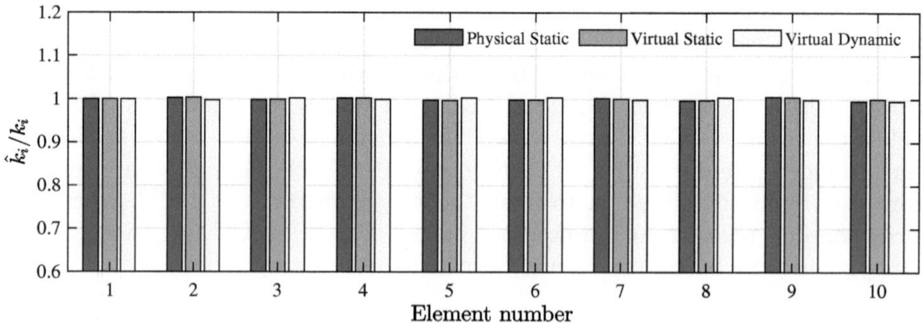

Fig. 2. Model updating results for the 10-DOF shear building with $\hat{k}$ and k being the converged estimate and true value of the stiffness components.

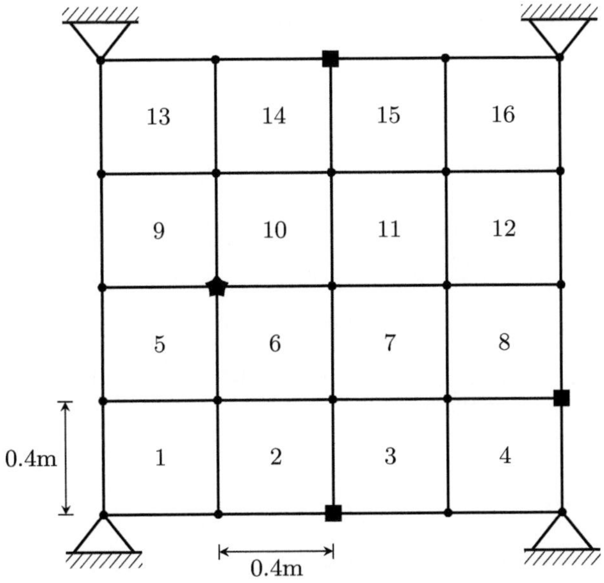

Fig. 3. Finite element plate simulation model with modulus of elasticity $E_i = E \; \forall i \in [2, 16]$ and $E_1 = 0.8E$. Out-of-plane input is applied at node 12 (⬟) and out-of-plane displacements are measured at node 3, 10, and 23 (■).

4.2 Plate Model

We consider the square finite element plate model, depicted in Fig. 3, which consists of 16 four-noded plate elements, where each node has 1 translational and 2 rotational degrees of freedom. The plate is assigned a material model corresponding to typical structural steel, and classical damping is assumed such that each mode is assigned a damping ratio of $\zeta_i = 2\%$ in open loop. The elements have a side length of 0.4m and thickness of 0.003m and are all assigned an initial modulus of elasticity of 200 GPa in the nominal models, while the elements of the simulation model is assigned a modulus of elasticity of $E_i = 200$ GPa, except for element 1 where $E_1 = 160$ GPa. The system is excited with white noise in node 12, and the resulting displacements are measured at node 3, 10, and 23 with a sampling frequency of 100 Hz for a duration of 5 min.

The model updating scheme converges to the results illustrated in Fig. 4. As in the previous example, the results from the virtual approach show to be comparable to the results obtained using the physical approach. The parameters are estimated with a maximum absolute error of 6.0% and 3.3% using the virtual approach with static and dynamic feedback, respectively, and 5.9% using the physical approach. The mean absolute percentage error is 3% for the physical approach and, respectively, 2.2% and 1.3% for virtual static and dynamic feedback.

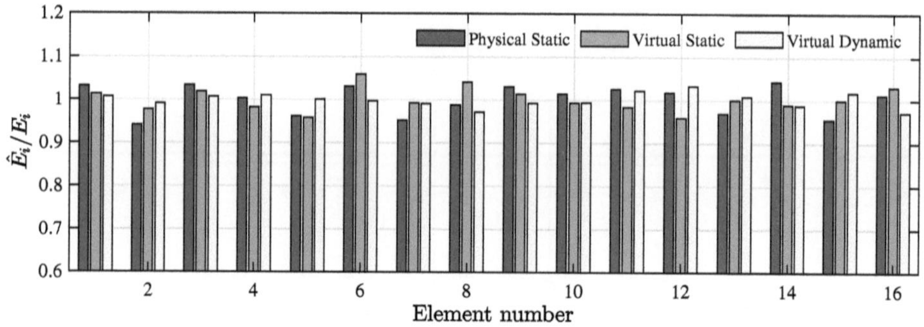

Fig. 4. Model updating results for the plate model with $\hat{E}$ and E being the converged estimate and true value of the moduli of elasticity of the elements.

5 Conclusion

The paper addresses model updating by use of closed-loop system formulations. In particular, we explore the applicability of a recently proposed virtual approach—based on processing of open-loop signals to form closed-loop systems—by comparison with physical operation in closed loop.

Numerical examination of a shear building and a finite element plate show the performance of the physical and virtual approaches to be comparable. As such, the two examples suggest the virtual approach to be a viable alternative to physical closed-loop testing; an alternative that eliminates the practical overhead associated with the latter.

Acknowledgement. The authors gratefully acknowledge the Danish Hydrocarbon Research and Technology Centre (DHRTC) for the financial support.

References

1. Aster, R.C., Borchers, B., Thurber, C.H.: Parameter Estimation and Inverse Problems, 2nd edn. Academic Press, Cambridge (2013)
2. Friswell, M.I., Mottershead, J.E.: Finite Element Model Updating in Structural Dynamics. Kluwer Academic Publishers, Dordrecht (1995)
3. Koh, B.H., Ray, L.R.: Feedback controller design for sensitivity-based damage localization. J. Sound Vib. **273**(1), 317–335 (2004)
4. Jiang, L.J., Tang, J.J., Wang, K.W.: An optimal sensitivity-enhancing feedback control approach via eigenstructure assignment for structural damage identification. J. Vib. Acoust. **129**(6), 771–783 (2007)
5. Nalitolela, N.G., Penny, J.E.T., Friswell, M.I.: A mass or stiffness addition technique for structural parameter updating. Int. J. Anal. Exp. Modal Anal. **7**(3), 157–168 (1992)
6. Cha, P., Gu, W.: Model updating using an incomplete set of experimental modes. J. Sound Vib. **233**(4), 583–596 (2000)

7. Bernal, D., Ulriksen, M.D.: Output feedback in the design of eigenstructures for enhanced sensitivity. Mech. Syst. Signal Process. **112**, 22–30 (2018)
8. Bernal, D.: Parameter estimation using virtual output feedback. Mechanical Systems and Signal Processing (to appear)
9. Bernal, D.: Parameter estimation using virtual dynamic output feedback. Mechanical Systems and Signal Processing (to appear)
10. Bernal, D., Ulriksen, M.D.: Virtual closed-loop parameter estimation. In: Proceedings of the International Conference on Structural Engineering Dynamics (ICEDyn), Viana do Castelo, Portugal (2019)
11. Ulriksen, M.D., Bernal, D.: On the use of complex gains in virtual feedback for model updating. In: Proceedings of the International Conference on Structural Engineering Dynamics (ICEDyn), Viana do Castelo, Portugal (2019)
12. Hansen, T.N., Jensen, M.S., Ulriksen, M.D., Bernal, D.: On the model order in parameter estimation using virtual compensators. In: Proceedings of the 13th International Conference on Damage Assessment of Structures (DAMAS 2019), Porto, Portugal (2009)
13. Trentelman, H., Stoorvogel, A.A., Hautus, M.: Control Theory for Linear Systems. Springer, Heidelberg (2001)
14. Ulriksen, M.D., Bernal, D.: On gain design in virtual output feedback for model updating. In: Proceedings of the 13th International Conference on Damage Assessment of Structures (DAMAS 2019), Porto, Portugal (2019)
15. Juang, J.N.: Applied System Identification. Prentice Hall, Upper Saddle River (1994)

On Gain Design in Virtual Output Feedback for Model Updating

Martin D. Ulriksen[1(✉)] and Dioniso Bernal[2]

[1] Department of Civil Engineering, Aalborg University, 6700 Esbjerg, Denmark
mdu@civil.aau.dk
[2] Center for Digital Signal Processing, Northeastern University, Boston, MA 02115, USA
D.Bernal@northeastern.edu

Abstract. The set of equations to be solved for parameter estimation in model updating has no unique solution when, as will often be the case in structural applications, the dimensionality of the model exceeds the number of target parameters estimated from experiments. One approach for enlarging the target space is to create closed-loop systems that, in addition, can be designed with pole sensitivities favorable for updating the model. The present paper will focus on designing gains for model updating using a recently proposed virtual implementation of output feedback, which allows computation of several closed-loop systems from a single open-loop realization and removes the constraint of closed-loop stability. The gains are designed through an eigenstructure assignment procedure, in which the model parameters of interest in the updating are divided into two different classes; one where the pole sensitivities with respect to the parameters are to be enhanced and one where they are to be reduced. A numerical example with a structural system is presented that demonstrates the merit of the proposed gain design procedure.

Keywords: Model updating · Output feedback · Virtual implementation · Gain design

1 Introduction

Parameter estimation for updating numerical models of structural systems is often resolved in an optimization setting, where some cost function expressing the discrepancy between experimental target poles and model-predicted ones is minimized [1]. Intrinsic deficiencies in this approach are that the structural system is never within the model space manifold and that the experimentally identified poles constituting the target vector are subject to estimation uncertainties [2]. What also hinder the applicability when strictly using poles to form the target vector are that the number of poles that can be identified is typically low and, in addition, that these have limited sensitivity to the parameters of interest.

M. A. Wahab (Ed.): DAMAS 2019, LNME, pp. 372–379, 2020.
https://doi.org/10.1007/978-981-13-8331-1_26

A recognized procedure for enlargement of the target space is to test the structure in question under known perturbations and/or changed boundary conditions [3–5]. Another approach, which avoids testing under modified structural conditions, is to operate in closed loop and increase the target space by designing multiple systems using different gains [6–8]. However, despite the noticeable merits of the approach, which also include allowing for eigenstructure assignment to tailor pole sensitivities, the applicability has thus far been hampered by the practical overhead associated with real-time testing in closed loop.

The applicability of closed-loop model updating has, however, recently been promoted by a proposed virtual implementation, where closed-loop eigencharacteristics are computed directly from processing of open-loop input-output data [9,10]. Besides eliminating the practical overhead associated with real-time operation, the virtual implementation also removes the constraint of closed-loop stability and allows computation of several closed-loop eigenstructures based on a single open-loop realization. The latter obviously implies that the target space can be readily increased by use of different gains, thus the task becomes the design and selection of these. Ultimately, the goal is to increase the Fisher information on the parameters to be updated, which can be achieved qualitatively by designing gains through optimizing some cost function promoting pole sensitivity or, as is currently being explored [11–14], quantitatively by simply generating an amount of random gains that highly overdetermine the set of equations to be solved in the updating.

In the present paper, we will follow the qualitative path of gain design for model updating, and we choose to divide the system parameters into three classes; (1) those with large uncertainties, (2) those with less but still notable uncertainties, and (3) those which are (almost) known exactly. While it is obvious that the parameters in the first class and the third class are, respectively, included and discarded in the model updating, a question opens up regarding the parameters in the second class. Whether or not to include these in the updating boils down to a tradeoff between the error introduced by discarding them (and hence treating them at their nominal values) and, if included, the increased size of the free parameter space to be handled in the optimization. In this study, we opt for the former and, in the gain design, minimize the sensitivity of the poles with respect to these parameters.

The paper is organized as follows: in Sect. 2, the basic principles of static output feedback, including the virtual implementation and eigenstructure assignment, are briefly presented. Sect. 3 discusses the design of gains and Sect. 4 outlines the model updating formulated as an optimization problem. The points made in the theoretical part of the paper will be demonstrated in a numerical example in Sect. 5, while some concluding remarks are provided in Sect. 6.

2 Output Feedback

We consider a linear and time-invariant structural system, $\mathcal{P}$, which is described in discrete time by the state-space formulation

$$x(k+1) = A_d x(k) + B_d u(k) \tag{1a}$$
$$y(k) = Cx(k), \tag{1b}$$

where $x(k) \in \mathbb{R}^n$, $u(k) \in \mathbb{R}^r$, and $y(k) \in \mathbb{R}^m$ are the state, input, and output vectors, while $A_d \in \mathbb{R}^{n\times n}$, $B_d \in \mathbb{R}^{n\times r}$, and $C \in \mathbb{R}^{m\times n}$ are the system matrices for which it is assumed that $\{A_d, B_d\}$ is controllable and $\{A_d, C\}$ observable. Also worth of explicit note is that Eq. 1b holds directly when measurements are displacements, velocities, or non-collocated accelerations, and it can also be used in the case of collocated acceleration measurements if the direct transmission term is subtracted from the measurements. One of the mentioned conditions will be assumed to hold in this paper.

Let $\mathcal{P}$ operate under the influence of static output feedback of the form

$$u(k) = -Gy(k) + v(k) \tag{2}$$

with some excitation $v(k) \in \mathbb{R}^r$ and gain $G \in \mathbb{R}^{r\times m}$, which we, for simplicity, restrict to be real from the outset. Substituting Eq. 2 into Eq. 1a yields the closed-loop formulation

$$x(k+1) = (A_d - B_d GC)\, x(k) + B_d v(k) \tag{3}$$

from which the closed-loop state matrix, $\bar{A} = A_d - B_d GC$, is defined.

2.1 Virtual Implementation

The transfer matrix of $\mathcal{P}$ in open loop, $H(z) \in \mathbb{C}^{m\times r}$, is defined as

$$H(z) = C(sI - A_d)^{-1} B_d, \tag{4}$$

so with the feedback law specified in Eq. 2, we establish

$$y(z) = H(z)\left(-Gy(z) + v(z)\right), \tag{5}$$

which results in the closed-loop transfer matrix

$$\bar{H}(z) = \left(I + H(z)G\right)^{-1} H(z). \tag{6}$$

It is appreciated that closed-loop eigenstructures can be identified directly from an open-loop realization. In fact, by using Eq. 6 with different gains, one can, in principle, generate as many closed-loop systems as required from just a single open-loop realization.

We close this part by noting that the virtual implementation follows directly when the identification of the open-loop system is conducted in frequency domain. If, however, a time-domain identification scheme is used, one must transform to z-domain to compute Eq. 6 and then return to time-domain to finish the identification. An approach for this, which is based on mapping observer Markov parameters to $H(z)$, is provided in [10]. Here, it is also brought to attention that the closed-loop system can have unstable poles, because these will be filtered when doing the inverse z-transformation to return to time-domain.

2.2 Pole and Eigenvector Placement

Let $\Lambda_{\mathcal{M}} = \{\lambda_1, \lambda_2, \ldots, \lambda_p\}$ denote a subset of $p \leq n$ poles that are to be placed, then, for $\lambda_j \in \Lambda_{\mathcal{M}}$, it follows from Eq. 3 that

$$(A_d - B_d G C - \lambda_j I)\, \psi_j = 0, \tag{7}$$

or, in partitioned form,

$$\begin{bmatrix} A_d - \lambda_j I & -B_d \end{bmatrix} \begin{bmatrix} \psi_j \\ b_j \end{bmatrix} = 0 \tag{8}$$

with $b_j = GC\psi_j$. The number of vectors that satisfy Eq. 8 equals the nullity of $\begin{bmatrix} A_d - \lambda_j I & -B_d \end{bmatrix}$ and is no less than the number of inputs, r.

Defining $\Psi = \begin{bmatrix} \psi_1\, \psi_2 \ldots \psi_p \end{bmatrix} \in \mathbb{C}^{n \times p}$, $\Phi = C\Psi \in \mathbb{C}^{m \times p}$, and $Z = [b_1 \ldots b_p] \in \mathbb{C}^{r \times p}$, it follows that

$$G\Phi = Z, \tag{9}$$

hence showing that the gain, G, is found by inversion when Φ has full rank and is square; with the latter obviously requiring that the number of placed poles, p, equals the number of outputs, m. Since transposition of a square matrix does not change its eigenvalues, operating with left-side eigenvectors and the appropriate transposes is also valid and allows placement of r poles.

3 Gain Design

Since $\bar{A} \in \mathbb{R}^{n \times n}$, the poles to be placed for the closed-loop system with gain G_i come in the self-conjugate subset

$$\Lambda_{\mathcal{M}}(G_i) = \{\lambda_1, \ldots, \lambda_l, \lambda_1^*, \ldots, \lambda_l^*\} \tag{10}$$

where superscript $*$ denotes complex conjugate and $p = 2l$. The number of poles selected for each gain can differ, but here we simplify the notation by taking l to be the same for all the gains.

The system parameters are gathered in $\theta = \{\theta_\alpha, \theta_\beta, \theta_\gamma\}$, where we, as described in Sect. 1, have defined three groups of which two of them, namely, $\theta_\alpha \in \mathbb{R}^{s_\alpha}$ and $\theta_\beta \in \mathbb{R}^{s_\beta}$, are included in the design of the gains. θ_α contains the parameters associated with large uncertainties and θ_β those with less but still notable uncertainties, thus the model updating is carried out in a setting where only the n_α parameters collected in θ_α are estimated.

Assume that q gains are gathered in the compound matrix

$$\mathcal{G} = \begin{bmatrix} G_1^T \ldots G_q^T \end{bmatrix}^T \in \mathbb{R}^{qr \times m}, \tag{11}$$

which is designed to maximize the sensitivity of

$$\Lambda_{\mathcal{M}} = \{\Lambda_{\mathcal{M}}(G_1), \ldots, \Lambda_{\mathcal{M}}(G_q)\} \in \mathbb{C}^{ql} \tag{12}$$

with respect to θ_α and minimize the sensitivity of $\Lambda_\mathcal{M}$ with respect to θ_β. Let

$$J = \begin{bmatrix} J_\alpha(G_1) & J_\beta(G_1) \\ \vdots & \vdots \\ J_\alpha(G_q) & J_\beta(G_q) \end{bmatrix} = \begin{bmatrix} \frac{\partial \Lambda_\mathcal{M}(G_1)}{\partial \theta_\alpha} & \frac{\partial \Lambda_\mathcal{M}(G_1)}{\partial \theta_\beta} \\ \vdots & \vdots \\ \frac{\partial \Lambda_\mathcal{M}(G_q)}{\partial \theta_\alpha} & \frac{\partial \Lambda_\mathcal{M}(G_q)}{\partial \theta_\beta} \end{bmatrix} \in \mathbb{C}^{ql \times (s_\alpha + s_\beta)} \tag{13}$$

be an extended Jacobian, which is partitioned into the sensitivities with respect to θ_α, gathered in $J_\alpha \in \mathbb{C}^{ql \times s_\alpha}$, and the sensitivities with respect to θ_β, gathered in $J_\beta \in \mathbb{C}^{ql \times s_\beta}$. For details on the computation of the sensitivities, the reader is referred to [8].

The Fisher information on, respectively, θ_α and θ_β is

$$\mathcal{F}_\alpha = J_\alpha^H \Sigma^{-1} J_\alpha \tag{14a}$$

$$\mathcal{F}_\beta = J_\beta^H \Sigma^{-1} J_\beta, \tag{14b}$$

where superscript H denotes conjugate transpose and where we have assumed normality and that the covariance on the poles, $\Sigma \in \mathbb{C}^{ql \times ql}$, is independent of the parameters. If Σ is taken as the identity, we see that the Fisher information is simply the dot product of the Jacobian by itself, which suggests that the gains can be designed from

$$\arg \max_{\mathcal{G}} ||J_\alpha||_* \tag{15a}$$

$$\arg \min_{\mathcal{G}} ||J_\beta||_*, \tag{15b}$$

as $||(\bullet)||_* = \text{trace}\left(\sqrt{(\bullet)^H(\bullet)}\right)$ is the nuclear norm of $(\bullet)$. A scalar cost function is conveniently formulated as

$$\begin{aligned} \arg \min_{\mathcal{G}} \quad & ||J_\alpha^{-1}||_* + ||J_\beta||_* \\ \text{subject to} \quad & \forall i \in [1, q] : ||G_i|| \leq \xi \end{aligned} \tag{16}$$

from which the compound gain, $\mathcal{G}$, is found under the noted constraint on the gains' norm. ξ must be selected such no undue error arises in the closed-loop pole estimates, which, according to Eq. 6, implies that $I + H(z)G_i$ must be well-conditioned. In the numerical example in Sect. 5, ξ is selected heuristically.

We note that the gain design, for necessary practicality, is carried out by use of a model of a structural reference state, which does not take into account the current realization of the system parameters. It is also worth mentioning that the outlined eigenstructure assignment procedure will, as elaborated in [8], typically yield a subset of unstable poles when one operates with homogeneous measurands. While this is obviously an intractable condition for physical real-time testing, eigenstructures with unstable poles are fully acceptable in the virtual implementation.

4 Model Updating Formulation

The parameters to be updated are $\theta_\alpha \in \mathbb{R}^{s_\alpha}$, and the updating is formulated as the following constrained optimization problem based on the use of q gains:

$$\begin{aligned} &\underset{\theta_\alpha \in \mathbb{R}^{s_\alpha}}{\arg\min} && \left\| \hat{\Lambda}_{\mathcal{M}} - \tilde{\Lambda}_{\mathcal{M}}(\theta_\alpha) \right\| \\ &\text{subject to} && \forall i \in [1,\, s_\alpha] : \tau_i \le \theta^0_{\alpha,i} \le \upsilon_i, \end{aligned} \tag{17}$$

where $\tau_i, \upsilon_i \in \mathbb{R}$ are lower and upper bounds on the ith nominal parameter, $\theta^0_{\alpha,i}$, while $\hat{\Lambda}_{\mathcal{M}} \in \mathbb{C}^{ql}$ and $\tilde{\Lambda}_{\mathcal{M}}(\theta_\alpha) \in \mathbb{C}^{ql}$ are, respectively, the estimated target poles and the corresponding model-predicted ones. Needless to say, the premise is to enforce $ql \ge s_\alpha$ such the system of equations to be solved in the optimization scheme is not underdetermined.

5 Numerical Examination

We consider the shear building model depicted in Fig. 1 and use the terms *nominal model* and *simulation model* to refer to, respectively, the model to be updated and the model used to simulate experiments. In the nominal model, all inter-story stiffnesses and floor masses are, respectively, 500 and 1 in some consistent set of units, and classical damping is assumed such each mode has a damping ratio of 2% in open loop. In the simulation model, perturbations are introduced such the floor masses are $\{1.02,\ 0.96,\ 1,\ 0.98,\ 1.04\}$ and the inter-story stiffnesses are $\{505,\ 503,\ 493,\ 400,\ 501\}$, where we note that the low value of the fourth inter-story stiffness could be due to, for example, structural damage.

In the simulations, noise excitation with unit standard deviation—low-pass-filtered such that only the first three modes are consistently excited—is applied as shown in Fig. 1, and the output is taken, with a sampling frequency of 100 Hz,

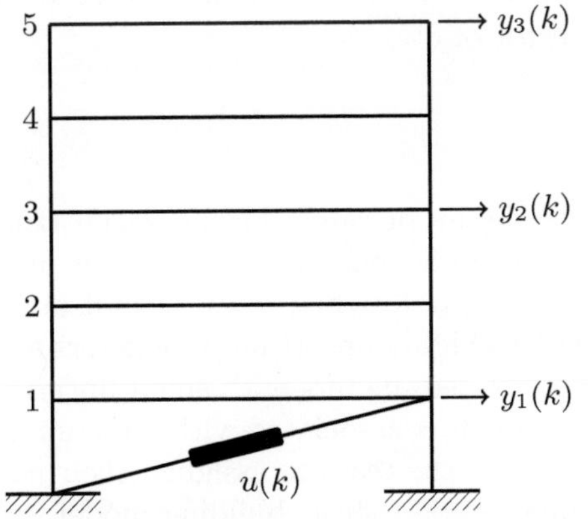

Fig. 1. Shear building with one input, $u(k)$, and three displacement outputs, $y_i(k)$.

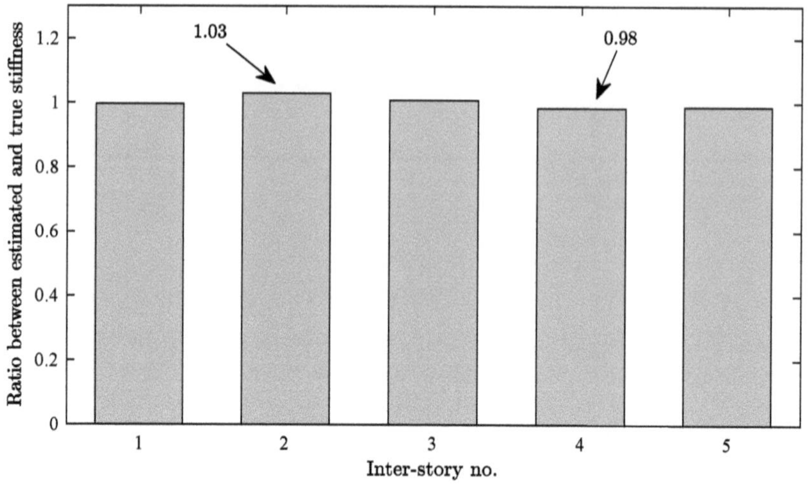

Fig. 2. Updating results for the inter-story stiffness in the shear building model.

as displacement response measured at floors 1, 3, and 5. The response is contaminated with 2% additive white noise, and the open-loop system identification is carried out using the Eigensystem Realization Algorithm [15].

In the gain design, the parameters θ_α and θ_β are composed of, respectively, the inter-story stiffnesses and the floor masses. We design two gains and select three poles from the identification of each of the resulting closed-loop systems, which provide the target vector $\hat{\Lambda}_{\mathcal{M}} \in \mathbb{C}^6$. The corresponding poles of the model, $\tilde{\Lambda}_{\mathcal{M}}(\theta_\alpha) \in \mathbb{C}^6$, are taken as those with the smallest discrepancy to the target poles. From this outset, the cost function defined in Eq. 17 is minimized with the constraints set to lower and upper bounds of 70% and 130% on the nominal inter-story stiffness parameters, θ_α^0. The minimization is conducted using the "fmincon" algorithm in MATLAB®, and it converges to the results presented in Fig. 2. Evidently, we estimate the inter-story stiffness parameters with a maximum absolute error of 3%.

6 Conclusion

This paper explores the design of gains through eigenstructure assignment in a recently proposed virtual implementation of static output feedback for parameter estimation. The gains are designed in an optimization setting, where pole sensitivities with respect to highly uncertain parameters are maximized and pole sensitivities with respect to parameters with small uncertainties are minimized. In this way, only the parameters associated with large uncertainties are included in the model updating while the rest are assigned their nominal values.

A numerical examination of a shear building model is conducted to demonstrate the governing concept. Here, the inter-story stiffness parameters are assumed to have large uncertainties (due to, for example, damage) and the floor masses small uncertainties. It is shown how the proposed gain design procedure allows for parameter estimation with a maximum error of 3% in a setting with open-loop output signals corrupted with 2% additive noise.

Acknowledgment. The first author gratefully acknowledge the Danish Hydrocarbon Research and Technology Centre (DHRTC) for the financial support.

References

1. Friswell, M.I., Mottershead, J.E.: Finite Element Model Updating in Structural Dynamics. Kluwer Academic Publishers, Dordrecht (1995)
2. Mottershead, J.E., Link, M., Friswell, M.I.: The sensitivity method in finite element model updating: a tutorial. Mech. Syst. Signal Process. **25**(7), 2275–2296 (2011)
3. Nalitolela, N.G., Penny, J.E.T., Friswell, M.I.: A mass or stiffness addition technique for structural parameter updating. Int. J. Anal. Exp. Modal Anal. **7**(3), 157–168 (1992)
4. Cha, P., Gu, W.: Model updating using an incomplete set of experimental modes. J. Sound Vib. **233**(4), 583–596 (2000)
5. Rade, D.A., Lallement, G.: A strategy for the enrichment of experimental data as applied to an inverse eigensensitivity-based FE model updating method. Mech. Syst. Signal Process. **12**(2), 293–307 (1998)
6. Koh, B.H., Ray, L.R.: Feedback controller design for sensitivity-based damage localization. J. Sound Vib. **273**(1), 317–335 (2004)
7. Jiang, L.J., Tang, J.J., Wang, K.W.: An optimal sensitivity-enhancing feedback control approach via eigenstructure assignment for structural damage identification. J. Vib. Acoust. **129**(6), 771–783 (2007)
8. Bernal, D., Ulriksen, M.D.: Output feedback in the design of eigenstructures for enhanced sensitivity. Mech. Syst. Signal Process. **112**, 22–30 (2018)
9. Bernal, D.: Parameter estimation using virtual output feedback. Mech. Syst. Signal Process. (to appear)
10. Bernal, D.: Parameter estimation using virtual dynamic output feedback. Mech. Syst. Signal Process. (to appear)
11. Ulriksen, M.D., Bernal, D.: On the use of complex gains in virtual feedback for model updating. In: Proceedings of the International Conference on Structural Engineering Dynamics (ICEDyn) 2019, Viana do Castelo, Portugal (2019)
12. Bernal, D., Ulriksen, M.D.: Virtual closed-loop parameter estimation. In: Proceedings of the International Conference on Structural Engineering Dynamics (ICEDyn) 2019, Viana do Castelo, Portugal (2019)
13. Jensen, M.S., Hansen, T.N., Ulriksen, M.D., Bernal, D.: Physical and virtual implementation of closed-loop designs for model updating. In: Proceedings of the 13th International Conference on Damage Assessment of Structures (DAMAS 2019), Porto, Portugal (2019)
14. Hansen, T.N., Jensen, M.S., Ulriksen, M.D., Bernal, D.: On the model order in parameter estimation using virtual compensators. In: Proceedings of the 13th International Conference on Damage Assessment of Structures (DAMAS 2019), Porto, Portugal (2019)
15. Juang, J.N.: Applied System Identification. Prentice Hall, Upper Saddle River (1994)

Damage Assessment in Beam-Like Structures Using Cuckoo Search Algorithm and Experimentally Measured Data

H. Tran-Ngoc[1,2], S. Khatir[1], G. De Roeck[3], T. Bui-Tien[2], and M. Abdel Wahab[1(✉)]

[1] Department of Electrical Energy, Metals, Mechanical Constructions, and Systems, Faculty of Engineering and Architecture, Ghent University, Ghent, Belgium
{Hoa.tran,magd.abdelwahab}@Ugent.be, khatir_samir@hotmail.fr

[2] University of Transport and Communications, Hanoi, Vietnam
ngochoa.utc@gmail.com, btthanh@utc.edu.vn

[3] Department of Civil Engineering, KU Leuven, B-3001 Louvain, Belgium
guido.deroeck@kuleuven.be

Abstract. This paper presents an approach for damage identification in a steel structure using Cuckoo Search (CS) algorithm. CS is an evolutionary algorithm based on global search techniques, which provides a higher opportunity for seeking the best solution and avoid local minima. A steel beam calibrated on experimental modal analysis is applied to assess the efficiency of the proposed algorithm. While a finite element (FE) model is created using MATLAB to estimate structural dynamic behavior, measurement is carried out using excitation sources of a hammer. Dynamic characteristics are selected as an objective function to minimize the discrepancy between the results of numerical model and measurements. The results show that the proposed algorithm can accurately identify damage location and extents in the considered structure.

Keywords: Damage identification · Evolutionary algorithm · Cuckoo search · Model updating

1 Introduction

Structural health monitoring (SHM) based on nondestructive testing has become a continuous interest for the scientific community over the past decades. Numerous successful applications of SHM based on nondestructive method as a tool for damage detection have been reported in the literature. Sampaio et al. [1] applied frequency-response-function (FRF) curvature method to identify damage in a lumped-mass system and a real bridge. The result indicated that the proposed method can accurately detect damage in the considered structures without the need for any modal identification. Kaveh et al. [2] presented an optimization algorithm called thermal exchange optimization for damage identification in a wide range of structures encompassing a 15-bar planar truss, a six-story steel shear frame, a 25-bar spatial truss and a 40-element

M. A. Wahab (Ed.): DAMAS 2019, LNME, pp. 380–385, 2020.
https://doi.org/10.1007/978-981-13-8331-1_27

beam. Both noise-free and noise modal data are evaluated and the damage location and severity are detected with high accuracy. Zare Hosseinzadeh *et al.* [3] used an evolutionary algorithm, namely cuckoo search for structural damage identification based on an objective function of static displacement using a flexibility matrix. The results indicated the good performance of the proposed approach. Villalba *et al.* [4] presented an approach for structural damage detection based on dynamic parameters and genetic algorithms using three chromosomes to codify each individual in the population. The objective function used dynamic characteristics consisting of both natural frequencies and mode shapes to identify damages in truss structures. Messina *et al.* [5] used a sensitivity and statistical-based method for damage identification in two truss structures based on the difference of natural frequencies between undamaged and damage cases.

Evolutionary algorithms comprise genetic algorithm (GA), particle swarm optimization (PSO), cuckoo search (CS) based on global search techniques to look for the best solution applied for a variety of optimization issues in recent years. For instance, Samir *et al.* [6] applied CS combined with proper orthogonal decomposition method and radial basis function to identify damage location and severity of Carbon Fiber Reinforced Polymer (CFRP) composite structures. The results indicated that the proposed approach can accurately detect damage in the considered structures even though the effect of noise is fully considered. Tran-Ngoc *et al.* [7] employed PSO and GA to identify uncertain parameters consisting of the stiffness of rotational springs at truss joints, the stiffness of springs under bearings, and Young's modulus of truss members in a large-scale truss bridge. Samir *et al.* [8] identified damage location and level in plate structures using CS coupled with Radial Basis Functions (RBF) and Proper Orthogonal Decomposition (POD) based on the objective function of the strains of the actual cracks.

In this paper, we employ CS algorithm to detect damage in a steel beam. CS is used for minimizing the difference between numerical and experimental modal analysis results based on an objective function of dynamic characteristics. In comparison with CS in terms of structural damage detection based on convergence speed and accuracy, another evolutionary algorithm, namely genetic algorithm (GA) is applied.

2 Cuckoo Search Algorithm

Cuckoo search (CS) is an evolutionary algorithm inspired by the reproduction strategy of cuckoo birds. As natural law, cuckoos use the nests of other birds to lay their eggs. In this case, the host birds may either throw away the egg or abandon the nest if they discover the parasite. CS algorithm is formed based on three idealized rules as presented below [9]:

1. The cuckoo birds choose a random nest to spawn their egg and spawn only one egg at the time.
2. The eggs with the highest quality will be passed to the next step.
3. While the number of host's nests is specified, there is a 0–100% probability that the host discovers the egg laid by a cuckoo. If the host birds detect cuckoo's eggs, they may throw away those eggs or abandon their own nest.

CS algorithm is based on Eq. (1) to seek the best solution [9]:

$$X_i^{t+1} = X_i^t + a \times levy(\lambda) \quad (1)$$

Where a is the step size that represents the scale of problem of interests, levy (λ) denotes step length. X_i^t is the previous solution. The levy (λ) is a random walk whose step length is obtained by applying Eq. (2) [9]:

$$Levy \sim u = t^{-\lambda}(1 \leq \lambda \leq 3) \quad (2)$$

3 Finite Element Model and Experimental Validation

A steel beam with free-free boundary condition as shown in Fig. 1 is employed to consider the efficiency of the proposed approach. The span length, the width, and the height of the steel beam are 0.6 m, 0.038 m, and 0.006 m, respectively. The Young's modulus of elasticity, the volumetric mass density, and Poisson's ratio are 2×10^{11} N/m^2, 7850 kg/m^3, 0.3, respectively. A finite element model is generated by using the MATLAB toolbox [10]. The beam includes 11 elements modeled using two-dimensional beam element with 4 degrees of freedom consisting of translations in the x, y-directions, and rotations around the z-direction at each node.

Fig. 1. The steel beam with free-free boundary condition

The experimental measurements are carried out under excitation sources of a hammer combined with PCB Accelerometers 356A15 located near the edge of the steel beam [11].

Damage was introduced at the middle position of the steel beam with crack length's 3 mm and 6 mm. Intact and damaged natural frequencies of first three modes of measurement as shown in Table 1.

Table 1. Experimental natural frequencies of intact and damaged cases [11].

Mode	Natural frequencies (Hz)		
	Intact case	Damaged case 1 (20%)	Damaged case 2 (38%)
1	526	523	514
2	1410	1409	1406
3	2751	2707	2673

CS and GA are used for damage identification in the considered structure. For CS, While Lévy exponent is 1.5, the probability of detecting the egg of parasite (p_a) is 0.25. For GA, crossover and mutation coefficients are 0.1 and 0.8, respectively. The number of population in both algorithms is 30. The objective function comprises natural frequencies and mode shapes. The best solution will be found when the difference between two consecutive iterations of the objective function is lower than 10^{-6} or the maximum number of iterations is 50.

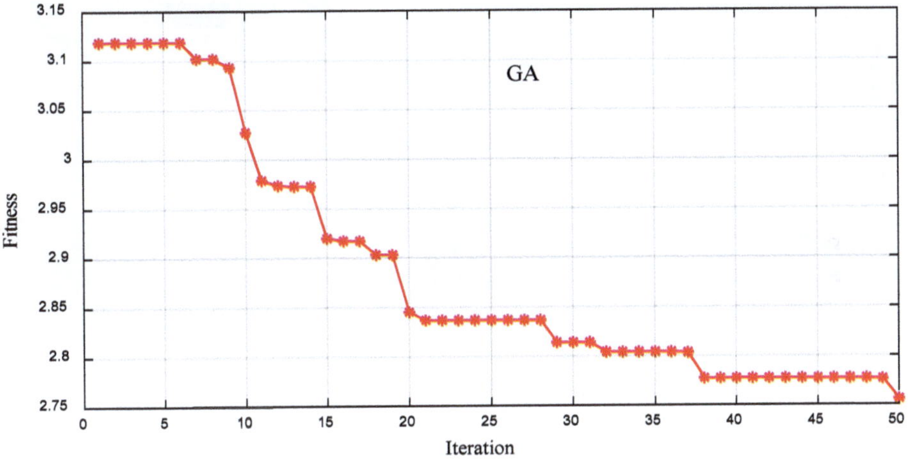

(a)

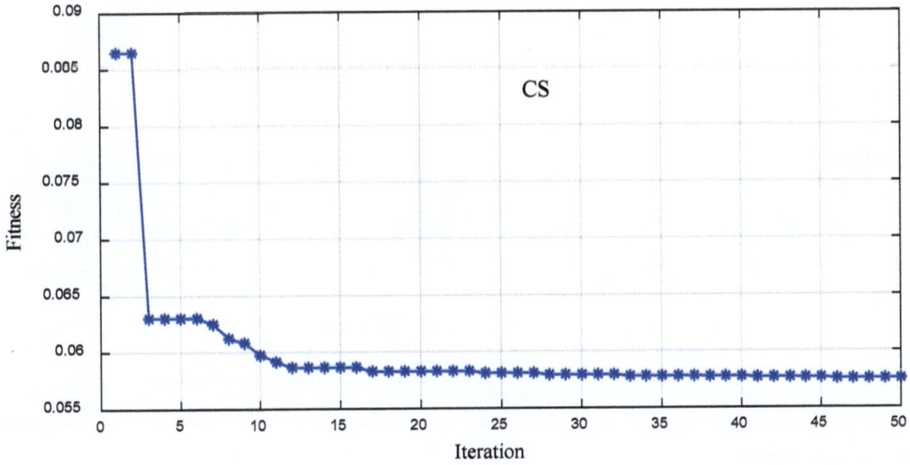

(b)

Fig. 2. Fitness tolerance; (**a**) GA; (**b**) CS

Figure 2 shows that CS can seek the best solution after 10 iterations, whereas GA needs approximately 38 iterations to look for the global best. Additionally, the tolerance of the objective function of CS is lower than that of GA at 0.06 and 2.8, respectively (Figs. 3 and 4).

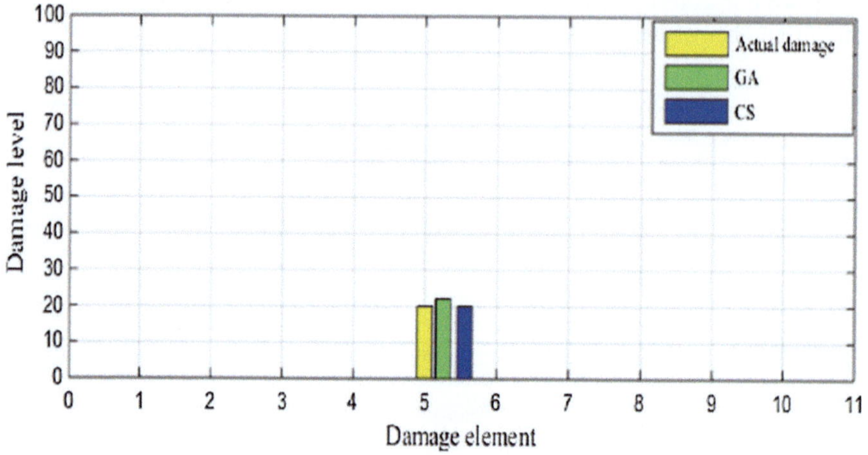

Fig. 3. Damage detection in the steel beam (20% of damage)

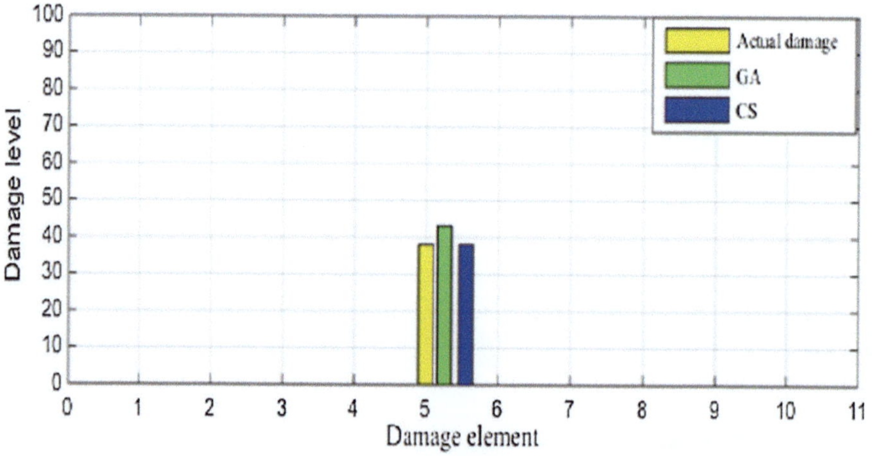

Fig. 4. Damage detection in the steel beam (38% of damage)

4 Conclusion

This paper presents an approach for damage detection in a steel beam based on cuckoo search algorithm combined with vibration measurements. Natural frequencies and mode shapes of first three modes are selected as an objective function. The results show that in most of the cases, both CS and GA successfully identify damage location in the steel beam. However, GA had some errors in determining damage severity of the considered structure. CS outperformed GA in terms of convergence speed and accuracy.

Acknowledgments. The authors acknowledge the financial support of VLIR-OUS TEAM Project, VN2018TEA479A103, 'Damage assessment tools for Structural Health Monitoring of Vietnamese infrastructures', funded by the Flemish Government. Moreover, the first author acknowledges the financial supports from University of Transport and Communications (UTC) under the project research "T2019- 02TĐ".

vliruos SHARING MINDS, CHANGING LIVES With the support of THE BELGIAN DEVELOPMENT COOPERATION .be

References

1. Sampaio, R., Maia, N., Silva, J.: Damage detection using the frequency-response-function curvature method. J. Sound Vib. **226**(5), 1029–1042 (1999)
2. Kaveh, A., Dadras, A.: Structural damage identification using an enhanced thermal exchange optimization algorithm. Eng. Optim. **50**(3), 430–451 (2018)
3. Zare Hosseinzadeh, A., Ghodrati Amiri, G., Koo, K.-Y.: Optimization-based method for structural damage localization and quantification by means of static displacements computed by flexibility matrix. Eng. Optim. **48**(4), 543–561 (2016)
4. Villalba, J., Laier, J.E.: Localising and quantifying damage by means of a multi-chromosome genetic algorithm. Adv. Eng. Softw. **50**, 150–157 (2012)
5. Messina, A., Williams, E., Contursi, T.: Structural damage detection by a sensitivity and statistical-based method. J. Sound Vib. **216**(5), 791–808 (1998)
6. Samir, K., Brahim, B., Capozucca, R., Wahab, M.A.: Damage detection in CFRP composite beams based on vibration analysis using proper orthogonal decomposition method with radial basis functions and cuckoo search algorithm. Compos. Struct. **1**(187), 344–353 (2018)
7. Tran-Ngoc, H., Khatir, S., De Roeck, G., Bui-Tien, T., Nguyen-Ngoc, L., Abdel Wahab, M.: Model updating for Nam O bridge using particle swarm optimization algorithm and genetic algorithm. Sensors **18**(12), 4131 (2018)
8. Khatir, S., Wahab, M.A.: Fast simulations for solving fracture mechanics inverse problems using POD-RBF XIGA and Jaya algorithm. Eng. Fract. Mech. **1**(205), 285–300 (2019)
9. Yang, X.S., Deb, S.: Cuckoo search via Lévy flights. In Nature & Biologically Inspired Computing, 2009. NaBIC 2009. World Congress on 2009 December 9, pp. 210–214. IEEE
10. Dooms, D., Jansen, M., De Roeck, G., et al.: StaBIL: A Finite Element Toolbox for Matlab. Version 2.0 User's Guide, 2010
11. Khatir, S., Wahab, M.A., Djilali, B., Khatir, T.: Structural health monitoring using modal strain energy damage indicator coupled with teaching-learning-based optimization algorithm and isogoemetric analysis. J. Sound Vib. **448**, 230–246 (2019)

State Evaluation of Centrifugal Compressor Unit Based on Parameter Distribution

Yuan Li[1,2], Zeyang Qiu[3(✉)], Ling Fan[2], Xiaolu Tan[2], and Tianyou Qiu[2]

[1] College of Energy and Power Engineering, Xi'an Jiaotong University, Xi'an, China
[2] SINOPEC Sichuan to Eastern China Transmission Gas Pipeline Co. Ltd., Dazhou, China
[3] College of Safety and Ocean Engineering, China University of Petroleum-Beijing, Changping, China
1164709426@qq.com

Abstract. To evaluate the state of centrifugal compressor unit, this paper presents a method based on Logistic Regression Model and Gaussian Model. First, the improved failure mode and effect analysis (FMEA) is used to select important unit parameters on the basis of subsystem of centrifugal compressor unit. And weights of selected important parameters are determined by means of analytic hierarchy process (AHP). Thus, the state evaluation index is established. Then, the distribution models of the parameters in state evaluation index are determined according to its characteristics. Meanwhile, the corresponding parameter distribution model is constructed based on the alarm range of each parameter. Finally, the quantitative evaluation model of centrifugal compressor unit is established using the evaluation value of each parameter distribution model instead of the expert experience score. We applied the model to a motor-driven centrifugal compressor unit in a gas transmission station, and the evaluation result demonstrates the model can evaluate the state of compressor unit accurately.

Keywords: Centrifugal compressor unit · Logistic regression model · Gaussian model · State evaluation

1 Introduction

The length of China's natural gas pipeline construction has exceeded 74,000 km besides the decommissioned and sealed pipelines by the end of 2016 [1]. As one of the most important power equipment for long-distance natural gas transmission pipelines, centrifugal compressor has been widely used due to its advantages of large displacement and stable operation, and plays an extremely important role in the natural gas transmission industry [2]. It has the characteristics of long continuous operation time, high working speed and so on, which requires auxiliary systems such as lubricating oil system and sealing system to ensure the operation.

M. A. Wahab (Ed.): DAMAS 2019, LNME, pp. 386–401, 2020.
https://doi.org/10.1007/978-981-13-8331-1_28

The essence of the state evaluation of centrifugal compressor unit is to realize the state evaluation of the unit through the indexes evaluation, thus it can avoid the occurrence of faults and cascading failures. Gao Guan ling and Zhang Lai bin et al. proposed a fault diagnosis and state evaluation method based on the fault characteristic frequency of centrifugal compressor, which combines fuzzy clustering and fuzzy closeness [3]. Yun song Lu concerns fault diagnosis of centrifugal compressor based on thermal parameters. An improved qualitative simulation (QSIM) based fault diagnosis method is proposed to diagnose the faults of centrifugal compressor in a gas-steam combined-cycle power plant (CCPP) [4]. Li [5] proposes a dynamic process monitoring method based on canonical variable analysis (CVA) and long short-term memory (LSTM), the results show that the proposed method can effectively detect process abnormalities and perform multi-step-ahead prediction of the system's behavior after the appearance of a fault. Sakthivel [6] presents the use of C4.5 decision tree algorithm for fault diagnosis through statistical feature extracted from vibration signals of good and faulty conditions.

There are two main problems in the state evaluation of centrifugal compressor group through analysis: First, the construction of state evaluation index mainly focuses on centrifugal compressor and driving machine, which ignore the importance of auxiliary systems, such as lubricating oil system and sealing system. Second, the state evaluation method is generally based on the traditional vibration evaluation method, which ignores the rich information contained in the process parameters, such as the inlet and outlet pressure and temperature of centrifugal compressor [7].

Therefore, in order to solve these two problems and realize accurate state evaluation of centrifugal compressor unit, this paper proposes a quantitative state evaluation method of centrifugal compressor unit based on parameter distribution. Firstly, the improved FMEA method is used to identify the common fault types and fault detection methods of the centrifugal compressor unit, and a group of monitoring parameters which can reflect the running state of the centrifugal compressor group is selected. The AHP method is used to calculate the weight of the selected monitoring parameters which completes the construction of the state evaluation index of the centrifugal compressor unit. Then, according to the characteristics of each monitoring parameter in the evaluation index and the normal operating range of the parameters, a corresponding parameter distribution model is established. Finally, according to the evaluation results of each parameter distribution model, the quantitative state evaluation of the centrifugal compressor unit is realized.

2 The Analysis Method Based on FMEA/AHP

2.1 The Improvement of FMEA Analysis Method

FMEA analysis method has an effect on other subsystems and complete systems when uses the form of tables to analyze the each subsystem fails. Based on this, we need to find ways to eliminate or control the impacts [8]. The traditional FMEA analysis can obtain the common failure types, failure positions, failure causes and impact consequences of equipment. In order to realize the construction of state evaluation index, this paper improves the traditional FMEA method and determines the corresponding fault detection methods for different fault types and fault causes. The analysis flow of the improved FMEA method can be shown in Fig. 1.

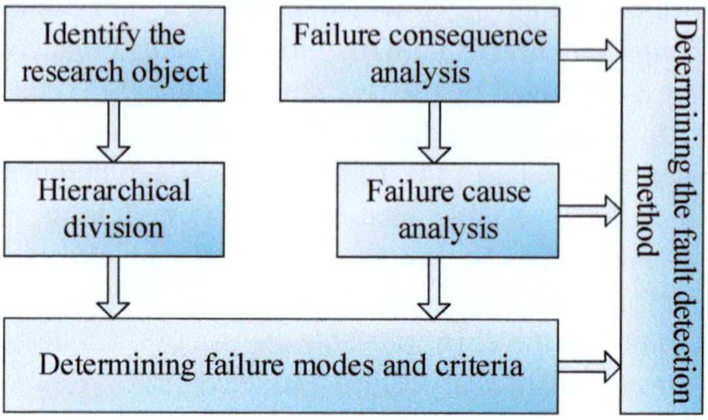

Fig. 1. Flow chart of improved FMEA.

2.2 The AHP Analysis Method

The AHP method transforms the index system into a hierarchical model, and then establishes a judgment matrix to determine the weight. Finally, consistency verification is carried out on the results to verify the rationality of the calculation [10].

The process of calculating weight by AHP method is as follows:

(1) Hierarchy model of indexes establishment
According to the top-down principle, a corresponding hierarchical structure model is constructed by dividing the indexes into structural levels. The hierarchy of AHP mainly includes the highest level (target level), the middle level and the lowest level.

(2) Judgment matrix construction
The indexes of the bottom layer are compared in pairs, and the judgment matrix of each index of the bottom layer is constructed according to certain rules. In this paper, the 1–9 scale method is used, and the scale size represents the relative importance of the two comparative indexes [11].

(3) Weight calculation
When the judgment matrix $\boldsymbol{A}$ meets the consistency requirement, the eigenvector corresponding to the maximum eigenvalue of the matrix $\boldsymbol{A}$ is normalized, and the result is the weight vector. In this paper, the sum-product method is used to calculate the weight. The specific calculation process is as follows:

① The judgment matrix $\boldsymbol{A}$ is normalized according to the Eq. (1) and the matrix obtained after normalization is denoted as $\boldsymbol{B}$.

$$b_{ij} = \frac{a_{ij}}{\sum\limits_{i=1}^{n} a_{ij}} \qquad (i,j = 1, 2, \cdots n) \tag{1}$$

② The matrix $\boldsymbol{B}$ is accumulated according to Eq. (2) and the result is recorded as vector $\boldsymbol{U} = (\boldsymbol{U_1}, \boldsymbol{U_2}\ldots, \boldsymbol{U_n})^{\mathrm{T}}$.

$$U_i = \sum_{j=1}^{n} b_{ij} \qquad (i = 1, 2, \cdots n) \tag{2}$$

③ Normalize the vector according to Eq. (3) and the calculation result is the corresponding weight vector, which is denoted as $\boldsymbol{W}$.

$$W_i = \frac{U_i}{\sum_{i=1}^{n} U_i} \qquad (i = 1, 2, \cdots n) \tag{3}$$

④ Calculate the maximum eigenvalue λ_{max} of the judgment matrix $\boldsymbol{A}$ according to Eq. (4).

$$\lambda_{\max} = \sum_{i=1}^{n} \frac{[AU]_i}{nU_i} \qquad (i = 1, 2, \cdots n) \tag{4}$$

(4) Consistency check
The judgment matrix $\boldsymbol{A}$ is prone to deviation caused by subjective factors, so consistency test is required to avoid deviation. According to Eqs. (5) and (6), the consistency ratio *CR* of the judgment matrix $\boldsymbol{A}$ is calculated, and if *CR* is less than 0.10, the consistency test is met; otherwise, the scale size between the comparison elements need to be readjusted.

$$CR = \frac{CI}{RI} \tag{5}$$

$$CI = \frac{\lambda_{\max} - n}{n - 1} \tag{6}$$

In the formula, *CI* is the consistency index of the matrix; *RI* is the average random consistency index, the reference is journal article [10]; λ_{max} is the maximum eigenvalue; n is the matrix order.

3 Distribution Model

3.1 Logic Regression Model

In data mining, logistic regression algorithm is a commonly used classification algorithm, especially for two types of classification problems [12]. The main idea is to study the correlation between input and output by means of probability estimation. Since the probability is used as the prediction result, the output of the model takes values in (0, 1), which can generally be expressed by Eq. (7).

$$P(y_i = 1|x_i) = \frac{1}{1 + e^{-(\alpha + \beta_1 x_1 + \beta_2 x_2 + \cdots + \beta_k x_k)}} \tag{7}$$

Where $\boldsymbol{x} = (x_1, x_2, \cdots, x_k)^T$ is an independent variable, k is a dimension; y_i is the actual observed response variable: when 1 is taken, it means that the event occurs; when 0 is taken, it means that the event does not occur; P_i is the probability of the occurrence of the ith event; And α and $\beta_1, \ldots \beta_k$ are the regression intercept and regression coefficient, respectively.

3.2 Gaussian Model

The Gaussian model is a normal model that can be used for multi-domain parameter prediction. The main idea is to accurately quantify the research object by Gaussian probability density function, and decompose each research object into several indicators based on Gaussian probability density function. The mathematical expression of the Gaussian model can be expressed by Eq. (8).

$$y = ae^{\frac{-(b-x)^2}{2c^2}} \tag{8}$$

Where x is an independent variable; y is a dependent variable that conforms to the Gaussian model; a, b, and c are the parameters estimated in the model, and are determined according to variable.

4 Evaluation Example

4.1 Construction of State Evaluation Index

State Parameter Filtering. In this paper, the centrifugal compressor unit of a gas transmission station is taken as an example for research, and the driving machine is a motor. Figure 2 is a schematic diagram of the process flow of the centrifugal compressor unit of the gas transmission station. According to the different functions, the unit is divided into five subsystems: centrifugal compressor body, motor, gear box, lubrication system and dry gas seal system for FMEA analysis. Tables 1, 2, 3, 4 and 5 shows the FMEA analysis results of the unit.

Compared with the traditional FMEA analysis, Tables 1, 2, 3, 4 and 5 adds different fault detection methods. At the same time, combined with the monitoring parameters of the on-site monitoring system of the centrifugal compressor group, a group of monitoring parameters that can reflect the running state of the centrifugal compressor group are selected according to different failure modes and characteristics, as shown in Table 6.

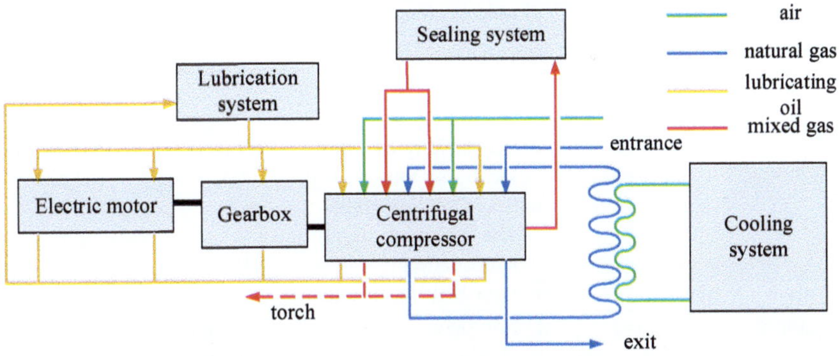

Fig. 2. Structure diagram of centrifugal compressor unit.

Table 1. FMEA for centrifugal compressor unit.

No.	Subsystem	Failure mode	Failure cause	Failure effect	Failure detection method
1	Shell	Excessive vibration and noise	① Poor alignment between bearing and motor ② Surge ③ Severe bearing wear ④ Rotor partial damage, out of balance ⑤ Foundation bolt looseness	Aggravation of mechanical damage, unit shutdown	Vibration detection, temperature detection, flow detection, pressure detection, empirical judgment
		Overheating	① Severe bearing wear ② The compressor displacement is larger than the actual displacement ③ Poor alignment between bearing and motor ④ Compressor internal friction	Aggravation of mechanical damage, unit shutdown	
2	Rotor	The rotor produces unbalance, excessive body vibration	① Design & manufacturing defect, non-performing installation ② The impeller wear or corrode badly ③ Plastic deformation stator & rotor are misalignment	Compressor performance reduction, unit shutdown	Vibration detection, change the load, comparison test
3	Bearing	Excessive bearing vibration,	① Rotating element balance is broken	Bearing wear intensively	Vibration detection, temperature detection,

(*continued*)

Table 1. (*continued*)

No.	Subsystem	Failure mode	Failure cause	Failure effect	Failure detection method
		The temperature is too high	② Eccentricity of rotor caused by serious wear of shaft and sealing ring ③ The lube with impurities or deterioration ④ Poor bearing installation		empirical judgment
4	Compressor Shaft	Bending deformation and crackle	① Improper shaft installation ② Poor alignment between bearing and motor ③ The axle material defect	Vibration intensifies, aggravation of mechanical damage, unit shutdown	Vibration detection
5	Mechanical Seal	Excessive leakage	① Improper installation of mechanical seal ② Serious worn bushings ③ Eccentricity of rotor caused by serious wear of compressor shaft or sealing ring ④ Poor alignment between compressor shaft and motor ⑤ Low sealing fluid pressure ⑥ Poor packing installation ⑦ Impurities on sealing surface	Aggravation of mechanical damage, unit shutdown	Temperature detection, difference-pressure detection

Weight Calculation. 17 monitoring parameters of the centrifugal compressor unit were selected using the FMEA analysis. In order to accurately realize the state evaluation of the unit, it is necessary to calculate the weight of each monitoring parameter and subsystem. In this paper, AHP method is used to calculate the weight, and the judgment matrix of centrifugal compressor group and its five subsystems is constructed. The corresponding weight is calculated according to the judgment matrix after the consistency check. The weight calculation results and judgment matrix are shown in Tables 7, 8, 9, 10 and 11. For convenience of presentation, each subsystem and 17 monitoring parameters are represented by the symbols in Table 6 in the following.

Table 7 shows the judgment matrix of the 5 subsystems of the centrifugal compressor group and their weights. Tables 8, 9, 10 and 11 show the judgment matrix of the centrifugal compressor body, motor, lubrication system and dry gas seal system and the weight distribution of the monitoring parameters of each subsystem respectively.

Table 2. FMEA for electric motor.

No.	Subsystem	Failure mode	Fault reason	Fault effect	Fault detection method
1	Rotor	Excessive motor vibration	Rotor eccentricity causes uneven magnetic force	Rotor damage, motor stop	Current detection, vibration detection
2	Bearing	Excessive bearing vibration and temperature	① Unbalance of the rotating part ② Abrasion of the shaft and seal ring causes the rotor to be eccentric ③ Lubricating oil has impurities or goes bad ④ Poor bearing installation ⑤ Insufficient or excessive lubrication	Increased bearing wear	Vibration detection, temperature detection, empirical judgment
3	Motor	Excessive vibration and overheating	① Serious bearing wear ② Internal friction of the motor	Mechanical damage, motor shutdown	Vibration detection, current detection, empirical judgment

Table 3. FMEA for gear case.

No.	Subsystem	Failure mode	Fault reason	Fault effect	Fault detection method
1	Gear	Snaggletooth	① Overload ② Poor quality of lubricating oil ③ Bearing damage	Excessive vibration, compressor unit shutdown	Vibration detection, temperature detection
		Fatigue damage	① The temperature of lubricating oil is too high and too low, which leads to poor lubrication ② The tooth surface receives excessive contact shear stress		
2	Bearing	Excessive bearing vibration and temperature	① The balance of the rotating part is destroyed ② Abrasion of the shaft and seal ring causes the rotor to be eccentric ③ Lubricating oil has impurities or goes bad ④ Poor bearing installation ⑤ Insufficient or excessive lubrication	Increased bearing wear	Vibration detection, temperature detection, empirical judgment

Table 4. FMEA for lubricating system.

No.	Subsystem	Failure mode	Fault reason	Fault effect	Fault detection method
1	Lubricating oil pump	Bearing temperature is too high	① Bearing wear ② Low oil flow	The lubrication effect is reduced, and the moving parts of the unit are intensified	Vibration detection, temperature detection, empirical judgment
		Excessive vibration	① Rotor imbalance ② Bearing wear ③ Pump internal rubbing		
2	lubricating oil	Temperature too high or too low	① Abnormal liquid level in high level oil tank ② Lubricating oil deterioration ③ Cooler failure	Lubrication effect decreases and wear of moving parts of the unit increases	Temperature detection

Table 5. FMEA for dry gas seal system.

No.	Subsystem	Failure mode	Fault reason	Fault effect	Fault detection method
1	Dry gas sealing system	Seal failure	① Rupture of dynamic ring and static ring ② The sealed cavity is mixed with impurities ③ Reuse of filter element ④ Sealing gas pressure is too low	Seal leakage, unit shutdown	Pressure detection, temperature check

4.2 Calculation of Parameter Distribution Model

Calculation of Logistic Regression Model. For monitoring parameters with only the upper limit of the alarm, such as vibration, temperature and other parameters, the smaller the value indicates the better the operating state of the unit, and the logistic regression model is used to indicate its distribution. In this paper, the one-dimensional logistic regression model is used to represent the distribution of each parameter. The mathematical expression is shown in Eq. (9).

Table 6. The construction of state evaluation index of centrifugal compressor unit.

No.	Subsystem	Monitoring parameter	Description
1	Compressor U_1	Bearing vibration C_1	It characterizes the vibration of the compressor
		Axial displacement C_2	It characterizes the operating state of the compressor shaft
		Inlet temperature C_3	It directly affects compressor power consumption and service life
		Output temperature C_4	It directly affects compressor power consumption and service life
		Import pressure C_5	It directly affects the safety and economic operation of the unit
		Output pressure C_6	It directly affects the safety and economic operation of the unit
2	Electric motor U_2	Bearing vibration M_1	It characterizes the vibration of the motor
		Electric current M_2	It characterizes the operating state of the motor
3	Gearbox U_3	Bearing vibration G_1	It characterizes the vibration of the motor
4	Lubrication system U_4	Lubrication filter differential pressure L_1	It characterizes filter performance and affects the lubrication of various components of the compressor unit
		Lubricating oil pump pressure L_2	It characterizes the lubricant pressure and affects the lubrication of the various components of the compressor unit
		Lubricating oil tank temperature L_3	It characterizes the temperature of the lubricant, which affects the quality of the lubricant and the lubrication of the various components of the compressor unit
5	Dry gas sealing system U_5	Dry gas seal pressure difference S_1	It affects the sealing performance of centrifugal compressors
		Dry gas seal gas pressure S_2	It affects the normal operation of the centrifugal compressor and affects the performance of the lubricating oil
		Dry gas seal filter pressure difference S_3	It characterizes filter performance and affects seal gas flow
		Exhaust temperature of secondary seal of drive end S_4	It affects the sealing performance of centrifugal compressors
		Exhaust temperature of secondary seal of non-drive end S_5	It affects the sealing performance of centrifugal compressors

Table 7. Weights result of centrifugal compressor unit.

U_1–U_5		U_1	U_2	U_3	U_4	U_5	Weights
	U_1	1	1	1	2	2	0.2500
	U_2	1	1	1	2	2	0.2500
	U_3	1	1	1	2	2	0.2500
	U_4	1/2	1/2	1/2	1	1	0.1250
	U_5	1/2	1/2	1/2	1	1	0.1250

Table 8. Weights result of centrifugal compressor.

C_1–C_6		C_1	C_2	C_3	C_4	C_5	C_6	Weights
	C_1	1	1	3	2	1/2	1/3	0.1364
	C_2	1	1	3	2	1/2	1/3	0.1364
	C_3	1/3	1/3	1	1/2	1/4	1/5	0.0519
	C_4	1/2	1/2	2	1	1/3	1/4	0.0812
	C_5	2	2	4	3	1	1/2	0.2304
	C_6	3	3	5	4	2	1	0.3637

Table 9. Weights result of motor.

M_1–M_2		M_1	M_2	Weights
	M_1	1	2	0.3333
	M_2	1/2	1	0.6667

Table 10. Weights result of lubricating system.

L_1–L_3		L_1	L_2	L_3	Weights
	L_1	1	1/2	1/2	0.2000
	L_2	2	1	1	0.4000
	L_3	2	1	1	0.4000

Table 11. Weights result of dry gas seal system.

S_1–S_5		S_1	S_2	S_3	S_4	S_5	Weights
	S_1	1	2	1	3	3	0.3130
	S_2	1/2	1	1/2	2	2	0.1765
	S_3	1	2	1	3	3	0.3130
	S_4	1/3	1/2	1/3	1	1	0.0988
	S_5	1/3	1/2	1/3	1	1	0.0988

$$P(x) = \frac{1}{1 + e^{-(\alpha + \beta x)}} \tag{9}$$

Where α and β are the distribution parameters of the parametric model; P characterizes the operational state of the parameter.

Figure 3 shows the logistic regression model distribution of the parameters. In this paper, the parameters and the state of the unit are divided into five grades: the state grades corresponding to the parameters $(0, x_0]$, $(x_0, x_1]$, $(x_1, x_2]$, $(x_2, x_3]$ and $(x_3, +\infty)$ are excellent, good, medium, poor and very poor respectively.

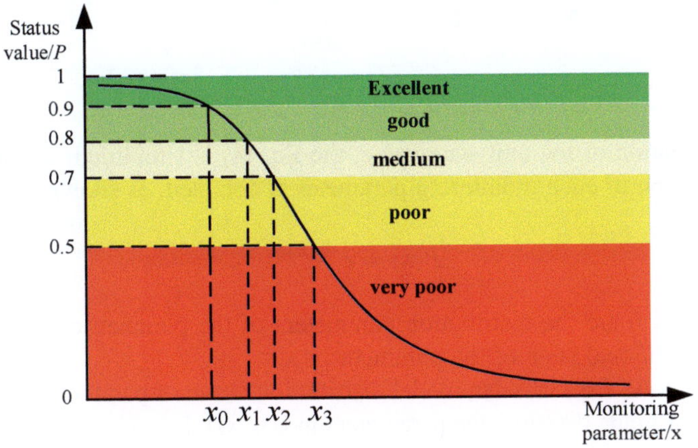

Fig. 3. Distribution curve of monitoring parameters (Logit regression model).

According to the operating characteristics of each parameter, the monitoring parameters in the state evaluation index that belong to the distribution of the logistic regression model include 9 parameters: compressor bearing vibration, compressor axial displacement, compressor outlet temperature, motor bearing vibration, gearbox bearing vibration, lubrication filter pressure difference, dry gas seal filter pressure difference, secondary seal exhaust temperature at the drive end, and secondary seal exhaust temperature at the non-drive end. Combined with the alarm values of the monitoring parameters, the distribution model parameters of the above 9 monitoring parameters are calculated with reference to Eq. (7), and the calculation results are shown in Table 12.

Calculation of Gaussian Model. The centrifugal compressor unit also has some monitoring parameters, such as inlet and outlet pressure, motor current and other parameters, which have a normal operating range. Exceeding this range means that the parameter is in an abnormal state, the parameter has the upper and lower limits of the alarm, and the Gaussian model is used to indicate its distribution. In order to conform

Table 12. Parameters of logit regression model.

Monitoring parameter	Model parameter		Monitoring parameter	Model parameter	
	α	β		α	β
Compressor bearing vibration	2.9655	−0.0424	Lubrication filter differential pressure	2.6629	−12.1043
Compressor axial displacement	2.9655	−4.2365	Dry gas seal pressure difference	2.6629	−12.1043
Compressor outlet temperature	4.8883	−0.0652	Exhaust temperature of secondary seal of drive end	6.4960	−0.0706
Motor bearing vibration	2.5419	−0.0212	Exhaust temperature of secondary seal of non-drive end S_5	6.4960	−0.0706
Gearbox bearing vibration	2.2595	−0.0222			

to the classification of the unit state value, the Eq. (8) is simplified, and the Gaussian model expression of each monitoring parameter is obtained, as shown in the Eq. (10).

$$P(x) = e^{-a(x-b)^2} \tag{10}$$

Where a and b are the distribution parameters of the parametric model; P characterizes the operational state of the parameter.

Figure 4 shows the Gaussian model distribution curve of parameters. According to the principle of grade division, the parameters in $[x'_1, x_1]$, $[x'_2, x'_1)$ and $(x_1, x_2]$, $[x'_3, x'_2)$ and $(x_2, x_3]$, $[x'_4, x'_3)$ and $(x_3, x_4]$, $(0, x'_4)$ and $(x_4, +\infty)$ respectively correspond to excellent, good, medium, poor and very poor state grades.

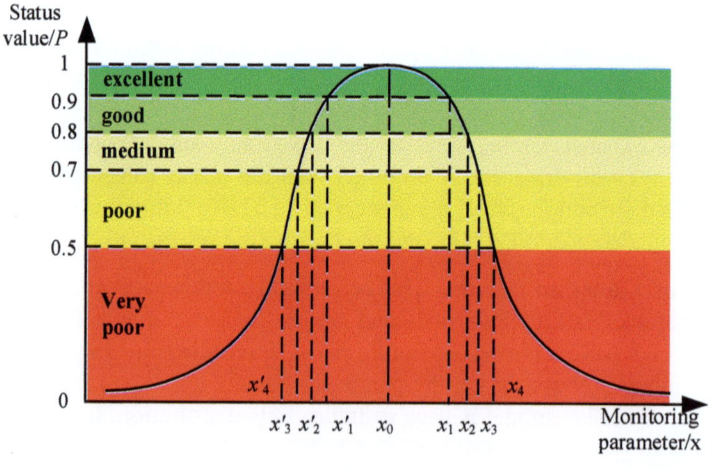

Fig. 4. Distribution curve of monitoring parameters (Gaussian model).

According to the operation characteristics of each parameter, the monitoring parameters in the state evaluation index, which belong to Gaussian model distribution, include 8 parameters: compressor inlet temperature, compressor inlet pressure, compressor outlet pressure, motor current, lubricating oil outlet pressure, lubricating oil tank temperature, dry gas seal air pressure difference and dry gas seal air pressure. According to Eq. (8), the distribution model parameters of 8 monitoring parameters are calculated, and the calculation results are shown in Table 13.

Table 13. Parameters of Gaussian model.

Monitoring parameters	Model parameter		Monitoring parameters	Model parameter	
	a	*b*		*a*	*b*
Compressor inlet pressure	0.0023	12.5000	Lubricating oil pump pressure	35.6673	0.7500
Compressor inlet pressure	0.9594	6.6500	Lubricating oil tank temperature	0.0003	57.5000
Compressor outlet pressure	0.3536	8.1000	Dry gas seal pressure difference	22.6334	0.2250
Motor current	0.0001	949.0000	Dry gas seal gas pressure	0.6931	9.1000

4.3 Evaluation Results

Taking the operation data of the 2 # centrifugal compressor unit of the gas transmission station at 16: 30 on August 6, 2017 as an example, the state evaluation of the compressor unit is implemented by the above method according to the established state evaluation index. The field test values and state values of each monitoring parameter are shown in Table 14.

According to the weight of each monitoring parameter in the corresponding subsystem, the state values of the five subsystems of the centrifugal compressor body, motor, gearbox system, lubrication system and dry gas seal system are obtained, as shown in Table 15.

According to the division principle of the state level, the centrifugal compressor unit and its five subsystems have excellent operating conditions and can operate for a long time. The centrifugal compressor unit was serviced on July 21, 2017, and the status evaluation results were consistent with the actual operating state of the compressor unit. However, the state value of the lubrication filter differential pressure parameter is 0.7483, and the state is poor. The filter was checked and found to be slightly blocked, and returned to normal after replacement.

Table 14. Running data of unit and its status values.

Monitoring parameters	Field test value	Sate values	Model parameter	Field test value	Sate values
Compressor bearing vibration	6.26 μm	0.9370	Lubrication filter differential pressure	0.13 MPa	0.7483
Compressor axial displacement	0.17 mm	0.9042	Lubricating oil pump pressure	0.73 MPa	0.9858
Compressor inlet temperature	14.8 °C	0.9879	Lubricating oil tank temperature	59.06 °C	0.9993
Compressor outlet temperature	41.9 °C	0.9063	Dry gas seal pressure difference	0.216 MPa	0.9982
Compressor inlet pressure	6.55 MPa	0.9905	Dry gas seal gas pressure	9.16 MPa	0.9975
Compressor outlet pressure	9.15 MPa	0.8182	Dry gas seal filter pressure difference	0.03 MPa	0.9089
Motor bearing vibration	13.27 μm	0.9056	Exhaust temperature of secondary seal of drive end	29.6 °C	0.9879
Motor current	976.84 A	0.9297	Exhaust temperature of secondary seal of non-drive end	24.8 °C	0.9914
Gearbox bearing vibration	6.05 μm	0.9033			

Table 15. Status values of centrifugal compressor unit and its subsystem.

System	Status value	System	Status value
Compressor	0.9018	Lubrication system	0.9437
Electric moto	0.9217	Dry gas sealing system	0.9685
Gearbox	0.9033	Centrifugal compressor unit	0.9207

5 Conclusion

In order to construct a comprehensive state evaluation index for the centrifugal compressor group, we use the improved FMEA and combine the monitoring parameters of the unit to select a group of monitoring parameters which can reflect the operation state of the centrifugal compressor. We calculated the weight of 17 monitoring parameters by AHP analysis method, and completed the construction of state evaluation index for centrifugal compressor group. According to the operating characteristics of each monitoring parameter, the corresponding parameter distribution model is determined. We determine the parameter values of each distribution model based on the normal operating range of the parameters, and complete the establishment of 17 monitoring parameter distribution models. Finally, taking the actual operation data of the centrifugal compressor unit of a gas station as an example, the unit was evaluated and the operating status of the unit was calculated. The state evaluation results are consistent with the actual operating state of the compressor unit, which proves the effectiveness of the method.

Acknowledgements. This paper is supported by SINOPEC Gas Company (No. 35150014-15-ZC0607-0002).

References

1. Gao, P., Tan, Z., Liu, G., et al.: New progress in China's oil and gas pipeline construction in 2016. Int. Pet. Econ. **25**(3), 26–33 (2017)
2. Cortinovis, A., Ferreau, H.J., Lewandowski, D., Mercangöz, M.: Experimental evaluation of MPC-based anti-surge and process control for electric driven centrifugal gas compressors. J. Process Control **34**, 13–25 (2015)
3. Jin-Qiu, H.U., Zhang, L.B., Liang, W., et al.: Quantitative HAZOP analysis for gas turbine compressor based on fuzzy information fusion. Syst. Eng. - Theory Pract. **29**(8), 153–159 (2009)
4. Lu, Y., Wang, F., Jia, M., et al.: Centrifugal compressor fault diagnosis based on qualitative simulation and thermal parameters. Mech. Syst. Signal Process. **81**, 259–273 (2016)
5. Li, X., Duan, F., Loukopoulos, P., et al.: Canonical variable analysis and long short-term memory for fault diagnosis and performance estimation of a centrifugal compressor. Control Eng. Pract. **72**, 177–191 (2018)
6. Sakthivel, N.R., Sugumaran, V., Babudevasenapati, S.: Vibration based fault diagnosis of monoblock centrifugal pump using decision tree. Expert Syst. Appl. **37**(6), 4040–4049 (2010)
7. Michal, T., Richard, M., Tomáš, S.: The analysis of the flow with water injection in a centrifugal compressor stage using CFD simulation. Meeting of Departments of Fluid Mechanics & Thermodynamics, 2017
8. Lin, Y., Liang, W., Qiu, Z., et al.: A new state evaluation method of oil pump unit based on AHP and FCE. In: 12th international conference on damage assessment of structures (2017)
9. Kang, J., Sun, L., Sun, H., et al.: Risk assessment of floating offshore wind turbine based on correlation-FMEA. Ocean Eng. **129**, 382–388 (2017)
10. Chen, D., Lü, L., Shang, M.S., et al.: Identifying influential nodes in complex networks. Fuel Energy Abs. **391**(4), 1777–1787 (2012)
11. Ooi, J., Promentilla, M.A.B., Tan, R.R., et al.: A systematic methodology for multi-objective molecular design via analytic hierarchy process. Process Saf. Environ. Prot. **111**, 663–677 (2017)
12. Saber, Q.O., Yahya, A.Z.: Feature selection using particle swarm optimization-based logistic regression model. Chemometr. Intell. Lab. Syst. (2018): S0169743918303800-

Crack Identification in Multi-Span Beams on Elastic Foundation by Using Transfer Matrix Method

Baran Bozyigit[1(✉)], Irem Bozyigit[2], Yusuf Yesilce[1], and M. Abdel Wahab[3]

[1] Department of Civil Engineering, Dokuz Eylul University, 35160 Buca, Izmir, Turkey
baran.bozyigit@deu.edu.tr
[2] Department of Civil Engineering, Ege University, 35040 Bornova, Izmir, Turkey
[3] Soete Laboratory, Faculty of Engineering and Architecture, Ghent University, Technologiepark Zwijnaarde 903, 9052 Zwijnaarde, Belgium

Abstract. In this study, an analytical based approach is proposed for detection of cracks in multi-span Euler-Bernoulli beams on Winkler foundation. Transfer matrix method (TMM) is used for both forward and inverse problems. The crack is modeled by means of a linear rotational spring. For the forward problem, natural frequencies of intact and cracked multi-span beam models are calculated via TMM for two different support arrangements and the results are compared with those obtained using finite element method (FEM). In the inverse problem, the crack location and crack length are calculated based on the natural frequencies obtained from FEM simulations using plots of rotational spring flexibilities versus crack location. The predicted crack properties are tabulated with actual data. It is seen that considering elastic foundation slightly decreases the accuracy of crack depth prediction for multi-span beams. However, the accuracy of localization of crack is not affected by Winkler foundation.

Keywords: Crack detection · Euler-Bernoulli beam · Natural frequency · Transfer matrix method · Winkler foundation

1 Introduction

The multi-span beams on elastic foundations are considered for various types of engineering problems especially in geotechnical and railroad engineering. The reduction of stiffness caused by cracks makes vibration analysis of damaged beam-like structures an important research area. There are some studies about vibrations of cracked beams on elastic foundation. Shin et al. [1] obtained natural frequencies of cracked single-span Euler-Bernoulli beams on elastic foundation. The results of considering Winkler foundation model were compared with those obtained from two-parameter (Pasternak) foundation model. Mirzabeigy and Bakhtiari-Nejad [2] proposed a semi-analytical approach for free vibrations of cracked beam on two-parameter elastic foundation. A single-span Euler-Bernoulli beam with elastically restrained ends was

M. A. Wahab (Ed.): DAMAS 2019, LNME, pp. 402–409, 2020.
https://doi.org/10.1007/978-981-13-8331-1_29

considered and the natural frequencies were calculated for general boundary conditions. Batihan and Kadioglu [3] investigated free vibrations of cracked beams on elastic foundation using Euler-Bernoulli beam theory (EBT) and Timoshenko beam theory (TBT). Winkler and Pasternak foundation models were considered and calculated natural frequencies were verified using previous studies. Free vibrations of cracked multi-span beams were studied in limited papers due to complicated formulations in comparison with cracked single-span beams. Liu et al. [4] obtained natural frequencies of cracked multi-span beams via a hybrid analytical/numerical method. Gillich et al. [5] performed a crack detection procedure based on the modal strain energy concept. In this study, a two span Euler-Bernoulli beam with a single crack was used.

Besides forward problem based on calculation of natural frequencies and mode shapes, the accurate detection of cracks is a very important topic to prevent destructive failure of structures. The crack modeling based on using massless rotational springs that divide beams into segments provide a straightforward solution for beam-like structures [6]. One of the effective crack identification procedure is combining measured natural frequencies with an analytical approach based on plotting rotational spring stiffness versus crack location [7–9]. The procedures of TMM are based on constructing a relation between two ends of structure by a chain multiplication of local transfer matrices of beam segments. Thus, transfer matrix formulations provide a straightforward solution for vibrations of cracked multi-span beams by adding a jump matrix that represents discontinuity of slope at crack location. The effectiveness of TMM on vibrations of beam-like structures with various types of boundary conditions were proven in previous studies [10–12].

In this study, a cracked two-span Euler-Bernoulli beam on Winkler foundation is considered. Two different support arrangements which are Fixed-Simple-Simple (F-S-S) and Simple-Simple-Simple (S-S-S) are used. Firstly, the natural frequencies of intact and cracked beam models are calculated and the results are presented using FEM package SAP2000. Then, using the calculated natural frequencies from FEM simulations, the crack is detected by the plots of rotational spring flexibility versus crack location. The results of cracked beam models without elastic foundation are also presented to show the effects of Winkler foundation on the accuracy of proposed crack detection approach.

2 Model and Formulation

The cracked two-span Euler-Bernoulli beams on Winkler foundation having two different support arrangements are presented in Fig. 1 where k_w is a spring constant of Winkler foundation, L^* is crack location, L is span length, x and y are beam axes and transverse displacement, respectively. The following assumptions are considered in this study:

1. The material of beam is isotropic.
2. The cross-section of beam is uniform.
3. The behavior of beam is linear elastic.
4. The crack remains open under bending.
5. The effects of damping are neglected.

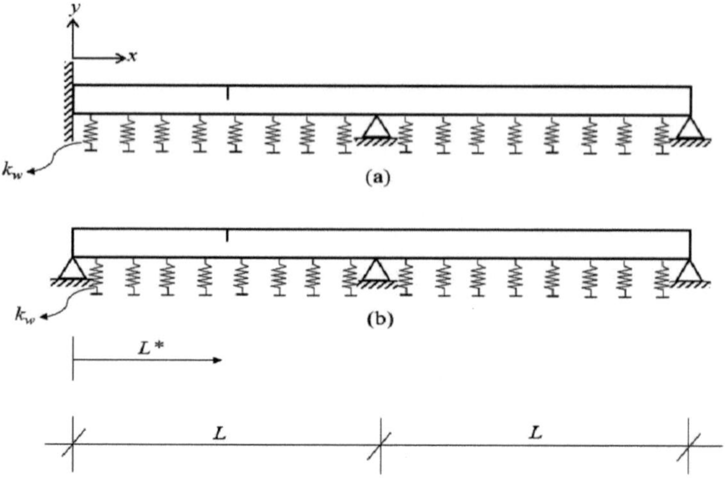

Fig. 1. Cracked multi-span beam on Winkler foundation with (a) F-S-S support arrangement and (b) S-S-S support arrangement

The crack modeling approach based on using rotational spring that divides beams into two segments are represented in Fig. 2 where C_R is spring flexibility that represents the reduction in bending stiffness at crack location, a and h are width and height of the beam, respectively.

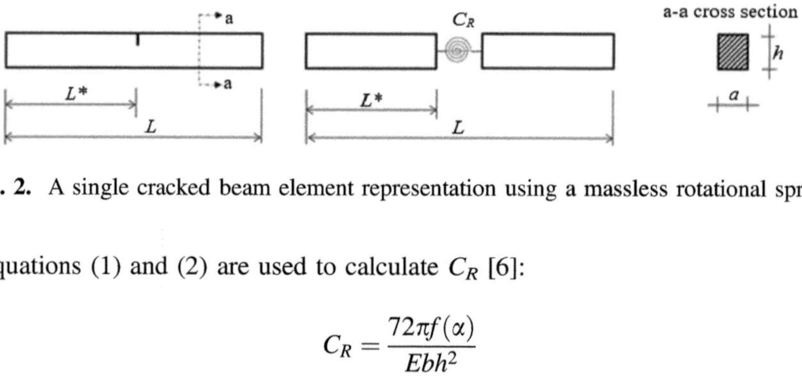

Fig. 2. A single cracked beam element representation using a massless rotational spring

Equations (1) and (2) are used to calculate C_R [6]:

$$C_R = \frac{72\pi f(\alpha)}{Ebh^2} \tag{1}$$

$$f(\alpha) = 0.6384\alpha^2 - 1.035\alpha^3 + 3.7201\alpha^4 - 5.1773\alpha^5 + 7.553\alpha^6 - 7.332\alpha^7 + 2.4909\alpha^8 \tag{2}$$

where α is crack ratio (h_c/h), h_c is crack length, E is elastic modulus and $f(\alpha)$ is local compliance function that can be obtained from linear elastic fracture mechanics [9].

The governing equation of motion of an Euler-Bernoulli beam on Winkler foundation is presented in Eq. (3) [1].

$$EI\frac{\partial^4 y(x,t)}{\partial x^4} + m\frac{\partial^2 y(x,t)}{\partial t^2} + k_w y(x,t) = 0 \tag{3}$$

where I is the area moment of inertia, m and t are mass per unit length and time, respectively. Equation (4) is obtained by applying separation of variables method with the assumption of harmonic motion.

$$\frac{d^4 y(z)}{dz^4} + \left(\frac{k_w L^4 - m\omega^2 L^4}{EI}\right) y(z) = 0 \tag{4}$$

where L is the beam length and $z = x/L$. It is assumed that the solution of Eq. (4) is in the following form:

$$y(z) = \{\bar{C}\} e^{isz} \tag{5}$$

By substituting Eq. (5) into Eq. (4), the transverse displacement function $y(z)$ and slope function $\theta(z)$ are obtained as in Eqs. (6) and (7), respectively.

$$y(z) = (\bar{C}_1 e^{is_1 z} + \bar{C}_2 e^{is_2 z} + \bar{C}_3 e^{is_3 z} + \bar{C}_4 e^{is_4 z}) \tag{6}$$

$$\theta(z) = (is_1 \bar{C}_1 e^{is_1 z} + is_2 \bar{C}_2 e^{is_2 z} + is_3 \bar{C}_3 e^{is_3 z} + is_4 \bar{C}_4 e^{is_4 z}) \tag{7}$$

Bending moment function $M(z)$ and shear force function $Q(z)$ are defined as:

$$M(z) = (R_1 s_1^2 \bar{C}_1 e^{is_1 z} + R_1 s_2^2 \bar{C}_2 e^{is_2 z} + R_1 s_3^2 \bar{C}_3 e^{is_3 z} + R_1 s_4^2 \bar{C}_4 e^{is_4 z}) \tag{8}$$

$$Q(z) = (R_2 i s_1^3 \bar{C}_1 e^{is_1 z} + R_2 i s_2^3 \bar{C}_2 e^{is_2 z} + R_2 i s_3^3 \bar{C}_3 e^{is_3 z} + R_2 i s_4^3 \bar{C}_4 e^{is_4 z}) \tag{9}$$

where $R_1 = EI/L^2$ and $R_2 = EI/L^3$.

3 Transfer Matrix Formulations

The state vector (Z) of left-hand side ($z = 0$) of beam element can be written as:

$$\{Z\}_{z=0} = [T_0]\{\bar{C}\} \tag{10}$$

where

$$[T_0] = \begin{bmatrix} 1 & 1 & 1 & 1 \\ is_1 & is_2 & is_3 & is_4 \\ R_2 i s_1^3 & R_2 i s_2^3 & R_2 i s_3^3 & R_2 i s_4^3 \\ R_1 s_1^2 & R_1 s_2^2 & R_1 s_3^2 & R_1 s_4^2 \end{bmatrix}; \{Z\}_{z=0} = \begin{Bmatrix} y \\ \theta \\ Q \\ M \end{Bmatrix}_{z=0}$$

The state vector of right-hand side ($z = 1$) of beam element is presented in Eq. (11).

$$\{Z\}_{z=1} = [T_1]\{\bar{C}\} \tag{11}$$

where

$$[T_1] = \begin{bmatrix} e^{is_1} & e^{is_2} & e^{is_3} & e^{is_4} \\ is_1e^{is_1} & is_2e^{is_2} & is_3e^{is_3} & is_4e^{is_4} \\ R_2is_1^3e^{is_1} & R_2is_2^3e^{is_2} & R_2is_3^3e^{is_3} & R_2is_4^3e^{is_4} \\ R_1s_1^2e^{is_1} & R_1s_2^2e^{is_2} & R_1s_3^2e^{is_3} & R_1s_4^2e^{is_4} \end{bmatrix}; \{Z\}_{z=1} = \begin{Bmatrix} y \\ \theta \\ Q \\ M \end{Bmatrix}_{z=1}$$

The vector $\{\bar{C}\}$ is rewritten as in Eqs. (12) and (13) by using Eqs. (10) and (11), respectively.

$$\{\bar{C}\} = [T_0]^{-1}\{Z\}_{z=0} \tag{12}$$

$$\{\bar{C}\} = [T_1]^{-1}\{Z\}_{z=1} \tag{13}$$

The relationship between the state vectors $\{Z\}_{z=0}$ and $\{Z\}_{z=1}$ is constructed as follows:

$$\{Z\}_{z=1} = [T_1]\,[T_0]^{-1}\{Z\}_{z=0} \tag{14}$$

$$\{Z\}_{z=1} = [T^*]\{Z\}_{z=0} \tag{15}$$

Where $[T^*] = [T_1]\,[T_0]^{-1}$ and $[T^*]$ is transfer matrix of beam element

The global transfer matrix of the whole vibrating system is obtained by a chain multiplication of transfer matrices of beam segments and an additional matrix representing the discontinuity of slope at crack location. The jump matrix $[C^*]$ for crack location is presented in Eq. (16) as:

$$[C^*] = \begin{bmatrix} 1 & 0 & 0 & 0 \\ 0 & 1 & 0 & C_R \\ 0 & 0 & 1 & 0 \\ 0 & 0 & 0 & 1 \end{bmatrix} \tag{16}$$

The natural frequencies of the cracked beam models are calculated by equating the determinant of reduced global transfer matrices to zero. The reduction of global transfer matrices are made according to boundary conditions.

For the inverse problem, the plots of C_R versus crack location are used for crack identification. The intersection of three plots of the first three natural frequencies is detected as crack location. After the localization of crack by spring flexibility, the length of crack is calculated using Eqs. (1) and (2). It should be noted that there may be two intersection points if the system is symmetrical by means of boundary conditions [9]. For reducing the error of graphical procedures and improve the accuracy of crack identification, the elastic modulus value must be updated. This calibration, which is also known as zero-setting, is made using Eq. (17).

$$E_n^{updated} = \left(\frac{\omega_{measured}}{\omega_{analytical}}\right)^2 E \qquad (17)$$

where E_n^{updated} represents updated elastic modulus, $\omega_{measured}$ and $\omega_{analytical}$ represent natural frequencies obtained from experiment or FEM simulation and natural frequency obtained from analytical based solution, respectively.

4 Numerical Case Study

The numerical example of this study is based on the following material and geometric properties: $L^* = 2.5$ m, $L = 5$ m, $a = h = 0.3$ m, $E = 30 \times 10^6$ kN/m^2 and $m = 0.023$ ts^2/m^2.

The first three natural frequencies of the multi-span beam models obtained from FEM simulations and TMM are presented in Tables 1 and 2 for intact and cracked beams, respectively.

Table 1. First three natural frequencies of intact multi-span beams

Support arrangement		Natural frequency (rad/sec)					
		ω_1	ω_2	ω_3	ω_1	ω_2	ω_3
		$k_w = 0$			k_w = 10000 kN/m		
F-S-S	TMM	137.1802	239.1329	509.5349	249.6430	315.5169	550.0719
	FEM	136.8258	236.7326	509.1148	249.6126	313.9646	549.6012
S-S-S	TMM	117.3048	183.0690	469.2189	239.5004	277.6946	513.5819
	FEM	117.2847	183.2211	469.1387	239.4594	277.7680	513.4942

Table 2. First three natural frequencies of cracked multi-span beams ($\alpha = 0.25$)

Support arrangement		Natural frequency (rad/sec)					
		ω_1	ω_2	ω_3	ω_1	ω_2	ω_3
		$k_w = 0$			k_w = 10000 kN/m		
F-S-S	TMM	136.5643	234.4648	508.3911	249.4991	313.9646	549.6012
	FEM	136.2115	232.1114	507.9719	249.2765	312.1875	549.1997
S-S-S	TMM	116.3947	182.1266	469.2189	239.0559	277.0742	513.5819
	FEM	115.1636	182.3378	469.1387	238.4277	277.3980	513.4941

It should be noted that FEM results of this study are obtained by dividing beams into 250 segments in SAP2000. For the crack modeling in SAP2000, the beam in first span is cut from crack location and a two joint link element is added at this section. Then, the spring stiffness is introduced as effective rotational spring stiffness value of two-joint link element. The other displacements of two-joint link element except rotation are restrained to reflect the jump of rotation due to crack. According to

Tables 1 and 2, the first three natural frequencies of intact multi-span beams are higher in comparison with cracked beam. Although, F-S-S support arrangement provides higher natural frequencies when compared to S-S-S support arrangement. Tables 1 and 2 imply that considering Winkler foundation increases natural frequencies significantly for both intact and cracked beams. The FEM results are in a very good agreement with results obtained from TMM.

The crack identification of multi-span beam models is performed using natural frequencies presented in Tables 1 and 2. The plots of spring flexibility values versus crack locations are presented in Fig. 3.

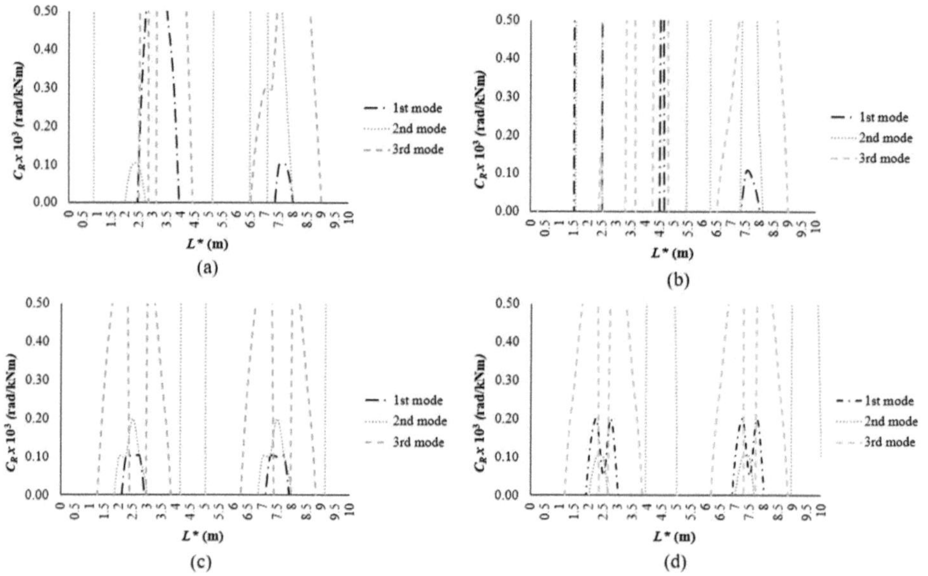

Fig. 3. Plots of C_R vs L^* for (a) F-S-S ($k_w = 0$), (b) F-S-S (k_w = 10000 kN/m), (c) S-S-S ($k_w = 0$) and (d) S-S-S (k_w = 10000 kN/m)

The result of crack identification procedure is presented in Table 3. According to Table 3, the accuracy of the proposed approach for detecting cracks in multi-span beams is very good. Table 3 shows that considering Winkler foundation does not affect the accuracy of prediction of location. However, the accuracy of prediction of crack length is slightly decreased by taking into account the Winkler foundation.

Table 3. Actual and predicted crack properties using proposed approach via TMM

Support arrangement	k_w (kN/m)	Crack properties			
		Actual length (cm)	Predicted length (cm)	Actual location (m)	Predicted location (m)
F-S-S	0	7.5	7.62	2.5	2.5
	10000	7.5	8.61	2.5	2.5
S-S-S	0	7.5	7.62	2.5	2.25
	10000	7.5	8.28	2.5	2.25

5 Conclusions

The analytical crack identification procedure based on relation between crack location and rotational spring flexibility that represents crack is applied to multi-span Euler-Bernoulli beams on Winkler foundation using transfer matrix formulations. A numerical case study including FEM simulations is presented. The results show that proposed approach can be used for detecting cracks in multi-span beams on elastic foundation. It is seen that considering Winkler foundation slightly affects the accuracy of crack length prediction.

References

1. Shin, Y., Yun, J., Seong, K., Kim, J., Kang, S.: Natural frequencies of Euler-Bernoulli beam with open cracks on elastic foundations. J. Mech. Sci. Technol. **20**, 467–472 (2006)
2. Mirzabeigy, A., Bakhtiari-Nejad, F.: Semi-analytical approach for free vibration analysis of cracked beams resting on two-parameter elastic foundation with elastically restrained ends. Front. Mech. Eng. **9**, 191–202 (2014)
3. Batihan, A.Ç., Kadioğlu, F.S.: Vibration analysis of a cracked beam on an elastic foundation. Int. J. Struct. Stab. Dyn. **16**, 1550006 (2016)
4. Liu, H.B., Nguyen, H.H., Xiang, Y.M.: Vibration analysis of a multi-span continuous beam with cracks. Appl. Mech. Mater. **256–259**, 964–972 (2013)
5. Gillich, G.R., Ntakpe, J.L., Abdel Wahab, M., Praisach, Z.I., Mimis, M.C.: Damage detection in multi-span beams based on the analysis of frequency changes. J. Phys: Conf. Ser. **842**, 012033 (2017)
6. Ostachowicz, W.M., Krawczuk, M.: Analysis of the effect of cracks on the natural frequencies of a cantilever beam. J. Sound Vib. **150**, 191–201 (1991)
7. Nandwana, B.P., Maiti, S.K.: Detection of the location and size of a crack in stepped cantilever beams based on measurements of natural frequencies. J. Sound Vib. **203**, 435–446 (1997)
8. Lele, S.P., Maiti, S.K.: Modelling of transverse vibration of short beams for crack detection and masurement of crack extension. J. Sound Vib. **257**, 559–583 (2002)
9. Kindova-Petrova, D.: Vibration-Based methods for detecting a crack in a simply supported beam. J Theor. Appl. Mech. **44**(4), 69–82 (2014)
10. Al Rjoub, Y.S., Hamad, A.G.: Free vibration of functionally Euler-Bernoulli and Timoshenko graded porous beams using the transfer matrix method. KSCE J. Civ. Eng. **21**, 792–806 (2017)
11. Tan, G., Liu, Y., Gong, Y., Shen, Y., Liu, Z.: Free vibration of the cracked non-uniform beam with cross section varying as polynomial functions. KSCE J. Civ. Eng. **22**, 4530–4546 (2018)
12. Lee, J.W., Lee, J.Y.: An exact transfer matrix expression for bending vibration analysis of a rotating tapered beam. Appl. Math. Model. **53**, 167–188 (2018)

Non Destructive Inspection of Corrosion in Rock Bolts Using an Ultrasonic Waveguide Approach

André Taras(✉) and Kaveh Saleh

Hydro Québec Research Institute, Varennes, QC J3X 1S1, Canada
taras.andre@ireq.ca

Abstract. Rock bolts are used frequently to consolidate and stabilize the rock mass in mining and civil engineering structures. Since their installations decades ago, there are concerns that corrosion might have set in, increasing the risk of a lack of holding strength.

In this study, preliminary measurements in rock gallery sites have shown that ultrasonic wave guide technic is able to determine the length of rock bolts. Then tests were undertaken on 4 m long bolts with different defects imbedded in a concrete block. They all respectively had different artificial defect to simulate corrosion or fracture in terms of cross-section reduction, length of loss of contact and crack. Besides depth of the defect, the following variables were added: the loss of contact length, the position of the defect and the length of bolt, with measurements taken for different conditions.

Based on a physical model for the wave energy propagation and reflection, an inspection method was developed to evaluate the effects of corrosion on section reduction, cracks and loss of contact. Results show a good correlation of the wave energy propagation model to the reflections of the different types of defects. Corrosion type defects (modelled by tapered reduction) were discriminated from crack defects through a signal analysis technic. Also, a reflection from a major defect was differentiated from an end of bolt reflection using the propagation model. Finally, a third approach was developed to identify the presence of a "ghost defect" and to determine its length of loss of contact and its influence on the end of bolt reflection. A thorough inspection method to determine anchor bolt corrosion has been put forth encompassing several inspection parameters.

Keywords: Ultrasonic · Guided waves · Rock · Bolt · Corrosion · Non destructive testing

1 Introduction

A rock bolt is a steel rebar that is introduced into a hole a few meters deep which was drilled into the rock mass. Then, the space between the bolt and the rock is injected with a grout or polymer.

Rock bolts have been used to reinforce mining and civil engineering structures such as: concrete structures, rock mass, mines, shafts, tunnels, hydro-electric galleries and

M. A. Wahab (Ed.): DAMAS 2019, LNME, pp. 410–432, 2020.
https://doi.org/10.1007/978-981-13-8331-1_30

foundations. They have been installed since the 1950's and the process of corrosion can jeopardize their structural integrity. They create serviceability and safety issues that incur important costs for their inspection, repair or replacement.

The main goal of this project is to evaluate the structural integrity of these rock bolts, based on the presence of corrosion in terms of section reduction and cracks, and in terms of the loss of bearing contact.

Several non-destructive technics have been evaluated among which "Impact hammer response", "piezoelectric shaker vibration" and "ultrasound guided waves", studies [1, 2]. Preliminary field tests, in a an underground water intake gallery of a dam at Hydro Québec, have shown the reliability of ultrasound guided wave technic to determine the length of rock bolts up to 3.5 m [3].

To evaluate the potential of ultrasonic guided waves to identify and characterize defects and cracks, Hydro Québec Research Institute (IREQ) built a concrete test bench. Imbedded in the concrete were eleven 4 m long and 25 mm diameter anchor bolts, each with a specific defects to simulate different degrees of simulated corrosion and cracks.

2 Wave Propagation Characteristics

The bolt acts as a waveguide imbedded in concrete, it channels the wave generated by the ultrasonic probe down its length. As it advances, it is attenuated due to material damping, to reflections and refraction at the ribs and to leakage. When the wave encounters a defect or the end of bolt, part of it is reflected.

An ultrasonic compression wave travels at a speed of around $C_P = 5900$ m/s in steel bolt as long as the wavelength is roughly 10 times smaller than the diameter of the bolt. Otherwise, it becomes a rod wave propagating at a bulk speed of around 5200 m/s.

Previous studies by Ervin et al. [4] and Shoji [5] were undertaken in a transmission mode on these types of bolts.

2.1 Reflection Characteristics

The ultrasonic wave can reflect back either from a defect along the bolt or from the end of the rod, in as much as it overcomes attenuation due to material damping, dispersion in the form of reflection or refraction and to "leakage" (Fig. 1).

As one zooms in on a reflection, one notices a succession of trailing groups of waves called "trails" (Fig. 2).

These "trails" are generated by mode conversion from parts of the compressional waves moving at grazing incidence along the surface of the bolt. They are partly reflected into a continuing grazing compression wave and a shear wave crossing at angle the section of the bolt. Also, leakage occurs through to the concrete occurs due to refraction of an incident waves coming at angle to the surface. Wave reflection and refraction is a continuous process and takes place behind the head compression wave front as it moves forward at speed "C_P" along the longitudinal axes of the bolt.

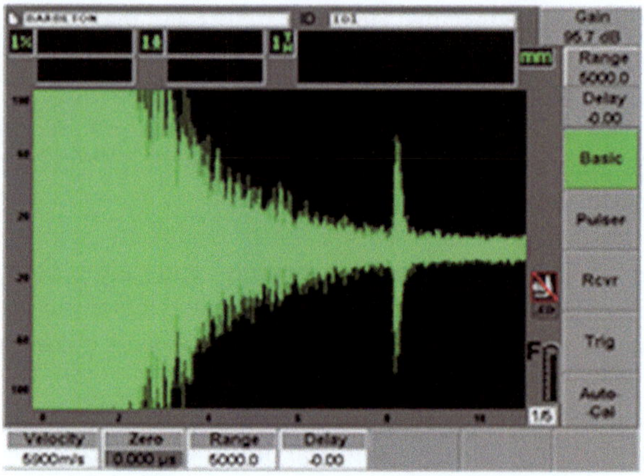

Fig. 1. End reflection of a 4 m long bolt, 25 mm in diameter, non-embedded

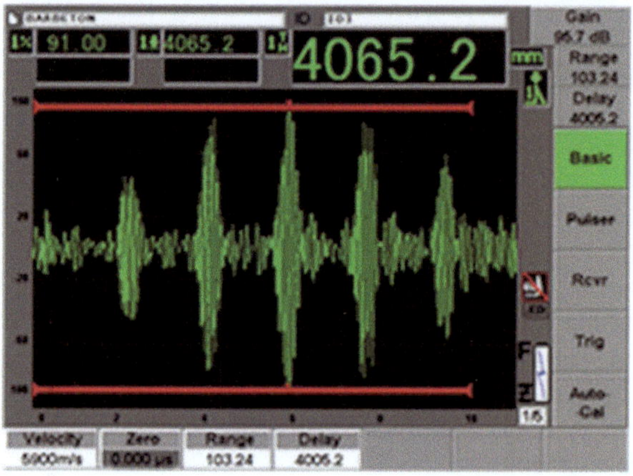

Fig. 2. Zoom on the reflection composed of trails in Fig. 1.

2.2 Ultrasonic Propagation Mode

Theory has shown that an ultrasonic probe emitting a wave with a frequency of emission at around 3.5 MHz and applied to a 25 mm diameter bolt induces an axially symmetric cross-sectional deformations concentrated at the center of the cross-section. It is designated by mode L(0,11) (Fig. 3). The frequency content of the propagating wave and its reflection have the same spectral frequency content as that of the pulse emitted (Fig. 4). Studies by "Erwin" [4] have indicated that the amplitudes of the lower frequencies in the spectrum decrease as uniform corrosion increases the mass loss, in a through transmission setup mode.

The L(0,11) deformation mode minimizes leakage losses according to the attenuation dispersion curves [6, 7]. This propagation mode is more suited to reach the end of the bolt and to be reflected back. On the other hand, it seems less optimized to

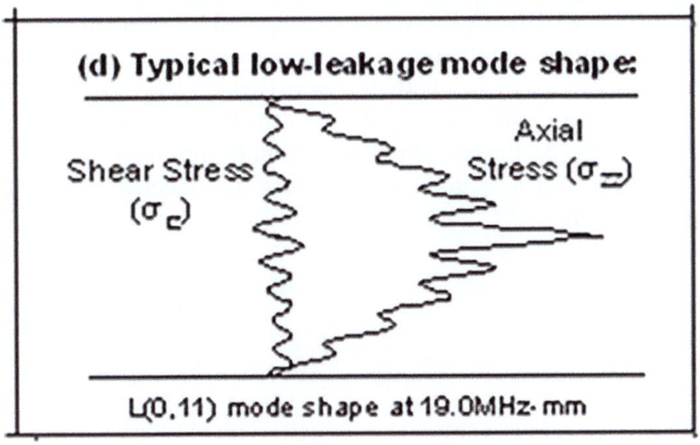

Fig. 3. An L(0,11) longitudinal propagation mode shape through a 25 mm cross-section [6]

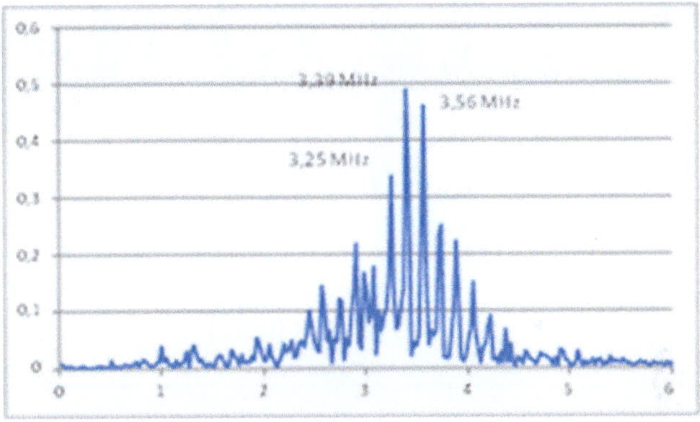

Fig. 4. Spectral frequencies of the end reflection presented in Fig. 1 and Fig. 2

characterize surface defects and de-bonding. A flat cross-section deformation mode of propagation L(0,1) would be more sensitive to surface defect, but would be more quickly attenuated through leakage. Its excitation frequency would be in the range of 100 to 500 kHz.

The central cross-section deformation mode L(0,11) was chosen for the measurements, given a wave pulse frequency of 3.5 MHz and a 25 mm diameter bolt. It was considered that mode conversion at the bolt surface would tend to redistribute across the section of the bolt the initial central deformation, for the subsequent trails of the reflection.

2.3 Wave "Energy" Propagation

The wave "energy" of a reflection was computed as the integration of the absolute value of the signal over the time length of the reflection that is visible above the noise level. The squared value of the signal was also computed but deemed to place too much emphasis on the maximum amplitudes.

The reflection over its time length is comprised of its wave front and all the successive trailing wave groups that are distinguishable above the noise level. As an illustrated example (Fig. 5), the end of bolt reflection is presented for a 3.5 m long bolt imbedded in concrete with a section reduction of 8 mm tapered over 15 cm on each side and situated at 1 m.

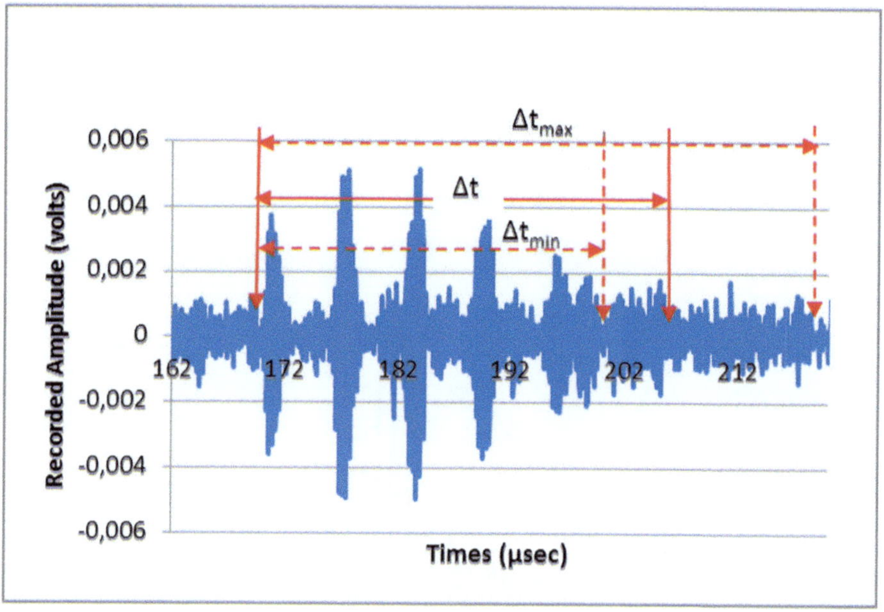

Fig. 5. End reflection of a 3.5 m long bolt imbedded in concrete, with 8 mm tapered reduction at 1 m

The recorded signal "a(t)" is an instrument voltage which is taken to be proportional to the compressive stress of the wave. The calculation procedure starts by defining the beginning of the wave front and the time length of the reflection. This time length is established based on how many trailing waves can be identified. This aspect can introduce 10% to 15% of variability in the interpretation of "Δt" ($\Delta t_{Min} \leq \Delta t \leq \Delta t_{Max}$). Then, the energy "*E*" of the reflected wave is obtained by adding the incremental area "a(τ)·Δτ" under the absolute value of the signal "|a(τ)|" over the "n" sampling points that constitutes the time length "Δt" of the reflection:

$$E = \sum\nolimits_{i=1}^{i=n} a_i \cdot \Delta\tau \tag{1}$$

where $n \cdot \Delta\tau = \Delta t$

2.4 Wave Attenuation in Rock Anchor Bolt

Wave attenuation is caused by material damping, leakage from the steel bolt to the concrete and reflection dispersion from the ribs, the filets and the end. Attenuation as a hole was modelled to be proportional to the amplitude of the wave energy propagating "E(x)" along the bolt. Energy loss was modelled with a length attenuation factor "λ" as follows:

$$E(x) = E_i \cdot e^{-\lambda \cdot (x - x_i)} \tag{2}$$

3 Experimental Set-up

Ultrasonic measurements were taken on 11 steel ribbed anchor bolts 25 mm in diameter and initially 4 m long. These bolts were imbedded in a concrete block (L = 3.80 m, H = 1.23 m, W = 1.21 m) and numbered (Fig. 6). Bolt lengths of 3.5 m, 2.5 m, and 2.0 m were obtained by drilling 15 cm core from the top of the concrete test bench down through each bolt to achieve the subsequent reduced lengths. Each bolt presented a specific characteristic; they are summarized in detail in Table 1. Bolt # 1 was the reference bolt without any defects. Bolts with cracks and tapered section reductions are presented in Fig. 7. Hollow bolts and constant section reduction are presented in Fig. 8. Hollow bolts 7 and 8 gave no ultrasonic reflection over 4 m, so they were neglected in the first stage of this study.

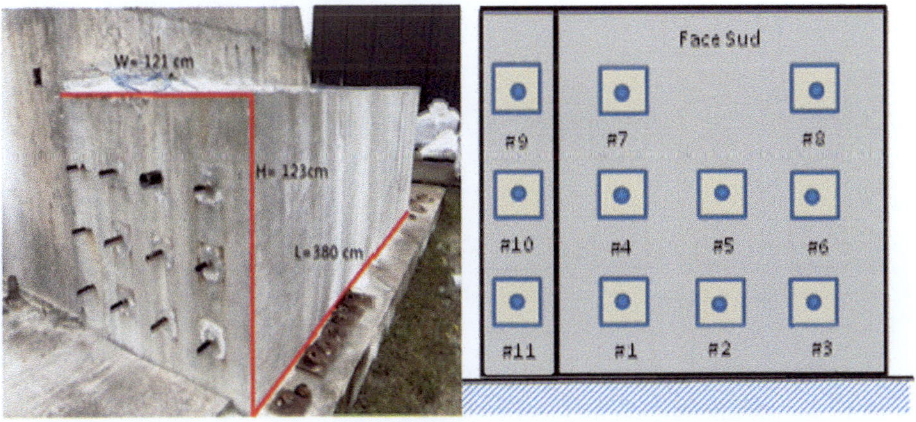

Fig. 6. Test bench and sketch of bolt position and numbering

Before imbedding the bolts in concrete, the defects were wrapped in a plastic sheet and the gap between the sheet and the reduced section of steel was filled with sand. The access was possible from both ends of the 4 m long bolts which produced two different measurement distances (1 m and 3 m) for the same defect. But, once the bolts were shortened through vertical coring the north end of the bolt became inaccessible to measurements.

Table 1. Descriptions of defects and distances of 11 bolts

Characteristics of Bolts	Bolt #	Length Δ of loss of contact	Reduction π of section	Distance between transducer and defect	
				South face	North face*
Reference bolt - without defect	1	–	–	–	–
Crack reduction 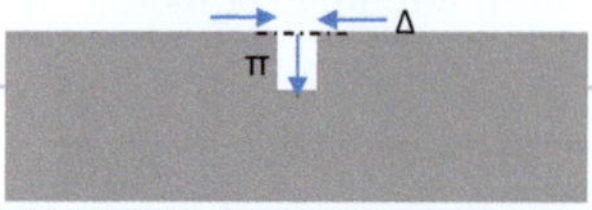	2	3 mm	8 mm	1 m	3 m
	3	3 mm	12 mm	1 m	3 m
Tapered reduction 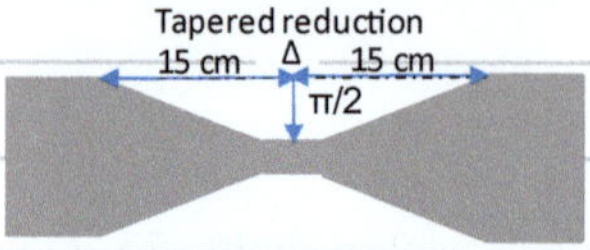	4	30 cm	4 mm	1 m **	3 m **
	5	30 cm	8 mm	1 m **	3 m **
	6	30 cm	12 mm	1 m **	3 m **
Hollow bolt without defect	7	–	–	–	–
Hollow bolt without defect	8	–	–	–	–
Loss of contact 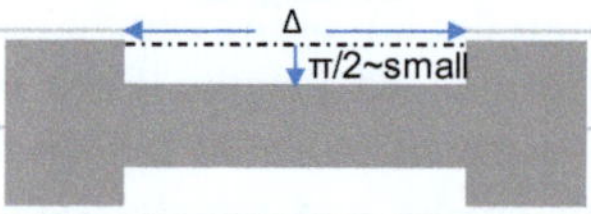	9	60 cm	4 mm	0.85 m	2.55 m
	10	90 cm	4 mm	0.85 m	2.225 m
	11	120 cm	4 mm	0.85 m	1.946 m
* only for 4 m long bolt	** distance from probe to the point of maximum section reduction				

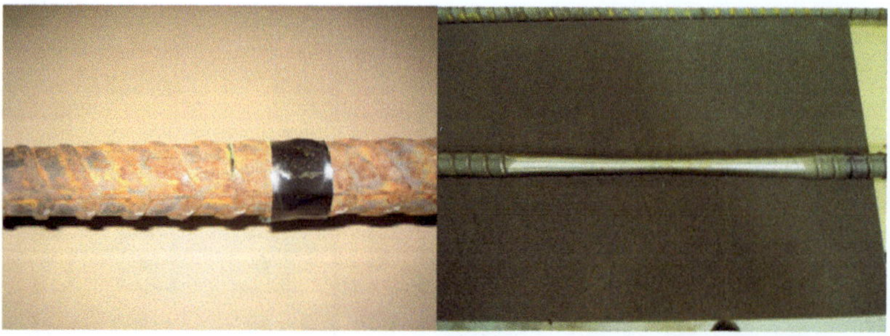

Fig. 7. Left, example of 8 mm crack; right, 12 mm tapered reduction over ± 15 cm

The ultrasonic measurements were undertaken with a ZETEK Z-SCAN UT. The Z_SCAN UT was used in a “full A-scan » capture mode, 300 V, 250 ns, a sampling frequency of 25 MHz and 2 to 10 MHz low and high filters. The ultrasonic emitter/receiver probe was an Olympus A181S - 3.5 MHz and 0.75 in diameter.

Tests on 1 m long bolts imbedded in concrete filled 15 cm diameter tubes were also undertaken to determine wave propagation parameters: “E_{IN} = 145 volts” the input ultrasonic wave pulse energy, “λ = 1.22” the coefficient of attenuation for leakage and material damping, “λ_0 = 0.88” the coefficient of attenuation for material damping only. These parameters were needed to be fed into the “wave energy” propagation model.

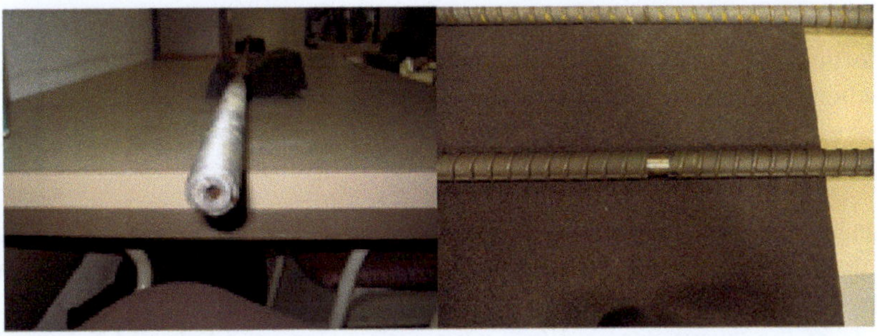

Fig. 8. Left, hollow centered bolt; right, loss of contact with radial reduction of 2 mm

4 Wave Energy Propagation Model for the Prediction of Defects

4.1 Wave Energy Propagation

The wave energy propagating through a bolt imbedded in concrete acting as a wave guide is formulated generically by expression (3) and is represented in Fig. 9. Based on conservation of energy, the energy of the generated ultrasonic wave pulse should match the sum of the energies of the following events: (a) the leakage energy "e_F" of the wave going back and forth along the contact length "L-Δ" of bolt as expressed in (2), (b) the reflected wave energy from the defect coming back to the sensor "E_D", (c) the reflected energy from the end of the bolt coming back to the sensor "E_B", (d) the reflected energy "E_{DR2}" at the defect of the returning wave energy reflected from the end of bolt. This energy balance is expressed as follows:

$$\begin{aligned} E_{IN} &= \int_0^x e_F(x)\cdot dx + \int_x^0 e_F(x)\cdot dx + E_D \\ &+ \int_{x+\Delta}^{L} e_F(x)\cdot dx + \int_L^{x+\Delta} e_F(x)\cdot dx + E_{DR2} + \int_x^0 e_F(x)\cdot dx + E_B \end{aligned} \tag{3}$$

One has to understand that the magnitudes of end and defect reflections behave in opposition, in the case where the two reflections coexist. Deeper is the section

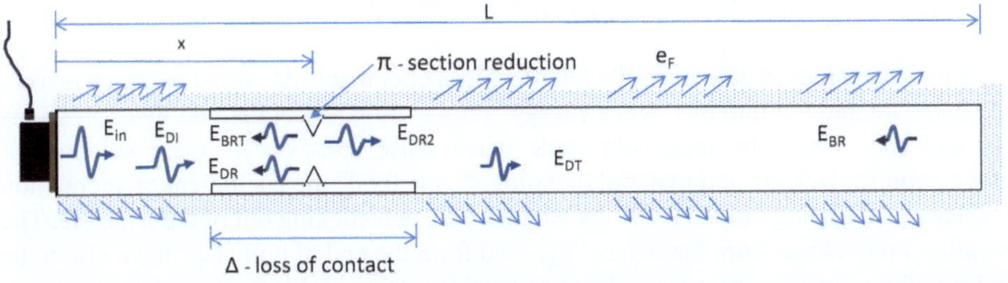

Fig. 9. Transmission, reflection, and leakage of the ultrasonic wave pulse through a bolt

reduction « π » of the defect, greater is the defect reflection "E_D" and smaller is the end reflection "E_B". On the other hand, greater is the length of loss of contact "Δ" with little section reduction and greater is the end reflection.

4.2 Wave Energy Attenuation

Wave attenuation is caused by material damping, leakage from refraction of the steel bolt to the concrete and dispersion due to reflection from the ribs, the filets, the bolt curvature and the end. Attenuation as a whole was modelled to be proportional to the amplitude of the "wave energy E(x)" propagating along the bolt and being reflected back. Energy loss "e_F" was modelled through a length attenuation factor "λ" as follows:

$$e_F(x) = \lambda \cdot E(x) \tag{4}$$

Knowing the attenuation mechanism, the energy "E(x)" at a point "x" can be expressed as the decay of the energy "Ei" at a starting point "xi" as follows:

$$E(x) = E_i \cdot e^{-\lambda \cdot (x - x_i)} \tag{5}$$

4.3 Guided Wave Energy Propagation Model

Using the decay principle, the measured defect reflection was expressed in terms of the coefficient of reflection "C_D". It represents an index of cross-section reduction. It was formulated as the ratio of the incoming wave energy to the defect and its reflected part. It is expressed in terms of the wave energy emitted from and reflected back (from the defect) to the sensor, as follows:

$$C_D = \frac{E_D}{E_{IN}} \cdot e^{2\lambda \cdot \left[x - \delta \cdot \left(1 - \lambda_0/\lambda\right)\right]} \tag{6}$$

Based on the energy balance expression (3), a mathematical expression was developed to predict the wave energy "E_B" reflected from the end of bolt, and returning back to the sensor, having gone through the defect one way and the other and having undergone attenuation:

$$E_B = E_{IN} \cdot (1 - C_D)^2 \cdot e^{-2\lambda \cdot \psi \cdot \left(L - \Delta\left(1 - \lambda_0/\lambda\right) - \delta\left(1 - \lambda_0/\lambda\right)\right)} \tag{7}$$

This model enables to determine the length of loss of contact "Δ". Input parameters have to be known and fed into the "wave energy" propagation model defined by expressions (6) and (7): "E_{IN} " the input ultrasonic wave pulse energy, "λ" the coefficient of attenuation for leakage and material damping, "$\lambda_0 = 0.88$" the coefficient of attenuation for material damping only and "δ" the length of the bolt sticking out of the concrete. The measured reflections from the defect "E_D" and from the end of bolt "E_B" have also to be fed into the model in order to evaluate section reduction and loss of contact.

A corrective function "Ψ" was introduced into the model to take into account nonlinear effects related to reflections and apparent loss of contact at section reductions.

A corrective function "Ψ" was obtained through a process of data correlation between predicted and measured values of "E_B". For an improved prediction, corrective functions "Ψ" were characterized separately on one hand for tapered section reduction (and loss of contact), and on the other for crack defects:

- for tapered section reduction "Ψ":

$$\psi = -a \cdot Ln\left(C_B \cdot \sqrt{C_D}\right) + b \tag{8}$$

- for cracks "Ψ":

$$\psi = c \cdot \left(C_B \cdot \sqrt{C_D}\right) + d \tag{9}$$

5 Prediction of Section Reduction

The coefficient of reflection of the defect « C_D » provided a good index for section reduction, as better alternative to the time interval between reflection "trails".

A regression function was established between the coefficient of reflection of the defect « C_D » and the reduction in diameter. It is independent of the distance of the defect and the length of bolt. The regression function is different for tapered section reduction and loss of contact than for cracks.

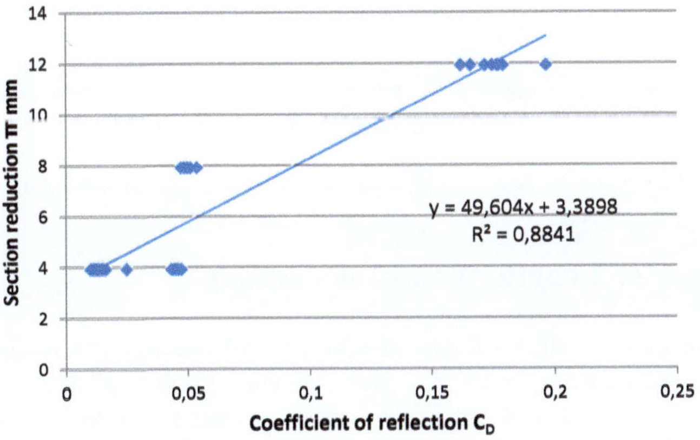

Fig. 10. Correlation between section reduction "π" and coefficient of reflection "C_D", for tapered reduction and loss of contact

The correlation for tapered section reduction (and loss of contact) between the reduction in diameter "π" and the coefficient of reflection "C_D" is linear with a regression coefficient of "$R^2 = 0.88$" (Fig. 10). It is based on three section reductions (4, 8 and 12 mm). It is of the form:

$$\pi = 49.6 \cdot C_D + 3.39 \tag{10}$$

Cracks distinguish themselves from the other types of defects with an abrupt reduction in section and virtually no length of loss of contact.

In predicting crack depth, the regression was established on only two section reduction (8 and 12 mm).

The correlation for cracks between the reduction in diameter "π" and the coefficient of reflection "C_D" is linear with a regression coefficient of "$R_2 = 0.94$" (Fig. 11). It is of the form:

$$\pi = 43.44 \cdot C_D + 1.80 \tag{11}$$

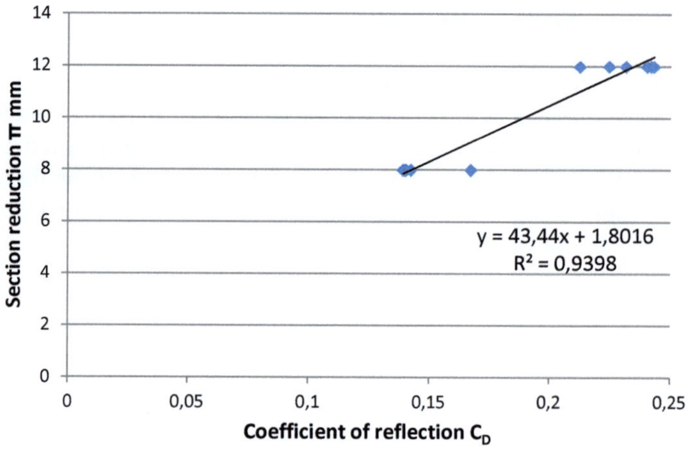

Fig. 11. Correlation between section reduction "π" and coefficient of reflection "C_D", for crack

6 Prediction of Length of Loss of Contact

The prediction of the length of loss of contact "Δ" requires the recording of both reflected wave energy from defect "E_D" and from end of bolt "E_B". It is evaluated using the expressions (6) and (7) of the wave energy propagation model in a convergence algorithm that reduces the spread between the measured and predicted values of end reflections « E_B » as the algorithm converges to the proper "Δ". The loss of contact correlation between actual and estimated values is a linear function with a coefficient of correlation of "$R^2 = 0.94$ (Fig. 12).

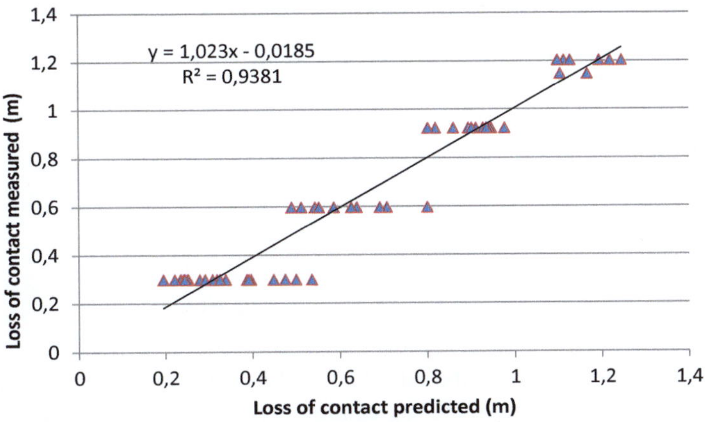

Fig. 12. Correlation between measured and predicted loss of contact "Δ"

7 Differentiating Cracks from Tapered Reductions

One has to be able to differentiate between a reflection from a tapered or a crack section reduction, in order to provide better prediction. Therefore, it was necessary to use the proper predictive functions (10) and (11) respectively according to the type of section reduction.

The signature of the reflections from cracks were differentiated from those of tapered reduction (and loss of contact). It was noticed that the trails that followed the head of the reflected wave diminished faster in time and monotonically for a crack than that for a tapered reduction reflection (Fig. 13).

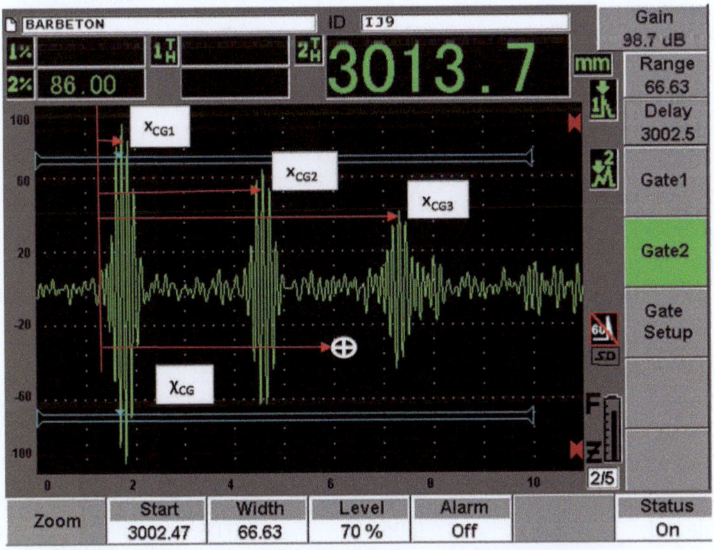

Fig. 13. Reflections from 12 mm crack at 3 m from probe

The procedure to discriminate cracks from tapered reduction involved recording the absolute values of each amplitude $|a_k|$ of the reflection and computing the lever arm "x_k" relative to the head of the reflection. Then a so-called center of gravity "χ_{CG}" was calculated based on the sum of the amplitudes; similar procedure was applied to the power of the amplitudes a_k^2:

$$\chi_{CG} = \frac{\sum_{k=1}^{m} x_k \cdot a_k}{\sum_{k=1}^{m} a_k} \text{ and } \frac{\sum_{k=1}^{m} x_k \cdot a_k^2}{\sum_{k=1}^{m} a_k^2} \tag{12}$$

It was found that by computing the ratio "ρ" of the centers of gravity of amplitude to amplitude square could improve the segregation between crack and tapered reduction:

$$\rho = \frac{\sum_{k=1}^{m} x_k \cdot a_k \Big/ \sum_{k=1}^{m} a_k}{\sum_{k=1}^{m} x_k \cdot a_k^2 \Big/ \sum_{k=1}^{m} a_k^2} \tag{13}$$

This computational procedure was applied to the signal of the reflection for the different types of reflections: end of bolt, tapered reduction, loss of contact and crack. The ratio "ρ" of the centers of gravity of amplitude to amplitude square are presented as example for a 3.5 m long bolt in Table 2. Crack reflections present a ratio "$\rho \leq 0.60$" as highlighted in yellow. Trend confirmed from the data of other bolt lengths.

Table 2. Center of gravities "χ_{CG}" and ratios "ρ" for different reflections on 3.5 m bolt

Bolt number	Reflection point	Distance of point (m)	Center of gravity amplitudes	Center of gravity amplitudes squarred	Ratio ρ of C.G.
2	8 mm crack	1	0.025	0.015	0.6
2	bolt end	3.5	0.051	0.052	1.02
2	12 mm crack	1	0.018	0.007	0.39
3bis	12 mm crack	1	0.018	0.007	0.39
3	bolt end	3.5	no reflection	no reflection	N/A
3bis	bolt end	3.5	no reflection	no reflection	N/A
4	4 mm tapered	1	0.065	0.062	0.95
4	bolt end	3.5	0.036	0.037	1.03
5	8 mm tapered	1	0.018	0.02	1.11
5	bolt end	3.5	0.037	0.034	0.92
6	12 mm tapered	1	0.055	0.049	0.89
6bis	12 mm tapered	1	0.089	0.098	1.10
6	bolt end	3.5	0.06	0.058	0.97
6bis	bolt end	3.5	0.065	0.063	0.97

8 Differentiating Defect from End of Bolt Reflections

It is important to determine whether the last reflection is coming from the end of the bolt or from a defect (crack or tapered reduction). This situation can arise whenever the defect is deep enough as a function of distance from the bolt tip so to block the transmission to and reflection from the end of the bolt. To determine the nature of the last reflection it is advised to check whether the distance of the reflection matches the length of bolt according to "design plans". If it is not the case, then one has to proceed by evaluating the source of the reflection.

8.1 Coefficient of End Reflection "C_{RB}"

To evaluate the source of the last reflection, a coefficient of end reflection "C_{RB}" was established. It was defined as the ratio of the incident wave energy "E_{IB}" to the reflected wave energy "E_{RB}" at the presumed source of reflection, as follows:

$$C_{RB} = \frac{E_{RB}}{E_{IB}} \tag{14}$$

Where $E_{RB} = E_B \cdot e^{+\lambda \cdot l}$ and $E_{IB} = E_{IN} \cdot e^{-\lambda \cdot l}$;

"E_{IN}" probe input wave energy; "E_B" measured bolt end reflected energy; "λ" attenuation coefficient; "l" distance recorded to source of reflection.

Incorporating expressions for "E_{RB}" and "E_{IB}" into (14) and taking into account the end of bolt "δ" sticking out of the rock, "C_{RB}" is expressed as:

$$C_{RB} = \frac{E_B}{E_{IN}} \cdot e^{2\lambda\left[l - 2\delta\left(1 - \left(\lambda_0/\lambda\right)\right)\right]} \tag{15}$$

If there is an intermediary reflection from a defect, then expression (13) is modified into:

$$C_{RB} = \frac{E_B}{E_{IN}} \cdot \frac{1}{\left[1 - C_D\right]^2} \cdot e^{2\lambda\left[l - 2\delta\left(1 - \left(\lambda_0/\lambda\right)\right)\right]} \tag{16}$$

8.2 Evaluating the End Reflection

A criterion on the source of the last reflection recorded was formulated based on the coefficient of end reflection "C_{RB}" to discriminate reflections of bolt end from those of major defect. This discrimination can be seen in the values of "C_{RB}" (Fig. 14).

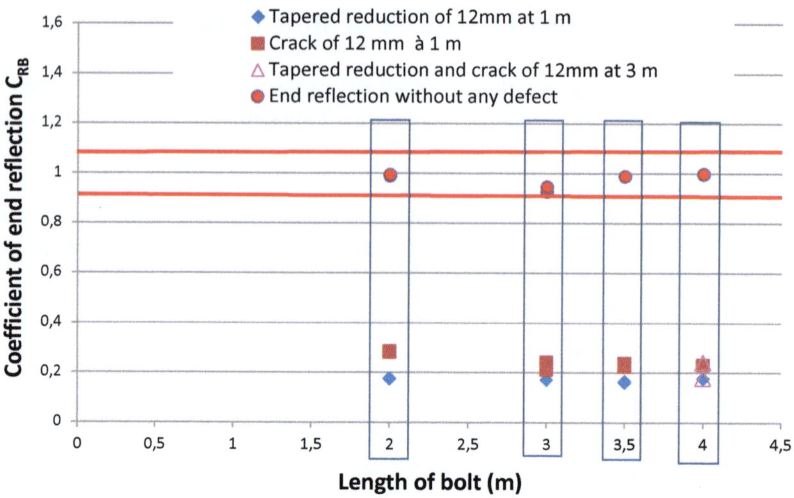

Fig. 14. Coefficient of end reflection "C_{RB}" for end of bolt and major defects

Total reflection from a bolt end without defect presents a coefficient of end reflection "$C_{RB} \sim 1.0$" close to unity. It is represented by red dots that represent end reflections for reference bolt #1 (without defect) that was sequentially shortened. These values vary within a range of "$0.90 < C_{RB} < 1.10$", identified by the two red lines.

Reflections that came from major defects showed low values of "C_{RB}" lying below a 0.30 threshold value, in the range of "$0.18 \leq C_{RB} \leq 0.30$". These major defects were characterized by 12 mm deep cracks (bolt #3) and 12 mm tapered section reduction (bolt #6). These low values of "C_{RB}" were used as a criterion to distinguish between reflections of end of bolt from that of major defects.

A greater uncertainty may arise when evaluating the end of bolt reflection. The possible presence of an undetected defect called "ghost defect" lying between the probe and the point of last reflection may increase the value of the coefficient of end reflection "C_{RB}", as explained in the next chapter.

When values of "C_{RB}" lye in the range of "$0.30 < C_{RB} < 0.90$", the test results indicate that the loss of contact from the presence of an undetected defect "ghost defect" artificially increases the value of the coefficient of end reflection "C_{RB}". If not for the presence of the "ghost defect", the "C_{RB}" values would otherwise lye in the range $0.18 \leq C_{RB} \leq 0.30$, characteristic of a major defect. Also, when values of "C_{RB}" lye above the level of "$C_{RB} = 1.10$", the presence of an undetected "ghost defect" enhances the coefficient of bolt end reflection "C_{RB}" from the range of "$0.90 < C_{RB} < 1.10$", characteristic of bolt end.

9 Identification of a "Ghost Defect"

9.1 Effects of a "Ghost Defect"

A "ghost defect" has no detected reflection recorded by the ultrasonic probe. If there is any reflection, it is so weak that it is buried in the noise level. It lies between the ultrasonic probe and the last reflection point. It is characterized by any given length of loss of contact for which the section reduction is so small that it doesn't give out measurable reflections. However, it has an effect on the reflection of the more distant defects or bolt end. The loss of contact of the "ghost defect" reduces the "leakage" of the wave energy propagating through and in turn raises the measured value "E_B" of the following reflection and consequently of "C_{RB}". Therefore, the "ghost defect" can be established whenever the coefficient of end reflection "C_{RB}" lies outside of the following limits:

- $C_{RB} \leq 0.30$, a major defect reflection;
- $0.90 \leq C_{RB} \leq 1.10$, bolt end reflection without defect.

"Ghost defects" were characterized in the present study by a 4 mm reduction in diameter set at 85 cm from the measured end and having the following profiles: a tapered section reduction over 15 cm on each side of the maximum reduction (bolt #4); a constant section reduction over three different lengths of loss of contact (bolts #9, 10, 11).

Tests were undertaken to evaluate the effect of the length of loss of contact on the coefficient of end reflection "C_{RB}" Fig. 15. Their effects were evaluated on the single reflection recorded that came from the end of bolt.

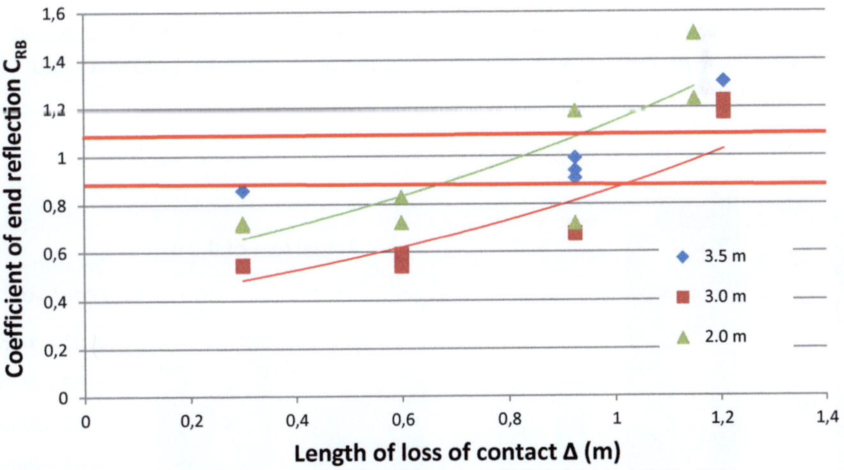

Fig. 15. Coefficient end reflection "C_{RB}" vs. loss of contact "Δ" for section reduction ≤4 mm

It was found that the "C_{RB}" values increased roughly by *0.7*, as the length of loss of contact increased from 0.3 m to 1.2 m; from an initial range of (0.5 to 0.8), due to variable bolt lengths, "C_{RB}" values reached levels of (1.2 to 1.5).

Therefore it is a possibility that a major defect with a coefficient "C_{RB} < 0.3" can be raised to a value in the range of 0.9 by the presence of a "ghost defects".

9.2 Length of Loss of Contact of a "Ghost Defect"

In order to estimate the length of loss of contact "Δ" of a "ghost defect", the expression (7) was rearranged to express "Δ". The coefficient of reflection "C_D" had to be determined.

To do so, the following hypothesis was formulated that the maximum diameter reduction for a defect to be undetected is no greater than "π = 4 mm". Hence, a conservative value for "C_D" can be obtained by applying this 4 mm reduction to expression (10) for tapered section reductions:

$$C_D = \frac{\pi - 3.39}{49.6} \tag{17}$$

Then, plugging the numerical value of "$C_D = 0.0123$" in expression (7) and rearranged it, the expression of "Δ" is obtained:

$$\Delta = \frac{L}{(1 - \lambda_0/\lambda)} - 2 \cdot \delta + \frac{1}{2 \cdot \lambda \cdot \Psi \cdot (1 - \lambda_0/\lambda)} \cdot Ln\left(1.02 \cdot \frac{E_B}{E_{IN}}\right) \tag{18}$$

The capacity of expression (18) to predict the length of loss of contact of a "ghost defect" was evaluated. The predicted lengths of loss of contact of the "ghost defect" were compared to the actual length of loss of contact of 0.3 m for two types of tapered reductions: the 4 mm and were added the 8 mm tapered section reductions Fig. 16. The actual length of loss of contact of "0.3 m" is indicated by a dotted line. Except for the 4 m bolts, there was good agreement between predicted and actual

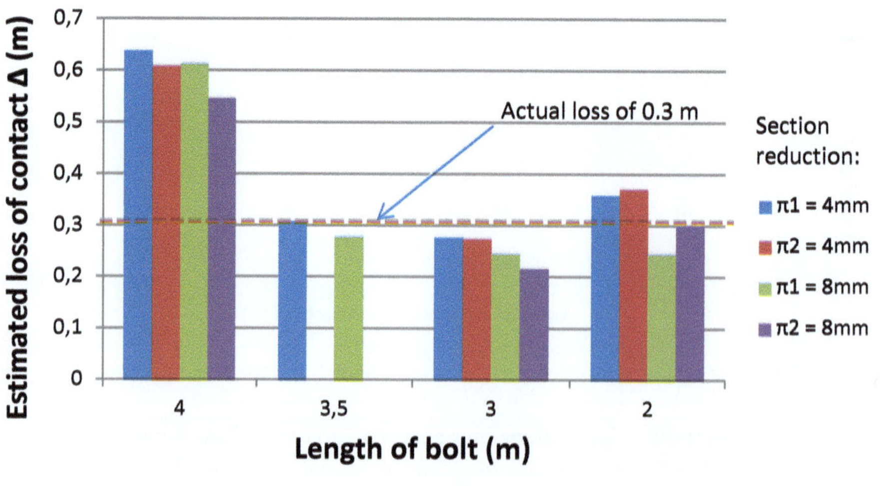

Fig. 16. Predicted loss of contact of "ghost defect" compared to actual length of 0.3 m for tapered reductions for different lengths of bolts

length of loss of contact. Measurements were repeated twice all through testing to be able to insure good measurements.

9.3 Procedure to Determine a "Ghost Defect"

In this procedure, the first step is to evaluate whether we have an end of bolt reflection. For such a case, the coefficient of end reflection "C_{RB}" has to lie within the range of "$0.90 \leq C_{RB} \leq 1.10$" or the point of reflection has to correspond to installation lengths.

If this first criteria is not met, the following step is to see whether the "C_{RB}" is lying in the range $C_{RB} \leq 0.30$", values characteristic of a major defect. In this the case, the reflection comes from a major defect which blocks the end of bolt reflection.

In the situation where the two previous steps have been dismissed, one should be confronted with "C_{RB}" value lying in one of the two ranges, either "$0.30 < C_{RB} < 0.90$" or "$C_{RB} > 1.10$". This is the indication of the presence of a "ghost defect" lying before the point of reflection. The first range characterizes the influence of the "ghost defect" on a major defect. The second range enhances the value of the coefficient of end reflection "C_{RB}" for the case of a bolt end reflection.

10 Inspection and Replacement Criteria

Ideally, bolt strength and corrosion should both be evaluated through a non-destructive procedure. Unfortunately, the non-destructive indices of the state of mechanical stress are often buried in the imprecision of the non-destructive measurement or within the effects of corrosion. However, corrosion is the main factor for the reduction of bolt holding strength along with the development of cracks. Therefore, by evaluating the corrosion and the presence of crack it becomes possible to evaluate the residual holding strength of the bolt, knowing in effect that ultimate strength of steel does not change.

10.1 Allowable Section Reduction

For "SDR" bolts used in this study, the bolt strength for a 25 mm diameter steel anchor is rated for a minimum ultimate tensile strength of "$F_U = 211$ *kN*" and for an elastic capacity of "$F_E = 156$ *kN*". At installation, the bolt is torqued to a service preload "F_S" of *65%* of a nominal strength "F_N" set by the manufacturer:

$$F_S = 65\% \cdot \sigma_N \cdot \frac{\pi \cdot D^2}{4} \tag{19}$$

The expression (19) corresponds to the stress relationship between for nominal bolt stress "σ_N" and service load stress "σ_S" at nominal bolt diameter "D":

$$\sigma_S = \sigma_N \cdot 0.65 \tag{20}$$

It can be argued that the 35% reserve in strength from nominal can be applied to the increase in stress from the service stress "σ_S" to the nominal stress "σ_N" due to a section reduction "d", while keeping the service load constant at *65%* of nominal load:

$$F_S = \sigma_N \cdot \frac{\pi \cdot d^2}{4} \tag{21}$$

Then the allowable reduced diameter "d" is obtained by setting the ratio of "F_S" (21) over "F_N" at 65%:

$$d = D \cdot \sqrt{\frac{F_S}{F_N}} \tag{22}$$

The maximum allowable diameter reduction "π_{Max}" is established as follows:

$$\pi_{Max} = D - d = D \cdot \left[1 - \sqrt{\frac{F_S}{F_N}}\right] = D \cdot \left[1 - \sqrt{0.65}\right] \tag{23}$$

The value of "π_{Max}" is rounded off to **"π_{Max} = *4.0 mm*"**.

10.2 Sufficient Holding Length of the Steel-Grout Interface

As a holding safety contingency, the steel-grout interface of the bolt can take up the holding strength of either the nut/plate or the expansion locking shell of the bolt, in the case they are too corroded. The holding strength of the steel-grout interface is based on the shear strength of the grout. This shear strength is of the order of **"**τ = *3 MPa***"**. For a given bolt diameter (D = 25 m), the necessary holding length "l_H" of the steel-grout interface due to shear can be calculated to match the elastic load strength "F_E = *156 kN*" of the bolt, in a conservative approach:

$$l_H = \frac{F_E}{\pi \cdot D \cdot \tau} \tag{24}$$

The value "l_H" is rounded off to the value **"**l_H = *0.70* m**"**.

10.3 Defining Acceptable Length of Loss of Contact

To evaluate the acceptable length of loss of contact "Δ" of the steel-grout interface, one has to consider the holding length of the steel-grout interface "l_H" in the context of a shear crack in the rock intersecting the bolt length.

If a bolt has no defect (section reduction or loss of contact) or corrosion of the nut/plate then the eventual presence of a shear crack in the rock will have no bearing on the inspection. The bolt will fill its holding function (unless the shear crack lies beyond the bolt length).

If there is presence of a defect along with that of a shear crack in the rock, they will both have a bearing on the acceptable length of loss of contact as a function of their relative positions to one another. A schematic of bolt with defect and shear cracks at two case study locations "A" and "B" in the rock is presented in Fig. 17.

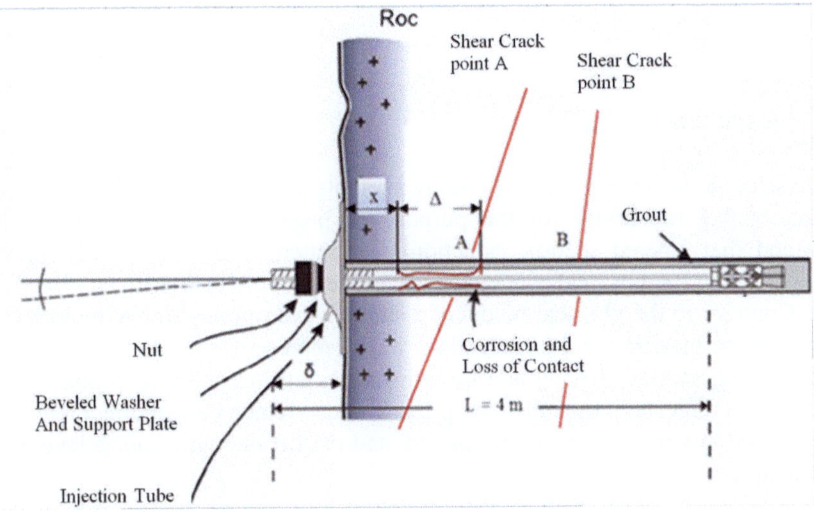

Fig. 17. Schematic of bolt, defect and two shear crack in the rock

If the shear crack in the rock intersects the bolt at location "A" (Fig. 17) along the bolt segment between the nut/plate and the defect, the contact length of the bolt beyond the defect can do little to prevent the block from falling, except for the holding strength of other adjacent bolts. In this case, it is pointless to define an acceptable length of loss of contact.

One has to understand that a defect has to be considered as an eventual source of failure. But the rock/grout around the defect can pick up the holding strength as a parallel link of the bolt grout rock holding system. Therefore, if a shear crack in the rock breaks the grout/rock holding link between the nut/bearing plate and the defect then failure occurs in terms of loss of holding strength of the bolt system.

On the other hand, if a shear crack in the rock runs across the segment of bolt beyond the defect, at "B" (Fig. 17), then there might be enough grout holding length "l_H". The minimum holding length "l_H" would have to lie beyond the defect but before the shear crack at "B" in the rock. As such the block of rock that would possibly slip away from location "B" is held by the minimum holding length "l_H" of grout to the bolt and its anchor shell deeper in the rock.

The families of shear cracks in the rock are known before installation of rock anchor bolts, as a geotechnical construction practice. Therefore, it is possible to estimate the position of intersection "lj" of the rock shear crack with the bolt.

The section reduction of a bolt due to corrosion can be mitigated by the holding capacity of grout which becomes a parallel holding link. The distance between the position of the defect "x" with its loss of contact "Δ" and that of the shear crack has to encompass the minimum holding length "l_H" of grout. This is expressed in the following formula:

$$l_J - x - \Delta \geq l_H, \text{ where } l_H = 0.70\,\text{m} \tag{25}$$

11 Discussion

The physical model developed in this study attempts to predict the actual measurements with reasonable reliability, for the purpose of inspection. However, it has to be understood that tapered section reduction is an attempt to represent the effects of corrosion. Also, the physical model for wave energy propagation is an oversimplification. Contrary to the physical model of a single wave running down the longitudinal axes, the probe measurement of reflection "trails" represents the contribution of a sum of different trajectories arriving at a same instances in time onto the probe.

The measurements of the reflection present some variation depending on the quality of the surface to which the probe is applied, and on the amount of noise level burying the reflection.

The degree of imprecision in the prediction was minimized by varying the following parameters: depth of cross-section reduction, length of loss of contact, position of defect and length of bolt. The physical model has acquired robustness by correlating it to the variation of these different parameters.

It was necessary to introduce into the physical model of the wave energy propagation a corrective function "Ψ" in order to obtain good correlation between the wave energy values of the measured reflections and the parameters of section reduction and loss of contact. The use of a corrective function compels us to revert in the future to numerical analysis of wave propagation to better understand the phenomenon of the reflection at a defect and the non-linearity's in the wave propagation and attenuation in order to improve the physical model.

Also, different factors can affected the measurements: signal saturation due to inappropriate instrument signal gains, sampling frequency of 25 MHz not high enough with respect to the probe central frequency of 3.5 MHz, noise masking the reflection, surface roughness, curvature or angle, bow along the length of the bolt, different lengths and diameter of bolt, bonding condition at interface grout-bolt, temperature, etc. Actual corrosion in bolts was not corroborated from data recorded in two field measurements undertaken in subterranean galleries. A tapered section reduction (over ± 15 cm) was adopted to simulate a corroded section. A real corroded section would attenuate more the trails of the reflection making it more difficult to measure corrosion at a distance.

The measured wave energy reflection was based on the integration in time of the amplitude of the signal (including the trails) above the noise level. Initial evaluation indicated that the integration of the amplitude over the square of the amplitude had a small edge. However, this evaluation did not encompass the different lengths of bolts. Therefore, we could go through the whole process of analysing the data and establishing the correlations with respect to the square of the amplitude of the signal.

To improve upon the prediction of the section reduction, we could have computed the time interval between trails and combined it with the coefficient of reflection of defect "C_D".

Spectral frequency analysis was not undertaken on the reflections. This analyses needs to be look at further. An initial analysis was started but not pursued. It seems promising in establishing a general rating on the degree of corrosion. Statistics on the spectral frequency distribution of a reflection could prove useful in quantifying the degree of corrosion.

12 Conclusion

This research has led to a successful application of a non-destructive technique based on ultrasonic wave propagation for the evaluation of defects in rock bolts. It has led to the development of a predictive physical model of the wave energy propagation and reflection applied to anchor bolts acting as waveguides. This model permitted to determine section reduction, loss of contact of a defect and the presence of "ghost defects". Also, the ultrasonic wave propagation technique used was shown to determine the length of bolts up to 4 m and the location of defects.

The ultrasonic non-destructive inspection method put forth presents a predictive reliability within an estimated range of the order of 15%. It shows a coherent prediction for any given change in the variables. The end reflections of 4 m bolts without defects seem the limit with the present instrumentation and signal analysis. Defects less than 4 mm in section reduction "ghost defects" presented no detectable reflections but would increase the end of bolt reflection.

Three innovative procedures were developed to identify and segregate three different types of anomalies. Corrosion type defects (modelled by tapered reduction) were differentiated from crack defects by computing a "center of gravity" of the amplitudes of trails of the reflection. Reflections from major defects were differentiated from reflections of end of bolt by formulating and computing a "coefficient of end reflection". Finally, an approach was developed to identify the presence of a "ghost defect" and to determine its length of loss of contact and the influence it has on the end of bolt reflection.

For improvements, one would have to consider the following actions: adjust the method on actual corroded bolts; undertake a numerical analysis study to explain the corrective function; undertake spectral frequency analysis and relate it the an index of section reduction; improve upon the prediction of section reduction by combining time interval between trails to the coefficient of reflection of defect "C_D"; investigate the results based on the square of the amplitude instead of the amplitude of the reflection.

We have not address the following; different diameter bolts, the state of mechanical stress in the bolt, effect of bolt end conditions, effect of temperature.

Nonetheless, a successful inspection method to determine anchor bolt corrosion has been put forth encompassing several inspection parameters.

References

1. Taras, A.: Mesures aux ultrasons et Impact-écho sur des boulons de consolidation ciblés à l'arrachement à Manic5 PA et des boulons creux à Manic2. Hydro Québec Report, Varennes, Québec (2014)
2. Taras, A.: Essais réalisés aux ultrasons de 2014 à 2015. Hydro Québec Report, Varennes, Québec, (June 2015)
3. Taras, A.: Mesures aux ultrasons sur des boulons de consolidation à Manic5 PA et Manic3. Hydro Québec Report, Varennes, Québec (Febuary 2015)
4. Ervin, B.: Monitoring or rebar embedded in mortar using high frequency guided ultrasonic waves. J. Eng. Mech. 9–16 (January 2009)
5. Shoji, M.: Guided Wave testing of cylindrical bars. In: 18th World Conference on Nondestructive Testing (2012)
6. Beard, M., Lowe, M., Cawley, P.: Inspection of steel tendons in concrete using guided waves. In: American Institute of Physics Conference (2003)
7. Beard, M., Lowe, M.: Non-destructive testing of rock bolts using guided waves. Int. J. Rock Mech. Min. Sci., 527–536, (2003)

Effects of Measuring Techniques on the Accuracy of Estimating Cable Tension in a Cable-Stay Bridge

Viet Long Ho[1,4](✉), Thanh Nam Hoang[5], Guido De Roeck[3], Tien Thanh Bui[2], and M. Abdel Wahab[4]

[1] University of Transport and Communications, Campus in Ho Chi Minh City, Ho Chi Minh, Vietnam
HoViet.Long@ugent.be
[2] University of Transport and Communications, Hanoi, Vietnam
[3] Department of Civil Engineering, KU Leuven, 3001 Leuven, Belgium
[4] Soete Laboratory, Faculty of Engineering and Architecture, Ghent University, Ghent, Belgium
[5] Department of Science and Technology, Ministry of Transport, Hanoi, Vietnam

Abstract. Nowadays, many methods have been employed to estimate the cable tension in a cable-stayed bridge. In these methods, the non-destructive test has a widespread application thanks to its feasibility and effectiveness. The vibration-based approach determines the cable tension based on natural frequency. In fact, there are many factors that can influence the tension values. Some of them are derived from objective factors, others from subjective factors. In this paper, a subjective issue, so-called measurement technique, is studied. A cable, without fillings between the HDPE duct and strands, is the main object of this study. In the first step, some scenarios related to the placement of sensors are conducted. The former considers the tension between different positions of sensors, directly and indirectly contact conditions with strands, under the ambient and pulling-rope excitation. The latter takes into account the directions, out-of-plane and in-plane, of sensor placement on the strands or HPDE duct. In the next step, the relationship between the force and frequency is utilized to identify the mean tension of the cable. In the finally step, results of cable tension estimation are compared with the real-time monitoring and finite element (FE) simulation. From the comparison, the effects of measuring technique in situ are discussed in the conclusion.

Keywords: Cable tension · Kien bridge · Non-destructive test · Vibration · Dynamic measurement · Finite element analysis

1 Introduction

One of the most important activities in cable-stayed bridge inspection is to estimate tension of cables from construction to operation time. The monitoring of tension reduction of cables is indispensable because the contribution of the stay cables to the

M. A. Wahab (Ed.): DAMAS 2019, LNME, pp. 433–445, 2020.
https://doi.org/10.1007/978-981-13-8331-1_31

overall resistance of the structure is extremely significant. After a long-term operation or an incident, the concern of the cable tension becomes more obvious and considerable. Based on the variation in tension, the managers have more clues to assess the behavior of the whole bridge then make a timely decision on bridge maintenance. Currently, Genoa bridge, the A-frame towers, which support the concrete-encased stay cables, collapsed in August 2018. Due to the effects of salted water, corrosion of the cables hidden inside the concrete can lead to damaged strands. The disaster could be prevented if the cables were received more attention.

Nowadays, many direct and indirect measurement methods are employed to estimate tension of cables in the operating stage. So far, researchers [4, 7, 8] have made use of these methods for estimating tension. Strain gauges and load cell (lift-off test) represent the direct measurement approach. These methods require a direct contact with strands. While the former is very sensitive to temperature changes and isolated epoxy-coat, the latter is often applied in construction stage due to complicated implement in the operating stage. In contrast, elasto-magnetic sensor and vibration-based method do not need a direct contact. However, a dedicated winding device is required to deploy the elasto-magnetic sensor. Besides that, increasing temperature due to long measurement time can disturb the measured data. Meanwhile, vibration-based approach can identify the tension from dynamic characteristics of cable [11]. With dominant advantages like simple and low-cost implementation in situ, the method is applied worldwide. Cunha et al. [10] and Peeters et al. [9] pointed out that the vibration-based approach is a suitable choice thanks to cheapness, quickness and reliability in evaluating the cable tension. Starting with the idea, in this study, the dynamic responses were generated under ambient load and pulling-rope excitation. The frequency values can be identified as much as possible from dynamic responses to warrant an accurate estimation. A linear regression of frequency will be achieved via the least-square method. Then a cable tension is estimated based on the identified fundamental frequency.

To achieve an acceptable estimation, two crucial factors, namely cable sag and bending rigidity have to be taken into account [3, 5, 8–10]. In the paper, two scenarios are conducted to consider the effect of cable sag and bending stiffness. Sensors were placed in two directions, in-plane (vertical direction) and out-of-plane (horizontal direction). In data processing stage, the authors applied the proposed equations in [3]. The results will be discussed in the next section.

The studies in Refs. [4–11] didn't mention how to measure dynamic response of cables without fillings between the strands and HPDE duct. Therefore, in the study, two other scenarios are carried to find out the effect of fillings. In the first case, sensors were placed on HPDE duct. The other, HPDE duct was removed and the sensors were directly placed on strands. For the purpose to evaluate the influence of fillings on cable vibration, a FE model of cable was constructed in regard to real measurement. The results were compared with the data from real-time monitoring and measured data.

The paper layout includes six sections. Section 1 is introduction of current study on tension estimation. Section 2 presents the relationship between frequencies and tension based on proposed equations in [3, 5–7]. Section 3 describes the specific measurement campaign and processing data of cable vibration. Section 4 presents a step-by-step simulation of cable vibration. Sections 5 and 6 present results and conclusion, respectively.

2 Theoretical Methodology of Vibration-Based Method

In the early stage, the simplified taut string theory was employed to estimate the tension force as shown in Eq. (1). From the equation, if we have some parameters for instance mass of cable per unit length *m*, free length of cable *L* and the fundamental frequency f_k, the tension can be identified as:

$$H = 4.m.L^2.f_k^2. \tag{1}$$

In fact, tension force of cable is affected by cable sag and bending rigidity as well as the way to identify the first frequency. From the obtained results, some authors proposed a modified fundamental frequency to achieve a more accurate force.

In 1987, Morse and Ingard started to study the effect of bending stiffness on frequency of cable. The frequency in Eq. (1) is recalculated as in Eq. (2):

$$f_k = \frac{k}{2L}\sqrt{\frac{H}{m}}.\left(1 + 2\sqrt{\frac{EI}{HL^2}} + \left(4 + \frac{k^2\pi^2}{2}\right).\frac{EI}{HL^2}\right). \tag{2}$$

Casas [7] proposed that a frequency f_k^* can be estimated by an average value of natural frequencies as in Eq. (3). In the equation, f_i refers the ith frequency and *k* is the number of modes.

$$f_k^* = \frac{1}{k}\sum_{i=1}^{k}\frac{f_i}{i}. \tag{3}$$

After that, in 1996 Zui et al. [8] developed a modified equation that considered the cable sad and bending stiffness. In 1998, based on the study of Zui et al., Mehrabi [3] proposed an improved method. In the method, two main issues were mentioned. One is the difference of end conditions. The other is difference of in-plane and out-of-plane vibration modes while the two non-dimensional parameters ξ and λ^2 for bending stiffness and sag extensibility were consider [3][1]

$$\xi = L\sqrt{\frac{H}{EI}}. \tag{4}$$

$$\lambda^2 = \frac{\left(\frac{m.g.L}{H}\right)^2.L.EA}{H.L_e}. \tag{5}$$

Where *EI* and *EA* are the flexural and axial stiffness, respectively, *g* is the gravitational acceleration and

[1] Interested readers are recommended to consult Ref. [3] for more details.

$$L_e \approx L.\left(1 + \frac{\left(\frac{m.g.L}{H}\right)^2}{8}\right). \tag{6}$$

Equation (2) was rewritten based on the relationship between the measured frequencies f_k and the ideal frequency f_{1s}, i.e.:

$$\left(\frac{f_k}{f_{1s}}\right) = k.\alpha. \tag{7}$$

$$\text{with: } \alpha = 1 + \frac{2}{\xi} + \left(4 + \frac{k^2\pi^2}{2}\right).\frac{1}{\xi^2}. \tag{8}$$

The authors in Ref. [6] used a minimization algorithm Gauss-Newton to fit Eq. (7) to the measured data f_k. After fitting procedure, the ideal frequency f_{1s} and non-dimensional parameters ξ are achieved. Finally, the cable tension H is determined as:

$$H = 4.L^2.f_{1s}.m. \tag{9}$$

In this paper, Eq. (9) will be applied to estimate the tension. Fitness diagrams of frequency are introduced in the next section.

3 In Situ Measurement of Cable Tension

3.1 Introduction to Kien Bridge

Kien bridge is a cable-stayed bridge in Vietnam with four lanes, two lanes for traffic road and the others for pedestrians and cyclists. Main bridge is a 3-span continuous prestressed concrete, with span 85 m, 200 m and 85 m, respectively. The approach bridge includes 12 simply-supported spans (see Fig. 1). The width of the bridge is varied from 16.7 m for main bridge to 15.1 m for approach bridge.

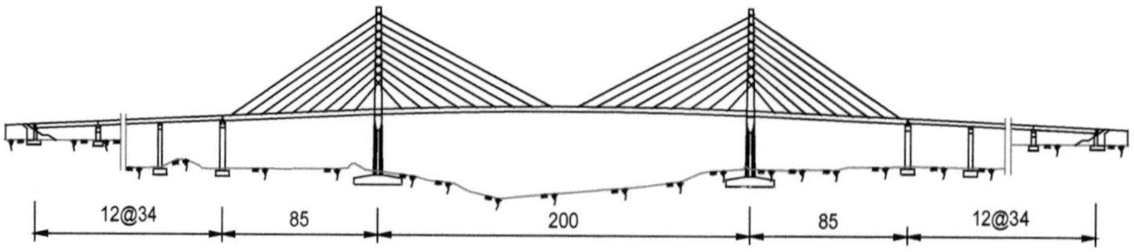

Fig. 1. Overview of Kien bridge

The 72 cables are diagonally stretched down from a H-pylon to support the main girder. Fan configuration is the arrangement of stay cables (see Fig. 2). The variation of the cable length is significant from 20 m to 103 m, at mid-span and pylon, respectively.

There are two types of tendon, one tendon consists of 37 strands of 15.2 mm diameter type, another has 48 strands. Each seven-wire strand is coated with high-density polyethylene (HDPE) as shown in Fig. 3. An interesting point is no filling material between protection duct and strands. Thanks to a periodic maintenance activity, authors have a great chance to study vibration of cables. In this case, dynamic response of cable will be considered when sensors are placed on protection ducts and on strands.

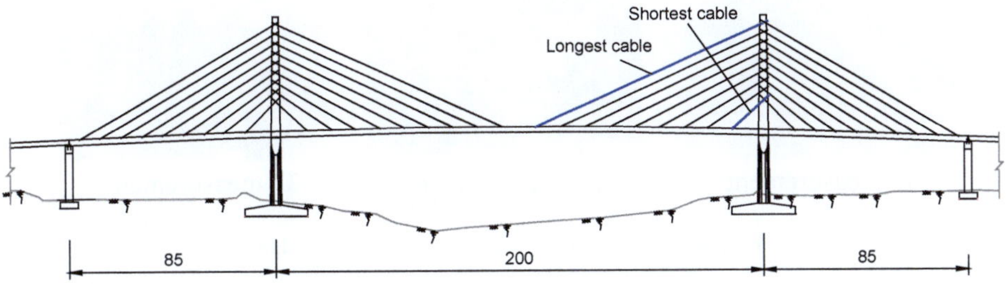

Fig. 2. Schematic drawing of fan-shaped stay cable arrangement and tested cables

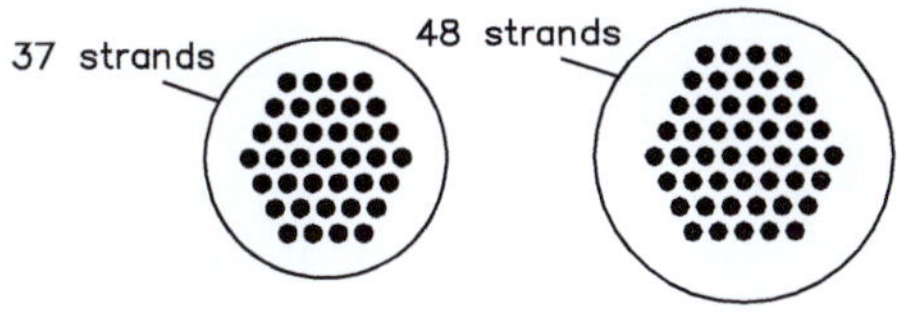

Fig. 3. Typical cross section of cable at anchorage

3.2 Field Measurement

In this measurement campaign, two cables were chosen to measure vibration, the longest and the shortest in the mid-span, as shown in Fig. 2. Each cable was measured three times with two scenarios. At first, vibration of cable stay was measured conventionally by measurement points on protection duct. Then, after removing the duct and anti-vandalism pipe, new measurement points were placed on exposed strands (see Fig. 4):

- Scenario 1: sensors were placed on the HPDE duct (1st measurement).
- Scenario 2: sensors were directly placed on strands, HPDE duct was removed and held by a triangle frame (2nd measurement) or supported by a wooden bar (3rd measurement).

Two one-dimensional accelerometers of PCB were employed for measuring the vibration responses of each cable in two directions, vertical (in-of-plane) and horizontal (out-of-plane). Positions of the mounting were chosen in the sidewalks because of accessibility. A clamp and magnetic bases were used to fix the position of sensors on the cable and on the duct (see Fig. 5).

1st measurement 2nd measurement 3rd measurement

Fig. 4. Measurement scenarios on Kien bridge.

a) On HPDE duct b) On strands

Fig. 5. Accelerometer placement. One accelerometer for out-of-plane (*y*-horizontal direction) modes and the rest one for in-plane modes (*z*-vertical direction)

Data acquisition was conducted at a sampling rate of 1651 Hz during fifteen minutes under ambient load, wind and vehicle passage and pulling-rope excitation. The data was collected and saved on a laptop. A Fast Fourier Transform (FFT) was employed to convert the measured signal to frequency domain, see Fig. 6. A frequency range of interest was chosen to perform from 0-20 Hz for clear observation.

From the frequency domain, higher modes can be achieved in pulling-rope excitation. The dynamic responses of in-plane modes are clearer as well. Besides that, in Fig. 6a, measured responses of sensor in *y*-direction is more dominant than in *z*-direction. This can be understandable because out-of-plane modes are not disturbed by the possible bridge modes. However, the dynamic responses in two directions are almost the same in Fig. 6b due to good excitation than ambient.

In the next step, the system model is performed by covariance-driven stochastic subspace identification method, so-called SSI-COV [1] and [2]. Stabilization diagram is constructed to estimate dynamic characteristics. Several criteria are input for a clear

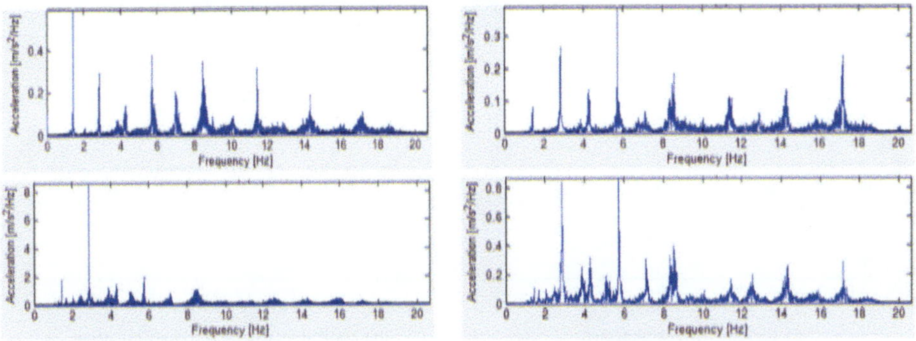

a) Ambient load (upper: out-of-plane, lower: in-plane)

b) Pulling-rope excitation (upper: out-of-plane, lower: in-plane)

Fig. 6. Frequency domain from measured data of longest cable

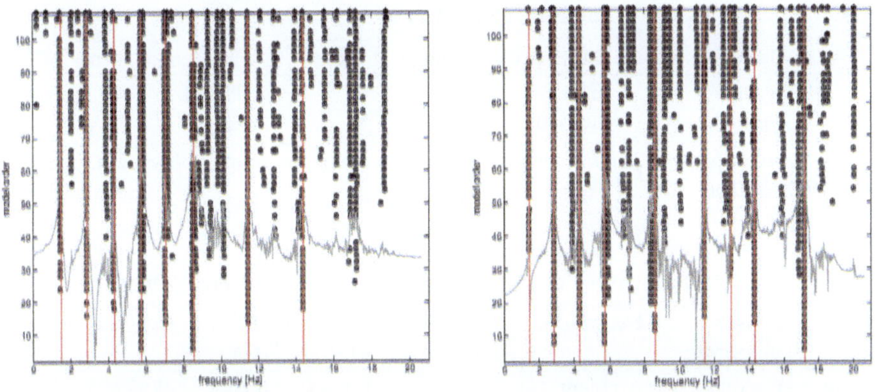

a) Under ambient load

b) Under pulling-rope excitation

Fig. 7. Stabilization diagram in an interval of 0–20 Hz. The stable poles are indicated with red dash lines. More out-of-plane modes are identified by pulling-rope.

stabilization diagram with 0.1%, 10% and 0.1% for frequency, damping factor and modal vectors, respectively. In Fig. 7, the stabilization diagram consists of clear vertical lines in an interval of 0 to 20 Hz.

Missing frequencies do not influence results too much if the number of identified frequencies is large enough. Then a deviation between two successive natural frequencies need to be calculated to assign mode order n to corresponding frequency (consulting Ref. [11] for a step-by-step procedure). With the obtained frequencies and corresponding order n, Eq. (7) was applied to find out the ideal frequency f_{1s}. Figure 8 shows the fitness between measured and calculated frequencies.

From as-built drawings and documents, the two left parameters m (kg/m) and free length of cable L (m) were determined as in Table 1. Finally, tension force was estimated by Eq. (9) and presented in Table 2. In the case of shortest cable, because of bad

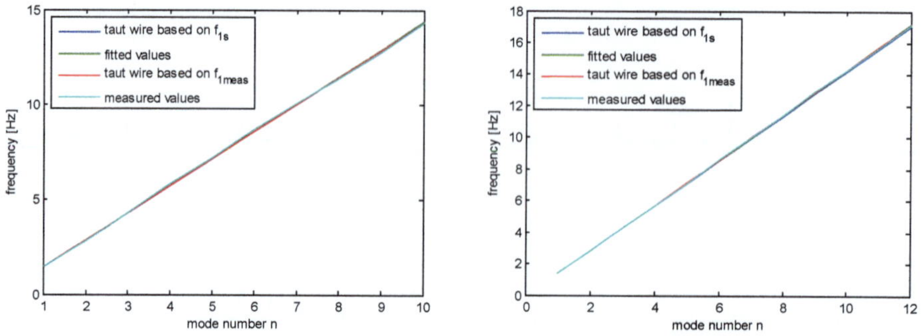

a) Under ambient load b) Under pulling-rope excitation

Fig. 8. Comparison between the measured and calculated frequencies of in-plane modes. The linear regression of frequency on the right-hand side is better than in the left-hand side thanks to the quality and number of identified modes

Table 1. Parameters of stay cables

Cable	No. strand	Mass per unit length m (kg/m) (*)	Free length L (m)
Longest	48	61.13	98.654
Shortest	37	49.019	14.507

(*): mass per unit length consisted of mass per unit length of cable as well as HDPE duct.

Table 2. Frequencies and tension forces of two measured cables

Cable	Sensor placement	Excitation	Sensor direction	f_{1s} (Hz)	Δ (%)	H (KN)	Δ (%)
Longest	On duct	Ambient	y	1.4284	0.154	4855.635	0.313
			z	1.4306		4870.85	
		Pulling rope	y	1.4219	0.007	4811.507	0.016
			z	1.4218		4810.727	
	On strand		y	1.4018	1.091	4676.704	2.186
			z	1.4171		4778.949	
			y	1.3968	0.995	4643.228	1.999
			z	1.4107		4736.062	
Shortest	On duct		y	8.3971		2909.654	
	On strand		y	8.2963		2840.173	

installation of sensors in z-direction, only measured data in y-direction was processed and shown in Table 2.

From Table 2 the maximum discrepancies of identified frequencies f_{1s} and estimated tensions of the longest cable between in-plane and out-of-plane modes are lower than 1.1% and 2.2%, respectively. The results reveal a good match between the two scenarios of sensor directions as well as under excitation load.

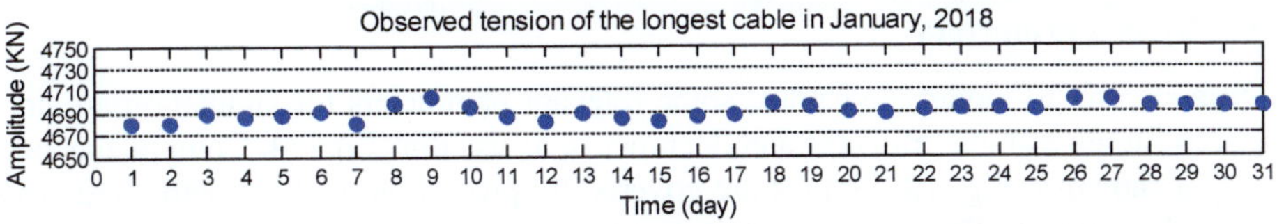

Fig. 9. Variation of tension force from 4670 kN to 4710 kN during January 2018 (average value of 4691 KN) (*source: The Road Engineering Center, Vietnam*).

From a real-time health monitoring system on the longest cable at the bridge, tension force was collected by the Road Engineering Center, Vietnam. The variation of force in the cable is plotted in Fig. 9. An average value of 4691 KN from real-time monitoring in January 2018 is used to compare with the measured tension. To facilitate the comparison, these measured data will be averaged according to the corresponding measurement scenario – on duct/on strands, as shown in Table 3. With the shortest cable, due to lack of real-time monitoring of shortest cable, frequency f_{1s} and measured tension force H are only compared for duct and on strands cases.

Table 3. Comparison of tension force in two scenarios

Cable	Sensors on	Frequency f_{1s} (Hz)		Tension H (kN)			
		Average	Δf	Average	ΔH	Realtime	ΔH
Longest	Duct	1.4257	1.36%	4837.180	2.73%	4691.097	3.11%
	Strands	1.4066		4708.736			0.38%
Shortest	Duct	8.3971	1.21%	2909.654	2.45%		
	Strands	8.2963		2840.173			

There is another issue during measurement, which is the influence of anti-vandalism pipe on estimating tension force. Especially, when the pipe is hung or held by a triangle frame or a wooden bar. Therefore, to have a general view of cable vibration in the paper, a FE model of cable is conducted in numerical model section. Then results obtained in Sects. 3 and 4 will be mentioned in the discussion section.

4 Numerical Model

In this step, a FE model of the cable is built to identify the behavior of cable when considering the mass of anti-vandalism pipe. The model does not simulate the gap between the strands and HPDE duct. It starts with an assumption that the strands and HPDE duct will behave as there are fillings between them. It means that they will have the same behavior even without fillings. Therefore, the HPDE duct will be assigned as a mass on strands. Positions of anti-vandalism pipe during the measurement also are considered as a varied mass along the strands. Hence, based on the differences between modelling and measured results, it can be pointed the effects of anti-vandalism pipe on vibration in the cables.

4.1 Input Parameters

Before starting to model the cables, some relevant parameters for simulation are collected from the as-built drawings and documents, as listed in Table 4. Components of a stay are shown in Fig. 10. Strands are covered by protection duct and anti-vandalism pipe in Fig. 10a. In Fig. 10b, strands are exposed after removing protection duct and anti-vandalism pipe by VSL Vietnam company.

Table 4. Input parameters in model

Component	α (degree)	*D-outer* (m)	*D-inner* (m)	Thick d (m)	Density (kg/m^3)	Mass m (kg/m)
Duct		0.2	0.17	0.015	950	8.282
Anti-vandalism pipe		0.7	0.68	0.01	7800	169.081
Shortest cable	24.89				7800	40.737
Longest cable	43.60				7800	52.848

a) Operation stage b) Removing the anti-vandalism pile

Fig. 10. Components of a stay

4.2 Modelling Scenarios

In general, when estimating the tension force as in Eq. (9), the mass unit per length is often calculated by the mass of strands and an equivalent mass of protection duct, ignoring the mass of anti-vandalism pipe. Therefore, to consider the real effect of mass of pipe, this paper proposes three scenarios, for which the details are shown in Fig. 11.

- Case study 1: This model only considers the mass of strands as well as the protection duct.

CS1: without anti-vandalsim pipe | CS2: anti-vandalism pipe in operational stage | CS3: Hanging anti-vandalism pipe

Fig. 11. Illustration of modelling scenarios

- Case study 2: This model considers the mass of anti-vandalism pile at the position in operation stage, together with the mass of strands and the protection duct.
- Case study 3: This model considers the mass of anti-vandalism pile at the position during the removing stage, together with the mass of strands and the protection duct. The anti-vandalism pile was hung at 1.5 m and 1.0 m height from the concrete base for longest and shortest cable, respectively as shown in Fig. 10b.

Prestressed load is very important in modelling. The load input in the model is an average value of 4691 KN as mentioned above, will be set up as the prestressed load of cable for longest cable. Unfortunately, there is no real-time monitoring on the shortest cable at bridge. Therefore, with studying purpose, measured tension of 2840.173 KN in Table 2, is considered as prestressed load. One end of the cable will be fixed, another is only allowed to move axially. The prestressed load is assigned at the movable end. To consider the prestress effects in a modal analysis, two load-steps are generated in the analysis. A static analysis will be performed first, then a modal analysis is conducted with prestress effects turned on.

With the boundary conditions and pre-stressed load as mentioned above, ANSYS 17 is employed to build the FE model. The first five frequency values of two types of cable are shown in Tables 5 and 6. A maximum discrepancy of frequency between CS1 and CS3 is also determined to emphasize the effect of anti-vandalism mass.

Table 5. The simulated frequency values and discrepancies of the longest cable, f and Δf (Hz)

Mode	Frequency f (Hz)			
	CS 1	CS2	CS3	Δf (CS1–CS3) (Hz)
1	1.405	1.405	1.404	0.01%
2	2.810	2.810	2.808	0.06%
3	4.217	4.217	4.211	0.14%
4	5.627	5.626	5.613	0.26%
5	7.041	7.038	7.011	0.43%

Table 6. The simulated frequency values and discrepancies of the shortest cable, *f* and *Δf* (Hz)

Mode	Frequency *f* (Hz)			
	CS 1	CS2	CS3	*Δf* (CS1–CS3) (Hz)
1	8.589	8.550	8.392	2.35%
2	17.438	17.087	15.817	10.25%
3	26.791	25.479	23.149	15.73%
4	36.866	33.980	32.446	13.62%
5	47.852	43.600	43.248	10.65%

From Table 5, the calculated results reveal that with the long cable, although the mass of anti-vandalism pipe reduces the magnitude of frequencies, its effect is still negligible. Even when the pipe is hung at a 1.5 m height from the concrete base, the changes of frequency still are very small, for instance, a maximum decrease of 0.43% of the fifth frequency from 7.0407 Hz to 7.0106 Hz, others are under 0.26%.

In contrast, the mass of anti-vandalism pile has a significant influence on the vibration of the short cable. The discrepancy of the fifth frequency between CS1 and CS3 is 10.65%, from 47.85 Hz to 43.25 Hz in Table 6, others are around 10.25% to 15.73%.

5 Results and Discussion

As mentioned above, the results obtained in two Sects. 3 and 4 are analyzed in this section to have an overview of cable vibration. Table 3. shows a similar downward trend of estimated tension value in two types of cable between placing on duct and on strands. The longest cable has a 2.73% decrease in tension, meanwhile the shortest reduces to 2.45%. Besides that, in comparison with real-time data, the deviation of tension from direct contact with strands is much closer than that on duct at 0.38% and 3.11%, respectively. From the simulation results in Table 5, the effect of anti-vandalism pipe on dynamic response of the longest cable is insignificant. Therefore, in the paper, it can be seen that the placing of sensor on strands achieves a more accurate result than on HPDE duct. In reality, the accuracy of measurement points on the protection duct to collect dynamic response of the longest cable is still acceptable with an error of 3.11%.

From Table 3, the deviation of tension value between the two scenarios is 2.45%. However, Table 6 shows a significant influence of anti-vandalism mass on the shortest cable, measuring vibration of the short cable still should be considered carefully. Especially in cable stay without filling material, more measurements need to be conducted to access accuracy of estimated tension.

6 Conclusions

This paper studied the vibration of two cables with varying lengths by means of different sensor placements. The study answered partly about the effects of fillings on dynamic response of cable via measured and simulation results.

In general, the mass of anti-vandalism pipe and protection duct influences the dynamic characteristics of cable. However, degree of influence does depend on the length of cable and filling material e.g. wax, grease, elastic epoxide products, etc., between the HPDE duct and the strands or not.

Acknowledgements. The authors acknowledge the financial support of VLIR-UOS TEAM Project, VN2018TEA479A103, 'Damage assessment tools for Structural Health Monitoring of Vietnamese infrastructures', funded by the Flemish Government.

Authors would like to express appreciation to colleagues: Nguyen Minh Hoa, Le Thanh Tung from UCT company for supporting the measurement.

vliruos SHARING MINDS, CHANGING LIVES With the support of THE BELGIAN DEVELOPMENT COOPERATION .be

References

1. Peeters, B.: System identification and damage detection in civil engineering. PhD thesis, Department of Civil Engineering, KU Leuven (2000)
2. Peeters, B., De Roeck, G.: Stochastic system identification for operational modal analysis: a review. J. Dyn. Syst. Meas. Control **123**(4), 659–667 (2001)
3. Mehrabi, A.B., Tabatabai, H.: Unified finite difference formulation for free vibration of cables. J. Struct. Eng. **124**(11), 1313–1322 (1998)
4. Cho, S., Yim, J., Shin, S.W., Jung, H.J., Yun, C.B., Wang, M.L.: Comparative field study of cable tension measurement for a cable-stayed bridge. J. Bridge Eng. **18**, 748–757 (2013)
5. Geier, R., De Roeck, G., Flesch, R.: Accurate cable force determination using ambient vibration measurements. Struct. Infrastruct. Eng. **2**(1), 43–52 (2006)
6. Van Gysel, E., De Roeck, G.: A report of estimation of cable forces and bending stiffness. 3rd July 2002
7. Casas, J.R.: A combined method for measuring cable forces: the cable-stayed Alamillo Bridge, Spain. Struct. Eng. Int. **4**(4), 235–240 (1994)
8. Zui, H., Shinke, T., Namita, Y.: Practical formulas for estimation of cable tension by vibration method. J. Struct. Eng. **122**(6), 651–656 (1996)
9. Peeters, B., De Roeck, G., Sa Caetano, E., Cunha, A.: Dynamic study of the Vasco da Gama Bridge. In: Proceedings of ISMA, pp. 545–554 (2002)
10. Cunha, A., Caetano, E., Calçada, R., Delgado, R.: Dynamic tests on Vasco da Gama cable stayed bridge. In: IABSE Conference on Cable Stayed Bridges-Past, Present and Future (1999)
11. Ho, V.L., Hoang, T.N., Bui, T.T.: On the measurement of cable tension of Kien bridge. In: Proceedings of ICSCE, pp. 412–417 (2018)

Finite Element Analysis of Pile Foundations Under Surface Blast Loads

Yasser E. Ibrahim[1,2](✉) and Marwa Nabil[2]

[1] Prince Sultan University, Riyadh, Kingdom of Saudi Arabia
ymansour@psu.edu.sa
[2] Zagazig University, Zagazig, Egypt

Abstract. Recent terrorist attacks on important buildings raised the significance of analysis of structures under blast loads in order to seek mitigations measures to have safer structures under such threats. Most of recent studies consider the superstructure only since it is the major part affected by the blast loads. However, the foundations may be affected severely as well. This can make the superstructure repair is impractical or inefficient. In this research, detailed finite element analysis using ABAQUS was used to study the effect of blast loads on the response of soil and nine reinforced concrete piles buried in the soil and connected by a 10 m × 10 m × 1.0 m reinforced concrete raft. The piles have a 0.6 m diameter and a length of 20 m. Drucker-Prager Cap model was used to model the soil behavior. The model accounts for soil hardening and softening and stress path dependence. The raft and piles were modelled using 8-node solid elements with reduced integration. The concrete damage plasticity model was used to model the reinforced concrete material for the pile and raft. The blast load was considered through several explosive weights of TNT at a height of 0.66 m above ground surface. The effect of standoff distance was studied through five different distances from the raft edge. Finally, observations and recommendations were provided to enhance the response of pile foundations under blast loads.

Keywords: Finite element analysis · Blast load · Pile foundation · Reinforced concrete

1 Introduction

Recently, many iconic buildings in the world have been targeted by terrorist attacks with explosions causing many causalities and significant damage in these buildings. This has led many researchers to study the effect of blast loads on structures and ways to mitigate such risks [1–6].

In addition to the superstructures, the foundations and underground structures can also be severely affected by the dynamic waves resulted from explosions that may take place at, or near the ground surface (above or under it). The dynamic waves generated from these above-ground surface explosions are transmitted and refracted at ground surface causing dynamic loads on the underground structures and foundation. Similarly, surface and underground explosions exert dynamic loads and may cause a crater [7–9].

M. A. Wahab (Ed.): DAMAS 2019, LNME, pp. 446–460, 2020.
https://doi.org/10.1007/978-981-13-8331-1_32

The severity of dynamic loads on underground structures and foundations from surface and above-ground surface explosions depends on many parameters such as the amount of charge of the explosives, the distance from the explosives, soil properties, structure orientation from the explosives and the structural aspects of the structural elements [10, 11].

There is a limited experimental work conducted to predict the pressure caused by conventional weapons on underground structures due to the complexity of the problem [10]. Accordingly researchers use computer packages to conduct finite element analysis of these structures under blast loads such as ABAQUS [12], LS-DYNA [13] and ANSYS [14]. Empirical equations were developed to calculate the pressure exerted on structures by air, surface, subsurface and under water explosives [7–9, 15–18]. Most of these equations were developed based on spherical explosive charges and were used to calculate different parameters, that affect the pressure on structures such as arrival time of explosion wave at a given location, its duration, peak pressure, impulse force, and particle displacement, velocity, and acceleration. The empirical equations were developed based on experimental investigations and most importantly analytical studies. Soils with air voids experience smaller peak pressure and considerable less peak particle velocity compared to saturated soils [19]. Peak stresses and accelerations are increased significantly for fully saturated clays, clay shales, and sandy clays, while the effect of water saturation is not considerable in case of granular soils with high relative density [11].

In this research, detailed finite element analysis using ABAQUS was used to study the effect of blast loads on the response of the soil as well as reinforced concrete piles and their connecting raft buried in this soil. The blast load effect was considered through different explosive weights of TNT charges at a height of 0.66 m above ground surface at a 5.0 m horizontal distance from the edge of the raft. The effect of standoff distance of the explosive charge was investigated through using a charge of 457 kg of TNT at a height of 0.56 m above ground surface at five different distances from raft edge; 2.5 m, 5.0 m, 7.5 m, 10.0 m and 15.0 m. Finally, observations and recommendations were provided to enhance the response of pile foundations under blast loads.

2 Structural and Soil Model

The soil block considered has the dimensions of 100 m × 100 m × 50 m so that no effect can be received through the reflection of artificial pressure from domain boundaries [20]. The model consisted of 9 reinforced concrete piles with 600 mm diameter and connected by a raft with a dimension of 10 m × 10 m and thickness of 1.0 m. The piles have length of 20 m. The distance between piles is 7 times the diameter (4.2 m). The explosion was considered at standoff horizontal distance of 5.0 m from the raft edge and at a vertical distance of 0.66 m (26 inches) above the ground surface. Figure 1 shows the arrangement of the piles connected by the raft and the location of blast charge.

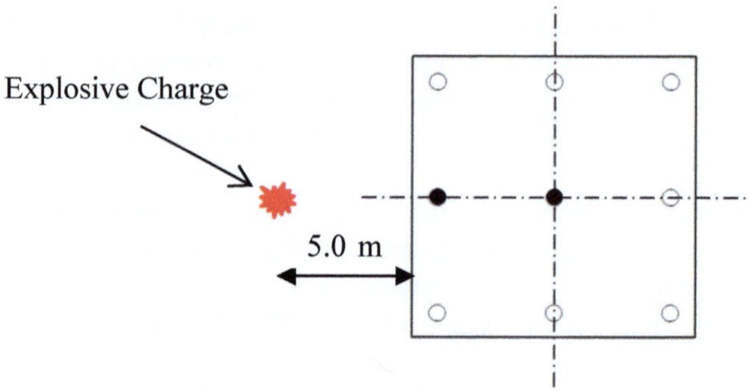

Fig. 1. Raft and piles (filled piles are those under consideration)

3 Finite Element Model

Lagrangian three-dimensional solid continuum elements were used to represent the soil block in the finite element analysis. The block was loaded by time-dependent surface pressure resulted from blast load. The analysis focused on the soil behavior including crater formation and pile foundation performance. Accordingly, air block was not considered in the model [21].

Drucker-Prager Cap model in ABAQUS was used to model the soil behavior. The model accounts for soil hardening and softening and stress path dependence [22]. The soil modulus of elasticity is 51.7 MPa with 0.45 Poisson ratio and 1920 kg/m^3 specific weight. The soil cohesion is 0.036 MPa and friction angle of 24°. The stress and corresponding plastic strain are given in Table 1.

Table 1. Stress-plastic strain of soil used in the analysis [23]

Stress (MPa)	Plastic strain
2.75	0.00
4.83	0.02
5.15	0.04
6.20	0.08

The raft and piles were modelled using C3D8/C3D8R elements, which are 8-node solid elements with reduced integration. The concrete damage plasticity model was used to model the reinforced concrete material for the pile and raft. This model uses the concept of isotropic damaged elasticity in combination with isotropic tensile and compressive plasticity to model the inelastic behavior of concrete. The stress-strain curves presented by Martin [24], which were developed based on the material properties of Chopra and Chakrabarti [25] were used in this model. Concrete strains at yield and at failure are 0.002 and 0.004, respectively. True stress and logarithmic plastic strain were obtained and used as input data to model plasticity within ABAQUS Package. Figure 2 shows the finite element model of the soil and pile foundation including the 9 piles and the raft.

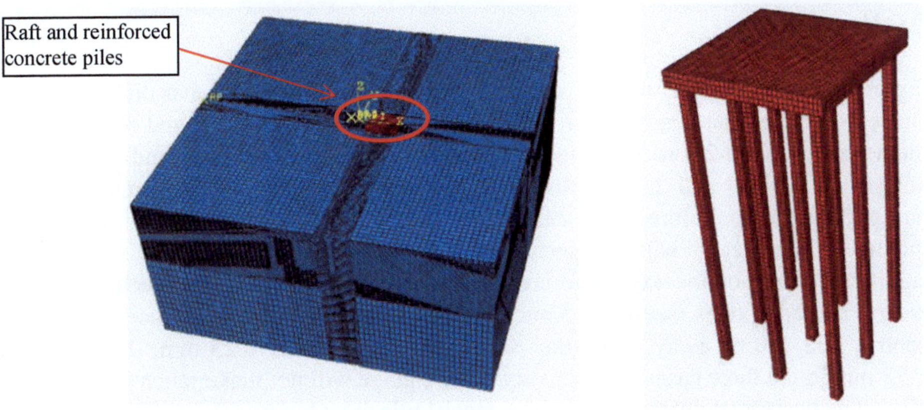

a) Soil and pile foundation model

b) Raft and reinforced concrete piles

Fig. 2. Finite element model using ABAQUS

4 Blast Pressure Loading

The CONWEP algorithm, which is available in ABAQUS/Explicit was used to model the dynamic load resulted from the explosive charge on the soil. This method is empirical with the condition that the distance from the explosive charge to the point of consideration should be greater than the radius of that charge. Three different explosive charges of TNT were used in the analysis; 45.4 kg, 227 kg and 457 kg. These cases are case 1, case 2 and case 3, respectively. All charges were assumed to have a spherical shape. The blast was assumed at a horizontal distance of 5.0 m from the edge of the raft and at a height of 0.66 m (26 inches) above the ground surface.

Five more cases were considered in order to study the effect of standoff distance on the soil and pile foundation response under blast load. An explosive charge of 457 kg of TNT was considered at a height of 0.56 m (22 inches) above ground surface at five different distances from raft edge; 2.5 m (case 4), 5.0 m (case 5), 7.5 m (case 6), 10.0 m (case 7) and 15.0 m (case 8). All cases are summarized in Table 2.

Table 2. Cases under consideration

Case no.	TNT charge	Distance from raft edge	Height above ground surface
1	45.4 kg	5.0 m	0.66 m
2	227 kg	5.0 m	0.66 m
3	457 kg	5.0 m	0.66 m
4	457 kg	2.5 m	0.56 m
5	457 kg	5.0 m	0.56 m
6	457 kg	7.5 m	0.56 m
7	457 kg	10.0 m	0.56 m
8	457 kg	15.0 m	0.56 m

5 Results

Figure 3 shows the crater shape for case1, case 2 and case 3. Case 1 has different crater shape from case 2 and case 3. The crater shape is identical for case 2 and case 3 with a diameter of around 2.0 m at ground surface. The depth of crater is around 2.30 m and 1.35 m for cases 3 and 2, respectively. There is a huge difference in terms of plastic energy dissipation in the three cases (see Fig. 4). The passive energy dissipation increases dramatically with the increase of the explosive charge. There is no much significant effect of the explosive charge weight on the vertical displacement of the central and edge piles (see Figs. 5 and 6). The maximum vertical displacement of the central pile (10 m away from the explosive charge) was 0.25 mm, 1.0 mm and 1.55 mm in the three cases, respectively. This change will not make much effect on the piles. The horizontal deformation of the central pile at its top was 0.7 mm, 1.6 mm and 1.9 mm, for the three cases respectively. This can explain the considerable high horizontal stresses on the pile caused by the lateral deformations in cases 2 and 3 compared to case 1 (see Figs. 7 and 8). As mentioned earlier, the small effect of the weight of the explosive charge on the vertical deformation of the central pile caused no significant difference in the vertical stresses developed in the central pile for the three cases (see Fig. 9). The maximum horizontal stress in the central pile was 0.05 MPa, 1.1 MPa and 1.7 MPa for case 1, case 2 and case 3, respectively. The maximum horizontal stress occurred at the top of the piles in the three cases. Considering the vertical stresses developed in the central pile, the maximum stress obtained was 1.4 MPa at the top of the pile, 2.2 MPa at a depth of 18 m from ground surface and, 2.5 MPa at the bottom of the pile, for cases 1, 2 and 3, respectively.

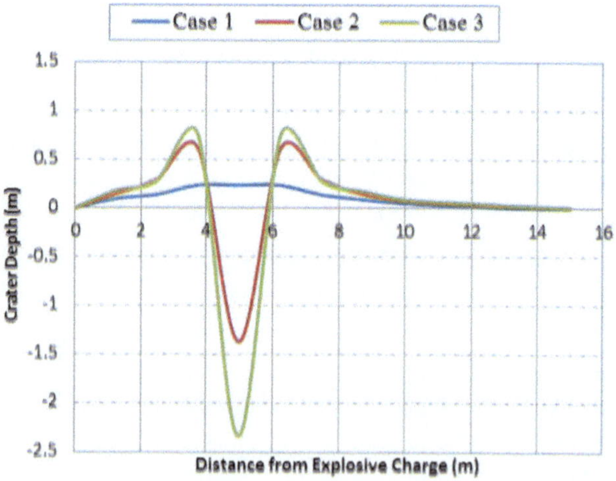

Fig. 3. Crater shape

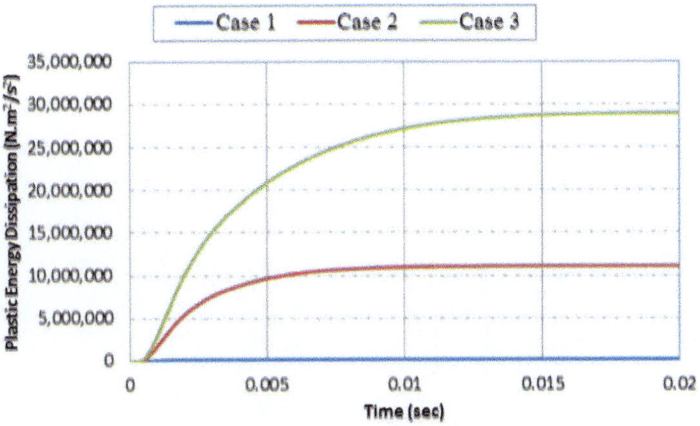

Fig. 4. Passive energy dissipation

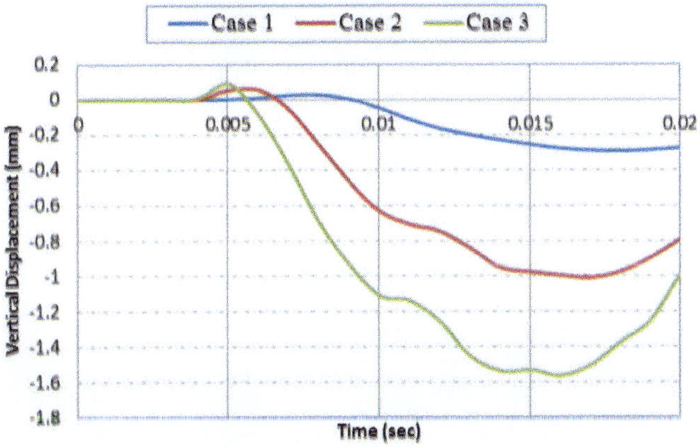

Fig. 5. Vertical displacement of central pile at its top

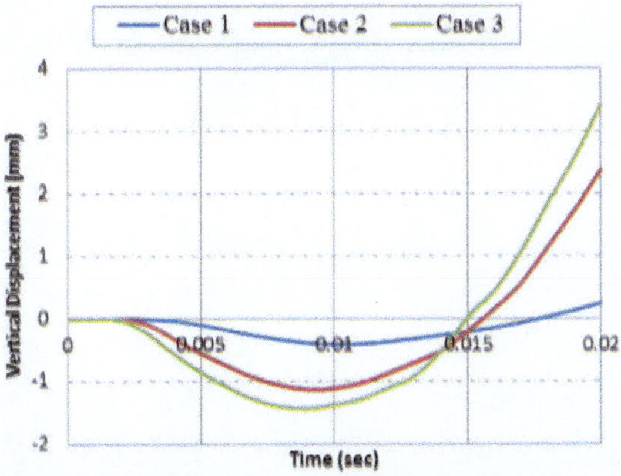

Fig. 6. Vertical displacement of edge pile at its top

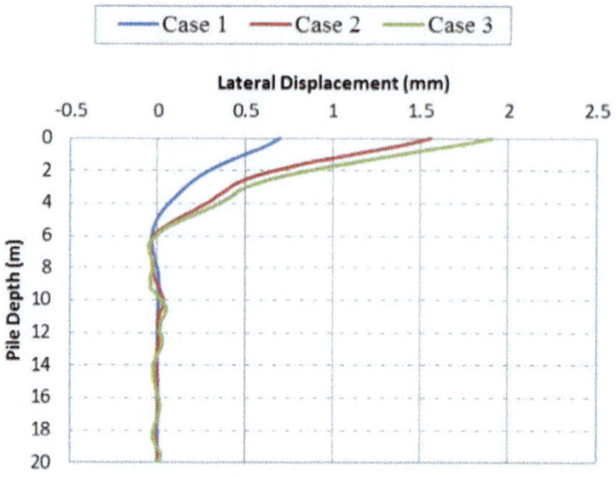

Fig. 7. Lateral displacement over the length of central pile

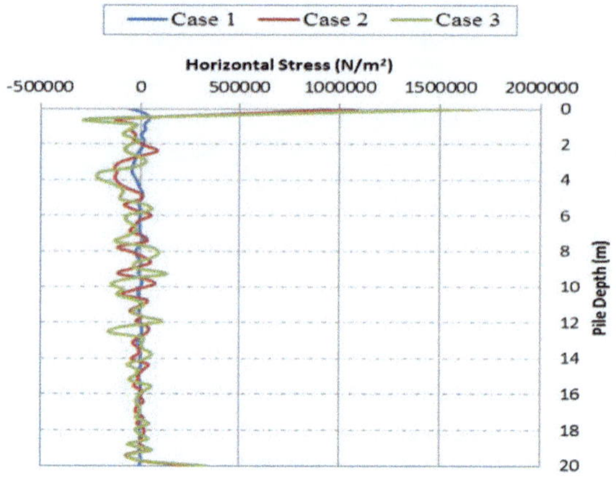

Fig. 8. Horizontal stresses over the length of central pile

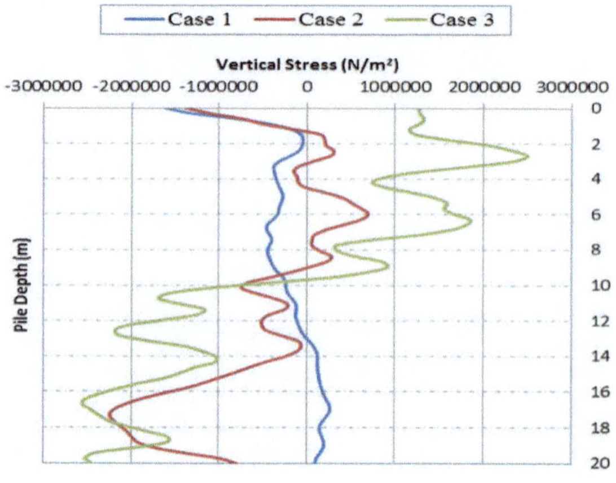

Fig. 9. Vertical stresses over the length of central pile

Considering the central pile on the left column of the piles (closest pile to the explosive charge; 5.8 m), the maximum lateral displacement at the top of the pile was around 2 mm, 7 mm and 10 mm for the three cases, respectively (see Fig. 10). These large values in case 2 and case 3 may not allow using the pile with its current configuration after the blast to support the superstructure. Compared to the horizontal deformations, the vertical deformations were less with smaller effect. The vertical deformations were 0.4 mm 1.2 mm and 3.9 mm and all occurred at the top of the pile.

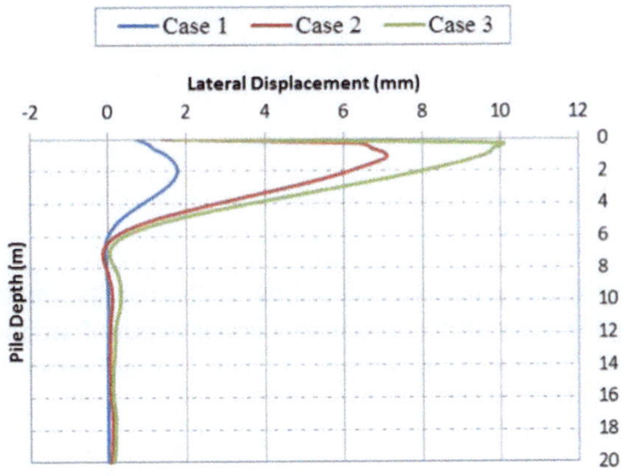

Fig. 10. Lateral displacement over the length of edge pile

The effect of standoff distance on the soil and pile foundations was investigated through considering a 457 kg of TNT at a height of 0.56 m above ground surface at different distances from the left edge of the raft. These standoff distances were 2.5 m, 5.0 m, 7.5 m, 10.0 m and 15.0 m. The results of studying these cases (case 4 through case 8) are shown in Figs. 11, 12, 13, 14, 15, 16, 17, 18, 19 and 20. Figure 11 shows the

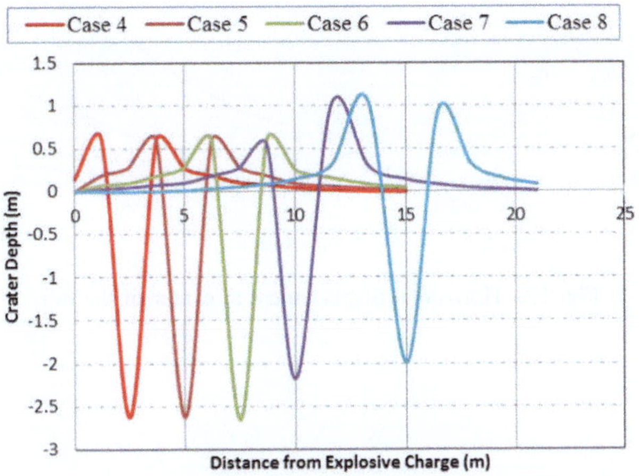

Fig. 11. Crater shape

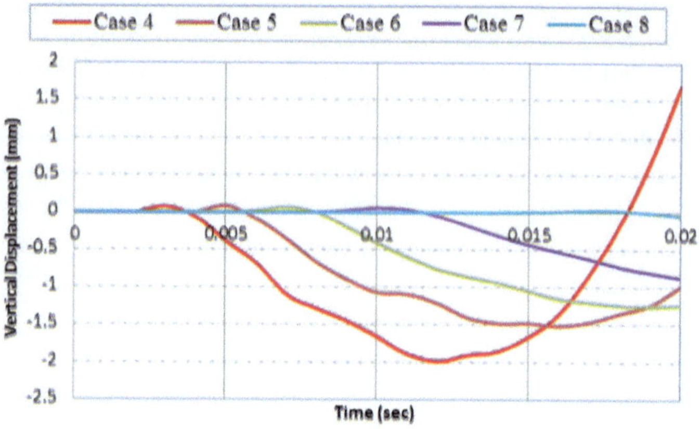

Fig. 12. Vertical displacement at center of the raft

crater shape developed in each case. According to the results obtained, similar crater shape was developed in cases 4, 5 and 6 with a crater depth of around 2.6 m below the ground surface and a diameter of around 2.0 m at ground surface. The distance of explosive charge to pile foundation has negligible effect on the crater shape in these cases. However, case 7 and case 8 experienced little different crater shape with a smaller crater depth below ground surface and higher height above ground surface. This may be attributed to the effect of pile foundation and its confinement effect on the soil.

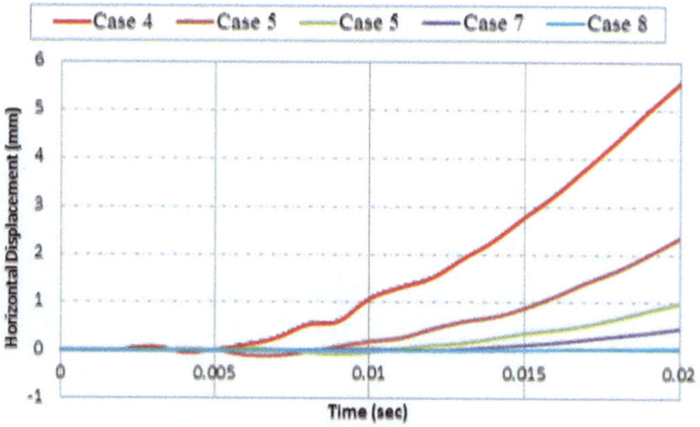

Fig. 13. Horizontal displacement at center of the raft

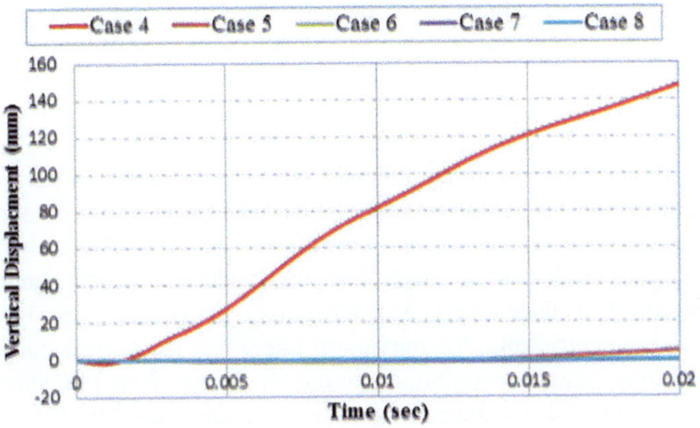

Fig. 14. Vertical displacement at edge of the raft

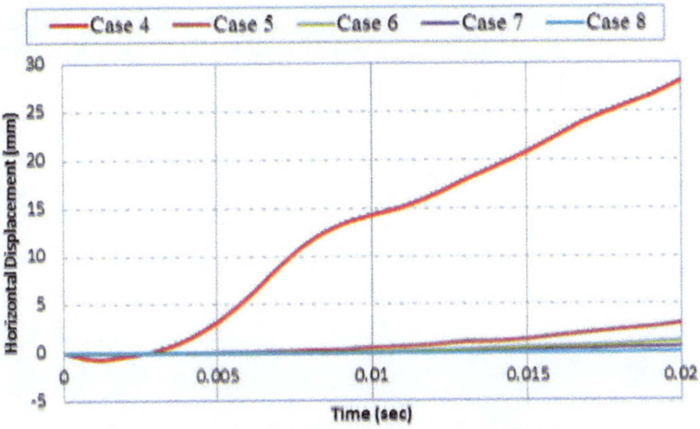

Fig. 15. Horizontal displacement at edge of the raft

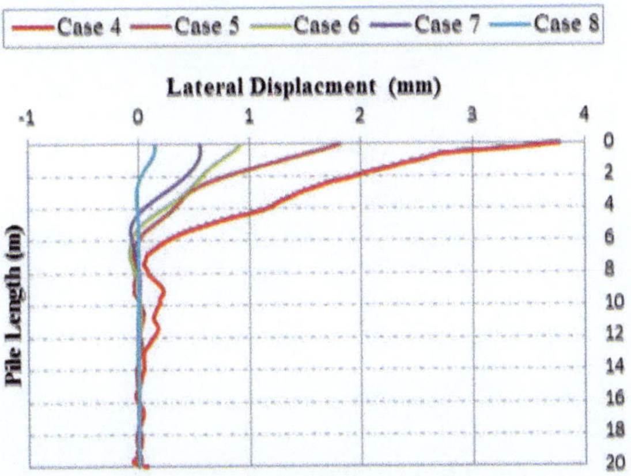

Fig. 16. Lateral displacement of central pile

Considering the deformations developed in the raft foundation caused by blast loading, the small standoff distance has a significant effect as appeared in case 4. In this case, the vertical and horizontal displacement of the center of the raft reached 2.0 and 5.5 mm (see Figs. 12 and 13). The edge of raft close to the explosive charge experienced much more deformations. The vertical displacement of the raft edge was around 145 mm upward while the horizontal displacement was around 28 mm away from the explosive charge. These high values cannot be allowed because it would affect the pile deformations and consequently, the equilibrium and stability of the superstructure if existed. Other cases experienced much less deformations, which can be tolerated, especially with standoff distance equal or more than 7.5 m (see Figs. 14 and 15).

Considering the central pile, the maximum lateral deformation occurred in the top of the pile with considerable deformations in the top 5.0 m of the pile length, while the rest of the pile length did not experience noticeable deformations. Cases 6, 7, 8 had lateral deformations less than 1.0 m (see Fig. 16). The maximum lateral stress in the central pile reached 0.9 MPa in case 4 and 1.7 MPa in case 5. However, in other cases, the maximum lateral stress was less than 0.5 MPa (see Fig. 17).

There is no trend in the effect of standoff distance on the vertical stress at the central pile. In all cases, the vertical stress was less than 4 MPa (see Fig. 18). Considering the edge pile, the lateral deformations in case 4 and case 5 had high values. The maximum lateral deformation in that pile was 35 mm and 10 mm, in case 4 and case 5, respectively. In all cases, the maximum lateral deformations were obtained in top 5 m of the pile length (see Fig. 19). The maximum vertical deformation in the edge pile was received in case 4, where the standoff distance was 2.5 m. In that case, the upward deformation was 35 mm in the top of the pile, while the maximum downward deformation was around 4 mm at depth of around 7 m. Much smaller vertical deformations were obtained in other cases (see Fig. 20).

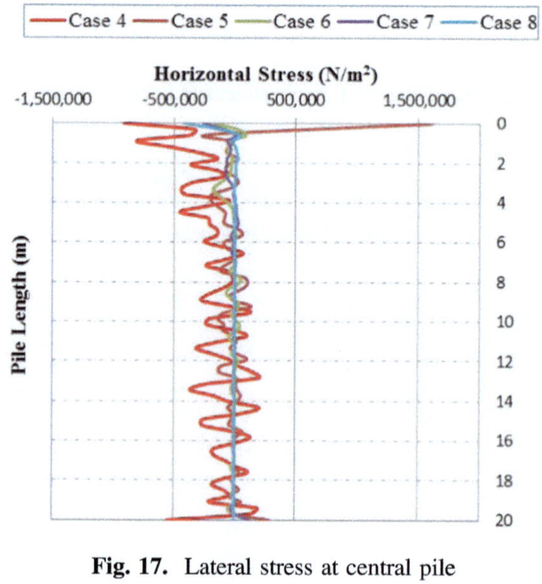

Fig. 17. Lateral stress at central pile

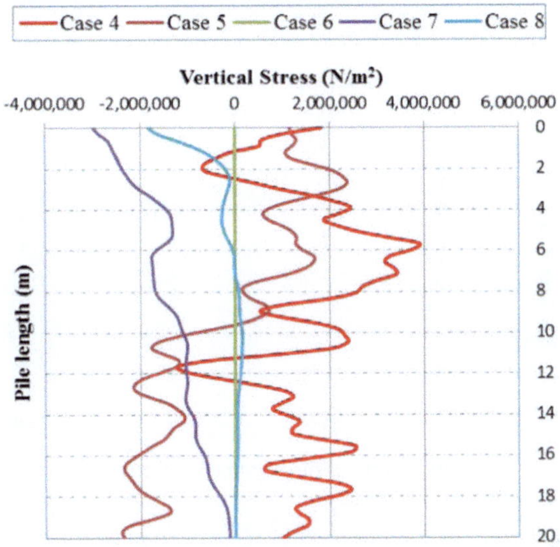

Fig. 18. Vertical stress at central pile

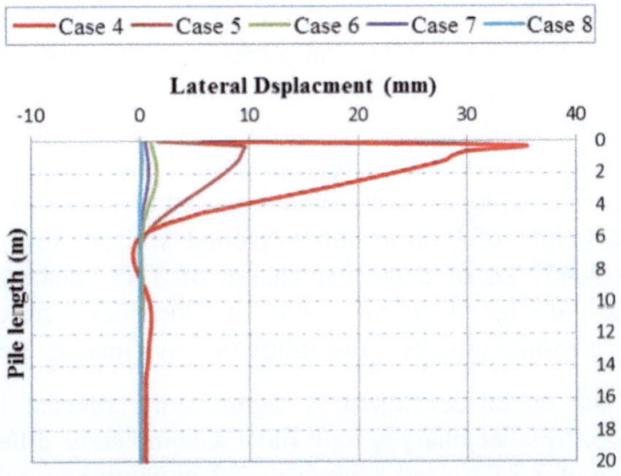

Fig. 19. Lateral displacement of edge pile

According to the previous results, the standoff distance has a considerable effect on the pile and raft response, when subjected to blast loads. Standoff distance of less than 7.5 m from the raft edge may cause high stresses and permanent deformations, that would make the use of these piles and raft foundation is impractical after such threat, or the instability of the whole structure is possible and highly expected.

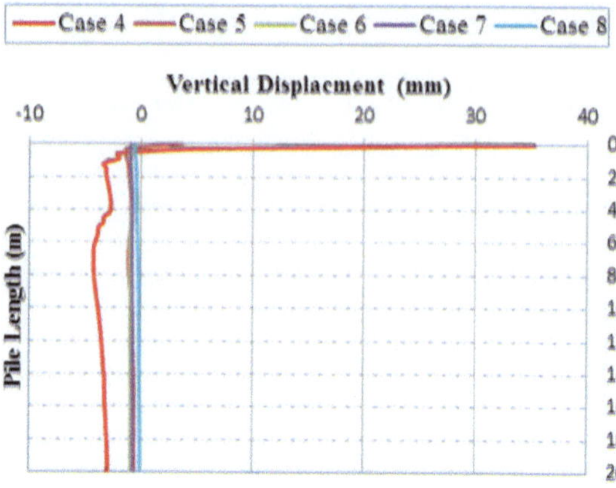

Fig. 20. Vertical displacement of edge pile

6 Conclusions

A finite element model was developed to study the performance of pile foundation under surface blast load. The model consisted of a soil block of a surface area of 100 m × 100 m and a depth of 50 m. A raft foundation of dimensions of 10 m × 10 m and a depth of 1.0 m was modeled with its top surface on the ground surface. The raft foundation rested on 9 reinforced concrete piles with a configuration of 3 rows and 3 columns with a length of 20 m each. The piles are 0.6 m in diameter and spaced at 4.2 m apart. Three weights of explosive charge of TNT were considered; 45.4 kg, 227 kg and 457 kg. The explosive charges were placed at a distance of 5.0 m from the raft edge and at a height of 0.66 m above ground surface. Five different standoff distances between a 457 kg of explosive charge of TNT, placed at 0.56 m above ground surface, and the edge of the raft connecting the piles were considered.

According to the results obtained, the following conclusions are obtained:

1. The shape of the crater developed is significantly affected by the weight of explosive charge. Smaller charges may have a completely different crater shape. A 457 kg charge of TNT caused a crater of 2.2 m depth and a diameter of 2 m at the ground surface.
2. Plastic energy dissipation increases dramatically with the increase of the weight of explosive charge.
3. High horizontal stresses are developed in piles due to the large lateral deformations in the top part of the piles, especially on the piles closest to the explosive charge of a specific weight. These large deformations can prevent piles from being used to support the superstructure, if the structure is not yet constructed or increases the structural damage if the blast occurs after the construction of the structure.
4. Standoff distance from raft edge has a considerable effect on the response of pile foundations subjected to blast loads. A minimum of 7.5 m standoff distance is recommended in order to protect the piles, raft and superstructure if already constructed.

The work will be extended to consider the effect of other parameters on the soil and pile foundation response under blast loads. These parameters include the soil type, pile configuration and material and axial load of piles when subjected to blast loads.

Acknowledgement. The authors would like to express their thanks to Prince Sultan University for funding the attendance of the conference and publishing the journal article.

References

1. Ismail, M., Ibrahim, Y., Nabil, M., Ismail, M.M.: Response of A 3-D reinforced concrete structure to blast loading. Int. J. Adv. Appl. Sci. **4**(10), 46–53 (2017)
2. Ibrahim, Y.E., Ismail, M.A., Nabil, M.A.: Response of reinforced concrete frame structures under blast loading. Procedia Eng. **171**, 890–898 (2017)
3. Fu, F.: Dynamic response and robustness of tall buildings under blast loading. J. Constr. Steel Res. **80**, 299–307 (2013)
4. Choi, J., Choi, S., Kim, J.J., Hong, K.: Evaluation of blast resistance and failure behaviour of prestressed concrete under blast loading. Constr. Build. Mater. **173**, 550–572 (2018)
5. Ritchie, C.B., Packer, J.A., Seica, M.V., Zhao, X.: Behaviour and analysis of concrete-filled rectangular hollow sections subject to blast loading. J. Constr. Steel Res. **147**, 340–359 (2018)
6. Hadianfard, M.A., Malekpour, S., Momeni, M.: Reliability analysis of H-section steel columns under blast loading. Struct. Saf. **75**, 45–56 (2018)
7. Baker, W.E.: Explosions in Air. University of Texas Press, Austin, TX (1973)
8. ASCE: Manual 42, Design of Structures to Resist Nuclear Weapons Effects. American Society of Civil Engineers, New York City, NY (1985)
9. Cooper, P.W.: Explosives Engineering. Wiley, Hoboken, NJ (1996)
10. DOD: Structures to Resist the Effects of Accidental Explosions. Unified Facilities Criteria (UFC) 3–340–02. Department of Defense, Arlington, VA (2008)
11. DOD: Fundamentals of Protective Design for Conventional Weapons. TM 5–855–1. Department of Defense, Arlington, VA (1986)
12. ABAQUS 6.14 User Documentation, Dessault Systems (2018)
13. ANSYS. AUTODYN User's Manual. ANSYS Inc., Canonsburg, PA (2018)
14. LS-DYNA User's Manual, Livermore Software Technology Corporation (2018)
15. Larcher, M.: Simulation of the Effects of an Air Blast Wave. JRC Technical Notes (JRC) 41337. Luxembourg: European Communities (2007)
16. Kinney, G.F., Graham, K.J.: Explosive Shocks in Air. Springer-Verlag, Berlin, Germany (1985)
17. Krauthammer, T., Altenberg, A.: Negative phase blast effects on glass panels. Int. J. Impact Eng. **24**, 1–17 (2000)
18. Smith, P.D., Hetherington, J.G.: Blast and Ballistic Loading of Structures. Butterworth-Heinemann, Oxford, England (1994)
19. Wang, Z., Lu, Y., Bai, C.: Numerical analysis of blast-induced liquefaction of soil. Comput. Geotech. **35**(2), 196–209 (2008)
20. Wilt, T., Ofoegbu, G., Hsiung, S.: Vulnerability Assessment of Buried Near-Surface Structures Subject to Ground Surface Blasts. Center for Nuclear Waste Regulatory Analyses, San Antonio, TX (2012)

21. NUREG/CR-7201: Characterizing Explosive, Effects on Underground Structures, Center for Nuclear Waste Regulatory Analyses, U.S. Nuclear Regulatory Commission Washington DC 20555-0001 (2015)
22. Huang, T.K., Chen, W.F.: Simple procedure for determining cap-plasticity-model parameters. J. Geotech. Eng. **116**(3), 492–513 (1990)
23. Nagy, N., Mohamed, M., Boot, J.C.: Nonlinear numerical modelling for the effects of surface explosions on buried reinforced concrete structures. Geomech. Eng. **2**, 1–18 (2010)
24. Martin, O.: Comparison of different Constitutive Models for Concrete in ABAQUS/Explicit for Missile Impact Analyses, JRC Scientific and Technical Reports. EUR 24151, European Commission, Joint Research Centre, Institute for Energy (2010)
25. Chopra, A.K., Chakrabarti, P.: The Koyna earthquake and the damage to the Koyna dam. Bull. Seism. Soc. Am. **63**(2), 381–397 (1973)

Recurrence Analysis for Damage Detection and Localization in Beam Structure

Joanna Iwaniec[1(✉)], Krzysztof Mendrok[1], Ángel J. Molina-Viedma[2], and Łukasz Pieczonka[1]

[1] Faculty of Mechanical Engineering and Robotics, Department of Robotics and Mechatronics, AGH University of Science and Technology, Mickiewicz Alley 30, 30-059 Krakow, Poland
jiwaniec@agh.edu.pl

[2] Departamento de Ingeniería Mecánica y Minera, Universidad de Jaén, Campus Las Lagunillas, 23071 Jaén, Spain

Abstract. In the paper, application of the **R**ecurrence **P**lots (RP) method to damage detection and localization in the beam structure is proposed. The RP is a well-established method of nonlinear data analysis dedicated to investigation into changes in dynamic behaviour of mechanical systems during their normal work. Interpretation of information provided by the recurrence plots can be supported by the estimation of the measures of **R**ecurrence **Q**uantification **A**nalysis (RQA).

The main emphasis of the paper is put on damage detection and localization in the aluminium beam subjected to burst random excitation provided by the electrodynamical shaker. In order to simulate the damage, the notch of known parameters and position was introduced to the system. System responses in the form of time series measured by vision methods (fast cameras) were analysed with the application of the RP and RQA methods. Obtained results proved high sensitivity of the applied methods to changes in dynamical system properties resulting from damage initialization.

Keywords: Beam structure · Recurrence plots · Damage detection and localization

1 Introduction

Nowadays, demand for mechanical systems reliability, durability, safety of operation and comfort of exploitation has been increasing. For economic reasons, shorter design time, cost reduction, longer exploitation period as well as minimization of necessary inspections and repairs are receiving increasing attention. Therefore the quest of development of reliable methods for early damage detection is still open.

For the purposes of system state diagnosing and structural health monitoring modal analysis is commonly used [1–3]. Although its application provides satisfactory results [4–6], it should be stressed that appearance and propagation of system faults invokes system nonlinear properties. For that reason, in the paper, application of the **R**ecurrence **P**lots (RP) [7–9] method to damage detection and localization in the beam structure is

M. A. Wahab (Ed.): DAMAS 2019, LNME, pp. 461–473, 2020.
https://doi.org/10.1007/978-981-13-8331-1_33

proposed. The RP is a well-established method of nonlinear data analysis dedicated to investigation into changes in dynamic behaviour of mechanical systems during their normal work. The main advantage of its application consists in the possibility of detecting transitions from periodic to chaotic behaviour (and the other way round), variability of parameters in time, transient states and cyclic processes on the basis of measured time histories of system dynamic responses. The method can be also used for the purposes of model-based diagnostics. Interpretation of information provided by the recurrence plots can be supported by the estimation of the measures of **R**ecurrence **Q**uantification **A**nalysis (RQA).

The main emphasis of the paper is put on damage detection and localization in the aluminium beam subjected to burst random excitation provided by the electrodynamical shaker. In order to simulate the damage, the notch of known parameters and position was introduced to the system. System responses in the form of time series measured by vision methods (fast camera) were analysed with the application of the RP and RQA methods. Obtained results proved high sensitivity of the applied methods to changes in dynamical system properties resulting from damage initialization.

The structure of the paper is as follows. Section 2 concerns the algorithms of the applied methods. Details of the identification experiment are provided in the Sect. 3. Results of damage detection and localization (beam model-based diagnostic) obtained with the application of the RP and RQA methods are discussed in the Sect. 4. Finally, the paper is concluded in the Sect. 5.

2 Applied Methods

The concept of recurrences was introduced by Henri Poincaré [8] in 1890. Based on the results of the Poincaré's research, Eckmann, Oliffson and Ruelle formulated the Recurrence Plots method (RP) [9], which was originally used for the purposes of visualisation of system trajectories in the higher dimensional phase spaces. Development of the Recurrence Quantification Analysis (RQA) [7, 10, 11], which provides the objective measures supporting interpretation of information contained in recurrence plots, has consolidated the RP method as an efficient tool for nonlinear data analysis. Nowadays, both RP and RQA methods find applications in detection of qualitative changes in the behaviour of dynamical systems [12]. Discussed methods are used in numerous fields of science, such as technique [13], astrophysics [14, 15], biology [16], chemistry [17], geology [18], neuroscience [19], cardiology [20] and economy [21].

2.1 Recurrence Plot (RP)

The idea of the RP method consists in detecting all the times when the phase space trajectory of the given dynamical system visits roughly the same area in the phase space. The recurrence of a state x at time instant i at a different time instant j is represented within a two-dimensional squared matrix $[R]$ (called recurrence matrix) by ones and zeros:

$$[R_{i,j}] = \Theta(\varepsilon - \| \{x_i\} - \{x_j\} \|), \quad i,j = 1, \ldots, N \tag{1}$$

where N denotes the number of considered states x_i, ε_i: a threshold distance, $\| \; \|$ a norm and $\Theta()$: the Heaviside function.

Graphical representation of the recurrence matrix $[R]$ is known in the literature under the term *recurrence plot*. In other words, the recurrence plot provides graphical representation of the matrix $[R_{i,j}]$ in the following form:

$$[R_{i,j}] = \begin{cases} 1: & \{x_i\} \approx \{x_j\} \\ 0: & \{x_i\} \neq \{x_j\} \end{cases}, \quad i,j = 1, \ldots, N \tag{2}$$

where $\{x_i\} \approx \{x_j\}$ are the points belonging to the neighbourhood of radius ε (defined according to the user-selected norm).

In order to estimate recurrence plot on the basis of the measured time response of the considered dynamic system it is necessary to specify values of three parameters: threshold ε, time delay τ and embedding dimension m. Values of these parameters have a significantly influence on the informative content of the estimated recurrence plot and, therefore, should be selected carefully. For instance, if the value of ε is too small, there may be almost no recurrence points and, what follows, the information concerning recurrence of the considered system states can be lost. On the other hand, if ε is too large, almost every point is a neighbour of every other point, which results in appearance of spurious structures on the estimated recurrence plot. In the literature various criterions for selection of threshold ε have been proposed. Nevertheless, in practical applications, the value of ε corresponding to the assumed RP point density is usually selected.

Proper time delay τ can be determined with the application of the auto-correlation or mutual information function [7, 22]. The mutual information function (MI) method, unlike the linear autocorrelation function approach, takes into account also nonlinear correlations:

$$MI = -\sum_{ij} p_{ij}(\tau) \ln \frac{p_{ij}(\tau)}{p_i p_j} \tag{3}$$

where p_i denotes probability of finding the value of system time history in the ith interval, p_{ij}: probability that observation in a given time instant t belongs to the ith interval and the observation in $(t + \tau)$ time instant to the jth interval.

In practical applications, as the 'appropriate' value of τ, the value corresponding to the first minimum of the mutual information function is assumed.

In order to determine the sufficient embedding dimension m, the false nearest neighbours method is frequently used [23]. In the original definition of the false nearest neighbours method [9] the L_2 norm is used and while the neighbourhood radius is selected in such a way that it contains a fixed amount of states. It is possible to modify this criterion in such a way that the recurrence point density has a fixed predetermined value. Such criterion is called **F**ixed **A**mount of Nearest **N**eighbours (FAN).

2.2 Recurrence Quantification Analysis (RQA)

Recurrence Quantification Analysis (RQA) is a method of nonlinear data analysis which quantifies the number and duration of recurrences of a dynamical system represented by its state space trajectory. Measures based on diagonal structures of the recurrence plot can be used for the purposes of finding chaos-to-order transitions while measures based on vertical structures can indicate chaos-to-chaos transitions [7, 19]. Definitions of the most popular RQA measures, such as the *RR* (recurrence rate), *DET* (determinism), *LAM* (laminarity), *L* (averaged length of diagonal lines), *TT* (trapping time), L_{max} (longest diagonal line), V_{max} (longest vertical line), *DIV* (divergence), *ENTR* (entropy) are provided in [7, 10–12].

3 Identification Experiment

Experimental research was carried out for the aluminium frame subjected to the burst random excitation provided by the electrodynamical shaker (Fig. 1). Dynamic response of the horizontal member was registered by two high speed cameras (model Phantom v9.1). Focus was set on the front surface of the member while images were recorded at the frame rate of 512 fps with 400 ms (1/2500 s) of exposure time. The acquisition was synchronized with the cameras.

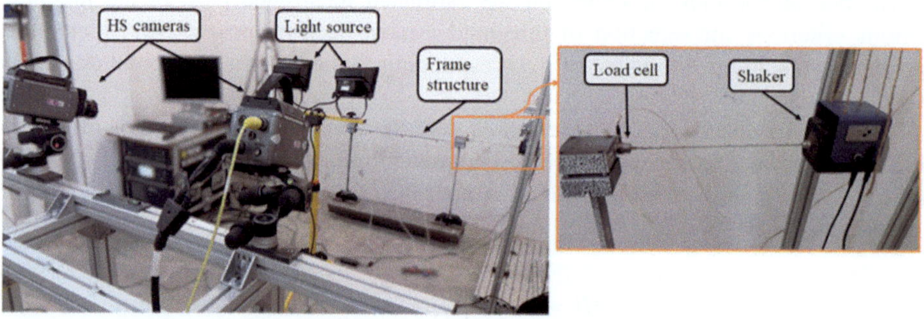

Fig. 1. Experimental setup used for 3D-DIC measurement under burst random excitation.

3D **D**igital **I**mage **C**orrelation (3D-**DIC**) technique was used for determining object deformation in a plane parallel to the image plane of the camera. The region of interest was divided into facets of 11 pixels size with the overlap step of 2 pixels, each of which represented a measurement point. Taking into account the predominant bending behaviour of the member, only vertical displacements were investigated. In order to reduce the redundant information and obtain 1D problem, a vector of spatial measurements was formulated by averaging along the beam thickness. As a result, the beam was modelled by 675 linearly equally distributed points, resulting in 1.49 mm spacing between measurement points.

In the course of the identification experiment, the controlled damage in the form of transversal notch in the upper face (located at 385 mm from the joint) was introduced.

The notch depth was gradually increased - initially the structure was undamaged. In the consecutive measurement sessions the notch depth equalled 0.8 mm, 1.5 mm, 2.5 mm and 4 mm. In each measurement session the network of measurement points remained unchanged.

4 Results of Damage Detection and Localization with the Application of the RP and RQA Methods

Damage of the considered structure is a local phenomenon. Therefore, it is expected that the sensors close to the damage are more heavily influenced than the sensores remote from the damage site. In order to be able to detect damage at an arbitrary location in a structure, densely distributed network of measurement points was assumed at the stage of experiment planning.

For the purposes of damage detection and localisation the RP method was applied to analysis of time histories of the considered system dynamic responses to burst random excitation. Interpretation of the informative content of the RPs estimated for the assumed measurement points was supported by the RQA analysis, which provides the measures of complexity and quantifies the small-scale structures visible on the estimated diagrams. The main idea of the carried out research consists in computing recurrence plot and RQA measures separately for the undamaged (reference) and current system state and, finally, comparing the RQA measures for both states [25–27].

In the course of the carried out research, the results of which are presented in this paper, the FAN definition of the neighbourhood was used. Phase space of the considered systems was reconstructed by delay embedding [11, 23]. For the purposes of consistency, in all the computations the threshold value $\varepsilon = 0.1$ was assumed. The smallest sufficient embedding dimension m was estimated with the application of the false nearest neighbours algorithm. Time delay τ was determined by searching for the first minimum of the mutual information function. All the computations were performed by means of the functions implemented in the CRP Toolbox for MATLAB [24].

In the paper, the results obtained for the selected measurement points are presented (Table 1). Distribution of these measurement points along the horizontal member of the considered structure is presented schematically in the Fig. 2.

Table 1. Measurement points for which the analysis results are presented in the paper.

Annotation of the measurement point used in the paper	p_1	p_2	p_3	p_4	p_5	p_6	p_7	p_8	p_9	p_{10}
Number of 'real' measurement point	10	110	210	310	400	410	420	430	510	630

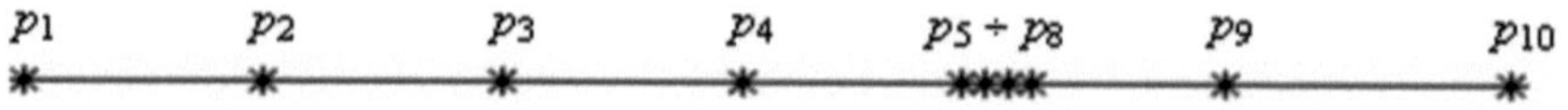

Fig. 2. Distribution of the measurement points denoted as $p_1 \div p_{10}$ along the horizontal member of the considered structure.

In order to asses sensitivity of the RP method to damage depth and location, the responses of the undamaged system and system with the introduced damages of depth $d_1 = 0.8$ mm and $d_2 = 4$ mm were considered. In the first step of analysis, the RP diagrams for each measurement point $p_1 \div p_{10}$ were estimated. In the course of the carried out analysis the FAN criterion was used. Example results – estimated on the basis of system responses measured in the points p_3 and p_7 are shown in the Fig. 3 and Fig. 4, respectively.

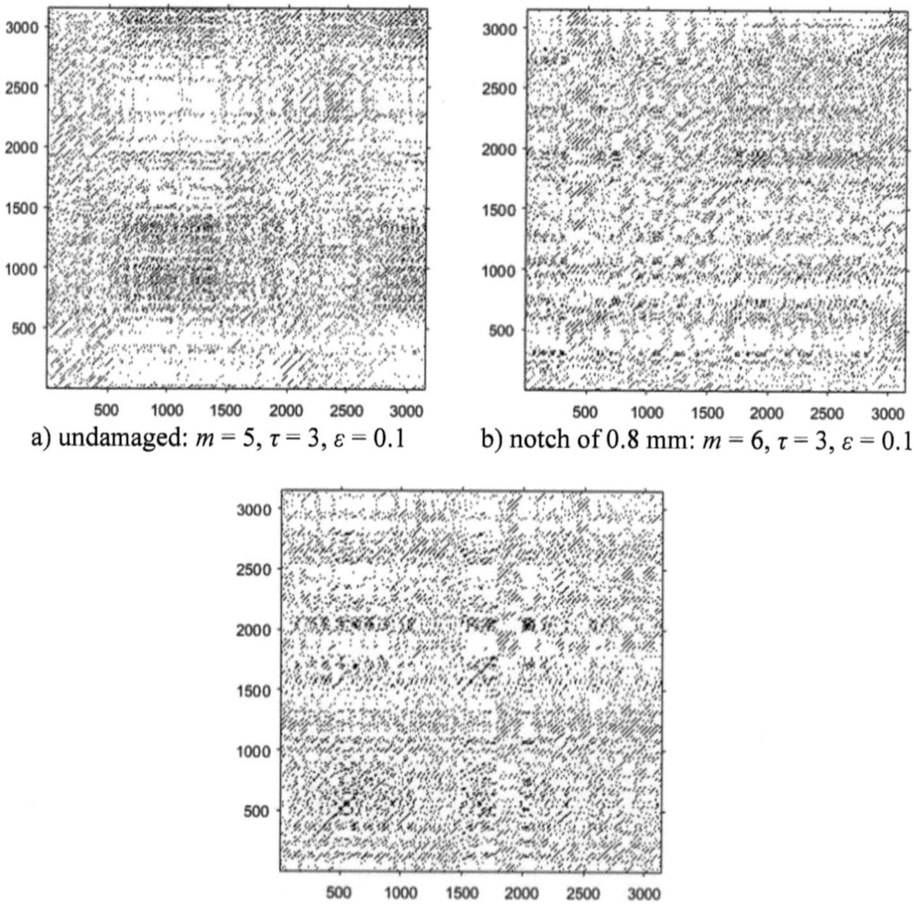

a) undamaged: $m = 5, \tau = 3, \varepsilon = 0.1$

b) notch of 0.8 mm: $m = 6, \tau = 3, \varepsilon = 0.1$

c) notch of 4 mm: $m = 6, \tau = 3, \varepsilon = 0.1$

Fig. 3. Recurrence plots determined on the basis of system responses measured in the point p_3 in the vertical direction, for undamaged system (a) and system with the introduced notch of 0.8 mm (b) and 4 mm (c) depth.

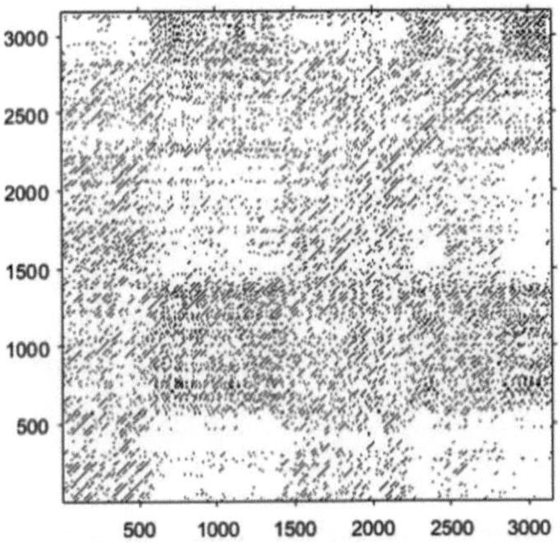

a) undamaged: $m = 6, \tau = 3, \varepsilon = 0.1$

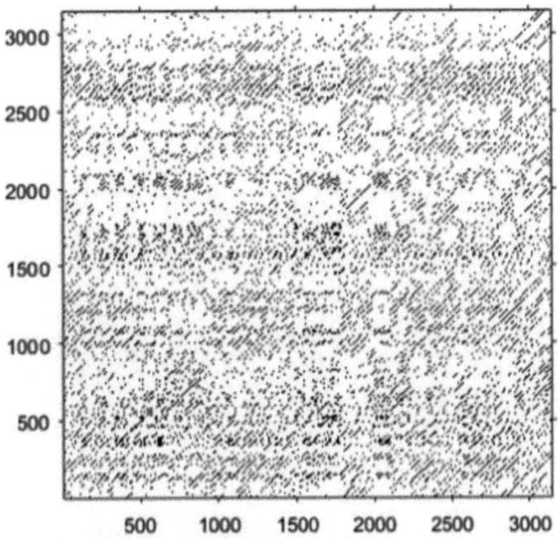

b) notch of 0.8 mm: $m = 6, \tau = 3, \varepsilon = 0.1$

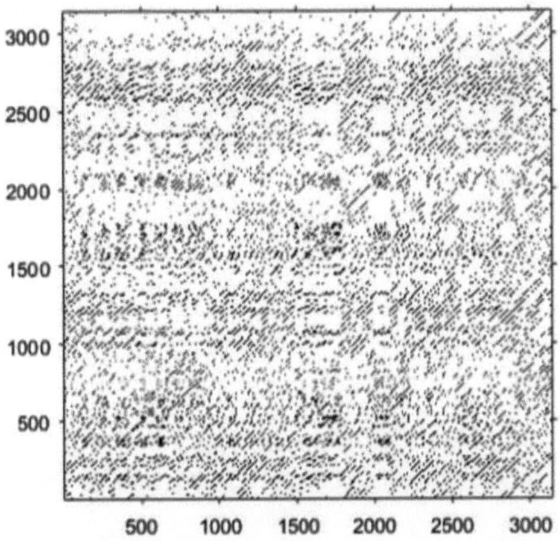

c) notch of 4 mm: $m = 6, \tau = 3, \varepsilon = 0.1$

Fig. 4. Recurrence plots determined on the basis of system responses measured in the point p_7 in the vertical direction, for undamaged system (a) and system with the introduced notch of 0.8 mm (b) and 4 mm (c) depth.

Recurrence plots determined for responses of the damaged system (with notch of $d_1 = 0.8$ mm and $d_2 = 4$ mm) differ significantly from the recurrence plots estimated for the undamaged system. In order to assess the observed differences qualitatively, the RQA analysis was carried out. Results of the RQA analysis for measurement points $p_1 \div p_{10}$ for undamaged structure and structure with the introduced damage (notch of increasing depth: 0.8 mm, 1.5 mm, 2.5 mm and 4 mm) are presented in the Tables 2, 3, 4 and 5.

Graphical representation of the results presented in the Tables 2, 3, 4 and 5 is provided in the Figs. 5, 6 and 7.

Results of the investigation into the variability of values of RQA measures (Figs. 5, 6 and 7) with respect to the damage depth and the distance of the considered

Table 2. Values of selected RQA measures for undamaged structure.

Point	p_1	p_2	p_3	p_4	p_5	p_6	p_7	p_8	p_9	p_{10}
	$m = 6$ $\tau = 10$	$m = 6$ $\tau = 10$	$m = 6$ $\tau = 10$	$m = 6$ $\tau = 3$	$m = 6$ $\tau = 2$	$m = 6$ $\tau = 2$	$m = 5$ $\tau = 2$	$m = 6$ $\tau = 3$	$m = 5$ $\tau = 3$	$m = 6$ $\tau = 10$
DET	0.6500	0.8221	0.7423	0.772	0.74602	0.7713	0.7814	0.7485	0.7827	0.7725
L	3.0200	3.6338	4.7517	7.8096	8.5729	8.7620	8.3275	8.4958	8.7115	3.3983
ENTR	1.3100	1.5715	1.4349	2.7785	2.8315	2.8612	2.8317	2.8397	2.8773	1.477
LAM	0.7300	0.8626	0.8136	0.5958	0.5661	0.6288	0.6917	0.6649	0.8355	0.8224
TT	3.3200	3.6126	2.8387	2.2331	2.1893	2.2559	2.3338	2.3158	2.9491	3.6029
T2	20.4200	26.6872	21.1805	14.9612	14.4878	15.4364	16.5905	16.1231	22.4036	24.7301

Table 3. Values of selected RQA measures for structure with introduced damage (notch of 0.8 mm depth).

Point	p_1	p_2	p_3	p_4	p_5	p_6	p_7	p_8	p_9	p_{10}
	$m = 5$ $\tau = 11$	$m = 5$ $\tau = 10$	$m = 6$ $\tau = 10$	$m = 6$ $\tau = 3$	$m = 6$ $\tau = 2$	$m = 6$ $\tau = 2$	$m = 5$ $\tau = 2$	$m = 6$ $\tau = 3$	$m = 5$ $\tau = 3$	$m = 6$ $\tau = 10$
DET	0.6800	0.8211	0.7749	0.6253	0.6617	0.6993	0.6636	0.6591	0.6701	0.7738
L	3.1500	3.5248	3.1892	3.7125	7.3307	7.0969	6.3135	3.9464	4.0599	3.3857
ENTR	1.4000	1.5247	1.4079	1.1977	2.6847	2.6621	2.5179	1.3115	1.4422	1.4431
LAM	0.7600	0.866	0.8191	0.5721	0.6434	0.7296	0.7047	0.7289	0.7938	0.8284
TT	3.4700	3.6982	3.0691	2.2691	2.344	2.539	2.5288	2.6857	3.2424	3.5831
T2	21.9900	27.2467	22.4049	14.7285	15.8987	17.9891	17.4903	18.464	22.2294	24.9069

Table 4. Values of selected RQA measures for structure with introduced damage (notch of 2.5 mm depth).

Point	p_1	p_2	p_3	p_4	p_5	p_6	p_7	p_8	p_9	p_{10}
	–	–	–	$m = 5$ $\tau = 10$	$m = 5$ $\tau = 3$	$m = 6$ $\tau = 3$	$m = 6$ $\tau = 3$	$m = 6$ $\tau = 3$	$m = 5$ $\tau = 3$	$m = 6$ $\tau = 2$
DET	–	–	–	0.8153	0.6898	0.7330	0.7193	0.7159	0.6831	0.7627
L	–	–	–	3.3181	4.3428	4.9141	4.8224	4.7411	4.2657	8.4597
ENTR	–	–	–	1.4546	1.5689	1.5624	1.5255	1.5060	1.5398	2.8030
LAM	–	–	–	0.8572	0.8223	0.8498	0.8372	0.8326	0.7733	0.6294
TT	–	–	–	3.2460	3.0919	3.2385	3.1456	3.1045	2.7176	2.2286
T2	–	–	–	24.6648	22.6008	24.2776	23.3515	22.9974	19.6112	15.3630

measurement point from the introduced damage location (point p_7) have proven that for points close to the damage location (Fig. 6):

- *DET* values are lower for all the damage depths than for the undamaged structure,
- increase in the depth of the introduced notch results in the decrease of the values of *L* and *ENTR* measures,

Table 5. Values of selected RQA measures for structure with introduced damage (notch of 4 mm depth).

Point	p_1	p_2	p_3	p_4	p_5	p_6	p_7	p_8	p_9	p_{10}
	$m = 7$ $\tau = 11$	$m = 7$ $\tau = 11$	$m = 5$ $\tau = 11$	$m = 6$ $\tau = 11$	$m = 7$ $\tau = 3$	$m = 7$ $\tau = 3$	$m = 7$ $\tau = 3$	$m = 6$ $\tau = 3$	$m = 7$ $\tau = 3$	$m = 6$ $\tau = 2$
DET	0.7748	0.8402	0.8111	0.7785	0.7177	0.7158	0.6998	0.6881	0.6581	0.7395
L	3.9800	4.0394	3.6172	3.4515	4.7081	4.7095	4.5739	4.4151	4.2858	7.9525
ENTR	1.6912	1.6472	1.5480	1.4698	1.4858	1.4905	1.4531	1.4496	1.3692	2.7771
LAM	0.8318	0.8767	0.8642	0.8431	0.8390	0.8383	0.8272	0.8193	0.7663	0.6482
TT	4.2331	4.0925	3.7397	3.4813	3.3105	3.2971	3.1683	3.1523	2.8119	2.3444
T2	27.5487	29.7693	27.3111	25.1025	24.2250	24.1344	23.1361	22.7332	19.8317	15.9678

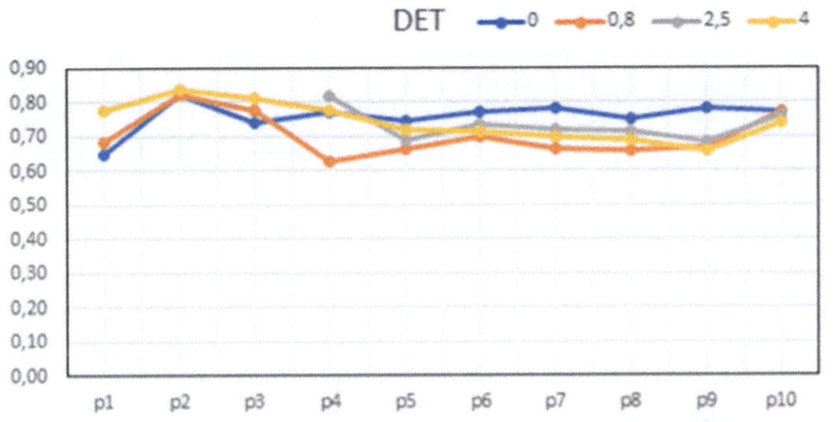

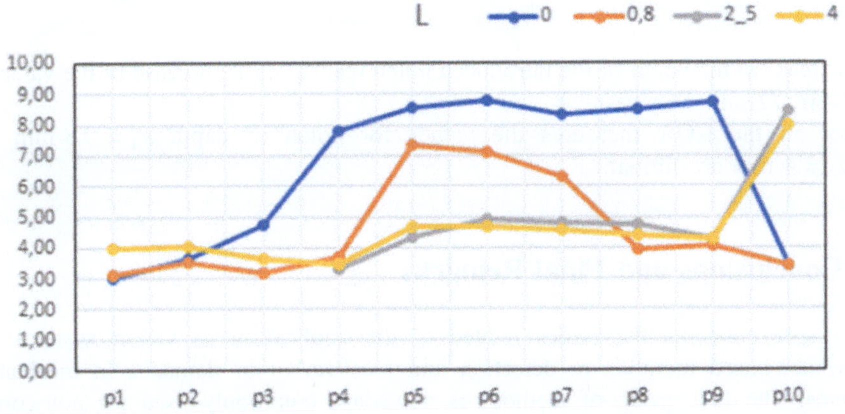

Fig. 5. Variability of RQA measures: *DET* and *L* with respect to damage (notch) depth.

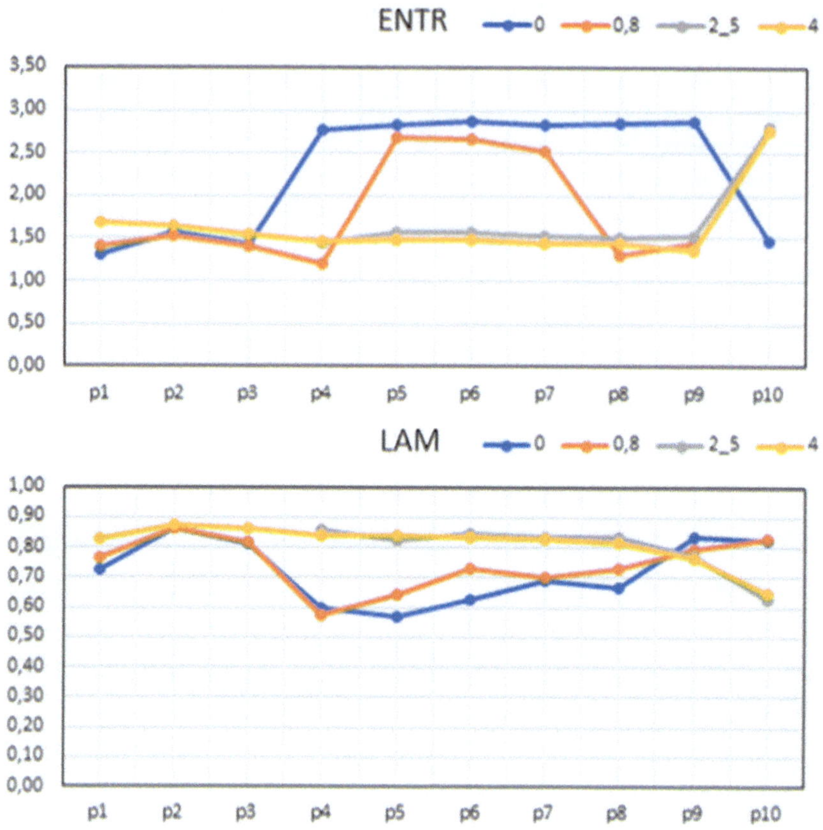

Fig. 6. Variability of RQA measures: *ENTR* and *LAM* with respect to damage (notch) depth.

- increase in the depth of the introduced notch results in the increase of the values of *LAM*, *TT* and *T2* measures,
- for all the RQA measures the values for notch of depth $d_1 = 0.8$ mm and $d_2 = 4$ mm are similar.

5 Conclusions and Final Remarks

The paper concerns the issues related to the application of vision systems and recurrence-based methods to detection and localization of damages in mechanical systems. The first group of methods is nowadays commonly used for non-contact measurements of system displacement time histories resulting from operational or controlled excitation. The second group of methods finds application in the analysis of nonlinear data from various fields of research, such as technique, astrophysics, biology, chemistry, geology, cardiology, neuroscience and economy.

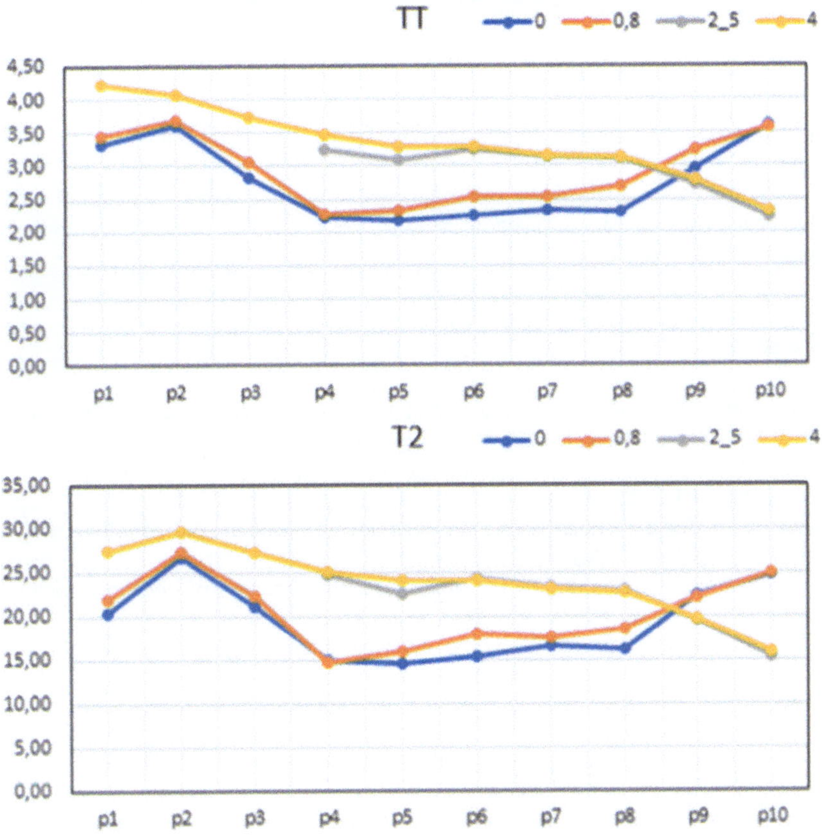

Fig. 7. Variability of RQA measures: *TT* and *T2* with respect to damage (notch) depth.

In the paper the authors discussed the idea of recurrence plots (RP) and Recurrence Quantification Analysis (RQA), briefly presented the most commonly used criterions and methods of embedding parameters (m, τ, ε) selection.

The research was carried out for the horizontal member of the aluminium frame subjected to the burst random excitation provided by the electrodynamical shaker. Dynamic responses of the horizontal member in the assumed measurement points were registered by two high speed cameras. 3D Digital Image Correlation (3D-DIC) technique was used for determining object deformation in a plane parallel to the image plane of the cameras. The process of damage propagation was simulated by the successive increasing the depth of the introduced fault. For each considered depth of damage (notch), measured time histories of system responses to burst random excitation were analysed with the application of RP and RQA methods. The smallest sufficient embedding dimension m was estimated with the use of the false nearest neighbours algorithm while the appropriate time delay τ was determined by searching for the first minimum of the mutual information function.

Obtained results proved that the RP and RQA methods are sensitive to changes in dynamical properties of the considered aluminium frame resulting from damage (notch) initialization and propagation.

Acknowledgments. The authors would like to acknowledge support from 16.16.130.942-KRIM subvention.

We would also like to acknowledge our collaborators: P. Kohut and K. Holak for their support with high speed cameras data acquisition; E. López-Alba and F.A. Díaz for their support and discussions on DIC procedures.

References

1. Heylen, W., Lammens, S., Sas, P.: Modal Analysis Theory and Testing. Katholieke Universiteit Leuven, Departement Werktuigkunde, Leuven (1997)
2. Maia, N.M.M., Silva, J.M.M.: Theoretical and Experimental Modal Analysis. Research Studies Press LTD, Taunton (1997)
3. Ewins, D.J.: Modal Testing: Theory, Practice and Application, 2nd edn. Research Studies Press, Hertfordshire (2000)
4. Doebling, S.W., Farrar, C.R., Prime, M.B.: A summary review of vibration-based damage identification methods. Shock. Vib. Dig. **30**(2), 91–105 (1998)
5. Cawley, P., Adams, R.D.: The location of defects in structures from measurements of natural frequencies. J. Strain Anal. **14**(2), 49–57 (1979)
6. Salawu, O.S.: Detection of structural damage through changes in frequency: a review. Eng. Struct. **19**(9), 718–723 (1997)
7. Marwan, N., Romano, M.C., Thiel, M., Kurths, J.: Recurrence plots for the analysis of complex systems. Phys. Rep. **438**, 237–329 (2007)
8. Poincaré, H.: Sur la probleme des trios corps et les equations de la dynamique. Acta Mathematica **13**, 1–271 (1890)
9. Eckmann, J.P., Oliffson, K.S., Ruelle, D.: Recurrence plots of dynamical systems. Europhys. Lett. **5**, 973–977 (1987)
10. Zbilut, J.P., Giuliani, A., Weber, C.L.: Recurrence quantification analysis and principal components in the detection of short complex signals. Phys. Lett. A **237**, 131–135 (1998)
11. Zbilut, J.P., Weber, C.L.: Embedding and delays as derived from quantification of recurrence plot. Physic. Lett. A **171**, 199–203 (1992)
12. Ngamga, E.J., Nandi, A., Ramaswamy, R., Romano, M.C., Thiel, M., Kurths, J.: Recurrences of strange attractors. PRAMANA J. Phys. **70**(6), 1039–1045 (2008)
13. Elwakil, A.S., Soliman, A.M.: Mathematical models of twin-T, wien-bridge and family of minimum component electronic chaos generators with demonstrative recurrence plots. Chaos, Solitons Fractals **10**(8), 1399–1411 (1999)
14. Kurths, J., Schwarz, U., Sonett, C.P., Parlitz, U.: Testing nonlinearity in radiocarbon data. Nonlinear Process. Geophys. **1**(1), 72–75 (1994)
15. Zolotova, N.V., Ponyavin, D.I.: Phase asynchrony of the north-south sunspot activity. Astron. Astrophys. **449**(1), L1–L4 (2006)
16. Manetti, C., Giuliani, A., Ceruso, M.A., Webber, C.L., Zbilut, J.P.: Recurrence analysis of hydration effects of nonlinear protein dynamics: multiplicative scaling and additive processes. Phys. Lett. A **281**(5–6), 317–323 (2001)

17. Giuliani, A., Manetti, C.: Hidden pecularities in the potential energy time series of a tripeptide highlighted by a recurrence plot analysis: a molecular dynamics simulation. Phys. Rev. E **53**(6), 6336–6340 (1996)
18. Marwan, N., Thiel, M., Nowaczyk, N.R.: Cross recurrence plot based synchronization of time series. Nonlinear Process. Geophys. **9**(3–4), 325–331 (2002)
19. Thomasson, N., Hoeppner, T.J., Webber, C.L., Zbilut, J.P.: Recurrence quantification in epileptic EEGs. Phys. Lett. A **279**(1–2), 94–101 (2001)
20. Zbilut, J.P., Koebbe, M., Loeb, H., Mayer-Kress, G.: Use of recurrence plots in the analysis of heart beat intervals. In: Proceedings of the IEEE Conference on Computers in Cardiology 1990, IEEE Computer Society Press, Chicago, pp. 263–266 (1991)
21. Hołyst, J.A., Zebrowska, M., Urbanowicz, K.: Observations of deterministic chaos in financial time series by recurrence plots, can one control chaotic economy? Eur. Phys. J. B **20**(4), 531–535 (2001)
22. Kantz, H., Schreiber, T.: Nonlinear Time Series Analysis. University Press, Cambridge (1997)
23. Cao, L.: Practical method for determining the minimum embedding dimension of a scalar time series. Phys. D **110**(1–2), 43–50 (1997)
24. Marwan, N.: Cross Recurrence Plot Toolbox for Matlab, Reference Manual, Version 5.15, Release 28.6. Available at: http://tocsy.pik-potsdam.de (2010)
25. Iwaniec, J., Uhl, T., Staszewski, W.J., Klepka, A.: Detection of changes in cracked aluminium plate determinism by recurrence analysis. Nonlinear Dyn. **70**(1), 125–140 (2012)
26. Iwaniec, J.: Investigation of selected mechanical systems by recurrence plots method. Int. J. Struct. Stab. Dyn. **13**(7), 1340008-1–1340008-10 (2013)
27. Iwaniec, J., Kurowski, P.: Experimental verification of selected methods sensitivity to damage size and location. J. Vib. Control **23**(7), 1133–1151 (2017)

Fault Detection for a Satellite-like Structure Using Sine Sweep Vibration Test Data

Gao Haiyang[1], Guo Xinglin[2(✉)], Xie Yicun[1], Yang Yanjing[1], and Ouyang Huajiang[2,3]

[1] Beijing Institute of Spacecraft Environment Engineering, Beijing 100094, People's Republic of China
eric415@126.com
[2] State Key Laboratory of Structural Analysis for Industrial Equipment, Dalian University of Technology, Dalian 116024, People's Republic of China
xlguo@dlut.edu.cn
[3] University of Liverpool, Liverpool L69 3BX, UK

Abstract. Sine sweep vibration test is one of the most important means for verifying the dynamic environmental adaptation and assessing the structural dynamic characteristics of a satellite structure, which should be generally carried out on the ground before launching. Many vibration-based techniques have been investigated and applied to civil structures and mechanical systems. However, very few studies have been reported on their applications to satellite structures. This paper makes fault detection of a satellite-like structure using sine sweep date. Firstly, the time and frequency domain acceleration responses on some specified points are measured to form the fault detection indicators respectively, which are derived from the variational modal decomposition technique. Then the indicators of three-dimensional tests are combined based on Dempster-Shafer's evidence theory. Finally, the proposed method is implemented in an experimental study on a two-story satellite-like structure. The result proves the effectiveness and feasibility of the proposed method. In addition, the effect of selection of data segments on the detection results is also discussed.

Keywords: Fault detection · Satellite structure · Sine sweep · Variational modal decomposition

1 Introduction

As we know that most satellites are unique and very difficult to repair in orbit. Therefore, high reliability is the basic requirement for the normal functioning of a satellite. Before operation on orbit, a satellite would be subjected to various types of mechanical environmental effects including vibration, noise, impact and acceleration. These are generally produced during transportation, loading and unloading, launch, flight and separation [1]. The above mentioned mechanical environment causes structural deformation, and may even damage the satellite and its components, such as mechanism and electronic equipment failure. The occurrence of these faults may affect the success of the flight mission or even lead to the termination of the entire flight

M. A. Wahab (Ed.): DAMAS 2019, LNME, pp. 474–486, 2020.
https://doi.org/10.1007/978-981-13-8331-1_34

mission. Consequently, the structural dynamic characteristics of the satellite should be strictly assessed before the scheduled launch.

Sine sweep vibration test is one of the most important means for verifying the dynamic environmental adaptation, in which the acceptance test, developed based on the satellite-launch vehicle coupling analysis, is a widespread adoption. To diagnose whether the structure is damaged during the acceptance test, two characteristic tests with the same excitation levels should be performed before and after the acceptance test. More specifically, the amplitude-frequency response curves of each measurement point from the two characteristic tests are compared to find out whether there is any frequency shift and amplitude change. However, this method can only make a qualitative judgment, and cannot provide the fault locations and failure modes, which often requires human judgments, and sometimes affects the duration of satellite development. More importantly, only the characteristics of the frequency domain data are used without the consideration of time domain characteristics that results in information omission, and thus affects the accuracy of fault localization. Although this method has solved some problems in practical applications in the past, it needs further theoretical investigation. Meanwhile, some enhanced signal processing techniques also should be employed to deal with nonlinear and non-stationary signals in order to improve the performance of the fault detection methods (Fig. 1).

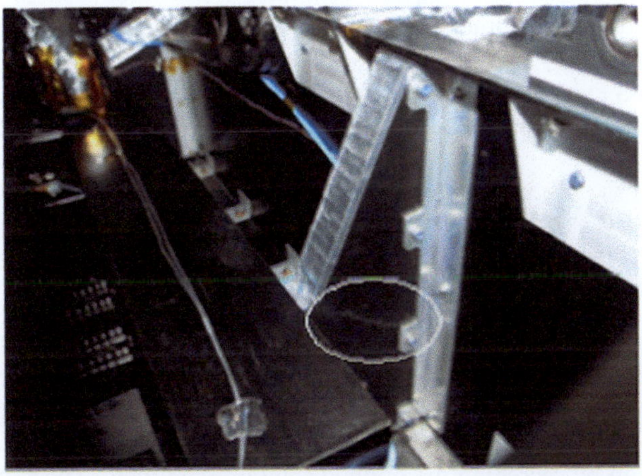

Fig. 1. The fracture of main satellite structure

A large number of vibration-based methods have been proposed and developed for structural damage detection and identification in civil engineering [2–5]. An overview of several damage location indicators built on modal parameters and their derivatives can be found in [6]. However, it is well known that these modal-based methods are more sensitive to noise contamination and require more sensors to obtain sufficient data. In contrast, some damage identification methods based on time-frequency domain analysis have attracted significant attention due to their strong capability of noise immunity. Some publications [7, 8] show the great advantages of various wavelet analysis techniques, and also indicate that the wavelet basis needs to be selected and cannot be replaced in the analysis process, which sometimes makes the damage identification task more difficult, especially when using the non-stationary signals with significantly varying frequency.

To be suitable for non-stationary signal analysis, some researchers studied and introduced the EMD (empirical mode decomposition) technique [5, 9, 10]. The IMFs (intrinsic mode functions) of the EMD and their derivatives are widely used for damage detection of mechanical systems. Xie et al. [10] presented a crack diagnosis method on a box structure based on empirical mode decomposition and cosine-based indicator, in which the sine sweep vibration responses in three directions were used to deduce damage detection indicators. It should be highlighted that the EMD has some problems associated with frequency mix and the signal boundary treatment uncertainty [11]. Besides, a disadvantage can be seen in [10] that it is inefficient and time consuming when decomposing a large-scaled time domain data of the sine sweep test. As an alternative to the EMD, the VMD (variational mode decomposition) [12] was recently proposed for the separation of composite real-valued time series into respective modes, which is less sensitive to noise and has the ability of avoiding frequency mix and boundary effect, compared with the EMD.

Motivated by the work of [10], a novel fault detection method is proposed based on sine sweep test data for fault detection of a satellite structure in this paper. In Sect. 2, the VMD technique and mean square definition were introduced to build the fault detection indicators using time and frequency domain acceleration responses, respectively. To better localize the fault the indicators of each direction test were combined by means of the D-S evidence theory. Section 3 presents an experimental study on a two-story satellite-like structure. Finally, some concluding remarks are made in Sect. 4.

2 Theory and Method

2.1 Variational Mode Decomposition

The VMD technique, as a multiple-scale signal processing method, can decompose a real signal $x(t)$ into a series of modes. Each mode is compact around a center frequency ω_k. The original signal is carried into the variational mode to acquire the center frequency and corresponding bandwidth, and then adaptively achieve the frequency domain separation by searching the optimal solution iteratively. The following scheme is proposed. (1) Hilbert transform firstly is applied to every mode u_k to obtain a unilateral frequency spectrum. (2) Then adding an exponential tuned $e^{-j\omega_k t}$ to each estimated center frequency and shift the frequency spectrum of every mode to the baseband. (3) Estimating the bandwidth of each mode by using Gaussian smoothness. In this way, the problem of decomposition can be defined as follows [12]:

$$\begin{cases} \min\left\{\sum_k \left\|\partial\left\{\left[\delta(t) + \frac{j}{\pi t}\right] * u_k(t)\right\}e^{-j\omega_k t}\right\|_2^2\right\} \\ \sum_k u_k = x(t) \end{cases} \tag{1}$$

In order to confirm the reconstruction constraint, a quadratic penalty term and Lagrangian multipliers λ are introduced. The problem can be rewritten in the following format.

$$
\begin{aligned}
L(\{u_k\},\{\omega_k\},\lambda) = \alpha \sum\nolimits_k \left\| \partial \left\{ \left(\delta(t) + \frac{j}{\pi t} \right) * u_k(t) \right\} e^{-j\omega t} \right\|_2^2 \\
+ \left\| x(t) - \sum\nolimits_k u_k(t) \right\|_2^2 + \left\langle \lambda(t), x(t) - \sum\nolimits_k u_t \right\rangle
\end{aligned}
\tag{2}
$$

where α is the balancing parameter of the data-fidelity constraint. Then the ADMM (Alternate Direction Method of Multipliers) is used to get the optimal solution upon the variational mode. And the modes u_k and ω_k can be derived iteratively. Detailed theory of VMD is given in [12]. Some previous publications indicate that the VMD is different from the EMD in dealing with the problem on frequency mix and the boundary effects [11].

2.2 VMD-Based Fault Detection Indicator

It is known that damages can cause a structural response change in the same excitation conditions, particularly at those locations close to the damage positions, which is attributed to the change of dynamic characteristics. Considering the acceleration responses $x_i^U(t)$ and $x_i^D(t)$ on the ith measurement point obtained from the undamaged structure and the damaged one. These responses are decomposed based on the VMD to gain n modes $u_1^U(t)$, $u_2^U(t)$, …, $u_n^U(t)$ and $u_1^D(t)$, $u_2^D(t)$, …, $u_n^D(t)$, respectively. Then the sum values of each mode can be calculated and organized as a vector in Eq. (3).

$$
\begin{aligned}
s_i^U &= \left[\sum\nolimits_{t_1}^{t_2} u_1^U \sum\nolimits_{t_1}^{t_2} u_2^U \cdots \sum\nolimits_{t_1}^{t_2} u_n^U \right] \\
s_i^D &= \left[\sum\nolimits_{t_1}^{t_2} u_1^D \sum\nolimits_{t_1}^{t_2} u_2^D \cdots \sum\nolimits_{t_1}^{t_2} u_n^D \right]
\end{aligned}
\tag{3}
$$

where t_1 and t_2 are the bounds of the selected time segment. The two vectors could be treated by normalization method. The cosine-based fault detection indicator on the ith measurement point can be written as follows.

$$
VFDI_i = \cos^{-1}(s_i^U \cdot s_i^D) \tag{4}
$$

2.3 Area-Based Fault Detection Indicator

During the test, the vibration system generates a sine sweep force which acts on the test specimen. Meanwhile, the frequency of excitation f_t is exported through the COLA (constant output level adaptor) signal. Although the response data of sine sweep vibration are deterministic periodic data, they are non-stationary. As the frequency varies linearly or logarithmically, the amplitude also changes. In addition, in the vibration test, since the satellite and its components are generally a complex system

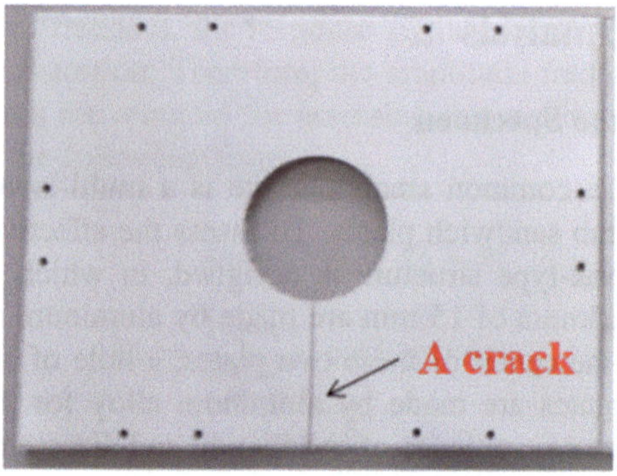

Fig. 3. A crack along with Y-axis on the top plate

Table 1. Experiment condition in X, Y and Z directions

Sweep rate (oct/min)	Frequency range (Hz)	Acceleration level (g)
4	5–500	0.1

signals in every direction vibration test. Point 1 is the closest to the crack. A 50 kN shaker is adopted to perform the sine sweep vibration test. The SD2560 vibration control system and LMS data acquisition system are employed. The sampling frequency is set to 3200 Hz. The tests were carried out in the order of X, Y and Z directions. For the construction of the *VFDI* vector, the mode number of VMD is set to 4, that is $n = 4$ in Eq. (3).

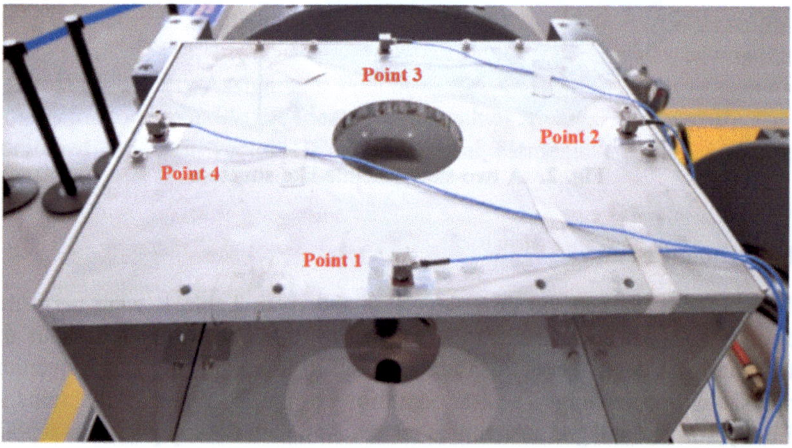

Fig. 4. The measurement point placement

3.3 Result and Discussion

For the test of X-direction, X is the direction of primary vibration and the responses in the X-axis is the maximal. Figure 5 shows the time domain acceleration series of Point 1 from the undamaged structure. During the test, the amplitude-frequency curves corresponding to the time series are simultaneously generated based on COLA signal using Eq. (5). Figure 6 shows the amplitude-frequency curve comparison of Point 1 between the undamaged structure and the damaged one. It can be seen in Fig. 6 that when the structure is damaged, the amplitude is higher than that of the undamaged structure in the frequency interval of about 5–30 Hz, which corresponds to the non-resonant regions of the structure. At the locations close to the resonant regions (100–300 Hz) of the structure, the increase and decrease of the amplitude values at certain natural frequency points appear random. Consequently, data of the non-resonant regions and resonant regions, and their corresponding time segments can be used to establish the *AFDI* and *VFDI*.

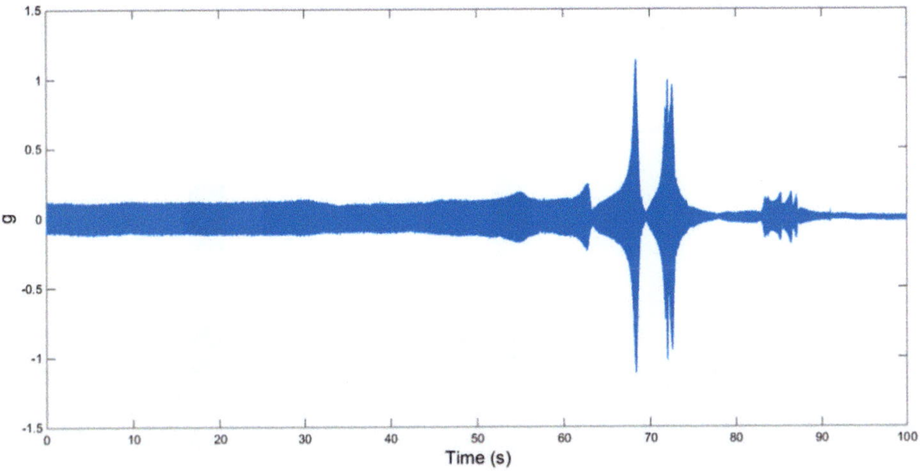

Fig. 5. X-axis response of Point 1 from undamaged structure in X-direction sine sweep test

In the same way, the test of the Y-direction and the Z-direction can be also carried out to provide more information for fault detection. After the combination of the *AFDI* and *VFDI* of each direction test, the sine sweep test of all three directions also can be combined by Eq. (7) to obtain the overall indicator.

Figures 7–9 show the fault indicators using the data in the non-resonant interval (5 Hz to 30 Hz) of each direction test. It can be seen that all the combined indicators show the crack is close to Point 1, in which the indicators of the X-direction test are more sensitive to the damage in the non-resonant range, compared with other direction tests.

Figures 10–12 illustrate the fault indicators using the data in the resonant interval (100 Hz to 300 Hz) of each direction test. It is obvious that the indicators of the

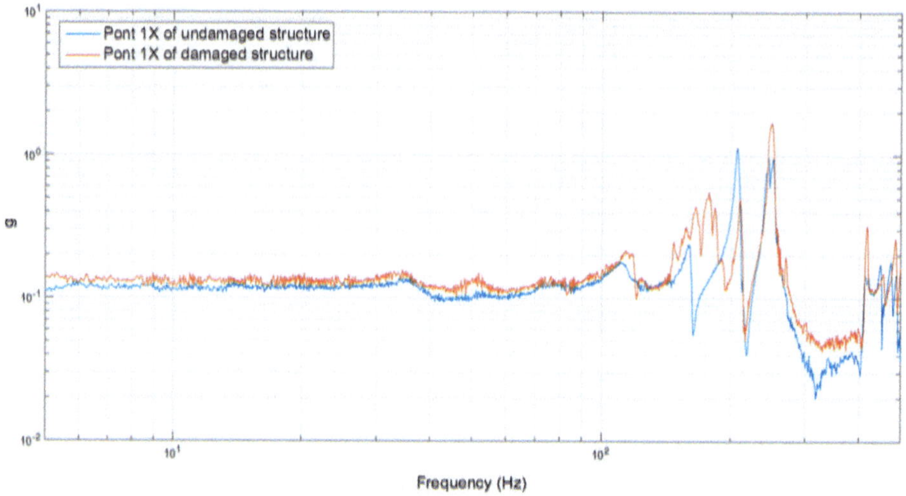

Fig. 6. The amplitude-frequency curve comparison of point 1 between the undamaged structure and the damaged structure

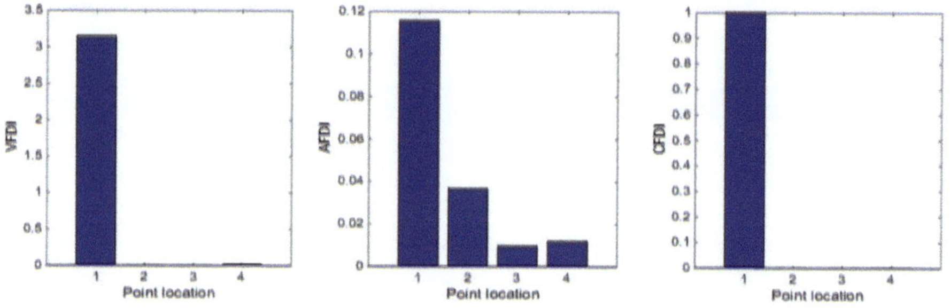

Fig. 7. Fault indicators of X-direction test using data in non-resonant interval

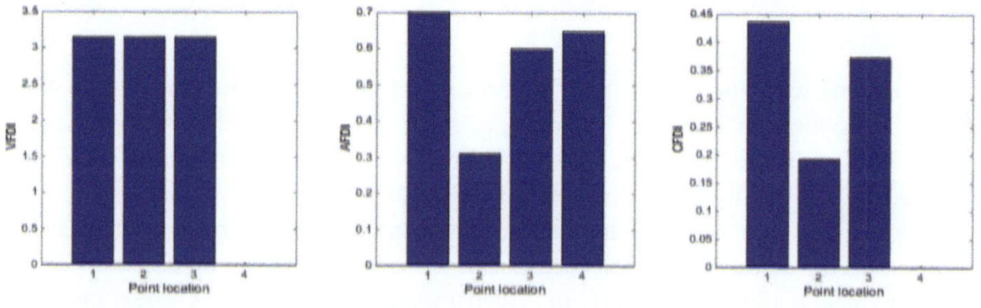

Fig. 8. Fault indicators of Y-direction test using data in non-resonant interval

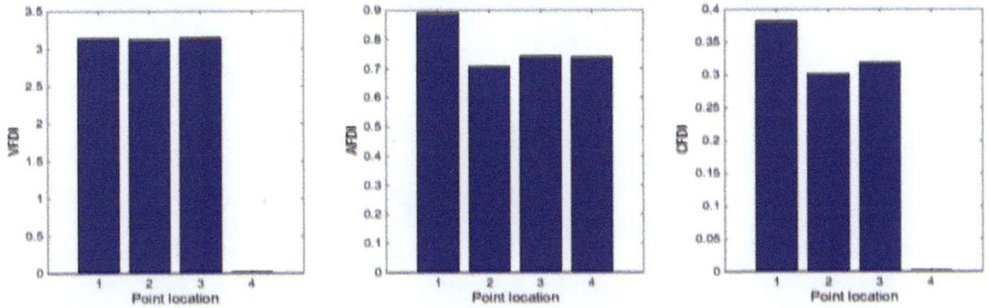

Fig. 9. Fault indicators of Z-direction test using data in non-resonant interval

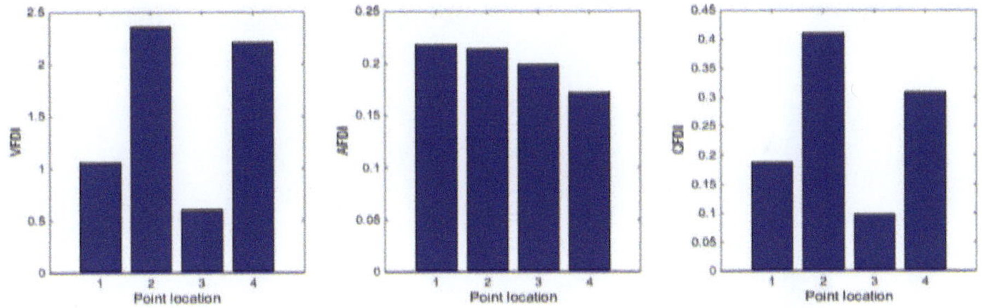

Fig. 10. Fault indicators of X-direction test using data in resonant interval

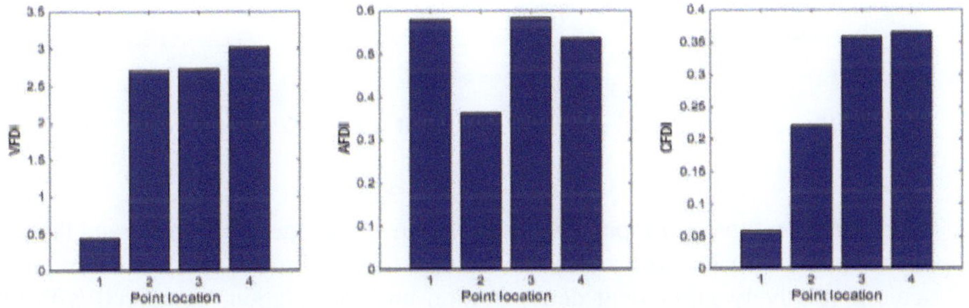

Fig. 11. Fault indicators of Y-direction test using data in resonant interval

Z-direction test are more sensitive to the damage by using the data segment in the resonant range. As shown in Fig. 13, the overall indicators can be obtained based on D-S evidence theory by using the *CFDI*s of all the three direction tests, which indicates that the most likely damaged location is close to Point 1.

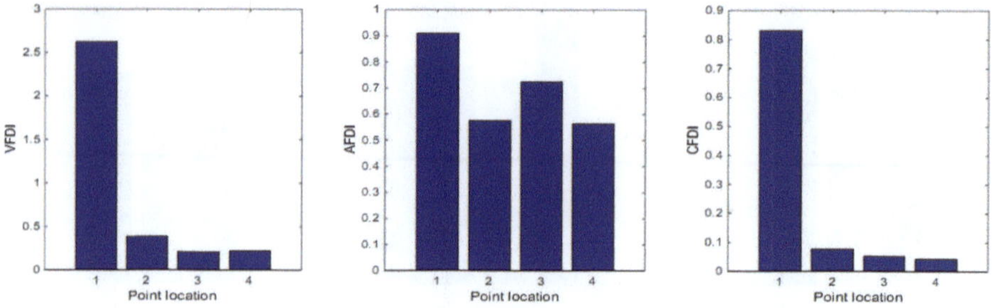

Fig. 12. Fault indicators of Z-direction test using data in resonant interval

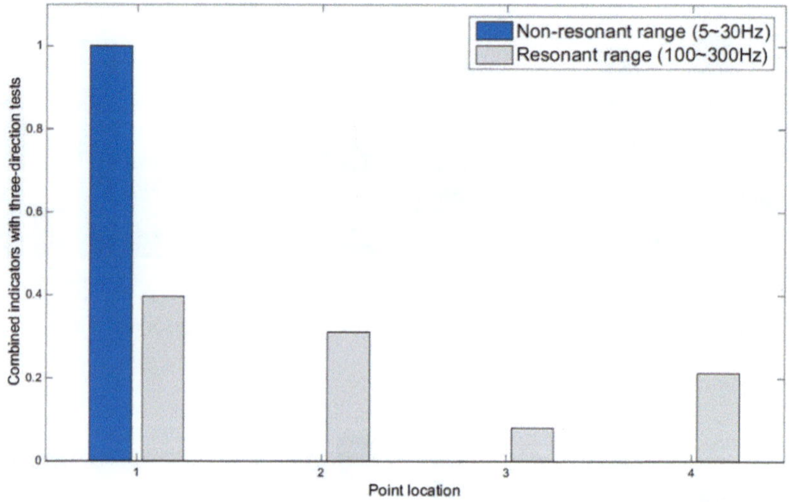

Fig. 13. The overall indicators combined with three direction test

The calculation time is compared in Table 2. An advantage can be observed that the proposed method by using data segments can significantly reduce the calculation time, and achieve relatively rapid fault determination and localization, compared with the EMD-based method.

Table 2. The comparison of calculation time

Segment selection	Frequency range	Calculation time of *VFDI*	Calculation time of *AFDI*
Non-resonant data	5–30 Hz	41.141 s	0.187 s
Resonant data	100–300 Hz	801.207 s	0.205 s
EMD-based method in [10]	5–500 Hz	29243.208 s	–

4 Conclusions

This paper puts forwards a fault detection method based on sine sweep test data. An experimental study on a two-story satellite-like structure demonstrate the effectiveness of the proposed method, and some conclusions can be drawn as follows: (1) Firstly, the variational mode decomposition technique can be used to deal with non-stationary sine sweep responses in time domain for fault detection. The fault indicator based on the VMD's modes is sensitive to local damage. (2) The data in time and frequency domains can be divided into several segments to reduce the time of variational mode decomposition. Different data segments have different sensitivities to the damage. (3) Responses from three direction tests can be used to increase useful information. Combined indicators, based on the Dempster-Shafer's theory, provide a clearer result on likely faulty regions, which remarkably improve the accuracy of crack localization.

Acknowledgement. This work is funded by the Open Fund of State Key Laboratory of Structural Analysis for Industrial Equipment, Dalian University of Technology (GZ18118).

References

1. Burke, W.R.: Spacecraft structures and mechanical testing. Esa Special Publication (1991)
2. Bayissa, W.L., Haritos, N.: Structural damage identification in plates using spectral strain energy analysis. J. Sound Vib. **307**(1), 226–249 (2007)
3. Li, J., Dackermann, U., Xu, Y., Samali, B.: Damage identification in civil engineering structures utilizing PCA-compressed residual frequency response functions and neural network ensembles. Struct. Control. Health Monit. **18**(2), 207–226 (2015)
4. Chang, K.C., Kim, C.W.: Modal-parameter identification and vibration-based damage detection of a damaged steel truss bridge. Eng. Struct. **122**, 156–173 (2016)
5. Han, J., Zheng, P., Wang, H.: Structural modal parameter identification and damage diagnosis based on Hilbert-Huang transform. Earthq. Eng. Eng. Vib. **13**(1), 101–111 (2014)
6. Frans, R., Arfiadi, Y., Parung, H.: Comparative study of mode shapes curvature and damage locating vector methods for damage detection of structures. Procedia Eng. **171**, 1263–1271 (2017)
7. Alamdari, M.M., Li, J., Samali, B.: Damage identification using 2-D discrete wavelet transform on extended operational mode shapes. Arch. Civ. Mech. Eng. **15**(3), 698–710 (2015)
8. Xiang, Z., Chen, R., Zhou, Q.: Damage identification using wavelet packet transform and neural network ensembles. Int. J. Struct. Stab. Dyn. In Press (2018)
9. Yi, S., Zhang, Y., Wang, Z.: Satellite fault diagnosis method based on predictive filter and empirical mode decomposition. J. Syst. Eng. Electron. **22**(1), 83–87 (2011)
10. Xie, Y., Fan, S., Yang, Y.: A new crack diagnosis method for box structures based on empirical mode decomposition. Spacecr. Environ. Eng. **32**(5), 489–495 (2015)
11. Ming, Z., Jiang, Z., Feng, K.: Research on variational mode decomposition in rolling bearings fault diagnosis of the multistage centrifugal pump. Mech. Syst. Signal Process. **93** (460), 460–493 (2017)
12. Dragomiretskiy, K., Zosso, D.: Variational mode decomposition. IEEE Trans. Signal Process. **62**(3), 531–544 (2014)

13. Bao, Y., Li, H., Ou, J.A.: Dempster-Shafer evidence theory-based approach for online structural health monitoring. Proc Spie **20**(9), 1191–1210 (2008)
14. Guo, T., Xu, Z.: Data fusion of multi-scale representations for structural damage detection. Mech. Syst. Signal Process. **98**, 1020–1033 (2018)

Debonding Detection in Reinforced Concrete Beams with the Use of Guided Wave Propagation

Beata Zima(✉)

Faculty of Civil and Environmental Engineering,
Department of Mechanics of Materials and Structures,
Gdańsk University of Technology, 80-233 Gdańsk, Poland
beata.zima@pg.edu.pl

Abstract. One of the most frequent damage of the reinforced concrete structures is debonding between steel bar and concrete cover. In the case of debonding occurrence not only the strength of the structure decreases, but also it is more vulnerable to corrosion damages. For this reason fast and effective methods of debonding detection in an early stage of its development need a significant boost. The paper presents analytical and experimental investigation of debonding detection in reinforced concrete beams using non-destructive method based on guided waves propagation. Concrete beams of rectangular cross-section consisting of four steel rebars with pre-existing debonding are investigated. Lack of adhesive connection between one rod and concrete cover is provided by introducing cellophane film of a small thickness (90 μm). The research is focused on detection of debonding with variable length on the basis of time-domain signals captured by piezoelectric sensors. Presented method of damage detection takes advantage of the time of flight of the reflections registered at the ends of the specimens.

Keywords: Debonding · Damage detection · Guided wave · Reinforced concrete

1 Introduction

One of the most frequently used material in civil engineering is reinforced concrete. Despite its durability and resistance to potential defects, it is constantly exposed to adverse environmental conditions, overloads or poor manufacturing. Various types of damages can occur in concrete, however, one of them is significantly dangerous for the state of the whole structure. Debonding between steel bar and concrete cover is difficult to detect, especially in an early stage of its development, when it is relatively small. It usually develops under concrete cover, which hinders its monitoring. A potentially useful non-destructive method for internal damage detection is guided wave propagation, which experienced a rapid development recent decades.

Guided waves allow to monitor large areas of tested structure with the use of small, light and relatively low-cost sensors [1]. An appropriate interpretation of captured signals allows to monitor the occurrence of structure severities. This approach have

M. A. Wahab (Ed.): DAMAS 2019, LNME, pp. 487–497, 2020.
https://doi.org/10.1007/978-981-13-8331-1_35

been investigated by many researchers and it was showed that elastic waves can be successfully applied in diagnostics of various types of structures with different damages (e.g. [2–5]). Because guided waves allow to inspect object of a considerable size without interfering with its integrity, they are highly attractive in civil engineering field when structural elements require continuous state assessment. The possibility of using waves in debonding detection was demonstrated in many papers. Na et al. [6] detected debonding in reinforced concrete blocks with varying length. Wu and Chang ([7, 8]) monitored amplitude and travel time of signals excited and registered at the bar, which was embedded in concrete beam. Li et al. [9] used a time reversal method for debonding detection in concrete beams under different CFRP bonding conditions. Guided waves in debonding detection in specimens of a circular-cross section were used by Zima and Rucka ([10, 11]). Reference-free method of determining debonding length in concrete beams with multiple debondings was proposed by Zima and Kędra [12]. It was shown the average wave velocity changes according to bonding and debonding lengths.

This study presents results of experimental investigation of wave propagation in reinforced concrete beams with partially debonded steel rods. The ultimate objective of the paper is to analyze how excitation frequency influences on effectiveness of damage detection. The investigations were conducted for five different beams with variable length of debonding and two different excitation frequencies.

2 Theoretical Background of Wave Propagation in Reinforced Concrete Specimens

Guided waves are dispersive waves and their propagation characteristics depend on specimen material and type of cross-section. The description of wave propagation phenomenon in beam with partial debonding between rod and concrete cover requires considering two types of cross-sections: single, debonded rod (Fig. 1a) and rectangular cross-section of the beam with four healthy connected rods (Fig. 1b).

2.1 Wave Propagation in Single Waveguide

When steel rod is debonded and it has no adhesive connection with the cover, it can be considered as single waveguide with circular-cross section made of elastic and isotropic material. The relation propagation velocity-frequency is then described by the dispersion equation given explicitly by Pochhammer [13] and Chree [14]:

$$\frac{2\alpha}{r}\left(\beta^2+k^2\right)J_1(\alpha r)J_1(\beta r)-\left(\beta^2-k^2\right)^2J_0(\alpha r)J_1(\beta r)+ \\ -4k^2\alpha\beta J_1(\alpha r)J_0(\beta r)=0, \tag{1}$$

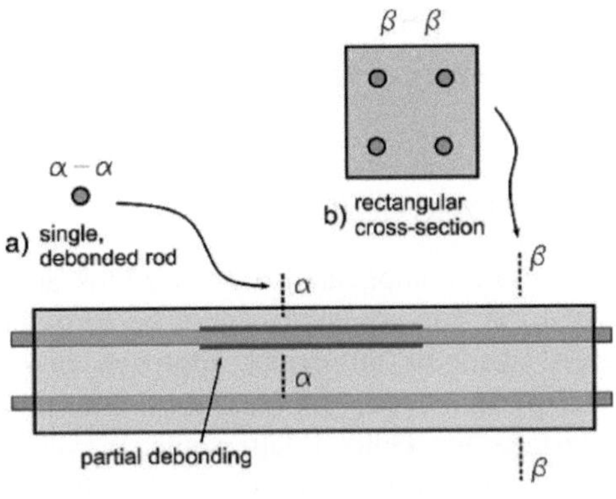

Fig. 1. Cross-sections distinguished in reinforced concrete beam with partial debonding between rod and concrete cover: (a) single, debonded rod and (b) rectangular cross-section with four healthy connected rods.

where parameters α and β depend on wavenumber k, angular frequency ω and propagation velocity of pressure wave c_P and shear wave c_S:

$$\alpha^2 = \frac{\omega^2}{c_P^2} - k^2,\ \beta^2 = \frac{\omega^2}{c_S^2} - k^2. \tag{2}$$

Functions J_0 and J_1 are Bessel's function of the first kind, while r is the radius of the rod. The solution of dispersion equation is usually presented as dispersion curves relating frequency and group velocity, which is defined as follows:

$$c_g = \frac{d\omega}{dk}. \tag{3}$$

Equation (1) was solved in MATLAB environment and the results are given in Fig. 2 in the form of dispersion curve. The calculations were performed for steel rod,

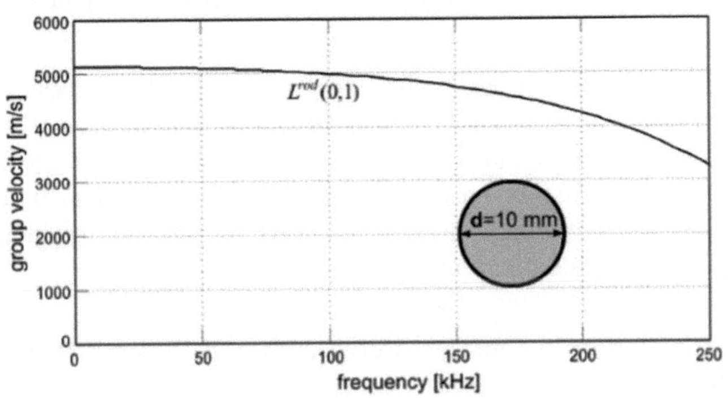

Fig. 2. Dispersion curve representing longitudinal mode for steel rod with circular cross-section (E = 207 GPa, ν = 0.3, and ρ = 7894 kg/m^3).

which is experimentally investigated in the further part of the paper. One can see that for frequency range 0–250 kHz only one longitudinal mode denoted as $L^{rod}(0,1)$ can be distinguished.

2.2 Wave Propagation in Concrete Beam

Unfortunately, in case of more complicated cross-sections analytical solution in the form of explicitly given dispersion equation cannot be formulated. Then, dispersion solution can be obtained using a numerical approach based on finite elements ([15–17]). In this study dispersion curves for cross-section of reinforced concrete beam the Graphical User Interface for Guided Ultrasonic Waves program (GUIGUW), written by Bocchini et al. [18] was used (see Fig. 3). GUIGUW allows to obtain the curves for the all wave mode families (longitudinal, flexural and torsional). However, in experimental investigation discussed later, waves in specimens are excited longitudinally. For this reason, only longitudinal wave modes should be taken into consideration. In addition, because of the significant number of the curves, the labels *L* and *F* denoting longitudinal and flexural modes were assigned only to the modes with the highest group velocity.

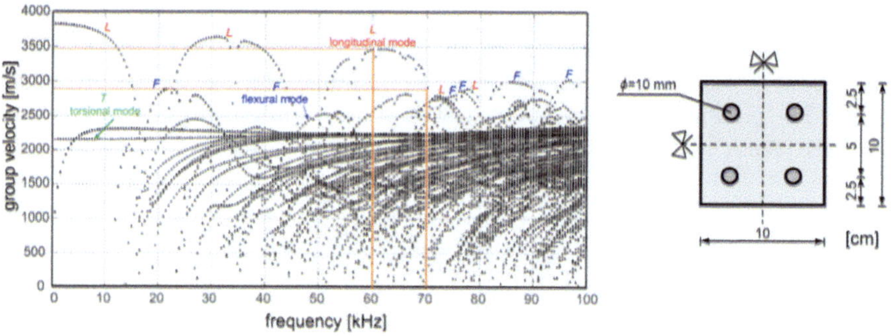

Fig. 3. Dispersion curves for reinforced concrete cross-section (steel: E = 207 GPa, ν = 0.3, and ρ = 7894 kg/m^3 and concrete: E = 30.05 GPa, ν = 0.2, and ρ = 2360 kg/m^3).

It can be seen that the number of possible wave modes for particular frequency is much higher for more complicated cross-section (compare Figs. 2 and 3), even though, dispersion curves for rectangular cross-section were plotted in narrower frequency range (0–100 kHz). The higher number of wave modes which can be generated can cause the increase of complexity of observed wave propagation phenomenon and in consequence hinders results interpretation.

2.3 Wave Propagation in Concrete Beam with and Without Partial Debonding

On the basis of dispersion solutions given in previous paragraph, differences between wave propagation in undamaged and damaged concrete beam are described [12].

Figure 4 presents scheme of wave propagation in beam without debonding. After longitudinal wave excitation (Fig. 4a) in beam several wave modes start propagating (Fig. 4b) and their number depends on excitation frequency (see Fig. 3). As a first travels mode with the highest velocity denoted as c_b. When it reaches the end of specimen reflects and travels back to the start end of the beam (Fig. 4c).

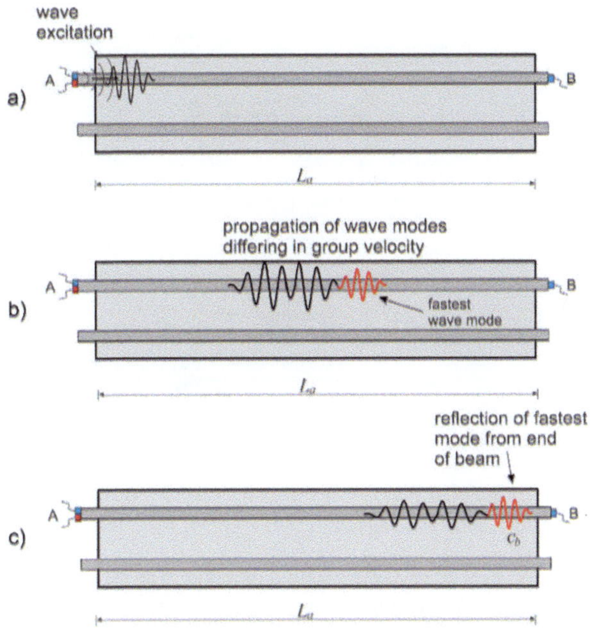

Fig. 4. Wave propagation in concrete beam: (a) wave excitation, (b) propagation of wave modes, (c) reflection from the end of specimen.

For specimen with total length L_a the registration times for the fastest mode at the ends A and B are as follows:

$$t_A = \frac{2L_a}{c_b}. \tag{4}$$

$$t_B = \frac{L_a}{c_b}. \tag{5}$$

Figure 5 presents scheme of wave propagation phenomenon for beam with debonding of length L_f [12]. After wave excitation along undamaged part a number of wave modes starts propagating. When they sequentially reach the start of debonding, they diffract and convert into longitudinal modes, which can exist in single debonded rod. According to dispersive solution in frequency range 0–200 kHz only one longitudinal mode can be generated (see Fig. 2), so the converted modes propagate along the rod with the same velocity denoted as c_f. After diffraction at the end of debonding they

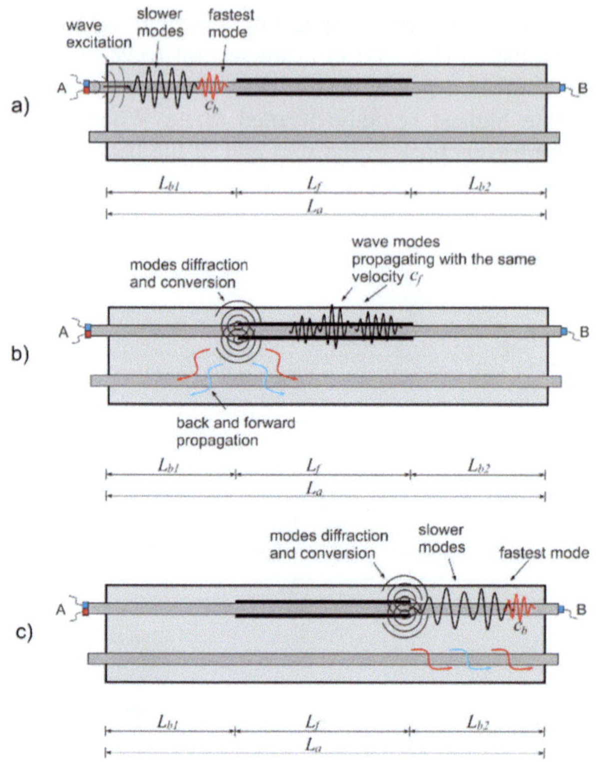

Fig. 5. Wave propagation in concrete beam with debonding: (a) wave excitation, (b) diffraction and mode conversion, (c) reflection from the end of specimen.

are again converted and travel with variable velocities in accordance with dispersive solution given in Fig. 3. In this case, the fastest mode is registered at the end B after the time of flight [12]:

$$t_a^B = \frac{L_f}{c_f} + \frac{L_b}{c_b} = \frac{L_a c_f - L_f\left(c_f - c_b\right)}{c_b c_f}, \quad (6)$$

where $L_b = L_{b1} + L_{b2}$. The relation (6) is a linear function with a slope:

$$\frac{-\left(c_f - c_b\right)}{c_b c_f}. \quad (7)$$

To register the first reflection at the end A the time needed is twice as long:

$$t_a^A = 2t_a^B = \frac{2L_a c_f - 2L_f(c_f - c_b)}{c_b c_f}. \tag{8}$$

The slope of the linear function (8) is:

$$\frac{-2(c_f - c_b)}{c_b c_f}. \tag{9}$$

In can be seen that the travel time depends on damage length L_f and the difference between velocities of the fastest mode in both, bonded and debonded part. When $c_f > c_b$ and the slope is negative, the increase of damage length is related with decrease of travel time, when $c_f < c_b$ and the slope is positive, the increase of damage length causes increase of travel time and when $c_f = c_b$ and the slope is zero, the occurrence of damage has no influence on travel time.

3 Experimental Investigation

3.1 Beam Model

The geometry of the experimental models is given in Fig. 6. The reinforced concrete beams of rectangular cross-sections consist of four steel rods with diameter of 1 cm, which were embedded longitudinally in concrete blocks.

The height and width of each beam was 10 cm, while the total length of the specimen was 50 cm. The material parameters were as follows: E = 207 GPa, ν = 0.3, and ρ = 7894 kg/m^3 for steel and for concrete: E = 30.05 GPa, ν = 0.2, and ρ = 2360 kg/m^3. Five different damage scenarios were taken into account (Fig. 6). Beam #A was undamaged. In beams #B, #C and #D deboning length was equal to 20, 30 and 40 cm, respectively, while in beam #E rod was totally debonded. The debonding was performed artificially by introducing cellophane film with total thickness of 90 μm.

3.2 Experimental Set up

Non-destructive experiment was conducted with the use of plate piezo actuators Noliac NAC2012 which generated and measured elastic waves at the ends of the rods embedded in concrete block. Waves were excited and acquired by the device PAQ-16000D.

The photo of experimental set up for the wave propagation experiments is given in Fig. 7. The excitation signal was a wave packet consisted of a ten-cycle sine function with a carrier frequency of 60 kHz and 70 kHz modulated by the Hanning window. The excitation frequency was selected on the basis of the difference between wave velocities c_b and c_f (see Figs. 2 and 3).

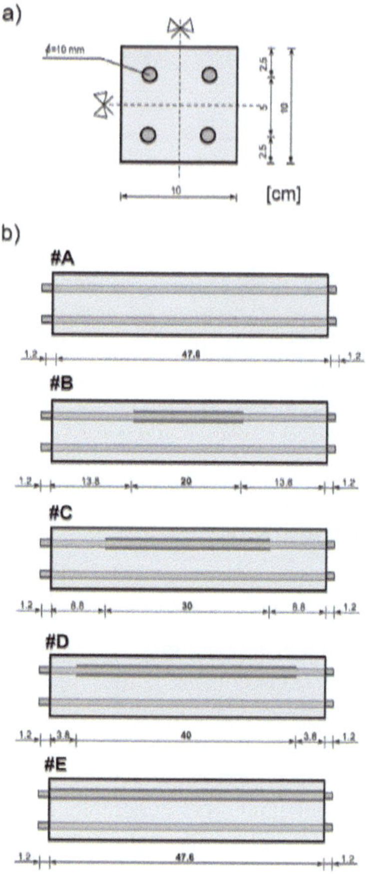

Fig. 6. Schemes of investigated specimens: (a) cross-section of the beam, (b) beams with variable debonding length (beam #A – 0 cm; #B – 20 cm; #C – 30 cm; #D – 40 cm and #E – totally debonded rod).

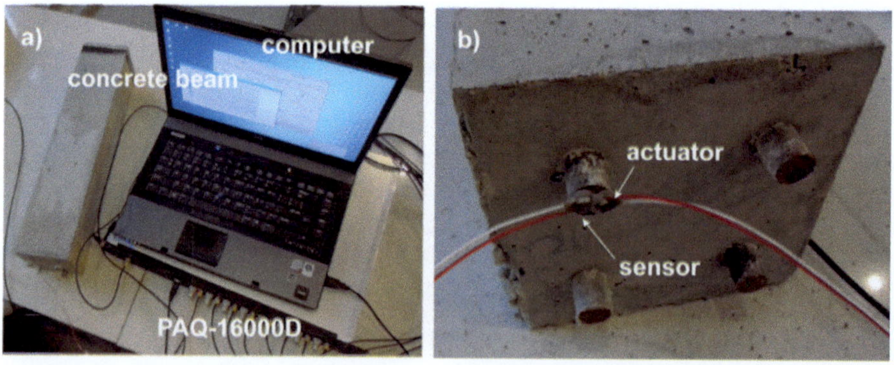

Fig. 7. Experimental set up: (a) equipment for guided wave propagation, (b) sensor and actuator attached to the reinforced concrete beam.

3.3 Results

The results of experimental investigation are presented in the form of time-domain signals given in Figs. 8 and 9 registered at the both ends of the beams with variable length of debonding. Figure 8 presents results obtained for frequency of 60 kHz, while in Fig. 9 results for 70 kHz are shown. In both cases first column contains normalized signals registered at the start of the beam and the second column contains normalized signals registered at the end of the beam.

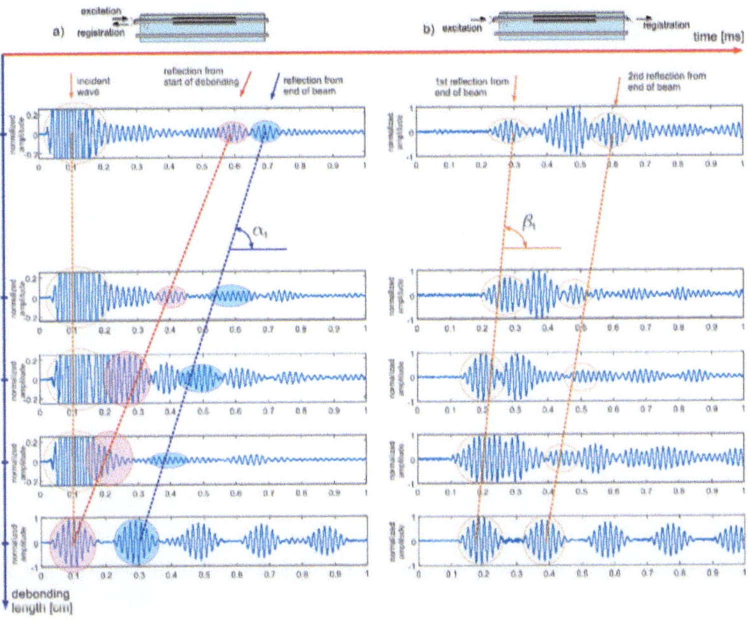

Fig. 8. Wave propagation signals registered at the (a) start and (b) end of the beams with variable debonding lengths for frequency of 60 kHz.

Characteristic wave packets were identified and indicated. Straight lines pass through incident waves and reflections from the end of the beam. The travel times of particular reflections differ for various debonding lengths: the longer debonding is, the reflection is registered earlier. For both frequencies (60 and 70 kHz) the velocity in concrete beam c_b is lower than velocity in free, debonded rod c_f (see Figs. 2 and 3).

According to Eqs. (6) and (8) travel time should decrease with increasing debonding length, what stays with good agreement with experimental results. In addition, in both figures, slopes of specific lines are denoted as α and β. Slopes α concerns slope of line connecting reflections from the end of the beam registered at the start of the specimen, while slope β concerns first reflections registered at the end of the beams. It can be seen that slopes α_1 and β_1 are greater than slopes α_2 and β_2, respectively. The reason for this fact is that frequency of 70 kHz is more sensitive to

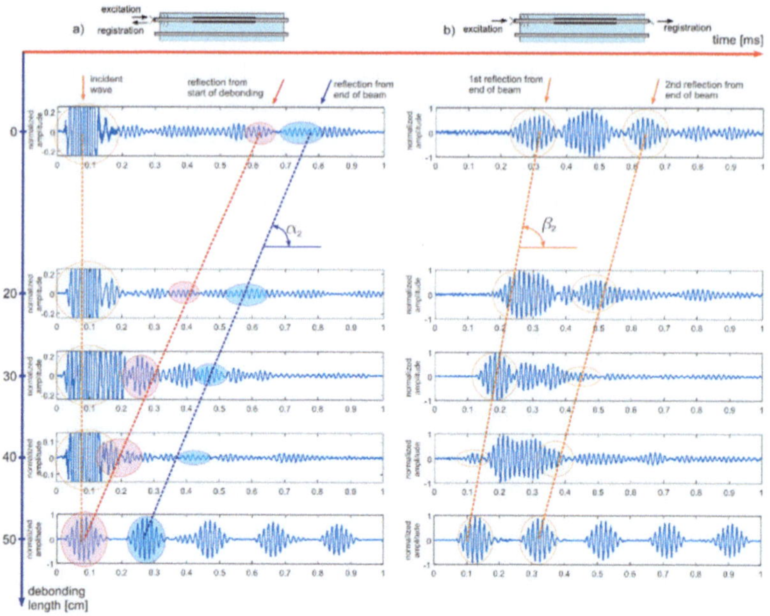

Fig. 9. Wave propagation signals registered at the (a) start and (b) end of the beams with variable debonding lengths for frequency of 70 kHz.

debonding detection, what means that a change in the debonding length has a greater influence on travel time than in case of excitation of 60 kHz. The sensitivity of particular frequency for debonding detection is characterized by a slope (Eq. (6) or (7)) which depends on the difference c_f–c_b and a product of c_f and c_b. In the analyzed case c_f for both frequencies is comparable and is equal to about 5000 m/s, while the velocity c_b is about 3500 and 2900 m/s for 60 and 70 kHz, respectively. The difference c_f–c_b is greater for frequency of 70 kHz, what results in greater slope and higher sensitivity to debonding detection.

4 Conclusions

This paper presents results of experimental investigation of wave propagation in reinforced concrete beams with partial debonding between rods and the cover. The following conclusions can be formulated:

- guided waves can be effectively used in debonding detection, even in an early stage of its development, when its size is not significant;
- the frequency and in consequence, wave propagation velocities of the fastest modes propagating in damaged and undamaged part have a crucial meaning in deboning detection;

- it is recommended to choose the excitation frequency for which the ratio between the difference c_f–c_b and a product of c_f and c_b is as high as possible – for such a case even a small change in debonding length will result in significant change in wave travel time, which facilitates debonding detection.

References

1. Giurgiutiu V.: Structural Health Monitoring with Piezoelectric Wafer Active Sensors (2008)
2. Kudela, P., Ostachowicz, W., Żak, A.: Damage detection in composite plates with embedded PZT transducers. Mech. Syst. Signal Process. **22**(6), 1327–1335 (2008)
3. Żak, A., Radzieński, M., Krawczuk, M., Ostachowicz, W.: Damage detection strategies based on propagation of guided elastic waves. Smart Mater. Struct. **21**, 1–18 (2012)
4. Grabowska, J., Palacz, M., Ostachowicz, W.: Damage identification by wavelet analysis. Mech. Syst. Signal Process. **22**(7), 1623–1635 (2008)
5. Staszewski, W.J., Lee, B.C., Traynor, R.: Fatigue crack detection in metallic structures with lamb waves and 3D laser vibrometry. Meas. Sci. Technol. **18**, 727–739 (2007)
6. Na, W.-B., Kundu, T., Ehsani, M.R.: Lamb waves for detecting delamination between steel bars and concrete. Comput.-Aided Civ. Infrastruct. Eng. **18**, 58–63 (2003)
7. Wu, F., Chang, F.-K.: Debond detection using embedded piezoelectric elements in reinforced concrete structures – part I: experiment. Struct. Health Monit. **5**, 5–15 (2006)
8. Wu, F., Chang, F.-K.: Debond detection using embedded piezoelectric elements in reinforced concrete structures – part II: analysis and algorithm. Struct. Health Monit. **5**, 17–28 (2006)
9. Li, D.S., Ruan, T., Yuan, J.H.: Inspection of reinforced interface delamination using ultrasonic guided wave non-destructive test technique. Sci. China Technol. Sci. **55**(10), 2893–2901 (2012)
10. Zima, B., Rucka, M.: Detection of debonding in steel bars embedded in concrete using guided wave propagation. Diagnostyka **17**(3), 27–34 (2016)
11. Zima, B., Rucka, M.: Guided ultrasonic waves for detection of debonding in bars partially embedded in grout. Constr. Build. Mater. **168**, 124–142 (2018)
12. Zima, B., Kędra, R.: Reference-free determination of debonding length in reinforced concrete beams using guided wave propagation. Constr. Build. Mater. **207**, 291–303 (2019)
13. Pochhammer, L.: Beitrag zur Theorie der Biegung des Kreiscylinders. Journal fur die reine und angewandte. Mathematik **81**, 33–61 (1876)
14. Chree, C.: The equations of an isotropic elastic solid in polar and cylindrical coordinates, their solutions and applications. Trans. Camb. Philos. Soc. **14**, 250–369 (1889)
15. Treyssède, F.: Numerical investigation of elastic modes of propagation in helical waveguides. J. Acoust. Soc. Am. **121**(6), 3398–3408 (2007)
16. Søe-Knudsen, A., Sorokin, S.: On accuracy of the wave finite element predictions of wavenumbers and power flow: a benchmark problem. J. Sound Vib. **330**(12), 2694–2700 (2011)
17. Liu, Y.J., Han, Q., Li, C.L., Huang, H.W.: Numerical investigation of dispersion relations for helical waveguides using the scaled boundary finite element method. J. Sound Vib. **333** (7), 1991–2002 (2014)
18. Bocchini, P., ASCE M., Marzani, A., Viola, E.: Graphical user interface for guided acoustic waves. J. Comput. Civ. Eng. **25**(3), 202–210 (2011)

On the Model Order in Parameter Estimation Using Virtual Compensators

Tommi Navntoft Hansen[1(✉)], Martin Skovmand Jensen[1], Martin Dalgaard Ulriksen[1], and Dionisio Bernal[2]

[1] Department of Civil Engineering, Aalborg University, 6700 Esbjerg, Denmark
{tnha14,mj13}@student.aau.dk, mdu@civil.aau.dk
[2] Center for Digital Signal Processing, Northeastern University, Boston, MA 02115, USA
D.Bernal@northeastern.edu

Abstract. Processing signals from open-loop system realizations can replace real-time operation using actuators in the design of closed-loop eigenstructures. One merit of the signal processing-based implementation is that it, in principle, allows virtual compensators of user-defined model order since the closed-loop systems are not to be realized during physical testing. The present paper explores the implication of the virtual compensator order in terms of the Fisher information on unknown parameters to be estimated in a model updating context. A numerical example with a structural system of engineering interest is presented that demonstrates the basic points outlined in the paper.

Keywords: Model updating · Output feedback · Eigenstructure assignment · Virtual implementation · Complex gains

1 Introduction

Calibrated numerical models are used extensively within structural and mechanical engineering for design, analysis, health monitoring, and control [1]. The calibrated models are often obtained through conventional updating schemes, in which the discrepancy between poles from a model, $\mathscr{M}$, and target poles estimated from the physical system, $\mathscr{P}$, is minimized [2]. An issue in this context—which will typically prevail when using poles as targets in structural and mechanical engineering applications—is that the system of equations to be solved in the minimization problem is ill-conditioned [3,4]. Proposals to increase the target space and, in this way, resolve the condition issue by testing the system under known perturbations have been made [5,6], but practicality is often limited. Another way to address the target space issue is to interrogate the structure in closed loop with different gains and, as such, increase the number of poles that can be identified. However, the closed-loop interrogation scheme has not yet had

M. A. Wahab (Ed.): DAMAS 2019, LNME, pp. 498–506, 2020.
https://doi.org/10.1007/978-981-13-8331-1_36

an important impact in applications, which, presumably, is due to the practical overhead associated with real-time operation.

Recently, it has been shown that the noted overhead can be eliminated by use of a virtual approach, in which multiple closed-loop systems can be computed based on a single open-loop realization [7–10]. The virtual implementation also removes stability and actuator design constraints [11], enables the use of complex gains because the control forces do not have to be physically delivered [10], and allows for a user-defined model order of the closed-loop system. In the present paper, focus will be on the last item, namely, to examine the feasibility of increasing the closed-loop system model order as an alternative approach to increase the target space. The theoretical outset in the examination is the Fisher information on the parameters to be updated; in particular, how the information increases as function of the model order and how this is reflected in the accuracy of the model updating.

The paper is organized as follows: Sect. 2 outlines the fundamental theory of dynamic output feedback, followed by a description of the virtual implementation in Sect. 3. Section 4 describes the concept of Fisher information, and subsequently, in Sect. 5, a model updating scheme based on minimizing the discrepancy between closed-loop target poles and model-based poles is outlined. In Sect. 6, a numerical example is presented to demonstrate the points made, and in Sect. 7 some concluding remarks are given to close the paper.

2 Output Feedback

Let a structural domain, $\mathscr{P}$, be described as a linear, time-invariant system in discrete time as

$$x(k+1) = A_d x(k) + B_d u(k) \tag{1a}$$

$$y(k) = Cx(k). \tag{1b}$$

Here, $x(k) \in \mathbb{R}^n$ is the state, $u(k) \in \mathbb{R}^r$ the control input, $y(k) \in \mathbb{R}^m$ the output, and $A_d \in \mathbb{R}^{n\times n}, B_d \in \mathbb{R}^{n\times r}$, and $C \in \mathbb{R}^{m\times n}$ the system matrices. It should be noted that when measurements are displacements, velocities, non-collocated accelerations or collocated accelerations where the direct transmission term has been subtracted, Eq. (1b) holds directly. It is assumed throughout this paper that one of the mentioned conditions is met.

In case of dynamic output feedback, u is the output of a discrete-time, finite-dimensional linear time-invariant system driven by y, thus

$$x_f(k+1) = A_f x_f(k) + B_f y(k) \tag{2a}$$

$$u(k) = C_f x_f(k) + D_f y(k) + v(k) \tag{2b}$$

for some excitation v and coefficient matrices $A_f \in \mathbb{C}^{q\times q}, B_f \in \mathbb{C}^{q\times m}, C_f \in \mathbb{C}^{r\times q}$, and $D_f \in \mathbb{C}^{r\times m}$. Augmenting Eq. (2a) with Eq. (1a), using Eqs. (1b) and (2b), yields

$$\begin{Bmatrix} x(k+1) \\ x_f(k+1) \end{Bmatrix} = \begin{bmatrix} A_d + B_d D_f C & B_d C_f \\ B_f C & A_f \end{bmatrix} \begin{bmatrix} x(k) \\ x_f(k) \end{bmatrix} + \begin{bmatrix} B_d \\ 0 \end{bmatrix} v(k), \tag{3}$$

which is referred to as the compensator. Defining $\tilde{x}(k) = \{x(k)^T \quad x_f(k)^T\}^T$, Eq. (3) can, as shown in [7,11], be rewritten as

$$\tilde{x}(k+1) = \tilde{A}\tilde{x}(k) + \tilde{B}\tilde{u}(k) \tag{4}$$

with the output

$$\tilde{u}(k) = \tilde{G}\tilde{C}\tilde{x}(k) + \begin{Bmatrix} 0 \\ v(k) \end{Bmatrix} \tag{5}$$

and

$$\tilde{A} = \begin{bmatrix} A_d & 0 \\ 0 & 0 \end{bmatrix}, \quad \tilde{B} = \begin{bmatrix} 0 & B_d \\ I & 0 \end{bmatrix}, \quad \tilde{C} = \begin{bmatrix} 0 & I \\ C & 0 \end{bmatrix}, \quad \tilde{G} = \begin{bmatrix} A_f & B_f \\ C_f & D_f \end{bmatrix}. \tag{6}$$

As shown in Eq. (4), using the feedback law in Eq. (5), the compensator can be viewed as a static output feedback system of order $n+q$ and with the system matrices $\tilde{A} \in \mathbb{R}^{(n+q)\times(n+q)}$, $\tilde{B} \in \mathbb{R}^{(n+q)\times(r+q)}$, $\tilde{C} \in \mathbb{R}^{(m+q)\times(n+q)}$ and $\tilde{G} \in \mathbb{C}^{(r+q)\times(m+q)}$.

3 Virtual Implementation

As seen in Sect. 2, implementation of the compensator to close the loop changes the model order of the system from n to $n+q$, and the inputs and outputs increase by q. The virtual implementation of the compensator model is achieved using the relation between the open- and closed-loop transfer matrices, $H(z)$ and $\mathscr{H}(z)$, which in a system controlled with positive static output feedback with the gain G is defined by [12]

$$\mathscr{H}(z) = (I - H(z)G)^{-1}H(z), \tag{7}$$

from which it follows that the closed-loop system can be identified from an open-loop realization. The open-loop transfer matrix can, in order to comply with the compensator model, be defined as [8]

$$H_C(z) = \begin{bmatrix} \frac{1}{z}I & 0 \\ 0 & H(z) \end{bmatrix}, \tag{8}$$

so by substituting Eq. (8) into Eq. (7), we get the compensator transfer matrix

$$\tilde{H}(z) = \begin{bmatrix} I - \frac{1}{z}A_f & -\frac{1}{z}B_f \\ -H(z)C_f & I - H(z)D_f \end{bmatrix}^{-1} \begin{bmatrix} \frac{1}{z}I & 0 \\ 0 & H(z) \end{bmatrix}. \tag{9}$$

Worth of explicit note is that when the identification of the open-loop system is conducted in the frequency domain, the implementation of the virtual compensator follows directly. If, however, the system is identified in the time-domain, one must transform to z-domain in order to calculate Eq. (9). An approach for this is provided in [8], and it is based on mapping observer Markov parameters to $H(z)$. Furthermore, it should be noted that the inverse z-transformation filters unstable poles, hence removing the system stability constraint required in physical testing.

4 Fisher Information on the Parameters

The amount of information that the closed-loop system carries on the system parameters to be updated, gathered in $\theta \in \mathbb{R}^s$, can be assessed, qualitatively, by use of the Fisher information matrix, $\mathcal{I} \in \mathbb{C}^{s \times s}$.

Let Y be an observable variable, carrying information on θ, then the Fisher information matrix can be expressed in terms of the likelihood function, f, with the entries

$$\mathcal{I}_{j,k} = -E\left(\frac{\partial^2 \ln f(Y;\theta)}{\partial\theta_j \partial\theta_k}\right) \quad \forall j,k \in [1,s], \tag{10}$$

where E is the expectation operator. Assuming $Y \sim \mathcal{N}(\mu(\theta), \Sigma)$, the Fisher information matrix can be expressed as [13]

$$\mathcal{I} = \mathcal{J}^H \Sigma^{-1} \mathcal{J}, \tag{11}$$

where superscript H indicates the conjugate transpose, $\Sigma \in \mathbb{C}^{s \times s}$ is the covariance matrix of Y, and $\mathcal{J} \in \mathbb{C}^{q+m \times s}$ is the Jacobian matrix containing the first-order derivatives of Y with respect to θ. If Y contains $q+m$ poles of the closed-loop system, which will be the case in this study, the Jacobian is given as

$$\mathcal{J} = \begin{bmatrix} \frac{\partial\lambda_1}{\partial\theta_1} & \frac{\partial\lambda_1}{\partial\theta_2} & \cdots & \frac{\partial\lambda_1}{\partial\theta_s} \\ \frac{\partial\lambda_2}{\partial\theta_1} & \frac{\partial\lambda_2}{\partial\theta_2} & \cdots & \frac{\partial\lambda_2}{\partial\theta_s} \\ \vdots & \vdots & \ddots & \vdots \\ \frac{\partial\lambda_{q+m}}{\partial\theta_1} & \frac{\partial\lambda_{q+m}}{\partial\theta_2} & \cdots & \frac{\partial\lambda_{q+m}}{\partial\theta_s} \end{bmatrix} \in \mathbb{C}^{q+m \times s} \tag{12}$$

when λ_i denotes the ith pole.

One gathers that the Fisher information can be used to assess the qualitative implication of the model order, q, as it is contented that an increase in Fisher information on the parameters will improve the estimation of these in model updating schemes. Another way to appreciate this is from the Cramér-Rao lower bound, $\mathcal{C} \in \mathbb{C}^{s \times s}$, which is defined as

$$\mathcal{C} = \mathcal{I}^{-1} \tag{13}$$

and composes a measure of the minimum covariance that any unbiased estimator of θ can achieve. In other words, $\mathcal{C}$ provides a lower bound on the variance by which we can estimate each component in θ, and one anticipates that this variance will decrease asymptotically as q increases.

5 Model Updating Using Virtual Compensator

The estimation of the parameters in $\theta \in \mathbb{R}^s$ is formulated as the constrained optimization problem

$$\begin{aligned} &\underset{\theta \in \mathbb{R}^s}{\text{argmin}} && ||\Lambda_{\mathcal{M}}(\theta) - \Lambda_{\mathcal{P}}|| \\ &\text{subject to} && \forall i \in [1,\text{s}] : \alpha_i \leq \theta_i \leq \beta_i, \end{aligned} \tag{14}$$

where $\Lambda_{\mathscr{P}} \in \mathbb{C}^{q+m}$ is a subset of poles estimated from the physical system, $\Lambda_{\mathscr{M}}(\theta) \in \mathbb{C}^{q+m}$ the corresponding model-predicted poles, and $\alpha_i, \beta_i \in \mathbb{R}$ lower and upper bounds on θ_i. The idea is to ensure that $q + m \geq s$, such that the system to be solved in the optimization scheme is (over)determined, which can be achieved by stacking identified poles from multiple closed-loop systems into a target vector [14], or by increasing the compensator order. In this paper, we opt for the latter and use the least square method described in [11] to design a single gain configuration to place $q + m \geq s$ poles, hence yielding the target vector

$$\Lambda_{\mathscr{P}} = \{\lambda_{\mathscr{P}_1}, \ldots, \lambda_{\mathscr{P}_{q+m}}\}^T \in \mathbb{C}^{q+m}. \tag{15}$$

Since the least square procedure uses arbitrary complex constants to collapse the eigenvector basis, different gain configurations yielding the same pole subset placement are achievable. One approach to investigate the error in the realization of the closed-loop poles, using the virtual dynamic approach with the gain $\tilde{G}$, is through the poles' sensitivity with respect to some parameter, g, of the gain, that is,

$$\frac{\partial \lambda_j}{\partial g} = \psi_j^T \frac{\partial \mathscr{A}_{cs}}{\partial g} \phi_j = \psi_j^T \tilde{B} \frac{\partial \tilde{G}}{\partial g} \tilde{C} \phi_j \tag{16}$$

where ψ_j and ϕ_j are the jth left and right eigenvectors of the compensator state matrix, $\mathscr{A}_{cs}$, presented in Eq. (3). In order to omit gains that cause undue error, we use the heuristic gain selection procedure proposed in [10], which circumvents calculation of the gain derivative, to select from a pool of candidates. In particular, we use the initial model of the closed-loop system to define the metric

$$\gamma_i = \|\Psi_{\mathscr{M}_i} \tilde{B}\| \|\tilde{C} \Phi_{\mathscr{M}_i}\|, \tag{17}$$

where $\Psi_{\mathscr{M}_i}$ and $\Phi_{\mathscr{M}_i}$ are the left and right eigenvectors of the closed-loop system model with gain $\tilde{G}_i$.

6 Numerical Examination

We consider the 10-DOF shear building depicted in Fig. 1 in the context of model updating for damage characterization. The example explores what influence the compensator order, q, has on the Fisher information and how this relates to the performance of the model updating when using a single closed-loop system.

The terms *simulation model* and *nominal model* are used to refer to, respectively, the model used for simulating the output for system identification and the numerical model to be updated. In this example, the simulation model is drawn from the manifold containing the virtual nominal model, which implies that in the absence of noise there is a set of parameters for which the model to be updated coincides with the simulation model. This will, of course, not be realizable in practice. The simulation model, which is assigned a 20% stiffness perturbation in the 6th floor, is excited by white Gaussian noise in the first floor, and the resulting displacements are measured in the 1st, 5th, and 9th floor for

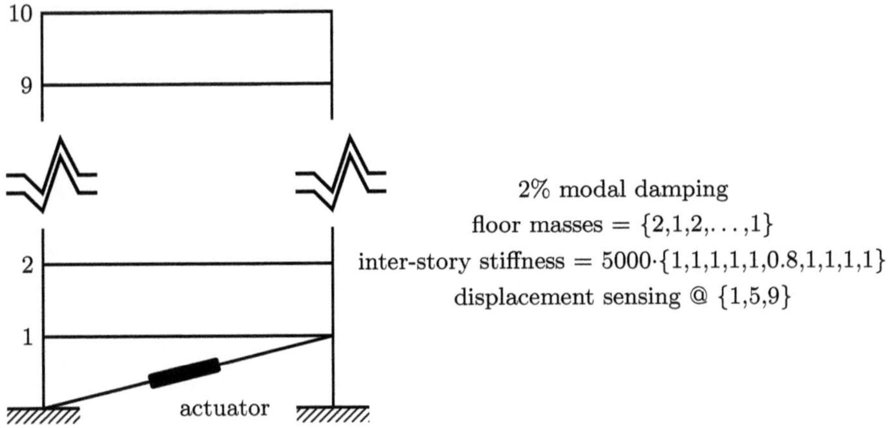

Fig. 1. 10-DOF shear building, where the inter-story stiffness and floor masses are in any consistent set of units, used for numerical illustrations.

a duration of 5 min with a sampling frequency of 100 Hz. The displacements are contaminated with 2% white Gaussian noise, and the open-loop system identification is performed using the Eigensystem Realization Algorithm [15].

We consider compensator orders of 7, 11, and 15, which result in target vectors with 10, 14, and 18 poles, hence making the optimization problem determined for $q = 7$ and overdetermined for $q = 11$ and $q = 15$. The gain is designed such that a subset of the closed-loop nominal model poles is assigned as $\{\lambda_1, \frac{3}{2}\lambda_1, \frac{4}{2}\lambda_1, \ldots, \frac{q+m+1}{2}\lambda_1\}$, with λ_1 being the first pole of the open-loop nominal model. To compute the Fisher information for each configuration of the compensator order, we estimate the covariance matrix in a Monte Carlo setting with 100 simulations and, subsequently, calculate the Fisher information using Eq. 11. Figure 2 presents the condition number of each Fisher informa-

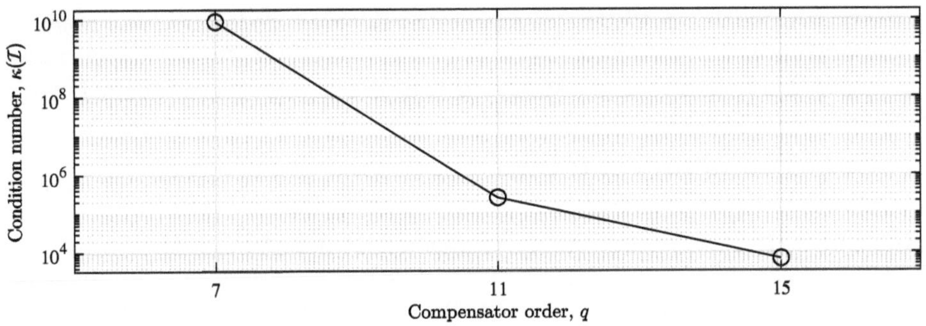

Fig. 2. Condition number of the Fisher information matrix for different configurations of the compensator order, q.

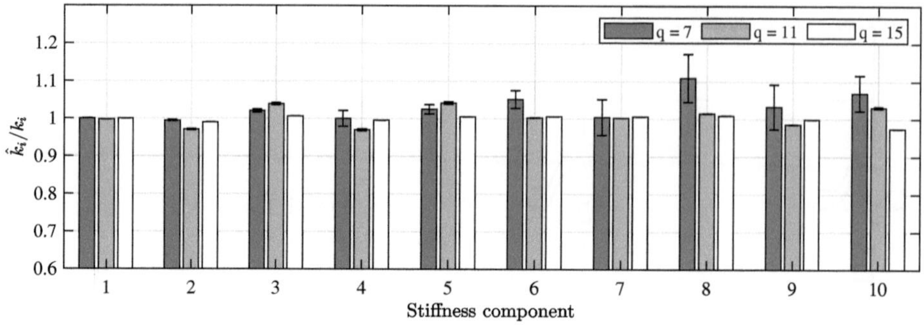

Fig. 3. Model updating results for the shear building for three configurations of q, with $\hat{k}_i$ and k_i being the converged estimate and true value of the stiffness components.

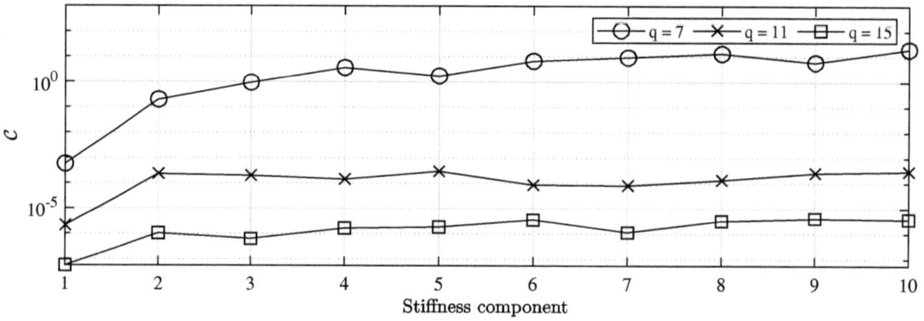

Fig. 4. Cramér-Rao lower bound on the individual parameters to be estimated for three compensator order configurations.

tion matrix, which decreases as q increases. Further numerical examination has confirmed that the decrease is asymptotic.

For the model updating, the target vectors are formed as $\Lambda_{\mathscr{P}} \in \mathbb{C}^{q+m}$ where the corresponding poles from the nominal models, $\Lambda_{\mathscr{M}}(\theta) \in \mathbb{C}^{q+m}$, are taken as the ones yielding the lowest discrepancy to the identified poles. The nominal models are updated for each realization of the target vectors by use of the "fmincon" algorithm in MATLAB® to solve the optimization problem in Eq. (14). It is assumed that no prior knowledge of the perturbation location exists, thus θ contains all the inter-story stiffness values in the shear building model.

The mean and variance of the converged estimates, which are normalized with respect to the true stiffness, are visualized in Fig. 3. The parameters are estimated with a mean absolute percentage error of 3.16%, 2.06%, and 0.37% for a compensator order of 7, 11, and 15, thus the Fisher information increase is reflected in the estimation accuracy. One gathers from Fig. 3 that the variance decreases as q increases, which is in agreement with the tendency of the Cramér-Rao lower bounds depicted in Fig. 4. For each model order, we also find reasonable correlation between which parameters that have the lowest Cramér-

Rao lower bounds and what is observed in terms of relative estimation variance for the parameters in the model updating.

7 Conclusion

The paper addresses model updating by use of a recently proposed virtual approach based on processing open-loop signals to form closed-loop compensators. In particular, we explore the feasibility of resolving the issue of ill-conditioned model updating by (over)determining the system of equations to be solved through an increase of the compensator order.

Model updating of a shear building in a Monte Carlo setting shows the mean of the estimated parameters to approach (asymptotically) the true values when increasing the compensator order. Furthermore, the variance of the estimated parameters and the condition number of the Fisher information matrix decrease asymptotically as the compensator order increases.

Acknowledgement. The authors gratefully acknowledge the Danish Hydrocarbon Research and Technology Centre (DHRTC) for the financial support.

References

1. Aster, R.C., Borchers, B., Thurber, C.H.: Parameter Estimation and Inverse Problems, 2nd edn. Academic Press, Cambridge (2013)
2. Friswell, M.I., Mottershead, J.E.: Finite Element Model Updating in Structural Dynamics. Kluwer Academic Publishers, Dordrecht (1995)
3. Koh, B.H., Ray, L.R.: Feedback controller design for sensitivity-based damage localization. J. Sound Vib. **273**(1), 317–335 (2004)
4. Jiang, L.J., Tang, J.J., Wang, K.W.: An optimal sensitivity-enhancing feedback control approach via eigenstructure assignment for structural damage identification. J. Vib. Acoust. **129**(6), 771–783 (2007)
5. Nalitolela, N.G., Penny, J.E.T., Friswell, M.I.: A mass or stiffness addition technique for structural parameter updating. Int. J. Anal. Exp. Modal Anal. **7**(3), 157–168 (1992)
6. Cha, P., Gu, W.: Model updating using an incomplete set of experimental modes. J. Sound Vib. **233**(4), 583–596 (2000)
7. Bernal, D.: Parameter estimation using virtual output feedback. Mechanical Systems and Signal Processing (to appear)
8. Bernal, D.: Parameter estimation using virtual dynamic output feedback. Mechanical Systems and Signal Processing (to appear)
9. Bernal, D., Ulriksen, M.D.: Virtual closed-loop parameters estimation. In: Proceedings of the International Conference on Structural Engineering Dynamics (ICEDyn), Viana do Castelo, Portugal (2019)
10. Ulriksen, M.D., Bernal, D.: On the use of complex gains in virtual feedback for model updating. In: Proceedings of the International Conference on Structural Engineering Dynamics (ICEDyn), Viana do Castelo, Portugal (2019)
11. Bernal, D., Ulriksen, M.D.: Output feedback in the design of eigenstructures for enhanced sensitivity. Mech. Syst. Signal Process. **112**, 22–30 (2018)

12. Trentelman, H., Stoorvogel, A.A., Hautus, M.: Control Theory for Linear Systems. Springer, London (2001)
13. Kay, S.M.: Fundamentals of Statistical Signal Processing: Estimation Theory. Prentice Hall, Upper Saddle River (1993)
14. Jensen, M.S., Hansen, T.N., Ulriksen, M.D., Bernal, D.: Physical and virtual implementation of closed-loop designs for model updating. In: Proceedings of the 13th International Conference on Damage Assessment of Structures (DAMAS 2019), Porto, Portugal (2019)
15. Juang, J.N.: Applied System Identification. Prentice Hall, Upper Saddle River (1994)

Smart Acoustic Band Structures

Wiktor Waszkowiak, Arkadiusz Żak, Magdalena Palacz(✉), and Marek Krawczuk

Gdansk University of Technology, Narutowicza 11/12, 80-233 Gdansk, Poland
mpalacz@pg.edu.pl

Abstract. Smart acoustic band structures exhibit very interesting and non-standard physical properties due to the periodic nature of their certain characteristic on different scale levels. They manifest mostly in their frequency spectra as so-called frequency band-gaps or stop-bands, what has a great impact on the behaviour of these structures in relation to the propagation of vibro-acoustic signals that can be transmitted through the structures in some precisely defined frequency bands. Properties of acoustic band structures are directly linked to their geometry on the level of the unit cell, which parameters determine structural dynamics of such structures on the macroscopic scale. Here the piezoelectric transducers play a significant role. The combined exploitation of active properties of acoustic band structures equipped with active piezoelectric elements, in order to filter or damp transmitted vibro-acoustic signals, allows for very effective their applications. In their paper, the authors present certain results of certain computer simulations by the time-domain spectral finite element method, related to 1-D smart active and passive acoustic band structures supplemented with experimental measurements.

Keywords: Acoustic band structure · Vibration

1 Introduction

Smart acoustic band structure as a periodic structure made up of sequentially arranged repetitive elements characterise certain unique dynamic properties. Those properties are utilised in many real engineering structures such as acoustic filters and dampers or isolation devices. The dynamic behaviour of periodic structures has been already intensively analysed by many researchers [1–6], however, there has not been found a numerical model of an active periodic structure based on the frequency domain spectral finite element method (FDSFEM) [7]. In this paper, the authors present a model of an active periodic beam based on FDSFEM and compare the results with a laboratory experiment. The aim of the presented work has been to examine the influence of parameters of the active periodical structure on the location and width of the forbidden bands.

M. A. Wahab (Ed.): DAMAS 2019, LNME, pp. 507–514, 2020.
https://doi.org/10.1007/978-981-13-8331-1_37

2 Numerical Model

The analysed structure has been presented in Fig. 1. It has been modelled as a sequence of 25 or 50 single unit cells (Fig. 2) consisting of an aluminium beam with permanently attached piezoelectric rectangular discs on the bottom and top.

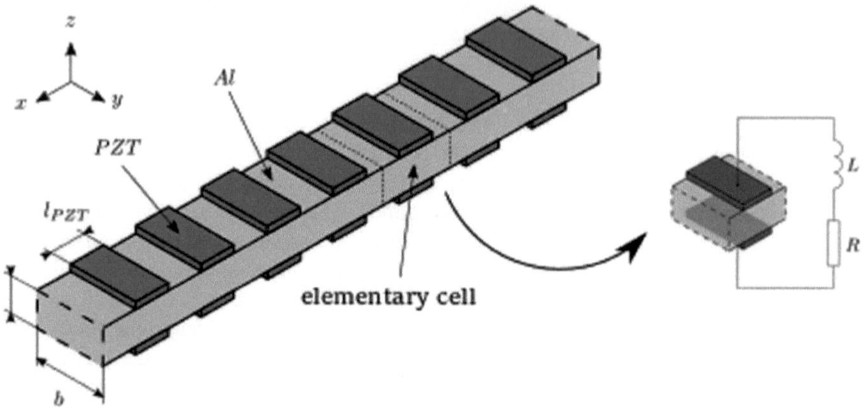

Fig. 1. The modelled smart acoustic band structure.

The displacement field of the analysed structure has been assumed for longitudinal and torsional displacement according to the elementary rod theory, for bending displacements – according to the Timoshenko theory. The mathematical formulae for displacement and strain fields have been given by the following equations:

$$\begin{cases} u(x) = u_0(x) - z\phi_y(x) - y\phi_z(x) \\ v(x) = v_0(x) - z\phi_x(x) \\ v(x) = v_0(x) - z\phi_x(x) \end{cases} \tag{1}$$

$$\begin{cases} \varepsilon_x = \frac{\partial u(x)}{\partial(x)} = \frac{du_0(x)}{dx} - z\frac{d\phi_y(x)}{dx} - y\frac{d\phi_z(x)}{dx} \\ \gamma_{xy} = \frac{\partial v(x)}{\partial x} + \frac{\partial u(x)}{\partial y} = \frac{dv_0(x)}{dx} - z\frac{d\phi_x(x)}{dx} - \phi_z(x) \\ \gamma_{xz} = \frac{\partial u(x)}{\partial z} + \frac{\partial w(x)}{\partial x} = \frac{dw_0(x)}{dx} - y\frac{d\phi_x(x)}{dx} - \phi_y(x) \end{cases} \tag{2}$$

In order to model a piezoelectric material, it has been necessary to use constitutive equations known from the literature [8], which couple the electric field and the mechanical stress field occurring in the piezoelectric material:

$$\begin{cases} \sigma = [D]\varepsilon - [e]^T E \\ q = [e]\varepsilon - [k^\varepsilon]E \end{cases} \tag{3}$$

Where: σ stands for stress vector, ε is the strain vector, E is the electrical field vector, q is the displacement vector, $[D]$ stands for the matrix of elastic coefficients for piezoelectric material, $[e]$ is the matrix of stress constants of the piezoelectric material and $[k^{\varepsilon}]$ is the matrix of dielectric constants of the piezoelectric material.

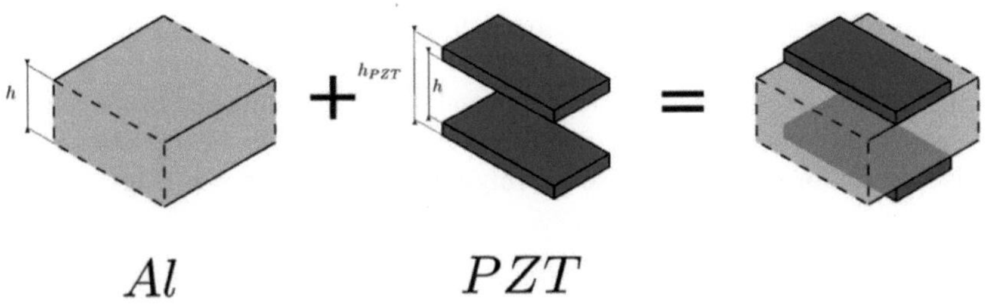

Fig. 2. Single periodic cell.

Numerical modelling requires proper determination of material properties of the element. Therefore the stiffness of piezoelectric material on the resonance frequency of the system in which this piezoelectric device works has been determined according to the numerical formula for an effective Young's modulus of piezoelectric material [6]. It uses the electrical capacitance of piezoelectric elements and the impedance of the resonance circuit depending on the resonance frequency of this circuit:

$$E_p^{SU}(\omega) = E_p^D\left(1 - \frac{k_{31}^2}{1 + i\omega C_p^{\varepsilon} Z^{SU}(\omega)}\right), \tag{4}$$

with E_p^{SU} meaning the effective young modulus of piezoelectric material in a closed circuit, E_p^D is the effective Young modulus in an open-circuit, k_{31}^2 means the electromechanical coupling coefficient, C_p^{ε} is the electrical capacitance of piezoelectric elements and Z^{SU} is the resonant circuit impedance.

3 Numerical Analysis

A single periodic cell has been defined as a combination of two materials: piezoelectric and aluminium (Fig. 2). A single resonance system of a piezoelectric crystal with a resistor and a coil has been proposed – as shown in Fig. 1. The dimensions of the pure single piezoelectric rectangular disc have been assumed as 0.01 × 0.02 × 0.001 [m]. The total analysed structure (as presented in Fig. 1) in this particular simulation case consisted of 50 single unit cells. The total geometrical dimensions of the periodic beam have been as the length 1 [m], the height 0.01 [m] and the width 0.02 [m]. The material parameters have been given from the datasheet delivered by the PZT (APC International Ltd., material type 721) and aluminium (E = 7200 [MPa], ρ = 2700 [kg/m^3]). The total beam has been analysed as a free-free element.

To solve the problem the FDSFEM has been utilized [7]. In this method, the solution is obtained in the frequency domain, and the transition to the time domain is performed through an inverted Fourier transform. The authors have been interested in frequency characteristics, that by the use of this method have been directly obtained by the solution of the set of system equations for every component of the fast Fourier Transform (FFT) of the excitation signal.

In this section exemplary results of the analysis of the influence of the single PZT circuit resonance frequency on the width of the active periodic beam, band gaps have been discussed. For that purpose, several figures have been shown (Figs. 3–5).

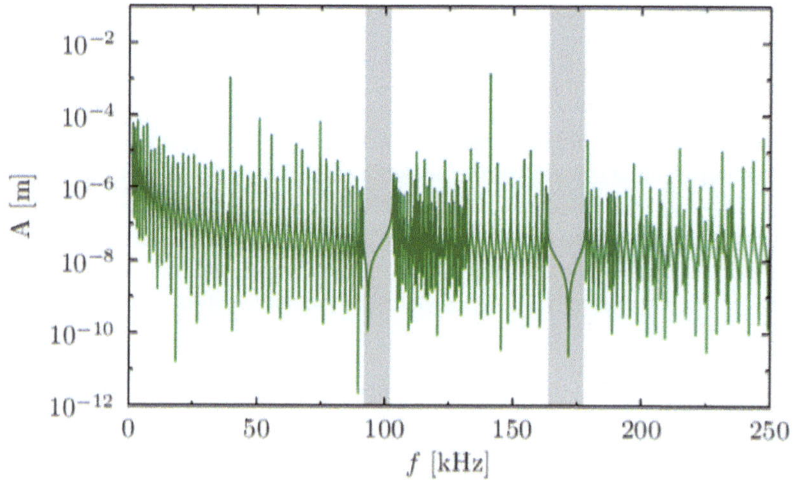

Fig. 3. Simulated FRF of the active periodic beam.

The graph that is shown in Fig. 3 represents the calculated FRF curves in the range from 0 to 250 [kHz] for a periodical beam in which the resonance circuits were switched off. It may be noticed, that there have occurred two characteristic band gaps (marked with grey colour), what is in good correspondence with the structural characteristics.

The following Fig. 4 presents the same range of simulated FRF curves, but in this case, the resonance frequencies of all the PZT cells have been assumed as 50 [kHz] (marked with a red line in the figure). It may be noticed that in comparison to the previous example an additional bandgap has occurred, which means, that active PZT unit cells definitely influence the periodic beam behaviour.

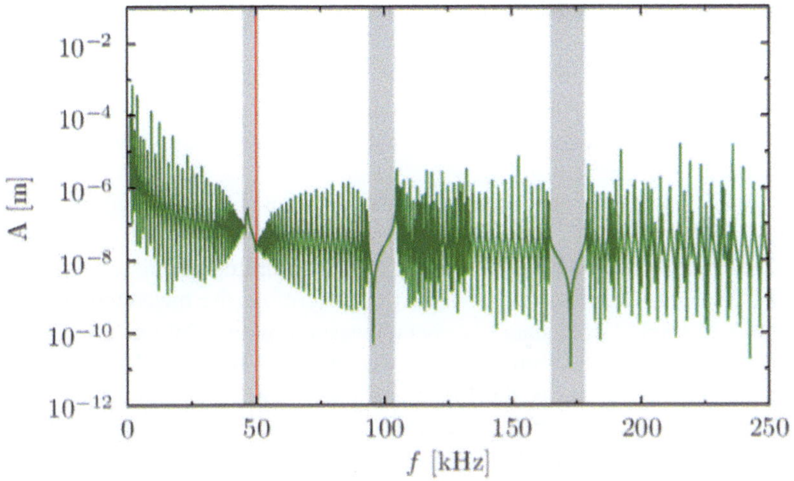

Fig. 4. Simulated FRF for the active periodic beam with PZT resonant frequency of 50 [kHz].

In Fig. 5 given in this section the resonance frequency of the PZT unit circuits has been set to 100 [kHz], what is the frequency of the first band gap for a periodic beam with inactive PZT elements. From the graph presented one can conclude that it is possible to increase the width of the band gap of the periodic beam with PZT active elements by active modification of the value resonance frequency of the PZT unit cells.

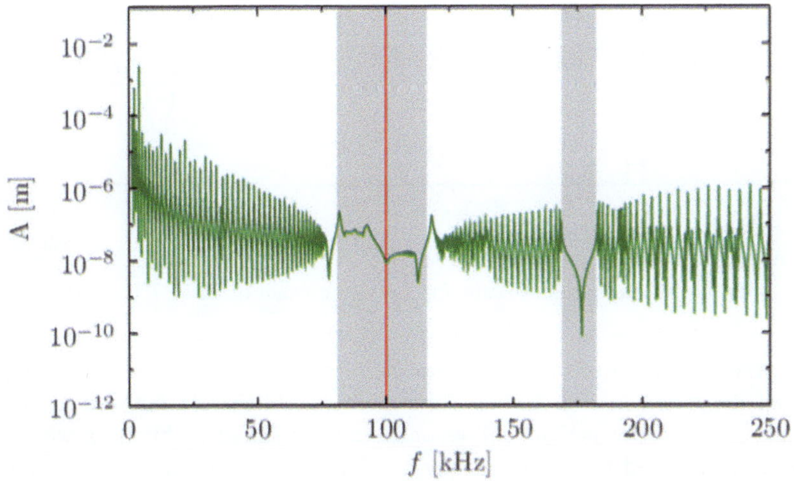

Fig. 5. Simulated FRF for the active periodic beam with PZT resonant frequency of 100 [kHz].

4 Experimental Verification

To verify the proposed numerical model an experimental set up has been prepared where the element tested was a periodic beam consisting of 25 PZT unit cells. The detailed picture of the analysed periodic beam with all the required electronic equipment is presented in Fig. 7.

To register the FRFs the prepared specimen has been excited with solenoid piezoelectric element (APC International, Ltd., cat# 42-1021) with a periodic chirp signal in the range from 20 [kHz] to 100 [kHz]. The sampling frequency has been set to 256 [kHz], the sampling time has been 64 [ms], therefore the measurement resolution has been 15,625 [Hz]. All the data have been measured with a Polytec PSV-400 SLDV.

The diagrams that have been shown in Fig. 6 represent two calculated cases. The one marked as blue represents the FRFs and their envelope obtained for a periodic beam with totally switched off PZT resonant cells. The red lines represent the FRFs measured for an analysed periodic beam with the 75 [kHz] resonance frequency of the PZT unit cells. It may be noticed that active PZT resonant cells attenuate a wider frequency range of natural bandgaps of the beam (the one attenuated in range of 50 [kHz]).

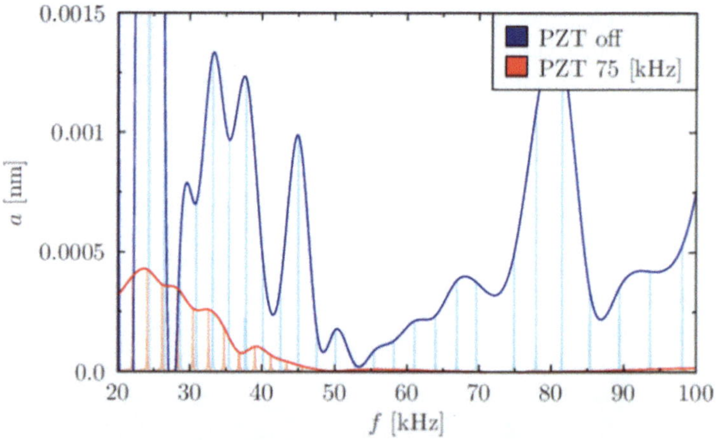

Fig. 6. The calculation results for a periodic beam of 25 PZT unit cells.

The experimental results registered for the same periodic beam (25 PZT unit cells) have been illustrated in Fig. 8, where the measured FRFs and their envelopes for a periodic beam with not active PZT resonant cells (marked as blue) as well as with resonant cells adjusted for the frequency of 75 [kHz] have been presented. From the results shown it can be concluded that in a real PZT periodic structure switching on the

Fig. 7. The analysed experimental specimen.

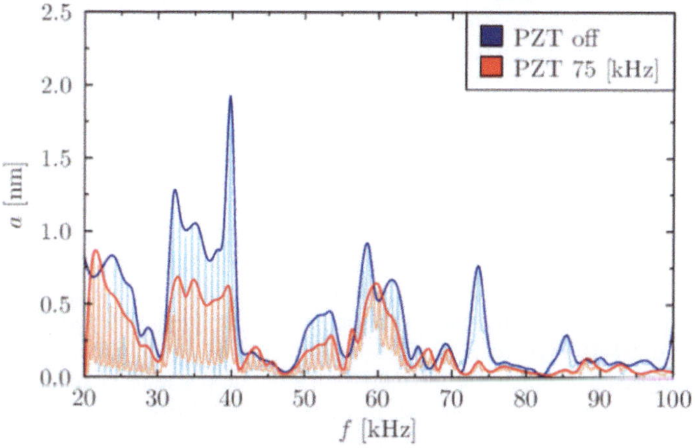

Fig. 8. The experimental results for a periodic beam of 25 PZT unit cells.

resonant circuits attenuates the vibration amplitudes in the whole analysed range. The most promising conclusion is that it has been possible to attenuate precisely the controlled resonant frequency of PZT periodic cells (75 [kHz]). This conclusion is very promising for future analysis.

5 Conclusion and Discussion

On the bases of the numerical and experimental results obtained it may be concluded that there is a possibility of active attenuation of a specific range of frequency spectrum in a periodic beam. In order to do it properly, it is necessary to control in a precise manner the parameters of resonant circuits. Test results obtained by the experiment lead to a statement that the proposed numerical model requires more attention and careful analysis as the calculation result do not coincide well with experimental measurements. This fact may rise from some numerical aspect of the model of piezoelectric parameters of the resonant circuit or from certain issues of the solution method chosen, as certain limits of the application of the FDSFEM already reported in the literature [9]. However, there is a chance to improve the proposed numerical model and this possibility shall be used by the authors of the research in their further work on the issue described in this paper.

References

1. Żak, A., Krawczuk, M., Palacz, M.: Periodic properties of 1D FE discrete models in high-frequency dynamics. In: Mathematical Problems in Engineering, ID9651430 (2016)
2. Baz, A.: Active control of periodic structures. Trans. ASME **123**, 472–479 (2001)
3. Wu, Z.J., Wang, Y.Z., Li, F.M.: Analysis on band gap properties of periodic structures of bar system using the spectral element method. Waves Random Complex Media **23**(4), 349–372 (2013)
4. Hussein, M.I.: Reduced Bloch mode expansion for periodic media band structure calculations. Proc. R. Soc. Lond. A: Math., Phys. Eng. Sci. **465**, 2825–2848. The Royal Society (2009)
5. Hagood, N.W., Flotow, A.: Damping of structural vibrations with piezoelectric materials and passive electrical networks. J. Sound Vib. **146**(2), 243–268 (1991)
6. Airoldi, L., Senesi, M., Ruzzene, M.: Piezoelectric superlattices and shunted periodic arrays as tunable periodic structures and metamaterials. In: Wave Propagation in Linear and Nonlinear Periodic Media, pp. 33–108. Springer, Vienna (2012)
7. Doyle, J.F.: Wave Propagation in Structures, 2nd edn. Springer, New York (1997)
8. Agnes, G.: Piezoelectric Coupling of Bladed-Disk Assemblies. In: Tupper Hyde, T. Proc. of Smart Structures and Materials Conference on Passive Damping, Newport Beach, CA, SPIE-Vol. 3672, pp. 94–103 (1999)
9. Palacz, M.: Spectral methods for modelling of wave propagation in structures in terms of damage detection—a review. Appl. Sci.-Basel **8**(7), 1–25 (2018)

Damage in Machineries

Development of a Novel Solution for Leading Edge Erosion on Offshore Wind Turbine Blades

William Finnegan[1,2(✉)], Tomas Flanagan[3], and Jamie Goggins[1,2]

[1] Civil Engineering, School of Engineering, National University of Ireland Galway, Galway, Ireland
william.finnegan@nuigalway.ie, jamie.goggins@nuigalway.ie

[2] MaREI Centre, Ryan Institute, National University of Ireland Galway, Galway, Ireland

[3] ÉireComposites Teo, An Choill Rua, Inverin, Co., Galway, Ireland

Abstract. In recent years, wind energy has become a leading source of renewable energy as the world strives to remove its reliance on fossil fuels. With the growing demand for wind energy, wind farms have begun to move offshore and the size of the average wind turbine has increased (up to 10 MW). However, as a result of these advances, additional challenges are presented – one of the most significant being leading edge erosion on wind turbine blades. This erosion requires additional maintenance, while lowering a turbine's annual energy production by up to 25%, which needs to be eliminated, or significantly reduced, if offshore wind energy is to become competitive within global energy markets. To this end, in this paper, the methodology proposed in LEAPWind, a new collaborative European research project, which aims to prevent blade leading-edge erosion by employing advanced composite materials and innovative manufacturing processes has been presented. An advanced thermoplastic-epoxy composite material is used to manufacture a leading edge component for a wind turbine blade. The critical technical stages, including material identification and characterisation, component design and manufacture have been discussed. Additionally, the details relating to de-risking of the novel technologies through mechanical and rain erosion testing, and full-scale operational trials on a 2.1 MW wind turbine, located in an onshore wind farm in Portugal, has been included. The results of this study, will not only have social and economic benefits, but also a significant environmental impact as it will allow for the manufacture of a more sustainable wind turbine blade.

Keywords: Advanced materials · Leading edge erosion · Structural integrity · Wind energy

M. A. Wahab (Ed.): DAMAS 2019, LNME, pp. 517–528, 2020.
https://doi.org/10.1007/978-981-13-8331-1_38

1 Introduction

As the world moves to a more sustainable way of living, the leading source of renewable energy is wind energy, where, by 2020, the global capacity is expected to nearly double to 650.8 GW [1]. As the wind energy industry grows, increasingly more wind farms are being developed offshore due to favourable social and environmental factors compared to onshore. With this development in the sector, wind turbine blades are now become much larger with the increased resource and the need for fewer turbines, where the average capacity of wind turbines installed in European waters has doubled, from 2 MW in 2000 to 4 MW in 2014 and SEIMENS Gamesa announced their 10 MW (193 diameter wind turbine) this year [2].

However, with larger turbines comes additional challenges - as blade tip speeds approach 500 km/hr, erosion of the leading-edge of the wind turbine blade begins to occur very early (within 5 years) of the blades design life (typically 20–25 years). This becomes an even more paramount issue for offshore wind as access to turbines is limited and the cost of maintenance increases ten-fold compared to onshore wind turbine blades. Additionally, this erosion has a knock on effect on the performance of the wind turbine as it can reduce the annual energy production of a wind turbine by between 2% and 25% [3].

A comprehensive review of the existing methods for addressing leading edge erosion of wind turbine blades was compiled in 2013 by Keegan et al. [4]. Currently, a number of leading edge protection methods that are designed to be applied to the completed wind turbine blade are available, which include tapes, paints and coatings [5]. Initially, the protective coatings were made from epoxy or polyester but over time, these rigid coatings were found to be inadequate and more ductile materials, such as polypropylene and polyurethane, were necessary. In recent years, manufacturers have moved towards multi-layered solutions, which can be designed to optimise performance and as a means of assessing the durability of the protection system. In general, leading edge protection methods can be divided into two categories: in-mould and post-mould solutions [4, 6]. The in-mould solutions are applied directly to the matrix substrate, using painting or spraying. These coating are typically rigid, brittle and have a high modulus, compared to the more flexible coatings, such as polyurethane [4], that are used for the post-mould solutions. The post-mould protective systems are typically multi-layer systems with the inclusion of filler and primer layers between the laminate substrate and surface coating. These methods provide additional protection from erosion during operation but usually require replace during the design life of the blade and this replacement becomes more regular in larger wind turbine blades. In addition, there are a number of products designed for maintenance of eroded wind turbine blade leading edge. These are usually in the form of spray foams or fillers, which are costly to apply even for onshore wind turbines.

In order to evaluate the effectiveness of these protective solutions, various testing methods and apparatus have been developed to simulate rain erosion. These can be classified into wheel and jet [7, 8], whirling arm [9–11], water jet [12, 13] and ballistic [14] testing apparatuses. The results showed that the whirling arm style rig is able to rank the tested materials in the same order as those that had been attached to an aircraft and flown is real-world conditions [15], which suggests that the whirling arm is the method that best represents real-life operational conditions.

This paper describes the methodology proposed in LEAPWind, a new collaborative European research project, which aims to prevent blade leading-edge erosion by employing advanced composite materials and innovative manufacturing processes. An advanced thermoplastic-epoxy composite material is used to manufacture a leading edge component for a wind turbine blade. The material identification and characterisation, using mechanical and rain erosion testing, along with demonstrator and full-scale component design, is described. An overview of the complementary finite element method model is also included in the paper. The technologies developed in the project will be demonstrated through operational trials on a 2.1 MW wind turbine, located in an onshore wind farm in Portugal, which is also discussed.

2 Methodology

2.1 Aim and Objectives

The overarching aim of this study is to eliminate leading edge erosion in wind turbine blades. This will be achieved during the design and manufacturing stages of blade production, where maintenance due to leading edge erosion will not be necessary during operation. Within this study a new leading edge blade component will be designed, manufactured and tested on a wind turbine to protect the wind blade from erosion, which will boost productivity, and increase blade life, while reducing maintenance costs.

However, in order to attain this aim, the following specific objectives must be achieved:

- To design a new leading edge blade component, which will protect the wind blade from erosion.
- To develop a commercial-scale process for manufacturing the leading edge blade component.
- To de-risk the technology by performing structural (mechanical and rain erosion) testing of a full-scale prototype leading-edge blade component.
- To develop a finite element prediction and analysis model of the new component, which can be used for future design and structural health monitoring.
- To perform full-scale operational tests on an existing wind turbine in Portugal using advanced sensory structural health monitoring techniques to assess the performance of the new blade component.

2.2 Methodology Overview

In order to achieve the aim and objectives of this study, a systematic project methodology of the main technical tasks has been established, which is shown graphically in Fig. 1. The initial tasks in the project will run in parallel in order to arrive at an optimum component design. Following this, a more linear approach will take place as the novel wind blade component is manufacture, mechanically tested and de-risked further through operational trials. These stages are discussed in further detail in Sect. 3. In parallel to this stages, commercialisation of the new product and dissemination of project finding and outcomes will be performed but these aspects are not discussed in this paper.

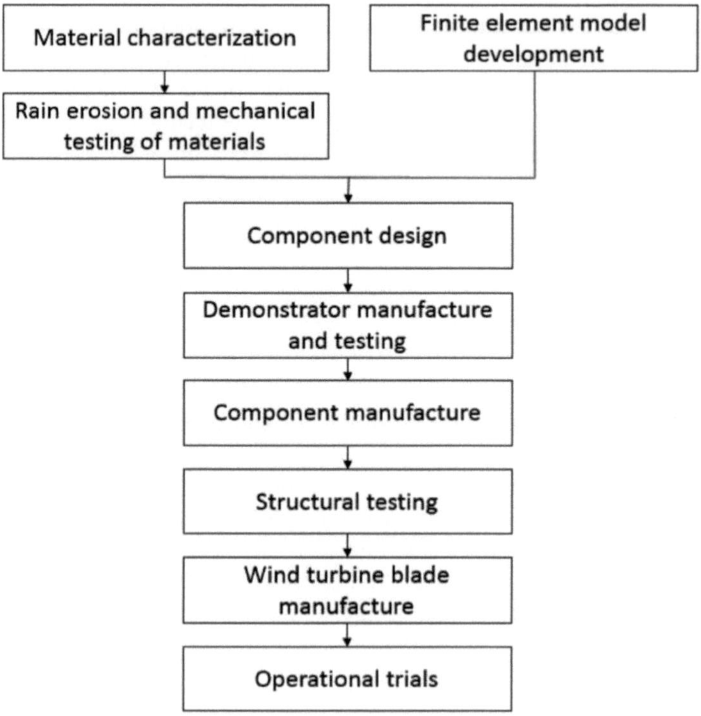

Fig. 1. Overview of the methodological approach employed to develop the new wind blade component

3 Methods and Discussions

3.1 Material Characterization

A key aspect of the study is to identify the most suitable material for protecting the leading edge of a wind turbine blade against erosion. Currently, there are a number of solutions for protecting against leading edge erosion, including tapes, coatings and shields. However, the proposed study intends to obtain a solution in the manufacture stage and eliminate the need for maintenance as a result of leading edge erosion. Based on the existing solutions, a material with properties of high tensile strength, high ductility and high elongation at break is required and, as a result, the following materials will be initially investigated:

- Polyurethane
- Polyphenylene sulphide (PPS)
- Polycarbonate

- Polyethylene terephthalate (PET)
- Polyamide 69 Nylon
- Polyamide 11.

The above list of materials, along with some baseline materials, will be structurally tested using mechanical (static and fatigue) testing and rain erosion testing, which is discussed further in the Sect. 3.2. The thermoplastic material selected will then be appended to the glass fibre reinforced powder epoxy, which makes up the main structural elements of the wind blade component. The thermoplastic material will effect a number of aspects of the design and, therefore, is one of the initial steps within the study.

3.2 Rain Erosion and Mechanical Testing of Materials

The materials detailed in Sect. 3.1 will be subjected to rain erosion testing in order to investigate their resistance to erosion in accordance to the testing standard ASTM G73-10 [16]. The tests will first be performed at the Whirling Arm Rain Erosion Rig (WARER) facility at University of Limerick (UL), which was developed in 2009 and has been extensively validated [9–11]. Following this, a set of specimens will be tested at a jet erosion testing facility to confirm the ranking of the selected materials.

Mechanical static and fatigue testing will be performed on the most promising materials when appended to glass fibre reinforced powder epoxy in order to investigate the effect on the mechanical properties of the resultant composite laminate. This static tensile and fatigue tension-tension testing was carried out in accordance with ASTM D3039 [17] and ASTM D3479 [18] using a 250kN Zwick test machine with wedge grips. Examples of material specimens that have been mechanically and rain erosion tested are shown in Fig. 2.

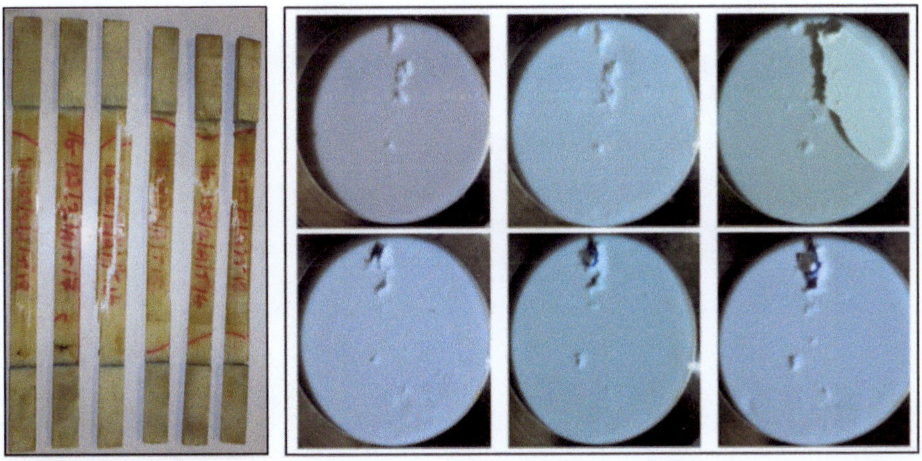

Fig. 2. Material specimens that have been mechanically tested (left) and rain erosion tested (right; [6])

The results of this testing campaign will form part of the decision process involved in the selection of the most suitable thermoplastic material, which is discussed in Sect. 3.1.

3.3 Finite Element Model Development

In parallel to the selection of the thermoplastic material that will form the outer costing of the leading edge component, where the structural part is made up of a glass fibre reinforced powder epoxy material, a finite element model of the component and materials will be developed.

The finite element model will be developed using ANSYS WorkBench [19], where it will combine a number of the ANSYS software packages, mainly DesignModeler, Composite PrepPost (ACP) and Mechanical. The development of this model will build on previous previous models for renewable energy systems [20–22]. This will allow for a complete design of the composite material part, including simulation of mechanical testing and a response displacement investigation. The model will be of the full-scale leading-edge blade component and will be used to predict the structural performance of the component in order to validate the model during in the experimental work performed, subsequently. A schematic of the new wind blade component that will be developed during this study is shown in Fig. 3, where it is labelled as "LEP". Additionally, the finite element model will also be used to model the blade component element demonstrators that will be produced and tested during the project. This will allow for the initial validation of the model and, therefore, improve its performance.

The finite element model will initially be used in the design of the wind blade component but it will also be used during the structural health monitoring of the component will in operation. This will allow the wind turbine operator to predict the performance and response of the component and feed into the on-site decision making processes. During the mechanical testing and operational trials within the project, any discrepancies between the model and the physical testing will be investigated and the model will be updated throughout the project in order to improve its reliability.

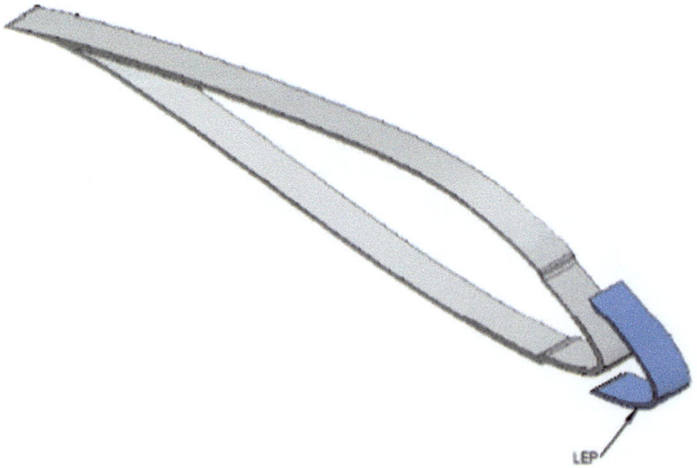

Fig. 3. Schematic of the new wind blade component that will be developed during this study, labelled as "LEP"

3.4 Leading Edge Component Design and Manufacture

The central research element of the study is the design and manufacture of a new leading edge blade component (a schematic of which is shown in Fig. 3).

The new leading edge blade component will be initially designed based on Suzlan's current design for the leading edge of their wind turbine blades. However, it will be adapted to ensure it has the required strength and stiffness to be incorporated into a wind turbine blade as a component, which is connected to the shear web of the blade's spar. The part will be mainly comprised of structure made from glass fibre reinforced powder epoxy material with an outer layer of the selected thermoplastic material, which is discussed in Sect. 3.1. The finite element model, which is discussed in Sect. 3.3, will also be used within the design of the new component.

Once the new leading edge blade component is design, manufacturing trials will take place in order to de-risk the more complicated aspects of the build. Included in this will be the manufacture and mechanical testing of a representative demonstrator, which will be used to further de-risk the component design, the manufacture process and the finite element model. This will ensure that any costly mistakes on the full-scale mould or other problems will be eliminated prior to the manufacture of the full-scale blade component.

The next stage in the production of the full-scale blade component is the manufacture of the mould tooling that the component will be built on. Firstly, a pattern for this mould is manufactured using a CAD model of the component outer profile that is inputted into a computer numerical control (CNC) machine. The accuracy in this CAD model, and subsequent CNC manufacture of the pattern, is paramount in the production of a high-performance part. Once the pattern is completed, the mould tooling will be cast on the pattern and it will be designed to provide the recommended manufacturing process window (vacuum level, temperature distribution along mould and through laminate thickness, heat-up time and power, dwell time and cooling rates). Testing of the heated mould tooling will be performed to ensure that it is fit for purpose and reliable for full scale, volume production, which ensures that the process can be completed in the future in as an efficient and environmentally friendly way as possible.

The full-scale blade component will be manufactured using the final blade component design and the mould tooling. To this end, rolls of glass-fibre will be impregnated with a powder epoxy and then kitted (or cut into plies). The outer layers of the component will be comprised of the selected thermoplastic material, which is discussed in Sect. 3.1. These outer layers will eliminate the risk of leading edge erosion during operation. The leading edge sub-components will be heated in an oven to 50 °C at which point they solidify but do not cross polymerise. Subsequently, these sub-components will be assembled together in the full-scale mould and cured to 180 °C under vacuum. Following the cure, the component will be de-moulded, visually inspected and non-destructive testing will be performed in order to ensure that it is of the highest standard and there are no defects.

3.5 Leading Edge Component Mechanical Testing

Mechanical testing of the new leading edge wind blade component will be performed in the Large Structures Test Cell at NUI Galway to ensure the component has sufficient structural integrity, where the facility and blade component testing is shown in Fig. 4. The wind blade component will be mounted on a steel testing fixtures within the Test Cell for the duration of the testing. This mechanical (static and fatigue) testing campaign will be performed to the relevant testing standards, DNV-DS-J102 [23] and IEC61400-23 [24], where actuators will be equally spaced along the length of the blade component to impose the loading on the component. The test loads applied during the mechanical testing campaign will be derived from actual loads applied to the wind turbine blades in operation.

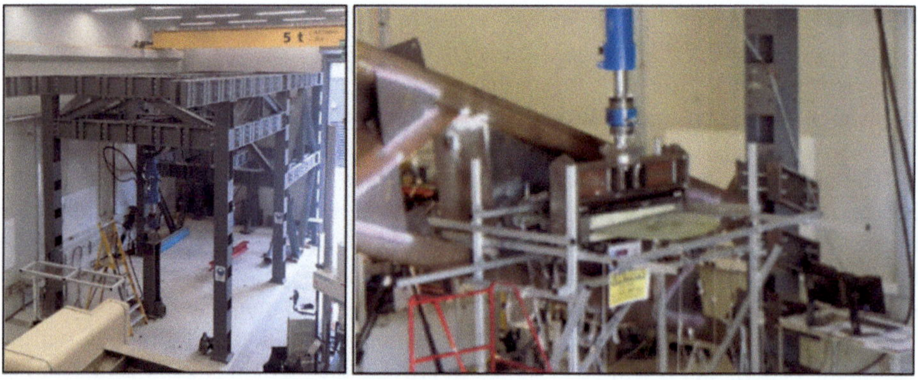

Fig. 4. The Large Structures Test Cell at NUI Galway, showing the test cell (left) and a blade component during a mechanical fatigue test (right)

During the static testing, the blade will be tested to the maximum predicted loads envisaged during operation, where loads will be applied in incremental steps of 5% of the maximum predicted operational load. The static testing will be monitored using a combination of strain gauges, LVDTs and a digital image correlation (DIC) strain measurement system. The data obtained during the physical test will be post-processed and will be compared with the results predicted by the finite element model. The key static testing cases are the flapwise and edgewise bending tests and the test data will confirm the stiffness distribution, natural frequency and strength of the blade. Following this, the fatigue testing will take place, where the blade component prototype will be tested in tension-tension fatigue loading within an R value of 0.1. Learnings from execution of the static test procedure will be inform the fatigue testing. The test loads will be based on operational learnings from the current SUZLON wind turbines, from coupon data, from static test results and from the updated finite element model. Up to 1,000,000 cycles of fatigue load cycles will be applied to the blade to demonstrate its fatigue strength and durability.

Once the mechanical testing campaign is complete, the results will be analysed and fed back into the design and manufacture of the commercial components. Additionally, the finite element model will be updated based on these results.

3.6 Operational Trials

A critical aspect in the development of a new wind turbine blade component is performing operational trials in a real-world scenario. This trial will evaluate the mechanical and erosion prevention performance of the blade component in real-world conditions.

Operational trials will take place in the Penamacor Wind Farm in Portugal that is being operated by Portugal OMS group, where there are 38 Suzlon S88/2100 (power 2,100 kW, diameter 88 m) wind turbines installed, which is shown in Fig. 5. This site has been selected as leading-edge erosion has been identified as a serious issue due the environmental condition at the site. The operational trials will be performed onshore as the conditions at the site and size of the turbines are comparable to offshore wind turbines, while reducing the cost of the trial and increasing access/monitoring capabilities due to the more controlled setting. However, ultimately, the technology will be deployed as a solution to leading-edge erosion in offshore wind turbine blades. Additionally, as part of this deployment, standard testing will be performed on the finished blades to DNVGL-CP-0424 [25].

Fig. 5. Penamacor Wind Farm in Portugal, where the operational trials for the new wind blade component will take place

Following the operational trials, the results will be analysed and reported to the project partners, the steering committee and the European Commission. Additionally, the results will be compared to the results from the physical testing and finite element models and this analysis will be included in the operational trials report. Following this, a concise summary of these results will be disseminated to key project stakeholders, the research community and members of the general public in the form of a peer-reviewed publication in a high impact journal.

4 Conclusion

In this paper, the methodology proposed in LEAPWind, a new collaborative European research project, which aims to prevent blade leading-edge erosion by employing advanced composite materials and innovative manufacturing processes has been presented. An advanced thermoplastic-epoxy composite material is used to manufacture a leading edge component for a wind turbine blade. The critical technical stages, including material identification and characterisation, component design and manufacture have been discussed. Additionally, the details relating to de-risking of the novel technologies through mechanical and rain erosion testing, and full-scale operational trials on a 2.1 MW wind turbine, located in an onshore wind farm in Portugal, has been included.

The immediate impact of this project will be the creation of a new wind energy product as the existing technologies are advanced from a technology readiness level of 6 to 9, which will have both economic and social benefits, including job creation. In a recent study, the levelised cost of offshore wind energy was shown to be €165/MWh [26]. The results from this project will lower this cost as maintenance costs will be significantly reduced by eradicating the need for repair due to leading-edge erosion, which is currently necessary every 5 years. Further environmental benefits will be generated as a more reliable wind turbine blade can be manufactured, increasing the sustainability of offshore wind energy, while contributing the UN Sustainable Development Goals [27].

Acknowledgements. This study was funded by the Executive Agency for Small and Medium sized Enterprises (EASME) in the European Commission through the LEAPWind project (Agreement no.: EASME/EMFF/2017/1.2.1.12/S1/06/SI2.789307). The first and last authors would like to acknowledge the support from Science Foundation Ireland (SFI), through the Marine and Renewable Energy Ireland (MaREI) research centre (Grant no. 12/RC/2302), and the Career Development Award programme (Grant No. 13/CDA/2200).

References

1. Conroy, J.: World wind power to almost double to 650GW by 2020. The Australian. 21 January 2015
2. SEIMENS Gamesa Newroom. http://www.siemensgamesa.com/en-int/newsroom/2019/01/new-siemens-gamesa-10-mw-offshore-wind-turbine-sg-10-0-193-dd. Accessed 1 Mar 2019

3. Budinski, K.G.: Guide to Friction, Wear and Erosion Testing. ASTM international, West Conshohocken, PA (2007)
4. Keegan, M.H., Nash, D.H., Stack, M.M.: On erosion issues associated with the leading edge of wind turbine blades. J. Phys. D: Appl. Phys. **46**, 383001 (2013)
5. Chen, J., Wang, J., Ni, A.: A review on rain erosion protection of wind turbine blades. J. Coat. Technol. Res. **16**, 15–24 (2018)
6. Cortes, E., Sanchez, F., O'Carroll, A., Madramany, B., Hardiman, M., Young, T.M.: On the material characterisation of wind turbine blade coatings: the effect of interphase coating-laminate adhesion on rain erosion performance. Materials **10**, 1146 (2017)
7. Cook, S.S.: Erosion by water-hammer. Proc. R. Soc. A: Math. Phys. Eng. Sci. **119**, 481–488 (1928)
8. Mann, B.S., Arya, V.: An experimental study to correlate water jet impingement erosion resistance and properties of metallic materials and coatings. Wear **253**, 650–661 (2002)
9. Tobin, E.F., Rohr, O., Raps, D., Willemse, W., Norman, P., Young, T.M.: Surface topography parameters as a correlation factor for liquid droplet erosion test facilities. Wear **328–329**, 318–328 (2015)
10. O'Carroll, A., Hardiman, M., Tobin, E.F., Young, T.M.: Correlation of the rain erosion performance of polymers to mechanical and surface properties measured using nanoindentation. Wear **412–413**, 38–48 (2018)
11. Young, T.M., Tobin, E.: Rain erosion testing of composite materials: the design of a laboratory test facility. In: 3rd International Supply Wings Conference, Frankfurt, Germany (2008)
12. Engel, O.G.: Mechanism of Rain Erosion Part 2: A Critical Review of Erosion by Water Drop Impact. Nation Bureau of Standards (1953)
13. Bourne, N.K., Obara, T., Field, J.E.: The impact and penetration of a water surface by a liquid jet. Proc. R. Soc. A: Math. Phys. Eng. Sci. **452**, 1497–1502 (1996)
14. Adler, W.F.: Influence of water drop distortion on impact damage. In: Proceedings of the Sixth Symposium on Electromagnetic Wind, Atlanta (1982)
15. Schmitt Jr., G.F.: Flight test-whirling arm correlation of rain erosion resistance of materials. Air Force Materials Laboratory, Wright-Patterson Air Force Base, OH (1968)
16. ASTM G73 – 10: Standard Test Method for Liquid Impingement Erosion Using Rotating Apparatus. ASTM International, West Conshohocken, PA, USA (2017)
17. ASTM, D3039 / D3039 M-17: Standard Test Method for Tensile Properties of Polymer Matrix Composite Materials. ASTM International, West Conshohocken, PA, USA (2017)
18. ASTM D3479 / D3479 M – 12: Standard Test Method for Tension-Tension Fatigue of Polymer Matrix Composite Materials. ASTM International, West Conshohocken, PA, USA (2017)
19. Lawrence, K.L.: ANSYS Workbench Tutorial Release 14. SDC publications, Mission, KS (2012)
20. Finnegan, W., Fagan, E., Flanagan, T., Doyle, A., Goggins, J.: Operational fatigue loadings on tidal turbine blades using computational fluid dynamics, Renewable Energy (Under Review)
21. Fagan, E.: Design of Fibre-Reinforced Polymer Composite Blades for Wind and Tidal Turbines. National University of Ireland Galway, Galway, Ireland (2017)
22. Finnegan, W., Goggins, J.: Linear irregular wave generation in a numerical wave tank. Appl. Ocean Res. **52**, 188–200 (2015)
23. DNV-DS-J102: Design and Manufacture of Wind Turbine Blades, Offshore and Onshore Wind Turbines. DNV GL Standard (2010)
24. IEC 61400-23:2014: Full-Scale Structural Testing of Rotor Blades. Wind Turbines, International Electrotechnical Commission Part 23 (2014)

25. DNVGL-CP-0424: Coatings for Protection of FRP Structures with Heavy Rain Erosion Loads. DNV GL Standard (2016)
26. Astariz, S., Vazquez, A., Iglesias, G.: Evaluation and comparison of the levelized cost of tidal, wave, and offshore wind energy. J. Renew. Sustain. Energy **7**, 053112 (2015)
27. UN Sustainable Development Goals. https://sustainabledevelopment.un.org.Accessed 1 Mar 2019

Fault Diagnosis of Shaft Misalignment and Crack in Rotor System Based on MI-CNN

Wang Zhao, Chunrong Hua(✉), Danyang Wang, and Dawei Dong

School of Mechanical Engineering, Southwest Jiaotong University, Chengdu 610031, China
hcrong@swjtu.cn

Abstract. Focusing on the difficulty in distinguishing the shaft misalignment and crack in a rotor system, a fault diagnosis method based on multi-input convolutional neural network (MI-CNN) is proposed in this paper. The time-domain vibration signals are directly taken as the input of a one-dimensional convolutional neural network, which are collected on the test bench in the conditions of health, shaft misalignment, crack and misalignment-crack coupling of the rotor system. Kernels of different sizes are adopted to extract the signal features of diverse dimensions at different input ends to fully use the information of the raw vibration signals, and then the extracted features from each input end are fused adaptively. Finally, the classification of shaft misalignment and crack of the rotor system is completed by softmax function. The results show that the intelligent diagnosis of shaft misalignment and crack in the rotor system can be realized effectively by the proposed method, and eventually the recognition rate reaches 99.42%, which has better accuracy and stability compared with other intelligent algorithms. The study achievements can provide a basis for intelligent fault diagnosis of rotating machinery.

Keywords: Rotor system · Misalignment · Crack · Convolutional neural networks · Fault diagnosis

1 Introduction

Rotors are widely used in power stations, compressors, aero-engines, and so on, while the faults of shaft misalignment and crack are very common for rotors, which will lead to performance deterioration and equipment malfunction. Serious axle-broken accidents will be happened once the fault is aggravated or the shaft misalignment and crack are coupled with each other, which will not only cause the outage of the whole system, resulting in immeasurable economic losses, but also pose a great threat to the security of the personnel involved, and even lead to catastrophic accidents. Therefore, it is of great significance to detect and distinguish the shaft misalignment, crack and misalignment-crack coupling in the rotor systems to maintain the stable operation of the equipments and protect the personal safety of the staff.

In recent decades, fault diagnosis of rotor systems based on vibration signals has been a research hotspot, which is sensitive to fault characteristics, convenient for testing and effective for on-line detection. Some scholars found that the 2X

M. A. Wahab (Ed.): DAMAS 2019, LNME, pp. 529–540, 2020.
https://doi.org/10.1007/978-981-13-8331-1_39

superharmonic component can be used as the characteristic property of the shaft misalignment fault for rotor systems [1, 2]. Sekhar and Prabhu [3] established a high-order finite element model of a rotor system and analyzed the effects of shaft misalignment on the 1X and 2X superharmonic components of the vibration signals at different speeds. Patel and Darpe [4] established a six-degree-of-freedom rotor system with a coupling using Timoshenko beam element, and analyzed the effects of parallel misalignment and angular misalignment of the coupling on torsional and longitudinal response by fast Fourier transform (FFT). However, the superharmonic components will also occur in cracked rotor signals [5], especially at 1/2 and 1/3 subcritical speeds, the 2X and 3X superharmonic components are particularly prominent [6]. Therefore, both misaligned and cracked rotors can produce superharmonic components in vibration signals, and it is difficult to distinguish them accurately only relying on the vibration characteristics in frequency domain. Moreover, the coupling of shaft misalignment and crack in rotor systems will produce more complex non-linear dynamic characteristics in the vibration response and increase the difficulty of the fault identification greatly.

Deep learning, as a branch of machine learning in the field of artificial intelligence, which aims to establish and simulate the learning mechanism of human brain to interpret data, has achieved great success in pattern recognition fields such as image processing [7, 8], speech recognition [9, 10], natural language understanding [11, 12] with its powerful automatic feature extraction ability. In 2013, Tamilselvan and Wang [13] applied the deep belief network (DBN) to fault diagnosis of aircraft engines for the first time, and achieved a higher recognition rate than the traditional algorithms, which attracted massive attention of scholars in this field. Convolutional neural network (CNN), as one of the most widely used and successful algorithms in deep learning, is a typical feedforward neural network. Compared with the traditional diagnosis methods, which excessively depend on expert diagnosis experience and consume a lot of time to extract fault features manually, CNN can directly train the network based on raw data and automatically learn signal features to achieve intelligent diagnosis of system health conditions. Yuan et al. [14] proposed a multi-mode convolutional neural network (M-CNN) based on multi-source heterogeneous monitoring data, and achieved the multi-classification fault diagnosis of a rotor system with one-dimensional vibration signals and two-dimensional infrared images as the model input simultaneously. Guo et al. [15] decomposed the vibration signals of a rotor system by continuous wavelet transform, and then used the wavelet coefficients to construct two-dimensional continuous wavelet transform scalogram (CWTS) as the input of CNN, and then verified the accuracy and robustness of the proposed method. The above researches mainly focus on the faults classification in rotor systems, such as imbalance, misalignment, rub-impact, and bearing seat looseness, but the identification of the crack fault and the misalignment-crack coupling fault has not yet been studied.

In this paper, focusing on the difficulty in distinguishing the faults of shaft misalignment and crack in a rotor system, a MI-CNN is designed to realize the intelligent fault diagnosis. A network structure is proposed and raw vibration signals are used as the input to fully exert the feature extraction ability of CNN. On this basis, the network is verified by experimental data at a rotor test bench, and the classification effects are compared with other intelligent algorithms. The rest of this paper is as follows. In Sect. 2, the convolutional neural network is introduced briefly and the model structure is proposed. In Sect. 3, the experiment verification and diagnosis results based on MI-CNN are present. And the last part summarizes and discusses the conclusions of this paper.

2 Convolutional Neural Network for Fault Diagnosis of Shaft Misalignment and Crack in Rotor System

2.1 Brief Introduction

The structure of convolutional neural networks tends to be flexibly adjusted according to different classification tasks, but conventional CNN usually includes five parts: input layer, convolutional layer, pooling layer, fully-connected layer, and output layer. The convolutional layer and the pooling layer, as the core parts of the CNN, are often used alternately and repeatedly to extract features from the input signals. The fully-connected layer is usually connected to the output layer to complete the classification process of the network.

As the main feature extraction layer, the convolutional layer performs convolution calculation on the input signals through multiple kernels with selected stride, the outputs are connected with the local receptive fields of the convolutional layer to realize sparse connection and each kernel keeps its parameters unchanged during the sliding on the signals, i.e. weight sharing. These two characteristics of the convolutional layer make the network translation invariant, and can reduce the amount of the network parameters and improve the training speed of the CNN significantly. The specific calculation of the convolutional layer is as follows:

$$x_j^l = f\left(\sum_i x_i^{l-1} * w_{ij}^l + b_j^l\right) \tag{1}$$

where x_j^l denotes the jth feature map generated by the layer l, x_i^{l-1} denotes the ith feature map in the layer l-1, w_{ij}^l denotes the connection weight between the jth kernel and the ith input feature map, b_j^l denotes the deviation of the jth kernel, $*$ denotes the convolution calculation, and $f(\cdot)$ denotes the activation function.

The activation function is often applied between the convolutional layer and pooling layer to express the extracted features non-linearly. The rectified linear unit (Relu) can avoid the problem of slow updating network parameters and the vanishing gradient in the process of back propagation to a large extent, which has simple gradient solution, and can speed up the convergence of the network. Therefore, Relu is selected as the activation function in this paper.

The feature information extracted from convolutional layer is often further processed by pooling layer to compress data and reduce the redundant information. At the same time, the process of the pooling can effectively improve the robustness of the extracted features, speed up the network operation and reduce the risk of overfitting. The pooling operation process is as follows:

$$x_j^l = \text{down}\left(x_j^{l-1}, s\right) \tag{2}$$

where $\text{down}(\cdot)$ is the down-sampling function, x_j^l is the output of the pooling layer, x_j^{l-1} is the feature map of the previous layer, s is the size of pooling layer. The pooling operation can be further classified into max pooling and average pooling, and the specific process is to take the maximum value or average value for each area of size s in the input map. In order to obtain the most prominent features of each signal segment, max pooling is selected in this paper.

The function of the fully-connected layer which is often connected to the final output layer, is to transform all the features learned by the network into a feature vector representation. For multi-classification tasks, the softmax function is commonly used in the output layer as the classifier. By mapping the output vector of the fully-connected layer into the interval (0, 1), the probability of each input being recognized as different labels is calculated to achieve the purpose of classification. The calculation of the softmax function is as follows:

$$P_n = \frac{\exp(x_i)}{\sum_{j=1}^{m} \exp(x_j)} \tag{3}$$

where n represents the number of labels, P_n represents the probability of the input x_i belonging to label n.

2.2 Model Design

In 2014, Simonyan and Zisserman [16] presented the famous VGGNet in the field of image recognition in the ILSVRC competition, and won the second and first place in image classification and target localization respectively. VGGNet reuses the 3 × 3 kernel continuously, and combines with the nonlinear activation layer alternately, so it has a deeper network structure and can extract deeper features in the signals. Combined with the network characteristics of VGGNet, reference [17] used successive small-size kernels and selected the wide kernel in the first convolutional layer of CNN to extract the medium and low frequency features of the signals for rolling bearings, and as a result, the influence of high-frequency noise is reduced to a certain extent. The proposed WDCNN is a 12-layer network with 5 convolutional layers, and has achieved good results in fault classification of bearings.

However, it is often unnecessary to design a relatively deep network structure for general mechanical fault diagnosis. In general, excessive stacking of small-size kernels causes network structure too deep, leading to excessive training parameters and the risk of network overfitting. In addition, kernels of the same size can only extract signal features of the same dimension, and continuous stacking fixed-size kernels will result in a single type of learned features. Consequently, for the fault identification of the rotor system in this paper, the number of convolutional layers and pooling layers of CNN are appropriately reduced on the premise of ensuring the classification accuracy, and the kernels of different sizes are adopted to process signals at different input ends to extract the features of different dimensions in order to make full use of the rich information of the raw signals, then the features learned by CNN at different ends are adaptively fused, and different fault types of the rotor system are distinguished by softmax classifier finally.

Focusing on the condition recognition of health, shaft misalignment, crack and misalignment-crack coupling of the rotor system, a MI-CNN is designed. Figure 1 is the detailed architecture of the MI-CNN, it can be seen from the figure that the one-dimensional raw signals are directly input to the CNN after overlapping sampling, and different processing is performed at three different input ends. The network structure of each input end includes 3 convolutional layers and 3 pooling layers. The first convolutional layer at different ends adopts the kernels of the same size and stride to extract features from the input signals, the second and third convolutional layers at each end use the same kernels and the size of the kernels at the three input ends are 3 × 1, 5 × 1, 7 × 1 respectively. Besides, the same padding is adopted in the first two convolutional layers, and the size of all the pooling layers is 2 × 1. After the third pooling layer, the various signal features of the network learned by the three input ends are merged and further processed through the fully-connected layer. Eventually, four types of the rotor signals are output by the softmax function. The specific parameters of each input end are shown in Table 1:

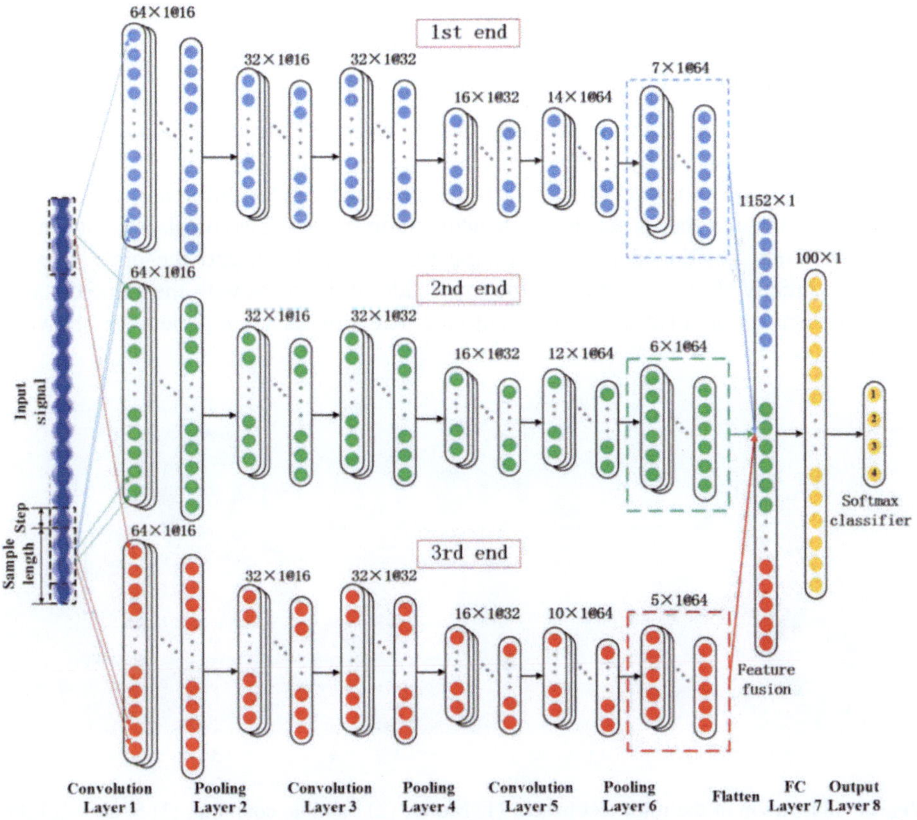

Fig. 1. Architecture of the proposed model

Table 1. Specific parameters of the proposed model

Layer	Type	Kernel size/stride in 1st end	Kernel size/stride in 2nd end	Kernel size/stride in 3rd end	Output size (1st/2nd/3rd)
1	Convolution	32 × 1/8 × 1	32 × 1/8 × 1	32 × 1/8 × 1	64 × 16/64 × 16/64 × 16
2	Pooling	2 × 1/2 × 1	2 × 1/2 × 1	2 × 1/2 × 1	32 × 16/32 × 16/32 × 16
3	Convolution	3 × 1/1 × 1	5 × 1/1 × 1	7 × 1/1 × 1	32 × 32/32 × 32/32 × 32
4	Pooling	2 × 1/2 × 1	2 × 1/2 × 1	2 × 1/2 × 1	16 × 32/16 × 32/16 × 32
5	Convolution	3 × 1/1 × 1	5 × 1/1 × 1	7 × 1/1 × 1	14 × 64/12 × 64/10 × 64
6	Pooling	2 × 1/2 × 1	2 × 1/2 × 1	2 × 1/2 × 1	7 × 64/6 × 64/5 × 64
-	Flatten	/	/	/	1152 × 1
7	Fully-connected	/	/	/	100 × 1
8	Softmax	/	/	/	4

3 Experimental Verification

In order to verify the validity of the proposed method and the reliability of the fault diagnosis of rotor systems, this section will conduct research through bench experiments.

3.1 Experimental Setup

Figure 2 shows the rotor test bench used in this paper, it is driven by a motor, and the rotor speed is regulated by a motor control system. Four eddy current displacement sensors collect the vibration signals at different speeds under different working conditions, and transmit the signals to a data acquisition system. Two displacement sensors are respectively positioned right above the two rotary discs, and the other two are arranged directly above the shaft on the inner side of the discs with the same distance from the discs on both sides. The relevant parameters of each component are shown in Table 2.

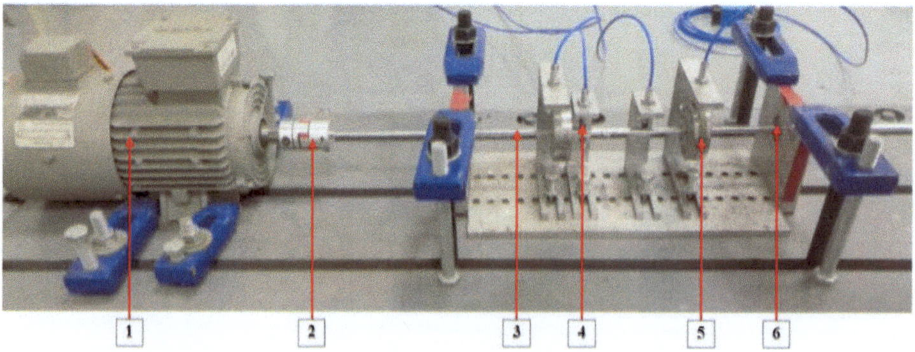

Fig. 2. Illustration of the rotor test bench. (1) Motor, (2) Flexible coupling, (3) Shaft, (4) Eddy current displacement sensor, (5) Rotary disc, (6) Bearing.

Table 2. Details of the rotor test bench

Components	Type	Components	Type
Driving motor	SIEMENSE 1.5 kW	Coupling	Flexible coupling
Motor control system	VFD-M 1.5 kW	Bearing	SKF 6300
Eddy current displacement sensor	ZA21-0803	The material of rotary shaft	40Cr
Signal acquisition system	DHDAS5922 N	The material of rotary disc	45# steel

3.2 Data Description

The vibration signals of the rotor system under four kinds of conditions are collected in the experiments, including health, shaft misalignment, crack and misalignment-crack coupling. The crack on the shaft is machined by wire cutting, with a width of 0.2 mm and a depth of 4 mm, and a gasket of 1 mm thickness is placed at the bottom of the motor to simulate the shaft misalignment.

The vibration signals of the rotor system at different rotating speeds within 10 Hz–26 Hz are collected. The specific operations are as follows: firstly, the motor control system is adjusted to a specific speed, and the rotor shaft is held rotating at this speed for 60 s to achieve system stability, and the vibration signals are collected for 20 s by the data acquisition system with a sampling frequency of 5 kHz, then the motor speed is changed and the same operations are repeated, finally, the above processes are performed for signals collecting of different fault types.

Considering the influence of the rotor speed on fault classification results, six groups of signals with different speeds are selected from the collected data under different conditions, and each group of signals is fused by intercepting data points of the same length, then the fused signals are normalized and the sample sets are formed by overlapping sampling. One fused shaft misalignment signal is shown in Fig. 3. In detail, each condition contains 1000 samples, and each sample contains 512 data points, therefore, the final data sets contain a total of 4000 samples, of which 70% are randomly selected as the training sets and the rest as the testing sets.

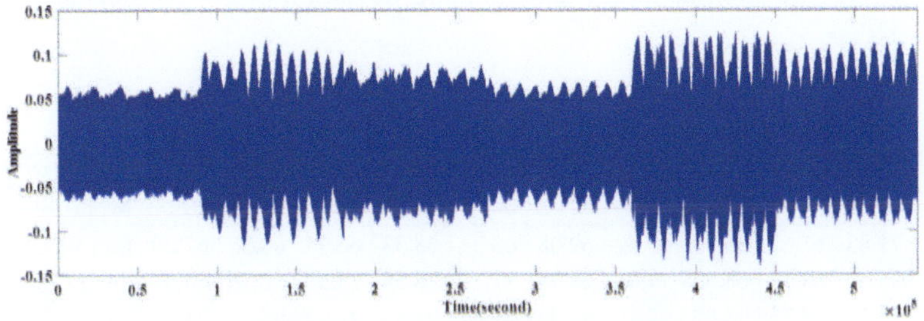

Fig. 3. The fused vibration signal

3.3 Results and Discussions

When training the network, Adam optimization algorithm is adopted with a learning rate of 0.001, the batch size is set to 128 and the epoch is 60, ten trials are conducted on each data set. The results of fault identification using MI-CNN is shown in Fig. 4, it can be seen that this method has achieved good results in identification of health, shaft misalignment, crack and misalignment-crack coupling of the rotor system. The training accuracy of 10 trials is greater than 99.6%, and the testing accuracy is not less than 98.89%.

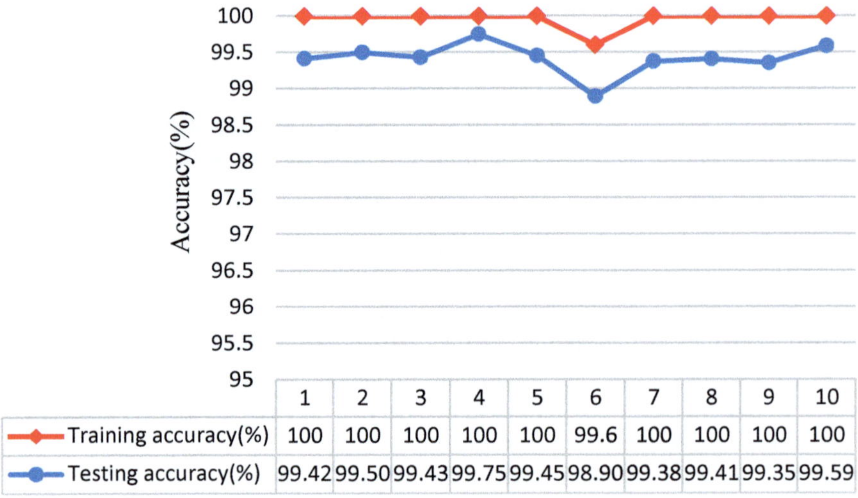

Fig. 4. Accuracy of the MI-CNN with training sets and testing sets

In order to verify the superiority of this method in fault diagnosis of the rotor system, three typical machine learning algorithms (K-Nearest Neighbor, Support Vector Machine and Random Forest) are utilized to compare the classification performance using the same data sets. The number of neighbors in KNN adopts the default value, SVM selects the polynomial kernel function, and the number of classifiers in RF is 250. The classification accuracy and standard deviation of the 10 trials are shown in Table 3.

Table 3. Comparison of classification accuracy of different methods (unit: %)

Methods	1	2	3	4	5	6	7	8	9	10	Mean	St-Dev
KNN	71.83	65.67	66.33	63.42	69.08	63.33	65.33	60.75	68.58	67.00	66.13	3.22
SVM	75.42	69.75	68.33	74.08	68.75	71.33	78.58	73.50	80.75	66.67	72.72	4.61
RF	89.25	92.75	93.83	97.17	92.92	91.67	91.83	91.92	95.25	95.58	93.22	2.30
MI-CNN	99.42	99.50	99.43	99.75	99.45	98.90	99.38	99.41	99.35	99.59	**99.42**	**0.22**

Note: StDev represents the standard deviation.

As can be seen from Table 3, compared with the traditional intelligent algorithms, the classification accuracy of the proposed method has been improved greatly, with an average recognition rate of 99.42%. Figure 5 is the classification results of the four methods, it is more intuitive that the recognition accuracy of MI-CNN is significantly higher than that of RF, SVM and KNN in distinguishing shaft misalignment and crack of the rotor system, and the dispersion degree of data is greatly reduced, indicating the stability of the algorithm is obviously improved. The comparison of the verification results shows that the fault diagnosis method based on MI-CNN can mine the hidden feature information of the raw time-domain signals, extract the deeper fault features automatically and establish more abstract and complex feature representation, and then realize the high-precision identification on the fault types of the rotor system.

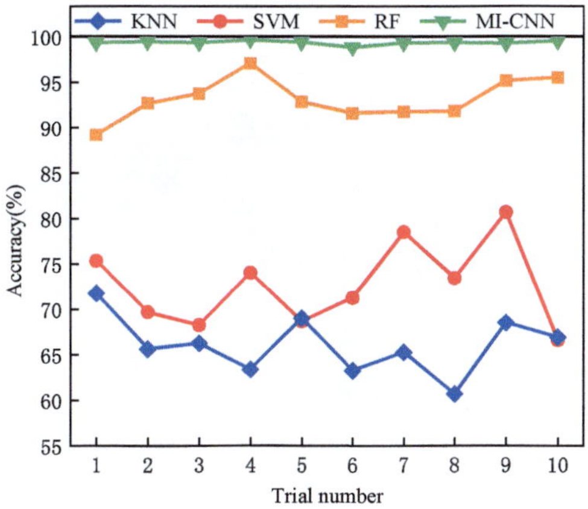

Fig. 5. Diagnosis results of the ten trials

In order to further study the influence of different fault signals on the recognition accuracy in the classification process, the results of KNN, SVM and RF are mapped as confusion matrices, as shown in Fig. 6. The vertical coordinate denotes the actual labels and the horizontal coordinate denotes the predicted labels, the values on the main diagonal represent the correct recognition rate of the signal in different condition, and the other values represent the probability that the signals are misdiagnosed as other conditions. It shows that the three algorithms can achieve high recognition rate for the healthy and cracked rotor system. For shaft misalignment signals, only RF successfully realizes the classification, while KNN and SVM both classify partial misalignment samples as the misalignment-crack coupling condition or the crack condition. However, for misalignment-crack coupling signals, all the three algorithms are difficult to achieve accurate identification, and some signal samples are misdiagnosed as the crack condition.

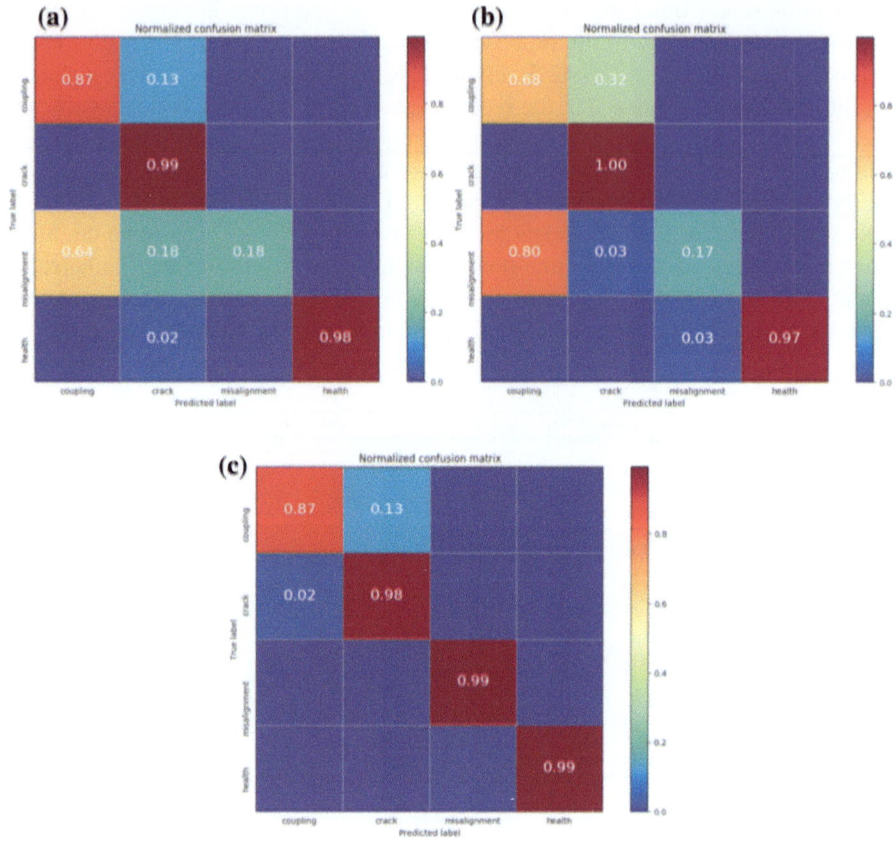

Fig. 6. The confusion matrix of (a) KNN (b) SVM (c) RF

In summary, the traditional intelligent algorithms can effectively recognize the conditions of health and crack of the rotor system, but it is difficult to identify the complex shaft misalignment and misalignment-crack coupling signals. Since a large number of misalignment samples are misclassified as the misalignment-crack coupling condition, and the coupling samples are misidentified as the crack fault by the traditional intelligent algorithms, it is demonstrated that the vibration signals of the misaligned and cracked rotor system have similar characteristics and mutual interference in classification, resulting in difficulty in distinguishing shaft misalignment, crack and misalignment-crack coupling signals effectively.

In addition, in order to verify the performance improvement of the MI-CNN proposed in this paper, the recognition results of the MI-CNN and the single-input CNNs are compared. Figure 7 shows the classification accuracy of the four networks and the calculated standard deviation based on the 10 random trials. Compared with different single-input CNNs, MI-CNN has higher accuracy and stability in identifying shaft misalignment and crack of the rotor system. In detail, MI-CNN has improved the classification accuracy by at least 1.03%, and the stability improvement is more significant with the error line shorter than the others several times. The results further show that the MI-CNN can make full use of the information in the raw vibration signals and improve the accuracy of fault diagnosis by fusing the various features extracted by kernels of different sizes.

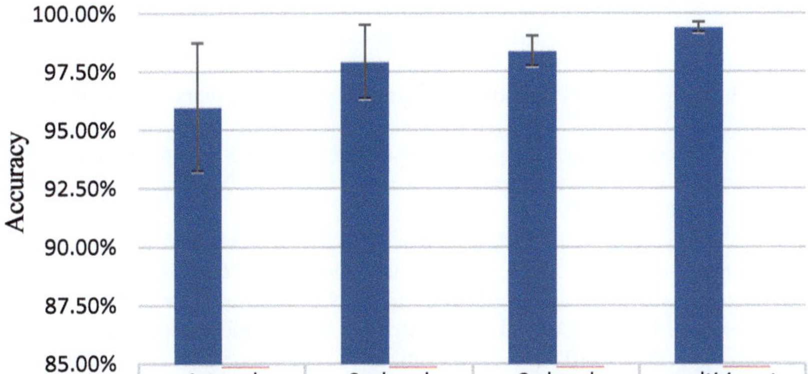

	1st end	2nd end	3rd end	multi-input
Accuracy	96.00%	97.93%	98.39%	99.42%
Variance	2.74%	1.57%	0.65%	0.22%

Fig. 7. Diagnosis results of MI-CNN and three single-input CNNs

4 Conclusions

In this paper, a recognition method for fault diagnosis of a rotor system based on MI-CNN is proposed, which can effectively distinguish the conditions of health, shaft misalignment, crack and misalignment-crack coupling of the rotor system.

1. Considering the full use of the raw vibration signals of the rotor system, a MI-CNN combined with different kernels is designed, and the relevant parameters of the network are determined.
2. The MI-CNN can effectively mine the hidden information in the raw vibration signals and the recognition rate of health, shaft misalignment, crack and misalignment-crack coupling of the rotor system reaches 99.42%, while it is difficult to distinguish different types of fault signals accurately based on the KNN, SVM, RF.
3. Compared with the recognition results of single-input CNNs with different kernels, this method improves the recognition accuracy by at least 1.03% and improves the stability of the network significantly.

Acknowledgments. This study was supported by the National Natural Science Foundation of China (No. 51875482).

References

1. Dewell, D.L., Mitchell, L.D.: Detection of a misaligned disk coupling using a spectrum analysis. J. Vib., Acoust., Stress Reliab. Des. **106**(1), 9–16 (1984)
2. Xu, M., Marangoni, R.D.: Vibration analysis of a motor-flexible coupling-rotor system subject to misalignment and unbalance, part I: theoretical model and analysis. J. Sound Vib. **176**(5), 663–679 (1994)
3. Sekhar, A.S., Prabhu, B.S.: Effects of coupling misalignment on vibrations of rotating machinery. J. Sound Vib. **185**(4), 655–671 (1995)
4. Patel, T.H., Darpe, A.K.: Vibration response of misaligned rotors. J. Sound Vib. **325**(3), 609–628 (2009)
5. Sinou, J.J.: Detection of cracks in rotor based on the 2 × and 3 × super-harmonic frequency components and the crack–unbalance interactions. Commun. Nonlinear Sci. Numer. Simul. **13**(9), 2024–2040 (2008)
6. Guo, C., Yan, J., Yang, W.: Crack detection for a Jeffcott rotor with a transverse crack: an experimental investigation. Mech. Syst. Signal Process. **83**, 260–271 (2017)
7. Krizhevsky, A., Sutskever, I., Hinton, G.E.: ImageNet classification with deep convolutional neural networks. In: 26th Annual Conference on Neural Information Processing Systems 2012, vol. 2, pp. 1097–1105. Neural Information Processing Systems Foundation, Nevada, USA (2012)
8. Russakovsky, O., Deng, J., Su, H., Krause, J., Satheesh, S., et al.: Imagenet large scale visual recognition challenge. Int. J. Comput. Vis. **115**(3), 211–252 (2014)
9. Graves, A., Mohamed, A.R., Hinton, G.: Speech recognition with deep recurrent neural networks. In: 2013 38th IEEE International Conference on Acoustics, Speech, and Signal Processing, pp. 6645–6649. Institute of Electrical and Electronics Engineers Inc, Vancouver, Canada (2013)
10. Li, Y.X., Zhang, J.Q., Pan, D., Hu, D.: A study of speech recognition based on RNN-RBM language model. J. Comput. Res. Dev. **51**(9), 1936–1944 (2014)
11. Cho, K., Merrienboer, B.V., Gulcehre, C., Bahdanau, D., Bougares, F., et al.: Learning phrase representations using RNN encoder-decoder for statistical machine translation. In: 2014 Conference on Empirical Methods in Natural Language Processing, pp. 1724–1734. Association for Computational Linguistics, Doha, Qatar (2014)
12. Yang, Z., Tao, D.P., Zhang, S.Y., Jin, L.W.: Similar handwritten Chinese character recognition based on deep neural networks with big data. J. Commun. **35**(9), 184–189 (2014)
13. Tamilselvan, P., Wang, P.: Failure diagnosis using deep belief learning based health state classification. Reliab. Eng. Syst. Saf. **115**, 124–135 (2013)
14. Yuan, Z., Zhang, L.B., Duan, L.X.: A novel fusion diagnosis method for rotor system fault based on deep learning and multi-sourced heterogeneous monitoring data. Meas. Sci. Technol. **29**(11) (2018)
15. Guo, S., Yang, T., Gao, W., Zhang, C.: A novel fault diagnosis method for rotating machinery based on a convolutional neural network. Sensors **18**(5) (2018)
16. Simonyan, K., Zisserman, A.: Very deep convolutional networks for large-scale image recognition. Computer Science (2014)
17. Zhang, W., Peng, G.L., Li, C.H., Chen, Y.H., Zhang, Z.J.: A new deep learning model for fault diagnosis with good anti-noise and domain adaptation ability on raw vibration signals. Sensors **17**(2) (2017)

Development and Tuning of a Simplified 1D Model for Generation of Transient States in Large Turbomachinery

Tomasz Barszcz[1(✉)], Piotr Czop[1], and Mateusz Zabaryłło[2]

[1] AGH University of Science and Technology, Krakow, Poland
{tbarszcz, pczop}@agh.edu.pl
[2] General Electric Sp. z o.o., Elblag, Poland
mateusz.zabaryllo@ge.com

Abstract. The Machine Learning methods require a large data sets for model training and it often causes a problem in a real application. Usage of a model based approach to generate the data can be a way to generate data in various degrees of malfunctions, e.g. with different sizes of unbalance.

The aim of this paper is to demonstrate the applicability of a 1D rotor-bearing model to reproduce the unbalance conditions during rotors coast-down operation. The model parameters adjustment case study is focused on an application of the model in order to reproduce rotor unbalance conditions of a 200 MW steam turbine. The rotor coast-down operation is considered to reduce external forces related to start-up or steady-state rotor operation. This allows to reduce and in turn clearly isolate unbalance conditions. The developed 1D model consists of first-principle rotor motion equations along the hydrodynamic bearing-support and foundation equations. The gray-box approach was applied to reduce the number of parameters required to be adjusted during system identification process. The rotor geometry and related mass-stiffness parameters were derived from the bearing-rotor assembly drawings while other phenomenological parameters were adjusted based on the measurements to obtain a good correlation with amplitude of vibrations at measurement locations along the shaft line.

Keywords: Rotordynamics · Simulation · Dynamic model · Gray-box rotor-bearing model

1 Introduction

Conventional turbogenerators constitute a major part of the power generation assets, and this situation will only slowly change towards renewable sources. Methods for fault detection and identification are an important direction of research. Typical rotating machinery malfunctions are listed according to the machinery part (rotor, seal, bearing, coupling, generator, foundation, etc.) in the area of diversified groups of machinery (power generation, production, auxiliary, etc.) (Eisemann 1998). The classification from practical viewpoint (Pareto analysis) can take into account frequency of malfunction occurrence (Table 1). Alignment, Balance, and incorrect Clearances

M. A. Wahab (Ed.): DAMAS 2019, LNME, pp. 541–554, 2020.
https://doi.org/10.1007/978-981-13-8331-1_40

Table 1. Fundamental malfunctions of steam turbines (Allianz Versicherungs 1978).

Location of failure	Frequency of occurrence %
Rotor blading	23.0
Bearings	15.0
Shaft seals balancing pistons	14.0
Rotor and wheels	13.0
Casing, bedplates and bolts	10.0
Strainers and valves	7.0
Governor	4.0
Nozzles and diaphragms	3.0
Gearwheels and gearing	3.0
Pipework and expansion joints	2.0
Other locations	6.0

(ABC) are the most common problems of rotating machinery regarding their severity therefore the effort should be focused on Early Warning (EW) diagnostics and increase of ABC malfunctions detectability.

The proposed approach towards detection of ABC malfunctions considers recent trends in parametric system identification providing possibilities for semi-automatic malfunction detection scenarios using physical interpretation of obtained results. A brief survey on damage detection and location methods is given in Friswell et al. (1997) outlining six categories, while discussion of typical rotating machinery malfunctions is provided by Muszynska (2005). Parametric methods are frequently used in modal analysis (Friswell et al. 1997; Heylen et al. 1998). Early attempts to parametric modeling and modal analysis in rotor dynamics are given in Mahrenholtz (2014) including rotor-bearing system identification from operational and experimental data (Krämer 2013), hydrodynamic bearing identification apart a rotor system using a first principal linearized model with adjustable parameters corresponding to stiffness and damping of the oil film, and advanced studies on hydrodynamic bearings stability (Frene et al. 1997).

Analysis of the technical state of a rotor system must be based on the data. For journal bearings the most popular source are eddy-current proximity probes, accompanied by auxiliary measurements, for instance rotational speed, expansions, thrust bearing position etc. There is a large literature in the field of rotor dynamics modelling and analysis. Fundamental works were given by Jeffcott, who laid the groundwork for the discipline. More recent books were published by Bently and Hatch (2003), Vance (1988) and Muszyńska (2005), to name only a few. Recent decade has witnessed rapid development of data analysis using methods from the field of Artificial Intelligence (AI) or its subset – Machine Learning (ML). As these methods require large data sets for training, numeric models offer a promising means to generate such data. The literature about rotor systems modelling is also very rich, with example works of Kicinski (2005), Pennacchi et al. (2010) and Bachschmid et al. (2009). ML methods can bring important benefits to turbo-generator users. First of all, it can assist experienced vibration experts. The automated methods can cover the vibration data at a much lower cost and the attention of a highly skilled expert can only be required when the algorithm will find an incorrect or a novel technical state. Interesting recent works in this area can be found in Fu et al. (2018), Li et al. (2016), Adamowicz and Zywica (2018).

The paper develops one of research problems stated in Barszcz and Zabaryłło (2019), where ML method for large turbomachinery was proposed. The goal of the paper is to present a simplified 1D model, which can be used for simulation of a transient state. The model will adopt the greybox approach, where the first principles physical model is tuned using the measurement data from a real object. In such an approach the model tuning is simplified and the ready model can be modified with seeded faults. The data from the modified model can be – in turn – applied to train the ML algorithms to detect real life faults.

The paper consists of 4 parts. Section 2 presents the fundamentals of the 1D model. Subsections present the model of the rotor, the bearing and model parameters. Section 3 presents the case study, where the model was tuned to the 200 MW steam turbine. The results of a model after balancing are presented. Finally, the Sect. 4 is a summary of the paper.

2 1D Gray-Box Rotor-Bearing Model

The model objective is to reproduce transient and steady-state operation scenarios of a 200 MW steam turbine equipped with hydrodynamic bearings. The model provides relative displacement of the bearing journals which are used in condition monitoring.

2.1 The Rotor Model

A bearing-rotor model has been developed using a transfer matrix approach (Krämer 2013) in order to provide simulation data. The rotor discrete finite components are shown in Fig. 1.

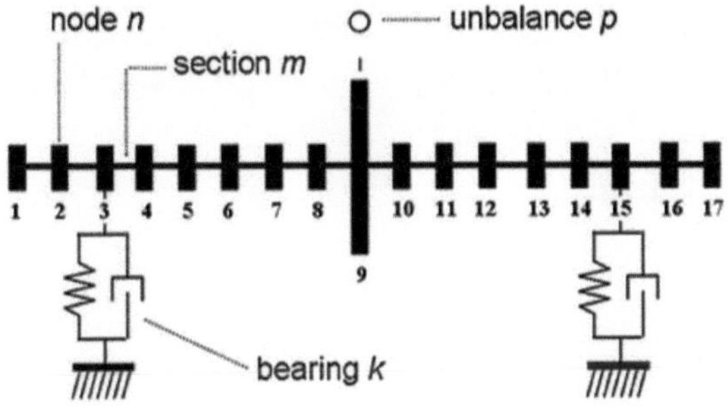

Fig. 1. Rotor model discrete representation.

The mechanical model is described using a matrix differential equation, where the subscripts denote particular inertia, gyroscopic, damping, and stiffness forces associated with the n-th node:

$$\mathbf{M}_{nxn}\mathbf{w} + \mathbf{G}_{nxn}\dot{\mathbf{w}} + \mathbf{D}_{nxn}\dot{\mathbf{w}} + \mathbf{K}_{nxn}\mathbf{w} + f(\dot{\mathbf{w}}, \mathbf{w}) = \mathbf{u} \tag{1}$$

where

$$\mathbf{w} = \begin{bmatrix} z \\ \phi \end{bmatrix}, \mathbf{u} = \begin{bmatrix} u_z \\ u_\phi \end{bmatrix} \tag{2}$$

and, it yields to the following matrix equation

$$\begin{aligned} &\mathbf{M}_{n,n}\begin{bmatrix} \ddot{z}_n \\ \ddot{\phi}_n \end{bmatrix} + \mathbf{G}_{n,n}\begin{bmatrix} \dot{z}_n \\ \dot{\phi}_n \end{bmatrix} + \mathbf{K}_{n,n-1}\begin{bmatrix} z_{n-1} \\ \phi_{n-1} \end{bmatrix} + \mathbf{K}_n\begin{bmatrix} z_n \\ \phi_n \end{bmatrix} + \mathbf{K}_{n+1}\begin{bmatrix} z_{n+1} \\ \phi_{n+1} \end{bmatrix} \\ &+ f\left(z_n, \dot{z}_n \phi_n, \dot{\phi}_n\right) = \begin{bmatrix} u_z \\ u_\phi \end{bmatrix} \end{aligned} \tag{3}$$

where $z_n = x_n + jy_n,\ \phi_n = \theta_n + j\varphi_n$. The expanded complex coordinates provide the matrix equation

$$\mathbf{M}_n\begin{bmatrix} \ddot{x}_n \\ \ddot{y}_n \\ \ddot{\theta}_n \\ \varphi_n \end{bmatrix} + \mathbf{G}_n\begin{bmatrix} \dot{x}_n \\ \dot{y}_n \\ \dot{\theta}_n \\ \dot{\varphi}_n \end{bmatrix} + \mathbf{K}_{n-1}\begin{bmatrix} x_{n-1} \\ y_{n-1} \\ \theta_{n-1} \\ \varphi_{n-1} \end{bmatrix} + \mathbf{K}_n\begin{bmatrix} x_n \\ y_n \\ \theta_n \\ \varphi_n \end{bmatrix} + \mathbf{K}_{n+1}\begin{bmatrix} x_{n+1} \\ y_{n+1} \\ \theta_{n+1} \\ \varphi_{n+1} \end{bmatrix} + f\left(z_n, \dot{z}_n \phi_n, \dot{\phi}_n\right) = \mathbf{u}_n \tag{4}$$

The above equation considers the coupled rotor motion in the translational and rotational coordinates as described in Krämer (2013). The particular matrices are formulated as follow:

$$\mathbf{M}_{n,n} = \begin{bmatrix} m_n & 0 & 0 & 0 \\ 0 & m_n & 0 & 0 \\ 0 & 0 & I_{Tn} & 0 \\ 0 & 0 & 0 & I_{Tn} \end{bmatrix}, \ \mathbf{G}_{n,n} = \begin{bmatrix} 0 & 0 & 0 & 0 \\ 0 & 0 & 0 & 0 \\ 0 & 0 & 0 & \Omega I_{Pn} \\ 0 & 0 & -\Omega I_{Pn} & 0 \end{bmatrix} \tag{5}$$

$$\mathbf{K}_{n,n-1} = \begin{bmatrix} -2k_{n-1} & 0 & -(kl)_{n-1} & 0 \\ 0 & -2k_{n-1} & 0 & -(kl)_{n-1} \\ (kl)_{n-1} & 0 & 1/3(kl^2)_{n-1} & 0 \\ 0 & (kl)_{n-1} & 0 & 1/3(kl^2)_{n-1} \end{bmatrix} \tag{6}$$

$$\mathbf{K}_{n,n} = \begin{bmatrix} 2(k_{n-1}+k_n) & 0 & -(kl)_{n-1}+(kl)_n & 0 \\ 0 & 2(k_{n-1}+k_n) & 0 & -(kl)_{n-1}+(kl)_n \\ -(kl)_{n-1}+(kl)_n & 0 & 2/3\left[(kl^2)_{n-1}+(kl^2)_n\right] & 0 \\ 0 & -(kl)_{n-1}+(kl)_n & 0 & 2/3\left[(kl^2)_{n-1}+(kl^2)_n\right] \end{bmatrix} \tag{7}$$

$$\mathbf{K}_{n,n+1}=\begin{bmatrix}-2k_n & 0 & (kl)_n & 0\\ 0 & -2k_n & 0 & (kl)_n\\ -(kl)_n & 0 & 1/3(kl^2)_n & 0\\ 0 & -(kl)_n & 0 & 1/3(kl^2)_n\end{bmatrix} \tag{8}$$

where rotor stiffness is

$$m_n = m_n^{disk} + (d_{n-1}\rho_{n-1}l_n + d_n\rho_n l_n)^{shaft},\quad k_n = \frac{6E_nI_n}{l_n^3} \tag{9}$$

The complete transfer matrix equation can be expanded of term g(ẇ, w) and is given as follow:

$$\mathbf{M}_{nxn}\mathbf{w} + \mathbf{G}_{nxn}\dot{\mathbf{w}} + \mathbf{D}_{nxn}\dot{\mathbf{w}} + \mathbf{K}_{nxn}\mathbf{w} + f(\dot{\mathbf{w}},\mathbf{w}) + g(\dot{\mathbf{w}},\mathbf{w}) = \mathbf{u}_n \tag{10}$$

where the excitation vector is defined as follow:

$$\mathbf{u}_n(t) = \Omega^2\begin{bmatrix} m_n e_n\, cos(\Omega t + \alpha_n)\\ m_n e_n\, sin(\Omega t + \alpha_n)\\ -(I_{pn} - I_{Tn})\,\gamma_n\, cos(\Omega t + \beta_n)\\ -(I_{pn} - I_{Tn})\,\gamma_n\, sin(\Omega t + \beta_n)\end{bmatrix} \tag{11}$$

The inertial excitation terms are neglected in the developed rotor model.

2.2 The Bearing Model

The physical bearing model is based on the Reynold's equation and allows to analyze the oil flow in a determined layer, comprising the balance equations for a fluid element and the equations of flow continuity. Reynold's equation (Frene et al. 1997) provides better insight in the dynamics of a rotor-bearing system especially under transient conditions when a nonlinear dynamic analysis needs to be carried out. The forces F_x and F_y generated by the fluid film are obtained based on the analytical solution of Reynold's equation for the short bearing approximation as follows:

$$f(z_n,\dot{z}_n) = \begin{bmatrix} F_x = -\mu\pi RL^3\left[\frac{\Omega y + 2\dot{x}}{2(c^2-x^2-y^2)^{3/2}} + \frac{3x(x\dot{x}+y\dot{y})}{2(c^2-x^2-y^2)^{5/2}}\right]\\ F_y = -\mu\pi RL^3\left[\frac{2\dot{y}-\Omega x}{2(c^2-x^2-y^2)^{3/2}} + \frac{3y(x\dot{x}+y\dot{y})}{2(c^2-x^2-y^2)^{5/2}}\right]\end{bmatrix} \tag{12}$$

The total force component in the x and y can be obtained in the following way:

$$f(z_n,\dot{z}_n,) = \begin{bmatrix} F_\beta cos\,\theta - F_\alpha sin\,\theta\\ F_\beta sin\,\theta + F_\alpha cos\,\theta\end{bmatrix},\quad x = c\cdot\beta\cdot cos(\alpha),\ y = c\cdot\beta\cdot sin(\alpha),$$
$$\beta = \frac{\sqrt{x^2+y^2}}{c}, y = arctan\left(\frac{y}{x}\right) \tag{13}$$

2.3 Model Parameters

The model was derived as a 19-node system of ordinary differential equations corresponding to the most significant rotor-bearing shaft line components. The model neglects the support rigidity and it is (Eq. 4) parameterized according to the geometry and physical properties of the reference 200 MW steam turbine (Table 2).

Table 2. The parameters of the shaft-bearing line assumed in the model based on the turbine operational manual.

Node no.	Section length [mm]	Outer diameter [mm]	Inner diameter [mm]	Bearing clearance [mm]	Bearing diameter [mm]	Bearing length [mm]
1	162	300	100	0,64	330	270
2	2323	600	100			
3	2,235	470	100			
4	180	330	100	0,65	360	290
5	2233	740	100			
6	1910	660	100			
7	215	360	100	0,75	360	290
8	1710	890	140			
9	220	450	0	0,88	450	358
10	1486	1270	140			
11	1798	1630	650			
12	1526	8500	1270			
13	220	450	0	0,89	460	298
14	1855	760	6			
15	270	400	0	0,89	400	500
16	1405	570	0			
17	4350	710	0			
18	1675	540	2			
19	245	400	0	0,25	400	400

3 Case Study

3.1 The Shaft-Bearing Line Parameters

The presented case study is focused on the turbine type TK200/13K215 (200 MW type of turbine) and the generator is denoted as GWW-200/GTHW 230 (200 MW type of generator). The actual parameter data were taken from a real 200 MW turbine in Poland. The graphical representation of the turboset is shown in Fig. 2 where the bearing are indicated with their numbers 1–9.

The turboset consists of 3 turbine's parts (1xHP – high pressure, 1xIP – intermediate pressure, 1xLP – low pressure) and 1 generator and 1 exciter. The turboset is

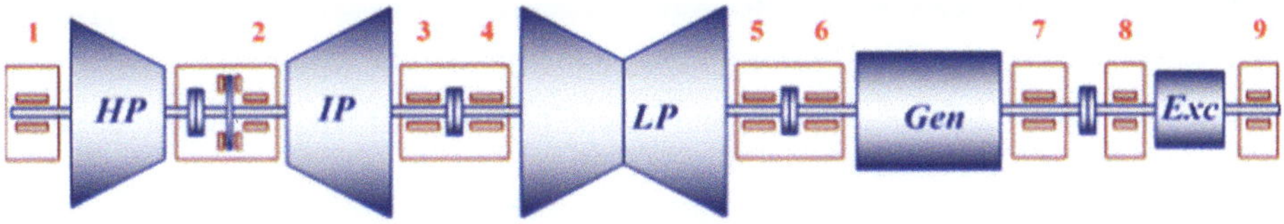

Fig. 2. Shaft-bearing line schematic representation

supported on 9 bearings which only first 7 are taken into account due to lack of probes mounted in the bearing no. 8 and no. 9. The critical speeds of the rotor-bearing system are indicated in Table 3.

Table 3. Critical speed from commercially used software

Critical	Rotational speed [rpm]	Affected rotor components
I	882	Generator 1. mode
II	2195	Turbine – High Pressure
III	2580	Generator 2. mode
IV	3415	Turbine – Low Pressure
V	3514	Turbine – Intermediate Pressure

3.2 Operational Measurement Data

The vibration measurements were taken by means of ADRE 408 by Bently Nevada (industrial data collector and first stage post-processing unit). The possible rearrangement of the measurement's system installed on the 200 MW machine type. The signals from relative probes are only taken into account. The orientation of the probes and the labels are as follows: each probe is oriented 45° from vertical axis (Fig. 3).

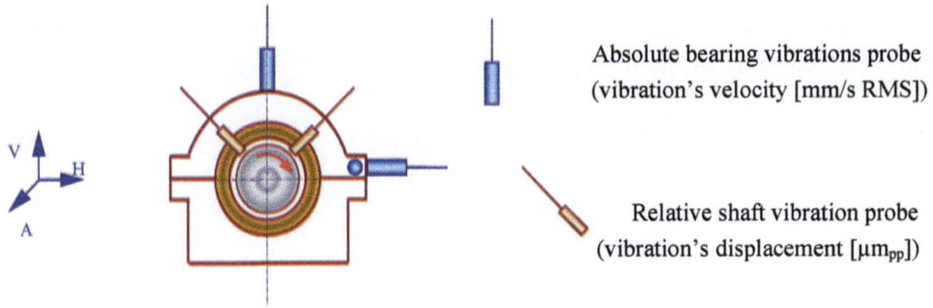

Fig. 3. Rotor model discrete representation

The Y probe is on the left and the X probe is on the right from the vertical axis seen from the bearing no. 1 to no. 9, respectively. The turbo-set rotates CW (clockwise) and orientation of the tacho-probe varies (depending on specific machine). The signals from Y and X probe are sampled based on two conditions:

- Δ_{time} – each and every n-seconds
- Δ_{rpm} – each and every ± N-rpm – revolutions per minute

Both parameters are been determined manually during of data collection configuration. Usually n = 20 ÷ 45 s, and N = 5 ÷ 20 rpm.

3.3 Model Parameters

The initial model conditions were assumed based on the simulation model of the rotor-bearing finite element model developed in house by GE R&D department. The bearing linearized stiffness and damping coefficients were obtained in a function of the rotor speed. The initial static and dynamic unbalance conditions are presented in Table 4.

Table 4. Static and dynamic unbalance conditions

Part affected	Unbalance type	Amount of unbalance [kg · mm]
HP	Static	74 (middle of rotor)
	Dynamic	32 (governor side)
		−32 (exciter side)
IP	Static	178 (middle of rotor)
	Dynamic	89 (governor side)
		−89 (exciter side)
LP	Static	316 (middle of rotor)
	Dynamic	158 (governor side)
		−158 (exciter side)
Generator	Static	496 (middle of rotor)
	Dynamic	248 (governor side)
		−248 (exciter side)

The rotor-bearing model is represented by a time-variant nonlinear adjustable-parameter ODE model in the form of set of nonlinear state-space equations formulated in the continuous-time domain (Fig. 4):

$$\begin{aligned} \frac{d}{dt}x(t) &= f(t, x(t), u(t), w(t); \theta) \\ y(t) &= h(t, x(t), u(t), v(t); \theta) \\ x(0) &= x_0 \end{aligned} \tag{14}$$

where vector f(.) is a nonlinear, time-varying function of the state vector x(t) and the excitation vector u(t), while vector h(.) is a nonlinear measurement function; w(t) and v

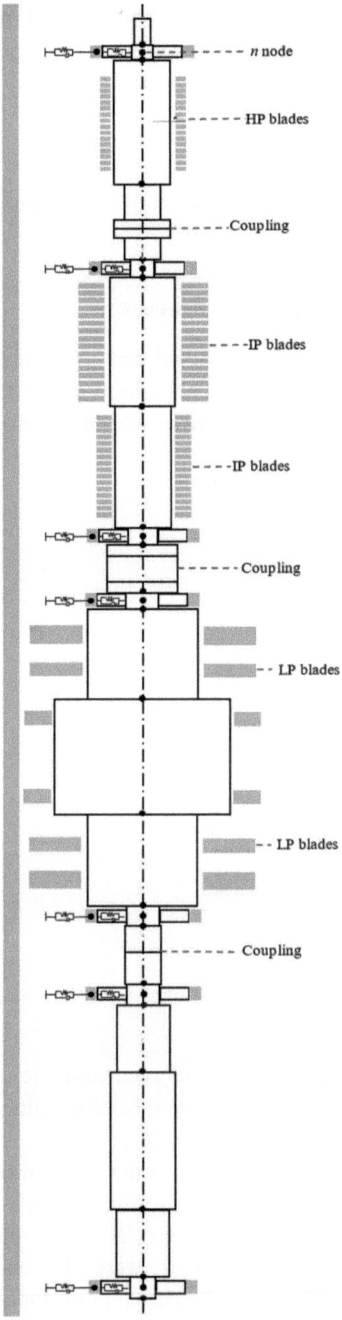

Fig. 4. Schematic representation of the reference shaft-bearing-foundation system used in the 19-node model.

(t) are sequences of independent random variables and θ denotes a vector of unknown parameters. The sum of squared errors is used as an error criterion. The model structure is defined by the function notation as follows:

$$ODE(ny, nu, nx) \tag{15}$$

where nu is the number of inputs, ny is the number of outputs, and nx is the number of the state variables.

3.4 Model Parameters Adjustment Process

The rotor coast-down operation was considered to reduce external forces related to start-up or steady-state rotor operation. This allows to reduce and in turn clearly isolate unbalance conditions. The rotor coast-down operation program applied during coast-down operation is presented in Fig. 5.

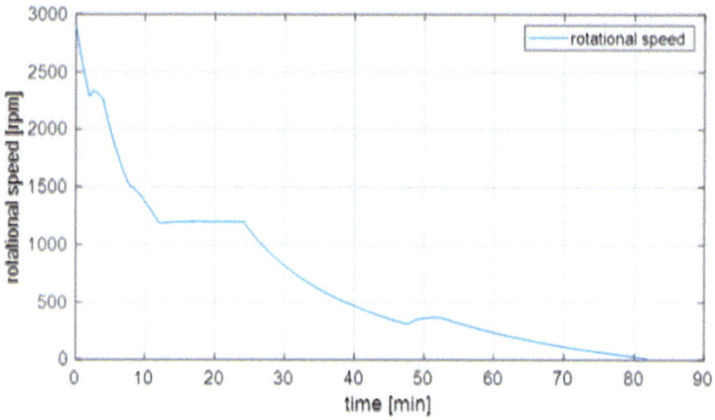

Fig. 5. The turboset was operated under transient coast-down conditions

The Bode and cascade plots were analyzed and the exported data were used to create a waterfall plot of the bearing relative vibration values expressed in μm-pp along the shaft's line vs. rotational speed (Fig. 6).

The Optimization Toolbox was used in order to minimize the error criteria during model adjustment process using Genetic Algorithms (GA). The target function values were the measured maximum values of the relative journal bearing vibration at each bearing section. The adjustable parameters were the unbalanced mass parameters, namely an equivalent unbalance mass. The model input is a vector of harmonic

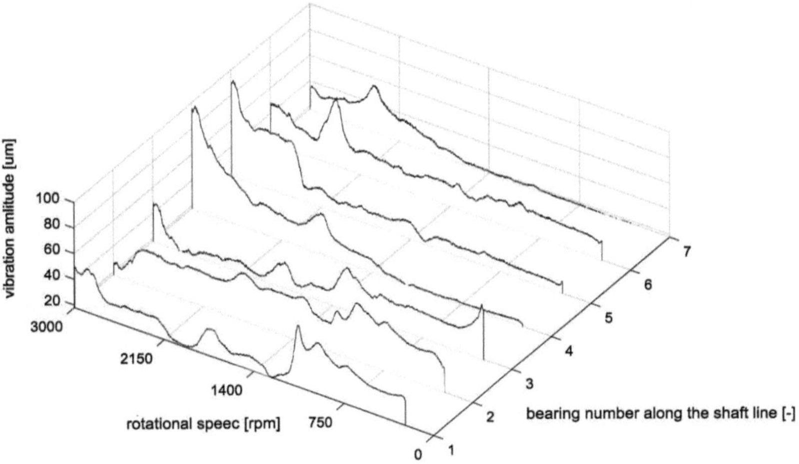

Fig. 6. Journal bearing vibration level along the shaft line vs. rotational speed.

excitation waveforms (Eq. 11) while the model output is a vector of relative displacement (vibration) signals of the shaft journal inside the hydrodynamic bearings. The 19-node model was applied according to the following structural number (Eq. 14):

$$ODE(14, 14, 152) \tag{16}$$

The model has 14 outputs (1 pair of a vertical/horizontal vibration signals along 7 bearings) and 152 states result from 19 sets of equations multiply by 4 equations per a set multiply by 2 states (velocity, displacement). The adjustment process is based on an optimization routine which allow to adjust a few model parameters starting from their initial values assigned based on the values from an operational manual or a turbine design documentation. The Global Optimization toolbox from Matlab package (MathWorks 2015) was used in order to conduct the error minimization between the measured and simulated maximum relative vibration signals at each bearing section. The algorithm parameters, such as initial population, are automatically generated. The optimal solution is numerically obtained. The optimization process is manually stopped after 100 epochs. Figure 7 shows the measured vibration amplitude vs. model response in a form of a waterfall plot, while Fig. 8 shows the error between the measured vibration amplitude vs. model response in a form of a waterfall plot.

Further improvement of the adjusting routine algorithm will be executed in order to increase accuracy at each bearing section (Table 5).

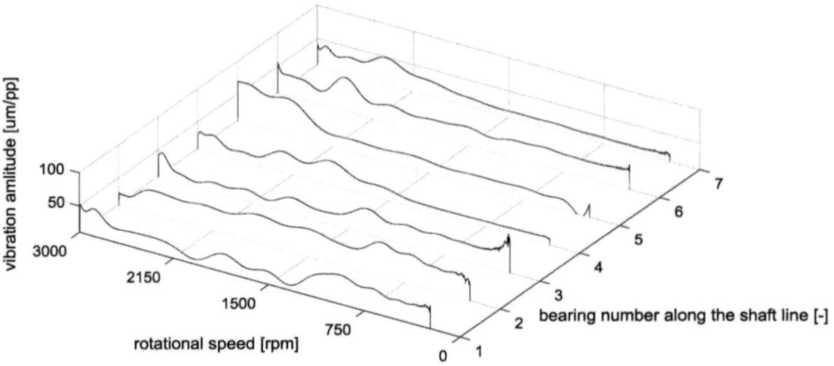

Fig. 7. Simulated journal bearing relative displacements (vibration) expressed in μm-pp reported along the rotor line vs. rotational speed

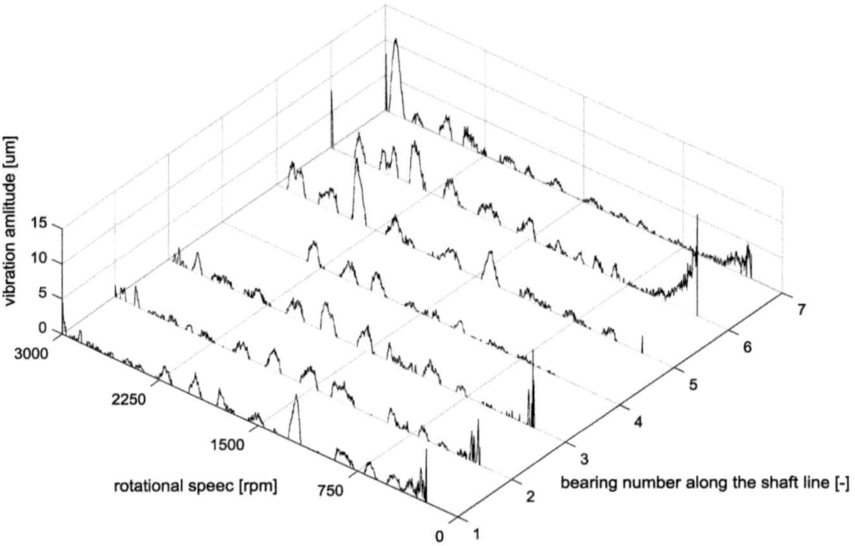

Fig. 8. Error plot of an absolute difference between the measured and simulated journal bearing relative displacements (vibration) expressed in μm-pp reported along the rotor line vs. rotational speed

Table 5. Error table of a percentage difference between the measured and simulated journal bearing relative displacements (vibration) expressed in μm-pp reported along the rotor line vs. rotational speed

Bearing no.	1	2	3	4	5	6	7
Error %	1.6	0.6	−2.7	20.4	−0.1	−0.4	−0.7

4 Summary

The 19-node 1D model of the rotor-bearing system was implemented in Matlab/Simulink environment. The 19-node model was successfully adjusted with relative error lest than 20% based on the operational vibration data acquired during coast-down turboset operation. The initial static and dynamic unbalanced condition were reproduced in the excitation model along the physical and geometrical properties of the rotor-bearing system. The model was adjusted using the global optimization approach and genetic algorithm available in Matlab. The model accuracy was evaluated and reported in the paper based on the accuracy plot. Further activities are required to improve the accuracy and involve more parameters, e.g. physical bearing parameters.

Nomenclature

$i = \sqrt{-1}$ – imaginary unit,
t - continuous time domain,
Ω - rotating speed [1/s],
L - total rotor length [m] or length of the journal [m],
x, y - lateral motion coordinates [m],
z - coordinate along the shaft axis [m],
θ, ϕ - dx/dz and dy/dz coordinates in angular motion [rad],
Z, φ - complex translational and rotational vibrations [m],
n - the number of degrees of freedom [-],
$w_{(nx1)}$ - response vector of translational and rotational vibrations,
$u_{(nx1)}$ - excitation vector of translational and rotational vibrations,
$M_{(nxn)}$ - inertia matrix of translational and rotational vibration,
$G_{(nxn)}$ - gyroscopic matrix of translational and rotational vibrations,
$D_{(nxn)}$ - external damping matrix of translational and rotational vibrations,
$K_{(nxn)}$ - stiffness matrix of translational and rotational vibrations,
$K_{s(nxn)}$ - support stiffness matrix of translational and rotational vibrations,
c - bearing clearance [m],
μ - dynamic viscosity [Pa·s],
θ - current angular position of the journal center [rad],
h - dimensional oil gap (oil film thickness) [m],
R - journal radius [m],
m - mass [kg],
I_T, I_P – transverse and polar inertia moment [kg·m2],
I - area moment [m4],
l - length of shaft section [m],
d - diameter of shaft section [m],
ρ - density [kg/m3],
E - Young's modulus,
F_x,F_y - bearing forces [N],
x, y - current position of journal center [m],

References

Adamowicz, M., Zywica, G.: Advanced gas turbines health monitoring systems. Diagnostyka **19** (2), 77–87 (2018)

Allianz Versicherungs, A.: Handbook of loss prevention. Springer, Berlin (1978)

Bachschmid, N., Pennacchi, P.E.L., Chatterton, S., Ricci, R.: On model updating of turbo-generator set. J. Vibroeng. **11**(3), 379–391 (2009)

Barszcz, T., Zabaryłło, M.: Concept of automated malfunction detection of large turbomachinery using machine learning on transient data. Diagnostyka **20**, 1641–6414 (2019)

Bently, D.E., Hatch'Charles, T.: Fundamentals of rotating machinery diagnostics. Mech. Eng. CIME **125**, 53–54 (2003)

Eisemann, R.: Machinery Malfunction Diagnosis and Correction: Vibration Analysis and Troubleshooting for Process Industries. Prentice Hall PTR, Upper Saddle River, NJ (1998)

Frene, J., Nicolas, D., Degueurce, B., et al.: Hydrodynamic Lubrication: Bearings and Thrust Bearings. Elsevier, Amsterdam (1997)

Friswell, M., Penny, J., Garvey, S.: Parameter subset selection in damage location. Inverse Probl. Eng. **5**, 189–215 (1997)

Fu, C., Ren, X., Yang, Y., et al.: An interval precise integration method for transient unbalance response analysis of rotor system with uncertainty. Mech. Syst. Sig. Process. **107**, 137–148 (2018)

Heylen, W., Lammens, S., Sas, P.: Modal Analysis Theory and Testing. Universiteit Leuven, Leuven, Belgium (1998)

Kiciński, J.: Dynamika wirników i łożzysk ślizgowych. Wydawnictwo IMP PAN (2005)

Krämer, E.: Dynamics of Rotors and Foundations. Springer, Berlin (2013)

Li, C., Cabrera, D., de Oliveira, J.V., et al.: Extracting repetitive transients for rotating machinery diagnosis using multiscale clustered grey infogram. Mech. Syst. Signal Process. **76**, 157–173 (2016)

Mahrenholtz, O.: Dynamics of Rotors: Stability and System Identification. Springer-Verlag Wien (2014)

MathWorks: Global Optimization Toolbox User's Guide. The MathWorks, Inc., Natick (2015)

Muszynska, A.: Rotordynamics. CRC Press, Boca Raton (2005)

Pennacchi, P., Vania, A., Chatterton, S.: Identification of mechanical faults in rotating machinery for power generation. In: 2010 IEEE International Symposium on Industrial Electronics, pp. 2109–2114 (2010)

Vance, J.M.: Rotordynamics of Turbomachinery. Wiley, New York (1988)

Static Strength Analysis of Centrifugal Compressor Balance Plate Based on Finite Element Method

Yuan Li[1,2], Zeyang Qiu[3](✉), Yu Zhang[2], Zhenyu Ding[2], and Ling Fan[2]

[1] Energy and Power Engineering College, Xi'an Jiaotong University, Xi'an, China
[2] SINOPEC Sichuan to Eastern China Transmission Gas Pipeline Co. Ltd., Beijing, China
[3] College of Safety and Ocean Engineering, China University of Petroleum, Beijing, China
1164709426@qq.com

Abstract. After the centrifugal compressor balance disc is installed in the final stage impeller, the rotor system can be protected because the axial force of the rotor is offset by the axial thrust generated by the pressure difference between the two sides. The balance plate plays an important role in the safe operation of the centrifugal compressor rotor system, and its strength is the guarantee for the safe operation of the centrifugal compressor. In order to analyze the reliability of the balance disc, this paper uses the static finite element method to analyze its strength. Considering the influence of centrifugal force, the stress and strain distribution of the balance plate under different rotating speeds are obtained, which plays a certain reference role for the design of the balance plate. The stress and strain distribution of the balance disc under the single action of centrifugal force is studied, and the influence of rotating speed on the strength of the balance disc is obtained.

Keywords: Centrifugal compressor · Balance disc · Strength analysis · Velocity

1 Introduction

The construction and development of long-distance natural gas pipelines are promoted because of the increasing demand for natural gas in the domestic market. Centrifugal compressors, as the core equipment of gas pipeline systems, have always been the focus of attention in the oil and gas industry [1]. Centrifugal compressor has the advantages of large displacement, stable operation, low failure rate, etc. Its operating conditions (such as inlet and outlet pressure, flow rate, etc.) often fluctuate under the influence of upstream and downstream pipelines, which is difficult to ensure the stable operation of the unit at the design point [2]. After the balancing disc of centrifugal compressor is installed on the last stage impeller, the axial thrust generated by the gas pressure difference between the two sides is used to partially offset the axial force of the

M. A. Wahab (Ed.): DAMAS 2019, LNME, pp. 555–565, 2020.
https://doi.org/10.1007/978-981-13-8331-1_41

rotor, which can protect the rotor system [3]. Due to the change of operation conditions of centrifugal compressor, the balance disc has to bear the effect of variable load and has higher requirements on strength [4]. Therefore, it is of great significance to carry out the strength analysis of the balancing disc under different working conditions to ensure the intrinsic safety of the centrifugal compressor during long-term operation.

In order to analyze the reliability of the balance disc, this paper uses static finite element method to analyze its strength. Taking PCL503 centrifugal compressor of a compressor station as an example, considering the influence of centrifugal force, the stress and strain distribution of the balance disc at different rotational speeds are obtained, which plays a certain reference role in the design of the balance disc. The stress and strain distribution of the balance disc under the single action of centrifugal force is studied, and the influence of rotational speed on the strength of the balance disc is obtained.

2 Static Finite Element Analysis

Static finite element analysis is used to solve displacement, stress and force caused by external load. Static analysis is very suitable for solving the problem that the influence of inertia and damping on the structure which is not significant [5]. Static analysis in ANSYS program can not only carry out linear analysis, but also nonlinear analysis, such as plasticity, creep, expansion, large deformation, large strain and contact analysis [6]. The main steps are as follows:

(1) Establishment of model of compressor and its key components.

The model is simplified without changing the mechanical properties of the compressor and its components. Three - dimensional modeling tool SOLIDWORKS is applied to model the simplified components, which are imported into ANSYS through PRAR interface.

(2) Determination of boundary conditions for key components.

Due to the complicated forces on the moving parts of the compressor, it is necessary to establish a simplified force model and a simplified motion model when applying boundary conditions before analysis [7].

(3) Finite element analysis on each key component.

Applying boundary conditions to the imported model, the finite element calculation results are obtained, and the failure modes are analyzed according to the results, then the results are compared with the actual situation of the compressor.

3 Static Analysis of Balance Disc

The material of the balance disc is Fv520B, Elastic modulus E = 194 GPa, Poisson's ratio $\mu = 0.3$, density $\rho = 7872\ kg/m^3$. The key interference fit is adopted between the inner surface of the balance disc and the outer circumferential surface of the shaft, and there is almost no displacement change between the balance disc and the main shaft. Therefore, it is considered that the two are rigidly connected to restrict the freedom degree of the inner ring surface of the balance disc [8]. The established three-dimensional model is appropriately simplified, and the meshed model is shown in Fig. 1.

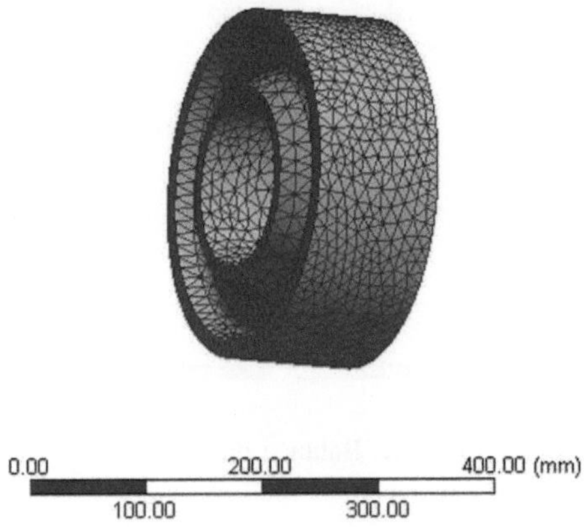

Fig. 1. Balancing disk grid division

The balancing disk is in contact with the impeller, so the pressure on the high pressure side should be applied to the part of the balancing disk that is not in contact with the impeller [9]. The pressure is approximately the compressor outlet pressure, and the pressure on the low pressure side is directly applied to the other end of the balancing disk. The greater the compressor rotational speed, the greater the centrifugal force on the balance disc [10].

Using the data of the compressor station at a certain time, the centrifugal compressor rotational speed is 5009 rpm, at which time the inlet pressure is 6.61 MPa and the outlet pressure is 7.45 MPa. According to the finite element analysis of ANSYS workbench, the stress deformation results are shown in Figs. 2 and 3:

The maximum stress of the balance disc is on the outside of the cavity formed by the high-pressure end and the impeller. The maximum stress is 61.589 MPa. The stress at the low-pressure end is the same as the stress at the high-pressure end but both are smaller than the high-pressure stress and are much smaller than the allowable stress of 1029 MPa. The maximum deformation of the balance disc is 0.01872 mm outside the high-voltage end. The deformation of the low-voltage end is the same as that of the high-voltage end, but the whole deformation is smaller than that of the high-voltage end. The thickness of the balance disc is smaller, which accords with the actual situation.

When the rotational speed, inlet pressure and outlet pressure are changed, the force on the balance disc is as follows:

When the compressor rotational speed is 5688 rpm, the corresponding inlet pressure is 6.58 MPa, and the outlet pressure is 7.83 MPa. According to the finite element analysis of ANSYS workbench, the stress deformation results are shown in Figs. 6 and 7 (Figs. 4 and 5):

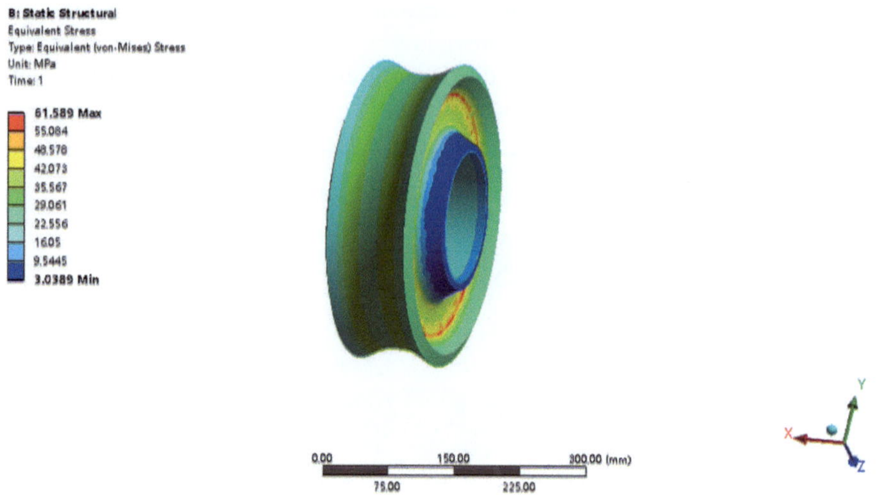

Fig. 2. Balance disc stress

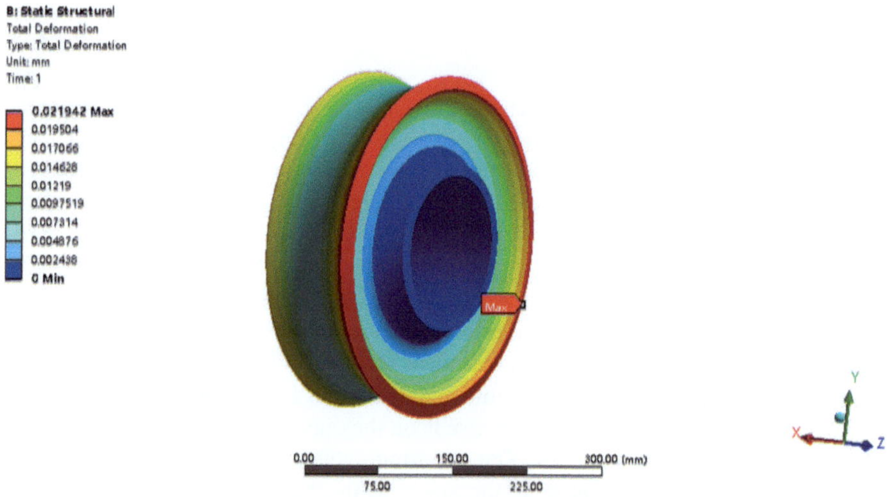

Fig. 3. Deformation of balance disc

The maximum stress of the balance disc is on the outside of the cavity formed by the high-pressure end and the impeller. The maximum stress is 72.284 MPa. The stress at the low-pressure end is the same as the stress at the high-pressure end but both are smaller than the high-pressure stress and are much smaller than the allowable stress of 1029 MPa. The maximum deformation of the balance disc is 0.021942 mm outside the high-voltage end. The deformation of the low-voltage end is the same as that of the high-voltage end, but the whole deformation is smaller than that of the high-voltage end. The thickness of the balance disc is smaller, which accords with the actual

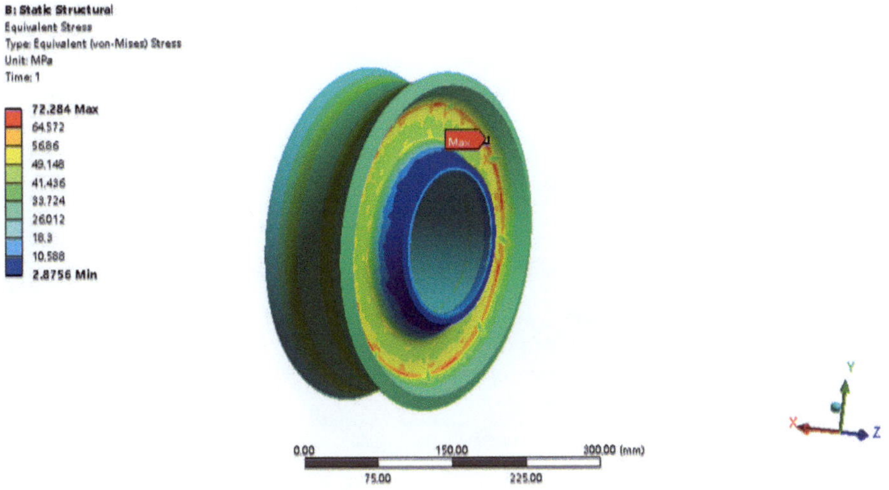

Fig. 4. Balance disc stress

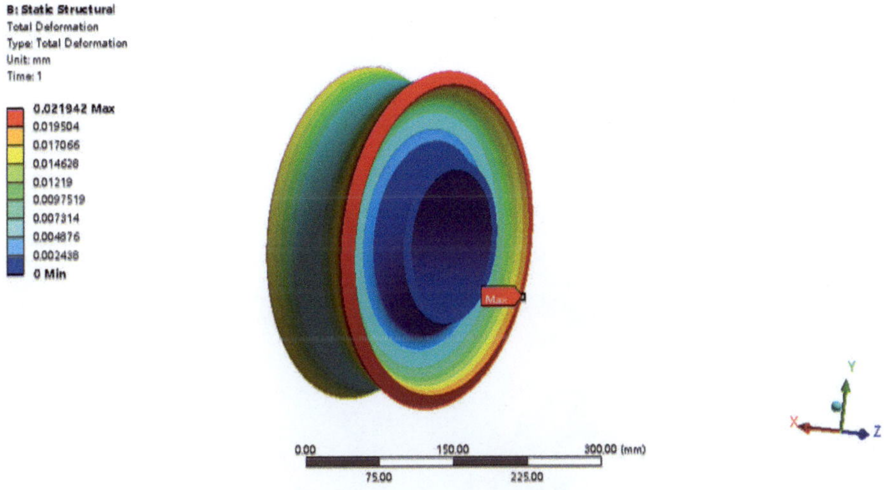

Fig. 5. Deformation of balance disc

situation. In this case, the maximum stress and the maximum deformation increase with respect to the case where the rotational speed is 5009 rpm.

When the compressor rotational speed is 6601 rpm, the corresponding inlet pressure is 6.52 MPa, and the outlet pressure is 7.24 MPa. According to the finite element analysis of ANSYS workbench, the stress deformation results are shown in Figs. 6 and 7:

The maximum stress of the balance disc is also on the outside of the cavity formed by the high-pressure end and the impeller. The maximum stress is 73.372 MPa. The stress at the low-pressure end is the same as the stress at the high-pressure end but both are smaller than the high-pressure stress and are much smaller than the allowable stress

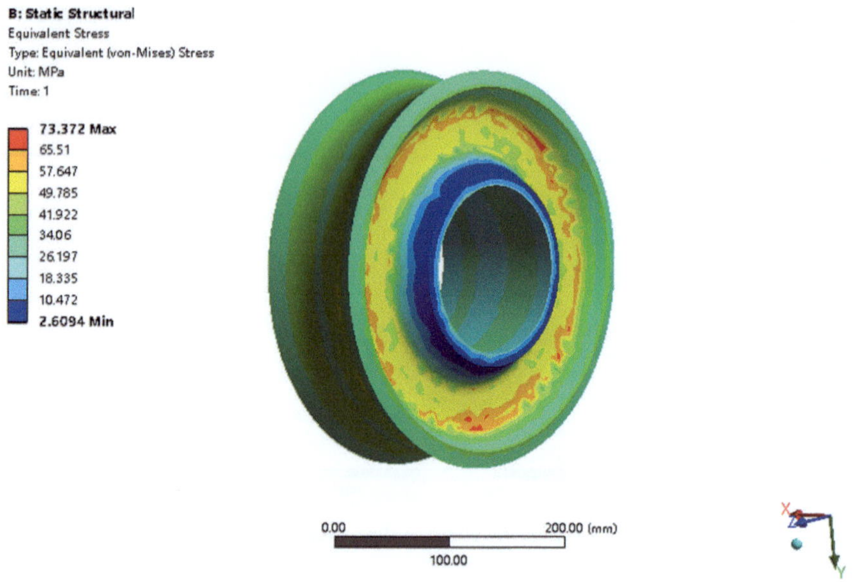

Fig. 6. Balance disc stress

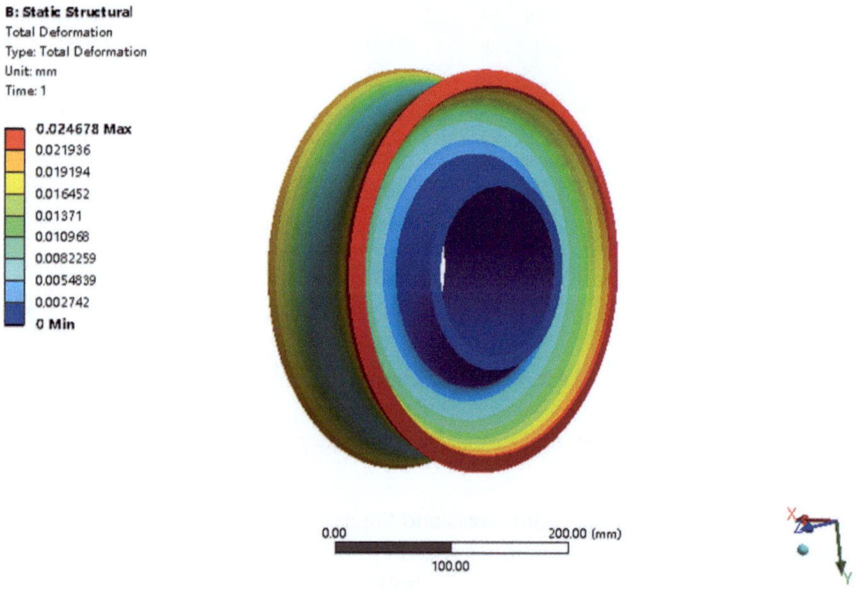

Fig. 7. Deformation of balance disc

of 1029 MPa. The maximum deformation of the balance disc is 0.024678 mm outside the high-voltage end. The deformation of the low-voltage end is the same as that of the high-voltage end, but the whole deformation is smaller than that of the high-voltage end. The thickness of the balance disc is smaller, which accords with the actual situation. In this case, the maximum stress and the maximum deformation increase with respect to the case where the rotational speed is 5009 rpm and 5688 rpm.

In order to study the influence of the rotational speed on the balance disc strength, the pressure on both sides of the balance disc is neglected, and only consider the influence of the rotational speed on the stress deformation of the balance disc.

When the rotational speed is 5009 rpm, the stress and deformation results of the balance disc are shown in the following Fig. 8. The maximum stress is 24.723 MPa and the maximum deformation is 0.0081481 mm. The stress and deformation trends of the high pressure end and the low pressure end are consistent and symmetrical (Fig. 9).

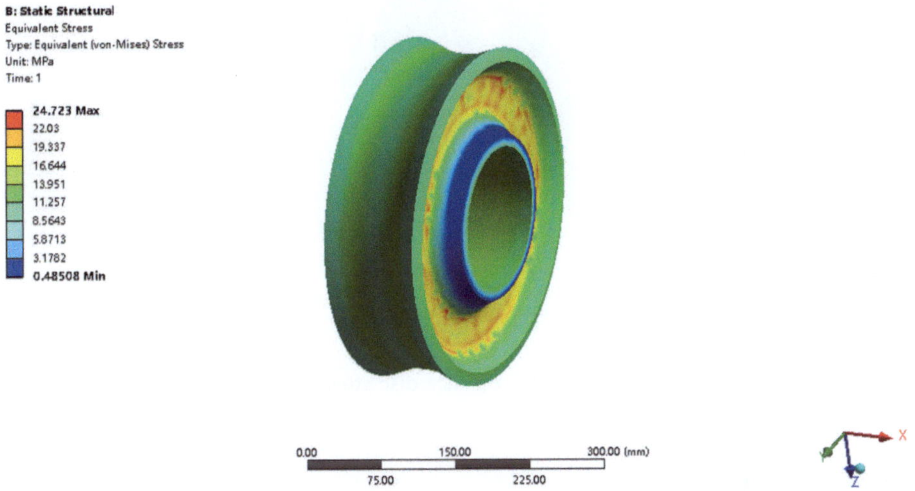

Fig. 8. Balance disc stress

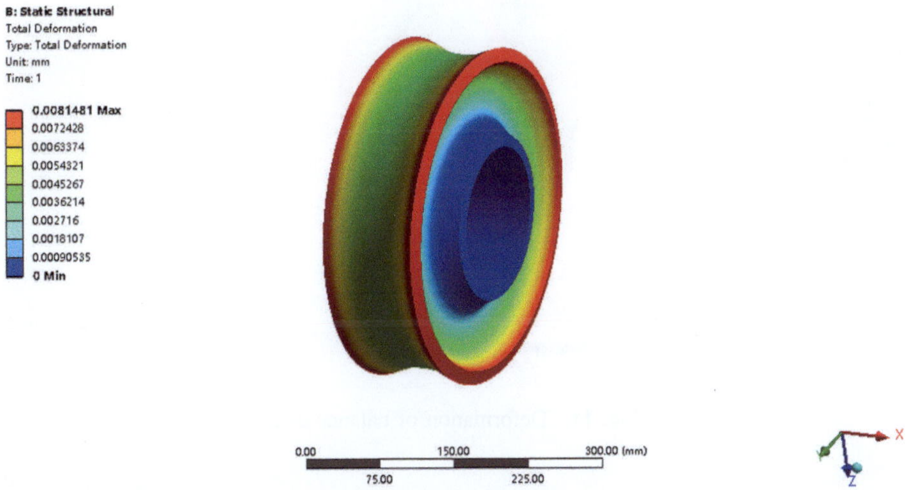

Fig. 9. Deformation of balance disc

When the rotational speed is 5688 rpm, the stress and deformation results of the balance disc are shown in the following Fig. 10. The maximum stress is 31.88 MPa and the maximum deformation is 0.010507 mm. The stress and deformation trends of the high pressure end and the low pressure end are consistent and symmetrical (Fig. 11).

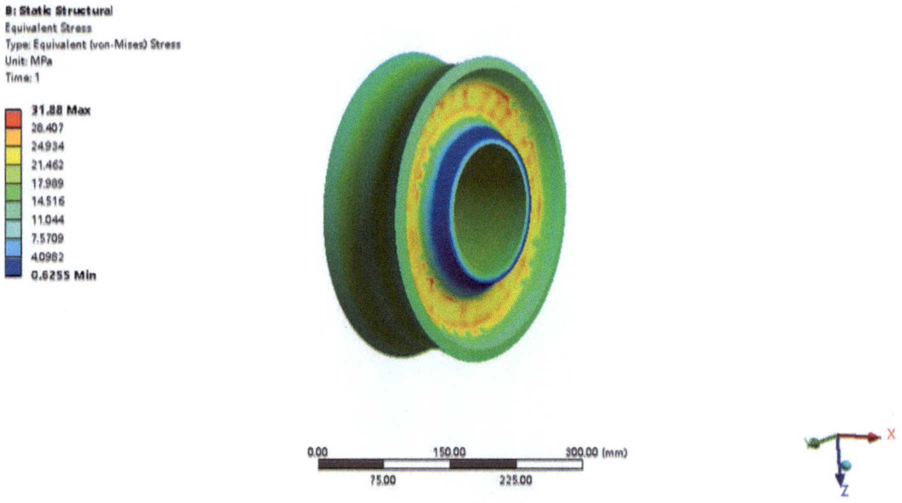

Fig. 10. Balance disc stress

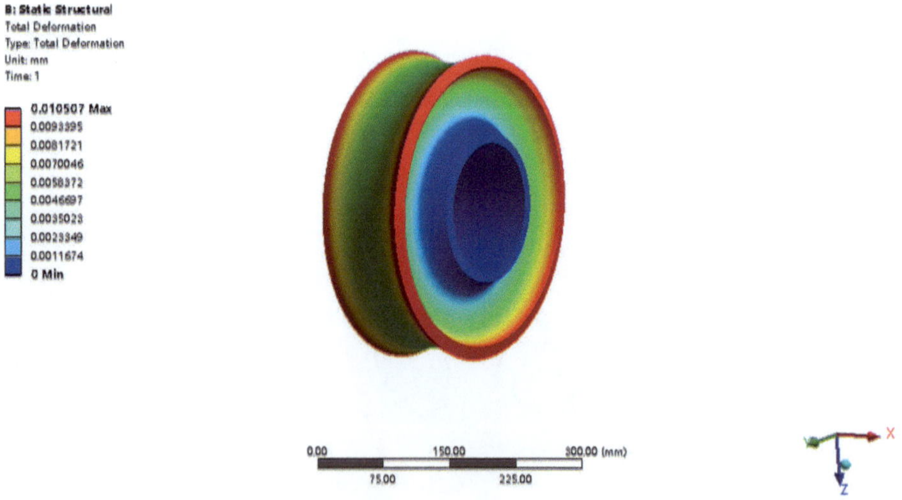

Fig. 11. Deformation of balance disc

When the rotational speed is 6601 rpm, the stress and deformation results of the balance disc are shown in the following Fig. 12. The maximum stress is 42.935 MPa and the maximum deformation is 0.014151 mm. The stress and deformation trends of the high pressure end and the low pressure end are consistent and symmetrical (Fig. 13).

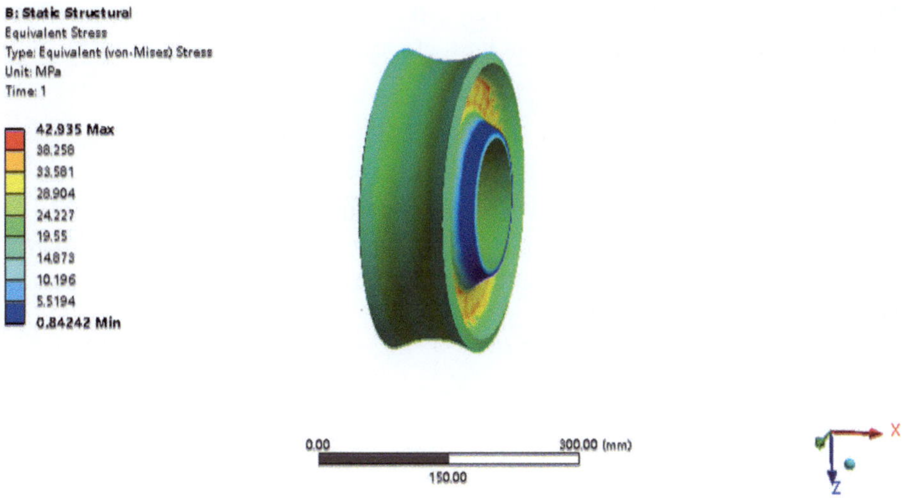

Fig. 12. Balance disc stress

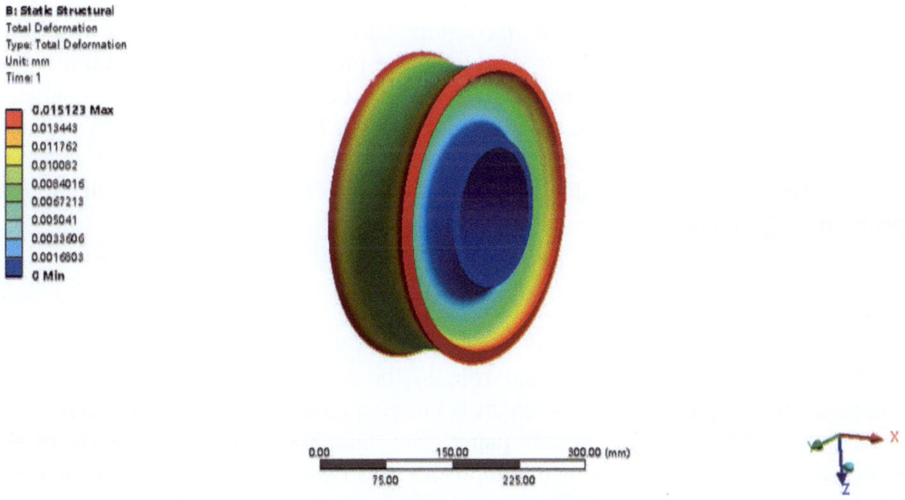

Fig. 13. Deformation of balance disc

When the rotational speed increases, the maximum stress and deformation of the balance disc increase as shown in Tables 1 and 2.

Table 1. The maximum stress of the balance disc at each speed.

Compressor rotational speed (rpm)	5009	5688	6601
Maximum stress consider stress (MPa)	61.589	72.284	73.372
Maximum stress under centrifugal force only (MPa)	24.723	31.88	42.935
Proportion only considers the maximum stress of rotating speed	40.1%	44.1%	58.5%

Table 2. The maximum deformation of the balance disc at each speed.

Compressor rotational speed (rpm)	5009	5688	6601
Maximum deformation consider stress (mm)	0.01872	0.021942	0.024678
Maximum deformation under centrifugal force only (mm)	0.0081481	0.010507	0.014151
Proportion only considers the maximum stress of rotational speed	43.5%	48%	57.3%

The maximum stress of balance disk is mainly affected by centrifugal force generated by rotational speed and axial pressure [11]. According to the results in the table, when the rotational speed is low, the pressure is the most important factor affecting the maximum stress of the balance disc, and when the rotational speed increases gradually, the proportion of the maximum stress of the balance disc generated by the rotational speed will also increase gradually. When the rotational speed is low, the pressure is the most important factor affecting the maximum deformation of the balance disc, and when the rotational speed increases gradually, the proportion of the maximum deformation of the balance disc generated by the rotational speed will also gradually increase. At the same time, when the rotational speed is high and reaches 6601 rpm, the maximum stress value generated is 73.372 MPa, which is much smaller than the allowable stress of the balance disc material of 1029 MPa. It does not affect the normal operation of the compressor.

4 Conclusion

The balance disc plays an important role in the safe operation of the centrifugal compressor rotor system, and its strength is the guarantee for the safe operation of the centrifugal compressor [12]. In this paper, the static finite element analysis of the balance disc of PCL503 centrifugal compressor in a compressor station is carried out, and the stress and strain conditions of the balance disc under different operating conditions are obtained, which can serve as a reference for the design of the balance disc. The stress and strain distribution of the balance disc under the single action of centrifugal force is further studied, and the influence of rotational speed on the strength of the balance disc is obtained.

Acknowledgements. This paper is supported by SINOPEC Gas Company (No. 35150573-16-ZC0607-0001).

References

1. Lin, Y.: Cause analysis and countermeasure of abnormal shutdown of centrifugal compressor for long distance natural gas pipeline. Pet. Chem. Equip. **20**(6), 92–94 (2017)
2. Zhang, P., Yang, J.: Thermal performance failure and avoidance of reciprocating natural gas compressor. Petrochem. Chem. Ind. Stand. Qual. **1**, 258 (2013)
3. Shi, W., Wei, Z.: Analysis of the generation and balance of the axial force of centrifugal rotating machinery. Chem. Eng. Equip. **12**, 38–39 (2008)
4. Research on static and dynamic strength and aerodynamic load of multistage impeller rotor system of compressor. Northeastern University (2012)
5. Wang, D., He, J., Rui, M.: Research and discussion on some problems in static elastoplastic analysis. Build. Constr. **5**, 96–99 (2007)
6. Zhang, X., Tang, Y., Li, H., et al.: Structural nonlinear simulation analysis and design based on ANSYS. Comb. Mach. Tools Autom. Mach. Technol. **10**, 29–32 (2007)
7. Geng, K., Du, S., Tang, M., et al.: Virtual simulation analysis of translational piston air compressor. Mech. Des. Manuf. **11**, 28–30 (2012)
8. Research on self-healing control of centrifugal compressor shaft displacement and improvement of sealing improvement and efficiency technology. Beijing University of Chemical Technology (2006)
9. Luo, Q., Si, D., Feng, J., et al.: Study on blade strength and vibration characteristics of centrifugal compressor based on fluid-structure coupling method. Veh. Engine **2**, 51–54 (2012)
10. Wang, W., Gao, J., Li, S., et al.: Improvement of axial force regulation and balance plate sealing of centrifugal compressors. J. Fluid Mech. **34**(7), 15–18 (2006)
11. Theoretical analysis and experimental study on the automatic balancing axial force of floating impeller. Lanzhou University of Technology (2008)
12. Research on static and dynamic strength and aerodynamic load of multistage impeller rotor system of compressor. Northeastern University (2012)

Modulation Signal Bispectrum Analysis of Motor Current Signals for Condition Monitoring of Electromechanical Systems

Funso Otuyemi[2], Haiyang Li[2], Fulong Liu[2], Jiongqi Wang[1,2(✉)], Fengshou Gu[2], and Andrew D. Ball[2]

[1] School of Mathematics and Big Data, Foshan University, Foshan 528000, People's Republic of China
wjq_gfkd@163.com
[2] Centre for Efficiency and Performance Engineering (CEPE), School of Computing and Engineering, University of Huddersfield, Huddersfield HD1 3DH, UK
u1369519@unimail.hud.ac.uk, {Haiyang.liu, f.gu, a.d.ball}@hud.ac.uk

Abstract. Induction motor is one of the most widely used prime drivers and electric energy consuming devices in industry. Accurate and timely diagnosis of faults in motors will help to maintain their operating under optimal status and avoid excessive energy consumption and severe damages to systems. In this study, instantaneous motor current and voltage signals (IMCVS) is analyzed by an advanced Modulation Signal Bispectrum (MSB) method to achieve accurate demodulations of Frequency Modulation (FM) and Amplitude Modulation (AM) by minimizing noise influence and enhancing modulation characteristics simultaneously. Firstly, the modulation effects due to motor faults and downstream mechanical components were modelled, thus finding the interaction between AM and FM effect and hence developed a scheme to use the signature of AM and FM jointly for accurate fault diagnosis. Then experimentations were carried out to verify the performance of the proposed scheme in detecting and diagnosing common mechanical faults including Shaft Misalignments(SM), Motor Rotor Broken Bar (BRB), Stator Resistance Imbalance (SRI) and compound BRB with SRI

Keywords: Modulation signal bi-spectrum · Electrical motor · Fault diagnosis

1 Introduction

Induction motors are one of the essential industrial equipment and used as a prime mover for almost all industries. About 60% of the industrial electric energy is converted by motors into mechanical energy for driving pumps, fans, gears, adjustable speed drives and machine tools. Especially, it is gaining even more prevalent with the rapid rise of electric vehicles in last 5 years. In addition, these machines equipped with motors are subject to electrical and mechanical faults which can be a significant financial loss due to production breakdowns and maintenance costs. Thus, considerable

M. A. Wahab (Ed.): DAMAS 2019, LNME, pp. 566–581, 2020.
https://doi.org/10.1007/978-981-13-8331-1_42

research efforts have been focused on topics like condition monitoring of induction motors to ensure motor's operational efficiency and reliability. Several of these electrical or mechanical faults may be as a result of prolonged activity times, harsh operating conditions, voltage or current unbalance, among other factors. Emerging faults can be as a result of the motor performance, which may result in remarkable financial losses [1, 2]. Therefore, condition monitoring is a recurrent research topic in recent years with the primary aim to detect faults in an early stage and avoid unwanted production situations [3]. The contribution of this paper is to present another dimension for detecting and diagnosing induction motor combined faults using electrical signals which will be examined under different operating conditions with a more advanced signal processing tool - modulation signal bispectrum (MSB), rather than previously used spectrum analysis methods as reviewed in [4, 5]

2 Theoretical Backgrounds

2.1 Modulations in Motor Drives

At the point when an induction motor working under a healthy, the electromagnetic relationship can be examined via one of the three phase current, for a simpler comprehension of the impact due to Broken Rotor Bar (BRB) fault. However, disregarding the higher order harmonics and inherent errors and referring to supply voltage signal, the current phase signal of one of the three phases can be illustrated as [6, 7]:

$$I_A = \sqrt{2}\cos(2\pi f_s t - \alpha_I) \tag{1}$$

where the magnetic flux in the motor stator is:

$$\emptyset_A = \sqrt{2\emptyset}\cos(2\pi f_s t - \alpha_I) \tag{2}$$

Also, the electrical torque exhibited by the interaction between the current and flux is presented as:

$$\mathrm{T} = 3P\emptyset I\sin(\alpha_I - \alpha_\emptyset) \tag{3}$$

where I and $\emptyset$ represent the Root Mean Square (RMS) of the supply current and flux linkage, respectively;α_I and $\alpha_\emptyset$ are the phase of the current and flux respectively;f_s represent the fundamental frequency of electrical supply; P is the number of pole pair.

2.2 Rotor Problem

According to [4, 5], motor signals from a faulty operation motor can be expressed in the form:

$$
\begin{aligned}
i_A^F &= i_A^{FS} + i_f \\
&= I\sin\left(2\pi f_s t - \alpha_\varphi - \varphi_z\right) \\
&\quad + I_F \cos[2\pi(f_s - f_F)t - \alpha_F] \\
&\quad + \frac{\sqrt{2}}{2} I_{FL} \sin\left[2\pi(f_s - f_F)t - \alpha_F - \varphi_z - \Delta\varphi_{ZL}\right] \\
&\quad I_{Fr} \sin\left[2\pi(f_s + f_F)t - 2\alpha_\varphi + \alpha_F - \varphi_z + \Delta\varphi_{Zr}\right]
\end{aligned} \tag{4}
$$

where

$$
I_{FL} = 2\pi(1+2s)f_s\Delta\varphi/(Z - \Delta Z);
$$

and

$$
I_{FL} = 2\pi(1+2s)f_s\Delta\varphi/(Z + \Delta Z);
$$

are the modulus of sideband components due to speed oscillations, which is caused originally by the BRB current. It shows that the current signal of the faulty case exhibits three new additional components. Two of them are the lower sideband components which are at the same frequency but with different phases, and the other is the upper sideband component with a phase different to the previous two. To observe the possible connection of these components, Eq. (4) can be expressed using phase shift relationships:

$$
\begin{aligned}
&\cos\left(\frac{\pi}{2} + \theta\right) = -\sin(\theta) \text{and } \cos(\pi + \theta) = -\cos(\theta) \\
&i_A^F = \sqrt{2} I \cos\left[2\pi\left(f_s t - \alpha_\varphi - \varphi_z + \frac{\pi}{2} + \pi\right)\right] \\
&\quad + \frac{\sqrt{2}}{2} I_F \cos[2\pi(f_s - f_F)t - \alpha_F] \\
&\quad + \frac{\sqrt{2}}{2} I_{FL} \cos\left[2\pi(f_s - f_F)t - \alpha_F - \varphi_Z - \Delta\varphi_Z + \frac{\pi}{2} + \pi\right] \\
&\quad + \frac{\sqrt{2}}{2} I_{Fr} \cos\left[2\pi(f_s + f_F)t - 2\alpha_\varphi + \alpha_F - \varphi_Z + \Delta\varphi_Z + \frac{\pi}{2}\right]
\end{aligned} \tag{5}
$$

by combining the two lower sidebands in Eq. (5), it can be given as

$$
\begin{aligned}
i_A^F &\approx \sqrt{2} I \cos\left(2\pi f_s t - \alpha_\varphi - \varphi_Z + \frac{\pi}{2} + \pi\right) \\
&\quad + \frac{\sqrt{2}}{2} I_L \cos[2\pi(f_s - f_F)t - \alpha_F - \delta] \\
&\quad + I_{Fr} \cos\left[2\pi(f_s + f_F)t - 2\alpha_\varphi + \alpha_F - I_F + \Delta\varphi_{Zr} + \frac{\pi}{2}\right]
\end{aligned} \tag{6}
$$

where the amplitude of the lower sideband in RMS value is:

$$I_L = \sqrt{I_F^2 + I_{FL}^2 - 2I_F I_{FI} \sin(\varphi_Z + \Delta\varphi_{ZI})} \tag{7}$$

and the phase is:

$$\cos(\delta) = \sqrt{2I_F} - \sqrt{2I_{FL}} \sin(\varphi_Z + \Delta\varphi_{ZL}) / \sqrt{2I_F^2 + 2I_{FL}^2 - 4I_F I_{FI} \sin(\varphi_Z + \Delta\varphi_{ZI})} \tag{8}$$

The above derivation and discussion can be concluded that BRB leads to additional current components which are the lower and upper sideband components at $(1 \pm 2)f_s$. However, the amplitudes of the sidebands are impacted by rotor inertia, changes in load and machine impedance. Specifically, the amplitude of original fault current i_f, appears in the lower sideband, and is suppressed due to speed oscillations. Therefore, this implies that the interference needs to be eliminated so as to acquire the amplitude and precisely obtain the amplitude of i_f imperative to foresee the severity of the fault. Furthermore, the phases of the sidebands not only narrate to the same factors as their magnitudes but also to the phase variation of the fault current. However, the phase combination between sidebands and the carrier can eliminate the phase variation to achieve reliable spectrum estimation.

2.3 Stator Winding Problem

Majority of motor stator winding faults occur due to the damage in turn protection brought about the short circuit in stator windings. This short circuits can be between inter-turns shorts of same phase, short between coils of the same phase, short between two phases and short between phases to earth. As indicated by an Institute of Electrical and Electronics Engineers (IEEE) and Electric Power Research Institute (EPRI) motor reliability study [7], stator faults are mostly responsible for 37% of the failures in an induction motor. Heretofore, it has not found an analytic expression for stator faults. According to [6], unbalance stator would produce addition modification at a frequency of considering an inter-turn short circuit in stator windings and assuming that the rotor structure is not modified, the distribution of the stator magneto motive force (MMF) in the air gap will be affected because of the current that will circulate through the short-circuited turns. Analysing the MMF of the affected coil group on the rotor currents, the following new frequencies can be obtained in these rotor currents [6, 7]

$$f_{sc} = f_s\{n/P(1 - s) \pm k\} \tag{9}$$

where f_s represent the fundamental supply frequency;

n represent any integer;

P represent the number of pole pairs;

k = 1, 3, 5…;

s = slip;

2.4 MSB Analysis

To improve the performance of characterizing motor current signals, a new variant of the conventional bispectrum, known as MSB [5, 19], is examined to show the mechanisms of its modulation extraction and noise reduction performances.

$$MSB_{(f_1,f_2)} = E\{X\ (f_2 + f_1)X\ (f_2 - f_1)X^*(f_2)X^*(f_2)\} \tag{10}$$

where f_1, f_2 and $(f_1 + f_2)$ are individual frequency components; X^*: the complex conjugate of $X(f)$; $E\{\}$: the statistical expectation operator.

However, both $(f_2 + f_1)$ and $(f_2 - f_1)$ are considered for the characterization of nonlinear coupling in modulation signals. Thus, if $(f_2 - f_1)$ and $(f_2 + f_1)$ are due to nonlinear coupling, bi-frequency MSB (f_1,f_2) will be shown, and therefore making it more accurate and effective to modulate the signal.

Hence, the phase of MSB from Eq. (10) above is:

$$\emptyset_{MSB}(f_1,f_2) = \phi(f_2 + f_1) + \phi(f_2 - f_1) - \phi(f_2) - \phi(f_2) \tag{11}$$

Thus, f_1 and f_2 are in coupling when their phases are illustrated as shown below;

$$\phi(f_2 + f_1) = \phi(f_2) + \phi(f_2) \tag{12}$$

$$\phi(f_2 - f_1) = \phi(f_2) - \phi(f_2) \tag{13}$$

Substituting Eq. (14) into Eq. (10), the total phase of MS bi-spectrum will be regarded as zero and its amplitude will be the product of the four magnitudes, which is the maximum of the complex product. Thus, a bi-spectral peak does appear at (f_2,f_1) If $(f_2 + f_1)$ and $(f_2 - f_1)$ are both due to nonlinear effect between f_1 and f_2, a bi-spectral peak will appear at bi-frequency (f_1,f_2). This is more accurate and apparent in representing the sideband characteristics of modulated signals. Hence, MSB phase for current signal is:

$$\emptyset_{MSB}(f_S,f_F) = 2\alpha_{\emptyset} + 2\alpha_F + 2_{\varphi} + \pi \tag{14}$$

Equation (16) shows that the bispectrum phase is connecting to only sideband components i.e. the angular position of faults and motor design and operating parameters but exclude the phase of the fundamental component. It means that MSB is independent of the angular position of the rotor or the start point of a signal acquired. This will allow an average of MSB estimation to be performed using a data set collected or framed at any time without the need of additional alignments signals such as shaft encoders. The average in turn will suppress random noise and unrelated components to obtain a reliable estimation of MSB of the hidden modulating signal [8].

Moreover, the excellent noise reduction capability of MSB will lead to a more accurate amplitude modulating component.

2.5 MSB Coherence

From Eq. (17), if $(f_1 + f_2)$ and $(f_1 - f_2)$ are both due to nonlinearity effect between f_1 and f_2, a bispectral peak will appear at bifrequency $B_{MS}(f_1, f_2)$ which is similar to a normalized form of MSB or modulated signal bicoherence and is given by the expression below:

$$b_{MS}^2(f_1, f_2) = \frac{|B_{MS}(f_1, f_2)|^2}{E|X(f_2)X(f_2)X^*(f_2)X^*(f_2)|^2 E|X(f_2 + f_1)X(f_2 - f_1)|^2} \quad (15)$$

The expression given above is more effective and precise in representing the sideband and characteristics of modulation signals component. Thus, the four component products will enhance signal component more and produces more vigorous detection results in random noise.

3 Experimental Methods

To evaluate the performance of MSB analysis in diagnosing induction motor faults, several electric current signals were measured from induction motor fault tests when the motor under different fault cases and operating conditions.

3.1 Test Systems

The test rig is made up of two primary parts, the mechanical system and the electrical control system, with additional measuring devices and data acquisition systems. To show the performance of MSB in analyzing current signals, a dataset is acquired in a motor rig under different simulated fault cases and operating condition. The electromechanical part of the system outlined in the Fig. 1 consists of a rated output power of 4 kW three phase AC induction motor at 1420 rpm as the prime driver (two-pole pairs), 28 rotor bars and 36 slots, and a load generator using flexible spider rubber couplings. Couplings' type is a hard rubber with three jaw in size of 150 mm and a transmission power up to 100 kW at 1500 rpm speed. The DC load generator is controlled by a DC variable drive that varies the armature current to provide the required loads to the AC motor. The operating speeds and loads are set and controlled by operators via a touch screen on the control panel system connected to a programmable logic controller (PLC) [8, 9], as shown in Fig. 2.

To investigate the characteristics of the current and voltage signals, a power supply measurement device was employed to measure the AC voltages, currents and power using Hall Effect voltage and current transducers and a universal power cell, which is independent of the controller to avoid any interruptions to the control process. During the experimental work, all of the data was recorded using a YE6232B high speed data acquisition system. This system has 16 channels, each channel with a 24-bit analogue-digital converter at a sampling frequency of 96 kHz. The three phase currents, voltages

Fig. 1. Test facility

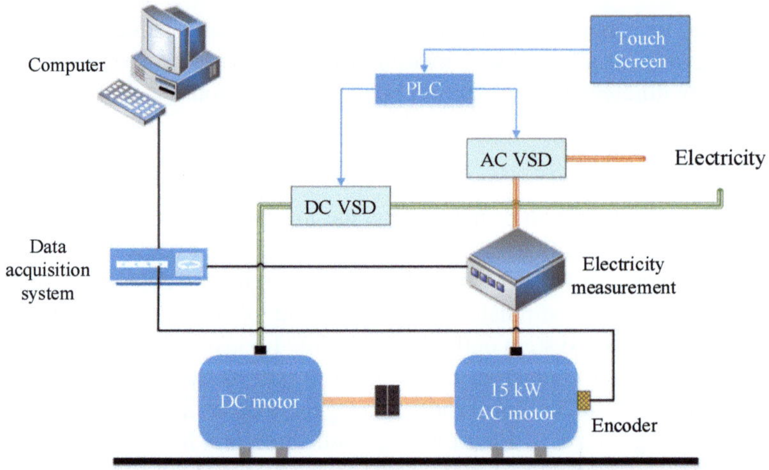

Fig. 2. Experimental test rig setup

and encoder pulse trains (for speed measurements) were acquired simultaneously under different testing cases when the system was in steady operations.

3.2 Fault Simulations

Based on the numerical study and theoretical analysis in [13, 14], it has been confirmed that MSB can capture AM characteristics in signals more accurately. To examine its performance in characterizing measured signals for fault detection and diagnosis, it is thus applied to the motor current signals measured from the induction motor with different common motor fault cases such as BRB and Stator Resistance Imbalance (SRI) under different operating conditions.

Broken Rotor Bar (BRB)
When there is a crack in the rotor bar, no current flows through it therefore causing two counter rotating magnetic fields at frequencies ± sf due to rotor asymmetry, there is a loss of current flow around the rotor, causing no magnetic flux around that bar. Thus, this phenomenon generates an asymmetry in the rotor magnetic field and produces a non-zero backward rotating field which rotates at the slip frequency speed with respect to the rotor. However, this effect creates harmonic currents in the stator windings, which are superimposed on the stator currents. These superimposed harmonics are used as signatures for BRB detection in Motor Current Signature Analysis (MCSA) techniques [10–12] In order to estimate the performance of MSB study, current and voltage signals were collected for five different motor formations: healthy motor (BL), half broken rotor bar ($\frac{1}{2}$ BRB),one broken rotor bar (1BRB), two broken rotor bar (2BRB) which are shown in Fig. 6, induced by drilling carefully into the bars along their height in such a way that the hole cut the bar completely to simulate the broken rotor bar fault (Fig. 3).

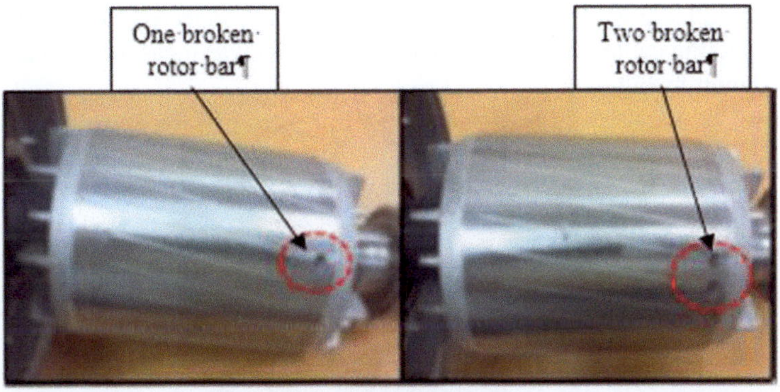

Fig. 3. One and two broken rotor bar (BRB) cases [14]

The occurrence of broken bars results in particular effects on the first-order sidebands (k = 1) which are significant to the diagnosis of BRB fault. While the right sideband $f_s(1+2ks)$ is due to the speed variation or ripple and the left sideband $f_s(1-2ks)$ is due to magnetic or electrical rotor asymmetry which is caused by the BRB. It has been deduced that the presence and amplitude of the sideband components are dependent on both the position, number of BRB with respect to the motor's speed and load [13, 14].

Stator Resistance Imbalance (SRI)
In this test, a small amount of motor resistance imbalance was introduced which would cause an increase in the winding temperature by a marginal amount. This imbalance in the stator winding causes the rotating flux to oppose the main flux unbalanced-voltage

operation, then creates a pulsating torque which will result in speed pulsation, vibration and consequently acoustic noise [16]. Also, it creates an imbalance in the stator circuit which reduces the magnetic flux generated by one of the stator windings. The effect of this has been studied for the reason of motor instability and winding temperature increase by comparing to normal running conditions [17].

According to Lee & Ching-Yin [16], voltage imbalance defined by the National Electric Motors Association (NEMA) is given by:

$$\text{Voltage imbalance} = (1 - \frac{3U_{min}}{\sum U_i}) \times 100 \tag{16}$$

Where U_{min} the lowest phase voltage among the three phases is, U_i is the voltage across each phase.

Therefore, the characteristics frequency for the stator current spectrum of an induction motor is determined by [19, 20]:

$$f_{usv} = (1 + 2\text{k})f_s \tag{17}$$

Where k is the order of harmonics i.e. 1, 2, 3,….

The resistance simulated fault which occurs in one winding inside the inductor motor thereby affecting only one motor phase in a star connected motor. This indicate that the connecting cable resistance from drive to motor was measured at 0.2 Ω for each phase, as shown in Fig. 4.

Fig. 4. Variable resistor bank

3.3 Test Procedure

This experiment work has been conducted with all test cases under five successive motor load increments: 0%, 20%, 40%, 60%, and 80% of full load condition. Each test repeated 3 times for obtaining reliable results, which results in 3 sets of data, each consisting of 5 load settings at a constant speed of 100% when the Variable Speed Drive (VSD) system which changes the frequency to change the speed of the motor

Table 1. Test profile under different speed and load

Operation duration	Speed	Load	Load (input in PLC)
3 min	100%	0%	0%
2 min		20%	10%
2 min		40%	20%
2 min		60%	30%
2 min		80%	40%

operating under sensorless control mode. In addition, each load condition lasts 2 min to ensure system is stabilized. Table 1 details the operation profile.

Thereafter, the tests were further carried out when the half BRB is compounded with a phase winding resistance increment of R_{fs} = 0.2 Ω, in which the resistance were realized by an external resistor bank (Fig. 4) connected into one of the three winding phases.

Furthermore, it is important to have a meaningful and consistent baseline data set from which to measure simulated faults. Therefore, the healthy baseline data was measured in three separate tests at the end, each with the same operating data applied as the faulty cases. The design of the test rig is such that a Programmable Logic Controller (PLC) provides a predictable load change to the AC driven motor system at precise, repeatable intervals. During the experimental work all the data was recorded using a YE6232B high speed data acquisition system. This system has 16 channels, each channel with a 24 bit analogue-digital converter at a sampling frequency of 96 kHz with a 30 s sample time was used for each test run. A total of 15 data sets were therefore obtained for the baseline data.

4 Results and Discussion

4.1 Characteristics of MSB Under BRB and SM

MSB coherence under BRB and SM

Figures 5 and 6 present a typical MSB coherence for a current signal under baseline and half broken bar under 20% & 80% load with half broken rotor bar with two distinctive peaks at bifrequency (24.72, 49.8) Hz and (24.9, 51.45) Hz This frequency does not show any connections to BRB which is difficult to identify the faults based on the spectra with the frequency resolution for the current signals. In all cases, the machine RPM (1x component) and its higher harmonics (2x, 3x,. ..) are available and no sideband recurrence at the fundamental recurrence identified with the slip recurrence has been believed to recognized the fault as suggested in [21, 22]. Amplitude demodulation at a frequency for any signal removes that frequency, but modulated frequencies can be clearly identified from the demodulated signal. Subsequently,

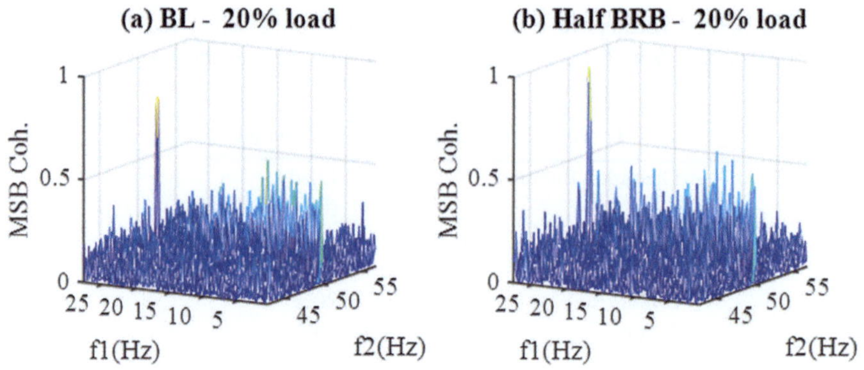

Fig. 5. MSB coherences for (a) healthy motor under 20% load and (b) half BRB motor with 20% load

amplitude demodulation at the mains frequency, 51.45 Hz, has been done for the phase current signals of the faulty rotor and stator conditions. The amplitude spectra of all the demodulated signals have demonstrated just a single peak at the RPM of the motor [23].

MSB magnitude under BRB and SM

Figures 7 and 8 show both two bispectra of current signals for a healthy induction motor and a BRB case, respectively. The difference of two bispectra for both the healthy cases and BRB are noticeable. First of all, they all have a high peak at bifrequency (24.9, 51.45) Hz and (1.09, 51.45) Hz as the load increases. The first one relates to the $2sf$s and can be relied on to detect and diagnose BRB, while the second one relates to rotor speed due to the speed oscillation. However, for the fault cases, a number of small peaks, arisen from the induction motor working cycle and supply frequency. These feature thus indicated the nonlinear coupling effects existing in the current signals, which made the difference between the normal and fault case.

4.2 Characteristics of MSB Under BRB, SRI and SM

MSB coherence under BRB with SRI and SM

Figures 9 and 10 present a typical MSB coherence for a current signal under baseline and half broken bar with 0.2 Ω SRI under 20% & 80% load with half broken rotor bar with two distinctive peaks at bifrequency (24.72, 49.8) Hz and (24.9, 51.45) Hz due to the fault severity This frequency does not show any connections to BRB which is difficult to identify the faults based on the spectra with the frequency resolution for the current signals which is predominated by background noise.

Figures 11 and 12 present typical MSB magnitude results from the current signal under half broken bars with a phase winding resistance increment (R_{fs} = 0.2 Ω) and 80% load with respect to sensorless control mode. As can be seen in Fig. 11, MSB shows two distinctive peaks at bifrequency (24.9, 51.45) Hz and (1.09, 51.45) Hz as the load increases. Clearly, the first one relates to the $2sf$s and can be relied on to detect

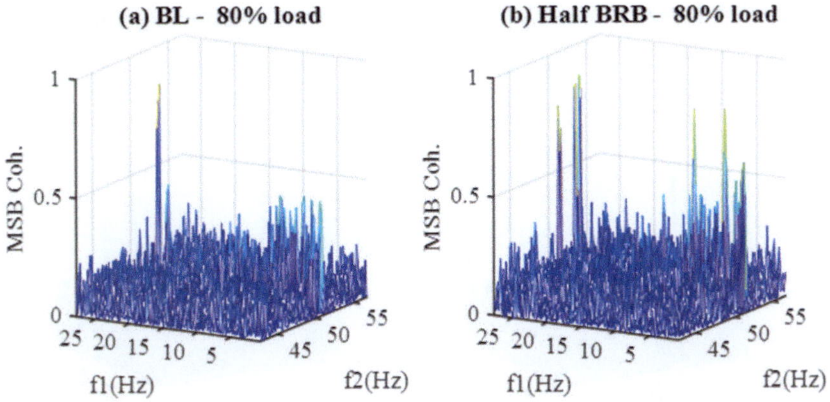

Fig. 6. MSB coherences for (a) healthy motor under 80% load condition and (b) half BRB motor with 80% load.

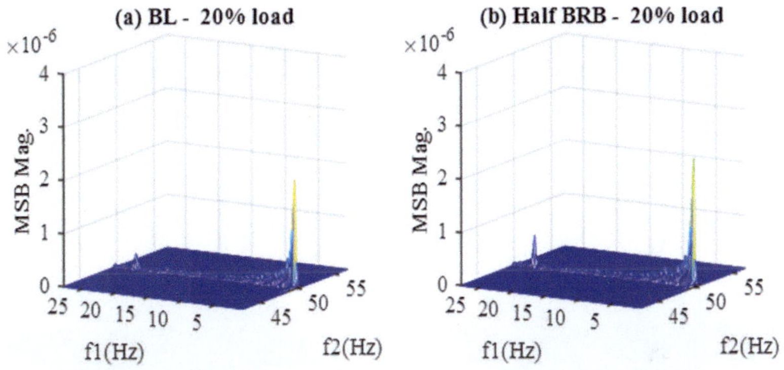

Fig. 7. MSB magnitudes of the stator phase current for (a) healthy motor under 20% load and (b) half broken rotor bar motor with 20% load

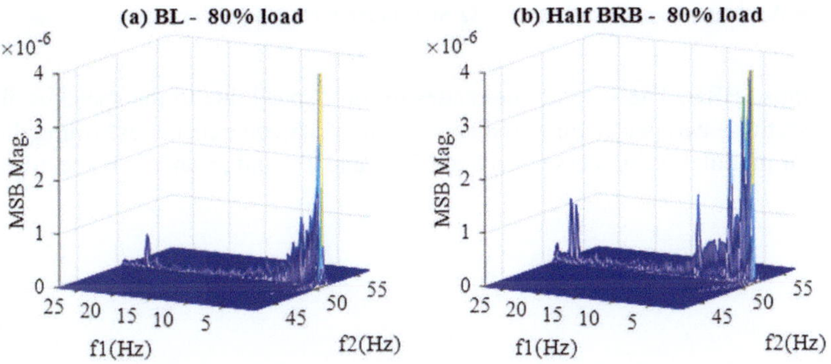

Fig. 8. MSB magnitudes of the stator phase current for (a) healthy motor under 80% load and (b) half broken rotor bar motor with 80% load.

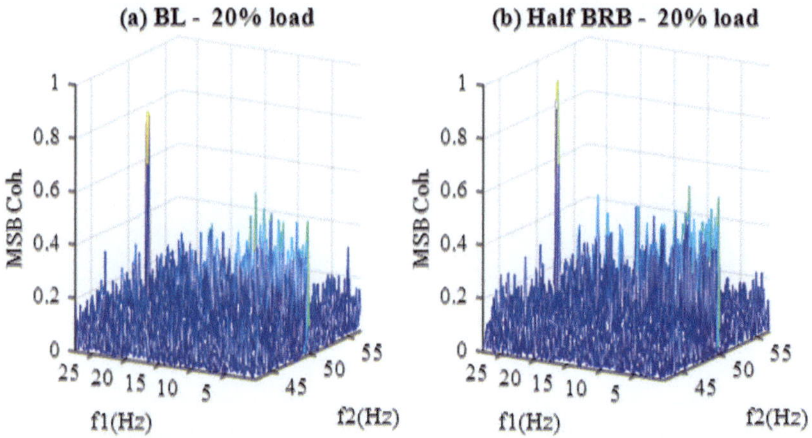

Fig. 9. MSB coherence of the stator phase current for (a) healthy motor under 20% load and (b) half broken rotor bar motor with 0.2 Ω SRI under 20% load.

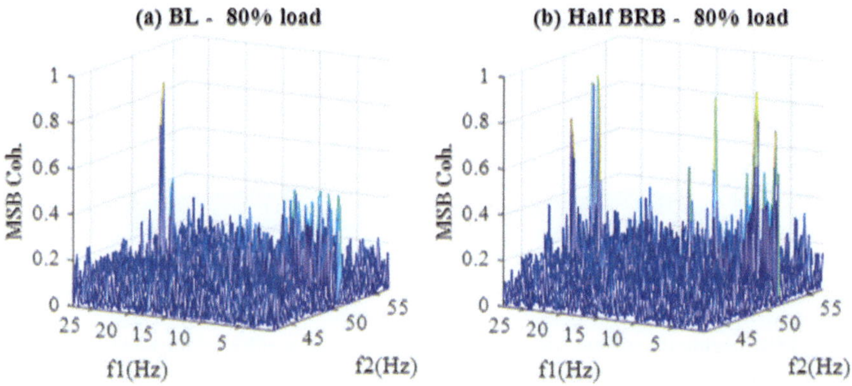

Fig. 10. MSB coherence of the stator phase current for (a) healthy motor under 80% load and (b) half broken rotor bar motor with 0.2 Ω SRI under 80% load.

and diagnose BRB. The second one relates to rotor speed due to the speed oscillation. Moreover, these two peaks are also distinctive in MSB coherence, confirming that they stem from modulation processes between $2sfs$ and fs, and fr and fs respectively.

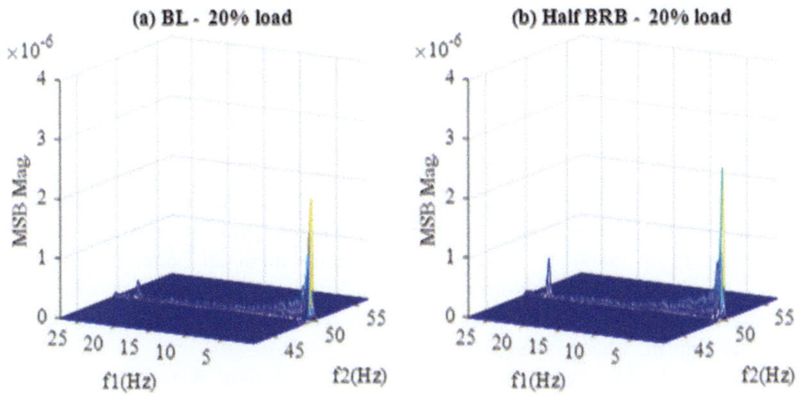

Fig. 11. MSB magnitudes of the stator phase current for (a) healthy motor with 20% load and (b) half BRB motor with 0.2 Ω SRI under 20% load

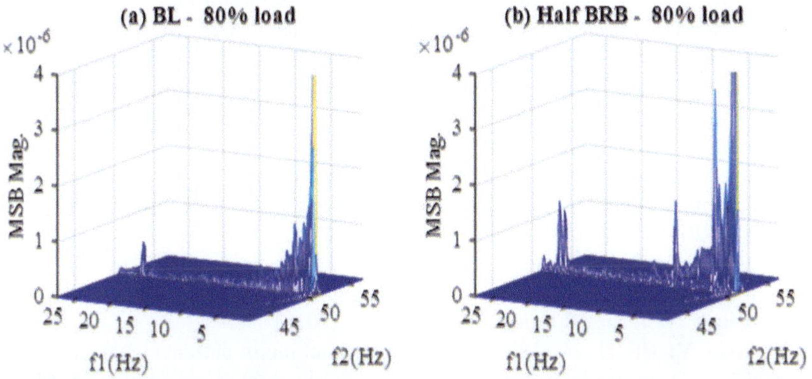

Fig. 12. MSB magnitudes of the stator phase current for (a) healthy motor under 80% load and (b) half BRB motor with 0.2 Ω SRI under 80% load

5 Conclusions

A method that can identify faults in a motor at an early stage and also capable of distinguishing between rotor faults and the stator faults is always important, so that remedial action can be carried out quickly. The induction motor phase current signal in the case of any fault is expected to contain a number of harmonic components related to the motor RPM and the mains frequency. Either the stator or the rotor may distort the sinusoidal response of the motor phase current signal which results in a number of harmonics of the motor RPM and the mains frequency. This has already been observed in the motor phase current spectra. It shows that the amplitude spectra are not able to detect the BRB and stator faults for the induction motor.

Therefore, MSB is evaluated using phase current signals to monitor stator faults. Based on the previous analysis, it can conclude that the MSB suggested in this paper has the capability to suppress noise influences to obtain an accurate estimation of the modulating components since it considered the phase effects of modulating components.

References

1. Jin, C., Ompusunggu, A.P., Liu, Z., Ardakani, H.D., Petré, F., Lee, J.: A vibration-based approach for stator winding fault diagnosis of induction motors: application of envelope analysis. Center for Intelligent Maintenance Systems Cincinnati United States (2014)
2. Drif, M., Cardoso, A.J.M.: Stator fault diagnostics in squirrel cage three-phase induction motor drives using the instantaneous active and reactive power signature analyses. IEEE Trans. Ind. Inform. **10**(2), 1348–1360 (2014)
3. Lamim Filho, P.C.M., Pederiva, R., Brito, J.N.: Detection of stator winding faults in induction machines using flux and vibration analysis. Mech. Syst. Signal Process. **42**(1–2), 377–387 (2014)
4. Bellini, A., Filippetti, F., Tassoni, C., Capolino, G.-A.: Advances in diagnostic techniques for induction machines. IEEE Trans. Ind. Electron. **55**, 4109–4126 (2008)
5. Mehrjou, M.R., Mariun, N., Marhaban, M.H., Misron, N.: Rotor fault condition monitoring techniques for squirrel-cage induction machine—a review. Mech. Syst. Signal Process. **25**, 2827–2848 (2011)
6. Sahraoui, M., Zouzou, S.E., Ghoggal, A., Guedidi S.: A new method to detect inter-turn short-circuit in induction motors. In: The XIX International Conference on Electrical Machines-ICEM 2010, pp. 1–6. IEEE (2010)
7. Ukil, A., Chen, S., Andenna, A.: Detection of stator short circuit faults in three-phase induction motors using motor current zero crossing instants. Electr. Power Syst. Res. **81**(4), 1036–1044 (2011)
8. Gu, F., Wang, T., Alwodai, A., Tian, X., Shao, Y., Ball, A.D.: A new method of accurate broken rotor bar diagnosis based on modulation signal bispectrum analysis of motor current signals. Mech. Syst. Signal Process. **50**, 400–413 (2015)
9. Gu, F., Shao, Y., Hu, N., Naid, A., Ball, A.: Electrical motor current signal analysis using a modified bispectrum for fault diagnosis of downstream mechanical equipment. Mech. Syst. Signal Process. **25**, 360–372 (2011)
10. Haram, M., Wang, T., Gu, F., Ball, A.D.: Electrical motor current signal analysis using a modulation signal bispectrum for the fault diagnosis of a gearbox downstream. J. Phys.: Conf. Ser. **364**(1), 012050. IOP Publishing, (2012)
11. Electrical motor current signal analysis using a modulation signal bi-spectrum for the fault diagnosis of a gearbox downstream. https://www.researchgate.net/publication/254495503_Electrical_Motor_Current_Signal_Analysis_using_a_Modulation_Signal_Bispectrum_for_the_Fault_Diagnosis_of_a_Gearbox_Downstream. Accessed 30 Jan 2019
12. Lane, M., Ashari, D., Gu, F., Ball, A.D.: Investigation of motor current signature analysis in detecting unbalanced motor windings of an induction motor with sensorless vector control drive. In: Vibration Engineering and Technology of Machinery, pp. 801–810. Springer, Cham (2015)
13. Shaeboub, A., Gu, F., Lane, M., Haba, U., Wu, Z., Ball, A.D.: Modulation signal bispectrum analysis of electric signals for the detection and diagnosis of compound faults in induction motors with sensorless drives. Syst. Sci. Control. Eng. **5**(1), 252–267 (2017)
14. Shaeboub, A., Lane, M., Haba, U., Gu, F., Ball, A.D.: Detection and diagnosis of compound faults in induction motors using electric signals from variable speed drives. In: Automation and Computing (ICAC), 2016 22nd International Conference on, pp. 306–312. IEEE (2016)

15. Shi, P., Chen, Z., Vagapov, Y., Zouaoui, Z.: A new diagnosis of broken rotor bar fault extent in three phase squirrel cage induction motor. Mech. Syst. Signal Process. **42**(1–2), 388–403 (2014)
16. Hamad, N., Brethee, K.F., Gu, F., Ball, A.D.: An investigation of electrical motor parameters in a sensorless variable speed drive for machine fault diagnosis. In: Automation and Computing (ICAC), 2016 22nd International Conference on, pp. 329–335. IEEE (2016)
17. Messaoudi, M., Sbita, L.: Multiple faults diagnosis in induction motor using the MCSA method. Int. J. Signal Image Process. **1**(3) (2010)
18. Shi, P., Chen, Z., Vagapov, Y., Zouaoui, Z.: A new diagnosis of broken rotor bar fault extent in three phase squirrel cage induction motor. Mech. Syst. Signal Process. **42**(1–2), 388–403 (2014)
19. Shaeboub, A., Lane, M., Haba, U., Gu, F., Ball, A.D.: Detection and diagnosis of compound faults in induction motors using electric signals from variable speed drives. In: Automation and Computing (ICAC), 2016 22nd International Conference on, pp. 306–312. IEEE (2016)
20. Lee, C.-Y.: Effects of unbalanced voltage on the operation performance of a three-phase induction motor. IEEE Trans. Energy Convers. **14**(2), 202–208 (1999)
21. Lane, M., Ashari, D., Gu, F., Ball, A.D.: Investigation of motor current signature analysis in detecting unbalanced motor windings of an induction motor with sensorless vector control drive. In: Vibration Engineering and Technology of Machinery, pp. 801–810. Springer, Cham (2015)
22. Lane, M., Shaeboub, A., Gu, F., Ball, A.D.: Investigation of reductions in motor efficiency and power factor caused by stator faults when operated from an inverter drive under open loop and sensorless vector modes. Syst. Sci. Control. Eng. **5**(1), 361–379 (2017)
23. Alwodai, A.: Motor fault diagnosis using higher order statistical analysis of motor power supply parameters. PhD diss., University of Huddersfield (2015)
24. Gu, F., Wang, T., Alwodai, A., Tian, X., Shao, Y., Ball, A.D.: A new method of accurate broken rotor bar diagnosis based on modulation signal bispectrum analysis of motor current signals. Mech. Syst. Signal Process. **50** (2015): 400–413
25. Treetrong, J., Sinha, J.K., Gu, F., Ball, Andrew: Bispectrum of stator phase current for fault detection of induction motor. ISA Trans. **48**(3), 378–382 (2009)
26. Henao, H., Razik, H., Capolino, G.-A.: Analytical approach of the stator current frequency harmonics computation for detection of induction machine rotor faults. Ind. Appl., IEEE Trans. **41**(3), 801–807 (2005)
27. Didier, G., Ternisien, E., Caspary, O., Razik, H.: A new approach to detect broken rotor bars in induction machines by current spectrum analysis. Mech. Syst. Signal Process. **21**, 1127–1142 (2007)

Diagnosis for Timing Gears Noise of a Diesel Generating Set

Wanyou Li[1], Chongpei Liu[1], Yunbo Hu[2], Shuwen Yu[2], BingLin Lv[3], and Yibin Guo[1(✉)]

[1] College of Power and Energy Engineering, Harbin Engineering University, Harbin, People's Republic of China
guoyibin@hrbeu.edu.cn
[2] Marine Design and Research Institute of China (MARIC), Shanghai, People's Republic of China
[3] China Shipbuilding Power Engineering Institute Co., Ltd, Shanghai, People's Republic of China

Abstract. There is an abnormal noise generated from the timing gears while practicing for the commissioning of a diesel generator set. As the load increases, this phenomenon becomes more apparent. In order to find the cause of the noise, a fault diagnosis model is established which considers not only the gears but also the effects of the associated structures. After eliminating the possible reasons such as the defects of timing gears, the abnormality of the valve system, fuel system controllers and generator, the injector problem is suspected to cause the fault. To verify this conjecture, the fuel injection pumps of the diesel generator set are inspected and the second injector is found to be blackened with some carbon deposition. By replacing the injector, the noise is gone. Hence, the fault diagnosis model can provide a new tool for future diagnosis of gear fault.

Keywords: Noise · Timing gears · Fault diagnosis model · Injector

1 Introduction

A marine diesel generator set which includes a diesel engine (5L21/31), a generator and a common base is studied in this paper, and the related parameters are listed in the Appendix. While during the commissioning of the set, a noise is generated from the timing gears as the diesel engine power increases to 75% of the rated load. With increasing of the load, this phenomenon becomes more apparent. Although the timing gears have been checked lots of times, including adjusting the center distance, tooth surface modification and replacing gears, however, the noise still remains.

There are intensive studies on the gear fault diagnosis available and many diagnostic methods have been proposed. Afia et al. [1] proposed a novel method named Autogram to detect the gear failures which can better reduce the influence of noise. Zhao et al. [2] presented a new method to eliminate the effects of spectrum smearing phenomenon and gear interferences. The simulated and experimental results showed that this method is reliable and effective to conduct the fault diagnosis without the angular resampling technique and tachometer. Cai and Li [3] proposed a new gear fault

M. A. Wahab (Ed.): DAMAS 2019, LNME, pp. 582–593, 2020.
https://doi.org/10.1007/978-981-13-8331-1_43

diagnosis approach based on S transform to separate the useful information from the noise. The validity of this method is verified through numerical simulation and experimental study of a gear with tooth crack. Based on kernel independent component analysis (KICA) and BP neural network (BPNN), Li [4] presented a new approach to address the problem that the vibration signals of gearbox are easily contaminated. Wang et al. [5] proposed alternately evolutionary digital filter (AEDF) to solve the issue of the lower convergence rate of the evolutionary digital filter (EDF), which can quickly detect the gear tooth spalling faults in a noisy environment. Based on SSTFA, Sun et al. [6] proposed a gear fault diagnosis method to extract the transient feature from the vibration signals of gears. The comparisons among the proposed method and some traditional time-frequency analysis methods showed that the proposed method has advantages on separating feature signals in strong noise.

The works mentioned above are mainly focused on exploring new methods to extract the useful information from gear fault signals, in fact, reducing the noise of whole engine is also a research focus in recent years. Yao et al. [7] proposed a new noise source identification method to separate and identify the piston slap noise and the combustion noise of a 6-cylinder diesel engine. The results showed that the independent noise obtained by the proposed method is more pure and accurate than that obtained by the EEMD-RobustICA-CWT method. Zhao et al. [8] built a boundary element model of the diesel engine to rapidly calculate the radiation noise, and the results showed that the noise can be reduced by optimizing the structure parameters such as oil sump, gear chamber cover and flywheel housing. Zhang [9] tested the noise of a diesel engine and torsional vibration of the shaft system, and found that the sound power level of noise will be reduced by 1 to 2 dB with installing dampers. Bhat et al. [10] investigated the influences of the fuel injection parameters on the combustion noise in a single-cylinder diesel engine. This investigation indicated that the split injection and injection timing can reduce noise effectively. Usuda et al. [11] developed the 1KD-FTV engine by using pilot injection optimization and high stiffness power plant structure, which reduced the noise of the engine.

Yet despite remarkable progress in gear fault diagnosis and reducing the noise of engine, there is little research has been devoted to reduce or eliminate the abnormal noise of timing gears. In this paper, a fault diagnosis model is proposed to determine the cause of the timing gears noise of a diesel generator set. After eliminating the possible reasons about the coupling of the gears with their associated structures, the injector problem is suspected to cause the fault. Following the indication of the results, the second injector is found to be blackened with some carbon deposition. By replacing this injector with new one, the noise is gone. This study shows that the fault diagnosis model can provide a new tool for future diagnosis of gear fault.

2 Analysis of the Noise

First, the noise is collected through a microphone which is placed 0.5 m far away from the gears. The test layout is shown in Fig. 1, and the test instruments are listed in Table 1. Note that the vibration measuring points will be introduced in the following

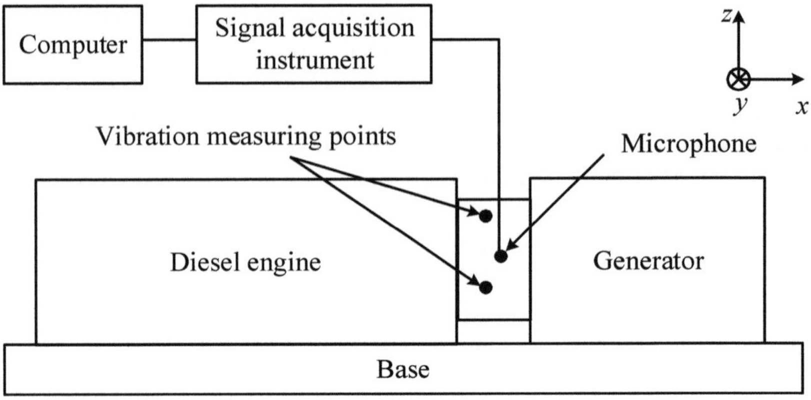

Fig. 1. Schematic diagram of noise test.

Table 1. Test instruments.

Instrument types	Numbers
B&K 3560C analysis system	1
B&K 4957 microphone	1
B&K 2258A-100 triaxial accelerometer	2

sections. The A-weighting and 1/3 octave are used in the data process, and the noise curves of 25%, 50%, 75% and 100% rated load are shown in Fig. 2.

The noise curves show that with increasing of the load, the sound pressure level (SPL) also shows an increasing trend. Under low loads, the fluctuations are small, about 3 dB. Under 100% rated load, the SPL fluctuates greatly, and the maximum fluctuation is up to 6 dB, as seen in Table 2.

In this study, the fluctuations of the noise are evaluated by defining "noise variance", as shown below

$$R=\frac{1}{N}\sum\left(Lp_i-\frac{1}{N}\sum(Lp_i)\right)^2 \tag{1}$$

Where R is the noise variance, Lp_i the noise sound pressure level at time i.

The noise variances of no load, 25%, 50%, 75% and 100% of rated load working conditions are listed in Table 3.

It can be seen that the noise variances show a general uptrend with load increasing, and it is particularly evident under 100% of rated load.

Furthermore, through frequency domain analysis of the noise of 100% rated load, as shown in Fig. 3, it can be found that the main characteristic frequency is 60.16 Hz, which is eight times of 7.5 Hz, half of the shaft frequency. The frequencies of other larger amplitudes are also the multiples of 7.5 Hz, such as 37.58 Hz and 30.08 Hz.

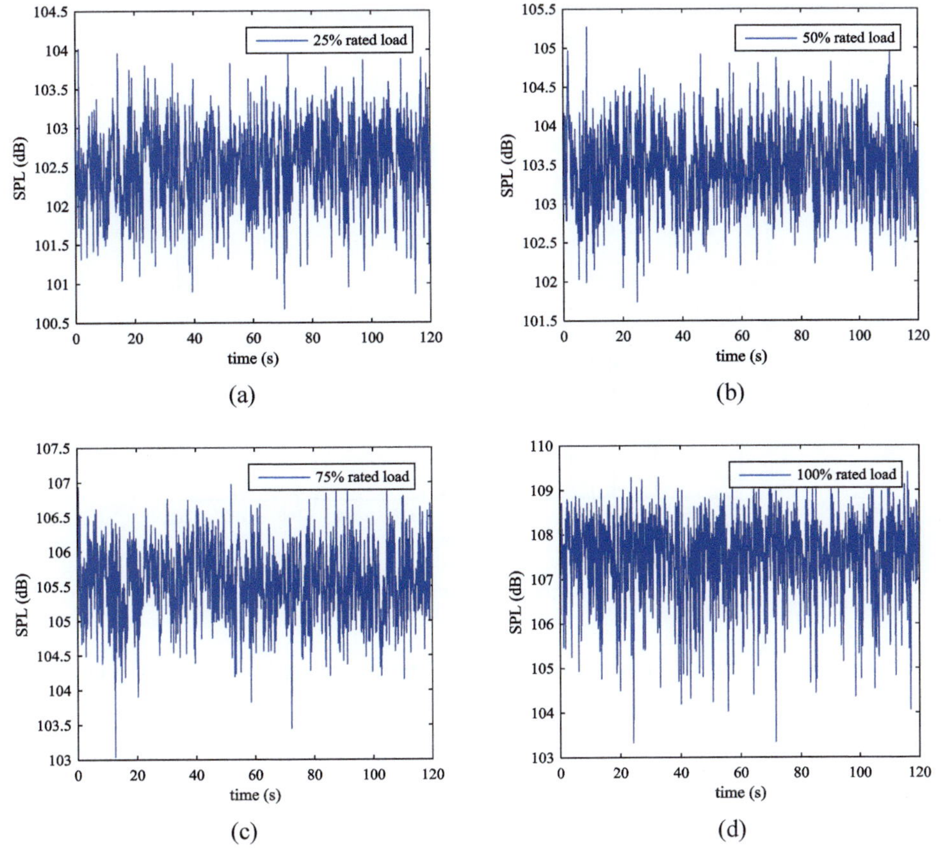

Fig. 2. Noise curves of (a) 25% rated load, (b) 50% rated load, (c) 75% rated load, (d) 100% rated load.

Table 2. Noise SPL at specific time under 100% rated load.

Time (s)	SPL (dB)	Time (s)	SPL (dB)
23.8	109.3	71.3	108.7
24.3	103.3	71.8	103.4
24.6	108.3	72.0	109.0

Table 3. Noise variances of variable working conditions.

Working conditions	No load	25% load	50% load	75% load	100% load
R	0.17	0.28	0.28	0.30	0.92

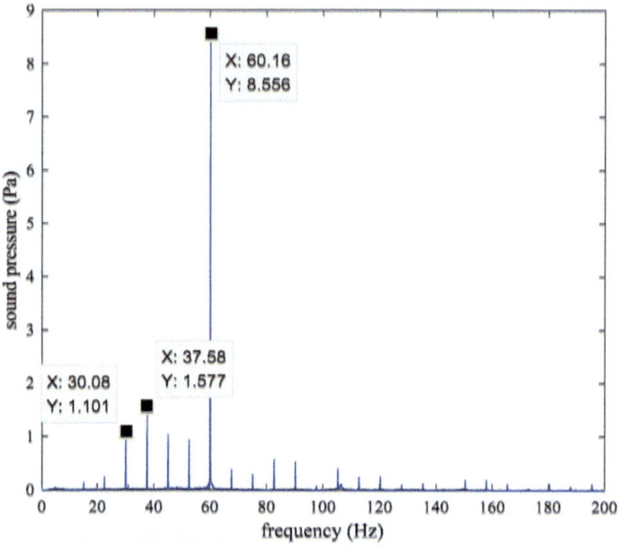

Fig. 3. Schematic diagram of vibration test.

To further verify the noise source, the vibration acceleration signals are collected through the acceleration sensors putting on the shell of the gears, as shown in Fig. 1. The noise and vibration signals in the range of 23.5s ~ 24.7s are shown in Fig. 4. The

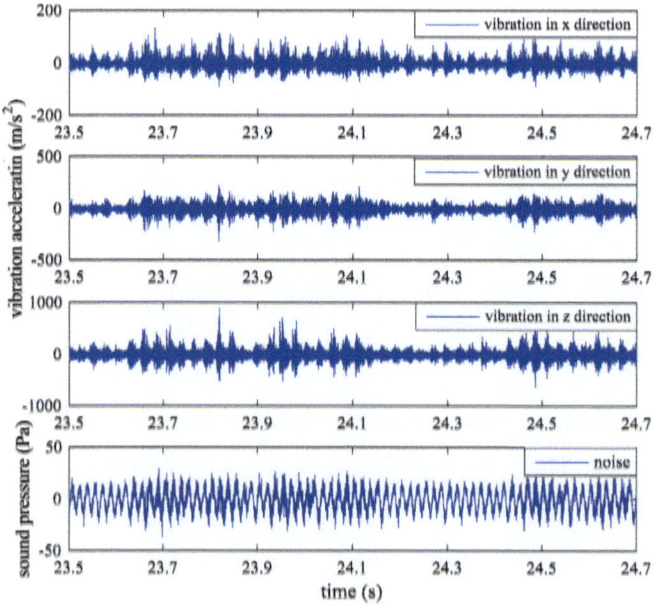

Fig. 4. Vibration and noise signals in time domain.

range of 23.5s ~ 24.0s is defined as time 1 while the range of 24.0s ~ 24.7s is defined as time 2. The comparison between the two ranges shows that the noise curve within time 1 is denser than that within time 2, and the amplitudes of the vibration accelerations within time 1 are larger than those within time 2.

The correlations between the noise and vibration accelerations are analyzed and the correlation curves are obtained. The curves of the measuring point 1 and point 2 are given in Fig. 5, and the maximum correlation values are listed in Table 4.

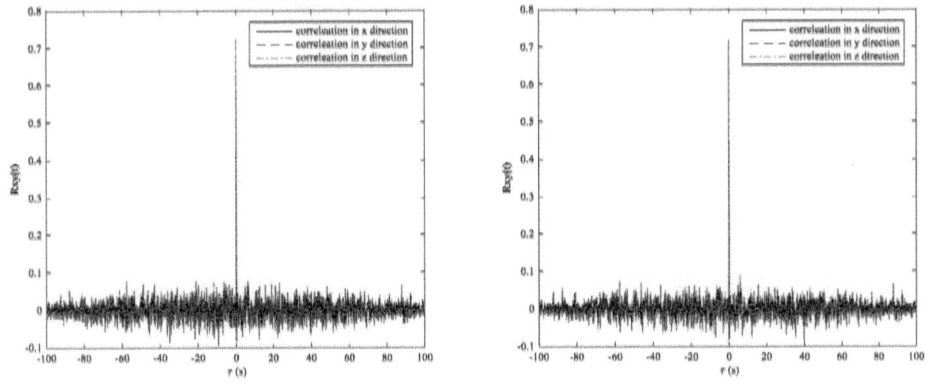

(a) (b)

Fig. 5. Correlation curves between noise and vibration accelerations in x, y and z directions (a) measuring point 1 (b) measuring point 2.

Table 4. Maximum correlation values between noise and vibration accelerations in x, y and z directions at the measuring point 1 and point 2.

Direction	x	y	z
Measuring point 1	0.628	0.653	0.725
Measuring point 2	0.706	0.665	0.721

In Fig. 5, τ is the time interval between the signals. Figure 5 and Table 4 show that the amplitudes of noise and vibration accelerations have the largest correlation 0.7 at $\tau = 0$, which means the variation laws of vibrations are consistent with the noise.

Through the analysis of the frequency domain of the noise of 100% rated load, it is known that the main characteristic frequencies are the multiples of 7.5 Hz, half of the shaft frequency. By analyzing the coherence of the noise and vibration accelerations, the correlations between them at x-direction are shown in Fig. 6. The results show that at the frequencies of 7.5 Hz and its multiplications, the signals show a high degree of coherence as the values are all greater than 0.9.

A high degree of the coherence between noise and vibration accelerations shows the noise is generated from the operation of timing gears.

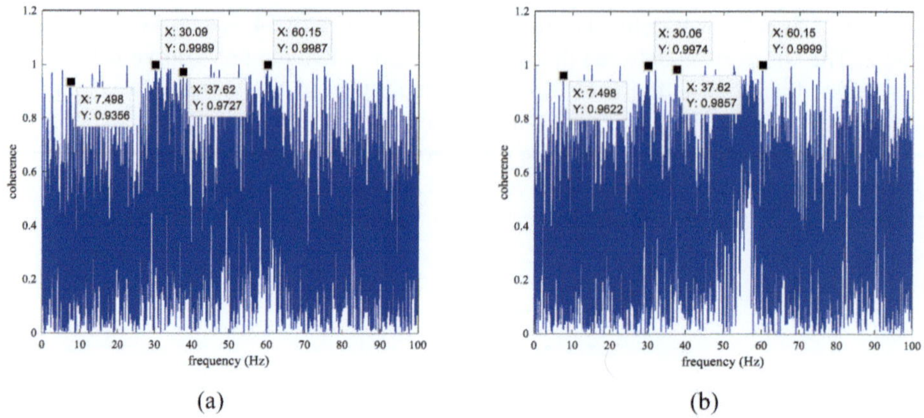

Fig. 6. Coherence of the noise and vibration accelerations at X direction of (a) measuring point 1 (b) measuring point 2.

3 Fault Reasoning Procedure

In this section, a fault diagnosis model is established to find the cause of the noise, as shown in Fig. 7. It considers not only the gears but also the effects of the associated structures, and all these possible causes will be checked one by one.

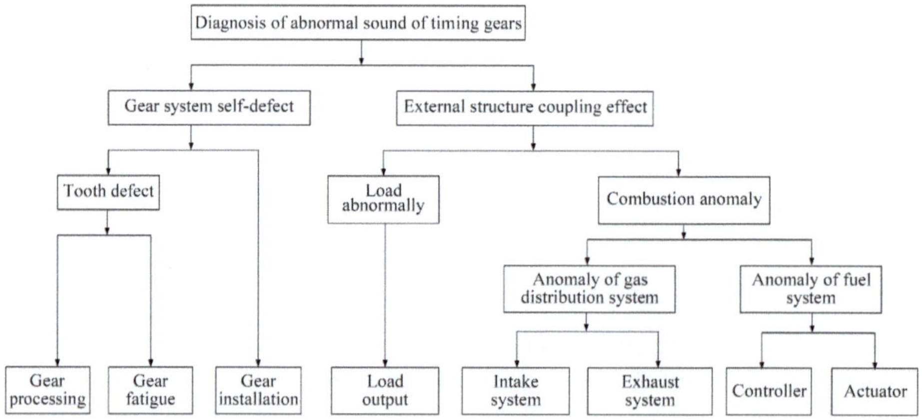

Fig. 7. Fault diagnosis model.

As already mentioned, it is ineffective to eliminate the noise by adjusting the center distance, tooth surface modification and replacing gears. Hence, the possibility of the gear system self-defect can be ruled out.

The operation parameters versus time of load can be obtained by detecting the current, voltage and power of the generator set, as shown in Fig. 8.

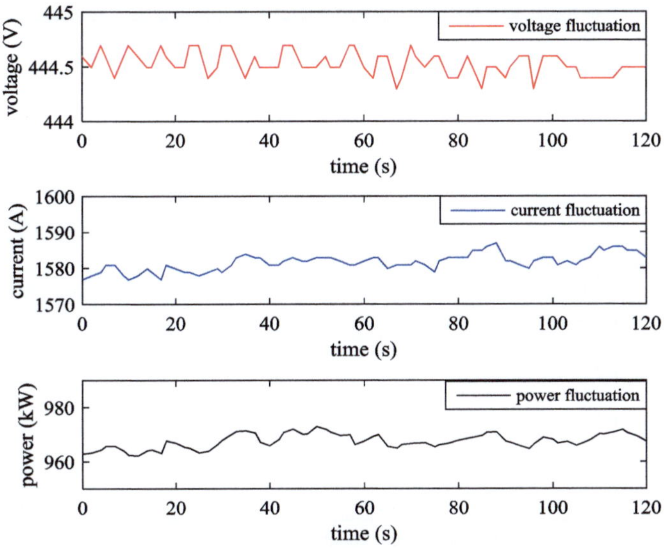

Fig. 8. Operation parameters versus time of the generator set.

It can be seen that the operation of the generator set is relatively stable in time domain. Hence, the possibility of the fault of generator set can be ruled out.

The running state of diesel engine mainly depend on the operation performance of gas distribution system and fuel system. No evident anomalies have been found by checking the structures of intake system and exhaust system. Hence, the possibility of the fault of gas distribution system can be ruled out.

The fuel system mainly includes controller and actuator, and PI control method is used to adjust the injection quantity. In the test, the PI parameters have been underwent several adjustments under 100% rated load, and the noise signals are also collected. The adjustments schemes are listed in Table 5, and the corresponding noise variances are shown in Fig. 9.

Table 5. Noise variances of different adjustments schemes under 100% rated load.

	P	I
Working condition 1	100%	100%
Working condition 2	50%	50%
Working condition 3	50%	150%
Working condition 4	150%	150%
Working condition 5	150%	50%

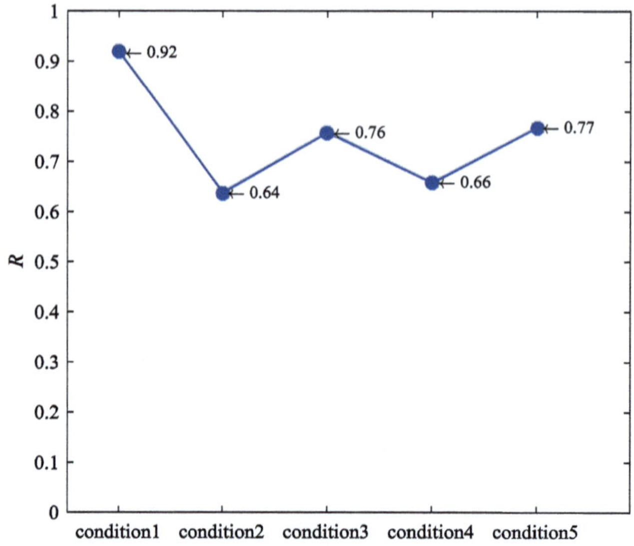

Fig. 9. Noise variances of different adjustments schemes under 100% rated load.

It can be seen that adjusting PI parameters can reduce the noise fluctuations in some degree, yet the effect is not remarkable. Hence, the PI parameters setting is not the main reason leading to the gears noise.

Through the above analysis, the possibilities of gear system self-defect, the generator set, gas distribution system and PI parameters of the controller have been ruled out. Hence, it can be speculated that the anomaly of injector may cause the gears noise. It is obvious that the anomaly of injector can affect cylinder combustion of the diesel engine and result the abnormal torsional vibration of shaft system, which will aggravate the timing gears vibration. Hence, the abnormal noise is generated here.

4 Problem Solved

By the analysis in Sect. 3, checking the fuel injection pumps to find that from right to left, the second injector is blackened with some carbon deposition, as shown in Fig. 10.

Replace this injector with a new one, and collect the noise signals under 100% rated load. The corresponding noise curves is shown in Fig. 11, and is compared with that before replacing. Table 6 shows the comparison the noise variance. It can be seen that the noise variances has decreased significantly, which verifies our conjecture.

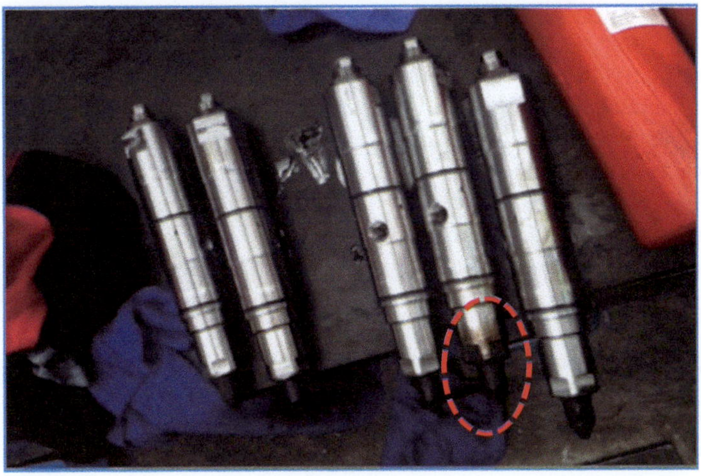

Fig. 10. Injectors.

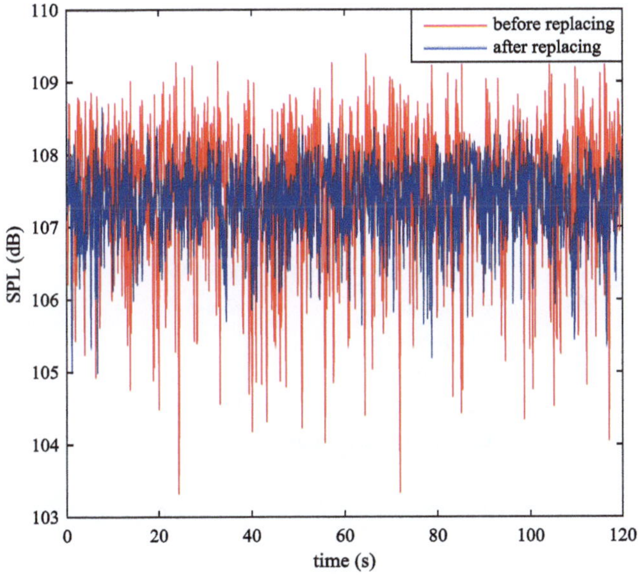

Fig. 11. Noise curves before and after replacing under 100% rated load.

Table 6. Noise variances before and after replacing under 100% rated load.

	Before replacing	After replacing
R	0.92	0.29

5 Conclusion

In this paper, a fault diagnosis model is established for noise generated by timing gears of a diesel generator set, which considers not only the gears but also the effects of the associated structures. After eliminating the possible reasons such as the defects of timing gears, the abnormality of the valve system, fuel system controllers and generator, the injector problem is suspected to cause the fault. The fuel injection pumps are inspected and the second injector is found to be blackened with some carbon deposition. By replacing the injector, the noise is gone. Hence, the proposed model provides a new way of thinking for the gear fault diagnosis.

Acknowledgement. The research work is supported by the National Natural Science Foundation of China (Grant No. 51805106), China Postdoctoral Science Foundation (Grant No. 2015M5 81433), Postdoctoral Science Foundation of Heilongjiang Province (Grant No. LBH-Z15038) and the Fundamental Research Funds for the Central Universities (Grant No. 3072019CFM0306).

Appendix

The parameters of the diesel generator used in this study are given in Table 7.

Table 7. Parameters of the diesel generator set.

Parameters	Values
RPM (@MCR)	900 rpm
Total power (@MCR)	1000 kW
Firing order	1-2-4-5-3
Number of strokes	4
Cylinder diameter	210.0 mm
Length of stroke	310.0 mm
Connecting rod ratio	0.223

References

1. Afia, A., Rahmoune, C., Benazzouz, D.: Gear fault diagnosis using Autogram analysis. Adv. Mech. Eng. **10**(12), 1–11 (2018)
2. Zhao, D., Li, J., Cheng, W., He, Z.: Generalized demodulation transform for bearing fault diagnosis under nonstationary conditions and gear noise interferences. Chin. J. Mech. Eng. **32**(1), 1–11 (2019)
3. Cai, J., Li, X.: Gear fault diagnosis based on time–frequency domain de-noising using the generalized S transform. J. Vib. Control **24**(15), 3338–3347 (2017)
4. Li, Z.C.: New detection method for gear faults based on kernel independent component analysis and BP neural network. Adv. Mater. Res. **909**, 371–374 (2014)

5. Wang, L., Ye, W., Shao, Y., Xiao, H.: A new adaptive evolutionary digital filter based on alternately evolutionary rules for fault detection of gear tooth spalling. Mech. Syst. Signal Process. **118**, 645–657 (2019)
6. Sun, R., Yang, Z., Chen, X., Tian, S., Xie, Y.: Gear fault diagnosis based on the structured sparsity time-frequency analysis. Mech. Syst. Signal Process. **102**, 346–363 (2018)
7. Yao, J., Xiang, Y., Qian, S., Wang, S., Wu, S.: Noise source identification of diesel engine based on variational mode decomposition and robust independent component analysis. Appl. Acoust. **116**, 184–194 (2017)
8. Zhao, Y.M., Cao, X.H., Guo, C.H., Ma, Q.Z.: Study on surface radiation noise of a diesel engine based on FEM. Adv. Mater. Res. **774–776**, 17–20 (2013)
9. Zhang, Z.G.: Experimental study on reducing diesel engine noise by using torsional vibration damper. Appl. Mech. Mater. **321–324**, 90–93 (2013)
10. Bhat, C.S., Meckl, P.H., Bolton, J.S., Abraham, J.: Influence of fuel injection parameters on combustion-induced noise in a small diesel engine. Int. J. Engine Res. **13**(2), 130–146 (2011)
11. Usuda, S., Otsuka, M., Nagata, M.: Noise and vibration reduction of newlydeveloped 3.0 l direct injection diesel engine. JSAE Rev. **23**(3), 285–289 (2002)

Adaptive Feature Selection for Enhancing Blade Damage Diagnosis on an Operational Wind Turbine

Artur Movsessian[1(✉)], David Garcia[1], and Dmitri Tcherniak[2]

[1] Department of Mechanical and Aerospace Engineering, University of Strathclyde, Glasgow, UK
artur.movsessian@strath.ac.uk

[2] Bruel & Kjaer Sound & Vibration Measurement, Skodsborgvej 307, 2850 Naerum, Denmark

Abstract. Monitoring wind turbine blades (WTB) is an important aspect when assessing the health of wind turbines. Structural Health Monitoring (SHM) systems enable continuous monitoring of the condition of WTB during operation. When SHM is coupled with advanced data analysis techniques, damage detection can be improved by customizing the methodologies to the structure being monitored. The work presented in this manuscript introduces an SHM methodology based on a Semi-supervised damage detection algorithm that uses preliminary findings to reinforce the selection of features used to identify damage. The methodology proposes a novel technique for feature extraction by sorting the acceleration values in each vibration response. Then, an adaptive feature selection algorithm is applied to identify the most sensitive characteristics of the feature for damage detection. This technique enhances the correlation between measurements of the same blade status and therefore the performance of the proposed SHM methodology. The methodology was implemented on real acceleration measurements on an operational Vestas V27 WTB. The results were compared with those from an alternative Semi-supervised methodology that considers only the measurements from the undamaged WTB. The comparison of the results demonstrated that the proposed adaptive feature selection algorithm enhances damage diagnosis.

Keywords: Feature selection · Damage detection · Structural Health Monitoring · Semi-supervised learning · Adaptive feature algorithm · Wind turbine blades

1 Introduction

Wind turbine blades (WTB) have grown both in size and complexity to enable the generation of more energy from a single wind turbine and this has made them a critical component of wind turbines. The design of a WTB is certified for lifetime period of 20 years [1]. However, this does not guarantee a failure-free operation and maintenance service is required. Structural Health Monitoring (SHM) systems are used for online, continuous and real-time early damage detection, thereby minimizing the risk of failure

M. A. Wahab (Ed.): DAMAS 2019, LNME, pp. 594–605, 2020.
https://doi.org/10.1007/978-981-13-8331-1_44

of WTB and reducing the downtime. These systems gather data from sensors mounted on the blade and transmit it to a supervisory control and data acquisition (SCADA) system. The main challenge is to detect and extract useful damage information hidden in the high frequent measurements collected by the SCADA system from the WTB.

A mature and robust technology for vibration-based SHM is built on recording vibration measurements such as acceleration signals. These sensors measure the dynamic responses of the WTB during testing and/or operation. Traditionally, such vibration measurements are used for physics-based modal analysis. However, they have been increasingly used in SHM data-driven approaches. Physics-based approaches are more appropriate for detailed examination of the behavior of a structure, whilst data-driven SHM applications are more suitable for automatic damage detection and localization [2]. The latter has a low computational demand when paired with advanced machine learning techniques making it an efficient method for online damage detection.

Depending on the data available about the structural damages, either supervised or unsupervised machine learning algorithms can be applied to detect damage on WTB. In [3], the authors successfully implemented a supervised Principal Component Analysis (PCA) -type algorithm on a laboratory case study with varying temperature and the potential presence of sprayed water and were able to select the optimal feature vector with high sensitivity to damage, thereby improving the overall damage detectability. In [4], supervised machine learning techniques such as Neural Networks, k-NN, and decision trees algorithms were applied to detect and diagnose delamination in a real WTB using experimentally obtained signals. The damage was successfully diagnosed by the chosen algorithms with Neural Networks performing the best. Despite the positive results of these studies, the main challenge remains to detect damage without having prior knowledge about its location or extent, e.g. during operation.

Supervised learning algorithms have been able to identify damage successfully. However, they require data to be available about the damage in the WTB to train the algorithms. This condition is rarely met during operation. Therefore, unsupervised learning algorithms seem more appropriate for SHM of real WTB. To overcome this limitation, unsupervised approaches have been applied. An unsupervised SHM methodology for WTB by using a single accelerometer was presented in [5]. The methodology is based in Singular Spectrum Analysis which is an extension of PCA for non-independent variables. The results demonstrated successful damage diagnosis for different damage sizes. Often semi-supervised methodologies are an alternative to unsupervised learning. In these cases, only the data from the reference state is labeled and used to train the algorithm. Semi-supervised methodologies do not require data from the damaged states and, therefore, are convenient in many applications where information about the damage is unknown or difficult to model [6]. A semi-supervised methodology was applied in [7], where accelerometers and a mechanical actuator were mounted on a WTB of a 225 kW Vestas V27 wind turbine. Artificial damage was introduced and increased from 15 cm to 45 cm. A semi-supervised learning methodology was able to successfully detect damage and track its growth.

To improve the performance of the SHM methodologies and therefore their reliability towards damage diagnosis, some studies have shown that by reducing the dimension of the data and extracting the features that enhance the damage detection is crucial. This has been studied in detailed in other fields of application as shown in [8], the authors highlighted the importance of feature selection in their empirical study. The aim was to identify the degradation of a slew bearing with a low rotational speed. The results showed that the selection of features was central to have reliable information about the condition of the bearing from vibration measurements and, therefore, to obtain a reliable health status classification. The health status classification can also be enhanced by adapting the features according to the performance of the methodology, as was demonstrated by [9]. The authors proposed an adaptive feature selection algorithm for SHM that accommodates signal perturbations caused by aging materials. This methodology enabled the reduction of the complexity and improved the performance of the algorithm by highlighting relevant segments of the features. It was tested on composite plates with simulated perturbations. The results showed that the standard algorithm accuracy decreased due to perturbations while the adaptive feature selection implementation was able to maintain successfully the level of correct performance.

This paper aims to improve damage detection and quantification in an operating WTB by developing an SHM methodology that implements an adaptive feature selection algorithm to enhance the detection of damage propagation. The algorithm uses the correlation between sorted vibration responses measured by accelerometers mounted on the blade of an operating Vestas V27 wind turbine aiming to find segments of the measurements that lead to early damage and propagation detection. Once damage symptoms are detected, the findings are used by the algorithm in an iterative and adaptive process to identify features in the vibration responses that are more sensitive to damage. This leads to better performance of the proposed SHM methodology and enhances the sensitivity to different damage scenarios. This methodology is more sensitive to damage propagation; thus, it eases the decision-making for building a more reliable SHM methodology.

The paper is organized as follows: Sect. 2 describes the implemented SHM methodology, Sect. 3 outlines the experiment setup and procedure, Sect. 4 summarizes the results, and, finally, Sect. 5 discusses the implications of the implemented SHM methodology and further opportunities for research.

2 Methodology

The SHM methodology presented in this paper aims to develop a semi-supervised learning algorithm to improve the damage diagnosis and reliability in the decision-making. The algorithm requires vibration measurements collected from sensors such as accelerometers installed on the structure under consideration create a reference state from which the new observations can be compared and eventually diagnosed.

The main steps of the methodology are summarized in Fig. 1 and include:

1. Trimmed discrete time-series measurements from each sensor (i.e. accelerometer) are sorted in a strictly increasing order;
2. Compute the covariance between the sorted vibration signals from the different accelerometer locations and obtain the feature vectors.
3. The feature vectors from the reference state are transformed by Principal Component Analysis (PCA);

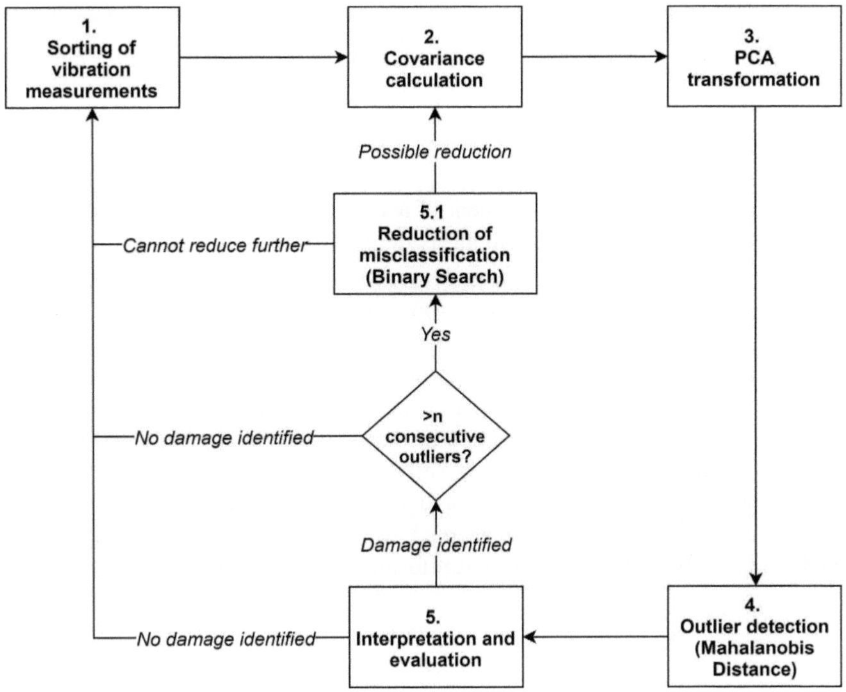

Fig. 1. Main methodological steps followed to identify and quantify the damage

4. A damage index based on the Mahalanobis distance between the feature vectors is calculated for damage diagnosis purposes;
5. The performance of the SHM methodology is evaluated by computing the corresponding confusion matrix against a defined threshold. If the damage indices are above the selected threshold, step 1 is repeated by feeding back the information obtained and hence adapt and enhance features towards the damage sensitivity. Then, new automatically adapted feature vectors are created to be used in the next steps of the methodology.

A detailed description of the procedure followed in these steps is provided in the following sections.

2.1 Data Collection and Sorting Vibration Measurements

Discrete time-series measurements gathered from the vibration measurements are organized in a matrix:

$$\boldsymbol{V}^k = \begin{pmatrix} v_{11} & v_{12} & \cdots & v_{1n} \\ v_{21} & v_{22} & \cdots & v_{2n} \\ \cdots & \cdots & \ddots & \vdots \\ v_{m1} & v_{m2} & \cdots & v_{mn} \end{pmatrix} = (v_{ij}) \in \mathbb{R}^{m \times n}, \quad (1)$$

where v_{ij} is the ith vibration measurement of accelerometer j for the kth realisation. The time series of a trimmed vibration signal is defined as:

$$\boldsymbol{v}_{*j} = \left[v_{1j}v_{2j}\ldots v_{mj}\right]^T \quad (2)$$

Given that this study uses the covariance between vibration measurements from sensors (i.e. accelerometers) to compare changes in response due to damage, it is important to account for possible delays in the readings of the sensors. A lag in the reading will certainly influence the covariance factor between the signals of the accelerometers and, thus, lead to wrong damage diagnosis. To avoid influenced or corrupted features, the data is transformed by sorting the absolute values of each trimmed signal $\boldsymbol{v}_{*j}$ strictly from minimum to maximum into $\boldsymbol{w}_{*j}$ such that:

$$\boldsymbol{w}_{*j} = \left[w_{1j}w_{2j}\ldots w_{mj}\right]^T \mathit{with}\ w_{ij} \leq w_{mj}\ \mathit{for\ all}\ i < j, \quad (3)$$

where $w_{ij} = |v_{\mathrm{ij}}|$. The resulting matrix of sorted signals is:

$$\boldsymbol{W}^k = \begin{pmatrix} w_{11} & w_{12} & \cdots & w_{1n} \\ w_{21} & w_{22} & \cdots & w_{2n} \\ \cdots & \cdots & \ddots & \vdots \\ w_{m1} & w_{m2} & \cdots & w_{mn} \end{pmatrix} = (w_{ij}) \in \mathbb{R}^{m \times n}\ \mathit{with} \quad (4)$$

The assumption made for this transformation is that decay in stiffness will be represented in a flatter slope at certain amplitudes, which would allow direct comparison of signals without the disturbance of lagged readings from the accelerometers. Sorting allows highlighting changes in certain amplitudes in consecutive order.

2.2 Covariance Matrix

To capture the behavior of the of the reference state, the covariance between signals is computed. The assumption is that the variance of a signal and the covariance between signals can be used as a reference to compare new measurement observations. Using the sorted vibration measurements $\boldsymbol{W}^k$, the covariance matrix is computed as follows:

$$\boldsymbol{S}^k = \begin{pmatrix} s_{11} & s_{12} & \cdots & s_{1n} \\ s_{21} & s_{22} & \cdots & s_{2n} \\ \vdots & \vdots & \ddots & \vdots \\ s_{n1} & s_{n2} & \cdots & s_{nn} \end{pmatrix} = (s_{jp}) \in \mathbb{R}^{n \times \mathrm{n}} \quad (5)$$

The variance-covariance between the sensors is given by:

$$s_{jp} = \frac{1}{n-1} \sum_{p=1}^{n} \sum_{j=1}^{n} \left(w_{ij} - \overline{w}_{*j}\right)\left(w_{ip} - \overline{w}_{*p}\right), \tag{6}$$

where s_{jp} is the covariance between sensor j and p and, for j = p, s_{jp} represents the variance. The mean of jth sensor measurements can be defined as:

$$\overline{w}_{*j} = \frac{1}{n} \sum_{1}^{n} w_{ij} \tag{7}$$

From the resulting covariance matrix $\boldsymbol{S}^k$, a feature vector $\boldsymbol{f}^k$ is constructed:

$$\boldsymbol{f}^k = \left[s_{11}s_{12}\ldots s_{1n}\cdots s_{22}s_{23}\ldots s_{2n}\cdots s_{33}s_{34}\ldots s_{3n}\cdots\cdots s_{nn}\right] \tag{8}$$

Where $\boldsymbol{f}^k := \{s_{ij} \in \boldsymbol{S}^k | j \geq i\}$ with $\dim(\boldsymbol{f}^k) = \frac{n(n+1)}{2}$. For further steps, the matrix consisting of all k realizations is defined as:

$$\boldsymbol{F} = \left[\boldsymbol{f}^1 \boldsymbol{f}^2 \ldots \boldsymbol{f}^k\right]^T \tag{9}$$

2.3 Principal Component Analysis

Principal Component Analysis (PCA) is a mathematical procedure that applies an orthogonal transformation to a set of possibly correlated variables and transforms it into a linearly uncorrelated set of variables known as principal components. This technique emphasizes variation and highlights potential strong patterns in a dataset. The first resulting principal component comprises most of the variability from the original dataset and each succeeding component as much of the remaining as possible. Although with this transformation the dimensionality of the dataset is reduced, most of the information in the original set is kept.

In this analysis, PCA is applied to reduce the dimensions of the dataset and highlight patterns, thereby minimizing the distances within clusters and maximizing the distance between them. Transforming $\boldsymbol{F}$ by applying PCA results in the following matrix:

$$\boldsymbol{F}_r = \left[\boldsymbol{f}_r^1 \boldsymbol{f}_r^2 \ldots \boldsymbol{f}_r^k\right]^T, \tag{10}$$

where r is the number of principal components and each $\boldsymbol{f}_r \in \mathbb{R}^{1 \times r}$ represents a new feature vector corresponding to an actuator hit with $r < dim(\boldsymbol{f}^k)$.

2.4 Mahalanobis Distance

The Mahalanobis distance (MD) measures the distance between two points for multiple variables. Unlike the Euclidean distance, the MD considers the correlation of the dataset. The most common use for the MD is to find multivariate outliers. Therefore, in this paper, the MD was used for the identification of damage and detection of outliers in the feature vectors resulting from the PCA transformation. The MD is calculated using the following equation:

$$\mathrm{d}(\boldsymbol{F}_r, \boldsymbol{Y}) = \sqrt{(\boldsymbol{F}_r - \boldsymbol{\mu}_{\boldsymbol{Y}})^T \boldsymbol{S}_{\boldsymbol{Y}}^{-1} (\boldsymbol{F}_r - \boldsymbol{\mu}_{\boldsymbol{Y}})} \tag{11}$$

where $\mathrm{d}(\boldsymbol{F}_r, \boldsymbol{Y})$ represents the Mahalanobis distance between the new feature vector $\boldsymbol{F}_r$ and the reference state created by the baseline matrix $\mathbf{Y}$, $\boldsymbol{\mu}_Y$ is the mean of the samples in $\mathbf{Y}$ and $\boldsymbol{S}_Y$ is the covariance between samples in $\mathbf{Y}$.

The threshold of the reference state during the training stage is represented by T_t. The threshold value T_t is estimated as follows:

$$T_t = max(d(\boldsymbol{F}_r, \boldsymbol{Y})), \tag{12}$$

where t is the tth percentile from the sorted MD values computed with Eq. 11. The values above this threshold are marked as potential outliers. This procedure allows for a factor $(1 - t)$ of outliers. An outlier or damage is defined as:

$$d_i\left(f_r^i, \mathbf{Y}\right) > T_t \tag{13}$$

2.5 Probability of Damage and Feedback of New Information

Once potential outliers have been identified, it is necessary to determine if the measurement is an actual outlier or damage. Since a single outlier cannot be defined as damage, it has been decided that n consecutive observations will be evaluated. Identifying n consecutive observations as outliers implies damage with a probability given by $(1 - t)^n$. Given a normal distribution of outliers results in a low probability of getting consecutive outliers. Identifying damage by the given probability the determined n consecutive outliers are treated as damage. In this sense, one of the novelties of the SHM methodology presented in this paper is to use the knowledge gained by the semi-supervised algorithm to improve the decision-making of the new measurements by adapting the algorithm to select the most relevant segments of the sorted vibration measurements.

Identifying the section of the sorted signal that is more sensitive to damage can contribute to maximizing the MD and therefore the damage index. This is based on the assumption that damage changes the slope of the sorted signals $\boldsymbol{W}^k$ in certain sections more than others. To identify the sections that are more sensitive to damage, a binary search is applied as illustrated in Fig. 2.

The sorted signal is divided into two equal parts and each half is evaluated as described in Sects. 2.2–2.4. If a section of the sorted signal that reduces misclassification and increases the MD between the n damaged observations and the undamaged observations it is chosen for the next stage of the binary search. The chosen section is divided again into two equal parts and the process repeats until no improvement in the classification is witnessed.

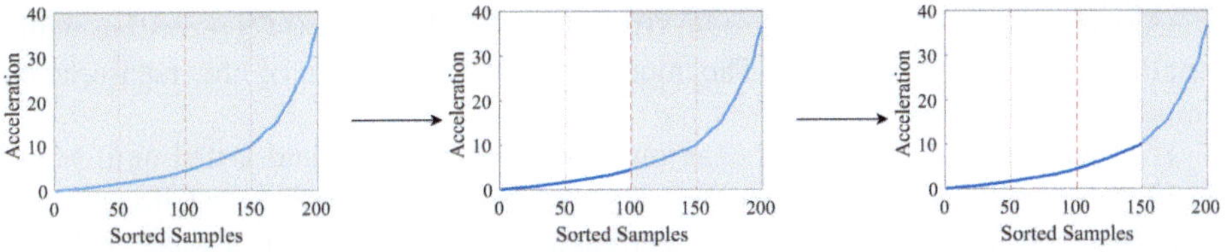

Fig. 2. Example of the binary search performed on sorted vibration measurements

New measurements from further actuator hits can now be evaluated using the section of the sorted vibration measurements that maximized the MD and minimized misclassification.

3 Experiment and Data

3.1 Setup and Procedure

The data for this study was generated during the experiment performed by [7]. The methodology was adjusted considering their conclusions selecting sensors mounted only on the edges. The object of study was a blade of the Vestas V27 wind turbine located on the campus of the Technical University of Denmark (DTU) during its operation. The V27 has a rotor diameter of 27 m, WTB length of 13.04 m and a rated power of 225 kW. Given its age, this wind turbine has only two operational regimens, namely 32 and 43 rpm. The data used to test the SHM methodology previously described was gathered during the operation mode of 32 rpm.

For the experiment, the blade was excited by an electro-mechanical actuator mounted 1 m away from the root on the surface of the blade. Driven by an electrical pulse, the actuator launches a plunger towards the surface of the blade. The blade's vibrations caused by the actuator's impact were measured by accelerometers mounted on the blade. Eight monoaxial piezoelectric accelerometers (Brüel & Kjær Type 4507B) were installed on the blade, as shown in Fig. 3. The accelerometers were

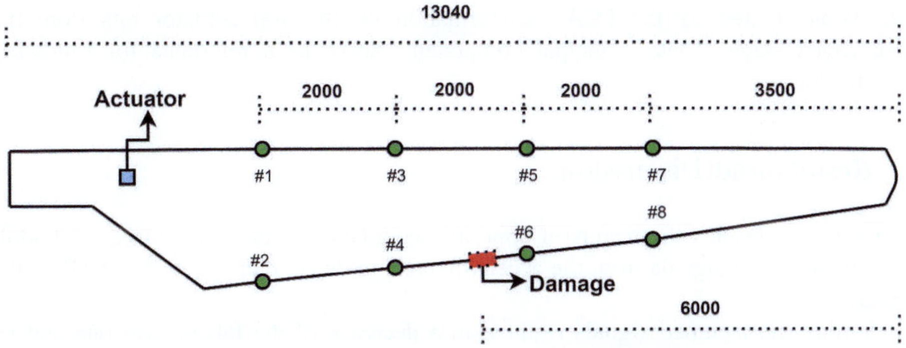

Fig. 3. Location of accelerometers mounted on the WTB from [7]

connected to a data acquisition system (Brüel & Kjær Type 3660-C) recording with a sampling frequency 16384 Hz. The more detailed description of the measurement campaign can be found in [7].

The measurement campaign began on November 28, 2014 and lasted until March 12, 2015. The actuator was triggered 12 times per hour (6 hits during the Christmas and New Year period) over the measurement campaign, i.e. data from a total of 24,693 actuator hits were recorded. The data acquisition system was programmed to record the data 10 s before the actuator impact and continue for another 20 s after the impact.

An artificial damage was introduced in the WTB on December 9 and then extended December 15 and January 6, by opening the trailing edge to simulate a crack of 15 cm, 30 cm, and 45 cm length, respectively. On January 19 the opening was repaired.

3.2 Data and Reference State

Each signal consists of 30 s of vibrational data sampled at 16,384 Hz frequency; this results in approximately 500,000 samples per sensor. A band-pass filter with the pass range 700–1200 Hz was applied to the time histories, similarly to [7]. The signals were trimmed to 200 samples; the samples right after the actuator impact were found the most useful to detect the damaged structure. The number of datasets available for each blade state for 32 rpm operational regime is summarized in Table 1.

Table 1. Summary of signals gathered during the operational mode of 32 rpm

	Undamaged	Damaged			Repaired	Total
		15 cm	30 cm	45 cm		
Signals	828[a]	66	117	105	500	1616

[a]400 signals(realizations) were used to create the reference state

A total of 828 datasets were recorded before damage was implemented in the WTB. To establish a reference for the Mahalanobis distance, 400 datasets were transformed into their principal components and used as a reference $\boldsymbol{F_r}$ and $\boldsymbol{Y}$ the following realizations in Eq. 11. Each following actuator hit was transformed into the reference dimension created by the PCA transformation on the 400 actuator hits from the undamaged stage. These principal components serve as a reference for following actuator hits.

4 Results and Discussions

This section presents the results of applying the adaptive, semi-supervised methodology on the sorted signals from the accelerometers to detect damage in the WTB of the Vestas V27.

Sorting the trimmed signals resulted in a decrease of the false alarm rate and an increase of the accuracy of detecting first damage as shown in Table 2. The results

Table 2. Comparison of results from different algorithms for damage detection during the operational mode of 32 rpm

Comparison	Total amount of observations	Initial (*cf.* Fig. 4a) (from [7])	Sorting	Variation Sorting vs. Initial	Sorting + Binary Search (*cf.* Fig. 4b)	Variation S + BS vs. Initial
False alarm	828	42	28	−33%	15	−64%
Overlap: Undamaged vs. 15 cm	66	61	25	−59%	11	−82%
Overlap: 15 cm vs. 30 cm	117	73	89	22%	1	−99%
Overlap: 30 cm vs. 45 cm	105	86	34	−60%	11	−87%
False alarm after repair	500	432	296	−31%	239	−44%

from training the algorithm with sorted signals show that approx. 33% less false alarms were registered. This technique prevents bias in the correlation factors between the features given that potential lags in signal response are prevented.

As a criterion to stop the binary search, the overlap between the first 5 consecutive outliers was considered and previous observations were compared. The smallest amount of overlaps identified during the binary search was after 3 iterations where the algorithm was trained with samples 150–200. Further iterations increased the error and, thus, were not considered. Due to sorting the values from lowest to highest, the range from 150 to 200 of these samples correspond to the highest accelerations recorded. Therefore, it can be concluded that sorting and trimming the vibration signals to 50 sorted samples by performing a binary search enable the identification of those responses more sensitive to damage and more efficient use of the information for damage detection and tracking purposes. By observing the results from the adapted algorithm presented in Table 2, it can be concluded that the newly selected feature from the vibration responses, provide a better separation between different damage stages. The results demonstrate a significant improvement of damage diagnosis, with up to 99% reduction of overlap between the different damage classes after adopting the feature and 33% less false alarms before adopting the feature. A graphical representation of the results is presented in Fig. 4.

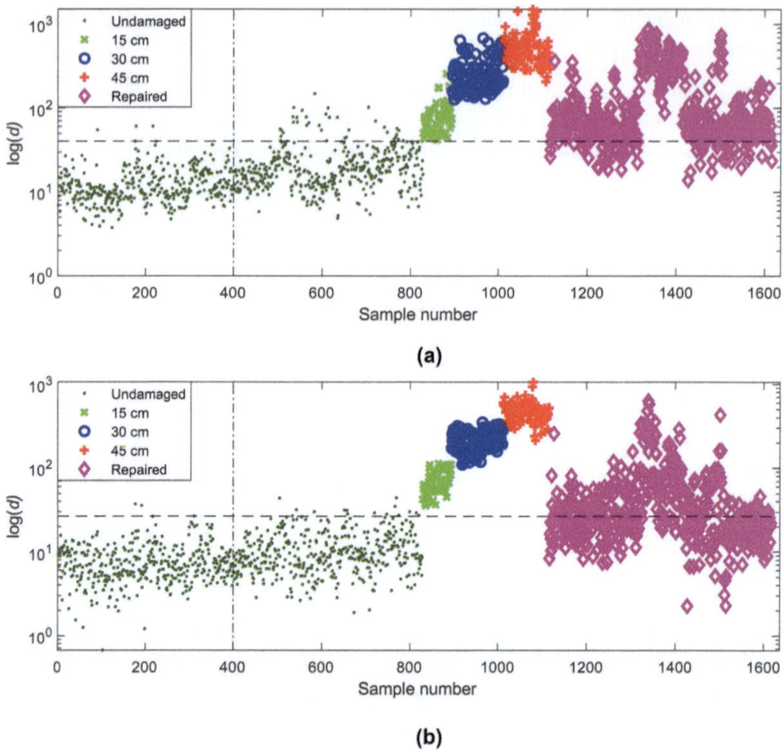

Fig. 4. Comparison of results from [7] (a) and this study (b)

5 Conclusions

The use of absolute sorted values to extract knowledge of the WTB through correlation showed a significant improvement. With additional adaption of the features after first damage was identified the response of the algorithm was more sensitive to the propagation of the damage. To the authors best knowledge sorted measurements were not used before for correlation and feature extraction, therefore the ongoing research is supposed to identify method's benefits and disadvantages. Binary search was successfully applied to the sorted measurements to identify the segments which reduced outliers and decreased the overlap between different damage scenarios. Nevertheless, further studies need to be carried out to evaluate different damage scenarios and how this SHM methodology performs on other WTB. Additional research on validation data is needed to be able to train the algorithm with ongoing observation, which is a constant improvement.

The focus of ongoing research is to build modules with a similar purpose as the binary search to implement in the main algorithm for constant improvement with the goal that SHM methodologies that are constantly reinforced by itself.

References

1. Coronado, D., Fischer, K., Fischer Bremerhaven, K.: Condition Monitoring of Wind Turbines: State of the Art, User Experience and Recommendations. Fraunhofer-IWES, Bremerhafen, Germany (2015)
2. Shahidi, G.: Physics-based and data-driven methods with compact computing emphasis for structural health monitoring (2016). https://preserve.lehigh.edu/etd/2801
3. Gómez González, A., Fassois, S.D.: A supervised vibration-based statistical methodology for damage detection under varying environmental conditions & its laboratory assessment with a scale wind turbine blade. J. Sound Vib. **366**, 484–500 (2016). https://doi.org/10.1016/j.jsv.2015.11.018
4. Jiménez, A.A., Gómez Muñoz, C.Q., García Márquez, F.P.: Machine learning for wind turbine blades maintenance management. Energies **11**, 1–16 (2018). https://doi.org/10.3390/en11010013
5. García, D., Tcherniak, D.: An experimental study on the data-driven structural health monitoring of large wind turbine blades using a single accelerometer and actuator. Mech. Syst. Sig. Process. **127**, 102–119 (2019). https://doi.org/10.1016/j.ymssp.2019.02.062
6. Chandola, V., Banerjee, A., Kumar, V.: Anomaly detection: a survey. ACM Comput. Surv. **41**(3), 15 (2009). https://doi.org/10.1145/1541880.1541882
7. Tcherniak, D., Mølgaard, L.L.: Active vibration-based SHM system: demonstration on an operating Vestas V27 wind turbine. In: 8th European Workshop On Structural Health Monitoring (EWSHM 2016). NDT.net, Bilbao, Spain (2016)
8. Caesarendra, W., Tjahjowidodo, T.: A review of feature extraction methods in vibration-based condition monitoring and its application for degradation trend estimation of low-speed slew bearing. Machines **5**, 21 (2017). https://doi.org/10.3390/machines5040021
9. Kessler, S.S., Agrawal, P.: Adaptive SHM methodology to accommodate ageing, maintenance and repair. In: Proceedings of the 6th International Workshop on Structural Health Monitoring, Stanford, CA (2007)

Condition Monitoring of Wind Turbines Using Adaptive Control Charts

Qinkai Han(✉) and Fulei Chu

The State Key Laboratory of Tribology, Tsinghua University, Beijing 100084, China
hanqinkai@hotmail.com, chufl@mail.tsinghua.edu.cn

Abstract. Due to the harsh working environments, variable speeds and alternating loads, wind turbines are likely to breakdown or suffer damage. Effective condition monitoring methods for wind turbines are essential for maintenance decisions which aim to reduce O&M costs. A typical supervisory control and data acquisition (SCADA) system records comprehensive wind turbine condition parameters, which would be fault informative. Thus, a framework for condition monitoring of wind turbines is introduced based on adaptive control charts and SCADA data. The adaptive exponential weighted moving average (AEWMA) is proposed for abnormal state alarm of wind turbines. Random forest (RF) is used for feature selection and regression prediction to establish the normal condition prediction model (NCPM) of wind turbine with fault-free SCADA data. The performance and robustness of various control charts are compared comprehensively. Compared with the exponential weighted moving average (EWMA) control charts, the AEWMA control chart behaves more sensitive to the abnormal state, and thus has more effective anomaly identification ability and better robustness.

Keywords: Computational geometry · Graph theory · Hamilton cycles

1 Introduction

Due to the harsh working environments, variable speeds and alternating loads, wind turbines are likely to breakdown or suffer damage. The high operation and maintenance (O&M) cost of wind turbines has significantly affected renewable energy utilization. Consequently, effective condition monitoring methods for wind turbines are essential for maintenance decisions which aim to reduce O&M cost [1]. Various signals, such as vibration, strain, acoustic emission, lubrication oil parameters, etc, have been utilized for condition monitoring of wind turbines [2]. However, these approaches require the installation of additional sensors and data acquisition devices, which increases the capital cost and wiring complexity of the wind turbine system. The supervisory control and data acquisition (SCADA) systems have been installed in most of modern wind turbines

M. A. Wahab (Ed.): DAMAS 2019, LNME, pp. 606–617, 2020.
https://doi.org/10.1007/978-981-13-8331-1_45

to monitor the operational performance. A typical SCADA system records comprehensive wind turbine condition parameters, including the temperatures (e.g., bearing temperature, oil temperature), wind parameters (e.g., wind speed, wind direction) and energy conversion parameters (e.g., output power, pitch angle, rotor speed), which would be fault informative. Since no additional sensors or data acquisition devices are needed, condition monitoring of wind turbines based on the SCADA data is a cost-effective approach to improve the reliability of wind turbines [2].

Building a model to predict the normal behavior of SCADA parameters is the first issue of condition monitoring of wind turbines. By using advanced SCADA data mining methods, various normal condition prediction models (NCPMs) have been developed to detect the significant changes in wind turbine behavior prior to anomaly occurrences [3,4]. After detailed comparisons based on the SCADA data collected at a large wind farm, they found that the random forest (RF) algorithm models provided the best accuracy [3]. For a given NCPM, the relationship between the input and output SCADA state variables of the wind turbine could be learned. Subsequently, the departure of the current turbine state from the NCPM could be measured online and yield a time series of residuals. The control charts from statistical process control (SPC) are a time-honored tool to monitor the residuals [5]. If the residuals are statistically different from a normal (or fault-free) reference, the process is said to be out of control and a alarm would be raised accordingly. In recent years, the NCPM combined with control charts has been increasingly used in fault diagnosis of wind turbines [6–10]. Most of these studies used the Shewhart type control charts, which have been proved to be pretty effective for detecting greater shifts [5]. However, they are slow in reacting to small and moderate shifts in the process mean. In that regard, the exponential weighted moving average (EWMA) control chart was developed to provide more sensitivity to small mean shifts [5]. Cambron et al. [11,12] first applied the EWMA control chart for condition monitoring of wind turbines. Through several applications on real SCADA data, the results shown that a shift of 3.4% in annual energy production over a period of five years could be detected in time to plan proper maintenance.

Although the EWMA control chart can provide greater sensitivity to small shifts, it is not as effective as the Shewhart chart when the shifts in the process mean level are relatively large due to what is known as the inertia problem [13]. In actual applications, such as the monitoring of wind turbines, the shift of the residuals from the NCPM is unknown, which might cause the EWMA control chart insufficient if the larger shift appears. To overcome the inertia problem, Cappizzi and Mastrotto [14] first presented an adaptive EWMA (AEWMA) by adaptively adjusting the weight on past observations according to a function of the prediction error. Later, Shu [15] extended the idea of AEWMA chart on monitoring process locations to the case of monitoring process dispersion. The AEWMA chart is a smooth combination of the Shewhart and EWMA charts, and thus can diminish the inertia problem. After examples on the capsules weights data and simulated data, both Cappizzi and Masarotto [14] and Shu [15] showed that the AEWMA control chart is able to offer an overall good detection performance against shifts of different sizes. However, in the condition monitoring of wind turbines, the residual data would be more complicated, and whether the AEWMA control chart could hold better performance than the EWMA chart is still unknown. To the authors' knowledge, the AEWMA control chart has not been used in the condition monitoring of wind turbines in the open literature.

Thus, a framework for condition monitoring of wind turbines is introduced based on AEWMA and SCADA data. The remainder of this paper is organised as follows. Section 2 introduces the proposed control charts, Sect. 3 describes feature selection and regression prediction on the SCADA data acquired from an operating wind turbine, Sect. 4 gives several condition monitoring examples and discusses the results. Finally some conclusions are summarized in Sect. 5.

2 AEWMA Control Charts

Monitoring data that obeys the same distribution is represented by $X_{i,1}, X_{i,2}, \cdots, X_{i,n}$, where $i = 1, 2, \cdots$ is the sampling time and n is the size of each sample. The mean and variance of the data are denoted by μ_0 and σ_0^2, respectively. When the process is out of control, the mean of the data becomes $\mu_1 = \mu_0 + \delta\sigma_0$, in which δ is the shift parameter. We define the mean of the sample data as $\bar{X}_i = \frac{1}{n}\sum_{j=1}^{n} X_{ij}$, and thus the EWMA statistics for monitoring mean shift of the sample data could be written as follows

$$Y_i = \lambda X_i + (1 - \lambda)Y_{i-1} \tag{1}$$

where λ is the smoothing parameter, and $0 < \lambda < 1$. Without loss of generality, we can let $Y_0 = 0$. Lucas and Saccucci [16] pointed out that for smaller value of λ, the EWMA statics can detect a smaller mean shift faster. When λ takes a greater value, the EWMA statistic would have a good sensitivity to the larger mean shift. Theoretically, the EWMA control charts can be customized to detect specific shifts in the process.

However, for the actual wind turbine monitoring data, the mean shift is usually fluctuated in a certain range. The designed value of λ is difficult to adapt to the change of the actual mean shift. To overcome the inertia problem, the AEWMA statistic is proposed by [14]

$$Y_i = Y_{i-1} + \phi(e_i) \tag{2}$$

in which $e_i = X_i - Y_i$ is the error term, and $\phi(\cdot)$ represents the score function. Note that for $e_i \neq 0$, the AEWMA statistic can be rewritten as

$$Y_i = w(e_i)X_i + (1 - w(e_i))Y_{i-1} \tag{3}$$

where $w(e_i) = \frac{\phi(e_i)}{e_i}$ is the equivalent smoothing parameter. Obviously, AEWMA statistic can adaptively adjust the weight of the estimate value at the past time (Y_{i-1}) according to the prediction error at the current time. Thus, it can balance the requirements of various mean shifts to the smoothing parameters. Yashchin

[13] suggested the Huber function as the score function, and its expression is given by

$$\phi(e) = \begin{cases} e + (1-\lambda)\gamma, & e < -\gamma; \\ \lambda e, & |e| \le \gamma; \\ e - (1-\lambda)\gamma, & e > \gamma; \end{cases} \tag{4}$$

where γ is the error limit. The static Y_i also obeys the same distribution with $\bar{X}$ and has the same mean value with $\bar{X}$. When the sampling size n is large enough, the variance of Y_i can be expressed as $\sigma_Y^2 \approx \sigma_{\bar{X}}^2$, where $\sigma_{\bar{X}}^2$ is the variance of $\bar{X}$, and we have $\sigma_{\bar{X}}^2 = \frac{\sigma_0^2}{n}$. Therefore, the upper control limit (UCL) and lower control limit (LCL) of the AEWMA control chart could be expressed as follows

$$\begin{aligned} UCL_{AEWMA} &= \mu_0 + k\sigma_{\bar{X}} \\ LCL_{AEWMA} &= \mu_0 - k\sigma_{\bar{X}} \end{aligned} \tag{5}$$

where k is the control limit parameter. From Eqs. (3–5), one can find that three parameters: λ, γ, k should be determined to obtain the control limits of AEWMA control charts. Determination of these parameters will be discussed in the following section. It is noticed that for $\gamma \to \infty$ we have $\phi(e_i) = \lambda e_i$ and $w(e_i) = \lambda$. In this case, the AEWMA statistic degenerate into EWMA statistic, and its control limits can be expressed as

$$\begin{aligned} UCL_{EWMA} &= \mu_0 + k\sigma_{\bar{X}}\sqrt{\frac{\lambda}{2-\lambda}} \\ LCL_{EWMA} &= \mu_0 - k\sigma_{\bar{X}}\sqrt{\frac{\lambda}{2-\lambda}} \end{aligned} \tag{6}$$

Three parameters, including λ, γ, k, should be determined for the AEWMA control charts. Obviously, selection of λ, γ plays a key role in the performance of AEWMA control charts. Generally, lower value of λ or greater value of γ should be selected for small mean shift, while greater value of λ or lower value of γ would be favorable for detecting large mean shift. Therefore, the design of AEWMA control charts is a multi-objective optimization problem. Cappizzi and Mastrotto [14] utilized the simulated annealing algorithm (SAA) for the parameter optimization of AEWMA control charts. However, the requirement of initial value of SAA is relatively high. Once the initial value significantly deviates from the optimal value, it is difficult to converge to the optimal value. In order to improve the convergence speed of SAA, Shu [15] proposed a "two step method". First, the AEWMA control chart is treated as a conventional EWMA control chart, and the optimal value of λ is obtained under certain ARL_0 using SAA. Then, on the premise of given value of λ, the value of γ is optimized. Detailed optimization procedure could be found in Ref. [15].

3 Feature Selection and Regression Prediction on Fault-Free SCADA Data

In this section, RF is used for feature selection and regression prediction to construct the NCPM of wind turbines with fault-free SCADA data. This study is aimed at monitoring and diagnosis of doubly fed wind turbines with rated power of 1.5 MW. Usually, the SCADA data of the unit include: output power, speed, torque, temperature, pitch angle, etc. The data record interval is 10 min. In order to correctly establish the NCPM of wind turbines, the anomaly data should be avoided as much as possible. By reading the record table of SCADA system, it was found that no anomaly was reported in the time period of 12/26/2013–2/12/2014. The wind turbine unit was built and connected to the grid in early 2012. In this time period, the unit has passed the initial running stage, and is in the stage of normal power generation. Therefore, the data segment is ideal for MRA to construct the NCPM of wind turbines. There are 45 variables recorded by the SCADA system. After excluding the lost data points and data points during the maintenance downtime, the total amount of data is 6135 points.

In our study, we choose the output power as the response variable and the remaining 44 variables as argument variables. But in the process of actual modeling, it is necessary to select a variable subset (feature selection) which has the best ability to explain the response variable to improve the regression and prediction accuracy of the NCPM. Before feature selection, the raw SCADA data should be standardized as follows

$$Z_{ij} = \frac{x_{ij} - \min(x_j)}{\max(x_j) - \min(x_j)} \tag{7}$$

where x_{ij} is the ith sample point of the jth variable. $\max(x_j)$ and $\min(x_j)$ are, respectively, denote the maximum and minimum values of the jth variable.

RF is an integrated machine learning method [17]. It uses random resampling technology bootstrap and node random splitting technology to construct multiple decision trees, and the final classification results are obtained by voting. RF has the ability to analyze the classification characteristics of complex interaction. It has good robustness for noise data and has a faster learning speed. Its variable importance measure can be used as a feature selection tool for high dimensional data. The core algorithm uses the RF package in R software, in which the parameter takes the value of $\sqrt{n}$ recommended by Breiman [17] (n is the number of features of the training data set). The number of trees is set to be $n_{tree} = 500$. Thus, the feature selection based on RF is carried out, and the top 15 features are shown in Table 1. Figure 1 also presents the comparison between the source SCADA data and regression prediction results. For the sake of simplicity, only four argument variables (the rotation torque, generator current, the average wind speed in 10 min and generator speed) on the response variable (output power of the wind turbine unit) are illustrated in the figure.

Table 1. Feature selection result based on RF.

Ranking	Features	Uncontrollable variable	Controllable variable
1	Rotation torque	+	
2	Generator phase A current	+	
3	Average wind speed in 10 min	+	
4	Generator speed	+	
5	Rotor speed	+	
6	Blade yaw angle		+
7	Generator temperature	+	
8	Gearbox bearing temperature	+	
9	Cabin angle		+
10	Generator phase A voltage	+	
11	Cabin speed	+	
12	Gearbox temperature	+	
13	Ambient temperature	+	
14	Cabin temperature	+	
15	Bearing temperature	+	

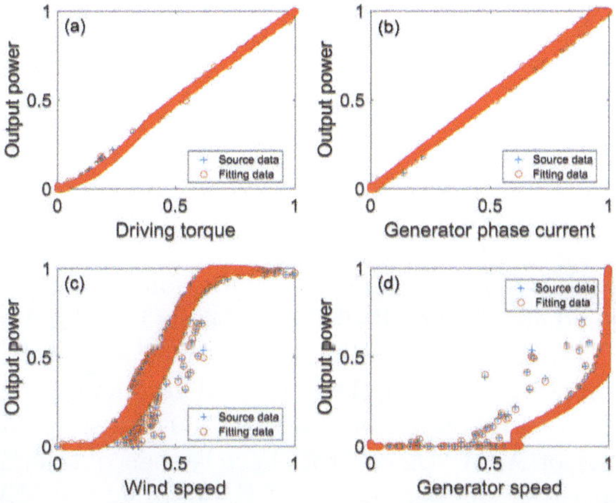

Fig. 1. Comparison between the source SCADA data and regression prediction results: (a) rotation torque; (b) generator current; (c) average wind speed in 10 min and (d) generator speed.

4 Condition Monitoring Examples

Based on the feature selection and regression prediction results, condition monitoring on the wind turbine unit is carried out. During the period of 12/1/2015–6/1/2016, there were three anomalies, namely, the generator brush worn, gearbox running hot in low generator stage and the shaft bearing overtemperature.

Table 2. Anomaly information for the wind turbine unit during the period of 12/1/2015–6/1/2016.

Anomaly no.	Anomaly description	Monitored period	Number of data points	Time of alarm log
1	Generator brush worn	12/3/2015 19:10–12/7/2015 6:30	500	12/6/2015 20:30 (alarm log: point no. 440)
2	Gearbox running hot in low generator stage	2/1/2016 00:00–2/4/2016 11:10	500	2/3/2016 22:10 (alarm log: point no. 421)
3	Shaft bearing overtemperature	5/28/2016 15:50–6/1/2016 3:10	500	5/31/2016 15:20 (alarm log: point no. 429)

Table 3. Optimization results of AEWMA and EWMA control charts for abnormal state alarm.

	ARL_0	Mean shifts	λ	γ	k
AEWMA	500	$\delta_1 = 0.4, \delta_2 = 4$	0.1	5.4750	1.3531
EWMA-1	500	$\delta = 0.4$	0.1	–	6.8110
EWMA-2	500	$\delta = 2$	0.2	–	5.9430
EWMA-3	500	$\delta = 4$	0.4	–	6.9531

The specific time of alarm log is shown in Table 2. For each anomaly, the number of monitored data points is 500, and the exact anomaly data point is also given in the table for comparisons.

By using the NCPM model obtained in the previous section, the output power of the unit before and after the fault (500 data points in Table 2) is predicted and then the residual is obtained by measuring the difference with the real output power. The mean and variance of the predicted residuals for three fault data are all less than 0.05 and 0.08. Given $ARL_0 = 500$ and shift range (0.4–4), the optimal parameters of AEWMA control chart is then obtained, and shown in Table 3. For comparison, the parameters of the optimal EWMA control charts corresponding to different shifts (0.4, 2, 4) are also given in the table.

As mentioned before, the out-of-control ARL is an important index to evaluate the performance of control charts. Figure 2 shows the variation of ARL_1 with mean shift in the range of 0–4 for the designed control charts in Table 3. Obviously, when the shift is zero, the out-of-control ARL is equal to the in-control ARL, i.e. $ARL_1 = ARL_0$. With the increase of shift, the value of ARL_1 gradually decreases. Under small shifts ($\delta < 1.5$), the value of ARL_1 of the AEWMA is lower than that of the EWMA control charts, especially for the EWMA control chart with larger smoothing parameters (EWMA-2 and EWMA-3). This means that the AEWMA behaves more sensitive and could give earlier warn of abnormal states than the EWMA control charts. When the shift becomes large enough

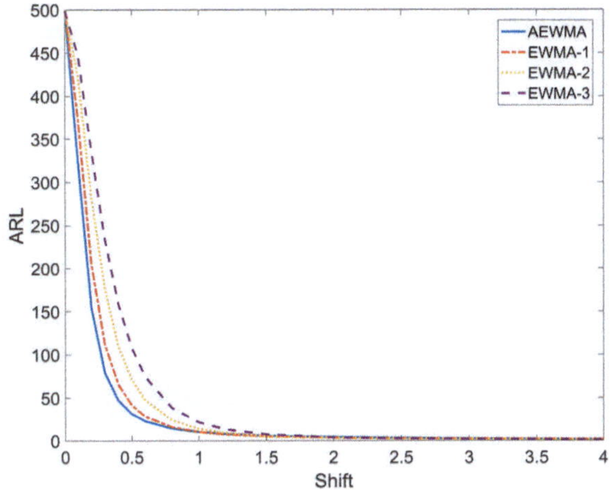

Fig. 2. Variation of ARL_1 with mean shift in the range of 0–4 for the designed control charts.

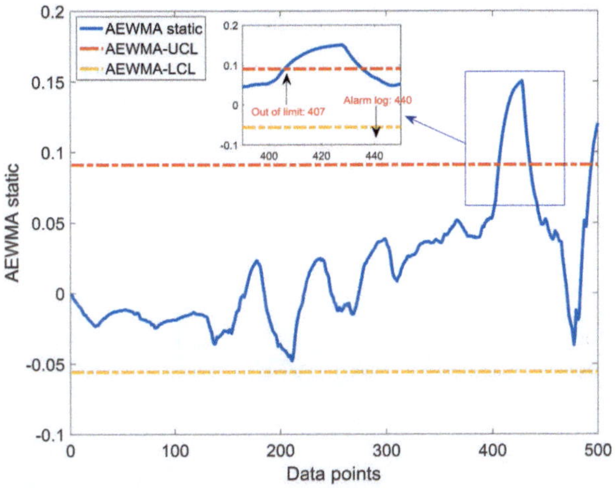

Fig. 3. Residuals with anomaly 1 monitored by AEWMA control chart.

($\delta > 2$), the difference of ARL_1 between the AEWMA and EWMA control charts is not distinct, indicating that under large shift the performance comparable to the EWMA control charts can still be maintained by the AEWMA. This is consistent with the theoretical expectation of the AEWMA.

The AEWMA and EWMA control charts are established for the output power residuals with fault 1, as shown in Figs. 3 and 4. Figures 5, 6, 7 and 8 are the results of control charts containing fault 2 and 3, respectively. It can be seen from the figures that the AEWMA control chart can effectively identify the abnormal state caused by the fault. Compared with the alarm log of SCADA system,

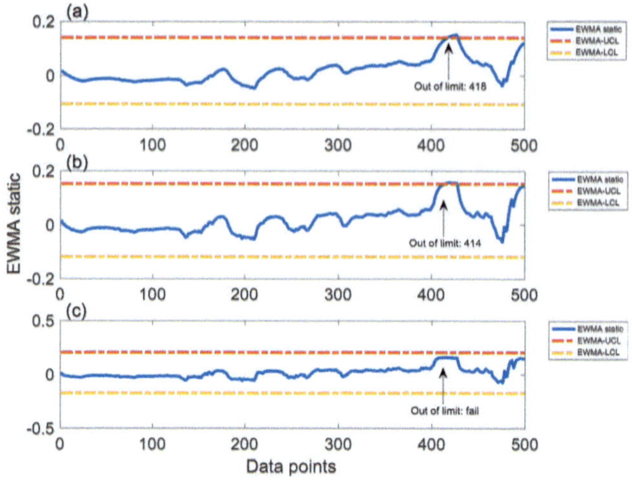

Fig. 4. Residuals with anomaly 1 monitored by various EWMA control charts: (a) EWMA-1; (b) EWMA-2 and (c) EWMA-3.

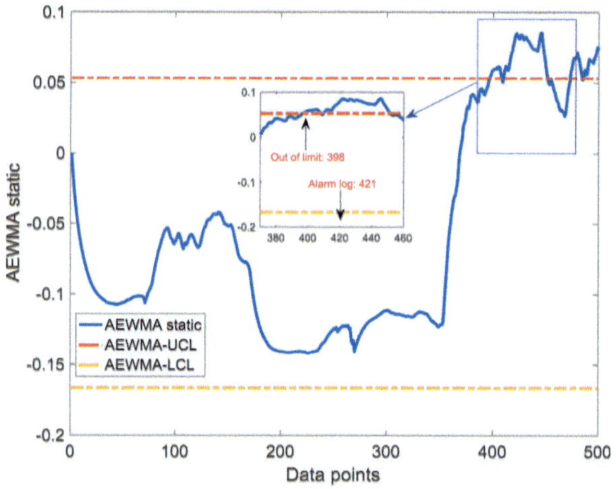

Fig. 5. Residuals with anomaly 2 monitored by AEWMA control chart.

the AEWMA control chart could send the alarm in early time. For fault 1 (see Fig. 3), one can see that the AEWMA alarm time is $(440 - 407) * 10 = 330\,\text{min}$ (about 5.5 h) ahead. For fault 2 and 3 (see Figs. 5 and 7), the time of advance is about $(421 - 398) * 10 = 230\,\text{min}$ (about 3.8 h) and $(429 - 407) * 10 = 220\,\text{min}$ (about 3.7 h), respectively. Thus, one can say that the alarm time of AEWMA control chart could be several hours ahead of the SCADA system, and in our study the maximum ahead of time appears in fault 1 (about 5.5 h).

Compared with the AEWMA control chart, the EWMA control charts behave less sensitive to fault and have poor robustness. For the EWMA-3 of fault 1 (see

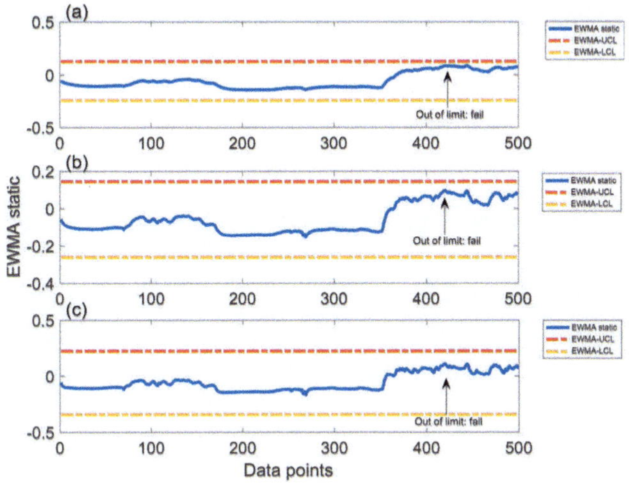

Fig. 6. Residuals with anomaly 2 monitored by various EWMA control charts: (a) EWMA-1; (b) EWMA-2 and (c) EWMA-3.

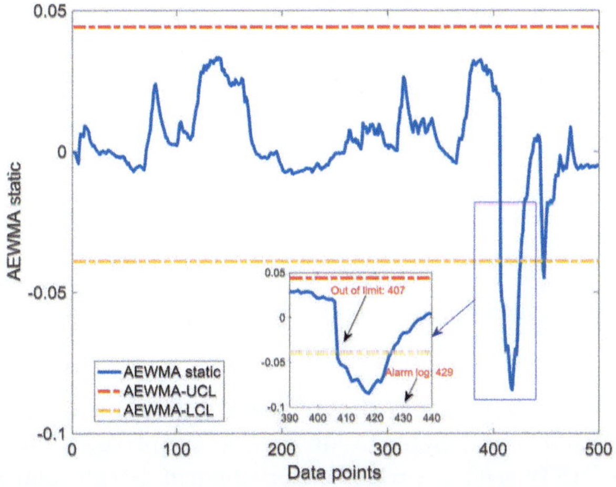

Fig. 7. Residuals with anomaly 3 monitored by AEWMA control chart.

Fig. 4(c)) and EWMA-1, EWMA-2, EWMA-3 of fault 2 (see Fig. 6(a, b and c)), the abnormal state is not identified during the monitoring period. Among the rest EWMA control charts, although the faults are alarmed earlier than the SCADA system, the alarm time still lags behind the AEWMA control chart. For fault 1, the EWMA-2 control chart (see Fig. 4(b)) sends the earliest alarm, about $(440 - 414) * 10 = 206$ min (about 4.3 h), which is still lagging behind the AEWMA control chart for $(5.5 - 4.3) = 1.2$ h. For fault 3, the EWMA-2 control chart (see Fig. 8(b)) has the best performance and the ahead of time

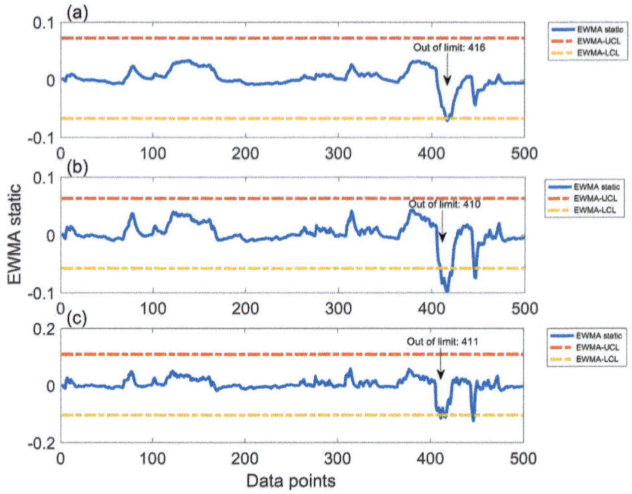

Fig. 8. Residuals with anomaly 3 monitored by various EWMA control charts: (a) EWMA-1; (b) EWMA-2 and (c) EWMA-3.

is $(429 - 410) * 10 = 190\,\mathrm{min}$ (about 3.2 h), which is still lagging behind the AEWMA control chart for $(3.7 - 3.2) = 0.5\,\mathrm{h}$.

From the above condition monitoring examples, one can say that compared with the EWMA control charts, the AEWMA control chart behaves more sensitive to the abnormal state. Thus, it can effectively identify the abnormal state and has better robustness. This is of great application value to the condition monitoring of practical wind turbine units.

5 Conclusions

A framework for condition monitoring of wind turbines is introduced based on adaptive control charts and SCADA data. The adaptive exponential weighted moving average (AEWMA) is proposed for abnormal state alarm of wind turbines. Random forest (RF) is used for feature selection and regression prediction to establish the normal condition prediction model (NCPM) of wind turbine with fault-free SCADA data. The performance and robustness of various control charts are compared comprehensively. Compared with the exponential weighted moving average (EWMA) control charts, the AEWMA control chart behaves more sensitive to the abnormal state, and thus has more effective anomaly identification ability and better robustness.

Acknowledgments. The research work described in the paper was supported by the National Science Foundation of China under Grant No. 11872222, and the State Key Laboratory of Tribology under Grant No. SKLT2019B09.

References

1. Garcia Marquez, F.P., Tobias, A.M., Pinar Perez, J.M., Papaelias, M.: Condition monitoring of wind turbines: techniques and methods. Renew. Energy **46**, 169–178 (2012)
2. Qiao, W., Lu, D.: A survey on wind turbine condition monitoring and fault diagnosis, part II: signals and signal processing methods. IEEE Trans. Ind. Electron. **62**(10), 6546–6557 (2015)
3. Kusiak, A., Verma, A.: A data-mining approach to monitoring wind turbines. IEEE Trans. Sustain. Energy **3**, 150–157 (2012)
4. Verma, A., Kusiak, A.: Fault monitoring of wind turbine generator brushes: a data-mining approach. J. Solar Energy Eng. **134**, 021001-1–021001-9 (2012)
5. Montgomery, D.: Introduction to Statistical Quality Control. Wiley, Hoboken (2007)
6. Marvuglia, A., Messineo, A.: Monitoring of wind farms' power curves using machine learning techniques. Appl. Energy **98**, 574–583 (2012)
7. Yampikulsakul, N., Byon, E., Huang, S., Sheng, S., You, M.: Condition monitoring of wind power system with nonparametric regression analysis. IEEE Trans. Energy Convers. **29**, 288–299 (2014)
8. Taslimi-Renani, E., Modiri-Delshad, M., Elias, M., Rahim, N.A.: Development of an enhanced parametric model for wind turbine power curve. Appl. Energy **177**, 544–552 (2016)
9. Wang, L., Zhang, Z., Long, H.: Wind turbine gearbox failure identification with deep neural networks. IEEE Trans. Ind. Inform. **13**, 1360–1368 (2017)
10. Yang, H., Huang, M., Lai, C., Jin, J.: An approach combining data mining and control charts-based model for fault detection in wind turbines. Renew. Energy **115**, 808–816 (2018)
11. Cambron, P., Lepvrier, R., Masson, C., Tahan, A., Pelletier, F.: Power curve monitoring using weighted moving average control charts. Renew. Energy **94**, 126–135 (2016)
12. Cambron, P., Masson, C., Tahan, A., Pelletier, F.: Control chart monitoring of wind turbine generators using the statistical inertia of a wind farm average. Renew. Energy **116**, 88–98 (2018)
13. Yashchin, E.: Estimating the current mean of a process subject to abrupt changes. Technometrics **37**, 311–323 (1995)
14. Capizzi, G., Masarotto, G.: An adaptive exponentially weighted moving average control chart. Technometrics **45**, 199–207 (2003)
15. Shu, L.: An adaptive exponentially weighted moving average control chart for monitoring process variances. J. Stat. Comput. Simul. **78**, 367–384 (2008)
16. Lucas, J., Saccucci, M.: Exponentially weighted moving average control schemes: properties and enhancements. Technometrics **32**, 1–29 (1990)
17. Breiman, L.: Random forests. Mach. Learn. **45**, 5–32 (2001)

A New Intelligent Fault Diagnosis Method and Its Application on Bearings

Yi Sun, Hongli Gao(✉), Liang Guo(✉), Xin Hong, Hongliang Song, Jiangquan Zhang, and Lei Li

School of Mechanical Engineering Research Center of Advanced Driving Energy-Saving Technology, Ministry of Education, Southwest Jiaotong University, Chengdu 610031, China
guoliang@swjtu.edu.cn

Abstract. Fault diagnosis is vital in manufacturing system, however, fault diagnosis is divided into three stages: signal preprocessing, feature extraction and fault classification, which destroys the relationship between each stage and causes a part of the loss of fault information. The feature extraction process depends on the experimenter's experience, and the recognition rate of the shallow diagnostic model does not achieve satisfactory results. In view of this problem, this paper proposes a method, the first step is converting raw signals into two-dimensional (2-D) images, the step can extract the features of the converted 2-D images and eliminate the impact of expert's experience on the feature extraction process. Next, an intelligent diagnosis algorithm based on convolutional neural network (CNN) is proposed, which can automatically complete the feature extraction and fault identification of the signal. The effectiveness of the method is verified by using bearing data. Test with different sample sizes and noise signals to analyze their impact on diagnostic capabilities. Compared with other mainstream algorithms, this method has a higher recognition rate and can meet the timeliness of fault diagnosis.

Keywords: Fault identification · Anti-noise · Raw signals · Convolutional neural networks

1 Introduction

With the development of modern industrial technology, mechanical and electrical equipment is developing large-scale, high-speed and intelligent. As a key component of mechanical and electrical equipment, the condition monitoring of rolling bearings has great significance to the stable operation of the equipment [1]. At present, intelligent fault diagnosis methods mainly include two main steps: feature extraction and fault classification, this kind of method has been widely studied and applied in the field of rotating machinery fault diagnosis. Many feature extraction methods have been put forward in the development of bearing fault diagnosis, such as short time Fourier transform [2], wavelet transform [3], empirical mode decomposition [4], singular value decomposition [5]. In fault classification, artificial intelligence methods such as

M. A. Wahab (Ed.): DAMAS 2019, LNME, pp. 618–628, 2020.
https://doi.org/10.1007/978-981-13-8331-1_46

clustering algorithm, genetic algorithm [6], support vector machine (SVM) [7], fuzzy inference and neural network [8–10] are commonly used.

There are two main problems in the above-mentioned methods: firstly, when the raw signal contains noise, it is difficult to extract fault features and requires expert experience; second, fault diagnosis is divided into three stages: signal preprocessing, feature extraction and fault classification, which destroys the relationship between each stage and causes a part of the loss of fault information. In this study, a new data preprocessing method is introduced to transform the raw signal data into two-dimensional (2D) gray images without any predefined parameters, the experience of experts can be eliminated as much as possible. Then, a new CNN is proposed to extract the features. The results show that this method is effective in fault diagnosis. The main contributions of this paper are summarized as the following three points:

(1) Features are automatically extracted from time-domain vibration signals, independent of fault diagnosis experience and signal processing technology;
(2) The deeper network structure, by combining low-level features to form more abstract high-level representations, to discover the distributed characteristics of the data;
(3) The new CNN model can extract features adaptively and achieve good results in the classification of noisy signals.

2 Methodology

With the continuous increase of equipment quantity, equipment monitoring points and acquisition frequency, a large number of diagnostic data has been obtained, which makes the field of fault diagnosis enter the era of "Big Data". CNN is one of the best classification algorithms with big data. The combination of CNN and big data has broad prospects.

2.1 The Difference Between CNN and Traditional Algorithms

Fault diagnosis is divided into three stages: signal preprocessing, feature extraction and fault classification, the expert experience is demanded, the algorithm design is time-consuming and the generality of the algorithm is poor, which can no longer meet the requirements of big data [11]. In order to solve above problems, the best method is to combine the three phases of fault diagnosis into one phase. This paper presents a fault diagnosis method based on CNN structure, the raw signal is used to diagnose directly which does not need to analyze the internal mechanism of mechanical equipment, avoiding the problem of information loss caused by preprocessing, as shown in Fig. 1.

2.2 Design of the CNN for Bearing Fault Diagnosis

The overall framework of DFCNN is shown in Fig. 2. In order to improve the classification accuracy and structural adaptability, the 3×3 convolution kernel is adopted, followed by max pooling layer to reduce the dimension of data.

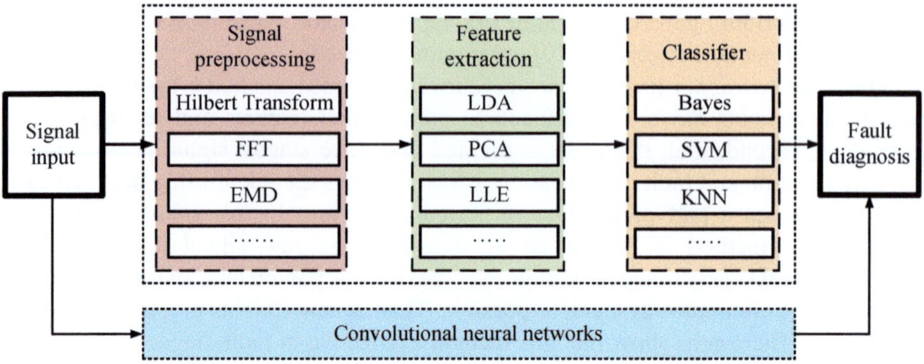

Fig. 1. Decomposition of the steps of bearing diagnosis algorithm.

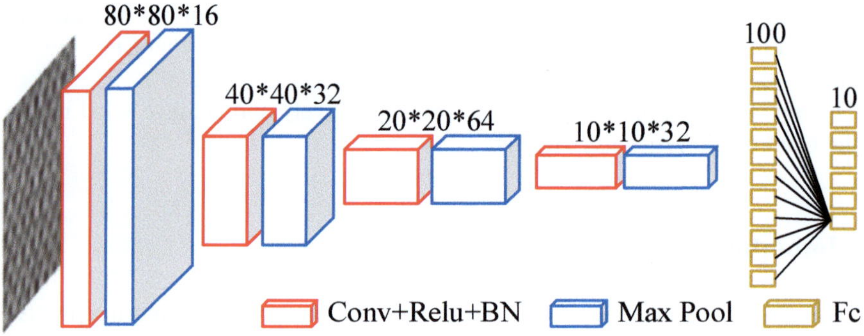

Fig. 2. The structure of the convolution neural network.

Convolutional operation. The convolutional layer uses a convolution kernel to perform convolution operations on the local region of the input signal and the corresponding characteristics are obtained. Weight sharing is the most important feature of the convolutional layer. In order to avoid over-fitting caused by excessive parameters and reduce the memory required for the system to train the network. When each convolution window traverses the entire image, the parameters of the convolution window are fixed. The specific convolution layer operation, as shown in formula (1).

$$y^{l(i,j)} = K_i^l * x^{l(r^j)} = \sum_{j'=0}^{W-1} K_i^{l(j')} x^{l(j+j')} \tag{1}$$

$K_i^{l(j')}$—the j weight of the i convolution kernel of the l layer
$x^{l(r^j)}$—the j of the l layer is a convoluted local area
W—the width of the convolution kernel

Activation operation. The data processed by the convolution layer, uses the activation layer to convert the logits value of each convolution output non-linearly. After processing the data, the linear separability of features will be enhanced. Select Relu (rectified linear unit) as the activation function, specific expression formula (2).

$$a^{l(i,j)} = f\left(y^{l(i,j)}\right) = \max\{0, y^{l(i,j)}\} \tag{2}$$

$y^{l(i,j)}$—the activation value of the convolution layer output y

Batch Normalization. In order to improve network training efficiency and network generalization ability, a batch normalization layer is connected after the activation layer. The BN layer process first subtracts the average value of mini_batch from the volume layer input, and then divides it by standard deviation. However, this will cause the input value to be limited to a small range, so after standardization, it needs to multiply by a scaling amount γ, plus an offset value β. The input of the L-bn layer is $y^1 = (y^{l(i,1)}, \ldots, y^{l(i,j)}, \ldots, y^{l(i,p)})$. The concrete expression of BN layer is shown in formula (3, 4)

$$\hat{y}^{l(i,j)} = \frac{y^{l(i,j)} - \mu_B}{\sqrt{(\sigma_B^2 + \varepsilon)}} \tag{3}$$

$$z^{l(i,j)} = \gamma^{l(i)} \hat{y}^{l(i,j)} + \beta^{l(i)} \tag{4}$$

$\gamma^{l(i)}, \beta^{l(i)}$—the scaling and offset of the BN layer
$z^{l(i,j)}$—the output of the BN layer
ε—a constant term that guarantees numerical stability

Pooling Layer. The pool layer is often connected behind the BN layer, the dimension of the feature map is reduced, at the same time, the invariance of the feature scale is maintained to some extent. The pooling layer is operated by the down sampling operation, and the main purpose is to reduce the parameters of the neural network. In this paper, the max-pooling layer is selected. The input feature window size is 2 × 2 and the step size is 2. As shown in the formula (5)

$$P^{l(i,j)} = \underset{(j-1)W+1 \leq t \leq jW}{max} \left\{a^{l(i,j)}\right\} \tag{5}$$

$a^{l(i,j)}$—the activation value of t neuron in l layer i frame
$P^{l(i,j)}$—the width of the pool area

2.3 Data Augmentation

In the field of computer vision, data enhancement is usually used to increase the number of training samples, thus improving the generalization of CNN [24]. The original vibration data is a one-dimensional time series signal, in order to train the deep learning algorithm, a large number of training samples are needed. Therefore, an overlapping training sample segmentation method is adopted to realize the expansion of training data. This process is shown in Fig. 3. Compared with a non-overlapping sample segmentation method, the overlapping sample segmentation method can keep the correlation between adjacent elements. At the same time, the number of samples participating in model training can be increased.

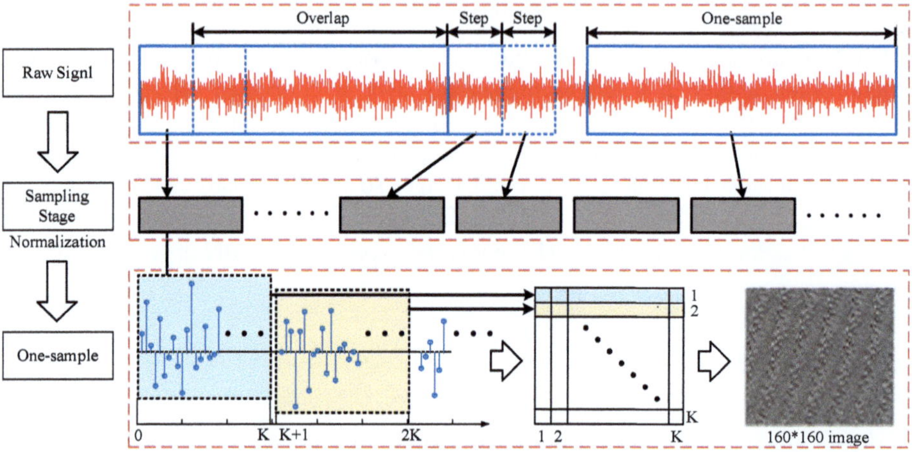

Fig. 3. Signal-to-image conversion process.

2.4 Signal-to-Image Conversion Method

The extraction of features depends on expert experience, and the extracted features are usually not universal. Considering the influence of features on classification results, this paper proposes an effective data preprocessing method. The original signal in time domain is converted into an image, this data preprocessing method can be implemented without any predetermined parameters.

The method is shown in Fig. 3. The raw signals in the time domain fills the pixels of the image sequentially. A segmentation signal of length K^2 is randomly obtained from the time-domain raw signals, and images with size of K * K are obtained by processing them. The intercepted signal is normalized to each line, ranging from 0 to 255, which is just the pixel strength of the gray image. The choice of 160 * 160 in this paper is depended on the volume of signal data. $\mathrm{L(i), i} = 1, \ldots, K^2$ denotes the value of the segment signal. $\mathrm{P(j, k), j} = 1, \ldots, \mathrm{K, k} = 1, \ldots, \mathrm{K}$ denotes the pixel strength of the image. The process is described as follows (6):

$$\mathrm{P(j, k)} = \frac{L(j \times K + k) - Min(L)}{Max(L) - Min(L)} \times 255 \tag{6}$$

3 Verification of Bearing Data Sets

In this section, the vibration signals of rolling bearings are analyzed, verify the feasibility and effectiveness of the proposed method, the robustness of this method under strong noise is discussed.

3.1 Data Set Description and Fault Type Definition

At present, fault diagnosis algorithms are developing rapidly, new theories are constantly put forward. In order to objectively evaluate the performance of this method, it is appropriate to use the third-party standard database and compare the mainstream methods. This experimental data set consists of open source data from the Case Western Reserve University (CWRU) Rolling Bearing Data Center, data acquisition system as shown in Fig. 4.

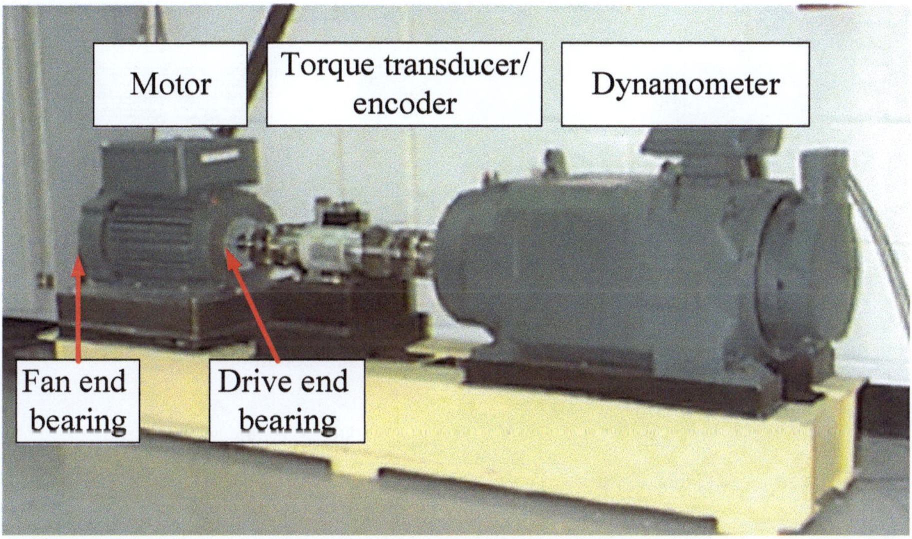

Fig. 4. CWRU bearing fault test-bed.

The rotation speed of the test motor is 1792 r/min and the sampling frequency is 12 kHz. In this paper, the vibration signals of bearings are divided into 10 failure modes, the 2-D graph is shown in Fig. 5. The four different types are respectively the normal state of the bearings, the inner race fault, the outer race and fault the ball fault. The fault depths are 18 mm, 36 mm and 54 mm. Vibration signals of each type of fault are divided into 200 samples, and a fault data set with 2000 samples, which is data set A. Under actual production conditions, the equipment may work under different loads (0 to 3HP). Data set E is established by data sets A, B, C and D. Table 1 is shown the information of the experimental data sets.

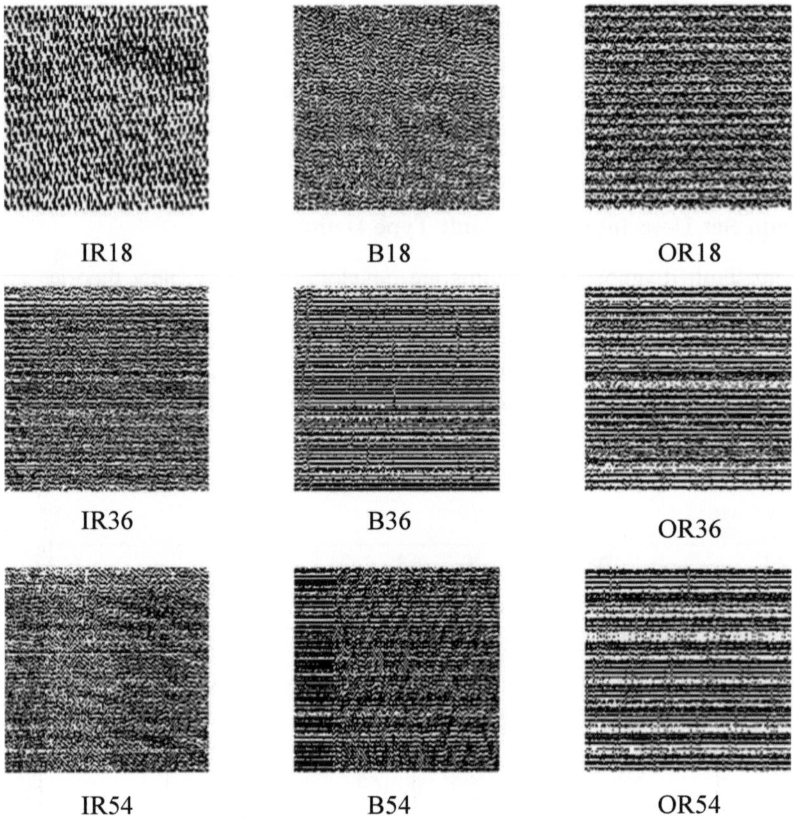

Fig. 5. Converted images on nine fault conditions.

Table 1. Description of bearing fault datasets.

Fault number	Fault category	Fault depth/mm	Data sets A/B/C/D	Data set E
1	B18	0.18	200	800
2	B36	0.36	200	800
3	B54	0.54	200	800
4	IR18	0.18	200	800
5	IR36	0.36	200	800
6	IR54	0.54	200	800
7	Normal	0	200	800
8	OR18	0.18	200	800
9	OR54	0.36	200	800
10	OR36	0.54	200	800

3.2 The Influence of Data Size on Training and Fault Identification

CNN has thousands of parameters needing to be trained. In order to avoid overfitting and enhance the generalization ability of CNN model, a large number of training samples are needed. The samples used in the data-driven diagnosis effect experiment are randomly selected from the Dataset E. The performances of CNN influenced by the data set size are observed respectively using the total amount of samples 100, 200, 500, 1000, 2000, 4000. The initial weights of the CNN model are randomly generated, and the maximum number of iterations is set to 100, and the learning rate of the network is 0.0008. In order to verify the stability of CNN, Each experiment repeats 10 times.

It can be seen from Fig. 6 that when the training sample increases, the diagnosis accuracy increases gradually, and the standard deviation of the 10 diagnostic accuracy decreases the same time. When training samples are relatively small, high diagnosis accuracy can still be achieved. The above result shows that the accuracy of CNN model increases with the quantity of input data, and the data feature extraction ability increases gradually, because the increase of training samples can improve the generalization of the model [37]. And the size of the training sample does not affect the time required to diagnose a signal sample. The time that consumed to diagnose a signal using the CNN with different sample numbers is shown on the right side of Fig. 6, which indicates that the CNN model diagnoses a signal in only 0.8572 ms, which can meet the real-time requirement of fault diagnosis.

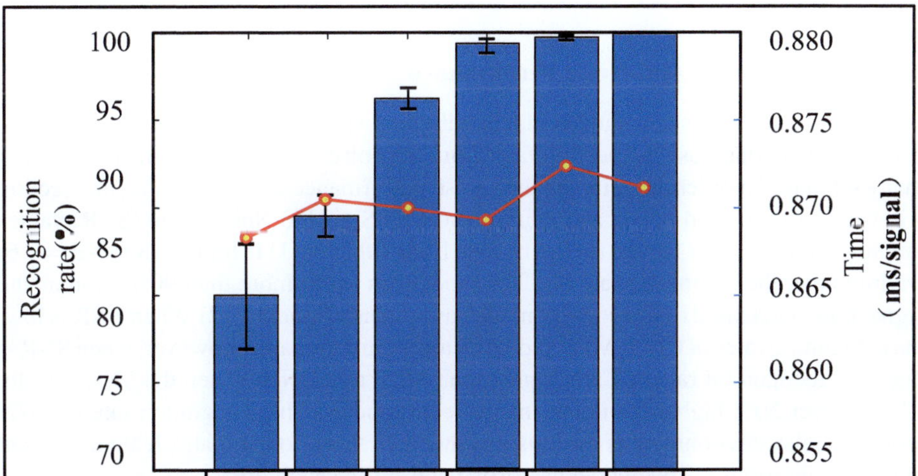

	100	200	500	1000	2000	4000
Recognition rate	0.8391	0.8726	0.9712	0.9856	0.9938	1
Time	0.8567	0.8571	0.8570	0.8569	0.8574	0.8572

Fig. 6. Number of different training samples, CNN's identification rate for test set E and time to diagnose individual signals.

In order to demonstrate the advancement of CNN, the diagnostic accuracy of some traditional methods is compared. The results are shown in Table 2. Compared with traditional methods, CNN has achieved better results. In document [12], the Multifractal is used to extract fault features and SVM classifier is used to classify fault signals. The classification accuracy of this method is 89.1%. In document [13], the time domain characteristics of faults are extracted by mathematical statistics. Then, six kinds of faults in three data sets are classified by multi-layer perceptron. The classification accuracy of this method are 95.7%, 99.6% and 99.4%, respectively. In the same data set, CNN gets 100% classification accuracy from the original data. In document [13], the method of directly diagnosing faults from original time domain signals by using DBN, the classification accuracy of this method is 98.8%. The accuracy rate of the method proposed in this paper reaches 100%, which is better than the existing methods.

Table 2. Description of bearing fault datasets.

					(%)
Method	A	B	C	D	E
Multifractal + SVM [12]	89.1	-	-		-
Time + MLP [13]	95.7	99.6	99.4		-
PSO-WKLFDA [14]	97.2	-	-		-
Time-domain signal + DBN [15]	100	100	100		98.8
Time-domain signal + CNN	100	100	100	100	100

3.3 Verification of Anti-Noise Performance

Noise from industrial sites is unavoidable, vibration signals are susceptible to contamination. How to diagnose the bearing fault from the noise signal has become the focus of many scholars. In order to verify the anti-noise performance of CNN model, this section will test its recognition rate in an environment with Signal to Noise Ratio (SNR) values ranging from −4 dB to 10 dB. As can be seen from Table 3: (1) In high SNR (SNR = −6) environment, the diagnostic rate of CNN model can reach more than 90%. And as the signal noise decreased, the recognition rate has increased quickly; (2) When SNR is low, the recognition rates of CNN, MLP and DNN are almost the same; however, when SNR is high, the recognition rate of CNN higher than MLP and DNN. When the SNR is 0 db, CNN is over 50% higher than other mainstream models. The recognition rate of CNN model in noisy environment is obviously higher than that of other mainstream intelligent diagnosis algorithms. This model has strong anti-noise capability.

Table 3. CNN's recognition rate of noise signal.

Method	SNR (dB)								
	−6	−4	−2	0	2	4	6	8	10
MLP	-	-	-	43.25	49.89	76.56	91.76	95.75	97.78
DNN	-	-	-	42.63	70.06	82.79	92.15	96.27	98.45
CNN	91.69	96.32	99.13	99.24	99.31	99.62	99.84	99.92	99.97

4 Conclusion

The method proposed in this paper, compared with the traditional algorithm, has the following three advantages: (1) The process of artificial feature extraction is no longer needed. The original vibration data are directly used as the input of CNN and the output of CNN is the result of fault identification. (2) Because of the deep structure of CNN, the essential characteristics of the original vibration signal are extracted by layer-by-layer nonlinearity, the powerful self-learning ability of deep learning has been fully utilized; (3) The algorithm has good robustness and generalization performance. Even if the signal contains noise, the fault identification rate is still high. However, it should be pointed out that there are still some shortcomings in this method: Compared with the shallow layer algorithm, the model of this algorithm needs a long training time; the selection of model parameters requires repeated tests. The next research focus should be to design an adaptive model parameter selection algorithm according to the nature of the signal. Improve the efficiency of the algorithm.

Acknowledgements. This research is supported Engineering Research Center of Advanced Driving Energy-saving Technology, Ministry of Education. The authors would also like to acknowledge the financial support from the National Natural Science Foundation of China (no. 51775452) and Fundamental Research Funds for Central Universities (A0920502051822-2).

References

1. Guo, L., Lei, Y., Xing, S., Yan, T., Li, N.: Deep convolutional transfer learning network: a new method for intelligent fault diagnosis of machines with unlabeled data. IEEE Trans. Ind. Electron. https://doi.org/10.1109/tie.2018.2877090
2. Lou, X., Loparo, K.A.: Bearing fault diagnosis based on wavelet transform and fuzzy inference. Mech. Syst. Sig. Proc. **18**(5), 1077–1095 (2004)
3. Yu, Y., YuDejie, Junsheng, C.: A roller bearing fault diagnosis method based on EMD energy entropy and ANN. J. Sound Vib. **294**(1), 269–277 (2005)
4. Qin, S.J.: Process data analytics in the era of big data. AIChE J. **60**(9), 3092–3100 (2014)
5. Brocki, Ł., Marasek, K.: Deep belief neural networks and bidirectional long-short term memory hybrid for speech recognition. Arch. Acoust. **40**(2), 191–195 (2015)
6. Cerrada, M., Zurita, G., Cabrera, D.: Fault diagnosis in spur gears based on genetic algorithm and random forest. Mech. Syst. Sig. Proc. **s70–71**, 87–103 (2016)
7. Shen, C., Wang, D., Kong, F., et al.: Fault diagnosis of rotating machinery based on the statistical parameters of wavelet packet paving and a generic support vector regressive classifier. Measur. J. Int. Measur. Confed. **46**(4), 1551–1564 (2013)
8. Linear feature selection and classification using PNN and SFAM neural networks for a nearly online diagnosis of bearing naturally progressing degradations. Eng. Appl. Artif. Intell. **42**, 67–81 (2015)
9. Zhang, L., Xiong, G., Liu, H., et al.: Bearing fault diagnosis using multi-scale entropy and adaptive neuro-fuzzy inference. Expert Syst. Appl.: An Int. J. **37**, 6077–6085 (2010)
10. Lei, Y., Jia, F., Lin, J., et al.: An intelligent fault diagnosis method using unsupervised feature learning towards mechanical big data. IEEE Trans. Ind. Electron. **63**, 3137–3147 (2016)

11. Guo, L., Lei, Y.G., Li, N.P., Yan, T., Li, N.B.: Machinery health indicator construction based on convolutional neural networks considering trend burr. Neurocomputing **292**, 142–150 (2018)
12. Du, W., Tao, J., Li, Y., Liu, C.: Wavelet leaders multifractal features based fault diagnosis of rotating mechanism. Mech. Syst. Sig. Proc. **43**(1–2), 57–75 (2014)
13. de Almeida, L.F., Bizarria, J.W.P., Bizarria, F.C.P., Mathias, M.H.: Condition-based monitoring system for rolling element bearing using a generic multi-layer perceptron. J. Vib. Control **21**(16), 3456–3464 (2015)
14. Particle Swarm Optimization; New Findings on Particle Swarm Optimization from University of Ulsan Summarized (Bearing defect classification based on individual wavelet local fisher discriminant analysis with particle swarm optimization). Computers, Networks & Communications (2016)
15. Zhao, G.Q., Ge, Q.Q., Zhao, X.Y., Peng, X.Y.: Research on fault feature extraction and diagnosis based on DBN. J. Instrum. Instrum. **37**(09), 1946–1953 (2016)

Study on Vibration Tracing and Vibration Reduction Technology of Reciprocating Compressor Pipeline

Yuan Li[1,2], Yang Lin[3(✉)], Ling Fan[2], Yu Zhang[2], and Yunfeng Chang[2]

[1] College of Energy and Power Engineering, Xi'an Jiaotong University, Xi'an, China
[2] SINOPEC Sichuan to Eastern China Transmission Gas Pipeline Co. Ltd., Dazhou, China
[3] College of Safety and Ocean Engineering, China University of Petroleum-Beijing, Changping, China
liny715@163.com

Abstract. Reciprocating compressor is the core equipment for storing natural gas in gas storage. The abnormal vibration of pipeline will lead to fatigue fracture of pipeline, loose and broken connecting parts, and even lead to leakage of high temperature and high pressure, flammable and explosive gas, causing fire in gas storage. and major accidents such as explosions. In this paper, the first-level outlet safety valve pipe with abnormal vibration of a gas storage reciprocating compressor is taken as an example. Through theoretical and experimental exploration, the vibration source tracing and vibration reduction technology of reciprocating compressor are studied. Based on the finite element analysis of the cause of vibration anomaly, the double-dynamic vibration absorber with vibration reduction is designed based on the fixed-point theory. After installing the vibration absorber, the finite element analysis results reduce the pipe vibration from 16.9 μm to 11.7 μm. Finally, the vibration reduction effect of the vibration absorber is verified by finite element analysis and field test, which provides certain guidance for pipeline vibration reduction.

Keywords: Reciprocating compressor · Pipeline vibration · Finite element analysis · Dynamic vibration absorber · Field test

1 Introduction

Natural gas demand is increasing day by day. In the natural gas supply system, gas storage plays an important role in peak regulation, so the safe and stable operation of gas storage is an important guarantee for the normal supply of natural gas [1]. The reciprocating compressor is the core equipment of natural gas storage in gas storage. Abnormal vibration of pipeline is a common problem in its operation [2]. Severe pipeline vibration will generate strong noise, which will harm the hearing of site operators. At the same time, it can cause fatigue damage of pipelines, loose and damaged connecting parts. Serious cases may lead to pipeline fracture, high temperature and

M. A. Wahab (Ed.): DAMAS 2019, LNME, pp. 629–638, 2020.
https://doi.org/10.1007/978-981-13-8331-1_47

high pressure, flammable and explosive gas leakage in the pipeline, and even cause major accidents such as fire and explosion in gas storage [3]. According to statistics, the direct property loss caused by abnormal vibration of pipelines in industrial production in the United States is nearly 10 billion dollars every year. According to the maintenance of compressor units in gas storage, the failure of compressor pipeline system is an important factor causing unplanned shutdown of compressor, accounting for 20–30%.

On the one hand, the violent vibration of compressor pipeline is caused by gas pulsation in the pipeline; on the other hand, it is caused by insufficient mechanical damping of pipeline [4]. It can be seen from the research status that: for the problem of insufficient damping of mechanical structure, the typical representative is dynamic vibration absorber [5]. The dynamic vibration absorber is composed of auxiliary mass, spring and damping, which belongs to the passive type of vibration damper to suppress resonance. The advantage of the dynamic vibration absorber lies in its small size, light weight and good vibration control performance. By reasonably optimizing the performance parameters, material selection and structural design of dynamic vibration absorber, the vibration energy of main vibration pipeline can be absorbed at the set frequency to reduce its forced vibration, and finally the vibration of pipeline can be suppressed [6]. At the same time, the dynamic vibration absorber can control the vibration of compressor pipeline effectively when the pipeline structure is changed in a small scale. Compared with traditional pipeline vibration reduction methods, such as optimized piping, adjusting support, setting orifice plate and buffer tank, etc., pipeline dynamic vibration absorber has engineering advantages of low cost, high reliability and easy operation. In 1909, H. Frahm took the lead in designing an undamped vibration absorber composed of only two parts, the mass element and the elastic element. In recent years, Nishihara et al. have made the analytic solutions of damped dynamic vibration absorbers more systematic and precise [7]. ASAMI et al. compared the vibration absorption effects for different types of vibration absorbers under the premise of same mass ratio [8]. Japanese scholar Back Households focused on practical engineering problems and studied the multi-mode vibration control method and active dynamic vibration absorber technology based on magnetic damping [9].

In this paper, the finite element method is used to trace the vibration of pipelines, and the causes of pipeline vibration are obtained. Then, according to the characteristics of pipeline vibration, the vibration reduction scheme of double dynamic vibration absorber is proposed, and the parameters of vibration absorber are determined. Finally, the finite element analysis of the pipeline after the installation of vibration absorber verifies the vibration reduction effect of vibration absorber.

2 Pipeline Vibration Traceability

Reciprocating compressor is large equipment which can improve the compressed gas pressure by transforming the mechanical energy of driving machine. As the weak link of gas transmission system, compressor has complex structure, including equipment body, sealing system, lubricating system, driving system and valves. Failure of unit subsystem or key components will also affect the normal operation of equipment [10]. As an auxiliary part of reciprocating compressor, safety valve pipeline is an indispensable part to ensure the reliable and safe operation of compressor.

The piping structure and composition of reciprocating compressors are very complex. Due to the different shapes and positions, the vibration causes are different, and the vibration state of pipelines is also different. At the same time, the design of dynamic vibration absorber needs to be customized according to the characteristics of pipeline. First of all, it is necessary to understand the actual vibration situation of unit and its pipeline in the production site, find out the location of pipeline with larger vibration, and then design the dynamic vibration absorber. Therefore, pipeline vibration test is the premise and foundation of dynamic vibration absorber design. This paper takes the first-level outlet safety valve pipe with abnormal vibration of reciprocating compressor in a gas storage as an example, and studies the vibration tracing and vibration reduction technology of reciprocating compressor pipeline through theoretical and experimental exploration.

The first-level outlet safety valve pipe of reciprocating compressor in a gas storage tank has excessive vibration, and the pipeline layout is shown in Fig. 1. Main parameters of the pipeline: the main pipe diameter is 168 mm, the vertical pipe height is 480 mm, the branch pipe length is 815 mm, and the flange, safety valve and other parts are approximately replaced by solid units. The pipe is made of No. 45 steel with a density of 7.9×10^3 kg/m^3, so the total mass of first-level outlet safety valve pipe model is calculated in SolidWorks software to be about 215 kg.

Fig. 1. The first-level outlet safety valve pipe.

In order to trace the source of vibration, the inherent frequency and natural mode of vibration for pipeline are solved. Specific steps are as follows:

(1) Material setup and meshing: The model of first-stage outlet safety valve pipeline established with SolidWorks is imported into ANSYS for modal analysis. The type of analysis unit is solid element. Material properties are defined as 45 steel

parameters, Passion ratio is 0.3, elastic modulus is 2.06×10^5 MPa, density is 7800 kg/m^3, and damping ratio is 0.005. The suitable precision is adjusted during the free meshing, and the model of first-stage outlet safety valve is meshed intelligently.

(2) Constraint condition: According to the distribution of pipeline support on site, constraints parallel to the pipeline section are imposed on the four support undersides of pipeline, that is, constraints other than the axial direction of pipeline.

(3) Modal analysis: Modal analysis is the main method in structural dynamics analysis. At present, modal analysis is mainly used to determine the cause of pipeline vibration [11]. Mode is the inherent property of structure, each mode has specifically resonant frequencies, damping values and mode shapes [12]. The Block Lanczos method is used to extract the pipeline modal when solving the pipeline modal. The first 10 order is selected for both eigenvalue and extended mode order, and the natural frequency range of calculated pipeline is set as 0 $\sim$ 1000 Hz.

Table 1. The maximum stress of the balance disc at each speed.

Modal order	1	2	3	4	5	6	7	8	9	10
Frequency (Hz)	16.3	42.5	112	135	186	235	240	327	361	371

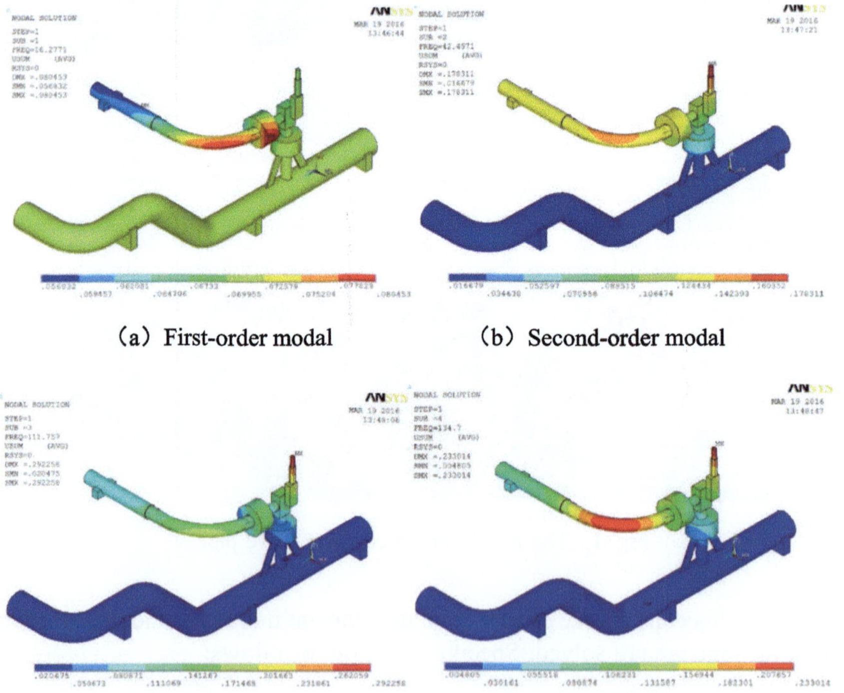

(a) First-order modal (b) Second-order modal

(c) Third-order modal (d) Fourth-order modal

Fig. 2. Modal mode diagram of the first-level exit safety valve.

The first ten natural frequencies of first-stage outlet safety valve pipeline are obtained through modal calculation, as shown in Table 1, and the first four modes are shown in Fig. 2. It can be concluded that the first-order natural frequency 16.28 Hz is relatively close to the basic frequency 16.6 Hz of compressor, while the second-order, third-order and fourth-order natural frequencies of pipeline are relatively far from the basic frequency and frequency multiplication. Therefore, it can be judged that the pipeline vibration of first-stage outlet safety valve is caused by the fact that the natural frequency of pipeline is close to the basic frequency of compressor, which causes structural resonance. From the vibration pattern diagram, it can be seen that the most violent vibration is concentrated at the elbow of branch pipeline.

3 Design of Dynamic Vibration Absorber

The whole pipe system or part pipe of reciprocating compressor is also a mechanical vibration system with natural frequency. The key parts of compressor are tested for vibration, the first-stage outlet safety valve pipeline with larger vibration is found, and the pipeline object of vibration reduction at the heavy point of dynamic vibration absorber is determined. A double dynamic vibration absorber for vibration reduction is designed based on set-point theory and the characteristics of pipeline vibration in this paper.

The set-point theory is to use the points on the frequency response curve which are independent of damping change to design the vibration control device for the vibration control system with damping [13]. The set-point theory is the theoretical basis for parameter design and optimization of double dynamic vibration absorber (DDVA).

According to the characteristics of first-stage outlet safety valve pipeline, the corresponding dynamic vibration absorber is designed to reduce the vibration of pipeline. DDVA has the characteristics of simple structure and good effect for vibration reduction. In this paper, DDVA is used to reduce the vibration of pipeline. DDVA is composed of approximately symmetrical structure, including holder, reinforcing plate, mass block, rubber cushion, pipe clamp, attachment bolt and other parts, as shown in Fig. 3.

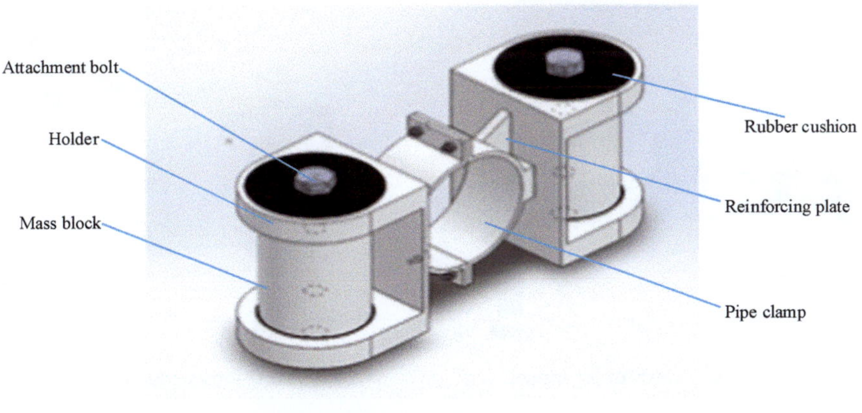

Fig. 3. Double dynamic vibration absorber structure model.

The shape of the mass block for DDVA is selected as a cylinder, which is convenient for installation and disassembly. According to the set-point theory, it is calculated that the mass of mass block is 3.225 kg, the design height is 100 mm, and the density of steel is 7.9×10^3 kg/m^3.

4 Analysis of Vibration Reduction Effect

In order to study the vibration absorption effect of DDVA, the simulation analysis is carried out on the pipeline installed with DDVA, and the vibration situation is compared before and after the installation of DDVA. The first 10 inherent frequencies of pipe after the additional DDVA are shown in Table 2, and the natural modes of first 4 orders are shown in Fig. 4.

Table 2. The first 10 inherent frequencies of pipe after the additional DDVA.

Modal order	1	2	3	4	5	6	7	8	9	10
Frequency (Hz)	17.6	24.5	25.3	35.2	54.3	57.3	71.0	71.9	72.8	75.9

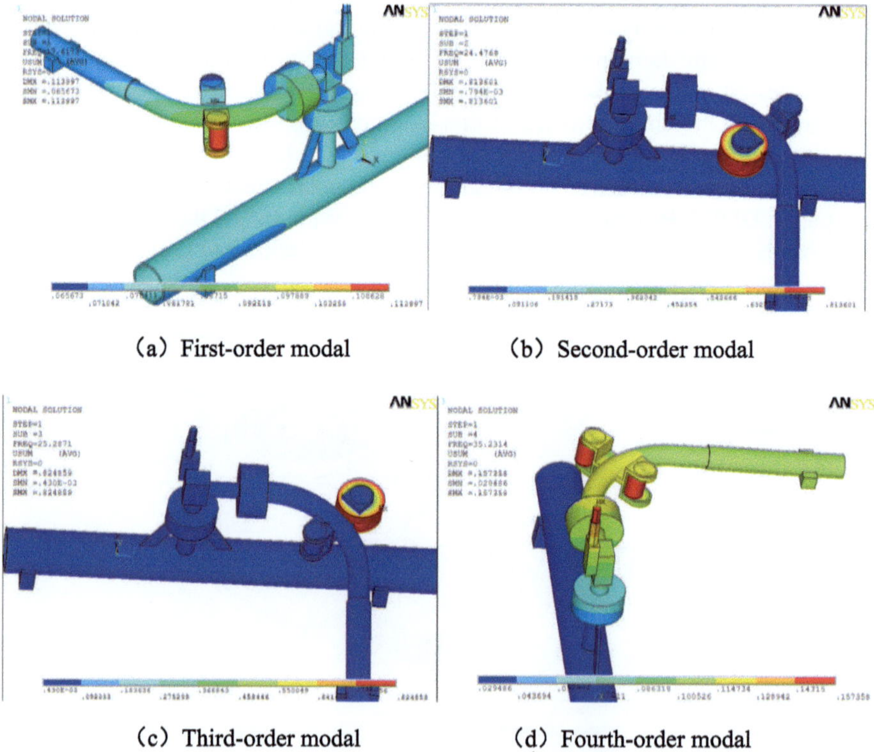

(a) First-order modal (b) Second-order modal

(c) Third-order modal (d) Fourth-order modal

Fig. 4. Pipeline mode after installation of the vibration absorber.

In order to compare the vibration of pipeline after the DDAV is installed, the transient analysis of vibration process is carried out by means of simulation method. When 16.6 Hz excitation force is applied to the model before installing DDAV, the maximum vibration amplitude of pipeline at the position of DDAV is 16.9 μm. When 16.6 Hz excitation force is applied to the model after installing DDAV, the maximum vibration displacement of pipeline at the position of DDAV is 11.7 μm. The comparison results are shown in Figs. 5 and 6.

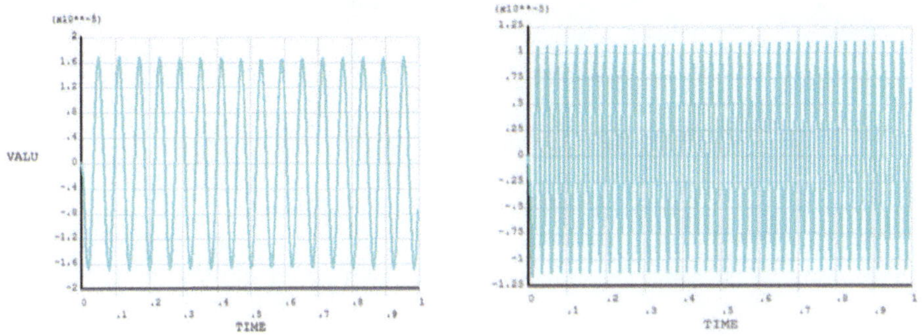

(a) Before installing DDAV (b) After installing DDAV

Fig. 5. Comparison of vibration displacement of elbow position.

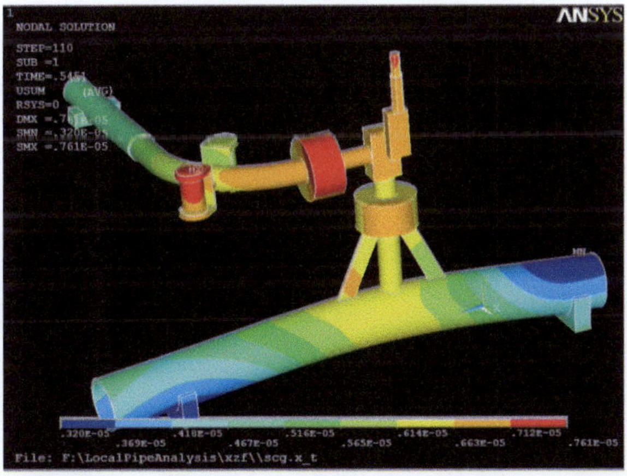

Fig. 6. Pipe displacement map after installing the vibration absorber.

In order to further verify the vibration reduction effect of DDAV, the vibration reduction technology of proposed DDAV is applied in the reciprocating compressor field, as shown in Fig. 7.

Fig. 7. Dynamic vibration absorber field test.

The results of vibration absorption are shown in Table 3 and Fig. 8:

Table 3. Dynamic vibration absorber field test results.

Test position	Direction	Before installation	After installation	Vibration reduction ratio
Support position	X	15.75	13.58	13.8%
	Y	12.42	11.68	6.0%
	Z	8.10	7.12	12.1%
Safety valve	X	11.04	6.75	38.9%
	Y	10.02	7.42	25.9%
	Z	9.85	8.74	11.3%

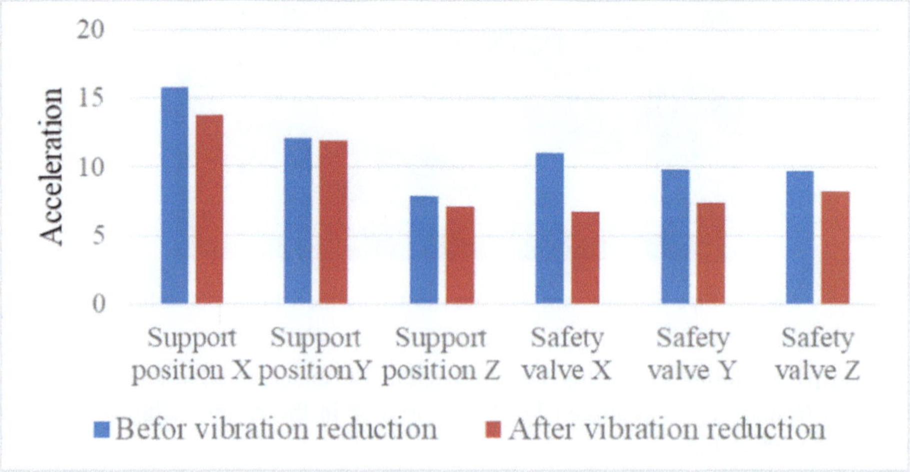

Fig. 8. Vibration displacement changes at various positions after installation DDAV.

The support position reduces vibration by 13.8% in the X direction, 6% in the Y direction and 12.1% in the Z direction. The safety valve reduces vibration by 38.9% in the X direction, 25.9% in the Y direction and 11.3% in the Z direction. The vibration in vertical and horizontal direction is obviously reduced, and the vibration absorber achieves good effect of vibration absorption and meets the requirement of overall vibration reduction.

The fault prediction value of the motor is in the defined normal range, which indicates that the motor works normally in the future. It is consistent with the actual situation, and verifies the correctness and reliability of the fusion prediction technology.

5 Conclusion

Taking the first-stage outlet safety valve pipe with abnormal vibration of reciprocating compressor in gas storage as the research object, the abnormal vibration of pipeline is analyzed by finite element method, and the reason for abnormal vibration is found to be mechanical resonance. A double dynamic vibration absorber for vibration reduction is designed based on set-point theory and the characteristics of pipeline vibration. The effect of DDAV is verified by finite element analysis and field test. The finite element analysis results show that the vibration of pipeline decreases from 16.9 μm to 11.7 μm after installing DDAV. The field test results show that the vibration of pipeline in all directions is obviously reduced, which provides some guidance for the vibration reduction of pipeline.

Acknowledgements. This paper is supported by SINOPEC Gas Company (No. 35150573-16-ZC0607-0001).

References

1. Wanyan, Q., Ding, G., Zhao, Y., et al.: Key technologies for salt-cavern underground gas storage construction and evaluation and their application. Nat. Gas Ind. B **5**(6), 623–630 (2018)
2. Bin, X.U., Feng, Q.K., Xiao-Ling, Y.U.: Study on pressure pulsation and piping vibration of complex piping of reciprocating compressor. Nucl. Power Eng. **29**(4), 79–83 (2008)
3. Liang, Z., Li, S., Tian, J., et al.: Vibration cause analysis and elimination of reciprocating compressor inlet pipelines. Eng. Fail. Anal. **48**, 272–282 (2015)
4. Zhao, B., Jia, X., Sun, S., et al.: FSI model of valve motion and pressure pulsation for investigating thermodynamic process and internal flow inside a reciprocating compressor. Appl. Therm. Eng. **131**, 998–1007 (2018)
5. Huang, X., Su, Z., Hua, H.: Application of a dynamic vibration absorber with negative stiffness for control of a marine shafting system. Ocean Eng. **155**, 131–143 (2018)
6. Liu, K., Jie, L.: The damped dynamic vibration absorbers: revisited and new result. J. Sound Vib. **284**(3), 1181–1189 (2005)

7. Nishihara, O., Asami, T.: Closed-form solutions to the exact optimizations of dynamic vibration absorbers (minimization of the maximum amplitude magnification factors). ASME J. Vib. Acoust. **124**(4), 576–582 (2002)
8. Asami, T., Nishihara, O., Baz, A.M.: Analytical solutions to H and H2 optimization of dynamic vibration absorbers attached to damped linear systems. Trans. Jpn. Soc. Mech. Eng. **67**(655), 597–603 (2002)
9. Back households, Deng Mingzhang: Dynamic vibration absorber and its application. Mechanical Industry Press, Beijing (2013)
10. Corvaro, F., Giacchetta, G., Marchetti, B., et al.: Reliability, Availability, Maintainability (RAM) study, on reciprocating compressors API 618. Petroleum **3**(2), 266–272 (2016)
11. Mikota, G.: Modal analysis of hydraulic pipelines. J. Sound Vib. **332**(16), 3794–3805 (2013)
12. Zhu, Y.Q., Zhang, H.Y., Zhan, P.X.: The vibration analysis of drive shaft based on SolidWorks. In: International Conference on Electric Information & Control Engineering (2011)
13. Wong, W.O., Fan, R.P., Cheng, F.: Design optimization of a viscoelastic dynamic vibration absorber using a modified fixed-points theory. J. Acoust. Soc. Am. **143**(2), 1064 (2018)

Research on Motor Fault Warning Technology Based on Second-Order Volterra Series

Yuan Li[1,2], Ning Ding[2], Zeyang Qiu[3(✉)], Song Yang[2], and Yongming Wang[2]

[1] Energy and Power Engineering College, Xi'an Jiaotong University, Xi'an, China
[2] SINOPEC Sichuan to Eastern China Transmission Gas Pipeline Co. Ltd., Beijing, China
[3] College of Safety and Ocean Engineering, China University of Petroleum, Beijing, China
1164709426@qq.com

Abstract. As important equipment for natural gas transmission stations, electric drive centrifugal compressor units are generally in continuous high-speed operation. Long-term high-load operation is easy to induce drive motor failure, resulting in huge economic losses and casualties. Compressor group condition monitoring, maintenance and repair are all closely related to motor fault diagnosis, making compressor group fault diagnosis very important. Therefore, the study of motor abnormal fault warning technology is of great significance for the prevention of centrifugal compressor accidents. In this paper, the motor fault warning model is established based on the multi-sensor data and second-order Volterra series. The obtained prediction data is compared with the fusion data collected by the sensor to obtain a range under the normal operation of the motor, that is, the numerical set [66, 67]. Predicting the development trend of motor running based on motor monitoring data. The model is used to predict the simulation data and motor data, and the results verify the correctness of the model.

Keywords: Motor · Information fusion · Fault warning technology · Second-order Volterra series

1 Introduction

The motor fault diagnosis itself is closely related to the mechanical system of the motor-related equipment. The diagnosis of rotating machinery is a larger area, but it is mainly for mechanical failures. The problem of motor fault diagnosis is more complicated because it involves both mechanical and electrical parts. At the same time, motor condition monitoring, maintenance and repair are closely related to motor fault diagnosis, making motor fault diagnosis very important. In 1987, Tavner P.J. and Penman J. firstly proposed the concept of motor condition monitoring, and began research on online monitoring and fault diagnosis of electric motors [1]. Schoen R.R. and others believe that motor bearing faults can be detected not only by vibration signals, but also by their current signals [2]. In the literature, Han compares the

M. A. Wahab (Ed.): DAMAS 2019, LNME, pp. 639–650, 2020.
https://doi.org/10.1007/978-981-13-8331-1_48

performance of Fourier transform and wavelet transform in dealing with motor fault signals, and replaces the traditional Fourier transform with wavelet transform [3]. In recent years, thanks to the rapid development of computer technology and artificial intelligence technology, motor fault diagnosis technology has gradually moved toward the path of intelligent development. In these emerging artificial intelligence methods, there are more artificial neural networks and fuzzy logic., support vector machines, genetic algorithms, and expert systems [4]. Although most of them are still only in the laboratory stage, these artificial intelligence methods, which have sprung up, have opened up new directions for motor fault diagnosis technology. In addition, multi-sensor information fusion technology, an emerging technology derived from military applications, is increasingly used in motor fault diagnosis [5].

Commonly used motor fault diagnosis methods are stator current analysis method, infrared diagnosis method, vibration analysis method, and noise diagnosis method [6]. Due to the existence of several interrelated working systems in the motor, the causes of faults and the signs of faults often show diversity, which makes it more difficult for motor fault diagnosis [7]. In order to diagnose the motor fault, it is essential to analyze the various fault mechanisms of the motor. According to the location of the fault, there are four common faults of the motor, such as rotor fault, stator fault, bearing fault and air gap eccentric fault [8].

The fault diagnosis problem is actually a decision problem based on pattern classification. The acquisition of the dataset of the pattern classification and the feature extraction are the key to the problem. As far as the classification technique is statistical or intelligent or purely analytical, it is based on the feature extraction of a specific dataset. Uncertainty theory is the basis of these methods in the solution of decision problems. Therefore, a motor fault diagnosis process can be seen as a process from the sensor feature decision.

In this paper, based on multi-sensor data and second-order Volterra series, an early warning model of motor fault is established, and the development trend of motor running is predicted based on motor monitoring data. The model is used to predict the simulation data and motor data, and the results verify the correctness of the model.

2 Volterra Series

For continuous nonlinear systems $y(t) = f(t, u(t))$, Volterra series model can be obtained by a series of convolution operations, as shown in Eq. (1).

$$\begin{aligned} y(t) = {} & h_0 + \int_{-\infty}^{\infty} h_1(\tau)u(t-\tau)d\tau + \\ & \int_{-\infty}^{\infty}\int_{-\infty}^{\infty} h_2(\tau_1,\tau_2)u(t-\tau_1)d\tau_1 d\tau_2 + \ldots + \int_{-\infty}^{\infty}\int_{-\infty}^{\infty}\int_{-\infty}^{\infty} h_3(\tau_1,\tau_2,\tau_3)u(t-\tau_1)u(t-\tau_2)u(t-\tau_3)d\tau_1 d\tau_2 d\tau_3 \\ & + \ldots + \int_{-\infty}^{\infty}\int_{-\infty}^{\infty}\cdots\int_{-\infty}^{\infty} h_n(\tau_1,\tau_2,\ldots,\tau_n)u(t-\tau_1)u(t-\tau_2)\ldots u(t-\tau_n)d\tau_1 d\tau_2\ldots d\tau_n \end{aligned} \quad (1)$$

The Volterra series model of discrete system is shown in Eq. (2)

$$
\begin{aligned}
y(k) = {} & h_0 + \sum_{\tau=0}^{\infty} h_1(\tau)u(t-\tau) \\
& + \sum_{t_1}^{\infty}\sum_{\tau_2}^{\infty} h_2(\tau_1,\tau_2)u(t-\tau_1)u(t-\tau_2) + \sum_{\tau_1}^{\infty}\sum_{\tau_2}^{\infty}\sum_{\tau_3}^{\infty} h_3(\tau_1,\tau_2,\tau_3)u(t-\tau_1)u(t-\tau_2)u(t-\tau_3) \\
& + \ldots + \sum_{\tau_1}^{\infty}\sum_{\tau_2}^{\infty}\cdots\sum_{\tau_\infty}^{\infty} h_\infty(\tau_1,\tau_2,\ldots,\tau_\infty)u(t-\tau_1)u(t-\tau_2)\ldots u(t-\tau_\infty)
\end{aligned}
\tag{2}
$$

The direct use of Volterra series model leads to dimensional disasters. Previous studies have shown that the nonlinear system in reality can be approximated by Volterra series model with finite series, and the infinite Volterra series model can be finite-truncated. The higher the reserved order is, the smaller the approximation error will be. But the computation increases exponentially as the order increases. Volterra series model of finite order is shown in Eq. (3).

$$
\begin{aligned}
y(k) = {} & h_0 + \sum_{\tau=0}^{\infty} h_1(\tau)u(t-\tau) \\
& + \sum_{t_1}^{\infty}\sum_{\tau_2}^{\infty} h_2(\tau_1,\tau_2)u(t-\tau_1)u(t-\tau_2) + \sum_{\tau_1}^{\infty}\sum_{\tau_2}^{\infty}\sum_{\tau_3}^{\infty} h_3(\tau_1,\tau_2,\tau_3)u(t-\tau_1)u(t-\tau_2)u(t-\tau_3) \\
& + \ldots + \sum_{\tau_1}^{\infty}\sum_{\tau_2}^{\infty}\cdots\sum_{\tau_p}^{\infty} h_p(\tau_1,\tau_2,\ldots,\tau_p)u(t-\tau_1)u(t-\tau_2)\ldots u(t-\tau_p) + e
\end{aligned}
\tag{3}
$$

Where e represents the truncation error, it means the difference between actual value and Volterra series model value. $h_p(\tau_1,\tau_2,\ldots,\tau_p)$ represents the Volterra kernel of p order.

Second-order Volterra series model is a commonly used model, which can satisfy the approximation accuracy in many cases and its computation is small. Set time series $x_i(i = 1,2\ldots N_x)$, the expression of Second-order Volterra series prediction model is shown Eq. (4).

$$
x'(n+1) = F[x(n)] = h_0 + \sum_{i=0}^{N_1-1} h_1(i)x(n-i) + \sum_{i=0}^{N_2-1}\sum_{j=o}^{N_2-1} h_2(i,j)x(n-i)x(n-j) \tag{4}
$$

The minimum embedding dimension m of the signal is obtained by using nearest neighbor method, Set $(N_1 = N_2 = m)$.

The input vector can be expressed as follows:

$$
X(n) = [1, x(n), x(n-1), \ldots, x(n-m-1), x^2(n), x(n)x(n-1), \ldots, x^2(n-m+1)]^T \tag{5}
$$

The prediction coefficient vector can be expressed as Eq. (6):

$$W(n) = [h_0, h(0), h(1), \ldots, h(m-1), h_2(0,0), h_2(0,1), \ldots, h_2(m-1, m-1)]^T \quad (6)$$

And Eq. (4) can be written as:

$$x'(n+1) = X^T(n)W(n) \quad (7)$$

The prediction coefficient vector $W(n)$ can be obtained by recursive least square method (RLS).

$$Q(0) = \delta^{-1}I \quad (8)$$

Where δ is a small normal number, I is unit matrix.

$$W(0) = 0 \quad (9)$$

$$G(n) = \frac{\lambda^{-1}Q(n-1)X(n)}{1 + \lambda^{-1}X^T(n)Q(n-1)X(n)} \quad (10)$$

Where λ is forgetting factor.

$$\alpha(n) = D(n) - W^T(n-1)X(n) \quad (11)$$

Where $D(n)$ is ideal output signal.

$$W(n) = W(n-1) + G(n)\alpha(n) \quad (12)$$

$$P(n) = \lambda^{-1}Q(n-1) - \lambda^{-1}G(n)X^T(n)Q(n-1) \quad (13)$$

By iterative operation of Eqs. (10)–(13), $W(n)$ can be obtained.
The phase space reconstruction of x_i is carried out.

$$X_{N_m \times m} = \begin{bmatrix} x_1 & x_{1+\tau} & x_{1+2\tau} & \cdots & x_{1+(m-1)\tau} \\ x_2 & x_{2+\tau} & x_{2+2\tau} & \cdots & x_{2+(m-1)\tau} \\ \vdots & \vdots & \vdots & & \vdots \\ x_{N_m} & x_{N_m+\tau} & x_{N_m+2\tau} & \cdots & x_{N_x} \end{bmatrix} \quad (14)$$

Where $N_m = N_x - (m-1)\tau$, τ is delay, m is embedded dimension.
The input sample and ideal output for training prediction coefficient can be obtained from Eq. (14). The input sample is shown in Eq. (15).

$$X = [X_1, X_2, \ldots, X_{N_x - m\tau}]^T = \begin{bmatrix} x_1 & x_{1+\tau} & x_{1+2\tau} & \cdots & x_{1+(m-1)\tau} \\ x_2 & x_{2+\tau} & x_{2+2\tau} & \cdots & x_{2+(m-1)\tau} \\ \vdots & \vdots & \vdots & & \vdots \\ x_{N_x - m\tau} & x_{N_x-(m-1)\tau} & x_{N_x-(m-2)\tau} & \cdots & x_{N_x-\tau} \end{bmatrix} \tag{15}$$

The ideal output is shown in Eq. (16).

$$Y = [x_{1+m\tau}, x_{2+m\tau}, \ldots, x_{N_x}]^T \tag{16}$$

The sample can be predicted by iterative algorithm.

$$x_n = f([x_{n-m\tau}, x_{n-(m-1)\tau}, \ldots, x_{n-\tau}]) \tag{17}$$

Where f is the prediction function corresponding to the second-order Volterra series prediction model.

3 Establishment of Motor Fault Warning Model

Faced with complex and diverse forms of multi-sensor information fusion, experts and scholars at home and abroad hold different views on multi-parameter fusion technology, but also reached a consensus that multi-sensor information fusion forms are basically divided into three types: data-level fusion, feature-level fusion and decision-level fusion [9].

Motor fault diagnosis is closely related to the motor-related mechanical system. Rotating fault diagnosis is a larger field, but it mainly focuses on mechanical faults. Fault diagnosis of rotating machines is a larger research field, but it mainly focuses on mechanical faults. Motor fault diagnosis is more complicated because it involves mechanical and electrical parts. At the same time, motor condition monitoring, maintenance and repair are closely related to motor fault diagnosis, so motor fault diagnosis is very important.

Fault diagnosis is actually based on the decision problem of pattern classification. The acquisition of the dataset of the pattern classification and the feature extraction are the key to the problem. Statistical classification techniques, intelligent classification techniques, and analytical classification techniques are all based on feature extraction of specific data sets. Uncertainty theory is the basis of these methods in the solution of decision problems. Therefore, a motor fault diagnosis process can be regarded as the process of sensor feature decision.

(1) Data level fusion

For the same kind of information of multiple sensors, the data can be directly fused, and the unprocessed sensor monitoring information can be directly combined. The advantage is that objective data is retained as much as possible to avoid data loss and omission, ensuring the highest precision, but its disadvantage is that the information processing is large and lacks real-time. At present, the challenges of multi-sensor data fusion mainly include four cases: data defects, data association, data inconsistency, and different data forms [10]. Its main application is the analysis, synthesis and interpretation of multi-source images. As shown in Fig. 1.

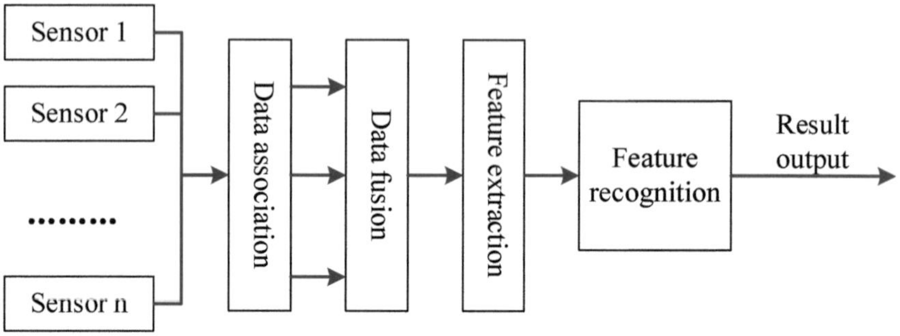

Fig. 1. Structure diagram of centrifugal compressor unit.

(2) Feature level fusion

On the one hand, the feature level fusion algorithm can increase the similarity between the features extracted from the sensor information and the important appearance of the corresponding environment, and on the other hand, can systematically create some additional combined feature information [11]. The effective information is extracted from different types of sensor data, and the feature vector is formed to represent and describe the original system, and then the data of different sensors are classified, analyzed of correlation, and fused with reference to the feature information. The advantage is not only to retain the main data information, but also improves the real-time of multi-sensor data analysis and maintains high precision, through a certain data processing method. However, the disadvantage is that some data may be lost, so the requirements for sensor information preprocessing are strict. The main scope of application is the fusion processing of homogeneous or heterogeneous multi-source information, multi-sensor target tracking. As shown in Fig. 2.

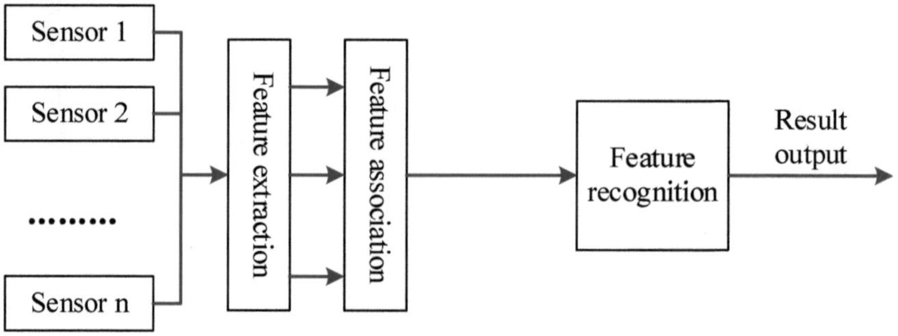

Fig. 2. Feature level fusion block diagram.

(3) Decision level fusion

The multi-sensor data is processed and analyzed first, and the decision results of each sensor are given. Then according to the credibility of each sensor decision and refer to the corresponding rules to comprehensively judge the decision results, and finally form a unified decision. This level of fusion can be directly based on the information of feature level fusion, starting from the actual situation and solving the problem of specific target decision. At this time, the effect of decision making is mainly affected by the feature level fusion information. At present, the algorithm of decision-level fusion mainly includes neural network algorithm, genetic algorithm and ant colony algorithm [12]. The advantages are small information demand, low transmission bandwidth, strong fault tolerance, and can be applied to heterogeneous sensor decisions; but the disadvantage is that the accuracy is lowered, the false judgment rate is increased, and the cost of data processing is relatively high. The results can provide a basis for command control and decision making. As shown in Fig. 3.

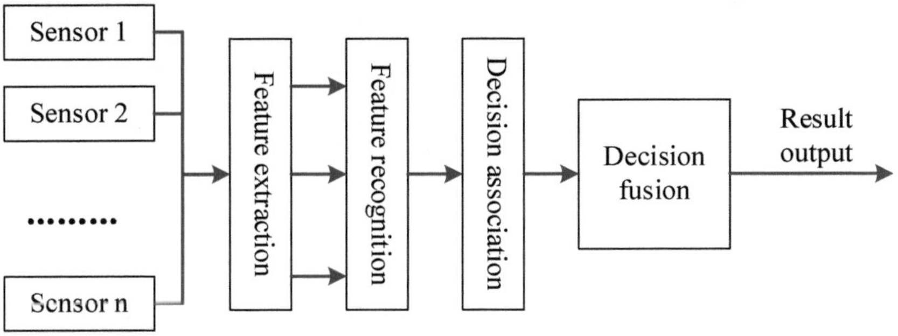

Fig. 3. Decision level fusion block diagram.

Multi-sensor information fusion compensates for the shortcomings of single sensor information by maximizing the correlation and complementarity of multiple homogeneous or heterogeneous sensor information. After comprehensive analysis, information about the equipment to be diagnosed is obtained with high accuracy and good comprehensiveness, and the accuracy and reliability of the whole system are improved. When a sensor or some sensors fail in the system, these sensors cannot transmit information that can characterize the state of the device to be diagnosed. However, due to the redundancy and correlation between the information of multiple sensors, the information fusion technology can make full use of other sensors to obtain relevant information and ensure that the normal operation of the system. Therefore, multi-sensor information fusion improves the reliability, accuracy and fault tolerance of the fault warning system.

All the information of sensor is extracted; the data information collected by the same type of sensor is fused. Then the feature data of the homologous information is extracted as the feature subspace. Different subspaces can be predicted by the multi-sensor signals in the second-order Volterra prediction model. By comparing with the limit value of the fault database, the fault identification of this kind of signal can be realized. At the same time, all subspaces constitutes a fused feature space. The predictive model is used to obtain the rotating equipment fault prediction model based on multi-sensor fusion. By comparing the fault features predicted by data fusion and multi-source information fusion with the fault mode library, the fault modes of the same multi-sensor information and different types of sensor information are identified to achieve the purpose of warning.

This paper designs a second-order Volterra fault warning system based on information fusion model combined with the characteristics of rotating equipment faults. The structure is shown in Fig. 4.

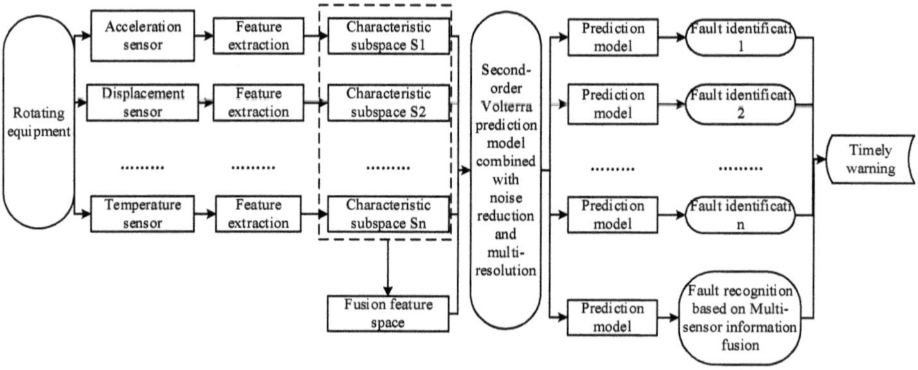

Fig. 4. Structure diagram of multi-sensor information fusion fault early warning system.

4 Application Examples

4.1 Simulation Signal Prediction Example

250 normal data is intercepted from a point on the X component of the Henon mapping, and the first 200 data are used as training samples (as shown in Fig. 5), and the next 50 data are predicted. Figure 6 shows the comparison of predicted and ideal outputs.

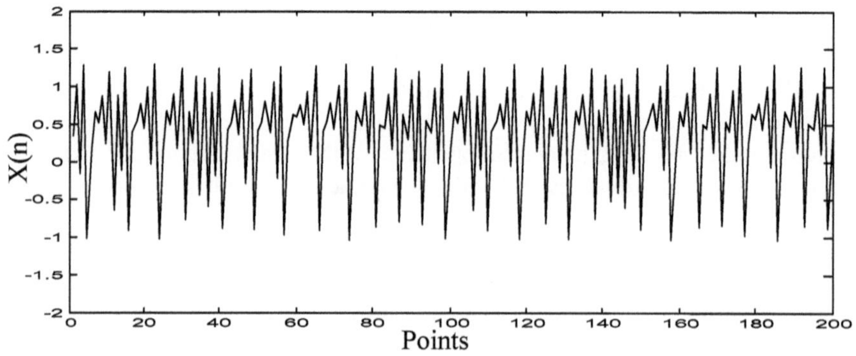

Fig. 5. Ideal Henon map X component.

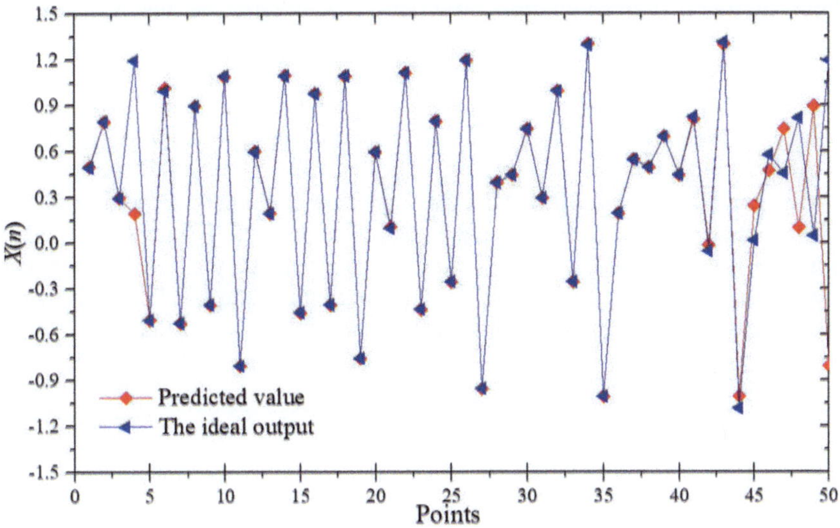

Fig. 6. Second-order Volterra series prediction results.

It can be seen from Fig. 6 that the second-order Volterra series can accurately predict the next 50 data of Henon time series samples, and the prediction relative error is almost zero. Only the last few prediction data errors are slightly larger.

4.2 Measured Signal Prediction Example

Now take an electric motor as the research object, collect the effective value of the vibration, temperature and axis displacement signals on the site, and feature fusion is carried out for subspace signals of the same type of sensors, and constitute the subspace of each source signal. The weight coefficient of each sensor in the homologous information is set by applying the weight algorithm. A single value is used to represent the signal change of the sensor and the state change of the motor.

The information of this kind of sensor, represented by a single value obtained, is taken as a new subspace, and the single value subspaces of different sources are weighted and fused. Finally, the eigenvalue data is fused into a fusion space which can represent the overall operation of the motor. The second-order Volterra prediction model is used to make the final prediction and achieve the final warning of compressor failure through pattern recognition.

The weighted method is used to fuse the data. According to the empirical analysis, the importance of the vibration sensor is basically the same in the vibration test. The weighting coefficient of each sensor is set to be the same. After the data is fused, 100 sample data representing vibration in the process of vibration operation were obtained, namely the running state of 50 min. The group of data can represent the overall state of motor vibration. The second-order Volterra prediction is performed by using the weighted fusion of the 100 sets of data. The 20 sets of data obtained by the prediction result, that is, the prediction of the vibration state of 10 min, can reveal the development trend of the vibration information. Similarly, the fusion space representing the axis displacement and temperature information is also obtained by a weighting method.

The vibration, axis displacement and temperature information data obtained by fusion are then weighted and fused. Based on the existing experience, according to the degree of contribution of various indexes to the state of the whole motor, the vibration weighting coefficient is set to 0.5, the axis displacement weighting coefficient is set to 0.25, and the temperature weighting coefficient is set to 0.25. Finally, a group of 100 data fusion space is obtained, and the next 20 sets of data are predicted by the second-order Volterra, so as to represent the development trend of the whole motor operation. The final predicted trend is shown in Fig. 7.

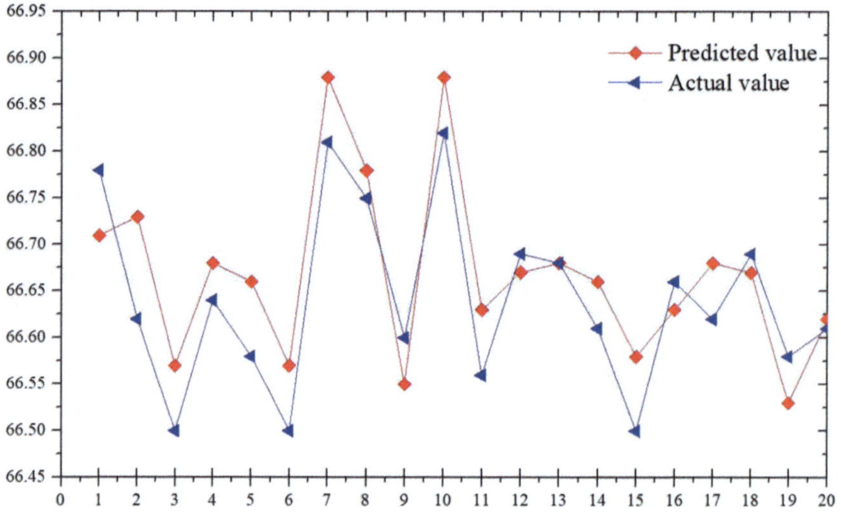

Fig. 7. Sensor information fusion prediction trend graph.

The obtained prediction data is compared with the fusion data collected by the sensor to obtain a range under the normal operation of the motor, that is, the numerical set [66, 67], and the amplitude range is set into the failure mode library. If the predicted value exceeds this range, the motor may fail in the following time period, so that early warning and maintenance can be provided to prevent the occurrence of equipment accidents. In the example, the fault prediction value of the motor is in the defined normal range, which indicates that the motor works normally in the future. It is consistent with the actual situation, and verifies the correctness and reliability of the fusion prediction technology.

5 Conclusion

In this paper, a motor fault warning model is established based on multi-sensor data and second-order Volterra series. At the same time, the development trend of motor operation is predicted according to motor monitoring data. In order to verify the accuracy of the model, the prediction of simulation data and measured data is carried out respectively: (1) 250 normal data are intercepted from a point on the X component of the Henon mapping. The first 200 data are used as training samples, and the last 50 data are predicted and verified. The relative error of the prediction is almost zero, and only the last few prediction data errors are slightly larger. (2) The measured data of vibration, axial displacement, temperature, pressure and flow rate of the motor are weighted and fused, and the second-order Volterra series is input for prediction, so as to obtain a range under normal operation of the motor, namely the numerical set [66, 67]. The fault prediction value of the motor is within the defined normal range, which indicates that the motor works normally in the future. It is consistent with the actual situation, and verifies the correctness and reliability of the fusion prediction technology.

Acknowledgements. This paper is supported by SINOPEC Gas Company (No. 35150014-15-ZC0607-0002).

References

1. Tavner, P.J., Hammond, P., Penman, J.: Contribution to the study of leakage fields at the ends of rotating electrical machines. Proc. IEE **125**(125), 1339–1349 (1978)
2. Schoen, R.R., Habetler, T.G., Kamran, F., et al. Motor bearing damage detection using stator current monitoring. IEEE Trans. Ind. Appl. **31**(6), 1274–1279 (2002)
3. Han, L., Hong, J., Wang, D.: Fault diagnosis of aero-engine bearings based on wavelet package analysis. J. Propuls. Technol. **30**(3), 327–328 (2009)
4. Cheang, T.S., Zhang, L.A: New prototype of diagnosis system of inner-faults for three-phase induction motors developed by expert system. In: International Conference on Electrical Machines & Systems, IEEE (2001)
5. Jafari, H., Poshtan, J.: Fault isolation and diagnosis of induction motor based on multi-sensor data fusion. In: Power Electronics, Drives Systems & Technologies Conference (2015)
6. Wang, G.W., Zhuang, J., Yu, D.H.: Research and application of manifold learning to fault diagnosis of reciprocating compressor. In: Seventh International Conference on Fuzzy Systems & Knowledge Discovery (2010)
7. Smeeton, P., Bousbaine, A.: fault diagnostic testing using partial discharge measurements on high voltage rotating machines. In: Universities Power Engineering Conference (2009)
8. Banerjee, T.P., Das, S.: Multi-sensor data fusion using support vector machine for motor fault detection. Inf. Sci. **217**(24), 96–107 (2012)
9. Jafari, H., Poshtan, J.: Fault isolation and diagnosis of induction motor based on multi-sensor data fusion. In: Power Electronics, Drives Systems & Technologies Conference (2015)
10. Frenay, B., Verleysen, M.: Classification in the presence of label noise: a survey. IEEE Trans. Neural Netw. Learn. Syst. **25**(5), 845–869 (2014)

11. Honda, K., Shrestha, A., Chinnachodteeranun, R., et al. Landslide early warning system for rural community as an application of sensor Asia. In: World Conference on Agricultural Information & It (2008)
12. Ghimire, R., Zhang, C., Pattipati, K.: A rough set theory-based fault diagnosis method for an electric power steering system. In: IEEE/ASME Trans. Mechatron. 1–1 (2018)

Topological Design of a Rotationally Periodic Wheel Under Multiple Load Cases

Lu Jiang[1,2], Wei Zhang[2], ChengWei Wu[2(✉)], LiPing Zhang[1], YiXiong Zhang[1], and ZhenYu Liu[1]

[1] National Key Laboratory of Science and Technology on Reactor System Design Technology, Nuclear Power Institute of China, Chengdu 610041, China
[2] State Key Laboratory of Structural Analysis for Industrial Equipment, Dalian University of Technology, Dalian 116024, China
cwwu@dlut.edu.cn

Abstract. This paper is dedicated to designing the overall structural topology for the lightweight design of an automobile wheel. A simplified two-dimensional finite element analysis (FEA) model for the wheel is established, in which the whole wheel structure is first defined as design domain during topology optimization. A rotationally periodic constraint is introduced to design the wheel into structural topology consisting of rotationally repetitive modules. Further, compliance-based topological design under multiple load cases within single module is carried out. In order to achieve a uniform deflection and stiffness distribution around the circumference of wheel, a weighted compliance under multiple load cases is taken as the objective function. In addition, some factors significantly affecting the structural topology are discussed.

Keywords: Wheel · Structural topology · Rotationally periodic structure · Multiple load cases

1 Introduction

The earliest design of wheel is a solid disk due to limitation of materials, structural design and manufacturing techniques. In modern industry, lightweight design has become one of hot research issues. The goal is mainly achieved either by selecting lightweight advanced material [1, 2] or by reducing material usage by introducing holes into structural topology [3, 4]. Though we have limited materials with characteristic of low weight, high strength and good formability, there is a huge space in designing structure using today's advanced design technique, e.g., shape or topology optimization [5–8] and manufacturing technique, e.g., 3D printing [9].

To achieve lightweight, Das [3] optimized the area between the rim and hub of wheel by topology optimization. Xiao et al. [10] proposed multi-objective topology optimization with objective function combining structural compliance and eigenfrequency. In those existing studies, design domains are all confined in partial region rather than overall structural topology. On the other hand, the wheel is a typical rotating structure and usually designed into structure consisting of rotationally repetitive modules [11]. Moses et al. [12] studied the topological design of periodic structures

M. A. Wahab (Ed.): DAMAS 2019, LNME, pp. 651–658, 2020.
https://doi.org/10.1007/978-981-13-8331-1_49

over single module. Zuo et al. [13] carried out stiffness optimization of wheel with rotationally periodic constraint. What they have in common is that structural compliances are all obtained under single static load. Actually, location of the reaction force is keeping moving when wheel is rotating, which means each location undergoes the same load condition in a rotation cycle. In other words, it has non-uniform deflection and stiffness around circumference. The single load case cannot accurately demonstrate the working characteristic.

Based on these problems, this paper is dedicated to designing the overall structural topology of a wheel and the overall wheel structure is first taken as design domain. To rationally simulate the rotating characteristic and achieve uniform stiffness distribution around circumference, multiple load cases are applied on wheel rim and weighted compliance of those load cases is taken as objective function. Effects of same factors affecting the final structural topology are studied.

2 Optimization Modeling

2.1 Optimization Algorithm

In the past three decades, various topology optimization algorithms have been proposed such as homogenization method [14], density method [15], evolutionary structural method [16] and level set method [17]. Among them, variable density method is most widely used in commercial software owing to its simplicity, convenience and high efficiency. In this paper, variable density method is used and solid isotropic material with penalization model (SIMP) is adopted:

$$[K]^e = \bar{\rho}^P [K_0]^e, \tag{1}$$

where $[K]^e$ and $[K_0]^e$ are respectively the penalized and original element stiffness matrix, $\bar{\rho}$ is the element relative density. P is the penalization factor larger than 1. The penalization factor is 5.

2.2 Stiffness Optimization Under Multiple Load Cases

The structure static behavior can be expressed as:

$$[K]\{u\} = \{P\}, \tag{2}$$

where $[K]$ is the structure global stiffness matrix, $\{u\}$ is the nodal displacement vector and $\{P\}$ is the nodal load vector. The structural compliance is commonly represented by:

$$C = \frac{1}{2}\{P\}^T\{u\} = \frac{1}{2}\{u\}^T[K]\{u\}, \tag{3}$$

which is an inverse measure of structural stiffness [18]. The wheel is simplified into two-dimensional model as shown in Fig. 1a. The parameters in Table 1 refer to the setting in the work of Zuo et al. [13]. By contrast, the overall wheel structure is defined as design domain and five bolt holes are introduced.

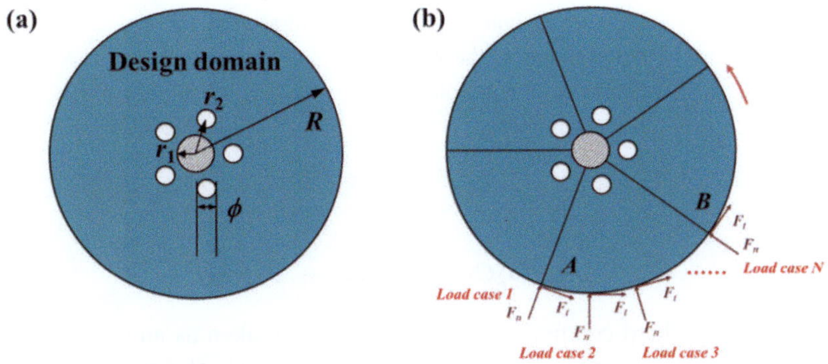

Fig. 1. Optimization model: (a) two-dimensional model of a wheel, (b) schematic of the wheel modeling under multiple load cases

The rotational periodicity of a wheel is a type of infinite periodicity. Single load case with infinite rotationally periodic constraint cannot be achieved as it will impose too strong constraints into design space and gives rise to infinite small bars and holes. However, topology optimization plays its essential role by introducing finite holes. Facing this contradiction, finite rotationally periodic constraint is added. Even with this constraint, wheel design is still unable to achieve a completely uniform stiffness as the single load case cannot simulate the rotating characteristic. To improve the uniformity, we cover the shortage of single load case by employing multiple load cases. Stiffness optimization under multiple load cases within a single module (see segment A-B in Fig. 1b) is studied. Five periods is taken as an example. The inner boundary is fixed and outer rim is subjected to external load. In reality, the external load is a complex contact problem [19]. However, the evaluation by single load case cannot ensure structural strength of wheel during a rotation cycle. The method proposed here maximizing the uniformity of stiffness distribution around circumference under multiple load cases can somehow improve the structural strength. The load case is treated as a moving single point force (number of load cases $N = 101$). These load cases are evenly distributed along rim and each load case consists of tangential and vertical forces.

The structural compliance, C_i, under load case i is given as follows:

$$C_i = \frac{1}{2}\{u\}_i^T[K]\{u\}_i. \tag{4}$$

The objective function is defined as the weighted compliance under all these load cases:

Table 1. Parameters of the wheel

Parameters	Description	Value
r_1	Inner radius of the wheel	30 mm
r_2	Distance of bolt hole center to central axis	60 mm
R	Outer radius of the wheel	235 mm
ϕ	Diameter of bolt hole	30 mm
E	Elastic modulus	210 GPa
μ	Poisson ratio	0.3
ρ	Mass density	7850 kg m^{-3}

$$C^* = \alpha_1 C_1 + \alpha_2 C_2 + \ldots + \alpha_i C_i + \ldots + \alpha_N C_N, \tag{5}$$

where α_i is the weighted coefficient. 101 load cases are taken as an example. Finally, the formulation of the topological design problem of wheel can be expressed as follows:

$$\text{Find.} \bar{\rho} \tag{6a}$$

$$\text{minimize.}\ C^* \tag{6b}$$

$$\text{subject to.}\ V/V_0 = \sum \rho_i v_i / V_0 \leq \delta, \quad i = 1, \ldots, n, \tag{6c}$$

where ρ_i and v_i represent the element density and volume, V_0 and V are respectively the initial volume and the volume after optimization. δ is a value less than 1 representing volume fraction and n is the total number of elements. The values of weighted coefficients are all specified as 1.

3 Optimization Results

3.1 Optimized Wheel Structural Topology

Topology optimization frequently encounters checkerboard pattern [20] and many methods have been developed such as perimeter control [21, 22], mesh independent filtering [23], density slope control [24] and minimum member size (MMS) control method [25]. Here, MMS control method is employed. It penalizes the formation of small members. A volume fraction δ = 0.6 and MMS control of 15 mm is studied. The result is expressed in different colors representing element relative densities (see Fig. 2a). The blue area is occupied by low density elements. In contrast, the red area with high density elements represents the best transmitting routine. Figure 2b shows structural topology after hiding the elements with density smaller than 0.5.

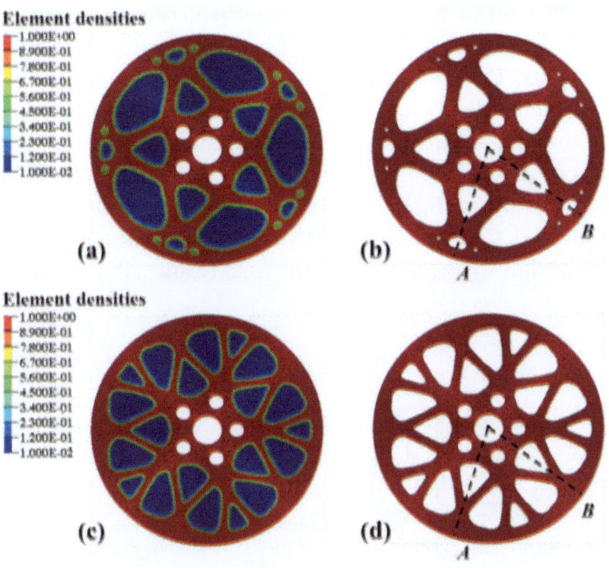

Fig. 2. The optimized wheel: (a)-(b) structural topology obtained under load cases consisting of only tangential force, (c)-(d) structural topology obtained under load cases consisting of both tangential force and vertical force

In fact, for a single load case consisting of tangential force, the stiffness topology optimization is reminiscent of the Michell-type solution. But here, additional restrictions, i.e., the rotationally periodic constraint and MMS control are introduced, and sum of weighted compliance under multiple load cases is taken as objective function. The results shown in Figs. 2a and b are obtained under tangential force. The final structural topology is composed of five rotationally repetitive modules. Each of them consists of a group of ribs and an outer ring. Apparently, this thick enough outer ring generated can help to carry the tangential forces around circumference in rotation cycle. By contrast, what is shown in Figs. 2c and d is obtained under multiple load cases consisting of both tangential force and vertical force. Comparing with Fig. 2b, two ribs are added into each module and connected to outer ring as shown in Fig. 2d. This can compensate for the vertical stiffness and hence helps to undertake the vertical forces around the circumference. Most of the material is distributed in the form of organized ribs connecting with a thick outer ring to undertake the external load.

3.2 Effect of MMS Control

MMS control directly affects the rib size in final structural topology. To study its influence, topological design under volume fraction $\delta = 0.6$ with different MMS controls is carried out. Figure 3a–d indicates that MMS control performs well in avoiding checkerboard pattern. However, a small MMS control gives rise to small size ribs in final structural topology, together with more holes appearing simultaneously. This will increase the complexity of design and difficulty in manufacturing process. By contrast, a big MMS control forces to generate thick ribs and reduce the number of small holes. Figure 3e gives the structural compliance around circumference under different MMS controls. The curve consisting of square symbols represents initial solid disk which shows a completely uniform stiffness distribution around circumference. However, after removing 40% of material, all optimized wheels show non-uniform stiffness distribution. In addition, it is seen that increasing the MMS control will enlarge the variation of the stiffness distribution and hence reduce the uniformity. This result can be explained that a larger MMS control actually will impose a stronger restriction into the design space, and hence the final result will deviate from the optimal solution.

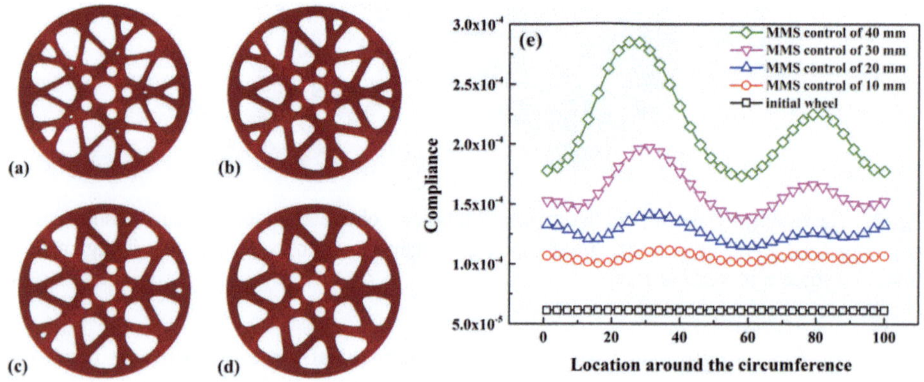

Fig. 3. Structural topologies obtained with MMS control of (a) 10 mm, (b) 20 mm, (c) 30 mm and (d) 40 mm. (e) structural compliances around the circumference

3.3 Effect of Material Usage

The volume fraction determines the remaining material. To study its influence, the wheel with different volume fractions is studied under MMS control of 15 mm as shown in Fig. 4. It can been that, with increasing volume fractions, more material joins the outer ring resulting in thicker outer ring, and also, the ribs grow thicker. The normalized compliance is defined as the ratio of objective function value of optimized wheel to initial wheel. With increasing volume fraction, structural compliance (stiffness) is significantly reduced (increased) at low material usage. However, increasing material usage after volume fraction of about 0.60 will no more contribute to remarkable improvement in stiffness. In short, the reduction rate of structural compliance decreases with increasing material usage.

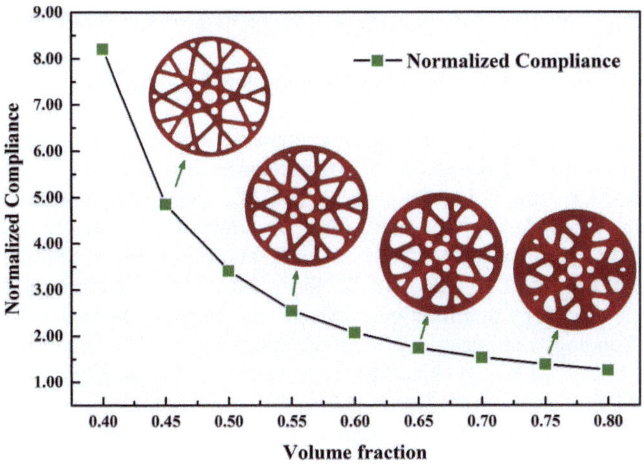

Fig. 4. Normalized compliances of the optimized wheels with increasing volume fractions

4 Conclusions

To explore the lightweight design of wheel, compliance-based topological design is presented in this paper. The overall wheel structure is taken as design domain by employing multiple load cases. By adding rotationally periodic constraint and taking weighted compliance of all load cases as objective function, optimized wheel is composed of repetitive modules which are composed of organized ribs and outer ring. The MMS control shows a remarkable effect in determining the size of ribs and outer ring among the final structural topology. Appropriate material usage needs to be designed to make effective use of the material. The work presented can provide novel guidance for further study of rotationally periodic wheel.

Acknowledgements. This work was supported by the National Natural Science Foundation of China (Grant No. 11772086), and Fundamental Research Funds for the Central Universities of China (Grant No. DUT18ZD302, DUT17ZD229).

References

1. Cole, G.S., MSherman, A.M.: Lightweight materials for automotive applications. Mater. Charact. **35**(1), 3–9 (1999)
2. Miller, W.S., Zhuang, L., Bottema, J., Wittebrood, A.J., De Smet, P., Haszler, A., Vieregge, A.: Recent development in aluminium alloys for the automotive industry. Mater. Sci. Eng., A **280**(1), 37–49 (2000)
3. Das, S.: Design and weight optimization of aluminum alloy wheel. Int. J. Sci. Res. Publ. **4** (6), 2250–3153 (2014)
4. Jiang, L., Wu, C.W.: Topology optimization of energy storage flywheel. Struct. Multi. Optim. **55**, 1917–1925 (2017)

5. Jiang, L., Zhang, W., Wu, C.W.: Shape optimization of energy storage flywheel rotor. Struct. Multi. Optim. **55**(2), 739–750 (2017)
6. Bendsøe, M.P., Sigmund, O.: Topology Optimization: Theory, Methods, and Applications. Springer Science & Business Media (2013)
7. Lin, P., Zhou, P., Wu, C.W.: Multi-objective topology optimization of end plates of proton exchange membrane fuel cell stacks. J. Power Sources **196**, 1222–1228 (2011)
8. Liu, B., Wei, M.Y., Ma, G.J., Zhang, W., Wu, C.W.: Stepwise optimization of endplate of fuel cell stack assembled by steel belts. Int. J. Hydrogen Energy **41**, 2911–2918 (2016)
9. Berman, B.: 3-D printing: the new industrial revolution. Bus. Horiz. **55**(2), 155–162 (2012)
10. Xiao, D.H., Zhang, H., Liu, X.D., He, T., Shan, Y.C.: Novel steel wheel design based on multi-objective topology optimization. J. Mech. Sci. Technol. **28**(3), 1007–1016 (2014)
11. Zuo, Z.H.: Topology optimization of periodic structures. RMIT University (2009)
12. Moses, E., Fuchs, M.B., Ryvkin, M.: Topological design of modular structures under arbitrary loading. Struct. Multi. Optim. **24**(6), 407–417 (2002)
13. Zuo, Z.H., Xie, Y.M., Huang, X.D.: Reinventing the wheel. J. Mech. Des. **133**(2), 024502 (2011)
14. Bendsøe, M.P., Kikuchi, N.: Generating optimal topologies in structural design using a homogenization method. Comput. Methods Appl. Mech. Eng. **71**(2), 197–224 (1988)
15. Bendsøe, M.P., Sigmund, O.: Material interpolation schemes in topology optimization. Arch. Appl. Mech. **69**(9–10), 635–654 (1999)
16. Xie, Y.M., Steven, G.P.: A simple evolutionary procedure for structural optimization. Comput. Struct. **49**(5), 885–896 (1993)
17. Wang, M.Y., Wang, X.M., Guo, D.M.: A level set method for structural topology optimization. Comput. Methods Appl. Mech. Eng. **192**(1), 227–246 (2003)
18. Liang, Q.Q., Steven, G.P.: A performance-based optimization method for topology design of continuum structures with mean compliance constraints. Comput. Methods Appl. Mech. Eng. **191**(13), 1471–1489 (2002)
19. Arslan, M.A., Kayabaşı, O.: 3-D rail-wheel contact analysis using FEA. Adv. Eng. Softw. **45** (1), 325–331 (2012)
20. Diaz, A., Sigmund, O.: Checkerboard patterns in layout optimization. Struct. Multi. Optim. **10**(1), 40–45 (1995)
21. Ambrosio, L., Buttazzo, G.: An optimal design problem with perimeter penalization. Calc. Var. Partial. Differ. Equ. **1**(1), 55–69 (1993)
22. Haber, R.B., Jog, C.S., Bendsøe, M.P.: A new approach to variable-topology shape design using a constraint on perimeter. Struct. Multi. Optim. **11**(1), 1–12 (1996)
23. Sigmund, O.: Morphology-based black and white filters for topology optimization. Struct. Multi. Optim. **33**(4–5), 401–424 (2007)
24. Petersson, J., Sigmund, O.: Slope constrained topology optimization. Int. J. Numer. Meth. Eng. **41**(8), 1417–1434 (1998)
25. Zhou, M., Shyy, Y.K., Thomas, H.L.: Checkerboard and minimum member size control in topology optimization. Struct. Multi. Optim. **21**(2), 152–158 (2001)

Effect of Loading Conditions in Fretting Fatigue on Wear Characteristics

S. Wang(✉) and M. Abdel Wahab

Laboratory Soete, Ghent University, Ghent, Belgium
Shengjie.Wang@UGent.be

Abstract. Fretting is a phenomenon that happens when there is a small slip amplitude between two contact surfaces. Fretting can cause damages like fretting wear, fretting fatigue, and fretting corrosion. These damages can affect each other in the real case, which is ignored in most research. In this paper, the effect of loading conditions such as axial stress, tangential stress and normal load in fretting fatigue on wear characteristics is analyzed in partial slip regime. The finite element fretting model of pad-on-specimen is designed to analyze the difference in wear characteristics between fretting fatigue models with different loading conditions. The results show that by increasing axial stress, the wear depth and wear width increase near the right contact edge and decrease near the right contact edge, while with the increment of the tangential load and normal load, the wear depth and width increase near both contact edges. Moreover, by increasing the normal load, the point of maximum depth moves closer to the contact center.

Keywords: Fretting wear · Fretting fatigue · Finite element method · Findley parameter

1 Introduction and Background

Fretting happens between two contact components when there is a small oscillatory movement between them [1]. Fretting can lead to wear and fatigue, which is sometimes detrimental for the contact parts [2, 3]. There are three fretting regimes: stick regime, mixed stick and slip regime, and gross slip regime, which are classified according to the relative slip amplitude [4, 5]. In the stick regime, limited debris is generated and no fatigue crack is observed. In the mixed stick and slip regime, fatigue is the dominant damage, meanwhile, wear also occurs. In the gross slip regime, wear is mostly observed. At present, many researchers have studied the fretting wear in gross slip regime, e.g. Refs. [6–11]. The initiation and propagation of fretting fatigue are analysed in Refs. [12–16] in partial slip regime. Most researchers analyse the fretting wear and fretting fatigue separately, which is not the real case.

The relations between fretting wear, fretting fatigue, and slip amplitude is shown in Fig. 1 [1]. In partial slip regime, fretting wear also occurs and may have some effect of fretting fatigue. The effect of fretting wear in partial slip regime is studied by many researchers. Ding et al. [17] studied fretting fatigue of the splined coupling and found that in partial slip regime, fretting wear can increase the fretting fatigue damage. To

M. A. Wahab (Ed.): DAMAS 2019, LNME, pp. 659–665, 2020.
https://doi.org/10.1007/978-981-13-8331-1_50

quantify the effect of wear profile on fretting fatigue, the parameter, D_{fret2}, was proposed by Ding et al. in Ref. [18]. Zhang et al. [19] studied the effect of fretting wear on fretting fatigue in the press-fitted shaft under rotating bending and concluded that the stress concentration at the contact edge was significantly relieved due to fretting wear. Shen et al. [20] found that in partial slip regime, the maximum contact pressure, and shear traction increased greatly due to the profile evolution of the contact surface induced by wear. Basseville et al. [21] analyzed the competition between fretting wear and crack initiation of Ti-6Al-4V by stress-based DV (Dang Van) model [22] and strain–stress based SWT (Smith–Watson–Topper) model [23]. The results showed that the DV parameter and SWT parameter decreased with the number of cycles. Furthermore, if the crack did not initiate at the beginning, the wear would become predominant.

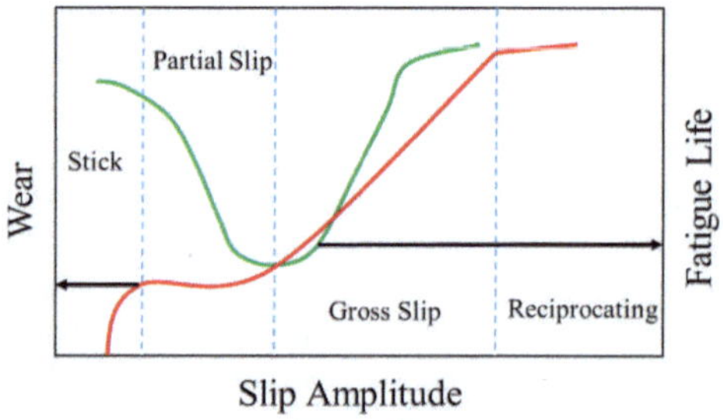

Fig. 1. The relations between fretting wear, fretting fatigue and slip amplitude

In common fretting fatigue model, axial stress, normal load and tangential load are applied, while in fretting wear model, the oscillatory displacement and the normal load are applied. Therefore, the effect of axial stress, tangential load and normal load on wear characteristics is analysed in this paper based on the fretting fatigue model in Ref. [24].

2 Model Details

The dimension and boundary conditions of the finite element (FE) model are shown in Fig. 2. The normal load, P, which is applied to the top centre of the pad. Axial pressure, σ_{axial}, is applied to the right side surface of the specimen. In the experiment, the left side of the specimen is fixed. To get the reaction pressure, $\sigma_{reaction}$, the following equation is used Ref. [24]:

$$\sigma_{axial} - \sigma_{reaction} = \frac{Q}{As} \tag{1}$$

Where Q is the tangential load, which can be measured in the experiment and As is the cross-section of the specimen. The thickness of the specimen is take as 4 mm.

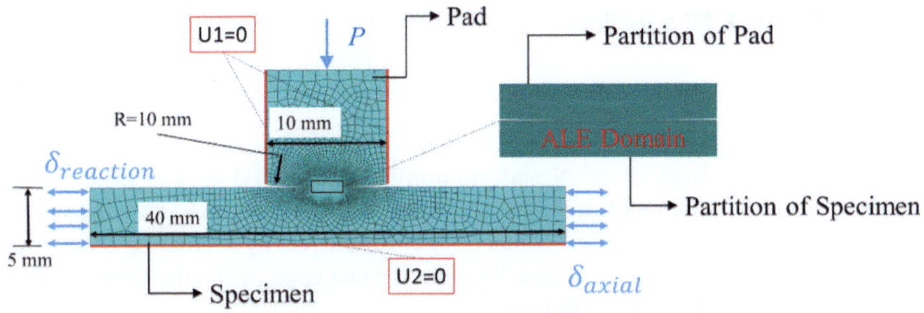

Fig. 2. Dimension and boundary condition of the FE model

The partition of the specimen is set as the ALE adaptive mesh domain to simulate the material removal of the contact surface. The loading condition is shown in Fig. 3. To simplify the simulation process, 1000 cycles is used as jump cycles. More details can be found in Ref. [24].

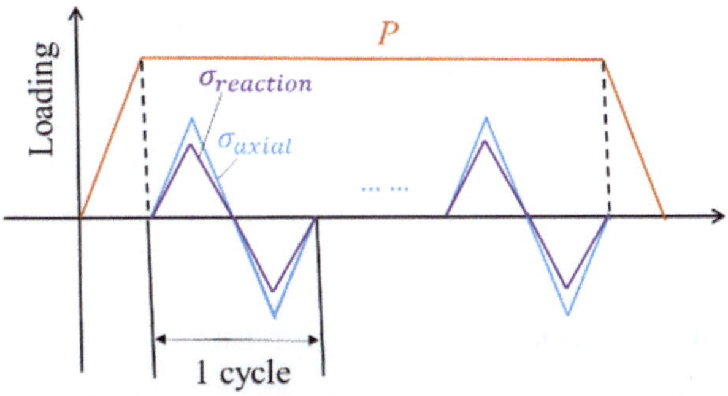

Fig. 3. Loading condition of the FE model

The material of the pad and the specimen is Ti-6Al-4V. Young's modulus and Poisson ratio are 121 GPa and 0.29 respectively. The friction coefficient between the pad and the specimen is 0.8 [25]. The energy wear coefficient K is 2×10^{-7} MPa^{-1}, which is based on dissipated energy wear model as in Ref. [20].

The contact pressure and shear stress are validated using the analytical solution and the simulation results from Ref. [24] when the material is aluminum alloy Al7075-T6. Both results show a good agreement with the reference results. Then, this FE model can be used to analyse the fretting wear and fretting fatigue in partial slip regime for the material Ti-6Al-4V.

3 Results and Discussion

3.1 Wear Profiles with Different Axial Stresses

The wear profiles after 5000 cycles with different axial stresses are shown in Fig. 4, when the tangential load is 150 N and the normal load is 500 N. The axial stress is applied to the right side, and therefore the wear near the right contact edge is more significant compared with the left contact edge. Moreover, by increasing axial stress, the wear depth and width increase near the right contact edge, while the wear depth and width decrease near the left contact edge.

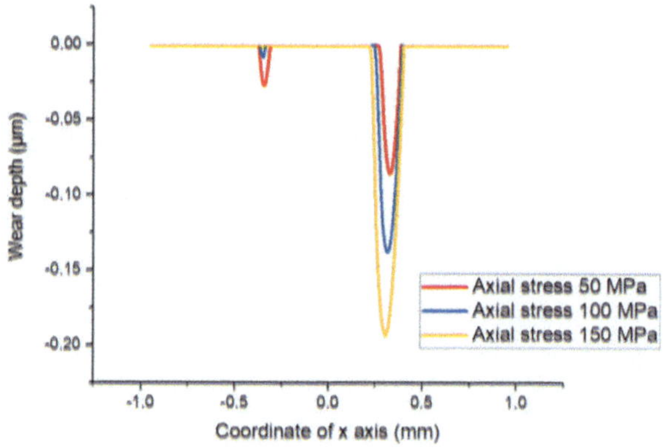

Fig. 4. The wear profiles after 5000 cycles with different axial stresses

3.2 Wear Profiles with Different Tangential Loads

The wear profiles after 5000 cycles with different tangential loads are shown in Fig. 5, when the axial stress is 100 MPa and the normal load is 550 N. The wear depth and wear width around both contact edge both with the increment of the tangential load.

3.3 Wear Profiles with Different Normal Loads

The wear profiles after 5000 cycles with different normal loads are shown in Fig. 6, when the axial stress is 100 MPa and the tangential load is 150 N. By increasing normal load, the wear depth and wear width around both contact edge increase. Compared with the wear profiles shown in Figs. 4 and 5, the point of the maximum wear depth at both sides moves closer to the contact centre with the increment of the normal load, as shown in Fig. 6.

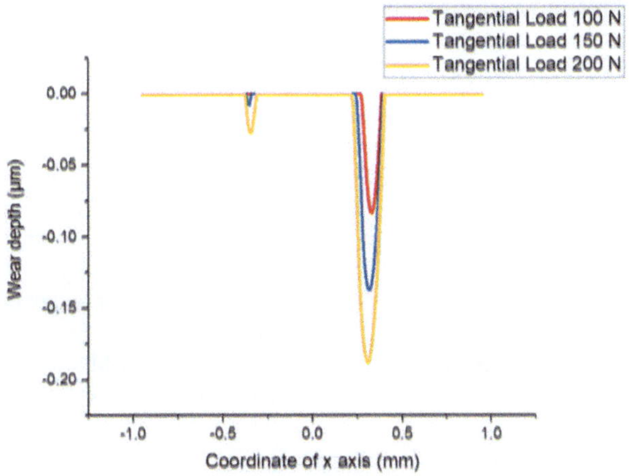

Fig. 5. The wear profiles after 5000 cycles with different tangential loads

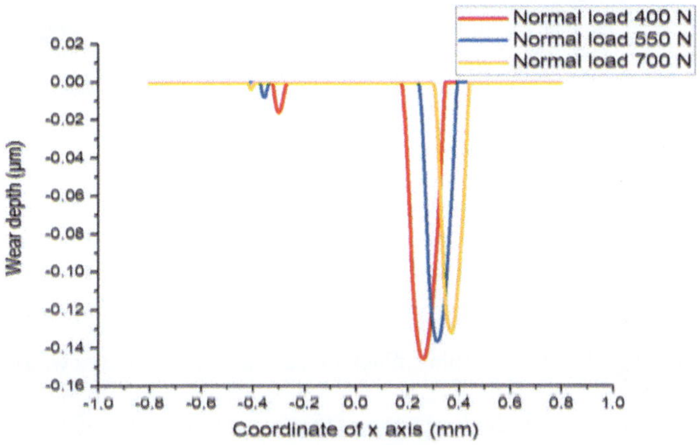

Fig. 6. The wear profiles after 5000 cycles with different normal loads

4 Conclusion

In this paper, the effect of loading conditions in fretting fatigue on wear characteristics is analysed. It can be concluded that by increasing the axial stress, the wear depth and wear width increase near the right contact edge and decrease near the left contact edge, while by increasing the tangential load and normal load, the wear depth and width increase near both contact edges. Moreover, the point of the maximum wear depth moves closer to the contact centre with the increment of the normal load.

References

1. Vingsbo, O., Söderberg, S.: On fretting maps. Wear **126**, 131–147 (1988)
2. Ovcharenko, A., Etsion, I.: Junction growth and energy dissipation at the very early stage of elastic-plastic spherical contact fretting. J. Tribol. **131**, 031602 (2009)
3. Fouvry, S., Kapsa, P., Vincent, L.: Quantification of fretting damage. Wear **200**, 186–205 (1996)
4. Yan, S., Zhang, D., Shirong, G.: Effect of fretting amplitudes on fretting wear behavior of steel wires in coal mines. Min. Sci. Technol. (China) **20**, 803–808 (2010)
5. Li, J., Lu, Y.: Effects of displacement amplitude on fretting wear behaviors and mechanism of Inconel 600 alloy. Wear **304**, 223–230 (2013)
6. Leonard, B.D., Ghosh, A., Sadeghi, F., Shinde, S., Mittelbach, M.: Third body modeling in fretting using the combined finite-discrete element method. Int. J. Solids Struct. **51**, 1375–1389 (2014)
7. Leonard, B.D., Patil, P., Slack, T.S., Sadeghi, F., Shinde, S., Mittelbach, M.: Fretting wear modeling of coated and uncoated surfaces using the combined finite-discrete element method. J. Tribol. **133**, 021601 (2011)
8. McColl, I., Ding, J., Leen, S.: Finite element simulation and experimental validation of fretting wear. Wear **256**, 1114–1127 (2004)
9. Pereira, K., Yue, T., Wahab, M.A.: Multiscale analysis of the effect of roughness on fretting wear. Tribol. Int. **110**, 222–231 (2017)
10. Yue, T., Wahab, M.A.: A numerical study on the effect of debris layer on fretting wear. Materials **9**, 597 (2016)
11. Yue, T., Wahab, M.A.: Finite element analysis of fretting wear under variable coefficient of friction and different contact regimes. Tribol. Int. **107**, 274–282 (2017)
12. Bhatti, N.A., Pereira, K., Wahab, M.A.: Effect of stress gradient and quadrant averaging on fretting fatigue crack initiation angle and life. Tribol. Int. **131**, 212–221 (2019)
13. Bhatti, N.A., Wahab, M.A.: Fretting fatigue damage nucleation under out of phase loading using a continuum damage model for non-proportional loading. Tribol. Int. **121**, 204–213 (2018)
14. Bhatti, N.A., Wahab, M.A.: Fretting fatigue crack nucleation: a review. Tribol. Int. **121**, 121–138 (2018)
15. Hojjati-Talemi, R., Wahab, M.A.: Fretting fatigue crack initiation lifetime predictor tool: Using damage mechanics approach. Tribol. Int. **60**, 176–186 (2013)
16. Hojjati-Talemi, R., Wahab, M.A., De Pauw, J., De Baets, P.: Prediction of fretting fatigue crack initiation and propagation lifetime for cylindrical contact configuration. Tribol. Int. **76**, 73–91 (2014)
17. Ding, J., Leen, S., Williams, E., Shipway, P.: Finite element simulation of fretting wear-fatigue interaction in spline couplings. Tribol. Mater. Surf. Interfaces **2**, 10–24 (2008)
18. Ding, J., Houghton, D., Williams, E., Leen, S.: Simple parameters to predict effect of surface damage on fretting fatigue. Int. J. Fatigue **33**, 332–342 (2011)
19. Zhang, Y., Lu, L., Zou, L., Zeng, D., Zhang, J.: Finite element simulation of the influence of fretting wear on fretting crack initiation in press-fitted shaft under rotating bending. Wear **400**, 177–183 (2018)
20. Shen, F., Hu, W., Meng, Q.: A damage mechanics approach to fretting fatigue life prediction with consideration of elastic–plastic damage model and wear. Tribol. Int. **82**, 176–190 (2015)
21. Basseville, S., Cailletaud, G.: An evaluation of the competition between wear and crack initiation in fretting conditions for Ti–6Al–4V alloy. Wear **328**, 443–455 (2015)

22. Dang-Van, K.: Macro-micro approach in high-cycle multiaxial fatigue. Advances in multiaxial fatigue, ASTM International (1993)
23. Smith, K.N.: A stress-strain function for the fatigue of metals. J. Mater. **5**, 767–778 (1970)
24. Kumar, D., Biswas, R., Poh, L.H., Wahab, M.A.: Fretting fatigue stress analysis in heterogeneous material using direct numerical simulations in solid mechanics. Tribol. Int. **109**, 124–132 (2017)
25. Paulin, C., Fouvry, S., Meunier, C.J.W.: Finite element modelling of fretting wear surface evolution: application to a Ti–6A1–4V contact. **264**, 26–36 (2008)

Damage Assessment in Fretting Fatigue Specimens with Micro-voids Using Critical Plane Approach

D. Infante-García[1(✉)], H. Miguelez[1], E. Giner[2], and M. Abdel Wahab[3]

[1] Department of Mechanical Engineering, Universidad Carlos III de Madrid, Avda. de La Universidad 30, 28911 Leganés, Madrid, Spain
dinfante@ing.uc3m.es

[2] Centre of Research in Mechanical Engineering – CIIM, Department of Mechanical and Materials Engineering, Universitat Politècnica de València, Camino de Vera, 46022 Valencia, Spain

[3] Soete Laboratory, Faculty of Engineering and Architecture, Ghent University, Technologiepark Zwijnaarde 903, 9052 Zwijnaarde, Belgium

Abstract. In this work, the influence of micro-voids on damage severity under fretting fatigue conditions is analysed for a regular and a random distribution of micro-voids. The finite element method is employed to obtain the stress field during a loading cycle. The well-known Smith-Watson-Topper and McDiarmid multiaxial fatigue criteria are assessed by means of critical plane analysis. Additionally, averaging methods are employed in order to consider the length scale for damage initiation. Furthermore, a direct comparison with the homogeneous case is presented. Different size and distribution of micro-voids are analysed. The numerical results show that the heterogeneity has a noticeable influence on the damage severity. In addition, the numerical model suggests that damage may firstly initiate at the upper edge of the micro-voids located close to the contact edge, generally leading to a mean increase of the damage severity. However, in some cases, the introduction of micro-voids reduces the stress intensity at the contact edge, and therefore, decreases the damage severity in the vicinity of the contact edges.

Keywords: Fretting fatigue · Micro-voids · Damage

1 Introduction

Fretting happens when two or more solids are in contact with some small relative displacement [1]. Fatigue lifetime of components under fretting conditions is significantly reduced compared to plain fatigue [2]. Most of the numerical models found in the literature usually take the material as perfectly homogeneous [2–6]. However, real materials are not completely homogeneous, and they may present some micro-porosity or defects due to manufacturing as in metal additive manufacturing [7]. In this way, the influence of the heterogeneity should be analysed for accurate prediction when using numerical methodologies [8].

M. A. Wahab (Ed.): DAMAS 2019, LNME, pp. 666–671, 2020.
https://doi.org/10.1007/978-981-13-8331-1_51

According to Forsyth [9], the damage process of fretting fatigue can be divided into two stages: crack initiation stage and crack propagation stage. In this way, numerical models usually consider both stages for lifetime prediction, calculating separately the lifetime of each stage [6]. Following the review of Bhatti and Wahab [10], methodologies employed for lifetime prediction of crack initiation can be classified into stress invariant approaches, critical plane analysis, fretting specific parameters and continuum damage mechanics. Linear elastic fracture mechanics and finite element models are usually employed to predict the lifetime during the crack propagation stage of an initial crack till complete fracture [3, 11].

Little information can be found regarding the influence of heterogeneity in the fretting fatigue behaviour. Kumar et al. [12] studied numerically the influence of voids in the stress state, contact pressure and shear tractions under fretting conditions through Direct Numerical Simulations. Erena et al. [13] analysed numerically the influence of the introduction of voids under fretting conditions in the absence of bulk stress on the damage severity using a critical plane approach and the Smith-Watson-Topper criterion. The authors predicted a decrease in the damage parameter for certain configurations of the micro-voids and suggested its employment as a future palliative. Lastly, Infante-García et al. [14] analysed the influence of voids in the crack initiation lifetime under fretting fatigue conditions.

In this work, the effect of micro-voids on the crack initiation damage severity under fretting fatigue conditions has been analysed numerically using numerical methods. Different micro-void distributions (normal distribution and random distribution), sizes and densities are analysed. A critical plane analysis with averaging methodologies is employed to assess the damage severity of crack initiation. The results are further compared to the homogeneous case to assess the influence of micro-voids.

2 Numerical Model

The experimental data published by Hojjati-Talemi et al. [4] is taken as reference. The test set-up consists of a dog-bone specimen subjected to cyclic axial load (σ_{axia}) and symmetrically clamped under constant normal load (P) to two cylinders of 50 mm radius, both made of AL2024-T3. The two cylinders are fixed with leaf springs to the platform, and therefore, giving rise to a cyclic tangential force (Q) between both components (see Fig. 1). Young's modulus (E), Poisson's ratio (v), yield strength (σ_y) and ultimate strength (σ_u) of AL2024-T3 are taken from [4].

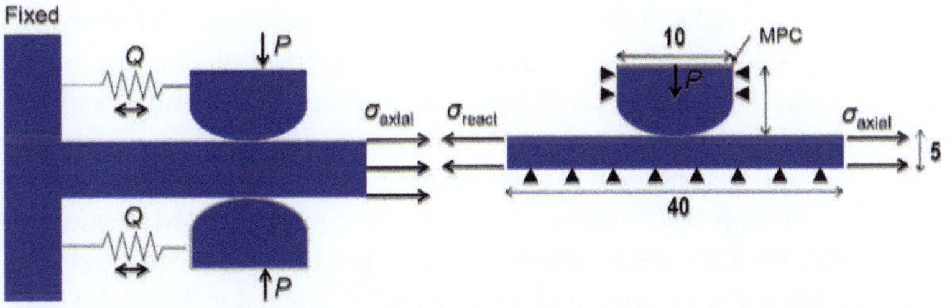

Fig. 1. Sketch of the test configuration (left) and numerical model (right).

The stress ratio of σ_{axial} and Q is 0.1 and −1, respectively. The loading case studied is: normal load equal to 135.75 N/mm, maximum tangential load equal to 38.79 N/mm and maximum axial load equal to 100 MPa. Tests were performed under partial slip conditions. The coefficient of friction between pad and indenter was measured by Hojjati-Talemi et al. in [4] resulting in 0.65.

The finite element method has been employed in order to calculate the stress state during a loading cycle in the specimen. Abaqus Standard © has been employed to develop the finite element models. A quasi-static problem is considered to obtain the stress field under fretting conditions along a cycle, since the influence of inertial effects is negligible in this problem. In addition, the material has been considered purely elastic. Q has been introduced through the definition of the cyclic reaction load (σ_{react}), analogous to [3, 4, 12]. Rotations are avoided by the application of the Multi Point Constraint (MPC) on the top edge of the indenter. The augmented Lagrange multipliers method has been employed to model friction. The problem is simplified due to symmetry and, solely, half of the specimen has been modelled in 2D under plane strain conditions. The x direction and y direction have been restricted on one side of the indenter and the bottom of the specimen, respectively. A structured mesh with an element size of 5 microns is employed in the contact region.

The crack initiation damage severity has been numerically assessed by means of critical plane analysis using the McDiarmid [15] (MD) and Smith-Watson-Topper [16] (SWT) parameters. The MD parameter can be calculated following Eq. 1, being $\Delta\tau_{f\text{-}1}$ the fatigue limit in torsion, $\Delta\tau_{max}$ the plane of maximum shear stress range and $\sigma_{n,max}$ maximum normal stress. The material constant $\Delta\tau_{f\text{-}1}$ can be estimated dividing the uniaxial fatigue limit by $\sqrt{3}$ [5]. On the other hand, the SWT parameter can be calculated following Eq. 2, being the plane normal to the orientation with maximum product of the amplitude of the first principal strain ($\Delta\varepsilon/2$) and first principal tensile stress (σ_{max}).

$$MD = \frac{\Delta\tau_{max}}{2} + \left(\frac{\Delta\tau_{f-1}}{2\sigma_u}\right)\sigma_{n,max} \tag{1}$$

$$SWT = \left(\sigma_{max}\frac{\Delta\varepsilon}{2}\right)_{max} \tag{2}$$

The damage parameters are averaged over a process area to take into account the length scale for crack initiation [17]. In this way, a semi-circular region centred in the maximum value, perpendicular to the free surface and with radius equal to $L/2$, being L the constant of Al-Haddad [18]. In this material, L is around 50 microns [4]. The micro-void shape is idealized to a circle. As commented in the introduction, two distribution of micro-voids are analysed: a normal distribution and a random distribution.

First, the specimen geometry has been divided into rectangular cells of 2 × 1 mm (100 cells). In the normal distribution, the voids are equally spaced inside the cells. In addition, two void densities are analysed: 4 voids per cell and 1 void per cell. On the other hand, 5 percentages of void surface (VS) are studied for 4 and 1 voids per cell (v.p.c.): 14%, 10%, 6%, 3.5% and 1.5%. The mesh has been refined to an element size of 2 microns in the cells contacting the indenter. An analogous method has been employed to develop the cases with random distribution. First, the specimen is again divided into rectangular cells of 2 × 1 mm. Then, a MatLab © script has been developed to define twelve randomly distribution cases with a VS of 1.5% and 4 v.p.c.. The twelve random cases are shown in Fig. 2.

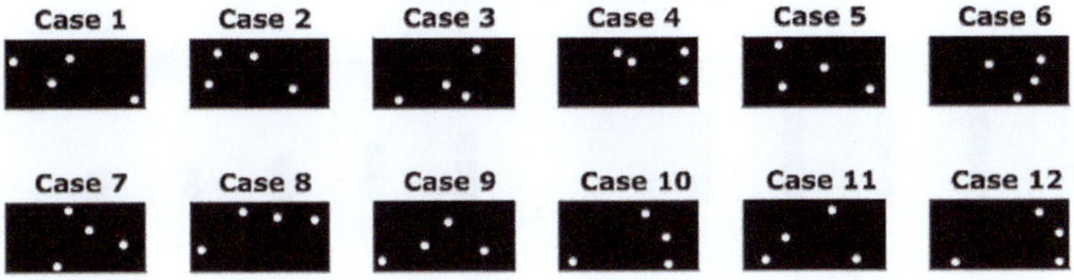

Fig. 2. Sketch of micro-void random studied distributions.

3 Results and Discussion

The introduction of the micro-voids in the specimen changes significantly the contact stress distribution [12], therefore, modifies the mechanics of the problem. The MD and SWT parameters are calculated in the centroid of each element belonging to the two contacting cells. In addition, the parameters are mapped over the geometry to show the damage distribution around the geometry, as shown in Fig. 3. In all studied cases, the damage is concentrated around the right contact edge of the contact and at the upper edge of the micro-void close to the right contact edge (see Fig. 3).

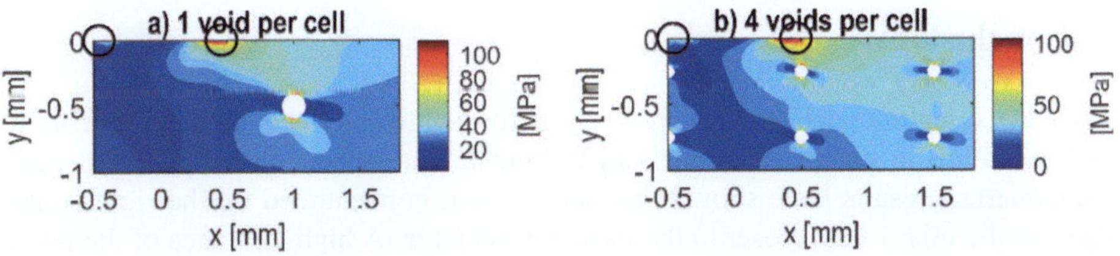

Fig. 3. Damage map of MD parameter for the case of 1 v.p.c. (left) and 4 v.p.c. right with 1.5% of void surface and normal distribution (circles represent contact edges).

The maximum of the parameters is found in one of the two hot-spots, depending on the case. For this reason, the damage parameter is averaged, as explained in the previous section, at the two hot-spots. In addition, the averaged damage parameter is normalized to the averaged damage parameter found when the specimen is considered homogeneous (without voids) and with the same nominal axial stress to make a fairer comparison. As shown in Fig. 4, the averaged damage parameter is higher than in the homogeneous case for all cases except for the micro-void with 1.5% VS. In addition, the damage severity is found to be higher at the micro-void spot than at the contact edge spot. Similar results are found for the case of 4 v.p.c., but with less damage severity. Furthermore, it has been found some cases with a decrease of the damage parameters at both hot-spots. As Erena et al. [13] pointed out, the loss of stiffness produced by the micro-voids may act in some configurations as stress relievers. Therefore, decreasing the damage severity at the contact edges.

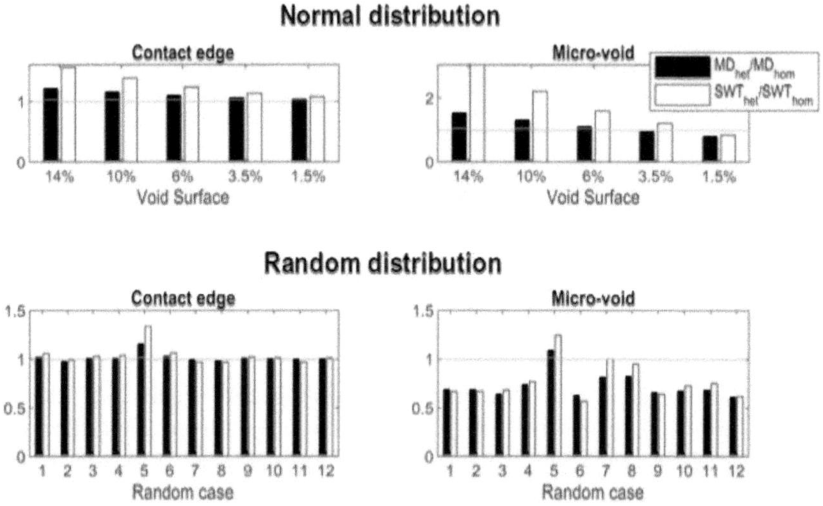

Fig. 4. Bar charts of the normalized MD and SWT averaged parameters for the normal distribution with 1 void per cell and random distribution with 4 void per cell and 1.5% of void surface.

4 Conclusions

In this work, a numerical analysis has been performed to assess the damage severity for crack initiation in numerical models of fretting fatigue specimens with micro-voids. The numerical results have shown that damage was concentrated on the right contact edge and the micro-void closer to the right contact edge. A high influence of the micro-void distribution, size and density has been found. Generally, an increase of the damage severity is produced due to the micro-void introduction. However, in some cases the micro-voids can be beneficial, acting as stress relievers.

Acknowledgements. The authors gratefully acknowledge the financial support given by the Spanish Ministry of Economy and Competitiveness and the FEDER program through the projects DPI2017-89197-C2-1-R, DPI2017-89197-C2-2-R and the FPI subprogram with the reference BES-2015-072070. The support of the Generalitat Valenciana, Programme PROMETEO 2016/007, is also acknowledged. The last author would like to acknowledge the financial support of the Research Foundation-Flanders (FWO), The Luxembourg National Research Fund (FNR) and Slovenian Research Agency (ARRS) in the framework of the FWO Lead Agency project: G018916 N 'Multi-analysis of fretting fatigue using physical and virtual experiments'.

References

1. Hills, D.A., Nowell, D.: Mechanics of fretting fatigue-Oxford's contribution. Tribol. Int. **76**, 1–5 (2013)
2. Hojjati-Talemi, R., Wahab, M.A., Giner, E., Sabsabi, M.: Numerical estimation of fretting fatigue lifetime using damage and fracture mechanics. Tribol. Lett. **52**, 11–25 (2013)
3. Pereira, K., Wahab, M.A.: Fretting fatigue crack propagation lifetime prediction in cylindrical contact using an extended MTS criterion for non-proportional loading. Tribol. Int. **115**, 525–534 (2017)
4. Hojjati-Talemi, R., Wahab, M.A., De Pauw, J., De Baets, P.: Prediction of fretting fatigue crack initiation and propagation lifetime for cylindrical contact configuration. Tribol. Int. **76**, 73–91 (2014)
5. Bhatti, N.A., Wahab, M.A.: A numerical investigation on critical plane orientation and initiation lifetimes in fretting fatigue under out of phase loading conditions. Tribol. Int. **115**, 307–318 (2017)
6. Szolwinski, M.P., Farris, T.N.: Observation, analysis and prediction of fretting fatigue in 2024-T351 aluminum alloy. Wear **221**, 24–36 (1998)
7. Gong, H., Rafi, K., Gu, H., Starr, T., Stucker, B.: Analysis of defect generation in Ti–6Al–4 V parts made using powder bed fusion additive manufacturing processes. Addit. Manuf. **1–4**, 87–98 (2014)
8. Chan, L.C., Lu, X.Z., Yu, K.M.: Multiscale approach with RSM for stress-strain behaviour prediction of micro-void-considered metal alloy. Mater. Des. **83**, 129–137 (2015)
9. Forsyth, P.: Proc crack Propag. Symp. Coll. Aeronaut. **1**, 76–94 (1961)
10. Bhatti, N.A., Wahab, M.A.: Fretting fatigue crack nucleation: a review. Tribol. Int. **121**, 121–138 (2018)
11. Marco, M., Infante-García, D., Díaz-Álvarez, J., Giner, E.: Relevant factors affecting the direction of crack propagation in complete contact problems under fretting fatigue. Tribol. Int. **131**, 343–352 (2019)
12. Kumar, D., Biswas, R., Poh, L.H., Wahab, M.A.: Fretting fatigue stress analysis in heterogeneous material using direct numerical simulations in solid mechanics. Tribol. Int. **109**, 124–132 (2017)
13. Erena, D., Vázquez, J., Navarro, C., Domínguez, J.: Voids as stress relievers and a palliative in fretting. Fatigue Fract. Eng. Mater. Struct. **41**, 2475–2484 (2018)
14. Infante-García, D., Giner, E., Miguélez, H., Wahab, M.A.: Numerical analysis of the influence of micro-voids on fretting fatigue crack initiation lifetime. Tribol. Int. (2019, in press). https://doi.org/10.1016/j.triboint.2019.02.032
15. McDiarmid, D.L.: A general criterion for high cycle multiaxial fatigue failure. Fatigue Fract. Eng. Mater. Struct. **14**, 429–453 (1991)
16. Smith, K., Watson, T.: Topper: a stress-strain function for the fatigue of metals. J. Mater. **5**, 767–778 (1970)
17. Taylor, D.: The theory of critical distances. Eng. Fract. Mech. **75**, 1696–1705 (2008)
18. El Haddad, M.H., Dowling, N.E., Topper, T.H., Smith, K.N.: J integral applications for short fatigue cracks at notches. Int. J. Fract. **16**, 15–30 (1980)

Preliminary Evaluation of Functional Coatings for Marine Based Renewable Energy Applications

M. Hegde[1], Y. Kavanagh[1], B. Duffy[2], and E. F. Tobin[1(✉)]

[1] Institute of Technology Carlow, Carlow, Ireland
edmond.tobin@itcarlow.ie
[2] CREST, Technical University of Dublin, Dublin, Ireland

Abstract. The reliability and maintenance of the tidal turbines is found to be a major problem due to cavitation erosion. The presence of hydrogen sulphide in the microbes leads to Microbially Induced Corrosion (MIC) in objects in the marine environment which can instigate biofouling to occur. Together, the synergistic effect of erosion, corrosion and fouling leads to reduced lifespan of the structural and operational components. Given the sheer scale of the marine renewable industry, which is estimated to reach around €9 billion by 2030, the effects of cavitation, corrosion and biofouling can cause large losses to the industry which will further spur significant costs in the operation of such off-shore technology. This is the main driver behind developing eco-friendly multi-functional sol-gel coatings for marine renewable applications. The present research is designed to investigate the two sol-gel coatings synthesized from organically modified silicon precursor 3-methacryloxypropyltrimethoxysilane (MAPTMS) mixed with zirconium (IV) propoxide. One of the coatings was modified using 1%v/v of hexamethyl di-isocyanate (HMDI) diluted in 60% ethanol (S65) whereas the second coating left unmodified (S6) diluted in 100% ethanol. The coatings were deposited on the aluminium (Al) panel using dip coater. The structure of the coatings was evaluated using ATR-FTIR. The coating properties such as hardness, adhesion and wettability were evaluated using pencil hardness, cross-cut adhesion, and water contact angle. Subsequently, the thermal and the chemical stability of the sol-gel coatings was also evaluated.

Keywords: Cavitation erosion · Renewable energy · Sol-gel coatings · Materials characterisation

1 Introduction

In the development of green energy systems, much of the investment and research is concentrated on the renewable energy resources. The energy extracted from the renewable resources is seen to be green as there is no release of CO_2. Marine renewable energy systems such as tidal turbines and wave energy harvesters convert energy from vast renewable resources [1]. Due to the marine environment, tidal turbines need to resist various harsh conditions such as cavitation erosion [2]. Cavitation erosion is the

M. A. Wahab (Ed.): DAMAS 2019, LNME, pp. 672–683, 2020.
https://doi.org/10.1007/978-981-13-8331-1_52

mechanism which occurs due to the growth and development of vapor bubbles in fluid systems due to rapid pressure drop. These bubbles implode and results in the formation of shock wave and micro jet. The micro jet collapses on the metal and results in cavitation [2, 3]. The cavitation erosion is a challenging erosive mechanism for the researchers as it occurs mainly in fluid-flow. For instance, pumps, water turbines, marine propellers, devices in the chemical and petrochemical industries, diesel engines and pipelines all suffer from cavitation erosion phenomena [4]. Additionally, due to the presence of Cl-ions in sea water, the corrosion of the metals is not minor. Due to this reason, there is a decrease in the operational life of the marine energy systems [5, 6]. Thereby, increasing the maintenance cost of the marine energy systems. Microbially Induced Corrosion (MIC) can also occur on objects in the marine environment due to the presence of hydrogen sulphide in microbes, allowing the instigation of biofouling to occur [7]. Together, the synergistic effect of erosion, corrosion and fouling leads to reduced lifespan of the structural and operational components. Given the sheer scale of the marine renewable industry, which is estimated to reach around €9 billion by 2030, the effects of cavitation, corrosion and biofouling can cause large losses to the industry which will further spur significant costs in the operation of such offshore technology [8].

Wide range of surface treatments are initiated to provide sustenance for the cavitation erosion resistance which incorporates vacuum diffusion, laser surface modifications and anodization. The use of these methods involves the requirement of expensive equipment's and plasma-assisted processes which are complicated [9]. A wide variety of metals and its alloys are employed as structural material in every major industrial sector such as power production [10], automotive, aerospace and marine industries. Specifically, aluminium is used largely as it's available in greater quantities [11]. Aluminium inherently has good corrosive resistance properties, as it forms organically a thin oxide layer on its surface when exposed to air [12]. Regardless, the self-protective efficacy is reduced immensely when exposed to corrosive environment [13–15]. Aiming to enhance the anti-corrosion property of aluminium when it is placed in harsh environments, many researches made use of chromate conversion coatings as it promotes excellent adhesion of organic coatings and thus acts as corrosion protector [16–18]. The compounds comprising of chromium were found to be hazardous to the environment. Thus, the chromate conversion process was banned by environmental regulations [19]. This being the case, there is a need for the development of much safer coatings which are eco- friendly and chromium free.

In comparison with the latest chromate substitute coatings, sol-gel coating method have shown positive results in protecting the aluminium metals from corrosion mechanisms [20]. The sol- gel process has the capability to develop coatings with low temperatures and provides uniform coatings with consistence thicknesses. Wide ranges of inorganic sol-gel coatings have been developed to enhance the anti-corrosion property of the aluminium metals. However, inorganic sol-gels have shown defects. For instance, microcracks, residual porosity and thickness issues. This can be resolved by using organic sol-gel coatings that are covalently attached to the silicon atoms [21].

The main objective of this research work is to investigate two sol-gel coatings synthesized from organically modified silicon precursor 3-methacryloxypropyltrimethoxysilane (MAPTMS) mixed with zirconium (IV) propoxide. One of the coatings was modified using 1%v/v of hexamethyl di-isocyanate (HMDI) diluted in

60% ethanol (S65) whereas the second coating left unmodified (S6) diluted 100% in ethanol. The structure of the coatings was evaluated using Attenuated Total Reflectance Fourier Transform Infra-Red spectroscopy (ATR-FTIR). The coating properties such as hardness, adhesion and wettability, was evaluated using pencil hardness, cross-cut tester and water contact angle. Subsequently, the thermal and the chemical stability of the sol-gel coatings was also evaluated.

2 Experimental

2.1 Materials

The constituents used in the preparation of the sol- gel coatings are outlined in Table 1. Q-panels made from aluminium were used for the coatings.

Table 1. Material preparations.

Materials	Grade	Manufacturers
3-(Trimethoxysilyl) propyl methacrylate	98%	Sigma- Aldrich
Zirconium(IV) propoxide solution	70 wt. % in 1- propanol	Sigma- Aldrich
Absolute ethanol	⩾99.7%	Sigma- Aldrich
Hexamethyl di-isocyanate (HMDI)	98%	Sigma- Aldrich

2.2 Sol-Gel Synthesis

The sol-gel coatings are synthesized by the generation of a stable and identical sols derived by mixing two hybrid precursors namely: an organically modified silicon precursor, 3-methacryloxypropyltrimethoxysilane (MAPTMS) and an organically modified zirconium complex prepared from the chelation of zirconium (IV) n-propoxide (ZPO). In the present work, two coatings are evaluated. All coatings are mixed in 50:50 ratio of precursor to solvent. The molar ratio of the coatings is summarised in Table 2.

Table 2. Molar ratios of sol-gel coatings.

Sample name	Precursors used	Solvent (% solids)
S6	MAPTMS-ZPO	100% diluted in ethanol
S65	MAPTMS-ZPO-1%v/v HMDI	60% diluted in ethanol

3 Methods

3.1 Pre-treatment of Aluminium Metal

The aluminium metal panels were pre-treated to remove any residual oil, oxide and mechanically distributed layers if present. In the current work, two cleaning processes were followed. This was done to evaluate, which cleaning procedure works the best on the sample material.

Process 1. The panels were first degreased with acetone and sonicated for 5 min. Subsequently, the panels were alkaline etched by immersing in 0.5%wt. NaOH @ 60 ° C for 2 min. The smut layer created by the alkaline etch was removed by immersing the panel in 5% HNO3 solution for 30 s at room temperature. The panels were cleansed with DI water after each step and dried in a fume hood for 24 h.

Process 2. The panels were initially degreased with isopropanol and thereafter etched using a commercial hydrofluoric acid aqueous solution (Novaclean 104, Henkel, Irl.) and a sulphuric acid aqueous solution (Novox 302, Henkel, Irl.). The panels were cleansed with DI water after each step and dried in a fume hood for 24 h.

3.2 Coating of the Al Samples

After the surface pre-treatment, the samples were dried for 24 h under the fume hood, the panels were dip coated using a Dip coater. The dip coating procedure involved the dip of the panel into the sol-gel solution with a rate of 40 mm/min. The panel remained in the solution for 1 min. The coated panel was cured at 120 °C for 1 h to achieve steadiness. The cured samples were allowed to cool down to achieve a glassy finish. The samples were now ready for characterisation.

4 Characterisation

4.1 Fourier Transform Infrared-Attenuated Total Reflectance (ATR-FTIR)

FTIR spectroscopy was used to identify the structure and functional groups of the sol-gel coatings. FTIR spectra of all coatings were recorded using a Perkin Elmer instrument operating in the ATR mode within the spectral range of 650–4000 cm^{-1} at room temperature.

4.2 Pencil Hardness Test

Pencil hardness test was to resolve the resistant of a coating when a friction from a sharp object is applied. Coated panel was placed on a firm horizontal surface. The pencil was held firmly against the film at a 45° angle and pushed away. Thirteen pencils (6H to 6B) were used to perform the test. The pencil hardness test was analysed by the level of hardness of the pencil which just doesn't scratch the surface. The hardness of the pencil was determined by the pencil hardness scale (Fig. 1). The test was performed using ISO 15184:2012.

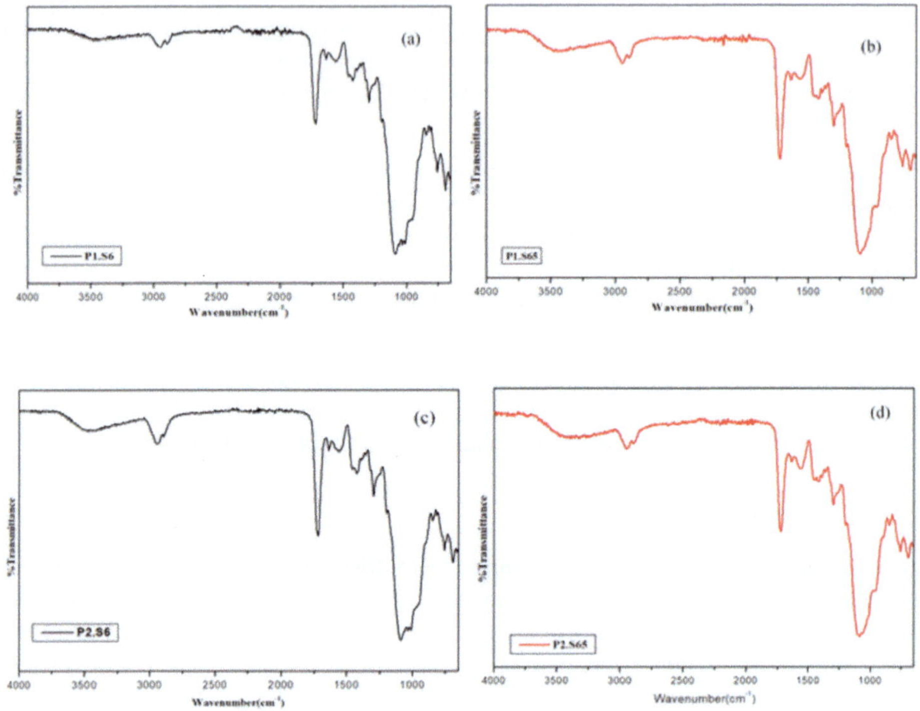

Fig. 1. ATR-FTIR spectra of (a) P1.S6, (b) P1.S65, (c) P2.S6, (d) P2.S65

4.3 Cross Hatch Adhesion Test

To ensure that the coatings have adhered to the Al substrates Cross-Cut Test was performed using standard ISO 2409:2013. A right-angle lattice pattern was cut into the coating using a blade 1 mm with 11 edge, a piece of tape was applied to the grid and removed to see the resistance of the coating to separation of the substrate. This was analysed using the magnifying glass/microscope.

4.4 Thermal Stability Test

The hydrophobic thermal stability test of the coatings was confirmed by placing samples in the vacuum oven at different temperatures (100 °C, 120 °C, 140 °C, 180 °C and 220 °C). After the set temperature of the oven, samples were exposed to the set temperature for 1 h. Finally, the samples were allowed to cool down to room temperature.

4.5 Chemical Stability Test

The chemical stability of the coatings was examined by immersing the sample in different pH solution (pH 2, 3, 4, 5, 9 & 12) for 1 h. The samples were then observed to determine the changes.

5 Results and Discussions

5.1 Pre-treatment of Aluminium Metal

Four samples were obtained. Two samples were pre-treated by performing process 1 and the other two samples pre-treated performing process 2. The samples pre-treated with commercial solutions (process 2) had a higher perceived reflectance when compared with that of process 1. The sample identities are abbreviated to P1 for process 1 and P2 for process 2.

5.2 ATR-FTIR

Figure 1 shows the ATR- FTIR spectra for the two pre-treated and prepared coatings (P1.S6, P1.S65 and P2.S6, P2.S65). The spectra show the presence of identical chemical vibrations owing to the use of same precursors and materials for the preparation of the coatings [11]. The peaks observed at 700–1700 cm^{-1} reveals the presence of MAPTMS [22]. The major bands were observed in the range between 850–1050 cm^{-1} which revealed the presence of Si-O-Si. The bands observed. The bands located at 1300–1650 cm^{-1} indicates the presence of Zr–OH and Zr–O–C bonds forming the zirconium complex. The bands situated at 1730, 2800, 3000 and 3200 cm^{-1} are assignable to C = O (stretching), C–H (stretching) and residual Si–OH and Zr–OH groups (stretching) [11].

5.3 Pencil Hardness Test

Two materials having different degrees of hardness or stiffness when forced against each other, one of them deteriorates while the other remains unaffected. The samples were found to have a scratch resistance of 4H as per ISO 15184:2012. This proves that, both S6 and S65 coated metal have good hardness property. The difference in pre-treatment method did not affect the property of hardness of sol gels (Fig. 2).

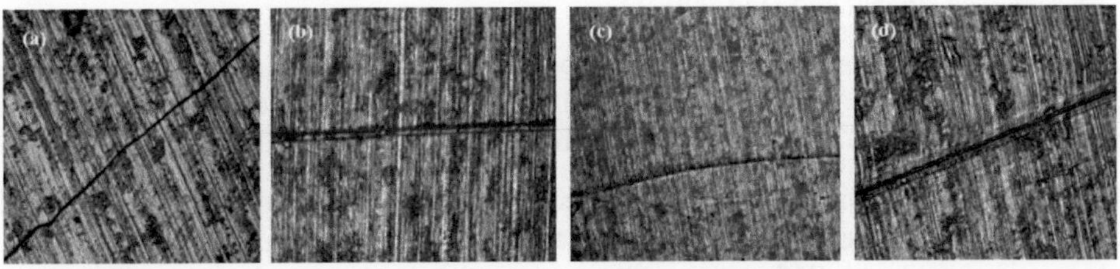

Fig. 2. Pencil hardness test for (a) P1. S6, (b) P1.S65, (c) P2.S6, (d) P2.S65

5.4 Cross Hatch Adhesion Test

The cross hatch adhesion test was performed using a blade with 11 grooves and 1 mm spacing. The coating is said to show good adhesion with the metal if the edges of the cut on the metal are smooth and none of the squares of the lattice are detached. The Al sample coated with S65 showed good adhesion (Fig. 3b and d). From Fig. 3a and c, there is a detachment of small flakes from the metal which clearly means that Al coated with S6 did not show good adhesion.

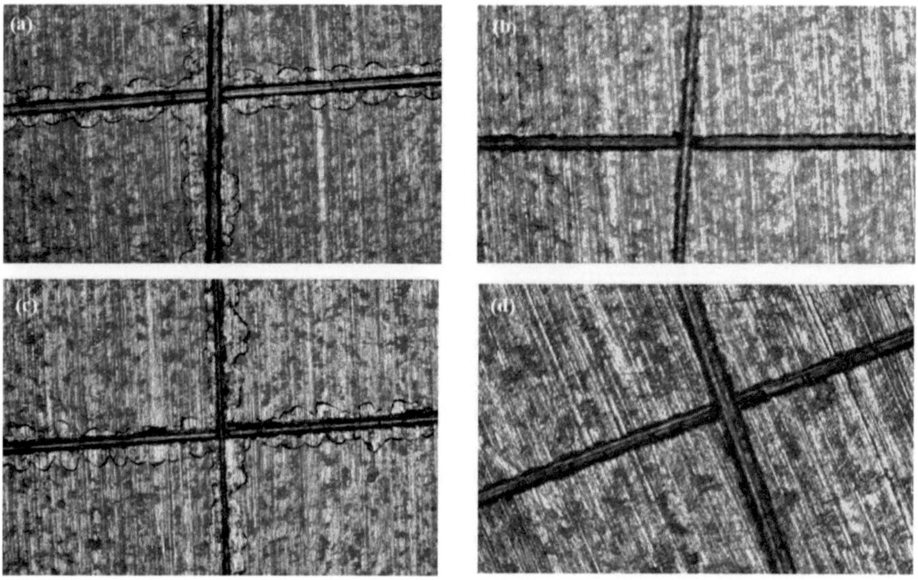

Fig. 3. Cross cut adhesion test for (a) P1. S6, (b) P1. S65, (c) P2.S6, (d) P2.S65

5.5 Thermal Stability Test

The thermal stability tests for Al metal with S6 and S65 coatings were observed at four different temperatures say, 100 °C,140 °C 180 °C and 220 °C.

100 °C for 1 h. At the end of heat treatment for an hour at 100 °C, color change was observed for the metal coated with S6 coatings (Fig. 4a) whereas the metal coated with S65 remained unchanged (Fig. 4b).

140 °C for 1 h. From Fig. 5(a), it was observed that, at 140 °C, there was insignificant degradation for the Al panel coated with S6 coatings and no major changes observed for the S65 coated metal (Fig. 5b).

180°C and 220°C for 1 h. The S6 coated Al metal exhibited major degradation of coating when observed after an hour of treatment at 180 °C and 220 °C (Fig. 6, 7(a)). The metal coated with S65 coating still remained the same even after thermally treated above the curing temperature (Fig. 6, 7(b)).

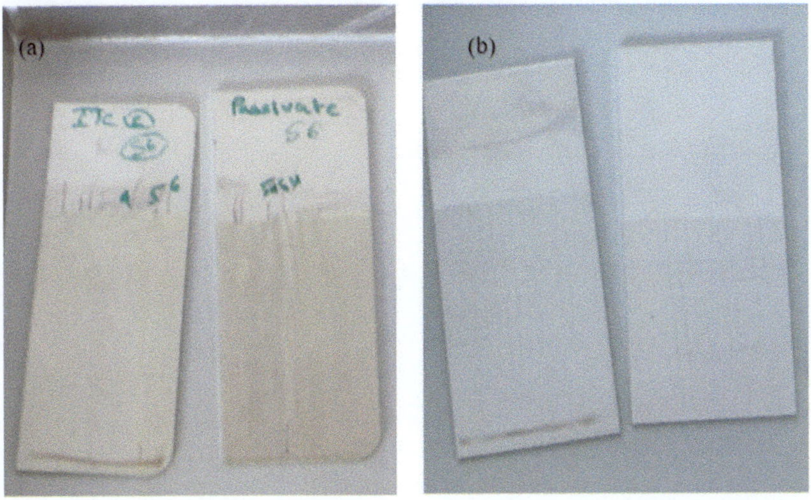

Fig. 4. Thermal stability test 100 °C for (a) P1.S6 & P2.S6, (b) P1.S65 & P2.S65

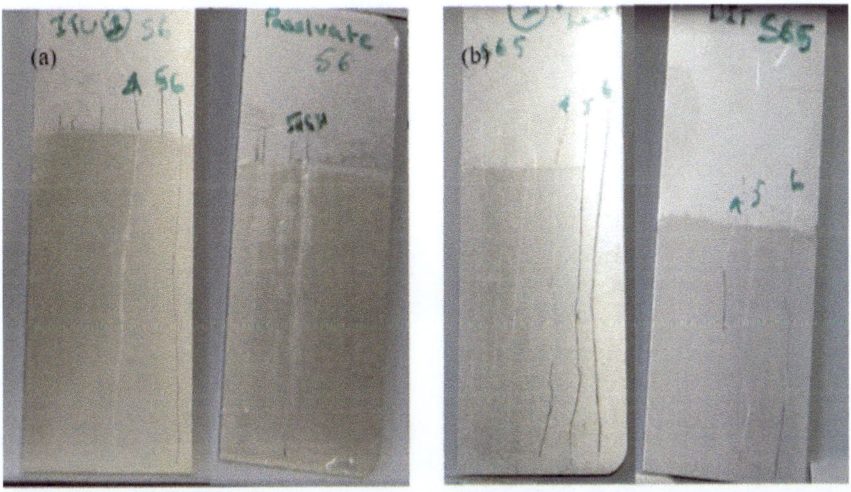

Fig. 5. Thermal stability test 140 °C for (a) P1.S6 & P2.S6, (b) P1.S65 & P2.S65

5.6 Chemical Stability Test

The coated Al samples were examined to test their chemical stability at different pH for 1 h. From Fig. 8, it was seen that the uncoated part of the metal did undergo some changes, whereas the metal coated with S6 and S65 sol-gel coatings did not undergo any changes and remained stable at different pH. Thereby, showing that S6 and S65 coatings are chemically stable.

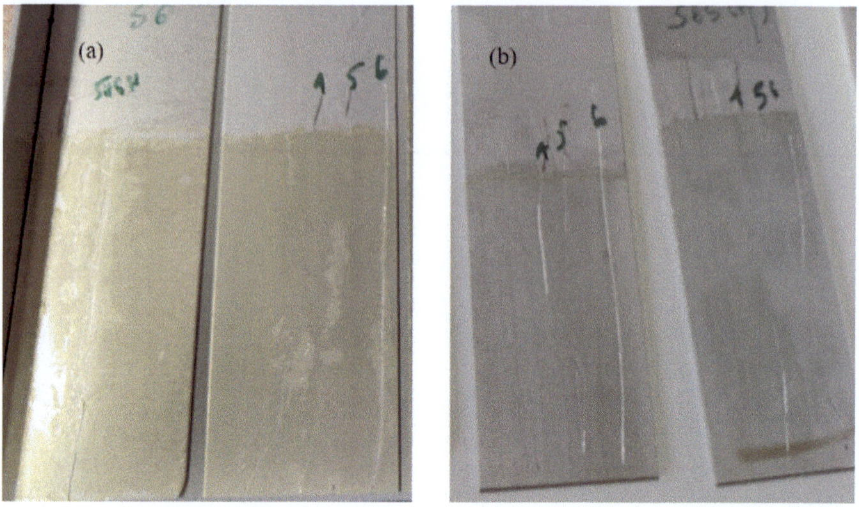

Fig. 6. Thermal stability test 180 °C for (a) P1.S6 & P2.S6, (b) P1.S65 & P2.S65

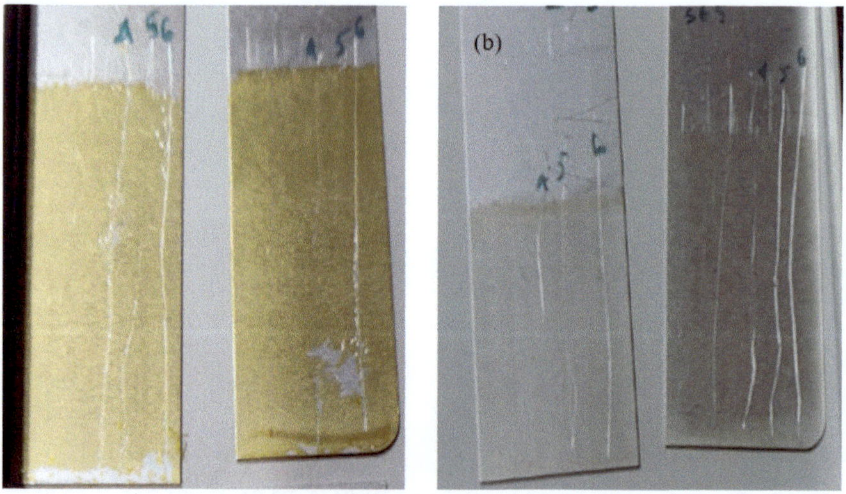

Fig. 7. Thermal stability test 220 °C for (a) P1.S6 & P2.S6, (b) P1.S65 & P2.S65

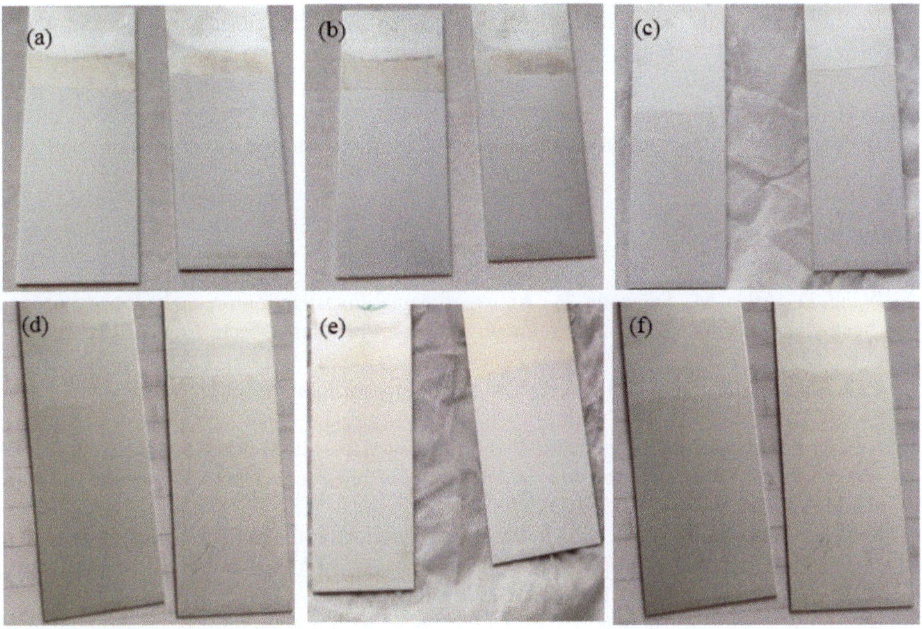

Fig. 8. Chemical stability test observed at different pH (a) pH 2, (b) pH 3, (c) pH 4, (d) pH 5, (e) pH 9, (f) pH 12

6 Conclusions

The primary objective of this study was to evaluate the structural and functional properties and the differences between the two sol-gel coatings coated on the Al metal panel. The sol-gel coatings synthesized from organically modified silicon precursor 3-methacryloxypropyltrimethoxysilane (MAPTMS) mixed with zirconium (IV) propoxide. One of the coatings was modified using 1%v/v of Hexamethyl di-isocyanate (HMDI) diluted in 60% ethanol (S65). The second coating was left unmodified (S6) and diluted in 50:50 in 100% ethanol. The structure of the coatings was evaluated using ATR-FTIR. The spectra showed that both the coatings were chemically bonded to the Al panel. The coating properties such as hardness, adhesion and wettability were evaluated using pencil hardness, cross-cut adhesion. The coatings showed negligible scratch when scratched with 4H-6H pencils and remained unaffected on using B-6B pencils. This proves that, both S6 and S65 coated metal possess good hardness property. The cross cut test proved that S65 sol-gel coating had excellent adhesive property when compared to S6 sol-gel coating. Both S6 and S65 sol gels demonstrated notable hydrophobicity. Subsequently, the thermal and the chemical stability of the sol-gel coatings was also evaluated. The S65 sols were thermally stable at all the four temperatures whereas S6 sols disintegrated at 180–220 °C but was chemically stable at different pH used.

To conclude, this work emphasized the structural, functional and mechanical properties of the sol-gel coatings. Additionally, in future the coatings will be tested conducting additional experiments to determine their cavitation erosion resistance properties.

Acknowledgments. The authors thank Dr Alicja Kaworek, CREST TU Dublin for her valuable help in developing coating and Dr Ray Benson, IT Carlow for his help and discussions on FTIR.

References

1. Wang, D., Atlar, M., Sampson, R.: An experimental investigation on cavitation, noise, and slipstream characteristics of ocean stream turbines. Proc. Inst. Mech. Eng. A **221**(2), 219–231 (2007)
2. Padhy Mamta, K., Saini, R.P.: A review on silt erosion in hydro turbines. Renew. Sustain. Energy Rev. **12**, 1974–1987 (2008)
3. Karimi, A., Martin, J.L.: Cavitation erosion of materials. Int. Mater. Rev. **31**, 1–26 (1986)
4. Wood, R.J.K.: Erosion–corrosion interactions and their effect on marine and offshore materials. Wear **9**(261), 1012–1023 (2006)
5. Park, C., Jong Kim, S.: Effect of stabilizer concentration on the cavitation erosion resistance characteristics of the electroless nickel plated gray cast iron in seawater. Surf. Coat. Tech. (2018). https://doi.org/10.1016/j.surfcoat.2018.08.098
6. Ryl, J., Wysocka, J., Slepski, P., Darowicki, K.: Instantaneous impedance monitoring of synergistic effect between cavitation erosion and corrosion processes. Electrochem. Acta **203**, 388–395 (2016)
7. Kip, N., van Veen, J.A.: The dual role of microbes in corrosion. ISME J. **9**(3), 542–551 (2015)
8. SQWenergy: Economic Study for Ocean Energy Development in Ireland. Technical Report, Sustainable Energy Authority of Ireland (2011)
9. Lin, C.J., Chen, K.C., He, J.L.: The cavitation erosion behavior of electroless Ni–P–SiC composite coating. Wear **261**, 1390–1396 (2006)
10. Aggarwal, L.K., Thapliyal, P.C., Karade, S.R.: Anticorrosive properties of epoxy-cardinol resin based paints. Prog. Org. Coat. **59**, 76–80 (2007)
11. Cullen, M., Morshed, M., O'Sullivan, M.: Correlation between the structure and the anticorrosion barrier properties of hybrid sol-gel coatings: application to the protection of AA2024-T3 alloys. J. Sol-Gel Sci. Technol. **3**(82), 801–816 (2017)
12. Akid, R., Gobara, M., Wang, H.: Corrosion protection performance of novel hybrid polyaniline/sol–gel coatings on an aluminium 2024 alloy in neutral, alkaline and acidic solutions. Electrochim. Acta **56**, 2483–2492 (2011)
13. Fontinha, I.R., Salta, M.M., Zheludkevich, M.L., Ferreira, M.G.S., Bacelar Figueira, R., Vaz Pereira, E., Silva, C.J.R.: Influence of pH on the corrosion protection of epoxysilica- zirconia sol-gel coatings applied on EN AW-6063 aluminium alloy. ECS Trans. **58**, 9–16 (2014)
14. Boukerche, I., Djerad, S., Benmansour, L., Tifouti, L., Saleh, K.: Degradability of aluminium in acidic and alkaline solutions. Corros. Sci. **78**, 343–352 (2014)
15. Deepa, P., Padmalatha, R.: Corrosion behaviour of 6063 aluminium alloy in acidic and in alkaline media. Arab. J. Chem. **10**, S2234–S2244 (2017). https://doi.org/10.1016/j.arabjc.2013.07.059
16. Metroke, T.L., Parkhill, R.L., Knobbe, E.T.: Passivation of metal alloys using sol-gel-derived materials—a review. Prog. Org. Coat. **41**(4), 233–238 (2001). https://doi.org/10.1016/S0300-9440(01)00134-5

17. Zhao, J., Xia, L., Sehgal, A., Lu, D., McCreery, R.L., Frankel, G.S.: Effects of chromate and chromate conversion coatings on corrosion of aluminum alloy 2024-T3. Surf. Coat. Technol. **140**, 51–57 (2001)
18. Kendig, M.W., Buchheit, R.G.: Corrosion inhibition of aluminum and aluminum alloys by soluble chromates, chromate coatings, and chromate-free coatings. Corrosion **59**, 379–400 (2003)
19. ECHA European Chemicals Agency, Chromium VI compounds – ANNEX XVII TO REACH – Conditions of restrictions (n.d.). https://echa.europa.eu
20. Bayer, I., Loth, E., Steele, A., Yeong, Y.H.: Adhesion strength and superhydrophobicity in polyurethane/organo clay nanocomposites. NSTI-Nanotech Conf. **1**, 399–417 (2011)
21. Pepe, A., Aparicio, M., Ceré, S., Durán, A.: Preparation and characterization of cerium doped silica sol–gel coatings on glass and aluminum substrates. J. Non-Cryst. Solids **348**, 162–171 (2004)
22. Rodic, P., Iskra, J., Milosev, I.: A hybrid organic–inorganic sol–gel coating for protecting aluminium alloy 7075-T6 against corrosion in Harrison's solution. J. Sol-Gel. Sci. Technol. **70**, 90–103 (2014)

An Implementation of Cyclic Cohesive Zone Models in ABAQUS and Its Applicability to Predict Fatigue Lives

K. Pereira(✉) and M. Abdel Wahab

Department of Electrical Energy, Systems and Automation,
Faculty of Engineering and Architecture, Ghent University, Ghent, Belgium
KyviadeFatima.ResendePereira@UGent.be

Abstract. This paper deals with an implementation of cyclic cohesive zone models (CCZM) in ABAQUS and its application for predicting fatigue lives. Firstly, a brief introduction on cyclic cohesive zone models is given, followed by a detailed discussion of its use in ABAQUS. Two user subroutines (USDFLD and UDMGINI) are written in junction with a python script in order to predict the damage accumulation in a cyclic loading condition. This damage variable is later on used to update the parameters of a cohesive zone model, allowing the simulation of material degradation in a XFEM (with cohesive segments) element. This implementation allows the user to compute fatigue damage in a cycle-by-cycle approach, being possible to model both crack initiation and propagation phases. In this paper, we present results of low cycle fatigue using this approach. The results are then compared with literature data, providing a verification of the implementation and showing the efficiency of this methodology for dealing with fatigue problems.

Keywords: Low cycle fatigue · Cyclic cohesive zone models

1 Introduction

The accurate prediction of fatigue life of mechanical components is of great importance and various numerical tools have played an important role in order to model and study fatigue. Most fatigue models rely on assumptions or simplifications that are not valid for cases of material non-linearity. For example, fatigue propagation phase has generally been modelled under the linear elastic fracture mechanics (LEFM) framework, which may no longer be applicable in situations where plasticity is present [1, 2]. In addition, empirical models for the crack growth law (such as Paris' Law) are often used to estimate propagation lives [3–5], which may not be suitable for fatigue problems where non-proportional loading is significant, for instance, fretting cases. An alternative approach that does not require the assumptions above is to treat failure using cyclic cohesive zone models (CCZM) [6–8]. A major advantage of CCZM is its ability to model both initiation and propagation phases in a unified way. It is an approach that can be used to model initiation phase for fretting phenomenon, as an another option to the common continuum damage mechanics [9–12] and critical plane approaches [13, 14].

M. A. Wahab (Ed.): DAMAS 2019, LNME, pp. 684–691, 2020.
https://doi.org/10.1007/978-981-13-8331-1_53

The aim of this paper is to present an implementation in ABAQUS that allows the user to simulate fatigue cracks using cyclic cohesive zones. Firstly, a brief introduction on cyclic cohesive zone models is given in Sect. 2, followed by a detailed description of its implementation in ABAQUS. Later, in Sect. 3, a low cycle fatigue case is analysed using the presented implementation and the results are compared with literature data. Finally, in Sect. 4, an overview of our main results and some conclusions are provided as well as directions for future researches.

2 Theoretical Background and Implementation

2.1 Cyclic Cohesive Zone Models

Cyclic cohesive zone models (CCZM) is an extension of monotonic cohesive zone models (CZM) for cyclic loading conditions. CZM is used to model failure without the necessity to add a predefined crack to the model [15]. Figure 1 illustrate the use of CZM to simulate failure, i.e., modelling the de-cohesion process between two cohesive surfaces on the bulk material. Initially, it is assumed that the material has no flaw or crack and there is no separation between the potential cohesive surfaces (where possibly the failure will occur). As the loading is monotonically increased, the material follows a local constitutive law (traction-separation law), based on the traction and separations transferred across the cohesive surfaces. This constitutive law is mainly divided into two regions, one reversible and one for damage accumulation. In the reversible and elastic region, separation increases as traction between surfaces increases, but once releasing the load, the surface tractions reduce to zero and surfaces are brought back to its initial condition, therefore, no damage is accumulated. The other part of the constitutive law models the process of failure, where degradation of the cohesive properties that maintain the undamaged material happens. It starts once the tractions reach an initial strength parameter ($T_{max,0}$) or separation reaches Δ_0 and it is defined by a softening state, in which the local material cohesive strength reduces as the load increases. Complete failure of the cohesive surfaces happens when cohesive strength reduces to zero or once separation reaches a critical value Δ_f.

Attempts have been made in the literature to use this type of monotonic cohesive zone for cyclic loading conditions, however they led to crack arrest instead of crack propagation. Therefore, this modelling has not been able to correctly predict fatigue problems. Aiming to increase its usability to handle fatigue problems, the cohesive models were further developed into cyclic cohesive zone models (CCZM), allowing for damage accumulation during unloading and reloading processes. CCZMs consider the effect of a damage evolution law in the traction-separation behavior of a cohesive model by making the cohesive strength T_{max} as a function of the current accumulated damage variable D.

$$T_{max} = T_{max,0}(1 - D) \tag{1}$$

where $T_{max,0}$ is the initial cohesive strength (of the undamaged material). The current damage variable D is obtained considering the damage evolution law (a function of the

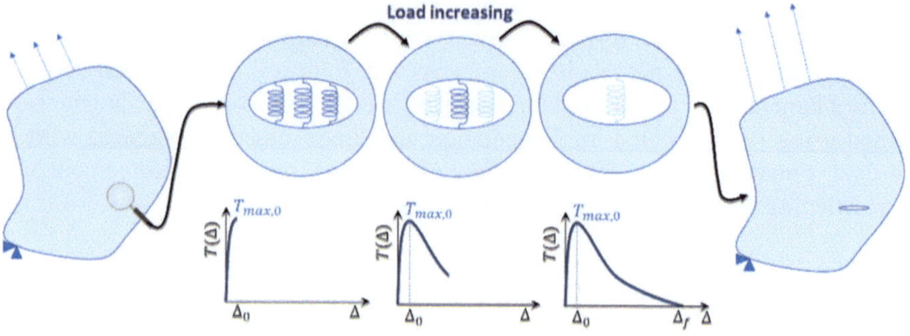

Fig. 1. Monotonic cohesive zone model: traction-separation curve.

effective tractions T and separations Δ at the cohesive surfaces and the damage variable), i.e., $\dot{D} = \dot{D}(T, \Delta, D)$.

In this work, the damage evolution law proposed by Roe and Siegmund [7] is used. It is based on the typical ideas of damage mechanics:

(a) The cohesive strength reduces with increasing damage;
(b) The monotonic cohesive zone traction-separation curve serves as an envelope (upper bound);
(c) The local damage endurance serves as a lower bound;
(d) Damage only starts if a deformation measure is greater than a critical value;
(e) The increment of damage is related to the increment of deformation (weighted by load level).

These requirements can be summarized in the following damage evolution equation:

$$\dot{D} = \frac{|\dot{\Delta}|}{\Delta_\Sigma}\left[\frac{T}{T_{max}} - \frac{\sigma_f}{T_{max,0}}\right]H(\Delta - \Delta_0) \tag{2}$$

where T and Δ are the effective tractions and effective separation, respectively. The parameters σ_f (fatigue endurance limit) and Δ_Σ (accumulated cohesive length) are the two additional material parameters introduced in the model in order to deal with damage accumulation in a cycle. Heaviside step function of the separation Δ is used to define the two regimes of the constitutive law. If separation is smaller than the critical value Δ_0, then there is no accumulation of damage and the model behaves elastic). The incremental damage $\dot{D}$ is considered proportional to the absolute value of increment of separation at the cohesive surfaces $|\dot{\Delta}|$.

In our implementation, we will extend ABAQUS XFEM with cohesive segments to account for cyclic damage accumulation. As it is not possible to work with the values of effective tractions and effective separation in this ABAQUS approach, to be able to incorporate the damage evolution law, Eq. (3) has to the adjusted for the stress and strains obtained from XFEM:

$$\dot{D} = \frac{|\dot{\varepsilon}|}{\varepsilon_{\Sigma}} \left[\frac{\sigma_{eff}}{T_{max}} - \frac{\sigma_f}{T_{max,0}} \right] \tag{3}$$

where $\dot{\varepsilon}$ is the incremental strain at the cohesive surfaces, σ_{eff} is the effective stress acting on the cohesive surface and ε_{Σ} is the accumulated cyclic strain. The incremental strain is given by:

$$\dot{\varepsilon} = \varepsilon_t - \varepsilon_{t-1} \tag{4}$$

where ε_t is the element strain at a time t and ε_{t-1} is the strain at an previous incremental time t−1.

The effective stress acting on the cohesive surface is given as a composition of the normal stress σ_n and tangential stress σ_t action on the cohesive surfaces:

$$\sigma_{eff} = \sqrt{\sigma_n^2 + \sigma_t^2} \tag{5}$$

Finally, the accumulated cyclic strain is given as a function of the cohesive parameters of Roe and Siegmund [7] model:

$$\varepsilon_{\Sigma} = \frac{K}{E} \Delta_{\Sigma} \tag{6}$$

where K is the cohesive stiffness, E is the Young's Modulus and Δ_{Σ} is accumulated cohesive length, typically adopted as five times the critical separation Δ_0.

2.2 ABAQUS Implementation

The damage evolution law was incorporated to the ABAQUS model using the user subroutine USDFLD to interpolate the material properties (cohesive strength T_{max} and Young's Modulus E) and a subroutine UDMGINI to define the crack orientation angle once the damage variable reaches unit. Figure 2 summarize the adopted implementation. A python code is written to read the model input file of one complete cycle loading (containing details of geometry, mesh details and material properties). ABAQUS solver is then called in junction with the two subroutines USDFLD and UDMGINI. After simulation of one cycle is complete, the python code extract the results and analyses the damage. If the damage variable is smaller than one, than another cycle is simulated. In order to do that, another input file is written using the RESTART option and the ABAQUS solver is again called in junction with the user subroutines. Once damage reaches one, the cycle on the python code is interrupted, the user subroutine UDMGINI is used to define the crack orientation angle and a crack that fully separate the element is generated in the model.

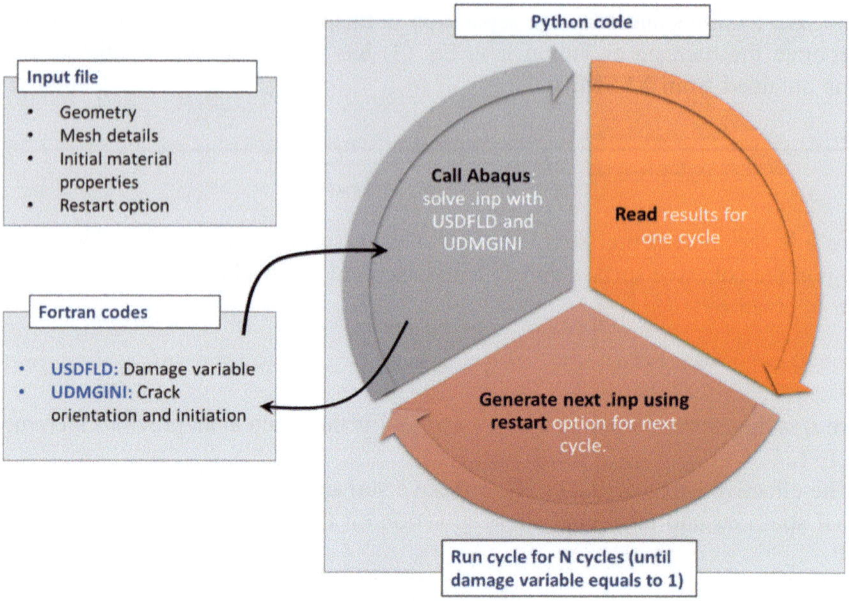

Fig. 2. Implementation flow chart.

3 Case Study: Low Cycle Fatigue

In order to verify the implementation of the cyclic cohesive model, the results of damage evolution of an element subjected to purely reversible shear stress is compared with literature data from Winter [6]. The model consisted of one XFEM element of unit size dimensions, here plane stress conditions were assumed. Loading and boundary conditions are summarized in Fig. 3.

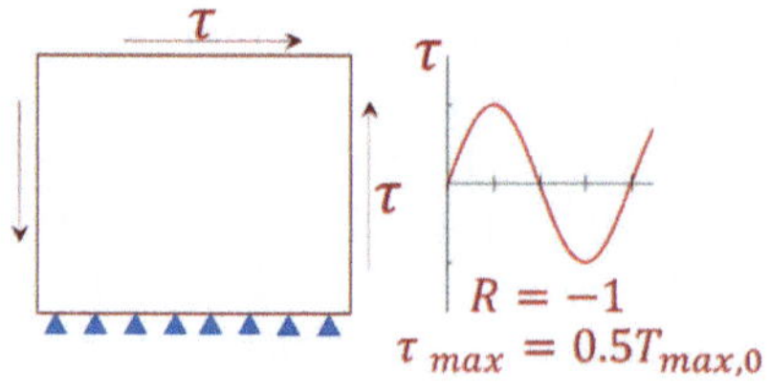

Fig. 3. Low cycle fatigue model: pure reversible shear stress condition.

In this model, the material properties used in the ABAQUS model (XFEM with cohesive segments in junction with a damage evolution law) were: $T_{max,0} = 20000$ MPa, Elastic modulus $E_0 = 10^6$ MPa, Fatigue endurance $\sigma_f = 0.0$, strain at failure $\varepsilon_f = 0.1$, effective cyclic strain $\varepsilon_\Sigma = 2\varepsilon_f$. A triangular shaped cohesive law is assumed between tractions and separations on the potential cohesive zone.

Firstly, a complete six loading cycles analysis was done using the already available implementation in ABAQUS (XFEM with cohesive elements). It is important to notice that this implementation consider a monotonic cohesive zone model in junction with a XFEM methodology [16]. As discussed in Sect. 2.1, it is expected that monotonic cohesive models are not able to predict damage accumulation under cyclic conditions and, therefore, are not a good option to model fatigue. We also used our implementation of a cyclic cohesive zone and run the same analysis.

The results comparing these two methodologies are presented in Fig. 4. The monotonic cohesive law is indeed unable to capture the cyclic damage accumulation that happened, and the behavior simulated is purely elastic. In our implementation the cohesive strength of the material reduced as the damage variable increased and the model was capable of adequately predict damage accumulation due to a cyclic loading condition.

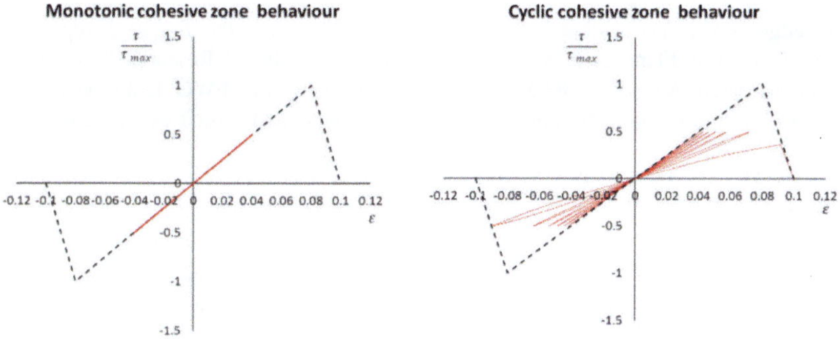

Fig. 4. Comparison between responses of a monotonic and cyclic cohesive zone model.

The results of damage accumulation as function of number of cycles were compared with literature data from Winter [6] and are shown in Fig. 5. Our implementation

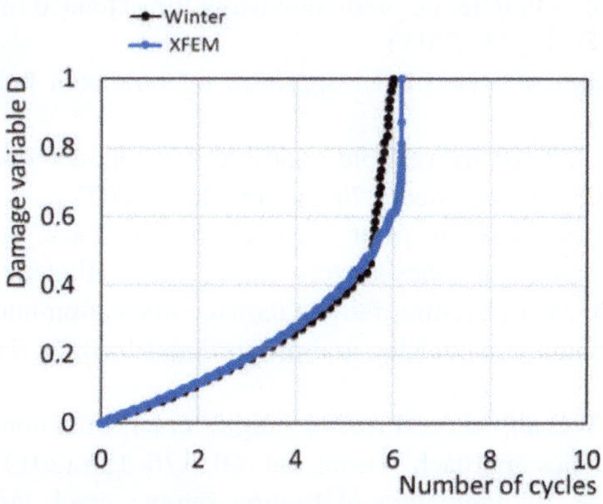

Fig. 5. Comparison between our implementation and literature data.

agrees well with literature data, revealing that it is able to correctly predict the evolution of damage in the element and the total number of cycles to failure.

4 Conclusions

This paper considers the implementation of a cyclic cohesive zone model in junction with an XFEM approach in ABAQUS. The results obtained showed that our implementation is able to accurately predict low fatigue cases with reasonable accuracy.

It is important to note that in order to use this implementation to predict cases of high cycle fatigue, an acceleration procedure should be developed, as it is extremely computational demanding to run a cycle-by-cycle analysis in those cases. The authors are currently working on this acceleration procedure and intend to present results in a near future.

Acknowledgements. The authors would like to acknowledge the financial support of the Research Foundation-Flanders (FWO), The Luxembourg National Research Fund (FNR) and Slovenian Research Agency (ARRS) in the framework of the FWO Lead Agency project G018916N 'Multi-analysis of fretting fatigue using physical and virtual experiments'.

References

1. Park, K., Paulino, G.H.: Cohesive zone models: a critical review of traction-separation relationships across fracture surfaces. Appl. Mech. Rev. **64**(6), 060802 (2011)
2. Elices, M., et al.: The cohesive zone model: advantages, limitations and challenges. Eng. Fract. Mech. **69**(2), 137–163 (2002)
3. Pereira, K., Wahab, M.A.: Fretting fatigue crack propagation lifetime prediction in cylindrical contact using an extended MTS criterion for non-proportional loading. Tribol. Int. **115**, 525–534 (2017)
4. Hojjati-Talemi, R.: Numerical modelling techniques for fretting fatigue crack initiation and propagation. In: Department of Mechanical Construction and Production, Ghent University (2014)
5. Giner, E., et al.: Fretting fatigue life prediction using the extended finite element method. Int. J. Mech. Sci. **53**(3), 217–225 (2011)
6. Winter, G.M.: A cohesive zone model approach to multiaxial fatigue. Purdue University (2009)
7. Roe, K., Siegmund, T.: An irreversible cohesive zone model for interface fatigue crack growth simulation. Eng. Fract. Mech. **70**(2), 209–232 (2003)
8. Li, H., Yuan, H., Li, X.: Assessment of low cycle fatigue crack growth under mixed-mode loading conditions by using a cohesive zone model. Int. J. Fatigue **75**, 39–50 (2015)
9. Bhatti, N.A., Wahab, M.A.: Fretting fatigue damage nucleation under out of phase loading using a continuum damage model for non-proportional loading. Tribol. Int. **121**, 204–213 (2018)
10. Hojjati-Talemi, R., Wahab, M.A.: Fretting fatigue crack initiation lifetime predictor tool: using damage mechanics approach. Tribol. Int. **60**, 176–186 (2013)
11. Hojjati-Talemi, R., et al.: Prediction of fretting fatigue crack initiation and propagation lifetime for cylindrical contact configuration. Tribol. Int. **76**, 73–91 (2014)

12. Hojjati-Talemi, R., et al.: Numerical estimation of fretting fatigue lifetime using damage and fracture mechanics. Tribol. Lett. **52**(1), 11–25 (2013)
13. Bhatti, N.A., Pereira, K., Wahab, M.A.: Effect of stress gradient and quadrant averaging on fretting fatigue crack initiation angle and life. Tribol. Int. **131**, 212–221 (2019)
14. Bhatti, N.A., Wahab, M.A.: A numerical investigation on critical plane orientation and initiation lifetimes in fretting fatigue under out of phase loading conditions. Tribol. Int. **115**, 307–318 (2017)
15. Roth, S., Hutter, G., Kuna, M.: Simulation of fatigue crack growth with a cyclic cohesive zone model. Int. J. Fract. **188**(1), 23–45 (2014)
16. Manual, A.: ABAQUS documentation version 6.13. Dassault Systems SIMULIA Corp., Providence, RI, USA (2013)

A 3S_BN Based Approach for the Quantitative Risk Assessment of Third-Party Damage on Pipelines

Xiaoyan Guo[1(✉)], Guozhi Zhang[1], Yunlong Wang[1], Laibin Zhang[2], and Wei Liang[2]

[1] State Key Laboratory of Safety and Control for Chemicals, SINOPEC Research Institute of Safety Engineering, Qingdao, China
guoxiaoyan5317@sina.com

[2] China University of Petroleum-Beijing, Beijing, China

Abstract. Third-party damage (TPD) is identified as the greatest threat to the safe operation of pipelines in different countries. The traditional TPD risk assessment methods cannot calculate the failure probability quantitatively and ignore the conditional dependence between risk influence factors (RIFs). In view of this, a theoretical system called 3S_BN is proposed based on Statistics, Scenario analysis, Safety barrier, and Bayesian network to realize the quantitative risk analysis of TPD. According to the 3S_BN, RIFs and their causal relationships are analyzed firstly. After that, a BN model including multi-state risk is developed and historical data and experts' opinions are used for the computation of conditional probability table. Besides, evidence theory is adopted in order to improve experts' belief. And then, the proposed approach is verified by a fire and explosion accident. Through sensitivity analysis and posterior probability reversal, the key influence factors and possible paths of the incident are confirmed. Finally, as discussed, the model can be applied to the TPD risk management of pipelines to continuously improve the safety level of pipelines.

Keywords: Third-party damage · Quantitative risk assessment · Bayesian theory · Safety barrier · Scenario analysis

1 Introduction

Third-party damage (TPD) is identified as the greatest threat to the safe operation of pipelines spread across the decades up to the present based on the investigation of accident records in different countries [1–3]. Take the explosion occurred in Mexican on Jan. 18, 2019 as an example, this TPD caused accident resulted in 98 deaths. Based on the findings of past accident investigation, it is identified that improper assessment of likely risk influence factors (RIFs) and correlations of these RIFs to the occurrence of a potential accident is the main causes of most incidents [4].

Modeling of TPD related pipe incidents used to be an active area of research. As discussed in the preliminary work of risk identification [5], researchers have been devoting themselves to analyzing the RIFs of it since the early 1990s. Variety

M. A. Wahab (Ed.): DAMAS 2019, LNME, pp. 692–707, 2020.
https://doi.org/10.1007/978-981-13-8331-1_54

approaches have been adopted to make a good understanding of the risk caused by TPD [6, 7]. Most of the research use qualitative and semi-quantitative methods and denied binary risk states, which is subjective and deviate from the practice [8].

However, the further quantitative analysis cannot be conducted due to the limit of data. Besides, RIFs of TPD is diverse as the TP itself is home to variations in its insecure relationship with the pipeline company and its random activities. New challenges arise because of the rapid urbanization process which brings more construction work and increases the risk of pipeline damage.

Nowadays, quantitative risk assessments (QRA) are increasingly used across industries to support risk assessment and management. A novelty theoretical system for the QRA of TPD is also required to calculate accident probability quantitatively and prepare for the subsequent dynamic risk assessment and early warning.

QRA approaches can date back to the domino theory proposed by Heinrich in 1931 [9] and the Swiss cheese model proposed by Reason [10]. Currently, based on these two models, researchers have developed Markov models, the logic mode, Reliability Block Diagram, the mixed model, and the Bayesian Network (BN).

The BN theory is demonstrated as the best choice for accident modeling and QRA due to its updating and experience learning mechanisms, especially dealing with multiple states problems [11, 12]. An in-depth study of BN has been conducted in the fields of medical diagnosis and treatment prognosis, engineering, finance, information technology and natural sciences [13].

In the industry of oil and gas (O&G), BN is widely used in the risk analysis of drilling incidents [14, 15], chemical process incidents [16–18]. In the accident modeling of O&G pipelines, BNs also play an effective role in clarifying causal relationships and accident probability calculation [19–21]. What's more, BN can be adopted for the dynamic risk assessment when giving real-time data for updating [22].

To the authors' knowledge, this is the first attempt to model TPD incidents using BN theory because researchers used to consider TPD as a whole cause of incidents instead of clarifying which kind of TPD. Even though the statistics of TPD caused incident does a favor to the QRA of TPD, it is still a difficult activity as the data is random and irregular. Therefore, a comprehensive approach by combing BN with other theories should be developed to overcome these problems.

In order to realize the QRA of TPD by using BN in the case of regular data limited, the following sections are organized as follows: Sect. 2 introduces the 3S_BN theoretical system developed in this paper. The BN model is provided in Sect. 3. Section 4 presents a case study to verify the proposed method, followed by the discussion of its possible application in Sect. 5. The conclusions are drawn and suggestions for further studies are given in Sect. 6.

2 The 3S_BN Theoretical System

As is shown in Fig. 1, a BN model will be developed when RIFs are identified and their causal relationships are clarified based on the findings of statistics, safety barrier analysis and scenario analysis. The state information provided by the pipeline monitoring system (PMS) can be used as the input of experience learning. The real-time information from the PMS can help to enhance experts' knowledge and provide new evidence to the BN model. The output results will present reference to risk decision making. Feedback messages from the decision making process can once again raise the risk perception of experts and prepare for a new round of probabilistic updates.

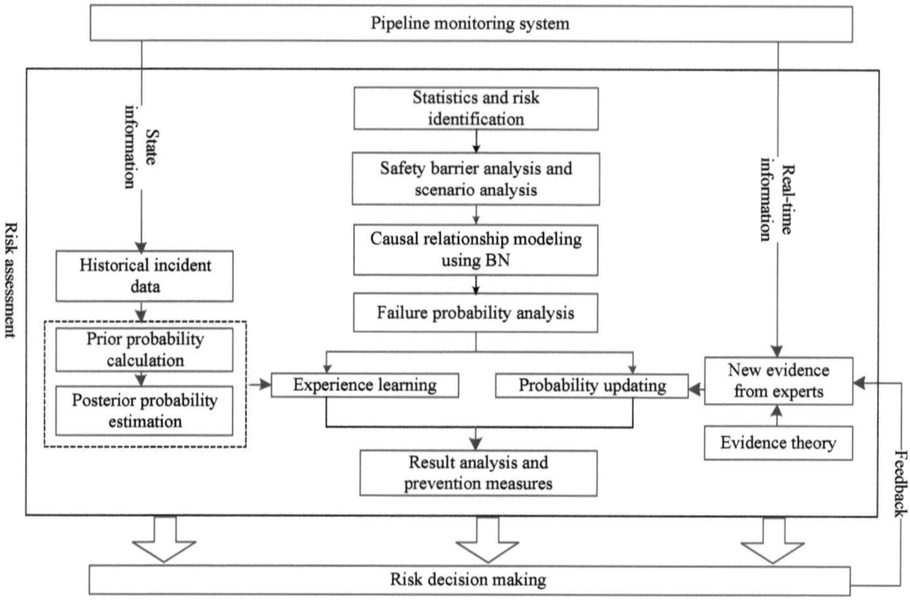

Fig. 1. Flowchart of the proposed quantitative risk assessment methodology

2.1 Safety Barrier

Safety barriers are defenses or safeguards that are used to prevent, control, or mitigate a threat from causing an incident [23]. Safety barriers can be used to prevent TPD from pipelines include surveillance systems (e.g. warning signs, patrol), optic fiber (OP) system, safety protection facilities, pipeline itself, emergency shutdown system (ESD), ignition proofing system, and emergency response measures.

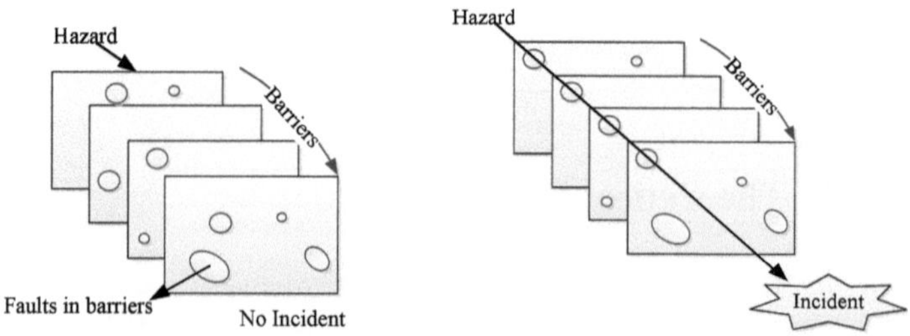

Fig. 2. Swiss cheese model adapted from Reason

As shown in the Swiss cheese model in Fig. 2, only when a threat penetrates through all of the "holes" (faults) in all of the barriers will an incident occur. These "holes" can be discovered by learning from incidents and capturing the risk variation on the basis of continuous surveillance of the pipeline.

According to statistics, when the current five types of barriers fail, it is easy to cause OP broken, pipeline damage or breakage (pipeline failure, PoP/L); when all the barriers fail, it is easy to cause a fire and explosion (F&E) incident.

2.2 Scenario Analysis

The incident scenario is a logical evolutionary sequence describing the incident triggered by specific RIFs. The triggering of different risk factors and the choice of response measures after triggering both have different effects on the consequences of incidents. The scenario analysis can help to understand the incident process and track the incident consequence. For a given incident scenario, based on field research and experts knowledge, we can identify the RIFs by reversing inference.

The incident scenario are identified as optical fiber broken, pipeline failure and F&E according to the statistics. A framework to analyze the evolutionary mechanism of TPD caused F&E is developed by conference to the basic framework made by Khan [24]. Similarly, frameworks for the scenario of pipeline failure and optical fiber broken can also be built respectively.

As is shown in Fig. 3, the RIFs of F&E incidents mainly include the condition of pipeline and ancillary facilities, pipeline surroundings, mechanical impact from TPs, pipeline transmission medium, and incident scenes.

2.3 Bayesian Network

BN is a directed graph that includes parent nodes and child nodes corresponding to random variables and direct arcs representing the influence links between nodes. A conditional probability table (CPT) is assigned to determine the conditional dependency between the linked nodes. Define a set of variables $U = X(X_1, X_2, X_3, \ldots X_n)$, whose joint probability distribution $P(U)$ can be described by the Eq. (1) [25].

$$P(U)=\prod_{i=1}^{n} P(X_i|Pa(X_i)) \tag{1}$$

Where X_i represents the random variables and $Pa(X_i)$ is the parent set of the variable X_i.

When giving a new evidence E, the posterior probability can be gained by the Eq. (2).

$$P(U|E) = \frac{P(U,E)}{P(E)} = \frac{P(U,E)}{\sum_U P(U,E)} \tag{2}$$

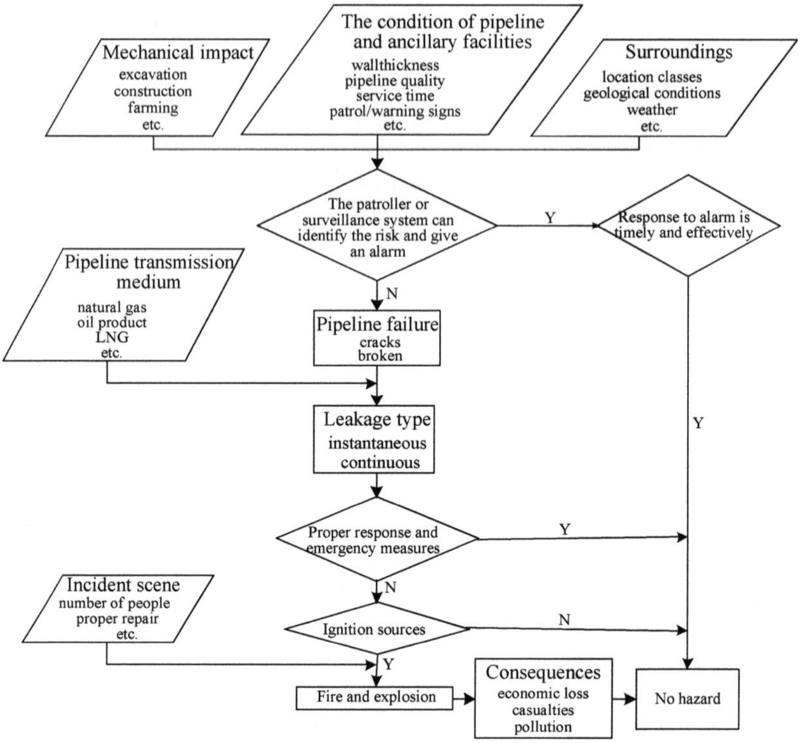

Fig. 3. Framework of fire/explosion scenarios due to TPD on pipelines

2.4 Methods of CPT Development

As far back as 1999, researchers [26, 27] have studied the approaches of developing CPTs in BN method in the situation of data limited. They found that expert judgment plays an important role in providing prior knowledge of BNs. Also, the work of this paper is carried out with the help of five experts who have profound academic background and field experience.

As a large number of probabilities are elicited from experts, issues, like avoiding different types of biases and ensuring consistency in the assessments should be highlighted. Hence, evidence theory [28] is adopted to minimize the biases, because it can directly express uncertainty under the condition of random data. The evidence theory uses sets to represent propositions and focus on the power set of the objective set.

Define a finite set $\Theta = \{a_1, a_2, \ldots a_N\}$, which is called the frame of discernment and means a combination of possible mutually exclusive solutions or assumptions of a proposition. All possible sets of Θ is described by 2^{Θ}, for example, if Θ contains N elements, then its power set 2^{Θ} contains 2^N elements, as is shown below.

$$2^{\Theta} = \{\emptyset, \{a_1\}, \ldots \{a_N\}, \{a_1, a_2\}, \ldots \{a_1, a_N\}, \ldots, \Theta\}$$

Define a set function m, which is the basic probability assignment (BPA) of Θ if m meets the conditions shown below [29]. For $\forall X \subseteq \Theta$, $m(X)$ is called the Mass function of X reflecting the trust to X.

$$m : 2^{\Theta} \rightarrow [0, 1]; \; m(\emptyset) = 0; \; \sum_{X \subseteq \Theta} m(X) = 1 \tag{3}$$

The combination rules of two Mass function $m(X)$ and $m(Y)$ are shown in the Eq. (4).

$$[m_1 \oplus m_2](E) = \begin{cases} \frac{1}{1-K} \sum\limits_{X \cap Y = E} m_1(X_i) \cdot m_2(Y_j) & E \neq \emptyset \\ 0 & E = \emptyset \end{cases} \tag{4}$$

Where, $\oplus$ is the direct sum; $K = \sum\limits_{X_i \cap Y_j = \emptyset} m_i(X_i) m_j(Y_j)$ is the conflict degree of evidences.

3 The QRA Model for TPD

In order to realize the QRA of the incidents caused by TPD, a BN model including three incident types of FoP/L, OP broken and F&E is developed. The states of the nodes are defined after the identification of the RIFs and their casual relationships by using safety barrier analysis and scenario analysis, as shown in Fig. 4. Table 1 presents a clear perspective of the definitions of nodes, their parent nodes and states.

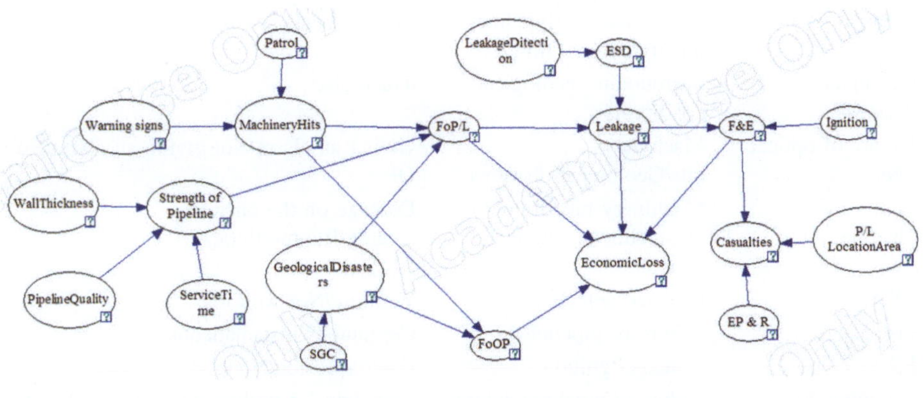

Fig. 4. The QRA model using BN for TPD

Machinery hit coming from activities of TPs, such as excavation, construction and farming, is the most common RIF of TPD. Whether this kind of damage will happen or not depends on the states of warning signs and patrollers which play an important role in damage prevention especially for those pipelines without OP system. Similarly, the OP system can also be seen as a proactive barrier, as it will give an alarm if the outside vibration is higher than the settled threshold value. It means that TPs' activities may be stopped before starting based on these early warning signals.

Given that these two barriers failed, OP cables may be broken, which can also cause an alarm. If workers respond in time, further damage will not happen. If not, the machinery will damage the pipeline coating or break the pipeline directly. Even though

Table 1. List of the states of RIFs and their parent nodes in BN for TPD caused incidents

Node	Parent node	States
Warning signs	–	Sufficient/Insufficient
Patrol	–	Working/Not working
Wall thickness	–	Thick/Thin
Pipeline quality	–	Good/Poor
Service time	–	Long time/Normal
Surrounding geological conditions	–	Good/Bad
Leakage detection	–	Working/Not working
Ignition	–	True/False
Pipeline location	–	Urban/Suburb
Emergency preparedness and responses	–	Proper/Poor
Machinery hits	Warning signs/Patrol	Excavation/Construction/Farming
Strength of pipeline	Wall thickness/Pipeline quality/Service time	Strong/Weak
Geological disasters	Surrounding geological conditions	True/False
Failure of optical fiber	Machinery hits/Geological disasters	Giving an alarm/Not giving an alarm/No OP
Failure of pipeline	Machinery hits/Strength of pipeline/Geological disasters	Damage on the outside coating/Rupture/Intact
ESD	Leakage detection	Working/Not working
Leakage	Failure of pipeline	Continuous/Instantaneous
F&E	Leakage/Ignition	Fire/Explosion/None
Economic loss	Failure of pipeline/Failure of optical fiber/Leakage/F&E	Less than 1.5 million dollar/1.5 million to 7.5 million dollar/More than 7.5 million dollar
Casualties	F&E/Pipeline location	Less than 10 people/10 to 30 people/More than 30 people

damage on the coating will not result in a direct leakage, if the pigging does not detect it, pipeline will be in a great risk of leak due to material failure.

When leaking, the ESD system will be implemented on the condition of leakage being detected successfully. Emergency preparedness and response including pipeline repair and evacuation has a great influence on the timely stop loss and operation recovery, accompanied by ignition prevention.

4 Case Study

The developed model is trained by using the statistic data in the previous paper considering different types of incident, such as leakage, F&E and OP broken. Here, an F&E incident case caused by TPD is conducted to verify the proposed approach. The input information acquired from experts is integrated by using evidence theory. Thus the conditional probability can be calculated by Eq. (1) in Sect. 2.3, followed by the analysis of the evaluation result.

4.1 CPT Development

An example of the probability distribution of "strength of pipeline" is given below in Tables 2 and 3. Where $P_n(1,2)$ represents the opinion of expert n on the states of SoP/L (Strong, Weak), and $S_0(1,2)$ represents the D-S results. The D-S results of other nodes can be obtained in the same way.

Table 2. The parent nodes of the node SoP/L and their states

Strength of pipeline (SoP/L): ①strong/②weak	Pipeline quality (PQ): ①good/②poor
	Service time (ST): ①long/②normal
	Wall thickness (WT): ①thick/②thin

4.2 Description of the Incident and the Analysis of Results

The third party did not report to the pipeline company before starting construction, which broke the pipeline and caused a leakage of crude oil. The leaked crude oil flowed into the municipal sewage pipe network, ignited by the flame and lead to an explosion. This incident resulted in direct economic loss of 301,754 dollars and no casualties.

There are four parallel pipelines buried in a shallow depth here with insufficient warning signs. Some of the signs were not located in an obvious place. The patrollers did not find any signs of construction so that the pipeline did not carry out supervision of work site. The leaked crude oil flowed into the municipal sewage pipe network

Table 3. Experts' judgments and the D-S results of the CPD of the node SoP/L

BN node				Experts' judgments					D-S results
	PQ	ST	WT	$P_1(1,2)$	$P_2(1,2)$	$P_3(1,2)$	$P_4(1,2)$	$P_5(1,2)$	$S_0(1,2)$
a	①	①	①	(0.75,0.25)	(0.75,0.25)	(0.65,0.35)	(0.72,0.28)	(0.78,0.22)	(0.788,0.212)
b	①	①	②	(0.60,0.40)	(0.66,0.34)	(0.70,0.30)	(0.70,0.30)	(0.76,0.34)	(0.698,0.302)
c	①	②	①	(0.80,0.20)	(0.78,0.22)	(0.72,0.28)	(0.75,0.25)	(0.85,0.15)	(0.829,0.171)
d	①	②	②	(0.68,0.32)	(0.67,0.33)	(0.60,0.40)	(0.60,0.40)	(0.70,0.30)	(0.647,0.353)
e	②	①	①	(0.30,0.70)	(0.33,0.67)	(0.20,0.80)	(0.25,0.75)	(0.35,0.65)	(0.298,0.702)
f	②	①	②	(0.20,0.80)	(0.28,0.72)	(0.15,0.85)	(0.20,0.80)	(0.30,0.70)	(0.231,0.769)
g	②	②	①	(0.35,0.65)	(0.30,0.70)	(0.25,0.75)	(0.30,0.70)	(0.40,0.60)	(0.325,0.675)
h	②	②	②	(0.35,0.65)	(0.27,0.73)	(0.28,0.72)	(0.33,0.67)	(0.30,0.70)	(0.332,0.668)

because of a delayed repair and the lack of facilities and materials for spilled oil sealing. The emergency plan lacks the information of the underground pipe network. The coordination efficiency of the emergency organization is not high at the beginning, which leads to the expansion of the accident.

The input information of the BN model is extracted from the report and the simulation results are shown in Fig. 5 and basically consistent with the facts. The pipeline rupture will cause instantaneous leakage in a probability of 80%. The probabilities leading to a fire and an explosion are 75% and 18% respectively. There is 74% probability of causing economic loss below 10 million dollars and 77% probability leading to less than 10 casualties.

The main paths to the F&E incident can be acquired by setting the posterior probability of this consequence as 1. There are three paths as is shown in Fig. 6, namely, (a) poor pipeline quality → low strength of pipeline → rupture → continuous leakage → F&E ← ignition, (b) leakage detection → ESD → leakage → F&E, (c) bad SGC → geological disasters → rupture → continuous leakage → F&E. Ignition has the greatest impact on the consequences of F&E incidents because the arrow connecting to the consequence node is the widest. Proper fire prevention measures must be taken to prevent disasters, such as using explosion-proof tools and prohibiting open flames.

The key RIFs of F&E incident can be recognized by sensitivity analysis, which are ignition, FoP/L, leakage detection, ESD and pipeline quality. Among them, ignition and FoP/L have a significant impact on the occurrence of F&E incident (the color of the nodes is darker), and should be paid attention to in risk management work (Fig. 7).

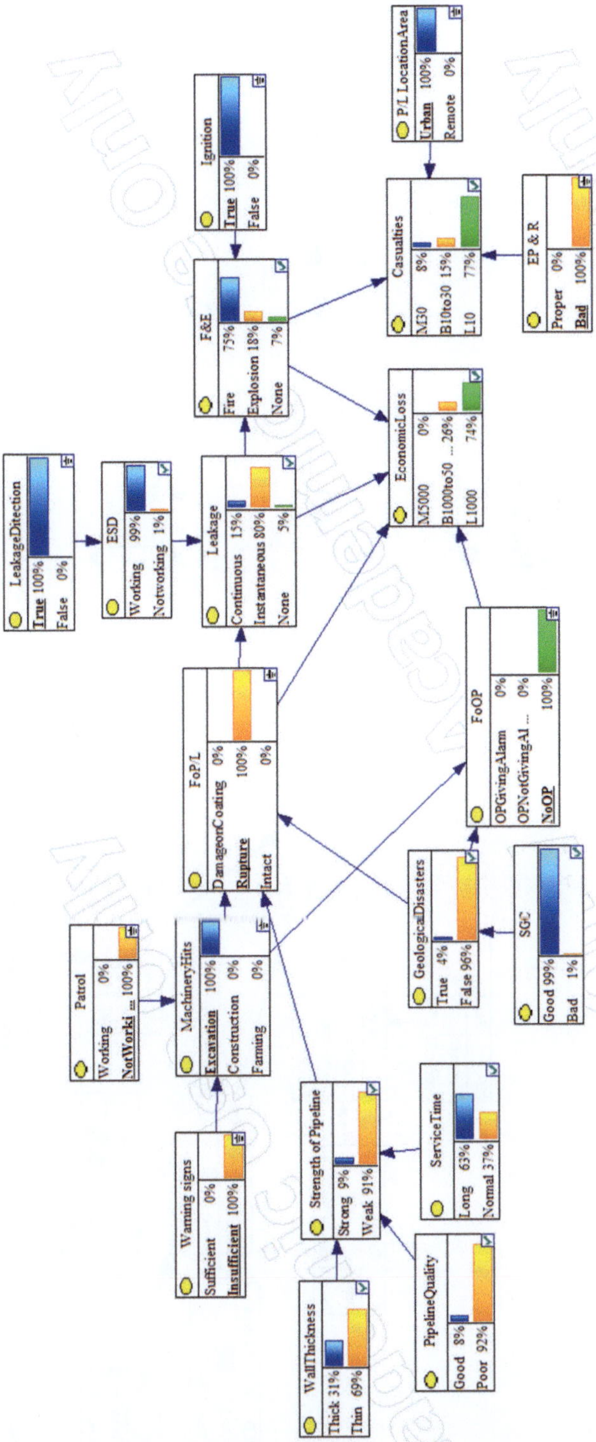

Fig. 5. Bayesian simulation results of F&E incident

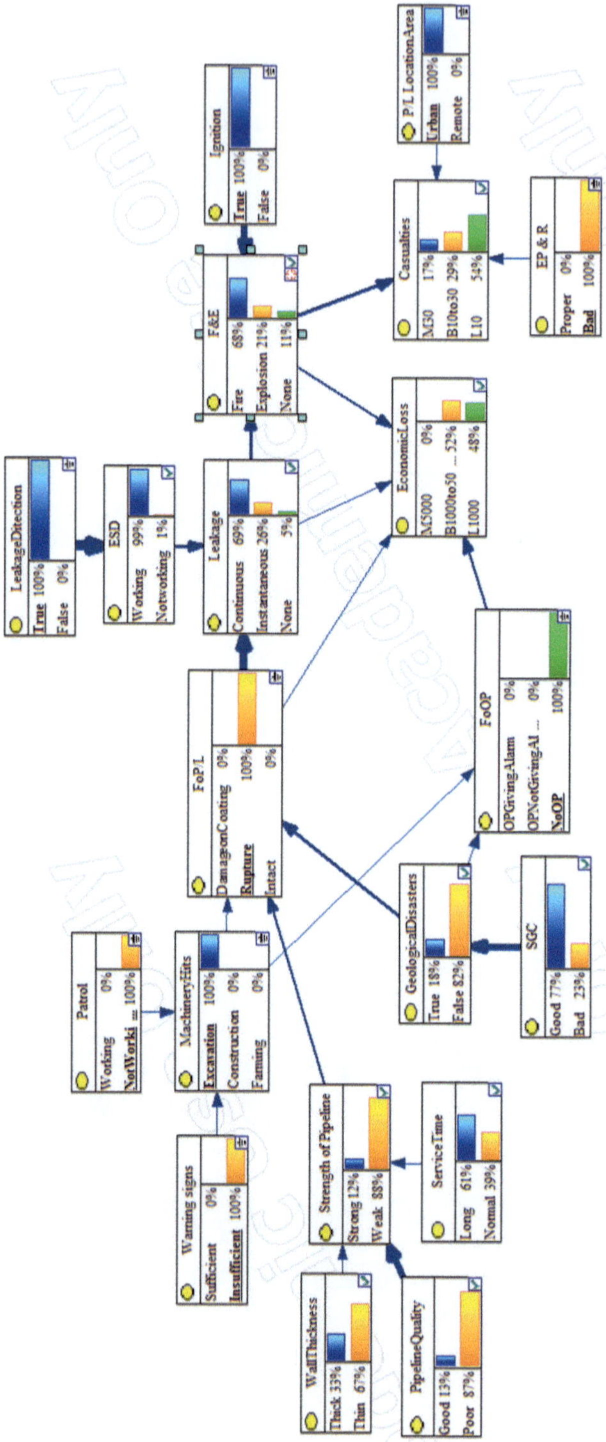

Fig. 6. Strength of influence analysis

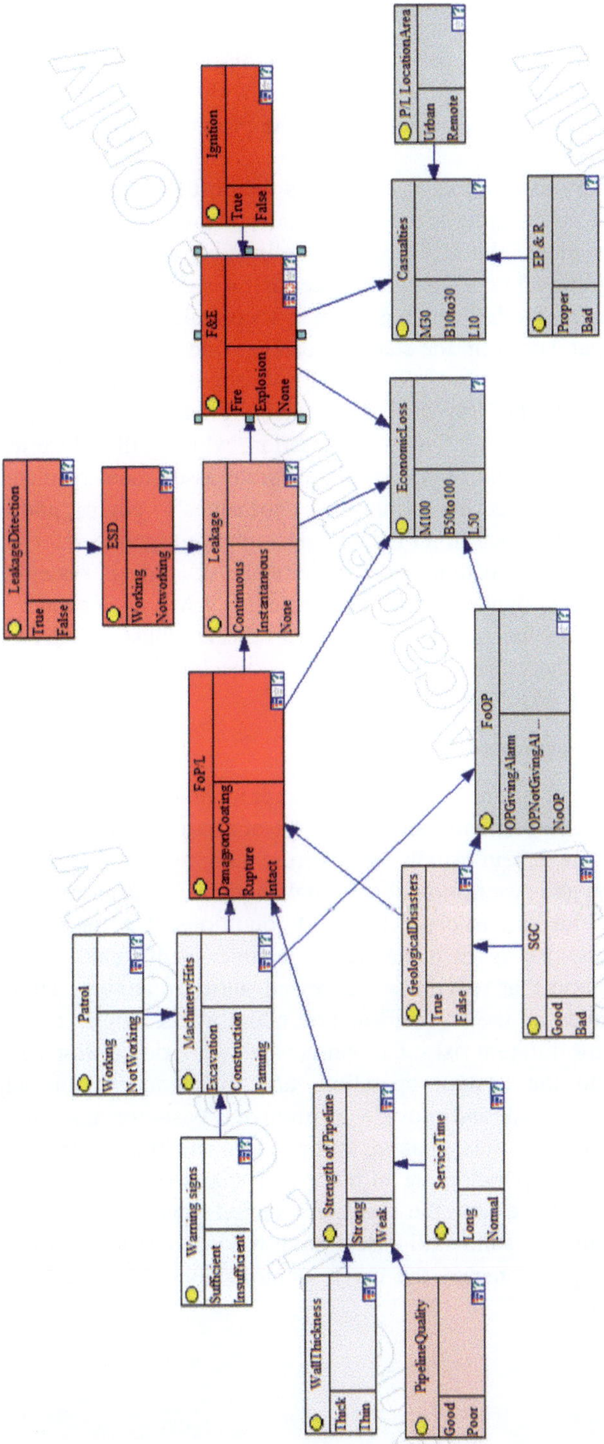

Fig. 7. Sensitivity analysis

5 Discussion

The QRA research in this paper is a continuation of the previous qualitative analysis using BN for TPD risk of pipelines. Compared with the commonly used FTA method in TPD related research, the advantages of the proposed approach can be reflected in the following aspects.

First, the safety barrier analysis identifies the post-incident RIFs to supplement the fault tree model and highlight the importance of proactive barriers in incident prevention. What's more, the RIFs can be defined in multi other than binary states. Second, the conditional dependence between RIFs is clarified by scenario analysis and presented by the BN model, which is more in accordance with the actual situation. And then, the historical data from the statistics can provide experience learning information for the BN model. While new information (evidence) is observed, the probability can be updated through the Bayesian mechanism. Finally, experts' risk awareness is increasing with the accumulation of information, which will reduce the uncertainty in the analysis procedure and lead to more accurate assessment results. Thus, potential incidents can be prevented by taking proper measures to control and reduce risk.

The proposed approach of TPD can help improve pipeline safety continuously by providing new risk information to the traditional risk analysis system [30]. Therefore, this approach enables dynamic risk management (DRM), as is shown in Fig. 8, which involves four main phases, namely, plan–do–check–adjust (PDCA cycle). The next section discusses the entire risk management mechanism, including the initial risk assessment and the PDCA cycle.

1. **Initial Risk Assessment**

The initial risk assessment involves the traditional analysis procedure, namely, risk identification, risk assessment, and evaluation of risk control measures.

Good risk management usually begins from a high-quality working plan with early risk identification and description of potential incidents. This process can be realized by learning from incidents and observing field conditions. The range of inadequacies in planning increases risk. Such inadequacies include insufficient work descriptions and relevant information that remains unaddressed during planning. Thus, a high-quality plan should provide all the key information about a task and be comprehensible with accepted levels for different risks. Comprehensive work descriptions containing all risk factors related to the design, operation, and maintenance can help increase the understanding of the risk and provide an improved basis for assessing such risk.

Managers should ensure that all human, organizational, and technical barriers involved in pipeline operation are in place and assess risks surrounding the pipeline. Given the assessment results, the managers should apply control measures to reduce risk. Good training and solid skill base among workers can help prevent many failures through proper operation and the early identification of incident precursors in the management process.

2. **PDCA Cycle**

Pipelines traversing high-population-density areas often involve a high risk of TPD. Design measures, such as changing the route of the pipeline or increasing the pipeline WT, can be performed to diminish risk. Surveillance systems can capture the change in risk caused by these modifications and the real-time signals caused by TP activities. These potential new risks and their possible effects should be added to the plan.

Risks surrounding the pipeline should be reassessed and additional risk control measures developed on the basis of the new risk to ensure that all procedures are in place. Risk control measures should be ensured as dynamic and changeable by comparing the new residual risk to the accepted risk level defined in the plan. Maintenance measures for fiber cables, pipelines, and supporting facilities should be conducted regularly on the basis of the performance evaluated from real-time data.

The PDCA cycle should be repeated throughout the entire service life of pipelines to continuously improve the safety standards by capturing new information and reassessing risks. In addition, the failure caused by TPD should be included in the pipeline company failure library in time to provide reference for subsequent risk assessment.

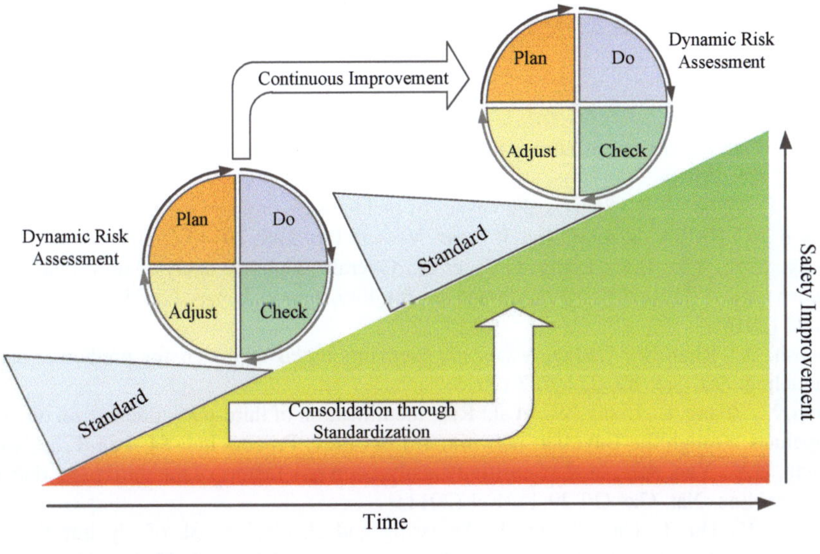

Fig. 8. Dynamic risk management with continuous safety improvement

6 Conclusion and Further Work

A 3S_BN theoretical system is proposed in this paper to overcome the shortages of traditional methods in the analysis of TPD risk and realize the QRA of TPD. A BN model including 3 incident types and 21 key RIFs is developed at first based on the findings of statistics, safety barrier analysis and scenario analysis. Then, information extracted from historical incidents and expert judgments makes up for the lack of regular data of TPD. Besides, the using of evidence theory improves the consistency and reliability of expert judgment which reflects the conditional dependence between RIFs more objectively. The case study results verify the logicality and reliability of the developed model. Meanwhile, 3 possible paths to an F&E and 5 key RIFs are acquired through the strength influence analysis and sensitivity analysis. The authors proposed that the 3S_BN approach can be applied in the DRMF and continuously improve the risk management level of TPD through probability updating.

TPD has a serious threat to the safe operation of O&G pipelines. The assessment of TPD risks has always been an important part of the overall pipeline risk management. This paper combines the Bayesian network with statistics, safety barrier analysis and scenario analysis to provide a novelty research idea for the prevention and mitigation of TPD risk. In the future, a further exploration using the dynamic risk assessment mechanism of BN is needed to predict TPD risk. This dynamic prediction mechanism will play an important role in the construction of intelligent pipeline system.

Acknowledgements. This research was supported by the National Key R&D Program of China (Grant No.2018YFC0809300) and the Major scientific and technological innovation projects of Shandong (Grant No. 2018YFJH0802). The authors are grateful for the valuable suggestions of reviewers.

References

1. EGIG: 10th Report of the European Gas Pipeline Incident Data Group (period 1970–2015), pp. 22–26 (2018)
2. Lam, C., Zhou, W.X.: Statistical analyses of incidents on onshore gas transmission pipelines based on PHMSA database. Int. J. Press. Vessels Pip. **145**, 29–40 (2016)
3. Peng, X.Y., Yao, D.C., Liang, G.C., et al.: Overall reliability analysis on oil/gas pipeline under typical third-party actions based on fragility theory. J. Nat. Gas Sci. Eng. **34**, 993–1003 (2016)
4. Baksh, A., Khan, F., Gadag, V., et al.: Network based approach for predictive accident modeling. Saf. Sci. **80**, 274–287 (2015)
5. Guo, X., Zhang, L., Liang, W., et al.: Risk identification of third-party damage on oil and gas pipelines through the Bayesian network. J. Loss Prev. Process Ind. **54**, 163–178 (2018)
6. Jiang, H.Y., Yao, A.L., et al.: Sensitivity analysis on risk factors of the third party damage in gas pipeline. Nat. Gas Oil **29**(1), 1–4 (2011)
7. Liang, W., Hu, J., Guo, C., et al.: Assessing and classifying risk of pipeline third-party interference based on fault tree and SOM. Eng. Appl. Artif. Intell. **25**(3), 594–608 (2012)
8. Li, J., Zhang, H., Han, Y., et al.: Study on failure of third-party damage for urban gas pipeline based on fuzzy comprehensive evaluation. PLoS One **11**(11), 112–127 (2016)
9. Hayhurst, E.R.: Industrial accident prevention, a scientific approach. Ind. Labor Relat. Rev. **4** (4), 609 (2011)
10. Reason, J.: Human Error, pp. 173–216. Cambridge University Press, Cambridge (1990)
11. Cai, B., Huang, L., Xie, M.: Bayesian networks in fault diagnosis. IEEE Trans. Industr. Inf. **13**(5), 2227–2240 (2017)
12. Lee, E., Park, Y., Shin, J.G.: Large engineering project risk management using a Bayesian belief network. Expert Syst. Appl. **36**(3), 5880–5887 (2009)

13. Jensen, F.V., Nielsen, T.D.: Bayesian Network and Decision Graphs. Springer Science & Business Media (2009). https://doi.org/10.1007/978-0-387-68282-2
14. Bhandari, J., Abbassi, R., Garaniya, V., et al.: Risk analysis of deepwater drilling operations using Bayesian network. J. Loss Prev. Process Ind. **38**, 11–23 (2015)
15. Wu, S.N., Zhang, L.B., Zheng, W.P., Liu, Y.L., Lundteigen, M.A.: A DBN-based risk assessment model for prediction and diagnosis of offshore drilling incidents. J. Nat. Gas Sci. Eng. **34**, 139–158 (2016)
16. Adedigba, S.A., Khan, F., Yang, M.: Process accident model considering dependency among contributory factors. Process Saf. Environ. Prot. **102**, 633–647 (2016)
17. Rathnayaka, S., Khan, F., Amyotte, P.: SHIPP methodology: predictive accident modeling approach. Part I: Methodology and model description. Process Saf. Environ. Prot. **89**(3), 151–164 (2011)
18. Baksh, A.A., Khan, F., Gadag, V., et al.: Network based approach for predictive accident modeling. Saf. Sci. **80**, 274–287 (2015)
19. Wang, W., Shen, K., Wang, B., et al.: Failure probability analysis of the urban buried gas pipelines using Bayesian networks. Process Saf. Environ. Prot. **111**, 678–686 (2017)
20. Yang, Y., Khan, F., Thodi, P., et al.: Corrosion induced failure analysis of subsea pipelines. Reliab. Eng. Syst. Saf. **159**, 214–222 (2017)
21. Zhang, C., Wu, J., Hu, X., et al.: A probabilistic analysis model of oil pipeline accidents based on an integrated Event-Evolution-Bayesian (EEB) model. Process Saf. Environ. Prot. **117**, 694–703 (2018)
22. Xin, P., Khan, F., Ahmed, S.: Dynamic hazard identification and scenario mapping using Bayesian network. Process Saf. Environ. Prot. **105**, 143–155 (2017)
23. Sklet, S.: Safety barriers: definition, classification, and performance. J. Loss Prev. Process Ind. **19**, 494–506 (2006)
24. Khan, F.I.: Use maximum-credible accident scenarios for realistic and reliable risk assessment. Chem. Eng. Prog. **97**(11), 56–64 (2001)
25. Li, X., Chen, G., Zhu, H.: Quantitative risk analysis on leakage failure of submarine oil and gas pipelines using Bayesian network. Process Saf. Environ. Prot. **103**, 163–173 (2016)
26. Cowell, R.G., Dawid, A.P., Lauritzen, S.L., et al.: Probabilistic Networks and Expert Systems, vol. 43, no. 1, pp. 289–320. Springer, Berlin (1999)
27. Ramoni, M., Sebastiani, P.: Parameter estimation in bayesian networks from incomplete databases. Intell. Data Anal. **2**(1–4), 139–160 (1998)
28. Wu, J., Zhou, R., Xu, S., et al.: Probabilistic analysis of natural gas pipeline network accident based on Bayesian network. J. Loss Prev. Process Ind. **46**, 126–136 (2017)
29. Hua, Z., Gong, B., Xu, X.: A DS–AHP approach for multi-attribute decision making problem with incomplete information. Expert Syst. Appl. **34**(3), 2221–2227 (2008)
30. Khan, F., Hashemi, S.J., Paltrinieri, N., et al.: Dynamic risk management: a contemporary approach to process safety management. Curr. Opin. Chem. Eng. **14**, 9–17 (2016)

Molecular Dynamics Simulation on Intergranular Crack Propagation Along $\sum 3$ Tilt Grain Boundary in Bcc Iron

Zhifu Zhao[1,2], Zhaoye Qin[1,2], Xueping Xu[1,2], and Fulei Chu[1,2](✉)

[1] State Key Laboratory of Tribology, Tsinghua University, Beijing 100084, People's Republic of China
chufl@mail.tsinghua.edu.cn

[2] Department of Mechanical Engineering, Tsinghua University, Beijing 100084, People's Republic of China

Abstract. As the major material of most machines and structures, bcc iron has a wide application in engineering. To well understand the fracture mechanism of bcc iron at atomistic scale, this work investigates the intergranular crack propagation behavior through carrying out molecular dynamics simulation, where two models containing $\sum 3$ $(1\bar{1}1)[110]$ and $\sum 3$ $(1\bar{1}2)[110]$ tilt grain boundaries are developed, respectively. For intergranular crack propagation along $\sum 3$ $(1\bar{1}1)[110]$ tilt grain boundary, nucleation and extension of twinning vertical to crack plane occur at crack left tip, but fast cleavage occurs at crack right tip. For intergranular crack propagation along $\sum 3$ $(1\bar{1}2)[110]$ tilt grain boundary, nucleation and extension of twinning inclined to crack plane occur at crack left tip, but stacking fault and phase transition from bcc to fcc occur at crack right tip. The intergranular crack propagation along $\sum 3$ tilt grain boundary presents strong directional anisotropy.

Keywords: Intergranular crack propagation · Molecular dynamics simulation · Directional anisotropy

1 Introduction

To well evaluate the service life of engineering components and structures and well satisfy the engineering requirement through improving the mechanical properties of materials, the way to investigate fracture has been developed from macroscopic and mesoscopic scales to atomistic scale. The concept that crack propagation is determined by the atomic motion at crack tip has been accepted by more and more researchers in recent years. Hence, it is necessary to investigate the crack propagation behavior at atomistic scale. To this aim, molecular dynamics simulation that can not only obtain the accurate trajectory of atomic motion but also obtain many microscopic details related to atoms is widely used [1].

At atomistic scale, the crystalline metals present strong anisotropy. The crack propagation behavior is closely correlated to the crystal orientation. For example, Spielmannova [2–4] proved that $(001)[110]$ crack could be cleaved fastly and

M. A. Wahab (Ed.): DAMAS 2019, LNME, pp. 708–717, 2020.
https://doi.org/10.1007/978-981-13-8331-1_55

accompanied by twinning formation and expansion but $(\bar{1}10)[110]$ crack could be blunted by dislocation nucleation and emission in bcc iron. Besides, the metals contain many fine grains. Hence, the crack propagation behavior is also significantly affected by grain boundary. According to the interaction between crack and grain boundary, the crack can propagate in transgranular and intergranular ways. Due to the different orientations of two monocrystal portions between grain boundary, the slip band and dislocation that are induced by crack propagation cannot pass through the grain boundary to continue moving at the other monocrystal portion, and the transgranular crack is resisted by grain boundary [5–7]. The resistance is controlled by the abilities of grain boundary sliding and misfit dislocation emission. Different from transgranular propagation way, the grain boundary can generally decrease the resistance against intergranular crack cleavage [8]. Hence, the intergranular crack propagation is more dangerous than transgranular one [9].

To identify the intergranular crack propagation behavior in crystalline metals, a few researchers have carried out molecular dynamics simulation on copper [8], aluminum [10], and nickel [11]. Gao [8] found that brittle cleavage was favored in one direction while dislocation emission was favored in the opposite direction for crack propagation along coherent grain boundaries in copper, which was consistent well with the analysis prediction based on Rice's model. However, for crack propagation along incoherent grain boundaries in copper, the dislocation emission was favored in both directions, which violated the analysis prediction based on Rice's model. Adlakha et al. [10] found the similar intergranular crack propagation behavior in aluminum which had a similar fcc crystal structure with copper. Besides, they also found that the grain boundary structure and the associated free volume could directly affect the crack growth rate, stress-strain behavior, and crack tip plasticity mechanism. Wu [11] found that temperature could significantly affect the intergranular crack propagation behavior and stress distribution at crack tip in nickel. With an increase in temperature, the intergranular propagation behavior could transform from twinning to slip band. Since the slip band was stronger to resist intergranular crack growth than twinning, the crack propagation could form an intergranular fracture at low temperature but not at high temperature.

Different from the above fcc crystalline metals, iron has bcc crystal structure at room temperature. Since bcc iron is widely used in engineering, it is necessary to identify its intergranular crack propagation behavior. According to the coincidence site lattice theory, the $\sum 3$ grain boundaries have better structural order and lower energy than general high angle grain boundaries. Hence, the $\sum 3$ grain boundaries are generally considered to be more resistance against intergranular crack cleavage. However, the intergranular crack propagation is not only determined by the intergranular cleavage but also determined by the plastic behaviors at crack tip. Through molecular dynamics simulation, this work mainly investigates the intergranular crack propagation behavior along $\sum 3$ tilt grain boundary in bcc iron, where two $\sum 3$ grain boundary structures are introduced and directional asymmetry effect is examined.

2 Modeling and Simulation

As shown in Fig. 1, two simulation models containing $\sum 3$ $(1\bar{1}1)[110]$ and $\sum 3$ $(1\bar{1}2)[110]$ tilt grain boundaries are developed. In Fig. 1(a), the *X*, *Y*, and *Z* axes at the upper portion of the model with $\sum 3$ $(1\bar{1}1)[110]$ tilt grain boundary are along $[\bar{1}12]$, $[1\bar{1}1]$, and $[110]$ crystal directions, and they are along $[\bar{1}1\bar{2}]$, $[\bar{1}11]$, and $[110]$ crystal directions at the lower portion. In Fig. 1(b), the *X*, *Y*, and *Z* axes at the upper portion of the model with $\sum 3$ $(1\bar{1}2)[110]$ tilt grain boundary are along $[\bar{1}11]$, $[1\bar{1}2]$, and $[110]$ crystal directions, and they are along $[\bar{1}1\bar{1}]$, $[\bar{1}12]$, and $[110]$ crystal directions at the lower portion. These two models have dimensions $L_X \times L_Y \times L_Z$ of 979.36 Å 642.99 Å × 24.23 Å and 979.41 Å × 643.52 Å × 24.23 Å, respectively. In each simulation model, a pre-existing central crack is created by deleting several layers of atoms. The pre-existing cracks in these two models have similar lengths l_{CX} of 61.7 Å and 57.9 Å, respectively.

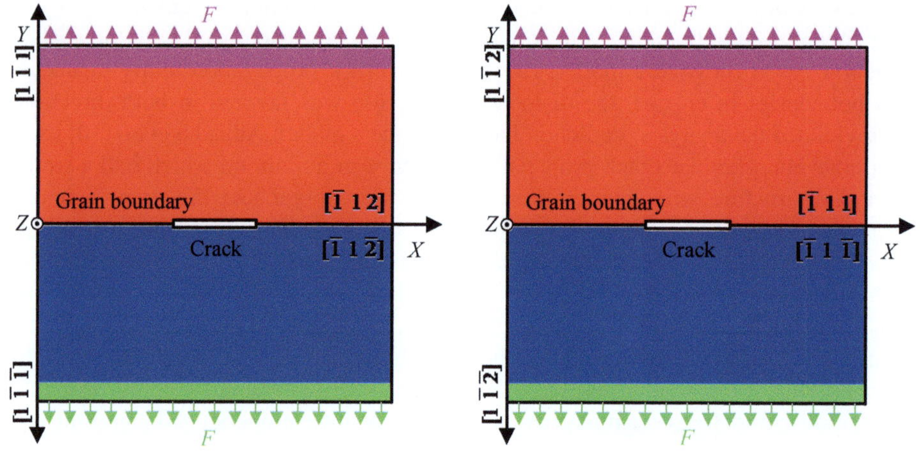

(a)$\sum 3$ $(1\bar{1}1)$ [110] (b) $\sum 3$ $(1\bar{1}2)$ [110]

Fig. 1. Molecular dynamics simulation.

Based on the precise embedded-atom potential of bcc iron [12], the molecular dynamics simulation is carried out by using LAMMPS [13]. Before simulation, the model is minimized by using conjugate gradient method and is fully equilibrated under microcanonical ensemble. After equilibration, the model is simulated under the same ensemble with time step of 0.001 ps. During simulation, the model is stretched by adding force on atoms in several layers along *Y* direction, as shown in Fig. 1. The tensile stresses applied on these two simulation models are same. Besides, free boundaries are applied along *X* and *Y* directions to freely shrink and stretch the models along these two directions, and periodic boundaries are applied along *Z* direction to avoid surface effect. Due to the large dimensions of these two models, the surface effect caused by free boundaries can be neglected.

3 Result Analysis

To reveal intergranular crack propagation behavior along $\sum 3$ tilt grain boundary, common neighbor analysis proposed by Helio [14] and software OVITO [15] are used to detect and visualize defects, respectively. Figure 2 presents the atomic distribution on central atomic plane along Z axis and the result obtained by common neighbor analysis for intergranular crack propagation along $\sum 3$ $(1\bar{1}1)[110]$ tilt grain boundary. In this figure, the pink atoms are located at grain boundaries, red atoms are located at crack surface, and green atoms have bcc crystal structure. As shown in Fig. 2(b), at 29.2 ps, the new surface atom is first observed at crack right tip. Based on Fig. 2(c), obvious cleavage can be observed at crack right tip, and no events can be observed at crack left tip. Hence, the intergranular crack first cleaves at right tip. At 30.6 ps, the block-like shear related to twinning forms but crack cannot cleave at left tip. Different from left tip, right one has moved 13.67 Å forward but the block-like shear related to twinning cannot occur. At 31.8 ps, the new surface atoms are observed at crack left tip, and left tip starts to cleave. After 31.8 ps, one can find that the twinning forms at crack left tip and it expands significantly. However, the crack left tip moves slowly forward. Different from left tip, the right one moves fastly forward. At 35.6 ps, block-like shear induced by atomic slip is observed at crack right tip. Then, this block-like shear develops as a twinning at 35.9 ps. With crack right tip further cleavage, the twining observed at 35.9 ps disappears and new twinnings form at the crack new right tip, as shown in Figs. 2(h) and (i). Based on the above illustration, one can conclude that the intergranular crack propagation behaviors at two tips are different. The crack left tip favors twinning nucleation and expansion, and the twinning nucleation and expansion impede crack fast cleavage. However, the crack right tip favors fast brittle cleavage, and the fast brittle cleavage impedes twinning nucleation and expansion. The fast brittle cleavage at crack right tip is easier than twinning formation at crack left tip. The crack left tip is more ductile than the right one.

Figure 3 describes the intergranular crack propagation behavior along $\sum 3$ $(1\bar{1}2)[110]$ tilt grain boundary through presenting the atomic distribution on central atomic plane along Z axis and the result obtained by common neighbor analysis. The pink and green atoms are same as those in Fig. 2, but red atoms in here are located at crack surface or nano-voids and blue atoms have fcc crystal structure. At 29.1 ps, the one-layer block-like shear inclined to crack plane can be observed at crack left tip. At 30.2 ps, the observed block-like shear has developed as a twinning inclined to crack plane and the crack starts to cleave at crack left tip, but no events can occur at crack right tip. After 30.2 ps, the inclined twinning at crack left tip expands significantly. At 30.7 ps, the stacking fault forms at crack right tip, as shown in Fig. 3(d). At stacking fault region, the phase transition from bcc to fcc occurs. After stacking fault formation, the new surface atom is observed at crack right tip at 32.2 ps. Then, the crack right tip starts to cleave. However, one can find that the crack right tip cleaves along the $(1\bar{1}0)$ plane instead of grain boundary, as shown in Figs. 3(f) and (g). This transformation of crack plane may be attributed to the formation of stacking fault in front of crack right tip. The stacking fault in front of crack right tip changes the atomic distribution, which can increase the free surface energy of the original crack plane. After crack cleaves

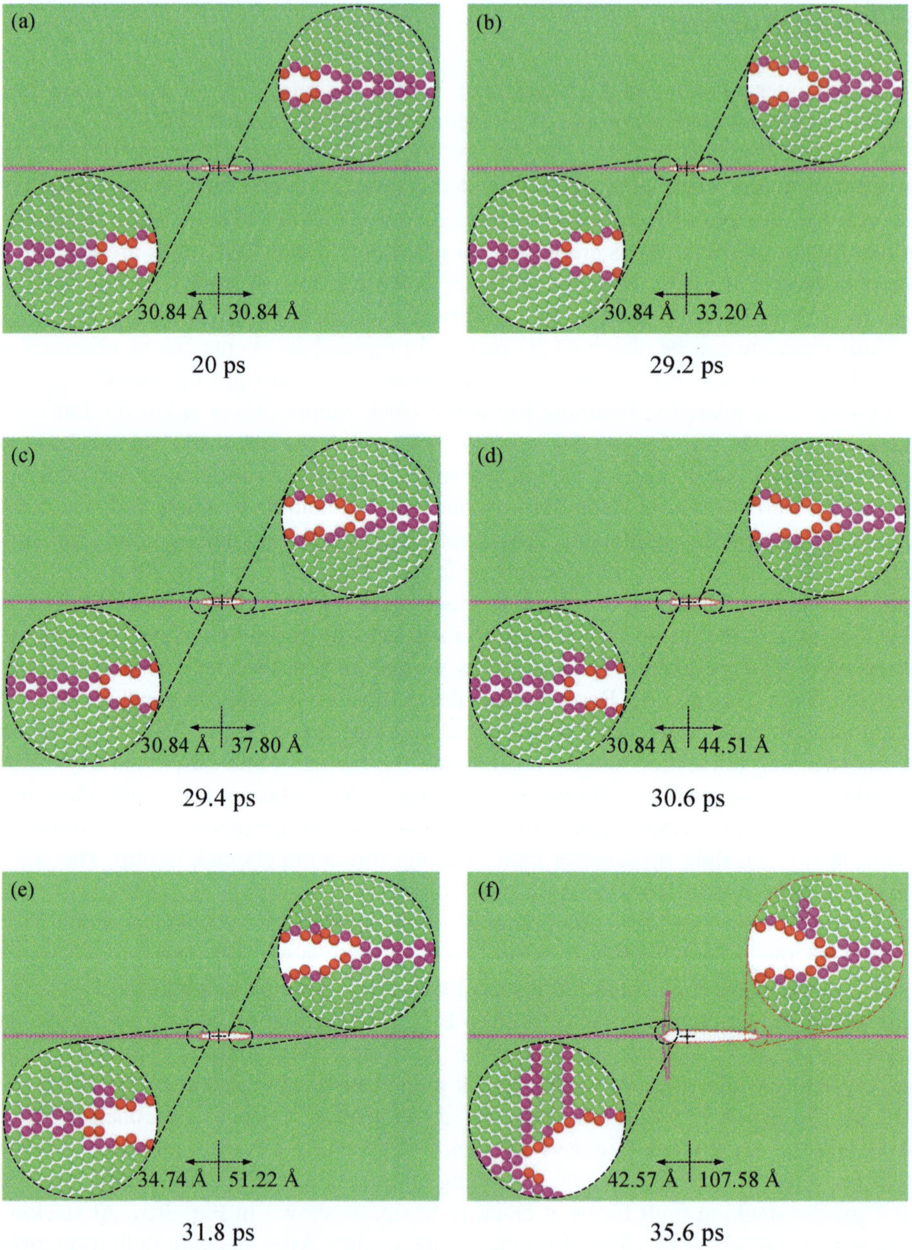

Fig. 2. Intergranular crack propagation behavior along $\sum 3$ $(1\bar{1}1)[110]$ tilt grain boundary, where the pink atoms are located at grain boundaries, red atoms are located at crack surface, and green atoms have bcc crystal structure.

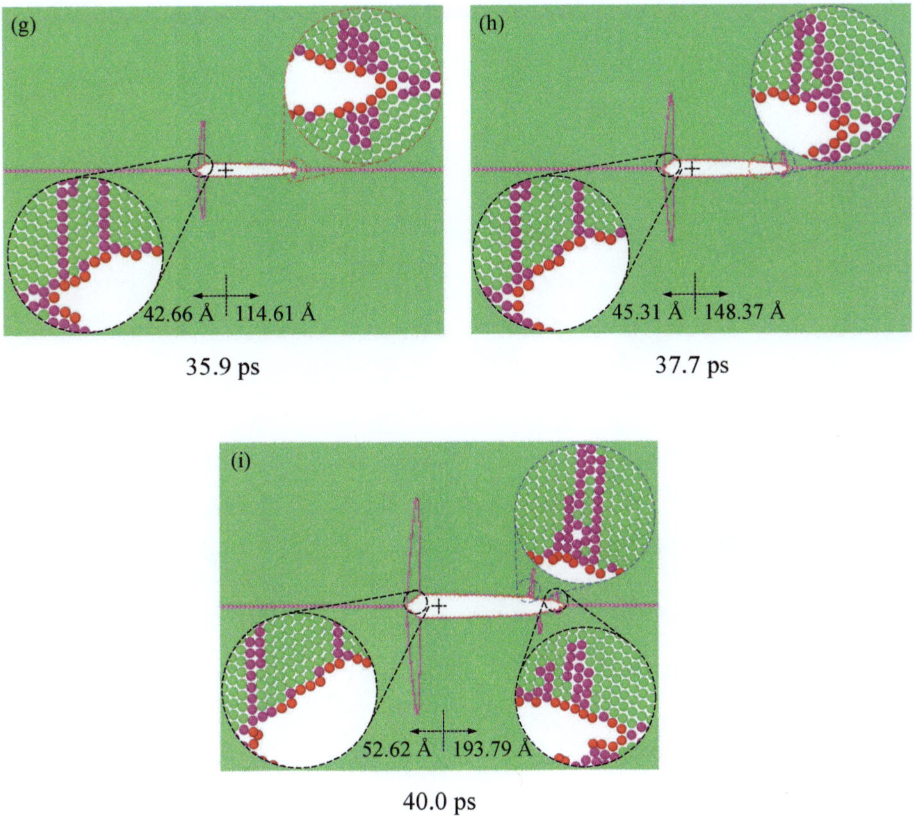

Fig. 2. (*continued*)

along the $(1\bar{1}0)$ plane, the stacking fault is located at the upper of crack right tip. Hence, the crack can transform back from $(1\bar{1}0)$ plane to the original crack plane, as shown in Figs. 3(g) and (h). In addition, nano-void forms at grain boundary, so crack can collect with the nano-void and then cleaves along the grain boundary again, as shown in Figs. 3(h) and (i). With crack right tip cleavage along grain boundary, the stacking fault expands obviously and more and more atoms transform from bcc to fcc, as shown in Figs. 3(j)–(l). In addition, there are more and more nano-voids, and they continue to collect with crack right tip. Different from crack right tip, the twinning expansion results in the deviation of crack plane from grain boundary at crack left tip, as shown in Figs. 3(j)–(l). Hence, the crack left tip is no longer at the centerline of the model. After the deviation becomes more obvious, the crack left tip will cleave in lower single crystal portion so that the left tip remains at the centerline of the model. Although the crack cleaves in lower single crystal portion, the crack plane is still parallel to the grain boundary plane. Comparing the distance that crack two tips move forward, one can find that the crack right tip cleaves more efficiently than the left one. Based on the above illustration, one can conclude that the intergranular crack propagation behaviors at two tips are different. The crack left tip favors nucleation and

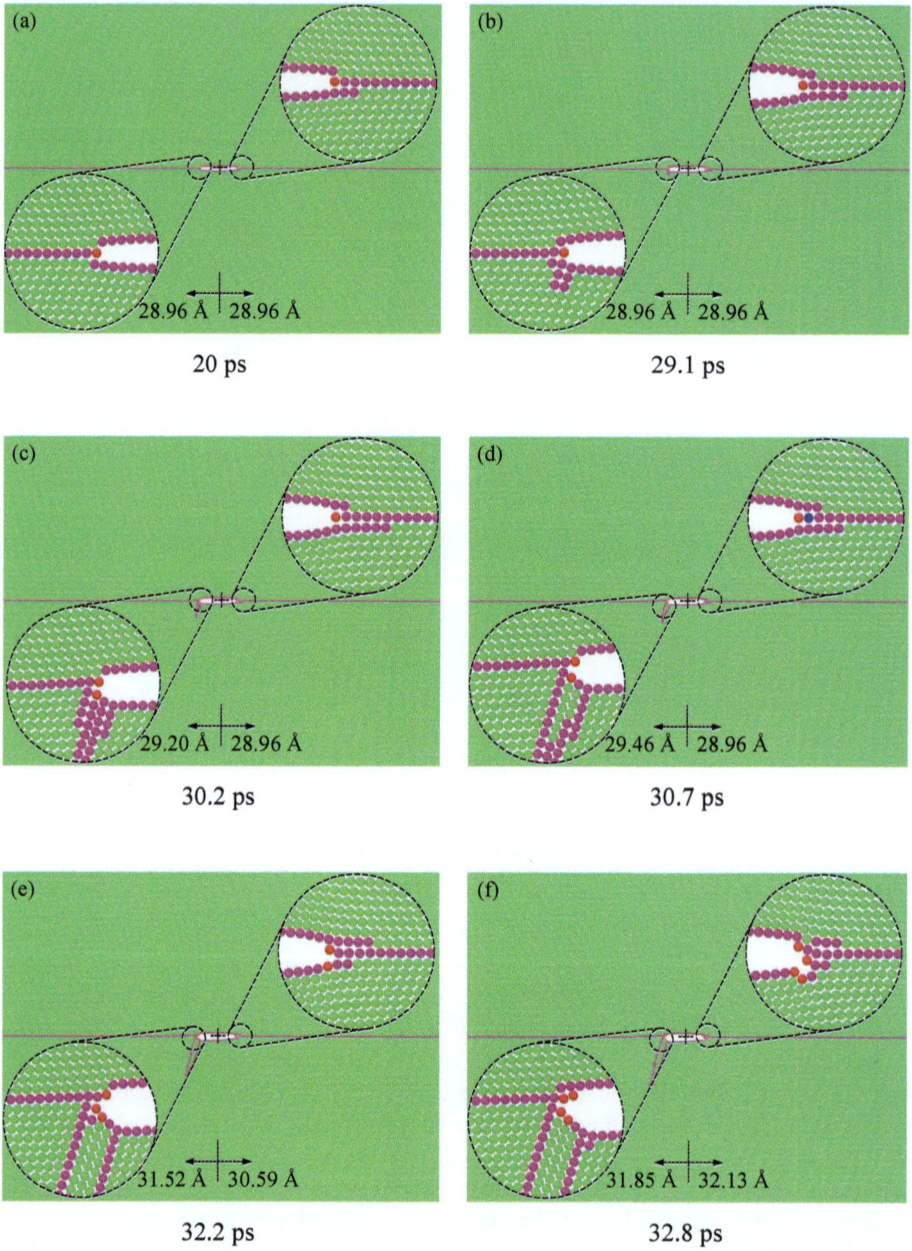

Fig. 3. Intergranular crack propagation behavior along $\sum 3$ $(1\bar{1}2)[110]$ tilt grain boundary, where the pink atoms are located at grain boundaries, red atoms are located at crack surface or nano-voids, green atoms have bcc crystal structure, and blue atoms have fcc crystal structure.

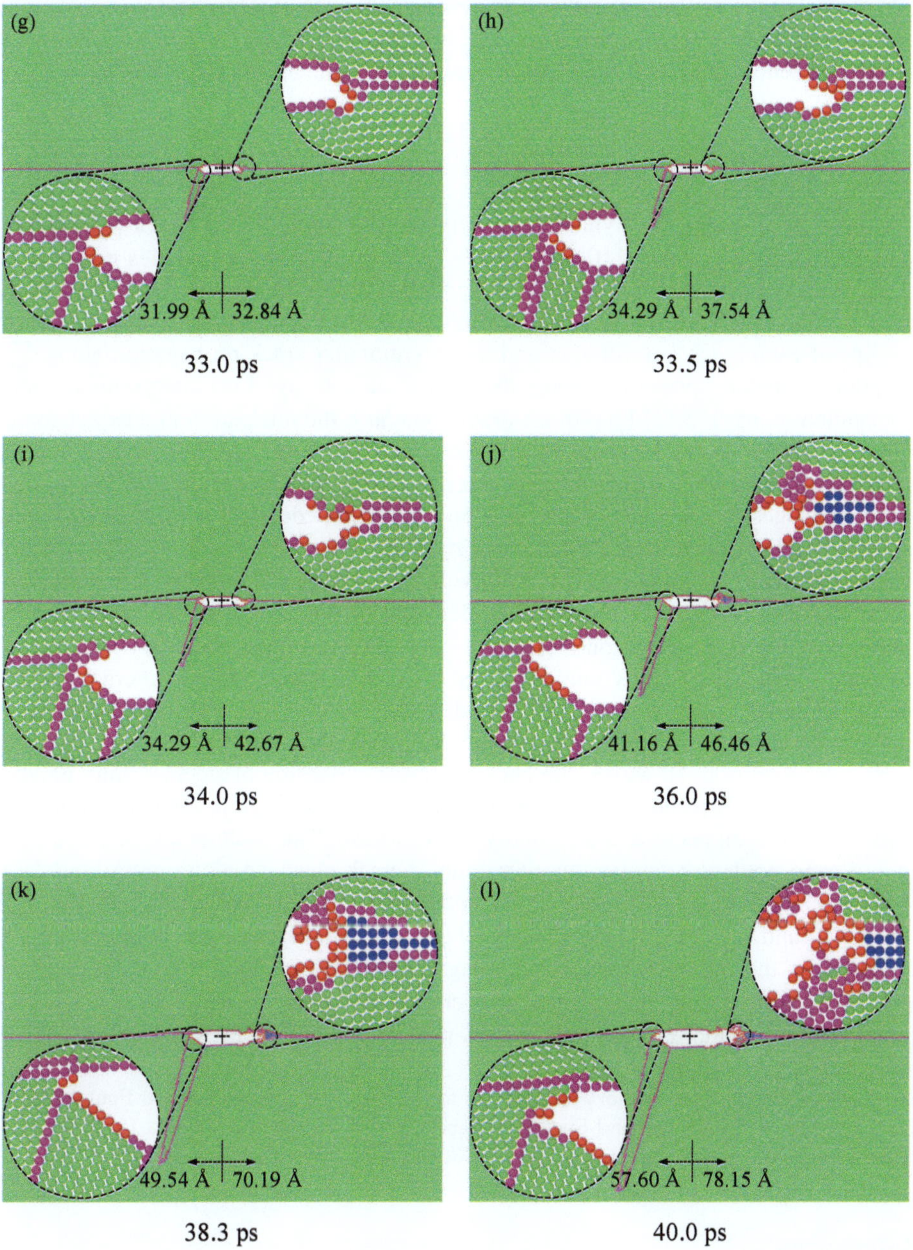

Fig. 3. (*continued*)

expansion of the twinning inclined to crack plane. The twinning nucleation and expansion impede crack fast cleavage and lead to the deviation of crack plane from grain boundary and the cleavage in lower single crystal portion. However, the crack right tip favors the nucleation and expansion of stacking fault. The coalescence of crack right tip and nano-voids accelerates the intergranular crack cleavage along grain boundary. Hence, the crack right tip cleaves more efficiently than the left one.

4 Conclusions

This work uses molecular dynamics to simulate the intergranular crack propagation behaviors along $\sum 3$ $(1\bar{1}1)[110]$ and $\sum 3$ $(1\bar{1}2)[110]$ tilt grain boundaries in bcc iron, where the common neighbor analysis is applied to track the evolutions of defect and crack surface. The results show that the intergranular crack propagation behaviors at two tips of each crack are different and the intergranular crack propagation along $\sum 3$ tilt grain boundary presents strong directional anisotropy. For intergranular crack propagation along $\sum 3$ $(1\bar{1}1)[110]$ tilt grain boundary, the nucleation and expansion of twinning vertical to crack plane are favored at crack left tip, and they impede left tip fast cleavage. The crack propagation at crack left tip is in a ductile way. However, fast brittle cleavage is favored at crack right tip, and it impedes twinning nucleation and expansion. The crack propagation at crack right tip is in a brittle way. Besides, the fast brittle cleavage at right tip is easier than twinning formation at left one. Different from $\sum 3$ $(1\bar{1}1)[110]$ tilt grain boundary, fast cleavage cannot occur at both crack tips for intergranular crack propagation along $\sum 3$ $(1\bar{1}2)[110]$ tilt grain boundary. The nucleation and expansion of the twinning inclined to crack plane are favored at crack left tip, and they impede crack fast cleavage. Besides, the twinning nucleation and expansion result in the deviation of crack plane from grain boundary and the cleavage in lower single crystal portion. However, the nucleation and expansion of stacking fault parallel to crack plane are favored at crack right tip. At staking fault region, the phase transition from bcc to fcc can occur and the nano-voids can form. The coalescence of crack right tip and nano-voids accelerates the intergranular crack cleavage along grain boundary. The crack right tip cleaves more efficiently than the left one. Although the $\sum 3$ $(1\bar{1}1)[110]$ and $\sum 3$ $(1\bar{1}2)[110]$ tilt grain boundaries have the same coincidence site lattice value, the intergranular crack propagation behaviors along these two tilt grain boundaries are different. The intergranular crack propagation along $\sum 3$ $(1\bar{1}1)[110]$ tilt grain boundary is more brittle than that along $\sum 3$ $(1\bar{1}2)[110]$ tilt grain boundary.

Acknowledgements. The authors are grateful to the National Natural Science Foundation of China (Grant No. 51335006) and State Key Laboratory of Tribology Tsinghua University Initiative Research Program (Grant No. SKLT2018C01) for supporting this research.

References

1. Frenkel, D., Smit, B.: Understanding Molecular Simulation from Algorithms to Applications. Academic Press, San Diego (2002)
2. Spielmannová, A., Landa, M., Machová, A., Haušild, P., Lejček, P.: Influence of crack orientation on the ductile–brittle behavior in Fe–3 wt.% Si single crystals. Mater. Charact. **58** (10), 892–900 (2007)

3. Prahl, J., Machová, A., Landa, M., Haušild, P., Karlík, M., Spielmannová, A., Clavel, M., Haghi-Ashtiani, P.: Fracture of Fe–3wt.% Si single crystals. Mater. Sci. Eng. A-Struct. Mater. **462**(1–2), 178–182 (2007)
4. Spielmannová, A., Machová, A., Hora, P.: Crack orientation versus ductile–brittle behavior in 3D atomistic simulations. Mater. Sci. Forum **567–568**, 61–64 (2008)
5. Terentyev, D., Gao, F.: Blunting of a brittle crack at grain boundaries: an atomistic study in BCC iron. Mater. Sci. Eng. A-Struct. Mater. **576**, 231–238 (2013)
6. Chandra, S., Kumar, N.N., Samal, M.K., Chavan, V.M., Raghunathan, S.: An atomistic insight into the fracture behavior of bicrystal aluminum containing twist grain boundaries. Comput. Mater. Sci. **130**, 268–281 (2017)
7. Kedharnath, A., Panwar, A.S., Kapoor, R.: Molecular dynamics simulation of the interaction of a nano-scale crack with grain boundaries in α–Fe. Comput. Mater. Sci. **137**, 85–99 (2017)
8. Cheng, Y., Jin, Z.H., Zhang, Y.W., Gao, H.: On intrinsic brittleness and ductility of intergranular fracture along symmetrical tilt grain boundaries in copper. Acta Mater. **58**(7), 2293–2299 (2010)
9. Musienko, A., Cailletaud, G.: Simulation of inter- and transgranular crack propagation in polycrystalline aggregates due to stress corrosion cracking. Acta Mater. **57**(13), 3840–3855 (2009)
10. Adlakha, I., Tschopp, M.A., Solanki, K.N.: The role of grain boundary structure and crystal orientation on crack growth asymmetry in aluminum. Mater. Sci. Eng. A-Struct. Mater. **618**, 345–354 (2014)
11. Wu, W.P., Li, N.L., Li, Y.L.: Molecular dynamics-based cohesive zone representation of microstructure and stress evolutions of nickel intergranular fracture process: effects of temperature. Comput. Mater. Sci. **113**, 203–210 (2016)
12. Mendelev, M.I., Han, S., Srolovitz, D.J., Ackland, G.J., Sun, D.Y., Asta, M.: Development of new interatomic potentials appropriate for crystalline and liquid iron. Philos. Mag. **83**(35), 3977–3994 (2003)
13. Plimpton, S.: Fast parallel algorithms for short-range molecular dynamics. J. Comput. Phys. **117**(1), 1–19 (1995)
14. Helio, T., Paulo, S.B., Jose, P.R.: Structural characterization of deformed crystals by analysis of common atomic neighborhood. Comput. Phys. Commun. **177**(6), 518–523 (2007)
15. Stukowski, A.: Visualization and analysis of atomistic simulation data with OVITO—The Open Visualization Tool. Model. Simul. Mater. Sci. Eng. **18**(1), 1–7 (2010)

Fatigue Crack Propagation in HSLA Steel Specimens Subjected to Unordered and Ordered Load Spectra

Jie Zhang[1,2](✉), Sven Trogh[2], Wim De Waele[2], and Stijn Hertelé[2]

[1] SIM vzw, Technologiepark 48, 9052 Zwijnaarde, Belgium
jie.zhang@ugent.be
[2] Soete Laboratory, Department of EEMMeCS, Ghent University, Technologiepark 46, 9052 Zwijnaarde, Belgium

Abstract. Prediction of fatigue crack propagation in metallic structures subjected to dynamic random load spectra, containing variable overloads and underloads, is challenging because of possible retardation and acceleration effects. In this paper, fatigue crack growth behaviour under random spectrum load is investigated on ESE(T) specimens made of DNV 460 steel, which is an HSLA steel widely used in the offshore industry. A reference spectrum composed of a sequence of random loads is transferred into various reduced and ordered spectra. Reduced spectra have been defined based on a peak-valley segmentation algorithm and on the deletion of non-damaging events. Ordered spectra consist of block loading sequences determined by rainflow counting methods. Specific control software has been developed that allows to execute the *K* (stress intensity factor) controlled experimental program and perform online crack growth measurement using a material compliance method. The different spectra are compared in terms of total crack extension and retardation in crack growth rate. Algorithms for crack growth simulation have been implemented in Abaqus using both existing and adapted plastic zone models. Numerical results are critically compared to the experimental data.

Keywords: Fatigue crack propagation · Load spectra · Variable amplitude · Retardation · Random load

1 Introduction

The most important concern in case of a cracked structure or component is its remaining useful life, or in other words how long a crack will grow until final failure. Fatigue crack propagation is a complex phenomenon as it is influenced by characteristics of the loading profile such as load sequences and amplitudes, load ratio, geometry of the structure, material resistance, etc. For constant amplitude loading, it has been successfully described with the help of fracture mechanics by means of well-known relationships between stress intensity factor range ΔK and fatigue crack growth distance per cycle da/dN, such as the Paris-Erdogan equation [1]. However, for real applications it is difficult to predict crack growth under complex and variable amplitude fatigue loading conditions, which may contain randomly distributed overloads and

M. A. Wahab (Ed.): DAMAS 2019, LNME, pp. 718–727, 2020.
https://doi.org/10.1007/978-981-13-8331-1_56

underloads. It is believed that a single overload contributes to a decrease of the crack growth rate, while an underload accelerates the rate [2]. The mechanisms contributing to the load interaction effects show that plastic straining or compressive residual stress in front of the crack tip inhibits crack opening [3]. However, load interaction effects of multiple overloads and underloads under a random load spectrum are more complex [4, 5] and result in different crack growth rates. Rainflow counting methods are widely used to process the complex load history, which allows to significantly reduce the complexity of a load spectrum by translating a random load into an ordered load spectrum with blocks of constant amplitude at the cost of losing the load interaction effect. In this paper, crack propagation under a random spectrum load and three differently ordered load spectra are investigated with the objective of comparing the influence of load ordering methods. Numerical simulations of fatigue crack propagation for the various load spectra, based on the Wheeler model, are performed and consider the plastic zone state in front of the crack tip to account for retardation. In the following sections, experiment details are elaborated firstly and then the theoretical backgrounds of numerical simulation and autocorrelation are explained. Comparison of prediction results and experimental data of crack propagation are critically analyzed, and conclusions are given at the end of the paper.

2 Experimental Details

2.1 Materials and Specimen Configuration

The material tested in this study is DNV F460, a high strength low alloy steel, which is widely used for offshore substructures. Its mechanical properties are listed in Table 1. Eccentrically-loaded single edge tension, ESE(T), specimens are used (Fig. 1). Before testing, the specimens were fatigue pre-cracked according to the ASTM standard E647 [6] with a minimum pre-crack length of 1.5 mm. To avoid the influence of the plastic zone generated by the pre-crack, the maximum stress intensity factor K of pre-cracking cannot exceed the initial maximum stress intensity factor value K_{max} of the effective fatigue test.

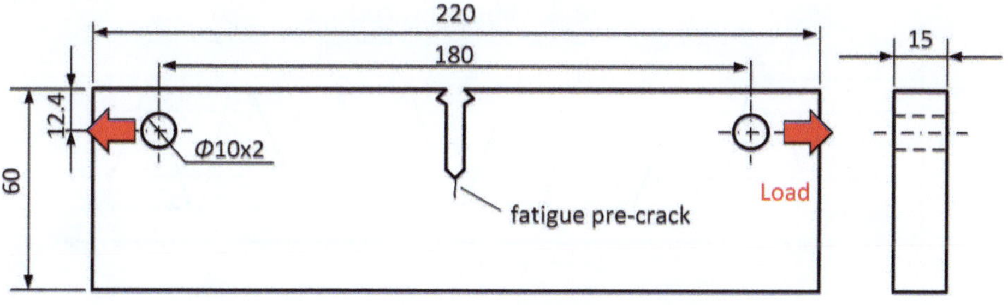

Fig. 1. ESE(T) specimen geometry configuration (mm)

Table 1. Mechanical properties of DNV F460 steel

Material	σ_y [MPa]	σ_{UTS} [MPa]	E [GPa]	ν	C	m	γ	ΔK_{th} [MPa $\cdot\sqrt{m}$]
DNV F460	560	635	210	0.28	3.6e−9	3.064	0.7	5.0

2.2 Load Profiles

K-controlled fatigue tests were performed using four different load spectra defined in a previous study [7]. 400 ΔK values with load ratios $R = 0.1$ are randomly shuffled into the first load profile. Based on this random profile, the second load profile is scaled down to 236 values by only considering the peaks and valleys of the random profile, illustrated in Fig. 2, which is referred hereafter as PaV profile. Assuming that no crack growth will occur when the driving force range is below the material threshold stress intensity factor range ΔK_{th}, the load profile is further reduced to 196 values. This third load profile is referred to as Reduced PaV profile. The rainflow counting method is applied to the PaV load profile to obtain a profile ordered in decreasing order of ΔK. Figure 3 demonstrates the so-obtained four load profiles. For each load profile, tests were performed by repeating the profile 2000 times. The details of the generation of four spectra are well-documented in reference [8].

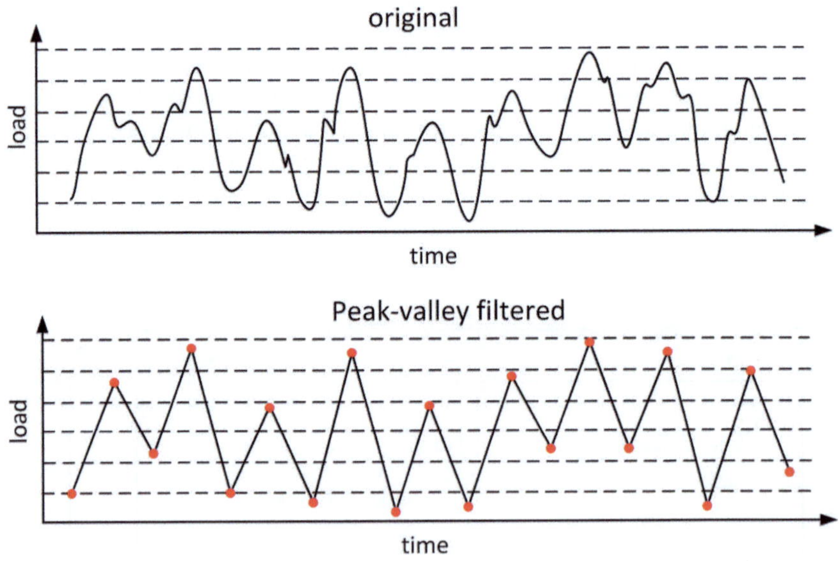

Fig. 2. Peak-valley filtering keeps only these data points (red dots) which represent reversals in slope

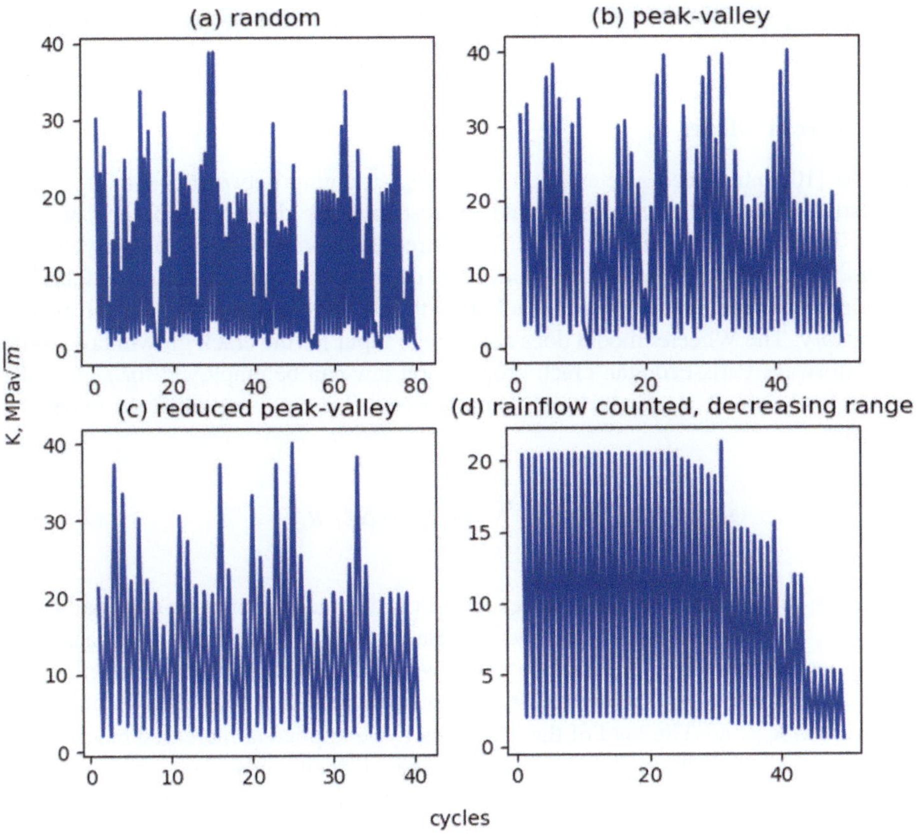

Fig. 3. Four types of load spectra profiles used in the testing: (a) random load; (b) peak-valley load; (c) reduced peak-valley load; (d) rainflow-counted load with decreasing order of SIF ranges

3 Theoretical Background

3.1 Autocorrelation Function

A mathematical autocorrelation function, defined as the cross-correlation of a signal with itself, is used to measure the degree of randomness of a load sequence. This function is given by the following Eq. (1) [9]:

$$\rho_k = \frac{1}{n-k}\sum_{i=1}^{n-k}\frac{(x_i - \mu)(x_{i+k} - \mu)}{\sigma^2} \quad (1)$$

where n is load spectrum length, k is lag of period, μ is the average value, and σ^2 is the variance of the load sequence. The autocorrelation is close to zero when the load spectrum is random, while an ordered load spectrum approaches to ± 1 as strong repetitions exist. Autocorrelation values of all four tested load sequences are discussed in Sect. 4.

3.2 Wheeler Model

Wheeler [10] introduced a retardation factor ϕ_{wh} (ranging between 0 and 1) into the calculation of fatigue crack growth rate, representing the degree of absence of interaction. The model is presented in Eq. (2), where n is the number of applied cycles and ΔK_i and R_i are the SIF range and load ratio of cycle i. Load interaction effects can potentially be simulated as the net crack growth rate is no longer independent of prior load history. The Wheeler model does not change input for the crack growth law, so the straightforward Paris-Erdogan crack propagation law can be employed for f, provided that the load ratio R_i is constant and its parameters C and m were obtained under that load ratio.

$$a_n = a_0 + \sum_{i=1}^{n} \phi_{wh,i} \cdot f(\Delta K_i, R_i) \tag{2}$$

Figure 4 illustrates the plastic zones ahead of the crack tip after an overload cycle, containing the relationship of all relevant elastic-plastic yield interfaces caused by subsequent fatigue cycles. In the case of one overload applied at the crack length a_{OL}, a plastic zone of size $r_{p,OL}$ is generated. $r_{p,OL}$ can be calculated from the maximum SIF of the overload $K_{max,OL}$. The load of the following cycle applied at increased crack length a, will cause a plastic zone of size r_p ahead of the advancing crack tip. As r_p is smaller than $r_{p,OL}$, the current plastic zone is fully embedded in the overload plastic zone for a certain number of cycles [10, 11]. The condition for retardation has been defined as the current elastic-plastic interface being embedded inside the overload elastic-plastic interface, namely $a_{OL} + r_{p,OL} > a + r_p$ [10]. Thus, the equivalent condition for retardation can be considered as $r_p^* > r_p$, whose r_p^* is defined as:

$$r_p^* = r_{p,OL} - (a - a_{OL}) \tag{3}$$

r_p^* represents the critical plastic zone size in the absence of retardation between two cycles. As the crack grows, the values of r_p^* and r_p approach each other and the load interaction effect will disappear. The retardation factor ϕ_{wh} can be obtained by Eq. (4):

$$\phi_{wh} = \begin{cases} \left(\frac{r_p}{r_p^*}\right)^{\omega} & r_p^* > r_p \\ 1 & r_p^* \le r_p \end{cases} \tag{4}$$

The shape exponent ω is determined experimentally for a given material and type of loading because of its dependence on material strength and overload ratio [10, 12, 13]. Typical reported values range from 1.0 up to 4.0.

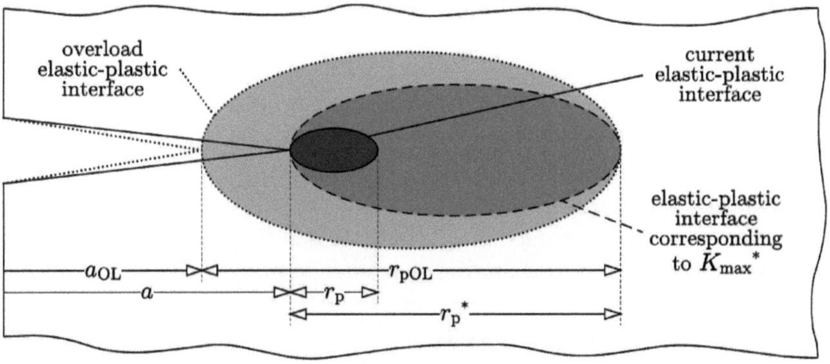

Fig. 4. Post-overload plastic zones considered in crack tip plasticity models [14]

3.3 Willenborg Model

Different from Wheeler's model, the Willenborg model assumes that retardation occurs due to the reduction of crack driving force by residual compressive stress generated in the overload plastic zone [11]. As reflected in the formula shown in Eq. (5), the Willenborg model modifies the input for the crack growth law, ΔK and R, to calculate the reduced crack growth rate indirectly.

$$a_n = a_0 + \sum_{i=1}^{n} f\left(\Delta K_{eff}^i, R_{eff}^i\right) \quad (5)$$

Since different effective load ratios R_{eff} are calculated by the Willenborg model, Paris law curves are required for different load ratios. Thus, models incorporating effects of R, such as the Forman equation or Walker equation are recommended [11, 14], although the Paris-Erdogan law can be adopted when a quantification of the load ratio effect is lacking or when the load ratio effect can be neglected [15].

The stress reduced K_{red} is defined in Eq. (6):

$$K_{red} = K_{\max}^* - K_{\max} \quad (6)$$

where K_{max}^* is the maximum SIF without retardation, K_{max} is the real maximum SIF. The effective SIF ranges and load ratio are calculated by reducing both K_{max} and K_{min} by K_{red} and are non-negative values, as shown in Eqs. (7) and (8):

$$R_{eff} = \begin{cases} \frac{K_{\min} - K_{red}}{K_{\max} - K_{red}} & K_{\min} > K_{red} \\ 0 & K_{\max} > K_{red} \geq K_{\min} \end{cases} \quad (7)$$

$$\Delta K_{eff} = \begin{cases} \Delta K & K_{\min} > K_{red} \\ K_{\max} - K_{red} & K_{\max} > K_{red} \geq K_{\min} \end{cases} \quad (8)$$

4 Results and Discussion

Fatigue crack growth tests using four different load spectra introduced higher were performed and numerical simulations were executed by a cycle-by-cycle fatigue analysis script based on Wheeler and Willenborg models. The predictions without considering the load interaction effect, using the classical Paris-Erdogan equation, are also included for comparison.

Firstly, the autocorrelation factors ρ_k are discussed. Figure 5 illustrates the autocorrelation values of the four load profiles as a function of the cycle lag parameter. The ρ_k of the random load profile fluctuates randomly in the range of -0.1 to 0.1. The variations are increasing with the degree of load complexity reduction in PaV and reduced PaV cases. This is caused by the increasing chance of identical load cycles when there are less peak-valley values in the load profile. Within logical expectations, for the rainflow counted load sequence, ρ_k is initially close to 1 and decreases with the increase of cycle lag due to the ordered load sequence. It is worthy to note that the autocorrelation curve of the rainflow counted load spectrum reaches its extreme negative value at a cycle lag value around 30. This corresponds approximately with the end of the first load block with the maximum load amplitude. Once the cycle lag k increases beyond the length of that first load block, the autocorrelation function slightly increases

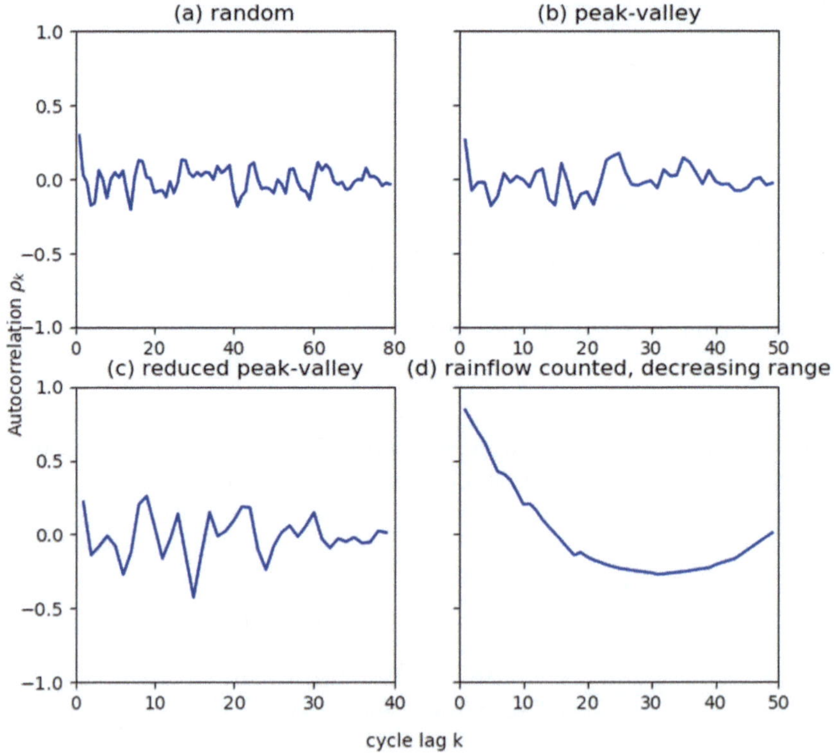

Fig. 5. Autocorrelation evolution in terms of cycle lag for four load spectrum profiles

again. The curve of the rainflow counted spectrum shows a rather smooth and regular behavior as compared to the other three which show scattered results as a function of cycle lags.

Figure 6 shows almost linear evolutions of normalized crack length *a/W* as a function of the number of cycles for all four load spectra. The reduction of the random load spectrum in three different ways produces a different number of cycles corresponding to the same normalized crack length. However, considering the final crack length, the three ordered load spectra yield almost identical values as compared to the random load spectrum. It shows that there is a tendency that the more ordered the load spectrum is and the faster the crack grow. This illustrates the possibility that random fatigue load testing can be significantly accelerated by using corresponding reduced load spectra.

The curves obtained for the PaV and rainflow counted spectra are almost identical which means that the general crack growth rates are similar. It is also observed that the load spectrum reduction does not noticeably influence the final crack extension. The difference between final crack extensions is limited to around 5%. Therefore, the damage accumulation processes for PaV and rainflow counted (decreasing range order) spectra can be considered as being similar.

Comparisons of experimental results and numerical predictions for the four load spectra are shown in Fig. 7. It is clearly demonstrated that predictions excluding the load interaction effect, curves denoted as 'no interaction', agree better with the real scenario in all four cases. It can be explained that influences from overload and underload events tend to cancel out each other, so that no significant net retardation or acceleration in crack growth rate can be observed. The Willenborg and Wheeler models give poor agreements with experimental results since both models fail to describe crack growth acceleration effects due to underloads.

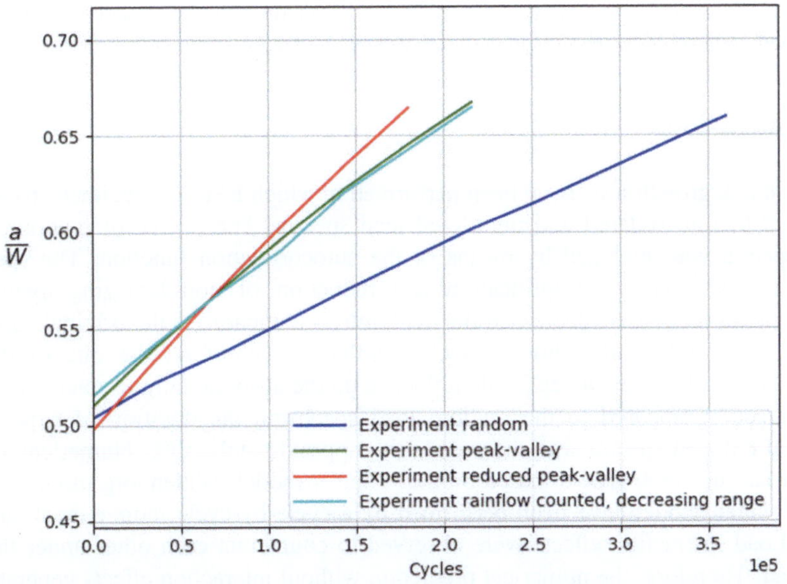

Fig. 6. Experimental results of fatigue crack growth for four different load spectra

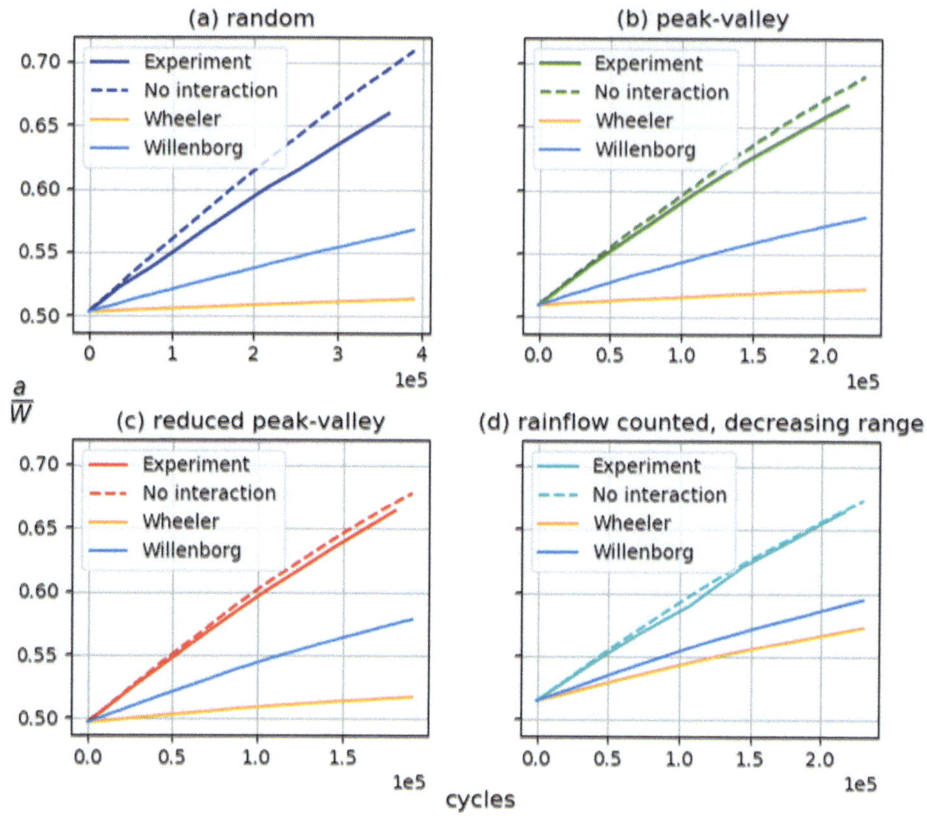

Fig. 7. Comparison of the numerical predictions based on Wheeler model and Willenborg model with experimental data for four different load spectra

5 Conclusions

Fatigue crack growth tests have been performed in which ESE(T) specimens have been subjected to four ordered and unordered load spectra. The degree of randomness of these spectra was analyzed by means of the autocorrelation function. The spectrum modifications, such as simplification and reduction of non-damaging cycles and ordering load sequences, decreased the randomness compared to the original spectrum. The experimental results showed a clear influence hereof on the rate of damage accumulation. However, a negligible influence on the absolute fatigue crack extensions was noticed. Compared to the random load spectrum, the duration of experiments based on reduced spectra was shortened with approximately 50%. Numerical simulations of fatigue crack growth based on the Wheeler model, Willenborg model, and the Paris-Erdogan model have been performed by a cycle-by-cycle fatigue crack analysis script. Load interaction effects were observed to counteract each other under the random load. Therefore, the numerical prediction without interaction effects generated the best agreement with the experimental data. Conversely, the plastic zone based retardation models of Wheeler and Willenborg had no added value for simulation of fatigue crack growth under random load spectra where both retardation and acceleration of crack growth rate occur.

Acknowledgements. The authors gratefully acknowledge the financial support via MaDurOS program (SBO DeMoPreCi-MDT) from SIM Flanders (Strategic Initiative Materials) and Vlaio (Flemish Agency for innovation and Entrepreneurship).

References

1. Paris, P., Erdogan, F.: A critical analysis of crack propagation laws. J. Fluids Eng. **85**(4), 528–533 (1963)
2. Skorupa, M.: Load interaction effects during fatigue crack growth under variable amplitude loading—a literature review. Part I: empirical trends. Fatigue Fract. Eng. Mater. Struct. **21** (8), 987–1006 (1998)
3. Skorupa, M.: Load interaction effects during fatigue crack growth under variable amplitude loading—a literature review. Part II: qualitative interpretation. Fatigue Fract. Eng. Mater. Struct. **22**(10), 905–926 (1999)
4. Abdelkader, M., et al.: Crack propagation under variable amplitude loading. Mater. Res. **16** (5), 1161–1168 (2013)
5. Bacila, A., et al.: Study of underload effects on the delay induced by an overload in fatigue crack propagation. Int. J. Fatigue **29**(9), 1781–1787 (2007)
6. ASTM E647: Standard test method for measurement of fatigue crack growth rates. In: Annual Book of ASTM Standards, Section Three: Metals Test Methods and Analytical Procedures, pp. 628–670 (2011)
7. Micone, N.: Development of testing methodologies for the analysis of variable amplitude fatigue and corrosion-fatigue of offshore steels. Ph.D. dissertation, Ghent University (2017)
8. Trogh, S.: Fatigue crack growth in components subjected to random load spectra. M.Sc. thesis, Ghent University (2018)
9. Post, N.L.: Reliability based design methodology incorporating residual strength prediction of structural fiber reinforced polymer composites under stochastic variable amplitude fatigue loading. Virginia Tech (2008)
10. Wheeler, O.E.: Spectrum loading and crack growth. J. Basic Eng. **94**(1), 181–186 (1972)
11. Willenborg, J., Engle, R.M., Wood, H.A.: A crack growth retardation model using an effective stress concept (1971)
12. Alawi, H.: Designing reliably for fatigue crack growth under random loading. Eng. Fract. Mech. **37**(1), 75–85 (1990)
13. Sheu, B.C., Song, P.S., Hwang, S.: Shaping exponent in wheeler model under a single overload. Eng. Fract. Mech. **51**(1), 135–143 (1995)
14. Gallagher, J.: A generalized development of yield zone models. DTIC Document (1974)
15. Muys, L.: XFEM-based cycle-by-cycle simulation of fatigue crack growth. M.Sc. thesis, Ghent University, p. 105 (2017)

Damage Detection in the Wind Turbine Blade Using Root Mean Square and Experimental Modal Parameters

Łukasz Doliński(✉), Marek Krawczuk, and Arkadiusz Żak

Department of Mechatronics and High Voltage Engineering, Faculty of Electrical and Control Engineering, Gdansk University of Technology, Gdansk, Poland
{lukasz.dolinski,marek.krawczuk, arkadiusz.zak}@pg.edu.pl

Abstract. The paper presents results of an experimental study related to a non-destructive diagnostic technique used for preliminary determination the location and size of delamination in composite coatings of wind turbine blades. The proposed method of damage detection is based on the analysis of the ten first mode shapes of bending vibrations, which correspond to displacements of rotor blades perpendicular to the rotor plane. Modal parameters depend on the physical properties of the structure. On the other hand, failures can affect these properties (e.g. locally reduce the stiffness of the structure). Monitoring of selected modal parameter can allow determination the technical condition of the structure. The main assumption of the presented method is a comprehensive analysis of the measured data by determination the root mean square value (RMS) for each measurement point from all forms of free vibration obtained from the experiment. As a result, information contained in all modes of vibrations that may indicate damage of the blade will be included in a single characteristic. The investigations were carried out on a scaled-down model of a wind turbine blade of a rotor diameter of 36 m. The modal parameters have been determined only experimentally using a Laser Doppler Scanning Vibrometer. Damage was simulated for three localizations by additional high stiffness elements fixed to the surface of the blade. The results of the research presented in this paper confirm the effectiveness of RMS calculation in detection damage using modes of vibrations.

Keywords: Modal testing · Non-destructive damage detection · Vibration based methods · Root Mean Square · Wind turbine blade

1 Introduction

Currently, wind energy is one of the most important elements of the electricity production system using renewable energy sources. The main drawbacks are lower efficiency compared to conventional energy sources and limited operating times due to unpredictable changes of weather conditions. Therefore, methods to improve their efficiency are intensively sought, for example by increasing their size or by locating wind farms in open waters, where wind conditions allow for continuous operation.

M. A. Wahab (Ed.): DAMAS 2019, LNME, pp. 728–742, 2020.
https://doi.org/10.1007/978-981-13-8331-1_57

However, this is connected not only to increased loads on individual turbine elements but also to extended operating time under loaded. These factors increase the risk of damage. Moreover, access to wind farms located on the open sea is limited, which undoubtedly also increases the costs of maintenance or periodic service inspections. That is why it is so important to use systems for monitoring and estimating the technical condition with special attention paid to the blades. The operation of the entire system depends on them. They convert wind energy into mechanical energy, which is transferred to a power generator. Due to the requirements for high strength and relatively low weight, the blades are usually made of laminates materials. It is related not only with the high price of blades but also with the occurrence of specific types of damage such as fibre failure or delamination [1]. The cost of the blades can account for 15–20% of the total turbine [2]. The causes of damage can be also specific: collision with a bird, icing, lightning strike [3], large temperature fluctuations, etc. Disassembling and replacing damaged parts at heights of several dozen meters is always very difficult, especially in the case of offshore wind turbine installations. For these reasons, intensive work is undertaken on non-destructive methods for monitoring wind turbine blades in order to detect and locate possible damage. In modern designs of wind turbines, diagnostic systems are often integral parts of rotor blade structures. The methods, which are the most commonly used for monitoring rotor blades are acoustic emission and ultrasonic wave propagation. Acoustic Emission examines the correlation between characteristic features of sound propagating within tested objects and their mechanical properties [4–7]. The ultrasonic method employs active systems for monitoring wind turbine rotor blades. In this method, the ultrasonic wave propagates in the material between transmitters and receivers. In the case of damage, the characteristics of the waves changes [8]. The measurement of strain is also used in rotor blades for monitoring. The sensors used in this method are typically foil strain gauges or fibre optical strain gauges [9]. There are also attempts to measure deformation using image correlation methods [10–12]. The next popular method to blade testing is thermography, which is a method to map the temperature distribution within the rotor blade using an infrared camera [13, 14].

Signal processing methods of diagnostic signals include nowadays wavelet transform [15, 16], fractal analysis [17], genetic algorithms [18], neural networks [19]. A common drawback of these methods is that the calculations are complicated and time-consuming, especially in the case of large amounts of collected measurement data. In these cases, statistical methods could be helpful. They are mainly used to isolate global features of signals or to determine the probability of occurrence of certain phenomenon. Determination of means in signal processing is usually associated with noise removal algorithms (median filter) and error determination. A special place among various means is occupied by the root means square (abbreviated RMS), which can be used to calculate the signal error, standard deviation or average power of a signal.

Obtained investigation results determine that damage changed the vibration spectrum of structure, which allow development of the non-destructive detection method of blades damage during normal running. The damage may significantly reduce the stiffness of structure which leads to the changes in natural frequencies and mode shapes. The main goal of this investigation was to develop a non-destructive method for damage detection and localization in a composite coat of a wind turbines blade using vibration parameters (free vibrations mode shapes) which depend on the physical properties of the structure.

2 Methodology

The results presented in this article are a continuation of the authors' work on the development of a fast and effective non-destructive diagnostic method in order to determine the location and size of damage of a wind turbine rotor blade [20]. In previous research, the authors presented calculations of the one-dimensional Continuous Wavelet Transform for bending forms of natural vibrations. The major conclusions of that research were that the method of detection and localization of imperfections of the blade coating is effective but only under certain conditions. These terms concern the quality and high accuracy of measurement signal data, as well as precise knowledge about the structure of the object under investigation. With noisy signals, the detection was very difficult. However, even in the case of low noise levels, the analysed signal required pre-processing. Further, each vibration form was analysed and interpreted separately. As a result, this method, despite its unquestionable advantages, is complicated and time-consuming. The authors note that proposed method based on the wavelet transform should be a second stage of the diagnostic investigation after some precalculation.

In this paper, the authors present a technique of preliminary detection and localisation of defects using statistic calculation. These calculations were carried out for experimentally determined modes of bending vibrations. A general research methodology is schematically depicted in Fig. 1.

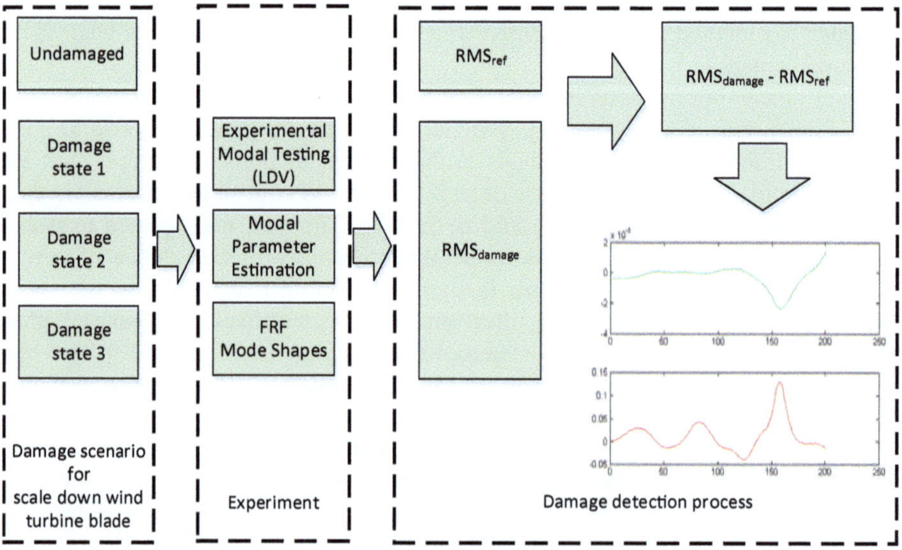

Fig. 1. General research methodology

The first step was experimental determination of the Frequency Response Function of the scaled down wind turbine blade, which was used to determine the values of resonant frequencies and corresponding modes of induced vibrations. Three scenarios of damage location of the blade coating were taken into account. The undamaged state was also measured. Data were obtained using laser vibrometry and analysed by calculating RMS.

2.1 Research Object

Investigations were carried out for a 1:10 scaled-down blade of a real wind turbine rotor of 36 m in diameter. The research object is schematically shown in Fig. 2.

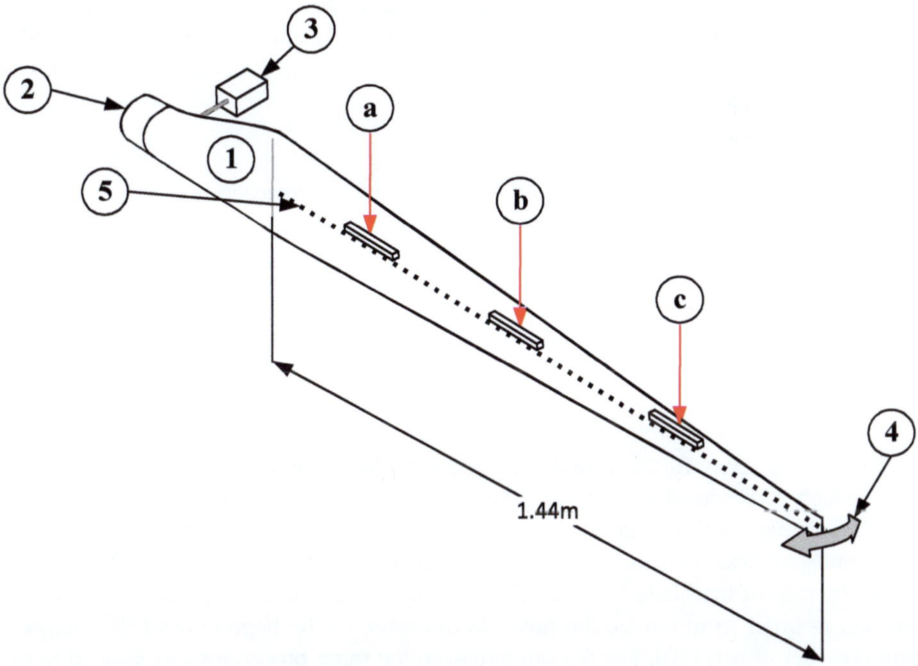

Fig. 2. A scheme of an experimental stand: 1 – rotor blade; 2 – blade fixing; 3 – modal shaker; 4 – the direction of the blade vibration; 5 – measurement points (dashed line); a, b, c – location of simulated damages.

The blade under investigation was 1.74 m in length and used a ClarkY aerodynamic profile, with no twist of the profile. Based on the design of the full-scale blade changes in the thickness of the blade coat were assumed. The blade was divided into three sections, and each section was characterised by a different coat thickness (number N of laminate layers). The total mass of the blade was about 2 kg. The geometrical features of the blade are summarised and presented in Table 1.

Table 1. Geometrical features of the blade

	Dimension [mm]
Total length	1740
Measured length	1440
Max width	160
Tip width	32
Thickness of spar	2.0
Thickness of section 1 (near blade root)	2.0
Thickness of section 2 (middle part of blade)	1.5
Thickness of section 3 (near blade tip)	1.0

A laminate coating of the blade was fabricated using custom lay-up moulding process. As components of the coating glass fibres and epoxy resin were used. The reinforcing fibres were symmetrically arranged as $[\pm 45°]_N$. Mechanical properties of the composite were presented in Table 2.

Table 2. Properties of composite material components

	Epoxy resin	Glass fibre
Young modulus, E	3.43 GPa	66.5 GPa
Poisson ratio, ν	0.35	0.23
Kirchhoff modulus, G	1.27 GPa	27 GPa
density, ρ	1250 kg/m3	2250 kg/m3

The current study focused on the influence of the defects located in the part of the blade which is responsible for the generation of the lifting force (aerodynamically active). This part begins at the widest point of the blade and ends at the tip. The length of the midspan was 1.44 m. Three positions of damage were considered. The closest one to the root of the blade (Fig. 2 - element a) was moved away from the start of the midspan in order to minimize the possible influence of the beginning of the analysed signal (border effect) [20]. For the same reason, the same procedure was applied to the damage located at the blade tip (element c). Halfway between the extreme locations, the third damage position was determined (element b). Further, for the damage position near the blade centre, the damage spanning over 12, 10 and 8 measurement points, corresponding to 6%, 5% and 4% of the total length measurement line, were considered. The measuring points are arranged in one line at the centre of the blade width (5). Measurement data were collected from 200 points. The object under investigation mounted on the testing stand is presented in Fig. 3a. An electrodynamics modal shaker (shown in Fig. 3c) was used to excite blade vibrations. Just above the measurement line, an additional high stiffness element was placed, which simulated damage. This additional element is shown in Fig. 3b.

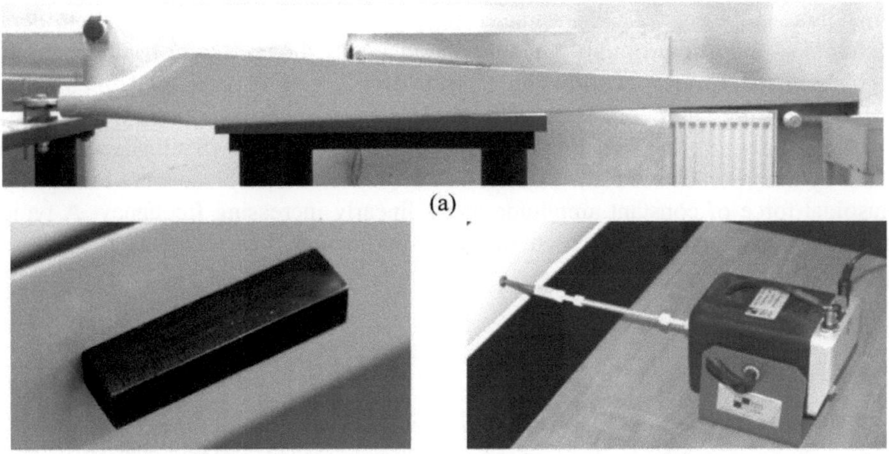

(a)

(b) (c)

Fig. 3. A view of the research object

The measurements were carried out using a PSV-400 Laser Scanning Vibrometer by Polytec Ltd., which enabled the authors for accurate and non-contact vibration measurements. The measurement stand used by the authors is presented in Fig. 4.

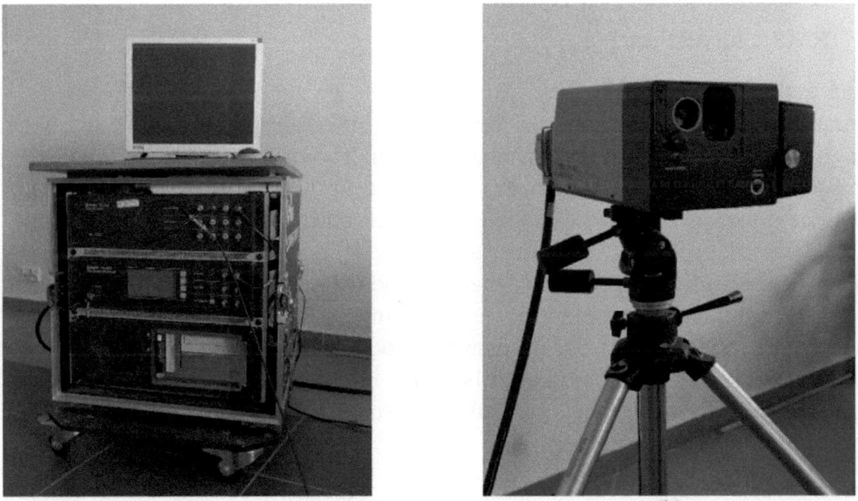

(a) (b)

Fig. 4. A detailed view of a: LSDV head (a), LDSV measurement unit (b)

The main and most important element of the vibrometer is a precise optical sensor, which is used to determine the velocities of vibrating objects. The device employs the principle of the Doppler effect to measure changes in the frequency of light reflected from vibrating objects. The vibrometer set also includes a vibrometer controller, a junction box and a control unit. The unit automatically moves a laser beam from point to point on the surface of vibrating objects over a grid of user predefined points. Therefore, during laboratory tests, appropriate measurement conditions were ensured, in order to maximize the signal level received by the vibrometer (the blade surface was coated by a retroreflective foil). The induced vibrations of the blade were excited by a sinusoidal force of constant amplitude and a linearly increasing frequency. A typical form of an excitation signal used during experimental measurements is shown in Fig. 5.

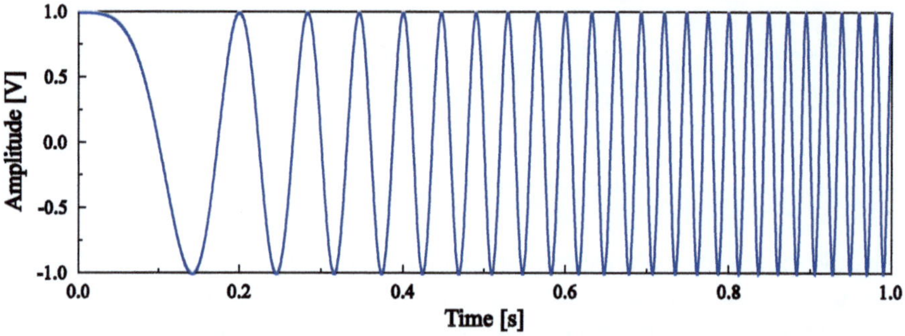

Fig. 5. A typical form of an excitation signal used during experimental measurements

2.2 RMS Calculation

In the authors' previous studies (using wavelets to analysis mode shapes), a high correlation was observed between the visibility of damage on scalograms obtained from given vibration forms and damage location. The damage was observed only if it coincided with local signal extremes. According to this, the higher the frequency of vibrations (more local extremes), the greater the chance of observing the damage.

Analysis of each vibration form separately is time-consuming and additionally requires pre-processing of source signals and adjustment of wavelet analysis parameters. The solution to this problem can be analysing the signals by determining the RMS for each measuring point from all available modes. This will allow the information contained in all forms of vibration to be taken into account in a single characteristic. Registered displacements on a particular measured point for all modes was used to determine the RMS value as:

$$RMS = \left(\sum_{k=1}^{N} f_k^2 / N \right)^{1/2} \tag{1}$$

where f_k^2 is a squared vibration amplitude of a sample (point) k, and N is a number of modes for RMS calculation.

Comparing the different forms of vibrations, it can be seen that their influence on the final result of the RMS is not the same. This is, of course, related to decreasing of the amplitudes of vibrations as their frequency increases. On the other hand, previous investigations proved that higher modes are more sensitive to damage. For these reasons, there was a need for data normalization, according to the formula:

$$f_{knorm} = \frac{f_k}{\max(|f(x)|)} \quad (2)$$

where f_{knorm} is a normalized vibration amplitude of a sample k, f_k is the vibration amplitude of a sample k and $\max(|f(x)|)$ is the maximum value from of all samples signal *f(x)*.

The normalization procedure equalizes the influence of the individual modes on the final result of RMS calculations.

3 Results and Discussion

Figure 6 shows a comparison of natural frequencies spectra for two damage state cases under consideration. The characteristic of each state is an averaged result of all vibration spectra obtained for each measurement point.

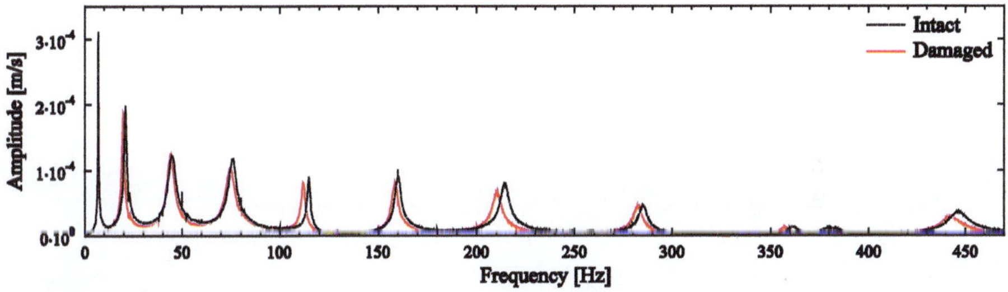

Fig. 6. Frequency Response Function for the undamaged and damaged blade under investigation

It can be seen from Fig. 6 that the position of peaks (the values of natural frequencies) is shifted for the damage case. However, taking into account the type of damage model used in the research (additional mass with high stiffness) such an effect of damage on the frequency spectra could be expected. Actually, small cracks and delamination do not necessarily cause changes in the vibration spectra. The literature indicates that analysis based on the value of vibration frequency is effective in the case of damage lengths greater than 15% of the total length of specimens [21, 22]. In addition, this type of analysis does not provide a direct and unambiguous information about the location and nature of damage. That is because the frequencies are global parameters and thus contain no direct information about the location and size of the damage. Therefore, it seems appropriate to include in calculations modes of vibrations, which are carriers of information from the entire blade length. The proposed method of analysis of vibration forms using RMS allows taking into account all forms at the same time. Figures 7, 8, 9 presents a summary of the results of RMS calculations for the blade with damage in all locations. The data has not been normalized. For comparison, a case of the RMS for undamaged blade has also been added. Vertical dashed lines mark the edge of additional mass. The length of simulated damage was 12 measurement points, corresponding to 6% of the total measurement line.

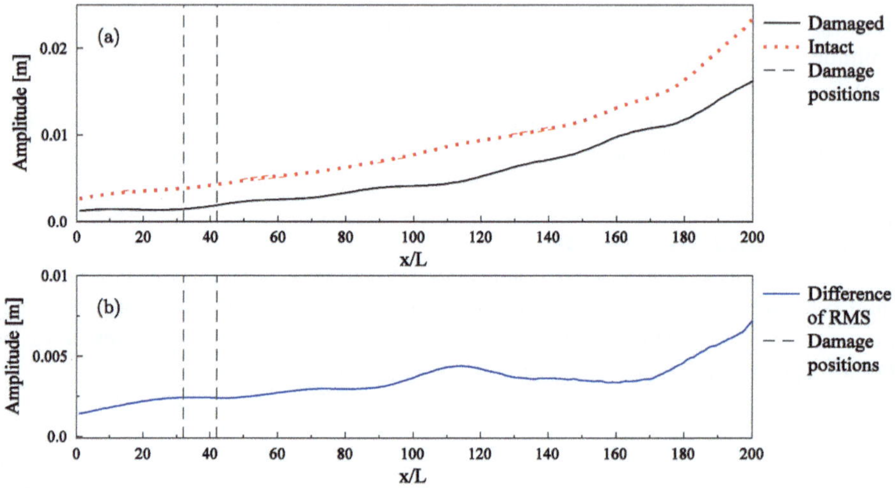

Fig. 7. Results of standard Root Mean Square calculation for the first case of damage localization: (a) comparison of the damaged and intact blade; (b) difference of RMS

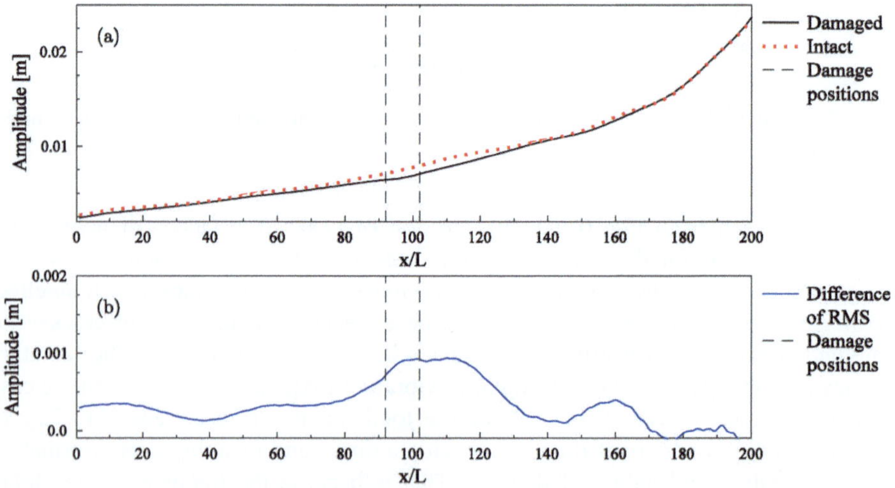

Fig. 8. Results of standard Root Mean Square calculation for the second case of damage localization: (a) comparison of the damaged and intact blade; (b) difference of RMS

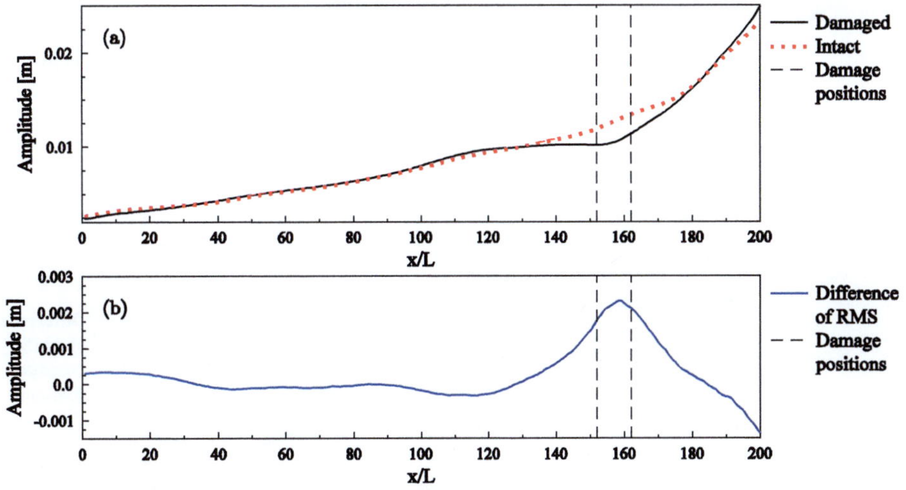

Fig. 9. Results of standard Root Mean Square calculation for the third case of damage localization: (a) comparison of the damaged and intact blade; (b) difference of RMS

The results obtained from the standard RMS calculations indicate a relationship between the sensitivity of the method and the location of the damage. Only in the most distant defect from the hub of the tested blade, it is possible to unambiguously indicate the location of the damage Fig. 9.

In the case of damage in the middle of the measured section, the result is inconclusive, and in the case of the first location, detection is completely impossible. That is because the effect of damage is relatively small there, where the cross-section of the blade has high stiffness values, as it is at the blade fixing. Changing the location of damage towards the tip of the blade results in an increase in the visible width of the damage zone. This is due to the fact that in this direction the stiffness of the blade decreases, due to a decrease in the cross-section of the blade as well as a decrease in the thickness of the blade coating. Thus, the relative impact of the damage on the cross-section will be stronger. The second conclusion is that damage is only visible if a reference state (undamaged) is taken into account in the calculation.

In order to improve the efficiency of the method and equalize the influence of all modes on the results of RMS calculations, the vibration form has been normalized. Figures 10, 11, 12 shows the results of these calculations.

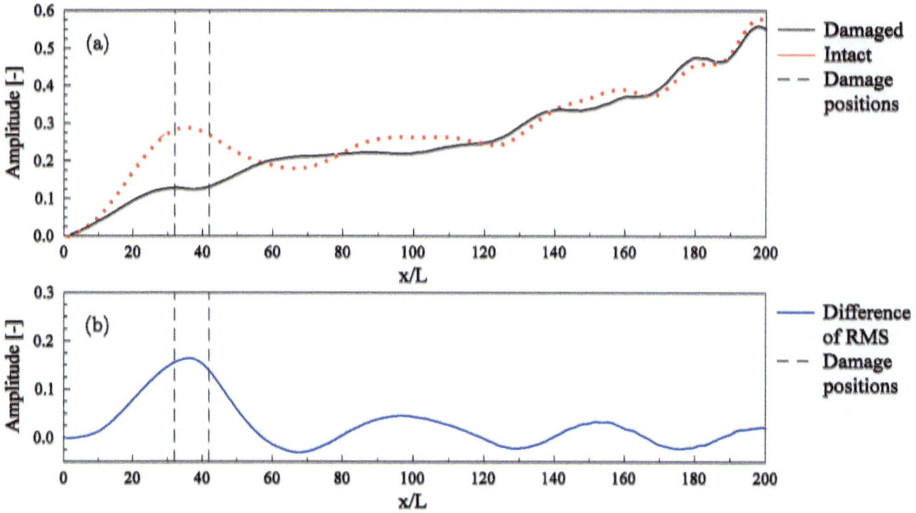

Fig. 10. Results of normalized Root Mean Square calculation for the first case of damage localization: (a) comparison of the damaged and intact blade; (b) difference of RMS

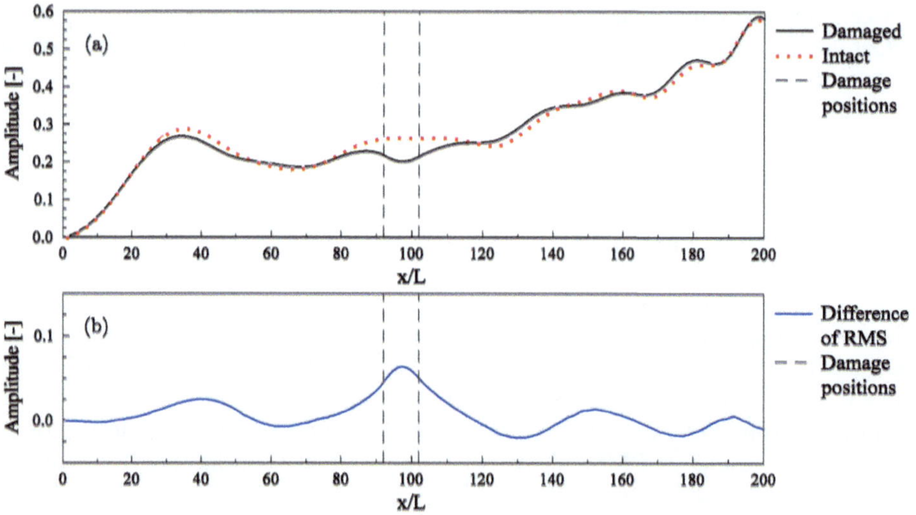

Fig. 11. Results of normalized Root Mean Square calculation for the second case of damage localization: (a) comparison of the damaged and intact blade; (b) difference of RMS

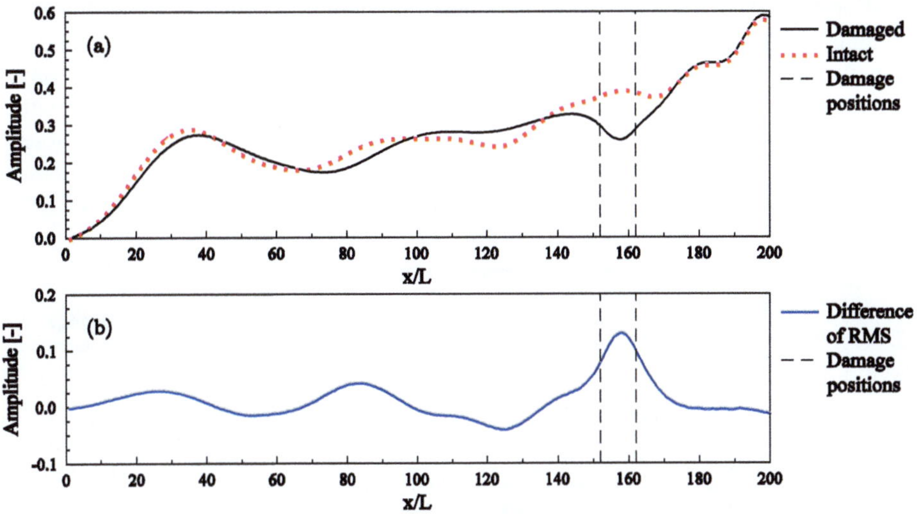

Fig. 12. Results of normalized Root Mean Square calculation for the third case of damage localization: (a) comparison of the damaged and intact blade; (b) difference of RMS

After the normalization procedure in all 3 cases, the difference in RMS between the damaged and reference state is clearly visible. In the case of second and third localization (Figs. 11, 12), it is even possible to directly determine the localization of damage. By determination the difference of RMS between damaged and undamaged state it is possible to detect the damage in all 3 positions unambiguously.

Figure 13 shows the results of calculation for three sizes of the damage zone close to the central location (6%, 5% and 4% of the total measurement line).

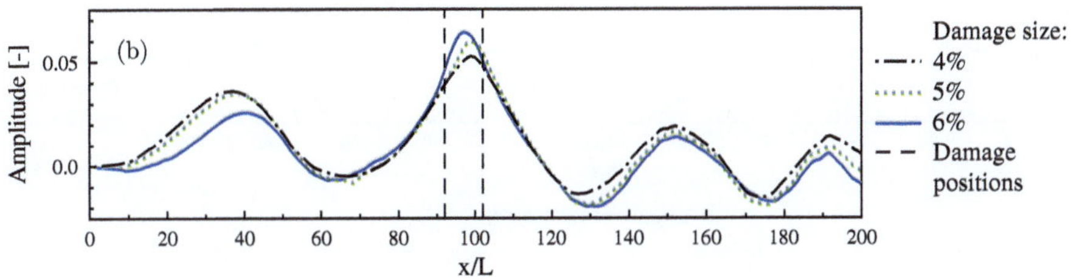

Fig. 13. The difference of normalized Root Mean Square calculation for the three sizes of the damage for the second case of localization

In real applications of blade condition monitoring systems it may be necessary to minimize the number of measurement points due to the measurement method used (e.g. piezoelectric transducers). A large number of measurement points means a large number of data contained in the signal, including information about the damage. On the other hand, it requires more time for measurement and data processing. The number of signal samples should not be smaller than the resolution of a finite signal length, i.e. the minimum number of samples needed to represent it. In the case of modes of vibration, the higher the frequency, the greater the number of points needed for the correct representation of the signal. Figure 14 (a–d) shows the results of the RMS calculation for the largest defect and a decreasing number of measurement points. The source signal with reduced density of measurement points to 50, 25, 10 and 5 measurement points were analysed. The undamaged state was considered in the calculation.

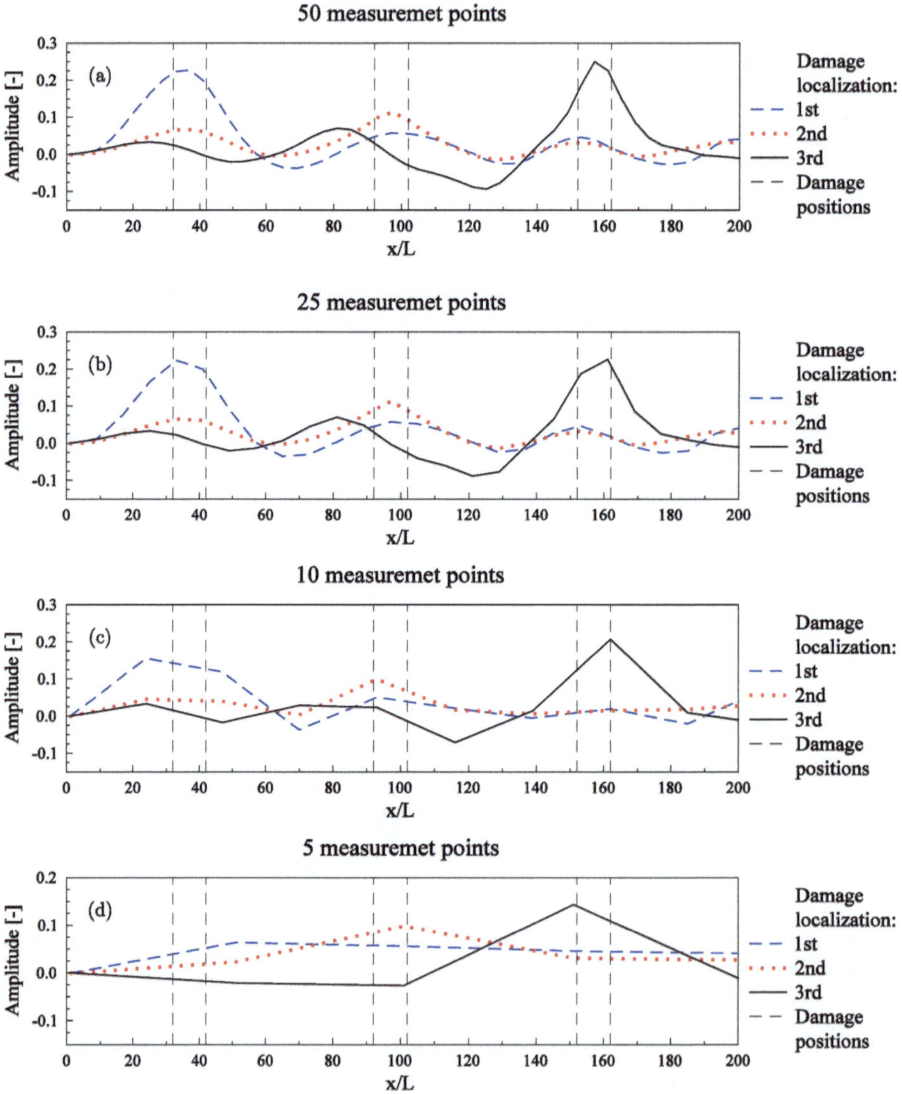

Fig. 14. The difference of normalized Root Mean Square calculation for the all case of damage localization and different number of measurement points: (a) 50 points; (b) 25 points; (c) 10 points; (d) 5 points

The results of the calculations show that to properly locate damage of the length of 6% of the measurement section, a signal of the length of 25 samples is sufficient (distance between consecutive samples was 4% of the total length of the measurement section). Therefore, in order to correctly determine the location of damage, the distance between subsequent samples should be smaller than the length of damage. However, as can be seen on Fig. 14c, even if the distance between subsequent samples is greater than the size of damage (10% of the measurement section length), it is possible to detect damage and determine its approximate location. For 5 measurement points, the representation of the vibration modes is insufficient, leading to ambiguous results.

4 Conclusions

The paper presents results of RMS calculation for modal responses in terms of natural frequencies and modes of vibrations for defect detection in laminated coatings of wind turbine blades. The modal parameters have been determined only experimentally using the Laser Doppler Scanning Vibrometer (LDSV). The investigations were carried out on a specially manufactured composite blade for several localizations of damage. Damage were simulated by additional high stiffness element fixed to the surface of the blade.

The results of the research presented in this paper confirm the effectiveness of RMS calculation in damage detection using modes of vibrations. Some of conclusions from obtained results can be summarized below:

- In order to extract damage information for all positions, it was necessary to normalize vibration forms.
- It is necessary to know the condition of the undamaged blade or its predecessor.
- With 200 measurement points and 10 forms of vibration, the method proved high effectiveness, allowing to determine the place of damage for all 3 examined cases.
- A study was also carried out to check the minimum number of measurement points by which it was possible to detect damage and to estimate the place of its occurrence.

However, the method does not allow for precise determination of damage limits, but it can be used for preliminary determination of the place of damage occurrence so that in the next step it is possible to apply a more accurate method to determine the exact size of damage and its nature.

References

1. Sørensen, B.F., Jørgensen, E., Debel, C.P., Jensen, F.M., Jensen, H.M., Jacobsen, T.K., Halling, K.M.: Improved design of large wind turbine blade of fibre composites based on studies of scale effects (phase 1) - summary report. Risø-R-1390(EN) report, Risø National Laboratory, Denmark (2004)
2. Ciang, C.C., Lee, J.R., Bang, H.J.: Structural health monitoring for a wind turbine system: a review of damage detection methods. Meas. Sci. Technol. **19**(12), 122001–122020 (2008)

injury. However, these methods have such disadvantages as expensive cost, inflammation or lacking of donors [6].

Replacement of the damaged articular cartilage with suitable artificial materials provides another option. Poly (vinyl alcohol) (PVA) hydrogel has been considered as a very interesting and promising material for articular cartilage replacement [7]. But the neat PVA hydrogels can not provide the sufficient bioactivity as the pores of PVA hydrogels are only a few micrometers [8]. To generate large pores in PVA hydrogels, the porogen such as polyethylene glycol [9], dichloromethane [10] and gelatin sponges [11] and the composites porogen consisting of poly (lactic-co-glycolic acid) and sodium chloride micro-particle [12] have been attempted.

In this communication, agarose (AG), a polysaccharide obtained from agar and used for a variety of life science applications, was used as a novel porogen to fabricate macroporous PVA hydrogel (mPVA). We also investigated the effects of preparation parameters (the duration of freezing, number of freezing-thawing cycles, the temperature of thawing) on the mechanical properties of mPVA.

2 Experiments

2.1 Materials

The PVA with a degree of polymerization of 1750 ± 50 was purchased from Sinopharm Chemical Reagent Co., Ltd., China. The agarose (G-10) was provided by Gene Company Ltd., Spain. The water used in experiments is deionized water.

2.2 Preparation of Macroporous PVA Hydrogels

PVA solution (15 wt.%) was prepared by dissolving the PVA in the deionized water and heated at 95 °C for 2 h under stirring. PVA/AG solutions were prepared by adding 4 wt.% AG powder (based on water) into the PVA solution (15 wt.%) with continuous stirring and the stirring was maintained 2 h at 95 °C until the mixtures became transparent. The mPVA were prepared by freezing-thawing method with different duration of freezing, number of freezing-thawing cycles and the temperature of thawing. Finally, AG used as porogen was removed by washing with deionized water.

2.3 Mechanical Test

Compression and tension experiments of hydrogel were carried out on 2 kN Sans Universal Testing Machine (Shenzhen SANS Testing Machine Co., Ltd., CMT-4204, China) [13–15]. The compression experiments were performed on the 4 mm × 4 mm × 4 mm hydrogels with a strain rate of 30%/min at room temperature. The tension experiments were performed on the 50 mm × 4 mm × 2 mm section of barbell shaped hydrogel samples with strain rate of 600%/min. Each experiment repeats at least three times.

2.4 Scanning Electron Microscopy (SEM)

The hydrogel slices were freeze-dried on a freeze drier (BK-FD10S, BIOBASE) for 16 h. Then the obtained hydrogel samples were coated with gold on a magnetron ion sputter metal coating device (Vacuum Device MSP-1s, Japan), and the surface morphology was examined on a scanning electron microscopy (SEM, FEI Quanta 200, FEI, USA).

3 Result and Discussion

3.1 Effect of Preparation Parameters on Mechanical Properties

The formation of macropores in the hydrogel will inevitably affect its mechanical properties. And it is necessary to maintain certain mechanical properties for artificial articular cartilage. The mechanical properties of mPVA can be modulated by adjusting the duration of freezing, number of freezing-thawing cycles and the temperature of thawing. The mechanical experimental results of hydrogels prepared with different parameters are presented in Figs. 1, 2 and 3. The stress values of tensile tests are

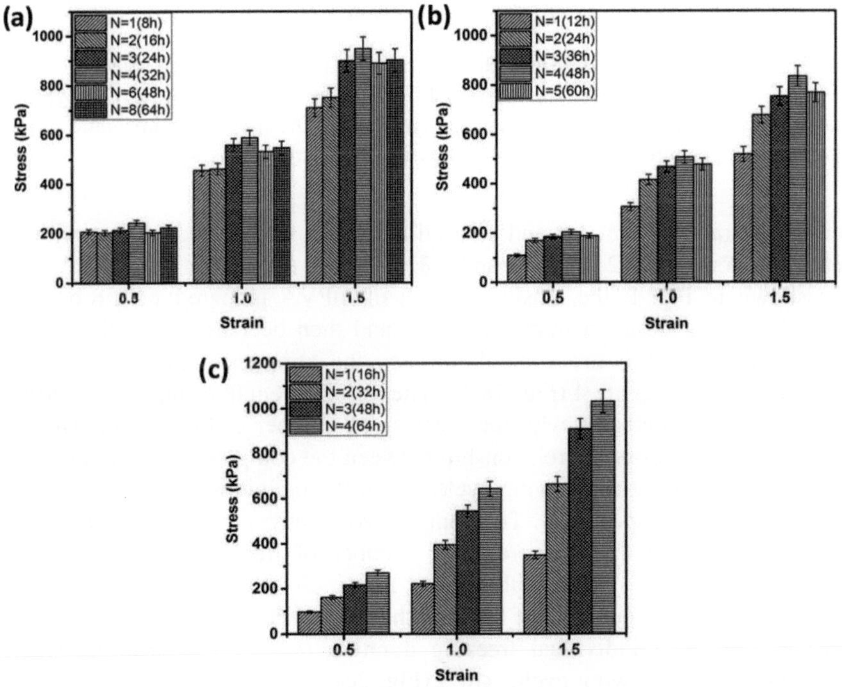

Fig. 1. Effect of individual freezing duration and number of freezing-thawing cycles on tensile strength of mPVA. The individual freezing durations are 8 h, 12 h and 16 h in (a), (b) and (c), respectively. N: freezing-thawing cycle time. The freezing temperature is −20 °C. The thawing temperature is 18 °C and the individual thawing time is 8 h.

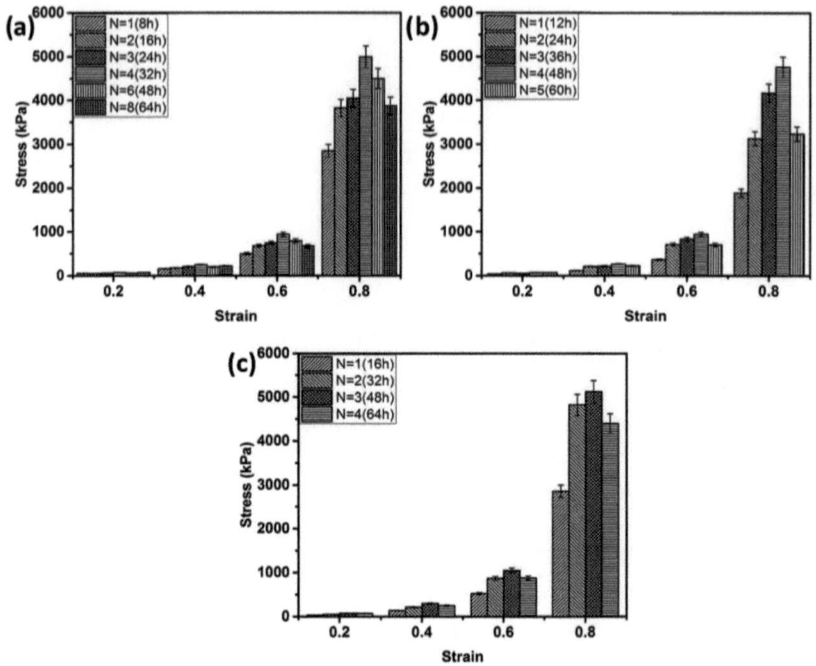

Fig. 2. Effect of individual freezing duration and number of freezing-thawing cycles on compression strength of mPVA. The individual freezing durations are 8 h, 12 h and 16 h in (a), (b) and (c), respectively. N: freezing-thawing cycle time. The freezing temperature is −20 °C. The thawing temperature is 18 °C and the individual thawing time is 8 h.

presented at strains of 0.5, 1.0 and 1.5 and the stress values of compression tests are presented at strains of 0.2, 0.4, 0.6 and 0.8.

As shown in Fig. 1, the tensile strength of mPVA prepared with 8 h and 12 h individual freezing duration firstly increases and then decreases with the number of freezing-thawing cycles, the strength reaches the strongest when the numbers of freezing-thawing cycles are 4 (Fig. 1a, b). The tensile strength of mPVA prepared with 16 h freezing time increases with the increase of number of freezing-thawing cycles (Fig. 1c). Figure 2 shows the relationship between the compressive strength of mPVA and the number of freezing-thawing cycles when the individual freezing durations are 8 h, 12 h and 16 h, respectively. The compressive strength of mPVA increases firstly and then decreases with the increase of number of freezing-thawing cycles. The compressive strength of mPVA with 8 h and 12 h individual freezing duration reaches the strongest when the numbers of freezing-thawing cycles are 4 (Fig. 2a, b), and that of mPVA with 16 h individual freezing duration reaches the strongest when the numbers of freezing-thawing cycles are 3 (Fig. 2c).

Figure 3 shows the effect of thawing temperature on the mechanical properties of mPVA. The thawing temperatures are designated as 18 °C and −3 °C. Other parameters, including freezing temperature (−20 °C), individual freezing duration (16 h), and individual thawing duration (8 h), are the same. Tensile strength increases with the

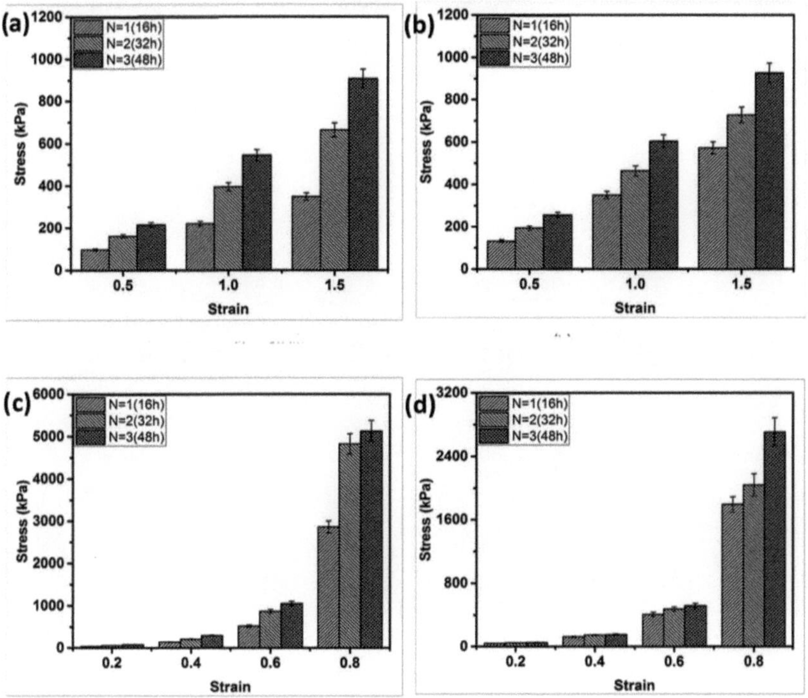

Fig. 3. Effect of thawing temperature on tension (a, b) and compression (c, d) strength of mPVA. The thawing temperature is 18 °C in (a), (b) and −3 °C in (c) and (d), respectively. N: freezing-thawing cycle time. The individual freezing duration is 16 h and the freezing temperature is −20 °C. The thawing time is 8 h.

increase of number of freezing-thawing cycles. This is true no matter the thawing temperature is 18 °C or −3 °C. But when the number of freezing-thawing cycles is fixed, the tensile strength of hydrogels at thawing temperature 18 °C is higher than that at thawing temperature −3 °C (Fig. 3a, b). From Fig. 3c, d, it can be found that the compressive strength of the hydrogel prepared with 18 °C thawing temperature is 5122 kPa (strain = 0.8, N = 3), and the compressive strength is only 2706 kPa (strain = 0.8, N = 3) when the thawing temperature is −3 °C.

In brief, the hydrogel with the best mechanical properties is obtained when the number of freezing-thawing cycles is 4 (for tension) or 3 (for compression), the individual freezing duration is 16 h and the thawing temperature is 18 °C.

3.2 Morphology Characterization

The micromorphology change of PVA hydrogels before and after the addition of AG is shown in Fig. 4. The honeycomb-like structure can be observed for the neat PVA hydrogel and the diameters of pores range from 2 to 3 μm (Fig. 4a). In contrast, macropores range in 20–200 μm are generated in PVA hydrogels when 4 wt.% AG is introduced (Fig. 4b), which means the macroporous PVA hydrogels are prepared

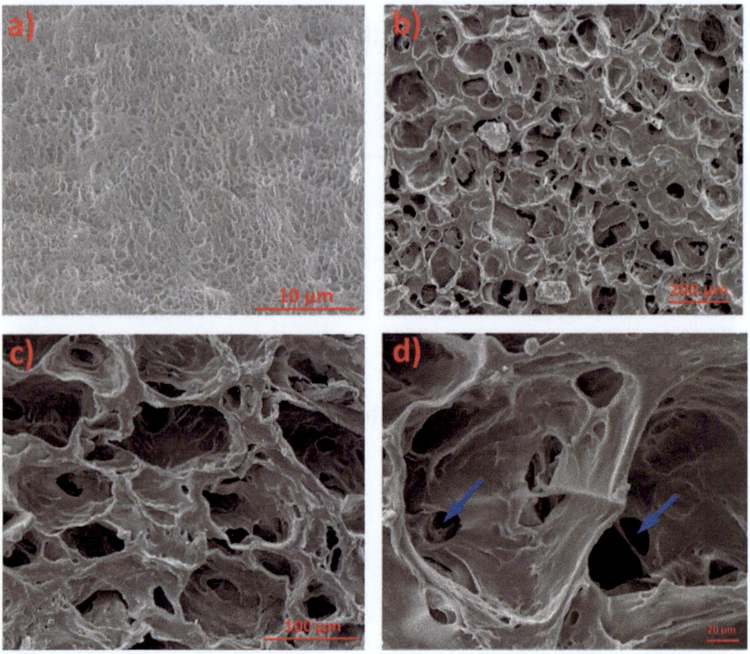

Fig. 4. SEM images of PVA hydrogels. (a): neat PVA hydrogel; (b): macroporous PVA hydrogel before removal of AG; (c): macroporous PVA hydrogel after removal of AG; (d): macroporous PVA hydrogel in (c) with high magnification.

successfully using AG as porogen. The AG particles exist in the pores, indicating that the introduced AG acts as the poregen. The AG particles in the pores need to be removed to leave space for the cell growth and proliferation. Fortunately, The AG particles can be removed simply by washing. The macroporous PVA hydrogel after the removal of AG is shown in Fig. 4c. Small pores (as indicated by the blue arrows) can also be observed on the walls of the original pores, as Fig. 4d shows, which means the macropores have good connectivity, which may facilitate the migration of cells, the diffusion of nutrients and the removal of waste products.

4 Conclusion

Macroporous PVA hydrogel is successfully prepared by using AG as porogen through freezing-thawing method. The pore size increases from less than 3 μm to 20–200 μm when AG is introduced, which meets the requirements of chondrocyte implantation. And the macropores have good connectivity, which may facilitate the migration of cells, the diffusion of nutrients and the removal of waste products. The mechanical properties of macroporous PVA hydrogels can be modified by adjusting the preparation parameters, and the proposed freezing-thawing parameters are 16 h individual freezing duration, 3 cycles of freezing-thawing and 18 °C thawing temperature.

Acknowledgments. This work was supported by grants from the National Natural Science Foundation of China (11772086, 51775541, 11572080, 51811530309), the Natural Science Foundation of Liaoning Province (201800935), and the Fundamental Research Funds for the Central Universities in China (DUT18ZD302).

References

1. Caplan, A.I.: Mesenchymal stem cells. J. Orthop. Res. **9**(5), 641–650 (1992)
2. Ito, Y., Fitzsimmons, J.S., Sanyal, A., Mello, M.A., Mukherjee, N., O'Driscoll, S.W.: Localization of chondrocyte precursors in periosteum. Osteoarthr. & Cartil. **9**, 215–223 (2001)
3. Bouwmeester, P.S.J.M., Kuijer, R., Homminga, G.N., Bulstra, S.K., Geesink, R.G.T.: A retrospective analysis of two independent prospective cartilage repair studies: autogenous perichondrial grafting versus subchondral drilling 10 years post-surgery. J. Orthop. Res. **20** (2), 267–273 (2002)
4. Hangody, L., Karpati, Z.: A new surgical treatment of localized cartilaginous defects of the knee. J. Orthop. Trauma **37**, 237–243 (1994)
5. Brittberg, M., Lindahl, A., Nilsson, A., Ohlsson, C., Isaksson, O., Peterson, L.: Treatment of deep cartilage defects in the knee with autologous chondrocyte transplantation. N. Engl. J. Med. **331**(14), 889–895 (1994)
6. Zeng, L., Yao, Y., Wang, D., Chen, X.: Effect of microcavitary alginate hydrogel with different pore sizes on chondrocyte culture for cartilage tissue engineering. Mater. Sci. Eng., C **34**(1), 168–175 (2014)
7. Ma, R., Xiong, D., Miao, F., Zhang, J., Peng, Y.: Novel PVP/PVA hydrogels for articular cartilage replacement. Mater. Sci. Eng., C **29**(6), 1979–1983 (2009)
8. Gutiérrez, M.C., García-Carvajal, Z.Y., Jobbágy, M., Rubio, F., Yuste, L., Rojo, F., Ferrer, M.L.: Poly (vinyl alcohol) scaffolds with tailored morphologies for drug delivery and controlled release. Adv. Func. Mater. **17**(17), 3505–3513 (2007)
9. Miao, T., Miller, E.J., Mckenzie, C., Oldinski, R.A.: Physically crosslinked polyvinyl alcohol and gelatin interpenetrating polymer network theta-gels for cartilage regeneration. J. Mater. Chem. B **3**(48), 9242–9249 (2015)
10. Scholten, P.M., Ng, K.W., Joh, K., Serino, L.P., Warren, R.F., Torzilli, P.A., Maher, S.A.: A semi-degradable composite scaffold for articular cartilage defects. J. Biomed. Mater. Res., Part A **97A**(1), 8–15 (2015)
11. Ng, K.W., Torzilli, P.A., Warren, R.F., Maher, S.A.: Characterization of a macroporous polyvinyl alcohol scaffold for the repair of focal articular cartilage defects. J. Tissue Eng. Regen. Med. **8**(2), 164–168 (2014)
12. Cao, Y., Xiong, D., Wang, K., Niu, Y.: Semi-degradable porous poly (vinyl alcohol) hydrogel scaffold for cartilage repair: evaluation of the initial and cell-cultured tribological properties. J. Mech. Behav. Biomed. Mater. **68**, 163–172 (2017)
13. Zhang, W., Liu, L.F., Xiong, Y.J., Liu, Y.F., Yu, S.B., Wu, C.W.: Effect of in vitro storage duration on measured mechanical properties of brain tissue. Sci. Rep. **8**, 1247 (2018)
14. Zhang, W., Zhang, R.R., Feng, L.L., Li, Y., Wu, F., Wu, C.W.: Mechanical response of brain stem in compression and the differential scanning calorimetry and FTIR analyses. ASME J. Appl. Mech. **83**(9), 091005 (2016)
15. Zhang, W., Zhang, R.R., Wu, F., Feng, L.L., Yu, S.B., Wu, C.W.: Differences in the viscoelastic features of white and grey matter in tension. J. Biomech. **49**(16), 3990–3995 (2016)

Damage in Composite Materials

Delamination Buckling of FRP: Experimental Tests and Theoretical Model

R. Capozucca(✉), E. Magagnini, and M. V. Vecchietti

Structural Section DICEA, Università Politecnica Delle Marche, Ancona, Italy
r.capozucca@univpm.it,
{e.magagnini,m.v.vecchietti}@pm.univpm.it

Abstract. In scientific literature many studies have focused on the aspects of debonding mainly due to bending and/or shear in steel/concrete beams strengthened with fiber reinforced polymers (FRPs). Few experimental investigations on delamination buckling process of fiber reinforced polymer (FRP) strip/sheet utilized as strengthening in structural damaged elements, beams or slab, have been carried out. This paper analyses about the experimental delamination buckling of two type of FRP: steel reinforced polymer (SRP) strips and basaltic reinforced polymer (BFRP) strips glued on the surface of slender homogeneous beam models. Several experimental tests have been done considering beam models made of homogeneous material marble assuming this material without tensile strength. The specimens strengthened with a double layer of FRP strips glued on both up and down surfaces were tested under increasing compressive force until the failure due to delamination buckling of strengthening. Analysis of experimental process has been developed considering an appraisal theoretical model. Finally, theoretical results with experimental data are compared and discussed.

Keywords: Homogeneous beams · Compression tests · SRP/BFRP strips · Delamination buckling

1 Introduction

The delamination of layer of fiber reinforced polymers (FRPs) used as strengthening in civil engineering structures, column or bending beams, under compression may buckle. This local instability process doesn't imply the ultimate load of structural element and usually the strengthened element can carry on in the post-buckling phase under loading. Delamination buckling of FRP layer causes separation of the added layer from the main structural element and this delamination may significantly influence the strength, stiffness and stability of a strengthened element typical of civil engineering as reinforced concrete/steel/timber beams.

The beginning of delamination in a strengthened columns or beams results from many sources such as defects, deterioration of bending material or local damages due to impact. The delamination buckling in beam/plate was investigated for many years from 1981 [1–8]. The delamination behavior of FRP strips in strengthened elements has received the attention of many researchers [4, 6]. Recently the analysis of damage due

M. A. Wahab (Ed.): DAMAS 2019, LNME, pp. 753–766, 2020.
https://doi.org/10.1007/978-981-13-8331-1_59

to delamination in CFRP composite has been developed also on the base of experimental vibration analysis [10, 11]. In this paper delamination buckling concerns the behavior of steel fiber reinforced polymer (SRP) and basaltic reinforced polymer (BFRP) strips of strengthening in column models of homogeneous material subjected to compression up buckling and post-buckling. This topic is increased in many civil engineering applications and it is mostly relevant in the structures of concrete and masonry damaged under seismic action where it is foreseen a strengthening with FRP that may interested under loading both tensile and compressive stresses [12].

The choice of marble as the material of beam models is convenient to have an almost perfect homogeneous brittle material, very different from concrete; such as substantially heterogeneous and porous, and thus with a higher capacity of bond for typical strengthening by externally bounded FRP lamina/strip.

In this paper compressive tests on slender column models of marble unreinforced and reinforced with SRP and BFRP strips are presented and experimental results have been discussed. Theoretical analysis based on simplified elastic model has been developed and obtained data have been compared and discussed with experimental results.

2 Experimental Tests and Results

2.1 Homogeneous Column Model Under Compression

The experimental column models were built by white marble of Carrara (Italy) (Fig. 1a) with following dimensions: thickness 6 mm, width 28 mm for a section of 168 mm^2 and length 300 mm. Young's modulus has been evaluated from bending experimental tests and it is equal to $E \cong 68.9$ kN/mm^2.

At the first phase, specimen of unreinforced marble column (UMC) was subjected to compression as shown in (Fig. 1b) under two steel plate measuring deflections, vertical displacements and vertical load P controlled by a cell of loading. In these tests three transducers (LVDTs) was utilizing to measure the lateral deflection at the midspan of column model and at the edges. Before loading two plates of neoprene were inserted between steel plates and edges of column model to improve the transfer of loading. In Table 1 the main experimental results are provided of UMC.

It is observed the displacement in midspan is occurred in horizontal direction until the specimen failure (Fig. 1c) under a load equal to P = 10.50 kN. The evaluated slenderness of column model, $\lambda = l_0/i$ with l_o: effective length of beam column and i = radius of inertia of section, was equal to $\lambda = 173.21$ assuming ideal hinges at the ends. Theoretical value of first Euler's load is equal to $P_E = 3.81$ kN.

In Fig. 2 the dimensionless diagram for the ratio P/P_E versus the deflection ratio δ/t is shown for the unreinforced column model (UMC). The first one is obtained dividing the experimental load, P, by the first Euler's load, P_E, the second is obtained dividing the horizontal displacement in midspan, δ, by thickness, t, of column model.

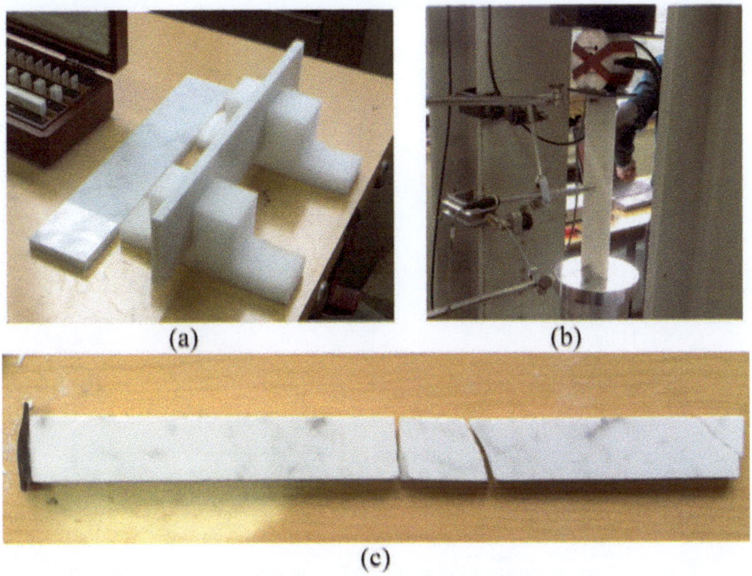

Fig. 1. (a) Specimen (UMC) of white marble; (b) set-up of compression test; (c) specimen after failure due to instability by compression test.

Table 1. Main results for unreinforced marble column model (UMC).

Compression load [kN]	Vertical displacement [mm]	Deflection at midspan [mm]
0.00	0	0
3.05	2.74	0.20
5.63	3.33	0.44
7.09	3.47	0.72
8.21	3.61	1.02
9.01	3.73	1.29
10.50	4.31	1.88

2.2 Strengthened Homogeneous Column Models

Reinforced marble columns (RMC) specimens strengthened with FRP strips are summarized and indicated in Fig. 3. Specimens with SRP/BFRP strips on both lateral surfaces - dimensions 300 mm·28 mm - were subjected to compression until failure to analyze the delamination process due to buckling of SRP/FRP strips strengthening on the compressed side of bending specimen. For both types of reinforcement the marble surfaces are treated; a primer, with the function of penetrating into the material across a microcavity, was used. Later a two-component epoxy adhesive, was employed like fiber matrix to guarantee impregnation and adhesion to the support. The main mechanical characteristics of composite materials SRP and BFRP were acquire through experimental tests on specimens of SRP/BFRP (Table 2) according to the regulations

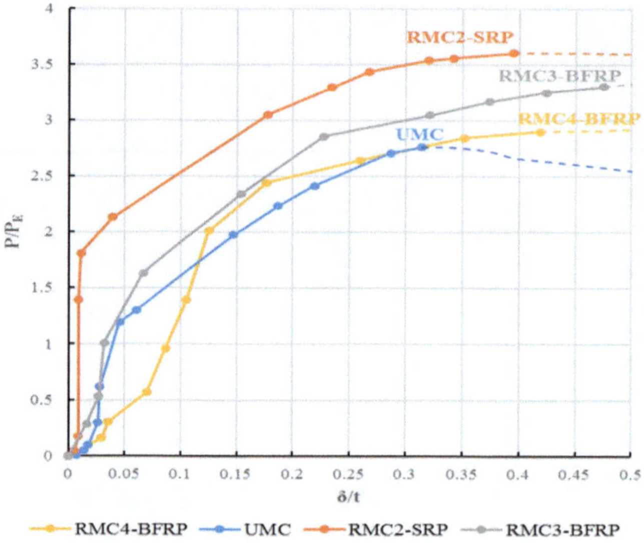

Fig. 2. Exp. diagram load ratio, P/P_E, vs displacement ratio, δ/t.

ASTM D 3039 [13]. In Figs. 4 and 5 the view of the different experimental failure by tensile tests of specimens of SRP e BFRP are shown.

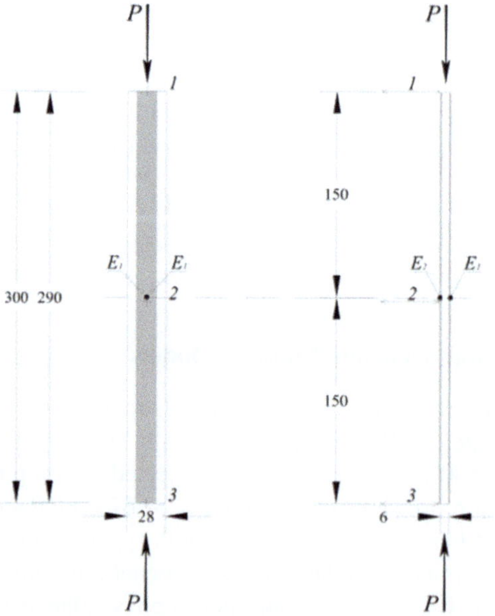

Fig. 3. Specimens with SRP/BFRP strips under compression.

Table 2. Mechanical exp. parameters of BFRP and SRP specimens.

Specimen	Composite material	Equivalent thickness t_e [mm]	Measured thickness t_m [mm]	Width b [mm]	Equivalent area A_{fib} [mm^2]
TB1	BFRP[1]	0.14	1.51	16.10	6.76
TB2	BFRP[1]	0.14	1.48	15.30	6.43
TA1	SRP[2]	0.48	3.30	21.04	4.81
TA2	SRP[2]	0.48	3.10	19.24	4.33
Specimen	Maximum Load F_{max} [N]	Normal stress σ_f [N/mm^2]	Equivalent strength σ_{fib} [N/mm^2]	**Ultimate strain*** ε_{FRP} [%]	Equivalent modulus E [GPa]
TB1	13820	568.5	2043.8	2.11	-
TB2	13842	611.3	2154.1	2.11	84.43
TA1	16910	243.5	3515.6	1.60	-
TA2	14267	239.2	3295.7	1.60	213.8

Fig. 4. View of experimental failure by tensile tests: SRP specimen – failure type LGM (Lateral, Gage, Middle).

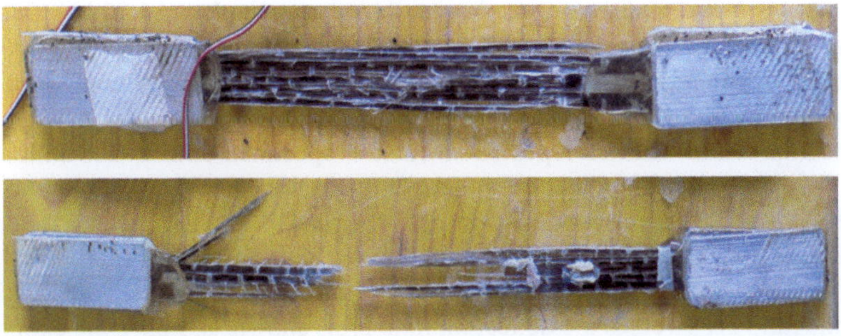

Fig. 5. View of experimental failure by tensile tests: BFRP specimen – failure type SGM (Long Splitting, Gage, Middle).

The instruments used during the compression tests on strengthened specimens was (Figs. 6 and 7): load of cell to measure compressive force, P; three LVDTs to record lateral deflections; one LVDT to measure the vertical displacements; two strain gauges (E_1-E_2) at the mid span of specimens on the lateral surface to measure strains of SRP/BFRP strips under compression tests both under bending of specimens due to the instability and post delamination buckling of FRP strips. In the Table 3 the tested specimens are shown with main experimental results. Tests were carried out applying incremental loads with increases of about 0.2 ÷ 0.3 kN. Data related to LVDTs and strain gauges, through software CATMAN, were recorded.

Fig. 6. Set-up of strengthened specimen with SRP strips (RMC1).

RMC1 was tested under compression until load equal to P ≅ 14.70 kN; failure was recorded on the edge of specimens. The RMC1 was strengthened at the edges and to improve the distribution of stresses at the ends multi-layered bases composed of lead and neoprene were interposed.

The compression test was repeated on RMC1′ specimen until a vertical load equal to P ≅ 14.2 kN, also in this case failure occurred on the local edge. The RMC2 specimen was subjected to compression until the load P ≅ 13.70 kN reaching failure for buckling (Fig. 7).

In Fig. 2 the dimensionless diagram for reinforced column model RMC2 with SRP strip of the ratio P/P_E versus the deflection ratio δ/t is shown. In Fig. 8 the experimental diagrams of the ratio, P/P_E, versus the strain measures recorded by two strain gauges, E1 - E2, during the compression test with the increase of load on RMC2 specimen until delamination buckling of SRP strip on the compressive side, are represented. In Fig. 8 the strain values on the tensile for bending are measured with strain gauges E1 while in the compressed SRP strip by E2.

RMC3 and RMC4 specimen were strengthened on both sides with BFRP strips with epoxy resin as matrix and subjected to compression test. The test was carried out with the same procedure as those without reinforcement and with reinforcement of SRP strips by applying progressively increasing loads. For both tests the detachment of the BFRP strips due to buckling of that located on the compressed side of the specimen,

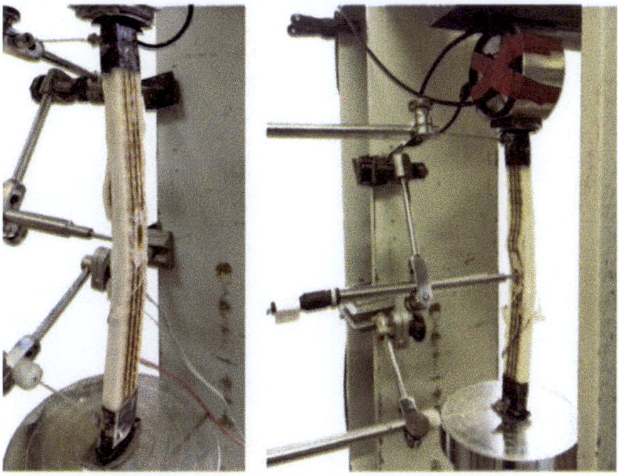

Fig. 7. View of delamination buckling of SRP strip on the compressive bending side (RMC2).

Table 3. Exp. results by compression tests on column specimens.

Exp. specimen	Type of strengthening	Exp. ultimate load [kN]	Exp. buckling load* [Kn]	Type of exp. failure
UMC	Unreinforced beam	10.50	4.56	Buckling failure
RMC1	SRP	14.70	-	Local edge failure
RMC1′	SRP	14.18	-	Local edge failure
RMC2	SRP	13.70	6.99	Buckling of SRP
RMC3	BFRP	12.65	5.28	Buckling of BFRP
RMC4	BFRP	11.10	5.78	Buckling of BFRP

*Exp. initial buckling load evaluated assuming a bilateral diagram of deflections.

also inflected due to exceeding the critical load value, was observed. For RMC3 the compression test was stopped until a value of load equal to 12.65kN. In Fig. 2 the experimental normalized diagrams of the load ratio, P/P_E, vs displacement ratio, δ/t, for specimen RMC3 and RMC4, are represented.

In Fig. 9 it is shown specimen RMC3 under compression until buckling and post-buckling with (Fig. 9c) delamination buckling of BFRP strip. In Fig. 9(d) the specimen RMC3 after test with detachment of BFRP strip on compressed side.

In the compression test on RMC4 specimen the compression load with consequent failure of the specimen was equal to $P \cong 11.1$kN. In Fig. 10 the failure due to detachment of compressed side BFRP strip and to the failure of the base specimen material.

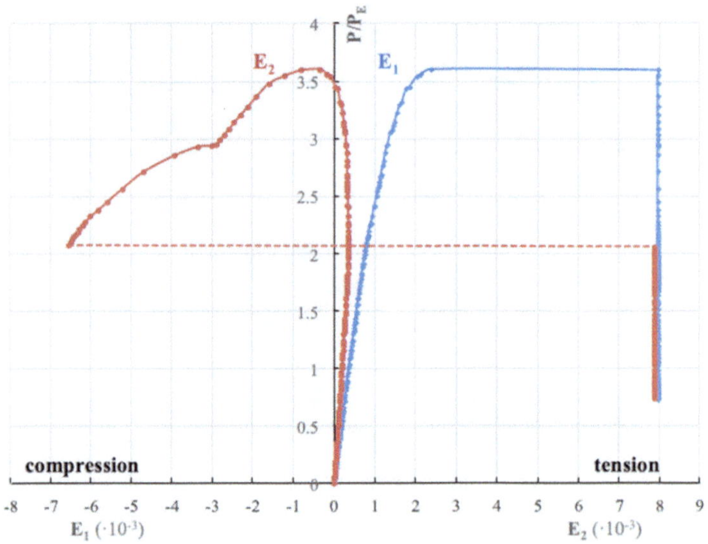

Fig. 8. Exp. diagram of the ratio P/P_E, versus strain values E1, E2 on the SRP strip, respectively, on the tensile side and compressive one (RMC2).

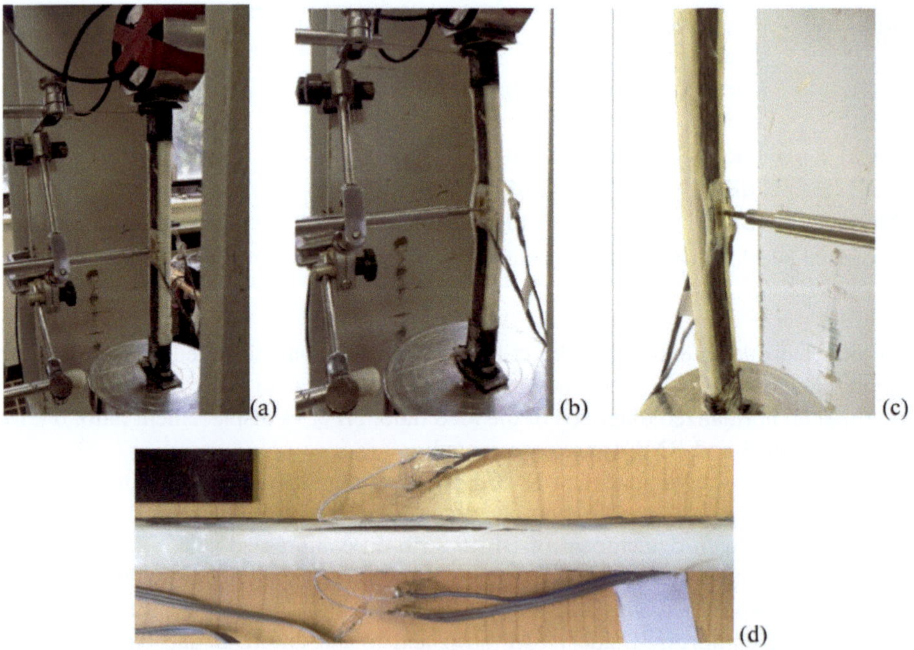

Fig. 9. (a) Specimen RMC3 under compression; (b) and (c) detachment of BFRP strip on compressed side of bending specimen for delamination buckling; (d) detachment of BFRP strip after test.

Fig. 10. View of detachment of BFRP strip on compressed side of RMC4 specimen and failure of marble beam.

In Figs. 11 and 12 the diagram of the compression load, P/P_E, vs the strain values measured, E1 and E2, during the tests, respectively, on RMB3 and RMC4 is shown. From the analysis of the dimensionless diagrams shown in Figs. 8, 11 and 12 the behavior of homogenous column models under compression clearly emerges. Actually, the strengthened specimens, once the critical load has been exceeded, start to in-stabilized and it involves the establishment of a double stress state, respectively of tension on a side and compression on the other.

By observing the experimental diagrams of P/P_E vs strains of FRP strips, for each kinds of specimens, it is possible to understand the starting point of the debonding process. After reached a critical value of compression strain equal to about $6.5 \div 710^{-3}$, the debonding mechanism due to compression starts with phenomenon of buckling of FRP strip/lamina. From a view of specimens at failure, debonding occurs involving a non-negligible thickness of the support, equal to about $1.5 \div 2$ mm, that propagates deep inside the marble columns. It is possible to conclude that under compression, columns reinforced with FRP strips have shown a weak behavior.

3 Analysis of Delamination Buckling with Analytical Model

The analysis of delamination buckling of FRP strip has been developed below using an approximal elastic model. The following assumptions were made: all materials are linear as in the above analysis of stress distribution; the plane sections remain plane after delamination buckling; the bond between adhesive and FRP strip remains up to displacement of FRP due to delamination buckling.

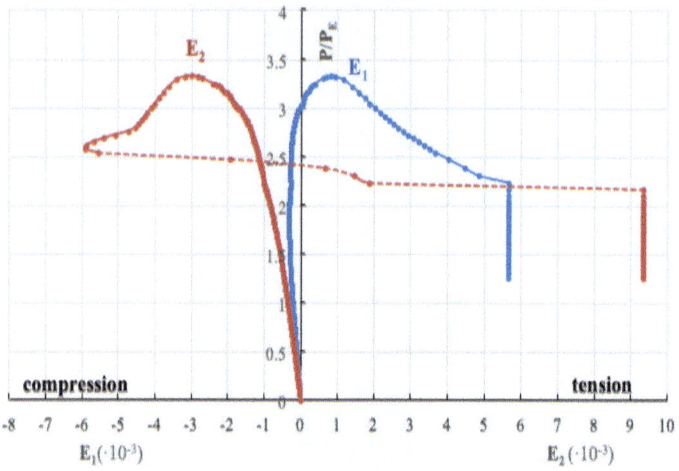

Fig. 11. Exp. diagram of the ratio, P/P_E, versus strain values (E1, E2) on the BFRP strip, respectively, on the tensile side (E1) and compressive (E2) one (RMC3).

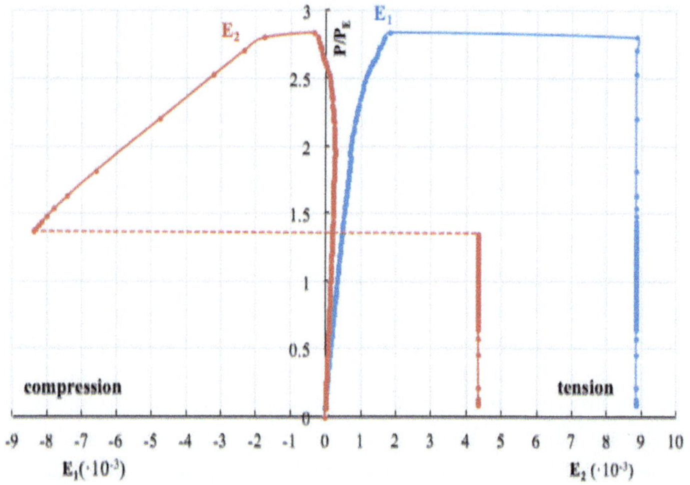

Fig. 12. Exp. diagram between the ratio, P/P_E, versus strain values (E1, E2) on the BFRP strip, respectively, on the tensile side (E1) and compressive (E2) one (RMC4).

In Fig. 13 the deflection of compressed FRP strip is shown with the adhesive layer assuming plane. The FRP strip is assumed as an elastic beam hinged at the ends, glued to the adhesive layer, of unknown length l.

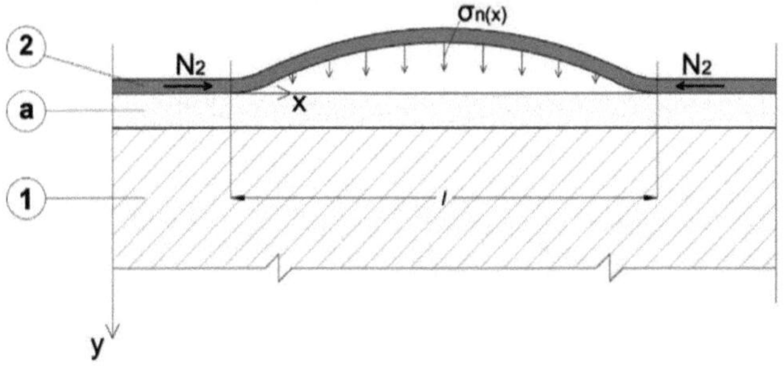

Fig. 13. Deflection of compressed FRP strip assuming adhesive layer plane.

The interfacial normal stress in the adhesive can be expressed as follows:

$$\sigma_n(x) = k_n \cdot w_a(x) \tag{1}$$

where k_n is the normal stiffness of the adhesive on unit length and can be written as:

$$k_n(x) = \frac{\sigma_n(x)}{w_a(x)} = \frac{\sigma_n(x)}{w_a(x) \cdot t_a} \cdot \frac{1}{t_a} = \frac{E_a}{t_a} \tag{2}$$

being

$$w_a(x) = w_2(x) \tag{3}$$

with w_a *(x)* the normal displacement of adhesive between the interface FRP strip and adhesive.

We consider that the elastic constant k_O is the value $k_O = k_n \cdot b$, being b the width of FRP, and the distribution of normal load on unit length may be expressed as:

$$p(x) = (k_n \cdot b) \cdot w_2(x) \tag{4}$$

On the basis of principal work, it is possible to write:

$$\frac{N_2}{2} \cdot \int w_2^{'}(x)dx = \frac{1}{2}\int K_O w_2^2(x)dx + \frac{1}{2}\int E_2 I_2 w_2^{''}(x)dx \tag{5}$$

Following Rayleigh's procedure [14], it is possible to assume a function $w_2(x)$ for the shape:

$$w_2(x) = w_0 \sin \pi \frac{x}{l} \tag{6}$$

and substitute it in Eq. (5) and solve for N_2 which is the critical buckling load for delamination buckling; in fact only when $N_2 = P_{E,FRP}$ the FRP strip is in equilibrium for deflected shape. We obtain:

$$\frac{N_2}{2}\cdot\int_0^l w_0(x)\cdot\frac{\pi}{l}\cdot\cos^2\frac{\pi x}{l}dx=\frac{1}{2}k_0\int w_2^2(x)dx+\frac{1}{2}E_2I_2\int\left(w_0\cdot\frac{\pi^2}{l^2}\right)^2\sin^2\frac{\pi x}{l}dx \tag{7}$$

The final expression may be obtained being in Eq. (3) both integrals square sines and cosines whose value is half the length *l/2*:

$$\frac{N_2}{2}\cdot\left(\frac{\pi}{l}\right)^2\cdot w_0^2\cdot\frac{l}{2}=\frac{1}{2}k_0\cdot w_0^2\cdot\frac{l}{2}+\frac{1}{2}E_2I_2\cdot\left(\frac{\pi}{4}\right)^4\cdot w_0^2\cdot\frac{l}{2} \tag{8}$$

and

$$P_{E,FRP}=k_0\cdot\left(\frac{l}{\pi}\right)^2+E_2I_2\cdot\left(\frac{\pi}{l}\right)^2 \tag{9}$$

The Eq. (9) says that in this case $P_{E,FRP}$ is greater than the Euler load which is augmented $k_0\cdot\left(\frac{l}{\pi}\right)^2$ due to adhesive layer.

Assuming $\beta=\sqrt[4]{\frac{k_0}{4E_2I_2}}$ it is possible to write:

$$\frac{P_{E,FRP}}{\sqrt{k_0E_2I_2}}=t^2+\left(\frac{1}{t}\right)^2 \tag{10}$$

with

$$t=\frac{\sqrt{2}}{\pi}\cdot\beta\cdot l \tag{11}$$

for $t=1$

$$P_{E,FRP}=2\cdot\sqrt{k_0E_2I_2} \tag{12}$$

which is the smallest load of delamination buckling for any length l. We obtain for the length:

$$l=\frac{\pi}{\sqrt{2}}\cdot\frac{1}{\beta}=\pi\cdot\sqrt[4]{\frac{E_2I_2}{k_0}} \tag{13}$$

From Eq. (12), it is possible to determine k_0 considering the strain's value in correspondence of the maximum load reached by the curves in the compressed zone (Figs. 8, 11, and 12). Multiplying the elastic stiffness, k_0, for the width of the FRP strip, b, it is possible to obtain a value of the adhesive's normal stiffness on unit length, k_n, equal to about 0.45 N/mm^3, for SRP, and 0.95 N/mm^3, for BFRP. For these values of k_n, we obtain a length of delamination equal to about 60 mm and 40 mm, respectively for SRP and BFRP strips. These results fit adequately the experimental evidences.

4 Conclusion

In this paper delamination buckling concerns the behavior of SRP and BFRP strips of strengthening in column models of homogeneous material subjected to compression up buckling and post-buckling. Investigation on slender column models of marble unreinforced and reinforced with SRP and BFRP strips is presented and experimental results have been discussed and compared with reference to theoretical analysis based on simplified elastic models. The main results of the investigation may be summarized as follows:

- Experimental compression tests confirmed that composite materials to brittle material bonded joints lose their capacity due to delamination buckling of FRP strips;
- FRP strips improve the structural response with strengthening of SRP strips;
- Theoretical analysis based on a simplified elastic model permits to control experimental results.

Notation

P exp. vertical compression load
P_E Euler's load
δ horizontal displacement at midspan
t thickness of specimens
E_1 tension strain on FRP strip
E_2 compression strain on FRP strip
E Elastic modulus of marble
σ_n interfacial normal stress
k_0 elastic stiffness
k_n normal stiffness of the adhesive on unit length
b width of the FRP strip
w_a normal displacement of the adhesive
$P_{E,FRP}$ critical buckling load for FRP delamination buckling
E_2 Elastic modulus of FRP strip
I_2 Moment of inertia of FRP strip

Acknowledgments. The Authors wish to thank all the technicians and students who worked to carry out the experimental tests. The experimental research was developed through founds provided by Polytechnic University of Marche, Italy.

References

1. Reddy, J.N., Barbero, E.J., Teply, J.L.: A plate bending element based on a generalized laminate plate theory. Int. J. Numer. Meth. Eng. **28**, 2275–2292 (1989)
2. Barbero, E.J.: On a generalized laminate theory with application to bending, vibration, and delamination buckling in composite laminates. Ph.D. dissertation, Virginia Polytechnic Institute and State University, Blacksburg, VA (1989)

3. Barbero, E.J., Reddy, J.N.: Modelling of delamination in composites laminates using a layer-wise plate theory. Int. J. Solid Struct. **28**(3), 373–388 (1991)
4. Kim, Y., Davalos, J.F., Barvero, E.J.: Delamination buckling of FRP layer in laminated wood beams. Compos. Strucures **37**(3–4), 311–320 (1997)
5. Capozucca, R.: Effects of mortar layers in the delamination of GFRP bonded to historic masonry. Compos. Part. B **44**, 639–649 (2013)
6. Kardomateas, G.A., Pelegri, A.A., Malik, B.: Growth of internal delamination under cyclic compression in composite plates. J. Mech. Phys. Solids **43**(6), 847–868 (1995)
7. Chai, H., Babcock, C.D., Knauss, W.G.: One dimensional modeling of failure in laminated plates by delamination buckling. Int. J. Solid Struct. **17**(11), 1069–1083 (1981)
8. Tounsi, A., Hassaine Daouadji, T., Benyoucef, S., Adda bedia, E.A.: Interfacial stresses in FRP-plated RC beams: effect of adherent shear deformations. Int. J. Adhes. Adhes. **29**, 343–351 (2009)
9. Lee, J., Gurdal, Z., Hayden Griffin Jr., O.: Layer-wise approach for the bifurcation problem in laminated composites with delaminations. AIAA J. **31**(2), 331–338 (1993)
10. Capozucca, R.: A reflection on the application of vibration tests for the assessment of cracking in PRC/RC beams. Eng. Struct. **48**, 508–518 (2013)
11. Katir, S., Brahim, B., Capozucca, R., Wahab, M.A.: Damage detection in CFRP composite beams based on vibration analysis using proper outraged method with radial basis function and Cuckoo Search algorithm. Compos. Struct. **187**, 344–353 (2018)
12. Capozucca, R.: Analysis of bond-slip effects in RC beams strengthened with NSM CFRP rods. Compos. Struct. **102**, 110–123 (2013)
13. ASTM D 3039/D 3039 M – 08: Standard test method for tensile properties of polymer matrix composite materials. American Standard of Testing and Materials (2008)
14. Den Hartog, J.P.: Advanced Strength of Materials. Mc Grace-Hill Book Company Inc., Cambridge (1952)

Preparation of N-Doped Carbon/Cobalt Ferrite Hybrid Nanocomposites for Lithium Ion Batteries Anodes

D. L. Dong, W. Zhang, J. L. Ma, and C. W. Wu(✉)

State Key Laboratory of Structural Analysis for Industrial Equipment, Dalian University of Technology, Dalian 116024, China
cwwu@dlut.edu.cn

Abstract. Transition metal oxides (TMOs) have much higher theoretical lithium storage capacities than that of the commercial graphite anode, but severe volume expansion (160–300%) and lower conductivity will result in poor cycle stability and low rate performance, which seriously hinder their applications. Developing inexpensive and efficient strategy to manufacture carbon/TMOs hybrids with good cycle stability and high rate performance is highly desired. Here, we report a facile approach to synthesize nitrogen-doped carbon/cobalt ferrite hybrids using the cheap renewable biomass, glucose and starch, as the carbon source. The resultant N-doped carbon/cobalt ferrite hybrid nanocomposites were investigated in terms of their morphology, structure, composition and electrochemical lithium storage performance. The results show that the hybrid composites possess porous structure and TMOs nanocrystals ($\sim$5.7 nm), which can not only effectively alleviate the sharp volume changes of cobalt ferrite (186%) during repeated lithiation/delithiation process, but also provide efficient electron and ion transport channels, demonstrating excellent rate performance and remarkable cycle stability. In particular, a high reversible capacity of 411 mAh/g was successfully maintained without decay over 200 cycles at a high current density of 3 A/g, exhibiting the potential as a promising lithium ion batteries anode material.

Keywords: Cobalt ferrite · N-doped carbon · Hybrids · Anode · Lithium ions batteries

1 Introduction

Although lithium-ions batteries (LIBs) have been widely used as power source in various devices owning to their high energy density, long cycle life and light weight, the energy and power densities of commercial LIBs cannot yet satisfy the ever-increasing demand of market. Developing advanced electrode materials with high special capacity and high rate performance is extremely urgent to boost the energy and power densities of LIBs [1]. Transition metal oxides (TMOs) have been identified as a potential alternative to commercial graphite anode owing to their natural abundance, environmentally benignity, safety voltage platforms and, most important, high special capacities (usually more than twice that of graphite, >700 vs 372 mAh/g) [2].

M. A. Wahab (Ed.): DAMAS 2019, LNME, pp. 767–774, 2020.
https://doi.org/10.1007/978-981-13-8331-1_60

Nonetheless, TMOs suffer from two bottlenecks that hinder their practical application. One is the rapid capacity fading caused by severe volume changes (160–300%) in lithiation/delithiation process, the other is the poor rate performance arising from their intrinsic low conductivity and sluggish reaction kinetics of conversion reaction [1, 3].

To tackle these problems, tremendous studies have been carried out. In brief, there are mainly three strategies, including downsizing the TMOs to nanoscale, controlling their morphology and structures, and building carbon/TMOs hybrid composite. Among them, building carbon/TMOs hybrid composite is the most attractive method to improve both cycle stability and rate capability of TMOs electrodes, since the induced carbon materials can not only buffer volume change of TMOs during lithiation/delithiation, but also provide high conductivity of whole electrodes [1, 4]. In the past decade, various carbon/TMOs hybrid composites have been developed for LIBs anodes. However, these materials usually involve expensive carbon sources (e.g. graphene, carbon nanotube, etc.), complex synthesis routes, toxic reagents, limiting their industrial production.

In this paper, we demonstrated the green and facile synthesis of carbon/cobalt ferrite hybrid nanocomposites with cheap and renewable biomasses as carbon sources. Two common biomasses (glucose and starch) were utilized to fabricate two hybrids with different lithium storage properties. We also discussed the mechanism of their different lithium storage properties by analyzing their morphology, structure and composition.

2 Experimental

2.1 Synthesis of CFO/C-G and CFO/C-S

The CFO/C-G and CFO/C-S hybrids were synthesized by hydrothermal method, which is commonly used to prepare nano-sized metal oxides and their complex [5–8]. In brief, 3.6 mmol $Fe(NO_3)_3{\cdot}9H_2O$, 1.8 mmol $Co(NO_3)_2{\cdot}9H_2O$ and 40 mmol Urea were added into 80 mL deionized water. Then, 1.8 g glucose or starch was added into solution with continuous stirring. Following this, the clear solution was transferred into 100 mL autoclave and maintained at 180 °C for 6 h. Finally, the obtained black precipitate was washed, dried and named as precursor-G or precursor-S. The CFO/C-G and CFO/C-S was prepared by annealing the precursor-G and precursor-S at 500 °C for 2 h under N_2 protection.

2.2 Characterization

The samples were characterized by field emission scanning electron microscope (FE-SEM, FEI NanoSEM 450), transmission electron microscope (TEM, FEI Tecnai F30), X-ray diffraction (XRD, PANalytical Empyrean X-ray diffractometer with Cu Kα radiation), Raman spectroscopy (Thermo Fisher DXR Microscope) and Fourier Transform Infrared Spectrometer (FTIR, Thermo Fisher). Electrochemical characterization was tested by Land CT 2001A with CR2025-type coin cells at room temperature

($\sim$20 °C), and all cells were made through a general method [9], except that the active materials are replaced by CFO/C-G or CFO/C-S.

3 Results and Discussions

The preparation route of hybrids is briefly illustrated in Fig. 1a. As can be seen, all reagents used here are common and cheap, and the synthetic process is green and efficient. SEM images (Fig. 1b–d) show that both precursor-G and CFO/C-G consist of many interconnected nanoparticles with loose porous structure. While particles of precursor-S (Fig. 1 d) is dispersed and their size is smaller compared with precursor-G. After calcination, these small nanoparticles were further fused together to form micron-sized secondary particles with tiny holes (Fig. 1f and g). Figure 2 shows different magnification TEM images of CFO/C-G (Fig. 1a–c) and CFO/C-S (Fig. 1 d-f). It can be seen that both CFO/C-G and CFO/C-S have plenty of TMOs nanoparticles evenly embedded in interconnected carbon skeleton. Nevertheless, the structures of two samples are obviously different, the former is loose branch-like, while the latter is relatively dense block. Actually, this morphological difference will significantly affect their electrochemical performance, which will be discussed below. Besides, the high

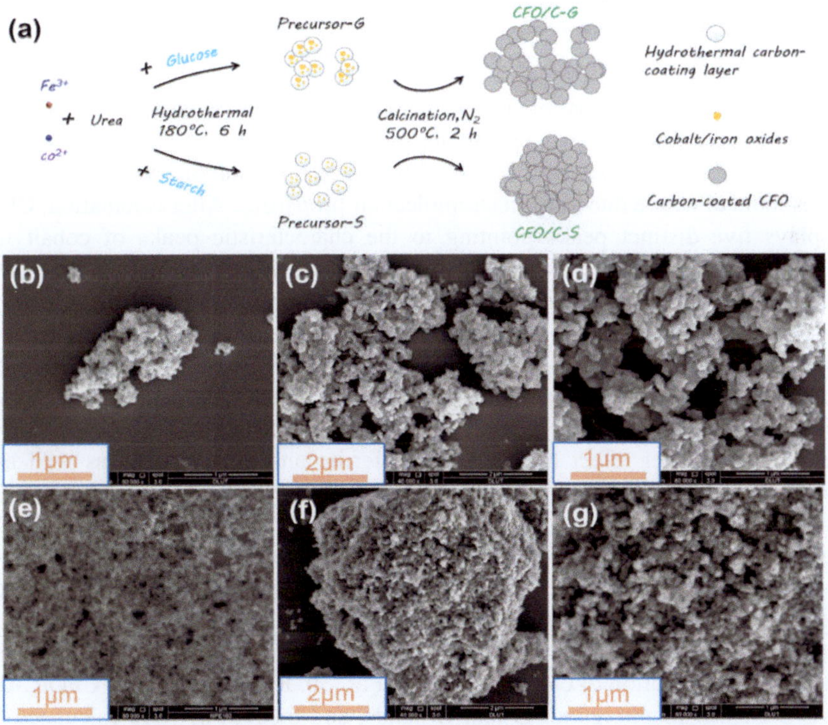

Fig. 1. (a) Schematic diagram for the synthesis of carbon/cobalt ferrite hybrids. SEM images of (b) precursor-G, (c) and (d) CFO/C-G, (e) precursor-S, (f) and (g) CFO/C-S.

resolution transmission electron micrograph (HR-TEM) image (Fig. 2c) depicts the lattice spacing of 0.247, 0.149 and 0.297 nm for the the CFO/C-G, ascribing to the (111) and (220) planes of cobalt oxides (CoO) and (220) plane of cobalt ferrite ($CoFe_2O_4$), respectively. In contrast, all facet stripes for CFO/C-S in Fig. 2g can be indexed to cobalt ferrite.

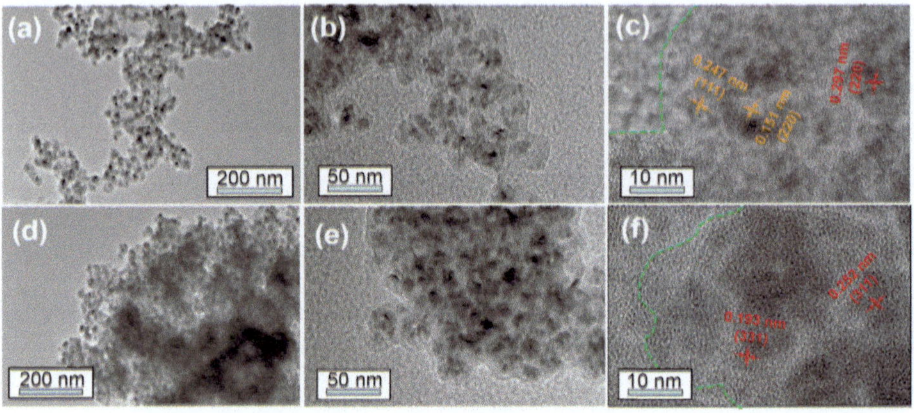

Fig. 2. TEM images of (a-c) CFO/C-G and (d-f) CFO/C-S.

To investigate evolution of TMOs in hybrids, XRD patterns were performed. Figure 3a shows that the two precursors have no obvious crystal structure, which indicates that both glucose and starch can significantly inhibit crystal growth, and starch is more effective due to its macromolecular properties. After calcination, CFO/C-S displays five distinct peaks pointing to the characteristic peaks of cobalt ferrite (JCPDS: 22-1086). As for CFO/C-G, its three humps can be associated with the overlapped signals of cobalt ferrite (JCPDS: 22-1086) and cobalt oxide (JCPDS: 70-2885). Apparently, glucose is not conducive to the formation of pure cobalt ferrite, because it is weakly reductive. Moreover, based on the Scherrer's formula, the average crystalline size of CFO/C-G and CFO/C-S were estimated to be about 5.7 nm and 10.2 nm, respectively. The obtained smaller crystalline size of CFO/C-G may be caused by the barrier effect between cobalt ferrite and cobalt oxide as well as the dispersion function of carbon skeleton, which can be manifested by the HR-TEM picture. However, the carbon has not been detected in XRD patterns, suggesting the carbon is low graphitized in hybrids. Then, Raman spectrum were carried out. As is known, D-band at $\sim 1360\ cm^{-1}$ indicates the defects and disorder of carbon, and the G-band at $\sim 1560\ cm^{-1}$ represents ordered sp^2 bond of graphite in Raman spectra [10]. Accordingly, the intensity ratio of D-band and G-band (I_D/I_G) can be utilized to assess the structural information of carbon. From the Fig. 3b, Raman spectra of CFO/C-G and CFO/C-S display unambiguous D and G bands, and their I_D/I_G were calculated to be about 2.25 and 2.33, respectively. Such high values reveal that the carbon existed in hybrids have a large number of defects, which facilitate the transport and storage of ions in the solid phase. Moreover, FTIR spectra indicate that both samples are doped with nitrogen (C = N, ~1300 cm^{-1}). This phenomenon will help to improve electrochemical lithium storage performance of two samples.

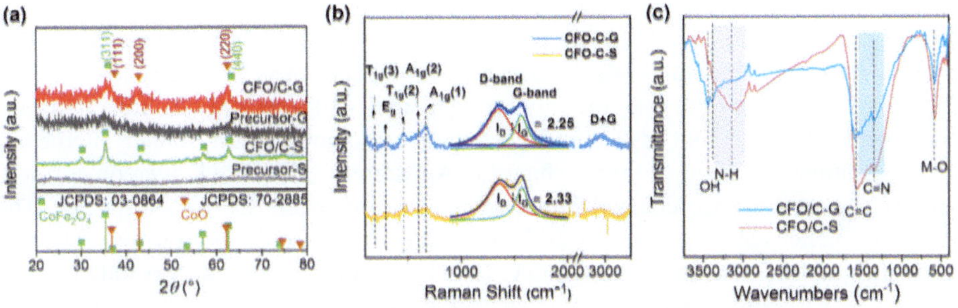

Fig. 3. The phase and composition characterization for the synthesized hybrids. (a) XRD patterns of hybrids. (b) Raman spectra (c) FTIR spectra of CFO/C-G and CFO/C-S.

The electrochemical performance of CFO/C-G and CFO/C-S electrodes was evaluated with CR2025 coin-type cells, where lithium foils were used as counter electrodes. Figure 4a and b show the rate performance of CFO/C-G and CFO/C-S. As can be seen, CFO/C-G (CFO/C-S) delivered average discharge capacities of 976.7 (1014.6), 838.3(848.7), 732.0 (659.0), 634.8 (520.1), 521.7 (391.0) and 315.2 (207.0) mAh/g at current densities of 0.1, 0.2, 0.5, 1, 2 and 5 A/g, respectively. When the current was reverted to 1 A/g, the CFO/C-G and CFO/C-S hybrids exhibited average discharge capacities of 716.6 and 512.5 mAh/g. The slightly higher capacities of CFO/C-S electrode at low current densities (<0.2 A/g) may be ascribed to the pure cobalt ferrite in CFO/C-S rather than mixture of cobalt ferrite/cobalt oxide, given that cobalt ferrite possess a higher theoretical capacity than that of cobalt oxide. On the contrary, once the current density is over 0.2 A/g, CFO/C-G shows much higher capacities than those of CFO/C-S. Considering that two hybrids have similar components, their conductivity should be similar, then their rate performance should be determined by the difference in ion transport (or diffusion) in the liquid phase, solid phase and electrochemical interface. Further, similar structured carbon in two hybrids may have a similar lithium diffusion rate. Accordingly, the poor rate performance of CFO/C-S is most likely caused by the greater concentration polarization arising from its compact structure, as proved by scanning and transmission electron microscopes above. In addition, high rate cyclic stability of CFO/C-G and CFO/C-S were also tested by galvanostatic charge-discharge. The results are presented in the Fig. 4c, in which CFO/C-G and CFO/C-S show discharge capacities of 412.3 and 344.5 mAh/g at 5th cycle, respectively. After 200 cycles, they retain the discharge capacities of 411.2 and

80.0 mAh/g, respectively. In fact, cobalt ferrite stores lithium through conversion mechanism, which can be described as follow [11],

$$CoFe_2O_4 + 8Li^+ + 8e^- \leftrightarrow Co + 2Fe + 4Li_2O \tag{1}$$

whereby the volume change during charge and discharge is calculated to be ~186% via the densities and molecular weights of reactants and products. In addition, fast reaction (the high rate) may also cause greater impact stress in electrode structure. Therefore, long-term high rate not only tests the ability of ions and electrons transport in electrode structure, but also reflects electrode structural stability under alternating stress stemmed from the volume changes. Apparently, CFO/C-G overcomes rapid volume changes during repeated lithiation/delithiation, while CFO/C-S did not. The excellent cyclic stability of CFO/C-G under high rate should be attributed by the following reasons. On one hand, CFO/C-G hybrid with smaller nanocrystals is more likely to resist volume changes, since smaller sized TMOs nanocrystals could induce relatively small local deformation to avoid excessive local stress in carbon skeleton, on the other hand, uniform carbon framework and loose porous structure of CFO/C-G hybrid can respectively bear and buffer volume changes during repeated lithiation/delithiation [12].

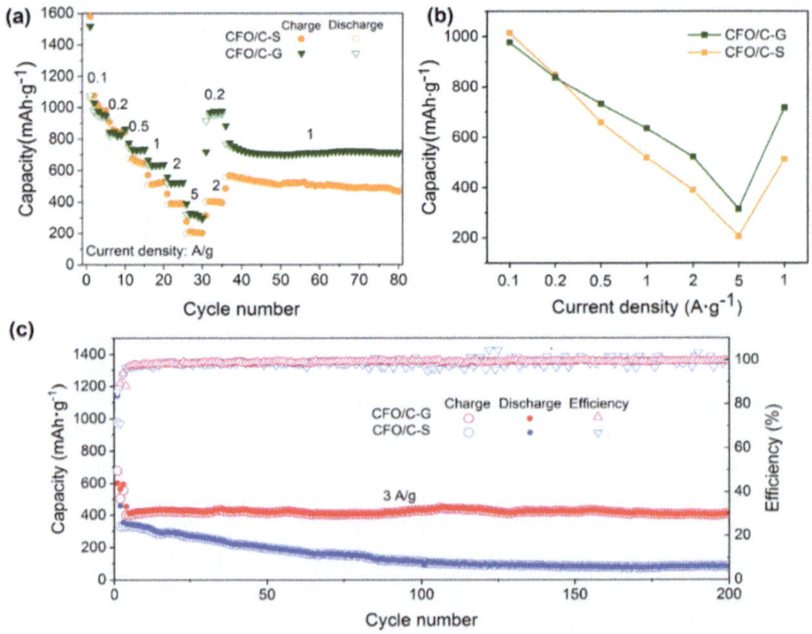

Fig. 4. Electrochemical performance characterization of CFO/C-G and CFO/C-S. (a) Rate performance of CFO/C-G and CFO/C-S for Lithium storage. (b) Average discharge capacities of CFO/C-G and CFO/C-S hybrids at various current densities. (c) High rate cycling stability of CFO/C-G and CFO/C-S.

4 Conclusions

One novel and efficient approach was proposed to manufacture N-doped carbon/cobalt ferrite hybrid nanocomposites using inexpensive and renewable biomass as carbon source. The structure, composition and electrochemical lithium storage property of hybrids synthesized with different biomasses (glucose and starch) were analyzed and compared. The results show that the hybrid (CFO/C-G) fabricated by using glucose as carbon source exhibits good rate performance and remarkable cycle stability as LIBs anodes. Specifically, it delivered capacities of 976.7, 838.3, 732.0, 634.8, 521.7 and 315.2 mAh/g at current densities of 0.1, 0.2, 0.5, 1, 2 and 5 A/g, respectively. And a high capacity of 411.2 mAh/g was maintained without decay over 200 cycles under 3A/g. Moreover, it has been demonstrated that small TMOs crystal size and loose porous structure are in favor of buffering sharp volume changes of TMOs during repeated lithiation/delithiation process. These evidences will provide a reference for rational construction of carbon/TMOs hybrid nanocomposites.

Acknowledgements. This work was supported by the National Natural Science Foundation of China (Grant No. 11772086), and Fundamental Research Funds for the Central Universities of China (Grant No. DUT18ZD302).

References

1. Zhao, Y., Wang, L.P., Sougrati, M.T., Feng, Z.X., Leconte, Y., Fisher, A., Srinivasan, M., Xu, Z.C.: A review on design strategies for carbon based metal oxides and sulfides nanocomposites for high performance Li and Na ion battery anodes. Adv. Energy Mater. **7** (9), 1601424 (2017)
2. Lee, S.H., Kotal, M., Oh, J.H., Sennu, P., Park, S.H., Lee, Y.S., Oh, I.K.: Nanohole-structured, iron oxide-decorated and gelatin-functionalized graphene for high rate and high capacity Li-ion anode. Carbon **119**, 355–364 (2017)
3. Zhao, Z.W., Wen, T., Liang, K., Jiang, Y.F., Zhou, X., Shen, C.C., Xu, A.W.: Carbon-coated Fe_3O_4/VO_x hollow microboxes derived from metal–organic frameworks as a high-performance anode material for lithium-ion batteries. ACS Appl. Mater. Interfaces. **9**(4), 3757–3765 (2017)
4. Li, P., Zhang, K., Park, J.H.: Dual or multi carbonaceous coating strategies for next-generation batteries. J. Mater. Chem. A **6**(5), 1900–1914 (2018)
5. Zhang, W., Zuo, X.D., Wu, C.W.: Synthesis and magnetic properties of carbon nanotube-iron oxide nanoparticle composites for hyperthermia: a review. Rev. Adv. Mater. Sci. **40**(2), 165–176 (2015)
6. Zhang, W., Zuo, X.D., Niu, Y., Wu, C.W., Wang, S.P., Guan, S., Silva, S.R.P.: Novel nanoparticles with Cr^{3+} substituted ferrite for self-regulating temperature hyperthermia. Nanoscale **9**(37), 13929–13937 (2017)
7. Zhang, W., Zuo, X.D., Zhang, D.M., Wu, C.W., Silva, S.R.P.: Cr^{3+} substituted spinel ferrite nanoparticles with high coercivity. Nanotechnol. **27**(24), 245707 (2016)
8. Zuo, X.D., Wu, C.W., Zhang, W., Gao, W.: Magnetic carbon nanotubes for self-regulating temperature hyperthermia. RSC Adv. **8**(22), 11997–12003 (2018)

9. Guo, S.S., Chen, Y.X., Shi, L.L., Dong, Y., Ma, J., Chen, X.H., Song, H.H.: Nitrogen-doped biomass-based ultra-thin carbon nanosheets with interconnected framework for high performance lithium-ion batteries. Appl. Surf. Sci. **437**, 136–143 (2018)
10. Du, L.Y., Xu, C.X., Liu, J.J., Lan, Y.Z., Chen, P.W.: One-step detonation-assisted synthesis of Fe_3O_4-Fe@BCNT composite towards high performance lithium-ion batteries. Nanoscale **9**(38), 14376–14384 (2017)
11. Wu, K.P., Liu, D.W., Tang, Y.: In-situ single-step chemical synthesis of graphene-decorated $CoFe_2O_4$ composite with enhanced Li ion storage behaviors. Electrochim. Acta **263**, 515–523 (2018)
12. Hu, J.K., Sun, C.F., Gillette, E., Gui, Z., Wang, Y.H., Lee, S.B.: Dual-template ordered mesoporous carbon/Fe_2O_3 nanowires as lithium-ion battery anodes. Nanoscale **8**(26), 12958–12969 (2016)

Random Vibration Based Robust Damage Detection for a Composite Aerostructure Under Assembly-Induced Uncertainty

Georgia Andriosopoulou, Andreas Mastakouris, Kyriakos Vamvoudakis-Stefanou, and Spilios Fassois(✉)

Department of Mechanical and Aeronautical Engineering, Stochastic Mechanical Systems and Automation (SMSA) Laboratory, University of Patras, 26504 Patras, Greece
{mead6698,mead6772}@upnet.gr, {vamvoudakisk,fassois}@mech.upatras.gr
http://www.smsa.upatras.gr

Abstract. The problem of random vibration based robust damage detection for a composite aerostructure under assembly-induced uncertainty is considered. The focus is on the exploration of the performance limits and critical assessment of two unsupervised Statistical Time Series (STS) type robust methods: a Multiple Model (MM) based method and a Principal Component Analysis (PCA) based counterpart. Three progressive damage scenarios are considered, the effects of which on the structural dynamics are 'minor' and almost completely 'masked' by assembly-induced uncertainty. The assessment, based on hundreds of experiments, suggests that damage as small as 10% reduction in the tightening torque of a single bolt is detectable in a reliable way. Both methods achieve remarkably good detection performance characterized by 100% correct detection for false alarm rates of 3.5% or higher, with the MM type method exhibiting a slight edge in performance over its PCA based counterpart! Overall the unsupervised STS type robust methods are shown to effectively detect 'minor' damage in the presence of significant assembly-induced uncertainty!

Keywords: Vibration based Structural Health Monitoring · Robust damage detection · Uncertainty · Multiple model methods · Principal component analysis based methods

1 Introduction

Random vibration based Structural Health Monitoring (SHM) has gained considerable popularity over the past several years as it may combine effectiveness, ease of use, and cost effective implementation with potential in-service operation without interrupting the structure's service cycle. The principle upon which this

M. A. Wahab (Ed.): DAMAS 2019, LNME, pp. 775–787, 2020.
https://doi.org/10.1007/978-981-13-8331-1_61

technology is founded is that incipient damage causes slight changes in the structural dynamics, which may, in turn, be detected via proper analysis of measured random vibration signals [1], [2, pp. 19–32].

A critical caveat of the technology is that changes in the dynamics may be due to factors other than incipient damage as well [3]. Such factors may include varying Environmental and Operating Conditions (EOCs), such as temperature, humidity, payload, as well as additional uncertainty sources, and their effects on the dynamics may be so significant as to totally 'mask' those of damage. This is a significant technology barrier, very often prohibiting important engineering applications [4].

As a result, considerable effort has been devoted to vibration based SHM methods that are *robust*[1] to varying EOCs, that is being capable of 'separating', to the extent possible, the effects of incipient damage from those of EOCs and other uncertainty [3,4]. This 'separation' is almost invariably accomplished via the 'implicit' or 'explicit' modeling of the structural dynamics under varying EOCs and uncertainty. *'Implicit'* methods include Principal Component Analysis (PCA) [5] and Factor Analysis (FA) [6] based methods, while *'explicit'* methods model the dynamics (for instance the 'healthy' structural dynamics in the context of damage detection) via explicit deterministic or stochastic modeling techniques. Deterministic techniques are, for instance, employed by the Functional Model (FM) based method [7–9], while stochastic techniques are employed by methods based on Gaussian [10] or non-Gaussian Random Coefficient (RC) models [11] and Multiple Model (MM) [4,12,13] representations.

Robust methods have been experimentally assessed under varying environmental (usually temperature and humidity), as well as varying operating (such as payload) conditions [3]. Yet, there is a sparsity of studies critically comparing robust methods in a statistically reliable way and exploring their performance limits via incipient damage of progressively varying magnitude, while other types of uncertainty, such as the practically important assembly-induced uncertainty, have yet to be systematically considered.

The goal of the present study thus is an innovative critical assessment of two, one explicit and one implicit, robust damage detection methods in a statistically reliable way, using hundreds of experiments, and exploring their performance limits with three progressive, 'minor', damage scenarios under assembly-induced uncertainty and uncontrolled ambient temperature variations.

The methods employed belong to the general class of *Statistical Time Series (STS)* type methods [14], which are characterized by various advantages such as the exclusive use of identified (data-based) and partial stochastic models of the dynamics (without the need for employing physics-based models). Moreover, they are *unsupervised*, implying that training is based on random vibration signals obtained only from the healthy structure, of the *model parameter based type*, implying that damage detection is based on changes in the parameters of a discrete-time stochastic structural model; an approach that has been repeatedly confirmed as providing high detection performance. *Multi-sensor versions*

[1] Such procedures are also referred to as (vibration) 'data normalization'.

(as opposed to more customary single-excitation single-response versions) of the methods are presently employed, with the aim being the achievement of the highest possible detection performance.

The *explicit* robust damage detection method employed is the U-MM-VARX (Un-supervised Multiple Model Vector AutoRegressive with eXogenous excitation) method [4]. As implied by their name, MM type robust methods use simple multiple model representations for roughly representing the healthy structural dynamics under uncertainty.

The *implicit* robust damage detection method considered is the U-PCA-VARX (Un-supervised Principal Component Analysis Vector AutoRegressive with eXogenous excitation) method [4]. PCA type robust methods implicitly represent the healthy structural dynamics, or part thereof, under uncertainty as a linear subspace implied by the maintained principal components.

The comparative critical assessment of the methods is based on an experimental procedure employing a lab-scale composite aerostructure, which represents the boom of an Unmanned Aerial Vehicle (UAV). Assembly consists of clamping the boom tightly at one end, simulating its connection to the fuselage, with its free end being attached to a mass representing the aircraft tail. Both connections are bolted, with the nominal (corresponding to the healthy state) tightening torques being equal to 2 Nm and 4 Nm, respectively. Three distinct damage scenarios are subsequently considered, each characterized by a specific reduction, of 10%, 30%, or 50%, in the tightening torque of its clamped (fuselage connection) end. It is noted that the tightening toque at its other end (tail connection) is not altered.

Assembly-induced uncertainty is introduced by disassembling and reassembling the structure after each experiment, with a total of 600 experiments carried out with the healthy structure and 70 with each of the three damage scenarios. As the experiments were performed over several days and at different times of the day, uncontrolled ambient temperature variations contribute some additional uncertainty as well.

The rest of the article is organized as follows: The experimental procedure is presented in Sect. 2, the versions of the robust damage detection methods employed are briefly outlined in Sect. 3, and the comparative performance assessment results are presented in Sect. 4. Concluding remarks are finally summarized in Sect. 5.

2 The Experimental Procedure

2.1 The Structure and the Test Rig

The experiments are based on a lab-scale composite aerostructure (Fig. 1), which represents the boom of a commercial Unmanned Aerial Vehicle (UAV). The structure has been manufactured by the Resin Transfer Molding method, and consists of several layers of woven and unidirectional fabric. Its cross-section is hollow square, with dimensions of $L \times W \times H = 600\,\text{mm} \times 65\,\text{mm} \times 65\,\text{mm}$, wall thickness of 3 mm, and corner radius of 8 mm.

Assembly consists of clamping the structure tightly at one end, simulating its connection to the fuselage, with its free end being attached to a mass representing the aircraft tail. Both connections are bolted, with the nominal (corresponding to the healthy state) tightening torques being equal to 2 Nm and 4 Nm, respectively.

During each experiment the structure is excited at its free end with a random white Gaussian force applied vertically at Point X (Fig. 1) via an electromechanical shaker. The vibration acceleration responses at points Y1, Y2 and Y3 are measured via 3 lightweight accelerometers. The measured force excitation and vibration response signals are driven through a signal conditioner into the data acquisition system. Details on the measured signals are provided in Table 1.

Table 1. Excitation and response signal characteristics

Excitation signal	Sampling frequency (Hz)	Operational bandwidth (Hz)	Total number of samples	Recording time (s)
White noise	4 654.5	3–2 327.25	120 000	25.8

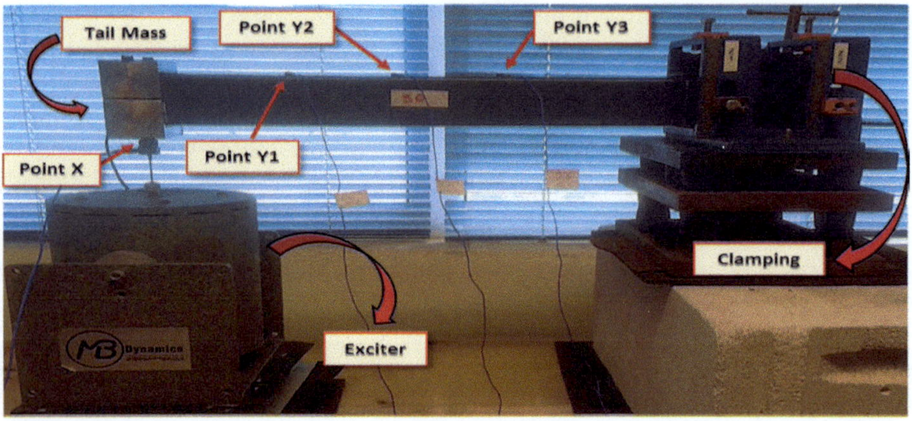

Fig. 1. Experimental set-up: a random force is applied at Point X and the resulting vibration acceleration signals are measured at Points Y1, Y2, Y3

2.2 The Damage Scenarios and the Experiments

Three distinct damage scenarios are considered, each characterized by a specific reduction, of 10%, 30%, or 50%, in the tightening torque of its clamped (fuselage connection) end. It is noted that the tightening toque at its other end (tail connection) is not altered. The specific tightening torques for each of the healthy and damage states are provided in Table 2, along with the corresponding number of performed experiments (600 with the healthy structure and 70 per each damage scenario). Assembly uncertainty is introduced by disassembling and re-assembling the structure after each experiment, while uncertainty due to ambient temperature variations is also present.

Table 2. The healthy and damaged scenarios

Structural state	Clamping torque (Nm)	Tail mass torque (Nm)	Number of experiments
Healthy	2	4	600
Damage 1	1.8	4	70
Damage 2	1.4	4	70
Damage 3	1	4	70

2.3 Assessment of the Effects of Each Damage Scenario on the Dynamics Under Assembly-Induced Uncertainty

The difficulty associated with effective vibration based damage detection may be demonstrated by examining the effects of each damage scenario on the dynamics under assembly-induced uncertainty. This is partially considered in Fig. 2, in which Welch-based [15, pp. 173–187] Point X to Point Y1 Frequency Response Function (FRF) magnitude estimates are shown for all 70 experiments performed for each damage scenario and the 600 experiments with the healthy structure.

Evidently, the effects of damage scenarios 1 and 2 on the dynamics are almost completely 'masked' by the variability caused by the assembly-induced uncertainty in the healthy case. Only for damage scenario 3, small deviations between the healthy and the damaged beam FRFs may be observed in the 600–1200 Hz range. As a consequence, damage detection is expected to be challenging for scenarios 1 and 2, and somewhat more straightforward for damage scenario 3.

3 The Damage Detection Methods

3.1 Preliminaries

The two damage detection methods employed belong to the general class of unsupervised Statistical Time Series (STS) methods and utilize stochastic Vector AutoRegressive with eXogenous excitation (VARX) models for representing the structural dynamics under specific environmental and operating conditions.

A $VARX(na, nb)$ model is of the form[2] [16]:

[2] Bold-face upper/lower case symbols designate matrix/column–vector quantities, respectively.

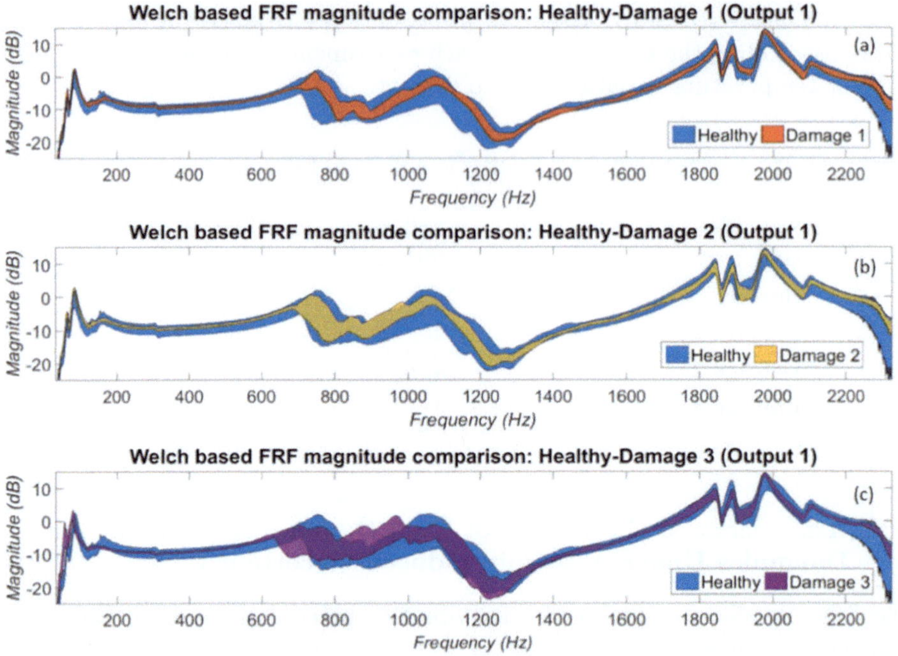

Fig. 2. Assessment of the effects of each damage scenario on the dynamics under assembly-induced uncertainty: Frequency Response Function magnitude estimates for 600 experiments with the healthy structure versus 70 experiments for each damage scenario [Response measurement Point Y1; Welch based estimates; Hamming windowing; 8 192-sample-long segment; 80% overlap; Matlab function: tfestimate.m]

$$\begin{gathered} \boldsymbol{y}[t] + \sum_{i=1}^{na} \boldsymbol{A}_i \cdot \boldsymbol{y}[t-i] = \sum_{i=0}^{nb} \boldsymbol{B}_i \cdot \boldsymbol{x}[t-i] + \mathbf{e}[t,\boldsymbol{\theta}] \\ E\left(\mathbf{e}[t,\theta] \cdot \mathbf{e}^T[t-\tau,\boldsymbol{\theta}]\right) = \boldsymbol{\Sigma}(\theta) \cdot \delta[\tau] \end{gathered} \tag{1}$$

where $t = 1, \ldots, N$ designates normalized (by the sampling period) discrete time, $\boldsymbol{x}[t] = [x_1[t] \ \ldots \ x_{nx}[t]]^T_{nx\times 1}$, $\boldsymbol{y}[t] = [y_1[t] \ \ldots \ y_{ny}[t]]^T_{ny\times 1}$ the excitation and response vector signals, respectively, and $\mathbf{e}[t] = [e_1[t] \ \ldots \ e_{ny}[t]]^T_{ny\times 1}$ the residual vector (one-step-ahead prediction error) that should be zero mean uncorrelated (white) over time, with covariance $\boldsymbol{\Sigma} = E\left(\mathbf{e}[t]\mathbf{e}^T[t]\right)$. $\boldsymbol{A}_i$ is the i-th $(ny \times ny)$ AR parameter matrix, $\boldsymbol{B}_i$ the i-th $(nx \times 1)$ X parameter matrix, while na and nb represent the AutoRegressive (AR) and eXogenous (X) orders, respectively. $E(\cdot)$ represents statistical expectation of the indicated quantity and $\boldsymbol{\theta}$ the $(n' := na \cdot ny^2 + (nb+1) \cdot nx) \times 1$ parameter vector comprising the elements of the AR, X, and innovations covariance matrices. The estimation of $\boldsymbol{\theta}$ is accomplished via methods based on the Ordinary Least Squares (OLS), the Weighted Least Squares (WLS), or the Maximum Likelihood criteria [15, pp. 318–320], [17].

3.2 The U-MM-VARX Method

This method uses a Multiple Model representation of the healthy dynamics under uncertainty. A brief overview is provided below, while for details the reader is referred to [4].

Baseline Phase. In the baseline phase a number, say p, of VARX models are obtained from a corresponding number of excitation–response vector signals for the healthy structure under various EOCs, and collectively constitute the multiple model representation, say $m_o := \{m_{o,1}, \ldots, m_{o,p}\}$, for the healthy dynamics.

Inspection Phase. Once a fresh set of excitation–response vector signals is obtained, with the structure being in an unknown health state, a new VARX model, say m_u, with parameter vector $\boldsymbol{\theta}_u$ and covariance matrix $\boldsymbol{\Sigma}_u$, is estimated. Then, a proper pseudo-distance is employed to determine a distance metric, say $D(m_0, m_u)$, between the current model m_u and the healthy representation m_o. Presently the distance metric is defined as the minimum of the pseudo-distances from m_u to the elements (individual models) in m_o, while the Kullback-Leibler (KL) divergence [18, pp. 756–758] is employed as the pseudo-distance between two conventional models.

Damage detection is then based on a decision making mechanism employing the distance metric $D(m_o, m_u)$ and a user specified threshold l_{lim}. The latter may be interpreted as a bound on the subspace containing the healthy structural dynamics under uncertainty, and therefore distance metric values lower or equal to the threshold indicate a healthy structure; else a damaged one:

$$\begin{array}{lcl} D(m_0, m_u) \leq l_{lim} & \longrightarrow & \text{Healthy Structure} \\ \text{Otherwise} & \longrightarrow & \text{Damaged Structure} \end{array} \qquad (2)$$

3.3 The U-PCA-VARX Method

This method employs the centered Principal Component Analysis (PCA) for representing a part of the healthy structural dynamics under uncertainty as a linear subspace implied by the maintained principal components [19]. A brief overview is provided below, while for details the reader is referred to [4].

Baseline Phase. In the baseline phase, p parametric VARX models are obtained from a corresponding number of vector excitation and vibration response signals with the healthy structure. The AR/X parameter covariance matrix, say $\boldsymbol{P}$, is then estimated. It should be noted that the number of experiments should be greater than the number of parameters in a single VARX model, that is $p > n'$, in order for the estimated covariance to be of reasonable accuracy. $\boldsymbol{P}$ is then decomposed as $\boldsymbol{P} = \boldsymbol{U}\boldsymbol{S}^2\boldsymbol{U}^T$ (Singular Value Decomposition, SVD), with $\boldsymbol{S}^2$ designating a diagonal matrix containing the (positive) eigenvalues of $\boldsymbol{P}$ in decreasing order, and $\boldsymbol{U}$ the matrix containing the corresponding eigenvectors of $\boldsymbol{P}$. Removing the first q columns (principal components) of $\boldsymbol{U}$ (which correspond to the largest eigenvalues that explain a certain high fraction, say γ

(%), of the total parameter vector variability) and retaining the $m = n' - q$ principal components, the (centered) parameter vector $\boldsymbol{\theta}$ may be transformed into the PCA coordinates as $\overline{\boldsymbol{\theta}}_m = \boldsymbol{U}_m^T \boldsymbol{\theta}$, with $\boldsymbol{U}_m^T$ designating the matrix composed of the remaining m columns of $\boldsymbol{U}$ and $\overline{\boldsymbol{\theta}}_m$ designating the projection of $\boldsymbol{\theta}$ into the m-dimensional transformed subspace.

Inspection Phase. Once a fresh set of excitation–response vector signals, with the structure being in unknown health state, is obtained, a fresh VARX model of the exact same orders as its baseline counterparts, is estimated and its AR/X parameters are centered by subtracting the sample mean values obtained in the baseline phase, thus leading to the parameter vector $\boldsymbol{\theta}_u$. This is subsequently projected into the m-dimensional PCA subspace as $(\overline{\boldsymbol{\theta}}_m)_u = \boldsymbol{U}_m^T \boldsymbol{\theta}_u$. As long as a suitable measure of it remains below a certain user-specified threshold l_{lim}, the structure is declared as healthy (see decision making interpretation in the previous subsection), else as damaged, that is:

$$\begin{array}{ccc} D = \|(\overline{\boldsymbol{\theta}}_m)_u\| \leqslant l_{lim} & \longrightarrow & \text{Healthy Structure} \\ \text{Otherwise} & \longrightarrow & \text{Damaged Structure} \end{array} \tag{3}$$

4 Damage Detection Performance Assessment

4.1 Preliminaries

Parametric Stochastic VARX Modeling. Parametric identification of the structural dynamics is based on N=12 000 sample-long excitation and vibration response signals. A scalar excitation ($nx = 1$) and vector (specifically trivariate, that is $ny = 3$) response signals are used, thus trivariate VARX models are adopted. The AR and X orders are selected equal, thus the models are of the VARX(n, n) type. Model estimation is based on the Ordinary Least Squares (OLS) method [15, pp. 318–320] (MATLAB function *arx.m*), with AR/X order selection based upon the trace of the residual covariance matrix, the Bayesian Information Criterion (BIC) [17], and frequency stabilization diagrams. The Samples Per (estimated) Parameter (SPP), defined as the total number of scalar signal samples over the total number of scalar model parameters, is also monitored, as it needs to be sufficiently high for purposes of statistical reliability; values of 20 or greater being typically preferred [17].

Based on the aforementioned criteria, a trivariate stochastic VARX(45,45) model of the healthy structural dynamics under fixed conditions is selected as adequate. Estimation details for an indicative estimation run (single data set) are provided in Table 3.

Baseline–Inspection Experiments and Method Details. 300 randomly selected experiments with the healthy structure are included in the 'baseline space', while an additional 300 experiments with the healthy structure and 210 with the damaged structure (70 per damage scenario) are included in the inspection phase; see Table 2.

Table 3. Parametric trivariate VARX estimation details

No. of scalar response signals	Scalar signal samples	Model	No. of model parameters	SPP[a]	BIC	Tr	Cond. No.[b]
Y1, Y2, Y3	48 000	VARX(45,45)	543	88.9	−26.68	0.0004	10^6

VARX model estimation details: Ordinary Least Squares (OLS). Matlab function: *arx.m*

[a]Samples per Estimated Parameter; [b]OLS inverted matrix

As mentioned in Sect. 3, PCA-based methods require greater number of baseline experiments than number of model scalar parameters ($p > n^*$). As the number of scalar parameters in the AR/X matrices is, for the selected trivariate VARX(45, 45) model, too high (543), a subset of 270 scalar parameters is selected for use in the context of the U-PCA-VARX method. This subset consists of the diagonal elements of the AR matrices (45 matrices $\times$ 3 = 135 scalar parameters) and the scalar parameters of the X (presently single column) matrices (45 matrices $\times$ 3 = 135). The value of γ is selected equal to 90.5% based on a sensitivity study. In the context of the U-MM-VARX method no corresponding constraint is present, hence the total number of estimated model scalar parameters (543) is employed. Details on the methods are provided in Table 4.

Table 4. Details on the U-MM-VARX and U-PCA-VARX damage detection methods

Method	No. of baseline experiments	No. of inspection experiments	No. of model parameters	γ (%)
U-MM-VARX	300	510	543	-
U-PCA-VARX	300	510	270	90.5

4.2 Damage Detection Performance

Damage detection results are presented in terms of the quantity D, as defined for each method, along with Receiver Operating Characteristic (ROC) curves which provide the true positive rate (percentage of correct damage detection) versus the false positive rate (percentage of false alarms) for varying decision threshold [20]. Ideally, for perfect performance, a ROC curve should go through the (0, 1) point.

Damage detection results are, separately for each damage scenario and for each of the two, U-MM-VARX and U-PCA-VARX, methods provided in Fig. 3. As observed in the left column of the figure, the D values corresponding to healthy inspection cases are, for each method, very well separated from those corresponding to damage. This is quite impressive in view of the minimal effects of damage on the dynamics under uncertainty, as presented in Sect. 2 (Fig. 2), and a prelude to excellent detection performance.

Indeed, very good performance is confirmed, separately for each damage scenario, via the ROC curves on the right column of Fig. 3. Specifically, for damage scenario 1, 100% correct detection is shown to be achieved for false alarm rates above 3% for the U-MM-VARX and above 4% for the U-PCA-VARX methods. For damage scenario 2, 100% correct detection is achieved for false alarm rates above 3% for the U-MM-VARX and above 2% for the U-PCA-VARX methods, respectively. Finally, for damage scenario 3, detection performance is perfect, characterized by 100% correct detection and 0% false alarm rates for both methods.

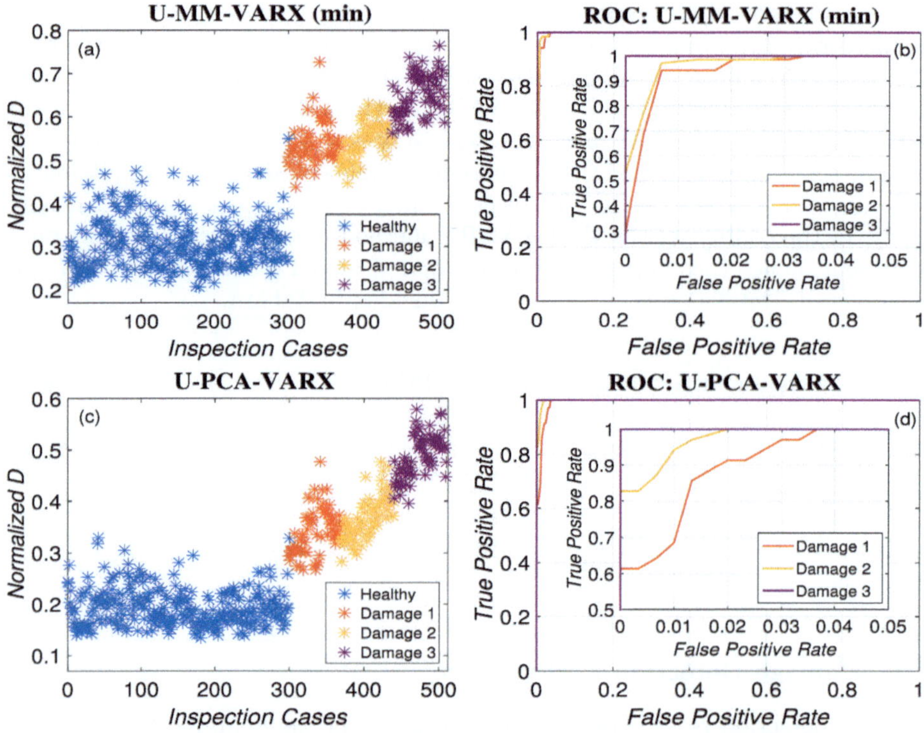

Fig. 3. Damage detection performance by each method and for each damage scenario: D quantity (left column) and ROC curve (right column) for the U-MM-VARX method (**a**, **b**), and the U-PCA-VARX method (**c**, **d**) (510 inspection experiments)

The aggregate (cumulative over all three damage scenarios) performance of each method is presented in Fig. 4 via ROC curves and AUC values (Area Under the Curve, which ranges from 0 for poor performance to 1 for excellent performance) [20]. The very good performance of both methods is evident, with 100% correct detection achieved for false alarm rates higher than 3.5%. Yet, the U-MM-VARX method exhibits a slight edge over its U-PCA-VARX counterpart, which is also reflected in a slightly higher (99.83% compared to 99.75%) AUC.

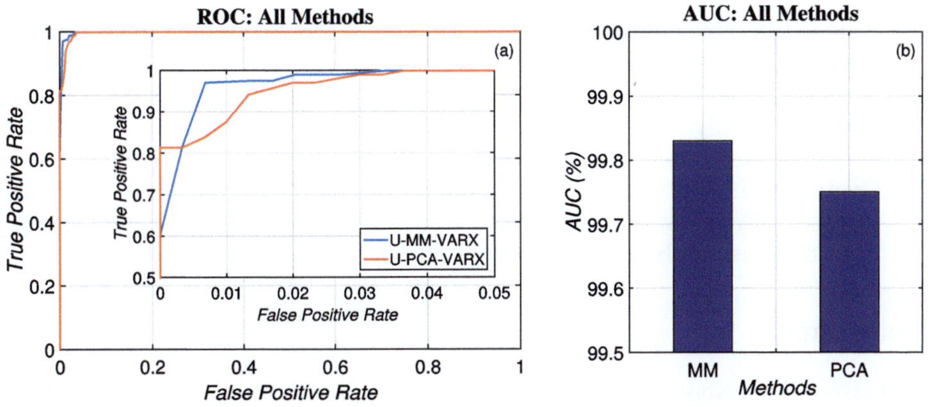

Fig. 4. Aggregate (cumulative over all damage scenarios) damage detection performance by each method: (**a**) ROC curves, and (**b**) AUC values (510 inspection experiments)

Overall, the obtained detection performance results reveal the capability of unsupervised STS type robust methods to effectively separate even slight damage-induced changes in the dynamics from their assembly-induced counterparts!

5 Concluding Remarks

The problem of random vibration based robust damage detection for a composite aero-structure under assembly-induced uncertainty was considered. Three damage scenarios, corresponding to 10%, 30%, or 50% reduction in the tightening torque of the structure's clamped (fuselage connection) end were employed, with effects on the dynamics shown to be 'minor' and largely 'masked' by assembly-induced uncertainty, thus leading to a challenging detection problem.

Two unsupervised Statistical Time Series (STS) type methods, employing Vector ARX (VARX) stochastic models, were selected for tackling the problem: An explicit U-MM-VARX method and an implicit U-PCA-VARX method. Their achievable detection performance was examined via 300 inspection experiments with the healthy structure and 70 inspection experiments under each damage scenario. The main conclusions of the study may be summarized as follows:

(a) Despite the 'minor' effects of each damage scenario on the structural dynamics, the unsupervised STS type robust methods achieved remarkably high aggregate (cumulative over all damage scenarios) performance reaching, for both methods, 100% correct detection for false alarm rates greater or equal to about 3.5%.
(b) Although the performance of the methods was grossly similar, the explicit U-MM-VARX method exhibited a slight edge over its implicit, U-PCA-VARX, counterpart.

(c) Among the three damage scenarios, scenario 3 (50% tightening torque reduction) was the easiest to detect, with detection performance being perfect. Scenario 2 (30% tightening torque reduction) was the second easiest, and scenario 1 (only 10% tightening torque reduction) was the least easy, but very good detection performance was still achieved!
(d) Overall, the results reveal the capability of the STS type robust methods to effectively separate even slight damage-induced changes in the dynamics from their assembly-induced counterparts!

References

1. Fassois, S.D., Kopsaftopoulos, F.P.: Statistical time series methods for vibration based structural health monitoring. In: Ostachowicz, W., Güemes, J. (eds.) New Trends in Structural Health Monitoring. CISM International Centre for Mechanical Sciences, vol. 542. Springer, Vienna (2013)
2. Deraemaeker, A.: Vibration based structural health monitoring using large sensor arrays: overview of instrumentation and feature extraction based on modal filters. In: Deraemaeker, A., Güemes, J., Worden, W. (eds.) New Trends in Vibration Based Structural Health Monitoring. CISM Courses and Lectures, vol. 520. Springer, Vienna (2010)
3. Sohn, H.: Effects of environmental and operational variability on structural health monitoring. Roy. Soc. – Philos. Trans.: Math. Phys. Eng. Sci. **365**(1851), 539–560 (2007)
4. Vamvoudakis-Stefanou, K.J., Sakellariou, J.S., Fassois, S.D.: Vibration-based damage detection for a population of nominally identical structures: unsupervised Multiple Model (MM) statistical time series type methods. Mech. Syst. Signal Process. **111**, 149–171 (2018)
5. Golinval, J.: Damage detection in structures based on principal component analysis of forced harmonic responses. Procedia Eng. **199**, 1912–1918 (2017)
6. Kullaa, J.: Vibration-based structural health monitoring under variable environmental or operational conditions. In: Deraemaeker, A., Worden, K. (eds.) New Trends in Vibration Based Structural Health Monitoring. CISM Courses and Lectures, vol. 520, pp. 1262-1269. Springer, Vienna (2010)
7. Worden, K., Sohn, H., Farrar, C.R.: Novelty detection in a changing environment: regression and interpolation approaches. J. Sound Vib. **258**(4), 741–761 (2002)
8. Hios, J.D., Fassois, S.D.: A global statistical model based approach for vibration response-only damage detection under various temperatures: a proof-of-concept study. Mech. Syst. Signal Process. **49**(1–2), 77–94 (2014)
9. Aravanis, T.-C.I., Sakellariou, J.S., Fassois, S.D.: Railway suspension fault detection under variable operating conditions via random vibration signals and the stochastic Functional Model based method. In: International Conference on Noise and Vibration Engineering (ISMA 2018), Leuven, Belgium (2018)
10. Vamvoudakis-Stefanou, K.J., Sakellariou, J.S., Fassois, S.D.: Random coefficient model based damage detection for a population of nominally identical structures: an exploratory study. In: International Conference on Noise and Vibration Engineering (ISMA 2016), Leuven, Belgium (2016)
11. Vamvoudakis-Stefanou, K.J., Fassois, S.D.: Vibration-based damage detection for a population of nominally identical structures via Random Coefficient Gaussian Mixture AR model based methodology. Procedia Eng. **199**, 1888–1893 (2017)

12. Vamvoudakis-Stefanou, K.J., Sakellariou, J.S., Fassois, S.D.: On the use of unsupervised response-only vibration-based damage detection methods for a population of composite structures. In: 8th European Workshop on Structural Health Monitoring (EWSHM), Bilbao, Spain (2016)
13. Vamvoudakis-Stefanou, K.J., Sakellariou, J.S., Fassois, S.D.: Assessment of a multiple model based parametric method for output-only vibration-based damage detection for a population of like structures. J. Phys: Conf. Ser. **628**(1), 012009 (2015)
14. Fassois, S.D., Sakellariou, J.S.: Time-series methods for fault detection and identification in vibrating structures. Roy. Soc. - Philos. Trans.: Math. Phys. Eng. Sci. **365**(1851), 411–448 (2007)
15. Ljung, L.: System Identification: Theory for the User, 2nd edn. Prentice-Hall, Upper Saddle River (1999)
16. Kopsaftopoulos, F.P., Fassois, S.D.: Scalar and vector time series methods for vibration based damage diagnosis in a scale aircraft skeleton structure. J. Theoret. Appl. Mech. **49**(3), 727–756 (2011)
17. Fassois, S.D.: Parametric identification of vibrating structures. In: Braun, S.G., Ewins, D.J., Rao, S.S. (eds.) Encyclopedia of Vibration, pp. 673–685. Academic Press (2001)
18. Press, W.H., Flannery, B.P., Teukolsky, S.A., Vetterling, W.T.: Numerical Recipes: The Art of Scientific Computing. Cambridge University, London (1986)
19. Yan, A.M., Kerschen, G., De Boe, P., Golinva, J.C.: Structural damage diagnosis under varying environmental conditions - Part I: a linear analysis. Mech. Syst. Signal Process. **19**(4), 847–864 (2005)
20. Fawcett, T.: An introduction to ROC analysis. Pattern Recognit. Lett. **27**(8), 861–874 (2006)

Random Vibration Damage Detection for a Composite Beam Under Varying Non-measurable Conditions: Assessment of Statistical Time Series Robust Methods

Tryfon-Chrysovalantis Aravanis, John Sakellariou, and Spilios Fassois(✉)

Department of Mechanical Engineering and Aeronautics,
Stochastic Mechanical Systems and Automation (SMSA) Laboratory,
University of Patras, 26504 Patras, Greece
{aravanis,sakj,fassois}@mech.upatras.gr
http://www.smsa.upatras.gr

Abstract. The problem of vibration-response-only damage detection for a composite beam under variable and non-measurable Environmental and Operational Conditions (EOCs) is considered via three unsupervised Statistical Time Series (STS) type robust detection methods. These include three versions of a novel Functional Model (FM) based method, a Multiple Model (MM) based method, and a Principal Component Analysis (PCA) based method. Performance assessment is based on hundreds of inspection experiments under temperature ranging from 0 to 28 °C and tightening torque ranging from 1 to 4 Nm. The results confirm the methods' high effectiveness, with a version of the FM based method and the MM based method achieving ideal performance, characterized by 100% correct detection rate for 0% false alarm rate.

Keywords: Robust damage detection · Structural Health Monitoring (SHM) · Vibration based methods · Unsupervised methods · Composite structures · Uncertainty

1 Introduction

Random vibration based Structural Health Monitoring (SHM) has rapidly progressed over the past several years, reaching high levels of technological maturity [1–3]. Statistical Time Series (STS) type methods [4], which employ corresponding models of the structural dynamics, are popular as they offer various advantages, including the exclusive use of data-based stochastic models which may be quite compact, only partially describing the dynamics.

Yet, a major challenge relating to effective damage diagnosis under variable Environmental and Operating Conditions (EOCs) still remains. The fundamental reason behind it has to do with the fact that variable EOCs may affect the underlying structural dynamics to a degree that may be similar or even greater

M. A. Wahab (Ed.): DAMAS 2019, LNME, pp. 788–803, 2020.
https://doi.org/10.1007/978-981-13-8331-1_62

than that caused by damage, and as such changes are at the core of damage diagnosis, the latter may become highly challenging and ineffective [5]. Overcoming this challenge requires the development of *robust* diagnosis methods, that are methods capable of 'separating', to the extent possible, the effects of variable EOCs from those of damage on the structural dynamics [5,6].

This generally requires the modeling of the considered dynamics under variable EOCs and uncertainty. Such modeling may assume various forms and be broadly classified—along with corresponding methods—as 'explicit' or 'implicit'. *'Implicit'* methods include Principal Component Analysis (PCA) [7] and Factor Analysis (FA) [8] based methods, while *'explicit'* methods model the dynamics (for instance the 'healthy' structural dynamics in the context of damage detection) via explicit deterministic or stochastic modeling techniques and include Multiple Model (MM) [9], Random Coefficient (RC) model based [10], and the newly introduced Functional Model (FM) based methods [11–15]. Although assessment of individual methods are available, systematic and critical comparative assessments are still scarce in the literature.

This study *aims* at contributing in this direction via the systematic and critical comparison of three distinct STS robust methods for damage *detection* which are based on *random vibration response signals*, that is:

(a) Versions of the Functional Model (FM) based method introduced by the authors and co-workers in a series of recent conference papers [12–15],
(b) a Multiple Model (MM) based method [9], and,
(c) a Principal Component Analysis (PCA) based method [9].

A new Residual Variance (FM-RV) based version of the FM based method is presently introduced and included in the assessment, along with two Residual Uncorrelatedness versions, one using the Portmanteau test (FM-RU-P version) [15] and one using the Peña-Rodríguez Test (FM-RU-PR version) [12].

All methods are *unsupervised* in nature, implying that random vibration signals only from the healthy structure are employed for their training in the baseline phase. Moreover, the variable EOCs are assumed *non-measurable* under the methods' normal (diagnostic) operation; yet, the FM method assumes their availability in the baseline (training) phase. Also, as only response signals are employed, the methods are based on transmittance type dynamics obtained through stochastic data-based parametric models of the AutoRegressive with eXogenous excitation (ARX) type [15–17].

It is also noted that the damage detection problem is tackled in a *batch mode*, implying that the methods operate on *short duration batches* of signal records collected periodically or on demand over time, with diagnostic decision making implemented at the end of a complete batch; not at each time instant, as it would be the case with *sequential* methods (for instance see [18]).

The comparative critical assessment of the methods is based on an experimental procedure employing a composite structure, which represents the topology of a commercial Unmanned Aerial Vehicle (UAV) boom that operates under variations in Environmental (temperature ranging from 0 to 28 °C) and Operating (tightening torque ranging from 1 to 4 Nm, simulating assembly variability)

Conditions (EOCs). Damage is simulated via the attachment of a small (12.6 g) mass on the beam, while additional uncertainty is introduced via the occasional attachment of a special adhesive tape. Damage detection is assessed in a systematic and statistically reliable way employing hundreds of inspection experiments, with the results presented in terms of Receiver Operating Characteristic (ROC) curves [19, pp. 34–35].

The rest of this article is organized as follows: The experimental set-up is presented in Sect. 2, the STS type robust damage detection methods are reviewed in Sect. 3, and their experimental assessment is presented in Sect. 4. Concluding remarks are finally summarized in Sect. 5.

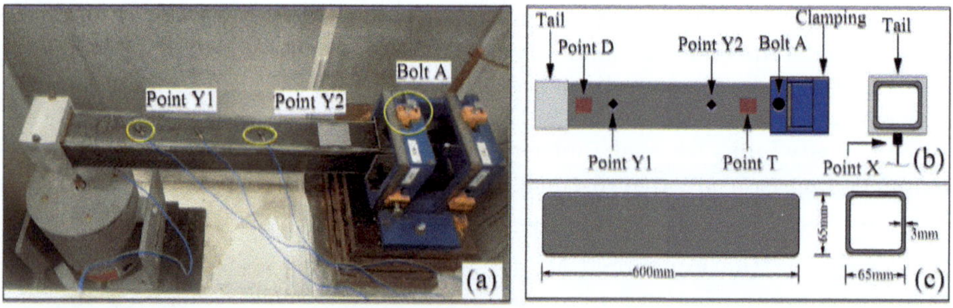

Fig. 1. The beam and the experimental set-up. **a** photo of the set-up. **b** schematic of the beam with the damage position (Point D), adhesive tape position (Point T), excitation position (Point X), and vibration measurement positions (Points Y1 and Y2). **c** geometrical details [12].

2 The Experimental Set-Up

The Structure, the Varying EOCs, and the Damage Scenario. The experiments are based on a lab-scale composite beam (further details in [12]), representing the topology of the main part of a commercial Unmanned Aerial Vehicle (UAV) boom. The beam is clamped at one end, simulating its connection to the fuselage, while its free end is attached to an aluminum mass representing part of the aircraft tail (Fig. 1(a)). The beam is placed in a freezer for temperature variation in the range [0–28] °C, while the tightening torque of Bolt A (Fig. 1(b)) is changed from 1 up to 4 Nm simulating assembly variability. The considered damage scenario is simulated via the attachment of a small, 12.6 g, mass at Point D (Fig. 1(b)) on the beam. Additional uncertainty is introduced via a piece of special (reinforced by plastic mesh) adhesive tape, placed (in certain experiments) at Point T on the surface of the beam (Fig. 1(b)). This serves to simulate potential material and/or manufacturing variability, such as variation in resin, fiber orientation, and so on, among nominally identical composite beams, thus exploring the methods' robustness to unknown uncertainty factors.

Table 1. Experimental details [12].

Structural state	Temperature range (°C)	Torque (Nm)	No. of exps
	Baseline (training) phase		
Healthy	Set A := [0° – 28°] with a step of 2°	1, 2, 3, 4	60*
	Inspection (diagnosis) phase		
Healthy (H)	A & {3°, 21°}	1	51◇
	A & {9°, 19°, 25°}	2	54◇
	A & {15°, 25°}	3	51◇
	A & {3°, 9°, 15°, 19°, 21°}	4	60◇
Healthy (H_1) (tape)	[0° 1° 3° 7° 9° 10° 14° ... 15° 17° 21° 23° 25° 27°]	1, 3, 4	117◇
Damaged (D) (12.6 g mass)	-//-	-//-	-//-

* 1 experiment per temperature and torque value.
◇ 3 experiments per temperature and torque value.
Sampling frequency $f_s = 4\,654.5$ Hz; signal length 2 500 samples (0.54 s).
BWD [5–2 327.25] Hz.

The Vibration Signals. Vibration experiments are performed using an electromechanical shaker applying a random, low frequency and band limited, white Gaussian force vertically at Point X, while the acceleration response signals at Points Y1 and Y2 on the beam are acquired through lightweight accelerometers (Fig. 1(a),(b)). Details on the experiments are provided in Table 1; also in [12]. Each measured vibration response signal is sample mean corrected and normalized by its own sample standard deviation.

Preliminary Analysis: Effects of Damage on the Uncertain Dynamics. Welch-based estimates [20, pp. 186–187] of the Transmittance Function (TF) [17] magnitude with the healthy and damaged beams under various EOCs are presented in Fig. 2. Significant variability is observed in the healthy dynamics, which, to a certain extent 'masks' the effects of damage.

3 The Statistical Time Series (STS) Type Robust Damage Detection Methods

Three Statistical Time Series (STS) type robust damage detection methods are employed and assessed. As aforementioned, they are all *unsupervised*, implying that signals obtained only from the healthy structure are employed in the baseline (training) phase, and do not require measurement of the EOC variability factors during their operation (in the inspection phase), although the first method assumes that such measurements are available in the preliminary baseline (training) phase. As the force excitation is assumed non-measurable, all three methods are based on the vibration response signals measured at Points Y1 and Y2, specifically on the corresponding transmittance dynamics. The three methods include: (a) The recently introduced Functional Model (FM) based method,

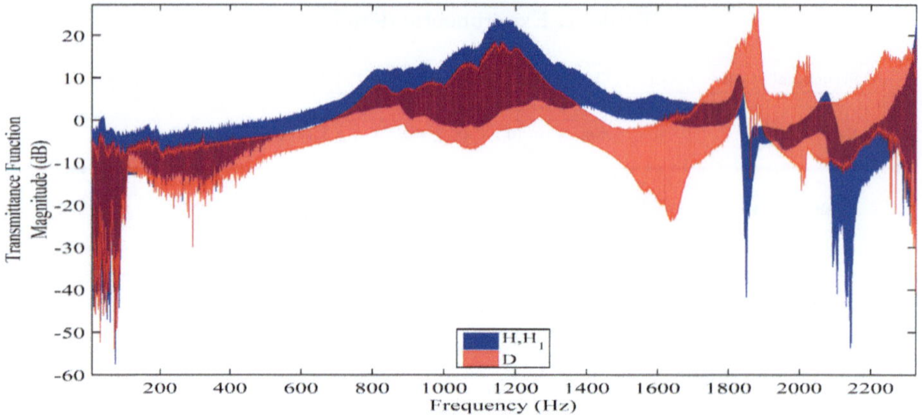

Fig. 2. Effects of damage on the transmittance dynamics under varying conditions: Welch-based transmittance function magnitude estimates based on 333 experiments with the healthy structure (H, H_1) and 117 experiments with the damaged structure. (Point Y1 to Point Y2 transmittance; Estimation based on $N = 100\,000$ sample long signals, Hamming windowing, segment length of 8 192 samples, 95% overlap, MATLAB function: `tfestimate.m`) [12].

with two versions making use of the Residual Uncorrelatedness (versions FM-RU-P and FM-RU-PR) and a third, new version, making use of the Residual Variance (version FM-RV); (b) a Multiple Model (MM) based method; and (c) a Principal Component Analysis (PCA) based method. Brief accounts of the methods are provided below.

3.1 The Functional Model (FM) Based Method

The cornerstone of the FM based method is the proper representation of the healthy structural dynamics under any EOCs in a parameter space, referred to as the '*healthy subspace*'. This subspace is constructed in the *baseline phase* using signal records obtained from controlled experiments (allowing—only in this phase—for measurement of the EOCs) and a data-based Functional Model (FM) [13–15,21].

Baseline (Training) Phase: Off-Line 'Healthy Subspace' Construction. The FM employed—for the healthy subspace construction—in this study is a Vector-dependent Functionally Pooled AutoRegressive with eXogenous excitation (VFP-ARX) model that incorporates the varying temperature and tightening torque through a 2-dimensional *operating (scheduling) parameter vector*[1] $\boldsymbol{k}$. The determination of this model is based on a total number of M controlled experiments under a specific combination of temperature and torque

[1] Vector/matrix quantities are designated by bold face lower/upper characters, respectively.

values. Thus, the *operating parameter vector* $\boldsymbol{k} = [k_l\, k_m]^T$ ($l = 1, 2, \ldots M_1$, $m = 1, 2, \ldots M_2$) is employed, where k_l and k_m designate variable temperature and torque, respectively, and the subscript discretization index.

The complete set of baseline experiments then provides $M = M_1 \times M_2$ random vibration response signal pairs $x_{\boldsymbol{k}}[t], y_{\boldsymbol{k}}[t]$ with each signal being N samples long. Based on this set, a VFP-ARX$(n_a, n_b)_p$ model of the form [22,23]:

$$y_{\boldsymbol{k}}[t] + \sum_{i=1}^{n_a} a_i(\boldsymbol{k}) \cdot y_{\boldsymbol{k}}[t-i] = \sum_{i=0}^{n_b} b_i(\boldsymbol{k}) \cdot x_{\boldsymbol{k}}[t-i] + e_{\boldsymbol{k}}[t] \tag{1a}$$

$$e_{\boldsymbol{k}}[t] \sim iid\, \mathcal{N}\big(0, \sigma_e^2(\boldsymbol{k})\big) \quad \boldsymbol{k} \in \mathbb{R}^2 \tag{1b}$$

$$a_i(\boldsymbol{k}) = \sum_{j=1}^{p} a_{i,j} \cdot G_j(\boldsymbol{k}), \quad b_i(\boldsymbol{k}) = \sum_{j=1}^{p} b_{i,j} \cdot G_j(\boldsymbol{k}), \quad \sigma_e^2(\boldsymbol{k}) = \sum_{j=1}^{q} s_j \cdot G_j(\boldsymbol{k}) \tag{1c}$$

is obtained. n_a, n_b designate the AR and X orders, respectively, *iid* identically independently distributed, $\mathcal{N}$ Gaussian distribution, $t = 1, 2, 3, \ldots$ the normalized (by the sampling period) discrete time, and $e_{\boldsymbol{k}}[t]$ the innovations (model residual) signal under conditions $\boldsymbol{k}$, which is assumed to be zero-mean, white (serially uncorrelated) with variance $\sigma_e^2(\boldsymbol{k})$ and potentially cross-correlated with their counterparts corresponding to different EOCs (different $\boldsymbol{k}$'s). As indicated by (1c), the AR and X parameters and variance $\sigma_e^2(\boldsymbol{k})$ are modeled as explicit functions of $\boldsymbol{k}$, by using p- and *q-dimensional* functional subspaces, respectively, spanned by the mutually independent functions $G_j(\boldsymbol{k})$. These form one functional subspace basis for the model parameters and one for the residual sequence variance, both consisting of bivariate polynomials obtained as tensor products from typical univariate polynomials such as Legendre, Chebyshev and so on (details in [22]). The constants $a_{i,j}$, $b_{i,j}$, s_j, designate the AR, X and $\sigma_e^2(\boldsymbol{k})$ coefficients of projection, respectively. $a_{i,j}$ and $b_{i,j}$ are estimated based on Ordinary Least Squares (OLS) [22]. Then the residual variance is estimated[2] as [22]:

$$\widehat{\sigma}_{e_o}^2(\boldsymbol{k}, \widehat{\boldsymbol{\theta}}) = \frac{1}{N} \sum_{t=1}^{N} e_{\boldsymbol{k}}^2[t, \widehat{\boldsymbol{\theta}}] \quad \forall\, \boldsymbol{k} = [k_l\, k_m]^T \; (l = 1, 2, \ldots M_1,\; m = 1, 2, \ldots M_2) \tag{2}$$

with the subscript o designating the structure under its *healthy* state (baseline phase) and $\boldsymbol{\theta} = [a_{1,1} \ldots a_{n_a,p}\; b_{0,1} \ldots b_{n_b,p}]^T$. Based on the above estimates and (1c), it may be written:

$$\begin{aligned}
\widehat{\sigma}_{e_o}^2(\boldsymbol{k}_1) &= s_1 G_1(\boldsymbol{k}_1) + s_2 G_2(\boldsymbol{k}_1) + \cdots + s_q G_q(\boldsymbol{k}_1) \\
\widehat{\sigma}_{e_o}^2(\boldsymbol{k}_2) &= s_1 G_1(\boldsymbol{k}_2) + s_2 G_2(\boldsymbol{k}_2) + \cdots + s_q G_q(\boldsymbol{k}_2) \\
&\vdots \\
\widehat{\sigma}_{e_o}^2(\boldsymbol{k}_M) &= s_1 G_1(\boldsymbol{k}_M) + s_2 G_2(\boldsymbol{k}_M) + \cdots + s_q G_q(\boldsymbol{k}_M)
\end{aligned}$$

[2] Estimators/estimates are designated by a hat.

Stacking the above in matrix form leads to:

$$\begin{bmatrix} \widehat{\sigma}^2_{e_o}(\boldsymbol{k}_1) \\ \widehat{\sigma}^2_{e_o}(\boldsymbol{k}_2) \\ \vdots \\ \widehat{\sigma}^2_{e_o}(\boldsymbol{k}_M) \end{bmatrix} = \begin{bmatrix} G_1(\boldsymbol{k}_1) & G_2(\boldsymbol{k}_1) & \dots & G_q(\boldsymbol{k}_1) \\ G_1(\boldsymbol{k}_2) & G_2(\boldsymbol{k}_2) & \dots & G_q(\boldsymbol{k}_2) \\ \vdots & \vdots & \ddots & \vdots \\ G_1(\boldsymbol{k}_M) & G_2(\boldsymbol{k}_M) & \dots & G_q(\boldsymbol{k}_M) \end{bmatrix} \cdot \begin{bmatrix} s_1 \\ s_2 \\ \vdots \\ s_q \end{bmatrix} \Leftrightarrow \boldsymbol{\sigma}_{e_o} = \boldsymbol{G} \cdot \boldsymbol{s} \quad (3)$$

based on which $\boldsymbol{s}$ is estimated as (^{T} designating matrix transposition):

$$\widehat{\boldsymbol{s}} = [\boldsymbol{G}^T \cdot \boldsymbol{G}]^{-1} \cdot [\boldsymbol{G}^T \cdot \boldsymbol{\sigma}_{e_o}] \quad (4)$$

The determination of the VFP-ARX model orders and its AR and X parameters functional subspace dimensionality p, for a given basis function family, is based on standard procedures using a Genetic Algorithm (GA) for the minimization of the Bayesian Information Criterion (BIC) and the Residual Sum of Squares (RSS), while model validation is based on formal verification of the residual sequence uncorrelatedness (whiteness) hypothesis corresponding to the M experiments used in model estimation. Full details on VFP-ARX model estimation and validation are provided in [22,23]. A similar procedure with a GA is also used for the determination of the residual variance functional subspace dimensionality q.

Inspection (Diagnosis) Phase: On-Line Damage Detection. Damage detection is achieved by examining—using a fresh random vibration signal pair[3] $x_u[t], y_u[t]$ obtained under unknown EOCs (that is unknown current value of $\boldsymbol{k}$)—whether or not the structural dynamics reside within the 'healthy subspace', so that the structure may or may not, respectively, be declared as healthy. This is essentially equivalent to examining whether or not the current signal pair $x_u[t], y_u[t]$ is 'consistent' with the available VFP-ARX model expressing the healthy subspace. This consistency examination may be realized in two steps [24]:

Step 1: Employ the VFP-ARX model of the baseline phase to estimate the unknown EOCs vector $\boldsymbol{k}$ that 'best' (according to a proper criterion) expresses the current signal pair. That is, given the current random vibration response signal pair $x_u[t], y_u[t]$, the estimates $\widehat{\boldsymbol{k}}$ and $\widehat{\sigma}^2_{e_u}(\widehat{\boldsymbol{k}})$ (innovations variance) are obtained using the equations of VFP-ARX model (1a) and (1c), as follows[4]:

$$\widehat{\boldsymbol{k}} = \arg\min_{\boldsymbol{k}} \sum_{t=1}^{N} e_u^2[t, \boldsymbol{k}], \quad \widehat{\sigma}^2_{e_u}(\widehat{\boldsymbol{k}}) = \frac{1}{N} \sum_{t=1}^{N} e_u^2[t, \widehat{\boldsymbol{k}}] \quad (5)$$

The estimate of $\boldsymbol{k}$ is obtained based on a GA algorithm followed by nonlinear refinement using Sequential Quadratic Programming [22,23].

[3] The subscript u designates the structure in an unknown health state.
[4] $e_u[t, \boldsymbol{k}]$ corresponds to the residual $e_{\boldsymbol{k}}[t]$ in (1a).

Step 2: 'Consistency' with the healthy subspace is confirmed through the *successful validation* of the model corresponding to $\widehat{\boldsymbol{k}}$. This may be achieved via two model residual based schemes, resulting in corresponding versions of the method: (a) A Residual Uncorrelatedness (RU) based version, and, (b) a Residual Variance (RV) based version that is presently postulated, motivated by [4].

(2a) The Residual Uncorrelatedness (RU) Based Versions. The residual $e_u[t, \widehat{\boldsymbol{k}}]$ whiteness testing, at a user selected risk level, may be achieved via two distinct tests, the Portmanteau Test or the Peña-Rodríguez Test, thus giving rise to the FM-RU-P or the FM-RU-PR versions of the method, respectively.

(i) The Portmanteau Test based (FM-RU-P) version: Residual whiteness is examined via the hypothesis test:

$$\begin{array}{ll} H_o : \rho[\tau] = 0 \quad \tau = 1, 2, \ldots, h & \text{(null hypothesis – healthy structure)} \\ H_1 : \rho[\tau] \neq 0 \quad \text{for some } \tau & \text{(alternative hypothesis – damaged structure)} \end{array} \tag{6}$$

where $\rho[\tau]$ designates the normalized autocovariance of the residual sequence at lag τ. The Q statistic below follows chi-square (χ^2) distribution with h degrees of freedom under the null (H_o) hypothesis of a valid model[5], that is [4]:

$$Q := N(N+2) \sum_{\tau=1}^{h} (N-\tau)^{-1} \widehat{\rho}^2[\tau] \sim \chi^2(h) \tag{7}$$

where h is the (user selected) maximum lag. The null hypothesis is then accepted, at a (user selected) risk level α (probability of rejecting H_o even though it is correct) as follows:

$$\begin{array}{rcl} Q < \chi^2_{1-\alpha}(h) & \Rightarrow & \text{(null hypothesis - healthy structure)} \\ \text{Else} & \Rightarrow & \text{(alternative hypothesis - damaged structure)} \end{array} \tag{8}$$

with $\chi^2_{1-\alpha}(h)$ designating the chi-square distribution's $(1-\alpha)$ critical point.

(ii) The Peña-Rodríguez Test based (FM-RU-PR) version: The partial autocorrelation $\pi_{e_u}[\tau]$ is examined via this testing procedure, according to which under the H_o hypothesis the D statistic below follows a standard normal distribution [25]:

$$D := (\lambda/\beta)^{-1/\varsigma} (\zeta/\sqrt{\lambda}) \left(Q^{1/\varsigma} - (\lambda/\beta)^{1/\varsigma} \left(1 - \frac{1}{2\lambda} \left(\frac{\varsigma - 1}{\varsigma^2} \right) \right) \right) \sim \mathcal{N}(0,1) \tag{9}$$

with:

$$Q = -N \sum_{\tau=1}^{h} \frac{h+1-\tau}{h+1} \log \left(1 - \frac{N+2}{N-\rho} \, \widehat{\pi}^2_{e_u}[\tau] \right)$$

[5] In which case the estimated innovations (residual) series $e_u[t, \widehat{\boldsymbol{k}}]$ is uncorrelated (white).

$$\beta = \frac{3(h+1)(h-2(n_a+n_b))}{2h(2h+1)-12(h+1)(n_a+n_b)}, \ \lambda = \frac{3(h+1)h^2}{2(2h(2h+1)-12(h+1))}$$

$$\zeta = \left(1 - \frac{2(h/2-(n_a+n_b))(h^2/(4(h+1))-(n_a+n_b))}{3(h(2h+1)/(6(h+1))-(n_a+n_b))^2}\right)^{-1}$$

where $\widehat{\pi}_{e_u}[\tau]$ designates the estimated partial autocorrelation of the residual series $e_u[t, \widehat{\boldsymbol{k}}]$ at lag $\tau = 1, 2, \ldots, h$.

Thus the following test is used at the α risk level as follows:

$$\begin{array}{ll} |\mathrm{D}| \leq Z_{1-a} \Rightarrow & H_o \text{ is accepted} \\ \text{Else} \qquad\quad \Rightarrow & H_1 \text{ is accepted} \end{array} \tag{10}$$

with Z_{1-a} designating the standard normal distribution' s $(1-a)$ critical point.

(2b) The Residual Variance (RV) Based Version. In this version damage detection is based on the fact that the variance $\sigma^2_{e_u}(\boldsymbol{k})$ becomes minimal, specifically equal to $\sigma^2_{e_o}(\boldsymbol{k})$ (see (2)), if and only if the current structure is healthy. Thus, the following hypothesis testing problem is constructed:

$$\begin{array}{ll} H_o : \sigma^2_{e_o}(\boldsymbol{k}) = \sigma^2_{e_u}(\boldsymbol{k}) & \text{(null hypothesis – healthy structure)} \\ H_1 : \sigma^2_{e_o}(\boldsymbol{k}) < \sigma^2_{e_u}(\boldsymbol{k}) & \text{(alternative hypothesis – damaged structure)} \end{array} \tag{11}$$

The F statistic below follows F distribution with $(N_u, N_o - d)$ degrees of freedom (N_o and N_u designate the number of samples used in estimating the residual variance in the healthy and current states, typically $N_o = N_u = N$; and d designates the dimensionality of vector $\boldsymbol{\theta}$) [4]:

$$\text{Under } H_o : \ F = \frac{\widehat{\sigma}^2_{e_u}(\widehat{\boldsymbol{k}})}{\widehat{\sigma}^2_{e_o}(\widehat{\boldsymbol{k}})} \sim F(N_u, N_o - d) \tag{12}$$

where $\widehat{\sigma}^2_{e_o}(\widehat{\boldsymbol{k}})$ is obtained via (1c) using the projection coefficients estimates as obtained by (4) and the q basis functions $G_j(\boldsymbol{k})$. The following test is then constructed at the α risk level:

$$\begin{array}{ll} F \leq f_{1-\alpha}(N_u, N_o - d) \Rightarrow & H_o \text{ is accepted (healthy structure)} \\ \text{Else} \qquad\qquad\qquad\quad \Rightarrow & H_1 \text{ is accepted (damaged structure)} \end{array} \tag{13}$$

with $f_{1-\alpha}(N_u, N_o - d)$ designating the corresponding F distribution's $1-\alpha$ critical point.

It is worth stressing that in the case where the structure is declared as healthy, the FM method also provides *precise estimates* of the current EOCs through $\widehat{\boldsymbol{k}}$.

3.2 A Multiple Model (MM) Based Method

In this method a Multiple Model (MM) representation is employed for modeling the healthy structural dynamics under variable EOCs. This representation consists of a set of conventional ARX-type models along with the Gaussian probability density functions of their estimated parameter vectors (also see [26]).

Baseline (Training) Phase. The M pairs of vibration response signals (also used in the FM based method) obtained under different EOCs are employed for the determination of the MM representation ('healthy set' of models) m_o, based on M transmittance $ARX(n_a, n_b)$ models [17], $m_{o,i}$ $(i = 1, \ldots, M)$, with corresponding parameter vectors $\{\boldsymbol{a}_o\} = \{\boldsymbol{a}_{o,1}, \ldots, \boldsymbol{a}_{o,M}\}$. Each estimated vector is (asymptotically, that is as the signal length $N \to \infty$) associated with a Gaussian probability density function, with mean equal to each point estimate $\boldsymbol{\alpha}_{o,i}$ and estimated covariance $\boldsymbol{\Sigma}_{o,i}$.

Inspection (Diagnosis) Phase. Once a fresh pair of vibration response signals is obtained from the structure in unknown health state, the objective is to decide whether or not its dynamics is adequately represented by the MM representation m_o, in which case the structure is declared as healthy, otherwise as damaged. Towards this end, a fresh transmittance function ARX model m_u (with parameter vector $\boldsymbol{a}_u$) and of same orders as those in m_o, is estimated. A distance metric $D(m_o, m_u)$ between m_o and m_u is then obtained (details in [9]), which is presently defined as:

$$D(m_o, m_u) := \sum_{k=1}^{M} d(m_{o,k}, m_u) \tag{14}$$

with $d(m_{o,k}, m_u)$ designating the Kullback–Leibler (KL) divergence (pseudo–distance) [27, pp. 756–758] between the standard ARX models m_o and m_u.

The structure is then declared as healthy if and only if $D(m_o, m_u)$ is smaller than a user–selected threshold; else it is declared as damaged.

3.3 A Principal Component Analysis (PCA) Based Method

This method (referred to as U-PCA-ARX) employs Principal Component Analysis (PCA) on the parameter vector of the transmittance ARX based representation of the structural dynamics [9].

Baseline (Training) Phase. Like in the previous method, M transmittance ARX(n_a, n_b) models [17], with corresponding parameter vectors $\boldsymbol{a}_o = \{\boldsymbol{a}_{o,1}, \ldots, \boldsymbol{a}_{o,M}\}$, are estimated and their sample mean and covariance matrix are then obtained. Each vector is centered by subtracting its sample mean. The Singular Value Decomposition of the covariance matrix is subsequently performed as $\boldsymbol{P} = \boldsymbol{U}\, \boldsymbol{S}^2\, \boldsymbol{U}^T$, where $\boldsymbol{S}^2$ is a diagonal matrix that contains the positive eigenvalues in decreasing order, while $\boldsymbol{U}$ is a real unitary matrix including the corresponding eigenvectors.

The first n eigenvectors of $\boldsymbol{U}$, which explain a certain, user–selected, fraction γ (%) of the total parameter vector variability, are dropped, assuming that they are mainly affected by the varying EOCs. The last m eigenvectors (presumably associated with damage) are stacked in a matrix $\boldsymbol{U}_m$ (containing m columns) that is used by the PCA to transform the (centered) parameter vectors $\{\boldsymbol{a}_o\} = \{\boldsymbol{a}_{o,1}, \ldots, \boldsymbol{a}_{o,M}\}$ ('healthy set' of parameters), into a reduced m-dimensional space [9]. It is noted that, as with all PCA–based methods, the detection performance may be significantly affected by the selected γ [9].

Inspection (Diagnosis) Phase. A new (transmittance) ARX model m_u of the same order as those of the baseline phase and parameter vector $\boldsymbol{a}_u$ is estimated based on a pair of vibration acceleration signals acquired from the current, unknown health state of the beam. The parameter vector's sample mean obtained in the baseline phase is then used to center $\boldsymbol{a}_u$, which is subsequently transformed, as $\bar{\boldsymbol{a}}_u = \boldsymbol{U}_m^T \boldsymbol{a}_u$, into the m-dimensional space. The Euclidean norm of $\bar{\boldsymbol{a}}_u$ is then obtained:

$$D := ||\bar{\boldsymbol{a}}_u||_{l_2} \tag{15}$$

and the structure is declared as healthy if and only if D is smaller than a user–selected threshold; else it is declared as damaged [9].

4 Experimental Assessment of the Methods

The experimental assessment of the three Statistical Time Series robust damage detection methods is based on 117 experiments with the damaged structure and 333 experiments with the healthy structure, all under various temperature and torque conditions which are different from those used in the baseline (training) phase (see Table 1). The comparative assessment of the methods' damage detection performance is based on Receiver Operating Characteristic (ROC) curves [19, pp. 34–35], each one representing the true positive rate (percentage of correct damage detections), versus the false positive rate (percentage of false alarms) for varying detection thresholds.

4.1 Baseline (Training) Phase

A transmittance (function) ARX model is initially obtained based on a pair of vibration-response signals from points Y1 and Y2 (Fig. 1(a)) under certain values of temperature (0°C) and tightening torque (2 Nm) from the structure under healthy state, and a standard identification procedure [20, pp. 203–205]. This includes model order selection based on the Bayesian Information Criterion (BIC) and the Residual Sum of Squares/Signal Sum of Squares (RSS/SSS), as well as model parameter estimation via OLS (MATLAB function: arx.m). This leads to an ARX(70,70) model characterized by zero delay, that is $b_0 \neq 0$ in the exogenous polynomial.

The Functional Model (FM) Based Method. Maintaining the AR and X orders and the zero delay for the VFP-ARX model, its AR and X parameters functional subspace is determined using $M = 60$ experiments and a GA based optimization procedure [12]. This procedure leads to a VFP-ARX(70,70)$_{30}$

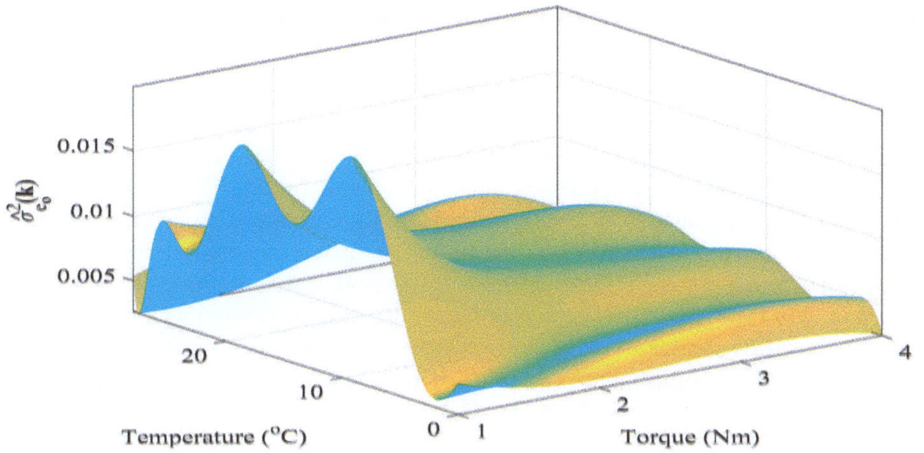

Fig. 3. Functional Model (FM) based method: innovations variance $\widehat{\sigma}^2_{e_o}(\boldsymbol{k})$ estimate as a function of temperature and torque.

model with functional subspace spanned by $p = 30$ bivariate Shifted Legendre polynomials. Then, the dimensionality of the residual variance functional subspace is selected as $q = 30$, using the 60 variance estimates (corresponding to the above experiments) obtained by (3), and bivariate Chebyshev type II polynomials. The functional dependence of the residual variance $\widehat{\sigma}^2_{e_o}(\boldsymbol{k})$ with respect to temperature and torque is presented in Fig. 3.

Table 2. Details on the detection methods.

The FM based method				
Selected model	*No. of train. experim.*	*Samples Per Param. (SPP)*	*Condition number*	*Whiteness testing maximum lag h*
VFP-ARX$(70, 70)_{30}$	60	70.92	1.73×10^7	Portmanteau: 10 Peña-Rodríguez: 210
The MM and PCA based methods				
Selected model	*No. of train. experim.*	*Samples Per Param. (SPP)*	*Condition number*	*No. of rejected components/γ (PCA based method)*
ARX$(70, 70)$	60	35.46	1.10×10^5	2 / 45.5%

The Multiple Model (MM) Based Method. Using the available vibration response signals (the exact same 60 experiments as previously reported), the 'healthy set' m_o that consists of 60 transmittance function ARX(70,70) models ($m_{o,i}$, $i = 1, \ldots, 60$) is constructed and the corresponding parameter vectors $\{\boldsymbol{a}_o\} = \{\boldsymbol{a}_{o,1}, \ldots, \boldsymbol{a}_{o,60}\}$ and covariance matrices are obtained. Estimation details are provided in Table 2.

The PCA Based Method. This is also based on the same 60 transmittance function ARX(70,70) models, $m_{o,i}$ ($i = 1, \ldots, 60$), previously reported. The obtained model parameters are centered by subtracting their sample mean, which leads to the 'healthy set' of parameters. The dimensionality of the sample covariance matrix is selected equal to 50×50, and it is constructed using the first 25 AR and first 25 X parameters from the 60 transmittance ARX models. Based on the fact that the number of variability (uncertainty) sources affecting the structural dynamics is two (temperature and tightening torque), the eigenvectors (principal components) which are removed are $n = 2$, which leads to $m = 48$ and $\gamma = 45.5\%$ (details in Table 2).

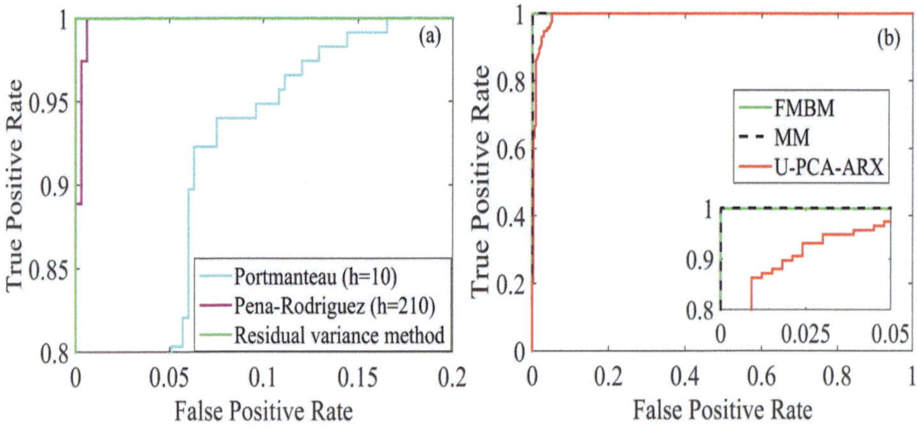

Fig. 4. Comparative damage detection performance assessment: ROC curves for the **a** three versions of the FM based method, and, **b** the FM based (RV version), the MM based, and the PCA based methods. (333 experiments with the healthy and 117 with the damaged structure.)

4.2 Inspection (Diagnosis) Phase

Based on the results of Fig. 4(a), it is evident that the Functional Model Residual Variance (FM-RV) version outperforms the two Residual Uncorrelatedness (FM-RU) versions, with the ROC curve being ideal (reaching the point (0,1)), implying the achievement of 100% correct detection rate with 0% false alarm rate. Among the two RU versions, the FM-RU-PR ($h = 210$) reaches somewhat lower, but still very good performance, while the FM-RU-P lags behind significantly.

A comparison of the best (FM-RV) FM method version with the MM and PCA based methods is provided in Fig. 4(b). Evidently, the MM based method achieves perfect performance as well, with the ROC coinciding with that of the FM-RV based method. On the other hand, the PCA based method achieves the lowest performance, characterized by 100% correct detection rate for false alarm rate greater than 8% (or 97% correct detection rate for 5% false alarm rate).

5 Concluding Remarks

The problem of vibration-response-only damage detection for a composite beam under variable and non-measurable Environmental and Operational Conditions (EOCs) was considered via three unsupervised, batch, Statistical Time Series (STS) type robust detection methods. These included a novel Functional Model (FM) based method in three distinct versions (FM-RU-P, FM-RU-PR, FM-RV), a Multiple Model (MM) based method, and a Principal Component Analysis (PCA) based method. Performance assessment was based on a total of 450 inspection experiments (333 with the healthy and 117 with the damaged structure), which were all excluded from the methods' baseline (training) phase. The main conclusions of the study may be summarized as follows:

(a) Despite the variations in Environmental (temperature ranging from 0 to 28 °C) and Operating (tightening torque ranging from 1 to 4 Nm) Conditions (EOCs), vibration-response-only damage detection was possible via the STS robust methods, with ideal performance—corresponding to 100% correct detection rate for 0% false alarm rate—achieved.
(b) Ideal performance was achieved by the Functional Model (FM) based and the Multiple Model (MM) based methods.
(c) The FM based method requires explicit knowledge of the precise EOC magnitudes in the baseline phase; this is not the case with the other methods.
(d) Of the three versions of the FM based method considered, the presently introduced, Residual Variance (RV) based version, achieved the best (ideal) performance.

Acknowledgement. This research was supported by Grant 56990000 from the Research Committee of the University of Patras via the 'K. Karatheodori' program.

References

1. Worden, K., Sohn, H., Farrar, R.: Novelty detection in a changing environment: regression and interpolation approaches. J. Sound Vibr. **258**(4), 741–761 (2002). https://doi.org/10.1006/jsvi.2002.5148
2. Figueiredo, E., Radu, L., Worden, K., Farrar, C.R.: A Bayesian approach based on a Markov-chain Monte Carlo method for damage detection under unknown sources of variability. Eng. Struct. **80**, 1–10 (2014). https://doi.org/10.1016/j.engstruct.2014.08.042
3. Deraemaeker, A.: Vibration based structural health monitoring using large sensor arrays: overview of instrumentation and feature extraction based on modal filters. In: Deraemaeker, A., Guemes, J., Worden, W. (eds.) New Trends in Vibration Based Structural Health Monitoring. CISM Courses and Lectures, vol. 520. Springer, Vienna (2010). https://doi.org/10.1007/978-3-7091-0399-9_2

4. Fassois, S.D., Sakellariou, J.S.: Statistical time series methods for SHM. In: Boller, C., Chang, F., Fujino, Y. (eds.) Encyclopedia of Structural Health Monitoring, pp. 443–472. Wiley, Chichester (2009). https://doi.org/10.1002/9780470061626.shm044
5. Sohn, H.: Effects of environmental and operational variability on structural health monitoring. Roy. Soc. - Phil. Trans. Math. Phys. Eng. Sci. **365**, 539–560 (2007). https://doi.org/10.1098/rsta.2006.1935
6. Zhang, Q.W., Fan, L.C., Yuan, W.C.: Traffic-induced variability in dynamic properties of cable-stayed bridge. Earth. Eng. Struct. Dyn. **31**, 2015–2021 (2002). https://doi.org/10.1002/eqe.204
7. Golinval, J.: Damage detection in structures based on principal component analysis of forced harmonic responses. Proc. Eng. **199**, 1912–1918 (2017). https://doi.org/10.1016/j.proeng.2017.09.449
8. Kullaa, J.: Vibration-based structural health monitoring under variable environmental or operational conditions. In: Deraemaeker, A., Worden, K. (eds.) New Trends in Vibration Based Structural Health Monitoring. CISM Courses and Lectures, vol. 520, pp. 1262–1269. Springer, Vienna (2010). https://doi.org/10.1007/978-3-7091-0399-9_4
9. Vamvoudakis-Stefanou, K.J., Sakellariou, J.S., Fassois, S.D.: Vibration-based damage detection for a population of nominally identical structures: unsupervised Multiple Model (MM) statistical time series type methods. Mech. Syst. Sig. Proc. **111**, 149–171 (2018). https://doi.org/10.1016/j.ymssp.2018.03.054
10. Vamvoudakis-Stefanou, K., Sakellariou, J., Fassois, S.: Random coefficient model based damage detection for a population of nominally identical structures: an exploratory study. In: The International Conference on Noise and Vibration Engineering, Leuven, Belgium (2016)
11. Hios, J.D., Fassois, S.D.: A global statistical model based approach for vibration response-only damage detection under various temperatures: a proof-of-concept study. Mech. Syst. Sig. Proc. **49**(1–2), 77–94 (2014). https://doi.org/10.1016/j.ymssp.2014.02.005
12. Aravanis, T.-C.I., Kolovos, S., Sakellariou, J.S., Fassois, S.D.: Random vibration-based damage detection for a composite beam under environmental and operational variability via a stochastic functional model based method. In: MATEC Web of Conference, vol. 188, no. 01003 (2018). https://doi.org/10.1051/matecconf/201818801003
13. Aravanis, T.-C.I., Sakellariou, J.S., Fassois, S.D.: Railway suspension fault detection under variable operating conditions via random vibration signals and the stochastic functional model based method. In: The International Conference on Noise and Vibration Engineering, Leuven, Belgium (2018)
14. Aravanis, T.-C.I., Sakellariou, J.S., Fassois, S.D.: On the problem of random vibration based fault detection in railway vehicle suspensions under variable and non-measurable operating conditions. In: The AVT-305 Research Specialists' Meeting on Sensing Systems for Integrated Vehicle Health Management for Military Vehicles, Athens, Greece (2018). https://doi.org/10.14339/STO-MP-AVT-305-08-PDF
15. Aravanis, T.-C.I., Sakellariou, J.S., Fassois, S.D.: Vibration based fault detection under variable non-measurable operating conditions via a stochastic functional model method and application to railway vehicle suspensions. In: The Surveillance 9 International Conference, FES, Morocco (2017)
16. Fassois, S.D.: Parametric identification of vibrating structures. In: Braun, S.G., Ewins, D.J., Rao, S.S. (eds.) Encyclopedia of Vibration, pp. 673–685. Academic Press, Cambridge (2001). https://doi.org/10.1006/rwvb.2001.0121

17. Poulimenos, A.G., Sakellariou, J.S.: A transmittance-based methodology for damage detection under uncertainty: an application to a set of composite beams with manufacturing variability subject to impact damage and varying operating conditions. Struct. Health Monit. **18**(1), 318–333 (2019). https://doi.org/10.1177/1475921718779190
18. Kopsaftopoulos, F., Fassois, S.D.: A vibration model residual-based sequential probability ratio test framework for structural health monitoring. Struct. Health Monit. **14**(4), 359–381 (2015). https://doi.org/10.1177/1475921715580499
19. Duda, R., Hart, P., Stork, D.: Pattern Classification, 2nd edn. Wiley, Hoboken (2000)
20. Ljung, L.: System Identification: Theory for the User, 2nd edn. Prentice-Hall, Upper Saddle River (1999)
21. Sakellariou, J.S., Fassois, S.D.: Functionally pooled models for the global identification of stochastic systems under different pseudo-static operating conditions. Mech. Syst. Sig. Process. **72–73**, 785–807 (2016). https://doi.org/10.1016/j.ymssp.2015.10.018
22. Sakaris, C.S., Sakellariou, J.S., Fassois, S.D.: A time series generalized functional model based method for vibration-based damage precise localization in structures consisting of 1D, 2D and 3D elements. Mech. Syst. Sig. Process. **74**, 199–213 (2016). https://doi.org/10.1016/j.ymssp.2015.07.014
23. Kopsaftopoulos, F., Fassois, S.D.: A functional model based statistical time series method for vibration based damage detection, localization, and magnitude estimation. Mech. Syst. Sig. Process. **39**(1–2), 143–161 (2013). https://doi.org/10.1016/j.ymssp.2012.08.023
24. Sakellariou, J.S., Fassois, S.D., Sakaris, C.S.: Vibration-based damage localization and estimation via the stochastic functional model based method (FMBM) - an overview. Struct. Health Monit. **17**(6), 1335–1348 (2018). https://doi.org/10.1177/1475921718793577
25. Peña, D., Rodríguez, J.: The log of the determinant of the autocorrelation matrix for testing goodness of fit in time series. J. Stat. Plan Infer. **136**(8), 2706–2718 (2006). https://doi.org/10.1016/j.jspi.2004.10.026
26. Vamvoudakis-Stefanou, K.J., Sakellariou, J.S., Fassois, S.D.: Unsupervised vibration-based damage detection methods for a population of nominally identical structures. In: 8th European Workshop on Structural Health Monitoring, Bilbao Spain (2016)
27. Press, W., Teukolsky, S., Vetterling, W., Flannery, B.: Numerical Recipes: The Art of Scientific Computing, 3rd edn. Cambridge University Press, New York (2007)

Damage Quantification in Composite Structures Using Autoregressive Models

Jessé A. S. Paixão[1(✉)], Samuel da Silva[1(✉)], and Eloi Figueiredo[2,3(✉)]

[1] Faculdade de Engenharia, Departamento de Engenharia Mecânica, Universidade Estadual Paulista - UNESP, Av. Brasil 56, Ilha Solteira, SP 15385-000, Brazil
jesseag.paixao@gmail.com, samuel.silva13@unesp.br

[2] Faculdade de Engenharia, Universidade Lusófona de Humanidades e Tecnologias, Campo Grande, 376, 1749-024 Lisboa, Portugal
eloi.figueiredo@ulusofona.pt

[3] CONSTRUCT, Institute of R&D in Structures and Construction, R. Dr. Roberto Frias s/n, 4200-465 Porto, Portugal
http://www.dem.feis.unesp.br/, https://www.ulusofona.pt

Abstract. When small damage is detected in its initial stage in a real structure, it is necessary to decide if the user must repair immediately or keep on safely monitoring it. Regarding the second choice, the present paper proposes a methodology for damage severity quantification of delamination extension in composite structures based on a data-driven strategy using autoregressive modeling approach for Lamb wave propagation. A pair of features is used based on the autoregressive (AR) model coefficients and residuals and a machine learning algorithm with Mahalanobis Squared Distance for outlier detection. The damage severity quantification is proposed through an experimentally identified smoothing spline trend curve between the damage index and its severity. The application of the methodology is demonstrated in a composite plate with various progressive damage scenarios. The proposed method proved to be able to identify and predict the localization and the damage index related to its respective extension of minimal simulated damage with promising accuracy.

Keywords: Damage quantification · Composite structures · Autoregressive models

1 Introduction

The use of composite materials in industrial applications has increased substantially in the last decades, due to their unique properties, such as high strength and stiffness combined with a low-density [1]. On the other hand, they have various and more complex types of damage such as matrix cracking, fiber debonding, and delamination [2]. Then, a drawback for the use of composite materials is to

M. A. Wahab (Ed.): DAMAS 2019, LNME, pp. 804–815, 2020.
https://doi.org/10.1007/978-981-13-8331-1_63

assure their reliability in service. In this scenario, structural health monitoring (SHM) techniques have been the focus of intensive research and development in recent years as a plausible solution, motivated by the potential of a substantial improvement in the safety of structures and economic benefits with maintenance cost reduction.

Damage identification methods can be decomposed in five levels that are related with: (1) detection, (2) localization, (3) classification, (4) quantification and (5) prognosis [3]. SHM methods based on guided waves and Lamb waves are the most widely used for damage identification [4]. They typically comprise the use of a network of piezoelectric elements (PZT) acting as both sensors and actuators to capture a baseline condition that after is related to an unknown condition to be classified. To ensure the reliability of damage identification is necessary to separate environmental and operational conditions from the structural changes associated with damages. Several works in the literature focused on damage detection and localization techniques in this context. However, a limited number of contributions have addressed the damage quantification level [5–8]. Ghrib et al. [9] presented a study of damage type classification and severity quantification in a composite structure with Support Vector Machine (SVM) using nonlinear model-based features to damage severity classification into the categories: "low," "mid" or "severe." Vitola Oyaga et al. [10] applied an approach for damage localization and quantification close to what will be employed in this work, using autoregressive models (AR). Nevertheless, it was based on vibration signals applied for civil structures and they did not verify a direct correlation among the index proposed and the damage size.

Thus, the primary purpose of this paper is to introduce a methodology for damage severity quantification of delamination size in composite structures based on a data-driven strategy. The paper is organized as follows: first, the proposed methodology of damage quantification is presented, where the damage identification using AR models is discussed concomitantly with the damage-sensitive features proposed. Next, the last step of the methodology of trend curve extrapolation to quantify the damage is detailed and discussed. Then, an experimental application of the methodology is demonstrated for a composite plate considering simulated damage. Finally, the results are discussed and further directions are suggested.

2 Quantification Methodology Proposed

Figure 1 illustrates the methodology proposed herein for damage quantification. The methodology can be separated into two steps: (1) learning and (2) test. First, a data set from the healthy state and small damage conditions known a priori is used to identify an AR model and to construct a trend curve between damage severity and the damage index. Next, this connection represented by a smoothing spline is used into the test step to quantify the damage severity of an unknown condition in a future state based on the curve extrapolation. As the methodology requires data from undamaged and damaged conditions, it is posed in the context of supervised methods [3].

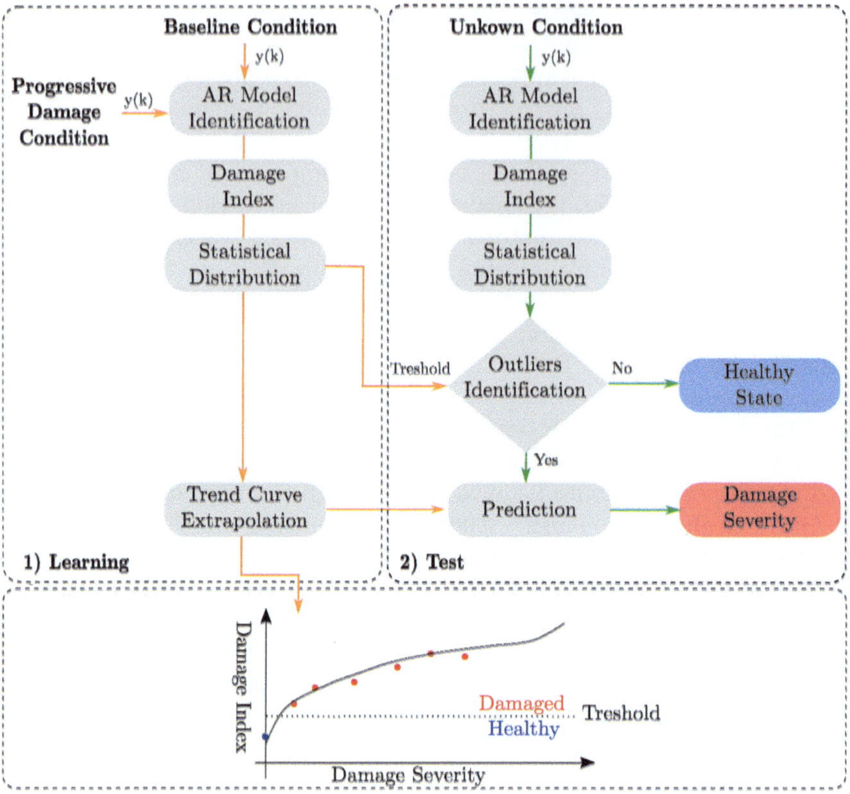

Fig. 1. Proposed methodology for damage quantification in composite structures.

2.1 Damage-Sensitive Index Using AR Models

An AR model in healthy condition using a time-series measured using a set of PZTs can be described by [11]:

$$A_{h_{ij}}(q)x_{ij}(k) = e_{h_{ij}}(k) \tag{1}$$

where $x_{ij}(k)$ is the output signal in i position caused by an excitation signal[1] applied in j spot in a sample time k, the healthy polynomial of the AR model is $A_{h_{ij}}(q) = 1 + a_1q^{-1} + \cdots + a_{n_a}q^{-n_a}$, where q^{-1} is a lag operator, i. e., $a_1q^{-1}x_{ij}(k) = a_1x_{ij}(k-1)$, and $e_{h_{ij}}(k)$ is the reference error prediction assumed to be a white noise. The order n_a can be found using Bayesian information criterion (BIC) and the parameters of the polynomial $A_{h_{ij}}(q)$ are identified using a simple least squares method with a focus in a one-step-ahead prediction [11–13]. Equation 1 can be used for monitoring an unknown situation signal $y_{ij}(k)$ in the path $i-j$ through:

[1] The input signal assumed here is a burst signal and it is not used to create the predictions models.

$$A_{h_{ij}}(q)y_{ij}(k) = \varepsilon_{ij}(k) \tag{2}$$

If the new error $\varepsilon_{ij}(k)$ has the same distribution (white noise) of the reference error, the system is in the healthy condition [14,15]. On another hand, if the error varies, probably it is induced by damage or environmental/operation variations. So, a new model must be identified:

$$A_{d_{ij}}(q)x_{ij}(k) = e_{d_{ij}}(k) \tag{3}$$

where $A_{d_{ij}}(q) = 1 + a_{d_1q}^{-1} + \cdots + a_{d_{n_a}}q^{-n_a}$ is the new polynomial in unknown condition. Important to note that it is assumed the same model order because there is an assumption of a small variation between the two states. Two features can be well applied to describe a damage detection procedure. First, to compute the ratio of variance of the error obtained:

$$\mathbf{X}_1 = \frac{\sigma(e_{d_{ij}})}{\sigma(e_{h_{ij}})} \tag{4}$$

and second using the parameters given by:

$$\mathbf{X}_2 = \frac{1}{n_a}\sum_{j=1}^{n_a}(a_j - a_{d_j})^2 \tag{5}$$

The Mahalanobis Squared Distance $\mathcal{D}^2$ is applied as the damage index for statistical outlier detection using a test matrix $\mathbf{Z}$ in a multivariate data set, including the two features presented previously [3]:

$$\mathcal{D}^2(\mathbf{Z}) = (\mathbf{Z} - \mu)\sum^{-1}(\mathbf{Z} - \mu)^T \tag{6}$$

where μ is the mean vector and $\sum$ is the covariance matrix assuming a training matrix formed by the indices in a defined baseline condition $\mathbf{X} = [\mathbf{X}_1 \quad \mathbf{X}_2]$.

To perform the detection is required the establishment of a threshold value to separate the damaged and healthy states. The strategy presented by Figueiredo et al. [16] is used in this study, where the threshold for outliers detection is defined by the most significant value of the $\mathcal{D}^2(\mathbf{Z})$ considering all signals corresponding to the safe condition.

2.2 Trend Curve Extrapolation

To establish a direct ratio among the damage index and its severity, it is proposed a trend curve fitting by a smoothing spline, where the data from undamaged and damaged states of the learning steps are used to define the curve, that can be extrapolated in order to predict the damage index and its respective severity in the next test step.

The damage severity (s) examined in this work corresponds to the area covered by simulated damage and can be measured for each state. As for each damage condition, it is considered a population of computed damage indices; then, its statistical model is used to relate to its particular severity. The statistical distribution of the damage indices is unknown a priori. Then, the kernel smoothing technique is used to estimate their probability density function (PDF) to obtain the mode of a set of damage index values in each damaged condition [17].

Suppose observed n pairs of damage index and severity $(\mathcal{D}_i^2, s_i)$, $i = 1..., n$, relating to the general smoothing spline regression [18]

$$\mathcal{D}_i^2 = f(s_i) + \xi_i \tag{7}$$

where ξ_i is an independent random error. A smoothing spline estimate f_p for f is defined as the minimizer of the penalized criterion [18].

$$\frac{1}{n}\sum_{i=1}^{n} \mathcal{D}_i^2 - f(s_i)^2 + (1-\delta)\int (\frac{\partial^2 f}{\partial s^2})^2 ds \tag{8}$$

where δ is a positive known as smoothing parameter. As long as the smoothing spline curve is estimated on the learning step of the methodology, it can be applied to predict an unknown damage size on the test step in a future state.

3 Experimental Application

Figure 2(a) illustrates a carbon-epoxy laminated with layup containing 10 plies unidirectionally oriented along 0° with four PZTs SMART Layers from Accelent Technologies, with 6.35 mm in diameter and 0.25 mm in thickness. A pitch-catch configuration is employed where the PZT 1 is used as an actuator. A five-cycle tone burst signal with 35 V of amplitude and center frequency of 250 kHz is applied. The outputs used are measured in PZT 2, PZT 3 and PZT 4 with a sampling frequency of 5 MHz and timespan of 200 µs. All signal generation and the acquisition was performed using the setup schematically described in Fig. 2(b), composed by a NI USB 6353 from National Instrument, a power amplifier EL 1225 from Mide QuickPack and an oscilloscope DSO7034B from Keysight, both controlled by Labview.

To simulate the damage reversibly, an industrial adhesive putty was inserted on the plate surface [19]. The additional mass introduced by the putty simulates local changes in the damping of the plate, which is an effect similar to the delamination in composites according to Lee et al. [19]. The damage severity of all states on the learning and test steps are presented in Table 1.

The experiments were conducted inside a temperature chamber SM-8 from Thermotron with a controlled temperature of 30 °C and placed in a free-free boundary condition in order to eliminate the effects from environmental and operational variability.

Therefore, in total, the structure was submitted to 12 conditions and, for each one, the experiments were performed 100 times to have enough data for statistical analysis. The signals used in the learning and test steps of the methodology were collected in different days and state conditions. The first condition corresponds to the healthy state (H30) used as a baseline condition. The next seven conditions

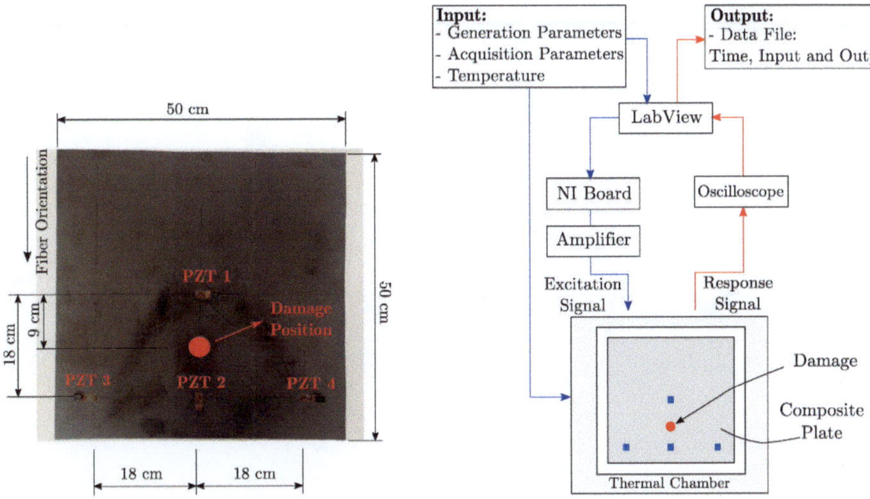

(a) Composite plate. (b) Scheme of experimental setup.

Fig. 2. Composite plate and schematic view of the experimental setup.

Table 1. Structural states examined.

Step	Learning								Test			
Structural state	H30	D1	D2	D3	D4	D5	D6	D7	DF1	DF2	DF3	DF4
Damage severity [mm^2]	0	490	707	962	1256	1963	2827	3848	2375	5026	5674	6361
Percentage of area covered [%]	0	0.19	0.28	0.38	0.50	0.78	1.13	1.54	0.95	2.01	2.27	2.54

correspond to the progressive damaged states (D1 to D7), used on the learning step to create the trend curve. Finally, the last four conditions correspond to the conditions of damage in future states of severity progression used on the test step (DF1 to DF4).

Figure 3 shows the consequence of the introduction of the progressive damage on the first arrival mode measured by PZT 2. An increase in the area covered by the damage is observed to be proportional to a reduction of the response signal amplitude. This phenomena is due to the nature of damage introduced, which adds local damping in the transducer path causing a higher attenuation of the wave [19]. Figure 4 presents the response signal acquired on all PZTs considering the baseline and a damaged condition D7, where it can be observed a more pronounced difference on the PZT 2 than on the others, as the damage is situated in a path along this transducer.

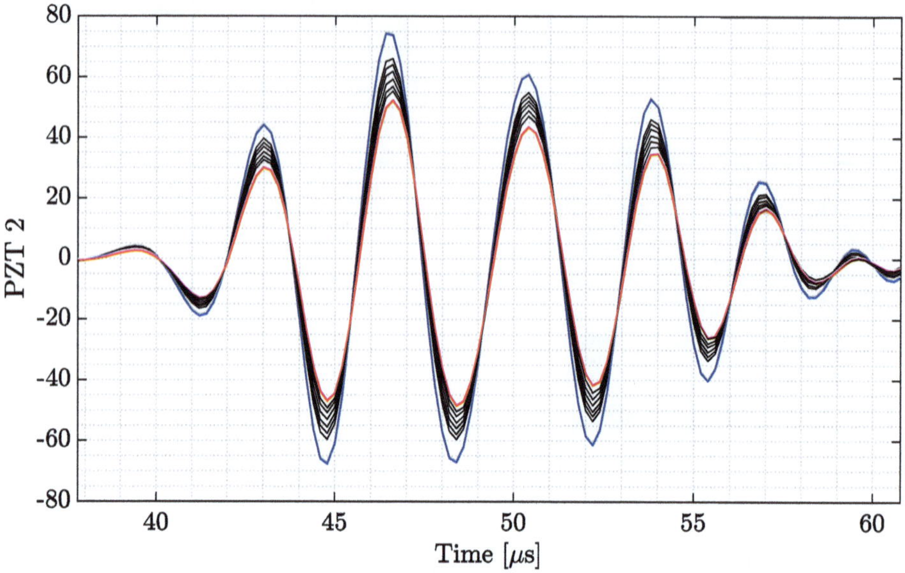

Fig. 3. Output time-series of first arrival mode for PZT 2 measured in mV for the structure in a healthy state (—) and progressive damage states. — represents the last state of damage D7 and — represents the progressive damage stages from D1 to D7.

Based on the BIC method, the model order was chosen as $n_a = 20$ for all PZTs. Figure 5 shows the response signals measured by the PZTs, the predicted signals with the estimated AR models and the residues. The residues are used as one of the features to interrogate the structure condition. As observed on the signals, the amplitude of residues is slightly more evident on the arrival modes. Then, the effectiveness of the damage detection is more dependent on the damage sensitivity from these modes.

Figure 6 shows the state-space of the two features, where it is observed a separation between the cluster states (points) for the healthy condition from the damaged one.

The proposed machine learning algorithm, based on the MSD for outlier detection, calculates the distance of the points from the test matrix to the centroid of the ellipse formed by the points from train matrix, which in this case corresponds to the features measured for the baseline condition. Figure 7 shows the damage index $\mathcal{D}^2$ calculated using the machine learning algorithm. It was employed 70% of the data from the damage features assuming the baseline condition. As can be noted in Fig. 7, the damage index manifested more accentuated for PZT 2 than PZTs 3 and 4, as the damage is positioned in the path between the transducers PZT 1 to PZT 2. The classification of the structural condition was performed based on the statistical model of PDF estimated for each condition, estimated using the kernel smoothing technique with cross-validation method to choose the smoothing parameter [17].

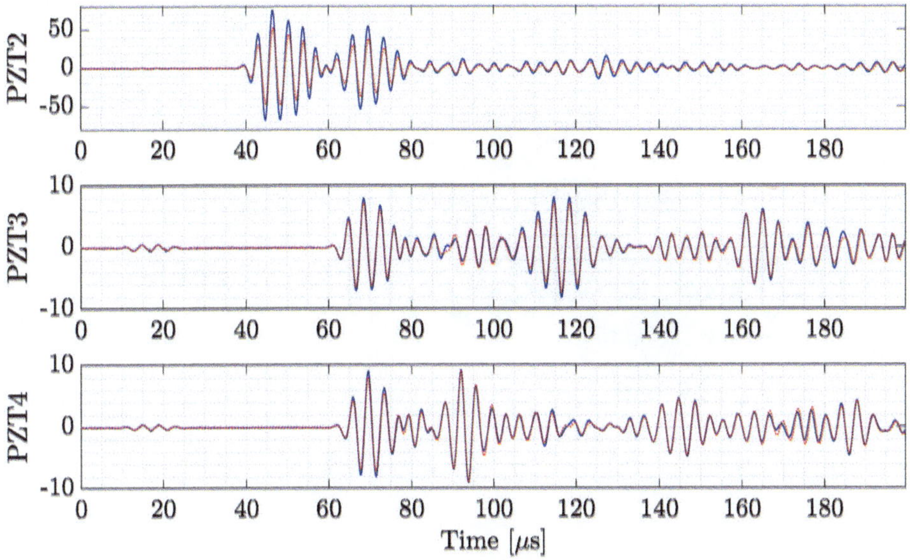

Fig. 4. Output time-series signal of the PZTs 2, 3 and 4 measured in mV for the structure in a healthy state (—) and the last state of damage D7 (....) in the learning step.

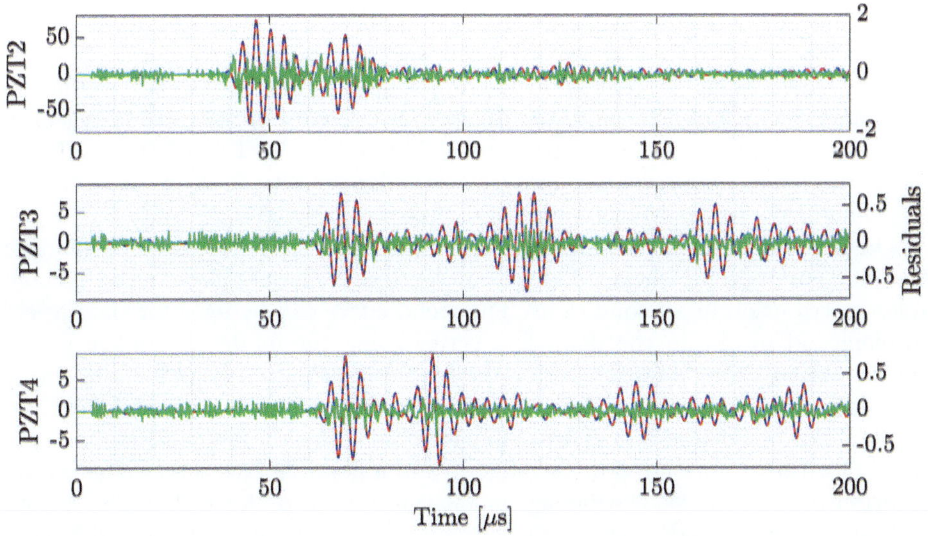

Fig. 5. Measured output signal in mV (—) compared with predicted signal using the AR reference model (- - -) and residuals in *mV* (—) for PZTs 2, 3 and 4 assuming the baseline condition.

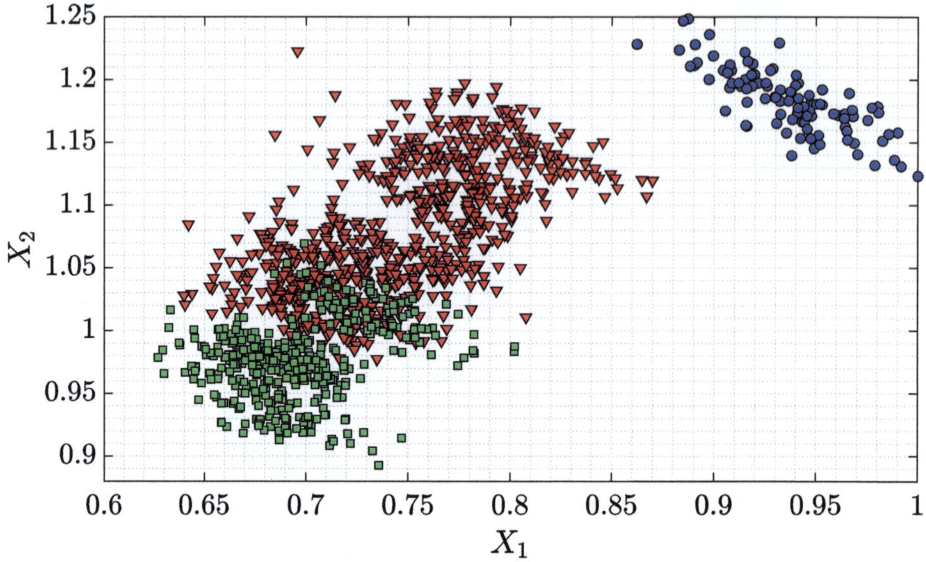

Fig. 6. Features for all structural conditions for the PZT 2. ● is the data set in the healthy condition, ▲ is the data set for progressive damage conditions used on the learning step and ■ is the data set for the damaged conditions used on the test step.

Based on a direct ratio among the mode of the statistical model estimated of the population of $\mathcal{D}^2$ index in each condition and the respective damage severity, a trend curve was fitted using a smoothing spline function to obtain a direct relationship between the two variables. Figure 8 shows the trend curve obtained using the smoothing fitted spline function and the boxplot of the population of damage index in each condition. The trend curve was obtained using only the pairs of data from the learning step (H30 and D1 to D7) of the mode for damage index in each condition and its particular severity. The extrapolation of the trend curve was performed until a damage severity of 8000 mm^2.

On the test step, the damaged conditions DF1 to DF4 were used to validate the trend curve extrapolation. The extrapolated curve is very close to the damage index distribution in all conditions. The trend curve can be used for the inverse problem and to obtain the damage severity using the mode of damage index. Table 2 presents the damage severity obtained using the trend curve and mode of damage index distribution compared with the measured one, for each condition, where it can be noticed a sufficient similarity presenting a low percentage error, mainly for the extrapolated damaged states in a possible future state before the occurrence. In this work, the damage quantification was performed only assuming the damage positioned in the path between the transducers PZT 1 and PZT 2 that corresponds the position where the small initial damage to be monitored is detected. However, each path of the PZT network has its trend curve, and it can be created on the learning step of the methodology.

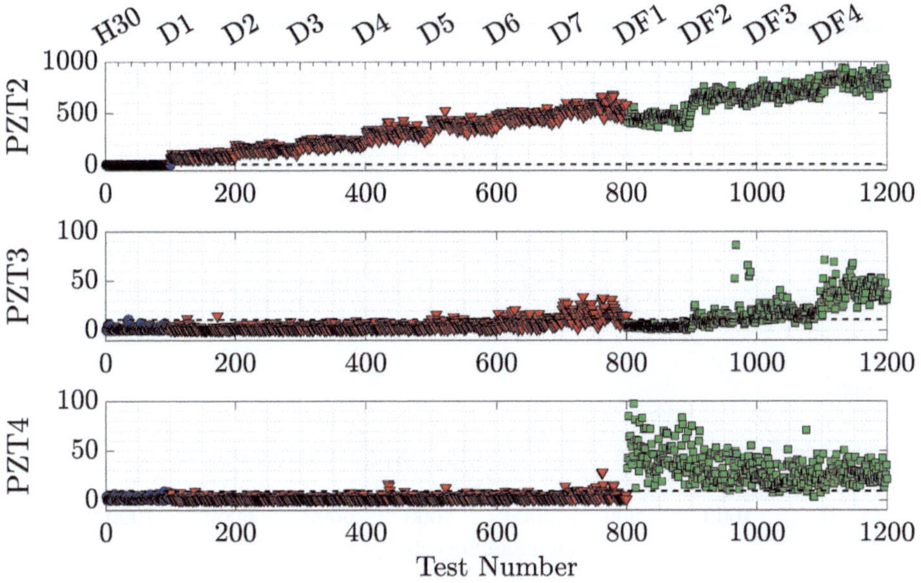

Fig. 7. Damage index $\mathcal{D}^2$ computed for all performed tests (learning and test steps) and the three PZTs. ● represents the index considering healthy condition (H30), ▲ represents the index for progressive damage conditions used on the learning test (D1 to D7) and ■ represents the index for the damaged conditions used on the test step (DU1 to DU4) and (- - -) is the threshold line considered for the outliers detection.

Table 2. Estimated damage severity for the damaged conditions on the test step

Structural state [mm^2]	DF1	DF2	DF3	DF4
Damage severity measured [mm^2]	2375	5026	5674	6361
Damage severity estimated [mm^2]	2892	5419	5625	6515
Error [%]	21.75	7.82	0.86	2.41

4 Final Remarks

The methodology presented in this paper, based on the use of AR models for damage quantification in composite structures, can predict the size and location of simulated damage with adequate precision. The set of features proposed was a combination of residuals and coefficients used along with a machine learning algorithm based on the Mahalanobis Squared Distance. This methodology was also able to extract information about the damage severity in a future state. The proposed methodology to obtain the relation between the damage index and its severity using smoothing spline fitting was validated in the test step by estimating the damage size, that presented a small percentage error. Additional research of the methodology is being carried about the influence of environ-

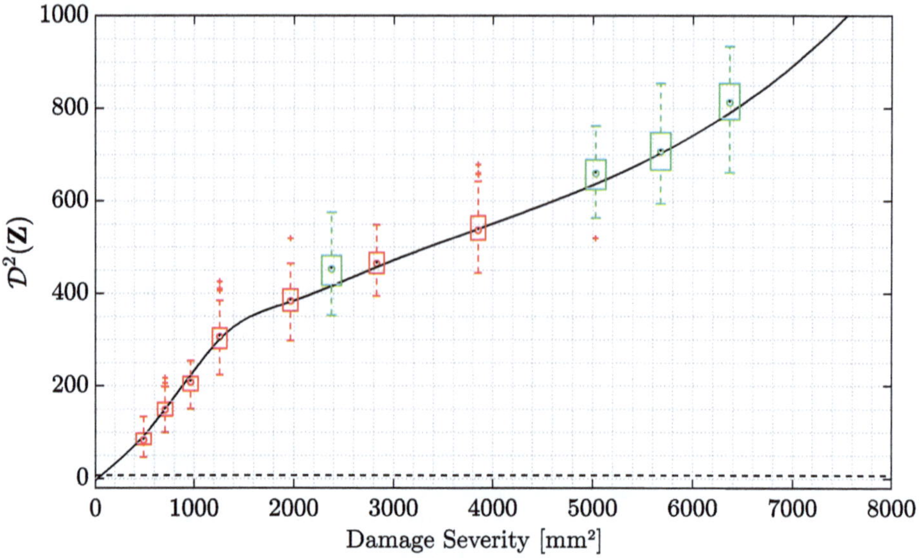

Fig. 8. Trend curve relating the damage severity to the damage index ($\mathcal{D}^2$) considering the PZT 2 and the boxplot of index population for the conditions of the learning test (—) and test step (—). The threshold line is represented by the dashed line (- - -).

mental/operational variability, that represents the central shortcoming to be overcome.

Acknowledgements. The authors are thankful for the financial support provided from São Paulo Research Foundation (FAPESP) grant numbers 2017/15512-8 and 2018/15671-1, and the Brazilian National Council for Scientific and Technological Development (CNPq) grant number 307520/2016-1.

References

1. Ghrib, M., Berthe, L., Mechbal, N., Rébillat, M., Guskov, M., Ecault, R., Bedreddine, N.: Generation of controlled delaminations in composites using symmetrical laser shock configuration. Compos. Struct. **171**, 286–297 (2017)
2. Talreja, R., Singh, C.V.: Damage and Failure of Composite Materials. Cambridge University Press, Cambridge (2012)
3. Nobari, A.S., Ferri Aliabadi, M.H.: Vibration-Based Techniques for Damage Detection and Localization in Engineering Structures. World Scientific (Europe) (2018). https://doi.org/10.1142/q0145
4. Su, Z., Ye, L.: Identification of Damage Using Lamb Waves: From Fundamentals to Applications, vol. 48. Springer, Heidelberg (2009)
5. Rebillat, M., Mechbal, N.: Damage localization in composite plates using canonical polyadic decomposition of lamb wave difference signals tensor. IFAC-PapersOnLine **51**(24), 668–673 (2018). https://doi.org/10.1016/j.ifacol.2018.09.647

6. Shiki, S.B., da Silva, S., Todd, M.D.: On the application of discrete-time volterra series for the damage detection problem in initially nonlinear systems. Struct. Health Monit. **16**(1), 62–78 (2017). https://doi.org/10.1177/1475921716662142
7. Memmolo, V., Pasquino, N., Ricci, F.: Experimental characterization of a damage detection and localization system for composite structures. Measurement **129**, 381–388 (2018). https://doi.org/10.1016/j.measurement.2018.07.032. http://www.sciencedirect.com/science/article/pii/S0263224118306274. ISSN 0263-2241
8. Mallouli, M., Ben Souf, M.A., Bareille, O., Ichchou, M.N., Fakhfakh, T., Haddar, M.: Damage detection on composite beam under transverse impact using the wave finite element method. Appl. Acoust. **147**, 23–31 (2019). https://doi.org/10.1016/j.apacoust.2018.03.022. http://www.sciencedirect.com/science/article/pii/S0003682X17305741. ISSN 0003-682X. Special Issue on Design and Modelling of Mechanical Systems Conference, CMSM 2017
9. Ghrib, M., Rébillat, M., des Roches, G.V., Mechbal, N.: Automatic damage type classification and severity quantification using signal based and nonlinear model based damage sensitive features. J. Process Control (2018). https://doi.org/10.1016/j.jprocont.2018.08.002. http://www.sciencedirect.com/science/article/pii/S0959152418301975. ISSN 0959-1524
10. Vitola Oyaga, J., Burgos, D.A.T., Vejar, M.A., Montero, F.P.: Structural damage detection and classification based on machine learning algorithms. In: Proceedings of the 8th European Workshop on Structural Health Monitoring (2016)
11. Kitagawa, G.: Introduction to Time Series Modeling. Chapman and Hall/CRC, Boca Raton (2010)
12. Figueiredo, E., Figueiras, J., Park, G., Farrar, C.R., Worden, K.: Influence of the autoregressive model order on damage detection. Comput.-Aided Civil Infrastruct. Eng. **26**(3), 225–238 (2010). https://doi.org/10.1111/j.1467-8667.2010.00685.x
13. Nardi, D., Lampani, L., Pasquali, M., Gaudenzi, P.: Detection of low-velocity impact-induced delaminations in composite laminates using auto-regressive models. Compos. Struct. **151**, 108–113 (2016). https://doi.org/10.1016/j.compstruct.2016.02.005. http://www.sciencedirect.com/science/article/pii/S0263822316300253. ISSN 0263-8223. Smart composites and composite structures In honour of the 70th anniversary of Professor Carlos Alberto Mota Soares
14. da Silva, S., Lopes Jr., V., Dias Jr., M.: Structural health monitoring in smart structures through time series analysis. Struct. Health Monit. **7**(3), 231–244 (2008). https://doi.org/10.1177/1475921708090561. http://shm.sagepub.com/content/7/3/231.abstract
15. Farrar, C.R., Worden, K., Todd, M.D., Park, G., Nichols, J., Adams, D.E., Bement, M.T., Farinholt, K.: Nonlinear system identification for damage detection. Technical report, Los Alamos National Laboratory (LANL), Los Alamos, NM (2007)
16. Figueiredo, E., Park, G., Figueiras, J., Farrar, C., Worden, K.: Structural health monitoring algorithm comparisons using standard data sets. Technical report, Los Alamos National Lab.(LANL), Los Alamos, NM (United States) (2009)
17. Bowman, A.W., Azzalini, A.: Applied Smoothing Techniques for Data Analysis: The Kernel Approach with S-Plus Illustrations, vol. 18. OUP, Oxford (1997)
18. Brien, C.M.: Smoothing splines: methods and applications by Yuedong Wang. Int. Stat. Rev. **80**(3), 475–476 (2012). https://doi.org/10.1111/j.1751-5823.2012.00196_6.x
19. Lee, J.-S., Park, G., Kim, C.-G., Farrar, C.R.: Use of relative baseline features of guided waves for in situ structural health monitoring. Journal of Intelligent Material Systems and Structures **22**(2), 175–189 (2011). https://doi.org/10.1177/1045389X10395643

Meso-Scale Damage Modeling of Hybrid 3D Woven Orthogonal Composites Under Uni-Axial Compression

Sohail Ahmed[1,2(✉)], Xitao Zheng[1,2], Tianchi Wu[1,2], and Nadeem Ali Bhatti[3]

[1] Department of Aeronautical Structural Engineering, School of Aeronautics, Northwestern Polytechnical University, Xi'an, People's Republic of China
s.ahmed@mail.nwpu.edu.cn

[2] Institute of Aircraft Composite Structures, Northwestern Polytechnical University, Xi'an, People's Republic of China

[3] Department of Electrical Energy, Metals, Mechanical Construction and Systems, Faculty of Engineering and Architecture, Ghent University, Ghent, Belgium

Abstract. In this study, a detailed predictive numerical modeling technique is developed for three different kinds of hybrid and non-hybrid architectures of 3D woven orthogonal composites. A unit cell, full finite element meso model, is developed to simulate the elastic and damage progression behavior under uni-axial compression. Material systems for 3D woven Composites used in this study are Carbon/epoxy, Kevlar epoxy, and Hybrid formulation. The geometry of the constituents of the unit cell, fiber tows, and matrix, was determined through the microscopic analysis of the woven structures. The unit cell model used an idealized geometry without incorporating the fiber undulation which is a pervasive effect during the weaving process of 3D woven composites. Abaqus standard solver and a UMAT code incorporating the damage model within Abaqus was used to compute the deformation of the unit cell under in-plane compression loading. Moreover, determining the effect of void contents on the mechanical properties of the material, Balshin's empirical equation was used to calibrate the input material parameters. The results from the numerical analysis are then compared with the experimental results and it is found that the predicted elastic modulus and compressive strength for all three types of architecture are in good agreement with the experimental results. Overall, the unit cell meso model furnished fairly accurate estimation of the compressive properties of 3D woven composites for different material systems with consideration of void contents.

Keywords: 3D woven composites · Hybridization · Compression · Unit cell · Numerical modelling

M. A. Wahab (Ed.): DAMAS 2019, LNME, pp. 816–826, 2020.
https://doi.org/10.1007/978-981-13-8331-1_64

1 Introduction

Three dimensional (3D) woven composites possess higher through-the-thickness (TTT) mechanical properties, delamination resistance, and impact damage tolerance compared to the laminated composites. However, the in-plane properties of these composites are compromised due to fiber crimping induced during the weaving process. Secondly, voids and resin rich pocket are likely to occur during the resin transfer molding process which further weakens the structure. Furthermore, the complex weave architecture makes it difficult for the designers to accurately predict the mechanical properties of 3D composites. These issues arise the need to mechanically characterize before their efficient and confident use in suitable applications. In the recent past, several researches have presented their research on experimental techniques as well as modelling techniques to characterize the 3D woven composites.

The tensile behavior for in-plane properties of 3D woven composites has been investigated by many researchers through experimental characterization [1–5]. It is commonly concluded that the tensile properties of a woven structure are mainly determined by the waviness of the fibers in the loading direction. Therefore, the structure with the least crimp in the loading direction exhibits higher tensile strength. The compressive properties of 3D woven composites were also reported by several authors [4, 6–8]. The compression testing results revealed that the failure of 3D woven composites under compression is mainly governed by fiber kinking. Since delamination is suppressed by the TTT fibers, fiber kinking becomes the dominant failure mechanism [8]. Cox et al. [7] suggested that the compressive properties can be improved by controlling the geometrical irregularities rather than increasing the fiber volume fractions of the preform.

In parallel to the experimental investigation, the predictive analytical and finite element (FE) models were also developed to characterize the elastic, plastic and failure behaviors of 3D woven composites. Analytical predictive models developed in the past are mainly an extension of the predictive models for 2D woven composites and/or are based on the Classical Laminate Theory (CLT) [9]. An analytical model to predict the macroscopic elastic properties of 2D and 3D composites based on orientation averaging (OA) method is proposed by Kreges et al. [10] in which the preform is discretized into small volume elements. The OA method, however, does not consider the manufacturing defects of the composite preforms like fiber undulation, voids in the matrix and fiber yarns etc. A solution to the problem is suggested by Cox et al. [11] by incorporating the tow waviness factor in the baseline OA method called modified OA method and presented the improved results for in-plane mechanical properties. Several FE modeling techniques have also been developed in the past, like unit cell (UC) model, binary model, and mosaic model. The UC modelling technique was used efficiently to conduct the numerical prediction of the elastic and strength of 3D woven composites by Tan et al. [12] and Drach et al [13]. Tan et al. [12] used the idealistic internal geometry of the composite preform to build the numerical model, however, Drach et al. [13] used the realistic geometry to build the UC for subsequent FE model.

A review of the experimental study and predictive models for mechanical behavior of 3D woven composites demonstrates that most of the research deals with the preform made from uniform reinforcement and uses idealistic geometry of the composite. Experimental investigation for hybridized 3D woven composites has been carried out recently by Muñoz et al. [14] on deformation and failure mechanism of carbon/glass hybrid materials. Pankow et al. [15] presented the experimental results on the tensile response of Carbon/glass/Kevlar 3D woven composite. However, the work on predictive models is very limited in the current literature hence arise the need of development of an efficient predictive model for hybridized 3D woven structure which can accurately predict the elastic parameters and strength.

Predictive FE models although time-consuming but cost-effective in comparison to the experimental techniques can be used to predict the effective properties of composite preforms having complex architecture with reliability and accuracy. However, the accuracy of the FE model mainly depends upon the assumptions to idealize the internal geometry of the composites, therefore, the internal architecture of the preform must be idealized on the basis of reasonable assumptions. The geometric model for representative volume element (RVE) and subsequent FE model for FE analysis is often created by employing major simplifications which include, modelling the fiber tows with rectangular cross-sections, ignoring the fiber tows undulation which takes place during the manufacturing process, ignoring voids within the matrix etc. These simplifications of composite RVE may lead to over-predict the mechanical properties. Therefore, in the present paper, an FE predictive model is presented in order to predict the stiffness and strength properties of hybrid 3D woven composite, utilizing more representative yet simple internal geometry of the composite preform with salient feature of voids volume consideration.

2 Composite Material and Experimental

The composite used in the present study is comprised of 3D woven orthogonal hybrid and non-hybrid preforms with TDE-86 matrix resin. Three different kinds of composite preforms were manufactured in terms of fiber tow materials. Configuration 1 (C-C) contains T700-12 K carbon fiber as warp and weft tows and T700-6 K as TTT yarns or z-yarns. Configuration 2 (C-K) formulates the hybrid architecture by using carbon fiber T700-12 K and aramid fibers Kevlar 49-4 as warp and weft fiber tows and kevlar 49-1 as z-yarns (Table 1). Configuration 3 (K-K) contains kevlar 49-4 aramid fiber as warp and weft tows and kevlar 49 as z-yarns. Composite preforms are manufactured with

Table 1. Preform specifications of carbon, carbon/Kevlar and Kevlar 3D woven composite panels

Architecture	Number of layers	Ratio C:K	Z-yarn	Warp tow	Weft tow	Fiber volume fraction
C-C	11	11:0	T700	T700 12 K	T700 2x6 K	52.3%
C-K	13	9:4	Kevlar 49	T700 12 K / Kevlar 49-4	T700 2x6 K / Kevlar 49-4	56.8%
K-K	13	0:13	Kevlar 49	Kevlar 49-4	Kevlar 49-4	57.2%

vacuum assisted resin transfer molding process (VARTM) in conjunction with 3D weaving process.

The compression experiments of 3D woven composites were conducted according to the ASTM D6641 test standard using electromechanical universal testing machine. Tests were carried out with the displacement control of 0.5 mm/min and the reaction force was continuously monitored during the test with a load cell of 200 KN. All the specimens are loaded in the weft direction. In order to obtain in-plane elastic modulus in longitudinal direction (weft direction) during compression, specimens were equipped with strain gauges on both faces along loading direction.

3 Methodology

The modelling methodology proposed in this study is based on the meso-scale analysis at RVE level of 3D woven composites with the aid of finite element based generalized micro-mechanics tools. After acquiring the internal geometry of the constituents of the composite system, it is a common practice to select an appropriate RVE from a composite preform which is assumed to be the smallest volume representative of the whole structure. Material properties for each constituent of the RVE are then assigned/calculated in collaboration with any material or geometric defect such as yarns waviness, voids etc. Internal geometry, material specification, and FE modeling are explained in the following sections.

3.1 Geometry of RVE

The constituents such as fibers and matrix and their construction in a composite material, affect the mechanisms operating in the composites during loading, elastic response, damage progression, failure modes and ultimately the strength. Generally, in a preform of 3D woven composites, the warp and weft tows are placed in-plane throughout the preform like [0/90] 2D laminates with little or no undulation. Moreover, they are bounded together by z-yarns which sweep along the warp fiber direction i.e. 0^0 from top to bottom to bind all the layers in the preform.

Three different architectures are analyzed in this study in terms of materials of the constituents and their geometry. The material specifications of all three architectures are given in Table 1. In order to calculate the exact geometric parameters for each type of preform, at first the specimens were cut in the warp and weft directions and sandpaper with roughness of 120 was applied to get precise location of the desired cross section. After that, sandpapers with the roughnesses of 320, 800 and 1200 were applied in ascending order to polish the surface and making it as smooth as possible. the microscopic analysis was conducted using Hirox KH-7700 digital microscope with the magnification scale of 10. The microscopic images of all three architectures are presented in Fig. 1. The calculated dimensions for RVE and its constituents are summarized in Table 2.

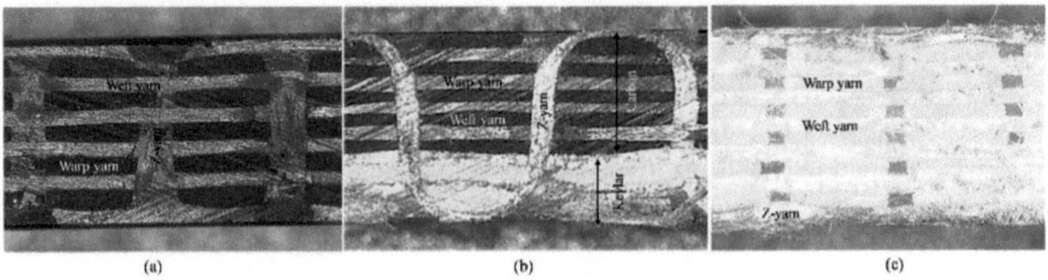

Fig. 1. Microscopic images for (a) carbon formulation, C-C (b) Hybrid formulation, C-K and (c) Kevlar formulation, K-K

Table 2. Dimensions of RVE

Weave type	Fiber Tow	Width (mm)	Height (mm)	RVE length (mm)	RVE width (mm)
C-C	Warp	1.85	0.32	4.86	4.8
	Weft	2.23	0.35		
	z-yarn	0.55	0.2		
C-K & K-K	Warp	2.12	0.32	4.54	4.98
	Weft	2.11	0.27		
	z-yarn	0.37	0.16		

The RVE for first architecture consists of 11 layers of warp and weft yarns with the warp tow insertion density at 4.27 ends/cm, weft tow insertion density at 4 picks/cm, and a Z yarn insertion of 1/end. All warp and weft layers used 12 K yarns with linear density of 800 Tex (g/1000 m) and z-yarns used 1 K yarns with linear density of 66 Tex for carbon composite. In the case of hybrid (C-K) and Kevlar (K-K) architecture, the RVE consists of 13 layers of warp and weft tows yarns with the warp tow insertion density at 4 ends/cm, weft tow insertion density at 4.4 picks/cm, and a Z yarn insertion of 1/end. The Kevlar yarns with linear density of 158 Tex are used.

3.2 Constituent Properties

Once the complete geometric model of the RVE including its constituents is created, the next step is to assign the correct material properties to each constituent. Considering the nature of warp, weft and TTT yarns as transversely isotropic, impregnated fiber tows were assumed basically carbon/epoxy and Kevlar/epoxy uni-directional laminae. Engineering elastic constants are estimated from Chamis model [16]. The failure strengths of the fiber tows are predicted using the rule of mixture method provided in [17]. Material properties, engineering elastic constants and strength parameters, for the constituent materials are listed in Table 3. The epoxy resin material is assumed as an isotropic material.

Table 3. The material properties of the constituents [18, 19]

	E_1 (GPa)	E_2 (GPa)	υ_{12}	υ_{23}	G_{12f} (GPa)	Strength
T700 [18]	230	18.2	0.27	0.3	36.6	4.9 GPa
Kevlar49 [19]	130	6.72	0.36	-	2.17	3.9 GPa
TDE86 [18]	3.5	3.5	0.35	0.35	1.3	80 MPa

3.3 Voids Consideration

Although one of the advantages of the VARTM process of manufacturing the composite preforms is the reduced possibility of voids formation, however, the voids have been observed in the intertow matrix regions of the composite preforms used in the current study. The presence of voids directly affects the mechanical properties of matrix epoxy and an indirect effect is also expected on the fiber tows properties due to the presence of the same matrix into the fiber tows. The voids are generally dispersed randomly throughout the material; therefore, the response of the matrix material is assumed to remain isotropic.

There are several approaches to measure the extent of voids in the material such as: density measurement and weighing of the constituents, quantitative microscopic examination of interior surfaces of the composite preform, using the indirect inference which is based on exposition of the material to ultrasonic or other radiations, and optical imaging techniques. In this paper, microscopic image analysis has been used to measure the void volume fraction present in the pure matrix.

Voids shape, distribution and content defines the function to which the mechanical properties are affected by the voids' presence. There are several methods which can determine the effect of void contents on the mechanical properties of a material [20] such as generalized rule of mixture, Eudier's model, minimum solid area method, and Balshin's empirical equation. In this study, the Balshin's empirical formula, as given by the following equation is used to define the homogenizing effect of void contents on the mechanical properties of the yarns and matrix.

$$\frac{M}{M_0} = (1 - \rho)^{1/J}$$

Where M is a specific property e.g. Young's modulus, yield strength etc. ρ is the void volume fraction and J is the scaling, fractal parameter which is controlled by the shape, size and distribution of the voids.

3.4 FE Model

FE model for the RVE, for three different architectures, were developed within FE software ABAQUS/standard. Partitioning-based modeling technique is used to generate the 3D models, which results in the combined material interfaces which give an advantage of avoidance of additional constraints between the overlapping regions of constituents. RVE is meshed by using solid elements with eight-node linear continuum brick elements with reduced integration (C3D8R). Enhanced stiffness relaxation method is used to prevent hour-glassing of the reduced integration elements [21]. Section based FE model of RVE and its constituents are shown in Fig. 2.

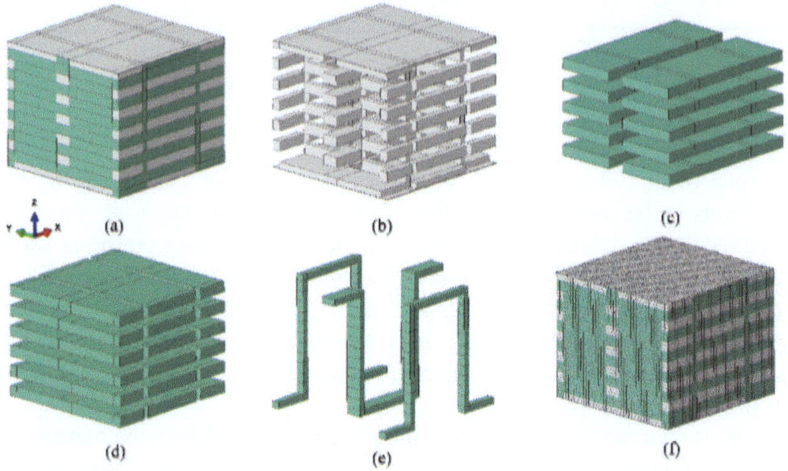

Fig. 2. FE model of complete RVE representing its constituents, warp layers, weft layers, Z-yarns and inter-tow matrix for C-C configuration

By utilizing the periodicity of 3D woven architecture, suitable periodic boundary conditions are specified on the four vertical faces of the RVE corresponding to $X = \pm L/2$ and $Y = \pm W/2$, The RVE faces located at $+H/2$ and $-H/2$ in the z-axis were set free for the sake of surface effect. Similarly, displacement boundary conditions are applied in coordination with periodic boundary.

3.5 Damage Modeling

Considering the hybrid formulation in terms of constitutive behavior of the RVE, two different kinds of materials are available namely transversely isotropic (fiber tows) and isotropic (matrix). Being the main load carrying capability of fiber tows as compared to the matrix material, the RVE is assumed to be failed when failure at fiber tows occurs. The damage criteria presented by Linde et al. [22] was deployed through user sub-routine UMAT in ABAQUS/standard [21] to evaluate the damage and constitutive behavior of the fiber tows. Detailed discussions regarding the derivation of the damage criteria are presented elsewhere [22].

4 Results and Discussion

A novel numerical simulation technique for 3D woven composites based on RVE is developed in this study. The predictive model is capable of considering the hybridization at inter and intra ply level and can take the effect of voids present in the matrix resin. Two different FE models were built for C-C and C-K weave types, in the first model the material is assumed to have no voids, in the second model the material properties were calculated considering the voids effect. The predicted compressive properties of all three configurations and experimental results are summarized in Table 4. As it can be seen that the models with no voids give promising results for C-K and K-K formulations however give overestimated results for the full carbon composite because the voids volume fraction in pure carbon composite is more than the hybrid and Kevlar composites. Therefore, the FE model with voids incorporation gives much more accurate results for carbon composite.

Table 4. Predicted compressive modulus and strength under void content effect in comparison with experiments

Architecture	Void volume fraction (%)	Modulus (GPa)	Experiment (GPa)	Strength (MPa)	Experiment (MPa)
C-C	0	65.97	55.23	642	438
	9.8	53.44		520	
C-K	0	46.97	45	422.78	397
	4.9	43.80		410	
K-K	0	30	33	117	124

Figure 3 shows the stress-strain curves for C-C, C-K and K-K configurations obtained from experimental and simulations results, it is observed that the curve for C-C behaves linearly until final failure which shows a brittle behavior however the curve for C-K shows the nonlinear behavior due the its hybrid nature, as low strength fibers breaks earlier than the high strength carbon fibers and K-K formulation shows the bi-linear behavior that is mainly because of the fiber kink band formation.

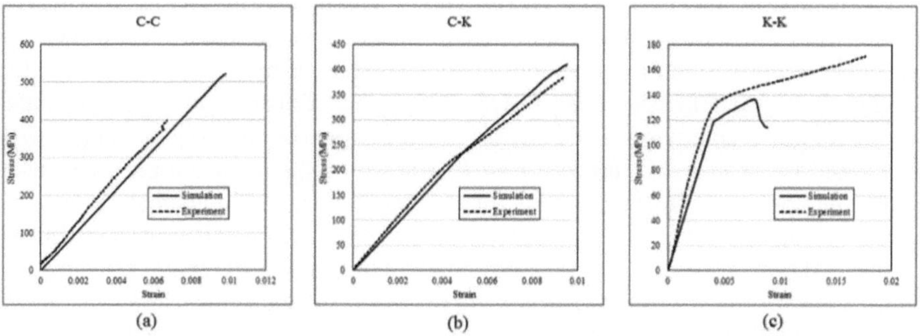

Fig. 3. A comparison of stress-strain date between simulation and experimental for different configurations

Figure 4 shows the progressive damage behavior of the fiber tows under the compression loading in weft direction. In the part (a) of the figure, the von-Mises distribution of stress in fiber tows is shown under elastic loading before the onset of any damage. It is observed that the main load bearing components in this case are weft fibers hence showing the maximum level of stresses. Part (b) and (c) depicts the longitudinal (fiber) failure index and transverse (matrix) failure index respectively. Fiber damage was observed in the weft fibers first and the matrix damage was occurred in the warp and the portion of TTT yarns in the warp direction. The damage in 3D woven composites onsets near the location of z-yarn and at the edges of the in-plane fiber tows orientated transversely to the loading direction which can also be observed from the simulations' results. Overall the model presented in this study considering the effect of voids, shows excellent agreement with experimental results, therefore, can prove a very useful tool in evaluating new and novel material combinations with inherent material defects.

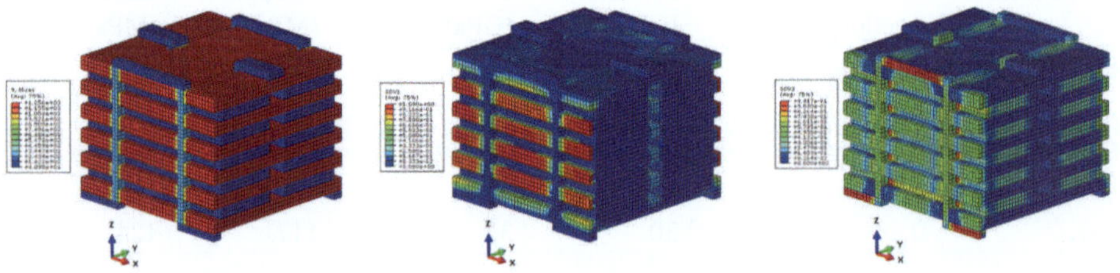

(a) (b) (c)

Fig. 4. Von-misses stress distribution and failure indices for C-C configuration (a) Von-misses stress distribution before onset of damage (b) fiber damage, (c) matrix damage during the compression loading in the weft direction

5 Conclusion

A meso-scale FE model for RVE is developed to simulate the elastic and damage behavior of 3 different formulations of 3D orthogonal woven composites. The void effect was incorporated to calculate the input properties of matrix and yarns and their behavior on overall stiffness properties was studied. The Balshin's empirical formula is used to determine the homogenized effect of void contents on the mechanical properties of the yarns and matrix. FE model for RVE is based on the idealized geometry where the main geometric parameters were extracted from the microscopic analysis of cross-sections of the composite preform in warp and weft directions. The predicted elastic modulus and compressive strength of all three configurations show a good correlation with the experimental results. Moreover, the FE model developed in this study can take the effect of voids present in the composite structure however is limited for only 3D orthogonal weaves.

Overall, the numerical predictive model presented in this study provided a fairly accurate estimation of the compressive properties of hybrid and non-hybrid 3D orthogonal woven composites and can be a useful tool for designers and engineers to design an optimum structure. The approach presented here can be used to understand

and quantify the effects of hybridization and voids content on the compressive response of 3D woven composites.

References

1. Ivanov, D.S., Lomov, S.V., Bogdanovich, A.E., Karahan, M., Verpoest, I.: A comparative study of tensile properties of non-crimp 3D orthogonal weave and multi-layer plain weave E-glass composites. Part 2: Comprehensive experimental results. Compos. Part A **40**(8), 1144–1157 (2009). https://doi.org/10.1016/j.compositesa.2009.04.032
2. Chou, S., Chen, H.C., Chen, H.E.: Effect of weave structure on mechanical fracture behavior of three-dimensional carbon fiber fabric reinforced epoxy resin composites. Compos. Sci. Technol. **45**(1), 23–35 (1992). https://doi.org/10.1016/0266-3538(92)90119-n
3. Stig, F., Hallström, S.: Assessment of the mechanical properties of a new 3D woven fibre composite material. Compos. Sci. Technol. **69**(11–12), 1686–1692 (2009). https://doi.org/10.1016/j.compscitech.2008.04.047
4. Cox, B.N., Dadkhah, M.S., Morris, W.L., Flintoff, J.G.: Failure mechanisms of 3D woven composites in tension, compression, and bending. Acta. Metall. Mater. **42**(12), 3967–3984 (1994). https://doi.org/10.1016/0956-7151(94)90174-0
5. Lomov, S.V., Bogdanovich, A.E., Ivanov, D.S., Mungalov, D., Karahan, M., Verpoest, I.: A comparative study of tensile properties of non-crimp 3D orthogonal weave and multi-layer plain weave E-glass composites. Part 1: Materials, methods and principal results. Compos. Part A **40**(8), 1134–1143. (2009). https://doi.org/10.1016/j.compositesa.2009.03.012
6. Kuo, W.S., Ko, T.H.: Compressive damage in 3-axis orthogonal fabric composites. Compos. A **31**(10), 1091–1105 (2000). https://doi.org/10.1016/s1359-835x(00)00066-x
7. Cox, B.N., Dadkhah, M.S., Inman, R.V., Morris, W.L., Zupon, J.: Mechanisms of compressive failure in 3D composites. Acta Metall. Mater. **40**(12), 3285–3298 (1992). https://doi.org/10.1016/0956-7151(92)90042-D
8. Mouritz, A.P., Cox, B.N.: A mechanistic interpretation of the comparative in-plane mechanical properties of 3D woven, stitched and pinned composites. Compos. A **41**(6), 709–728 (2010). https://doi.org/10.1016/j.compositesa.2010.02.001
9. Ansar, M., Xinwei, W., Chouwei, Z.: Modeling strategies of 3D woven composites: a review. Compos. Struct. **93**(8), 1947–1963 (2011). https://doi.org/10.1016/j.compstruct.2011.03.010
10. Kregers, A.F., Teters, G.A.: Determination of the elastoplastic properties of spatially reinforced composites by the averaging method. Mech. Compos. Mater. **17**(1), 25–31 (1981). https://doi.org/10.1007/BF00604878
11. Cox, B.N., Dadkhah, M.S.: The macroscopic elasticity of 3D woven composites. J. Compos. Mater. **29**(6), 785–819 (1995). https://doi.org/10.1177/002199839502900606
12. Tan, P., Tong, L., Steven, G.P.: Behavior of 3D orthogonal woven CFRP composites. Part II. FEA and analytical modeling approaches. Compos. Part A **31**(3), 273–281. (2000). https://doi.org/10.1016/s1359-835x(99)00071-8
13. Drach, A., Drach, B., Tsukrov, I.: Processing of fiber architecture data for finite element modeling of 3D woven composites. Adv. Eng. Software **72**, 18–27 (2014). https://doi.org/10.1016/j.advengsoft.2013.06.006
14. Muñoz, R., Martínez, V., Sket, F., González, C., LLorca, J.: Mechanical behavior and failure micromechanisms of hybrid 3D woven composites in tension. Compos. A **59**, 93–104 (2014). https://doi.org/10.1016/j.compositesa.2014.01.003

15. Pankow, M., Yen, C.F., Rudolph, M., Justusson, B., Zhang, D., Waas, A.: Experimental investigation on the deformation response of hybrid 3D woven composites. In: 53rd AIAA/ASME/ASCE/AHS/ASC Structures, Structural Dynamics and Materials Conference 20th AIAA/ASME/AHS Adaptive Structures Conference 14th AIAA, pp. 1572. April 2012. https://doi.org/10.2514/6.2012-1572
16. Chamis, C.C.: Simplified composite micromechanics equations for hygral, thermal, and mechanical properties. SAMPE Quar **15**(3), 14–23 (1984)
17. Daniel, I.M., Ishai, O.: Engineering Mechanics of Composite Materials. Oxford University Press, Oxford (1994)
18. Lu, H., Guo, L., Liu, G., Zhang, L.: A progressive damage model for 3D woven composites under compression. Int. J. Damage Mech. (2018). https://doi.org/10.1177/1056789518793994
19. Kulkarni, S.V., Rosen, B.W., Rice, J.S.: An investigation of the compressive strength of Kevlar 49/epoxy composites. Composites **6**(5), 217–225 (1975). https://doi.org/10.1016/0010-4361(75)90417-6
20. Ji, S., Gu, Q., Xia, B.: Porosity dependence of mechanical properties of solid materials. J. Mater. Sci. **41**(6), 1757–1768 (2006). https://doi.org/10.1007/s10853-006-2871-9
21. Abaqus, V.: 6.14 Documentation, Dassault Systemes Simulia Corporation, p. 651 (2014)
22. Linde, P., Pleitner, J., de Boer, H., Carmone, C.: Modelling and simulation of fibre metal laminates. In: ABAQUS Users' Conference, pp. 421–439 (2004)

Experimental Study on the Stiffness Evolution and Residual Strength of a Pre-damaged Structural Component Made from SMC CFRP Material

Stefan Sieberer[1(✉)], Susanne Nonn[1,2], and Martin Schagerl[1,2]

[1] Institute of Structural Lightweight Design, Johannes Kepler University Linz, Altenberger Strasse 69, 4040 Linz, Austria
stefan.sieberer@jku.at

[2] Christian Doppler Laboratory for Structural Strength Control of Lightweight Constructions, Johannes Kepler University Linz, Altenberger Strasse 69, 4040 Linz, Austria

Abstract. The mechanical characterisation of pre-damaged structural components made from sheet moulding compound (SMC) carbon-fibre reinforced polymer (CFRP) is still an open research question. This study shows the stiffness evolution of an automotive suspension part after damage initiation for a cyclic fully-reversed loading sequence with incrementally increased displacement amplitude. The cyclic test is followed by a quasi-static tensile test to estimate the residual strength of the component. Visual assessment of a highly stressed area is performed via 3D-DIC for the static and dynamic tests. As a result from the load and displacement measurements, the decrease in structural stiffness for the fatigue test is obtained and described by a logarithmic law. The residual strength and stiffness in the quasi-static tensile test gives insight into how pre-damaged structural parts can be used for limited durations (e.g. limp home mode in a passenger vehicle after light impact).

Keywords: CFRP · SMC · Crack propagation · Stiffness degradation

1 Introduction

In the assessment of fractures in metals, formation of cracks and their propagation is mainly governed by high stress levels, surface roughness, defects, and the local microstructure (see e.g. [1,2]). For specimen testing of discontinuous carbon long-fibre reinforced polymers (CFRP), however, a mesostructure influence is reported as further driving force for crack initiation. The mesostructure is formed by chopped strands or tows of a characteristic length of about 10 to 50 mm [3–5]. Thus, the finished component, made from sheet moulding composite (SMC) or bulk moulding composite (BMC) is comprised of small patches of

M. A. Wahab (Ed.): DAMAS 2019, LNME, pp. 827–836, 2020.
https://doi.org/10.1007/978-981-13-8331-1_65

material with defined fibre orientation. It is well-studied, e.g. in [6–8], that stress concentrations form at the interface between these elements. During the fatigue life, cracks are likely to form at stress concentrations originating either from the part geometry or the mesostructure interfaces, dependent on local conditions. A number of studies on the topic on specimen level were reviewed in [9], but no clear indication of a dominant mechanisms for fatigue damage initiation in discontinuous CFRP exists. Furthermore, applying these findings to structural parts made from SMC or BMC has been shown for some automotive components [10, 11], but little is known about the propagation of cracks on component level in structures made from such materials.

Investigations into stiffness degradation during fatigue life of SMC material have been conducted by specimen testing on short- and long-fibre discontinuous reinforced polymers by a number of authors. In [6], Caprino reviewed stiffness degradation of discontinuous glass- and carbon-fibre specimens, and a modulus degradation following a power law was proposed. The test data, however, included tensile cyclic testing only, and the influence of creep on the modulus was stated, but could not be quantified. It was furthermore mentioned that carbon fibres show much lower stiffness degradation compared to glass fibres before fracture of the specimen. Launay *et al.* [12] have shown a similar stiffness reduction for specimen made from glass short-fibre reinforced polymers under tensile-tensile fatigue testing. In their study, the stiffness loss was related to inelastic energy density W_{in}, and an exponential law describing the stiffness loss in relation to W_{in} was proposed. The underlying mechanism for the degradation is microcracking [6, 13], and the degradation is reportedly more pronounced in tensile than in compressive loading conditions. In [13], a fully-reversed SMC-R test with glass-fibre reinforced material is reported, yielding continuous stiffness degradation until fracture of the specimen. The degradation appears to follow logarithmic laws in two regimes, up to 10,000 cycles with a larger gradient, and after that with a lower gradient. Whether test frequency dependent influences were an additional influence in these results was not discussed, however the frequency of 10 Hz could yield some additional degeneration from specimen dissipative heating. Despite these experiments on specimen level, for SMC composite structural components, the stiffness degradation after crack initiation is not yet studied.

In this contribution, an automotive structural suspension part from a material characterisation and damage initiation study was used for investigations into the stiffness evolution during crack propagation and a final strength evaluation. The findings yield insight into the capability of the structure to endure load application during the damage progression stage and the strength reserve in this regime.

The paper is organised as follows: the test set-up of the component is described in Sect. 2 and the results for stiffness degradation and residual strength are presented in Sect. 3. A discussion is given in Sect. 4 before the main findings are summarised in Sect. 5.

2 Test Set-Up

Figure 1 shows the experimental set-up. The component is fixed to the ground at hinge A, where all degrees-of-freedom are restricted. Hinge B is guided in the horizontal plane, and is free to rotate about the hinge axis in the plane. The vertical actuation at B is applied via a Zwick-Roell BPS-LH0025 hydraulic cylinder controlled by a CATS Control Cube system, yielding bending-torsional loading in the component. For the cyclic tests, a displacement-controlled set-up was chosen. From preliminary Finite Element analysis, the highest stressed areas were identified and previous fatigue test showed that formation of visible cracks initiated in these areas. For monitoring purposes, a Correlated Solutions digital image correlation (DIC) system was used to measure strains and crack formation on one side of the structure, and an industrial digital camera was positioned to monitor the opposite side as only one DIC system was available.

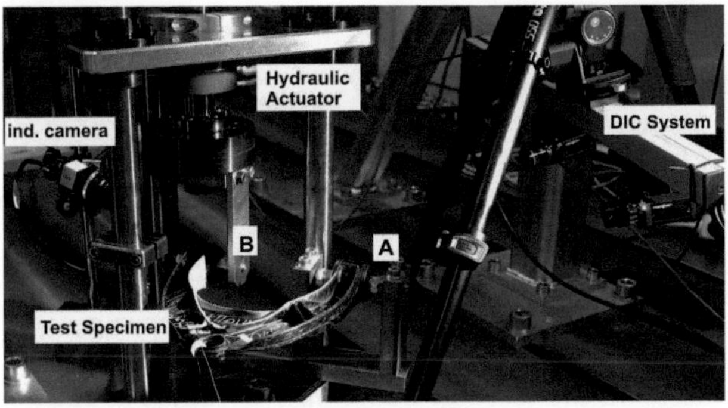

Fig. 1. Component with test equipment and measurement devices on the hydraulic test rig.

Table 1 shows the test schedule, starting with Phase 1, where a standard, fully-reversed fatigue test is performed. Phase 1 was performed to obtain a component at the end of its fatigue life, and is not focussed on in this publication. Following this, in Phase 2 a stepwise increase in the amplitude of the fully reversed cyclic testing was realised with approximately 500 cycles at each displacement level. The amplitude was increased until a set limit of 35 mm was reached, for which a larger number of cycles was recorded. Stiffness and damage evolution was monitored for each step. Finally in Phase 3, the residual stiffness and strength of the component was obtained by means of a quasi-static test with stepwise increase in positive displacement. At each displacement step, the structure was allowed to relax and the cylinder force at constant deflection was monitored.

Damage increase was visually assessed with both available inspection systems, as well as by monitoring the stiffness of the component via a load cell mounted between the top alignment guide and the component. No influence

Table 1. Overview of load conditions for the tested component and the reference component.

Phase	Test mode	Max. amplitude in mm	Normalised load
Phase 1	Fully-reversed cyclic	15	0.09
Phase 2	Fully-reversed cyclic	35	0.32
Phase 3	Quasi-static tension	64	0.37
Reference	Quasi-static tension	100	1.00

of a horizontal force component on the vertical force measurement is expected because of internal compensation in the cell. For calculation of stiffness of the structure, the load cell and cylinder position transducer readings were used. This gives valid results for the large amplitude cyclic and static displacements as parasitic influence of the hydraulic stiffness can be neglected.

3 Experiment Results

In this section, the stiffness evolution in Phase 2 is presented first, followed by the residual strength results from a quasi-static tensile test in Phase 3.

3.1 Stiffness and Crack Evolution - Phase 2

The stiffness in direction of the actuation can be calculated from the force and position transducers on the test rig. Using data obtained at the peak and trough only, and therefore assuming linear elastic behaviour, the stiffness during the cyclic test can be obtained. All reported stiffness values are normalised to the stiffness at the start of Phase 2.

Figure 2 shows the stepwise increase in cyclic amplitude over the number of cycles on one vertical axis, and the evolution of the stiffness over the number of cycles on a secondary vertical axis for Phase 2 cyclic testing. It is noteworthy that with each amplitude step, the stiffness abruptly increases before gradually falling. For the last stage with a larger number of cycles at the final amplitude, the stiffness drops from 107% to about 93% of the start value.

Describing the stiffness reduction with a logarithmic law $E = a + k \log N$ where E is the stiffness, N is the cycle count, and a and k are fitting parameters, the regression lines as shown in Fig. 2 are obtained. For all these lines, the gradient $k = 0.240$ was used.

From the rapid deterioration of stiffness in Phase 2, a large damage progression rate is expected. Figure 3 shows the area-of-interest for crack initiation and progression covered by the DIC. Area 1 and Area 2 are monitored for crack growth, whilst the control area shows a low-stress, crack-free area. In the graph, the correlation error, an indirect measure of potential crack initiation and

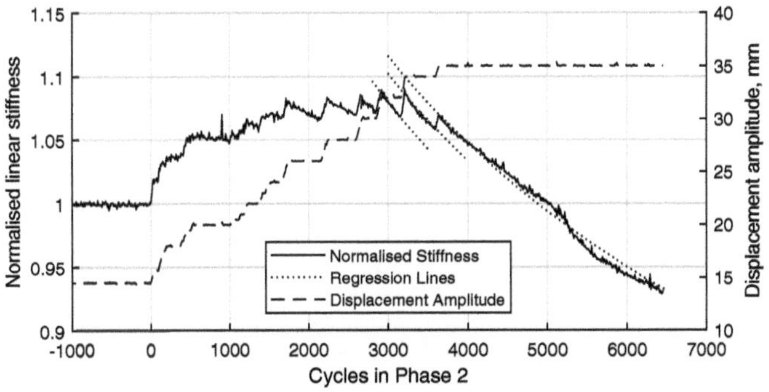

Fig. 2. Stiffness evolution on increase of the amplitude of the fully reversed cyclic testing.

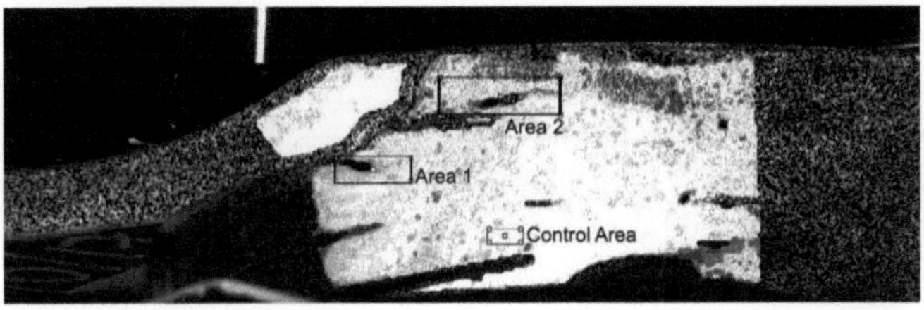

Fig. 3. Correlation error over the measurement field on the component. Bright areas indicate low correlation error, black indicates areas of highest correlation error. Areas 1 and 2 are crack initiation areas, and the control area is used to normalise the correlation error.

progression, is shown. Cracks are defined as the complete loss of correlation in the DIC measurements, indicating that the surface cohesion has been severely disturbed.

The crack initiation and growth in Phase 2 is visualised in Fig. 4. The crack length is measured at each visually observed increase of the crack length.

For Area 1, a visible crack is formed on the first increase in cyclic amplitude, and continuous crack growth is observed until approx. 4500 cycles. The sharp rise from 3500 to 4500 cycles can be explained by additional cracks forming near the primary crack and merging to one crack. In Area 2, a small crack was already formed during Phase 1, and is growing continuously with the increasing cyclic amplitude. Only smaller growth is present in the final cycling at maximum amplitude. However, in this last stage of Phase 2, further cracks are starting to form and the component becomes visibly damaged over large parts of the structure.

The crack formulation on the opposite side shows similar crack growth behaviour, but the crack length cannot be easily measured as to the limitations of using a single low resolution camera to monitor this side only.

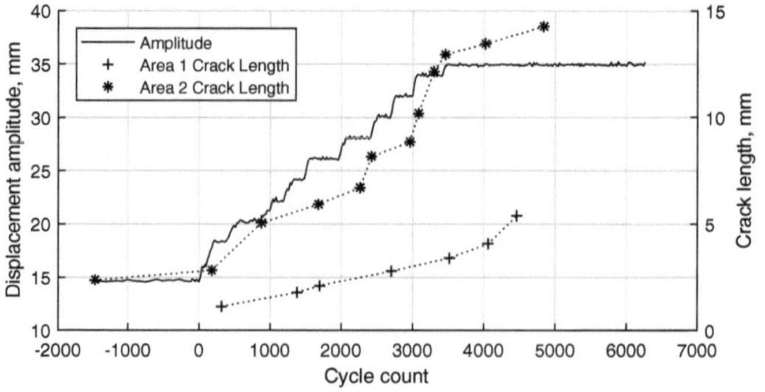

Fig. 4. Crack length and displacement amplitude over the Phase 2 cyclic testing of the component.

3.2 Residual Strength

Following the cyclic testing, a static strength test was performed on the pre-damaged component. The displacement amplitude was increased in steps, leaving a time period of at least 30 s between steps to allow for saving of data gained from additional transducers. During this period, the tensile force behaviour was observed.

Figure 5 shows the reference figure for the correlation error, taken at 30 mm displacement amplitude, and a correlation plot at 64 mm displacement. The large correlation error indicated as dark areas in the area-of-interest and the increased crack length can be seen.

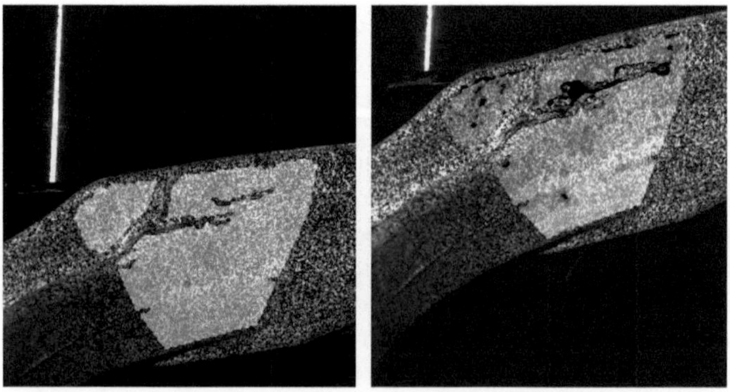

(a) 30mm (b) 64mm

Fig. 5. DIC images at 30 mm (reference) and 64 mm displacement. The increase in correlation error (dark areas) and crack length is clearly visible.

Figure 6 shows the displacement and normalised force over time. In addition to the visible damage inspection by DIC and camera, the force reduction can indicate the progression of damage. Also, the residual stiffness of the component can be estimated. For comparison, the experimental results from a tensile test on a pristine component, and a linear load-displacement curve obtained from the secant modulus at the end of the constant-load Phase 1 cyclic testing are added to the figure.

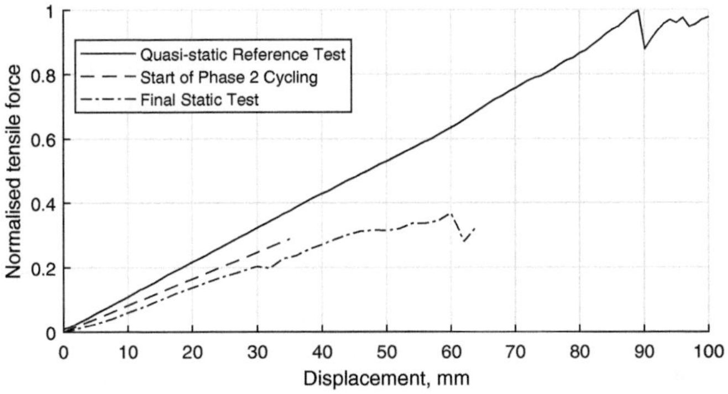

Fig. 6. Tensile test results of a pristine component, stiffness value (linear approximation) at the start of the increased load cycling, and residual tensile test results.

Despite the visible damage accumulated over the cyclic testing, the residual strength was in the region of 37% of the strength of the reference component, obtained at 68% of the displacement. The tensile testing of the damaged component was stopped after a tensile force peak was reached. However, the component did not exhibit full-fracture and, after unloading, elastic forces led to crack closing and little residual plastic deformation was observed.

4 Discussion

4.1 Stiffness Degradation

Because only one specimen in the damaged state was tested, no general hypothesis on post-damage stiffness degradation of carbon-fibre SMC material can be made from the observations in this study. A more in-depth investigation into the stiffness degradation behaviour of components made from these discontinuous long-fibre reinforced polymers and how to relate this to results from specimen testing is ongoing, and could also lead to better insight to crack propagation behaviour.

Figure 2 shows that in Phase 2, at each increase in displacement amplitude, the calculated stiffness surges. Such a phenomenon could be originating from a diminished influence of existing cracks on the observed dynamic stiffness in the fully-reversed cycle. As long as the cracks are open, the current stiffness of the component is small compared to the initial stiffness, and rises to a value in the range of the initial stiffness after the cracks located at zones in compression are closed. The stiffness calculated from maximum and minimum displacement and load values is thus a bilinear function and can be described by $E_{\mathrm{dam}} = bE_{\mathrm{co}} + (1 - b)E_{\mathrm{init}}$ where E_{dam} is the stiffness of the damaged part, E_{co} and E_{init} are the stiffnesses for the open-crack and closed-crack ranges, respectively, and b is the fraction of open-crack to close-crack range. With each step-up in amplitude, the value b drops and the combined stiffness E_{dam} increases sharply.

It can be imagined that the crack depth could be estimated for a simple component under uni-axial loading conditions from the stiffness change on increase of the cyclic displacement amplitude. For the component under investigation, this was not investigated as to the multitude of crack locations at this point (see e.g. Fig. 7(d)).

4.2 Crack Propagation

Very early into Phase 2 testing, a crack is recorded in Area 1, which is the area of highest stress in the component. However, the first crack on the DIC side is initiated in Area 2, where significantly lower stresses are obtained from Finite Element results. This crack could originate from the reported mesostructure influence, i.e. stress concentrations at fibre bundle ends or a unfavourable fibre alignment yielding matrix-dominated material properties for this area.

Figure 7 shows that opposite of the DIC-monitored side, a crack forms already in Phase 1 and in the equivalent of Area 1. The crack is initially formed through the first 8% of the Phase 1 test duration, and does not grow significantly over the rest of Phase 1. This indicates some stress re-distribution after the crack has been formed. Over Phase 2, the initial crack grows substantially and merges with further cracks that appeared in the area. Note that the white surface coating applied on the non-DIC side in Fig. 7(d) emphasises the crack expansion.

For the DIC-monitored side, there is crack growth during Phase 3 (see Fig. 5), however, no further crack growth is observed on the non-DIC side at maximum displacement of the quasi-static test. After removing the tensile load, all crack close almost fully. In the regions where no surface coating was applied to the component, visual detection of cracks is very difficult. When inspecting such components in an industrial or workshop environment, the detection method must be chosen with consideration in order to avoid erroneous inspection results.

4.3 Residual Strength and Implications for Post-damage Use

From the residual strength test, the reserve for use after damage-initiation can be estimated. For the component under consideration, the cyclic testing was performed at much lower load than the ultimate static load. This is because in the highest-stressed regions, the onset of fatigue damage occurs at load levels far below the ultimate load. This strength reserve in the pristine component yields the potential of the damaged component to fulfil a large number of cycles at loads higher than the load for 'safe fatigue life' of the component. In the final stage of Phase 2 testing, the displacement amplitude was higher by a factor of

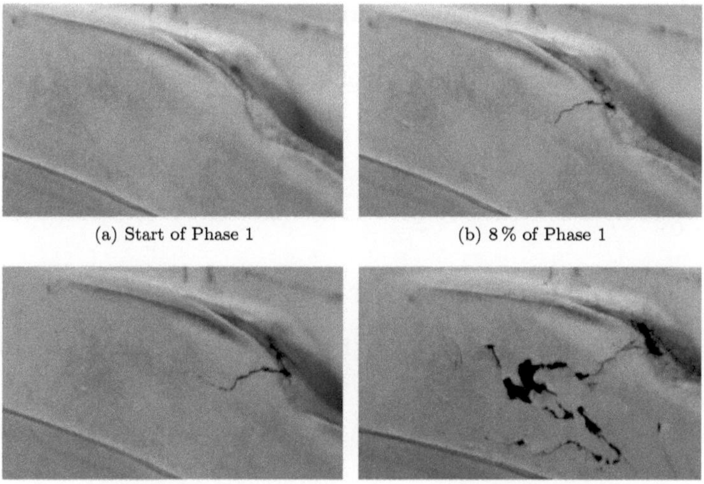

(a) Start of Phase 1
(b) 8 % of Phase 1
(c) End of Phase 1
(d) End of Phase 2

Fig. 7. Crack formation and propagation on the non-DIC side.

approx. 2.3 compared to Phase 1 loading, and a significant number of cycles was endured by the component. The following tensile test indicated a safety factor of Phase 1 load against the residual strength of approx. $S_f = 3$. This indicates that despite the damage (e.g. from an impact or light road accident), there is sufficient fatigue life and residual strength in discontinuous long-fibre SMC CFRP components to avoid sudden collapse in a pre-damaged component.

5 Conclusions

In this study, the stiffness evolution and residual strength of one pre-damaged component made from SMC CFRP were investigated. The main findings are:

- Small-scale damage is observed at the end of the initial fully-reversed cyclic test, and the component can be considered at the end of its fatigue life (i.e. cracks are forming).
- On increase of the cyclic displacement amplitude, the stiffness of the component initially rises sharply before gradually declining. This holds true for all amplitude steps.
- The gradual reduction closely follows a logarithmic law.
- There is large-scale crack propagation in the increased-amplitude cyclic test, leading to a heavily damaged component.
- A final quasi-static tensile test yields the residual strength of the component. There is still a sufficient safety factor between the load in initial cycling and the residual strength.

It can be concluded from the component behaviour, that despite heavy pre-damage and above-critical cyclic loading, a sudden collapse under sub-critical, normal operational loads, is unlikely and post-damage operation over a limited amount of time (limp-home mode) is potentially possible. However, identification of the damage can be difficult because of residual elastic forces closing existing cracks, and suitable inspection methods are necessary to avoid non-detection of damage in such components.

References

1. Radaj, D., Vormwald, M.: Ermüdungsfestigkeit: Grundlagen für Ingenieure. Springer, Berlin (2007)
2. Forschungskuratorium Maschinenbau: FKM Richtlinie Rechnerischer Festigkeitsnachweis für Maschinenbauteile, 6 Auflage. FKM (2012)
3. Feraboli, P., Peitso, E., Cleveland, T., Stickler, P.B.: Modulus measurement for prepreg-based discontinuous carbon fiber/epoxy systems. J. Compos. Mater. **43**, 1947–1965 (2011)
4. Li, Y., Pimenta, S., Singgih, J., Ottenwelter, C., Nothdurfter, S., Schuffenhauer, K.: Understanding and modelling variability in modulus and strength of tow based discontinuous composites. In: 21st International Conference on Composite Materials, ICCM 21, Xi'an, China, 20–25 August 2017, p. 12 (2017)
5. Wan, Y., Takahashi, J.: Multi scale internal geometry analysis and mechanical modelling of randomly oriented strands. In: 21st International Conference on Composite Materials, ICCM 21, Xi'an, China, 20–25 August 2017, p. 12 (2017)
6. Harris, B. (ed.): Fatigue in Composites. Woodhead Publishing Ltd., Cambridge (2003)
7. Qian, C., Harper, L.T., Turner, T.A., Warrior, N.A.: Notched behaviour of discontinuous carbon fibre composites: comparison with quasi-isotropic non-crimp fabric. Compos. Part A **42**(3), 293–302 (2011)
8. Feraboli, P., Peitso, E., Cleveland, T., Stickler, P.B., Halpin, J.C.: Notched behavior of prepreg-based discontinuous carbon fiber/epoxy systems. Compos. Part A **40**, 289–299 (2009)
9. Mortazavian, S., Fatemi, A.: Fatigue behavior and modeling of short fiber reinforced polymer composites: a literature review. Int. J. Fatigue **70**, 297–321 (2015)
10. Zago, A., Springer, G.S.: Fatigue lives of short fiber reinforced thermoplastics parts. J. Reinf. Plast. Compos. **20**, 606–620 (2001)
11. Zago, A., Springer, G.S.: Constant amplitude fatigue of short glass and carbon fiber reinforced thermoplastics. J. Reinf. Plast. Compos. **20**, 564–595 (2001)
12. Launay, A., Marco, Y., Maitournam, M.H., Raoult, I., Szmytka, F.: Cyclic behavior of short glass fiber reinforced polyamide for fatigue life prediction of automotive components. Proc. Eng. **2**, 901–910 (2010)
13. Pritchard, G. (ed.): Reinforced Plastics Durability. Woodhead Publishing Ltd., Cambridge (1999)

Delamination Detection via Reconstructed Frequency Response Function of Composite Structures

A. Alsaadi(✉), Yu Shi, and Yu Jia

The University of Chester, Chester CH2 4NU, England, UK
a.alsaadi@chester.ac.uk
https://www1.chester.ac.uk/group-user/5790/44

Abstract. Online damage detection technologies could reduce the weight of structures by allowing the use of less conservative margins of safety. They are also associated with high economical benefits by implementing a condition-based maintenance system. This paper presented a damage detection and location technique based on the dynamic response of glass fibre composite laminate structures (frequency response function). Glass fibre composite laminate plates of $200 \times 200 \times 2.64$ mm, which had a predefined delamination, were excited using stationary random vibration waves of 500 Hz band-limited noise input at ≈ 1.5 g. The response of the structure was captured via Micro-ElectroMechanical System (MEMS) accelerometer to detect damage. The frequency response function requires data from damaged structures only, assuming that healthy structures are homogeneous and smooth. The frequency response of the composite structure was then reconstructed and fitted using the least-squares rational function method. Delamination as small as 20 mm was detected using global changes in the natural frequencies of the structure, the delamination was also located with greater degree of accuracy due to local changes of frequency response of the structure. It was concluded that environmental vibration waves (stationary random vibration waves) can be utilised to monitor damage and health of composite structures effectively.

Keywords: Frequency response function (FRF) · Structural health monitoring (SHM) · Structural integrity · Damage assessment

1 Introduction

Early detection of damage in a structural system is of interest to all industries, as such there has been an extensive research to adapt various damage detection techniques [1]. In general, damage can be detected using five closely related methods that include Condition monitoring (CM), statistical process control

M. A. Wahab (Ed.): DAMAS 2019, LNME, pp. 837–843, 2020.
https://doi.org/10.1007/978-981-13-8331-1_66

(SPC), damage prognosis (DP), and structural health monitoring (SHM) [2–5]. CM can be used to monitor a certain condition in machinery, especially rotating machinery. SPC is used to monitor a process that may lead to a structural failure. DP is used to assess damage in structures and predict the remaining life of structures. SHM systems offer online health monitoring of structures, this is particularly important as it has significant advantages for both manufacturers and business owners [6]. Adapting SHM in practice changes a maintenance programme from time-based maintenance systems to condition-based maintenance systems. In order for an SHM system to be effective, the system should monitor natural global properties, such as dynamic or electric properties. There are many SHM systems are available, one of which is vibration-based SHM system [7,8]. This system adopts the dynamic properties of a structure as a tool to detect and locate damage. The fundamental premise of vibration-based SHM systems is that damage causes a reduction in structural stiffness, mass, or damping properties of a system, which, in turn alter dynamic properties of the system.

Vibration-based SHM systems can be divided into two types, modal-based and non-modal based damage identification systems [9]. In the modal-based SHM system, such modal-frequency approaches and modal-shape approaches, the damage is detected using direct comparison of modal parameters, such as mode shapes, strain mode shapes, and frequency curves or wave parameters, such as amplitude and/or phase angle. On the other hand, in the non-modal based SHM system, such as frequency response function, damage is detected using direct comparison of frequency response function and its derivatives. FRF technique relies on the fact that damage will degrade the stiffness of the structure, hence the FRF curve changes [10]. FRF technique is also less prone to error as it extracts the data from the response of the structure directly.

Few types of signals can be used to excite a structure, such as stationary or non-stationary signals [11]. In spite of the fact, that all signals in nature are non-stationary because their statistical characteristics change with time, however, from an engineering prospect this change is very slow, therefore, it makes a practical sense to assume signals are stationary [12]. Stationary vibration signals can be divide into two types that are deterministic or random vibration [13]. Random vibration are met more often in nature rather than deterministic vibration. It is important to mention that it is much more difficult to interrogate structural response data collected from a structure was excited using random vibration signals than deterministic vibration signals.

This paper presents a vibration-based structural health monitoring system to detect and locate damage based on the response of a glass fibre reinforced polymer (GFRP) laminate structure to a stationary random vibration signal.

2 Methodology

2.1 Experimental Set up

The GFRP laminate panels were manufactured using twill weave 2×2 VTC 401 with aerial weight of $295\,\mathrm{gm}^{-2}$ (SHD Composites, UK). The pre-preg was cut into sheets and hand laid onto a steel plate that is covered by PTFE sheet to prevent sticking. A predefined delamination in the form of PTFE strips of various lengths 20 mm (S1), and 50 mm (S2) were inserted between plies during the lamination as shown in Fig. 1. The pre-preg stack was enveloped in a vacuum bag and placed in an oven to cure according to the manufacturer recommendations, the curing cycle shown in Fig. 2. Once the curing cycle ends, the panels were taken of the oven and the panels were trimmed and a hole of 5 mm was drilled in the middle of the panels to be used to install the panel on the electrodynamic shaker.

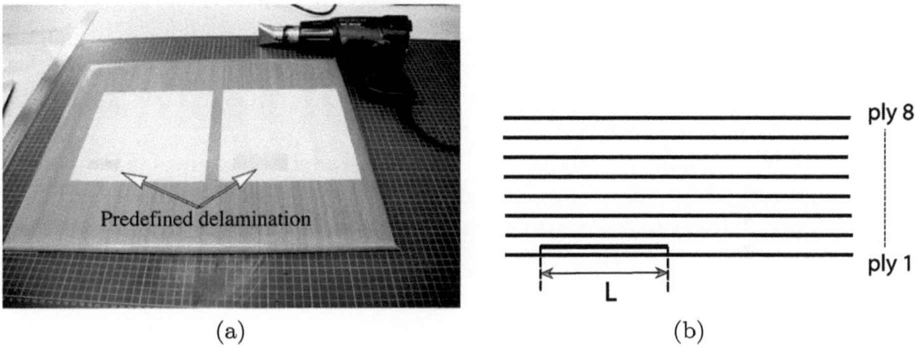

Fig. 1. (a) shows the composite lamination process, and (b) the location of predefined lamination through-thickness of a laminate panel.

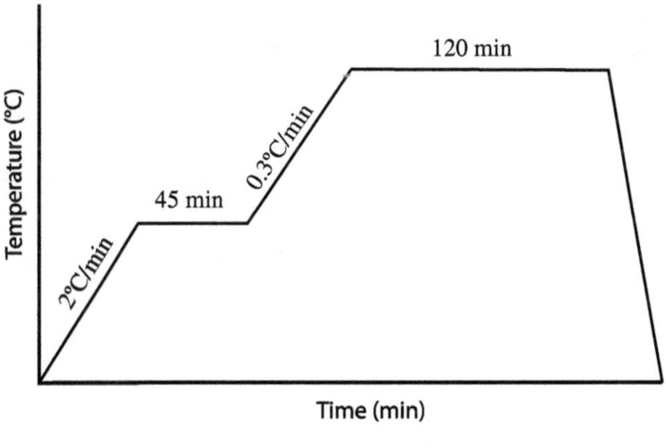

Fig. 2. Curing cycle of VTC 401 glass fibre prepreg.

2.2 Excitation Signal

The panels were excited using electrodynamic shaker (LDS 406, *Brüel & Kajer*) as shown in Fig. 3. A stationary vibration signal was used to excite the GFRP panels as shown in Fig. 4 and the response of the structure was collected using an accelerometer (ADXL325Z, Analog Devices, USA). The signal was sampled according to Nyquist theorem, however, 5 times the highest frequency was used rather than 2 times, therefore, the data was collected at a sampling frequency of 2.4 kHz. The data was collected using oscilloscope DSOX-2004A (Agilent Technologies, USA), the oscilloscope has a record length of 2000 points. To increase signal to noise ratio (SNR) 20 set of time - domain signals were acquired and averaged using the global averaging technique. To ensure the high quality signal further signal processing techniques were undertaken in MATLAB using different filters, such as normalisation, trend remover, and outliers. The signals were then smoothed using Savizky - Golary filter.

In order to estimate the frequency response function of the GFRP panels Welch's method was used via predefined function *"modalfrf"* in MATLAB [14]. The processed signals both excitation and response signals were imported into the function along with a time window of 1000. This time window was used so as to reduce the sidelobes by minimising their rate of roll-offs.

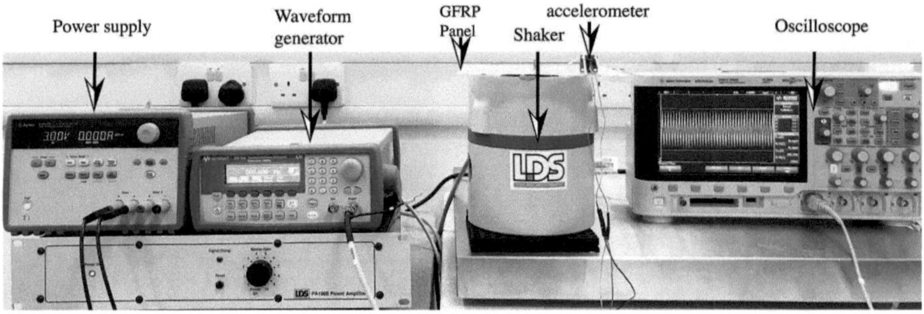

Fig. 3. Experimental set up to measure the response of GFRP laminate panels.

3 Results and Discussions

Figure 5 shows the response of damaged GFRP panel (S1) in terms of the time-domain. The dissimilarity between excitation acceleration in Fig. 4 and response acceleration in Fig. 5 can be attributed to two main factors that are the presence of damage as well as the response of GFRP laminate panels. To investigate this issue further, Welch's method was used to estimate the frequency response function as show in Fig. 6. It can be seen in the Fig. 6 that the damage changes response of the structure. Ozdemir et al. [15], proposes a fitting technique, least-square rational function (LSRF) fitting method was used to reconstruct the

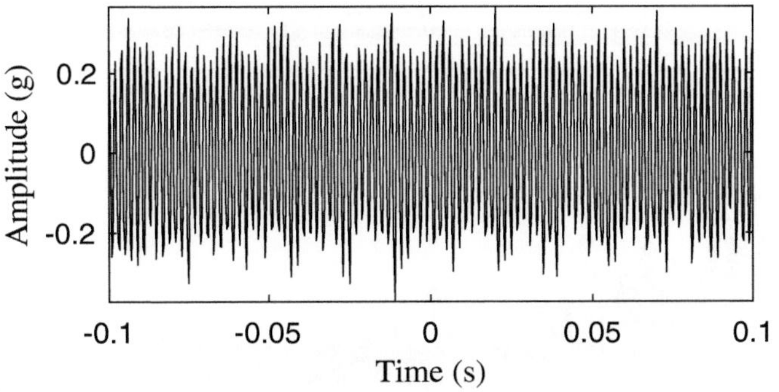

Fig. 4. Excitation signal, signal to noise ratio is 12.57 dB.

relationship between the frequency and the dynamic flexibility of the GFRP laminate panels as shown in Fig. 7. LSRF can be used to fit nonuniform frequency data.

To distinguish between peaks the phase was plotted using a predefined function *"modalfit"* in MATLAB. It can be seen in Fig. 7 that there are few peaks at the frequency band between 0 to 80 Hz, those are attributed to the initial structural dynamics as a response to the initial boundary conditions. Once the structure has been settled after a short period of time, the structure began to respond to excitation signal in a more systematic way as can be seen in Fig. 7 at frequency band between 80 to 800 Hz. It can be seen in the frequency response function in Fig. 7 that damage caused a frequency shit as well as changes in other modal parameters, such as amplitudes. In ongoing and future work, the authors will map the correlation of the frequency changes to delimitation characteristics. Furthermore, nonlinear profiles will be used alongside FRF to reveal nonlinear structural behaviour of the composite laminates as delimitation develops.

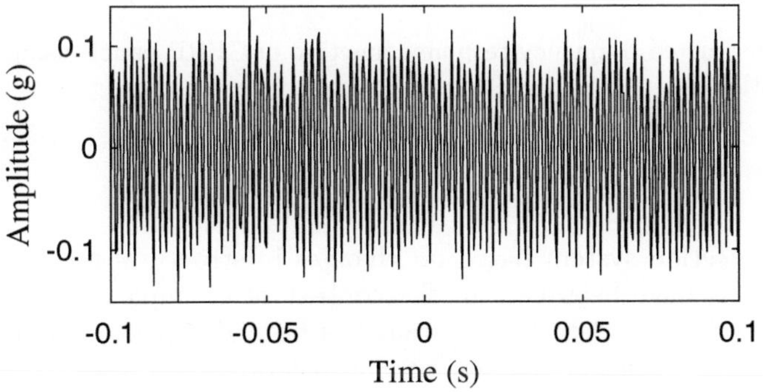

Fig. 5. Response signal of damaged GFRP panel (S1), signal to noise ratio is 9.998 dB.

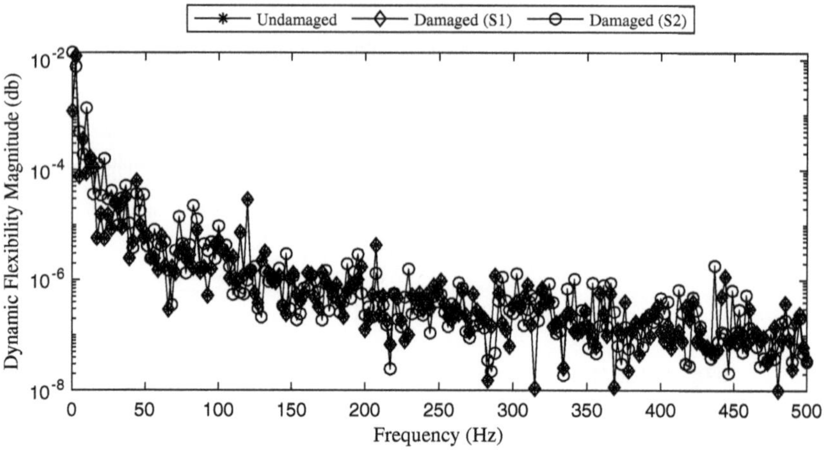

Fig. 6. Frequency response function of GFRP panels using Welch's method.

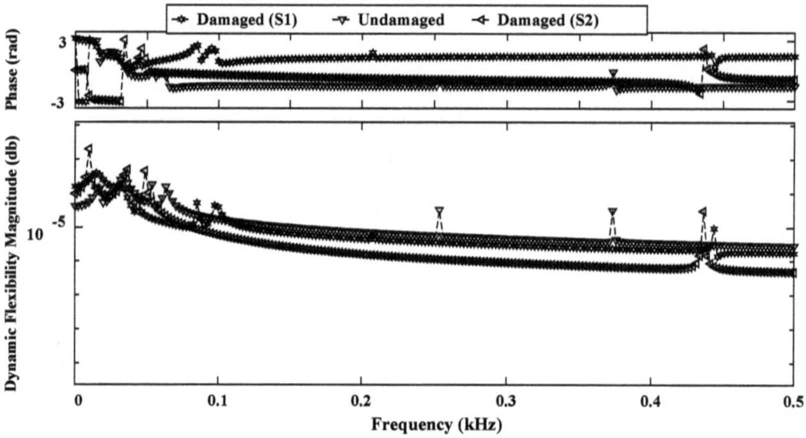

Fig. 7. Reconstructed frequency response function of GFRP panels using least square rational function.

4 Conclusion

A damage detection system based on changes in structural response of GFRP laminate panels have been cast in the context of a simple frequency response function. The damage detection system has been designed, manufactured and customised so as to detect damage in woven fabric glass fibre composite laminate panels. It has been found that small damage as small as 20 mm can cause changes in frequency response of the structure and change the modal parameters consequently.

References

1. Balageas, D., Fritzen, C.P., Güemes, A. (eds.): Structural Health Monitoring, vol. 90. Wiley, Hoboken (2010)
2. Worden, K., Farrar, C.R., Manson, G., Park, G.: The fundamental axioms of structural health monitoring. Proc. R. Soc. A: Math. Phys. Eng. Sci. **463**(2082), 1639–1664 (2007)
3. Liu, W.Y., Tang, B.P., Han, J.G., Lu, X.N., Hu, N.N., He, Z.Z.: The structure healthy condition monitoring and fault diagnosis methods in wind turbines: a review. Renew. Sustain. Energy Rev. **44**, 466–472 (2015)
4. Farrar, C.R., Lieven, N.A.: Damage prognosis: the future of structural health monitoring. Philos. Trans. R. Soc. A: Math. Phys. Eng. Sci. **365**(1851), 623–632 (2006)
5. Sohn, H., Czarnecki, J.A., Farrar, C.R.: Structural health monitoring using statistical process control. J. Struct. Eng. **126**(11), 1356–1363 (2000)
6. Chang, F.K.: Structural health monitoring: a summary report on the first international workshop on structural health monitoring. In: Proceedings of the 2nd International Workshop on Structural Health Monitoring, 18–20 September 1997. Stanford University, Stanford, California, 1999 September
7. Niezrecki, C.: Structural health monitoring & damage detection, vol. 7
8. Doebling, S.W., Farrar, C.R., Prime, M.B., Shevitz, D.W.: Damage identification and health monitoring of structural and mechanical systems from changes in their vibration characteristics: a literature review (No. LA-13070-MS). Los Alamos National Lab., NM, USA (1996)
9. Alamdari, M.M., Li, J., Samali, B.: FRF-based damage localization method with noise suppression approach. J. Sound Vib. **333**(14), 3305–3320 (2014)
10. Peng, Z.K., Lang, Z.Q., Billings, S.A.: Crack detection using nonlinear output frequency response functions. J. Sound Vib. **301**(3–5), 777–788 (2007)
11. Broch, J.T.: Mechanical vibration and shock measurements (1980)
12. Trampe Broch, J.: Mechanical Vibration and Shock Measurements, 2nd edn. Bruel & Kjaer, Naerum (1984)
13. Rao, S.S., Yap, F.F.: Mechanical Vibrations, vol. 4. Prentice Hall, Upper Saddle River (2011)
14. Welch, P.: The use of fast fourier transform for the estimation of power spectra: a method based on time averaging over short, modified periodograms. IEEE Trans. Audio Electroacoust. **15**(2), 70–73 (1967)
15. Ozdemir, A., Gumussoy, S.: Transfer function estimation in system identification toolbox via vector fitting. IFAC-PapersOnLine **50**(1), 6232–6237 (2017)

Free Vibration of Angle-Ply Laminated Micro-plates Using Isogeometric Analysis and Modified Couple Stress Theory

Cuong-Le Thanh[1,2], Samir Khatir[1], and M. Abdel Wahab[1(✉)]

[1] Department of Electrical Energy, Metals, Mechanical Constructions and Systems, Faculty of Engineering and Architecture, Ghent University, Ghent, Belgium
{Le.ThanhCuong, Magd.AbdelWahab}@UGent.be
[2] Faculty of Civil Engineering and Electricity, Open University, Ho Chi Minh City, Vietnam

Abstract. Based on new modified couple stress and isogeometric analysis, this paper presents a non-classical Reissner-Mindlin plate theory model for free vibration of angle-ply laminate micro-plate. This study extends the new modified couple stress for complicated geometrical structures with internal cutouts. The governing equations for size-dependent free vibration of angle-ply laminate micro-late obtained from the Galerkin weak form are solved by using isogeometric analysis. Various numerical examples are examined to verify the convergence and the accuracy of proposed model. In addition, the numerical results show the influences of material length scale parameter, fiber orientation and BCs of micro-plates. Only one material length scale parameter is investigated to predict the size effects. By increasing the material length scale parameter, the size-dependent behavior leads to an increase in non-dimensional frequencies and critical buckling loads of angle-ply laminate micro-plates.

Keywords: IGA · Size-dependent · Free vibration

1 Introduction

Based on establishing a new symmetric couple stress moment tensor and asymmetric couple stress curvature tensor for the size effect constitutive relationships for anisotropic materials, MCST was first extended by Chen et al. [5]. It was used to study the size effect analysis of composite laminated micro-plate under the name of new modified couple stress theory (NMCST). Moreover, Chen and Li [6] proposed a general NMCST with three material length scale parameters for anisotropic materials. This model can be applied to laminated micro-plate by using two material length scale parameters, and it can be indistinguishable from MCST for isotropic materials. Chen et al. [7] studied static bending of composite laminated micro-plates with various BCs using three-node triangular finite element method. Yang and Chen [8] developed Reddy shear deformation field for capturing the size effects of composite laminated micro-plates. A global-local theory based on composite laminated Reddy micro-plate was presented by Chen and Wang [9] to analyze the static bending of cross-ply simply supported

M. A. Wahab (Ed.): DAMAS 2019, LNME, pp. 844–852, 2020.
https://doi.org/10.1007/978-981-13-8331-1_67

laminated micro-plates. Yang and He [10] presented a size dependent model for composite laminated micro-plate that satisfy the continuity conditions of transverse shear stresses at interface layers.

The review of size effect models in this study indicates that the studies of small scale effects focus on analytical solutions that are limited to simply supported BCs, bending load and geometry. Therefore, some size effect finite element models have been developed such as the spline finite strip method, meshfree method, isogeometric analysis. Based on isogeometric analysis (IGA) proposed by Hughes and co-workers [11], some IGA numerical size effect models were established for bending, free vibration and buckling of micro- and nano-plates [12–15]. To the best of author's knowledge, the small scale free vibration analysis of composite laminated angle-ply micro-plates with new modified couple stress and isogeometric finite element method are still absolutely absent in the literature.

The rest of the paper is organized as follows. In Sect. 2, we present the basic equations in the modified couple-stress theory and the governing equations for free vibration problem. Numerical examples are provided in Sect. 3, and conclusions are drawn in Sect. 4.

2 Preliminaries

2.1 New Modified Couple Stress Theory

The new modified couple stress was established by Chen and Li [6], in which the couple stress moment tensor and the curvature tensor were asymmetric. By applying the NMCST, the virtual strain energy in the body V for the strain and curvature tensor can be written as:

$$\delta U = \int_V (\sigma_{ij}\delta\varepsilon_{ij} + m_{ij}\delta\chi_{ij})\mathrm{d}V \tag{1}$$

in which, the strain and asymmetric curvature tensor are:

$$\varepsilon_{ij} = \frac{1}{2}\left(\frac{\partial u_i}{\partial x_j} + \frac{\partial u_j}{\partial x_i}\right);\ \chi_{ij} = \frac{\partial \theta_i}{\partial x_j} \tag{2}$$

and σ_{ij}, m_{ij}, respectively, are the stress tensor and the couple stress moment tensor which are defined as follow [6]:

$$\sigma_{ij} = C_{ijkl}\varepsilon_{kl};\ m_{ij} = \left(\ell_i^2 G_i \chi_{ij} + \ell_j^2 G_j \chi_{ji}\right) \tag{3}$$

where C_{ijkl} are elasticity constants, σ_{ij} are the stress tensor, $\theta_i = 1/2curl(u_i)$ is the component of rotation vector. In addition, G is the shear modulus and ℓ is the material length scale parameter.

2.2 Plate Constitutive Relations

The displacement field of an arbitrary point in plate domain $V = \Omega x(-h/2, h/2)$ of Reinner-Mindlin plate theory is expressed as follows:

$$u_x(x,y,z) = u_0(x,y) + z\beta_x;\ u_y(x,y,z) = v_0(x,y) + z\beta_y;\ u_z(x,y) = w_0(x,y) \quad (4)$$

in which, u_0, v_0, w_0 are displacement components at the middle surface according to x, y, and z axes, respectively, while β_x, β_y are the rotation components in y-z and x-z surfaces.

Similar to the conventional finite element method, the NURBS basis function which was presented in previous work [12] is employed to build a finite approximation of displacement field as:

$$\mathrm{u}^h = \sum_I \mathbf{N}_I \mathbf{d}_I \quad (5)$$

where $\mathbf{d}_I = \{ u_{0I} \quad v_{0I} \quad w_{0I} \quad \beta_{xI} \quad \beta_{yI} \}^T$ is the vector of degree of freedom associated with the control point I.

Based on the discrete Galerkin weak form for free vibration analysis [12], the matrix form of global equilibrium equation for free vibration of angle-ply laminate microplates is established such as:

$$(\mathbf{K} - \omega^2 \mathbf{M})\mathbf{d} = 0 \quad (6)$$

where $\mathbf{K} = \mathbf{K}^u + \mathbf{K}^\theta$ and $\mathbf{M}$, respectively, are the global stiffness matrix and mass matrix. These matrices can be defined in a clear form as:

$$\mathbf{K}^u = \int_\Omega \left[(\mathbf{B}^u)^T \mathbf{D}_u^b \mathbf{B}^u + (\mathbf{B}^s)^T \mathbf{D}_u^s \mathbf{B}^s\right] \mathrm{d}\Omega;\ \mathbf{K}^\theta = \int_\Omega (\mathbf{B}^\chi)^T \mathbf{D}_c \mathbf{B}^\chi \mathrm{d}\Omega;\ \mathbf{M} = \int_\Omega \mathbf{R}^T \mathbf{I}_m \mathbf{R} \mathrm{d}\Omega \quad (7)$$

in which

$$\mathbf{B}_I^u = \sum_{I=1}^{m \times n} \{\mathbf{B}_I^m\ \mathbf{B}_I^{b1}\}^T;\ \mathbf{D}_u^b = \begin{bmatrix} \mathbf{A}^u & \mathbf{B}^u \\ \mathbf{B}^u & \mathbf{D}^u \end{bmatrix};\ \mathbf{D}_u^s = \sum_{k=1}^{N} \int_{-h_k/2}^{h_k/2} \begin{bmatrix} \bar{C}_{44}^k & \bar{C}_{45}^k \\ \bar{C}_{45}^k & \bar{C}_{55}^k \end{bmatrix} dz\ \mathrm{I}_m = \begin{bmatrix} \mathrm{I}_1 & \mathrm{I}_2 \\ \mathrm{I}_2 & \mathrm{I}_3 \end{bmatrix}$$

$$\mathbf{R}_I = \{\mathbf{R}_I^1\ \mathbf{R}_I^2\}^T;\ (\mathrm{I}_1, \mathrm{I}_2, \mathrm{I}_3) = \sum_{k=1}^{N} \int_{-h_k/2}^{h_k/2} \rho(1, z, z^2) dz \quad (8)$$

$$\left(\mathrm{A}^u,\mathrm{B}^u,\mathrm{D}^u\right)=\sum_{k=1}^{N}\int_{-h_k/2}^{h_k/2}\left(1,z,z^2\right)\begin{bmatrix}\bar{C}_{11}^k & \bar{C}_{12}^k & \bar{C}_{16}^k\\ \bar{C}_{12}^k & \bar{C}_{22}^k & \bar{C}_{26}^k\\ \bar{C}_{16}^k & \bar{C}_{26}^k & \bar{C}_{66}^k\end{bmatrix}dz;$$
$$\mathbf{D}_c=\sum_{k=1}^{N}\int_{-h_k/2}^{h_k/2}\begin{bmatrix}Q_{11}^{kc} & 0 & 0 & Q_{14}^{kc}\\ 0 & Q_{22}^{kc} & Q_{23}^{kc} & 0\\ Q_{31}^{kc} & Q_{32}^{kc} & Q_{33}^{kc} & Q_{34}^{kc}\\ Q_{41}^{kc} & Q_{42}^{kc} & Q_{43}^{kc} & Q_{44}^{kc}\end{bmatrix}dz \tag{9}$$

$$\mathbf{B}_I^m=\begin{bmatrix}N_{I,x} & 0 & 0 & 0 & 0\\ 0 & N_{I,y} & 0 & 0 & 0\\ N_{I,y} & N_{I,x} & 0 & 0 & 0\end{bmatrix};\ \mathbf{B}_I^{b1}=\begin{bmatrix}0 & 0 & 0 & N_{I,x} & 0\\ 0 & 0 & 0 & 0 & N_{I,y}\\ 0 & 0 & 0 & N_{I,y} & N_{I,x}\end{bmatrix}$$
$$\mathbf{B}_I^s=\begin{bmatrix}0 & 0 & N_{I,x} & N_I & 0\\ 0 & 0 & N_{I,y} & 0 & N_I\end{bmatrix};\mathbf{B}_I^{\chi}=\frac{1}{2}\begin{bmatrix}0 & 0 & N_{I,xy} & 0 & -N_{I,x}\\ 0 & 0 & -N_{I,xy} & N_{I,y} & 0\\ 0 & 0 & N_{I,yy} & 0 & -N_{I,y}\\ 0 & 0 & -N_{I,xx} & N_{I,x} & 0\end{bmatrix} \tag{10}$$

where the transform material constants $\bar{C}_{ij}^k$ are defined in [16] and

$$\begin{aligned}&Q_{11}^{kc}=2G_{13}^k\ell_b^2m^4+2G_{13}^k\ell_b^2m^2n^2;\ Q_{14}^{kc}=2G_{13}^k\ell_b^2mn\left(m^2+n^2\right);\ Q_{22}^{kc}=2G_{13}^k\ell_b^2n^4+2G_{13}^k\ell_b^2m^2n^2\\&Q_{23}^{kc}=2G_{13}^k\ell_b^2mn\left(m^2+n^2\right);\ Q_{41}^{kc}=Q_{42}^{kc}=Q_{31}^{kc}=Q_{32}^{kc}=G_{13}^k\ell_b^2mn\left(m^2+n^2\right)\\&Q_{43}^{kc}=Q_{33}^{kc}=G_{13}^k\ell_b^2m^2\left(m^2+n^2\right);\ Q_{44}^{kc}=Q_{34}^{kc}=G_{13}^k\ell_b^2n^2\left(m^2+n^2\right)\end{aligned} \tag{11}$$

where ℓ_b is the material length scale parameter of fiber rotation in $y-z$ plane.

3 Numerical Examples

In this section, a three-layer laminated composite plate with complicated cut out under SSSS and SCSC (S: Simply supported; C: Clamped) boundary conditions (BCs) is studied. The geometry and the mesh of cubic element of the plate is chosen as in Fig. (1). The material properties are consider as: $E_1/E_2=2.45$, $\nu_{12}=0.23$, $G_{12}/E_2=0.48$ and density $\rho=8000\,\mathrm{kg/m^2}$. The frequencies are normalized by $\varpi=\left(\rho h\omega^2a^4/D_0\right)^{1/2}$ with $D_0=E_1h^3/12(1-\upsilon_{12}\upsilon_{21})$. For verification of the accuracy of the proposed approach, Table 1 shows a comparison between the presented non-dimensional frequencies of laminated composite square plate for length to thickness ratio $L/h=1000/6$ and those of other references' solutions including the meshfree moving Kriging interpolation [17], IGA with level sets [18], IGA with multi-patches [19]. In can be observed that the obtained frequencies of the first six mode shape are in good agreement with other numerical solutions. Additionally, Table 2 shows an accuracy of present method through comparison between the present solution and the

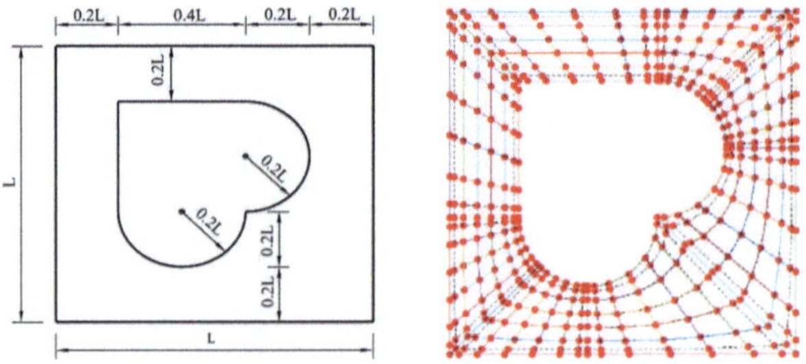

Fig. 1. A laminated composite square plate with heart cut-out: model geometry (left) and mesh of cubic element (right)

Table 1. Non-dimensional frequencies of a square SSSS three layer angle-ply laminated plate with heart cutout

Ply angle	Method	Mode					
		1	2	3	4	5	6
(15,−15,15)	MKI [17]	18.323	31.472	37.617	63.077	66.538	86.486
	IGA-level sets [18]	19.100	32.149	36.458	57.573	63.361	84.776
	IGA-8 patches [19]	18.912	32.045	36.004	56.345	63.370	83.629
	Present	19.062	32.019	36.174	57.385	62.848	84.665
(30,−30,30)	MKI [17]	20.310	33.987	39.898	58.111	69.699	92.099
	IGA-level sets [18]	20.606	33.997	37.610	59.797	65.688	88.809
	IGA-8 patches [19]	20.316	33.933	37.074	58.484	65.895	87.973
	Present	20.552	33.842	37.272	59.576	65.039	88.805
(45,−45,45)	MKI [17]	20.987	34.897	39.269	63.375	69.017	96.588
	IGA-level sets [18]	21.313	34.801	38.289	60.897	66.885	91.601
	IGA-8 patches [19]	20.982	34.848	37.559	59.325	67.518	91.220
	Present	21.249	34.632	37.918	60.654	66.154	91.991

Table 2. Non-dimensional frequencies of a square SCSC three layer angle-ply laminated plate with heart cut out

Ply angle	Method	Mode					
		1	2	3	4	5	6
(15,−15,15)	IGA-level sets [18]	29.909	39.816	56.158	70.114	81.783	102.274
	Present	29.918	39.495	56.168	69.974	81.356	102.233
(30,−30,30)	IGA-level sets [18]	31.898	41.453	59.480	73.628	85.336	107.280
	Present	31.903	41.094	59.477	73.456	84.805	107.065
(45,−45,45)	IGA-level sets [18]	33.863	43.227	63.869	77.595	89.129	106.308
	Present	33.870	42.849	63.844	77.391	88.658	106.636

numerical method established based on IGA with level sets [18]. As can be seen, for SCSC BCs, the obtained frequencies agree very well with those of Yin et al. [18].

Next, the size-dependent effect on free vibration of laminated composite with heart cutout is investigated. For simplicity, we assumed the material length scale parameter $\ell_b = \ell$. By changing the length scale ratio ℓ/h, the comparison of the first non-dimensional frequencies of microplate under SSSS and SCSC is displayed in Tables 3 and 4, respectively. It is noted that the NMCST is degenerated into classical model as $\ell/h = 0$. The results are generated for the ratio $\ell/h = 0, 0.2, 0.4, 0.6, 0.8, 1$. It can be observed that the increase in ratio ℓ/h leads to an increase in non-dimensional frequencies. It is worth commenting that this behavior is due to the increasing in the stiffness of plate as the ratio ℓ/h becomes higher. The highest frequencies are obtained for $\ell/h = 1$. Moreover, the first ten mode shapes of laminate (30,−30,30) microplate under simply supported BCs for $\ell/h = 1$ are plotted in Fig. 2.

Table 3. Comparison of non-dimensional frequencies of a square SSSS three layer angle-ply laminated plate with heart cut out

Ply angle	ℓ/h	Mode					
		1	2	3	4	5	6
(15,−15,15)	0	19.062	32.019	36.174	57.385	62.848	84.665
	0.2	20.658	34.899	39.103	61.195	68.200	90.911
	0.4	24.613	41.936	46.302	70.547	80.871	106.913
	0.6	29.712	50.557	55.911	82.755	95.815	124.278
	0.8	35.297	59.458	66.981	96.439	110.67	139.028
	1	41.086	68.379	78.779	110.88	125.09	155.177
(30,−30,30)	0	20.552	33.842	37.272	59.576	65.039	88.805
	0.2	22.073	36.545	40.202	63.414	69.884	94.659
	0.4	25.847	43.203	47.424	72.782	81.613	109.642
	0.6	30.745	51.675	56.816	84.870	96.265	126.681
	0.8	36.184	60.875	67.335	98.360	112.01	143.141
	1	41.912	70.429	78.473	112.67	128.32	161.088
(45,−45,45)	0	21.249	34.632	37.918	60.654	66.154	91.991
	0.2	22.739	37.264	40.847	64.513	70.740	97.395
	0.4	26.438	43.790	48.041	73.905	81.989	111.225
	0.6	31.252	52.250	57.270	85.938	96.503	127.684
	0.8	36.627	61.643	67.473	99.322	112.686	145.686
	1	42.323	71.547	78.241	113.56	129.882	165.053

Table 4. Comparison of non-dimensional frequencies of a square SCSC three layer angle-ply laminated plate with heart cut out

Ply angle	l/h	Mode					
		1	2	3	4	5	6
(15,−15,15)	0	29.918	39.495	56.168	69.974	81.356	102.233
	0.2	32.603	43.215	61.507	75.908	88.693	110.408
	0.4	39.176	52.321	74.635	90.577	105.964	127.691
	0.6	47.539	63.825	91.118	109.50	125.637	149.080
	0.8	56.655	76.135	108.436	130.25	144.761	174.912
	1	66.120	88.608	125.565	151.75	163.846	202.965
(30,−30,30)	0	31.903	41.094	59.477	73.456	84.805	107.065
	0.2	34.222	44.589	64.077	78.919	91.093	113.176
	0.4	40.022	53.242	75.560	92.431	106.312	128.281
	0.6	47.620	64.364	90.339	109.93	125.055	149.204
	0.8	56.120	76.550	106.405	129.32	144.902	173.908
	1	65.124	89.241	122.970	149.68	165.344	200.784
(45,−45,45)	0	33.870	42.849	63.844	77.391	88.658	106.636
	0.2	35.825	46.076	67.511	82.215	93.864	112.710
	0.4	40.786	54.161	76.783	94.163	106.980	128.314
	0.6	47.440	64.772	89.121	109.74	124.276	149.306
	0.8	55.050	76.673	103.133	127.26	143.721	173.177
	1	63.245	89.319	118.157	145.98	164.411	198.739

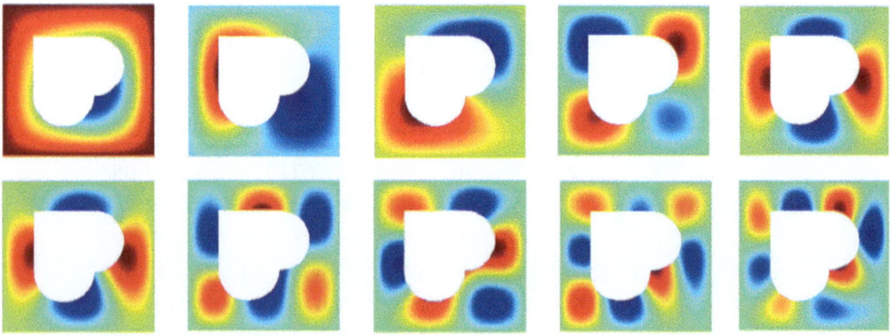

Fig. 2. First ten mode shapes a SSSS square three layer angle-ply laminated plate with complicated cut out. $\ell/h = 1$ (30,−30,30).

4 Conclusions

Based on new modified couple stress theory, the isogeometric analysis associated with Reinner-Mindlin plate theory is proposed to study the size-dependent effect on free vibration of angle-ply laminated composite microplate. The small scale phenomenon is captured by using one material length scale parameter. The accuracy of present method is examined by comparing the obtained results with those of references' solutions. Some benchmark results for free vibration of laminated composite microplate are presented, which can be used as reference data for future size-dependent analysis. The inclusion of the material length scale parameter causes an increase in the stiffness of plate that leads to a higher frequency than that of classical model and the highest result is obtained for $\ell/h = 1$.

Acknowledgements. The authors acknowledge the financial support of VLIR-OUS TEAM Project, VN2017TEA454A103, 'An innovative solution to protect Vietnamese coastal riverbanks from floods and erosion', funded by the Flemish Government.

1

References

1. Fleck, N.A., Muller, G.M., Ashby, M.F., Hutchinson, J.W.: Strain gradient plasticity: theory and experiment. Acta Metall. Mater. **42**(2), 475–487 (1994)
2. Stölken, J.S., Evans, A.G.: A microbend test method for measuring the plasticity length scale. Acta Mater. **46**(14), 5109–5115 (1998)
3. Chong, A.C.M., Lam, D.C.C.: Strain gradient plasticity effect in indentation hardness of polymers. J. Mater. Res. **14**(10), 4103–4110 (2011)
4. Lam, D.C.C., Yang, F., Chong, A.C.M., Wang, J., Tong, P.: Experiments and theory in strain gradient elasticity. J. Mech. Phys. Solids **51**(8), 1477–1508 (2003)
5. Chen, W., Xu, M., Li, L.: A model of composite laminated Reddy plate based on new modified couple stress theory. Compos. Struct. **94**(7), 2143–2156 (2012)
6. Chen, W., Li, X.: A new modified couple stress theory for anisotropic elasticity and microscale laminated Kirchhoff plate model. Arch. Appl. Mech. **84**(3), 323–341 (2014)
7. Wanji, C., Shengqi, Y., Xiaopeng, L.: A study of scale effect of composite laminated plates based on new modified couple stress theory by finite-element method. J. Multiscale Comput. Eng. **12**(6), 507–527 (2014)
8. Yang, S., Chen, W.: On hypotheses of composite laminated plates based on new modified couple stress theory. Compos. Struct. **133**(Supplement C), 46–53 (2015)
9. Chen, W., Wang, Y.: A model of composite laminated Reddy plate of the global-local theory based on new modified couple-stress theory. Mech. Adv. Mater. Struct. **23**(6), 636–651 (2016)
10. Yang, W., He, D.: Bending, free vibration and buckling analyses of anisotropic layered micro-plates based on a new size-dependent model. Compos. Struct. **189**, 137–147 (2018)
11. Hughes, T.J.R., Cottrell, J.A., Bazilevs, Y.: Isogeometric analysis: CAD, finite elements, NURBS, exact geometry and mesh refinement. Comput. Methods Appl. Mech. Eng. **194**(39), 4135–4195 (2005)

12. Thanh, C.-L., Phung-Van, P., Thai, C.H., Nguyen-Xuan, H., Abdel Wahab, M.: Isogeometric analysis of functionally graded carbon nanotube reinforced composite nanoplates using modified couple stress theory. Compos. Struct. **184**(Supplement C), 633–649 (2018)
13. Thanh, C.-L., Vu-Huu, T., Phung-Van, P., Nguyen-Xuan, H., Wahab, M.A.: Size-dependent analysis for FG-CNTRC nanoplates based on refined plate theory and modified couple stress. In: International Conference on Numerical Modelling in Engineering. Springer (2018)
14. Nguyen, H.X., Nguyen, T.N., Abdel-Wahab, M., Bordas, S.P.A., Nguyen-Xuan, H., Vo, T. P.: A refined quasi-3D isogeometric analysis for functionally graded microplates based on the modified couple stress theory. Comput. Methods Appl. Mech. Eng. **313**, 904–940 (2017)
15. Ansari, R., Norouzzadeh, A.: Nonlocal and surface effects on the buckling behavior of functionally graded nanoplates: an isogeometric analysis. Phys. E: Low-Dimens. Syst. Nanostructures **84**, 84–97 (2016)
16. Reddy, J.N.: Mechanics of Laminated Composite Plates and Shells Theory and Analysis, 2nd edn. CRC Press, New York (2004)
17. Bui, T.Q., Nguyen, M.N., Zhang, C.: An efficient meshfree method for vibration analysis of laminated composite plates. Comput. Mech. **48**(2), 175–193 (2011)
18. Yin, S., Yu, T., Bui, T.Q., Xia, S., Hirose, S.: A cutout isogeometric analysis for thin laminated composite plates using level sets. Compos. Struct. **127**, 152–164 (2015)
19. Shojaee, S., Izadpanah, E., Valizadeh, N., Kiendl, J.: Free vibration analysis of thin plates by using a NURBS-based isogeometric approach. Finite Elem. Ts Anal. Des. **61**, 23–34 (2012)

Damage Assessment of Laminated Composite Plates Using a Modified Cornwell Indicator

S. Tiachacht[1(✉)], M. Slimani[1], S. Khatir[2], A. Behtani[1], L. Mansouri[1], A. Bouazzouni[1], and M. Abdel Wahab[2]

[1] Laboratory of Mechanics, Structure, and Energetics (LMSE), Mouloud Mammeri University of Tizi-Ouzou, B.P.N°17 RP, 15000 Tizi Ouzou, Algeria
Tiachachtsamir@gmail.com
[2] Department of Electrical Energy, Metals, Mechanical Constructions and Systems, Faculty of Engineering and Architecture, Ghent University, Ghent, Belgium

Abstract. This paper presents an effective approach based on modal analysis for damage identification of three-layer (0°/90°/0°) laminated composite plates. Free vibration analysis of simply supported laminated composite plates using Finite Element Method (FEM) combined with Cornwell indicator (CI) and Modified Cornwell Indicator (MCI) is presented. The indicators are determined using modal analysis information extracted from a finite element code in Matlab and used to locate potential damage elements more effectively, after eliminating most of the healthy elements. A single and multiple damages are introduced in the laminate plate. The obtained results indicate that the proposed approach can identify the damage sites using both indicators. MCI can estimate the extent of damage with higher precision than CI.

Keywords: Laminated composite plates · Free vibration · Damage detection · Modified Cornwell indicator

1 Introduction

Structural health monitoring (SHM) literature using vibration data based on damage identification methods is widely used for composite materials. However, composite structures are easily affected by certain types of damage such as cracks or delamination. Therefore, damage identification for composite materials at an earlier stage in various engineering topics is a practical important aspect of the structural assessment. Non-Destructive Testing (NDT) is used particularly in the aerospace topic for delaminations identification in composite structures. Shi et al. [1] developed damage sensitivity matrix based on Mean Square Error (MSE) changes in structure elements for damage quantification. Yan et al. [2, 3] presented two damage indicators based on the MSE and generalized flexibility technique using natural frequencies and mode shapes. The obtained results showed good accuracy for locating and quantifying the presence of damage.

M. A. Wahab (Ed.): DAMAS 2019, LNME, pp. 853–862, 2020.
https://doi.org/10.1007/978-981-13-8331-1_68

Khatir et al. [4] presented an approach based inverse problem using different optimization techniques coupled with a reduced model based on Proper Orthogonal Decomposition and Radial Basis Function (POD-RBF) for a crack identification using experimental data. Li et al. [5] presented a generalized flexibility matrix change to detect the location and severity. Villalba and Laier [6] used a self-adaptive multi-chromosome GA for damage detection and localization in the structure. The results showed that the proposed approach could detect the damage with higher accuracy through different damage scenarios. Kulikov and Plotnikova [7] presented a method of 3D sampling surfaces to analyze the modal analysis of the simply supported piezo-electric plate. Milazzo [8] enhanced a refined equivalent single layer (ESL) formulation into the finite element method for free vibration analysis of smart laminated composites. Benedetti and Milazzo [9] studied the effects of advanced models for smart multilayered plates based on Reissner's mixed variational theorem on free vibrations and static analysis of magneto-electroelastic multilayer plates. Several numerical works were also studied for free vibration analysis of the piezo-laminated plates using FEM [10–13]. Damage identification in rectangular plates using spectral strain energy distribution was presented by Tufoi et al. [14]. Recently, several methods for damage identification in structures using modal data directly by considering the modal deformation energy before and after damage. Among these methods, Cornwell et al. [15] proposed a damage detection and localization indicator based on the variation of strain energy. The Modified Cornwell indicator (MCI) is used for truss and 3D structure for damage identification by Tiachacht et al. [16].

In this study, CI and MCI indicators are applied to laminated composite using FEM. Damage identification is investigated in the cases of single and multiple damages to define which method is the best to predict damage location and severity.

2 Method Description

FE models of laminated composites with different layer and orientation were presented in Ref. [17]. Cornwell et al. [15] presented his indicator for damage detection and localization based on the variation of strain energy. The combined MCI [16] with laminated composited plates is presented in our study.

2.1 MCI

The potential energy is written in the following:

$$E_{pot} = \frac{1}{2}\int_{0}^{L} EI\left(\frac{\partial u}{\partial x}\right)^{2} dx \tag{1}$$

Where E is the elastic modulus, I is the second moment of inertia, L is length, and u is deformation. For $\phi_i(x)$ proper mode deformation, the beam is subdivided into N elements, the energies associated with each element j due to the ith mode is:

$$(E_{pot})_i = \frac{1}{2}\int_0^L EI\left(\frac{\partial \phi_i(x)}{\partial x}\right)^2 dx \quad (2)$$

Considering the deformation, kinetic and dissipation energy of each j element and the total deformation energy of the beam for the ith eigen mode, the energy fractions are then given by:

$$FU_{ij} = \frac{(E_{pot})_{ij}}{(E_{pot})_i} \quad (4)$$

The expressions of these energies analogous to Eqs. (2) to (4) are written for the structure with defects as:

$$(E_{pot})_i^* = \frac{1}{2}\int_0^L (EI)^*\left(\frac{\partial \phi_i^*(x)}{\partial x}\right)^2 dx \quad (5)$$

$$(E_{pot})_{ij}^* = \frac{1}{2}\int_{a_j}^{a_{j+1}} (EI)_j^*\left(\frac{\partial \phi_i^*(x)}{\partial x}\right)^2 dx \quad (6)$$

$$FU_{ij}^* = \frac{(E_{pot})_{ij}^*}{(E_{pot})_i^*} \quad (7)$$

Where the symbol "*" in Eqs. (5) to (7), refers to the damaged structure. Modified Cornwell Indicator (MCI) is thus given by:

$$\beta_j = \frac{FU_{ij} - FU_{ij}^*}{\left(FU_{ij} - FU_{ij}^*\right)_{\max}} \quad (9)$$

Where
FU_{ij}: Fraction of deformation energy for the undamaged case.
FU_{ij}^*: Fraction of deformation energy for the damaged case.

3 Numerical Examples

The material parameters of a layer used here are: $E_1/E_2 = 40$; $G_{12} = G_{13} = 0.6E_2$; $G_{23} = 0.5E_2$; $\upsilon_{12} = 0.25$. The eigenproblem is solved using MATLAB to obtain the natural frequencies and mode shapes of laminated composite plates. The plate is discretized using a mesh of 8 × 8 (64 elements) see Fig. 1. In order to compare with the published results of Ferreira and Fasshauer [18], Liew [19], the same shear correction factors and nondimensionalized natural frequencies are also employed, i.e. shear correction factor $K_s = \pi^2/12$ and nondimensionalised natural frequency: $\bar{\omega} = \omega(b^2/\pi^2)\sqrt{\rho h/D_0}$ with $D_0 = E_2h^3/12(1 - \nu_{12}\nu_{21})$. The first four mode shapes are presented in Fig. 2.

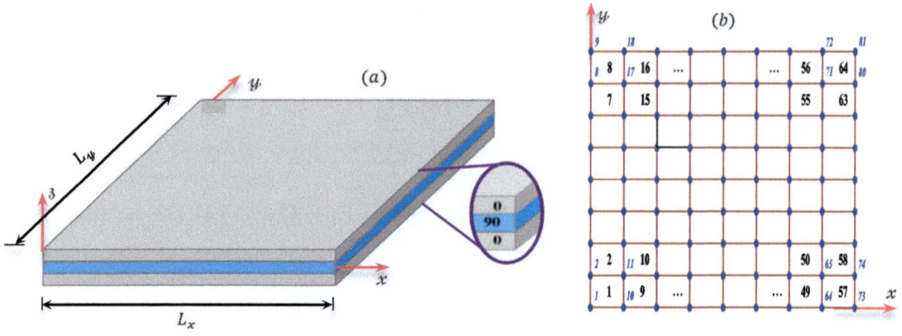

Fig. 1. Mesh of 8 × 8 for a laminated plate

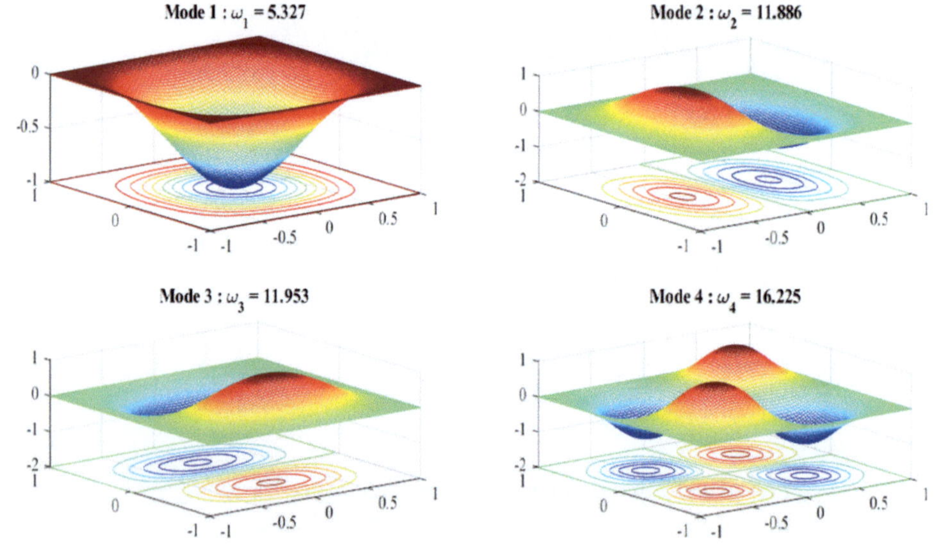

Fig. 2. Mode shapes for simply supported three-ply $[0°/90°/0°]$ square laminated plate with a grid of 8 × 8.

In order to validate the proposed damage assessment technique, five damage scenarios are considered, in which single damage, as well as, multiple damage cases, are studied as shown in Table 1. Table 2 presented the natural frequencies of undamaged and damaged laminate plate structure scenarios. Five damaged scenarios are simulated by reducing the global stiffness of individual elements as presented in the following equation:

Table 1. Damage scenarios.

Damage scenarios	Damaged element					
D1	Element	16	–	–	–	–
	% Reduction in stiffness	15%	–	–	–	–
D2	Element	28	29	–	–	–
	% Reduction in stiffness	5%	5%	–	–	–
D3	Element	8	16	24	–	–
	% Reduction in stiffness	15%	5%	10%	–	–
D4	Element	32	40	63	64	–
	% Reduction in stiffness	10%	5%	5%	10%	–
D5	Element	6	7	8	60	61
	% Reduction in stiffness	15%	10%	5%	10%	15%

$$K = \sum_{e=1}^{nele} (1 - \alpha)K^e \tag{10}$$

Where K is the global stiffness matrix of damaged structures, K^e is the stiffness matrix of eth element, respectively and α represents the damage ratio.

3.1 Case 1: Damage Scenario D1

In the first scenario, it is assumed that the 16th element of the plate is 15% damaged. The obtained results using both Indicators are shown in Fig. 3.

3.2 Case 2: Damage Scenario D2

In the second scenario, it is assumed that the 28th element of the laminated plate is 5% damaged and the 29th element is 5% damaged. The obtained results using CI and MCI are shown in Fig. 4.

3.3 Case 3: Damage Scenario D3

In the third scenario, it is assumed that each of the 8th, 16th, and 24th of the laminated plate are 15%, 5%, and 10% damaged, respectively. The obtained results are shown in Fig. 5.

3.4 Case 4: Damage Scenario D4

In the fourth scenario, four damaged elements 32, 40, 63 and 64 with loss of rigidity 10%, 5%, 5% and 10%, respectively, are considered. The obtained results are shown in Fig. 6.

Table 2. Natural frequencies of the undamaged and damaged laminated plate.

Mode	Undamaged f[Hz]	Damage scenario									
		D1		D2		D3		D4		D5	
		f[Hz]	%	f[Hz]	%	f[Hz]	%	f[Hz]	%	f[Hz]	%
1	5.328	5.321	0.124	4.768	10.504	5.265	1.175	5.058	4.238	5.005	6.053
2	11.886	11.871	0.126	11.332	4.662	11.701	1.560	11.382	1.018	11.162	6.088
3	11.953	11.938	0.123	11.571	3.199	11.908	0.375	11.831	1.257	11.852	0.850
4	16.226	16.174	0.317	15.967	1.596	15.954	1.672	16.022	2.161	15.854	2.293
5	25.528	25.515	0.052	22.928	10.184	25.412	0.456	24.976	0.872	24.568	3.761
6	25.704	25.680	0.093	24.551	4.485	25.560	0.561	25.480	1.741	25.615	0.347
7	28.054	27.984	0.251	27.606	1.598	27.661	1.402	27.566	0.628	27.468	2.088
8	28.196	28.123	0.260	27.783	1.468	28.063	0.473	28.019	1.929	27.904	1.038
9	36.656	36.528	0.348	35.780	2.390	36.353	0.825	35.949	0.693	35.851	2.197
10	51.393	51.381	0.024	44.369	13.667	51.368	0.050	51.037	5.056	50.536	1.668

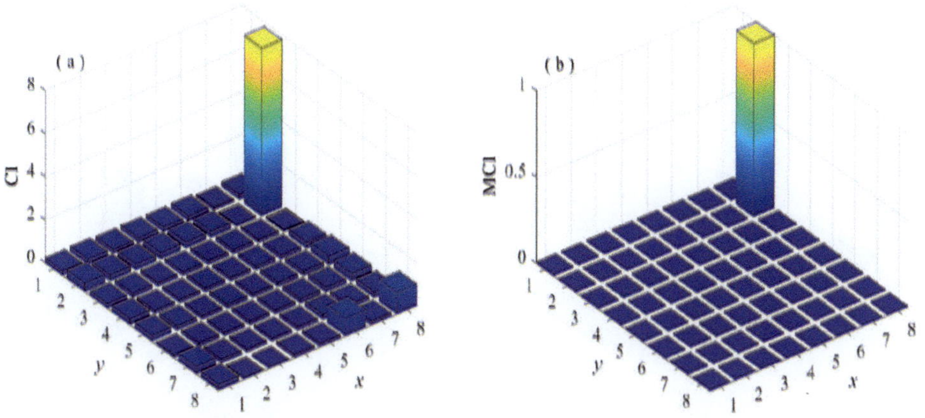

Fig. 3. Damage detection using (a) CI and (b) MCI for damage scenario 1.

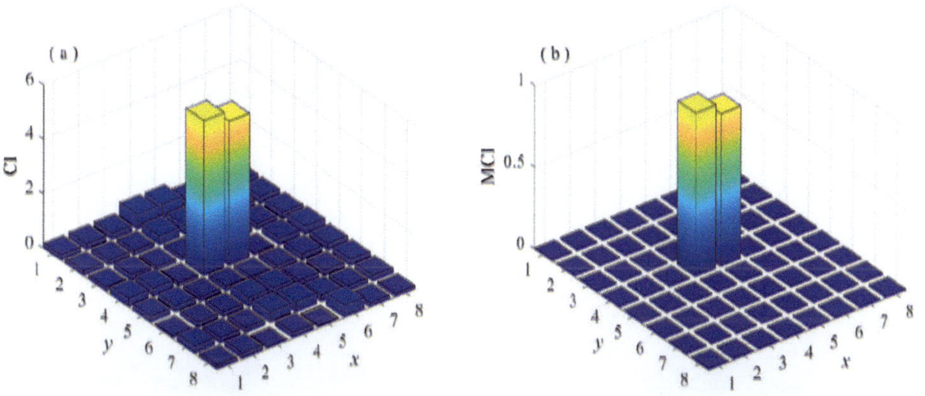

Fig. 4. Damage detection using (a) CI and (b) MCI for damage scenario 2.

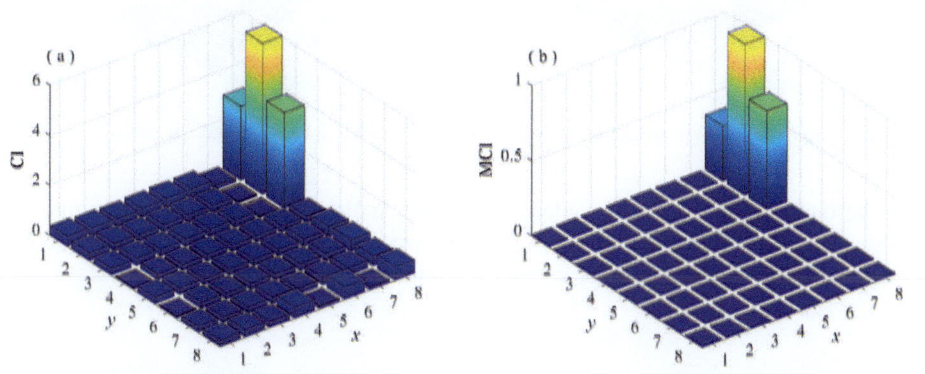

Fig. 5. Damage detection using (a) CI and (b) MCI for damage scenario 3.

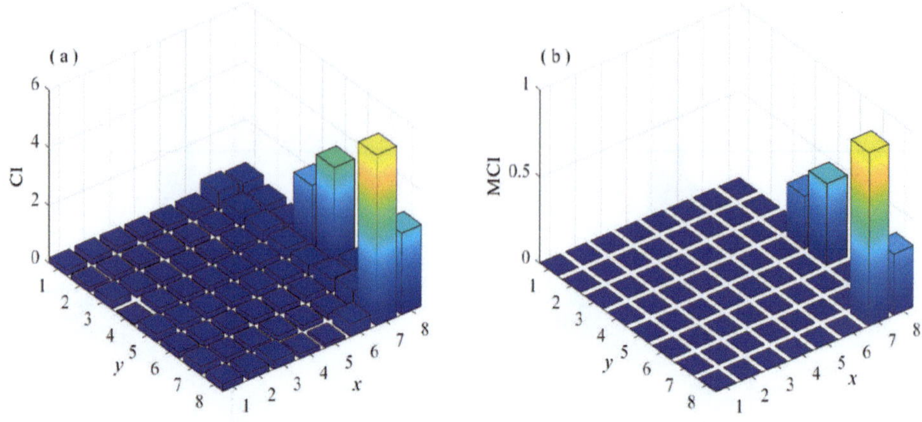

Fig. 6. Damage detection using (a) CI and (b) MCI for damage scenario 4.

3.5 Case 5: Damage Scenario D5

In the fifth scenario, five damaged elements 6, 7, 8, 60 and 61 with loss of rigidity 15%, 10%, 5%, 10% and 15%, respectively, are considered. The obtained results are shown in Fig. 7.

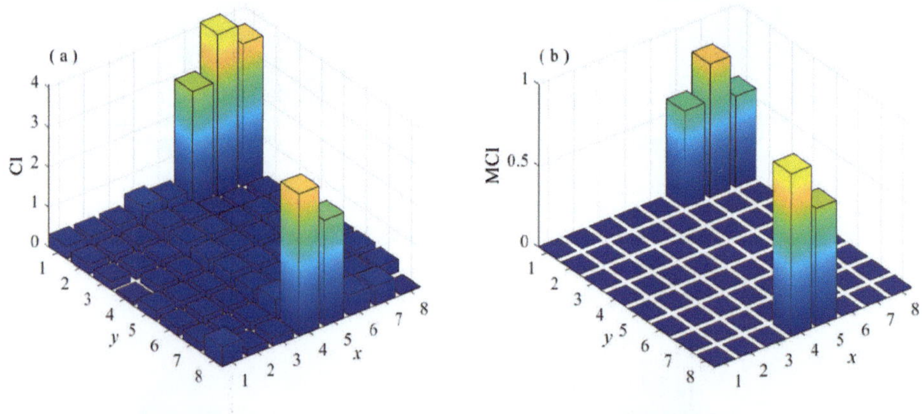

Fig. 7. Damage detection using (a) CI and (b) MCI for damage scenario 5.

The damage localization problem using both indicators are conducted to find out the suspiciously damaged elements in a composite plate. The results show that MCI predicts the damage with higher accuracy compared with CI.

4 Conclusions

In this paper, CI and MCI indicators are used for damage detection in laminated composites plates structures in free vibration simply supported three-ply $[0°/90°/0°]$ square laminated plates based on Finite Element Method (FEM). Different damage scenarios based on single and multiple damages are considered in each example. However, by comparing the obtained results, the result of MCI was more accurate in predicting the damage location and severity than CI.

References

1. Shi, Z., Law, S., Zhang, L.M.: Structural damage detection from modal strain energy change. J. Eng, Mech. **126**(12), 1216–1223 (2000)
2. Yan, W.-J., Huang, T.-L., Ren, W.-X.: Damage detection method based on element modal strain energy sensitivity. Adv. Struct. Eng. **13**(6), 1075–1088 (2010)
3. Yan, W.-J., Ren, W.-X.: Closed-form modal flexibility sensitivity and its application to structural damage detection without modal truncation error. J. Vib. Control **20**(12), 1816–1830 (2014)
4. Samir, K., Brahim, B., Capozucca, R., Wahab, M.A.: Damage detection in CFRP composite beams based on vibration analysis using proper orthogonal decomposition method with radial basis functions and Cuckoo Search algorithm. Compos. Struct. **187**, 344–353 (2018)
5. Li, J., et al.: A generalized flexibility matrix based approach for structural damage detection. J. Sound Vib. **329**(22), 4583–4587 (2010)
6. Villalba, J., Laier, J.E.: Localising and quantifying damage by means of a multi-chromosome genetic algorithm. Adv. Eng. Softw. **50**, 150–157 (2012)
7. Kulikov, G.M., Plotnikova, S.V.: Benchmark solutions for the free vibration of layered piezoelectric plates based on a variational formulation. J. Intell. Mater. Syst. Struct. **28**(19), 2688–2704 (2017)
8. Milazzo, A.: Refined equivalent single layer formulations and finite elements for smart laminates free vibrations. Compos. B Eng. **61**, 238–253 (2014)
9. Benedetti, I., Milazzo, A.: Advanced models for smart multilayered plates based on Reissner Mixed Variational Theorem. Compos. B Eng. **119**, 215–229 (2017)
10. Akhras, G., Li, W.: Stability and free vibration analysis of thick piezoelectric composite plates using spline finite strip method. Int. J. Mech. Sci. **53**(8), 575–584 (2011)
11. Wankhade, R.L., Bajoria, K.M.: Free vibration and stability analysis of piezolaminated plates using the finite element method. Smart Mater. Struct. **22**(12), 125040 (2013)
12. Rao, M., et al.: Finite element modeling and analysis of piezo-integrated composite structures under large applied electric fields. Smart Mater. Struct. **25**(5), 055044 (2016)
13. Kpeky, F., et al.: Linear and quadratic solid–shell finite elements SHB8PSE and SHB20E for the modeling of piezoelectric sandwich structures. Mech. Adv. Mater. Struct. **25**(7), 559–578 (2018)
14. Tufoi, M., et al.: Damage identification in rectangular plates using spectral strain energy distribution. In: Health Monitoring of Structural and Biological Systems 2015. International Society for Optics and Photonics (2015)
15. Cornwell, P., Doebling, S.W., Farrar, C.R.: Application of the strain energy damage detection method to plate-like structures. J. Sound Vib. **224**(2), 359–374 (1999)

16. Tiachacht, S., et al.: Damage assessment in structures using combination of a modified Cornwell indicator and genetic algorithm. Eng. Struct. **177**, 421–430 (2018)
17. Ferreira, A.J.: MATLAB codes for finite element analysis: solids and structures, vol. 157. Springer Science & Business Media, Berlin (2008)
18. Ferreira, A., Fasshauer, G.: Analysis of natural frequencies of composite plates by an RBF-pseudospectral method. Compos. Struct. **79**(2), 202–210 (2007)
19. Liew, K.M., Ng, T.Y., Zhao, X., Reddy, J.N.: Harmonic reproducing kernel particle method for free vibration analysis of rotating cylindrical shells. Comput. Methods Appl. Mech. Eng. **191**(37–38), 4141–4157 (2002)

The Sensitivity of Modal Strain Energy for Damage Localization in Composite Stratified Beam Structures

A. Behtani[1(✉)], S. Tiachacht[1], S. Khatir[2], M. Slimani[1], L. Mansouri[1], A. Bouazzouni[1], and M. Abdel Wahab[2]

[1] Laboratory of Mechanics, Structure, and Energetics (LMSE), Mouloud Mammeri University of Tizi-Ouzou, B.P.N°17 RP, 15000 Tizi Ouzou, Algeria
Amar.behtani@ummto.dz, Behtani2007@hotmail.com
[2] Department of Electrical Energy, Metals, Mechanical Constructions and Systems, Faculty of Engineering and Architecture, Ghent University, Ghent, Belgium

Abstract. In the present paper, a modal analysis of undamaged and damaged Carbon Fibre Reinforced Polymer (CFRP) beam elements with fixed and free ends and a cross-ply (0°/90°/0°) rectangular beam with clamped boundary conditions at both ends is investigated for damage identification. The presence of single and multiple damages in composite structures based on changes in the dynamic properties of the structure are considered. The Modal Strain Energy (MSE) damaged indicator based on frequency and mode shape combined with composite structure is studied. To enhance the results, the sensitivity of MSE using mode shape is investigated by varying the number of modes. The results showed that MSE using frequency much better than mode shape. Moreover, MSE with the first mode is much better than other modes.

Keywords: CFRP beam · Cross-ply · Dynamic properties · Free vibration · Damage detection

1 Introduction

A literature review of damage detection and structural health monitoring based on vibration characteristics has been published in recent years. From this literature review, a large number of proposed methods for damage identification based on modal analysis were presented by Doebling et al. [1, 2]. The existing of damage identification study was based on modal curvature and investigated the changes in indicator value between the undamaged and damaged structures [3]. A comparison between different techniques proposed by Rytter [4], which consists of four stages. The first stage is based on detection, the second stage is based on localization, the third stage is based on assessment, and finally, the fourth stage, which is the consequence of damage, predicts the remaining life of the structure in a certain state of damage.

Yang and Liu [5] presented a method based on the best eigenvector to solve the incomplete measurement problem. Khatir et al. [6] presented a new approach for fast

M. A. Wahab (Ed.): DAMAS 2019, LNME, pp. 863–874, 2020.
https://doi.org/10.1007/978-981-13-8331-1_69

simulation to predict the damage identification based on snapshot matrix using Proper Orthogonal Decomposition with Radial Basis Function (POD-RBF). The combined FEM with optimization techniques as an inverse problem for damage identification using modal analysis as objective function was introduced by Khatir et al. [7–9]. Valdes and Soutis [10] presented an application based on frequency in damage identification. The response showed that the modal analysis with higher frequencies is much better for the identification of delamination in the cantilever composite beams. Yoon et al. [11] presented a procedure of damage detection in structural stiffness using the data obtained by damaged structure, the mode shapes of the undamaged structure are smooth without irregularity and using a curve fitting technique, the first mode shape has approximated this procedure. Chiang and Lai [12] presented an application based on residual force combined with an analytical solution for damage identification. The damage detection and localization of the presented approach were used successfully to locate structural damage. Kahl and Sirkis [13] used the concept of the residual force, a so call subspace rotation damage identification approach. Tiachacht et al. [14] presented a modified indicator for damage quantification of 2D and 3D structures to identify the location of damages and their severities. For more accurate to predict the severity, the modified indicator was used as an objective function with a Genetic Algorithm (GA). The obtained results showed that the proposed indicator was accurate and efficient. Capozucca and Bonci [15] presented an experimental free vibration of CFRP cantilever beam elements were investigated by dynamic tests. Damage of CFRP specimens was locally due to double notches close to the fixed end with different reduction of section and bending stiffness. Theoretical and experimental analysis of double notches of Carbon Fiber Reinforced Polymer (CFRP) laminate elements under free vibration of CFRP laminate specimens were experimentally investigated using a mechanical apparatus capable of simulating hinge conditions at the edges of simply supported laminate beams presented in Ref. [16].

This work presents an MSE formulation based on frequencies and mode shapes for detecting and locating damage in clamped-free stratified beam structure discretized using FEM. The obtained results show that the MSE with frequencies is much better than with the mode shapes.

2 Damage Indicators

In this part, we present two damage indicators for damage identification. The first indicator is MSE indicator based on frequencies. The second indicator is the MSE based on mode shapes.

2.1 Modal Strain Energy Indicator

The eigenvalue equation of a structure with n degrees of freedom can be written as:

$$([K] - \lambda_i[M])\{\phi\}_i = 0 \tag{1}$$

where $[K]$ and $[M]$ are the rigidity and mass matrices of the structure, respectively.

λ_i and ϕ_i are its eigenvalues and eigenvectors.

We define below the modal strain energies MSE_i^{es} and MSE_i^{ed}, respectively, for the healthy and damaged finite element structure:

$$MSE_i^{es} = \frac{1}{2}\{\phi\}_i^T[K]^e\{\phi\}_i;\; MSE_i^{ed} = \frac{1}{2}\{\phi\}_i^{dT}[K]^e\{\phi\}_i^d \quad (2)$$

where n is the number of finite elements of the structure, m is the number of modes and $[K]^e$ is the stiffness matrix of the structure elementary sound.

The total world energy in the structure can be calculated by adding the MSE's of all elements, then we can write:

$$MSE_i^s = \sum\nolimits_{e=1}^{n} MSE_i^{es};\; MSE_i^d = \sum\nolimits_{e=1}^{n} MSE_i^{ed} \quad (3)$$

To normalize MSE_i^{es} and MSE_i^{ed}, we divide each elemental energy by the total energy, so that the normalized energy is written:

$$NMSE_i^{es} = \frac{MSE_i^{es}}{MSE_i^s};\; NMSE_i^{ed} = \frac{MSE_i^{ed}}{MSE_i^d} \quad (4)$$

After normalization of the *MSE*, we can choose the first m modes as effective parameters, and we can write

$$MNMSE^{es} = \frac{\sum_{i=1}^{m} MSE_i^{es}}{m};\; MNMSE^{ed} = \frac{\sum_{i=1}^{m} MSE_i^{ed}}{m} \quad (5)$$

The damage indicator based on modal strain energy (*MSEBI*) is defined as follows:

$$MSEBI^e = \frac{MNMSE^{es} - MNMSE^{ed}}{MNMSE^{es}} \quad (6)$$

To locate the damage, we just take the nonzero *MSEBI* indicators and the corresponding elements, knowing that the indicator of healthy elements is zero.

2.2 Modified Modal Strain Energy Indicator

In this part, we will follow the same steps as before except that we will use the frequency response in the calculation of the damage indicator instead of the eigenvectors.

The frequency response of the healthy and damaged structure is given by:

$$\begin{cases} \{X(w)\} = \left\{\left(-w^2[M] + [K]^{-1}\right)\right\}\{F(w)\} \\ \{X_d(w)\} = \left\{\left(-w^2[M]_d + [K]_d^{-1}\right)\right\}\{F(w)\} \end{cases} \quad (7)$$

The modified modal strain energy (*MMSEBI*) is defined as follows:

$$MMSEBI^e = \frac{MMNMSE^{es} - MMNMSE^{ed}}{MMNMSE^{es}} \quad (8)$$

For the calculation of this indicator, Eq. (2) becomes:

$$MMSE_i^{es} = \frac{1}{2}\{X\}_i^T[K]^e\{X\}_i; \; MMSE_i^{ed} = \frac{1}{2}\{X\}_i^{dT}[K]^e\{X\}_i^d \quad (9)$$

3 Numerical Results and Discussion

Two examples are considered here to study the performance of the present two methods. In the first example, Carbon Fibre Reinforced Polymer (CFRP) beam with fixed and free ends, is studied and in the second example a cross-ply (0°/90°/0°) rectangular beam with clamped boundary conditions at both ends is considered. Six damaged scenarios are simulated by reducing the global stiffness of individual element as presented in the following equation:

$$K = \sum_{e=1}^{nele}(1 - \alpha)K^e \quad (10)$$

where K is the global stiffness matrix of damaged structures, K^e is the stiffness matrix of eth element, respectively and α represents the damage ratio.

3.1 CFRP Cantilever Beam

In the first example, the mechanical and geometrical properties of CFRP beam specimens are shown in Table 1. In Fig. 1 the setup of the free vibration test for a CFRP cantilever beam specimen is shown with strike position of an impact hammer and accelerometer. The discretized finite element CFRP cantilever beam in 11 elements is presented in Fig. 2. The natural frequencies of damaged and undamaged are presented in Table 2. Table 3 presented the damage cases of CFRP.

Table 1. Experimental parameters of CFRP beam [15]

Width $b(mm)$	Thickness $h(mm)$	Length $L(mm)$	Young's modulus $E_f(kN/mm^2)$	Density $\rho(Ns^2/mm^4)$
20.253	1.7	200	~ 133	1.376×10^{-9}

Fig. 1. Experiment setup of the CFRP beam test [15]

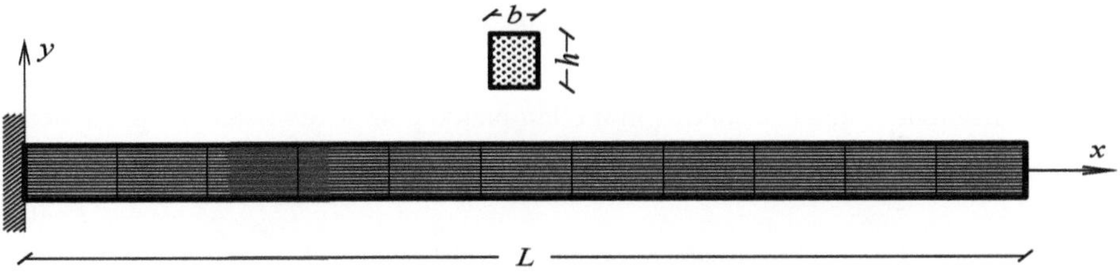

Fig. 2. CFRP cantilever beam discretized in 11 elements

Table 2. Natural frequencies of damaged and undamaged CFRP beam

$f[Hz]$	FEM	Undamaged beam CFRP [15]	Er %	Damage					
				Case 1	Er_1 %	Case 2	Er_2 %	Case 3	Er_3 %
f_1	67.49	67.49	0.005	63.28	−6.23	66.75	−1.09	63.98	−5.20
f_2	422.84	423.00	−0.038	405.60	−4.11	406.69	−3.86	396.86	−6.18
f_3	1183.46	1184.53	−0.090	1150.23	−2.90	1128.32	−4.75	1108.38	−6.43
f_4	2318.15	2321.23	−0.13	2271.28	−2.15	2281.90	−1.69	2184.15	−5.91
	$Er = 100 \times \frac{f_{FEM} - f_{Exp}}{f_{Exp}}$			$Er_i = 100 \times \frac{f_{Case_i} - f_{Exp}}{f_{Exp}}; i = 1 \cdots 3$					

Table 3. Damage cases for the CFRP beam

Scenario	1	2		3		
Element no.	1	5	8	1	5	9
Damage ratio	0.30	0.20	0.30	0.20	0.30	0.30

3.2 A Three Cross-Ply (0°/90°/0°) Rectangular Beam

The second example is a cross-ply (0°/90°/0°) rectangular beam with clamped boundary conditions at both ends, as depicted in Fig. 3. The geometrical and mechanical parameters of the beam are given as length $L = 0.2$ m, width $b = 0.02$ m, thickness $t = 0.02\,\text{m}$, $E_1 = 40\,\text{N/m}^2$, $E_2 = 1\,\text{N/m}^2$, $G_{12} = G_{13} = 0.6E_2$, $G_{23} = 0.5E_2$, $\nu_{12} = 0.25$, and the thickness of each layer is $t/3$. The composite beam is discretized into 16 beam elements (see Fig. 3) using first-order shear deformation theory (FSDT) as presented in Ref. [17]. Each element has two nodes with three DOFs (two translational displacements in the horizontal and vertical directions, and one rotation) per node.

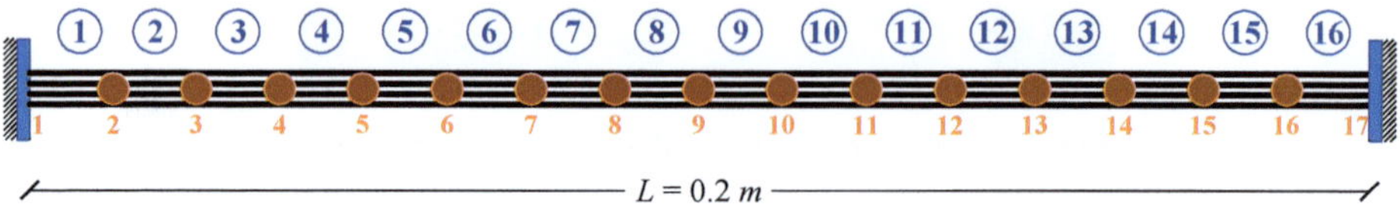

Fig. 3. Node and element numbering of the discretized composite beam [18].

Three different damage scenarios in the laminated composite beam are presented in Table 4 and natural frequencies for damaged and undamaged cases in Table 5.

Table 4. Three different damage scenarios in the laminated composite beam.

Scenario	4	5		6		
Element no.	1	5	16	1	2	9
Damage ratio	0.30	0.20	0.30	0.20	0.30	0.30

Table 5. Natural frequencies of damaged and undamaged for a three cross-ply (0°/90°/0°) beam.

$f[Hz]$	FEM	Unreduced model [18]	Er %	Unreduced model (FOBT) [17]	Damage					
					Case 4	Er_4%	Case 5	Er_5 %	Case 6	Er_6 %
f_1	20.309	19.125	6.191	19.051	20.214	5.694	20.21	5.67	20.169	5.458
f_2	40.552	38.983	4.025	–	40.547	4.011	40.48	3.84	40.528	3.963
f_3	62.821	61.861	1.552	–	62.817	1.546	62.81	1.53	62.640	1.260
f_4	84.811	85.374	−0.659	–	84.802	−0.670	84.79	−0.68	84.733	−0.751
f_5	106.793	109.741	−2.686	–	106.782	−2.696	106.73	−2.74	106.696	−2.775

4 Results and Discussion

Three damage scenarios are studied based on single and multiple damages for Carbon Fiber Reinforced Polymer (CFRP) beam elements with fixed and free ends and a cross-ply (0°/90°/0°) rectangular beam with clamped boundary conditions at both ends. The

Modal strain energy based on frequency ($MSECR_f$) and modal strain energy based mode shape (MSECR) are studied to predict the damages locations and their severities. Different modes are investigated using MSECR to enhance the results.

4.1 CFRP Cantilever Beam

4.1.1 Damage Scenario 1 (D1)

Figures 4 and 5.

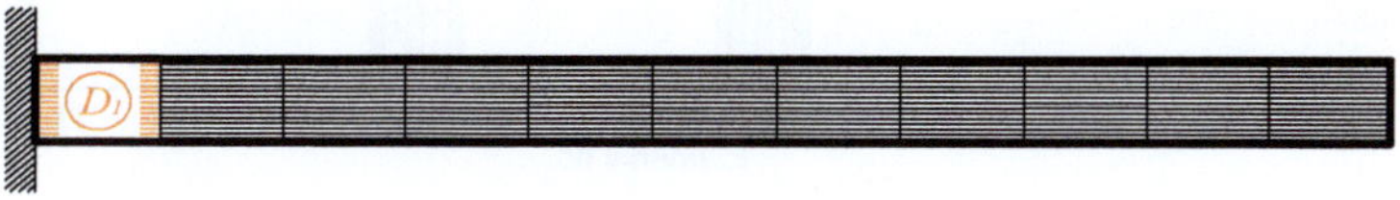

Fig. 4. Single element damage (Element 01) – damage case 1

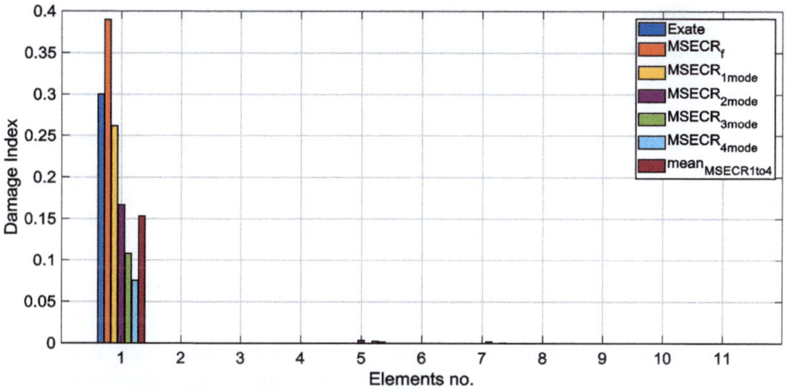

Fig. 5. Damage location based on elements – case 1.

4.1.2 Damage Scenario 2 (D2)

Figures 6 and 7.

Fig. 6. Two element damage (Element 05 and 08) – damage case 2

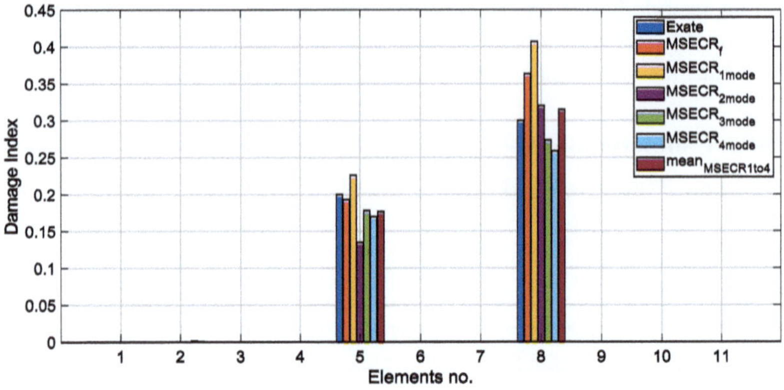

Fig. 7. Damage location based on elements – case 2.

4.1.3 Damage Scenario 3 (D3)

The obtained results show that the MSECR$_f$ performs much better than MSECR for different modes as shown in Table 6, Figs. 8 and 9.

Table 6. Damage cases for the CFRP beam using both indicators

	Case 1	Case 2		Case 3		
Elements	1	5	8	1	5	9
Exacted	0.3	0.2	0.3	0.2	0.3	0.3
MSECR_f	0.39	0.19	0.36	0.16	0.32	0.32
Error	30%	3%	21%	20%	8%	8%
MSECR using 1 mode	0.26	0.23	0.41	0.12	0.28	0.28
Error	13%	13%	36%	38%	6%	7%
MSECR using 2 modes	0.17	0.13	0.32	0.10	0.3	0.25
Error	45%	33%	7%	50%	0%	17%
MSECR using 3 modes	0.11	0.18	0.27	0.05	0.27	0.34
Error	64%	11%	9%	77%	10%	14%
MSECR using 4 modes	0.08	0.17	0.26	0.04	0.25	0.21
Error	75%	15%	14%	79%	18%	29%
Mean (MSECR 1 to 4 modes)	0.15	0.18	0.31	0.08	0.27	0.27
Error	49%	11%	5%	61%	9%	10%

Fig. 8. Multiple element damage (Element 01, 05 and 09) – damage case 3

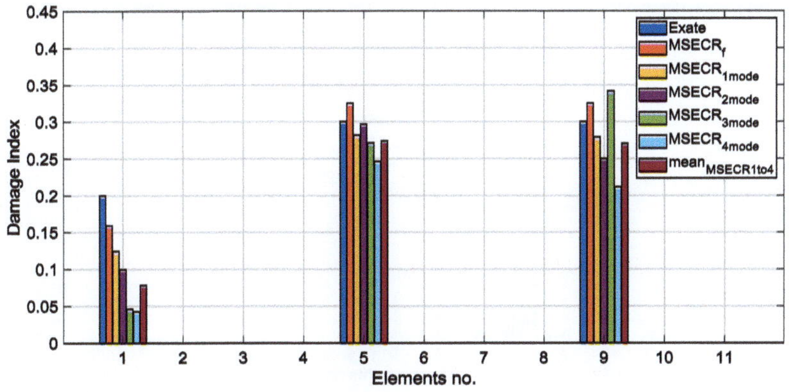

Fig. 9. Damage location based on elements – case 3.

4.2 A Three Cross-Ply (0°/90°/0°) Rectangular Beam

4.2.1 Damage Scenario 1 (D1)

Figure 10.

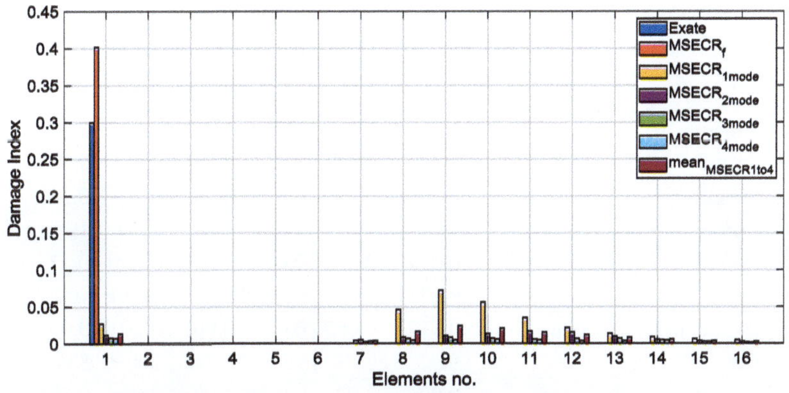

Fig. 10. Damage location based on elements – case 4.

4.2.2 Damage Scenario 2 (D2)

Figure 11.

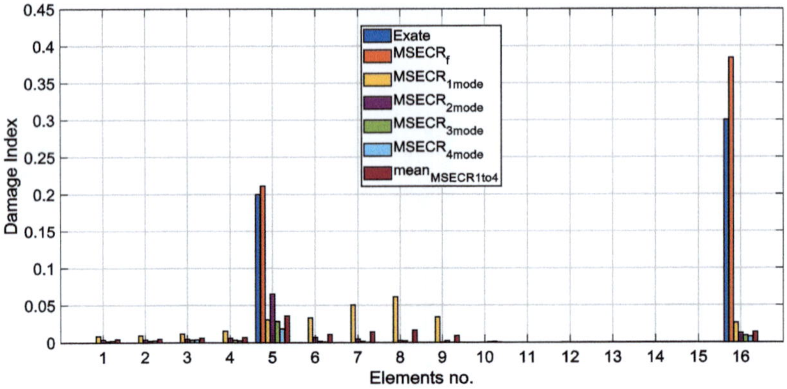

Fig. 11. Damage location based on elements – case 5.

4.2.3 Damage Scenario 3 (D3)

As shown in Table 7, the $MSECR_f$ method can localize damaged elements cautiously. According to Figs. 10, 11 and 12, MSECR with the first mode is much better than other modes and shows less computation time.

Table 7. Damage cases for the cross-ply (0°/90°/0°) beam using both indicators

	Case 4	Case 5		Case 6		
Elements	1	5	16	1	2	9
Exacted	0.3	0.2	0.3	0.2	0.3	0.3
MSECR_*f*	0.40	0.21	0.38	0.19	0.36	0.36
Error	34%	5%	28%	6%	19%	19%
MSECR using 1 mode	0.027	0.031	0.026	0.000	0.000	0.421
Error	91%	85%	91%	100%	100%	40%
MSECR using 2 modes	0.012	0.065	0.012	0.001	0.001	0.051
Error	96%	67%	96%	100%	100%	83%
MSECR using 3 modes	0.008	0.028	0.009	0.003	0.006	0.078
Error	97%	86%	97%	99%	98%	74%
MSECR using 4 modes	0.008	0.018	0.008	0.005	0.010	0.044
Error	97%	91%	97%	97%	97%	85%
Mean (MSECR 1 to 4 modes)	0.014	0.036	0.014	0.002	0.004	0.149
Error	95%	82%	95%	99%	99%	50%

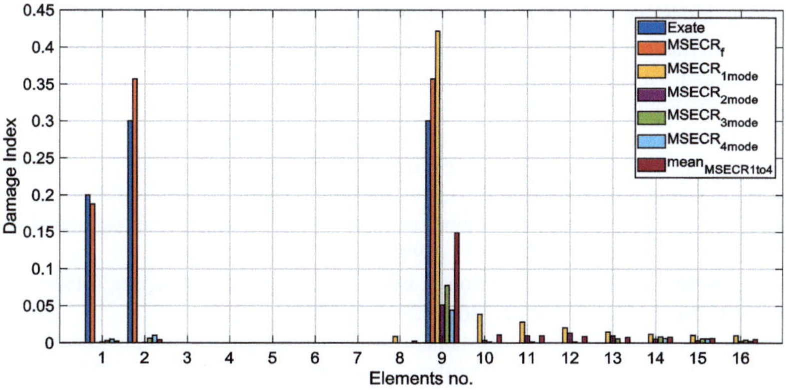

Fig. 12. Damage location based on elements – case 6.

5 Conclusion

The present study focuses on the identification of the location of damage in CFRP cross-ply rectangular beams using Modal Strain Energy based on frequency and mode shape indicators. Several damage scenarios are also considered for each type of structure. The sensitivity of MSE using mode shape is investigated by varying the number of modes. The results showed that the MSE using frequencies much better than that using mode shapes. Moreover, MSE with first mode shape is much better than other modes.

References

1. Doebling, S.W., et al.: Damage identification and health monitoring of structural and mechanical systems from changes in their vibration characteristics: a literature review (1996)
2. Zou, Y., Tong, L., Steven, G.P.: Vibration-based model-dependent damage (delamination) identification and health monitoring for composite structures—a review. J. Sound Vib. **230** (2), 357–378 (2000)
3. Wang, Y.-L.: New damage localization indicator based on curvature for single-span beams. Struct. Eng. Mech. **51**(6), 1037–1046 (2014)
4. Doebling, S.W., Farrar, C.R., Prime, M.B.: A summary review of vibration-based damage identification methods. Shock Vib. Digest **30**(2), 91–105 (1998)
5. Yang, Q., Liu, J.: Structural damage identification based on residual force vector. J. Sound Vib. **305**(1), 298–307 (2007)
6. Khatir, S., et al.: Damage detection and localization in composite beam structures based on vibration analysis. Mechanics **21**(6), 472–479 (2016)
7. Khatir, S., et al.: 1884. Numerical study for single and multiple damage detection and localization in beam-like structures using BAT algorithm. J. Vibroengineering **18**(1) (2016)
8. Khatir, S., et al.: Genetic algorithm based objective functions comparative study for damage detection and localization in beam structures. In: Journal of Physics: Conference Series. IOP Publishing (2015)

9. Khatir, A., et al.: Damage detection and localization on thin plates using vibration analysis. In: 23rd International Congress on Sound and Vibration (ICSV23). International Institute of Acoustics and Vibration (2016)
10. Valdes, S.D., Soutis, C.: Delamination detection in composite laminates from variations of their modal characteristics. J. Sound Vib. **228**(1), 1–9 (1999)
11. Yoon, M., et al.: Local damage detection using the two-dimensional gapped smoothing method. J. Sound Vib. **279**(1), 119–139 (2005)
12. Chiang, D.-Y., Lai, W.-Y.: Structural damage detection using the simulated evolution method. AIAA J. **37**(10), 1331–1333 (1999)
13. Kahl, K., Sirkis, J.: Damage detection in beam structures using subspace rotation algorithm with strain data. AIAA J. **34**(12), 2609–2614 (1996)
14. Tiachacht, S., et al.: Damage assessment in structures using combination of a modified Cornwell indicator and genetic algorithm. Eng. Struct. **177**, 421–430 (2018)
15. Capozucca, R.: Vibration of CFRP cantilever beam with damage. Compos. Struct. **116**, 211–222 (2014)
16. Capozucca, R., Bonci, B.: Notched CFRP laminates under vibration. Compos. Struct. **122**, 367–375 (2015)
17. Khdeir, A., Reddy, J.: Free vibration of cross-ply laminated beams with arbitrary boundary conditions. Int. J. Eng. Sci. **32**(12), 1971–1980 (1994)
18. Dinh-Cong, D., Dang-Trung, H., Nguyen-Thoi, T.: An efficient approach for optimal sensor placement and damage identification in laminated composite structures. Adv. Eng. Softw. **119**, 48–59 (2018)

A Comparative Study of the Behavior of Glass Fiber-Reinforced Polyester Composite Laminates Under Static Loading

L. Mansouri[1(✉)], D. Arezki[1], S. Khatir[2], A. Behtani[1], S. Tiachacht[1], M. Slimani[1], and M. Abdel Wahab[2]

[1] Laboratory of Structural Mechanics and Energetics (LMSE), Mouloud Mammeri University of Tizi Ouzou, BP17RP, Tizi Ouzou, Algeria
mansouri_lila@yahoo.fr, lilla.mansouri@ummto.dz
[2] Soete Laboratory, Faculty of Engineering and Architecture, Ghent University, Technologiepark Zwijnaarde 903, 9052 Zwijnaarde, Belgium

Abstract. Glass fiber-reinforced polyester matrix composites are increasingly used because of their many advantages, namely, lightness, cost, strength, and ease of processing. Thus, they can meet sometimes special needs that conventional materials cannot meet. Nevertheless, their properties depend on the nature of the plies used, e.g. UD, mat, woven, and the stacking sequence. This work aims to study the mechanical behavior of several composites stratified under static loading and to identify the different modes of damage leading to rupture. This paper mainly studies the influence of the shape of the reinforcement on rigidity and static behavior. In addition, we are interested in the evolution of shear stress as a function of the distance between supports.

Keywords: Woven laminates · Mechanical behavior · Stiffness · Damage

1 Introduction

Composite materials occupy a growing place in our daily universe that is the automobile, naval and aeronautic construction. Their penetration in these sectors requires a better knowledge of their mechanical behavior in static rather than in dynamics. The development of these composites relies on a better knowledge of their behavior as well as, the mechanical properties of its constituents, i.e. (their geometrical, nature and type, etc.), to avoid damage and offer good products adapted to the desired properties.

In recent decades, several studies on the fatigue behavior of glass-reinforced organic matrix composite materials have been conducted. However, it remains difficult to reach a better understanding of many damage mechanisms that can lead to complete failure. The breaking mechanisms of composite materials can be classified into four main groups: fiber/matrix shear, fiber fracture, matrix cracking, and delamination. Philippidis and Vassilopoulos [1] also modified the Tsai-Hill criterion to develop a deterministic model of fatigue life based on the static failure criterion in the cyclic load. Damage to composite structures is due to many mechanisms acting at different scales. Khatir et al. [2] presented a composite damage identification based on experimental

M. A. Wahab (Ed.): DAMAS 2019, LNME, pp. 875–886, 2020.
https://doi.org/10.1007/978-981-13-8331-1_70

analysis using model reduction, optimization techniques, and modal analysis. Capozucca and Magagnini [3] presented a static and dynamic analysis of damaged composites using experimental and numerical analyzed. A semi-continuous approach presented by Pascal et al. [4] of the five-beam woven composite material. Gillich et al. [5] presented an algorithm to assess damages in composite structures, based on how natural frequencies of weak-axes bending vibration modes change due to damage.

Composite materials can be degraded in various ways by the action of their environment, and these degradations naturally affect their mechanical behavior. Mansouri et al. [6] presented the effect of hygrothermal aging in different media on the mechanical behavior of short fiber /woven composite laminates of 4 layers of fiberglass fabrics (matt 300 g/m^2, matte 450 g/m^2, taffeta 800 g/m^2) and unsaturated polyester resins. The obtained results showed that the immersion in a humid environment over time causes physical and chemical degradation phenomena (hydrolysis of the resin and the fiber/matrix interface and fiber degradation), these phenomena could be at the same time caused a loss of strength and ductility. Maozhou et al. [7] studied the effect of moisture infiltration on flexural fatigue of laminated composites. The authors developed accelerated tests to study the correlation between composite fatigue and moisture diffusion effects. The effect of the fiber orientation angle of composite material based on the phenomenon of humidity diffusion presented by Boukhoulda et al. [8] using an analytical solution.

The purpose of this study is twofold. Firstly, it experimentally examines the mechanical behavior of the glass reinforced reinforcement and polyester matrix, consisting of 04 folds. Static tests are performed on various laminated composites to analyze the effect of reinforcement type on behavior and on different types of damage up to failure. Secondly, we are interested in following the evolution of the shear stress versus the distance between supports by changing the geometry of the test.

2 Material Characteristics

A laminated composite is primarily a stack of several folds oriented in different directions, which it is important to know its behavior. Therefore, we used in our study, different laminates consisting of 04 plies of fiberglass type E in the form of mats and fabrics (Mat 350 g/m^2, Mat 450 g/m^2 and Taffeta 800 g/m^2) and a thermosetting resin made of unsaturated polyester. The staking sequences are presented in Table 1.

Table 1. Type of stratification used

	Designation	Number of folds
LOT 1	Mat 300 g/m^2 + Mat 450 g/m^2 + Taffeta 800 g/m^2 + Mat 300 g/m^2	04
LOT 2	Mat 300 g/m^2	04
LOT 3	Mat 450 g/m^2	04
LOT 4	Taffeta 800 g/m^2	04

In this case, the laminate plates used to cut the specimens are obtained by the same manufacturing process as that used for fishing vessels, namely contact molding. The process is manual and we used the stratification shown in Fig. 1.

Fig. 1. Stratification used

3 Experimental Conditions

Three-point bending tests are carried out according to the recommendations of French standards NFT57-105. The tests make use of a universal machine model IBERTEST, instrumented with a load capacity of 100 KN with speeds ranging from 10 to 500 mm/min. The central cylinder has a radius 5 mm, and the cylindrical supports have a radius of 2.5 mm. The geometry of the test is shown in Fig. 2.

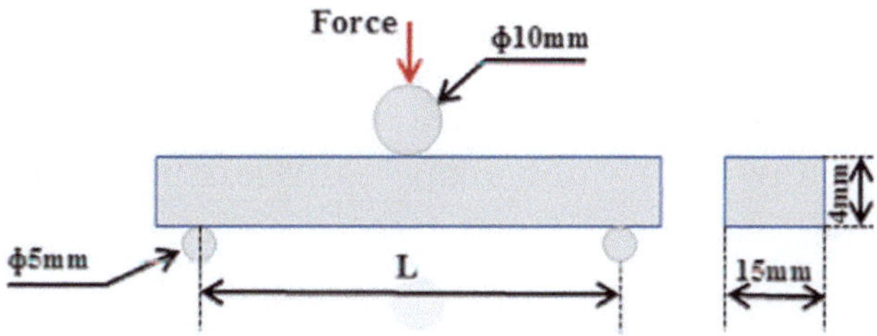

Fig. 2. Geometry of the bending test.

The purpose of these tests is to identify the mechanical behavior until the failure of different laminates in two directions from the following resistance formulas:

$$\sigma = \frac{3PL}{2bh^2} \tag{1}$$

$$\varepsilon = \frac{6\mathrm{Yh}}{\mathrm{L}^2} \tag{2}$$

$$E_{app} = \frac{L^3}{4bh^3}\frac{\Delta P}{\Delta Y} \tag{3}$$

$$\tau_{max} = \frac{3P}{2bh} \quad (4)$$

where, σ normal stress, ε relative deformation, E_{app} apparent modules, τ_{max} shear stress, P strength, b width, h thickness, L distance between supports and Y Displacement.

During the bending stress, the stress field is not homogeneous and the applied force generates not only normal tensile stresses - compression (1), but also shear stresses (4). Hence, we are interested in following the evolution of the shear stress with the variation of the distance between supports by changing the geometry of the test. In practice, for a large enough L/h value, the effects of the shear can be neglected. To follow the evolution of the shear stress, we performed bending tests at different distances L = 40, 50, 64, 80 and 100 mm.

4 Results and Discussion

We present the results of the various experimental tests carried out as part of our study. The results obtained are summarized in Table 2 and presented in Fig. 3.

Table 2. The mechanical characteristics of different batch

Material	LOT 1	LOT 2	LOT 3	LOT 4
$E_{app}[Gpa]$	6.32 (0.67)	3.34 (0.73)	4.50 (0.43)	15.63 (0.54)
$\sigma_e[Mpa]$	120 (26.04)	89.01 (21.15)	125.61 (15.08)	240.25 (39.13)
$\sigma_r[Mpa]$	177.29 (20.89)	115.93 (22.80)	147.81 (7.64)	263.32 (37.91)
$\varepsilon_e[\%]$	0.019 (0.004)	0.026 (0.001)	0.028 (0.004)	0.016 (0.001)
$\varepsilon_r[\%]$	0.039 (0.002)	0.04 (0.004)	0.035 (0.003)	0.02 (0.00)

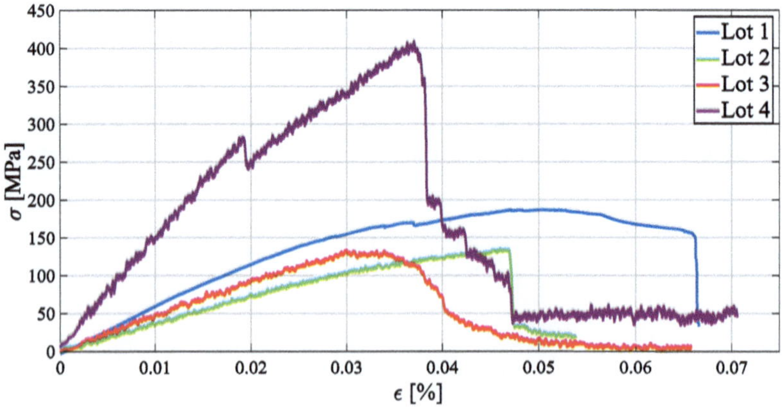

Fig. 3. Stress-strain curve in bending

All specimens exhibit similar behavior for the same lot. The bending curve of the mechanical properties shows that the behavior of the laminates is influenced by the type of the stack, where we find that the behavior of the material of batches 1 and 3 is ductile compared with that of batches 2 and 4, which is fragile. We note that the mechanical properties are different from one batch to another as shown in Fig. 4.

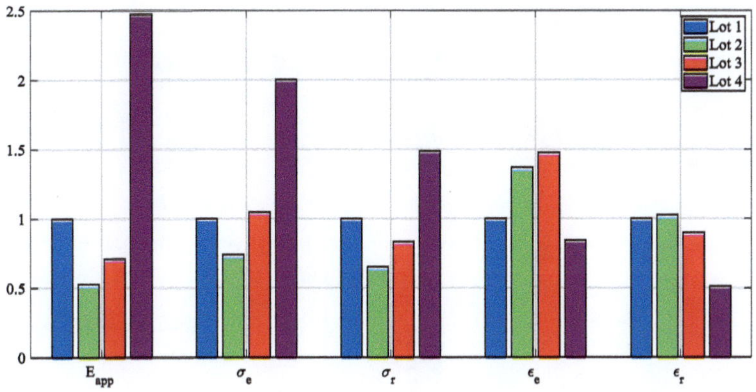

Fig. 4. Comparison between different batch's characteristics.

The comparison of the results obtained (see Fig. 4) shows that the modulus and the maximum stress of batch 4 are clearly higher than other batches. However, the opposite is to be noted concerning the deformation, which proves that the taffeta is not very deformable. In addition, batch 1 contains different type of layers of intermediate values.

4.1 Bending Results in Two Directions

The results obtained according to the direction of the stresses are presented in Figs. 5, 6, 7 and 8. The different figures show the evolution of the mechanical properties as a function of the direction of fibers. We find that the mechanical properties between the two senses are almost different for all lots. According to the literature, the fibers in the mats are arranged in a random manner, which leads to isotropy of the mechanical properties. On the other hand, according to the results obtained, we notice a dispersion of the mechanical characteristics. This is due mainly to the existence of defects (air bubbles) and to the process of obtaining materials (molding in contact).

4.2 Influence of the Distance Between Supports

The evolutions of the mechanical properties obtained as a function of the length to thickness ratio are summarized in Tables 3 and 4 and represented by the curves in Figs. 9 and 10.

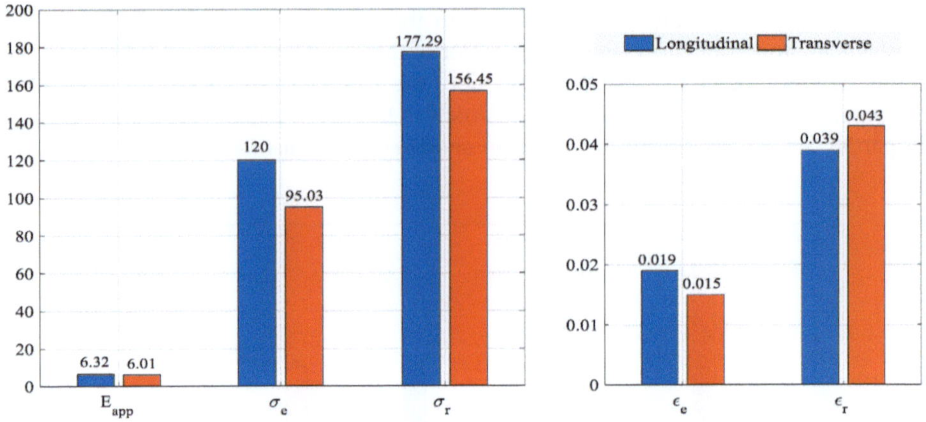

Fig. 5. Comparison of the mechanical properties according to the direction of fibers of lot 1.

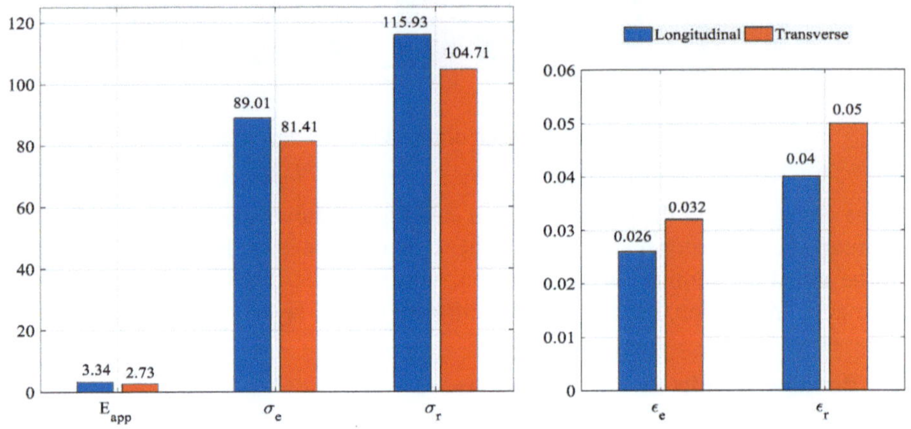

Fig. 6. Comparison of the mechanical properties according to the direction of fibers of lot 2.

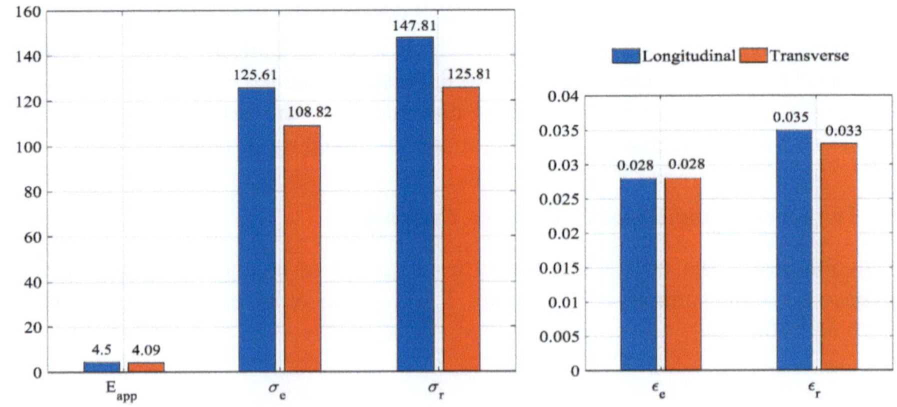

Fig. 7. Comparison of the mechanical properties according to the direction of fibers of lot 3.

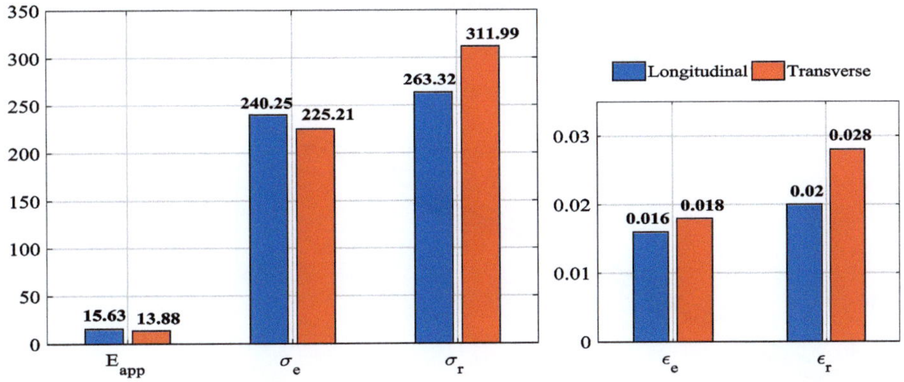

Fig. 8. Comparison of the mechanical properties according to the direction of fibers of lot 4.

According to the results obtained, we noticed that all the specimens exhibit similar behavior for the same material with a dispersion of the mechanical characteristics. The overall mechanism of rupture of our test pieces differs according to the type of the fold.

Table 3. Mechanical characteristics of lot 3.

Length to thickness ratio $[mm]$	$E_{app}[Gpa]$	$\sigma_e[Mpa]$	$\sigma_r[Mpa]$	$\varepsilon_e[\%]$	$\varepsilon_r[\%]$	$\tau_{max}[Mpa]$
40	4.63 (0.45)	80.53 (24.16)	148.91 (6.16)	0.018 (0.007)	0.041 (0.001)	8.21 (0.37)
50	5.51 (1.77)	142.59 (21.86)	179.40 (37.36)	0.027 (0.006)	0.036 (0.005)	7.13 (0.57)
64	5.83 (2.23)	138.15 (33.32)	171.74 (29.74	0.025 (0.003)	0.036 (0.005)	5.34 (0.57)
80	2.87 (1.77)	108.52 (3.29)	133.11 (10.18)	0.028 (0.001)	0.039 (0.001)	3.67 (0.15)
100	4.11 (0.16)	100.76 (2.56)	127.78 (2.52)	0.024 (0.002)	0.038 (0.002)	2.79 (0.09)

Table 4. Mechanical characteristics of lot 4.

Length to thickness ratio $[mm]$	$E_{app}[Gpa]$	$\sigma_e[Mpa]$	$\sigma_r[Mpa]$	$\varepsilon_e[\%]$	$\varepsilon_r[\%]$	$\tau_{max}[Mpa]$
40	9.55 (0.72)	210.14 (43.62)	223.06 (49.65)	0.022 (0.003)	0.032 (0.015)	9.47 (2.12)
50	12.09 (0.46)	216.23 (30.88)	249.02 (4.11)	0.018 (0.002)	0.03 (0.007)	8.47 (0.23)
64	11.71 (0.51)	90.33 (34.13)	249.98 (42.08)	0.016 (0.003)	0.032 (0.007)	6.55 (1.16)
80	11.09 (0.43)	207.29 (33.51)	236.43 (27.71)	0.018 (0.002)	0.034 (0.001)	4.85 (0.6)
100	13.18 (0.54)	209.85 (19.74)	238.44 (16.82)	0.016 (0.001)	0.026 (0.005)	3.91 (0.24)

According to the results presented in Figs. 9 and 10 (force versus displacement curves), it can be seen that the behavior of material 1 is semi-fragile.

For lot 4 (see Fig. 10), we noticed a nonlinear zone associated with the appearance of plastic deformation and the beginning of the damage where we heard cracks, which

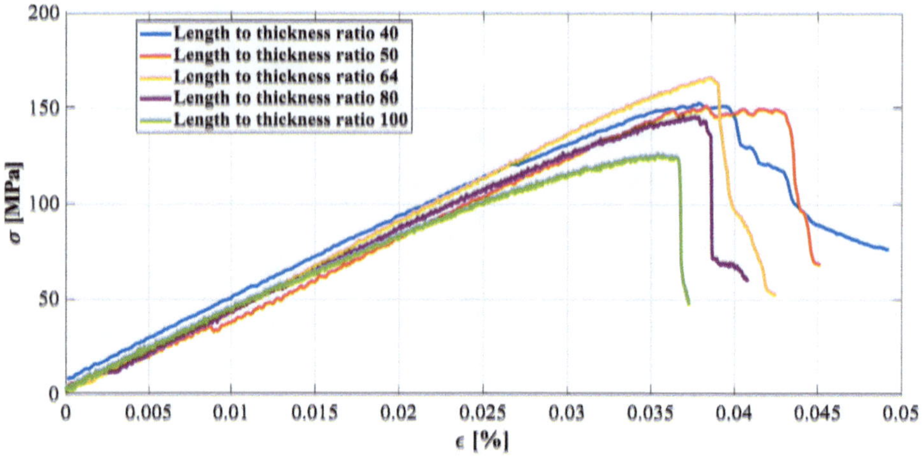

Fig. 9. Stress-strain curves in bending for lot 3.

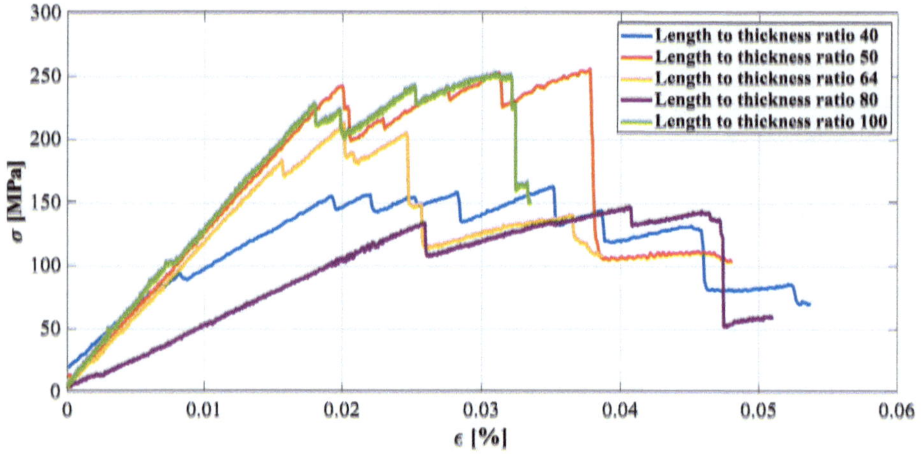

Fig. 10. Stress-strain curves in bending for lot 4.

correspond to the delamination inter folds. In order to characterize the effect of the distance between supports and the type of folds effect, a comparative study of the mechanical properties is conducted.

Figures 11 and 12 show the dimensionless evolution of mechanical properties as a function of length to thickness ratio. It is noted that all the mechanical properties are influenced by the variation of the length to thickness ratio. We note, for both materials, a sudden decrease in the shear stress with the variation of the length to thickness ratio, which shows that the shear is inversely proportional to the distance between supports. For sufficiently large values of *L/h*, the effects of shear can be neglected.

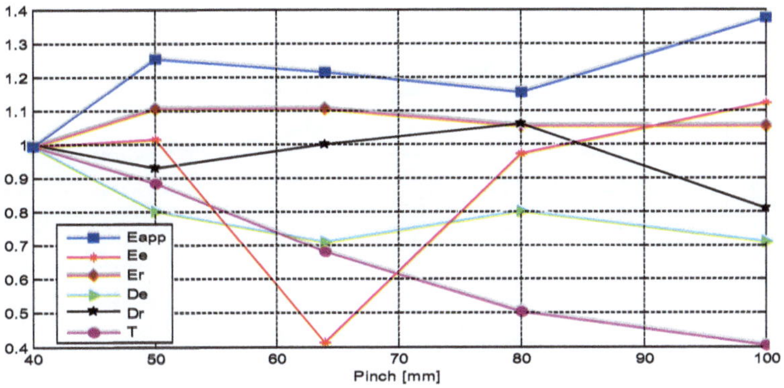

Fig. 11. Evolution of the mechanical properties versus deflection of lot 3.

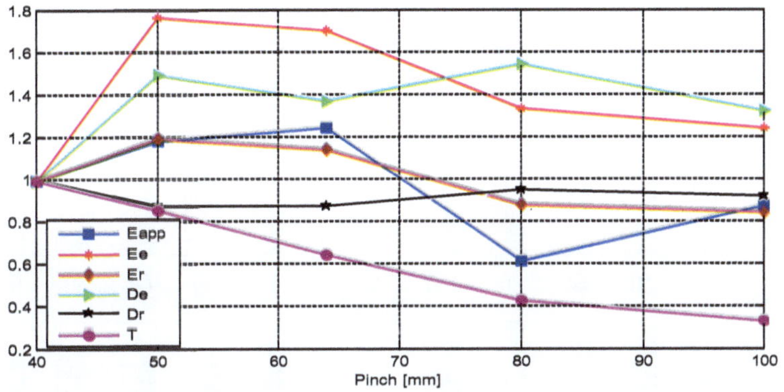

Fig. 12. Evolution of the Mechanical characteristics versus length to thickness ratio of lot 4.

4.3 Influence of Type of Folds

The first remark to extract from the above figures is that the behavior between the two materials is different and the stresses of the specimens of the laminate lot 4 are significantly higher than those of the specimens of the laminate lot 3.

In Fig. 14, we observe that the modulus and stress of the specimens of lot 4 are higher than those of the specimens of lot 3. This difference can be explained by the fiber content. Lot 3 contains 31.15% fiber, whereas lot 4 contains 63.16%. An increase in the volume fraction of the fibers increases the modulus of elasticity according to the law of the mixtures and Young's modulus. Nevertheless, the opposite is to be noted concerning the breaking strain and the elastic stress for L = 64 mm.

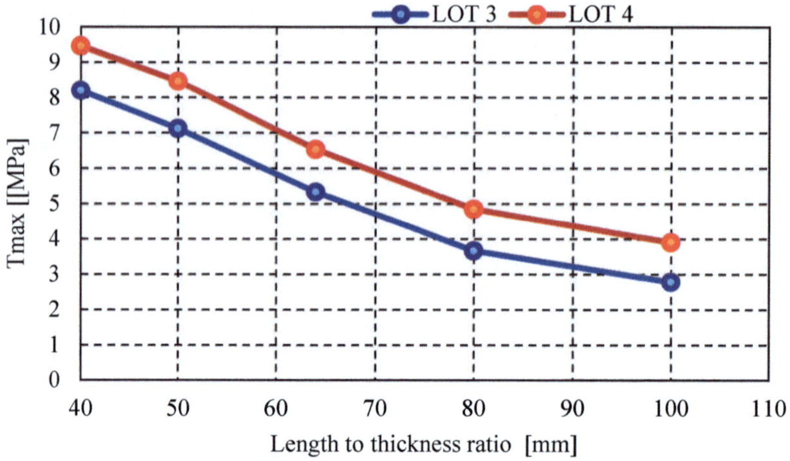

Fig. 13. Evolution of the shear stress versus length to thickness ratio of lot 3 and lot 4.

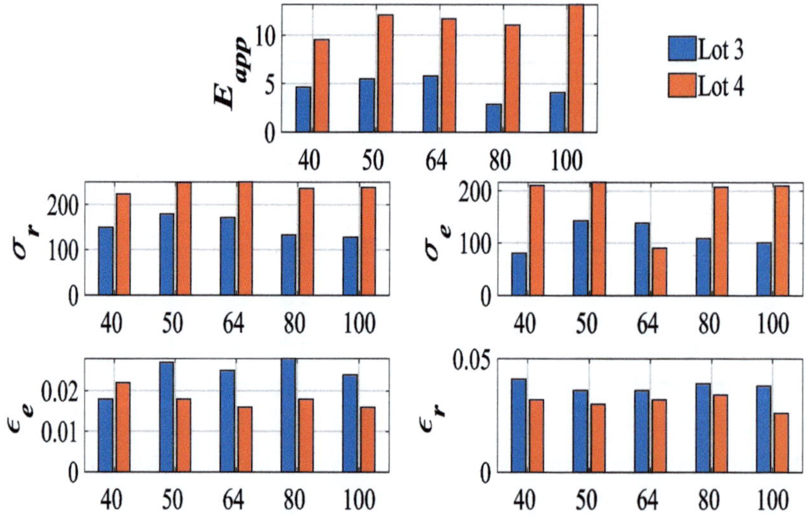

Fig. 14. Evolution of the mechanical characteristics versus length to thickness ratio of lot 3 and lot 4.

5 Observation of Fracture Surfaces Using SEM

In order to better understand the mechanisms of damage and the differences between them, the failure of bending-stressed specimens for the two composites were analyzed under a scanning electron microscope (SEM). Two batches of composite specimens subjected to bending were observed. The images of Figs. 15 and 16 were taken with different magnification for specimens that did not break during bending tests on the most damaged area. We observe several types of damage, which is mainly mechanism of micro-cracking of the matrix, de-cohesion of the interface and fiber breakage. According to the micrographs obtained, for both materials, the breaking of the matrix by cracking is observed. At the interface between the fibers and the matrix, there is formation of micro-voids and loosening of fibers.

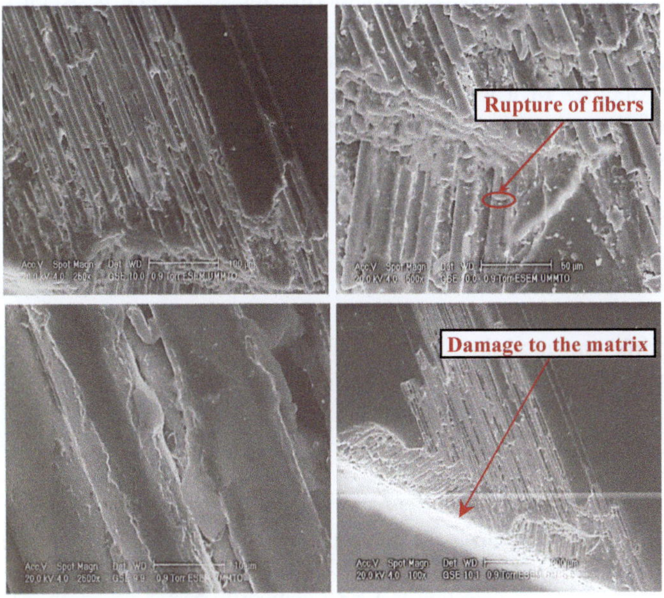

Fig. 15. Observation of fracture faces of a fiberglass-reinforced composite mat.

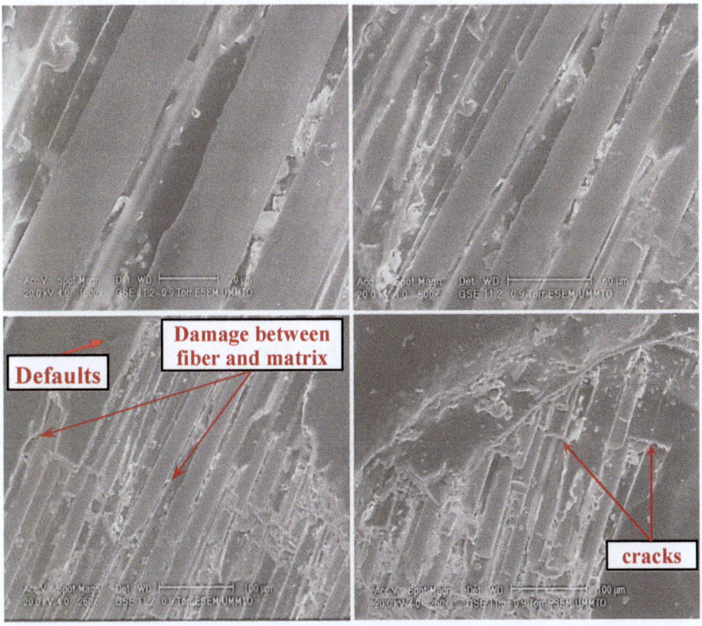

Fig. 16. Observation of fracture faces of composite fiberglass woven.

6 Conclusions

The analysis of all the results obtained highlights the main parameters that must be integrated in order to correctly describe the behavior of the composites. We highlight the influence of several parameters considering the difference between the characteristics of the constituent elements.

This study aims to compare the influence of the type of reinforcement on the mechanical properties of a woven laminate. It is particularly interesting for the characterization of the resistance and the mechanisms of rupture. The cracking of a fold differs according to its position in the stack, more particularly if it is on the inner or outer layers.

Strong interest is focused on the effect of the stacking sequences and the distance between supports on the mechanical properties of the composite material and their evolutions. The analysis of all the results shows that the mechanical characteristics evolve with the variation of the distance between supports. The designer possesses the ability to modify and modulate the mechanical behavior by changing geometry proportion of constituents and the orientation of the fibers.

References

1. Philippidis, T., Vassilopoulos, A.: Fatigue strength prediction under multiaxial stress. J. Compos. Mater. **33**(17), 1578–1599 (1999)
2. Samir, K., et al.: Damage detection in CFRP composite beams based on vibration analysis using proper orthogonal decomposition method with radial basis functions and cuckoo search algorithm. Compos. Struct. **187**, 344–353 (2018)
3. Capozucca, R., Magagnini, E.: Experimental vibration response of homogeneous beam models damaged by notches and strengthened by CFRP lamina. Compos. Struct. **206**, 563–577 (2018)
4. Pascal, F., et al.: Impact damage prediction in thin woven composite laminates–part I: modeling strategy and validation. Compos. Struct. **190**, 32–42 (2018)
5. Gillich, G.-R., Praisach, Z.-I., Negru, I.: Damages influence on dynamic behaviour of composite structures reinforced with continuous fibers. Mater. Plast. **49**(3), 186–191 (2012)
6. Mansouri, L., et al.: Effect of hygrothermal aging in distilled and saline water on the mechanical behaviour of mixed short fibre/woven composites. Compos. Struct. **207**, 816–825 (2019)
7. Meng, M., et al.: Moisture effects on the bending fatigue of laminated composites. Compos. Struct. **154**, 49–60 (2016)
8. Boukhoulda, B., Adda-Bedia, E., Madani, K.: The effect of fiber orientation angle in composite materials on moisture absorption and material degradation after hygrothermal ageing. Compos. Struct. **74**(4), 406–418 (2006)

Damage Detection in Laminated Composite Plates Based on Local Frequency Change Ratio Indicator

S. Khatir[1(✉)], S. Tiachacht[2], C. Le Thanh[1,5], T. Khatir[3], R. Capozucca[4], and M. Abdel Wahab[1]

[1] Department of Electrical Energy, Metals, Mechanical Constructions, and Systems, Faculty of Engineering and Architecture, Ghent University, 9000 Ghent, Belgium
Khatir_samir@hotmail.fr, Samir.khatir@ugent.be
[2] Laboratory of Mechanics, Structure, and Energetics (LMSE), Mouloud Mammeri University of Tizi-Ouzou, B.P.N°17 RP, 15000 Tizi-Ouzou, Algeria
[3] Institute Of Science and Technology, University Centre Salhi Ahmed, Naama, Algeria
[4] Struct. Section DICEA, Università Politecnica Delle Marche, Ancona, Italy
[5] Faculty of Civil Engineering and Electricity, Open University, Ho Chi Minh City, Vietnam

Abstract. This paper presents an application based on Local Frequency Change Ratio (LFCR) for damage assessment of three-layer (0°/90°/0°) laminated composite plates. The indicator is used to help locating single and multiple potential damaged elements. The obtained results indicate that even when increasing damaged elements, LFCR indicator can detect the damage accurately. For more accuracy to prove that the LFCR is much better to identify the damage location in laminated composite, we introduced white Gaussian noise with different levels. The obtained results indicate that even under measurement noise level 2%, the LFCR can identify the actual damage with high precision.

Keywords: LFCR · Vibration analysis · Laminated composite · Noise

1 Introduction

Laminated composite plates are widely investigated in practice, in civil and mechanical engineering. A new approach for crack identification in Carbon Fiber Reinforced Polymer (CFRP) composite using vibration analysis based on model reduction was presented by Khatir et al. [1]. This approach was based on the inverse problem using a snapshot matrix used to build matrix data collected from experiments. Moreover, Genetic Algorithm (GA) and Cuckoo Search algorithm were used to compute the inverse problem. Analytical solution of double notch crack with different boundary condition was used and the results were validated experimentally by Capozucca et al. [2, 3]. Khatir et al. [4] proposed a damage identification technique in beam-like structures based on frequencies, normalized Modal Strain Energy Damage Indicator (nMSEDI),

M. A. Wahab (Ed.): DAMAS 2019, LNME, pp. 887–898, 2020.
https://doi.org/10.1007/978-981-13-8331-1_71

IGA and Finite Element Method (FEM) combined with Teaching-Learning-Based Optimization Algorithm (TLBO). The proposed indicator was used as an objective function to enhance the results. Dahak et al. [5] presented a technique based on normalized frequencies in cantilever beam based on experimental data. Numerical simulation and experimental study on crack identification based on the piezoelectric ceramic lead zirconate titanate impedance technology were investigated by Wu et al. [6]. Modal Strain Energy method and an improved differential evolution algorithm for damage detection in applied for a cross-ply (0°/90°/0°) laminated composite beam and a cross-ply (0°/90°/0°) square laminated composite plate with multiple damaged elements was investigated by Vo-Duy et al. [7]. Tiachacht et al. [8] presented a modified indicator for damage quantification of 2D and 3D structures to identify the location of damages and their severities. The modified indicator was used as an objective function with GA to predict the severity of damage. The obtained results showed that the proposed indicator was accurate and efficient. Zenzen et al. [9, 10] presented an approach based on the inverse problem using frequency response function and frequency response as an objective function using different optimization techniques. Experimental vibration analysis of homogeneous beam damaged by notches and strengthened by CFRP was presented by Capozucca and Magagnini [11]. Gillich and Praisach [12] presented an algorithm for damage identification in beam-like structure using the power spectrum and time–frequency analysis. The proposed algorithm has been validated experimentally.

This manuscript is organized as follows. In Sect. 2, the Local Frequency Change Ratio (LFCR) is presented. Section 3 presents the numerical application of the laminated composite. Results and discussion are presented in Sect. 4. In Sect. 5, the effect of noise is studied and finally, conclusions are presented in Sect. 5.

2 Local Frequency Change Ratio Indicator

This method has been proposed by Shi et al. [13]. The local frequency of the jth element to the ith mode before and after the occurrence of the damage is calculated by using the ith undamaged and damaged mode shape, respectively.

$$LF_{ij}^{u} = \frac{\phi_{ui}^{T} K_j \phi_{ui}}{\phi_{ui}^{T} M_j \phi_{ui}} \text{ and } LF_{ij}^{d} = \frac{\phi_{di}^{T} K_j \phi_{di}}{\phi_{di}^{T} M_j \phi_{di}} \tag{1}$$

It is recognized that the local damage only reduce the local stiffness of the structure and the mode shape ϕ_i is not sensitive to a small change in the local stiffness except at or near the damage domain. The Local Frequency of the jth element, LF_{ij}, will have a large change when there is a damage in the jth element, and the LF_{ip} of the other element $p(p \neq j)$ will only change a little. Thus, LFCR can be a meaningful indicator to localize the damage location and is defined as:

$$LFCR_{ij} = \frac{|LF_{ij}^{d} - LF_{ij}^{u}|}{LF_{ij}^{u}} \tag{2}$$

The terms in $LFCR_{ij}$, are calculated for all the elements and for every measured mode shape. If a damage exits in element p, the value of $LFCR_{ip}$, will be larger than the value of other element j in the same set with $j \neq p$. The location of damage can be identified by examining those values of $LFCR_{ij}$, which are larger than others.

3 Results and Discussion

3.1 Numerical Example

The eigenvalue problem is solved using a MATLAB code to obtain the natural frequencies and mode shapes of laminated composite plates. The material parameters of a layer used here are: $E_1/E_2 = 40$; $G_{12} = G_{13} = 0.6E_2$; $G_{23} = 0.5E_2$; $\upsilon_{12} = 0.25$. The plate is discretized using a mesh of 8 × 8 (64 elements) as shown Fig. 1. In order to compare with the published results of Ferreira and Fasshauer [14] and Liew [15], the same shear correction factors and non-dimensional natural frequencies are also employed, i.e. shear correction factor $K_s = \pi^2/12$ and non-dimensional natural frequency: $\bar{\omega} = \omega\left(b^2/\pi^2\right)\sqrt{\rho h/D_0}$ with $D_0 = E_2h^3/12(1 - \nu_{12}\nu_{21})$. The first four mode shapes are shown in Fig. 2. A six damaged scenarios are simulated by reducing the global stiffness of individual element as presented in the following equation:

$$K = \sum_{e=1}^{nele}(1 - \alpha)K^e \tag{3}$$

where K is the global stiffness matrix of damaged structures, K^e is the stiffness matrix of eth element, respectively and α represents the damage ratio.

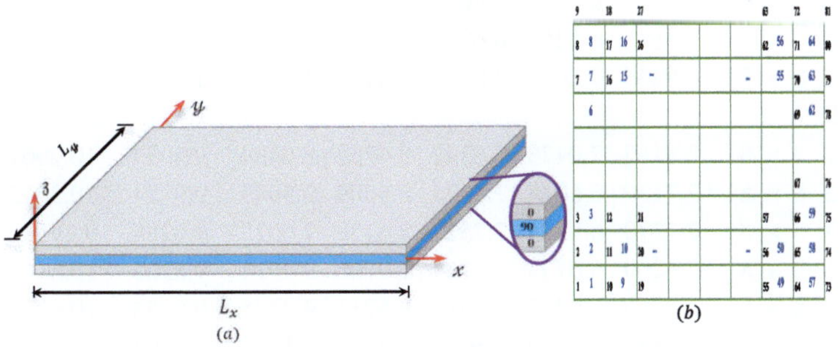

Fig. 1. Simply supported three-ply $[0°/90°/0°]$ rectangular laminated plate with a grid of 8 × 8.

All the damage scenarios are presented in Table 1 and the frequencies of each scenario are presented in Table 2.

Table 1. The damage scenarios of the laminate plate structure.

Scenarios	Damaged element & Reduction in stiffness						
Scenario 1	Element Nbr	1	–	–	–	–	–
	Reduction in stiffness [%]	10	–	–	–	–	–
Scenario 2	Element Nbr	3	12	–	–	–	–
	Reduction in stiffness [%]	20	15	–	–	–	–
Scenario 3	Element Nbr	4	9	16	–	–	–
	Reduction in stiffness [%]	15	15	20	–	–	–
Scenario 4	Element Nbr	9	10	35	43	–	–
	Reduction in stiffness [%]	20	25	25	20	–	–
Scenario 5	Element Nbr	6	7	17	25	33	–
	Reduction in stiffness [%]	30	35	30	35	20	–
Scenario 6	Element Nbr	24	32	33	40	48	56
	Reduction in stiffness [%]	35	35	35	35	35	35

Table 2. Naturel frequencies of undamaged and damaged laminate plate structure.

Intact	Grid	Mode sequence number							
		Undamaged							
		1	2	3	4	5	6	7	8
Ref [16]	11 x 11	2.3618	6.6252	6.6645	9.447	14.287	14.3846	16.1347	16.2051
Ref [14]	19 x 19	2.3728	6.6869	6.7991	8.3924	9.6042	13.9864	14.0793	15.8732
Ref [15]		2.367	6.6331	6.6691	9.4676	14.2921	14.3915	16.1009	16.1009
Exact [17]		2.3618	6.6252	6.6845	9.447	14.2869	16.3846	16.1347	16.2051
FEM	8 × 8	2.8235	7.0670	7.8433	9.4901	10.1453	10.2625	12.6285	12.7133
Damaged									
Scenario 1	8 × 8	2.8162	7.0473	7.8342	8.4762	8.9484	9.8977	10.0961	12.2751
Scenario 2	8 × 8	2.7530	6.6088	7.8024	8.6309	9.8943	9.9724	10.2013	12.3480
Scenario 3	8 × 8	2.7970	6.8888	7.7523	8.4158	9.0023	9.6288	9.8955	11.6305
Scenario 4	8 × 8	2.6200	6.6733	7.2264	7.9706	9.4914	9.8203	10.8168	11.0221
Scenario 5	8 × 8	2.7766	6.9084	7.5014	8.6211	9.2325	9.8053	10.6823	11.9798
Scenario 6	8 × 8	2.7853	7.0320	7.3998	8.4977	9.5626	9.8503	11.6734	12.0826

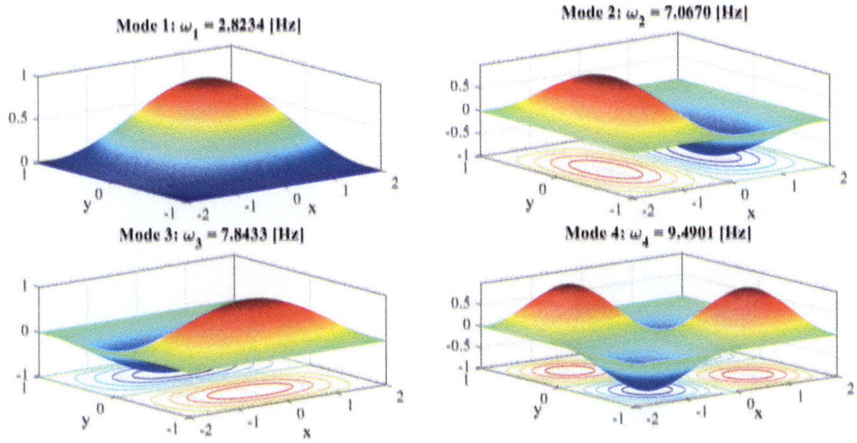

Fig. 2. Mode shapes for simply supported three-ply [0°/90°/0°] rectangular laminated plate with a grid of 8 × 8.

Figure 3 shows the error between the different damage Scenarios frequencies and the exact frequencies [17].

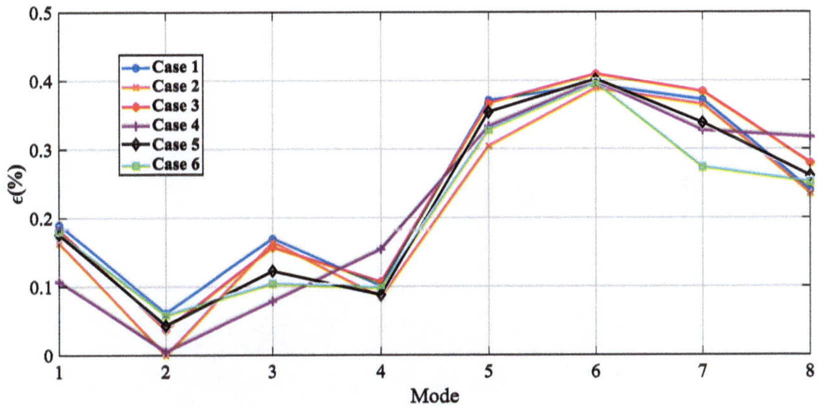

Fig. 3. Simply supported three-ply [0°/90°/0°] rectangular laminated plate: errors of the nondimensional fundamental frequency with different damage scenarios $\left(\in = {|\omega_{Exact} - \omega_{\#}|}/{\omega_{Exact}}\right)$.

3.2 Damage Scenarios

This section is conducted to investigate the capability of LFCR for damage identification by studying numerical example of laminated composite plates with single and multiple damages.

In the first scenario, it is assumed that the first element in the laminated composite plate is 10% damaged. The obtained result using the LFCR indicator is presented in Fig. 4.

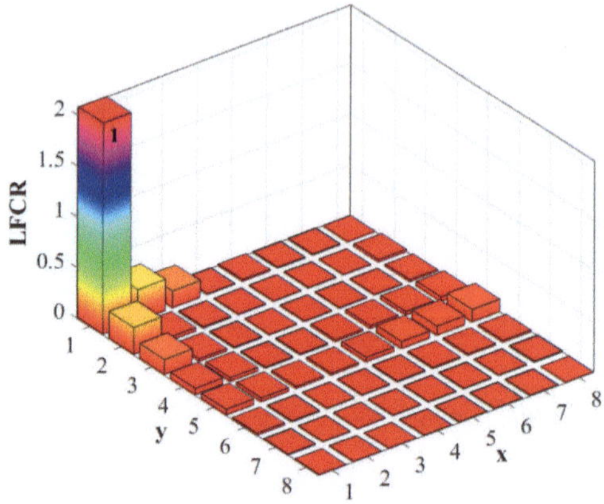

Fig. 4. Damage location for single damage - Scenario 1.

In the second scenario, two elements are affected, i.e. 3 and 12, with a reduction in stiffness 20% and 15%, respectively. The obtained result is presented in Fig. 5.

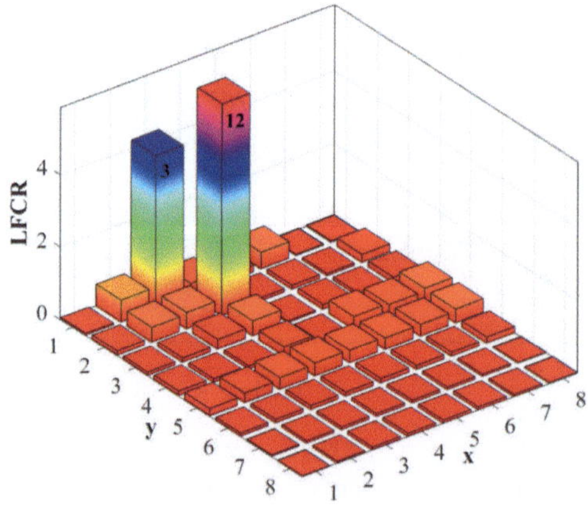

Fig. 5. Damage location for two damages - Scenario 2.

In the third scenario, three damaged elements, i.e. 4, 9 and 16, with a reduction in stiffness of 15%, 15%, and 20%, respectively. The obtained result is presented in Fig. 6.

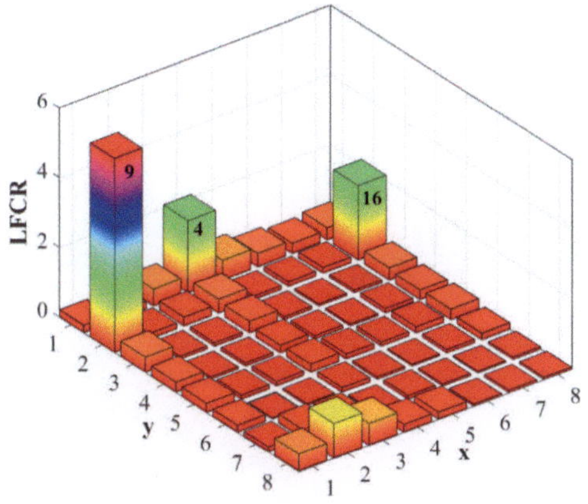

Fig. 6. Damage location for three damages - Scenario 3.

In the fourth scenario, four damaged elements, i.e. 9, 10, 35 and 43, with a reduction in stiffness of 20%, 25%, 25% and 20%, respectively. The potential of damaged elements obtained by LFCR is presented in Fig. 7.

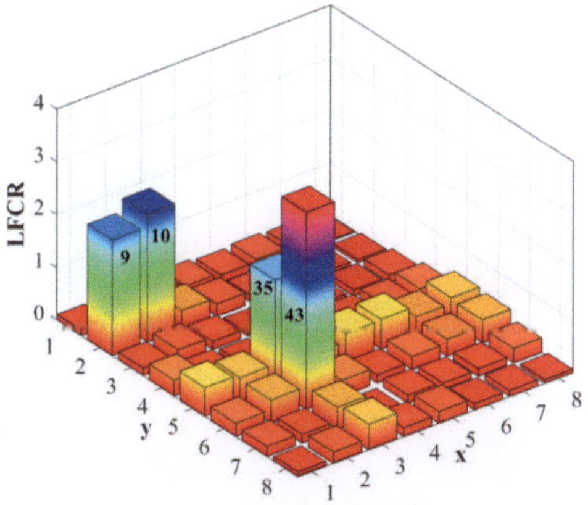

Fig. 7. Damage location for four damages - Scenario 4.

In the fifth scenario, LFCR investigated to predict five damaged elements, i.e. 6, 7, 17, 25 and 33, with a reduction in stiffness of 30%, 35%, 30%, 35% and 20%, respectively. The obtained potential of damaged elements is presented in Fig. 8.

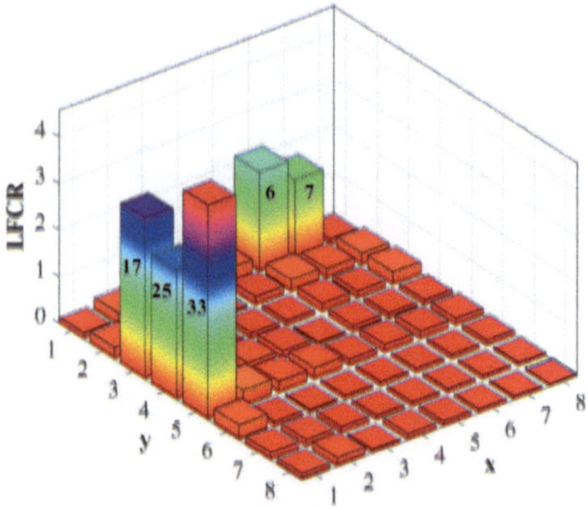

Fig. 8. Damage location for five damages - Scenario 5.

In the last scenario, also LFCR is investigated to predict six damaged elements, i.e. 24, 32, 33, 40, 48 and 56, with a reduction in stiffness of 35% for all damaged elements. The obtained potential of damaged elements is presented in Fig. 9.

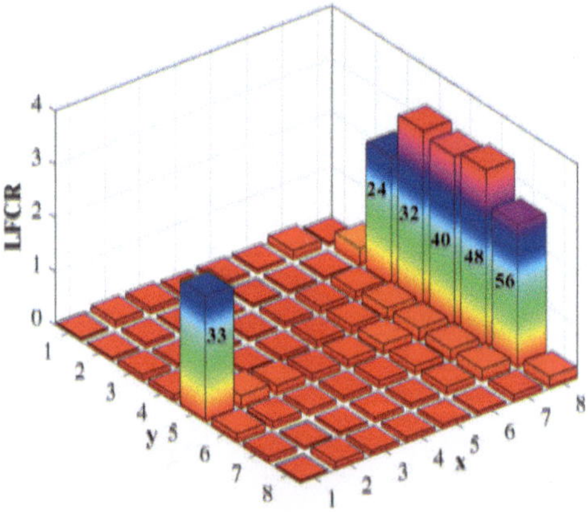

Fig. 9. Damage location for six damages - Scenario 6.

The obtained results from Sect. 4 show that the LFCR can predict the damage location correctly even when increasing the number of damaged elements. Moreover, the LFCR found difficulties to predict the quantification of damage.

4 Effect of Noise

To investigate the effect of noise, we introduced white Gaussian noise with different level in the modal data using two scenarios with single and multiple damages.

In the first case, it is assumed that the first element in the laminated composite plate is 10% damaged with 2%, 4%, and 6% noise level. The LFCR can predict the location of damaged elements in the case of a noise level of 2%. The obtained results are presented in Fig. 10.

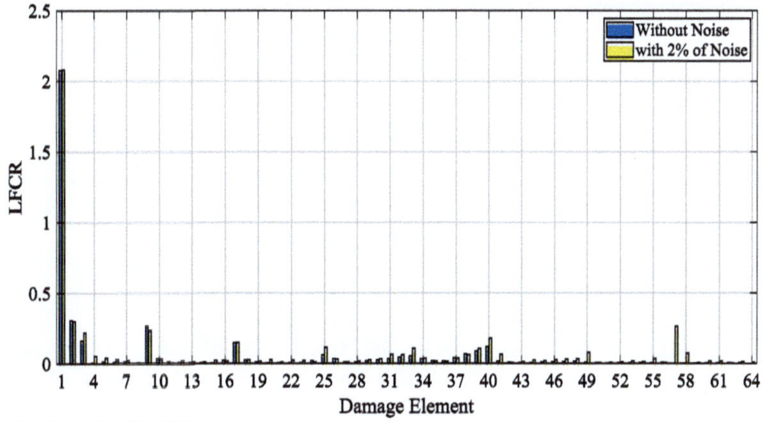

(a) Case 1 with 2%

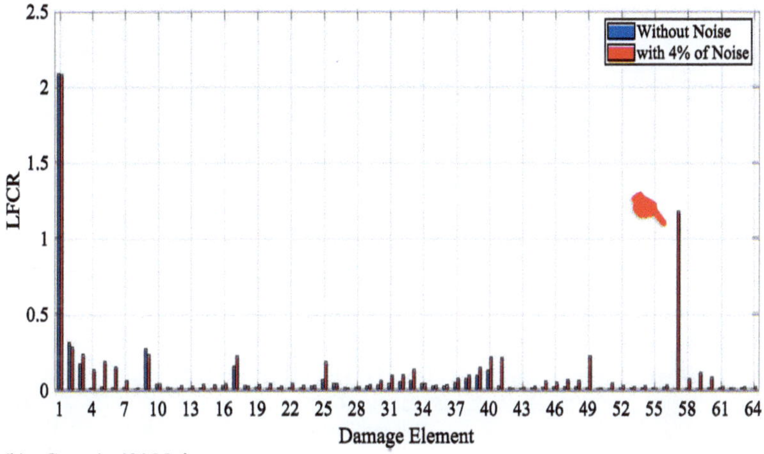

(b) Case 1: 4% Noise

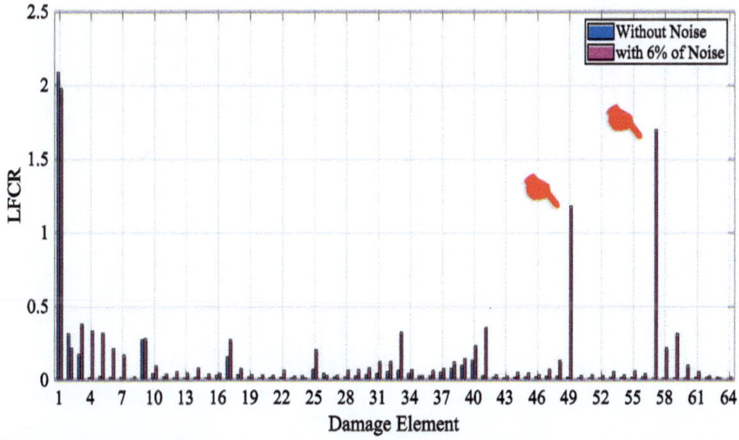

(c) Case 1: 6% Noise

Fig. 10. Damage identification with noise level 2%, 4% and 6% for single damage - case 1

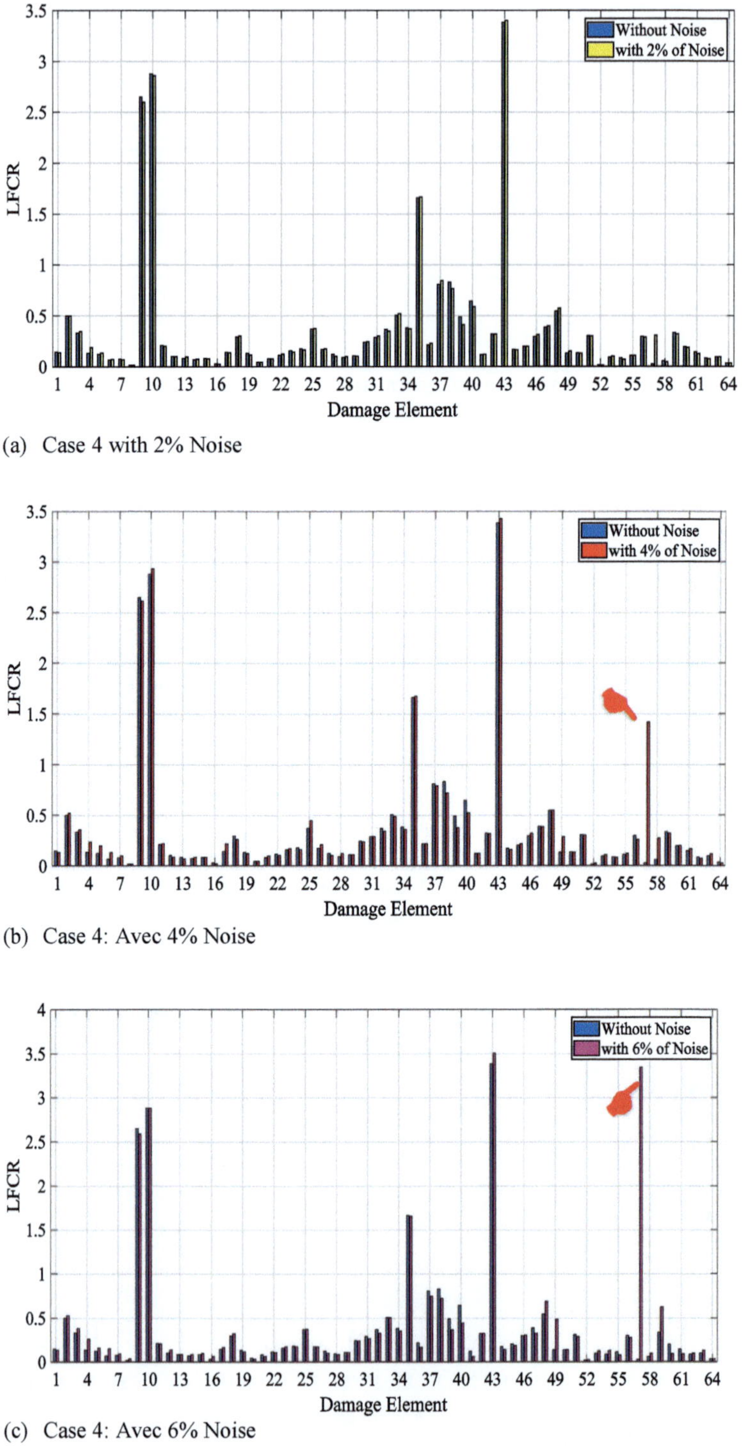

Fig. 11. Damage identification with noise level 2%, 4% and 6% for four damages - case 4

In the second case, four damaged elements 9, 10, 35 and 43 are considered with a reduction in stiffness of 20%, 25%, 25%, and 20%, respectively and by introducing noise level with 2%, 4%, and 6%. In the case of noise level, 2% the LFCR can predict the damaged element correctly. Moreover, when we increase the noise level more than 4% the indicator cannot predict the right damaged elements. The obtained results are presented in Fig. 11.

When we introduce the noise in both cases, LFCR can predict the damage accurately in the case of noise level less than 4%. For the noise level 4%, the suspected damage elements for both cases are not well identified in scenario 1, as well as, in the scenario of four damages, scenario 4.

5 Conclusion

An effective method based on Local Frequency Change Ratio (LFCR) has been presented for damage assessment of three-layer (0°/90°/0°) laminated composite plates. Furthermore, white Gaussian noise with different level is considered. The following conclusion can be drawn:

1- The LFCR provides superior accuracy for locating the right damage location in the scenarios of single and multiple damaged element.
2- By introducing noises, the results showed that LFCR predicts the damage with more accuracy for damaged elements in the case noise level 2%. However, some damaged elements are wrongly identified when we increase the noise level more than 2%.

References

1. Samir, K., et al.: Damage detection in CFRP composite beams based on vibration analysis using proper orthogonal decomposition method with radial basis functions and cuckoo search algorithm. Compos. Struct. **187**, 344–353 (2018)
2. Capozucca, R.: Vibration of CFRP cantilever beam with damage. Compos. Struct. **116**, 211–222 (2014)
3. Capozucca, R., Bonci, B.: Notched CFRP laminates under vibration. Compos. Struct. **122**, 367–375 (2015)
4. Khatir, S., et al.: Structural health monitoring using modal strain energy damage indicator coupled with teaching-learning-based optimization algorithm and isogoemetric analysis. J. Sound Vib. **448**, 230–246 (2019)
5. Dahak, M., Touat, N., Benseddiq, N.: On the classification of normalized natural frequencies for damage detection in cantilever beam. J. Sound Vib. **402**, 70–84 (2017)
6. Wu, Y., et al.: Numerical and experimental study on crack identification based on the piezoelectric ceramic lead zirconate titanate impedance technology. J. Intell. Mater. Syst. Struct. **30**(11), 1706–1716 (2019)
7. Vo-Duy, T., et al.: A two-step approach for damage detection in laminated composite structures using modal strain energy method and an improved differential evolution algorithm. Compos. Struct. **147**, 42–53 (2016)

8. Tiachacht, S., et al.: Damage assessment in structures using combination of a modified Cornwell indicator and genetic algorithm. Eng. Struct. **177**, 421–430 (2018)
9. Zenzen, R., et al.: A damage identification technique for beam-like and truss structures based on FRF and Bat Algorithm. Comptes Rendus Mécanique **346**(12), 1253–1266 (2018)
10. Zenzen, R., et al.: Structural health monitoring of beam-like and truss structures using frequency response and particle swarm optimization. In: Numerical Modelling in Engineering. Springer (2018)
11. Capozucca, R., Magagnini, E.: Experimental vibration response of homogeneous beam models damaged by notches and strengthened by CFRP lamina. Compos. Struct. **206**, 563–577 (2018)
12. Gillich, G.-R., Praisach, Z.-I.: Modal identification and damage detection in beam-like structures using the power spectrum and time–frequency analysis. Signal Process. **96**, 29–44 (2014)
13. Shi, Z., Law, S., Zhang, L.: Two stages damage detection in structure based on modal data. In: Proceedings of the 15th International Modal Analysis Conference (1997)
14. Ferreira, A., Fasshauer, G.: Analysis of natural frequencies of composite plates by an RBF-pseudospectral method. Compos. Struct. **79**(2), 202–210 (2007)
15. Liew, K.M., Ng, T.Y., Zhao, X., Reddy, J.N.: Harmonic reproducing kernel particle method for free vibration analysis of rotating cylindrical shells. Comput. Methods Appl. Mech. Eng. **191**(37–38), 4141–4157 (2002)
16. Ngo-Cong, D., et al.: Free vibration analysis of laminated composite plates based on FSDT using one-dimensional IRBFN method. Comput. & Struct. **89**(1–2), 1–13 (2011)
17. Reddy, J.N.: Mechanics of Laminated Composite Plates and Shells: Theory and Analysis, CRC press (2004)

The Impact of the Selected Exploitation Factors on the Adhesive Joints Strength

Anna Rudawska[1(✉)], Izabela Miturska[1], Jakub Szabelski[1], M. Abdel Wahab[2], Dana Stančeková[3], Nadežda Čuboňová[3], and Radovan Madleňák[4]

[1] Faculty of Mechanical Engineering, Lublin University of Technology, Nadbystrzycka 36 Str, 20-618 Lublin, Poland
{a.rudawska, i.miturska, j.szabelski}@pollub.pl
[2] Faculty of Engineering and Architecture, Ghent University, Technologiepark Zwijnaarde 903, 9052 Zwijnaarde, Belgium
magd.abdelwahab@ugent.be
[3] Faculty of Mechanical Engineering, University of Žilina, Univerzitna 1, 010 26 Žilina, Slovakia
{dana.stancekova, nadezda.cubonova}@fstroj.uniza.sk
[4] Faculty of Operation and Economics of Transport and Communications, University of Žilina, Univerzitna 1, 010 26 Žilina, Slovakia
radovan.madlenak@fpedas.uniza.sk

Abstract. The aim of the present article is to determine the impact of the selected exploitation factors on strength of the wooden elements' adhesive joints. Two types of adhesive joints were subject to experimental tests: butt adhesive joints and cross lap adhesive joints. Two types of adhesives were used: Prefere 6312 (a single-component PVAc adhesive for wood of D3 quality) and a two-component epoxy adhesive (Epidian 57/TFF/100:22). The exploitation factor under analysis was resistant to different temperature values above and below freezing. Six different variants of the joints' ageing temperature value were used. The adhesive joints have been aged for 2 weeks (2 variants) and 3 months (3 variants). One ageing variant also included thermal shock (−40 °C/ +60 °C), for which the ageing time was 2 months. A climatic/temperature chamber ESPEC SH 661 and a thermal shock chamber ESPEC TSE – 1 were used for ageing. Strength tests of the adhesive samples were carried out on the testing machine Zwick/Roell Z150, with accordance to the PN-EN 15870 standard. Based on the tests, it was observed that the adhesive joints of the wooden elements showed different dependencies between the strength and the ageing variant depending on the type of adhesive used. The results emphasise the importance of the temperature impact on the adhesive joints strength.

Keywords: Adhesive joints · Adhesive · Wood · Ageing · Strength

M. A. Wahab (Ed.): DAMAS 2019, LNME, pp. 899–913, 2020.
https://doi.org/10.1007/978-981-13-8331-1_72

1 Introduction

The strength of the adhesive bonded-joints depends on numerous factors: structural, technological, material and exploitation ones [1–6]. Among the basic constructional factors having a significant impact on the strength of adhesive joints, the most important of them is the type of joints [3, 4, 7, 8]. Adhesive bonds may be of different shapes, which are demonstrated on the example of butt and lap joints [1]. The structural factors also include the adherends and adhesives thicknesses and the lap length [1, 3, 6, 9–12]. The technological factors include for example: the method of surface treatment [13–18] and temperature, pressure and curing time [5, 19–22]. Material factors are also of great importance. They include, among others, type and characteristics of the material and type of adhesive [1, 2, 6, 7, 19, 23, 24]. The last group of factors impacting the adhesive joints' strength includes the exploitation ones [24–34]. These are the conditions in which the adhesive bonds are exploited.

An example of such factor may be temperature that acts on the adhesive bonds for a specified period [5, 35]. The issues related to the temperature's impact on the adhesive joints' strength were described in numerous publications [36–39].

The present article describes the impact of the various temperature values, both positive and negative ones, on the wooden elements' adhesive joints. Both time of the adhesive joints' exposure to temperature and the adhesive type were taken into consideration in the analysis.

2 Methodology

2.1 Adherend Characteristics

The experimental part is related to the strength tests conducted on the samples made of one of the most popular pine type, i.e. common pine (Scotch pine). The common pine is a type of tree that tolerates numerous kinds of soil, which is why it grows in almost whole Europe and Asia. It is also very popular in Poland as it represents ca. 70% of the whole forest area. It occurs in almost all environmental conditions, from the sandy areas, through the clayey ones to the peat lands [40].

2.2 Adhesive Joints

In order to conduct the strength tests, 480 samples made of pine wood were prepared. 240 adhesive joints were created, including 120 adhesive butt joints and 120 adhesive cross lap joints. The adhesive joints enumerated above are presented in Table 1.

When analysing the wooden samples' shape, it should be pointed out that each sample is a bit different when it comes to dimensions. There are very low dimensional deviations that oscillate at the accuracy level of hundredths part of millimetre. Thus, the dimensions that are the closest to the real dimensions were assumed for the tests' needs. The joints' shapes and dimenssions are shown in Fig. 1.

Table 1. Adhesive joints types used in the experimental tests.

Type and number of joints	Main view of the adhesive joint
Butt joints / 120 pieces	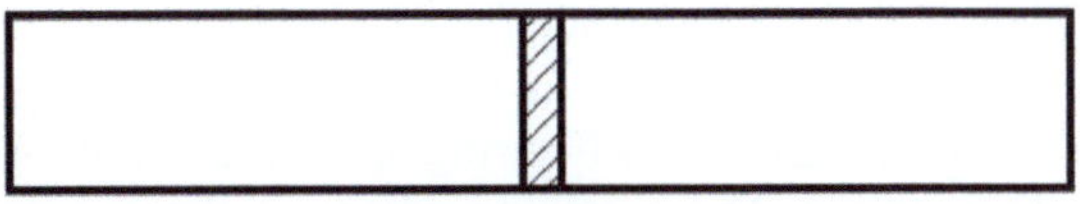
Cross lap joints / 120 pieces	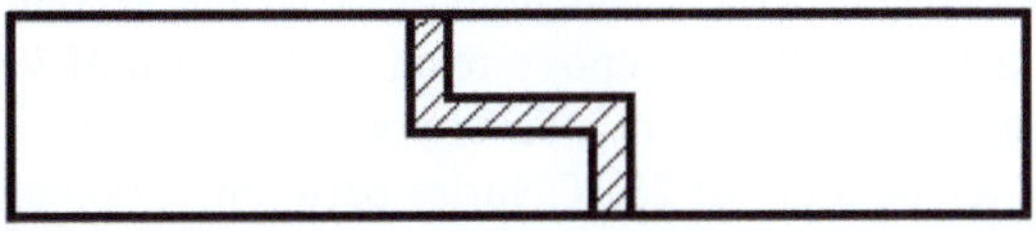

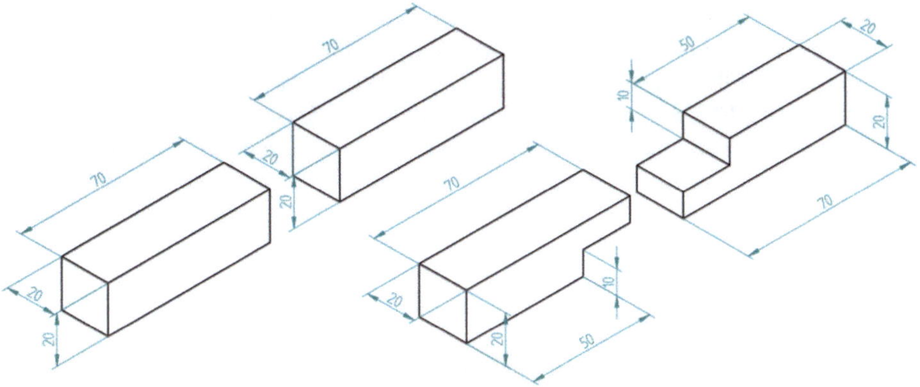

Fig. 1. Shape and dimension of the elements samples used for experimental tests.

2.3 Adhesives Characteristics

Strength tests were conducted with use of the types of adhesives: a single-component adhesive based on water dispersion PVAc and a two-component epoxy adhesive. They were used to stick the wooden samples together in order to obtain the adhesive joints.

Prefere 6312 is a single-component, wood-oriented adhesive based on water dispersion PVAc [41]. It is also used for other materials, e.g. wood-metal bonding. Such solutions, though, often cause corrosion because Prefere 6312 is acidic, i.e. its PH varies between 2.8 and 3.2, which causes reaction with metal. Moreover, this adhesive is water resistance-tested, is in class D3 and is characterised by high thermal resistance. It is of white colour and viscous, i.e. its viscosity range is 15000 ± 3000 MPas. The adhesive's life is about 6 months, provided it is stored in appropriate conditions (temperature between 10 and 20 °C, tight containers). The adhesive layer should be about 100–200 g/m^2, but it mostly depends on the surface, material type and its absorptivity. Prefere 6312 gains initial hardness after just 6–7 min at the temperature of ca. 20 °C. A factor that is necessary for curing is feed force, whose value should be between 0.2–0.6 N/mm^2 (2–6 kP/cm^2). As a consequence of such feed force, the adhesive is usually squeezed out of the joint, which provides information whether the feed force's value is appropriate. The time of exerting the feed force should be between 5 and 10 min. However, it mostly depends on the adhesive's thickness. Prefere 6312 gets its highest strength after 24 h.

The second adhesive a two-component epoxy adhesive made from Epidian 57 epoxy resin [42] and TFF curing agent [43] in the stoichiometric ratio 100:22 (Epidian 57/TFF/100:22).The main ingredient of the second adhesive compound used in the tests is the Epidian 57 epoxy resin in the form of viscous liquid of yellowish to light-brown colour. The resin's density at 20 °C is 1.14 – 1.17 g/cm^3, whereas its viscosity at the temperature of 25 °C varies between 13000 and 19000 MPas. Epidian 57 epoxy resin dissolves in alcohols, ketones, esters, aromatic hydrocarbons. It does not dissolve in water. It is necessary to add a selected curing agent in appropriate proportions to the resin. Thanks to this, the cured adhesives will show high pull-off strength.

The TFF curing agent (the Mannich type base) is used for curing of the epoxy compounds dedicated for construction works where air humidity is high and temperature is relatively low. The TFF curing agent' viscosity at the temperature of 25 °C amounts to 10000 MPas, whereas its density at 20 °C varies from 1.15 to 1.20 g/cm^3.

In order to prepare the adhesive, epoxy resin was mixed with the TFF curing agent in the proportion 100:22, i.e. 22 g of curing agent per 100 g of epoxy resin. In such proportion, the break tensions' values are about 50 to 60 MPa. Bending strength varies between 80 to 90 MPa, whereas compressive strength is about 55 MPa.

2.4 Conditions of Making Joints

The wooden samples were made on a milling machine with use of a milling cutter with a diameter of 70 mm. The samples of desired dimensions were then grinded because after being cut with a circular saw, the burrs occurred on some samples' surfaces. Sandpaper with 180 grit was used in order to remove them. The wooden samples' surfaces were cleaned out of wood and dust filing that occurred during the operations conducted beforehand. To this end, an air blow gun connected to a compressor was used.

After having prepared the surface of the elements to be joined, the adhesives were prepared and applied. A half on the prepared samples was joined with a single-component adhesive Prefere 6312, which was mixed with a stick before the application. The second part of the wooden samples was joined with the Epidian 57/TFF/100:22 epoxy adhesive compound. Both components (epoxy resin and curing agent) were mixed mechanically in the stoichiometric ratio 100:22, with use of a horseshoe mixer on a bench drill. The mixing speed was 460 rpm and the mixing time was 2 min. The mixing process was aimed at obtaining the best bonding properties possible.

The next stage of the bonding process was to apply the adhesive on the surfaces to be joined. The adhesive was distributed manually on both surfaces of the joined elements. It was done evenly, with one stroke of a brush.

After the adhesive was applied, the wooden samples were stuck together in order to obtain an adhesive joint. When sticking the samples together, the adhesive got out of the joined surface in many cases. Its excess was removed with a cloth.

A special fixing device [44] was used to exert the feed force during curing and to set the right position of the elements. It enabled to fix 48 joints at the same time. The feed during curing was exerted with use of the press bolts. The press bolts' feed force was ca. 250 MPa. The adhesive joints' curing time was 48 h. The curing process of all adhesive joints was conducted at ambient temperature and air humidity ca. 31–35%.

2.5 Description of the Ageing Conditions

The process of the adhesive joints' ageing was realised in six variants as presented in Table 2.

Table 2. The ageing process parameters and designation of the ageing.

Ageing variant	V1	V2	V3	V4	V5	V6
Temperature, °C	−20 °C	+6 °C	+20 ± 1 °C	+50 °C	+80 °C	−40 °C/+ 60 °C
Ageing time, weeks/months	3 months	3 months	3 months	2 weeks	2 weeks	2 months

The ageing process was conducted in the climatic-temperature chamber ESPEC SH 661- product of ESPEC, Klimatest, Poland [45] (which is characterised by the ability to work in the temperature range –60 °C to 150 °C and air humidity 30% to 95%) and the thermal shock chamber ESPEC TSE – 1 (product of ESPEC, Klimatest, Poland [45]), as well as in the laboratory with constant temperature (20 ± 1 °C).

In the climatic chamber where the temperature was of 50 °C, the air humidity for variant 4 (V4) was of 95%, whereas for variant 5 (V5) it was of 80%.

The sixth ageing variant (V6) was related to the use of thermal shocks in a compact, 2-zone thermal shock chamber. It was aimed at obtaining the information on the impact of sudden thermal changes on the adhesive joints' strength. The first zone of the thermal shock chamber, i.e. the hot zone, had the set temperature of 60 °C, whereas the second zone (cold) had the set temperature of –20 °C. Thanks to a mobile basket of 11 litre capacity, which is an element of the thermal shock chamber, it was possible to move the adhesive joints from one temperature to another. The ageing time at 60 °C was 15 min. After that the mobile basket went down to the cold chamber (–20 °C), where the stabilisation process occurred. When putting the hot adhesive joints into the cold chamber, its temperature increased to –12 °C. The stabilisation process was then about waiting until the temperature in the cold zone would decrease to –20 °C. It lasted from 2 to 5 min. After that, the adhesive joints were left at this temperature for 15 min more (Fig. 2). The number of cycles was 3300, whereas the time spent by the adhesive joints in the thermal shock chamber was 2 months.

In addition, one reference sample, which was not subject to the ageing process, was made. It was subject to the strength tests after curing. The samples, together with the exploitation factors, were placed in the containers made of chemically inert polymers.

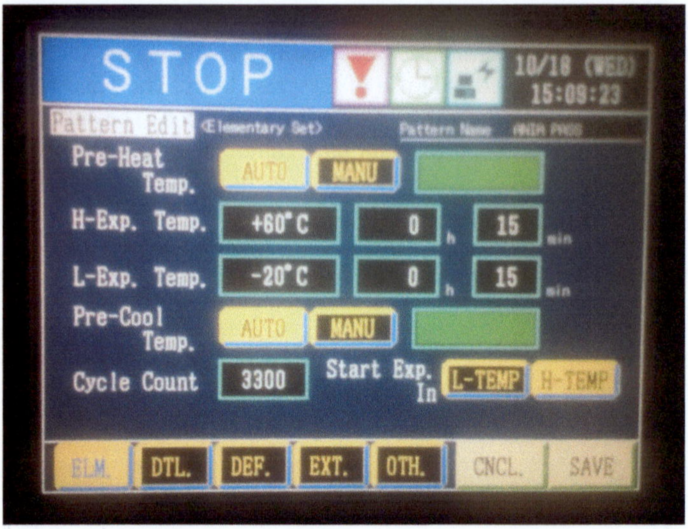

Fig. 2. Diagram from the thermal shocks chamber ESPEC TSE – 1.

2.6 Strength Tests Characteristics

After the ageing process, in accordance with the specific ageing variant, the strength tests were conducted. The tensile strength tests were carried out with accordance to the PN-EN 15870 standard [46] on the testing machine Zwick/Roell Z150.

3 Experimental Results

3.1 Adhesive Butt Joints Strength

The obtained results were divided into three groups with reference to various ageing times in specific temperature conditions. Three diagrams presenting the impact of the ageing conditions and the adhesive type on the adhesive butt joints were prepared. Figure 3 shows the impact of the ageing temperature on the adhesive butt joints' strength in relation to the adhesive type (variants: V1, V2 and V3).

Taking into consideration the temperature of –20 °C (V1 variant), it may be observed that the adhesive joints made with use of the Prefere 6312 adhesive showed higher strength, which was equal to 5.32 MPa. The adhesive joints made with use of Epidian 57/TFF/100:22 adhesive showed average strength of 4.71 MPa. The difference between them is 0.52 MPa, which is 10% of the strength value.

In case of the samples aged at the temperature of 6 °C (V2 variant), it may be observed that the adhesive joints made with use of the Prefere 6312 adhesive showed higher strength, which was equal to 6.31 MPa. A lower strength value was observed for the adhesives joints made with use of the Epidian 57/TFF/100:22 adhesive compound; it amounted to 4.70 MPa, which was 75% of the highest strength value.

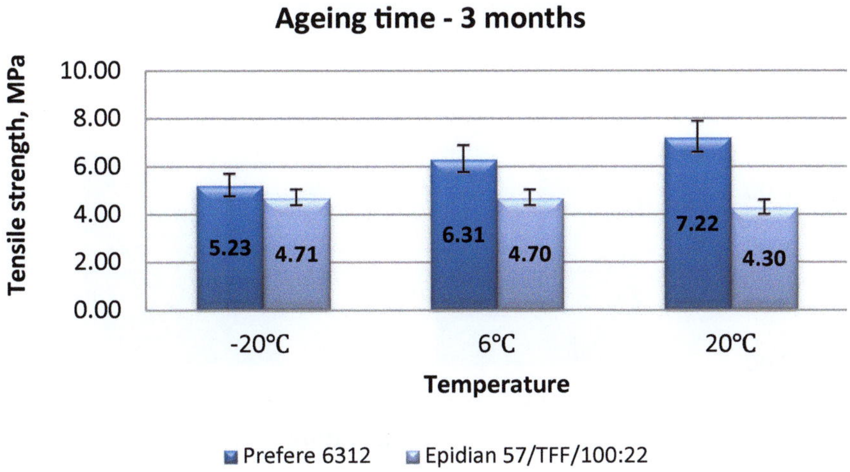

Fig. 3. The impact of the ageing temperature on the adhesive butt joints' strength in relation to the adhesive type (variants: V1, V2 and V3).

The last variant (V3) is related to the adhesive joints aged at 20 °C. As in the previous cases, the adhesives joints made with the Prefere 6312 adhesive showed higher strength, whose value amounted to 7.22 MPa, whereas the lowest strength value was of 4.30 MPa. The difference between them was of 2.92 MPa, which means that the adhesive joints made with use of Prefere 6312 adhesive were 40% stronger than those made with the Epidian 57/TFF/100:22 adhesive compound.

It is worth mentioning that the strength values 7.22 MPa and 4.30 MPa are also the highest and the lowest strength values obtained for all adhesive butt joints that were aged at three temperature variants, i.e.: –20 °C (V1), 6 °C (V2) and 20 °C (V3).

Figure 4 shows the impact of elevated temperature, i.e. 50 °C (V4) and 80 °C (V5) on the adhesive butt joints' strength in relation to the adhesive type. The ageing time in these variants was 2 weeks.

When analysing the impact of the ageing temperature 50 °C (the W4 variant) on the adhesive butt joints' strength, it may be observed that the highest strength was obtained by the adhesive joints made with the Prefere 6312 adhesive and it was of 6.90 MPa. The strength obtained by adhesive joints made with the Epidian 57/TFF/100:22 adhesive compound was 3.02 MPa lower than of the samples joined with Prefere 6312 adhesive and it amounted to 3.88 MPa. In percentage terms, it turns out that the adhesive joints made with use of Prefere 6312 adhesive are 44% stronger than those made with the Epidian 57/TFF/100:22 adhesive compound.

Taking into consideration the temperature of 80 °C (V5 variant), it may be observed that the adhesive joints made with use of the Epidian 57/TFF/100:22 adhesive compound showed the highest strength, which was equal to 7.80 MPa. In case of the adhesive joints made with use Prefere 6312 adhesive, the strength value was of 5.52 MPa, which is 2.28 MPa lower than 7.80 MPa. In percentage terms it means that the adhesive joints made with use with Epidian 57/TFF/100:22 adhesive compound are 29% stronger than those made with Prefere 6312 adhesive. Thus, it may be assumed

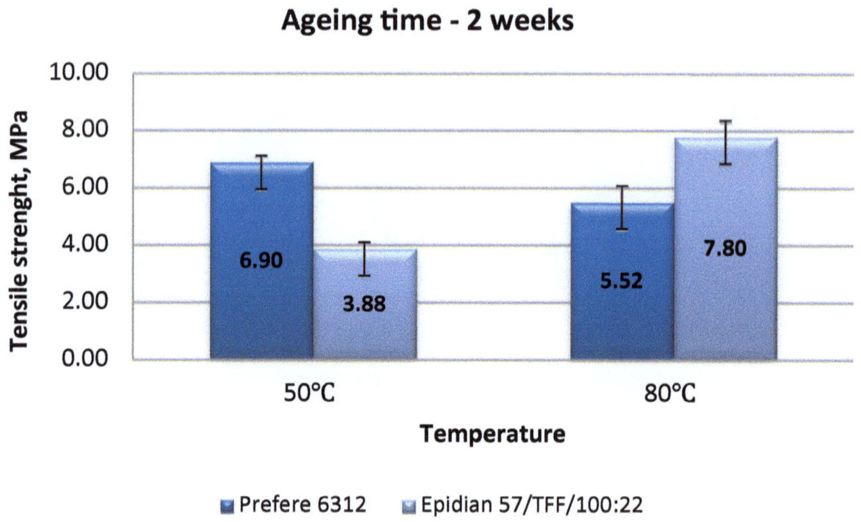

Fig. 4. The impact of the ageing temperature (V4 and V5 variants) on the adhesive butt joints' strength in relation to the adhesive type.

that the epoxy adhesive should be used when the temperature of environment is above 80 °C, whereas when the temperature is of 50 °C, the Prefere 6312 adhesive gives better results.

Figure 5 shows a diagram presenting the impact of a constant temperature of –40 °C to +60 °C for 2 months (3300 cycles) on strength of the adhesive butt joints made with two types of adhesives.

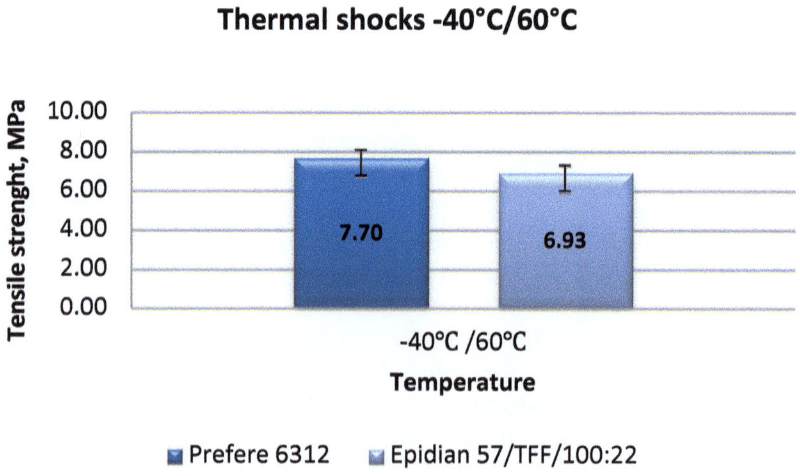

Fig. 5. Strength of the adhesive butt joints that were subject to thermal shocks (−40 °C/60 °C) - variant V6

When interpreting results of the strength tests presented in Fig. 5, it may be observed that the adhesive joints made with use of the Prefere 6312 adhesive were stronger than those made with use of the Epidian 57/TFF/100:22 adhesive compound - its value amounted to 7.70 MPa. It is different than the strength value obtained by adhesive joints made with the Epidian 57/TFF/100:22 adhesive compound, which amounted to 6.93 MPa; the difference between these two values was 0.77 MPa. In percentage terms it turns out that the adhesive joints made with the single-component adhesive based on water dispersion PVAc are 10% stronger than those made with use of the epoxy adhesive.

3.2 Adhesive Cross-Lap Joints' Strength

Figure 6 shows the comparison of strength values obtained by the adhesive joints ageing for 3 months at different temperatures, i.e. –20 °C (V1), 6 °C (V2) and 20 °C (V3).

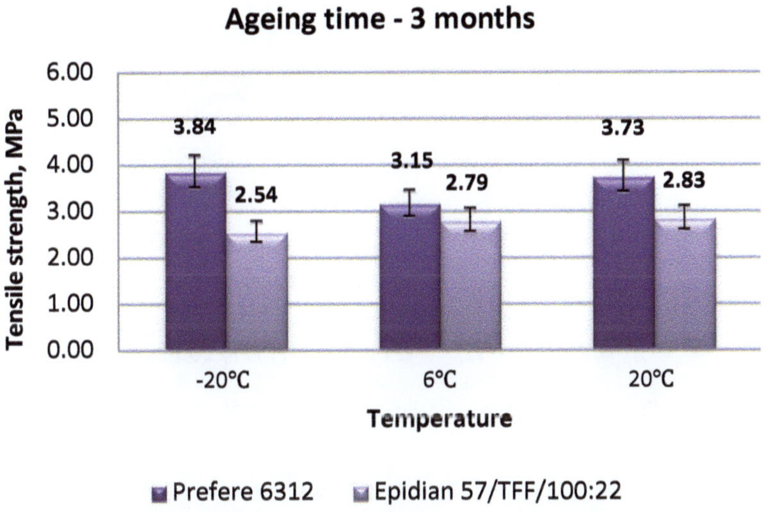

Fig. 6. Strength of the adhesive cross-lap joints aged at three temperature variants: −20 °C, 6 ° C and 20 °C

Taking into consideration the temperature of –20 °C (V1 variant), it may be observed that the adhesive joints made with use of the single-component adhesive based on water dispersion PVAc showed higher strength, which was equal to 3.84 MPa. The adhesive joints made with use of the epoxy adhesive showed the average strength of 2.54 MPa. The difference between them is 10%.

A similar dependence and the difference between the obtained strength values was observed for V2 variant - ageing at 6 °C.

The last variant (V3) in this group is related to the adhesive joints aged at 20 °C ± 1 °C. The adhesive joints made with the Prefere 6312 adhesive showed higher strength, whose value amounted to 3.73 MPa, whereas the lowest strength value was of 2.83 MPa. The difference between them was of 0.90 MPa, which means that the adhesive joints made with use of Prefere 6312 adhesive are 25% stronger than those made with the 57/TFF/100:22 adhesive compound.

It is worth mentioning that the strength values 3.84 MPa and 2.54 MPa are also the highest and the lowest strength values obtained for all adhesive cross-lap joints that were aged at three temperature variants, i.e.: –20 °C, 6 °C and 20 °C for 3 months.

Figure 7 shows results of the strength tests obtained by the adhesive cross-lap joints made with use of both single- and two-component adhesives, aged at 50 °C (V4) and 80 °C (V5) for 2 weeks.

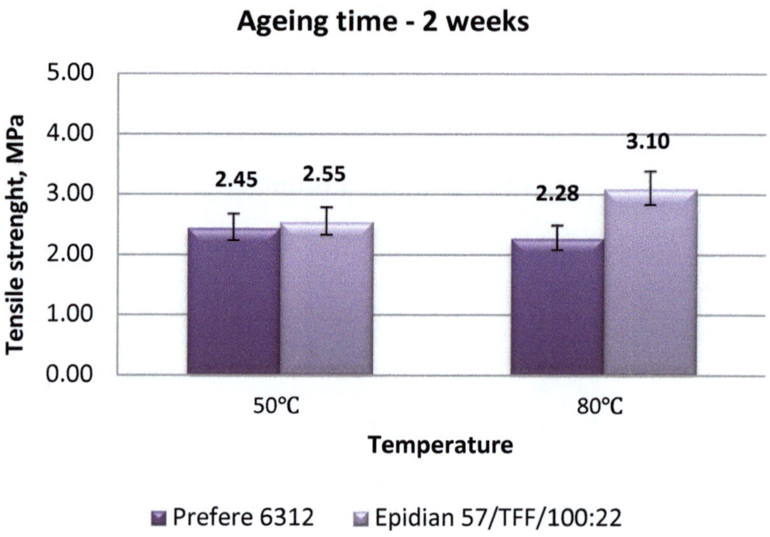

Fig. 7. Strength of the adhesive cross-lap joints aged at 50 °C (V4) and 80 °C (V5).

When analysing the impact of the elevated temperature (ageing variants: V4 and V5) on the adhesive-bonded cross-lap joints' strength, it may be observed that:

(1) In case of the adhesive joints made with a single-component adhesive based on water dispersion PVAc, lower strength was obtained by the samples from V5 variant (ageing at 80 °C), although the difference was not very significant (ca. 7%).
(2) Adhesive cross-lap joints made with use of the two-component epoxy adhesive were characterised by higher strength after being subjected to ageing at higher temperature, i.e. 80 °C (V5). The difference between strength values of the adhesive joints for variants V4 and V5 was ca. 18%.
(3) When analysing the impact of the ageing temperature 50 °C (V4 variant), it may be observed that the highest strength was obtained by the adhesive joints made with the Epidian 57/TFF/100:22 adhesive compound and it was of 2.55 MPa. The strength of the adhesive joints made with the Prefere 6312 adhesive was 0.10 MPa, which gives a 4% difference.

(4) Taking into consideration the temperature of 80 °C (V5 variant), it may be observed that the adhesive cross-lap joints made with use of the Epidian 57/TFF/100:22 adhesive compound showed higher strength, which amounted to 3.10 MPa. In percentage terms, it means that the adhesive joints made with use with Epidian 57/TFF/100:22 adhesive compound are 26% stronger than those made with Prefere 6312 adhesive.

Figure 8 shows the strength values of the adhesive cross-lap joints that were subject to thermal shocks for 2 months.

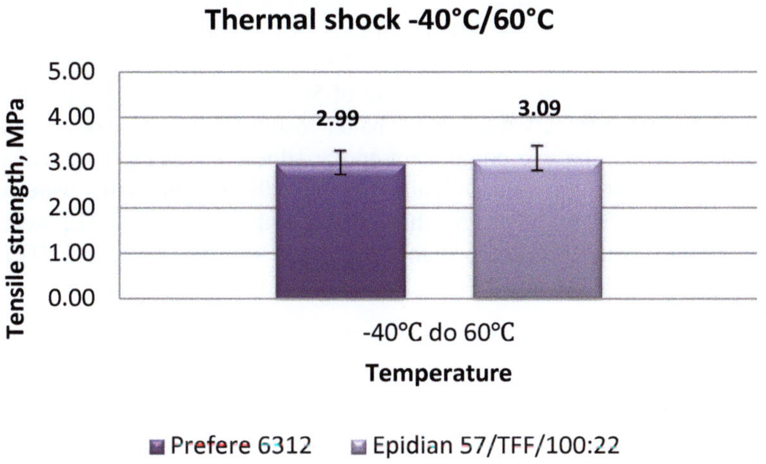

Fig. 8. Strength of the adhesive cross-lap aged in the thermal shock conditions (V6).

When interpreting the results of the strength tests presented in Fig. 8, it may be observed that the adhesive joints made with use of the epoxy adhesive were stronger than those made with use of the adhesive compound based on water dispersion PVAc - its value amounted to 3.09 MPa. It is 3% different from the strength value obtained by adhesive joints made with the Prefere 6312 adhesive. Thus, it may be assumed that the type of adhesive in such ageing conditions does not influence the analysed adhesive joints' strength.

4 Summary

Based on the test results the following conclusions may be drawn:

1. Results of the strength tests conducted on the adhesive butt joints showed that:
(i) Among 3 ageing variants at the temperature –20 °C (V1), 6 °C (V2) and 20 °C (V3) for 3 months, in case of the adhesive joints made with a single-component adhesive based on water dispersion PVAc, the most favourable ageing temperature was 20 °C (V3), whereas the least favourable temperature was –20 °C (V1). Among 3 ageing

variants at the temperature of –20 °C (V1), 6 °C (V2) and 20 °C (V3) for 3 months, in case of the adhesive joints made with a single-component adhesive based on water dispersion PVAc, the most favourable ageing temperature was 20 °C (V3), whereas the least favourable temperature was –20 °C (V1). It stems from the fact that each adhesive behaves differently at a specific temperature. Moreover, it needs to be pointed out that the single-component adhesive based on water dispersion PVAc turned out to be more effective than the two-component epoxy adhesive, as it let obtain higher strength for the adhesive joints aged in all 3 variants for 3 months.
(ii) Among two ageing variants at elevated temperature, i.e. 50 °C (V4) and 80 °C (V5), the most favourable ageing temperature for the adhesive joints made with the single-component adhesive based on water dispersion PVAc was 50 °C whereas the least favourable temperature was 80 °C. In case of the adhesive joints made with the Epidian 57/TFF/100:22 adhesive compound, the result was quite the opposite. In terms of the strength parameters at the temperature of 50 °C, the single-component adhesive based on water dispersion PVAc showed better results. When it comes to the ageing temperature of 80 °C, the two-component epoxy adhesive was more effective.
(iii) In terms of the strength values obtained by the adhesive joints that were subject to thermal shocks (–40 °C/60 °C - V6), those made with the Prefere 6312 were characterised by slightly better properties than those made with the Epidian 57/TFF/100:22 adhesive compound. However, these values are at similar level.

2. Results of the strength tests conducted on the adhesive cross-lap joints showed that:

(i) Among 3 ageing variants at the temperature –20 °C (V1), 6 °C (V2) and 20 °C ± 1 °C (V3) for 3 months, in case of the adhesive joints made with a single-component adhesive based on water dispersion PVAc, the most favourable ageing temperature was –20 °C (V1), whereas the least favourable temperature was 6 °C (V2). In case of the adhesive joints made with the Epidian 57/TFF/100:22 adhesive compound, the most favourable ageing temperature was 20 °C (V3), whereas the least favourable temperature was –20 °C (V1). It stems from the fact that each adhesive behaves differently at a specific temperature. Moreover, it needs to be pointed out that the Prefere 6312 adhesive was more effective than the Epidian 57/TFF/100:22 adhesive compound, as in all ageing variants (V1, V2 and V3) the strength of the adhesive joints made with this adhesive is higher.
(ii) Among two temperature values, i.e. 50 °C and 80 °C, the most favourable ageing temperature for the adhesive joints made with the single-component adhesive based on water dispersion PVAc was 50 °C (V4), whereas the least favourable temperature was 80 °C (V5). In case of the adhesive joints made with the Epidian 57/TFF/100:22 adhesive compound, the most favourable ageing temperature was 80 °C, whereas the least favourable temperature was 50 °C. Moreover, it needs to be pointed out that the Epidian 57/TFF/100:22 adhesive compound was more effective than the Prefere 6312 adhesive, as it showed better bonding properties in all exploitation conditions where the ageing time was 2 weeks.
(iii) In case of the adhesive joints subject to thermal shocks (−40 °C/60 °C), the Epidian 57/TFF/100:22 adhesive compound showed slightly better bonding properties than the Prefere 6312 adhesive. However, strength of the adhesive cross-lap joints

made with both the epoxy adhesive and the single-component adhesive based on water dispersion PVAc is at comparable level.

3. While conducting strength tests of adhesive-bonded joints, almost all joints were damaged. Only in two cases, the joined material was damaged.
4. Both adhesives used to join the wooden samples show completely different behaviour at different temperatures, which proves that exploitation conditions have enormous impact on the adhesive's characteristics and the adhesive joints' strength.

To conclude, it needs to be stated that the temperature plays a vital role during the ageing (exploitation) process of the adhesive joints. Based on the tests it also turns out that it is necessary to adjust the adhesive type to the ageing temperature. It will permit us obtain a specific joints' strength depending on the ageing temperature. Results of the tests strengths showed a significant impact of the selected exploitation conditions (temperature and exploitation time) on the adhesive joints' strength.

References

1. Adams, R.D., Comyn, J., Wake, W.C.: Structural Adhesive Joints in Engineering, 2nd edn. Chapman & Hall, London (1997)
2. Wahab, M.A.: Joining Composites with Adhesives. DEStech Publications Inc, Lancaster (2016)
3. Adams, R.D.: Adhesive Bonding. Science, Technology and Applications. Woodhead Publishing Limited, Camridge (2010)
4. Kinloch, A.J.: Adhesion and Adhesives. Science and Technology. Springer, Dordrecht (1987)
5. Mays, G.C., Hutchinson, A.R.: Adhesives in Civil Engineering. Cambridge University Press, Cambridge (1992)
6. da Silva, L.F.M., Carbas, R.J.C., Critchlow, G.W., Figueiredo, M.A.V., Brown, K.: Effect of material, geometry, surface treatment and environment on the shear strength of single lap joints. Int. J. Adhes. Adhes. **29**, 621–632 (2009)
7. Brockmann, W., Geiß, P.L., Klingen, J., Schröder, B.: Adhesive Bonding. Materials, Applications and Technology. Wiley-Vch Press, Weinheim (2009)
8. Packham, D.E. (ed.): Handbook of Adhesion. Longman, New York (1992)
9. Rudawska, A.: Influence of the thickness of joined elements on lap length of aluminium alloy sheer bonded joints. Adv. Sci. Technol. Res. J. **9**, 35–44 (2015)
10. Arenas, J.M., Narbón, J.J., Alía, C.: Optimum adhesive thickness in structural adhesives joints using statistical techniques based on Weibull distribution. Int. J. Adhes. Adhes. **30**, 160–165 (2010)
11. States, D.N., deVries, K.L.: Geometric factors impacting adhesive lap joint strength and design. Int. J. Adhes. Adhes. **26**, 89–107 (2012)
12. Adams, R.D., Harris, J.A.: The influence of local geometry on the strength of adhesive joints. Int. J. Adhes. Adhes. **7**, 69–80 (1987)
13. Rudawska, A., Reszka, M., Warda, T., Miturska, I., Szabelski, J., Stančeková, D., Skoczylas, A.: Milling as a method of surface pre-treatment of steel for adhesive bonding. J. Adhes. Sci. Technol. **23**, 2619–2636 (2016)

14. Rudawska, A., Bociąga, E., Olewnik-Kruszkowska, E.: The effect of primers on adhesive properties and strength of adhesive joints made with polyurethane adhesives. J. Adhes. Sci. Technol. **31**, 327–344 (2017)
15. Rudawska, A., Danczak, I., Müller, M., Valasek, P.: The effect of sandblasting on surface properties for adhesion. Int. J. Adhes. Adhes. **70**, 176–190 (2016)
16. Ebnesajjad, S., Ebnesajjad, C.: Surface Treatment of Materials for Adhesive Bonding, 2nd edn. William Andrew Inc, Norwich, New York (2013)
17. Prolongo, S.G., Ureña, A.: Effect of surface pre-treatment on the adhesive strength of epoxy-aluminium joints. Int. J. Adhes. Adhes. **29**, 23–31 (2009)
18. Hass, P., Kläusler, O., Schlegel, S., et al.: Effects of mechanical and chemical surface preparation on adhesively bonded wooden joints. Int. J. Adhes. Adhes. **51**, 95–102 (2014)
19. Rudawska, A.: Pressure during curing and the strength of 2024, 2017A and 1050 aluminium alloy sheet adhesive joints. Adv. Sci. Technol. Res. J. **9**(26), 96–103 (2015)
20. Lee, H.L., Neville, H.: Handbook of Epoxy Resins. McGraw-Hill, New York (1988)
21. Petrie, E.M.: Epoxy Adhesive Formulations. McGraw-Hill Professional (2006)
22. May, C.A.: Epoxy Resins, Chemistry and Technology, 2nd edn. Marcel Dekker, New York (1988)
23. Rudawska, A., Głogowska, K., Vitenko, T., Stančeková, D., Čuboňová, N., Kasperek, D.: The impact of selected technological and material parameters on the strength of adhesive steel sheets joints. Adv. Sci. Technol. Res. J. **11**, 8–16 (2017)
24. Rudawska, A., Wahab, M.A., Barta, D., Pukalskas, S.: The effect of technological and structural factors on the strength of polyethylene adhesive joints. In: Wahab M.A. (ed.) Proceedings of the 7th International Conference on Fracture Fatigue and Wear, FFW 2018, 9–10 July 2018, Ghent University, Belgium. Lecture Notes in Mechanical Engineering, Springer, Singapore, 2019, pp. 241–257 (2019)
25. Shaw, S.J.: Adhesives in demanding applications. Polym. Int. **41**, 193–207 (1996)
26. Loh, W.K., Crocombe, A.D., Abdel Wahab, M.M., Ashcroft, I.A.: Environmental degradation of the interfacial fracture energy in an adhesively bonded joint. Eng. Fract. Mech. **69**, 2113–2128 (2002)
27. Datla, N.V., Ameli, A., Azari, S., Papini, M., Spelt, J.K.: Effects of hygrothermal aging on the fatigue behavior of two toughened epoxy adhesives. Eng. Fract. Mech. **79**, 61–77 (2012)
28. Patil, O.R., Ameli, A., Dalta, N.V.: Predicting environmental degradation of adhesive joints using a cohesive zone finite element model based on accelerated fracture tests. Int. J. Adhes. Adhes. **76**, 54–60 (2017)
29. Ameli, A., Dalta, N.V., Azari, S., Papini, M., Spelt, J.K.: Prediction of environmental degradation of closed adhesive joints using data from open-faced specimens. Compos. Struct. **94**, 779–786 (2012)
30. Wylde, J.W., Spelt, J.K.: Measurement of adhesive joint fracture properties as a function of environmental degradation. Int. J. Adhes. Adhes. **18**, 237–246 (1998)
31. Gledhill, R.A., Kinloch, A.J.: Environmental failure of structural adhesive joints. J. Adhes. **6**, 315–330 (1974)
32. Gledhill, R.A., Kinloch, A.J., Shaw, S.J.: A model for predicting joint durability. J. Adhes. **11**, 3–15 (1980)
33. Nam, B.J., Jang, S.Y.: A Study on the Degradation properties of DGEBA/TETA epoxy system for restoration of ceramics by temperature. J. Korean Conserv. Sci. Cult. Prop. **31**, 373–386 (2015)
34. Crocomble, A.D., Hua, Y.X., Loh, W.K., Wahab, M.A., Ashcroft, I.A.: Predicting the residual strength for environmentally degraded adhesive lap joints. Int. J. Adhes. Adhes. **26**, 325–336 (2006)

35. Sousa, J.M., Correira, J.R., Firmo, J.P., Cabral-Fonseca, S., Gonilha, J.: Effects of thermal cycles on adhesively bonded joints between pultruded GFRP adherends. Compos. Struct. **202**, 518–529 (2018)
36. Çolak, A., Çoşgun, T., Bakirci, A.E.: Effects of environmental factors on the adhesion and durability characteristics of epoxy-bonded concrete prisms. Constr. Build. Mater. **23**, 758–767 (2009)
37. Han, X., Crocombe, A.D., Anwar, S.N.R., Hu, P.: The strength prediction of adhesive single lap joints exposed to long term loading in a hostile environment. Int. J. Adhes. Adhes. **55**, 1–11 (2014)
38. Poeller, M., Chung, D.D.L.: Effect of heating on the structure of an adhesive joint, as indicated by electrical resistance measurement. J. Adhes. **79**, 549–557 (2003)
39. da Silva, L.F.M., Adams, R.D.: Joints strength prediction for adhesive joints to be used over a wide temperature range. Int. J. Adhes. Adhes. **27**, 362–379 (2007)
40. Common pine. http://drzewa.nk4.netmark.pl/atlas/sosna/sosna_zwyczajna/sosna_zwyczajna.php. Last Accessed 22 Jan 2019
41. Prefrere 6312 adhesive. https://kadimex.com.pl/portfolio/kleje-dla-przemyslu-drzewnego/. Last Accessed 22 Jan 2019
42. Epoxy resins. https://ciechgroup.com/produkty/chemia-organiczna/zywice/zywice-epoksydowe/. Last Accessed 15 Jan 2019
43. Curing agents. https://ciechgroup.com/produkty/chemia-organiczna/zywice/utwardzacze/. Last Accessed 15 Jan 2019
44. Rudawska, A.: Fixing-fixing holder, especially adhesive joint. P.423717. Patent Office News
45. Klimatest. http://www.klimatest.eu/katalog/. Last Accessed 29 Jan 2019
46. PN-EN 15870. Adhesives determination of tensile strength of butt joints

The Influence of the Packing Material Type on the Adhesive Joints Strength of the Paperboard Packages

Anna Rudawska(✉) and Arkadiusz Gola

Faculty of Mechanical Engineering, Lublin University of Technology,
Nadbystrzycka 36 Str, 20-618 Lublin, Poland
{a.rudawska, a.gola}@pollub.pl

Abstract. This article presents selected issues related to the impact of the material factors on the adhesive joints' strength of the paperboard packages and strength of various types of the packaging materials: Alaska Plus GC2 (grammage 235 g and 280 g) and Crescendo C1S (grammage 300 g). The bonded material consisted of different types of paperboard used in production of packages, which were subject to the surface modification process by applying a varnish coat and, in case of one material, to the lamination process. The adhesive joints were made with use of a cold adhesive produced by SALTADIS (identification number: SI 5240). It is water-soluble and may be used for a variety of industrial applications. The adhesive joints were made with use of different packaging materials and different technological modifications of these materials' surface on an automated production line. It was in order to determine which material and which surface modification method has an impact on both the strength of the adhesive joints and the packaging material itself. Based on the packaging materials' properties, it was observed that together with the increase of the basis weight, the material's strength also increases. In addition, the lamination process increases the material's strength as well.

Keywords: Adhesive joints · Strength · Packing materials · Surface modification

1 Introduction

Packages are very important for the current consumers. They are widely used in the food, pharmaceutical and mechanical industries. They are the first element the consumers pay attention to and that encourages them to buy a given product. Due to this fact it is of great importance to make sure that the package is as durable as possible, protects the product and is visually aesthetic [1–3].

The paper materials have been the most popular material types used in the packaging industry for many years. The main advantages of the paper packages are: low mass, easy printing, good mechanical properties and easy utilisation. It is also important to adjust the packaging material to the product that will be placed in a final package. The packaging materials come in a variety types and grammages, which have impact on the final package's strength [4–6].

M. A. Wahab (Ed.): DAMAS 2019, LNME, pp. 914–925, 2020.
https://doi.org/10.1007/978-981-13-8331-1_73

However, it needs to be taken into consideration that the packaging materials have numerous drawbacks, such as: susceptibility to moisture, water, light and oxidation [1, 5, 7]. It is easy to overcome them, though, by modifying the material with use of different surface modification operations [8–10]. A proper method of the surface preparation before bonding needs to be selected so that the adhesive joints have good adhesion and cohesion [1, 11, 12].

The paper products aimed at the packaging production include: (i) coated paper: coated with liquid materials: wax, fusible masses, paraffin or polymer, whereas the coats themselves penetrate the paper to a lesser extent; metallised or laminated papers, (ii) paperboard - single- and multi-ply, the grammage of such paper ranges from 161 to 315 g/m^2 and (iii) cardboard - with the grammage above 315 g/m^2. Unfortunately, due to the specifics of paper as a material, it is characterised by strong water absorption, as well as gas and fat permeability. However, all these drawbacks are successfully eliminated thanks to using various surface modification operations, such as laminating or coating with different coats [1–3, 9].

The aim of the packaging materials' modification is not only to improve the visual properties of their surface, but also to add some specified properties, such as: (i) increased mechanical strength of the base paper (e.g. tensile strength), (ii) increased mechanical strength of the surface (e.g. shear strength), (iii) increased surface resistance to the aggressive environmental factors, such as: moisture, water, light, oxidation [10].

The selected methods of the surface modification that increase its strength are the following [1, 2, 5, 13]:

(i) Filming - also referred to as laminating; it consists of covering the paper plane with a film, also known as a laminate. It increases the mechanical strength and moisture resistance. This process may be cold or hot. There is a wide range of various films with different properties, whereas there are the following types of film: matte, glossy, scratch-resistant, barrier, metallised, holographic, structural, or, the latest and the most popular one, the soft touch.

(ii) Varnishing – it consists of covering the paper plate with a varnish coat. It is aimed at making the printing paper base more attractive and of better quality. The best substances for varnishing are the UV stoving varnishes that create a layer on the surface that protects it from print erasing and increases its mechanical strength.

(iii) Laminating - a similar process to filming; it consists of bonding a thin layer of paper onto a much thicker paper, such as cardboard, corrugated board, etc. Laminating enables to increase the packaging materials' strength and resistance to the environmental factors or to change their surface structure.

There are two types of joining of packing materials in the packaging production: stapling and bonding [1, 5, 10, 14, 15]. Stapling consists of bonding the material's elements with staples with use of a tool called a long reach stapler (manual, pneumatic, electric). As for bonding, it may be single-point or multi-point. A multi-point bonding consists of applying the adhesive by the adhesive nozzles in several points at the same time during the production process. A single-point bonding (linear) is when a roller immersed in the adhesive rolls at high speed and applies the adhesive on the material surface.

The aim of the present article is to determine the impact of the material and technological factors of the selected packaging materials on the strength of adhesive joints. The object of the research was the samples of the adhesive joints of the packing materials and the samples of the materials themselves that were subject to the surface modification process. It was in order to determine which material and which surface modification method has an impact on both the strength of the adhesive joints and the packaging material itself.

2 Methodology

2.1 Packaging Materials' Characteristics

Three types of the packaging materials made of different types of paperboard were used in the tests. The packing material types and their grammages were presented in Table 1.

Table 1. Packaging material types, designation and grammages [16, 17].

No.	Type of material	Designation	Grammage of paperboard
1	Alaska Plus	GC2	235 g
2	Alaska Plus	GC2	280 g
3	Crescendo	C1S	300 g

Alaska Plus GC2 is a low-grammage paperboard used for production of the packages mainly for pharmaceutical products, cosmetics, chocolates and sweets, as well as dehydrated food. This material has all necessary certificates, such as ISEGA, PZH or Robinson Test. This paperboard is suitable for the following types of printing: offset, rotogravure, flexographic, digital, as well as for different types of technological improvements with use of: UV varnish, dispersive varnish, hot and cold stamping, perforation, holographic film, soft touch film, or Braille stamping.

The second type of material used in the tests was paperboard Crescendo C1S 300 g. Thanks to a unique fibre mix, this material is characterised by high stiffness and strength. Double-coated surface protects the material from the aggressive environmental factors. Thanks to all these properties, this paperboard is mainly used for production of the luxurious packages.

2.2 Description of the Package Types Subject to Tests

The packages that were subject to tests were the consumer packages that serve to pack the individual products, mostly used in retail. The materials that the packages were made of the following materials: Alaska Plus GC2 (grammage 235 g and 280 g) and Crescendo C1S (grammage 300 g). Figure 1 shows the consumer package made of the material Crescendo 300 g, which had been covered with a matt varnish and laminated. Figure 1b shows a bearing package made of Alaska Plus 235 g. Its surface was covered with the UV varnish.

a) b)

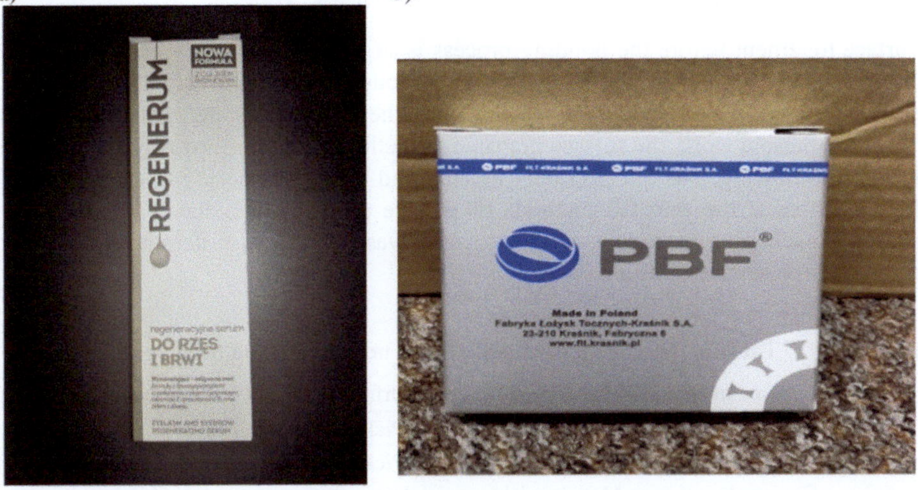

Fig. 1. Examples of the consumer packages: (a) a cosmetic package, (b) a bearing package

2.3 Description of the Packaging Materials' Surface Modification

Three types of different materials were used in the tests: Alaska Plus 235 g, 280 g and Crescendo 300 g. Different method of the surface modification was used for each material. The packing material types and the surface modification methods were presented in Table 2.

Table 2. Packaging materials' surface modification.

No.	Type of material	Designation	Type of modification
1	Alaska Plus	GC2	• UV varnish
2	Alaska Plus	GC2	• Matt varnish
3	Crescendo	C1S	• Matt varnish • Laminating

At the first stage the packaging materials were printed and varnished, and then immediately cured with use of the UV lamps. Printing and varnishing was conducted with use of the Heidelberg Speedmaster XL 75–5 L machine. The Crescendo 300 g material was additionally laminated with a polymer matt film.

The next stage included die-cutting of the desired shape from the packaging material's surface. A die-cutter BOBST COMMERCIAL was used to this end.

2.4 Surface Treatment Before the Bonding Process

Surface treatment before the bonding process is of great importance, regardless of the bonded material type. In this case this stage consisted of removing the impurities. However, due to the material type - paperboard, the possibilities are extremely limited.

In the present work, the surface preparation methods used for the analysed materials were the following: the "picking out" method and the atmospheric plasma activation. Juxtaposition of the selected methods of surface preparation before bonding of the analysed packaging materials, whose surface was subject to different modification types, was presented in Table 3.

Table 3. Packaging materials' surface treatment before bonding.

Type of material	Type of modification	Surface treatment of paperboard
Alaska Plus	• UV varnish	"Picking out" method
Alaska Plus	• Matt varnish	"Picking out" method
Crescendo	• Matt varnish • Laminating	Atmospheric plasma activation

Special adhesives are used to bond the material covered with the UV varnish. However, they are much more expensive. It is also possible to remove the varnish layer from the bonded surface with a mill. The samples were prepared with the "picking out" method, which consists of putting a polymer sheet with some cut-out points (Fig. 2) where there should be no varnish, on the cylinder transporting the varnish. Thanks to that the varnish does not interfere with the bonded surface's structure and the material itself becomes stronger.

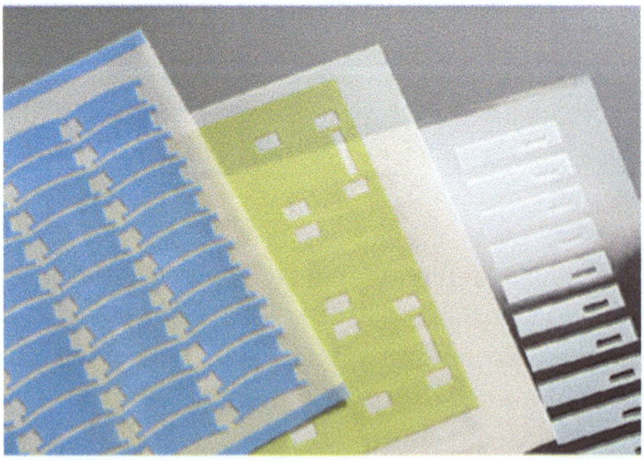

Fig. 2. A polymer sheet [18]

The laminated samples' surface for bonding was subject to the atmospheric plasma activation. A high voltage is applied to the gas at atmospheric pressure until the ignition of plasma. A stream of plasma is produced by a nozzle with use of the compressed air.

The reactive plasma particles cause activation and purification of the surface the stream is directed at [16]. The Plasmatreat PT-200 unit was used for the atmospheric plasma activation.

2.5 Making the Paperboard Packages with Use of the Bonding Technology

The process of making the adhesive joints of the packaging materials included four stages.

1. The first stage - preparing the surface of the packaging materials,
2. The second stage - covering the bonded elements' surface with the adhesive,
3. The third stage - bonding the elements,
4. The fourth stage - curing.

All the stages were performed at an automated cycle in a batch production with use of the BOBST Alpina 75 folder- bonding machine. The first stage was described in the point 2.4 herein.

The adhesive used for bonding was produced by SALTADIS. Its identification number was: SI 5240. It is a cold water-based adhesive for industrial applications. It is aimed at production of packages with metallised or laminated base paper. Its remarkable property is a resistance to both very high and very low temperatures. The basic physical and chemical properties of the SI 5240 adhesive were presented in Table 4.

Table 4. Basic physical and chemical properties of the SALTANDIS SI 5240 adhesive [19].

Properties	Description/Value
Physical state	Liquid
Colour	White
Smell	Barely perceptible
pH	4–8
Density at 20 °C	1.07 g/cm^3
Viscosity at 20 °C	610 mPa/s
Ignition point	Not defined
Boiling point	Not defined
Solubility in water	Soluble

The adhesive was applied at single-point with use of a roller, which was immersed in the adhesive and was applying the adhesive on the surface by rolling (stage 2). It is important to make sure that the length and width of the adhesive path is the same as the length and width of the lap.

The fourth stage was curing. Thanks to the fact that the adhesive used for the tests was a cold one, this process was performed at ambient temperature. Another important element of the bonding process was clamp - it needed to be selected in such a way so that the adhesive layer does not flow off from the bonded surface. The packages containing the adhesive joints were being cured for 3 days. The curing process was conducted at ambient temperature of 20 ± 3 °C and air humidity of 50%–55%.

2.6 Samples' Preparation for the Strength Tests

Two types of samples were used in the strength tests:

A. Samples of the adhesive joints of the selected packaging materials,
B. Samples of the packaging materials.

The elements were bonded with the lap joints. The samples with the dimensions of 104 mm x 57 mm and the lap length of 10 mm were cut out of the prepared packages. In total 10 samples of each type were prepared: A – 10 samples of the adhesive joints of the selected packaging materials and B – 10 samples of the packaging material of each type. A digital cutting machine SEPYA-UC was used for accuracy.

Figure 3 presents the samples of the adhesive joints of the selected materials and the samples of the material itself. Figure 4 shows the lap joints' scheme with the average dimensions. Figure 5 presents the scheme of the packaging material's sample.

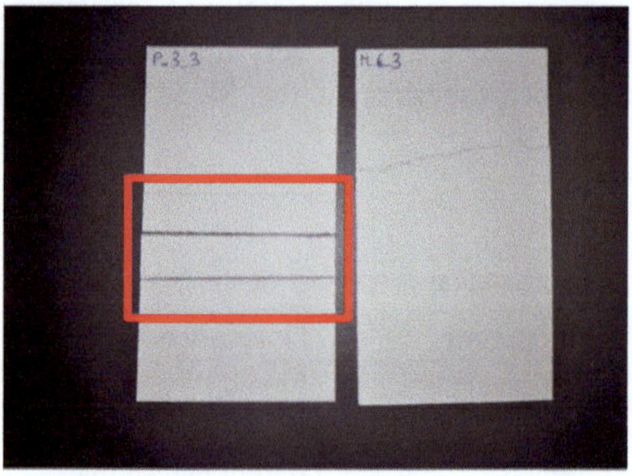

Fig. 3. On the right (A sample) - a sample with the adhesive joint highlighted in red, on the left (B sample) - a sample of the packaging material.

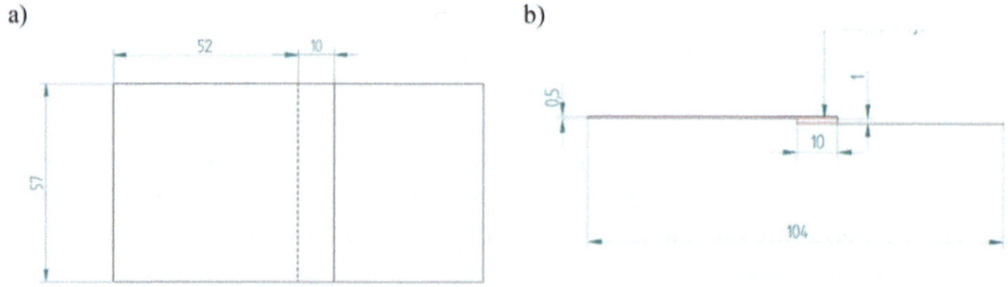

Fig. 4. Adhesive lap joint scheme: (a) top view, (b) said view

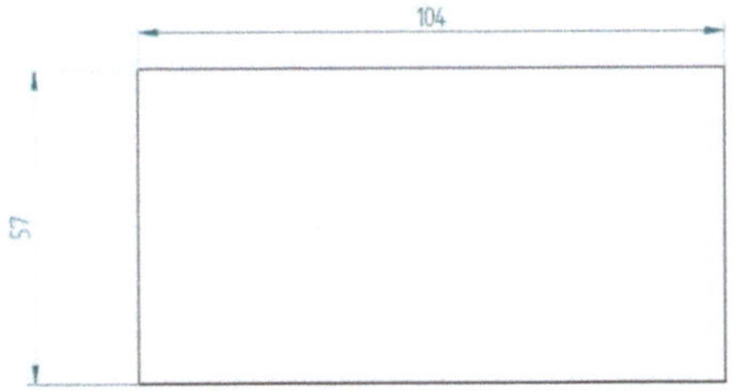

Fig. 5. Scheme of the packaging material's sample - top view

The strength tests of the adhesive joints were performed on the testing machine Zwick/Roell Z150, according to the norm DIN EN 1465. The crosshead speed was 5 mm/min.

3 Test Results

3.1 Adhesive Joints' Strength Test Results

Figure 6 shows the comparison of the adhesive joints' shear strength test results. Based on the data showed in Fig. 6, it may be observed that the highest shear strength was obtained by the adhesive joint samples made of the material Alaska Plus 235 g covered with the UV varnish. Their strength was 0.590 MPa, whereas the difference between the highest and the lowest value was 66%. The lowest strength value was 0.200 MPa and it was obtained by the adhesive joint made of the material Crescendo 300 g, which had been laminated. Based on the obtained results, it may be assumed that the atmospheric plasma activation did not activate the bonded surface effectively enough.

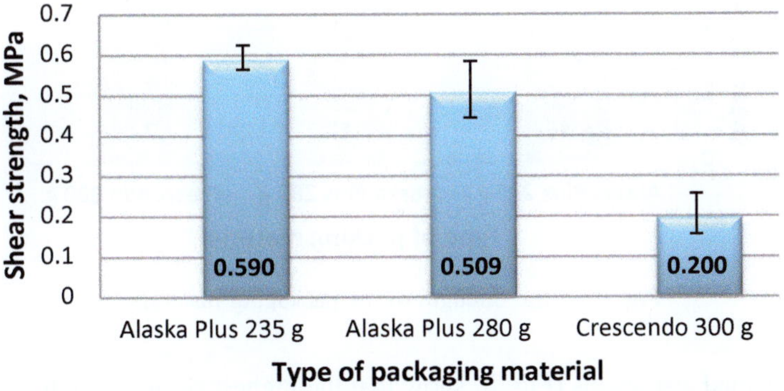

Fig. 6. Shear strength of the adhesive joints of different packaging materials.

Figure 7 presents the comparison of the obtained results of elongation of the adhesive joints made of the selected packaging materials.

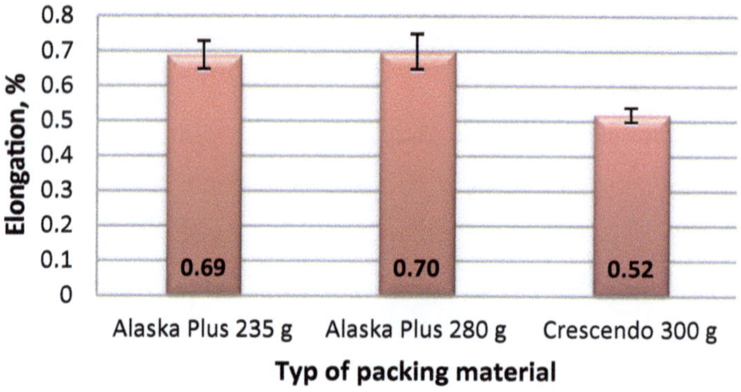

Fig. 7. Elongation of the adhesive joints made of the selected packaging materials.

Based on the results showed in Fig. 7, it may be observed that the highest elongation of the adhesive joints was obtained by the samples made of the material Alaska Plus 280 g - it amounted to 7%. It is also worth noticing that elongation for the adhesive joints of Alaska Plus 235 g was similar and amounted to 6.9%. The difference between the highest and the lowest elongation value was 25%.

3.2 Packaging Materials' Strength Test Results

Figure 8 shows the comparison of the packaging materials' shear strength test results.

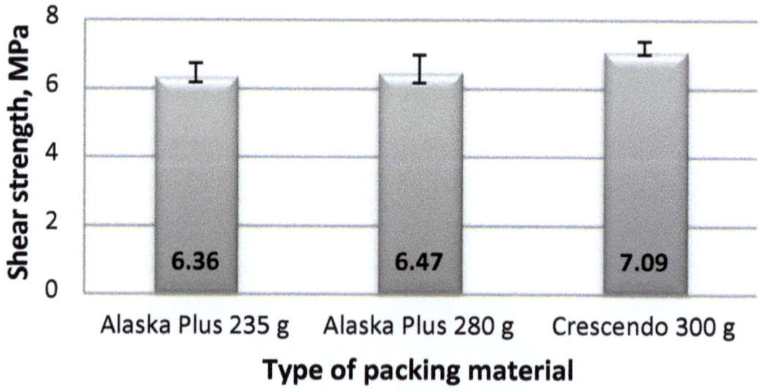

Fig. 8. Shear strength of the packaging materials.

The obtained test results (Fig. 8) show that the highest shear strength was obtained by the samples made of the material Crescendo 300 g. Their strength was 7.09 MPa, whereas the difference between the highest and the lowest value was 10%.

Based on the test results, it may be observed that both laminating and high grammage increased the material's strength. It may be assumed that together with the increase of the grammage, the strength of the packaging material increases as well.

Figure 9 presents the comparison of the obtained results of elongation of the selected packaging materials.

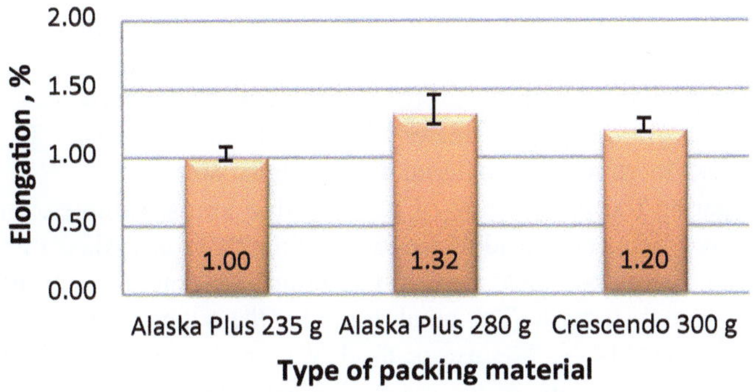

Fig. 9. Elongation of the selected packaging materials.

Based on the results presented in Fig. 9, it may be observed that the highest elongation was obtained by the samples of the material Alaska Plus 280 g and it was of 1.32%. The difference between the highest and the lowest elongation value of the packaging materials' samples was 24%.

3.3 Comparative Analysis of the Material and the Adhesive Joints

Figure 10 presents the comparison of the break force for the samples of the packaging materials and their adhesive joints.

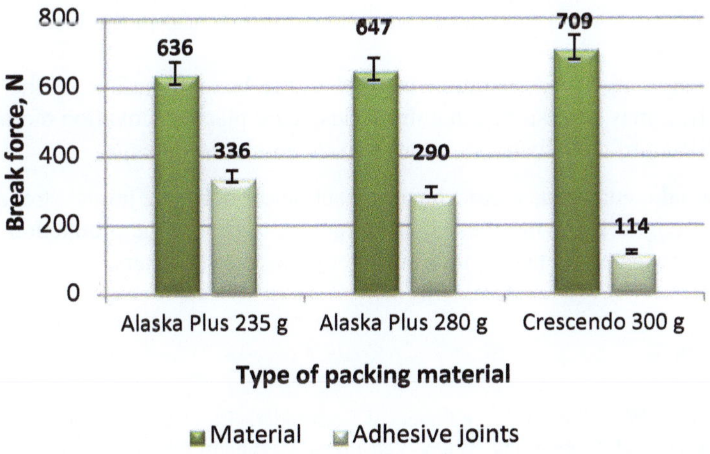

Fig. 10. Break force for the samples of the packaging materials and the adhesive joints of the analysed materials.

When analysing the results presented in Fig. 10, it may be observed that among all the samples, the material Crescendo 300 g was characterised by the strongest force, which was needed to break the material samples - it amounted to 709 N. At the same time, the samples of the adhesive joints of the material Crescendo 300 g showed the weakest break force, which was 114 N. The difference between these values was 595 N, i.e. 83%.

Based on the packaging materials' properties, it may be observed that together with the increase of the grammage, the material's strength also increases. In addition, the lamination process increases the material's strength as well.

When comparing the break force of the adhesive joints' samples, there is a considerable difference between the adhesive joints of the materials Alaska Plus 235 g and Crescendo 300 g. It may be stated the laminating process causes a significant decrease in the adhesive joints' strength. The break force for the adhesive joints of the material Crescendo 300 g was 114 N. The use of the UV varnish resulted in the break force's increase for the adhesive joints of the material Alaska Plus 235 g, which amounted to 336 N. The difference was 222 N, i.e. 66%.

4 Summary

The object of the tests was the samples of the packaging materials subject to the surface modification and the adhesive joints made of these materials. The variable factors were the materials and the modifications of their surfaces.

Based on the strength tests results, the following conclusions may be drawn:

- type of the material has an impact on the adhesive joints' strength,
- the strength of the packaging material increases together with the increase of the grammage,
- laminating increases the packaging material's strength,
- the use of the UV varnish has an impact on the increase in the break force of the adhesive joints,
- when comparing the preparation method of the bonded surface of the packaging materials, it may be assumed that the atmospheric plasma activation did not activate the bonded surface effectively enough in the analysed cases.

To conclude, numerous factors have impact on the adhesive joints' strength and the material's strength. Knowing these factors enables to design the production process of the paperboard materials taking into consideration, among others, their strength.

References

1. Brockmann, W., Geiß, P.L., Klingen, J., Schröder, B.: Adhesive Bonding. Materials, Applications and Technology. Wiley-Vch Press, Weinheim (2009)
2. Emblem, A., Hardwidge, M.: Adhesives for packaging. In: Emblem, A., Emblem, H. (eds.) Packaging Technology. Fundamentals, Materials and Processes, pp. 294–381. Woodhead Publishing, Cambridge (2012)

3. Adhesives in packaging. Symposium, 15th December 1992. https://www.iom3.org/fileproxy/307714. Accessed 2 Mar 2019
4. Aznar, M., Vera, P., Canellas, E., Nerín, C., Mercea, P., Störmer, A.: Comparison of the adhesives used in food packaging multilayer materials and migration studies from packaging to food. J. Mater. Chem. **21**, 4358–4370 (2011)
5. Vera, P., Aznar, M., Mercea, P., Nerín, C.: Study of hotmelt adhesives used in food packaging multilayer laminates. Evaluation of the main factors affecting migration to food. J. Mater. Chem. **21**, 420–431 (2011)
6. Rudawska, A., Čuboňova, N., Pomarańska, K., Stančeková, D., Gola, A.: Technical and organizational improvements of packaging production process. Adv. Sci. Technol. **30**, 82–192 (2016)
7. Stoermer, A., Franz, R.: Migresives: a research project on migration from adhesives in food packaging materials in support of European legislation an standardization. Food Addit. Contam. **26**, 1581–1591 (2009)
8. Kropp, M., Behr, A.: Innovations in IC packaging adhesives. https://electroiq.com/2005/08/innovations-in-ic-packaging-adhesives/. Accessed 2 Mar 2019
9. Chen, H., Jiang, B., Cai, Z.-Q.: Preparation and properties of paper–plastic laminating adhesive used for medical packaging materials. Polym. Adv. Technol. **26**, 1065–1069 (2015)
10. Toenniessen, M.: Packaging materials. 10. Adhesives for food packaging applications. International Life Science Institute. ILSI Europe Report Series, 2018. http://ilsi.eu/wp-content/uploads/sites/3/2018/12/2018-Packaging-materials_10_Interactif.pdf. Accessed 2 Mar 2019
11. Ebnesajjad, S., Ebnesajjad, C.: Surface Treatment of Materials for Adhesive Bonding, 2nd edn. William Andrew, Norwich (2013)
12. Rudawska, A., Reszka, M., Warda, T., Miturska, I., Szabelski, J., Stančeková, D., Skoczylas, A.: Milling as a method of surface pre-treatment of steel for adhesive bonding. J. Adhes. Sci. Technol. **30**, 2619–2636 (2016)
13. Ashley, R.J., Cochran, M.A., Allen, W.: Adhesives in packaging. Int. J. Adhes. Adhes. **15**, 101–108 (1995)
14. Pertie, E.M.: Handbook of Adhesives and Sealants, 2nd edn. McGraw-Hill Professional, New York (2007)
15. Schumacher, K.-H.: New water-based adhesives for flexible food packaging chemical design, performance and toxicological safety. J. Appl. Packag. Res. https://scholarworks.rit.edu/japr/vol8/iss1/7/. Accessed 2 Mar 2019
16. https://www.internationalpaper.com/Apps/Alaska-Plus/files/en/. Accessed 1 Mar 2019
17. https://www.papyrus.com/MEDIA_CustomProductCatalog/m9260255_16_5810_Datasheet_CrescendoC1S_PEFC_DE_WR_en.pdf. Accessed 2 Mar 2019
18. http://www.pxp.com.pl/images/glowne/39.jpg/. Accessed 2 Mar 2019
19. https://saltadis.com/oferta/wg-zastosowania/kleje-zamykania-opakowan/. Accessed 2 Mar 2019

Studies of Fibre Reinforced Polymer Samples with Embedded FBG Sensors

Magdalena Mieloszyk(✉), Katarzyna Majewska, and Wieslaw Ostachowicz

Institute of Fluid Flow Machinery, Polish Academy of Sciences,
14 Fiszera Street 80-231, Gdansk, Poland
mmieloszyk@imp.gda.pl

Abstract. The paper presents analyses of fibre reinforced polymer (FRP) samples with embedded fibre Bragg grating (FBG) sensors. As a reinforcement glass or carbon fibres in a form of textiles with different grammage and bundles fibre arrangement were used. All samples were manufactured using infusion method. FBG sensors are applied for analyses of strain distribution inside sample due to static point loads. Samples internal structures were analysed using THz spectroscopy. Analyses of possibility of detection and localisation of fibre optics lying under textile or embedded into carbon/glass FRP are performed. For carbon FRP the maximal depth of inspection is determined. The THz wave propagation depth depends on conductive carbon fibres arrangement in non-conductive resin matrix.

Keywords: Fibre Bragg grating sensor · Fibre reinforced polymers · Structural health monitoring · Embedded sensors · THz spectroscopy

List of Abbreviation

1, 2, 3, ..., N_{n-2}, N_{n-1}, N_n	static point location columns
A-F	static point location rows
B1, B2, B3	measurements point in CFRP sample B
carbon B, carbon C, carbon D	carbon textile or CFRP sample
CFRP	carbon fibre reinforced polymer
D1, D2	measurements point in CFRP sample D
E1	measurement point in GFRP sample E
F	free
FBG	fibre Bragg grating
FBG1, FBG2	fibre Bragg grating sensors
FO	fibre optic
FRP	fibre reinforced polymer
FWHM	full width half maximum
GFRP	glass fibre reinforced polymer
glass E	glass textile or GFRP sample
NDT	non-destructive technique
S	embedded
SHM	structural health monitoring

M. A. Wahab (Ed.): DAMAS 2019, LNME, pp. 926–936, 2020.
https://doi.org/10.1007/978-981-13-8331-1_74

1 Introduction

Fibre reinforced polymer (FRPs) are nowadays commonly used in many industrial branches: e.g. aviation [1], marine [2] or civil engineering [3]. Therefore, FRP elements are working in different environmental and loading conditions. Safety and reliability requirements results in development of a variety of non-destructive techniques (NDTs) and structural health monitoring (SHM) systems.

One of NDTs based on electromagnetic waves propagation is THz spectroscopy. The method can be applied to analyses of internal structures of non-conductive material. THz radiation can be used in analyses of materials that affect minimum one of three THz wave parameters: refractive index, absorption coefficient or wave scattering [4]. The technique can be used for inspection of internal structures of e.g: tissues [5], paper [6] or glass fibre reinforced polymer (GFRP) laminates [7]. Carbon fibre reinforced polymers (CFRP) are highly reflective for electromagnetic waves in THz range of frequencies. CFRP contains conductive carbon fibres and non-conductive polymeric matrix. Therefore, in some cases it is possible to detect defects that are within few hundreds of microns from an element surface [8]. THz spectroscopy can be applied for measurements of misprocessed (e.g. missing cleaning steps, contaminations, improper amount of topcoat) coatings on CFRP panels [9], burn damage on CFRP laminate surface [10] or impact damage in honeycomb sandwich panel skin [11].

SHM systems contains different types of sensors that are mounted on structures or embedded into element material during manufacturing process [3,12]. One of the sensor types are Fibre Bragg Grating (FBG) sensors. They are a great tool for SHM of composite elements due to their small size and weight, high multiplexing capabilities, high corrosion resistance [13] as well as possibilities of embedding into a composite structure [1]. FBG sensors are typically applied to strain [3] or temperature [13] measurements. Embedded FBG sensors can be applied for impact load localisation in GFRP sandwich panel [14] or damage detection (e.g. delamination [15]).

The paper is organised as follow. Firstly, FRP samples with embedded FBG sensors are described. Next, analyses of embedding process influences on FBG sensors behaviour as well as THz spectroscopy application for textiles and FRP samples are presented and discussed. Finally, some conclusions are drawn.

2 Experimental Investigation

The experimental investigation is divided into two parts. The first one is related to FBG sensors, while the second to THz spectroscopy analyses.

2.1 Samples

The experimental investigations were performed on FRP samples (GFRP and CFRP) with embedded FBG sensors. The samples were manufactured using infusion method. Inside every sample two FBG sensors were embedded between

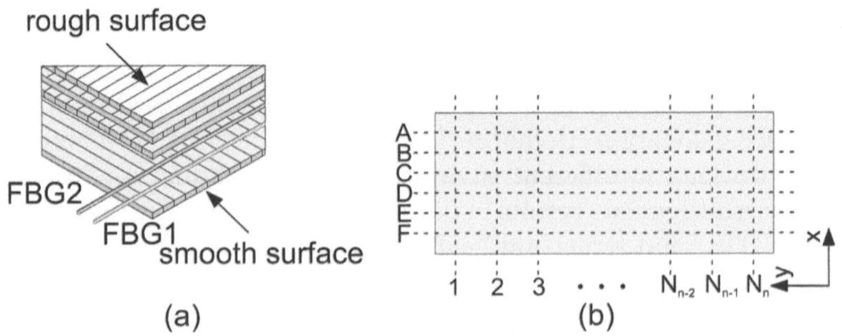

Fig. 1. Schema of: (a) layers and FBG sensors alignment in samples, (b) static point load locations on sample surface; FBG1, FBG2 – FBG sensors.

the 1st and the 2nd layer counting from the bottom surface. One (denoted as FBG1) with gage length equal to 10 mm and the other (denoted as FBG2) with gage length equal to 1 mm. The scheme of a sample part with marked FBG sensors and surfaces roughness is presented in Fig. 1(a). The samples contains glass/carbon fibre reinforcement in a form of textiles that photographs are presented in Fig. 2.

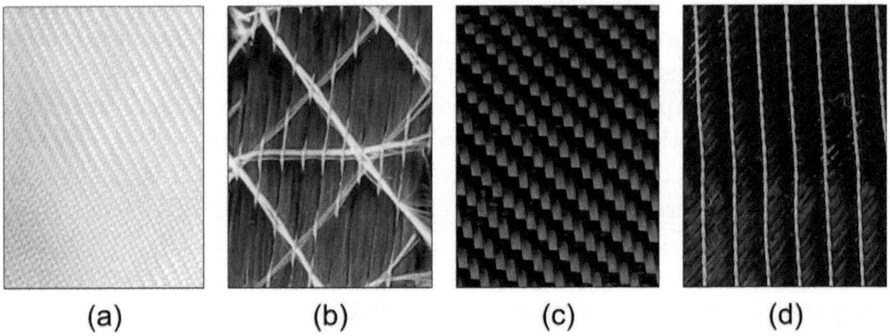

Fig. 2. Textiles: (a) glass E, (b) carbon B, (c) carbon C, (d) carbon D.

2.2 FBG Sensors

The experimental investigation is concerned on comparison between behaviour of FBG sensors embedded into GFRP (glass E) and CFRP (carbon C) samples. Analyses were performed on GFRP sample (70 mm × 240 mm × 1 mm) and CFRP sample (70 mm × 270 mm × 2 mm) made out of textiles with similar parameters (2 × 2 weave, grammage 200 g/m^2).

In Fig. 3 a comparison of FBG sensors spectra for both samples is presented. It can be observed that the full width at half maximum (FWHM) of FBG1 (10 mm) sensors is ca. 13 times smaller than FWHM of FBG2 (1 mm). Probably

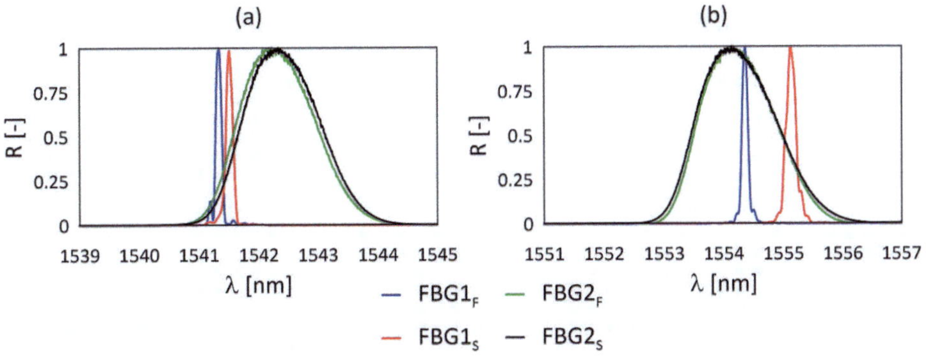

Fig. 3. Comparison of FBG sensors spectra for sample: (a) GFRP, (b) CFRP; FBG1, FBG2 – sensors, F – free, S – embedded.

due to this embedding process influence is neglected for FBG2 sensors regardless sample reinforcement material. For FBG1 sensors embedding process results in Bragg wavelength increase – local tension occurrence.

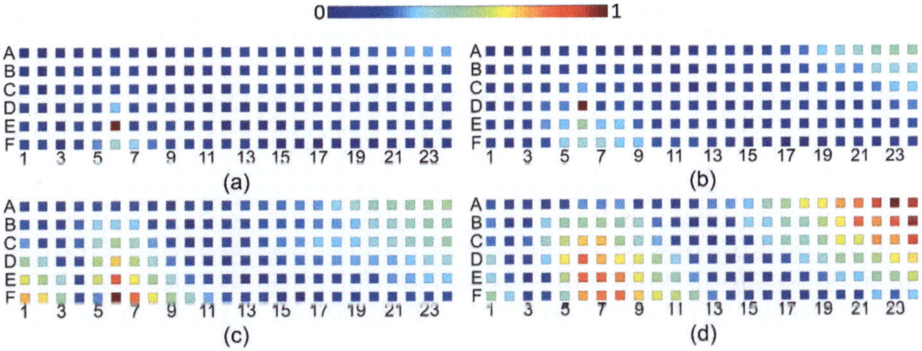

Fig. 4. Sensitivity of FBG sensors embedded into GFRP sample (rough side): (a) FBG1 and 0.1 N, (b) FBG2 and 0.1 N, (c) FBG1 and 0.2 N, (d) FBG2 and 0.2 N.

Next, a comparison between embedded FBG sensors sensitivity on static point load (0.1 N, 0.2 N) was performed for the samples in a form of simply supported beams. The investigations were performed on both sides (smooth and rough) of the samples. On every surface a grid of equally distributed points with a 10 mm distance between them was established. The points are arranged in rows denoted by letters A-F and columns denoted by numbers as it is presented in Fig. 1(b). In Figs. 4, 5, 6 and 7 the calculated sensitivity distribution of each sensor individually is shown. It is presented in a form of normalised strain values in particular points in relation with the highest strain value received for a particular sensor. The measurements were performed using interrogator (si425-500, Micron Optics) with measurement frequency equal to 250 Hz.

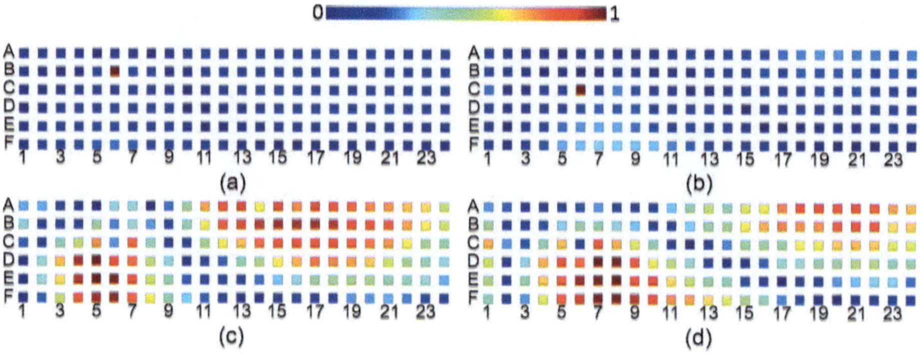

Fig. 5. Sensitivity of FBG sensors embedded into GFRP sample (smooth side): (a) FBG1 and 0.1 N, (b) FBG2 and 0.1 N, (c) FBG1 and 0.2 N, (d) FBG2 and 0.2 N.

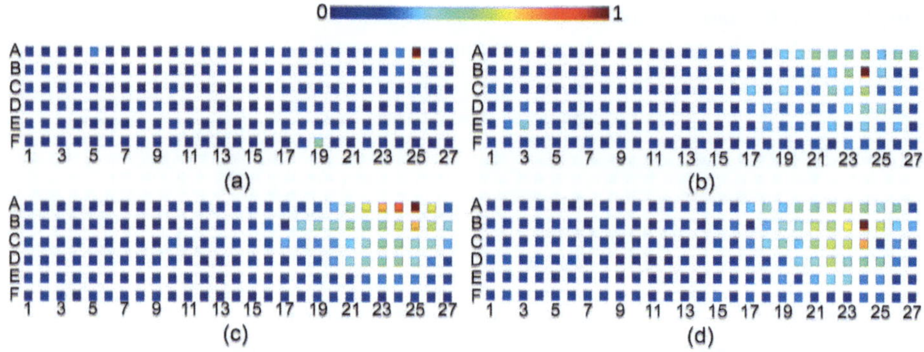

Fig. 6. Sensitivity of FBG sensors embedded into CFRP sample (rough side): (a) FBG1 and 0.1 N, (b) FBG2 and 0.1 N, (c) FBG1 and 0.2 N, (d) FBG2 and 0.2 N.

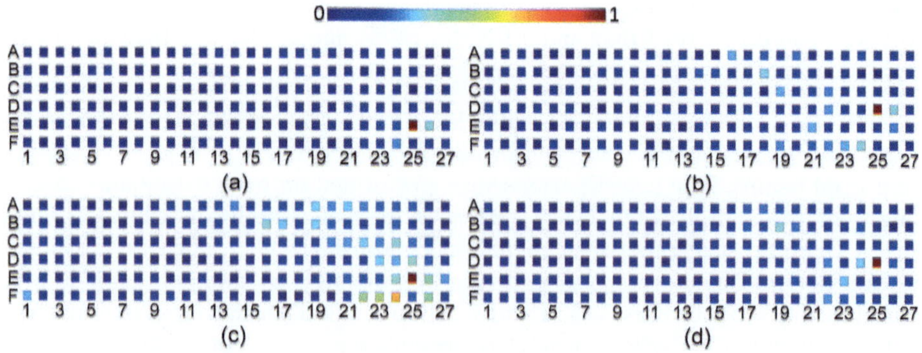

Fig. 7. Sensitivity of FBG sensors embedded into CFRP sample (smooth side): (a) FBG1 and 0.1 N, (b) FBG2 and 0.1 N, (c) FBG1 and 0.2 N, (d) FBG2 and 0.2 N.

In the rough surface of the GFRP sample fibre optics with FBG1 and FBG2 sensors are located along E and D lines, respectively. While for smooth side the fibre optics with FBG1 and FBG2 sensors are located along B and C lines, respectively. For static point load 0.1 N for both sides (smooth and rough) and sensors (FBG1 and FBG2) the highest response is visible exact on every sensor location. Responses with smaller strain level are also noticeable for FBG2 sensor in areas: close to the sensor location and the upper right corner. The responses are higher for rough than smooth surface. Higher static point load value (0.2 N) results in enlargement of FBG sensors sensitivity areas. For both sensors (FBG1 and FBG2) it is not limited to close neighbourhood of the sensors and covers both close area of the sensors and upper right corner of the sample. The region sizes are wider for smooth side of the sample.

In the rough side of the CFRP sample fibre optics with FBG1 and FBG2 sensors are located along A and B lines, respectively. While for smooth side the fibre optics with FBG1 and FBG2 sensors are located along E and D lines, respectively. For both static point load values (0.1 N and 0.2 N) the highest responses are visible exact on sensors locations, regardless surface roughness. For all cases the FBG sensors sensitivity areas in CFRP sample are significantly smaller than for GFRP. For static point load 0.2 N the area is limited to the neighbourhood of sensors. The observable differences are related to stiffness differences between glass and carbon fibres.

2.3 THz Spectroscopy

The experimental investigation was performed using the THz spectrometer (TPS Spectra 300 THz Pulsed Imaging and Spectroscopy from TerraView®) in reflection mode. During the measurements materials/samples were laying on metal table. The analyses were performed on three fibre textiles (Fig. 2) and finished samples with reinforcement made form the same textiles. It allowed to compare interactions between THz waves and glass/carbon fibres. The used glass textile was the same as during investigations with FBG sensors – material/sample

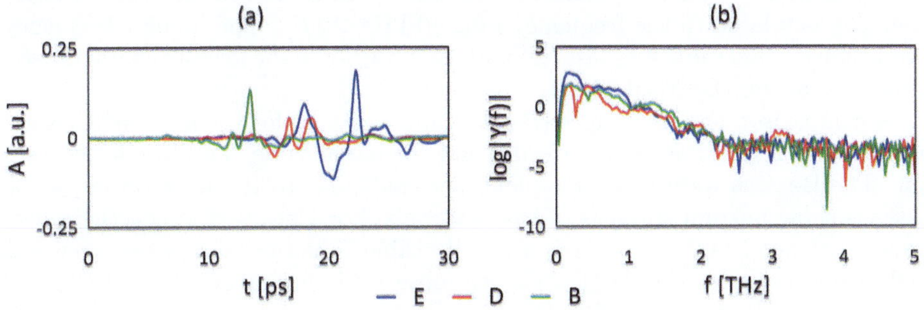

Fig. 8. Comparison of THz signals in: (a) time, (b) frequency domains; E – glass textile, D, B – carbon textile.

denoted as E. Two carbon fibre textiles were also analysed: denoted as D (biaxial $\pm 45^\circ$ with grammage $600\,\mathrm{g/m^2}$) and denoted as B (unidirectional with grammage $600\,\mathrm{g/m^2}$).

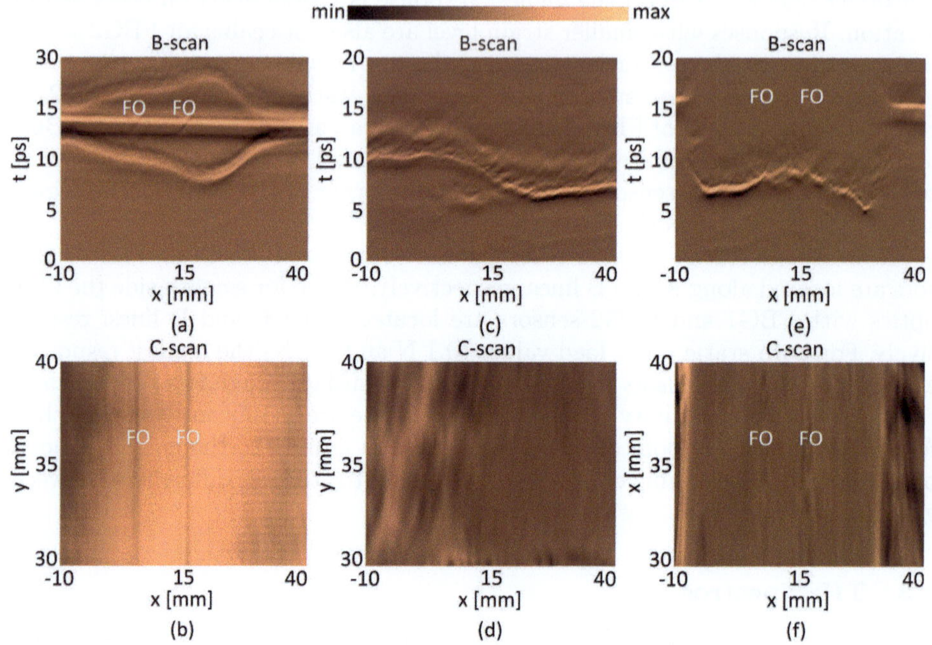

Fig. 9. THz images for textiles: (a)-(b) glass E, (c)-(d) carbon D, (e)-(f) carbon B; FO-fibre optic.

A comparison of THz signals and spectra for three materials are presented in Fig. 8. For all textiles reflections from the material surfaces are visible, while for glass (E) additionally reflection from the metal table (22.3 ps) is observable. The measured frequency spectra are similar for all materials and the majority of energy is related to the frequency range 0 THz–2 THz. The highest frequency amplitude is observable for glass (E) for 0.18 THz, while the lowest (characteristic drop) for carbon (D) for 0.27 THz.

A comparison among images (B-scans and C-scans) for the same textiles are presented in Fig. 9. Between materials and the metal table two fibre optics were put. The B-scans were made for plane perpendicular to the fibre optics, while C-scans show bottom surfaces of the materials. For glass textile (E) the reflections from metal table (ca. 14 ps) are well visible. The fibre optics occurrence is well visible for glass textile (E), slightly observable for carbon textile (B) and unnoticeable for carbon textile (D).

A comparison of THz signals and spectra for FRP samples made out of the three materials are presented in Fig. 10. For all samples reflections from

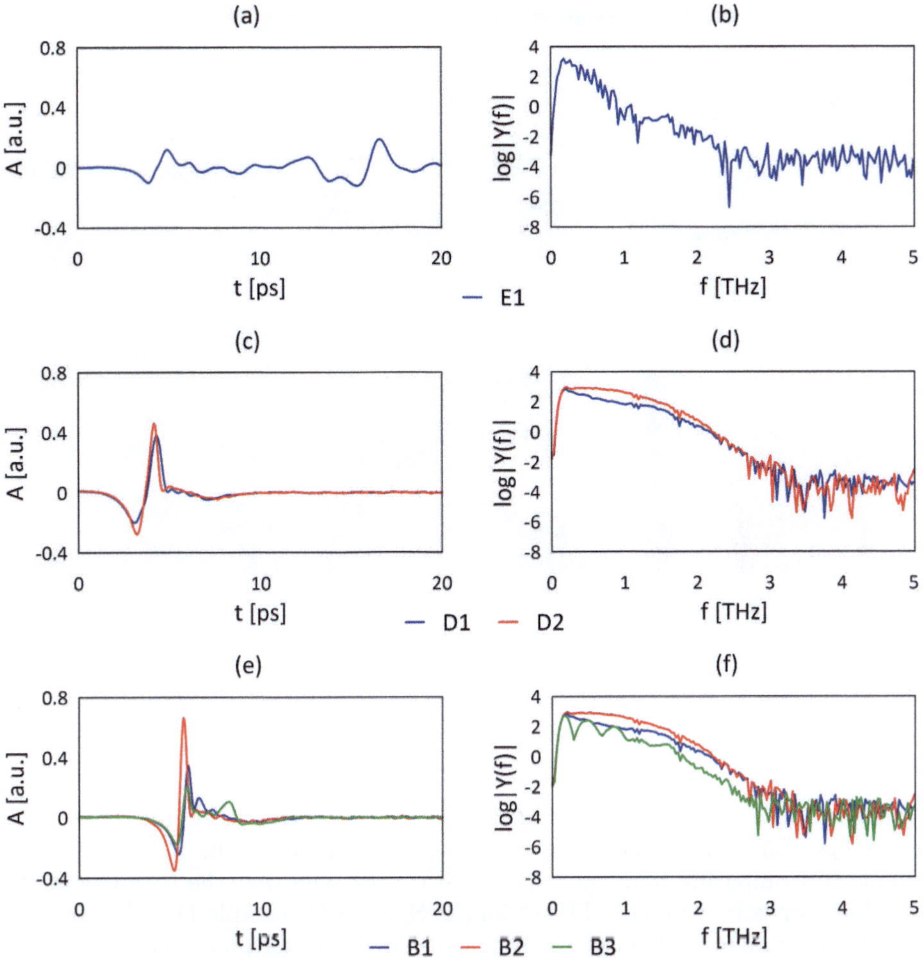

Fig. 10. Comparison of THz signals in: (a), (c), (e) time, (b), (d), (e) frequency domains; E – GFRP, D, B – CFRP.

the upper surfaces are visible (ca. 5 ps). For the GFRP sample (E1) also reflections from the internal layers, bottom (12.8 ps) surface and metal table (16.7 ps) can be distinguished. For CFRP sample D a comparison of signals for carbon fibre bundles (D2) and between the bundles on a connecting thread (D1) are presented. The observed differences are neglected. THz waves do not propagate throughout material, but are reflected from its surface. For CFRP sample B three points were analysed: carbon fibre bundle (B2), between bundles on a connecting thread (B1), and on a crossing of the connecting treads (B3). For such material differences between measured THz signals are observable. The signal amplitudes for reflections from the upper surface decrease form the highest (B2) to the lowest (B3). It shows that for points B1 and B3 a part of THz waves

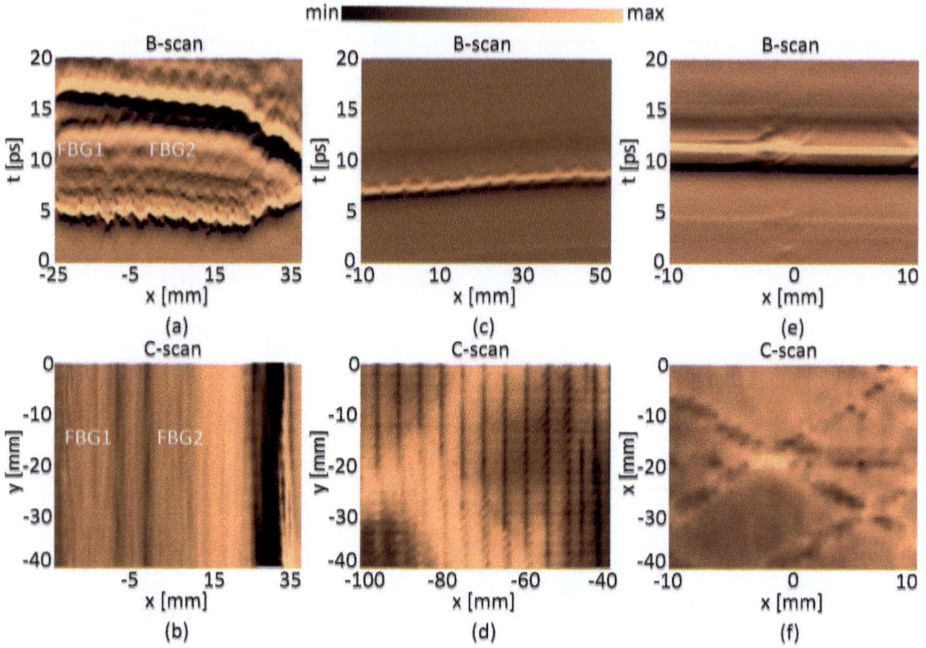

Fig. 11. Comparison of THz images for samples: (a)-(b) GFRP, (c)-(d) CFRP D, (e)-(f) CFRP B.

propagates also into material. Additional reflections are visible for 6.6 ps (B1) and 8.2 ps (B3). They can be linked with reflections from the 2nd layer of the sample B. Comparing frequency spectra it can be seen that the majority of the signal energy is in a range 0 THz–2.5 THz. Spectra for sample D and points B1 and B2 are similar to themselves. While the spectrum for point B3 have similar shape to the previous one, but some amplitude distortions also occur.

A comparison among images (B-scans and C-scans) for the FRP samples are presented in Fig. 11. The B-scans were made for plane perpendicular to the embedded fibre optics. The C-scan for GFRP was determined for embedded fibre optics plane inside material. The C-scan for CFRP sample D was determined for the upper surface, while for the CFRP sample B the image was made for 6.6 ps related to the reflection for point B1. GFRP laminate layers as well as embedded fibre optics are well visible in B-scan for GFRP sample. The fibre optics location between the 3rd and the 4th layer of the sample counting from the rough surface can be determined – compare with Fig. 1(a). For both CFRP samples embedded fibre optics are not visible. Images present the samples upper surfaces pattern – carbon fibre reinforcements and connecting treads.

3 Conclusions

The experimental investigations performed on FRP samples (GFRP and CFRP) with embedded FBG sensors were presented. The analyses contained two parts. The first one was related to FBG sensors, while the second to THz spectroscopy analyses.

A comparison of embedded FBG sensors spectra for both samples shows strong differences related to spectra shape and FBG sensors sensitivity on embedding process. Additionally, a comparison among sensitivity of FBG sensors embedded into samples on static point load (0.1 N, 0.2 N) was performed. It showed that for the first load (0.1 N) the highest strain response amplitudes are visible on exact sensors locations, regardless fibre reinforcement material. For the second load (0.2 N) the sample stiffness differences stronger influences on embedded sensors responses and the sensitivity areas covers almost all sample (GFRP) or ca. One third closer to the embedded sensors (CFRP).

In the experimental investigation using the THz spectrometer a comparison among glass and carbon fibres (in a form of textiles and samples reinforcement) influences on THz wave propagation was presented. For glass textile and GFRP sample is possible to determine fibre optics locations. For carbon textiles fibre optics visibility strongly depends on material pattern. THz waves propagation throughout CFRP material was also observable, but it strongly depends on the carbon fibre alignment. Therefore, THz waves are totally reflected from carbon fibre bundles and partially from connection between them. The deepest THz wave penetration was observed for connecting thread crossing, where it was possible to observe THz wave reflection from the next layer of the sample.

The authors wish to continue their work in this area of research in the future to overcome all the problems and difficulties encountered during the line of work presented in this paper.

Acknowledgement. The research was supported by the project entitled: The influence of temperature and moisture interaction effect on anisotropic structures: from theory to experimental investigation (2016/23/B/ST8/03088) granted by National Science Centre in Poland.

References

1. Kahandawa, G.C., Epaarachchi, J., Wang, H., Lau, K.T.: Use of FBG sensors for SHM in aerospace structures. Photonic Sens. **3**, 203–214 (2012)
2. Ferreira, P., Caetano, E., Pinto, P.: Real-time flying shape detection of yacht sails based on strain measurements. Ocean Eng. **131**, 48–56 (2017)
3. Gebremichael, Y.M., Li, W., Boyle, W.J.O., Meggitt, B.T., Grattan, K.T.V., McKinley, B., Fernando, G.F., Kister, G., Winter, D., Canning, L., Luke, S.: Integration and assessment of fibre bragg grating sensors in an all-fibre reinforced polymer composite road bridge. Sens. Actuators A: Phys. **118**, 78–85 (2005)
4. Mittleman, D.: Sensing with Terahertz Radiation. Springer, Heidelberg (2013)

5. Woodward, R.M., Cole, B.E., Wallace, V.P., Pye, R.J., Arnone, D.D., Linfield, E.H., Pepper, M.: Terahertz pulse imaging in reflection geometry of human skin cancer and skin tissue. Phys. Med. Biol. **47**, 38–53 (2002)
6. Peccianti, M., Fastampa, R., Conte, A.M., Pulci, O., Violante, C., Łojewska, J., Clerici, M., Morandotti, R., Missori, M.: Terahertz absorption by cellulose: application to ancient paper artifacts. Phys. Rev. Appl. **7**, 064019 (2017)
7. Mieloszyk, M., Majewska, K., Ostachowicz, W.: Application of THz spectroscopy for localisation of fibre optics embedded into glass fibre reinforced composite. Compos. Struct. **209**, 548–560 (2019)
8. Redo-Sanchez, A., Laman, N., Schulkin, B., Tongue, T.: Review of terahertz technology readiness assessment and applications. J. Infrared Millimeter Terahertz Waves **34**, 500–518 (2013)
9. Ospald, F., Zouaghi, W., Beigang, R., Matheis, C., Jonuscheit, J., Recur, B., Guillet, J.P., Mounaix, P., Vleugels, W., Bosom, P.V., Gonzalez, L.V., Lopez, I., Edo, R.M., Sternberg, Y., Vandewal, M.: Aeronautics composite material inspection with a terahertz time-domain spectroscopy system. Opt. Eng. **53**, 031208 (2013)
10. Karpowicz, N., Dawes, D., Perry, M.J., Zhang, X.C.: Fire damage on carbon fiber materials characterized by THz waves. In: Proceedings of SPIE, vol. 6212, p. 62120G (2006)
11. Im, K.H., Kim, S.K., Chiou, C.P., Jung, J.A.: Characterization of terahertz waves on foreign materials of composite materials. In: AIP Conference Proceedings, vol. 1949, p. 230023 (2018)
12. Oromiehie, E., Prusty, B.G., Compston, P., Rajan, G.: Characterization of process-induced defects in automated fiber placement manufacturing of composites using fiber Bragg grating sensors. Struct. Health Monit. **17**, 108–117 (2018)
13. Udd, E., Spillman Jr., W.B.: Fiber Optic Sensors: An Introduction for Engineers and Scientists. Wiley, Hoboken (2011)
14. Majewska, K., Mieloszyk, M., Ostachowicz, W.: Elasto-acoustic wave source localization in composite plate-like structure using muti-rosetts sensing. In: Proceedings of IWSHM, pp. 731–738 (2017)
15. Kakei, A.A., Islam, M., Leng, J., Epaarachchi, J.A.: Use of an elasto-plastic model and strain measurements of embedded fibre Bragg grating sensors to detect mode I delamination crack propagation in woven cloth (0/90) composite materials. Struct. Health Monit. **17**, 363–378 (2018)

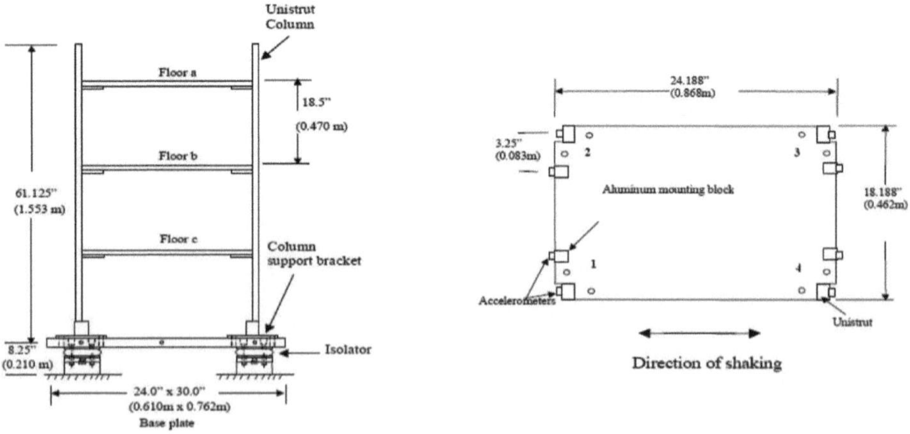

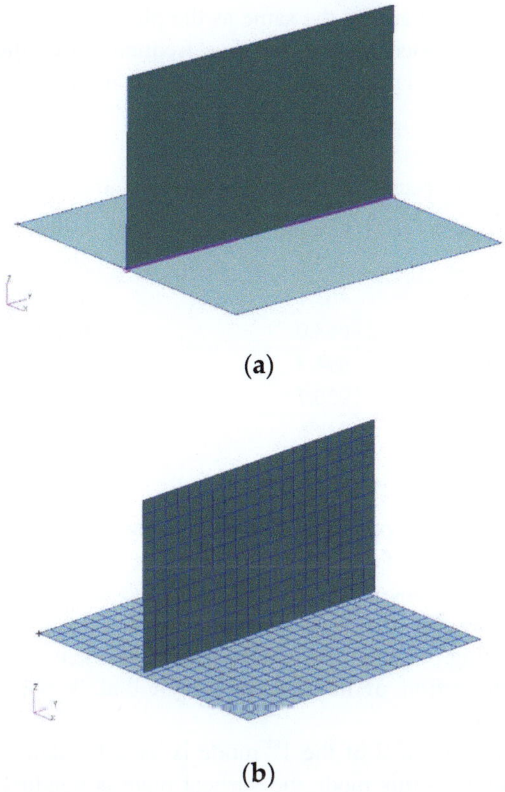

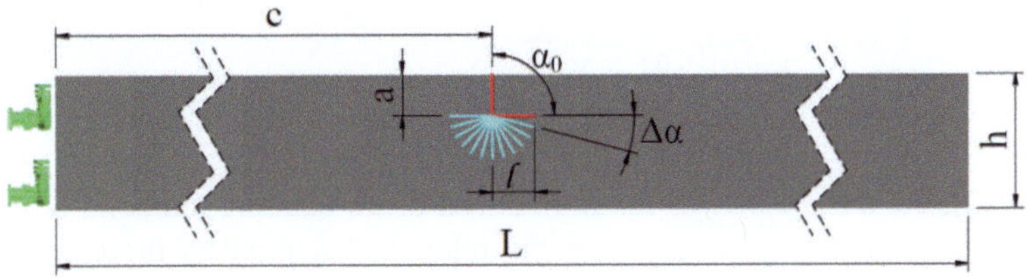

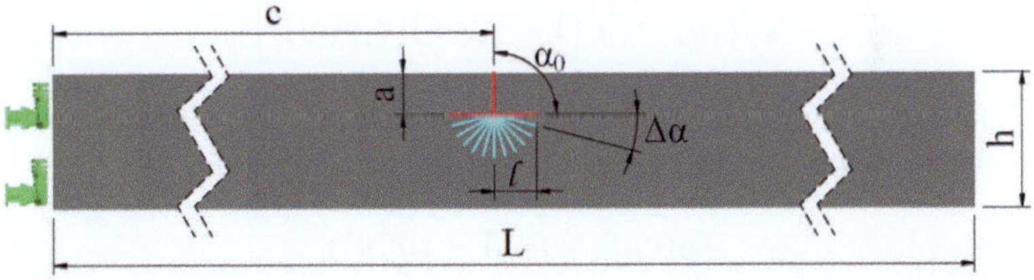

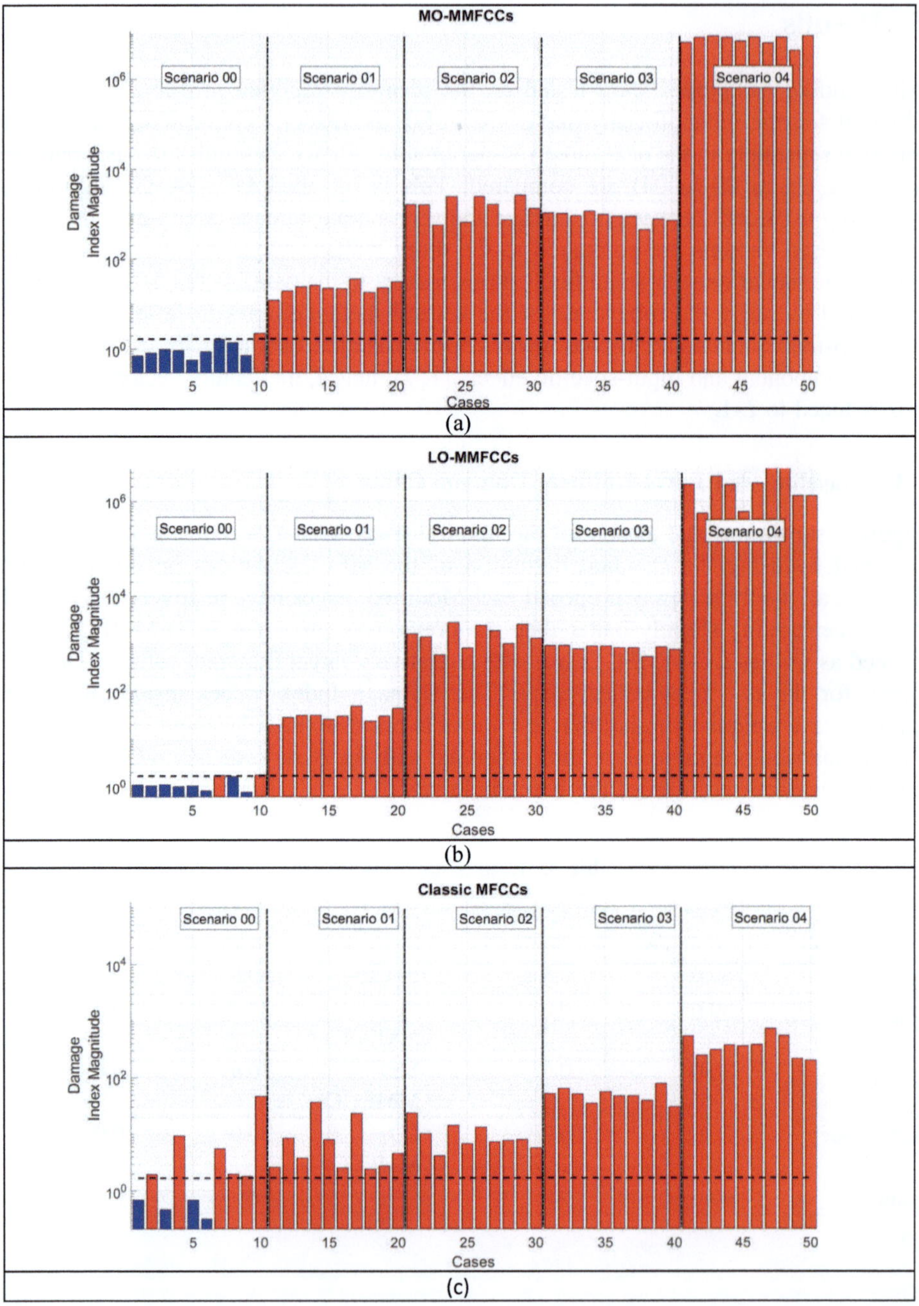

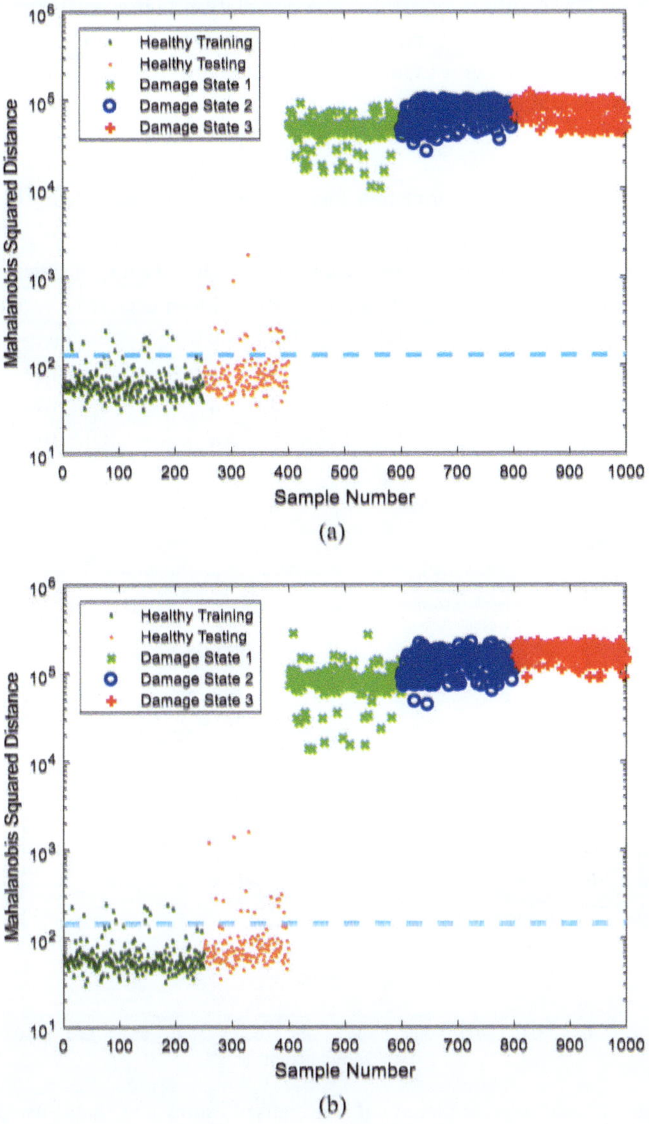

4.3 Damage Scenario 2 (D2)

For the second scenario, the results of damaged elements 6 and 14 with a loss of rigidity 50% and 30% respectively, are presented in Fig. 3. TMAC-1 is used as an objective function with PSO.

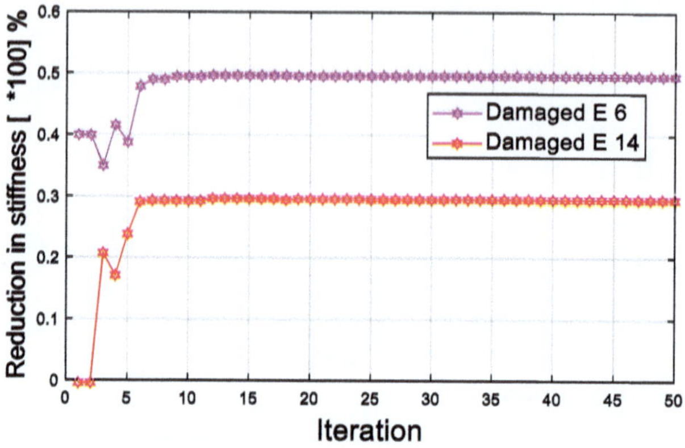

(a) Reduction in stiffness

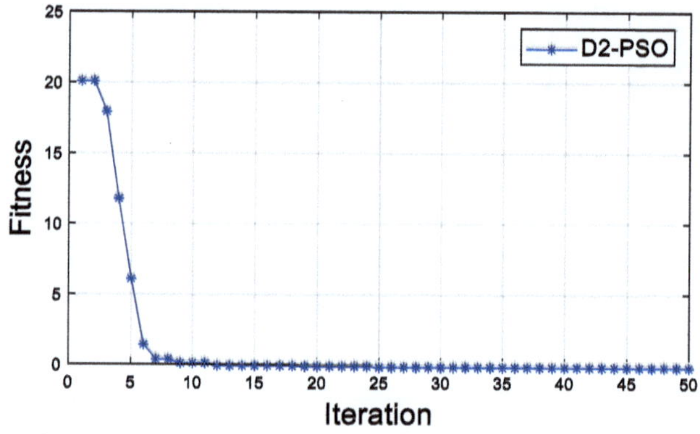

(b) Fitness

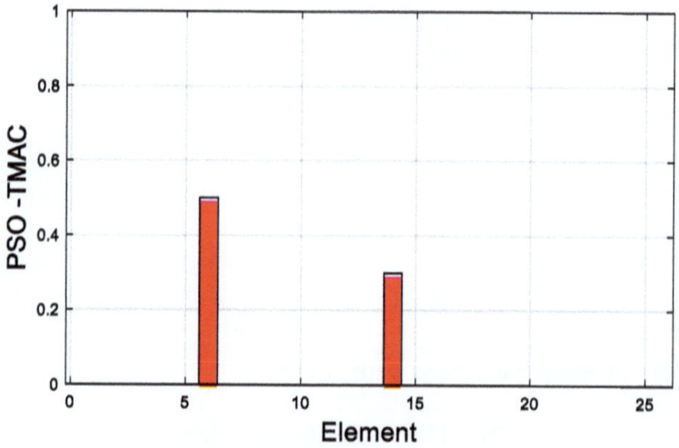

(c) Damaged element with severity

Fig. 3. PSO-TMAC for damaged scenario 2 (D2)

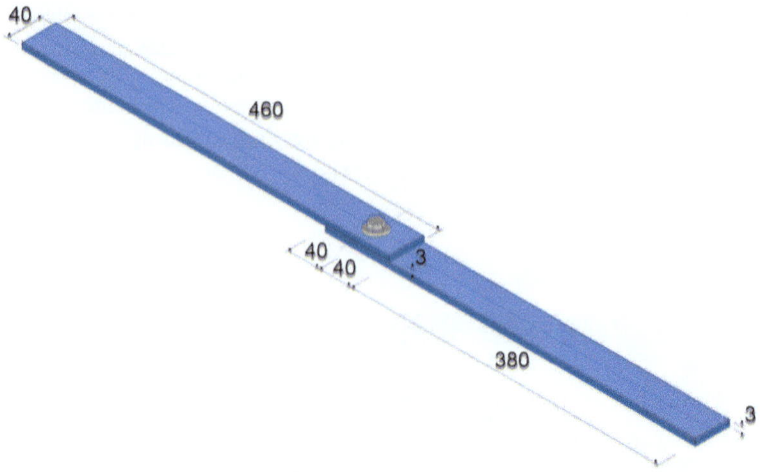

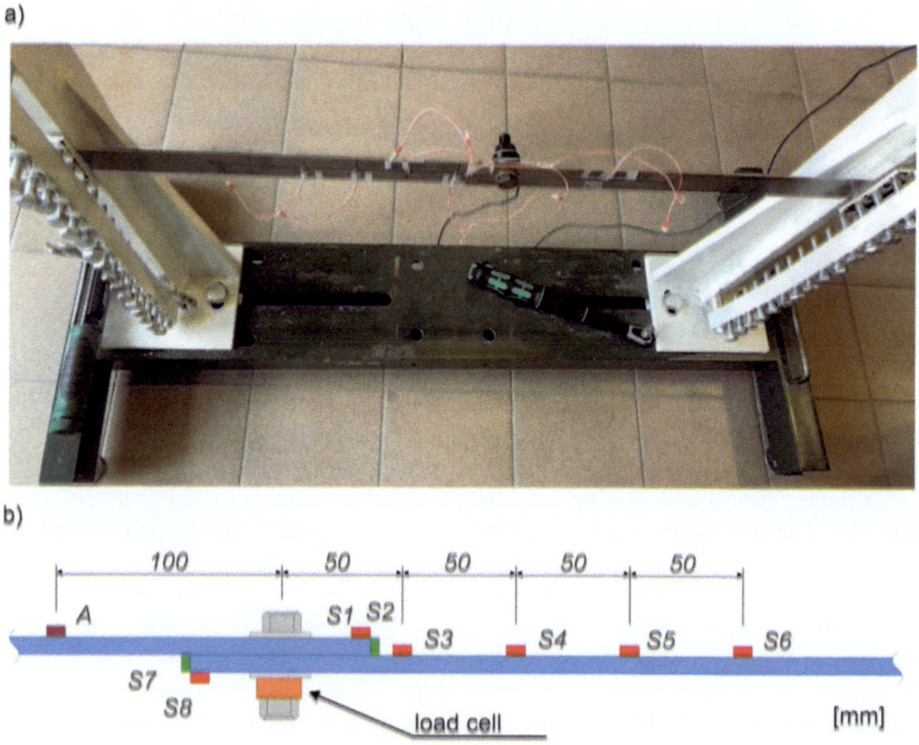

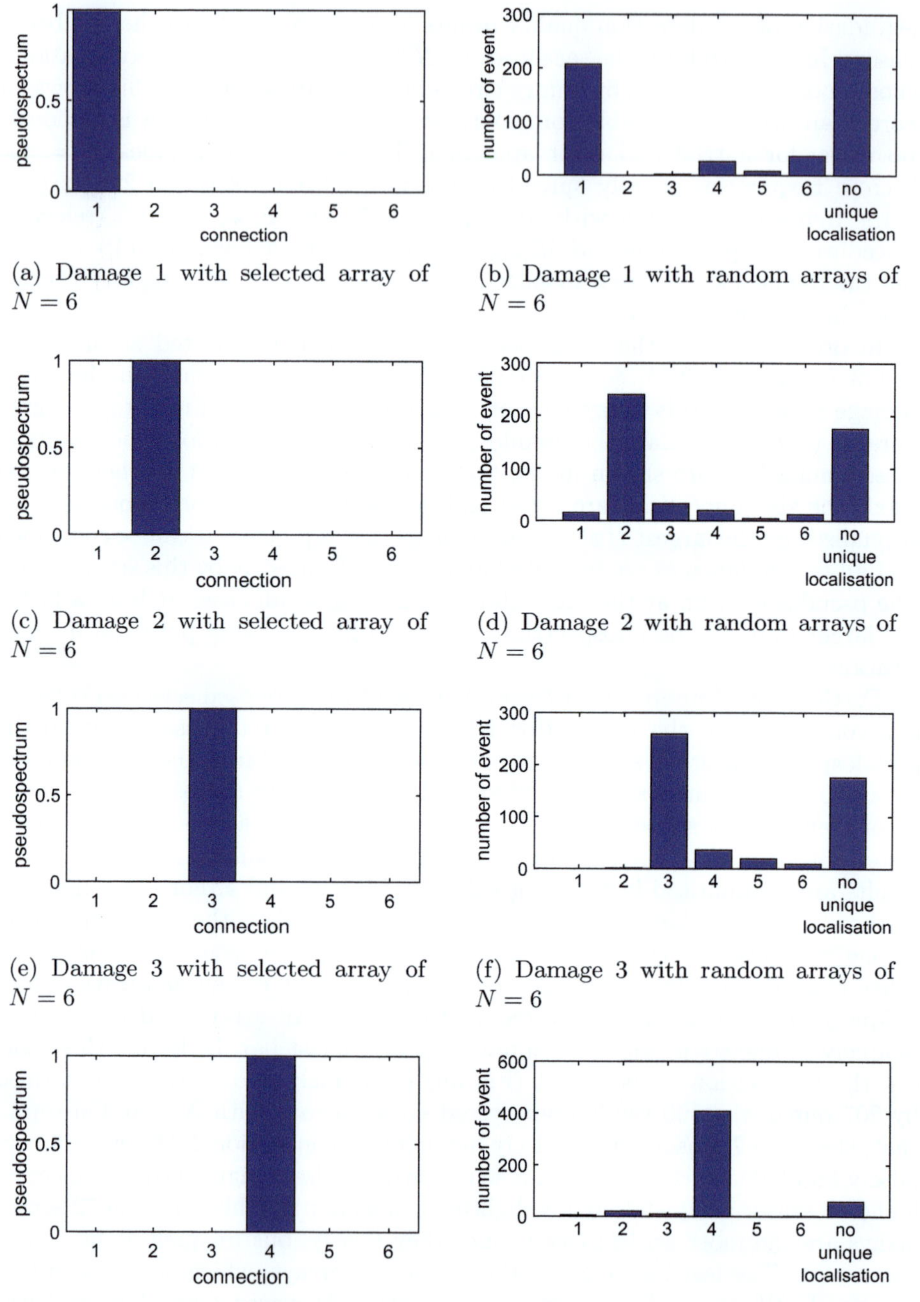

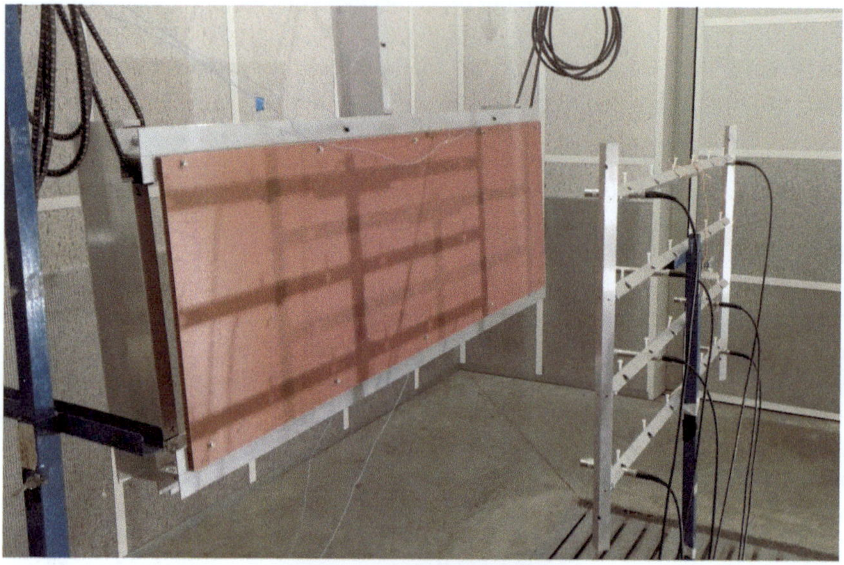

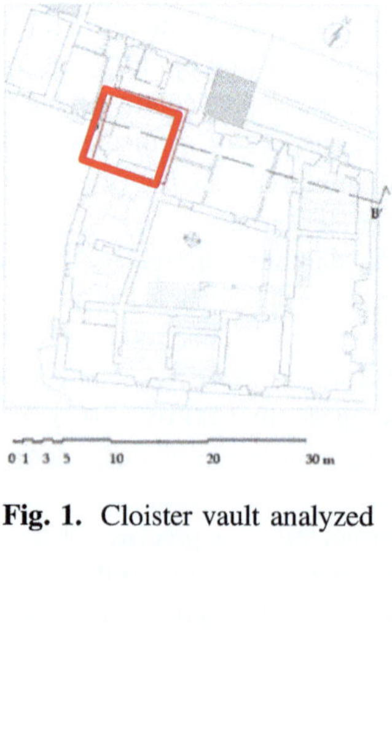

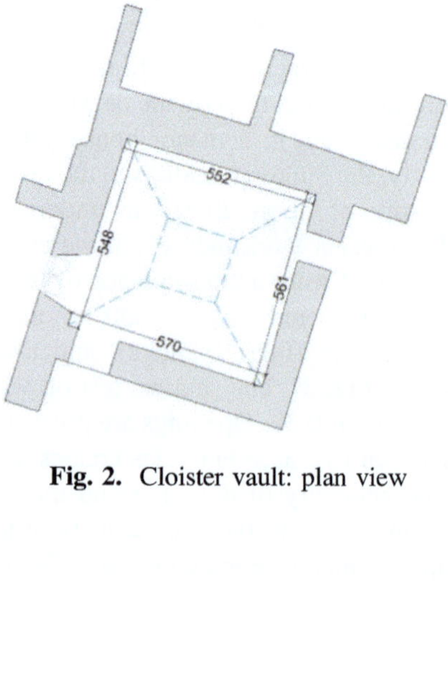

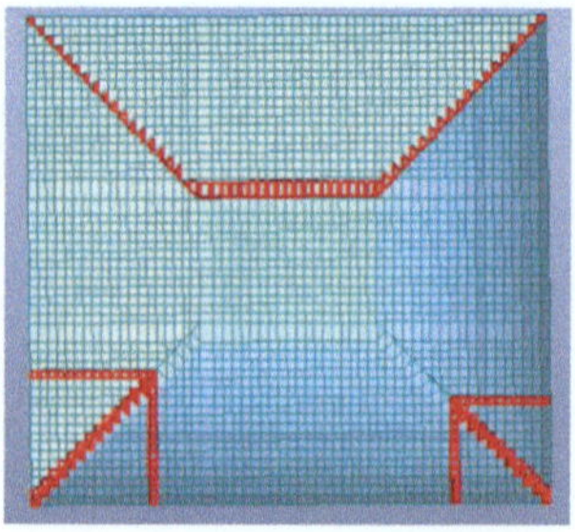

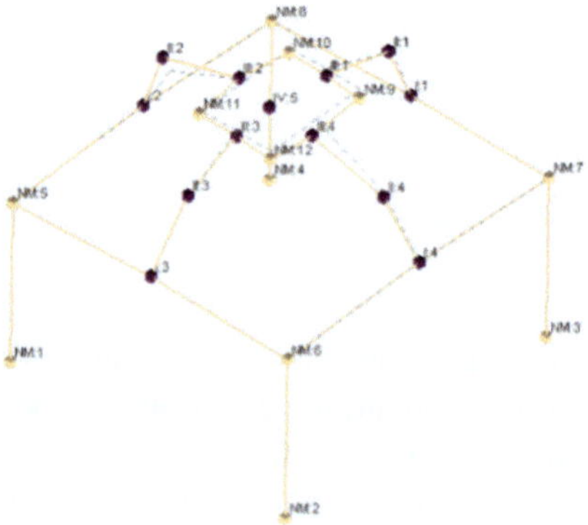

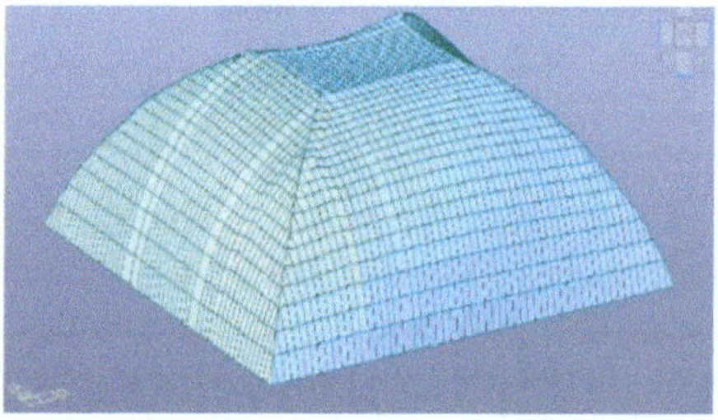

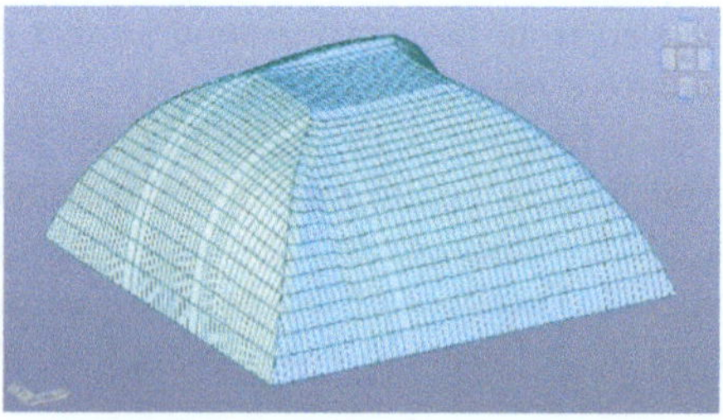

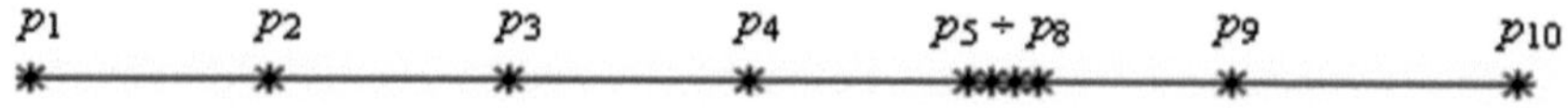

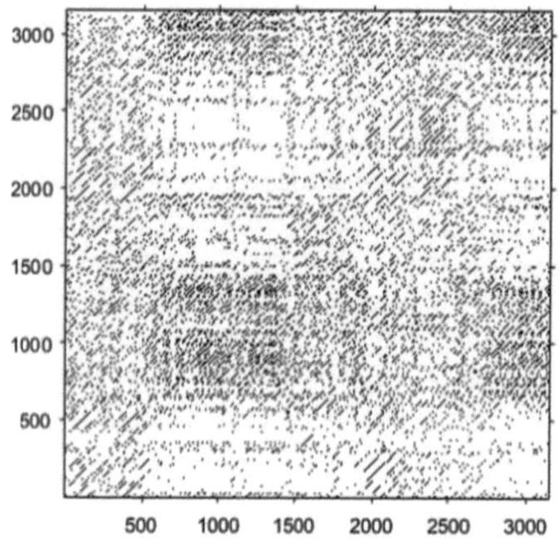

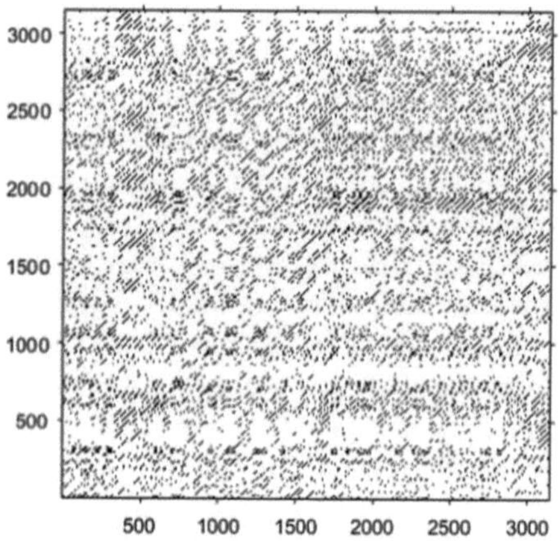

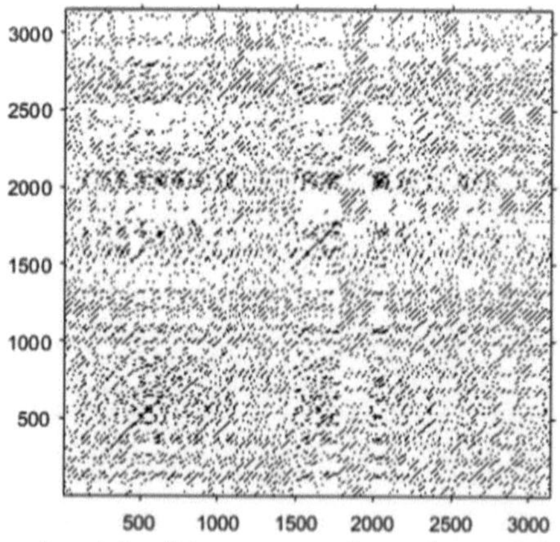

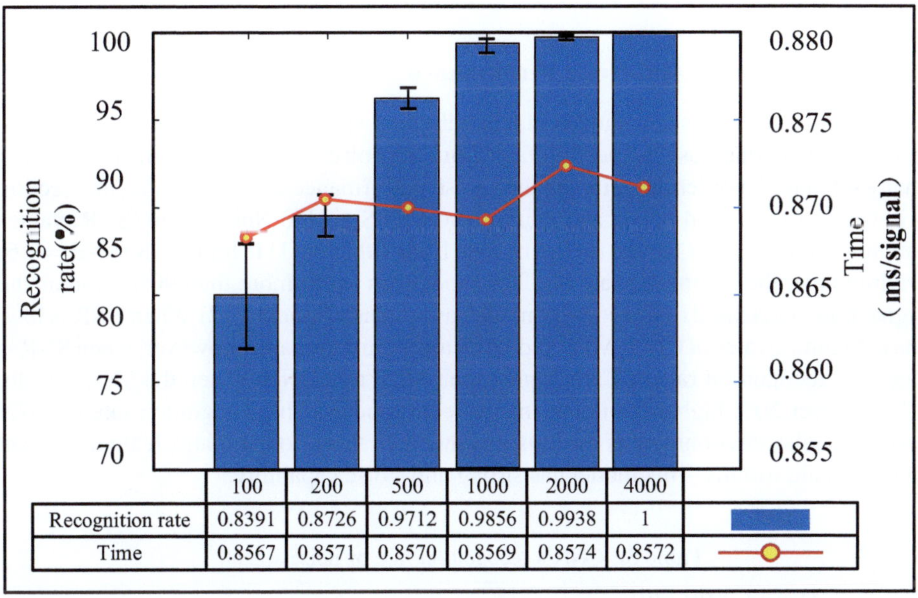

captured by using one material length scale parameter. The accuracy of present method is examined by comparing the obtained results with those of references' solutions. Some benchmark results for free vibration of laminated composite microplate are presented, which can be used as reference data for future size-dependent analysis. The inclusion of the material length scale parameter causes an increase in the stiffness of plate that leads to a higher frequency than that of classical model and the highest result is obtained for $\ell/h = 1$.

Acknowledgements. The authors acknowledge the financial support of VLIR-OUS TEAM Project, VN2017TEA454A103, 'An innovative solution to protect Vietnamese coastal riverbanks from floods and erosion', funded by the Flemish Government.

1 vliruos SHARING MINDS, CHANGING LIVES

With the support of THE BELGIAN DEVELOPMENT COOPERATION .be

References

1. Fleck, N.A., Muller, G.M., Ashby, M.F., Hutchinson, J.W.: Strain gradient plasticity: theory and experiment. Acta Metall. Mater. **42**(2), 475–487 (1994)
2. Stölken, J.S., Evans, A.G.: A microbend test method for measuring the plasticity length scale. Acta Mater. **46**(14), 5109–5115 (1998)
3. Chong, A.C.M., Lam, D.C.C.: Strain gradient plasticity effect in indentation hardness of polymers. J. Mater. Res. **14**(10), 4103–4110 (2011)
4. Lam, D.C.C., Yang, F., Chong, A.C.M., Wang, J., Tong, P.: Experiments and theory in strain gradient elasticity. J. Mech. Phys. Solids **51**(8), 1477–1508 (2003)
5. Chen, W., Xu, M., Li, L.: A model of composite laminated Reddy plate based on new modified couple stress theory. Compos. Struct. **94**(7), 2143–2156 (2012)
6. Chen, W., Li, X.: A new modified couple stress theory for anisotropic elasticity and microscale laminated Kirchhoff plate model. Arch. Appl. Mech. **84**(3), 323–341 (2014)
7. Wanji, C., Shengqi, Y., Xiaopeng, L.: A study of scale effect of composite laminated plates based on new modified couple stress theory by finite-element method. J. Multiscale Comput. Eng. **12**(6), 507–527 (2014)
8. Yang, S., Chen, W.: On hypotheses of composite laminated plates based on new modified couple stress theory. Compos. Struct. **133**(Supplement C), 46–53 (2015)
9. Chen, W., Wang, Y.: A model of composite laminated Reddy plate of the global-local theory based on new modified couple-stress theory. Mech. Adv. Mater. Struct. **23**(6), 636–651 (2016)
10. Yang, W., He, D.: Bending, free vibration and buckling analyses of anisotropic layered micro-plates based on a new size-dependent model. Compos. Struct. **189**, 137–147 (2018)
11. Hughes, T.J.R., Cottrell, J.A., Bazilevs, Y.: Isogeometric analysis: CAD, finite elements, NURBS, exact geometry and mesh refinement. Comput. Methods Appl. Mech. Eng. **194** (39), 4135–4195 (2005)

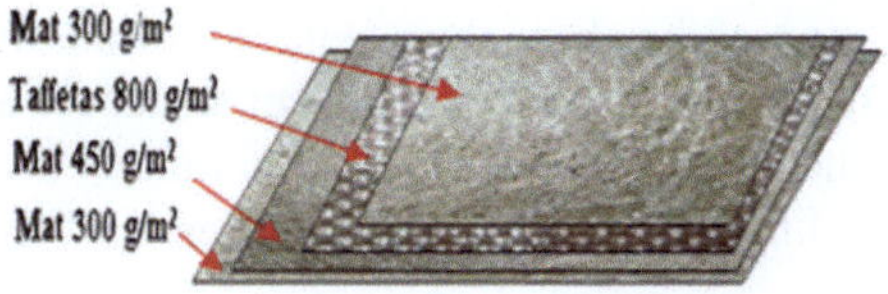

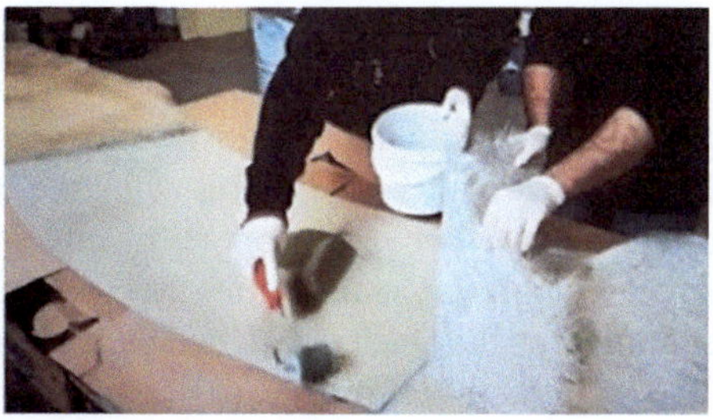

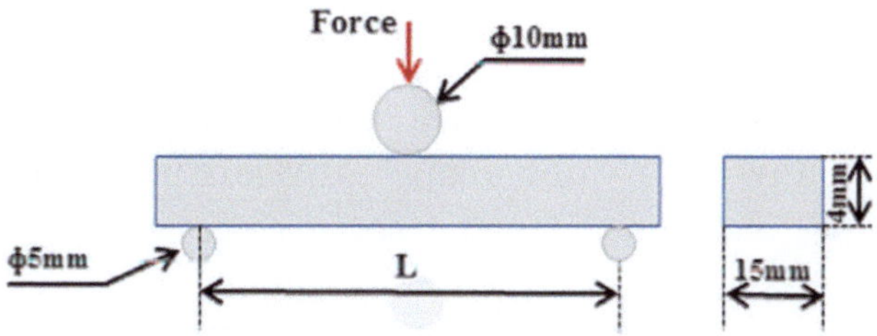

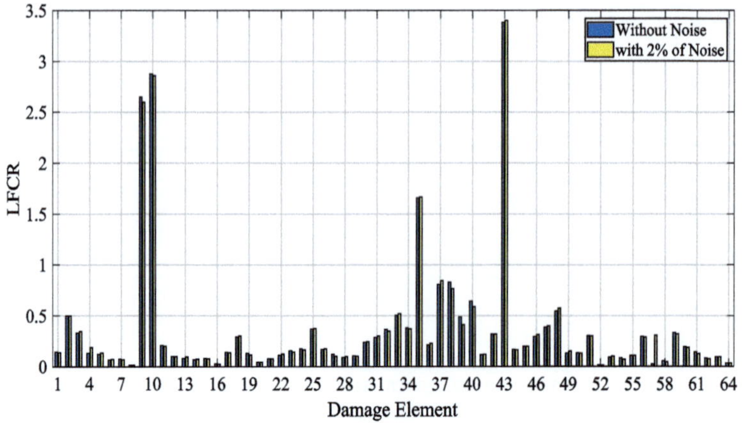

(a) Case 4 with 2% Noise

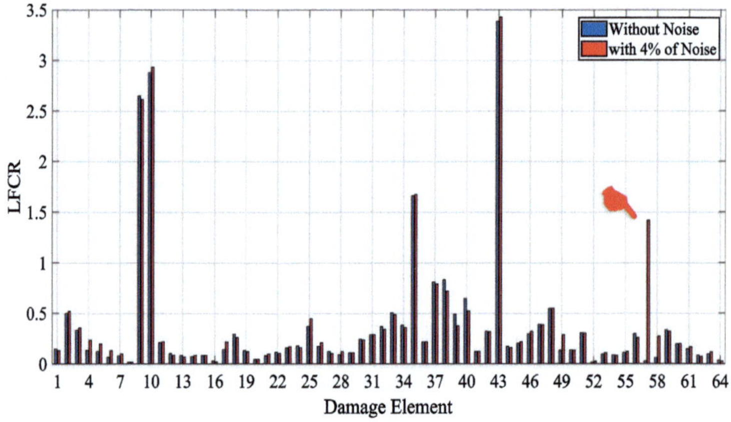

(b) Case 4: Avec 4% Noise

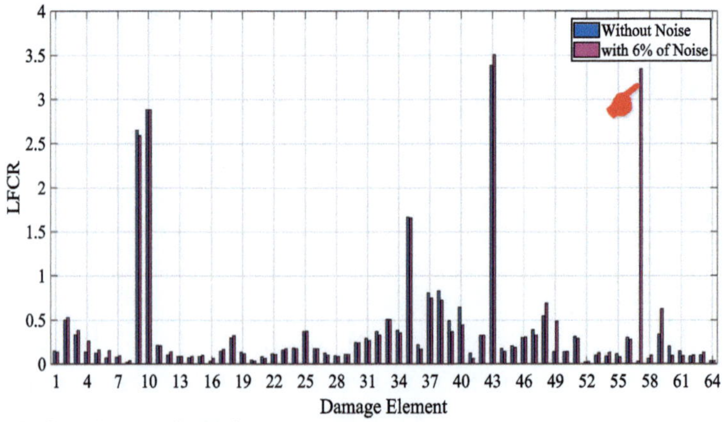

(c) Case 4: Avec 6% Noise

Fig. 11. Damage identification with noise level 2%, 4% and 6% for four damages - case 4

The Preparation of Smart Magnetic Nanoparticles for Intracellular Hyperthermia

XiaoGang Yu[1], RenPeng Yang[1], ChengWei Wu[1], Wei Zhang[1(✉)], DongFeng Deng[2], XuXin Zhang[2], and YanZhao Li[2]

[1] State Key Laboratory of Structure Analysis for Industrial Equipment, Department of Engineering Mechanics, Dalian University of Technology, Dalian 116024, China
wei.zhang@dlut.edu.cn
[2] Affiliated Zhongshan Hospital of Dalian University, Dalian 116001, China

Abstract. Magnetic induction hyperthermia is a new "green" cancer therapy method, which can destroy the tumor cells and undamage healthy cells. However, the difficulty of *in vivo* temperature rise control limits its clinical application. In this paper, we prepare a kind of smart magnetic nanoparticles for magnetic induction hyperthermia. The smart nanoparticles can self-control the temperature at 59.9 °C in the hyperthermia due to the low Curie temperature of nanoparticles. The formation mechanism of the smart magnetic nanoparticles is investigated. On the other hand, owing to the low thermal conductivity of the cell membrane, it is reasonable to believe that the intracellular hyperthermia is superior to extracellular hyperthermia. As such, we also discuss the feasibility of intracellular hyperthermia using the obtained smart magnetic nanoparticles. The results indicate that the temperature changes of the cell can meet the requirement of hyperthermia temperature when a single cell internalizes 2 pg of the smart magnetic nanoparticles.

Keywords: Smart magnetic nanoparticles · Intracellular hyperthermia · Magnetic induction hyperthermia

Hyperthermia therapy is a type of medical treatment in which body tissue is exposed to a suitable high temperature environment to damage and kill cancer cells. It is based on the biological principle that normal cells usually possess higher heat resistance and resilience to temperature than tumor cells, and the cancerous cells can be selectively destroyed by increasing the temperature to a desired temperature range (41 °C–46 °C), while ensuring healthy cells are unharmed [1]. Magnetic nanoparticles are widely used as the heat generation media [2, 3]. Sanz *et al.* [4] conducted a series of hyperthermia experiments by utilizing PEI-magnetic nanoparticles and human neuroblastoma cells, and compare the effect of exogenous heating source hyperthermia with the magnetic induction hyperthermia (MIH) at the same target temperatures. The results show that for the same target temperature, MIH induces a larger decrease in cell viability than the corresponding exogenous heating source hyperthermia. In other words, compared with exogenous heating source hyperthermia, MIH can destroy cancer cells more effectively [5]. However, owing to the difficulties of *in vivo* temperature rise control and temperature monitoring, it is still a challenge for the clinical application of MIH. To

M. A. Wahab (Ed.): DAMAS 2019, LNME, pp. 937–943, 2020.
https://doi.org/10.1007/978-981-13-8331-1_75

overcome the difficulties, a smart method of self-regulating heat temperature is proposed, which is taking the advantage of the Curie temperature (T_c) of the magnetic media. The T_c of magnetic media can be tuned by varying chemical composition through doping suitable amount of magnetic or nonmagnetic elements [6]. Here, through doping suitable amount of Zn^{2+} and Cr^{3+}, we synthesized a kind of smart magnetic nanoparticles (SMNPs) for MIH, the T_c of which is 60.8 °C. In order to enhance the curative effect of MIH, the intracellular hyperthermia is an alternative method to heat and destroy cancer cells [7]. Because the cell membrane possesses a low thermal conductivity, it may cause a lower temperature in the intracellular space. As a result, it is reasonable to believe that the intracellular hyperthermia is superior to extracellular hyperthermia as intracellular heating can overcome the thermal barrier [8, 9]. In this paper, we proposed a single cell model, established a heat conduction differential equation, and discussed theoretically the feasibility of intracellular hyperthermia by employing the SMNPs synthesized. The calculated results show that when a single cell internalizes 2 pg of the SMNPs, the temperature changes of the cell can meet the requirement of hyperthermia temperature.

1 Experimental

1.1 Synthesis of Smart Magnetic Nanoparticles

The SMNPs were synthesized by hydrothermal method. $ZnCl_2$ ($\geq$98%), $CoCl_2 \cdot 6H_2O$ ($\geq$99%), $CrCl_3 \cdot 6H_2O$ ($\geq$99%) and $FeCl_3 \cdot 6H_2O$ ($\geq$99%) were dissolved in 80 mL deionized water at a stoichiometric ratio of 0.54:0.46:0.6:1.4 to form a transparent metal salts solution of 0.61875 mol L^{-1}. Then 150 mL NaOH solution of 1 mol L^{-1} was added dropwise into the transparent metal salt solution under magnetic stirring at room temperature to form a precursor. Following this, the formed precursor solution was sealed in an autoclave and heated to 350 °C and maintained for 6 h. Then, the system was cooled naturally to room temperature, and the obtained black magnetic suspension was washed with deionized water and ethanol until neutral, and then dried at 80 °C for 8 h in vacuum drying chamber to obtain the $Zn_{0.54}Co_{0.46}Cr_{0.6}Fe_{1.4}O_4$ nanoparticles.

1.2 Characterization

The SMNPs were dispersed in ethanol and one drop of the mixture was placed on the copper grid until the complete evaporation of ethanol. The transmission electron microscopy (TEM, FEI Tecnai G2 F30, USA) was carried out to obtain the morphology and size distribution of the SMNPs. The thermogravimetric analysis (TGA, Mettler-Toledo TGA 851) was employed to record the mass-temperature curve of the SMNPs in the temperature range of 25–350 °C with a heating rate of 5 °C per min. The T_c of the SMNPs could be obtained by taking the derivative of the mass-temperature curve [10]. The magnetic heating experiment was conducted to determine the specific absorption rate (SAR) of the SMNPs, in which the SMNPs were dispersed into deionized water to produce a suspension with a concentration of 50 mg mL^{-1}. Once placed in an alternating magnetic field, the temperature of the suspension rose and the time-dependent temperature curve can be obtained by measuring the temperature of suspension per minute. The SAR of the SMNPs was calculated by employing the initial slope of the time-dependent temperature curve [11].

2 Results and Discussion

2.1 The Formation Mechanism

The formation process of the $Zn_xCo_{1-x}Cr_yFe_{2-y}O_4$ nanoparticles consists of three stages. The first stage is the dissolution of $ZnCl_2$, $CoCl_2{\cdot}6H_2O$, $CrCl_3{\cdot}6H_2O$ and $FeCl_3{\cdot}6H_2O$ in deionized water to form a transparent metal salts solution [12].

$$ZnCl_2 \rightarrow Zn^{2+} + 2Cl^-$$

$$CoCl_2 \cdot 6H_2O \rightarrow Co(H_2O)_6^{2+} + 2Cl^-$$

$$CrCl_3 \cdot 6H_2O \rightarrow Cr(H_2O)_6^{3+} + 3Cl^-$$

$$FeCl_3 \cdot 6H_2O \rightarrow Fe(H_2O)_6^{3+} + 3Cl^-$$

The second one is the co-precipitation process, which is the reaction of metal ions solution and precipitator (NaOH solution). Through the dropwise addition of the NaOH solution, the Zn^{2+}, $Co(H_2O)_6^{2+}$, $Cr(H_2O)_6^{3+}$ and $Fe(H_2O)_6^{3+}$, the mixed solution first becomes $Zn(OH)_n^{2-n}$, $Co(OH)_n^{2-n}$, $Cr(OH)_n^{3-n}$ and $Fe(OH)_n^{3-n}$. With the increase in the amount of NaOH, the precipitates of $Zn(OH)_2$, $Co(OH)_2$, $Cr(OH)_3$ and $Fe(OH)_3$ are formed, and then become $Zn(OH)_3^-$, $Co(OH)_3^-$, $Cr(OH)_4^-$ and $Fe(OH)_4^-$ eventually [12–15].

$$Zn^{2+} + nOH^- \rightarrow Zn(OH)_n^{2-n} \xrightarrow{OH^-} Zn(OH)_2{\downarrow} \xrightarrow{OH^-} Zn(OH)_3^-$$

$$Co(H_2O)_6^{2+} + nOH^- \rightarrow Co(OH)_n^{2-n} \xrightarrow{OH^-} Co(OH)_2{\downarrow} \xrightarrow{OH^-} Co(OH)_3^-$$

$$Cr(H_2O)_6^{3+} + nOH^- \rightarrow Cr(OH)_n^{3-n} \xrightarrow{OH^-} Cr(OH)_3{\downarrow} \xrightarrow{OH^-} Cr(OH)_4^-$$

$$Fe(H_2O)_6^{3+} + nOH^- \rightarrow Fe(OH)_n^{3-n} \xrightarrow{OH^-} Fe(OH)_3{\downarrow} \xrightarrow{OH^-} Fe(OH)_4^-$$

The third one is the hydrothermal treatment process, in which the dehydration reaction takes place on $Co(OH)_3^-$ and yields crystallized CoO. This CoO serves as the basis for the formation of $Zn_xCo_{1-x}Cr_yFe_{2-y}O_4$ nanoparticles by either heterogeneous nucleation or homogenous nucleation [14].

$$xZn(OH)_3^- + (1-x)Co(OH)_3^- + yCr(OH)_4^- + (2-y)Fe(OH)_4^-$$
$$\rightarrow Zn_xCo_{(1-x)}Cr_yFe_{(1-y)} + 3OH^- + 4H_2O$$

The formation mechanism of $Zn_xCo_{1-x}Cr_yFe_{2-y}O_4$ nanoparticles synthesized by hydrothermal method can be outlined in Fig. 1.

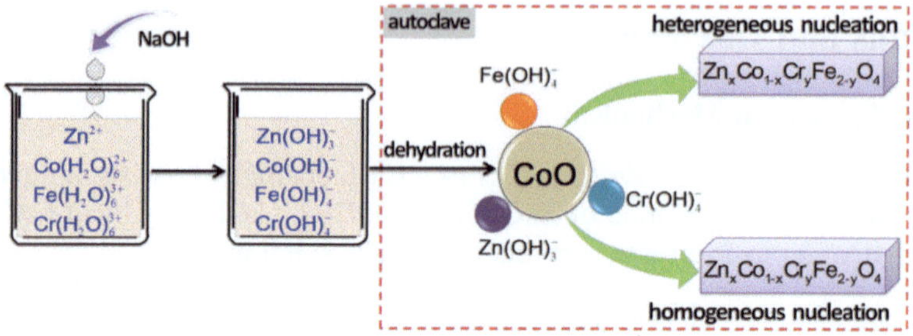

Fig. 1. The formation mechanism of $Zn_xCo_{1-x}Cr_yFe_{2-y}O_4$ nanoparticles synthesized by hydrothermal method.

2.2 Physical Properties of the SMNPs

As shown in Fig. 2a, the morphology image of the SMNPs shows that the nanoparticles are regular tetragonal structure and the size is in the range of 10–55 nm. The size distribution histogram is given in the inset of Fig. 2a, which describes that the average size of the SMNPs is 27.1 nm. Figure 2b shows the mass-temperature curve of the SMNPs, whose maximum value of the first derivative is considered to be the T_c. As displayed in the inset of Fig. 2b, the T_c is 60.8 °C. Figure 2c exhibits the temperature variety curve of the SMNPs suspension under the alternating magnetic field with an intensity of 32 kA m^{-1} and a frequency of 100 kHz. Utilizing the calculation formula [10], the SAR of the SMNPs can be calculated to 23.24 W g^{-1}.

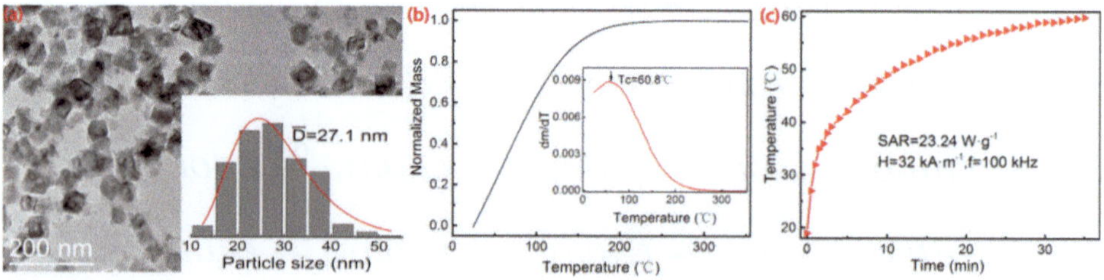

Fig. 2. (a) The morphology image and size distribution of the $Zn_{0.54}Co_{0.46}Cr_{0.6}Fe_{1.4}O_4$ nanoparticles. (b) The mass-temperature curve of the SMNPs and its first derivative. (c) The time-dependent curve of the SMNPs suspension under the alternating magnetic field of 32 kA m^{-1}, 100 kHz.

2.3 Intracellular Hyperthermia

To discuss the feasibility of intracellular hyperthermia, a case of a single nanoparticle in a single cancer cell is considered (as shown in Fig. 3), where D_p is the diameter of the nanoparticle. Applying the classical Fourier law for heat transfer analysis around the magnetic nanoparticle, the heat conduction differential equation can be established as the Formula 1 [9].

$$\frac{\partial \theta}{\partial \tau} = a[r^2 \frac{\partial}{\partial r}(r^2 \frac{\partial \theta}{\partial r})] \tag{1}$$

where a, equal to 1.3×10^{-7} m^2 s^{-1}, is the thermal diffusivity of the surrounding domain of the nanoparticle, θ is the temperature change, r is the radius, and τ is heating time.

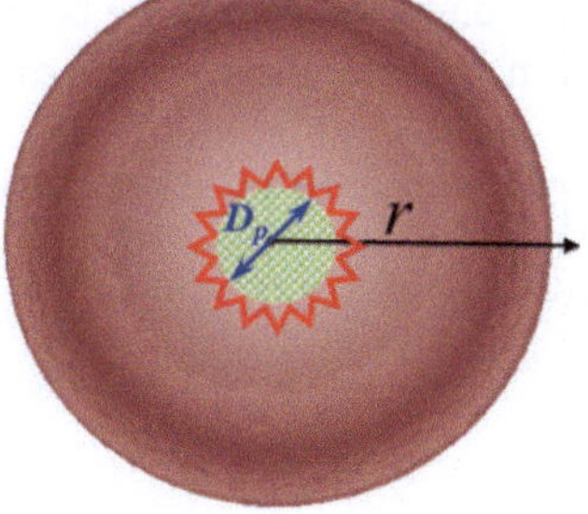

Fig. 3. The single cell model of a single spherical nanoparticle.

Taking the cancer cell as a finite heat conductor, the boundary condition can be expressed as the Formula 2.

$$Q\tau = \frac{\lambda}{u} \int_0^R 4\pi r^2 \theta dr \tag{2}$$

where $Q = q\rho\pi D_p^3/6$, q is the heating efficiency of the SMNPs, ρ is the density of the SMNPs, R is the radius of the cancer cell, $\lambda = 0.64$ W·m^{-1} K^{-1} is the thermal conductivity of the body solutions [9]. Solving the Eqs. (1) and (2), the temperature change θ can be obtained as the Formula 3.

$$\theta(r,t) = \frac{q\rho D_p^3 a\pi}{6\lambda} \int_0^t \frac{1}{(4\pi a\tau)^{3/2}} \frac{1}{f(\tau)} e^{-\frac{r^2}{4a\tau}} d\tau \tag{3}$$

where $f(\tau) = erf(\frac{R}{\sqrt{4a\tau}}) - \frac{2R}{\sqrt{4\pi a\tau}} e^{-\frac{R^2}{4a\tau}}$.

Assuming heating time t and cancer cell radius are 1800 s and 10 μm, respectively, and taking the parameters of the SMNPs prepared, $q = 23.24 \times 10^3$ W kg^{-1}, $\rho = 5.26 \times 10^3$ kg m^{-3}, $D_p = 27.1$ nm into Formula 3, θ_p (θ at $r = D_p/2$, the edge of the nanoparticle) is got to 1.1×10^{-4} °C, which can be neglected. But when the concentration of the nanoparticles internalized by cancer cells reaches 2 pg cell^{-1}, θ_p

will be up to 4.1 °C, which can meet the requirement of hyperthermia temperature. It is noted that the internalized SMNPs of 2 pg $cell^{-1}$ is much less than 80 pg $cell^{-1}$ that one cell can normally internalize [16]. As a result, intracellular hyperthermia using the developed SMNPs is feasible.

3 Conclusion

The formation mechanism of the SMNPs prepared by hydrothermal method is proposed. $CoCl_2$ $6H_2O$ converts to $Co(OH)_3^-$ with the addition of excessive alkali. After $Co(OH)_3^-$ being dehydrated to yield crystallized CoO, CoO serves as the basis for the formation of SMNPs by either heterogeneous nucleation or homogenous nucleation. The results of numerical calculation based on a finite boundary condition show that the SMNPs prepared can be used in intracellular hyperthermia.

Acknowledgments. This work was supported by grants from the National Natural Science Foundation of China (51775541, 11572080, 51811530309), and the Fundamental Research Funds for the Central Universities in China (DUT18ZD302).

References

1. Zhang, W., Wu, C.W., Ravi, S.R.P.: Proposed use of self-regulating temperature nanoparticles for cancer therapy. Expert Rev. Anticancer Ther. **18**(8), 723–725 (2018)
2. Zhang, W., Stolojan, V., Silva, S.R.P., Wu, C.W.: Raman, EELS and XPS studies of maghemite decorated multi-walled carbon nanotubes. Spectrochim. Acta Part A Mol. Biomol. Spectrosc. **121**(3), 715–718 (2014)
3. Zhang, W., Zuo, X.D., Wu, C.W.: Synthesis and magnetic properties of carbon nanotube-iron oxide nanoparticle composites for hyperthermia: a review. Rev. Adv. Mater. Sci. **40**(2), 165–176 (2015)
4. Sanz, B., Calatayud, M.P., Torres, T.E., Fanarraga, M.L., Ibarra, M.R., Goya, G.F.: Magnetic hyperthermia enhances cell toxicity with respect to exogenous heating. Biomater. **114**, 62–70 (2017)
5. Calatayud, M.P., Soler, E., Torres, T.E., Campos-Gonzalez, E., Junquera, C., Ibarra, M.R., Goya, G.F.: Cell damage produced by magnetic fluid hyperthermia on microglial BV2 cells. Sci. Rep. **7**(1), 8627 (2017)
6. Shaterabadi, Z., Nabiyouni, G., Soleymani, M.: Physics responsible for heating efficiency and self-controlled temperature rise of magnetic nanoparticles in magnetic hyperthermia therapy. Prog. Biophys. Mol. Biol. **133**, 9–19 (2017)
7. Hedayatnasab, Z., Abnisa, F., Daud, W.M.A.W.: Review on magnetic nanoparticles for magnetic nanofluid hyperthermia application. Mater. Des. **123**, 174–196 (2017)
8. Gordon, R.T., Hines, J.R., Gordon, D.: Intracellular hyperthermia a biophysical approach to cancer treatment via intracellular temperature and biophysical alterations. Med. Hypotheses **5**(1), 83–102 (1979)
9. Rabin, Y.: Is intracellular hyperthermia superior to extracellular hyperthermia in the thermal sense? Int. J. Hyperth. **18**(3), 194–202 (2002)

10. Zhang, W., Zuo, X., Niu, Y., Wu, C., Wang, S., Guan, S., Silva, S.R.P.: Novel nanoparticles with Cr^{3+} substituted ferrite for self-regulating temperature hyperthermia. Nanoscale **9**(37), 13929–13937 (2017)
11. Prasad, A.I., Parchur, A.K., Juluri, R.R., Jadhav, N., Pandey, B.N., Ningthoujam, R.S., Vatsa, R.K.: Bi-functional properties of Fe3O4@YPO4: Eu hybrid nanoparticles: hyperthermia application. Dalton Trans. **42**(14), 4885–4896 (2013)
12. Zhao, D., Wu, X., Guan, H., Han, E.: Study on supercritical hydrothermal synthesis of CoFe2O4 nanoparticles. J. Supercrit. Fluids **42**(2), 226–233 (2007)
13. Zhang, Y., Yang, Z., Yin, D., Liu, Y., Fei, C., Xiong, R., Shi, J., Yan, G.: Composition and magnetic properties of cobalt ferrite nano-particles prepared by the co-precipitation method. J. Magn. Magn. Mater. **322**(21), 3470–3475 (2010)
14. Zhang, W., Zuo, X., Zhang, D., Wu, C., Silva, S.R.P.: Cr^{3+} substituted spinel ferrite nanoparticles with high coercivity. Nanotechnol. **27**(24), 245707 (2016)
15. Zuo, X., Wu, C., Zhang, W., Gao, W.: Magnetic carbon nanotubes for self-regulating temperature hyperthermia. RSC Adv. **8**, 11997–12003 (2018)
16. Dias, C.S., Hanchuk, T.D., Wender, H., Shigeyosi, W.T., Kobarg, J., Rossi, A.L., Tanaka, M.N., Cardoso, M.B., Garcia, F.: Shape tailored magnetic nanorings for intracellular hyperthermia cancer therapy. Sci. Rep. **7**(1), 14843 (2017)

Author Index

M. A. Wahab (Ed.): DAMAS 2019, LNME, pp. 945–947, 2020.
https://doi.org/10.1007/978-981-13-8331-1